SOIL STRESS-STRAIN BEHAVIOR: MEASUREMENT, MODELING AND ANALYSIS

SOLID MECHANICS AND ITS APPLICATIONS
Volume 146

Aims and Scope of the Series

The fundamental questions arising in mechanics are: *Why?, How?,* and *How much?* The aim of this series is to provide lucid accounts written by authoritative researchers giving vision and insight in answering these questions on the subject of mechanics as it relates to solids.

The scope of the series covers the entire spectrum of solid mechanics. Thus it includes the foundation of mechanics; variational formulations; computational mechanics; statics, kinematics and dynamics of rigid and elastic bodies: vibrations of solids and structures; dynamical systems and chaos; the theories of elasticity, plasticity and viscoelasticity; composite materials; rods, beams, shells and membranes; structural control and stability; soils, rocks and geomechanics; fracture; tribology; experimental mechanics; biomechanics and machine design.

The median level of presentation is the first year graduate student. Some texts are monographs defining the current state of the field; others are accessible to final year undergraduates; but essentially the emphasis is on readability and clarity.

For a list of related mechanics titles, see final pages.

Soil Stress-Strain Behavior: Measurement, Modeling and Analysis

A Collection of Papers of the Geotechnical Symposium in Rome, March 16–17, 2006

Edited by

HOE I. LING

Columbia University, New York, NY, USA

LUIGI CALLISTO

University of Rome "La Sapienza", Rome, Italy

DOV LESHCHINSKY

University of Delaware, Newark, DE, USA

and

JUNICHI KOSEKI

University of Tokyo, Japan

Springer

A C.I.P. Catalogue record for this book is available from the Library of Congress.

ISBN 978-94-017-7666-0
ISBN 978-1-4020-6146-2 (eBook)
DOI 10.1007/978-1-4020-6146-2

Published by Springer,
P.O. Box 17, 3300 AA Dordrecht, The Netherlands.

www.springer.com

Printed on acid-free paper

CONTENTS

PREFACE

This Publication is an outgrowth of the Proceedings for the *Geotechnical Symposium in Roma*, also known as *Tatsuoka Symposium*, which was held on March 16 and 17, 2006 in Rome, Italy. The Symposium was organized to celebrate the 60th birthday of Prof. Tatsuoka. The occasion also provided a chance to honor Prof. Tatsuoka for his research achievement. Prof. Tatsuoka collaborated with many international researchers, and thus the most beautiful and historical city of Rome naturally became an ideal location for his friends, colleagues and former students from different parts of the world to meet and celebrate this special occasion. The generosity of the University of Rome "La Sapienza" directed all roads to Rome by providing the venue for the Symposium.

Prof. Tatsuoka retired from the University of Tokyo at the end of March 2004 following a 30-year distinguished career in teaching, research and professional service. During his tenure at the University of Tokyo, he published over 300 papers and graduated about 30 PhD and 25 MS students. Prof. Tatsuoka continues his research and teaching at the Tokyo University of Science. Thus, the Symposium also congratulated his new endeavor.

The Symposium focused on the recent developments in the stress-strain behavior of geomaterials, with an emphasis on testing and applications, including soil modeling, analysis and design. The Symposium was declared open by Prof. Bucciarelli, Dean of the School of Engineering at the University of Rome "La Sapienza," followed by addresses by Prof. Manassero, on behalf of the Italian Geotechnical Society and Dr. Cazzuffi, the President of the International Geosynthetic Society. The Symposium included a Special Lecture delivered by Prof. Tatsuoka and five Keynote Lectures delivered by Profs. Lo Presti (Italy), Jardine (UK), Di Benedetto (France), Shibuya (Japan) and Leshchinsky (US). A total of 90 papers were solicited and the overwhelming response did not allow all papers to be presented despite the shortening of the time allocated for each presentation. The individuals from the organizing institutions volunteered to give the slots of presentation to the young researchers and their cooperation was much appreciated. Due to the capacity of the lecture hall, the total number of participants was restricted to 120.

The Organizing Committee extended their appreciations to many individuals who assisted in the Symposium: the members of the Local Hosting Committee, the Scientific Committee, the reviewers, and the authors who contributed so much of their time and efforts in preparing the papers. The attendance of Mrs. Yoko Tatsuoka was especially appreciated.

Local Hosting Committee

Luigi Callisto, University of Rome "La Sapienza"
Sebastiano Rampello, University of Rome "La Sapienza"
Filippo Santucci de Magistris, University of Molise
Alessandro Flora, University of Napoli Federico II

Scientific Committee

Andrew Whittle, MIT (Chair), USA
A. Anandarajah, Johns Hopkins University, USA

Ronald Borja, Stanford University, USA
David Frost, Georgia Tech, USA
Satoshi Goto, Yamanashi University, Japan
Marte S. Gutierrez, Virgina Tech, USA
Sam C.-C.Huang, National Chi-Nan University, Taiwan
R. Lancellotta, Technical University of Turin, Italy
Radoslaw L. Michalowski, University of Michigan, USA
Yoshiyuki Mohri, NRIAE, Japan
Juan Pestana, University of California at Berkeley, USA
S. Rampello, University of Rome "La Sapienza", Italy
M.S.A. Siddiquee, Bangladesh University of Technology
Kazuo Tani, Yokohama National University, Japan
Jonathan T.H. Wu, University of Colorado-Denver, USA
Jerry A. Yamamuro, Oregon State University, USA
Conference Advisor: Raimondo Betti, Columbia University, USA

The presentations were grouped under 8 different sessions, each led by a Session Chair and a Discussion Leader, as listed below:
Behavior of Granular Materials (Tanaka, Flora*)
Behavior of Clays and Viscous Properties (Verdugo, G.M.B. Viggiani)
Modified Soils and Soil Mixtures (Lizcano, Hirakawa)
Cyclic/Dynamic Soil Behavior (Santucci de Magistris*, Yasuda*)
Soil Liquefaction (Pradel*, Kohata)
Soil Constitutive Models and Numerical Analysis (Siddiquee, Maugeri*)
Soil Reinforcement and Earth Retaining Structures (Mohri, Uchimura)
Buried Structures, Slopes and Other Geotechnical Issues (Rampello, Taylor)
[*: members in charge of paper review]

The technical competency of the Chairs and Leaders stimulated the discussions and greatly improved the standard of this Symposium.

The Banquet was held in the evening of the first day of Symposium at the *Palazzo Brancaccio*. This function provided an unforgettable memory to all participants.

The Symposium was financially supported in part by the University of Rome "La Sapienza", TREVI Corporation, and DMS. Their generosity allowed the Organizers to arrange for a minimum possible registration fee, and free registration to students who also attended the banquet at a reduced rate. These supports were gratefully acknowledged.

The preliminary preparation of the Symposium was done while the first editor was at Columbia University, but most of the works related to the Symposium were accomplished while he was on sabbatical at Harvard University. The support and friendship of Prof. James Rice, Dr. Renata Dmowsky, and the Division of Engineering and Applied Sciences was highly appreciated.

Last but not least, the secretarial assistance of Ms. Chieko Nohara, the invaluable help from the young researchers at Montedoro, editorial assistance including the cover design by Emi Ling, and the work of the editorial staff at the Springer (Ms. Nathalie Jacobs and Anneke Pot in particular), made this Volume possible.

Hoe I. Ling, Columbia University
Luigi Callisto, University of Rome "La Sapienza"
Dov Leshchinsky, University of Delaware
Junichi Koseki, University of Tokyo

August 2006

Acknowledgments:

TREVI

FOREWORD
The Italian Geotechnical Society

Due to unavoidable duties the president of the Italian Geotechnical Society, Prof. Alberto Burghinoli, cannot be here today.

Therefore, as a member of the Italian Geotechnical Society board and on the behalf of its President, I would like to welcome all the delegates of this Symposium.

The symposium has been organized for celebrating, in the occasion of 60th birthday of Prof. Fumio Tatsuoka, his past outstanding scientific career within the Tokyo University of Science and for inspiring, in some way, his future activities within other research institutions.

I would like to thank Prof. Tatsuoka for being here with us and the whole organizing staff of the symposium for the brilliant idea to locate such an important event in Rome under the auspicious of the Italian Geotechnical Society (AGI).

I would also remember that two other important events took place just yesterday in Rome, that are the board meeting of the International Society for Soil Mechanics and Geotechnical Engineering (ISSMGE) and the board meeting of the International Geosynthetics Society (IGS).

These occasions make the city of Rome and, I would say, the Italian Geotechnical Society, the focal points of the international geotechnical community at least for this week and we are very proud about this.

Therefore, it is a real pleasure and honour for me to extend the welcome to the president of ISSMGE Prof. Pedro Seco Pinto, to the president of IGS Daniele Cazzuffi (that is at home) and to the related society boards.

I also hope to have other occasions in Italy in the near future for organizing this kind of event and for promoting scientific and professional progress within the Geotechnical Engineering field.

Finally, I whish for all the delegates, coming worldwide, a pleasant and fruitful work.

Prof. Mario Manassero
AGI Board Member

The International Geosynthetics Society

It's really an honour for me to introduce this Roma Geotechnical Symposium organised at the University "La Sapienza" on 16 and 17 March 2006 to celebrate Prof. Tatsuoka's 60th birthday.

I met Fumio for the first time almost twenty years ago in Vienna, at the Hofburg palace, where in early April 1986 the 3rd International Conference on Geotextiles took place, and I was immediately impressed by his comprehensive view of the different aspects of geotechnical engineering and also by his personal attitude towards informal international contacts.

Then, in October 1992, I was invited from Prof. Tatsuoka to give a lecture at the First Seiken Symposium he organised at the University of Tokyo and I was experiencing his peculiar charisma as teacher and as researcher among his colleagues and particularly among his students : this charisma was able to attract dozens of foreign students coming from all over the world to his university to conduct their researches under his coordination.

Presently, Fumio is Professor of Geotechnical Engineering at the Department of Civil Engineering, at the Tokyo University of Science, while from 1977 to 2004 he was Associate Professor and then Professor of Geotechnical Engineering at the University of Tokyo.

Among his various research interests in the different fields of geotechnics, I could mention at least the following: laboratory testing methods for geomaterials, including clays, sands, gravels, soft rocks and geosynthetics; deformation and strength characteristics of geomaterials; foundation engineering, including bearing capacity of shallow foundations; ground improvement by cement-mixing and soil reinforcing with geosynthetics.

Fumio was awarded several times, both in Japan and overseas. Among his international awards, I have to quote at least the following: the IGS Award from the International Geosynthetics Society (1994), the Hogentoglar Award from ASTM (1996 and 2003), the Mercer Lectureship jointly from the IGS (International Geosynthetics Society) and from the ISSMGE (International Society for Soil Mechanics and Geotechnical Engineering) on "Geosynthetic-Reinforced Soil Retaining Walls as Important Permanent Structures" (1996) and the best paper Award of the "Ground Improvement" Journal (1997).

Fumio was also Editor of two books in English ("Permanent Geosynthetic-Reinforced Soil Retaining Walls", Balkema, 1994 and "Reinforced Soil Engineering, Advances in Research and Practice", Marcel Dekker, 2003) and Author or Co-Author of more than 300 technical papers published in "Soils and Foundations", "Geotechnical Testing Journal", "Géotechnique", "Geotechnical Testing Journal", "Journal of Geotechnical and Geoenvironmental Engineering", "Ground Improvement" , "Geosynthetics International", " Geotextiles and Geomembranes" and others relevant journals.

Being here in Roma, I like to mention that Prof. Tatsuoka is also a member of the Advisory Board of the Italian Geotechnical Journal ("Rivista Italiana di Geotecnica"), together with myself and with other Italian and international experts.

Fumio has been and also presently is very active in major learned Japanese and international societies : from 2001 to 2005 he was Vice President for the Asia Region of the International Society for Soil Mechanics and Geotechnical Engineering; from 2003 to 2005 Vice President of the Japanese Geotechnical Society and from 2005 to 2006 Vice President of the Japanese Society for Civil Engineers, while this year he was elected President (till 2008) of the Japanese Geotechnical Society.

But let me concentrate on the International Geosynthetics Society, where Fumio was elected Vice President in 2002 till 2006: in that period, as IGS President, I had the unique opportunity to work closely with him, with a day-by-day fruitful communication, that resulted in an impressive growth and outreach of the International Geosynthetics Society.

I am very proud to report that next September in Yokohama, Japan, at the end of the 8th International Conference on Geosynthetics, Prof. Tatsuoka will be officialy appointed as the new IGS President for the period 2006-2010, therefore being not only the first Japanese ,but also the first Asian, President of the International Geosynthetics Society.

I am really confident that we will continue to work closely as we did in the last four years, with our new rules, himself as the new president and myself as the immediate past president, but both having in mind a progressive consolidation of the IGS in the different countries around the world and also looking for a better understanding of the geosynthetics discipline among the entire geotechnical community.

But I could not conclude my introduction without mentioning the full dedication of Fumio to his wonderful family, his wife and his two daughters recently married: all the times I'm asking him about his family, he's starting to relax and he's always saying: "I'm very very busy, but my entire free time is for them". I'm sure that this full support of his family has helped Fumio a lot in order to allow him to achieve the impressive results of his academic and professional career.

Finally, let me sincerely congratulate the main organisers of this symposium,in strict alphabetical order namely Luigi Callisto, Dov Leshchinsky, Hoe Ling and Junichi Koseki, for being able to convey to Roma a very good participation from all over the world, both in terms of number and also of quality of the papers.

Ad majora !

Daniele Cazzuffi

IGS President

CESI SpA, Milano, Italy

International Society for Soil Mechanics and Geotechnical Engineering

It is for me a great honour and privilege, to write this letter to introduce Prof. Fumio Taksuoka following the request of the Organizing Committee of Geotechnical Symposium in Rome, 16-17 March 2006, to celebrate Professor Tatsuoka 60th Birthday.

Professor Fumio Taksuoka does not need any introduction as he is well known by the international geotechnical community. He is a man of prodigious energy and fine intellect. We are indebted for his outstanding contribution for the advancement of knowledge in the areas of stress strain strength testing of geomaterials, retaining walls, geosynthetics and soil dynamics.

Professor Taksuoka has authored/co-authored over 150 journal and conference proceedings papers in the area of geotechnical engineering. The impressive list of prestigious journals includes Soils and Foundations, Journal of Geotechnical Engineering (ASCE), Testing Journal ASTM, Geotextiles and Geomembranes and Geosynthetics International._

He is the Current President of Japanese Geotechnical Society (JGS) and the Current Vice President of International Geosynthetics Society (IGS). He has served as Vice President of ISSMGE for Asia (2001-2005) and Chairman of TC29 Laboratory Stress Strain Strength Testing of Materials.

Professor Taksuoka is often invited to be State-of-the-Art or Keynote Speaker at international conferences of geotechnical engineering and we always listen his lectures with great interest and pleasure, as they are challenge and open new avenues of research.

Dr. Taksuoka has received several awards and honors due his distinguished achievements.

I would like to highlight from Prof. Fumio Taksuoka outstanding curriculum: i) his solid scientific background and research contributions for the advancement of knowledge of soils mechanics and geotechnical engineering; ii) his excellent lecturing and teaching ability and immense contribution to improve the engineering geotechnical education level in developed countries.

I wish Fumio the best success in his professional and family life.

Pedro Sêco e Pinto

PEDRO SÊCO E PINTO
President
International Society for Soil Mechanics and
Geotechnical Engineering

August 2006

INTRODUCTION: PROFESSOR FUMIO TATSUOKA

Dov Leshchinsky
University of Delaware, USA

The most gratifying reward for a professional is recognition of achievements by his own peers. A forum such as the Symposium in Roma must be the ultimate recognition. This 2-day symposium was held to honor Professor Fumio Tatsuoka's achievements over a 34 years period. It brought together over 200 experts in geotechnical engineering and soil mechanics from over 20 countries spanning over four continents. Many of the attending experts are leaders in their field of specialization. Such an international event is extremely rare and it reflects the impact Professor Tatsuoka have on the theory of soil mechanics as well as the practice of geotechnical engineering.

Professor Tatsuoka received his Doctor of Engineering from the University of Tokyo in 1972. Following his graduation, he worked for 5 years as a research engineer at the Public Works Research Institute, Ministry of Construction, in Chiba (now Tsukuba). Professor Tatsuoka established a productive, state of the art soils lab in the Institute of Industrial Science, University of Tokyo, between 1977 and 1991. While in the Institute, Professor Tatsuoka carried out numerous experimental lab and field studies with the assistance of dedicated graduate students. An example of his productivity at the Institute is the development of a unique system of geosynthetic reinforced soil walls. This is the only reinforced wall system used in critical applications by Japan Rail. This economical wall system exhibited an outstanding performance during the Kobe earthquake.

Professor Tatsuoka moved to the main campus of the University of Tokyo, serving as a Professor of Geotechnical Engineering (1997-2004). Once again, Professor Tatsuoka built a state of the art experimental facility thus enabling him to continue producing high-level research in an on-campus environment. In 2004 Professor Tatsuoka decided to move to Tokyo University of Science, a place which enabled him to continue doing research until the age of 65. This movement entailed a reconstruction of an advanced soil lab for the third time. Rebuilding three productive soils labs within about 25 years, at Professor Tatsuoka's level, indicates an unusual stamina and passion for geotechnical research.

Professor Tatsuoka's active research spans over several areas. It includes elemental tests, model tests, and full scale field tests and analyses. His advanced testing includes soft rock, clay, sand and gravel. These tests include studying the stress-strain-time characterization of geomaterials. His research is also concerned with the behavior and bearing capacity of shallow foundations. The applied aspect of his research includes ground improvement using cement-mixing as well as the effective use of geosynthetic reinforcement in walls and slopes considering severe seismic conditions.

Professor Tatsuoka's research gained domestic and international recognition as is indicated by the many awards he has received from organizations such as the International Geosynthetic Society, American Society of Testing and Materials, and the

Japanese Society of Geotechnical Engineering. For his important contributions to advancing the application of geosynthetics, he received the prestigious Mercer Lecture award. As a result of his productive research, Professor Tatsuoka was invited to deliver numerous keynote lectures in major international events.

Professor Tatsuoka has been extremely busy in activities that can be broadly categorized as a service to the profession. He has served as the Secretary, Vice President, and President of the Japanese Geotechnical Society in 1988-1993, 2003-2005, and 2006-2008, respectively. He is the Vice President of the International Geosynthetics Society (2002-2006) and was the Vice President of the Asian Region of the International Society of Soil Mechanics and Foundation Engineering (2001-2005). Professor Tatsuoka served as Editor in Chief of *Soils and Foundations* (1995-1999). Currently he is on the editorial board of several archived journals such as the *Geotechnical Testing Journal, Geotextile and Geomembranes, Geosynthetics International, Ground Improvement, Italian Geotechnical Journal*, and *Mechanics of Cohesive-Frictional Materials*.

Professor Tatsuoka has advised over 30 PhD and 25 MS students. Such numbers have already impacted the profession as many of his students hold professional key positions in industry and governmental agencies as well as serve as academics in four continents. An energetic advisor, rigorous research and highly-motivated graduate students have produced close to 400 technical papers in English authored and coauthored by Professor Tatsuoka. These publications appear in *Soils and Foundations*, *Geotechnical Testing Journal*, *Geotechnique*, Journal *of Geotechnical and Geoenvironmental Engineering*, *Ground Improvement Journal*, *Geosynthetic International*, and various conferences.

Generally, an engineer will accept a reasonably conservative analysis in design, even if the science is undermined. Conversely, a scientist would have a hard time accepting a solution where science is vaguely or incorrectly used. Professor Tatsuoka is a rare combination of excellent engineer and scientist, attempting to bridge sometimes conflicting attitudes. His record shows an impact in both basic and applied research. He is sufficiently flexible to produce needed practical solutions, proven as safe, for geotechnical structures although the theory is not yet fully understood. At the same time he is fascinated with soil behavior at the very basic level where there are no immediate applications for such research. Perhaps Professor Tatsuoka is propelled by the desire to simultaneously satisfy challenges in both engineering and science. Such perspective will surely continue in keeping him busy in a productive manner without an end in sight. Professor Tatsuoka is truly an extraordinary researcher at a global scale.

PHOTOS
Banquet at ***Palazzo Brancaccio***

[from left]
- Ora Leshchinsky
- Hoe Ling
- Herve Di Benedetto
- Fumio Tatsuoka
- Yoko Tatsuoka
- Simonetta Cola
- Laura Tonni
- Dov Leshchinsky

- Viviana Yumbaca
- Logan Brant
- Jieh-Jiuh Wang
- Jui-Pin Wang
- Taro Uchimura
- Tomokazu Ise
- Min-hao Wu

- Filippo Santucci de Magistris
- Giuseppe Modoni
- Alessandro Flora
- Stefania Lirer
- Paola Caporaletti
- Daniela Boldini
- Fabiana Maccarini
- Efisio Erbi
- Enzo Fontanella

- Shunichi Sawada
- Yukihiro Kohata
- Keiko Sawada
- Yuri Yasuda
- Susumu Yasuda
- Noriyuki Horii
- Rieko Horii
- Kimio Umetsu
- Nohara Chieko
- Michie Torimitsu

- Saomoto Hidetaka
- Tadao Enomoto
- Tadatsugu Tanaka
- You-seoung Kim
- Yoshiyuki Mohri
- Lin Wang
- Junichi Koseki
- Katsuhiro Uemoto

- Salvatore Miliziaono
- Marco D'Elia
- Takashi Matsushima
- Ivo Sterba
- Claudio Di Prisco
- Renato Lancellotta
- Alexander Puzrin

- Marte Gutierrez
- Arcesio Lizcano
- Daniel Pradel
- Radoslaw Michalowski
- Mounir Bouassida
- Neil Taylor
- Osamu Kusakabe
- Ms. Kusakabe

- Takashi Kiyota
- Yoshikazu Kobayashi
- Daiki Hirakawa
- Warat Kongkitkul
- Luca Lovati
- Giulia Sforzi
- Ilaria Giusti

- Alessandra Verona
- Maria Rossella Massimino
- Michele Maugeri
- Pedro Pinto
- Diego Lo Presti
- Renzo Pallara
- Giancarlo Verona
- Glenda Abate

- David Muir Wood
- Angelo Amorosi
- Augusto Desideri
- Giulia Viggani
- Sebastiano Rampello
- Ernesto Cascone
- Luigi Callisto
- Helen Muir Wood

- Samuel Kofi Ampadu
- Goran Arangelovski
- Lalana Kongsukprasert
- Jiro Kuwano
- Richard Jardine
- Jayne Jardine
- Satoru Shibuya
- Cecilia Ampadu

- Beatrice Baudet
- Cristiana Ferreira
- Nadia Benahmed
-Antonio Viana da Fonseca
- David Nash
- Erdin Ibraim
- Kenny Sorensen

- Satoshi Yamashita
- Tara Lohani
- Askar Zhusupbekov
- Valentina Zhusupbekova
- Damien Pham Van Bang
- Antoine Duttine
- Junko Kawaguchi
- Takayuki Kawaguchi

- Munaz Ahmed Noor
- Kabirul Islam
- Hoe Ling
- Mohammed Siddiquee
- Sarwar Yasin

Participants not photographed in Banquet:
Ezaoui Alan*, Gioacchino Altamura*, Jianliang Deng*, Eleonora Di Mario*, Sonia Fortuna*, Martin Lings*, Lorenzo Marini*, Kenichi Matsushima, Irene Mensi, Giuseppe Mortara*, Anna d'Onofrio, Angelina Parlato, Nunziante Squeglia*, Ramon Verdugo*
(*: photographed in front cover)

Special and Keynote Lectures

Prof. Fumio Tatsuoka

Prof. Diego C.F. Lo Presti

Prof. Herve Di Benedetto

Prof. Richard J. Jardine

Prof. Satoru Shibuya

Prof. Dov Leshchinsky

Soil Stress-Strain Behavior: Measurement, Modeling and Analysis
Geotechnical Symposium in Roma, March 16 & 17, 2006

INELASTIC DEFORMATION CHARACTERISTICS OF GEOMATERIAL

Fumio Tatsuoka
Department of Civil Engineering
Tokyo University of Science, Noda City, Chiba, Japan
E-mail: tatsuoka@rs.noda.tus.ac.jp

ABSTRACT

The inelastic strain characteristics of geomaterial are analysed in the framework of a non-linear three-component model while based on a number of laboratory stress-strain test results. The followings are shown. Inelastic strain increments develop by plastic yielding that is controlled by viscous effect and inviscid cyclic loading effect. Inelastic strain increments that develop by these different factors cannot be linearly summed up. The concept of double yielding consisting of shear and volumetric yielding mechanisms is relevant to describe the plastic yielding characteristics of geomaterial. Shear yielding is dominant with dense granular materials while volumetric yielding with soft clay. Three basic viscosity types, *Isotach*, *TESRA* and *Positive & Negative*, have been observed with different geomaterial types subjected to shearing. The viscosity type is controlled by geomaterial type in terms of grading characteristics, particle shape and particle crushability. Inviscid cyclic loading effect is analysed in relation to plastic yielding and viscous effect. The ageing effect on the inviscid shear yielding characteristics and its interactions with the viscous effect are examined and modelled. Three different types of time effect (i.e., delayed dissipation of excess pore water pressure, viscous effect or delayed development of plastic strain, and ageing effect) are involved in a complicated way in soft clay consolidation. Related some fundamental issues are analysed in the framework of the three-component model in the case of Isotach viscosity.

1. INTRODUCTION

To predict the pre-failure deformation of ground and embankment as well as displacement of structure, accurate evaluation of both elastic and inelastic strains of geomaterial (i.e., soil and rock) is essential. This has been one of the classical topics of geotechnical research. A number of long-lasting studies on the elasto-plastic stress-strain behaviour were crystallised into several constitutive models, including the Cam Clay model (Schofield & Wroth, 1968) and a series of their modified versions (e.g., Muir-Wood, 1990). A great amount of studies on the small strain deformation characteristics under dynamic loading conditions was performed in the field of soil dynamics related to geotechnical seismic design and ground vibration problems. More recently, research on the elastic deformation characteristics has revived in the study on the stress-strain properties under static monotonic as well as cyclic loading conditions. A number of SOA papers were published on this subject for the last two decades, including those by the

Hoe I. Ling et al. (eds.), Soil Stress-Strain Behavior: Measurement, Modeling and Analysis, 1–108.

author and his colleagues (Tatsuoka & Shibuya, 1991; Tatsuoka & Kohata, 1995; Tatsuoka et al., 1995, 1999a&b). It is known that, only when based on very small strains measured accurately, the elastic deformation property of a given soil mass evaluated by static tests becomes consistent with the one evaluated by dynamic tests, including the wave propagation measurement, performed under otherwise the same conditions. A good agreement is obtained particularly when the static tests are performed at a relatively high strain rate after some long duration of drained sustained loading and when a comparison is made with finer soils. After having well understood the elastic deformation characteristics, it becomes possible to accurately evaluate the in-elastic deformation characteristics. Perhaps it is now the time to revisit the issue of inelastic deformation characteristics of geomaterial with the ultimate goal of being capable of predicting the stress-strain behaviour for given arbitrary loading histories (**Fig. 1.1**).

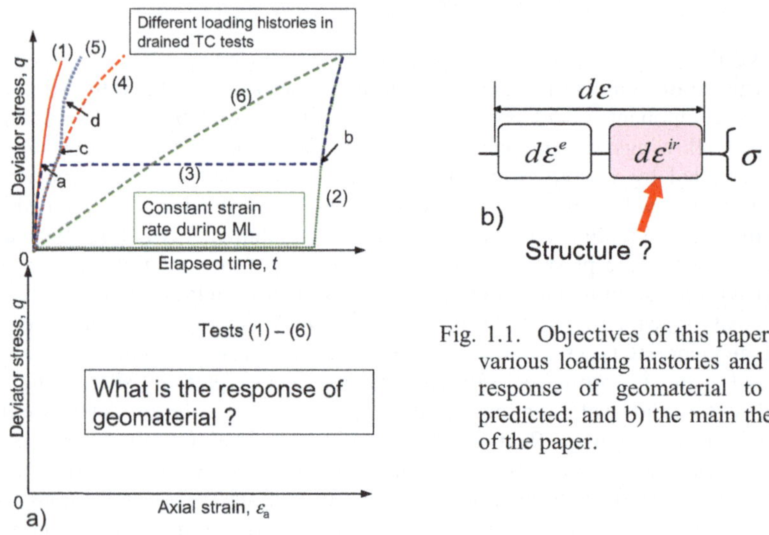

Fig. 1.1. Objectives of this paper; a) various loading histories and the response of geomaterial to be predicted; and b) the main theme of the paper.

Three major causes for the development of inelastic (or irreversible) strain increment, $d\varepsilon^{ir}$, are:

1) plastic (i.e., inviscid) yielding;
2) viscous deformation (i.e., delayed plastic yielding); and
3) inviscid cyclic loading effect.

These three factors are all affected by ageing effect. Then, the question is whether an given irreversible strain increment, $d\varepsilon^{ir}$, can be separated into three independent components presented by three components connected in series as illustrated in **Fig. 1.2**. The answer is no, as discussed in this paper.

In this paper, the following topics, among other important ones, are sketched:

1) some basic issues of plastic yielding characteristics of geomaterial;
2) the viscous property of geomaterial;
3) ageing effect;
4) inviscid cyclic loading effect; and

5) some fundamental issues in one-dimensional clay consolidation, as one of the most practical issues in which all the three issues, 1), 2) and 3), are important.

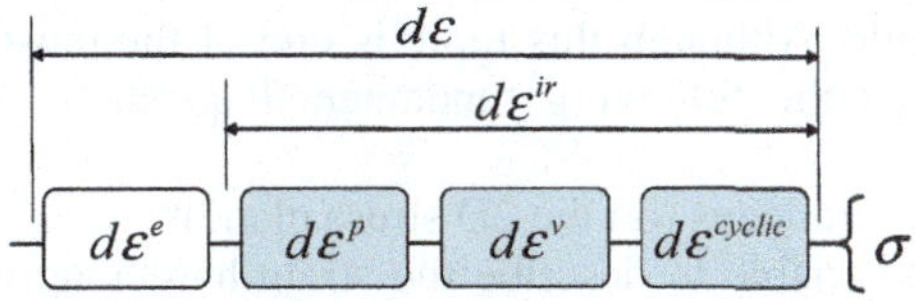

Fig. 1.2. Extended Maxwell model (a strain-additive model)

2. PLASTIC YIELDING CHARACTERISTICS

2.1 *Some basic issues*

In this chapter, the stress-strain behaviour of an elasto-plastic material, free from any viscous effect, that is also free from inviscid cyclic loading effect and ageing effect is discussed. **Fig. 2.1** illustrates the response of such an elasto-plastic material as described above when subjected to different loading histories in drained triaxial compression (TC), for example. The material exhibits a unique stress-strain curve in all of the tests 1 – 6, without showing any creep deformation during sustained loading and any effect of strain rate during monotonic loading (ML).

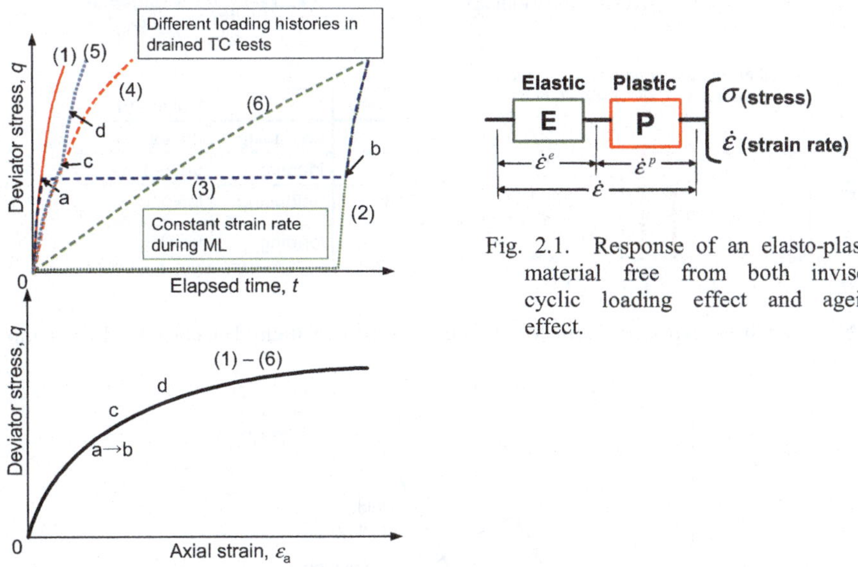

Fig. 2.1. Response of an elasto-plastic material free from both inviscid cyclic loading effect and ageing effect.

The development of in-elastic (or irreversible) strain increments, which are plastic strain increments in the present case, is associated with irreversible changes in the fabrics of geomaterial (i.e., plastic yielding). The plastic yielding characteristics for different loading stress paths can be conveniently described by yield surfaces in the three-dimensional stress space and yield loci on the two-dimensional stress plane that consecutively develop with yielding. They are one of the major constitutes of the classical elasto-plastic constitutive models with the other two being the hardening function and the flow rule. Although this topic is one of the most classical ones in Soil Mechanics, it seems that the following fundamental questions are still relevant even today:

1. What is the shape of yield locus (on the 2D stress plane)?
2. What is the relevant variable to describe the strain-hardening associated with plastic yielding?
3. How are the effects of recent stress path on the shape of yield locus?
4. How are the loading rate effects (i.e., viscous effect and ageing affect) on the shape of yield locus?

These topics are discussed below.

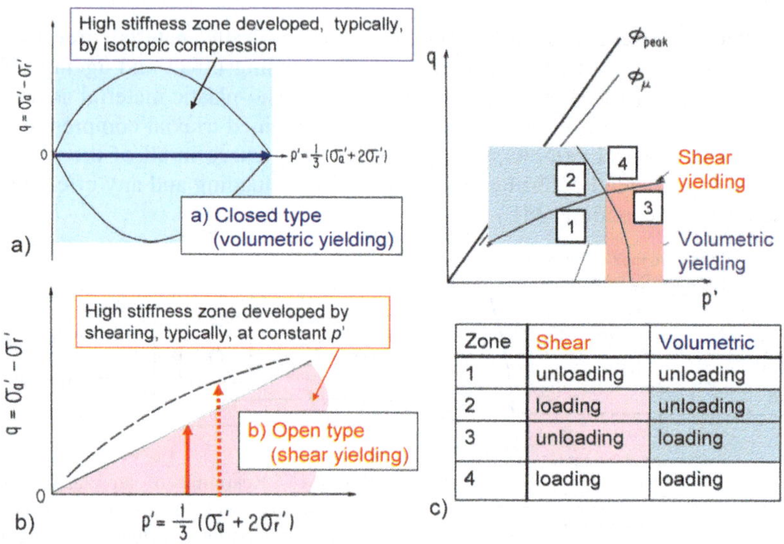

Zone	Shear	Volumetric
1	unloading	unloading
2	loading	unloading
3	unloading	loading
4	loading	loading

Fig. 2.2. Two basic types of plastic yield locus and their relation (Tatsuoka & Molenkamp, 1983).

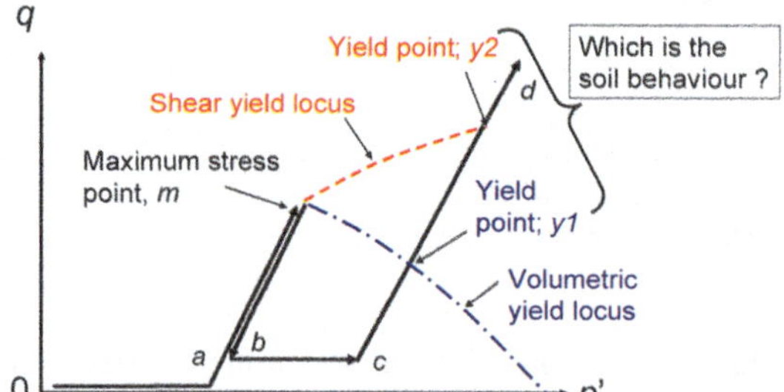

Fig. 2.3 Typical stress paths to examine the plastic yielding property of sand, first employed by Poorooshasb et al. (1967) & Poorooshasb (1971).

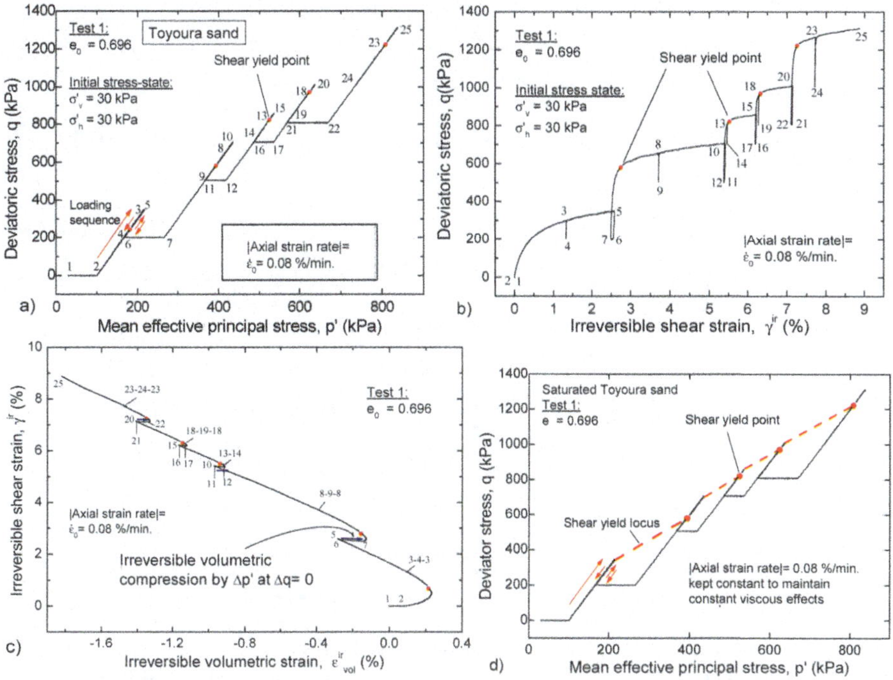

Fig. 2.4. Drained TC test on Toyoura sand to evaluate shear yield locus (Nawir et al., 2003b); a) stress paths; b) q - γ^{ir} relation; c) irreversible strain path; and d) observed shear yield locus segments.

2.2 *Shape of yield surface (or yield locus)*

Fig. 2.2a shows schematically the closed-form yield locus on the q - p' plane in the case of triaxial test. This type of yield surface has been employed most widely in constitutive modelling of geomaterial since the Cam Clay model was proposed (Schofield & Wroth, 1968). This type of yield locus is particularly relevant to highly compressive soil, such as soft clay. On the other hand, a different type of yield locus that is open in the p' axis direction (**Fig. 2.2b**) has been proposed to explain the yield characteristics of low-compressive soil, such as dense granular material (e.g., Stroud, 1971; Poorooshasb et al., 1967; Poorooshasb, 1971; Tatsuoka & Ishihara, 1974; Tatsuoka, 1980; Tatsuoka & Molenkamp, 1983). For example, when loaded into zones 2 & 3 from point A in **Fig. 2.2c**, different responses are obtained by following these two types of yield locus. The relevance of these two types of yield locus can be examined by performing drained TC tests along such a stress path as $a \rightarrow m \rightarrow b \rightarrow c \rightarrow y1 \rightarrow y2 \rightarrow d$ shown in **Fig. 2.3**.

The stress paths similar to the one presented in Fig. 2.3 and the relationships between the deviator stress, $q = \sigma_v' - \sigma_h'$, and the inelastic (or irreversible) shear strain, $\gamma^{ir} = \varepsilon_v^{ir} - \varepsilon_h^{ir}$, from two typical drained triaxial compression (TC) tests on dense Toyoura sand are shown in **Figs. 2.4** and **2.5**. The specimens were rectangular prismatic (18 cm high x 11

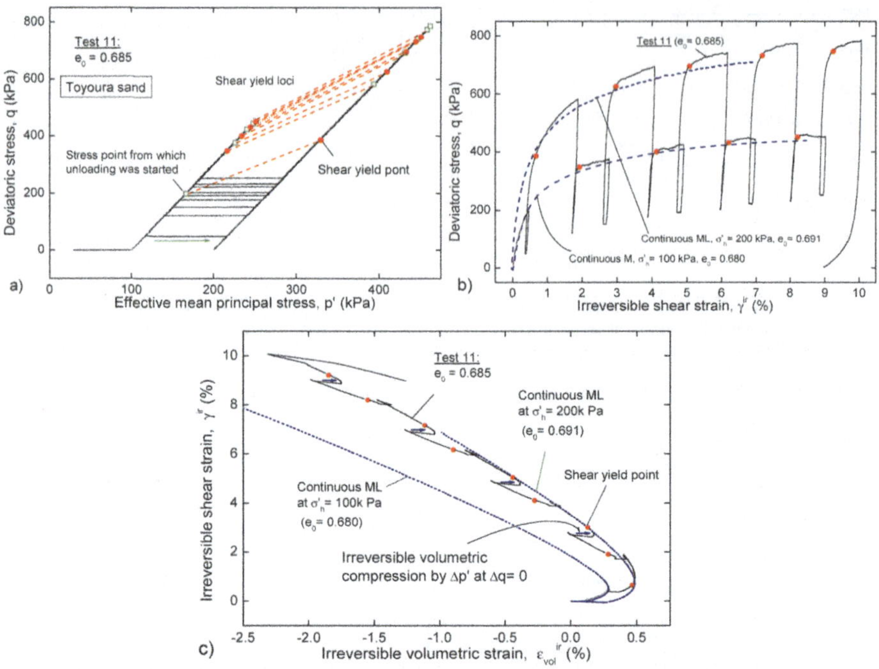

Fig. 2.5 A drained TC test on Toyoura sand to evaluate the shear yield locus (Nawir et al., 2003b); a) stress paths and observed shear yield locus segments; b) $q - \gamma^{ir}$ relation; and c) irreversible strain path.

cm x 11 cm) with well-lubricated top and bottom ends. Vertical and lateral LDTs were used to locally measure vertical and lateral strains free from, respectively, bedding errors at the top and bottom specimen ends when changing the effective vertical (axial) stress, σ'_v, and membrane penetration effects at the specimen lateral faces when changing the effective lateral stress, σ'_h. Irreversible strains were obtained based on the hypo-elastic model described in **Appendix A**. In the test described in Fig. 2.5, two stress paths at two different confining pressures were traced repeatedly while increasing the maximum value of q at the respective confining pressure. In Figs. 2.5b, the q - γ^{ir} relations from two continuous ML tests performed at these two confining pressures are also depicted as a reference. In these tests described in Figs. 2.4 and 2.5, to maintain the viscous effect constant as much as possible in the course of primary loading, unloading, reloading and so on, the absolute value of axial strain rate was always kept constant. The yield points shown in Fig. 2.4b and other similar figures were defined as the points of maximum curvature in a full-log plot of q - γ^{ir} relation. **Fig. 2.6** summarises the shear yield locus segments obtained from these tests. In this figure, results from other similar tests are also included, in which the viscous effect was changed in the course of testing by using a strain rate during reloading ($c \rightarrow d$) different from the one during primary loading ($a \rightarrow m$) and performing drained sustained loading at the maximum stress point m (Fig. 2.3) (see Fig. 3.8 for example). It may readily be seen from Figs. 2.4, 2.5 and 2.6 that the open-type yield locus (Fig. 2.2b) is appropriate in this case. As this yield locus type was

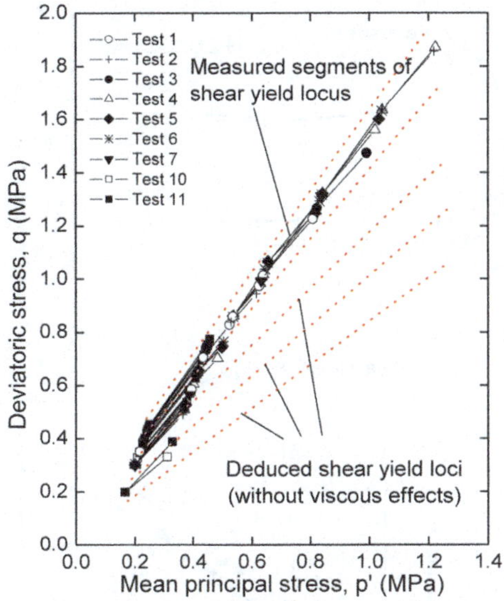

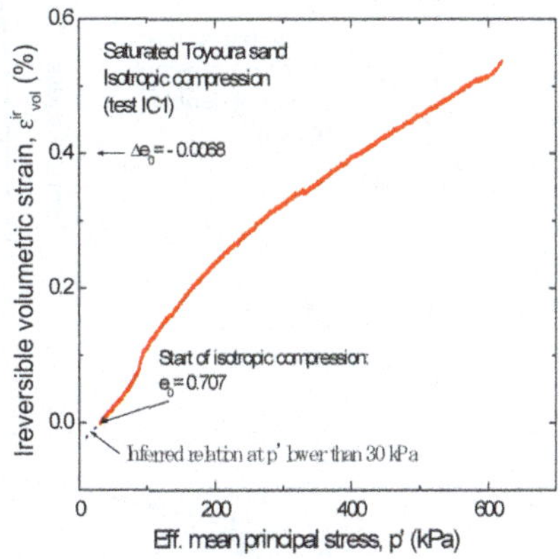

Fig. 2.6 (left) Summary of shear yield locus segments, some including various amounts of viscous effect, drained TC tests on Toyoura sand (Nawir et al., 2003b).
Fig. 2.7 (right) Compression of dense Toyoura sand (Tatsuoka et al., 2004a).

defined based on the shear deformation characteristics, it will herein be called the shear yield locus.

It is also to be noted from Figs. 2.4a and 2.5b that small but noticeable negative irreversible shear strain increments, $(\Delta\gamma^{ir})_{unloading}$, take place when the shear strain increment exceeds some limit during TC unloading (i.e., along stress path *m*→*b* in Fig. 2.3). In fact, the elastic-limit shear strain is very small, of the order of 0.001 %, with unbound granular material. The yielding point observed during the process of TC reloading becomes less clear with an increase in $(\Delta\gamma^{ir})_{unloading}$. These facts indicate that the stress-strain behaviour during unloading and reloading while the stress state is below the respective instantaneous shear yield point may not be perfectly elastic, showing the relevance of kinematic yielding models. So, the shear yielding characteristics that can be described by this type of yielding locus may be called "large-scale shear yielding".

2.3 *Double yielding*

Despite that the shear yielding described by the shear yield locus (Fig. 2.2b) is relevant to sand subjected to shear loading, as described above, it is also true that, even with dense sand, the volumetric yielding (Fig. 2.2a) also becomes important for some other loading histories. First of all, noticeable irreversible volumetric strains take place not associated with irreversible shear strains when subjected to an isotropic increase in *p*' as shown in **Fig. 2.7**. The volumetric yielding in such a case cannot be explained by the shear yield loci depicted in Figs. 2.2b, 2.4a, 2.5a and 2.6.

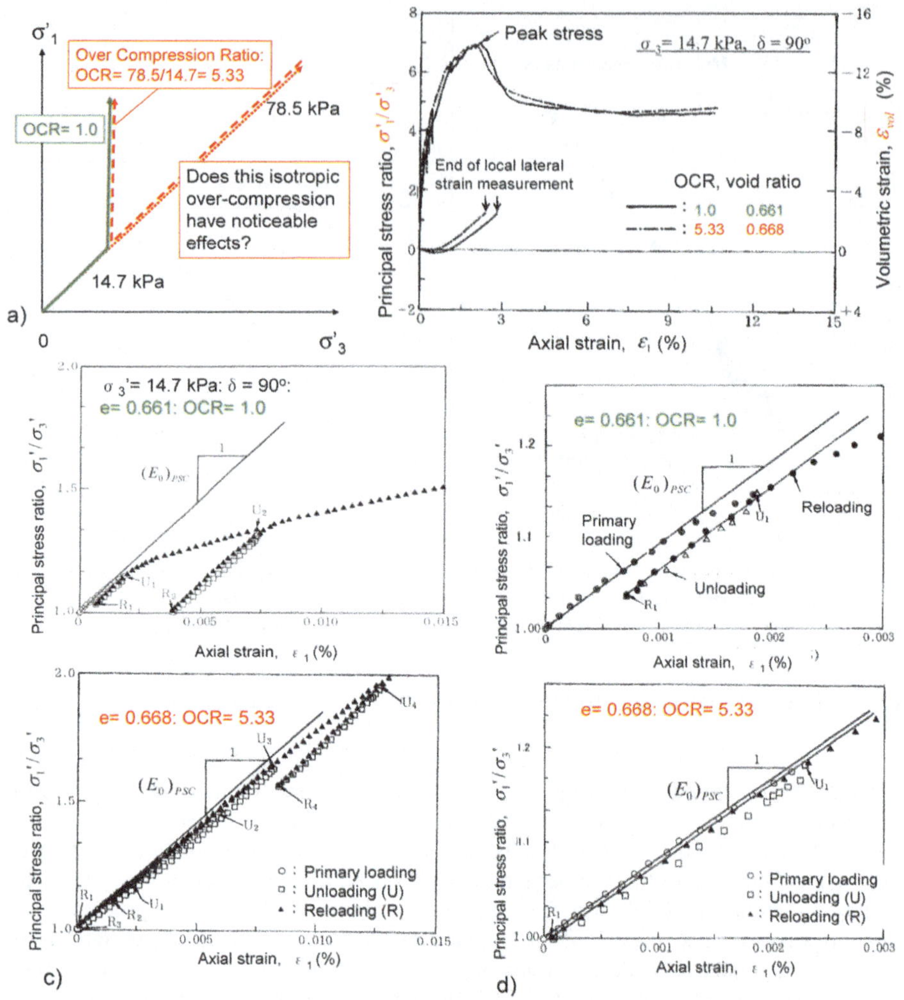

Fig. 2.8 Effects of isotropic OC on the stress-strain behaviour of dense Toyoura sand in drained PSC (Park &Tatsuoka, 1994): a) stress paths; b) overall stress-strain relations; c) behaviour at small strains; and d) behaviour at very small strains (to continue).

Moreover, by isotropic over-consolidation (OC), the stress-strain behaviour of sand when subsequently subjected to shear loading at lower confining pressure becomes noticeably different from the one when sheared normally consolidated. For example, **Fig. 2.8a** shows the stress paths applied to two similar dense saturated specimens of Toyoura sand. The irreversible volumetric strain that took place by isotropic over-compression from 14.7 kPa to 78.5 kPa was small, of the order of 0.01 %. For this reason, as seen from **Fig. 2.8b**, the effect of the OC loading history on the overall stress - strain behaviour in drained PSC, in particular after the onset of dilative behaviour, is negligible. On the other hand, noticeable effects can be seen on the stress - strain behaviour at small strains equivalent to the one that took place during the isotropic OC history: i.e., at this small strain level, the sand became less contractive (Fig. 2.8b) and noticeably stiffer (**Figs. 2.8c** and **d**), while the stress-strain relation became noticeably more linear (**Figs. 2.8e** and **f**).

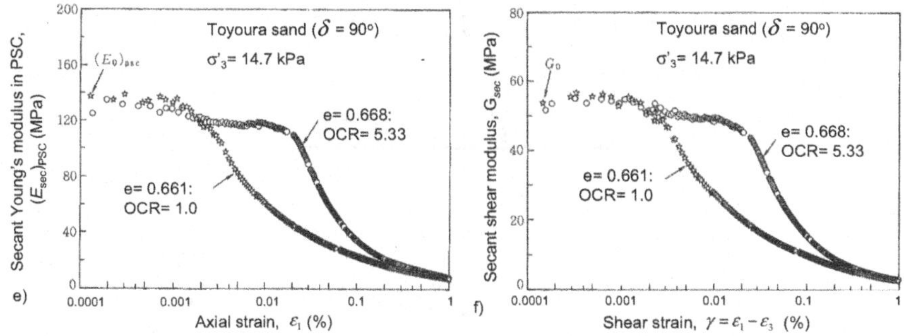

Fig. 2.8 (continued) e) Stiffness – logε_1 relations; and f) shear modulus – logγ relations.

Isotropic OC history does not increase the initial elastic modulus of sand in this case (Fig. 2.8d), but it develops the volumetric yield locus. Tatsuoka (1973) and Tatsuoka and Molenkamp (1983) reported that, in drained TC tests on relative loose sand, pre-isotropic compression history has significant effects on the stress-strain behaviour during the subsequent TC at lower pressure, developing volumetric yield loci (Fig. 2.2a) in the similar way as with soft clay. That is, as shown in **Fig. 2.9a**, isotropic OC histories A'-C-A' and A'-D'-A' were applied to two specimens of loose Fuji river sand. The stress-strain behaviour in drained TC when OCR= 2 and 3 are shown in **Fig. 2.9b**. The stress ratio (q/p') - shear strain (γ) curves of these two OC specimens have been shifted along the shear strain axis so that that the q/p' - γ curves after q/p' becomes larger than a certain value fit the one of the NC specimen. Obvious shear yielding starts at points C and D along the respective q/p' - γ curve. Similar starts of yielding may be seen in the q/p' - volumetric strain (ε_{vol}) relations (n.b., ε_{vol} is defined zero at point A' before applying the OC history). Curves C-C' and D-D' depicted in Fig. 2.9a can be considered as segments of the volumetric yield loci that have been developed by the isotropic OC histories A'-C-A' and A'-D'-A'. The yield locus segments A-A' and B-B'-B" were obtained by other similar TC tests. To depict the yield locus segment B-B'-B", it was assumed that the stress - strain behaviour after the stress state has entered the dilative zone is not affected by isotropic OC history. It may be seen from Fig. 2.9a that the volumetric yield loci developed by isotropic OC histories are totally different from the shear yield loci (Fig. 2.2b). This conclusion was confirmed by Ishihara and Okada (1978) and more recently by Kiyota et al. (2005) (**Fig. 2.10**).

Based on the fact that pre-shearing loading histories develop open-type shear yield loci (Figs. 2.4, 2.5 and 2.6), while isotropic OC histories develop the closed-type volumetric yield loci (Figs. 2.9 and 2.10), Tatsuoka and Ishihara (1974) and Tatsuoka and

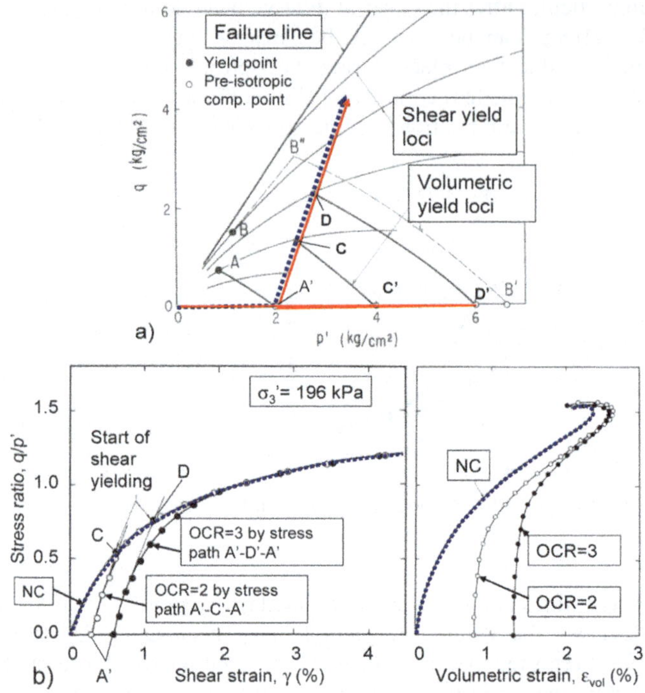

Fig. 2.9 Drained TC tests on loose Fuji river sand to evaluate the relationship between shear and volumetric yield loci (Tatsuoka & Molenkamp, 1983); a) stress paths; and b) stress – strain relations.

Molenkamp (1983) suggested a double yielding (or double hardening) concept to describe the yielding characteristics of sand subjected to general stress paths. According to this concept, the yielding of sand consists of shear yielding and compression (or volumetric) yielding (**Fig. 2.11**). In fact, several double yielding elasto-plastic models for sand were proposed (Lade & Duncan, 1975; Lade, 1976; Vermeer, 1978; Molenkamp, 1980; Vermeer & Neher, 1999; Schanz et al., 1999; among others).

In summary, both types of yielding (i.e., shear and volumetric) are relevant, either or both becoming important depending on imposed stress histories as well as geomaterial type. Generally, the shear yielding is dominant with dense sand and gravel, while the volumetric yielding is dominant with soft clay. Then, the issue of interactions between the two types of yielding becomes important. The interaction, if it takes place, is basically negative in the sense that the fabrics produced by shear yielding are altered by subsequent volumetric yielding and vice versa. For example, the shear yield point becomes less clear with an increase in the irreversible volumetric strain increment and an associated negative irreversible shear strain increment that take place during the intermediate isotropic loading (along stress path *b*→*c* in Fig. 2.3), applied before the restart of TC loading at a higher pressure level. In Fig. 2.5b, for example, the shear yield points seen during drained

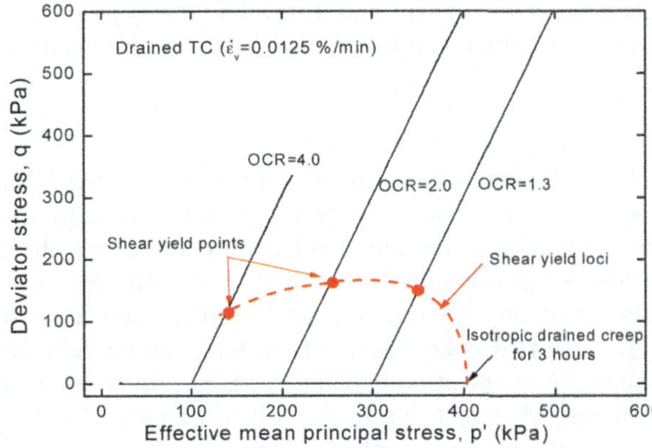

Fig. 2.10 Volumetric yield loci in drained TC observed with loose Silica No. 8 sand (Kiyota et al., 2005).

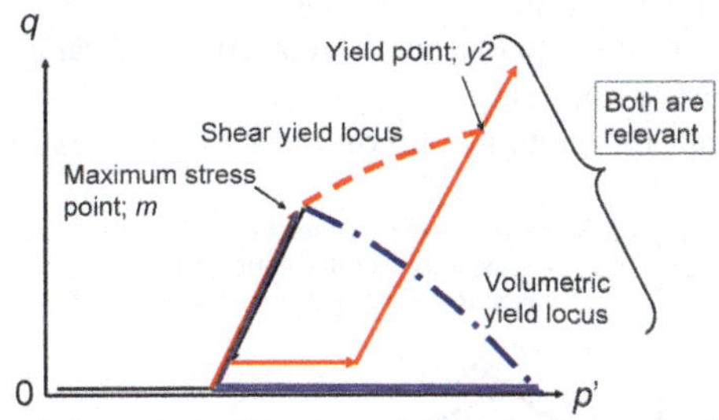

Fig. 2.11. Co-existence of shear and volumetric yielding loci (i.e., double yielding).

TC loading that is restarted at a higher confining pressure are less clear than those observed when TC reloading is restarted after some isotropic unloading. It is considered that the fabrics that are formed by shear loading up to a given maximum point (i.e., point *m* in Fig. 2.3) is altered to some extent by irreversible volumetric strain increments and associated negative shear strain increments that take place subsequently (Tatsuoka & Molenkamp, 1983). This complicated interaction between the two types of yielding could be one of the important research topics if the geomaterial model should become more realistic (or geomaterial-like).

2.4 Parameter to describe the inviscid yielding characteristics

It is convenient if the current yield locus and associated strain-hardening parameter that controls the process of yielding are independent of precedent stress paths. With the Cam Clay model, the void ratio is the state parameter in that the void ratio at a given stress state in the course of yielding is unique, independent of precedent stress paths. This assumption is based on the results from TC tests on clay reported by Rendulic (1936) and later confirmed by Henkel (1960). Therefore, it is called "the Rendulic assumption" (Gens, 1986). The use of inelastic volumetric strain as the strain-hardening parameter made the Cam Clay systematic and workable in the numerical analysis of many boundary-value problems. However, in the rigorous sense, the void ratio is not the state parameter, but it is stress path-dependent (e.g., Henkel & Sowa, 1963; Gens, 1986; Nakai,

1989; Yasin & Tatsuoka, 2000). Yasin and Tatsuoka (2003) proposed a more relevant strain-hardening parameter, which is not very different from irreversible shear strain, for shear yielding.

2.5 *Effects of recent stress path on yield locus*

According to the classical elasto-plastic theory, the current yielding locus is controlled basically by precedent stress paths, in particular recent ones. In this respect, some researchers proposed a yield locus for clay that has been anisotropically consolidated to a certain stress state that is considerably different from the one for clay which has been initially isotropically consolidated and then brought to the same stress state by shearing, as illustrated in **Fig. 2.12**. However, the trends of stress-strain behaviour that are seen during subsequent shearing of anisotropically and isotropically consolidated specimens of clay could become largely different due to the following two other factors that are not taken into account in the classical elasto-plastic theory;

a) different viscous effects for different strain rate histories, which is very important even when tracing the same stress paths; and
b) effects of precedent stress paths on the current void ratio (i.e., the non-Rendulic behaviour, as discussed above).

These issues, in particular, the first one, are discussed in the next chapter.

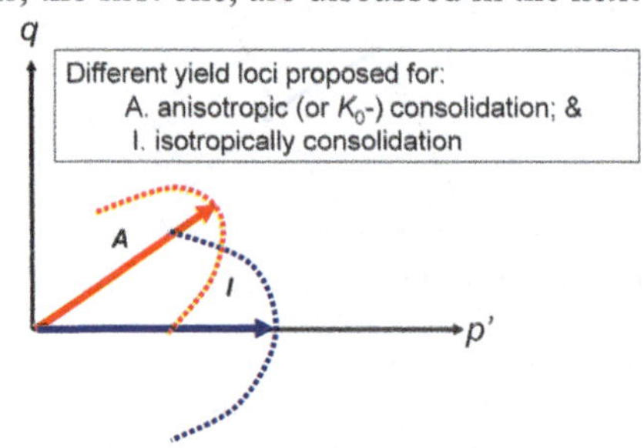

Fig. 2.12 Different yield loci for different recent stress paths assumed in some existing elasto-plastic models..

3. VISCOUS EFFECTS ON YIELDING CHARACTERISTICS

3.1 *A brief review*

The engineering importance of viscous effect on the strength and deformation of soft clay has been recognized via several time-dependent phenomena including "so-called secondary consolidation". The deformation of clay is always affected by the viscous property and it is necessary to take into account the viscous effect when predicting not only the secondary consolidation but also the primary consolidation in a given clay deposit (e.g., Tanaka, 2005a & b). On the other hand, the study on the viscous property of unbound granular material has been relatively limited (e.g., Murayama et al., 1984; Mejia et al., 1988; Yamamuro & Lade, 1993; Lade & Liu, 1998, 2001; Nakamura et al., 1999; Howie et al., 2001; Kuwano & Jardine, 2002), probably due to less engineering importance. However, there are a number of engineering problems for which proper

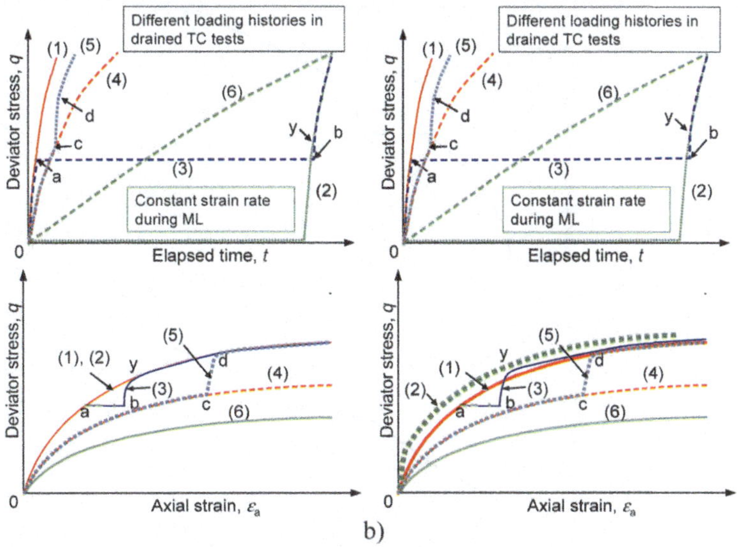

Fig. 3.1 Response of an elasto-viscoplastic material having Isotach viscosity without ageing effect; a) shear yielding only; and; b) double yielding.

understanding of the viscous property of unbound granular material is important (e.g., Tatsuoka et al., 2000, 2001; Jardine et al., 2005).

Fig. 3.1a illustrates the response against different loading histories of an elasto-viscoplastic material having the Isotach viscosity property (explained later in this paper) in the case where only the shear yielding is active without ageing effect. Unlike elasto-plastic materials, different stress-strain curves are observed for different loading histories due to the viscous property, controlled by instantaneous irreversible strain rate (because of Isotach viscosity). It is to be noted that the same stress-strain curves are observed in tests 1 and 2, because, in test 2, no ageing effect develops and no volumetric yielding takes place during drained sustained loading at $q=0$. **Fig. 3.1b** illustrates the response when the double yielding (Fig. 2.11) is relevant without ageing effect. In this case, different stress-strain curves, in particular at small strain levels, are observed between tests 1 and 2 due to effects of volumetric creep during drained sustained loading at $q=0$ in test 2. **Fig. 3.2** presents results from a pair of CD TC tests on loose sand in which drained sustained loading was performed for short and long durations (i.e., 3 and 180 minutes) under isotropic stress conditions before the start of drained TC loading, like tests 1 and 2 in Fig. 3.1b. It may be seen that the effects of volumetric creep under isotropic stress conditions on the stress-strain behaviour during the subsequent drained TC loading become more important with an increase in the duration of drained sustained loading at $q=0$.

3.2 *Viscous effects on yielding characteristics*

In test 3 illustrated in Figs. 3.1a and b, the stress-strain behaviour exhibits high stiffness for a relatively large stress range (b→y) immediately after ML is restarted at the original

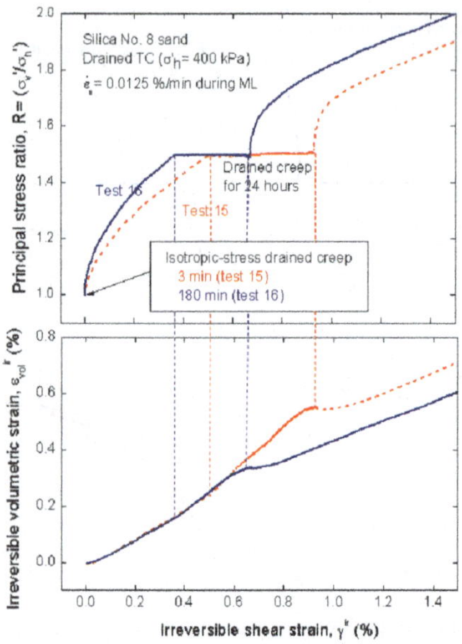

Fig. 3.2. Effects of drained creep at initial isotropic stress state on the subsequent stress-strain behaving in drained TC of loose sand (Kiyota et al., 2005; Kiyota &Tatsuoka, 2006).

constant strain rate following a drained creep stage for some duration a→b. This trend of behaviour can be seen typically from the test data presented in Fig. 3.2. This phenomenon should be called "apparent ageing effect" caused by the viscous property of sand, because any significant ageing effect is not involved in this test.

It is illustrated in **Fig. 3.3** that the shape and size of the current yield locus, within which the stress-strain relation exhibits high stiffness, depends on not only recent stress paths but also recent strain rate histories. Suppose that stress point *S* is reached by relatively fast ML along two different stress paths *1* and *2*. The shape of high stiffness stress zones that are observed when ML is restarted at a relatively high strain rate after drained creep for a relatively short duration at stress point *S* should be largely different, biased toward the directions of precedent stress paths (stress paths *1* and *2* in this case). With an increase in the duration of drained creep loading at stress point *S*, the high stiffness stress zone observed when ML is restarted at a relatively high strain rate becomes more similar for different precedent stress paths. This is because that the effect of creep strains taking place at stress point *S* on the stress-strain behaviour during the subsequent ML becomes more important than that of the precedent stress paths before having reached stress point *S* and associated strain histories. The data supporting the above are presented below.

Figs. 3.4a and **b** show the results from a series of undrained TC tests on isotropically reconsolidated specimens of reconstituted Fujinomori clay. The specimens were prepared by one-dimensionally consolidating well de-aired slurry having an initial water content that was twice of the liquid limit w_L. Consolidation normal stress equal to 70 kPa was applied for a duration three times longer than that at the end of primary consolidation in a

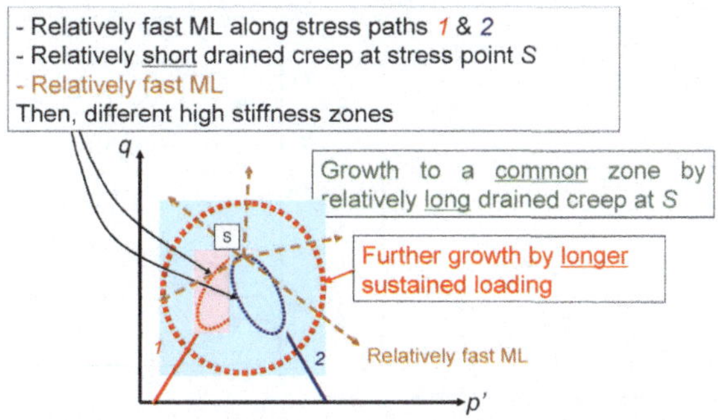

Fig. 3.3 Growth of yield locus by drained creep in the case free from ageing effect (modified from Tatsuoka et al., 1999a).

large oedometer with an inner diameter of 20 cm and an inner height of 60 cm. Specimens 7, 16 and 19 were subjected to drained sustained loading for two days at stress point *S* (at an effective stress ratio $K=\sigma'_h/\sigma'_v= 0.5$) during otherwise undrained ML at a constant axial strain rate equal to 0.05 %/min. It may be seen from Fig. 3.4a that a noticeable high stiffness stress zone developed by drained creep at stress point *S*. As seen from Fig. 3.4b, the initial Young's modulus when ML is restarted is very high, which is actually very close to the quasi-elastic Young's modulus measured by applying unload/reload cycles with a small axial strain amplitude of the order of 0.001 %. The size of this high stiffness zone increases with an increase in the duration of drained creep loading.

The behaviours of specimens 8 and 16 presented in Fig. 3.4 are reproduced in **Fig. 3.5**. In the other test (test 9) described in this figure, the specimen was anisotropically consolidated to stress point *S*, from which undrained TC at a constant strain rate was started without an intermission of drained creep at point *S*. Such a high stiffness stress zone as observed with specimen 16 cannot be seen in test 9. Furthermore, in tests 14, 16 and 28 described in **Fig. 3.6**, the same stress point (*S*) was reached via three different stress paths. After drained creep for two days at stress point *S*, these three specimens exhibit a very similar high stiffness stress zone. A similar result for specimens consolidated anisotropically at a higher stress ratio is reported by Tatsuoka et al. (2000). It is readily seen from the test results presented above that the stress-strain behaviour, in particular the initial one at small strains, is strongly controlled by the immediately precedent drained creep deformation history, even more than recent stress paths. Different behaviours seen when approaching the peak stress state between specimens 8 and 9 in Fig. 3.5 and among specimens 14, 16 and 28 in Fig. 3.6 are due to the dependency of void ratio on consolidation stress path (i.e., a higher void ratio when consolidated more anisotropically; i.e., the non-Rendulic behaviour; Tatsuoka et al., 2000).

The yield locus defined in terms of instantaneous effective stresses can be affected by arbitrary viscous effects, which are different for different strain histories. In fact, despite that it is subtle, some variance may be seen among the experimentally obtained segmental shear yield loci presented in Fig. 2.6. This is not an experimental scatter, but it is due

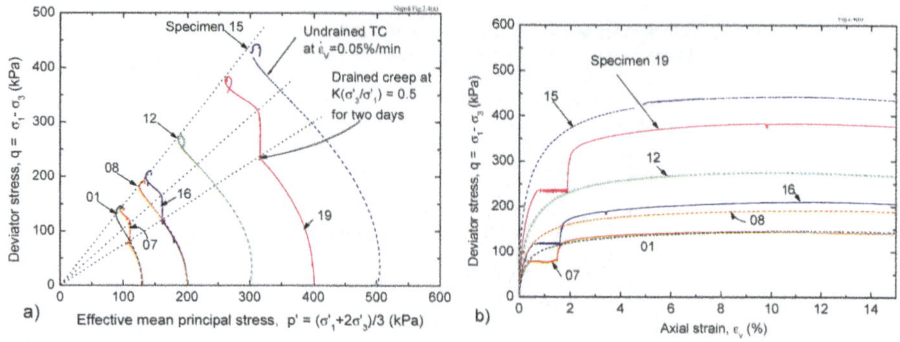

Fig. 3.4 Effects of drained creep on the behaviour during subsequent undrained TC on reconstituted Fujinomori clay (Momoya, 1998; Tatsuoka et al., 2000); a) effective stress paths; and b) q - ε_v relations.

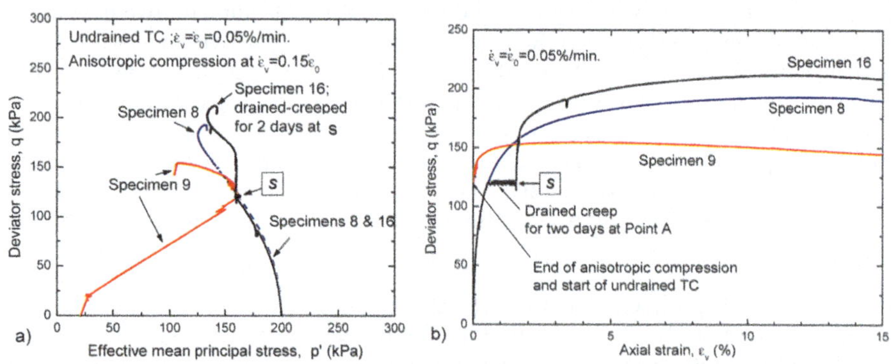

Fig. 3.5 Effects of recent stress path and drained creep on the behaviour during subsequent undrained TC on reconstituted Fujinomori clay (Momoya, 1998; Tatsuoka et al., 2000); a) effective stress paths; and b) q - ε_v relations.

mainly to different effects of the viscous property of the tested sand among different loading histories. To describe this trend of rate-dependent yielding behaviour, a constitutive model incorporating the viscous effect, such as the non-linear three-component model described in **Fig. 3.7a** (Di Benedetto et al., 2002: Tatsuoka et al., 2002), becomes necessary. This model is the simplified version of the general non-linear three-component model, **Fig. 3.7b** (Di Benedetto & Hameury, 1991; Di Benedetto & Tatsuoka, 1997; Di Benedetto et al., 2004). The essence of the simplified non-linear three component model is given in **Appendix B.** It seems that the simplified version is sufficient in most cases with geomaterial. In the case of multi-dimensional stress space, the (effective) stress, σ, is decomposed into the inviscid and viscous components, σ^f and σ^v, as illustrated in **Fig. 3.7c.**

On the premise that objective inviscid shear yield loci exist for component P (Fig. 3.7a), independent of current irreversible strain rate and strain history, Nawir et al. (2003b) and Tatsuoka et al. (2004a) examined the viscous effect on the shape of shear yield locus

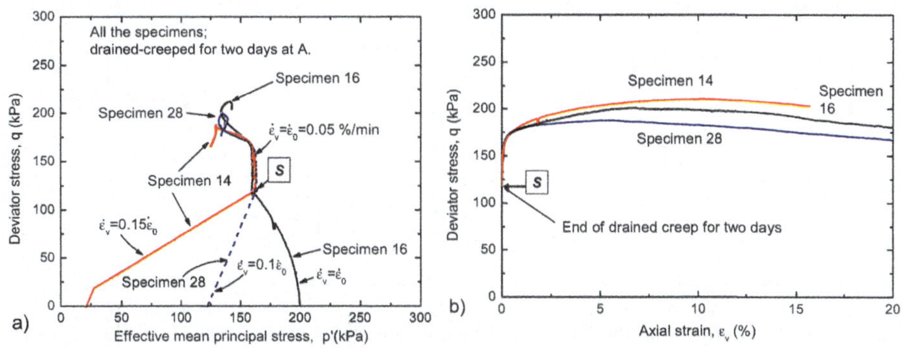

Fig. 3.6 Effects of recent stress path and drained creep on the behaviour during subsequent undrained TC behaviour on reconstituted Fujinomori clay (Momoya, 1998; Tatsuoka et al., 2000); a) effective stress paths; and b) q - ε_v relations.

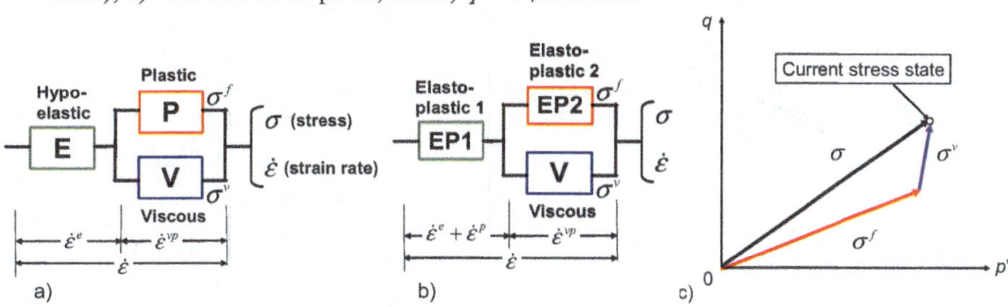

Fig. 3.7 Non-linear three-component rheology model; a) a simplified version (Di Benedetto et al., 2002; Tatsuoka et al., 2002); and b) the original version (Di Benedetto & Hameury, 1991); and c) vector summation of stress components.

defined in terms of measured effective stresses (i.e., σ in Fig. 3.7a). **Fig. 3.8a** shows the stress path employed in a typical drained TC test on Toyoura sand performed to this end. At two maximum stress points, drained creep loading was performed for five hours, followed by TC unloading, an increase in p' and then TC reloading at an increased confining pressure, σ_c', as c→d in **Fig. 3.9**, at a strain rate larger by a factor of ten than during the primary TC loading to reach the previous maximum stress point, as a→m in Fig. 3.9. In other cases, drained TC loading was restarted at an increased σ_c' at a strain rate that was the same as, or smaller than, the value during the primary TC loading (a→m). It may be seen from Figs. 3.8a and b that the shear yield point observed during TC reloading became higher by drained creep loading at the previous maximum stress point and by an increase in the strain rate during TC reloading when compared with the case without these two procedures. On the other hand, the shear yield point became lower by a decrease in the strain rate during TC reloading.

To quantify the viscous effect on the shape of shear yield locus, Nawir et al. (2003b) and Tatsuoka et al. (2004a) introduced the following equation referring to **Fig. 3.9:**

$$q = c \cdot p'^{\beta_s} \qquad (3.1)$$

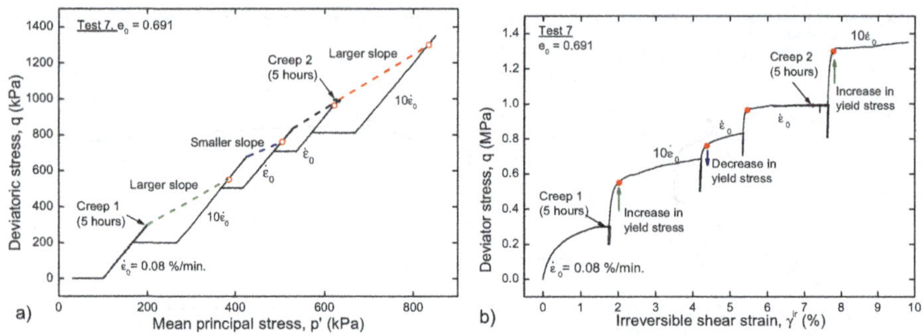

Fig. 3.8 Viscous effects on the shear yielding characteristics in drained TC on Toyoura sand (Nawir et al., 2003b); a) stress paths; and b) q - γ^{ir} relations.

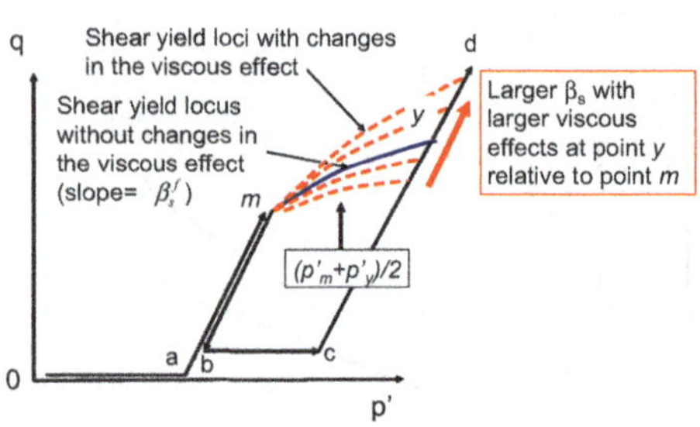

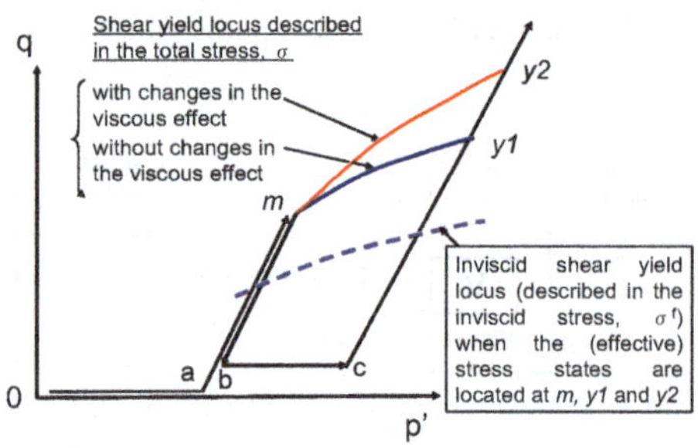

Fig. 3.9 (left) Illustration of viscous effects on shear yield locus segment
Fig. 3.10 (right) Inviscid shear yield locus and yield loci described in the (effective) stresses.

where β_s is the parameter obtained as:

$$\beta_s = \frac{\log(q_y / q_m)}{\log(p'_y / p'_m)} \qquad (3.2)$$

where the subscripts m and y mean, respectively, the maximum shear stress state (m) and the measured shear yield point (y). The test results showed that the value of β_s is not unique, but it depends on the loading history as follows:

a) The value of β_s becomes larger as the strain rate during TC reloading (at a greater confining pressure) is higher than the value during primary TC loading.
b) The value of β_s becomes larger as the duration of drained creep loading applied at the maximum stress state before TC unloading increases.
c) The effects of the two factors above on the β_s value are additive to each other. As a result, the value of β_s can become even larger than unity.
d) The value of β_s becomes smaller when the strain rate during TC reloading (at a greater confining pressure) is lower than the value during primary TC loading.

According to the three-component model, the observed shear yield locus is described in the total stress, σ, while the inviscid shear yield locus is in σ^f. Therefore, the difference between the shear yield loci described in terms of σ and σ^f at the same moment represents the instantaneous viscous effect, which is controlled by the instantaneous irreversible strain rate (and others). On the other hand, we can consider that the viscous effect was kept rather constant along the respective yield locus segment depicted in Figs. 2.4a and 2.5a. The yield locus segment in this case is depicted as $m - y1$ in **Fig. 3.10** and the value of β_s, denoted as β_s^f, is equal to around 0.84 in this case for Toyoura sand in TC. When the viscous effect when the total stress state is located at point *y2* is larger than the one when the total stress state is at point *m*, shear yield point *y2* is located above shear yield point *y1*. It is assumed that any yield locus segment along which the viscous effect is kept constant, as segment $m - y1$, is proportional to the inviscid shear yield locus described in terms of inviscid stresses, σ^f, as:

$$q^f = c \cdot (p'^f)^{\beta_s^f} \qquad (3.3)$$

$$\beta_s^f = \frac{\log(q_y^f / q_m^f)}{\log(p'^f_y / p'^f_m)} \qquad (3.4)$$

A set of broken curves depicted in Fig. 2.6 represent shear yield loci for different values of $(q/p_a)/(p'/p_a)^{\beta_s^f}$ according to Eq. 3.3 with β_s^f = 0.84. It may be seen that the formulated shear yield loci are consistent with the general trend of the measured yield locus segments. A deviation of the slope of respective measured segment of shear yield locus from the nearby shear yield locus expressing Eq. 3.3 with β_s^f = 0.84 is due to different viscous effects between the maximum stress point (*m*) and the shear yield stress point (*y*).

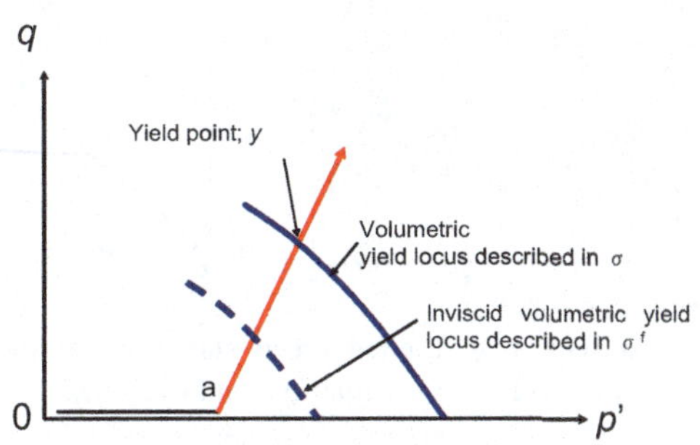

Fig. 3.11 Yield point determined by volumetric yielding, observed during shearing.

3.6 *Viscous effect in the framework of double yielding*

It may be seen from Fig. 3.2 that the shear yielding characteristics change by drained creep at the initial isotropic stress state. This trend of behaviour is illustrated in **Fig. 3.11**: i.e., while sand is subjected to drained creep at isotropic stress state *a*, the inviscid volumetric yield locus (described in σ^f) develops. Then, large-scale yielding starts at the stress point *y* where the volumetric yield locus (described in σ) intersects with the stress path (described in σ) during ML TC at a certain strain rate. In this sense, the effect of drained creep at the initial isotropic stress state has the same effects as isotropic-stress OC history (Fig. 2.10).

3.7 *Summary*

Both shear and volumetric yielding mechanisms are relevant to the yielding of geomaterial and their relative importance depends on the geomaterial type and given conditions. In any case, both shear and volumetric yield loci are affected by viscous effects. Kiyota et al. (2005) argued that undrained and drained creep deformations observed during sustained loading at a constant deviator stress and their relation can be properly understood only when referring to both shear and volumetric yielding mechanisms. In the following, only viscous effects on the shear yielding mechanism are discussed. The viscous effects on the volumetric yielding mechanism are not discussed more due to a lack of experimental data that are necessary for meaningful discussions.

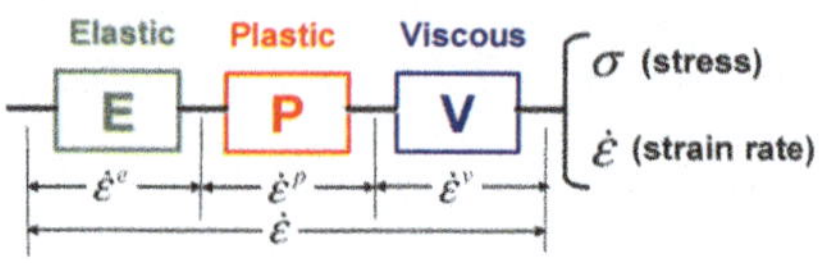

Fig. 4.1 Elasto-viscoplastic model described by an extended Maxwell model (a strain-additive model).

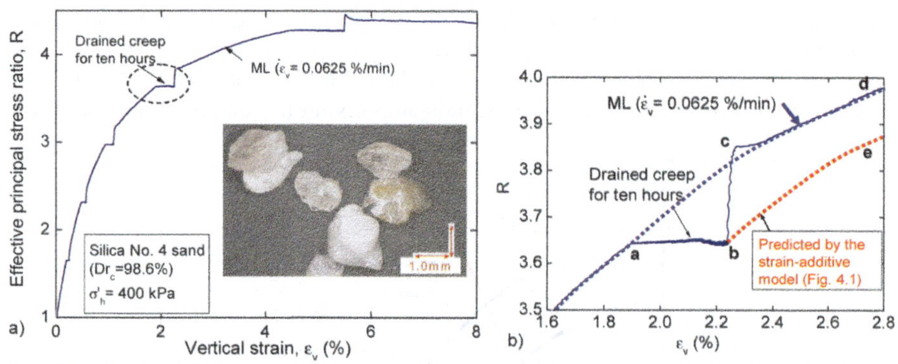

Fig. 4.2 Drained TC test with five drained creep stages on air-dried silica No. 4 sand (Enomoto et al., 2006); a) overall $R-\varepsilon_v$ relation; and b) a close-up.

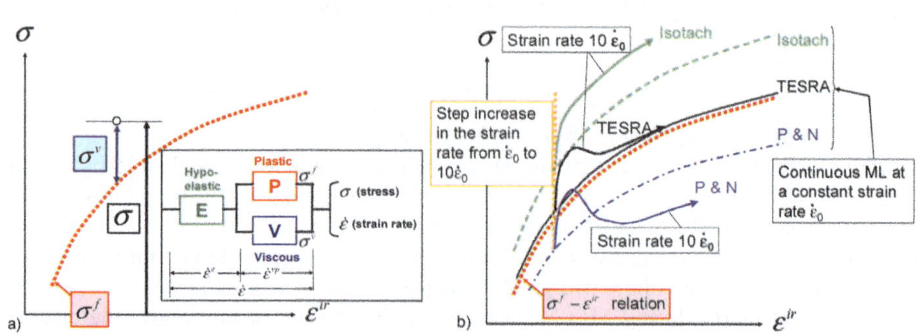

Fig. 4.3 Structure of the simplified non-linear three-component model: a) reference stress - strain relation; and b) stress - strain relations for three types of viscosity.

4. VISCOUS EFFECTS ON SHEAR YIELDING CHARACTERISTICS

4.1 *A basic issue in the visco-plastic formulation*

Fig. 4.1 illustrates a constitutive model in which components of elastic, plastic and viscous strain rates (or increments) are separated and then connected in series. Although several researchers employed this strain-additive model, this model cannot describe realistically the stress-strain behaviour of geomaterial when subjected to arbitrary loading histories (Tatsuoka et al., 2000, 2001). **Fig. 4.2a** presents typical data showing the above, which was obtained from a drained TC test on an air-dried specimen of silica No.4 sand. In this test, drained sustained loading lasting for ten hours was performed at five stages during otherwise ML at a constant strain rate. As seen from **Fig. 4.2b**, when ML is restarted at the original strain rate after the respective drained sustained loading stage, the stress-strain relation first exhibits a very high stiffness for a wide range of stress (*b* – *c*). After exhibiting clear large-scale shear yielding at point *c*, the stress - strain relation tends to rejoin the one that would have been obtained if ML had continued without an intermission of drained sustained loading (*a* – *d*). This result indicates that the 'plastic' and 'viscous' strains are not independent of each other. It is known that 'plastic' and 'viscous' strains should be represented by a single irreversible, or visco-plastic, strain, as expressed by the non-linear three-component model (i.e., the elasto-viscoplastic model; Fig. 3.7a). According to the three-component model, the major part of the 'plastic' strain increment that is to take place during stress range *b* – *c* has already taken place as the 'viscous' strain increment during drained sustained loading *a* – *b*. On the other hand, when based on the strain-additive model (Fig. 4.1), plastic shear yielding starts when the deviator stress, *q*, starts increasing upon the restart of ML from point *b* in the same way as when *q* increases from point *a* during the primary ML, which results in unrealistic behaviour *b* – *e* without showing a high stiffness zone *b* – *c* in Fig. 4.2b. Another simple example showing that the strain-additive model is not relevant is that creep strain does not stop developing as long as the stress is kept constant. We can conclude therefore that any type of strain-additive model (Fig. 4.1) is not able to properly describe the creep behaviour as well as post-creep behaviour, in particular, of geomaterial.

4.2 *Different viscosity types in shear yielding*

Fig. 4.3a illustrates the inviscid stress and irreversible strain relation (i.e., σ^f - ε^{ir} relation) of the simplified non-linear three-component model (Fig. 3.7a). The stress, σ, is obtained by adding the viscous stress component σ^v to σ^f at the same value of ε^{ir}. So far, the following three types of viscous property have been found by laboratory stress-strain tests on a wide variety of geomaterials:

1) *Isotach viscosity*: This is the most classical type of viscosity, which has been observed with rather coherent types of geomaterial, such as sedimentary soft rock, highly plastic clay and well-graded angular granular gravelly soil (only pre-peak). The specific type, called the new Isotach, was proposed by Tatsuoka et al. (1999c, 2001, 2002), Di Benedetto et al. (2002) and Tatsuoka et al. (2002) to describe this type of viscosity (**Appendix B**). As illustrated in **Figs. 4.4** and **4.5**, under the loading conditions (i.e., as far as $\dot{\varepsilon}^{ir}$ is kept positive), the current viscous stress component, σ^v, is a unique function of instantaneous values of ε^{ir} and its rate, $\dot{\varepsilon}^{ir}$, irrespective of precedent strain history and the sign of instantaneous $\dot{\sigma}$. Therefore, a unique σ - ε^{ir} relation is obtained by ML at a given constant strain rate $\dot{\varepsilon}$.

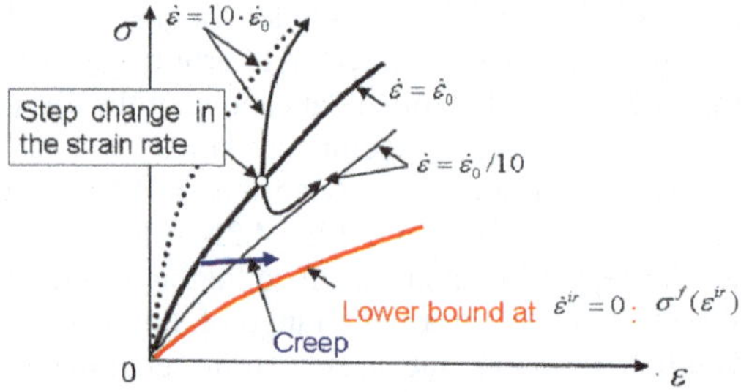

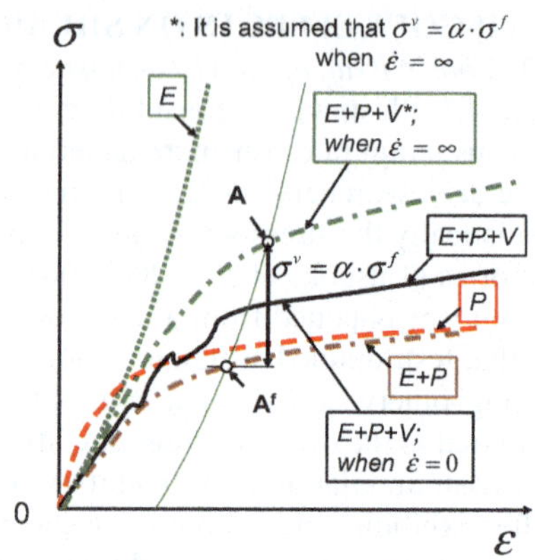

Fig. 4.4 (left) Stress-strain behaviour for different loading histories according to the Isotach viscosity (Di Benedetto et al., 2002; Tatsuoka et al., 2002).
Fig. 4.5 (right) Structure of the three-component model (Isotach viscosity).

2) *TESRA viscosity*: Based on the results from drained plane strain compression (PSC) tests of poorly-graded angular granular materials (i.e., Toyoura and Hostun sands), Di Benedetto et al. (2002) and Tatsuoka et al. (2002) proposed this new type viscosity, called the TESRA viscosity (**Appendix B**). TESRA stands for "temporary or transient effects of strain rate and strain acceleration", which means that, even under the loading conditions, the effects of $\dot{\varepsilon}^{ir}$ and its rate (i.e., irreversible strain acceleration) on the σ^v value become temporary and, therefore, the current σ^v value becomes a function of not only instantaneous values of ε^{ir} and $\dot{\varepsilon}^{ir}$ but also by recent history of strain rate. Then, the σ - ε^{ir} relation for ML at a given constant strain rate $\dot{\varepsilon}$ becomes non-unique. As explained below, these peculiar trends of stress-strain behaviour are due to such a feature that the viscous stress increment that develops by increments of ε^{ir} and $\dot{\varepsilon}^{ir}$ decays with an increase in ε^{ir} during subsequent loading.
3) *Positive and negative (P & N) viscosity*: This is the viscosity type that was found most recently (Enomoto et al., 2006; Kawabe et al., 2006; Duttine et al., 2006). The viscous stress component, σ^v, consists of two components as::

$$\sigma^v = \sigma^v_{TESRA}{}^* + \sigma^v_{NI} \qquad (4.1)$$

where $\sigma^v_{TESRA}{}^*$ is a kind of TESRA viscous stress component, of which the increment is positive upon a step increase in $\dot{\varepsilon}^{ir}$; and σ^v_{NI} is the negative Isotach component, which is always negative and so is its increment upon a step increase in $\dot{\varepsilon}^{ir}$.

Three broken curves illustrated in **Fig. 4.3b** show the different σ - ε^{ir} relations during continuous ML at a certain constant strain rate (denoted by $\dot{\varepsilon}_0$) when the viscosity is the respective three types. The σ - ε^{ir} curve when the viscosity is of Isotach type is located consistently above the reference curve (i.e., the σ^f - ε^{ir} curve). By assuming a proportionality of the viscous stress component to σ^f, as represented by Eq. B5

Table 4.1 Summary of viscosity type of geomaterial.

Viscosity type (θ) / Influencing factors	Isotach → Intermediate → TESRA → P & N (θ =1) (0< θ <1.0) (θ =0) (θ <0: could be less than –1.0)
Particle shape (in case of stiff particles)	More angular → More round
Grading characteristics	Better graded → More poorly graded
Particle size (if saturated)	Smaller (*clay*) → Larger (*sand/gravel*)
Particle crushability	More crushable ?? → Less crushable ??
Inter-particle bonding	Stronger → Weaker → Null (e.g., *rock/cement-mixed soil /bituminous-mix*) (*unbound granular materials*)
Strain level	Pre-peak → Post-peak (in particular, at residual state)
Inter-particle contact point	More stable (better bound, better interlocking & larger co-ordination numbers) → Less stable (less bound, less interlocking & smaller co-ordination numbers)
- *Deformation by cyclic loading*	*Smaller* → *Larger*
- *Creep deformation*	*Larger* → *Smaller*

(Appendix B), the σ value during for ML at a constant strain rate is always proportional to the σ^f value at the same ε^{ir}. The σ - ε^{ir} curve when the viscosity is of TESRA type is located only slightly above the σ^f - ε^{ir} curve and the difference disappears as the tangent modulus of the σ^f - ε^{ir} curve becomes zero. The σ - ε^{ir} curve when the viscosity is of positive and negative (P & N) type is located consistently below the σ^f - ε^{ir} curve. The difference between the values of σ and σ^f at the same ε^{ir} is the viscous stress component, σ^v. As the tangent modulus of the σ^f - ε^{ir} curve becomes zero, as $\dot{\varepsilon}^{ir}$ is nearly constant, the TESRA viscosity component σ^v_{TESRA} * of σ^v (Eq. 4.1) becomes zero and then σ^v becomes the same as the negative Isotach component, σ^v_{NI}. For this reason, the location of the σ - ε^{ir} curve becomes <u>lower</u> than the σ^f - ε^{ir} curve to a larger extent with an <u>increase</u> in the strain rate.

Now, the general Isotach viscosity is defined as:

$$\sigma^v_{G.I.} = \theta \cdot \sigma^f \cdot g_v(\dot{\varepsilon}^{ir}) = \theta \cdot (\sigma^v)_{isotach} \qquad (4.2)$$

where $g_v(\dot{\varepsilon}^{ir})$ is the viscosity function (Eq. B4); and θ is the parameter, which can have a value in a range from a certain negative value, which is less than - 1.0 as shown later, to 1.0. When θ is equal 1.0, $\sigma^v_{G.I.}$ becomes the ordinary Isotach viscous stress component, σ^v (Eq. B5). When θ is a negative constant, $\sigma^v_{G.I.}$ becomes negative (i.e., σ^v_{NI} of Eq. 4.1). We will also assume a proportionality of $\sigma^v_{G.I.}$ to σ^f, as the ordinary Isotach viscous stress component, σ^v (Eq. B5).

It is illustrated in Fig. 4.3b that, when the strain rate is increased stepwise during otherwise ML at a constant total strain rate, $\dot{\varepsilon}$, the viscous stress σ^v exhibits a sudden

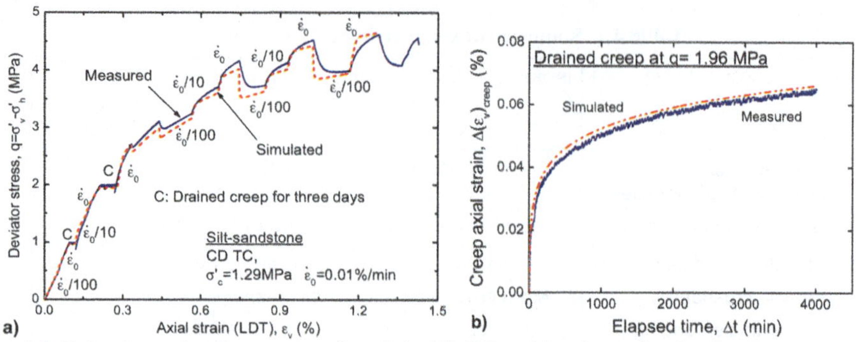

Fig. 4.6 Behaviour of sedimentary soft rock in CD TC and its simulation by the three-component model (Isotach viscosity) (Hayano et al., 2001): a) q - ε_v relation; and b) time history of axial strain during a drained creep stage.

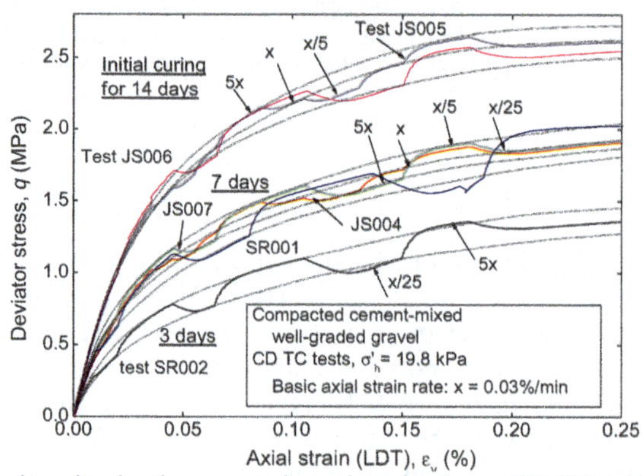

Fig. 4.7 Isotach viscosity in the pre-peak regime in three CD TC tests on specimens of compacted cement-mixed gravel cured for different durations: smooth thin solid curves denotes inferred stress - strain curves for different constant strain rates at the respective curing period (Kongsukprasert & Tatsuoka, 2005).

jump and so does the total stress σ. In this case, for a step increase in $\dot{\varepsilon}$ from $\dot{\varepsilon}_0$ to $10\dot{\varepsilon}_0$, the same amount of the positive stress jump is assumed with the three types of viscosity. Note that, when $\dot{\varepsilon}$ is increased stepwise, the change in the irreversible strain rate, $\dot{\varepsilon}^{ir}$, is not stepwise but it changes gradually, because the elastic strain rate, $\dot{\varepsilon}^{e}$, first changes associated with changes in the total stress, σ. In all the test results available to the author in which the P & N viscosity was observed (as shown later in this paper), the stress jump upon a step increase in the strain rate, which is equal to $\Delta\sigma^{v} = \Delta\sigma^{v}_{TESRA} * + \Delta\sigma^{v}_{NI}$ (Eq. 4.1), is positive. This fact means that, for the same strain rate increase, the value of $\Delta\sigma^{v}_{TESRA}$ * (positive) is larger than the absolute value of $\Delta\sigma^{v}_{NI}$ (negative).

The subsequent trend of the σ - ε^{ir} behaviour is different depending on the viscosity type. With the Isotach viscosity, the stress jump is persistent and the σ - ε^{ir} curve soon rejoins the one that is obtained by continuous ML at the constant strain rate after a step

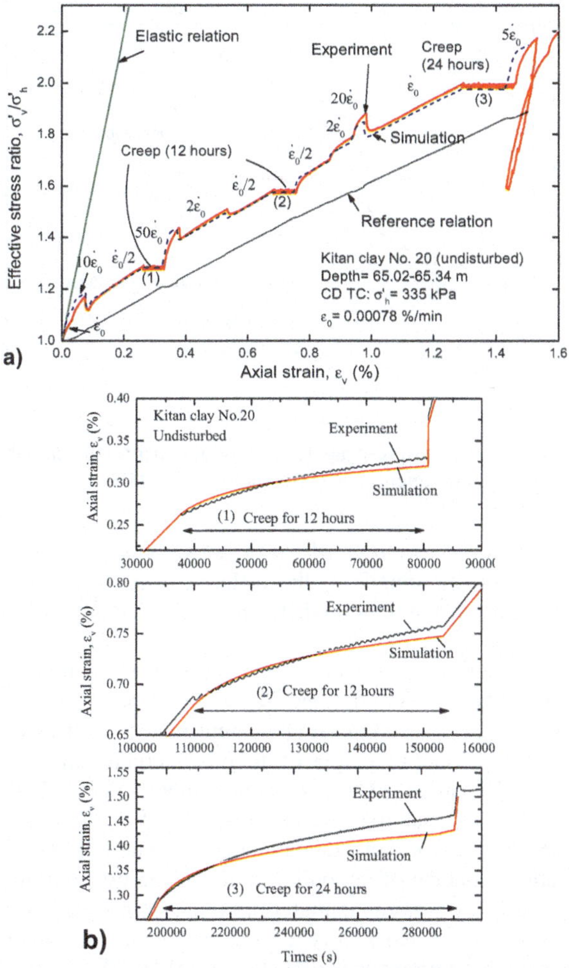

Fig. 4.8 Isotach viscosity in the pre-peak regime of undisturbed Pleistocene clay in CD TC and its simulation by the three-component model (Komoto et al., 2003); a) R- ε_v relation; and b) time histories of axial strain during drained creep stages.

increase. With the TESRA viscosity, the increase in the stress is temporary or transient and starts decaying as ε^{ir} increases with the σ - ε^{ir} curve approaching the σ^f - ε^{ir} curve. With the P & N viscosity, $\Delta\sigma^v_{TESRA}$* decays with an increase in ε^{ir} as the TESRA viscosity. Subsequently, the σ - ε^{ir} curve tends to rejoin the one that is obtained by continuous ML at the constant strain rate after a step increase, located below the σ - ε^{ir} curve that would be obtained if ML had continued at a constant strain rate without a step increase. As discussed below and as listed in **Table 4.1**, there is a distinct dependency of the viscosity type on particle shape, degree of uniformity, particle size (perhaps, linked to the effect of particle shape), inter-particle bonding and strain value, among other factors.

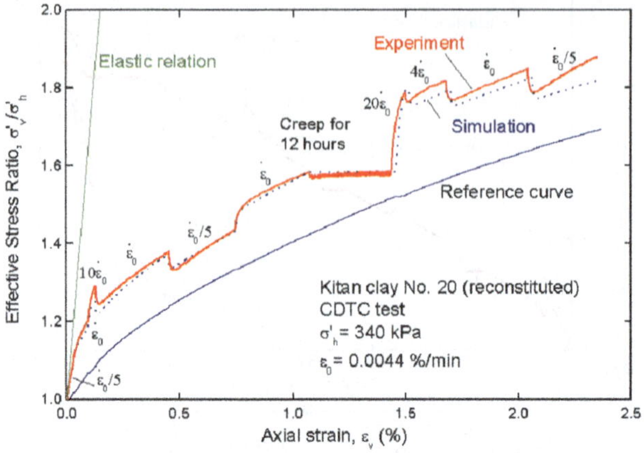

Fig. 4.9 Isotach viscosity in the pre-peak regime of reconstituted clay in CD TC and its simulation by the three-component model (Komoto et al., 2003).

4.3 *Different viscosity types observed in laboratory shear tests*
Some more discussions on the details of the different viscosity types are given below based on results from stress-strain tests on different types of geomaterials.

Isotach viscosity: **Figs. 4.6** through **4.9** show typical results showing the Isotach viscosity in the pre-peak regime obtained from drained TC tests on two types of bound geomaterial (i.e., sedimentary soft rock and compacted cement-mixed well-graded gravelly sand cured for different durations) and two types of unbound geomaterial (i.e., undisturbed and reconstituted samples of Pleistocene clay). The experimental data that are available to the author show a general trend that the viscous property of better bound materials is of Isotach type, in particular before the damage to the bounding by shearing becomes significant. The viscous property of unbound rather plastic clay having higher plasticity indexes and better interlocked granular materials (e.g., compacted well-graded angular gravelly soils) tends to be of Isotach type. The results of simulation based on the simplified non-linear three-component model (Fig. 3.7a) are also shown in Figs. 4.6, 4.8 and 4.9. It is important to note that, Fig. 4.8b, the time history of creep axial strain during the respective drained sustained loading stage is well simulated by using the parameters that were determined from the behaviour during continuous ML and those upon step changes in the strain rate.

Fig. 4.10 illustrates effects of recent loading histories on the size of high-stiffness zone in the case of Isotach viscosity. The development of such a high-stiffness zone as illustrated is due solely to the viscous effect (so nothing due to the ageing effect). The size of high stiffness zone increases with an increase in the duration of sustained loading from the one in test 1 toward the one in test 2 and also with an increase in the strain rate during ML reloading from the one in test 1 toward the one in test 3. For this reason, when loaded at a larger strain rate, older soil deposits tend to have a larger high-stiffness stress zone and the stress-strain behaviour tends to become more linear and reversible.

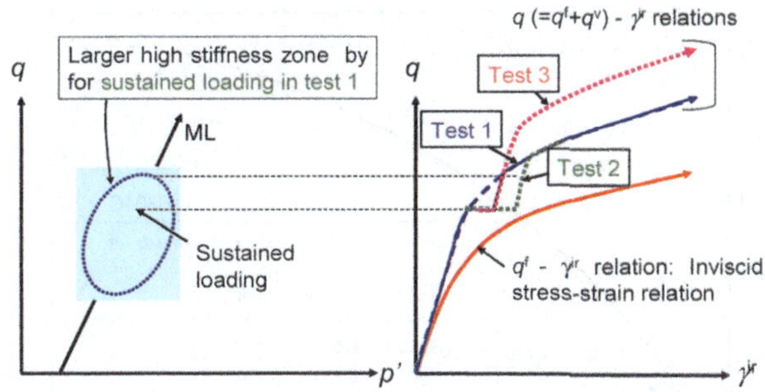

Fig. 4.10 Effects of recent loading histories and strain rate after the restart of ML on the size of high-stiffness zone (illustrated in the case of Isotach viscosity).

TESRA viscosity: With respect to the rate-dependency of the stress-strain behaviour of unbound granular materials not having significant particle crushability, basically the following two groups of test results can be found in the literature:

1) The stress-strain relation is rather independent of strain rate in ML drained TC tests.
2) Noticeable creep strain develops when subjected to drained sustained loading.

With respect to term 1), Yamamuro and Lade (1993) reported that, in the undrained TC tests on sand at elevated pressures, the stress-strain behaviours at different strain rates are noticeably different showing a trend of Isotach viscosity. This trend of behaviour is considered due to the effects of particle crushing, of which the amount increases with an increase in the elapsed time. In the following, the viscous property of unbound granular material without significant effects of particle crushing is discussed.

The trends of behaviour 1) and 2) described above apparently contradict each other, but both trends were also observed in recent experiments using the same apparatus and the same type of sand performed by the same persons. **Fig. 4.11** shows the relationships between the effective principal stress ratio, $R=\sigma'_v/\sigma'_h=\sigma'_1/\sigma'_3$, and the shear strain, $\gamma=\varepsilon_v-\varepsilon_h=\varepsilon_1-\varepsilon_3$ from drained PSC tests at $\sigma'_h=$ 392 kPa on saturated Hostun sand. The dense specimens were subjected to continuous ML at constant vertical strain rates, $\dot{\varepsilon}_v$, that were different by a factor of up to 500. Noticeable strain rate effects may be seen only in the initial stress-strain behaviour immediately after the start of ML following anisotropic compression (**Fig. 4.11b**). The effects of strain rate on the overall stress-strain behaviour are negligible (**Fig. 4.11a**). In another test, the specimen was subjected to drained sustained loading and stress relaxation during otherwise ML at a constant strain rate (**Fig. 4.12**). Despite that any significant and systematic strain rate-dependency cannot be seen in the continuous ML tests at constant but different strain rates (Fig. 4.11), the sand exhibits noticeable creep strain and stress relaxation in this test (Fig. 4.12).

The key to understand the apparently contradicting trends of rate effect seen from Figs. 4.11 and 4.12 is the fact that the viscous stress increment that has developed by a step change in the strain rate decays with an increase in the irreversible strain during the subsequent ML, as seen from **Fig. 4.13**. Di Benedetto et al. (2002) and Tatsuoka et al. (2002) showed that, referring to Eq. B7 (Appendix B), the current viscous stress at a

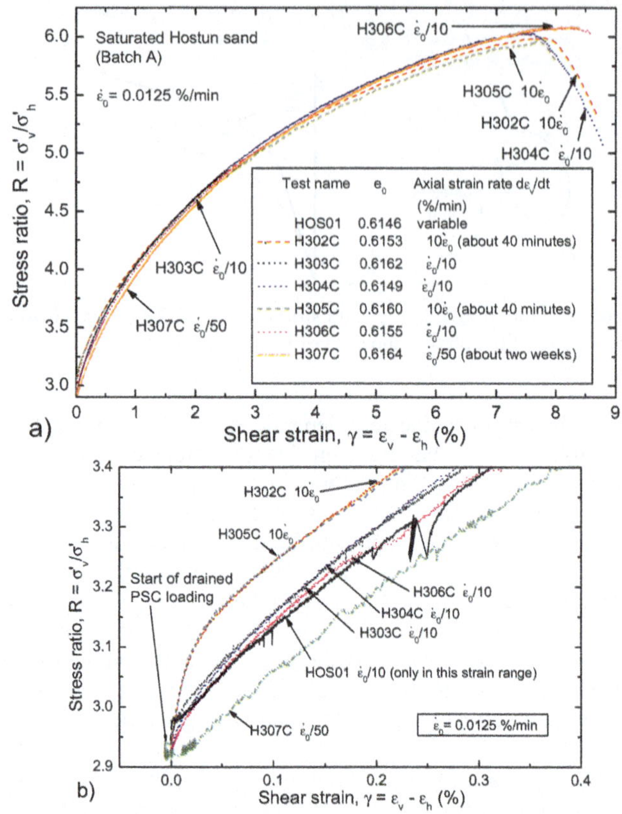

Fig. 4.11 Stress - strain behaviours during continuous ML at different strain rates of Hostun sand in CD PSC (σ'$_h$= 392 kPa) (Matsushita et al., 1999; Di Benedetto et al., 2002); a) overall R - γ relations; and b) initial stress – strain relations at small strains.

given stage σ^v can be obtained by integrating the viscous stress increment, $[d\sigma^v]_{(\tau,\varepsilon^{ir})} = [d\sigma^v]_{(\tau)} \cdot g_{decay}(\varepsilon^{ir} - \tau)$ with respect to the irreversible strain (not time), where $[d\sigma^v]_{(\tau)}$ is the viscous stress increment that developed by an irreversible strain increment, $d\varepsilon^{ir}$, and/or an increment of irreversible strain rate, $d\dot{\varepsilon}^{ir}$, at a respective moment when $\varepsilon^{ir} = \tau$ during a given loading history. $[d\sigma^v]_{(\tau)}$ has decayed with an increase in the irreversible strain from τ to the current value, ε^{ir}, by a factor of $g_{decay}(\varepsilon^{ir} - \tau)$, where $g_{decay}(\varepsilon^{ir} - \tau)$ is the decay function (Eq. B8). It is known that an exponential equation, $g_{decay}(\varepsilon^{ir} - \tau) = r_1^{\varepsilon^{ir} - \tau}$ (**Fig. 4.14**), is relevant, where r_1 is a positive constant less than unity. When r_1 is equal to 1.0, $g_{decay}(\varepsilon^{ir} - \tau) \equiv 1.0$ (i.e., no decay) and the viscosity type returns to the Isotach viscosity. As r_1 becomes smaller, the stress - strain relations during ML at different strain rates tend to collapse at a faster rate into a unique one (i.e., the reference relation, the σ^f - ε^{ir} relation). Then, as illustrated in **Fig. 4.15**, the current value of σ^v becomes dependent on not only instantaneous values of ε^{ir} and $\dot{\varepsilon}^{ir}$ but also recent strain history. Therefore, the value of σ^v at a given value of ε^{ir} could become the same for the different instantaneous values of $\dot{\varepsilon}^{ir}$. Creep deformation develops and stress relaxation

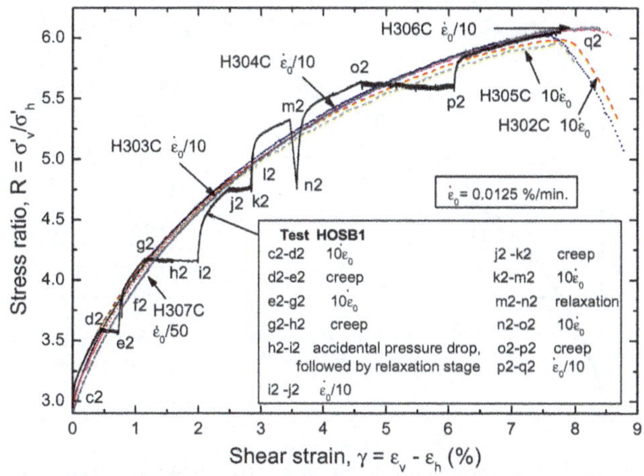

Fig. 4.12 Stress – strain behaviour in a test with creep and stress relaxation stages compared with those for ML at different strain rates of Hostun sand in CD PSC (σ'_h= 392 kPa) (Matsushita et al., 1999; Di Benedetto et al., 2002).

takes place as the process in which both $\dot{\varepsilon}^{ir}$ and σ^{v} decrease with time. That is, during a sustained loading stage at a given fixed value of σ, the creep strain continues increasing as far as the inviscid stress, σ^{f}, continues increasing satisfying the condition that $\dot{\sigma}$ (= 0) = $\dot{\sigma}^{f}$ (> 0) + $\dot{\sigma}^{v}$ (< 0) and $\dot{\varepsilon}$ (> 0)= $\dot{\varepsilon}^{e}$ (= 0) + $\dot{\varepsilon}^{ir}$ (> 0). On the other hand, during a stress relaxation stage at a given fixed value of ε, the total stress, $\sigma = \sigma^{f} + \sigma^{v}$, continues decreasing as far as the inviscid stress, σ^{f}, increases satisfying the condition that $\dot{\sigma}$ (< 0)= $\dot{\sigma}^{f}$ (> 0) + $\dot{\sigma}^{v}$ (< 0) and $\dot{\varepsilon}$ (= 0)= $\dot{\varepsilon}^{e}$ (< 0) + $\dot{\varepsilon}^{ir}$ (> 0). Note that, during sustained loading and stress relaxation, except for a certain initial stage, the viscous stress, σ^{v}, becomes negative while its absolute value continues increasing at a decreasing rate. That is, as illustrated in Fig. 4.15, during arbitrary loading history, the viscous stress σ^{v} could be either positive, zero or negative depending on recent strain history.

Fig. 4.16 compares the measured stress - strain relation of Hostun sand from a drained PSC test (σ'_h = 392 kPa) and the one simulated by the three-component model (TESRA viscosity). It may be seen that the decay property of the viscous stress is well-simulated. **Fig. 4.17** shows results from two drained PSC tests performed on saturated dense Toyoura sand specimens having very similar initial void ratios. It may be seen that the creep strain at the end of the respective sustained loading stage for 24 hours is considerably larger when the sustained loading starts during otherwise ML at a higher strain rate, despite that the respective sustained loading process starts from nearly the same stress and strain states in the two tests. Furthermore, upon the restart of ML at the original constant strain rate following a sustained loading stage, the stress-strain behaviour exhibits a high stiffness zone for a larger stress range in the test in which the total creep strain is larger. As seen from Fig. 4.17, these peculiar trends of behaviour can be explained by the TESRA-type model, incorporating the decay function (Eq. B8). It may also be seen that, in the simulation of sustained loading, the σ^{v} value becomes negative soon after the start of sustained loading.

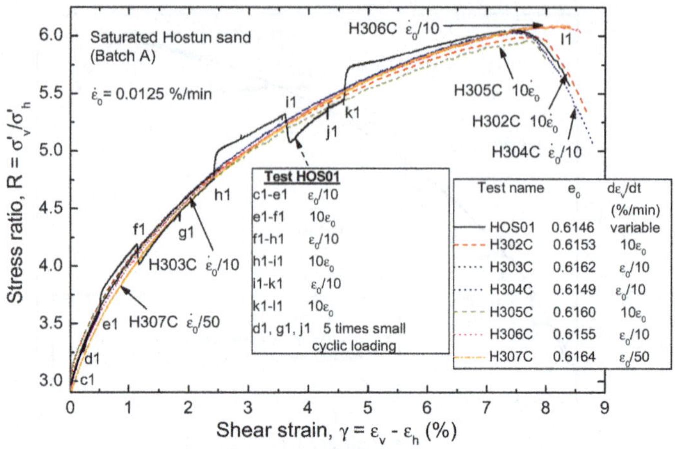

Fig. 4.13 Stress - strain behaviour in CD PSC tests (σ'_h= 392 kPa) with step changes in the strain rate compared with those from ML at different strain rates of Hostun sand (Matsushita et al., 1999; Di Benedetto et al., 2002).

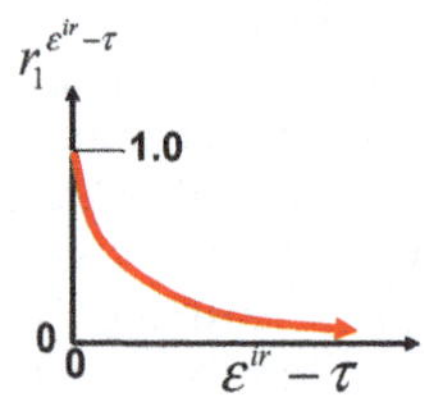

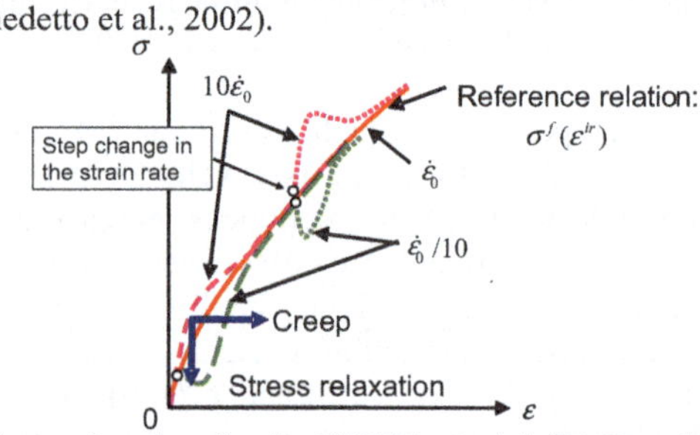

Fig. 4.14 (left) Exponential type decay function for the TESRA model (Di Benedetto et al., 2002; Tatsuoka et al., 2002).

Fig. 4.15 (right) Stress-strain behaviour according to the TESRA viscosity (Di Benedetto et al., 2002; Tatsuoka et al., 2002).

Positive and negative (P & N) viscosity: This viscosity type, which is most peculiar among all the viscosity types that have been observed, was found very recently by performing drained TC tests on three types of granular materials consisting of relatively round and rigid particles at the Tokyo University of Science. Referring to **Fig. 4.18**, Albany silica sand is a fine silica sand from Australia, corundum A is an artificial stiff round material (Al_2O_3) and Hime gravel is a natural fine gravel from a river bed in the Yamanashi Prefecture, Japan. The other types of sands that are also described in Fig. 4.18 are referred to in this paper. **Figs. 4.19** through **4.21** show results from a series of ML drained TC tests performed at different constant strain rates on air-dried dense specimens of these three types of granular materials. It may be seen that the strength decreases with an increase in the strain rate, of which the trend is stronger in the order of Albany silica sand, corundum A and Hime gravel. The dry densities (or void ratios) of the specimens in the respective case are very similar. It has been confirmed that a small variation of dry density (or void ratio) among different specimens in the respective case cannot explain

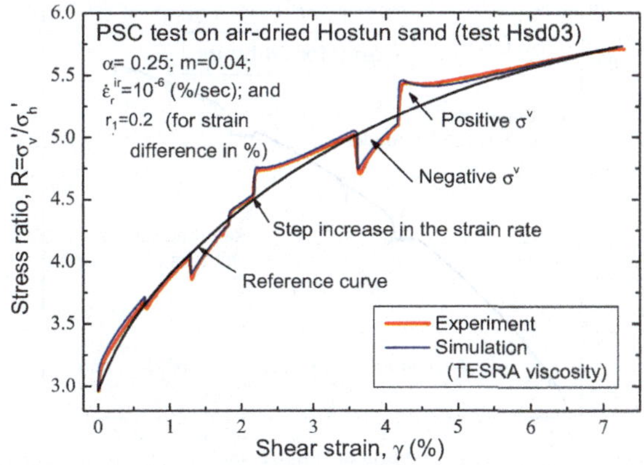

Fig. 4.16 Simulation by the three-component model (TESRA viscosity) of the stress - strain behaviour of Hostun sand in drained PSC (σ'_h= 392 kPa) (Di Benedetto et al., 2002).

this negative effect of strain rate on the stress - strain behaviour. This peculiar trend of behaviour is opposite to the one of the Isotach viscosity (as seen from Figs. 4.6 through 4.9), for which the strength increases with an increase in the strain rate. Calling this classical type of Isotach viscous property the positive Isotach viscosity, we can call the one seen in Figs. 4.19 through 4.21 the negative Isotach viscosity.

Fig. 4.22a illustrates the σ - ε^{ir} relations during continuous ML at different constant strain rates when solely the negative Isotach viscosity is active, showing the trend of behaviour seen from Figs. 4.19, 4.20 and 4.21. Relation A→B→C presented in **Fig. 4.22b** shows a sudden decrease in the stress upon a stepwise increase in the strain rate during otherwise ML at a constant strain rate, which is observed if solely the negative Isotach viscosity is active. On the other hand, actual relatively round and rigid granular materials exhibit the behaviour like A→B→D→C shown in **Fig. 4.23** upon a step increase in the strain rate during otherwise ML at a constant strain rate. This trend of an immediate positive stress jump upon a step increase in the strain rate is similar to those with the Isotach viscosity and the TESRA viscosity. However, subsequently, the stress starts decreasing at a relatively high rate from an increased value toward a value lower than the one that would be observed if ML had continued at the same strain rate without an increase in the strain rate. The test results showing these trends of behaviour described above are presented in **Figs. 4.24**, **4.25** and **4.26**. It may also be seen from these figures that the stress suddenly decreases upon a step decrease in the strain rate and subsequently the stress starts increasing at a high rate toward a value higher than the one that would have been observed if ML had continued at the same strain rate without a decrease in the strain rate. As stated earlier, these trends of viscous effects can be described by Eq. 4.1 while following the three-component model (Fig. 3.7a). As seen from these figures, upon a step increase in the strain rate, first the stress suddenly increases. A positive stress jump, $\Delta\sigma^v$, upon a step increase in the strain rate indicates that the absolute value of the increment of TESRA viscosity stress component, $\Delta\sigma^v_{TESRA}$*, by a sudden change in the strain rate is larger than that of the corresponding increment of negative Isotach viscosity stress component, $\Delta\sigma^v_{NI}$. The same conclusion can be obtained from the stress - strain

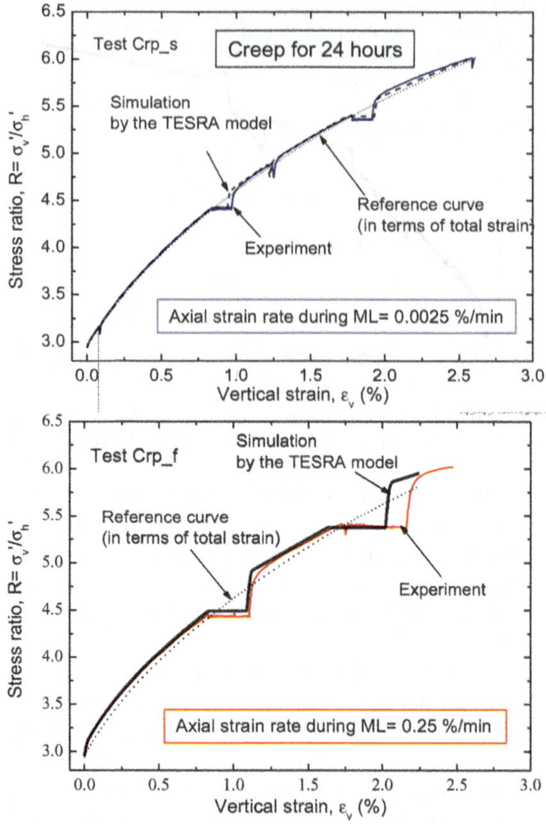

Fig. 4.17 Simulation by the three-component model (TESRA viscosity) of the stress – strain behaviour of saturated Toyoura sand (e= 0.74) in CD PSC (σ'_h= 392 kPa) (Tatsuoka et al., 2002).

behaviour upon a step decrease in the strain rate presented in Figs. 4.24 through 4.26. More specific structure of Eq. 4.1 is discussed later in this paper.

4.4 *Summary of various viscous types*

Fig. 4.27 schematically summarizes the different viscosity types of geomaterial that are described in the precedent sections. In this figure, for simplicity, it is assumed that the σ - ε^{ir} relation before a step change in the strain rate by a factor of 10 is the same for the different viscosity types. The changing rate of stress upon a step change in the irreversible strain rate is defined as the rate-sensitivity coefficient, β , as follows (Tatsuoka, 2004; Di Benedetto et al., 2004; Tatsuoka et al., 2006):

$$\beta = \frac{\Delta\sigma/\sigma}{\log\{(\dot{\varepsilon}^{ir})_{after}/(\dot{\varepsilon}^{ir})_{before}\}} \tag{4.3}$$

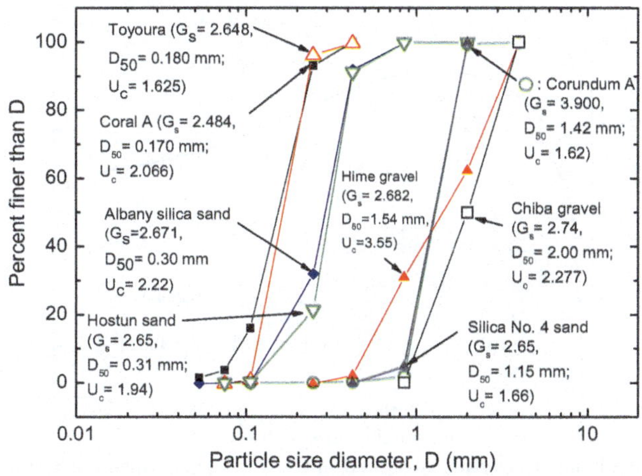

Fig. 4.18 Grading curves of various types of granular materials referred to in this paper (Enomoto et al., 2006; Kawabe et al., 2006).

where $\Delta\sigma$ is the stress jump upon a step change in the strain rate (more rigorously, upon a step change in $\dot{\varepsilon}^{ir}$ at a fixed value of ε^{ir}), which is equal to a jump in σ^v, $\Delta\sigma^v$; σ is the total stress when the strain rate is stepwise changed, which is equal to the instantaneous value of $\sigma^f + \sigma^v$ in the case of Isotach viscosity; and $(\dot{\varepsilon}^{ir})_{after}$ and $(\dot{\varepsilon}^{ir})_{before}$ are the irreversible strain rates after and before a step change, where $(\dot{\varepsilon}^{ir})_{after}/(\dot{\varepsilon}^{ir})_{before} = 10$ in the illustration in Fig. 4.27. For simplicity, the same value of β is assumed for the different viscosity types. The actual β value is a function of soil type, as discussed by Tatsuoka et al. (2006) and Enomoto et al. (2006, this volume).

In addition to the Isotach, TESRA and P&N viscosity types, the intermediate type is depicted in Fig. 4.27, for which the stress decays toward a value higher than the one that would have been observed if ML had continued at the same strain rate without a step increase in the strain rate. This type of behaviour, which is in between those of the Isotach and TESRA viscosity types, was observed in undrained TC tests on reconstituted specimens of relatively low PI clay (Fujinomori clay; Tatsuoka et al., 2002).

4.5 *Transition of viscosity type with strain*

Fig. 4.28 shows a result from a drained TC test on compacted moist well-graded gravelly soil (AhnDan et al., 2006). Despite that it is subtle, the decay rate of the viscous stress gradually increases with an increase in the strain. A more obvious transition in the viscosity type with an increase in the strain can be seen in the data from drained TC tests on compacted cement-mixed gravelly soil specimens aged for different durations presented in **Fig. 4.29** (see Fig. 4.7). In this case, the viscosity in the pre-peak regime is obviously of Isotach type, while it is of TESRA type in the post-peak regime. It seems that an increase in the damage to bonding at inter-particle contact points with an increase in the strain is related to this transition of viscosity type. Tatsuoka et al. (2002) reported that, in the pre-peak regime in CUTC tests on reconstituted Fujinomori clay, the viscosity type changes from the Isotach at the initial small stage towards to the intermediate one.

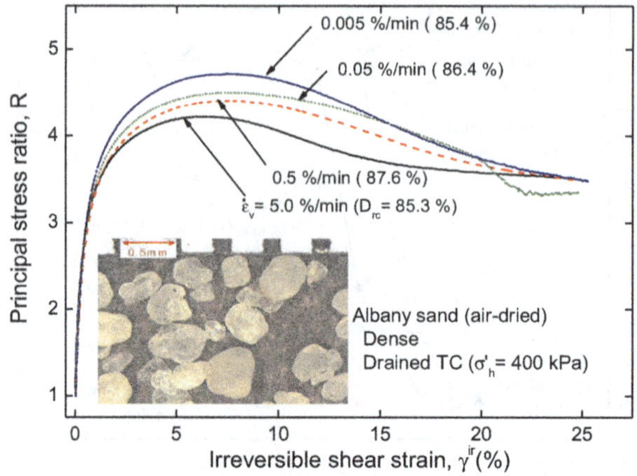

Fig. 4.19 Negative Isotach viscosity observed in CD TC tests at different constant strain rates on air-dried dense Albany silica sand (stiff round particles with D_{50}= 0.30 mm, U_c=2.2, FC= 0.1 %; Enomoto et al., 2006).

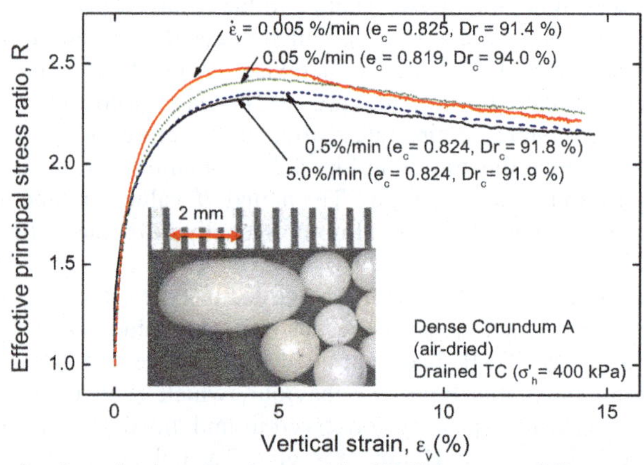

Fig. 4.20 Negative Isotach viscosity observed in CD TC tests at different constant strain rates on air-dried dense corundum (stiff round particles with D_{50}= 1.42 mm, U_c=1.62, FC= 0.0 %; Enomoto et al., 2006; Kawabe et al., 2006).

It is quite difficult to evaluate in details the transition of viscosity type that may take place around the peak stress state or in the post-peak strain-softening regime and to examine the viscosity type at the residual state by means of PSC tests (e.g., Oie et al., 2003) and TC tests. For this reason, a series of direct shear tests were performed on several different types of poorly graded sands (Duttine et al., 2006). The rigid shear boxes had a total height equal to 12 cm and a cross-section of 12 cm times 12 cm. The shear displacement rate was stepwise changed many times during otherwise ML at a constant shear displacement rate. **Fig. 4.30** shows results from two typical tests on dense air-dried Toyoura and Hostun sands. It may be seen from **Fig. 4.30b** that the viscosity type of Toyoura and Hostun sands is of TESRA type in the pre-peak regime as have been

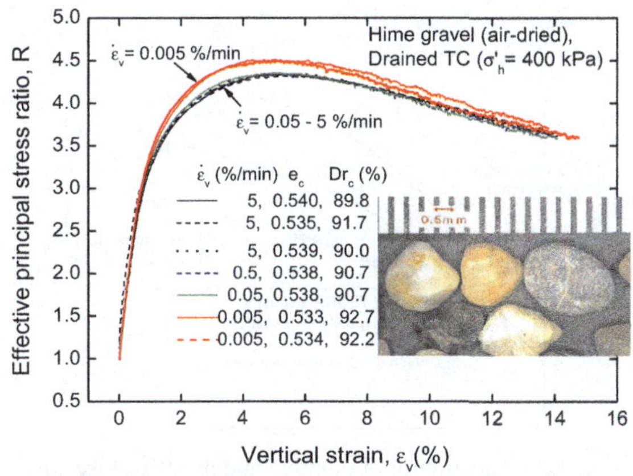

Fig. 4.21 Negative Isotach viscosity observed in CD TC tests at different constant strain rates on air-dried dense Hime gravel (stiff round particles with D_{50}= 1.54 mm, U_c=3.5, *FC*= 0.0 %; Enomoto et al., 2006 & Kawabe et al., 2006).

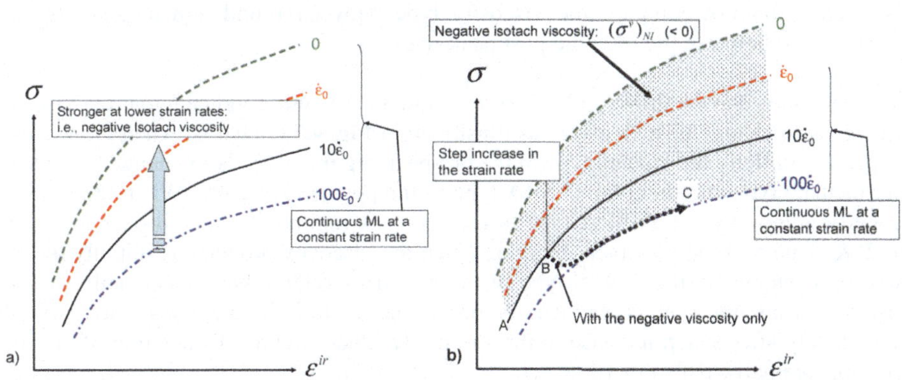

Fig. 4.22 Illustration of purely negative Isotach viscosity; stress-strain behaviour during a) ML at different constant strain rates; and b) when the strain rate is stepwise increased.

observed in drained PSC and TC tests (see Figs. 4.13 and 4.16). The viscosity type changes to the positive and negative (P & N) type during the post-peak strain-softening regime (**Fig. 4.30c**) and the viscosity type at the residual state is obviously the P & N type (**Fig. 4.30d**). Despite that the data are not presented here, in drained direct shear tests on Albany silica sand, the viscosity type is already the P & N type in the pre-peak regime as shown in Fig. 4.24. In the post-peak regime, the viscosity type is still of the P & N type while instable behaviour (i.e., a sudden temporary large drop in the stress) takes place occasionally, in particular immediately after a step increase in the strain rate.

The observation of P & N viscosity in the post-peak regime in drained direct shear tests on granular materials described above is not the first, but this trend has been observed with other types of round granular materials by other researchers (i.e., Ottawa sand in direct shear by Mair & Marone, 1999; glass beads in simple shear by Chambon, et al.,

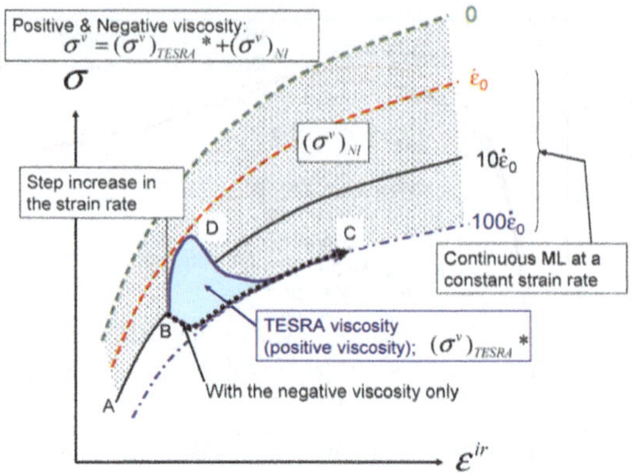

Fig. 4.23 Illustration of positive and negative viscosity.

2002). However, they tested only poorly-graded round granular materials and evaluated only their rate-dependency in the post-peak regime by means of direct or simple shear tests. They did not compare the viscosity type between round and angular granular materials and between the pre- and post peak regimes.

The test results shown above and others available to the author indicate that the viscosity type tends to change in such a way as illustrated in **Fig. 4.31**. That is, when the viscosity property is initially the Isotach type in the pre-peak regime, it tends to change towards the intermediate type and then the TESRA type in the post-peak regime. When the viscosity property is initially the TESRA type in the pre-peak regime, it tends to change towards the P & N type in the post-peak regime. When the viscosity property is initially the P & N type, it remains at the P & N type in the post-peak regime but occasionally showing unstable behaviour (i.e., a sudden temporary large drop in the stress) in particular immediately after a step increase in the strain rate. These trends of transition of viscosity type are summarized also in Table 4.1.

It seems that the viscous property at inter-particle contact points is basically the Isotach type, which likely affects the trend of viscous stress - strain behaviour of a given mass of geomaterial via the following two mechanisms having opposite effects on the strength and stiffness of a given mass of geomaterial:

a) The shear load - shear displacement relation at inter-particle contact points becomes stiffer and stronger with an increase in the shear displacement rate at the inter-particle contact point resulting from an increase in the global strain rate of a given mass of geomaterial.
b) The inter-particle contact points becomes more stable, thereby so does the global stress–strain behaviour of a given mass of geomaterial, by more creep compression at inter-particle contact points resulting from lower global strain rates.

Due to mechanism a), any geomaterial mass exhibits a sudden increase in the shear stress upon a step increase in the global shear strain rate (i.e., the trend of positive viscosity). On the other hand, it seems that the importance of the effects of mechanism b) on the

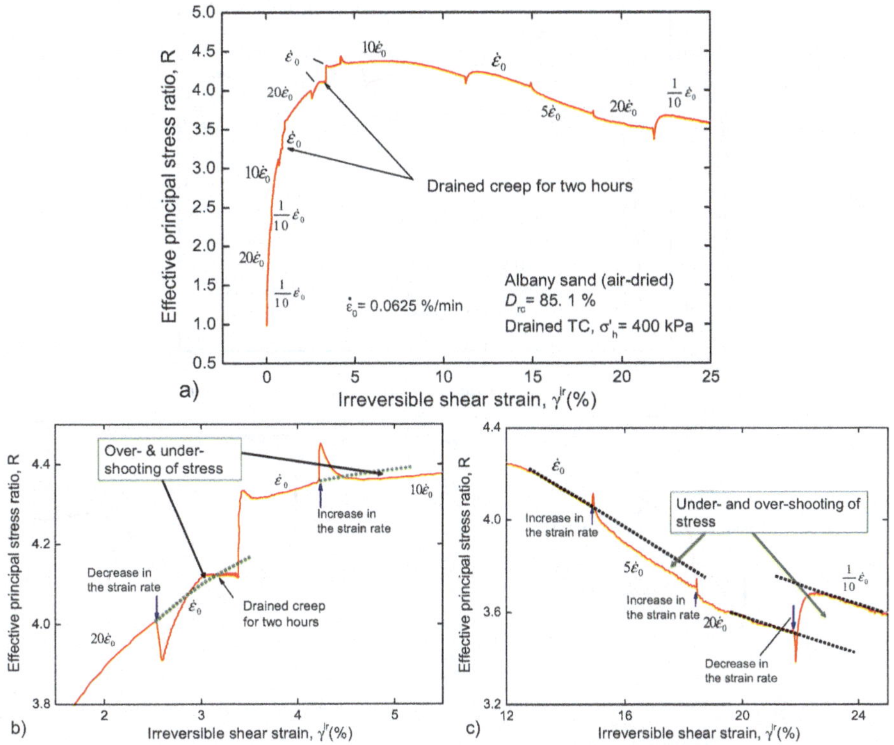

Fig. 4.24 Positive and negative viscosity observed in a CD TC test on Albany silica sand (Enomoto et al., 2006); a) overall behaviour; and b) and c) close-ups.

stability of a geomaterial mass depends on the geomaterial type. The particles become more stable against sliding and rotation with an increase in the inter-particle bonding as well as the interlocking strength as particles become more angular and as the coordination number (i.e., the average number of inter-particle contact points per particle) increases. Then, the global stability of fabrics should become less sensitive to a slight change in the stability at inter-particle contacts points. In this case, the effects of mechanism b) become insignificant relative to the effects of mechanism a) and the stress jump upon a step change in the strain rate becomes more persistent, resulting into the Isotach type viscosity. On the other hand, with relatively uniform round unbound materials, the stability of particles against sliding and rotation is very low. In addition, the effects of mechanism b) on the global stability of the fabrics become more significant relative to the effects of mechanism a). Then, the stress jump upon a step change in the strain rate becomes less persistent, resulting in decay in the viscous stress component with strain, and the global stability of the fabrics during ML decreases with an increase in the strain rate (i.e., the trend of negative Isotach viscosity). It seems that essentially no dependency of the stress - strain relation during ML on the strain rate in the case of TESRA viscosity may be a result of balancing of the effects of two mechanisms a) and b).

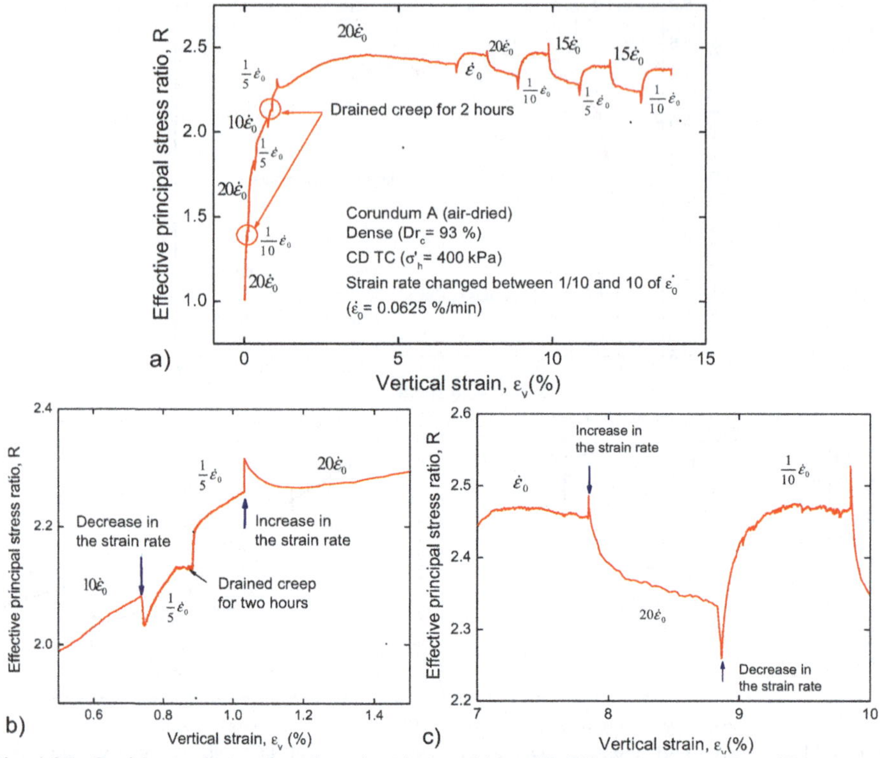

Fig. 4.25 Positive and negative viscosity observed in a CD TC test on corundum (Enomoto et al., 2006); a) overall stress – strain behaviour; and b) and c) close-ups.

4.6 *General expression for various viscosity types*

A wide variety of viscosity type and the transition of viscosity type with an increase in the irreversible strain described in the precedent sections can be described by the following general expression. First, referring to Eq. 4.2, Eq. 4.1 can be re-written as:

$$\sigma^v = \sigma^v_{TESRA}{}^* + \sigma^v_{NI} = \sigma^v_{TESRA}{}^* + \sigma^v_{G.I.} = \sigma^v_{TESRA}{}^* + \theta(\varepsilon^{ir}) \cdot (\sigma^v)_{isotach} \qquad (4.4)$$

where $\theta(\varepsilon^{ir})$ is the viscosity type parameter that distinguishes different viscosity types, which is a function of ε^{ir}. To express all the possible viscosity types by a single equation, it is proposed to express $\sigma^v_{TESRA}{}^*$ as:

$$\sigma^v_{TESRA}{}^* = \{1 - \theta(\varepsilon^{ir})\} \cdot (\sigma^v)_{TESRA} \qquad (4.5)$$

Then, we obtain the current viscous stress (at $\dot{\varepsilon}^{ir}$) as:

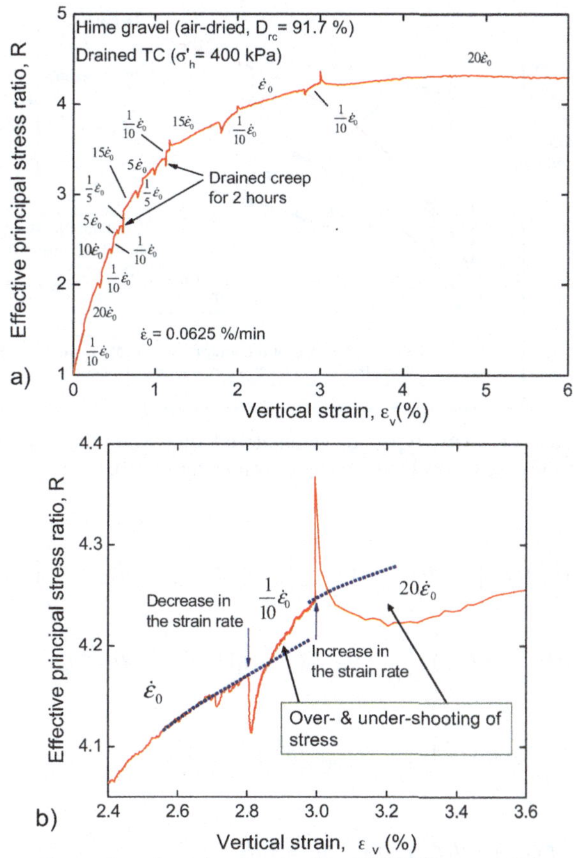

Fig. 4.26 Positive and negative viscosity observed in a CD TC test on Hime gravel (Kawabe et al., 2006); a) overall stress − strain behaviour; and b) a close-up.

$$[\sigma^v]_{(\varepsilon^{ir})} = \{1-\theta(\varepsilon^{ir})\}\cdot[(\sigma^v)_{TESRA}]_{(\varepsilon^{ir})} + \theta(\varepsilon^{ir})\cdot[(\sigma^v)_{isotach}]_{(\varepsilon^{ir})} \qquad (4.6)$$

It seems that the viscosity type parameter, $\theta(\varepsilon^{ir})$, generally decreases with an increase in the irreversible strain, as illustrated in **Fig. 4.32**. Hirakawa et al. (2003) first used Eq. 4.6 with a constant value of θ to simulate the rate-dependent load - strain behaviour of polyester (PET) geogrid reinforcement observed in tensile tests (described in the next chapter). For geomaterials, the possible largest value of $\theta(\varepsilon^{ir})$ is 1.0 (for the conventional Isotach viscosity), $\theta(\varepsilon^{ir}) = 0$ for the TESRA viscosity, and $\theta(\varepsilon^{ir})$ is negative for the P & N viscosity with the possible smallest value being lower than -1.0. Some typical combinations of the values of $1-\theta(\varepsilon^{ir})$ and $\theta(\varepsilon^{ir})$ are presented is Fig. 4.32 and **Table 4.2.**

If the first term, $(\sigma^v)_{TESRA}$, and the second term, $(\sigma^v)_{isotach}$, of Eq. 4.6 have the same viscosity function, $g_v(\dot{\varepsilon}^{ir})$, Eq. 4.6 becomes:

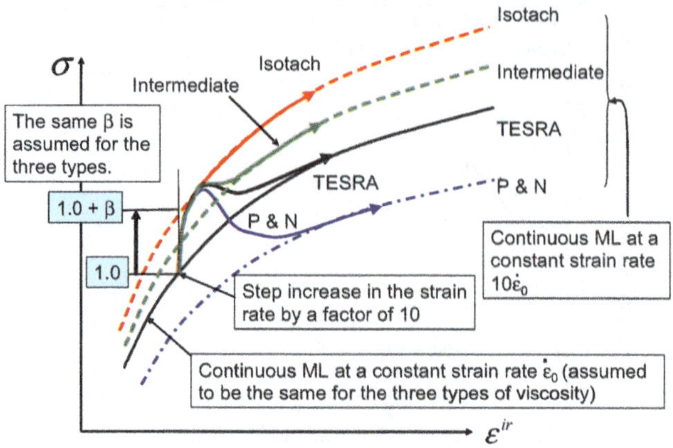

Fig. 4.27 Different viscosity types of geomaterial and definition of the rate-sensitivity coefficient (in the case of strain rate change by a factor of 10).

$$\begin{aligned}[\sigma^v]_{(\varepsilon^{ir})} &= \{1-\theta(\varepsilon^{ir})\}\cdot \int_{\tau=\varepsilon_1^{ir}}^{\varepsilon^{ir}} \left[d\{\sigma^f \cdot g_v(\dot{\varepsilon}^{ir})\}\right]_{(\tau)} \cdot r_1(\varepsilon^{ir})^{\varepsilon^{ir}-\tau} + \theta(\varepsilon^{ir})\cdot \int_{\tau=\varepsilon_1^{ir}}^{\varepsilon^{ir}} \left[d\{\sigma^f \cdot g_v(\dot{\varepsilon}^{ir})\}\right]_{(\tau)} \\ &= \int_{\tau=\varepsilon_1^{ir}}^{\varepsilon^{ir}} \left[d\{\sigma^f \cdot g_v(\dot{\varepsilon}^{ir})\}\right]_{(\tau)} \cdot [\{1-\theta(\varepsilon^{ir})\} r_1(\varepsilon^{ir})^{\varepsilon^{ir}-\tau} + \theta(\varepsilon^{ir})] \\ &= \int_{\tau=\varepsilon_1^{ir}}^{\varepsilon^{ir}} \left[d\{\sigma^f \cdot g_v(\dot{\varepsilon}^{ir})\}\right]_{(\tau)} \cdot g_{decay.g}(\varepsilon^{ir}, \varepsilon^{ir}-\tau)\end{aligned} \tag{4.7a}$$

$$g_{decay.g}(\varepsilon^{ir}, \varepsilon^{ir}-\tau) = \{1-\theta(\varepsilon^{ir})\}\cdot r_1(\varepsilon^{ir})^{\varepsilon^{ir}-\tau} + \theta(\varepsilon^{ir}) \tag{4.7b}$$

where $[\sigma^v]_{(\varepsilon^{ir})}$ is the current viscous stress when the irreversible strain is equal to ε^{ir}; $g_{decay.g}(\varepsilon^{ir}, \varepsilon^{ir}-\tau)$ is the generalised decay function, which decreases from 1.0 towards $\theta(\varepsilon^{ir})$ with an increase in the strain increase, $\varepsilon^{ir}-\tau$ (**Fig. 4.33a**); and $r_1(\varepsilon^{ir})$ is the generalised decay parameter, which could be a positive constant equal to, or less than, 1.0, or could decrease with an increase in ε^{ir} from a certain positive value to another lower positive value (Tatsuoka et al., 2002). The following comments on $r_1(\varepsilon^{ir})$ are important:

1) The functional form, $r_1(\varepsilon^{ir})$, means that the parameter r_1 is a function of the instantaneous irreversible strain, ε^{ir}, for which the current viscous stress, $[\sigma^v]_{(\varepsilon^{ir})}$, is to be obtained. Therefore, the value of $r_1(\varepsilon^{ir})$ is kept constant in the integration with respect to the past irreversible strain, τ, in Eq. 4.7a.
2) When the parameter r_1 decreases with an increase in ε^{ir} for which the current viscous stress, $[\sigma^v]_{(\varepsilon^{ir})}$, is to be obtained (**Fig. 4.33b**), the decay function, $g_{decay.g}(\varepsilon^{ir}, \varepsilon^{ir}-\tau)$, for a fixed value of $[d\{\sigma^f \cdot g_v(\dot{\varepsilon}^{ir})\}]_{(\tau)}$ that has taken place at a given moment in the past (when $\varepsilon^{ir} = \tau$) changes as ε^{ir} increases in the way that the decay rate becomes larger with an increase in ε^{ir} (see Fig. 5 of Tatsuoka et al., 2002).

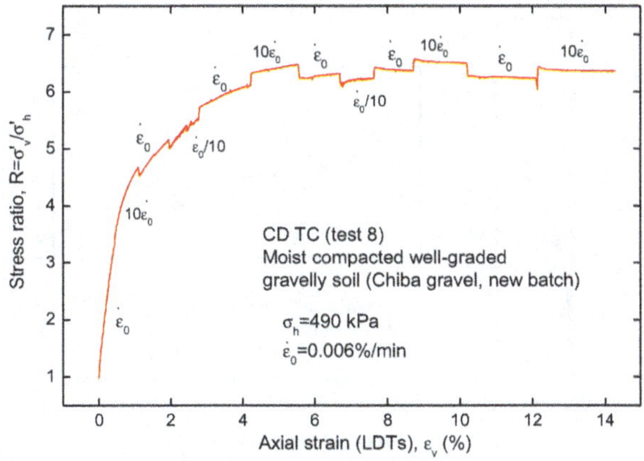

Fig. 4.28 Transition of viscous type from Isotach in the pre-peak regime to weak TESRA in the post-peak regime, CD TC test on compacted well-graded gravel (angular, D_{max}= 38 mm, D_{50}= 3.5 mm & U_c= 12.75; specimen: 30 cm-dia.& 60 cm-high; AhnDan et al., 2006).

3) If it is assumed that the parameter r_1 is a function of τ (i.e., $r_1(\tau)$), that is, if the general decay function becomes $g_{decay.g}(\tau, \varepsilon^{ir}-\tau)$ instead of $g_{decay.g}(\varepsilon^{ir}, \varepsilon^{ir}-\tau)$, the decay characteristics becomes different from the one expressed by Eq. 4.7b. Although this assumption is mathematically possible, the integration of Eq. 4.7a becomes much more complicated. Moreover, Eq. 4.7a, which uses $r_1(\varepsilon^{ir})$, can be approximated into an incremental form as Eq. B10 and this approximation is necessary to be incorporated into a FEM code (Tatsuoka et al., 2002). However, it is not the case when using $r_1(\tau)$. Moreover, the test results support the functional form of $r_1(\varepsilon^{ir})$.
4) Comments 1), 2) and 3) above are also relevant to the parameter $\theta(\varepsilon^{ir})$.
5) Although Eq. 4.7 can be one of other possible specific cases of Eq. 4.6, it is true that Eq. 4.7 can express all the three basic viscosity types, the Isotach, TESRA and P & N, and the intermediate type (between the Isotach and TESRA types) as well their transition as illustrated in Fig. 4.31.

The stress jump that is observed upon a step change in the strain rate, which is before the decay becomes meaningful (i.e., while $r_1^{\varepsilon^{ir}-\tau}$ is nearly equal to 1.0), is obtained as follows according to Eq. 4.7:

$$\begin{aligned}\Delta\sigma^v &= \{1-\theta(\varepsilon^{ir})\}\cdot\Delta\{\sigma^f\cdot g_v(\dot{\varepsilon}^{ir})\} + \theta(\varepsilon^{ir})\cdot\Delta\{\sigma^f\cdot g_v(\dot{\varepsilon}^{ir})\} \\ &= \Delta\{\sigma^f\cdot g_v(\dot{\varepsilon}^{ir})\} = (\Delta\sigma^v)_{isotach}\end{aligned} \tag{4.8}$$

So, the observed value of β (Eq. 4.3) becomes:

$$\beta = \{1-\theta(\varepsilon^{ir})\}\cdot\beta_{isotach} + \theta(\varepsilon^{ir})\cdot\beta_{isotach} = \beta_{isotach} \tag{4.9}$$

where $\beta_{isotach}$ is the β value when the viscosity is of the ordinary Isotach type.

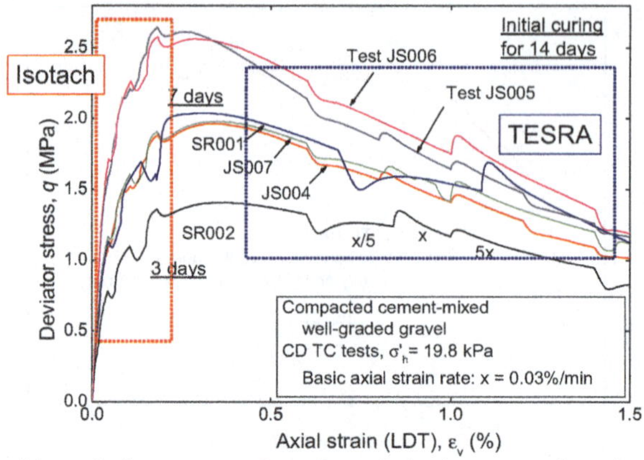

Fig. 4.29 Transition of viscous type from Isotach in the pre-peak regime to TESRA in the post-peak regime, compacted cement-mixed gravel (Kongsukprasert & Tatsuoka, 2005).

Then, the viscosity function, $g_v(\dot{\varepsilon}^{ir})$ for the Isotach viscosity can be defined based on the measured value of $\beta = \beta_{isotach}$. After the stress jump upon an increase in ε^{ir} has fully decayed during the subsequent ML at a constant strain rate, Eq. 4.9 becomes:

$$\beta_{residual} = \theta(\varepsilon^{ir}) \cdot \beta_{isotach} = \theta(\varepsilon^{ir}) \cdot \beta \tag{4.10}$$

where $\beta_{residual}$ is the rate-sensitivity coefficient after full decay of a stress jump (Enomoto et al., 2006; **Fig. 4.34**). Eq. 4.10 means that $\theta(\varepsilon^{ir})$ can be evaluated as $|\beta_{residual}|/\beta$ as illustrated in Fig. 4.32. That is, $\beta_{residual}$ is equal to $\beta =$ "$\beta_{isotach}$ for the Isotach viscosity"; a positive value less than β for the intermediate viscosity; zero for the TESRA viscosity; and a negative value for the P & N viscosity. As shown in Fig. 4.32, when $\theta(\varepsilon^{ir})$ is negative, the magnitude of $1-\theta(\varepsilon^{ir})$ becomes larger than $|\theta(\varepsilon^{ir})|$. It may be seen from Figs. 4.24c and 4.25c that $\theta(\varepsilon^{ir})$ can become smaller than – 1.0 with $|\theta(\varepsilon^{ir})|$ being larger than 1.0. The lower limit of $\theta(\varepsilon^{ir})$ with granular materials is not known.

4.7 *Summary*

The actual viscous property of geomaterial is much more complicated than has been considered and modelled in the previous research. Yet, the viscosity types described above may not cover all the possible types. For example, crushed concrete aggregate, consisting of stiff and strong coarse core gravel particles covered with a relatively soft and weak mortar layer exhibits a peculiar rate-dependency as shown in **Fig. 4.35**. The viscosity type in the pre-peak regime is apparently the P & N type, while it is transformed to an intermediate viscosity type around the peak stress state. This is a reversed transition of viscosity type compared with those described above. It seems that, due to its peculiar particle composition, the effects of mechanism b) described before are significant in the pre-peak regime, where the crushing of surface mortar layer is significant. For this reason, slower loading results in more compressive deformation at inter-particle contact points, which makes the inter-particle contact points more stable, resulting to a stronger response

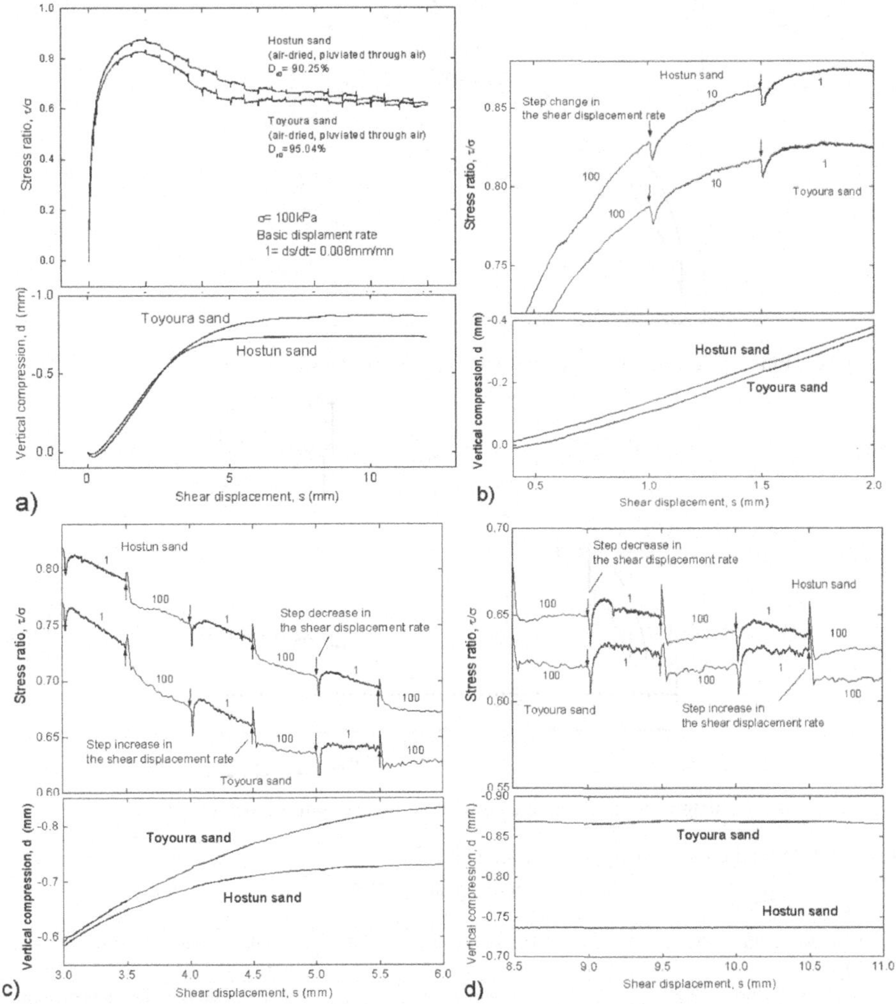

Fig. 4.30 Transition of viscosity type from TESRA in the pre-peak regime to P & N in the post-peak regime in direct shear tests on air-dried dense Toyoura and Hostun sands (Duttine et al., 2006); a) overall stress ratio– shear displacement – vertical displacement behaviour; and b) - d) close-ups.

of the specimen (i.e., a mass of crushed concrete aggregate). This is a trend opposite to the case of undrained TC on ordinary sand at elevated pressure, in which the whole particle is rather homogeneous (e.g., Yamamuro & Lade, 1993). It seems that the effects of mechanism a) become important around the peak stress state and dominant in the post-peak regime, where stiff and strong coarse core particles are in contact with each other supporting most of the applied load. For this reason, it seems that the viscous property becomes an intermediate type, as seen with ordinary well-graded granular materials consisting of particles having a low crushability.

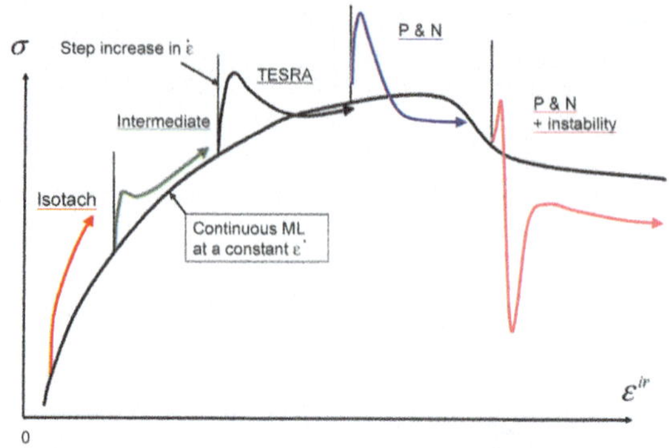

Fig. 4.31 Transition of viscosity type with an increase in the strain.

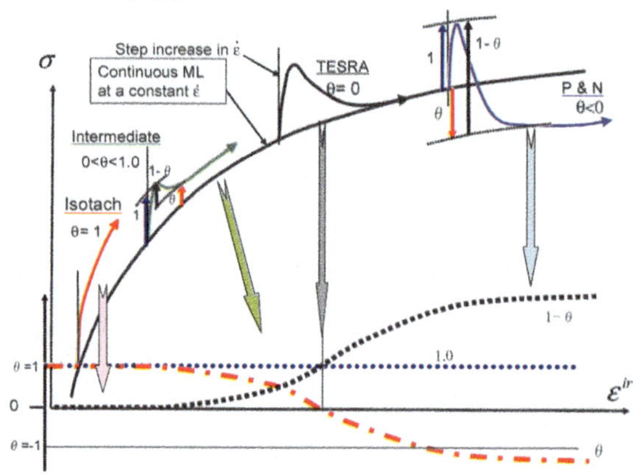

Fig. 4.32 Likely changes in the parameter θ with an increase in the strain.

Table 4.2 Meaning of viscosity type parameter $\theta(\varepsilon^{ir})$.

Viscosity type / Coefficient	Isotach	Intermediate	TESRA	P & N	
$\theta(\varepsilon^{ir})$ (for Isotach viscosity term)	1.0	Positive value (e.g., 2/3)	0.0	Negative value (e.g., -1.0)	Negative value (e.g., -2.0)
$1-\theta(\varepsilon^{ir})$ (for TESRA viscosity term)	0.0	Positive value (e.g., 1/3)	1.0	Positive value (e.g., 2.0)	Positive value (e.g., 3.0)

Another important issue, which was not touched upon above, is the viscous effect on the flow rule (i.e., the relationship between the irreversible volumetric and shear strain rates). It may be seen from Fig. 4.30 that the relationship between the vertical compression and the shear displacement of sand in drained direct shear is insensitive to step changes in the strain rate, unlike a high rate-sensitivity of the shear stress, τ. It seems that the flow rule is controlled by the instantaneous inviscid stress, σ^f, independent of the viscous stress component, σ^v, as assumed by the three-component model (Fig. 3.7a). On the other hand, the over-stress model (Perzyna, 1963), which basically exhibits the Isotach viscosity, describes the flow rule in terms of the stress, $\sigma = \sigma^f + \sigma^v$. More discussions on this important topic are however beyond the scope of this report.

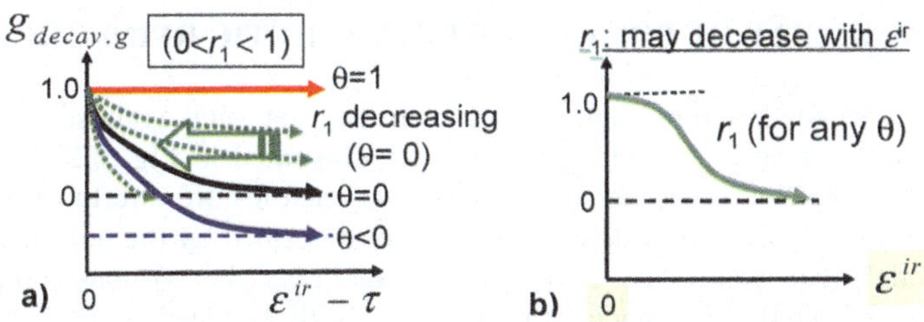

Fig. 4.33 Generalised decay function and generalised decay parameter r_1 (in the case when decreasing from 1.0 towards 0.0)

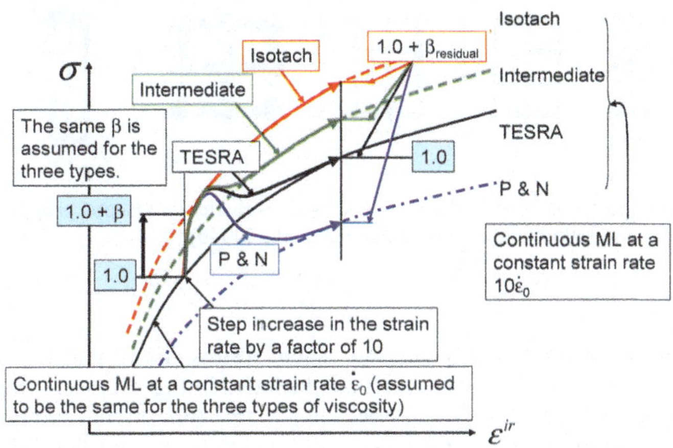

Fig. 4.34 Definition of β and $\beta_{residual}$ (in the case of strain rate change by a factor of 10).

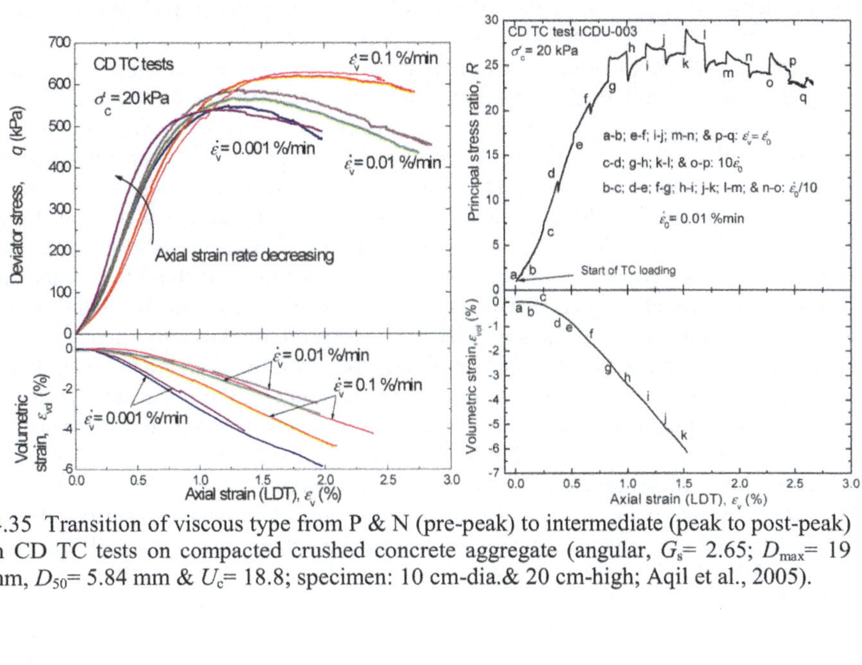

Fig. 4.35 Transition of viscous type from P & N (pre-peak) to intermediate (peak to post-peak) in CD TC tests on compacted crushed concrete aggregate (angular, G_s= 2.65; D_{max}= 19 mm, D_{50}= 5.84 mm & U_c= 18.8; specimen: 10 cm-dia.& 20 cm-high; Aqil et al., 2005).

5. VISCOUS PROPERTY OF POLYMER GEOSYNTHETIC REINFORCEMENT

5.1 *General*

The tensile load - tensile strain relations of polymer geogrids, such as those described in **Fig. 5.1**, are known to be highly rate-dependent as illustrated in **Fig. 5.2**. It is shown below that, despite that the raw materials and micro-structures are utterly different, the viscous property of polymer geogrids can be properly described by the non-linear three-component model that has been developed to describe the viscous property of geomaterial (Fig. 3.7a) only by replacing the stress, σ, with the tensile load, T (**Fig. 5.3**).

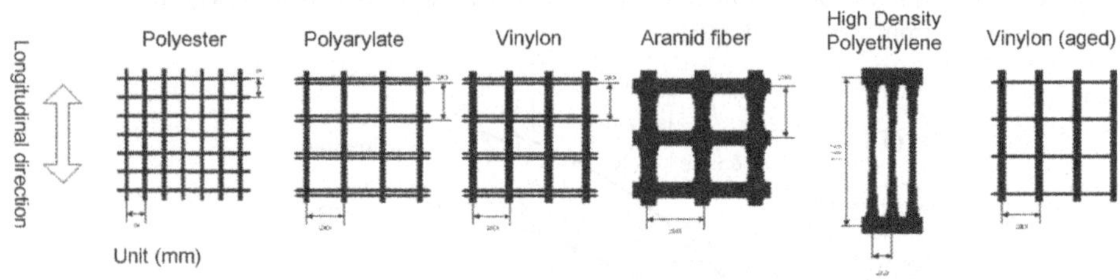

Fig. 5.1 Polymer geogrids used in the tensile tests to evaluate the viscous property (Hirakawa et al., 2003; Kongkitkul et al., 2004; Tatsuoka et al., 2004b).

5.2 *Isochronous model*

Historically in geosynthetics engineering, such results from tensile loading tests performed at different constant strain rates, as illustrated in Fig. 5.2, were first interpreted by the isochronous concept (**Fig. 5.4a**). According to this concept, the current tensile load, T, is a unique function of the instantaneous tensile strain, ε, and the time, t, that has elapsed since the start of loading when T= 0. All the T and ε states at the same t are represented by a single isochrone. When following this concept, the time, t, necessary to reach a certain T and ε state, such as point b in **Fig. 5.4b**, is the same (i.e., isochronous) when tracing different loading paths, such as continuous ML at a relatively low strain rate, $\dot{\varepsilon}$ (O→b); ML at a relatively high $\dot{\varepsilon}$ followed by sustained loading (O→a→b); and ML at a relatively high $\dot{\varepsilon}$ followed by load relaxation (O→c→b). Suppose that ML is restarted at the original $\dot{\varepsilon}$ during the primary loading (O→a) after sustained loading (a→b). As we cannot go back to the past, when following the isochronous concept, we must traverse isochrones for the elapsed times longer than the one at point b (**Fig. 5.5a**). Then, the ultimate tensile rupture strength decreases with an increase in the duration of sustained loading. However, the actual behaviour is utterly different: i.e., the ultimate strength is controlled by $\dot{\varepsilon}$ at tensile rupture, while it is independent of the duration of sustained loading, as illustrated in **Fig. 5.5b**. Several isochronous theories have been proposed also to simulate the rate-dependent stress-strain behaviour of geomaterial (in particular the secondary consolidation of clay) and some of these theories are still widely in use in geotechnical engineering practice. This issue is discussed later in this paper.

5.3 *Simulation by the non-linear three-component model*

The trend of rate-dependent load–strain behaviour of polymer geogrids observed in tensile tests can be simulated by the non-linear three-component model illustrated in Fig. 5.3. **Fig. 5.6a** shows the results from continuous ML tensile tests at different $\dot{\varepsilon}$ values of

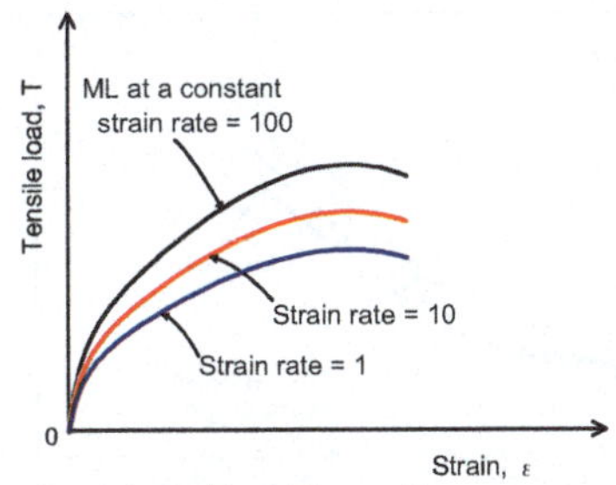

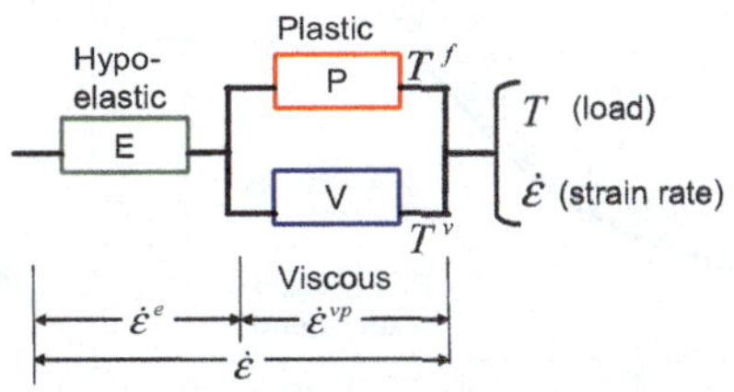

Fig. 5.2 (left) ML tensile rupture tests at different strain rates of polymer reinforcement.
Fig. 5.3 (right) Non-linear three-component model to simulate rate-dependent tensile load - tensile strain behaviour of geogrids (Hirakawa et al., 2003; Kongkitkul et al., 2004).

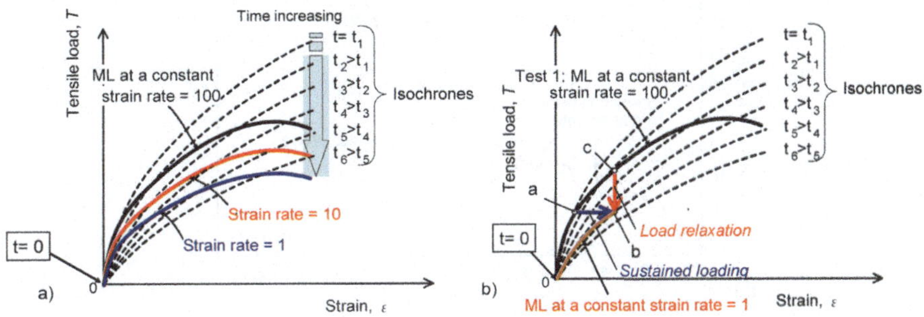

Fig. 5.4 Isochronous concept; a) interpretation of tensile test results; and b) the same time (i.e., isochronous) to reach point b via different loading paths (Tatsuoka et al., 2004b).

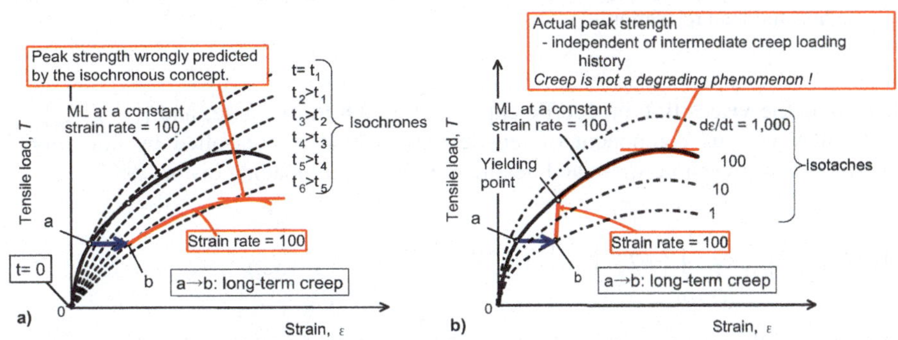

Fig. 5.5 Post-creep behaviour; a) the isochronous concept; and b) the Isotach viscosity (Tatsuoka et al., 2004b).

polyester (PET) geogrid and their simulation based on the non-linear three-component model. The viscous load component is predicted based on Eq. 5.1, which is the origin of the general expression of the viscous property of geomaterial, Eqs. 4.6 and 4.7.

$$[T^v]_{(\varepsilon^{ir})} = (1-\theta)\cdot[(T^v)_{TESRA}]_{(\varepsilon^{ir})} + \theta\cdot[(T^v)_{isotach}]_{(\varepsilon^{ir})} \qquad (5.1)$$

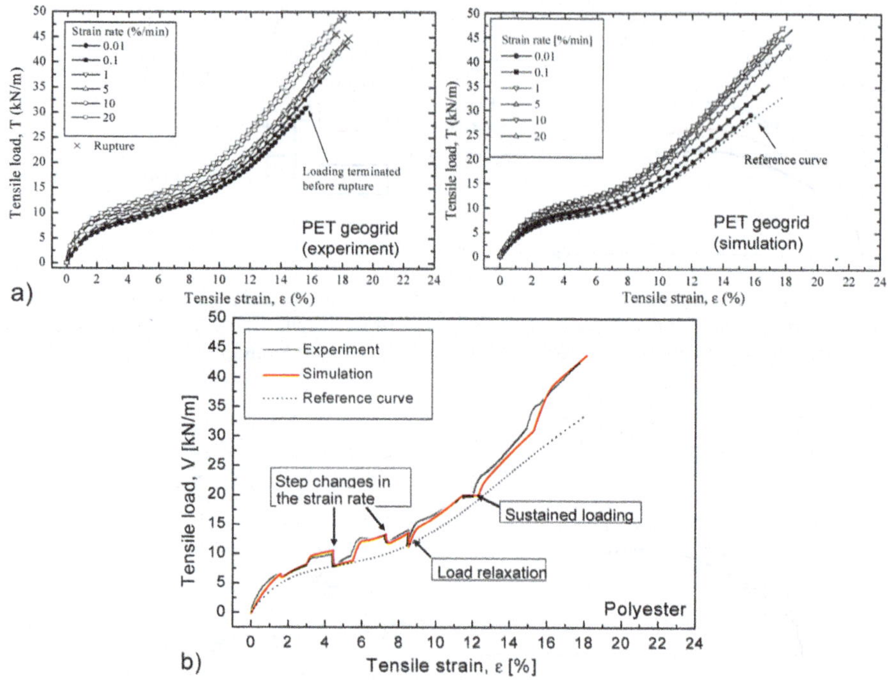

Fig. 5.6 Results from tensile tests of polyester (PET) geogrid and their simulation by the non-linear three-component model (intermediate viscosity type) (Hirakawa et al., 2003); a) ML tests and their simulation; and b) a test with step changes in the strain rate, sustained loading and load relaxation.

where θ is the viscosity type coefficient, which is 0.8 in this case; and $[(T^v)_{TESRA}]_{(\varepsilon^{ir})}$ is the TESRA viscous component of tensile load, which is controlled by not only the instantaneous irreversible strain and its rate but also recent loading history as:

$$[(T^v)_{TESRA}]_{(\varepsilon^{ir})} = \int_{\tau=\varepsilon_1^{ir}}^{\varepsilon^{ir}} \left[dT_{iso}^v\right]_{(\tau)} \cdot \left\{r_1\left(\varepsilon^{ir}\right)\right\}^{\left(\varepsilon^{ir}-\tau\right)} \qquad (5.2)$$

where $\left\{r_1\left(\varepsilon^{ir}\right)\right\}^{\left(\varepsilon^{ir}-\tau\right)}$ is the decay function: and $r_1\left(\varepsilon^{ir}\right)$ is the decay parameter that decreases with an increase in ε^{ir}. In the present case, r_1 decreases from 1.0 to 0.15. $\left[dT_{iso}^v\right]_{(\tau)}$ is the increment of the Isotach type viscous tensile load component given as:

$$T_{iso}^v = T^f\left(\varepsilon^{ir}\right) \cdot g_v\left(\dot{\varepsilon}^{ir}\right) \qquad (5.3)$$

where $g_v\left(\dot{\varepsilon}^{ir}\right)$ is the viscosity function (Eq. B4). The other details of the experiment and the model simulation are reported in Hirakawa et al. (2003). **Fig. 5.6b** compares measured and simulated T - ε relations when $\dot{\varepsilon}$ is stepwise changed several times and

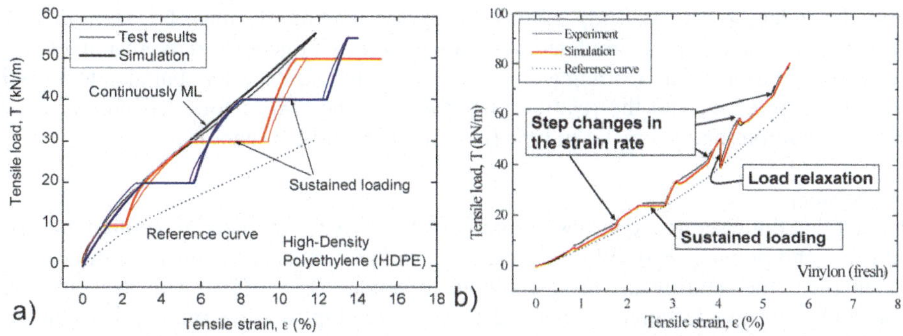

Fig. 5.7 Experimental results from tensile tests of geogrid and their simulation by the non-linear three-component model (Isotach viscosity); a) load-control tests on HDPE (Kongkitkul et al., 2004) and b) strain-controlled tests on Vinylon (Hirakawa et al., 2003).

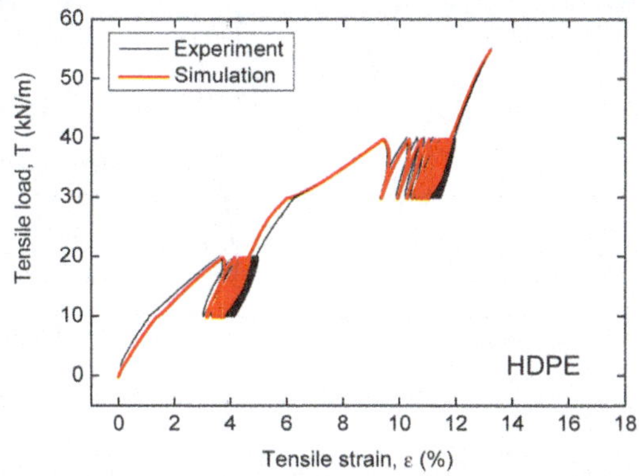

Fig. 5.8 Experimental results from tensile tests (load-controlled) of HDPE geogrid and their simulation by the non-linear three-component model (Isotach viscosity) (Kongkitkul et al., 2004).

sustained loading and load relaxation are performed during otherwise ML at a constant strain rate. It may be seen that the three-component model (Fig. 5.3) can simulate very well the whole trends of rate-dependent behaviour of the geogrid observed in the experiments as in the case of geomaterials.

Fig. 5.7 shows similar comparisons between measured and simulated T - ε relations of other types of polymer geogrids that exhibit the Isotach viscous property. **Fig. 5.8** shows the results from tensile tests (load-controlled) of HDPE geogrid in which cyclic loading was applied during otherwise ML at a constant load rate. The simulation of the test result by the non-linear three-component model (the Isotach type) is also presented in this figure. In the model simulation, the residual strain that develops during cyclic loading is due solely to the viscous property, like creep strains that develop during sustained loading at a constant tensile load. That is, specific hysteretic T^f - ε^{ir} relations were introduced to describe cyclic loading behaviour that do not exhibit any residual strain by specific rate-independent (i.e., inviscid) effects of cyclic loading. It may be seen that the experimental result is well simulated. As shown later, with geomaterials, residual strains that develop during cyclic loading are not due totally to viscous effect but also to inviscid cyclic loading effect, and the relative importance of the two factors depends on cyclic loading conditions and geomaterial type (in particular, particle shape).

5.4 *Summary*

The fact that the viscous properties of totally different types of material, geomaterial (soil and rock) and polymer geogrid reinforcement, can be simulated by the same constitutive model (i.e., the non-linear three-component model, Fig. 3.7a) indicates its high applicability to a wide range of material (e.g., a tire chip mass, Nirmalan & Uchimura, 2006). Moreover, with polymer geogrids, residual strains that develop during a given cyclic loading stage are due fully to the viscous property.

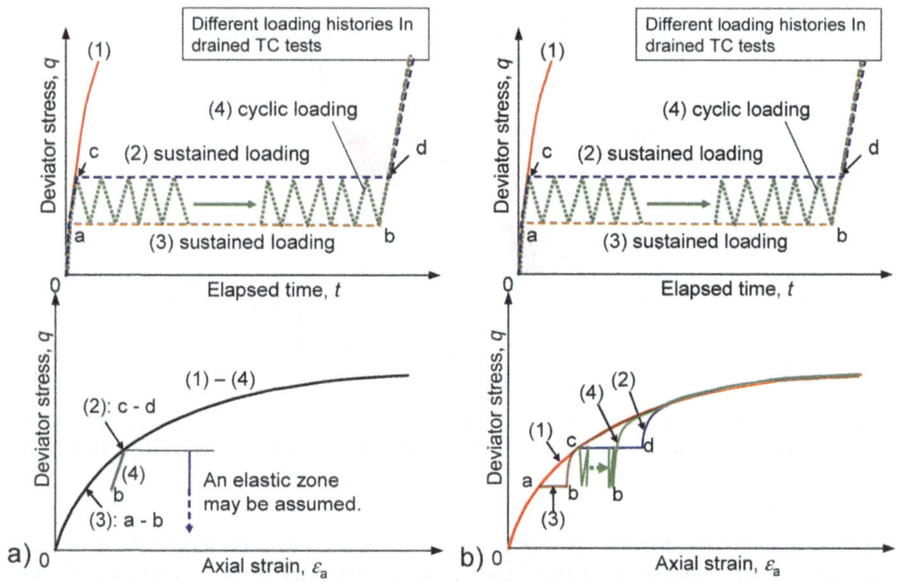

Fig. 6.1 Response of: a) an elasto-plastic material free from both inviscid cyclic loading effect and ageing effect; and b) an elasto-viscoplastic material free from both inviscid cyclic loading effect and ageing effect.

6. STRESS-STRAIN BEHAVIOUR DURING CYCLIC LOADING

6.1 *General*

Residual strain may accumulate when subjected to cyclic shear stresses for a given duration. By introducing into an elasto-plastic model a kinematic hardening yielding framework in which the purely elastic stress zone is small and always moving being dragged by the current stress state, plastic strain can continue developing even when cyclic shear stresses are repeatedly applied along the same stress path. However, this methodology has the following inherent drawbacks:

1. In actuality, the residual strain developed by cyclic loading is not totally different in nature from the one developed by sustained loading at a fixed stress state (i.e., creep strain). So, these two types of residual strain are linked to each other and somehow inter-changeable (as shown below).
2. For the same reason, the residual strain developed in the course of cyclic loading cannot be free from the viscous effect.

The pure effect of cyclic loading, which will herein be called "inviscid cyclic loading effect", can be accurately evaluated only when taking into account the viscous effect.

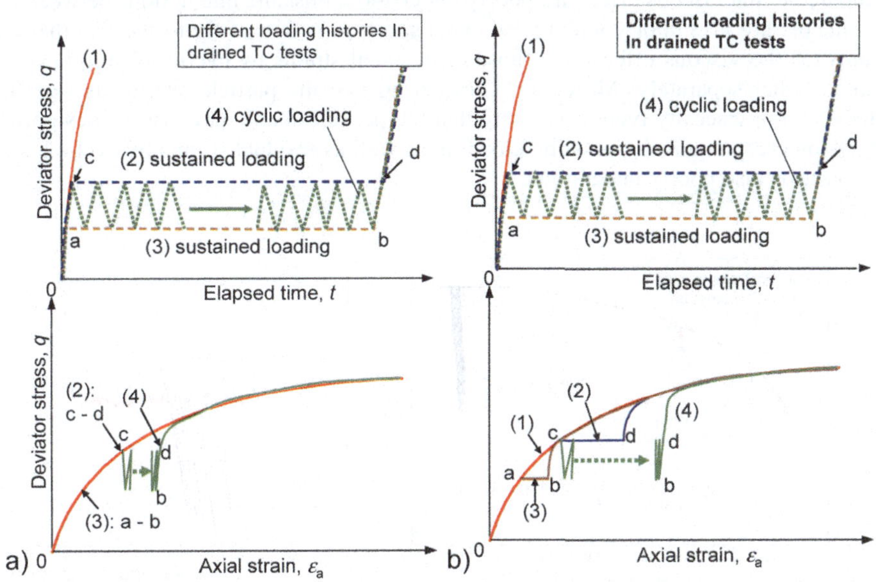

Fig. 6.2 Response of: a) an elasto-plastic material with significant inviscid cyclic loading effect while free from ageing effect: and b) an elasto-viscoplastic material with significant inviscid cyclic loading effect while free from ageing effect.

An elasto-plastic material that does not incorporate inviscid cyclic loading effect exhibits no accumulation of residual strain when subjected to cyclic loading along a fixed stress path (test 4; **Fig. 6.1a**). On the other hand, even when free from inviscid cyclic loading effect, an elasto-viscoplastic material exhibits residual strain which accumulates with time when subjected to cyclic loading along a fixed cyclic stress path (test 4; **Fig. 6.1b**). The nature of this residual strain is the same as creep strain observed when subjected to sustained loading under fixed stress conditions (tests 2 & 3). This is the case with the residual strains observed with a polymer geogrid subjected to cyclic loading (Fig. 5.8). In Fig. 6.1b, the maximum deviator stress, q_{max}, during a cyclic loading stage in test 4 is the same as the sustained deviator stress in test 2. Then, the residual strain at the end of the cyclic loading stage in test 4 is smaller than the one at the end of the sustained loading stage for the same duration in test 2. This is because the deviator stress, q, is cyclically unloaded from q_{max} during the cyclic loading stage, which decreases the creep strain rate when compared to the sustained loading at $q = q_{max}$. On the other hand, an elasto-plastic material showing inviscid cyclic loading effect exhibits residual strain that accumulates during cyclic loading along a fixed stress path (test 4; **Fig. 6.2a**), compared with no creep strain in tests 2 and 3. Then, as illustrated in **Fig. 6.2b**, if the inviscid cyclic loading effect is significant, an elasto-viscoplastic material exhibits residual strain at the end of a given cyclic loading stage in test 4 that may be larger than the one at the end of sustained loading at $q = q_{max}$ that lasts for the same duration as the cyclic loading stage (test 2).

The conditions under which the inviscid cyclic loading effect becomes important when compared to the viscous effect are poorly understood. Possible interactions between these two factors are also poorly understood. This situation is due partly to the fact that creep strains (as the viscous effect) and those on residual strains by cyclic loading have been studied rather separately. Moreover, considering that the particle shape has significant effects on the viscosity type, it is likely that the particle shape may also have significant effects on creep strains by sustained loading as well as residual strains by cyclic loading. This issue is also poorly understood.

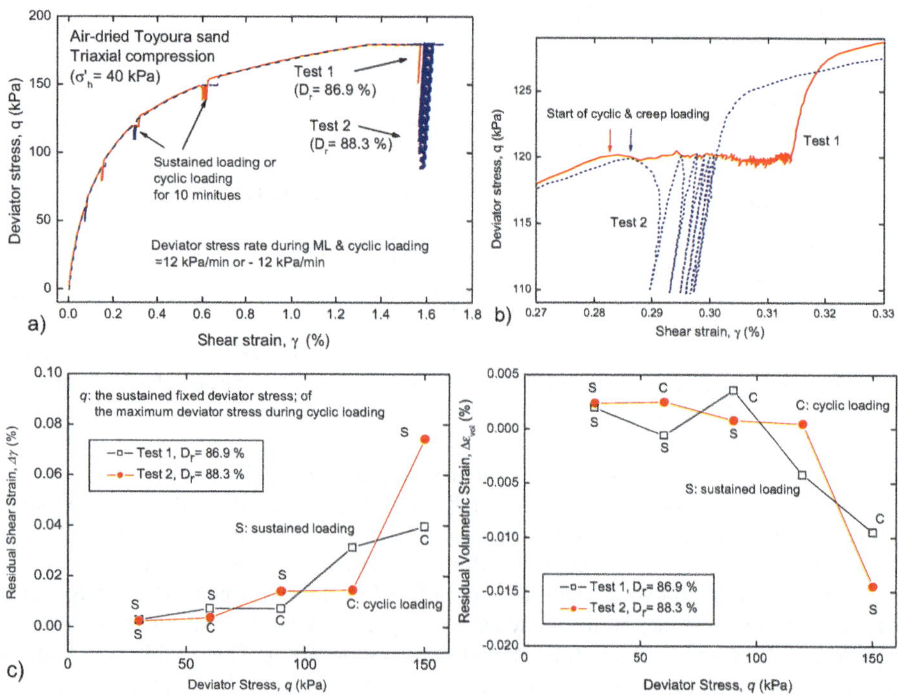

Fig. 6.3 Comparison of residual strains by sustained loading and cyclic loading in a pair of drained TC tests on air-dried Toyoura sand (Ko et al., 2003): a) overall stress-strain behaviour; b) a close-up; and c) residual shear and volumetric strain increments by sustained and cyclic loading plotted against maximum deviator stress during cyclic loading or fixed deviator stress during sustained loading.

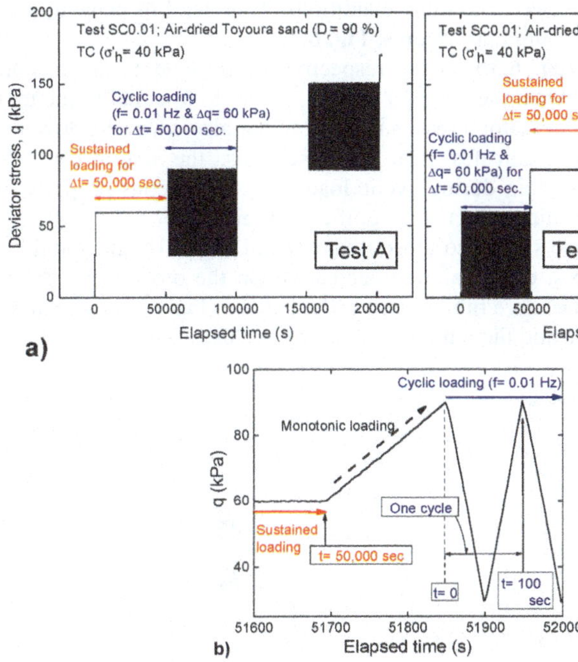

Fig. 6.4 Loading histories employed in a pair of TC tests on Toyoura sand to evaluate the importance of inviscid cyclic loading effect (Hayashi et al., 2005, 2006); a) overall loading histories; and b) part of loading history of test A.

6.2 *Viscous effect and inviscid cyclic loading effect in drained triaxial tests on granular materials*

The issues indicated above are examined below based on results from a series of drained triaxial tests in which sustained loading and cyclic loading histories were applied under otherwise the same conditions to a single type of granular material (i.e., Toyoura sand) as well as different types of granular materials having different particle shapes. Several different cyclic stress amplitudes and different numbers of loading cycles, among other cyclic loading parameters, were employed in these tests.

Fig. 6.3 shows typical results from a pair of stress-controlled triaxial tests at a fixed confining pressure (σ'_h = 40 kPa) on two dense specimens of air-dried Toyoura sand prepared by the air-pluviation method. Cyclic deviator stresses with a relatively small amplitude were applied with q_{max} during cyclic loading being the same with the fixed deviator stress during the corresponding sustained loading stage. The behaviours during sustained and cyclic loading histories at different shear stress levels were compared by applying these two loading histories alternatively but at different sequences to a pair of very similar specimens. The following trends of behaviour may be seen from Fig. 6.3:

1) Due likely to a relatively small amplitude of cyclic deviator stress applied in these tests, the residual strain developed by sustained loading is always larger than that by

cyclic loading under otherwise the same conditions. This trend of behaviour can also be clearly noted from the summary figures, **Fig. 6.3c**.

2) As may be seen from **Fig. 6.3b**, at the respective cyclic loading stage, noticeable residual strain develops when the deviator stress is closer to or equal to the maximum stress not only when the deviator stress is increasing but also when the deviator stress is decreasing. This fact indicates that, in these tests, the major cause for the development of residual strain during cyclic loading is the viscous property of sand, but the inviscid cyclic loading effect is insignificant, if any.

These trends of behaviour are similar to those illustrated in Fig. 6.1b and should become more relevant as the cyclic stress amplitude decreases. On the other hand, the inviscid cyclic loading effects become more important, as illustrated in Fig. 6.2b, with an increase in the cyclic stress amplitude and the number of loading cycles, as shown below.

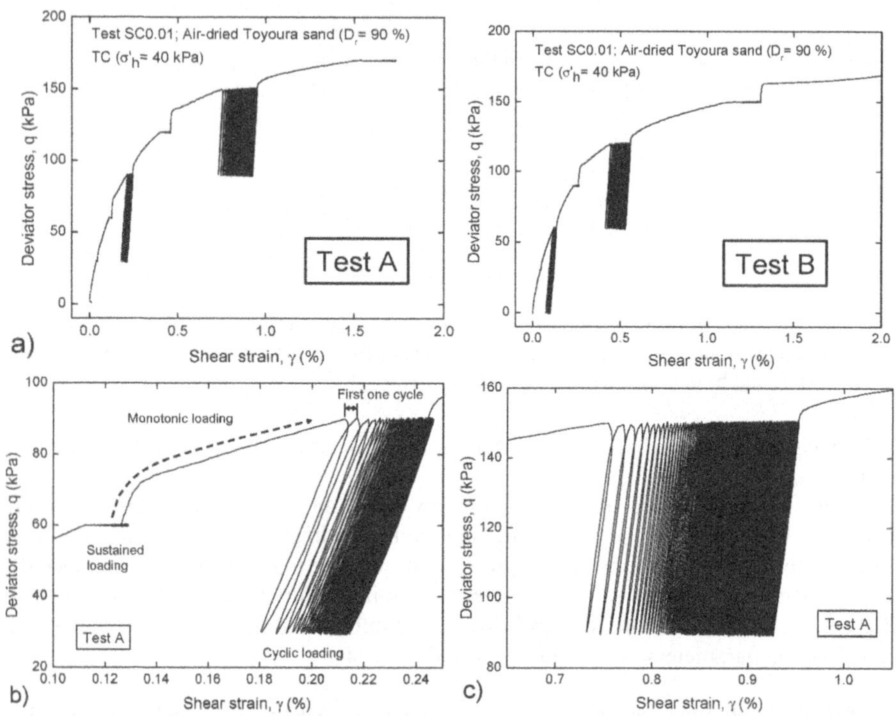

Fig. 6.5 Results from the tests using loading histories described in Fig. 6.4: a) overall stress ratio – shear strain relations; and b) & c) close-upped stress - strain relations from test A.

Figs. 6.4 and **6.5** show, respectively, the loading histories and the results from a similar series of triaxial tests (σ'_h = 40 kPa) on air-dried Toyoura sand. Both the cyclic deviator stress amplitude and the total number of loading cycles are significantly larger in these tests than in the tests described in Fig. 6.3. A close-upped stress - strain relation presented in Fig. 6.5b (test A) corresponds to the closed-upped time history of deviator stress presented in Fig. 6.4b. **Fig. 6.6** compares the residual shear strains developed by cyclic

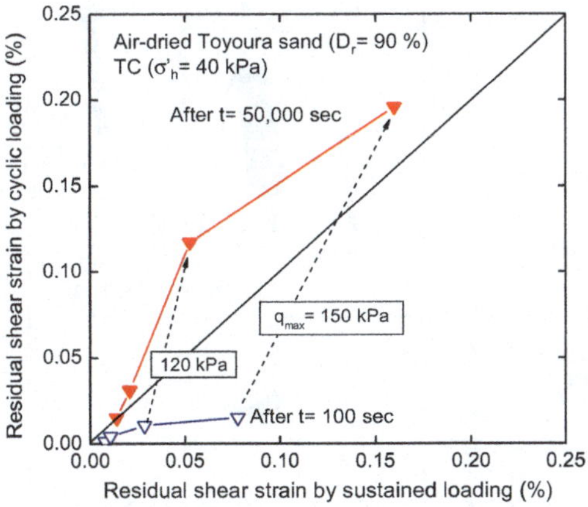

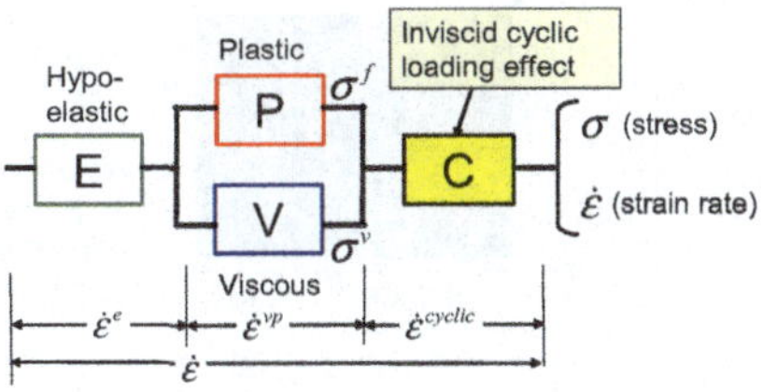

Fig. 6.6 (left) Comparison between residual strains by cyclic and sustained loading histories for short and long durations from TC on Toyoura sand (Hayashi et al., 2005, 2006).
Fig. 6.7 (right) A strain-additive model to incorporate inviscid cyclic loading effect.

and sustained loading histories for a short duration (i.e., the first 100 seconds or the first one cycle, Fig. 6.4b) and for a long duration (i.e., the whole 50,000 seconds or the whole 500 cycles) obtained from this pair of TC tests. The four data points of the respective relation were obtained at four deviator stress levels, equal to q (sustained stress) = q_{max} (the maximum stress during cyclic loading) = 60 kPa, 90 kPa, 120 kPa and 150 kPa (see Fig. 6.5a). It may be seen from Fig. 6.6 that, for the first 100 seconds (i.e., the first one unload/reload cycle), the residual shear strain by sustained loading is consistently larger than the one by cyclic loading at any deviator stress level. This test result is consistent with the one presented in Fig. 6.3. However, after a duration of 50,000 seconds (i.e., after 500 unload/reload cycles), the residual shear strain by cyclic loading becomes much larger than the one by sustained loading applied for the same duration. This result indicates that the inviscid cyclic loading effect continues for a longer duration than the viscous effect and, for a given loading duration, its importance relative to the viscous effect increases with an increase in the number of loading cycles. This point is re-confirmed below by other test results.

6.3 *Interactions between viscous effect and inviscid cyclic loading effect*

It is shown above that the inviscid cyclic loading effect on the residual strain of granular materials, which is different from the viscous effect, cannot be ignored when the cyclic deviator stress amplitude exceeds some limit and becomes more important with an increase in the number of unload/reload cycles. On the other hand, if the viscous effect and inviscid cyclic loading effect on the residual strain characteristics are totally independent of each other, such a strain-additive model as illustrated in **Fig. 6.7** may be relevant. That is, the strain increment that develops by the inviscid cyclic loading effect in component C is independent of the one taking place by the visco-plastic property in components connected in parallel, P+V. This point is examined below.

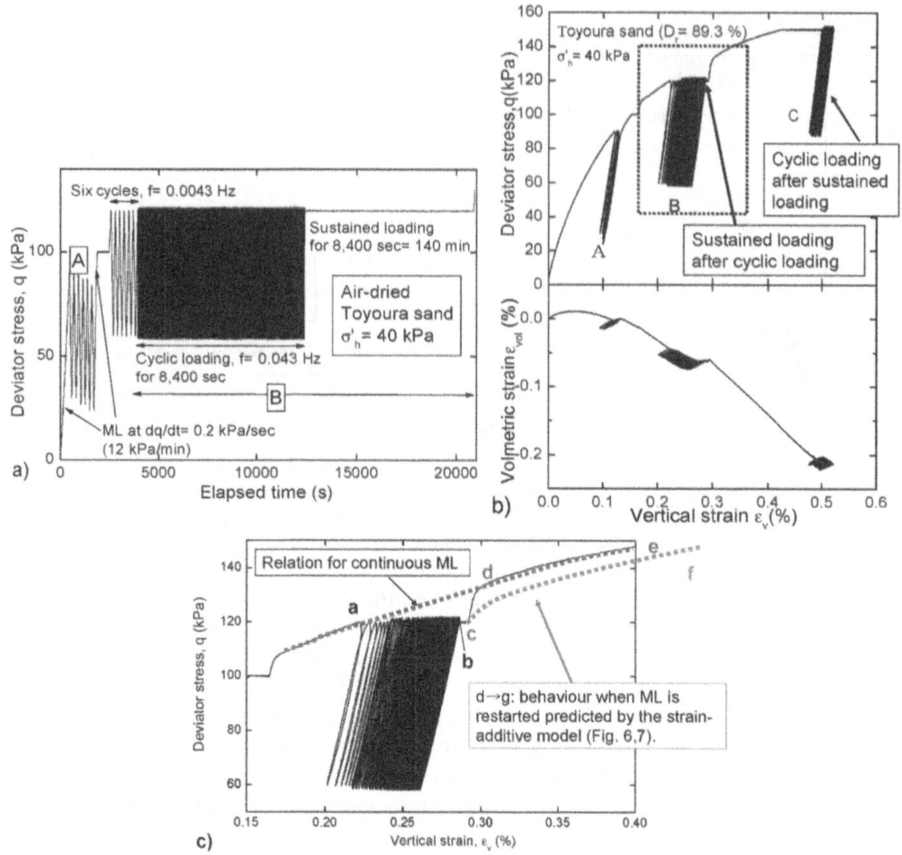

Fig. 6.8 TC tests on air-dried dense Toyoura sand to evaluate the relationship between residual strains by sustained and cyclic loading histories (Hayashi et al., 2005); a) loading history; b) overall stress - strain behaviour; and c) a close-up of stage B.

The importance of inviscid cyclic loading effect is re-confirmed while the relevance of the model presented in Fig. 6.7 is examined below based on results from another series of TC tests performed on air-dried Toyoura sand employing loading histories that combine the following two types of loading histories:

1) Before applying the main loading history shown below, pre-loading history consisting of six unload/reload cycles was applied, during which the maximum deviator stress, q_{max}, was kept constant or slightly increased or decreased. For loading stage B presented in **Fig. 6.8a**, the q_{max} value was kept constant.
2) Cyclic loading with a number of cycles equal to 360 cycles for 8.400 seconds was followed by sustained loading for 8,400 seconds keeping the sustained deviator stress the same as q_{max} during the precedent cyclic loading stage. Loading stage B presented in Fig. 6.8a is typical of the above. In the other tests, the sequence of sustained and cyclic loading histories was reversed.

Fig. 6.8 shows the results from one of these TC tests. In this test, six unload/reload cycles were applied keeping q_{max} constant before the start of loading stage B, where cyclic loading for 8,400 seconds was followed by sustained loading for 8.400 seconds. In other tests, q_{max} was increased by a factor of 1.05 or 1.10 or decreased by a factor of 0.975 or 0.95 or 0.925 during the precedent six unload/reload cycles. **Figs. 6.9a, b** and **c** compare the q - shear strain relations when q_{max} was kept constant; increased by a factor of 1.05; and decreased by a factor of 0.95 during the precedent six unload/reload cycles.

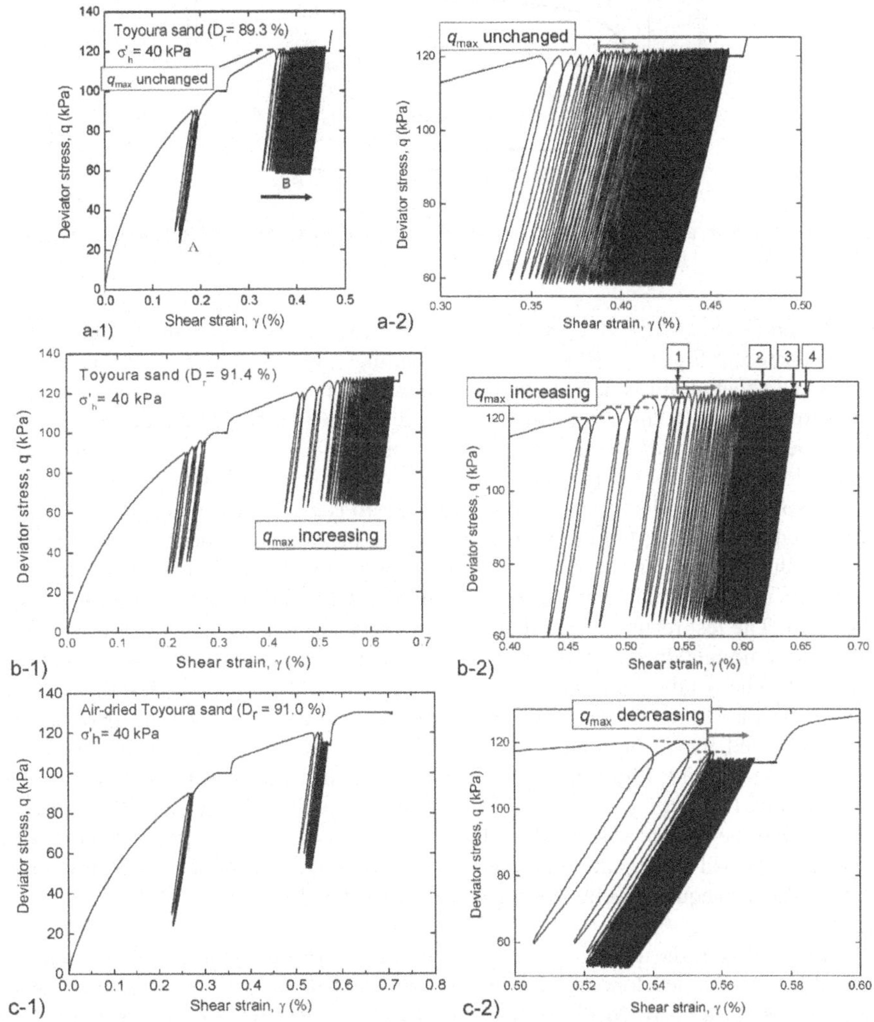

Fig. 6.9 Effects of change in q_{max} on the residual strain by subsequent cyclic loading followed by sustained loading in TC (σ'_h= 40 kPa) on Toyoura sand (Hayashi et al., 2005); q_{max} was: a) unchanged; b) slightly increased; and c) slightly decreased.

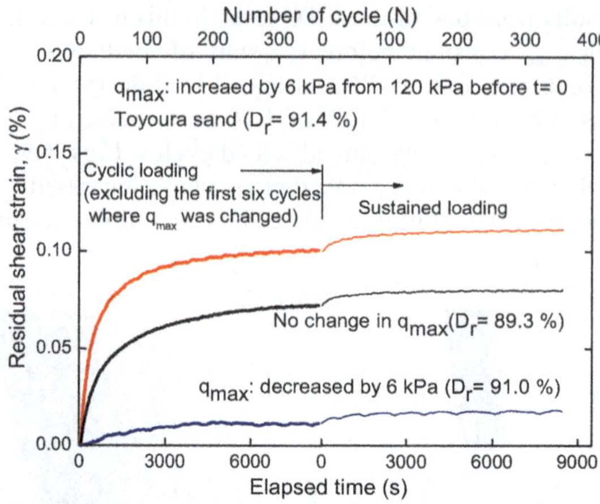

Fig. 6.10 Effects of a slight change in q_{max} on the residual shear strain by subsequent cyclic loading followed by sustained loading in TC on Toyoura sand (Hayashi et al., 2005).

Fig. 6.10 summarizes the time histories of residual shear strain from these three typical tests (presented in Fig. 6.9), in which cyclic loading was followed by sustained loading. The residual shear strain is defined zero at the start of the cyclic loading stage (as shown by an arrow in Figs. 6.9a-2, b-2 and c-2). Large effects of a slight change in q_{max} during the precedent six unload/reload cycles on the residual shear strain that took place during the six unload/reload cycles as well as the subsequent cyclic and sustained loadings may be seen. Note that the changes in q_{max} are very small (i.e., 6 kPa compared to the neutral value, equal to 120 kPa). **Fig. 6.11** summarizes the residual shear strains that have developed during the first cyclic loading stage (denoted by C) and those by the end of the subsequent sustained loading stage (denoted by C + S) obtained from the tests described in Fig. 6.10. The test results from other similar tests in which the q_{max} value was changed by different amounts during the precedent six unload/reload cycles are also summarized in Fig. 6.11. The results from another set of TC tests, similar to those described in Figs. 6.9 and 6.10, that were performed by reversing the loading sequence (i.e., first sustained loading followed by cyclic loading) under otherwise the same test conditions are also presented in this figure. **Fig. 6.12** shows results from one of these tests in which q_{max} was increased from 120 kPa by a factor of 1.05 during the precedent six unload/reload cycles applied before the start of sustained loading for 8,400 seconds. The residual shear strains that have developed during the first sustained loading stage are denoted by S and those by the end of the subsequent cyclic loading stage denoted by S + C in Fig. 6.11.

Fig. 6.13 compares the time histories of residual strain during a cyclic loading stage followed by a sustained loading stage (Fig. 6.9b) and those during a sustained loading stage followed by a cyclic loading stage (Fig. 6.12). In these tests, q_{max} was increased by a factor of 1.05 during the precedent six unload/reload cycles. The following trends of behaviour may be seen from Fig. 6.13:

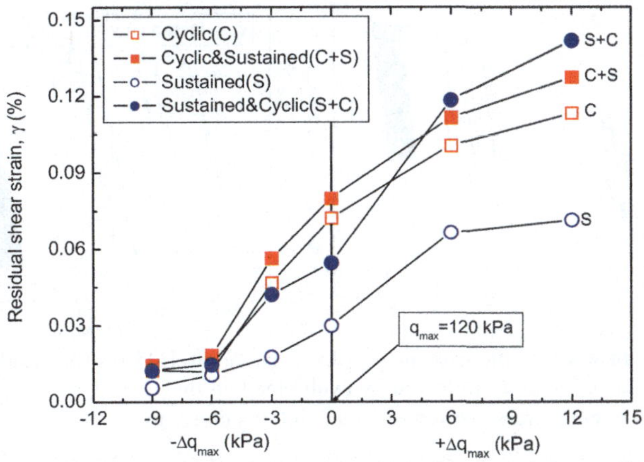

Fig. 6.11 Summary of effects of changes in q_{max} on the residual strain by cyclic and sustained loading in TC tests on Toyoura sand (σ'_h= 40 kPa) (Hayashi et al., 2006).

1) At the initial stage until the elapsed time becomes about 300 seconds (starting from initial points 1 and 5), the increasing rate of residual strain during sustained loading is larger than the one during cyclic loading. The opposite becomes true after the elapsed time becomes longer. Eventually the residual strain at the end of cyclic loading (at point 3) becomes much larger than the one at the end of sustained loading (at point 6). Moreover, after relatively large residual strain has taken place during the first sustained loading stage (5-6), noticeable residual strain still develops during the subsequent cyclic loading stage (6-7). These facts indicate significant inviscid cyclic loading effects that cannot be ignored in these tests.
2) Significant viscous effects can be seen from the development of relatively large residual strain during the initial sustained loading (5-6). A slight increase in the residual strain rate immediately after the start of the subsequent sustained loading (3-4), following the initial cyclic loading (1-2-3), may be due to an increase in the creep strain rate because q is not unloaded from q_{max} during the subsequent sustained loading. These results indicate significant viscous effects that cannot be ignored on the residual strain that develops during cyclic loading.
3) The residual strains that have developed due to both viscous and inviscid cyclic loading effects by the end of these two reversed loading sequences (i.e., at points 4 and 7) are rather similar. This trend of behaviour can be confirmed from Fig. 6.11: that is, the total residual strains indicated by the data points denoted by S + C and C + S are similar for any change in q_{max} during the precedent six unload/reload cycles.

These facts reconfirm the importance of both viscous effect and inviscid cyclic loading effect on the development of residual strain during cyclic loading.

6.4 *Examination of the strain-additive model (Fig. 6.7)*

Fig. 6.8c shows a close-up of the deviator stress ratio - axial strain relation at loading stage B indicated in Figs. 6.8a and b. If the strain-additive model (Fig. 6.7) is relevant, the residual strain that takes place at stage a→b→c is a linear summation of the

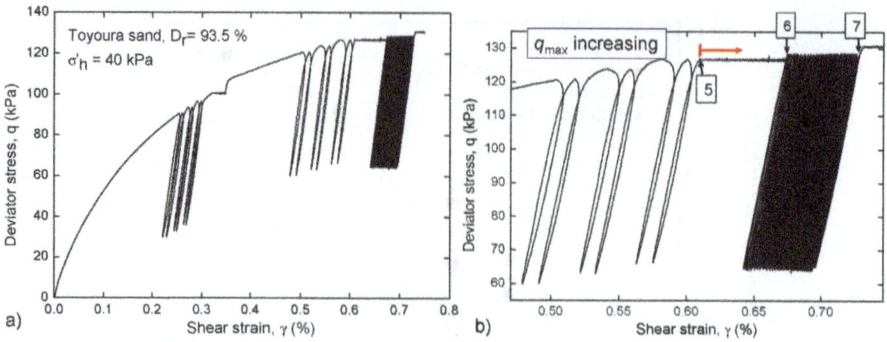

Fig. 6.12 Effects of a slight increase in q_{max} by a factor of 1.05 on the residual strain by subsequent sustained loading followed by cyclic loading in TC on Toyoura sand (Hayashi et al., 2005); a) overall stress − strain relation; and b) a close-up.

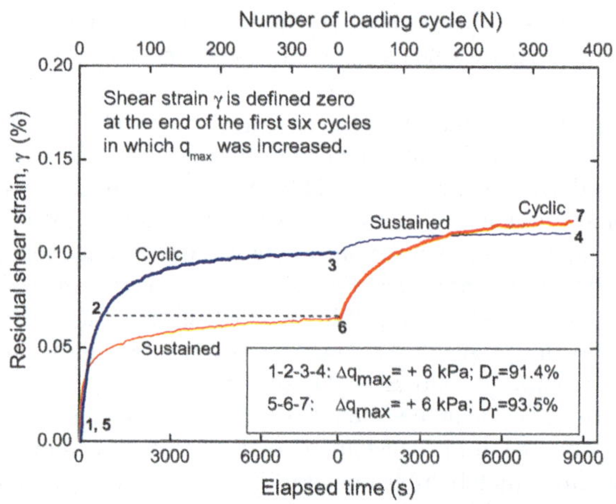

Fig. 6.13 Time history of residual strain by cyclic loading followed by sustained loading compared to the one by sustained loading followed by cyclic loading (q_{max} increased by a factor of 1.05 before the start of loading), TC on Toyoura sand (Hayashi et al., 2005).

components that take place in components in parallel V+P and component C. As the residual strain that has taken place in component C does not contribute to the strain-hardening process of component P, the stress–strain relation after ML is restarted at the original loading rate should become like c→f. This behaviour is different from the actual behaviour c→d→e, where clear yielding takes place at point d and then the stress-strain relation rejoins the one that is obtained by continuous ML at the original loading rate.

In addition, it may be seen from Figs. 6.9 and 6.10 that a slight change in q_{max} during the precedent six unload/reload cycles results into a significant change in the increasing rate of residual shear strain during the six unload/reload cycles as well as the subsequent cyclic loading stage. This trend of behaviour can be confirmed by the relation denoted by C in Fig. 6.11: i.e., the residual shear strain by the subsequent cyclic loading increases at

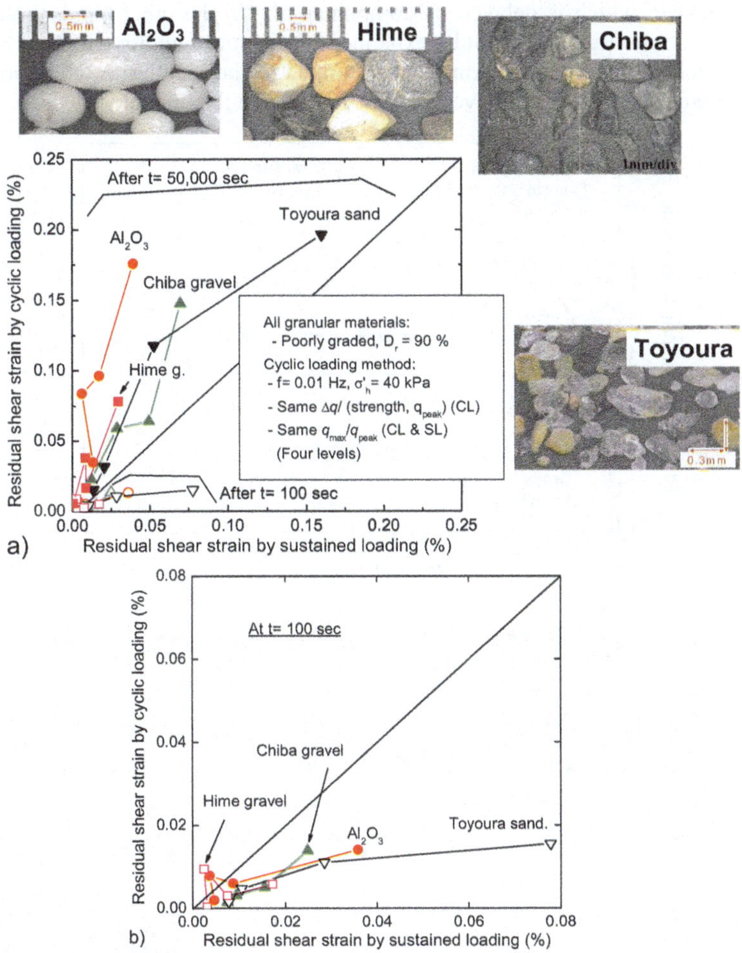

Fig. 6.14 Effects of particle shape on the relative largeness between residual strains by sustained and cyclic loading histories in TC on sands (Hayashi et al., 2006): comparison after; a) 100 seconds and 50,000 seconds; and b) 100 seconds.

a high rate with an increase in q_{max} during the precedent six unload/reload cycles, while it becomes nearly zero when q_{max} is decreased by only 6 kPa (i.e., 5 % of the neutral shear stress, 120 kPa). These trends of behaviour can be interpreted according to the non-linear three-component (Fig. 3.7a). That is, when q_{max} is increased (i.e., when σ is increased), component P yields by an increase in the inviscid stress, σ^f, while the viscous stress, σ^v, of component V increases, resulting in an increase in the viscous effect. These two factors increase the irreversible strain rate during the subsequent cyclic loading stage. On the other hand, a decrease in q_{max} suppresses the yielding of component P and decreases σ^v, resulting in a decrease in the residual strain rate during the subsequent cyclic loading stage. In short, the residual strain rate during cyclic loading is controlled by the yielding of component P, which is delayed by component V. On the other hand, these trends of behaviour described above cannot be explained by the strain-additive model (Fig. 6.7), in which the behaviour of component C is not linked to the yielding of component P. The introduction of such a link as above may just complicate this model.

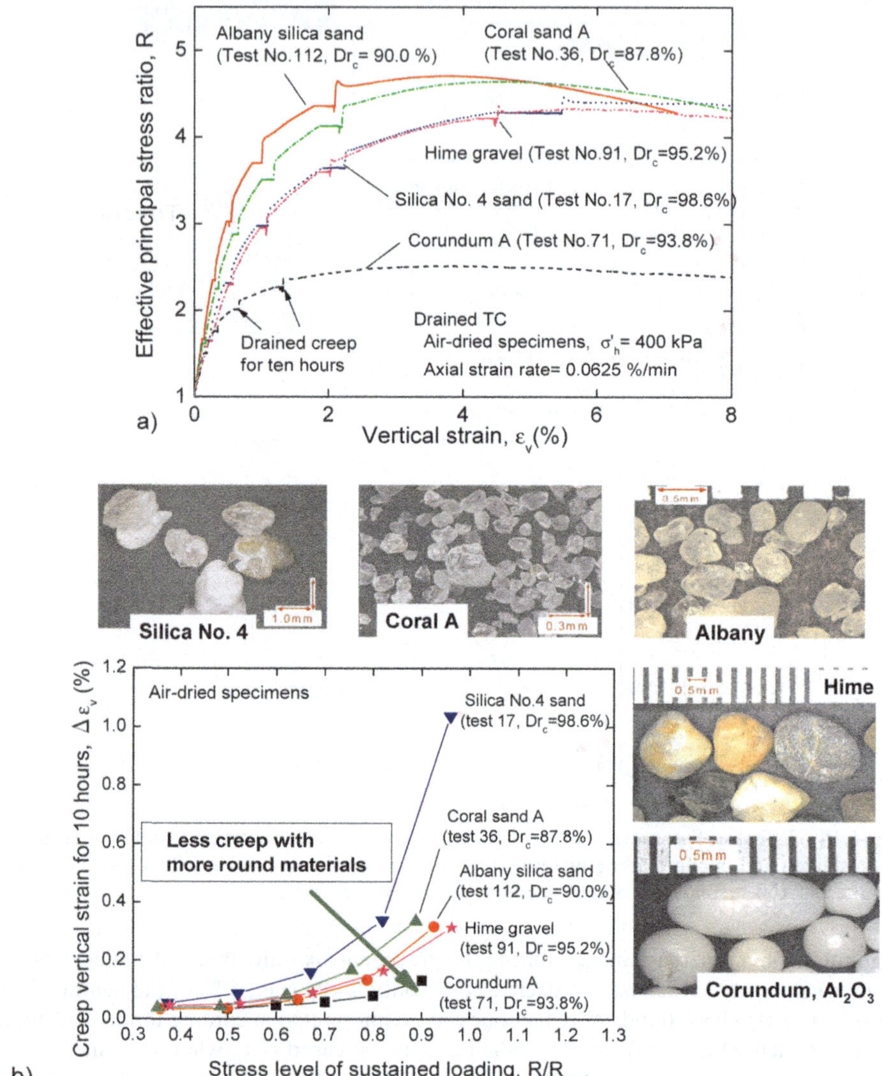

Fig. 6.15 Effects of particle shape on the residual strain in TC on dense granular materials (Enomoto et al., 2006); a) overall stress – strain behaviour; and b) creep vertical strain versus sustained load level.

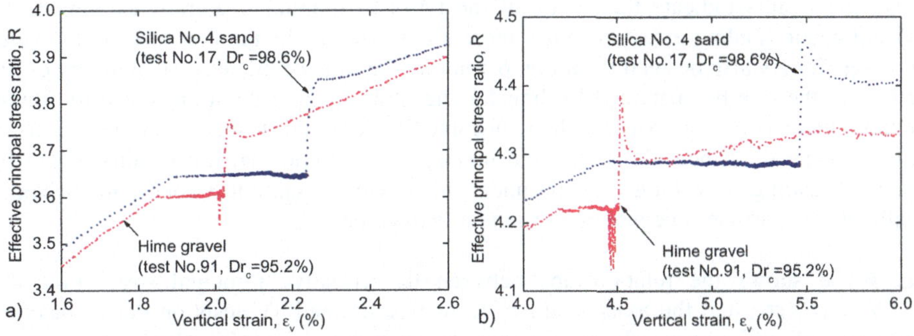

Fig. 6.16 Comparison of creep strain during a sustained loading for 10 hours between silica No.4 sand and Hime gravel, from Fig. 6.15a.

Moreover, the effects of a slight change in q_{max} during the precedent six unload/reload cycles on the residual strain rate during the subsequent cyclic loading stage described above (i.e., relation C in Fig. 6.11) are similar to those on the creep strain rate during the subsequent sustained loading stage (i.e., relation S). Similarly, as seen from Fig. 6.13, the residual strain (6-7) during the cyclic loading applied subsequently to the first sustained loading is noticeably smaller than the one during the first cyclic loading (1-2-3). The residual strain during the sustained loading (3-4), applied subsequently to the first cyclic loading (1-2-3), is significantly smaller than the one during the first sustained loading (5-6). That is, the residual strain by a cyclic loading history is strongly affected by the residual strain that has taken place in advance by sustained loading, and vice versa. These facts also indicate that the residual strains by cyclic and sustained loading histories have a common basis. Therefore, we can conclude that the strain-additive model (Fig. 6.7) is not relevant but the inviscid cyclic loading effect should be incorporated in some way into component P and perhaps also component V of the three-component model (Fig. 3.7a).

6.5 *Effects of particle shape on viscous effect and inviscid cyclic loading effect*

Fig. 6.6 compares the residual strains developed by cyclic and sustained loading histories for short and long durations obtained from a pair of TC tests on Toyoura sand. **Fig. 6.14** shows data from other similar TC tests performed on several similarly poorly-graded granular materials having different particle shapes, added to those presented in Fig. 6.6. The grading curves of these granular materials are presented in Fig. 4.18. The sustained and cyclic deviator stresses in the same ratios to the compressive strength, q_{peak}, from the respective drained TC test were applied in these tests. It may be seen from **Fig. 6.14b** that, except for two data points at low deviator stress levels of Hime gravel and corundum A, the residual shear strain by a single unload/reload cycle applied for a duration of 100 seconds is noticeably smaller than the one by sustained loading applied for the same duration at the fixed deviator stress that is the same as the maximum deviator stress during the cyclic loading. It may be seen that the general trend of behaviour is very similar among these granular materials having different particle shapes. On the other hand, it may be seen from **Fig. 6.14a** that the residual strain by 500 unload/reload cycles applied for a duration of 50,000 seconds (i.e., 13.89 hours) is noticeably larger than the one by the corresponding sustained loading applied for the same duration.

These test results indicate that, with all the types of granular materials examined, the inviscid cyclic loading effect becomes more significant as the number of loading cycles increases. It may also be seen from Fig. 6.14a that the ratio of the residual strain by cyclic loading to the one by sustained loading for the same loading duration (equal to 50,000 seconds) increases as the particle shape becomes more round. It seems that the effects of particle shape described above are due mainly to the trend that the residual strain by sustained loading for a long duration under otherwise the same test conditions becomes smaller as the particles become more round as shown below.

Fig. 6.15a shows the relationships between the effective principal stress ratio $R = \sigma'_v/\sigma'_h = \sigma'_1/\sigma'_3$ and the axial strain ε_v from five drained TC tests on dense air-dried specimens of different poorly-graded granular materials having similar coefficients of uniformity (Fig. 4.18) but having different particle shapes. The relative densities of these specimens are similar. In all the tests, drained sustained loading for ten hours were performed several times during otherwise ML at the same constant axial strain rate (0.0625 %/min.). **Fig. 6.15b** shows the relationships between the residual axial strain at the end of the respective sustained loading and the sustained stress level, R/R_{peak}, obtained from the test results presented in Fig.6.15a, where R is the principal stress ratio at which the respective sustained loading was performed and R_{peak} is the peak R value in the respective TC test. The following trends of behaviour may be seen from Fig. 6.15b:

1) The residual strain increases with an increase in the sustained load level, R/R_{peak}, with all the types of granular materials examined.
2) The residual strain at the same R/R_{peak} decreases as the particle shape becomes more round.

The magnitude of residual strain has no systematic link to the value of R at which sustained loading was performed among these different materials. For example, the residual strain at R= 4.29 of Silica No.4 sand is largest among this series of TC tests, whereas the residual strains at similar R values of Albany silica sand, Coral sands and Hime gravel are much smaller. This point can be seen very well from **Fig. 6.16**. That is, for the same sustained deviator stress compared to nearly the same peak strength, the residual strain of Silica No. 4 sand is much larger than that of Hime gravel. A number of small unload/reload cycles that took place during the sustained loading stage in the test on Hime gravel is considered to have insignificant effects on the residual strain rate because of a very small cyclic stress amplitude. Moreover, the general trend of behaviour is not significantly affected by some scatter in the relative densities among the different TC tests. For example, as shown from Fig. 6.15, the creep strain of Silica No. 4 sand (an angular sand), which has a D_r= 98.6 %, is consistently larger than the respective value at the same R of Hime gravel (a round gravel) having a smaller D_r, equal to 95.2 %.

The trends of the effects of particle shape on residual strains developed by cyclic and sustained loading histories are summarized in Table 4.1. It seems that, with round stiff particles, inter-particle contact points are relatively stable when subjected to fixed shear and normal loads. This mechanism increases the stability of the whole fabrics against sustained loading. On the other hand, inter-particle contact points become less stable when subjected to cyclic shear loads than when subjected to fixed stresses and this mechanism becomes more important with more round particles, increasing the ratio of residual strain by cyclic loading to that by sustained loading with round stiff granular materials. It is likely that these particle shape effects on residual strains by sustained and cyclic loading histories are linked to those on the viscosity type and the viscosity type transition. However, the details of this link are not known.

Table 7.1 Two components of "the time effects"

	Phenomenon	Mechanism	Parameter	
e.g., cement-mixed soil	Ageing effect	Time-dependent changes in the material property+	Time with a fixed origin (t_c)	
	Apparent ageing Loading rate effect (creep, stress relaxation, etc.)	Material viscosity	Strain rate, $\dot{\varepsilon}^{ir}$	e.g., air-dried Sand*

+ Positive ageing: e.g., cementation
Negative ageing: e.g., weathering

* excluding geological effects

7. AGEING EFFECTS

7.1 General

The "so-called time effect" on the stress-strain behaviour of geomaterial consists of the following two different components (**Table 7.1**; Tatsuoka et al., 2000, 2001, 2003a):

1) Loading rate effect due to the material viscosity, as discussed in details in the precedent chapters.
2) Ageing effect due to time-dependent changes in the material properties, which is discussed in relation to the viscous effect in this chapter.

It should be noted that the basic parameters to describe these two components, which are due to different mechanisms, should be different. For the ageing effect, we can use the time having the origin that is defined as zero when the ageing effect starts developing. For the viscous effect, the irreversible strain rate is the basic parameter, while the general time cannot be the basic parameter, although some existing constitutive models for geomaterials can be categorised into isochronous theories using the general time. These two components of time effect should be properly and separately taken into account to correctly describe and predict the stress - strain behaviour of geomaterial when they are both important. The ageing effects on the elastic property, the inviscid stress-strain behaviour and the viscous property (i.e., the ageing effects on the properties of three components, E, P and V in Fig. 3.7a) may be different.

7.2 *Interactions between viscous and ageing effects developing during sustained loading*

Fig. 7.1 illustrates the development of a high stiffness zone by ageing effects additive to viscous effects by sustained loading at stress state S. This is the case of test 3 illustrated

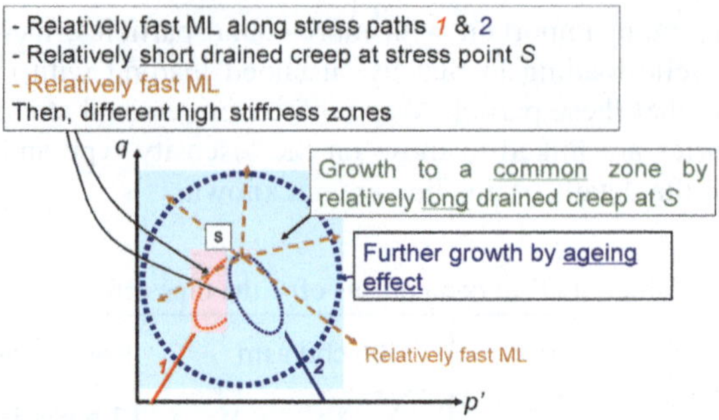

Fig. 7.1 Ageing effect on the size of high stiffness zone (revised from Tatsuoka et al., 1999a).

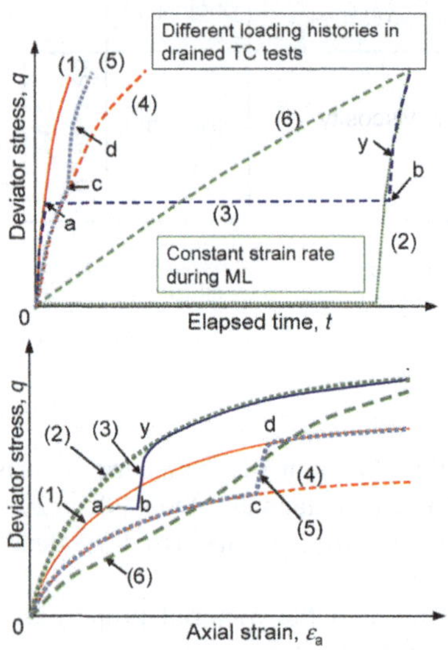

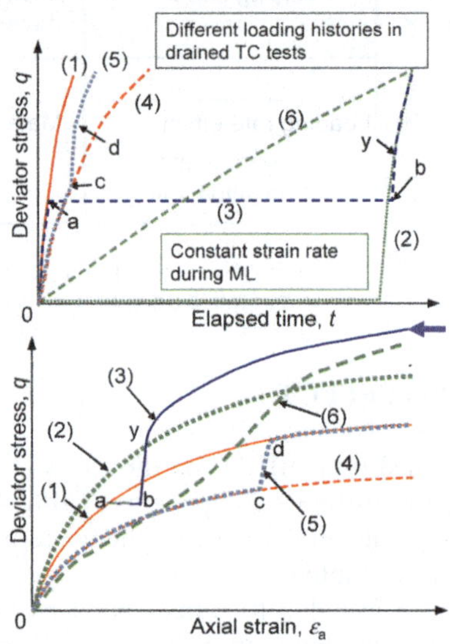

Fig. 7.2 (left) Response of an elasto-viscoplastic material with ageing effect (no interaction between Isotach viscosity and ageing effect).

Fig. 7.3 (right) Response of an elasto-viscoplastic material with ageing effect (positive interaction between Isotach viscosity and ageing effect).

in **Figs. 7.2** and **7.3**. Without ageing effects, a smaller high stiffness zone develops only by viscous effects, as illustrated in Figs. 3.1 and 3.3.

In Figs. 7.2 and 7.3, loading history until point *b* is different between tests 2 and 3. In these tests, yield point *y* appears at the end of high-stiffness zone upon the restart of ML from point *b*. The yield deviator stress, *q*, at yield point *y* in test 3 relative to the stress-strain relation in test 2 is different between the two cases illustrated in Figs. 7.2 and 7.3. In Fig. 7.2, the viscous and ageing effects are always independent and the stress - strain relation after yield point *y* in test 3 rejoins the corresponding one in test 2, showing that the ultimate strength for the same total time since the start of ageing effect and at the same strain rate is the same, independent of precedent loading history. On the other hand, in Fig. 7.3, the viscous and ageing effects interact with each other in a positive way. In this case, the stress - strain relation after yield point *y* in test 3 is located above the corresponding one in test 2, showing that the ultimate strength for the same ageing period and at the same strain rate is larger in test 3 due to longer ageing at higher deviator stresses. As shown below, both cases are possible depending on geomaterial type.

Fig. 7.4 shows the first case (no interaction between ageing and viscous effects, Fig. 7.2)

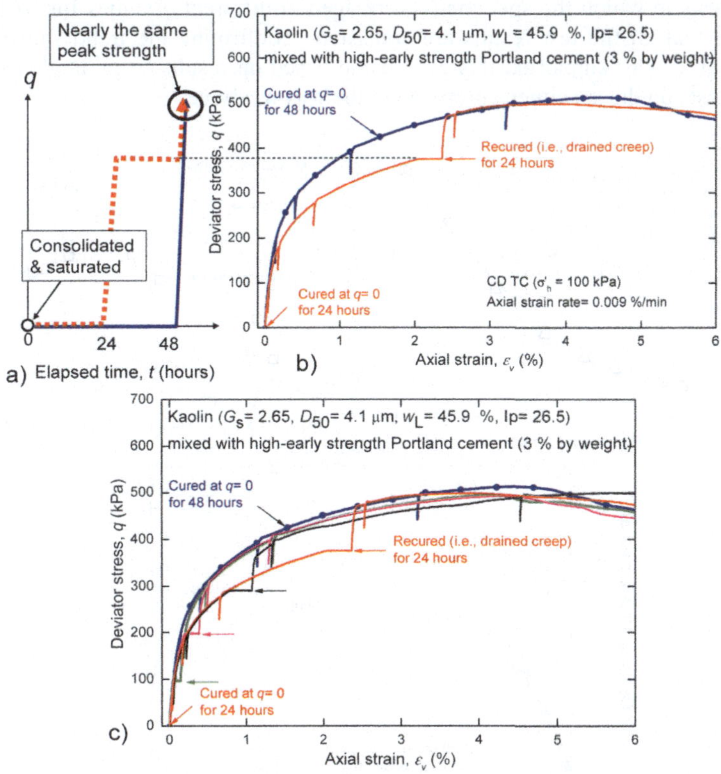

Fig. 7.4 Nearly no interaction between ageing and viscous effects in CD TC on a relatively weakly cemented geomaterial (Komoto et al., 2003): a) loading histories; b) comparison between two tests; and c) comparison among all the tests.

obtained from a series of CD TC tests on relatively weakly cemented kaolin. Dry powder of kaolin (G_s= 2.65; D_{50}= 4.1 μm; w_L= 45.9 %; and I_p= 26.5) was first mixed with air dried powder of high-early strength Portland cement (3 % by weight) without water and compacted to produce TC specimens of 100 mm in height and 50 mm in diameter. After isotropic compression to 100 kPa, the specimens were made saturated to start the development of inter-particle bonding due to cement hydration. The ageing time was defined zero at this moment. As shown in **Fig. 7.4a**, in the first test, a saturated specimen was aged for 48 hours under the isotropic stress conditions before the start of drained TC loading toward the ultimate failure. In the second test, another saturated specimen was aged for 24 hours before the start of drained TC loading. When the deviator stress, *q*, became 375 kPa, the specimen was aged for another 24 hours before the restart of ML at the original strain rate. It may be seen from **Fig. 7.4b** that the behaviour of these two specimens are like those in tests 2 and 3 illustrated in Fig. 7.2. That is, a high-stiffness zone developed for a large stress range upon the restart of ML in the second test. However, no specific effects of ageing at q= 375 kPa were observed on the stress-strain behaviour after yielding in the second test. **Fig. 7.4c** shows the results from these and other three tests in which the specimens were aged at different q values for 24 hours. It may be seen that all the test results are consistent, confirming the trend shown above. Sorensen et al. (2006) supported this conclusion based on results from another series of drained TC tests on the specimens prepared in the way as above.

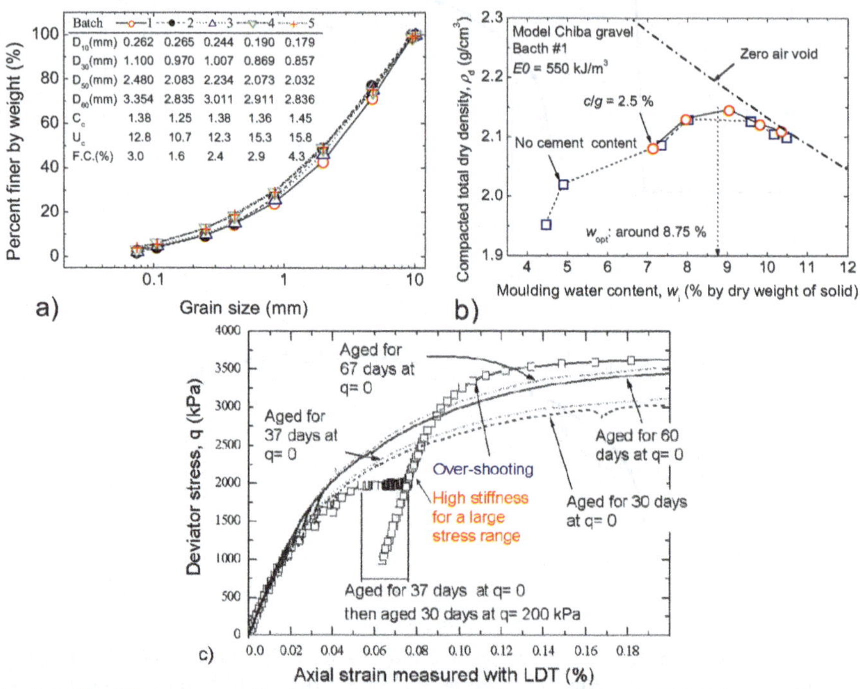

Fig. 7.5 Significant interaction between ageing and viscous effects in CD TC on a compacted moist cement-mixed gravel (σ'_h= 19.7 kPa and an axial strain rate of 0.03 %/min; Kongsukprasert & Tatsuoka, 2004 & 2005); a) grading curves; b) compaction curve; and c) typical set of test results.

Fig. 7.5 shows the second case (illustrated in Fig. 7.3), obtained from a series of CD TC tests on well-compacted moist cement-mixed well-graded gravelly soil, which was much denser and stronger than the cement-mixed kaolin described above. In this case, positive interactions between the viscous and ageing effects are significant. The gravel soil, which has the grading characteristics presented in **Fig. 7.5a**, was mixed with normal Portland cement (2.5 % by weight) at the optimum water content shown in **Fig. 7.5b** to a dry density equal to 2.0 g/cm^3. Rectangular prismatic TC specimens (95 mm x 95 mm x 190 mm) were prepared by compaction. **Fig. 7.5c** shows results from six drained TC tests (σ'_h = 19.7 kPa) on specimens that were aged in different ways. Five specimens were aged at zero deviator stress (i.e., q= 0) for 30 days, 37 days, 60 days and 67 days before the start of drained TC at an axial strain rate of 0.03 %/min toward the ultimate failure. The last specimen was aged first at q= 0 for 37 days and then at q= 200 kPa for other 30 days during otherwise ML at a constant strain rate. It may be seen that the last specimen exhibits an ultimate strength that is noticeably larger than the one aged at q= 0 for the same total duration (i.e., 67 days). A number of similar tests were performed as shown in **Fig. 7.6**. All the test results show that the ultimate peak strength for the same total duration of ageing at the same strain rate becomes larger when aged longer at higher shear stresses (Fig. 7.6a). This trend of behaviour indicates that the strength for the same total duration of ageing can become larger when inter-particle bonding develops after the fabrics have been modified by shear strains to resist more efficiently against larger shear stresses. It is not known why the above is the case with densely compacted granular materials but not with weakly cemented kaolin.

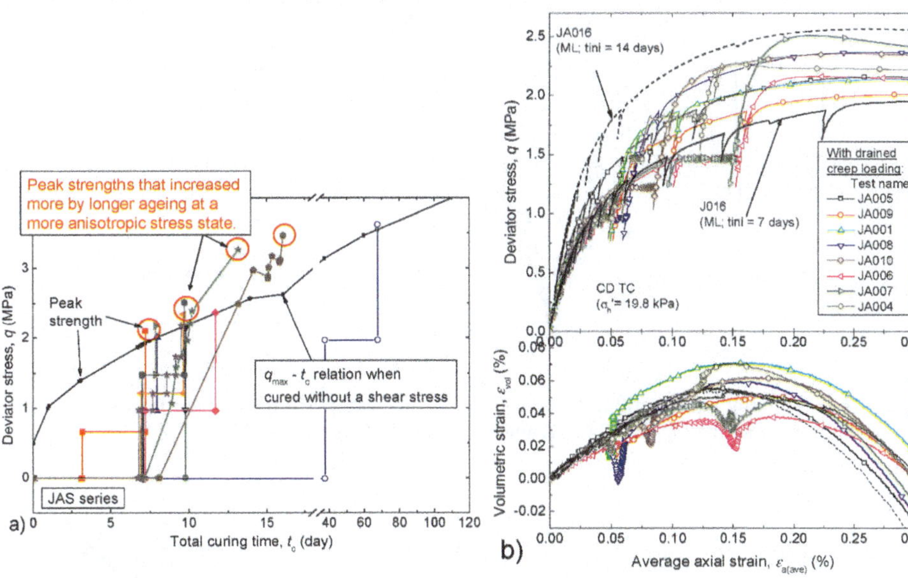

Fig. 7.6 Significant interactions between ageing and viscous effects in CD TC tests on a compacted moist cement-mixed gravel (σ'_h= 19.7 kPa and axial strain rate of 0.03 %/min; Kongsukprasert & Tatsuoka 2005); a) loading histories; b) overall stress - strain behaviour (except for two tests at extreme low strain rates, presented in Fig. 7.7).

In two tests among those indicated in Fig. 7.6a, continuous ML was performed at extremely low strain rates with the total curing period when the peak strength was mobilized being 14 days. The stress-strain relations are presented in **Fig. 7.7**. It may be seen that the peak strength in these tests are noticeably higher than the one from a ML test performed at a much higher strain rate on a specimen that was aged at q= 0 for the same total ageing period, 14 days. In one of these two tests, the strain rate was stepwise increased or decreased during otherwise ML at a constant strain rate. As illustrated in **Fig. 7.8a** and as seen from the test result presented in **Fig. 7.8b**, upon a step decrease in the strain rate, the deviator stress, q, suddenly decreases due to the viscous effect, which is followed by an increase in the increasing rate of q by which the q value subsequently over-shoots the stress-strain curve that would be obtained if the strain rate had not been decreased. The reversed behaviour takes place when the strain rate is stepwise increased. These peculiar trends of behaviour are due to viscous effects that are combined with changes in the developing rate of ageing effect per strain by changes in the strain rate. This test result shows clearly that both viscous and ageing effects on the stress - strain behaviour should be incorporated in constitutive models for geomaterials, as shown in the next section, to describe the stress - strain behaviour when both effects are important.

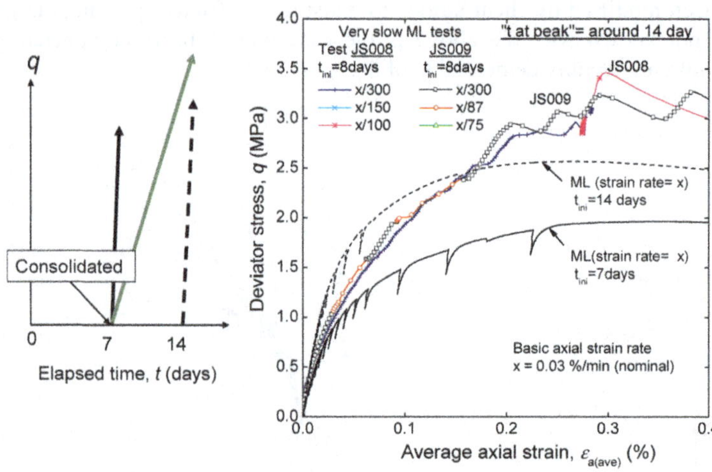

Fig. 7.7 Continuing ageing effects in very slow CD TC tests performed under the test conditions described in Fig. 7.6a (Kongsukprasert & Tatsuoka, 2005).

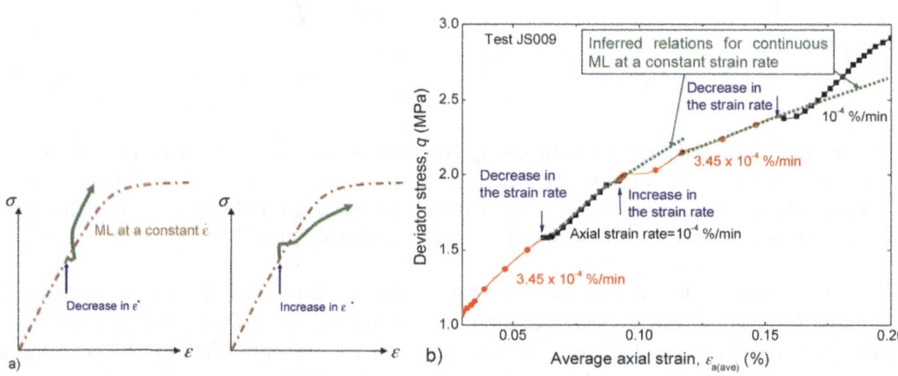

Fig. 7.8 Peculiar behaviour due to continuing ageing effect upon a step change in the strain rate during otherwise ML at a constant strain rate; a) illustration; and b) close-up of the test result presented in Fig. 7.7.

7.3 *Simulation of simultaneous viscous and ageing effects*

The method to incorporate the ageing effects into the non-linear three-component model (Fig. 3.7a) is discussed in this section. For simplicity, no interaction between the viscous and ageing effects is assumed (i.e., the case of Fig. 7.2: discussions taking into this interaction are given in Tatsuoka et al., 2003a). Furthermore, possible ageing effects on the viscosity function (Eq. B4) are ignored. **Figs. 7.9a** and **b** show the loading histories of considered four ML tests and their stress-strain curves obtained by simulations taking into account the positive ageing effect as well as the viscous effect, similar to those illustrated in Fig. 7.2 (in the case of Isotach viscosity). The ultimate strength is the same in tests *c* and *d* due to no interaction between the viscous and ageing effects. The ultimate strength in test *b* is smaller than the one for the same total duration of ageing in tests *c* and *d* due to a significantly lower strain rate at failure in test *b*.

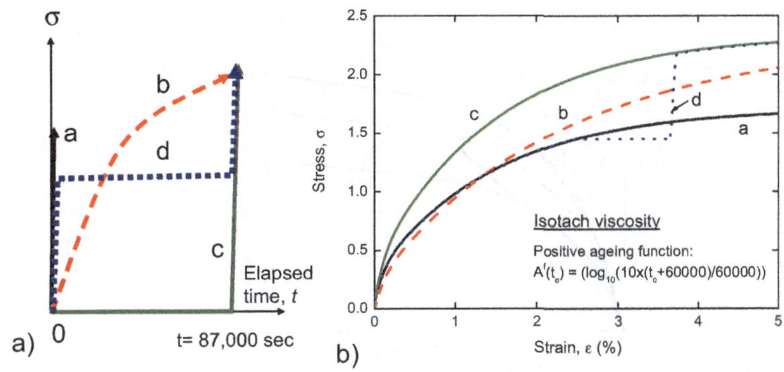

Fig. 7.9 Simulation of positive ageing effect (in case of no interaction between Isotach viscosity and ageing effect).

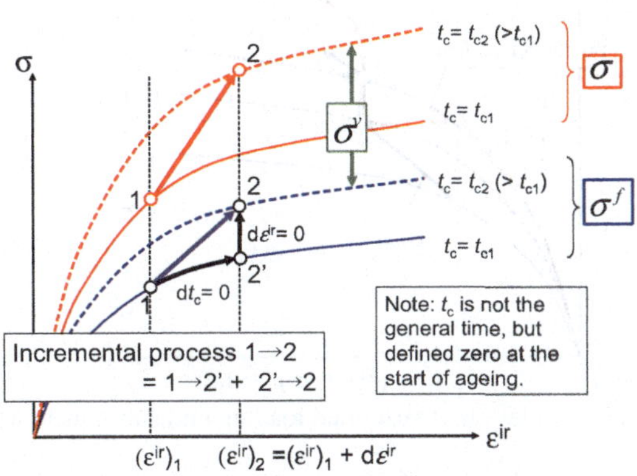

Fig. 7.10 Essence of the simulation of simultaneous ageing and viscous effects in case of no interaction between Isotach viscosity and ageing effect (Tatsuoka et al., 2003a).

Fig. 7.10 illustrates the essence of this simulation (in the case of Isotach viscosity). Suppose that we have reached state *1*, where the elapsed time since the start of ageing effect is equal to t_{c1} and the irreversible strain is equal to $(\varepsilon^{ir})_1$. Suppose that we are moving towards state *2*, where $t_c = t_{c2} = t_{c1} + dt_c$ and $\varepsilon^{ir} = (\varepsilon^{ir})_2 = (\varepsilon^{ir})_1 + d\varepsilon^{ir}$, by increasing t_c by amount of dt_c and ε^{ir} by amount of $d\varepsilon^{ir}$. The strain-hardening process of the inviscid stress, σ^f, during this incremental process *1*→*2* consists of the following two sub-processes:

Sub-process *1*→*2'*: ε^{ir} increases by amount of $d\varepsilon^{ir}$ at a fixed time (i.e., $dt_c = 0$). Despite that the ageing effect does not develop during this sub-process, the slope *1-2'* is the one that has increased by ageing effects until state *1*.

Sub-process *2'*→*2*: t_c increases by amount of dt_c without an increase in ε^{ir} (i.e., $d\varepsilon^{ir} = 0$). This sub-process represents pure on-going ageing effects, which increase with an increase in dt_c.

During loading process *1*→*2*, ε^{ir} is always increasing and, therefore, the value of σ^f is always equal to its yield stress, $(\sigma^f)_y$.

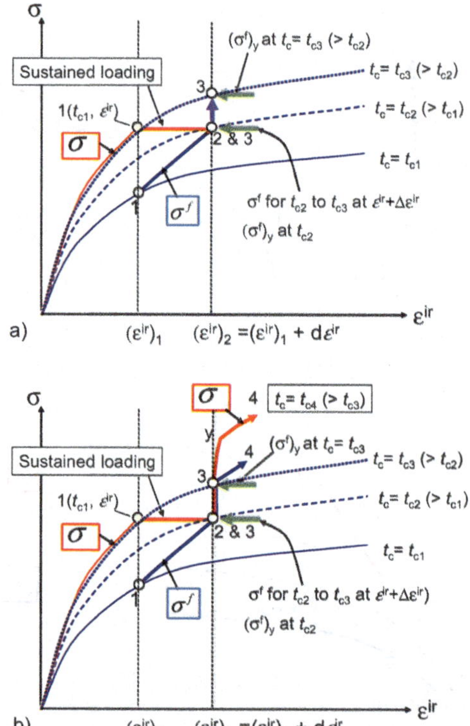

Fig. 7.11 Essence of the simulation of sustained loading and subsequent ML.

Figs. 7.11a and **b** illustrate the sustained loading process at a fixed total stress, σ, (*1*→*2*→*3*), followed by the restart of ML at a constant strain rate (*3*→*4*). During the first stage *1*→*2* described in Fig. 7.11a, the creep strain increases until the viscous stress σ^v becomes zero while t_c increases from t_{c1} to t_{c2} and ε^{ir} increases from $(\varepsilon^{ir})_1$ to $(\varepsilon^{ir})_2$. When state *2* is reached, the value of σ^f becomes the same as the constant sustained stress, σ. The value of σ^f increases not only by an increase in the irreversible strain but also by an increase in the elapsed time, t_c. Therefore, at the latter stage *2→3* of the subsequent sustained loading, only the elapsed time, t_c, increases from t_{c2} to t_{c3} without an increase in ε^{ir}, while the yield stress for σ^f, $(\sigma^f)_y$, continues increasing toward the value at point *3* solely by ageing effects. As ε^{ir} does not increase, the σ^f value remains the same as the value at point *2*. As shown in Fig. 7.11b, when ML is restarted at a constant strain rate from state *3*, the yielding of σ^f does not take place until it reaches the current yield stress $(\sigma^f)_y$ at state *3*. Once the yielding of σ^f starts, process *3→4* starts and the viscous stress σ^v is re-activated. The total stress σ increases accordingly, exhibiting a large high-stiffness stress zone *2→y*, as in test *d* illustrated in Fig. 7.9b.

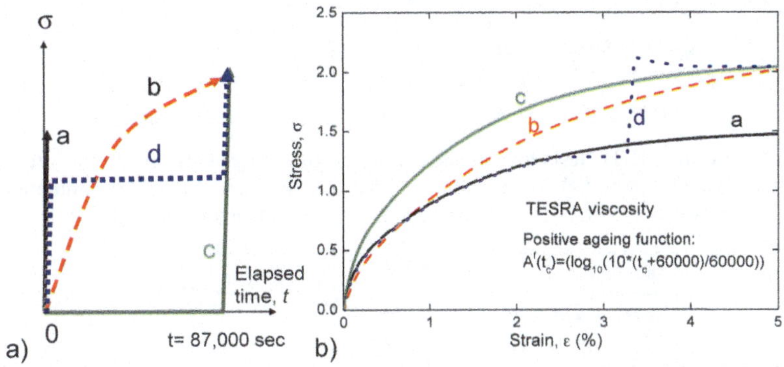

Fig. 7.12 Simulation of positive ageing effect (no interaction with TESRA viscosity; Tatsuoka et al., 2003a).

Fig. 7.12 shows a simulation of positive ageing effects when the viscosity is of TESRA type under otherwise the same conditions as the one shown in Fig. 7.9. In this case, due to the TESRA viscosity, the stress in test *d* temporarily overshoots the stress-strain relation in test *c* after ML is restarted at the original strain rate following a drained sustained loading stage. Due also to the TESRA viscosity, the ultimate strength in test *b* (at an extremely low strain rate) is the same as the one for the same total ageing period in tests *c* and *d* (at a much higher strain rate).

Fig. 7.13a shows the results from three CD TC tests on lightly compacted cement-mixed sand (Kongsukprasert et al., 2001). Natural sand from Aomori (G_s= 2.80, U_c= 3.0 & D_{max}= 2 mm) was mixed with Portland cement (4.36 % by weight) at water content of about 22.5 %, close to the optimum water content. The specimens were compacted to a dry density of 1.23 - 1.24 g/cm^3 and cured under the atmospheric pressure at constant water content for 11 days. The specimens, moist as prepared, were isotropically consolidated at 200 kPa and aged for 20 hours (tests A11APSC and C11APSC) and 92 hours (test Cc11APSC) before the start of ML drained TC (σ_3'= 200 kPa) at $\dot{\varepsilon}_a$= 0.03 %/min. Only in test A11APSC, the specimen was cured again at an anisotropic stress state for 72 hours during otherwise ML at constant $\dot{\varepsilon}_a$. The effects of curing at an anisotropic stress state seen in this figure are essentially the same as the one seen in test *d* illustrated Fig. 7.12. **Fig. 7.13b** shows the simulation of the test results presented in Fig. 7.13a assuming the TESRA viscosity with r_1= 0.001. The observed trends of behaviour are well simulated. It may be seen that a large temporary over-shooting seen in the test (Fig. 7.13a) can be interpreted as the TESRA viscosity in the simulation (Fig. 7.13b).

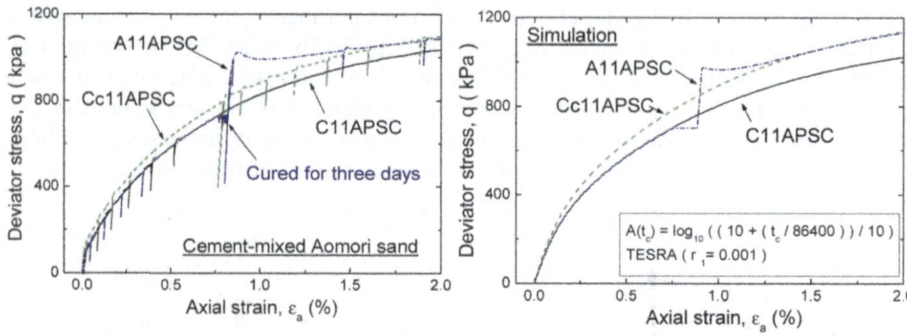

Fig. 7.13 a) Two ML CD TC tests to evaluate the effects of curing at an anisotropic stress state (after the stresses corrected for a scatter of dry density); and b) model simulation (no interaction between TESRA viscosity and ageing effect: Tatsuoka et al., 2003a).

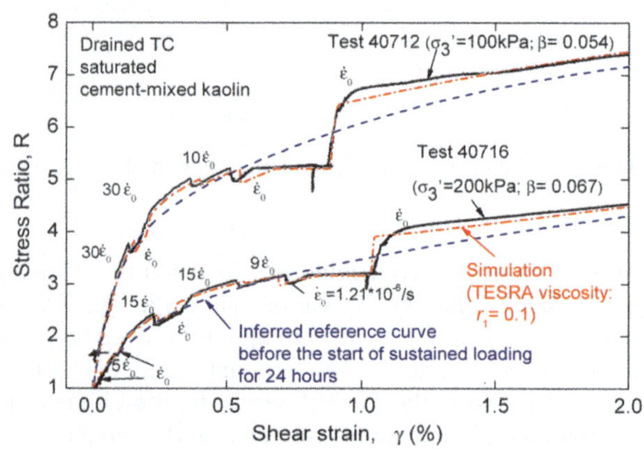

Fig. 7.14 Two ML CD TC tests on saturated cement-mixed kaolin and their simulation (no interaction between TESRA viscosity and ageing effects: Deng & Tatsuoka, 2006).

Deng and Tatsuoka (2006, this volume) reports the simulation of the results from two drained TC tests on saturated cement-mixed kaolin in which the strain rate was stepwise changes many times and sustained loading for 24 hours was performed one time during otherwise ML at a constant strain rate (**Fig. 7.14**). The specimens were prepared by mixing air-dried kaolin powder with air-dried power of high-early-strength Portland cement (3 % by weight) without water. The specimens were made saturated taking 2 days after σ_3' became 100 kPa. In the simulation, the TESRA viscosity type was assumed. Significant ageing effect can be observed in the stress-strain relation immediately after the restart of ML at a constant strain rate following a sustained loading stage. It may be seen that the viscous and ageing effects are both well simulated.

In summary, the key for realistic simulations of the stress-strain behaviour of geomaterial when both viscous and ageing effects are simultaneously important is the introduction of the yield stress, $(\sigma^f)_y$, for the inviscid stress, σ^f. Here, $(\sigma^f)_y$ is a function of not only the irreversible strain (as in the case when free from ageing effects) but also the time that has elapsed since the start of ageing. Depending on loading history, the value of σ^f can become smaller than $(\sigma^f)_y$ even when unloading process is not involved (typically during sustained loading).

The simulation when the ageing effect is negative (e.g., in the case of weathering) can be made by following the same framework as in the case where the ageing effect is positive described in this chapter (Tatsuoka et al., 2003a). Further study is necessary to simulate cases where interactions between the viscous and the ageing effects are significant.

8. 1D CONSOLIDATION OF CLAY

General

One-dimensional consolidation of a saturated clay deposit, as illustrated in **Fig. 8.1**, is one of the most classical soil mechanics problems. Despite so many years of studies by so many researchers, it seems that this problem is still not fully understood. This situation is due partly to an extremely high complexity of this problem due to the involvement of the three different time'-dependent factors listed in **Table 8.1**. The viscous and ageing effects, in particular among these three factors, are not simple to understand as discussed in the preceding chapters. Despite the above, they should be properly taken into account to understand the so-called primary and secondary consolidation processes of a saturated clay deposit. **Table 8.2** lists different combinations of these three factors. Considering a high complexity of the problem, the discussions which follow (Tatsuoka & Tani, 2006) start from the simplest case among those listed in Table 8.2 and then proceeds towards

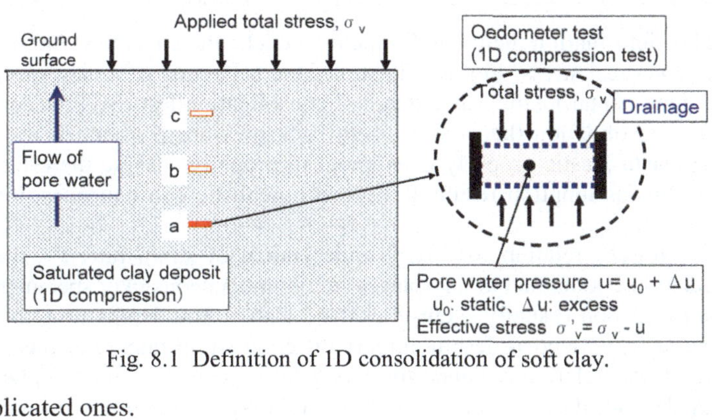

Fig. 8.1 Definition of 1D consolidation of soft clay.

more complicated ones.

8.2 *Elasto-plastic model (without ageing effect)*

Fig. 8.2 illustrates the linear e - log σ'_v relation assumed in the Terzaghi clay consolidation theory and many other recent elasto-plastic models developed for the

Table 8.1 Three 'time' -dependent factors in clay consolidation.

'Time'-dependent factor	Basic mechanism	Parameter for modelling
1. Delayed dissipation of Δu	Flow of pore water and compression of clay	Time (t^*) defined zero at the start of dissipation of Δu; $T=c_v t^*/H^2$ in the Terzaghi theory
2. Rate-dependent behaviour	Material viscosity	
3. Ageing effect	Time-dependent change in strength, stiffness ...	Time (t_0) defined zero at the start of ageing

Table 8.2 Different combinations of three factors listed in Table 8.1.

	$e - \log\sigma'_v$ behaviour	Delayed dissipation of Δu	Ageing effect	Note
1	Definitions of elasticity, plasticity and so on	No	No	Applicable only to loading conditions, not able to explain effects of OC
		Yes	No	
2	Elasto-plastic	No	No	Fully drained conditions
		Yes	No	Terzaghi theory; and many other recent models
3	Elasto-visco-plastic	No	No	Fully drained conditions
		Yes	No	- Consolidation of a clay deposit for a relatively short period
4	Elasto-visco-plastic	No	Yes	Fully drained conditions
		Yes	Yes	- Long-term sedimentation - Consolidation of young cement-mixed soil

analysis of clay deformation including Cam clay model, where e is the void ratio and σ'_v is the effective vertical stress. **Fig. 8.3** illustrates the behaviour of a clay specimen when subjected to incremental loading in a standard consolidation test (SCT) according to the Terzaghi clay consolidation theory (n.b., zero back pressure is assumed in this figure). Although the elasto-plastic property is essential to properly analyse the effects of over-consolidation, this assumption results in several unrealistic consequences, including the following:

1) As the e - $\log\sigma'_v$ relation for ML is independent of strain rate, $\dot{\varepsilon}_v$, the same e - $\log\sigma'_v$ relation is obtained at any places in a homogeneous clay specimen subjected to 1D consolidation (usually 2 cm-thick) and also at any depths in a corresponding homogeneous clay deposit despite largely different strain rates. For this reason, it is considered that a relation obtained for a very thin specimen can be applied to a very thick clay deposit, despite that the average strain rate is extremely different between these two cases due to largely different times until the end of so-called primary consolidation (i.e. until Δu becomes zero).
2) Any secondary consolidation (i.e., any long-term drained creep deformation) does not take place after the end of primary consolidation.

3) When ML is restarted after long-term sustained loading under the condition of $\Delta u = 0$, the e - $\log\sigma'_v$ behaviour follows the same relation as the one during the original primary loading without exhibiting any initial high-stiffness stress zone.

It is to be noted that a straight e - $\log\sigma'_v$ relation cannot to be extended until σ'_v becomes close to 0 and until e becomes close to zero, but the straight relation is only part of an actually reversed-S-shaped relation for a wide range of σ'_v and e as illustrated in **Fig. 8.4**. Imai (1981) reported curved e - $\log\sigma'_v$ relations starting from extremely low

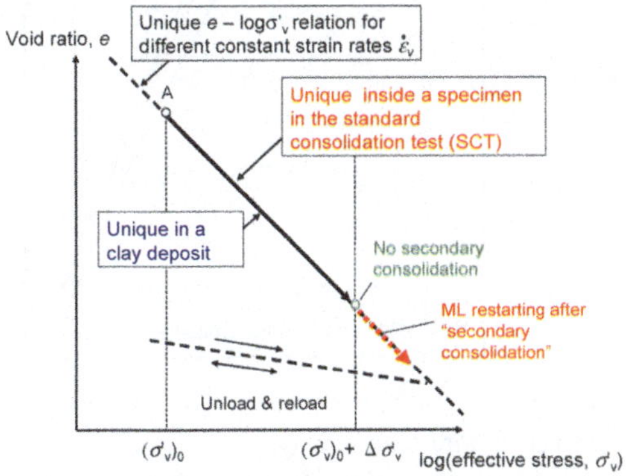

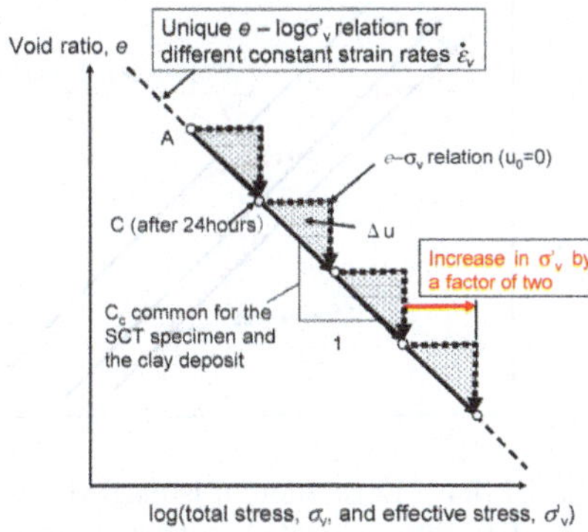

Fig. 8.2 (left) Stress – strain behaviour assumed in the Terzaghi clay consolidation theory and many other elasto-plastic models.

Fig. 8.3 (right) Behaviour in a standard consolidation test (SCT) according to the Terzaghi consolidation theory.

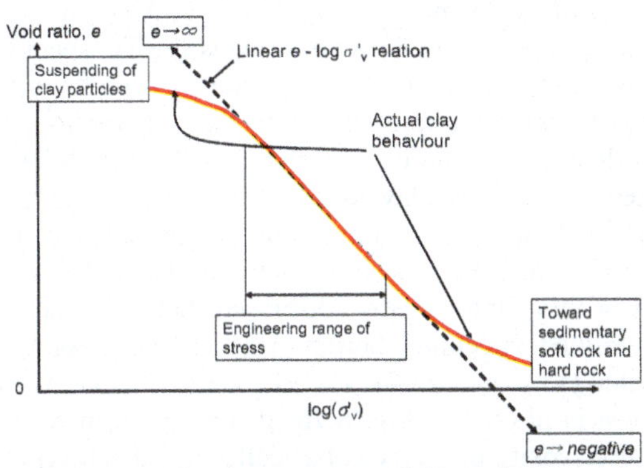

Fig. 8.4 Reasons for an apparent straight $e - \log\sigma'_v$ relation.

σ'_v values when clay sedimentation starts from slurry.

8.3 *Elasto-viscoplastic model (without ageing effect)*

Isotach behaviour: **Fig. 8.5** illustrates the e - $\log\sigma'_v$ behaviour in 1D compression of clay under fully drained conditions (i.e., always and everywhere essentially $\Delta u = 0$) when the Isotach type viscosity is relevant. Despite that other types of viscosity as discussed in Chapter 4 are likely to exist also in the 1D compression of clay, only the Isotach viscosity is herein considered, as this type of viscosity is most often observed in 1D compression

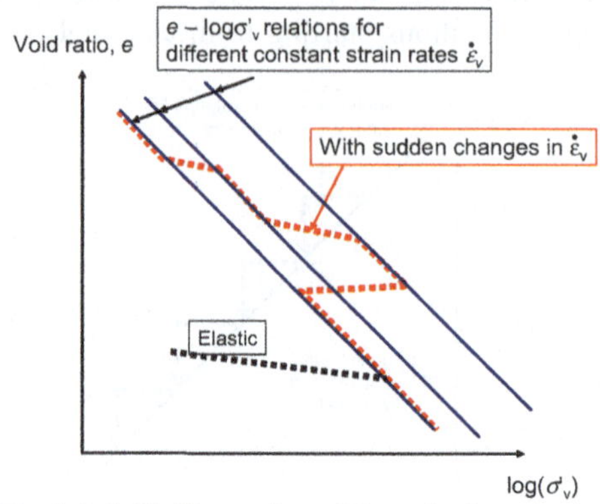

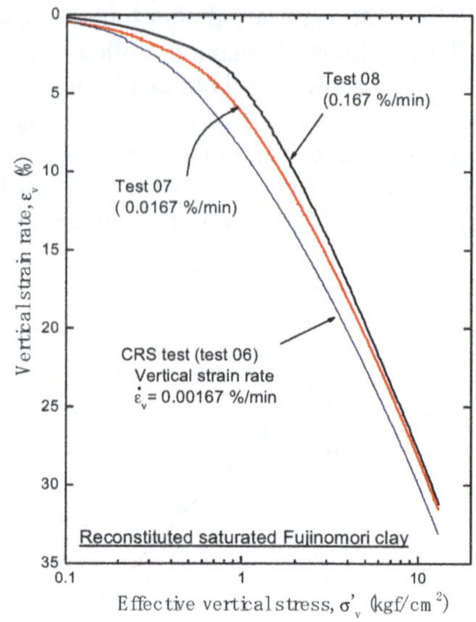

Fig. 8.5 (left) Illustration of Isotach viscosity in 1D compression of clay.
Fig. 8.6 (right) A trend of Isotach viscosity in CRS tests on saturated clay (Momoya, 1998).

of soft clay (e.g., Imai & Tang, 1992; Imai, 1995; Leroueil & Marques, 1996; Tanaka, 2005a & b). **Fig. 8.6** shows the ε_v - $\log\sigma'_v$ relations from constant-rate-of-strain (CRS) 1D compression tests performed at three largely different strain rates on saturated reconstituted specimens of Fujinomori clay (w_L= 62 %; PI= 33; D_{50}= 0.017 mm; & U_c~ 10). The specimens were prepared in the same way as the TC specimens prepared for the tests described in Figs. 3.4 through 3.6. Although the ε_v - $\log\sigma'_v$ curves after the start of yielding obtained at different strain rates shown in Fig. 8.6 are not perfectly parallel, they are well separated, indicating a trend of Isotach viscosity. **Fig. 8.7a** shows results from a CRS test on saturated Fujinomori clay in which the strain rate was stepwise changed several times and drained sustained loading was performed one time during otherwise ML at a constant strain rate. In this and other similar CRS tests on saturated clay specimens described in this paper, it was confirmed that the excess pore water pressure measured at the undrained specimen bottom was always negligible compared to the instantaneous σ'_v value (Li et al., 2004). The σ'_v - ε_v relation, like those from PSC and TC tests, from this test is plotted in **Fig. 8.7b**. It may be seen from Fig. 8.7 better than from Fig. 8.6 that the viscosity property is basically of Isotach type. It may also be seen from Fig. 8.7b that the differences among the σ'_v values at different strain rates increase with an increase in the stress level. According to the three-component model (Fig. 3.7a), this result means that the viscous stress component, σ^v, is always proportional to the instantaneous inviscid stress, σ^f, as seen in the PSC and TC test results.

Fig. 8.8 shows the result from another CRS 1D compression test that also exhibits a trend of Isotach viscosity. It is to be noted that the specimen was prepared by compacting air-dried clay powder and the specimen was kept air-dried throughout the test. This and other many similar test results show that, even without pore water, clay exhibits significantly

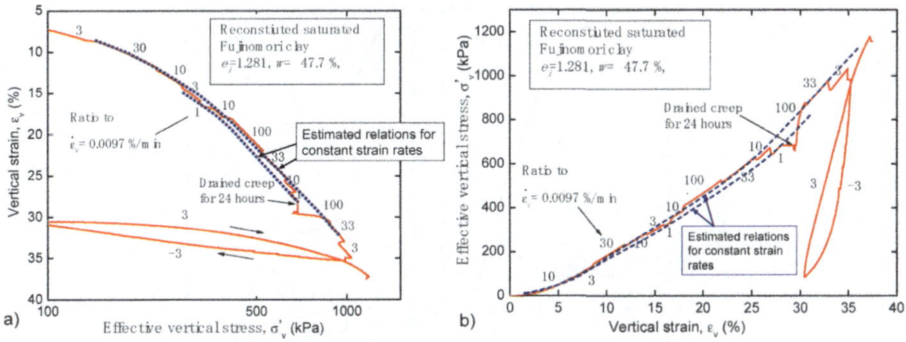

Fig. 8.7 Isotach viscosity in a CRS test on saturated clay (Li et al., 2004; Acosta-Martínez et al., 2005); a) $\varepsilon_v - \log \sigma'_v$ relation; and b) $\sigma'_v - \varepsilon_v$ relation.

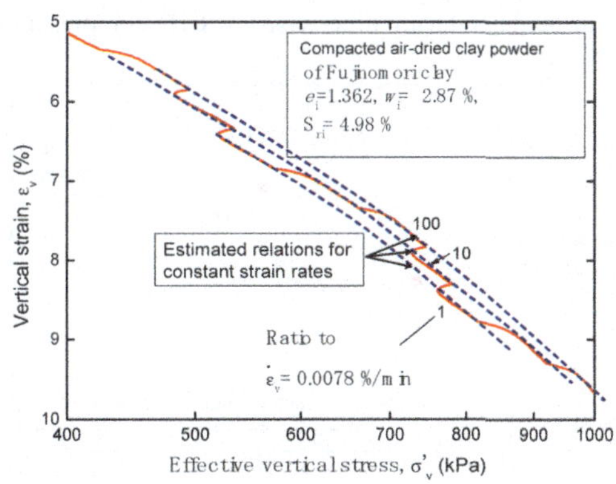

Fig. 8.8 Isotach viscosity in a CRS test on compacted air-dried clay powder (Li et al., 2004).

viscous behaviour in the same way as granular materials (i.e., unbound sand and gravel) (Li et al., 2004; Tatsuoka, 2004). They also showed that, despite that the ML stress - strain behaviour of clay changes significantly by saturation, the viscous property of saturated clay specimen is basically the same as the one of compacted oven- and air-dried clay powder specimens.

Fig. 8.9 shows the results from another CRS test on saturated Fujinomori clay in which a drained creep loading test was performed for 30 days during otherwise ML at a constant strain rate. It may be seen that a significant creep strain took place, followed by a large high-stiffness stress zone immediately after the restart of ML at the original strain rate. Then, the stress - strain relation slightly over-shoots the one that would have been obtained if ML loading had been continued without an intermission of 30 day sustained loading. The following two possible causes for this over-shooting can be conceived:

1) *A weak trend of TESRA viscosity of Fujimonori clay*, supported by the following facts:
   a) A trend of decay of viscous stress was observed at relatively large strains in CU TC tests on Fujinomori clay (Tatsuoka et al., 2002).
   b) The stress - strain curves from three CRS tests at different strain rates on saturated

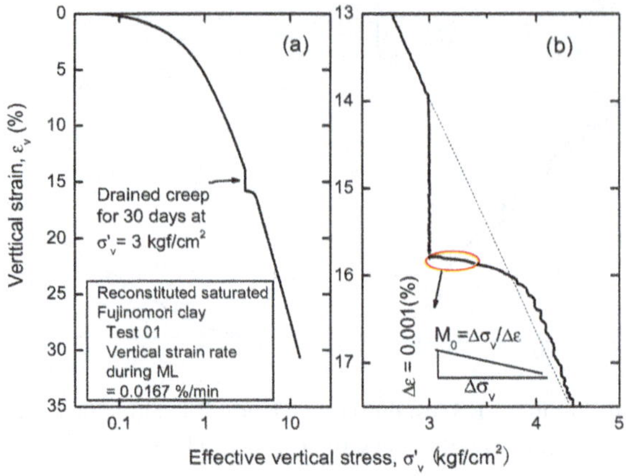

Fig. 8.9 Drained creep for 30 days during otherwise CRS 1D compression (Momoya, 1998).

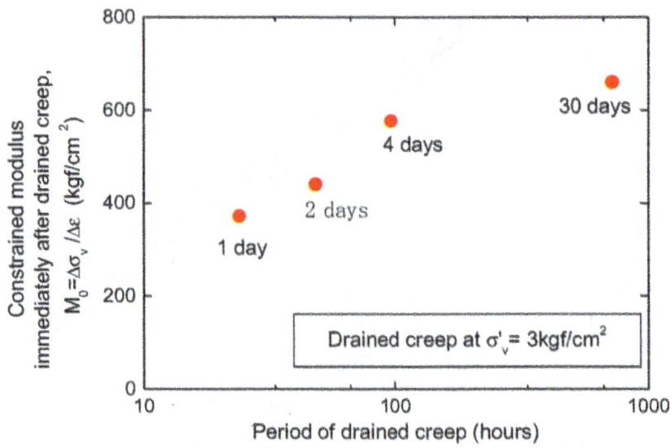

Fig. 8.10 Increase in constraint modulus, M_0, at small strains by drained creep for different periods (Momoya, 1998).

Fujinomori clay presented in Fig. 8.6 tend to converge into a single relation, despite that it is very gradual.

c) By examining very carefully the data from a CRS compression test on saturated Fujinomori clay presented in Fig. 8.7, despite that it is subtle, a weak trend of decay of viscous stress may be seen.

2) *Damage by subsequent straining to ageing effects that developed during drained sustained loading stage*. The initial constraint modulus, M_0, observed at small strains immediately after the restart of ML increased with an increase in the duration of drained creep loading (**Fig. 8.10**). The increase was much larger than the one that can be explained by a decrease in the void ratio during sustained loading. It is also known that the elastic shear modulus, G_0, of saturated clay measured by the resonant-column test or the Bender Element method increases with time during drained sustained loading for a relatively short period to an extent much larger than the one that can be explained by a decrease in the void ratio (e.g., Anderson & Woods, 1975: Shibuya et al., 2001). This extra increase in the M_0 and G_0 values may be due to stabilization at

inter-particle contact points between clay particles and associated stabilization of fabrics not accompanying a decrease in the void ratio (i.e., a sort of ageing effect). It seems that, unless the ageing effect becomes significant as produced in a geological time scale or strong artificial bonding, this type of ageing effect can be easily damaged by subsequent irreversible straining and its effect on stresses at larger strains, including the peak strength in TC tests, may become insignificant. More discussion on this issue is, however, beyond the scope of this paper.

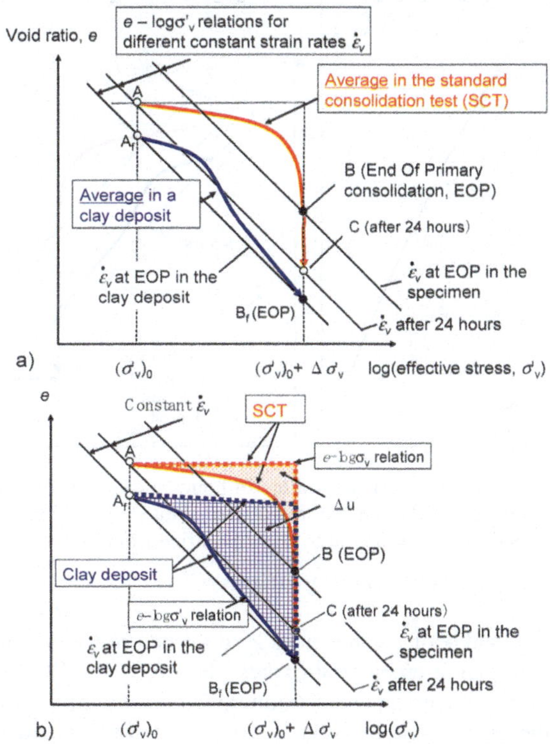

Fig. 8.11 Average behaviours of clay in a SCT and in a clay deposit when following the Isotach viscosity (Imai, 2006) (EOP is assumed at point A_f); a) e - $\log\sigma_v'$ relations; and b) dissipation process of excess pore water pressure.

Behaviour in standard consolidation tests (SCTs): In a SCT, the total vertical load is stepwise increased by a factor of two every 24 hours. As illustrated in **Fig. 8.11a**, the dissipation rate of excess pore pressure, Δu, is not constant with time at a given point and is different at different points in the SCT specimen and so is the strain rate. Only the average e - $\log\sigma'_v$ relation for a given SCT specimen is illustrated in Fig. 8.11a. For the same reason, the strain rate is not constant with time at a given point and is different at different points in a full-scale clay deposit. In Fig. 8.11a, only the average e - $\log\sigma'_v$ relation in a clay deposit when subjected to a sudden increase in the total vertical stress, σ_v, on the ground surface is presented. It is assumed that the average e and σ'_v state when a step increase in σ_v is made is at the end of primary consolidation (EOP). Because of utterly larger drainage lengths, the dissipation rate of Δu in a full-scale clay deposit is substantially lower than in a SCT specimen, which results in a substantially lower average strain rate (**Fig. 8.11b**: n.b., zero back pressure is assumed in this figure). According to the Isotach viscosity, the average e - $\log\sigma'_v$ relations in a SCT specimen

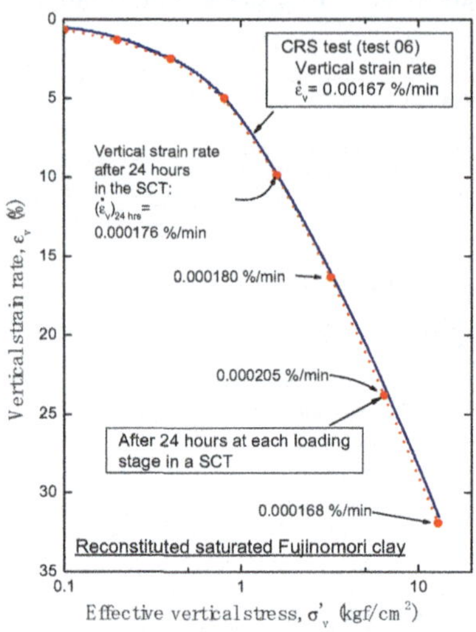

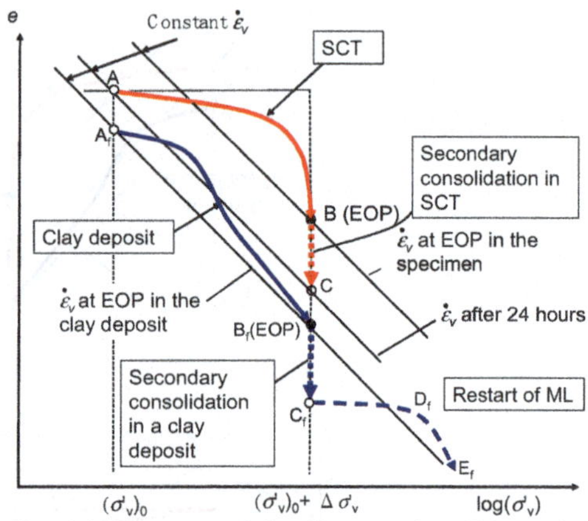

Fig. 8.12 (left) Comparison between states after 24 hours at each loading step in a SCT and the behaviour in a very slow CRS test, saturated reconstituted clay (Momoya, 1998).

Fig. 8.13 (right) Secondary consolidation in clay (modified from Fig. 8.11).

and any full-scale clay deposit are utterly different.

In a SCT, the average value of Δu becomes zero far before the end of respective loading step for 24 hours, and large creep strain takes place after the EOP until 24 hours (Fig. 8.11). For this reason, the average creep strain rate at the end of respective loading step is very low. When the viscosity is of Isotach type, the average e and $\log\sigma'_v$ states at the ends of a series of 24 hour-loading stages in a SCT become essentially the same as the e - $\log\sigma'_v$ relation from a CRS test performed at the average strain rate at the ends of 24-hour loading stages in the SCT. **Fig. 8.12** compares data points of the ε_v and $\log\sigma'_v$ states after 24 hours (i.e., at the ends of respective loading stage) from a SCT and the continuous ε_v - $\log\sigma'_v$ relation from a very slow CRS test performed at $\dot{\varepsilon}_v$ equal to 0.00167 %. The specimens were saturated reconstituted Fujinomori clay. It may be seen that the test results from the CRS test and the SCT are rather consistent when taking into account the strain rate effects on strain-strain behaviour. That is, as the strain rate $\dot{\varepsilon}_v$ = 0.00167 % in the CRS test is larger by a factor of about ten than the strain rates at the ends of 24 hour-loading stages in the SCT, the continuous ε_v - $\log\sigma'_v$ relation from the CRS test is located on the right of the relation from the SCT (despite that it is slight). A slight difference may be due to that the ε_v - $\log\sigma'_v$ relations in these two tests are already close to the reference relation for loading (explained later) and, therefore, the effects of strain rate have become very small.

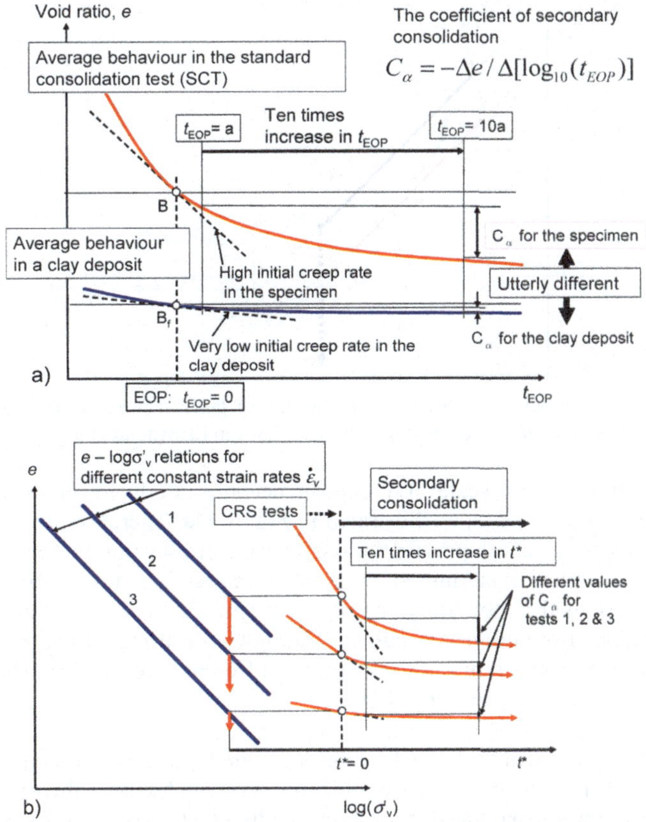

Fig. 8.14 Non-objectivity of C_α; a) comparison between a SCT and a clay deposit; and b) drained creep tests during otherwise CRS tests at different strain rates.

Secondary consolidation: In geotechnical engineering practice, the coefficient of secondary consolidation (defined below) obtained from 1D creep tests in the laboratory is often used to predict the rate of residual compression during the secondary consolidation stage of a clay deposit:

$$C_\alpha = -\Delta e / \Delta[\log_{10}(t_{EOP})] \tag{8.1}$$

where t_{EOP} is the time that has elapsed since the EOP (i.e., t_{EOP}= 0 at the EOP); and $-\Delta e$ is the decrease in the void ratio by an increase in t_{EOP} by a factor of ten from a certain moment. However, the coefficient of secondary consolidation, C_α, defined as above is not objective and therefore not the material property, but it depends on the loading duration required to reach the EOP since the start of loading, which increases with an increase in the drainage length. That is, when counting the time since the moment when step load is applied, secondary consolidation starts substantially later in a full-scale clay deposit than in a SCT (**Fig. 8.13**). For this reason, the average strain rate at the EOP is substantially lower in a full-scale clay deposit than in a SCT. Therefore, the values of C_α

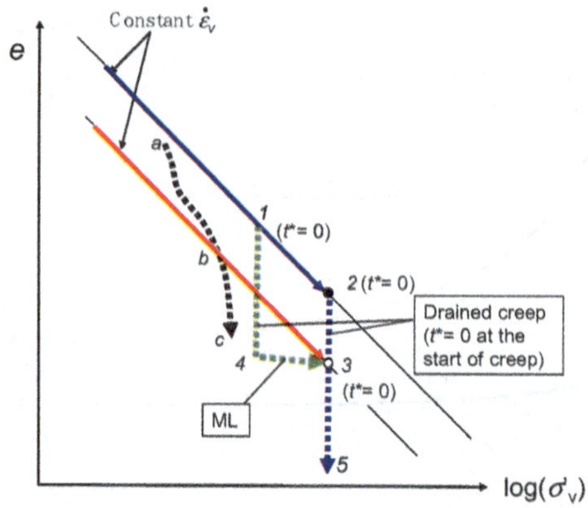

Fig. 8.15 Several fundamental problems with the use of general time' as the parameter to describe the behaviour of elasto-visco-plastic geomaterial (Tatsuoka et al., 2000).

defined by Eq. 8.1 for a full-scale clay deposit become utterly lower than the values determined from laboratory tests, as illustrated in **Fig. 8.14a**. Therefore, it is not straightforward, or very complicated, to predict the secondary consolidation rate in the field clay deposit from C_α values determined by 1D compression tests with secondary consolidation stages in the laboratory. Even when limited to the laboratory tests, the values of C_α obtained from drained sustained loading tests starting during otherwise CRS loading at largely different strain rates are largely different (**Fig. 8.14b**). The time t^* in Fig. 8.14b is equivalent to t_{EOP} in Fig. 8.14a.

Indeed, the use of C_α defined by Eq. 8.1 means a certain type of isochronous concept. As stated earlier in Chapter 5, it is not possible for any isochronous theory to properly predict the stress - strain behaviour for arbitrary loading histories. More specifically, it is not possible to define the origin of the time t^* (i.e., t_{EOP}) when the secondary consolidation starts in the objective way, as it depends on the loading history as well as drainage length. For example, suppose that several loading histories shown in **Fig. 8.15** are applied to a clay element under fully drained conditions. In one test, secondary consolidation (i.e., drained sustained loading) starts from point *1* during otherwise a CRS test at a relatively high strain rate. In this test, t^*= 0 may be defined at point *1*. Suppose that secondary consolidation has continued for a very long duration until point *4*. Then, ML is restarted at the original strain rate to reach point *3*, from which secondary consolidation is then restarted. In another test, secondary consolidation is started from point *2* after CRS loading has continued until point *2*. In this case, t^*= 0 may be defined at point *2*. Subsequently, stress point *3* is reached after some duration that is much shorter than the one that is needed to reach point *3* from point *1* in the first test. Furthermore, in the third test, secondary consolidation is started from point *3* during otherwise ML at a substantially lower strain rate. In this case, t^*= 0 may be defined at point *3*. It can be readily seen from the above that the value of t^* at point *3* is utterly different among these three tests. Finally, in a full-scale clay deposit, the strain rate may change arbitrarily without any duration for which σ'_v is kept constant, as path $a \rightarrow b \rightarrow c$ in Fig. 8.15. In this case, it is not possible to define the moment when t^*= 0.

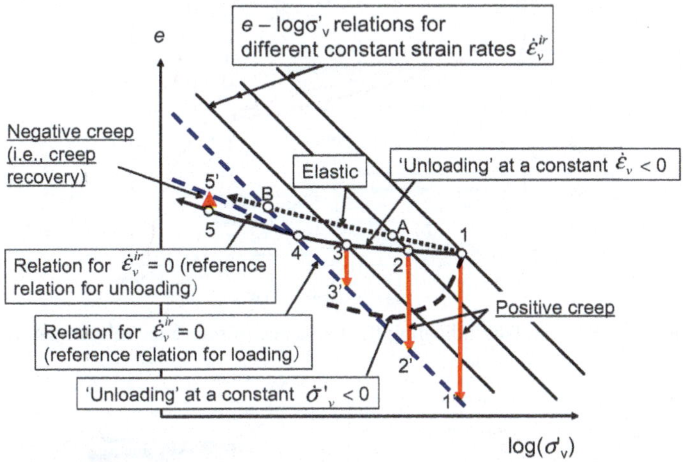

Fig. 8.16 Isotaches for zero strain rate and the reference curves for loading and unloading.

Toward constitutive formulation in the framework of the non-linear three-component model: To simulate the stress - strain behaviour when the stress is not only increasing but also decreasing as well as when the irreversible strain rate is not only increasing (i.e., loading condition) but also decreasing (i.e., unloading condition), it is necessary to develop a sort of general elasto-viscoplastic model. In the following, discussions on this issue for the 1D compression case are made. To simulate the rate-dependent stress-strain behaviour of geomaterial subjected to general stress paths, including TC and PSC loading at a fixed confining pressure and 1D compression, such a model as above should incorporate inviscid yielding characteristics based on a double-hardening concept (Tatsuoka et al., 2003b). This topic is however beyond the scope of this paper.

The discussions that follow are based on the three-component model (Fig. 3.7a), in which the stress and strain, σ and ε, are the effective vertical stress, σ_v', and the vertical strain, ε_v. Then, by referring to Eq. B11, the viscous stress component is obtained as:

$$(\sigma'_v)^V(\varepsilon_v^{ir}, \dot{\varepsilon}_v^{ir}) = (\sigma'_v)^f(\varepsilon_v^{ir}) \cdot g_v(\dot{\varepsilon}_v^{ir}) \qquad (8.2)$$

In **Fig. 8.16**, three linear e - $\log\sigma'_v$ relations from ML CRS tests at three different strain rates are depicted. When the total strain, ε_v, is reduced at a certain rate, relation 1→2→3→4 (from point 1 until point 4), which satisfies the conditions that $\dot{\varepsilon}_v$ (< 0) = $\dot{\varepsilon}_v^e$ (< 0) + $\dot{\varepsilon}_v^{ir}$ (> 0) may be obtained. This tnloading' relation is slightly different from the truly elastic relation 1-A-B, for which $\dot{\varepsilon}_v$ (< 0) = $\dot{\varepsilon}_v^e$ (< 0) and $\dot{\varepsilon}_v^{ir}$ = 0. Then, the stress - strain relations at zero $\dot{\varepsilon}_v^{ir}$ (i.e., the reference curves) for loading and unloading can be introduced. The reference curves represent the elasto-plastic stress-strain relations during loading an unloading of components E +P (c onnected in series) of the three-component model (Fig. 3.7a). For loading conditions (i.e., when $\dot{\varepsilon}_v^{ir}$ is kept positive), there is only one reference curve (i.e., the reference curve for loading), which is assumed to be straight and parallel to the relations from ML CRS tests at constant strain rates in this figure.

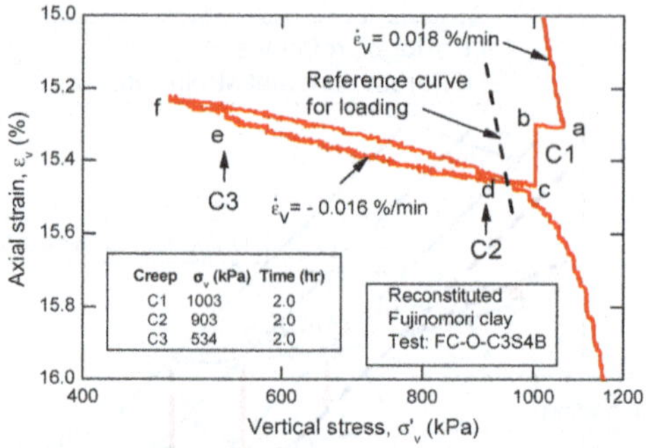

Fig. 8.17 Typical CRS test data indicating the existence of reference relation (Acosta-Martńez et al., 2005).

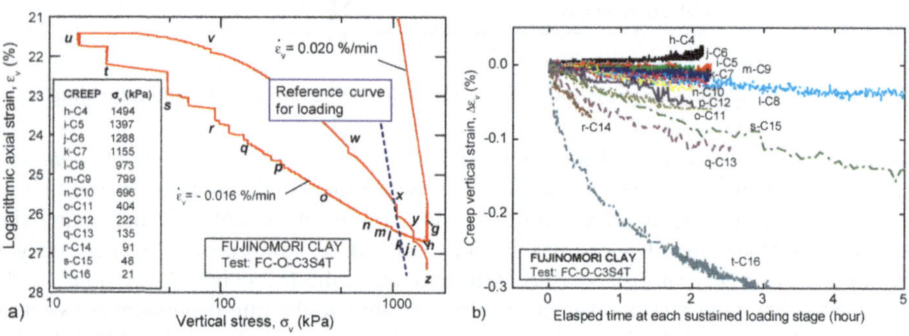

Fig. 8.18 Typical CRS test data indicating the existence of reference relation (Acosta-Martńez et al., 2005); a) overall stress-strain relation; and b) time histories of vertical strain at selected stages during otherwise global ŧnloading'.

There are an infinite number of reference curves for unloading (i.e., when $\dot{\varepsilon}_v^{ir}$ is kept negative) starting from different points along the reference curve for loading. In this figure, the one starting from point 4 (i.e., relation 4→5') is depicted. When the total strain, ε_v, is reduced at a certain rate from point 4, relation 4→5 is obtained, for which $\dot{\varepsilon}_v$ (< 0) = $\dot{\varepsilon}_v^e$ (< 0) + $\dot{\varepsilon}_v^{ir}$ (< 0). Note that the stress - strain relation obtained by ŧnloading' of σ'_v at a negative constant rate is largely different from the ŧnloading'relation for a negative constant rate of $\dot{\varepsilon}_v$, 1→2→3→4→5.

According to the Isotach viscosity concept, the irreversible strain rates at points 2 and 3, located on the right of the reference curve for loading, are still positive (i.e., under loading conditions in terms of the sign of $\dot{\varepsilon}_v^{ir}$) and equal to the values for the CRS e - $\log\sigma'_v$ curves on which points 2 and 3 are located. For this reason, the positive creep strain develops when drained sustained loading starts from points 2 and 3, which continues until reaching points 2' and 3', located on the reference curve for loading. On the other hand, the irreversible strain rate during sustained loading starting from point 4 is zero, as this point is located on the reference curve for loading. The irreversible strain rate at point 5 is negative, as this point is located on the left of the reference curve for loading. When drained sustained loading starts from point 5, the negative creep strain

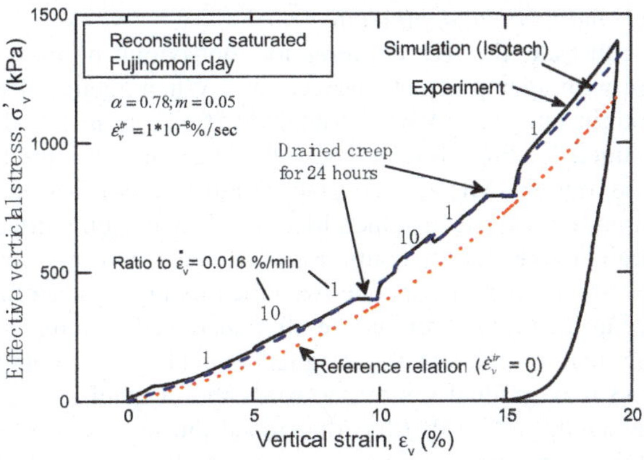

Fig. 8.19 Simulation of the 1D compression test result of saturated clay by the three-component model incorporating a reference relation for loading (Acosta-Martínez et al., 2005).

develops and continues until reaching point 5', located on the reference curve for unloading starting from point 4. This phenomenon is known as creep recovery.

The rate-dependent stress - strain behaviour described above can be seen from the data presented below. **Fig. 8.17** shows the results from a CRS test, in which drained sustained loading was started from points *b*, *d* and *e* during otherwise monotonic 'unloading' at a constant negative $\dot{\varepsilon}_v$. The reference curve for loading depicted in Fig. 8.17 was determined from the signs of the creep strain rates observed when drained creep loading was started from different stress points. That is, when drained sustained loading was started from point *b*, which was reached by 'unloading ' from a CRS loading condition, considerable positive creep strain took place as denoted by C1. When started from point *d*, reached after more 'unloading', the creep strain rate became very small and negative. When started from *e*, reached by further 'unloading', noticeable negative creep strain started. **Fig. 8.18** shows results from another CRS test, in which drained sustained loading was started during otherwise not only ML loading at a constant positive $\dot{\varepsilon}_v$ (as denoted by *g*) but also monotonic 'unloadi ng' at a constant negative $\dot{\varepsilon}_v$. It may be seen from **Fig. 8.18b** that the negative creep strain rate increases when drained sustained loading was started from more unloaded states. It may be seen from the above that the reference stress -strain relation for loading can be inferred from such tests described above in which drained sustained loading tests are started from differently unloaded stress states, rather than performing extremely long-term sustained loading tests.

Fig. 8.19 shows an example of simulation of a typical 1D compression test of saturated clay in which the strain rate was stepwise changed many times and drained sustained loading was performed two times during otherwise ML at a constant strain rate by the three-component model (Fig. 3.7a) incorporating a reference relation for loading. It may be seen that all the details of the Isotach type viscous behaviour are well simulated. Other examples of similar simulations are reported in Li et al. (2004) and Acosta-Martinez et al. (2005).

8.4 *Elasto-viscoplastic model (with ageing effect)*

Figs. 8.20a and **b** illustrate the 1D compression behaviour of clay having Isotach viscosity without and with ageing effect, respectively. When ageing effect is not active (Fig. 8.20a), stress-strain curve 3→4→5 is obtained if ML is restarted at the original strain rate after sustained loading (1→3). The reloading curve rejoins the original one (1→2) obtained by continuous ML at a constant strain rate without an intermission of secondary consolidation (or drained sustained loading). When ageing effect is active (Fig. 8.20b), the stress-strain curve during continuous ML at a constant strain rate (1→2) gradually deviates toward a larger stress zone from the one in the case of no ageing effect. When ML is restarted at the original strain rate after sustained loading (1→3), the stress-strain curve 3→4→5 that overshoots the original one (1→2) is obtained. The over-shooting relation 4→5 may gradually move towards the original relation (1→2) as the strain increases if the ageing effect that has developed during sustained loading 1→3 is gradually damaged by subsequent straining. However, as the amount of ageing effect at the same stress level is different between the relations 1→2 and 4→5, relation 4→5 would not rejoin relation 1→2.

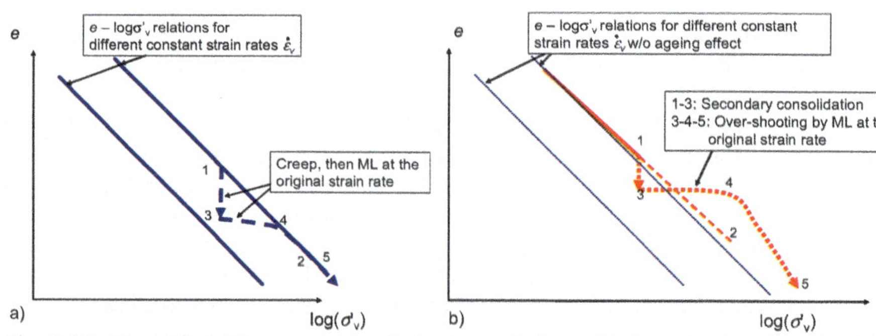

Fig. 8.20 Simplified 1D compression behaviour of clay with Isotach viscosity; a) without ageing effect; and b) with ageing effect.

Considering that it is nearly impossible to examine the ageing effect, as described above, taking place in a natural clay deposit, Deng and Tatsuoka (2005) performed a series of 1D compression tests on cement-mixed clay accelerating the ageing effect. A typical test result is presented in **Fig. 8.21**. The specimen (6 cm in diameter and 2 cm in height) was made by compacting air-dried kaolin powder mixed with air-dried powder of high-early-strength Portland cement (3 % by weight). The specimen was subsequently made saturated to start ageing effects when $\sigma'_v = 0$ and then cured for one day before the start of 1D compression. During otherwise ML at a constant strain rate, the strain rate was stepwise changed many times and drained sustained loading was performed one time for 24 hours. Upon a step decrease in the strain rate, the stress suddenly decreased, which was due to the viscous effect. During the subsequent ML at a decreased constant strain rate, however, the stress - strain relation gradually deviated from the one that would have been obtained if the ageing effect had not been active. Eventually, the stress-strain relation overshot the one that would have been obtained if ML had been continued without a step decrease in the strain rate. It can be inferred that, if the strain rate had been

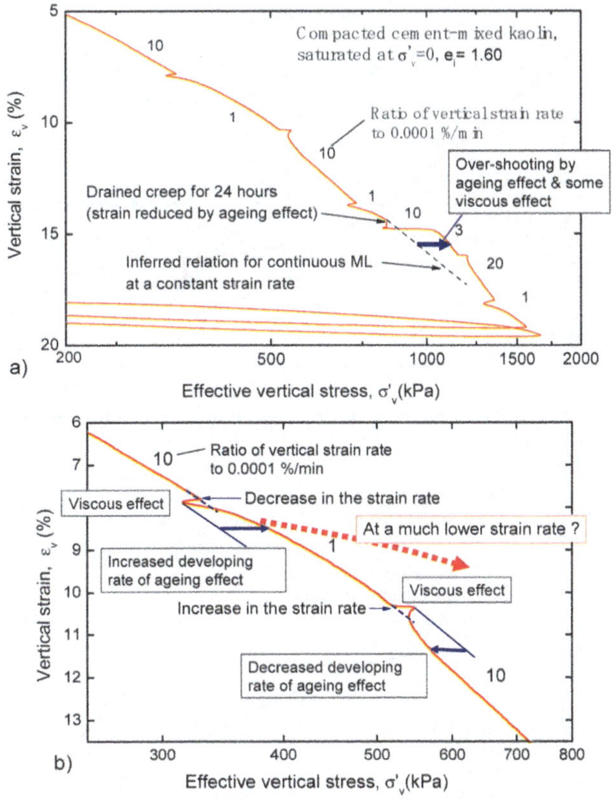

Fig. 8.21 1D compression of saturated cement-mixed clay (with continuing ageing effects) (Deng & Tatsuoka, 2005); a) overall stress-strain behaviour; and b) close-up.

decreased to a much smaller value, the stress - strain relation during the subsequent ML would have shown a larger stress increase as illustrated in **Fig. 8.21b**. The opposite trend of behaviour can be seen after a step increase in the strain rate. These trends of behaviour are similar to those described in Fig. 7.8. Moreover, the stress - strain behaviour became very stiff for a large stress zone when ML was restarted at a constant strain rate after sustained loading for 24 hours, similar to the one seen in Fig. 7.5c. The amount of creep strain during the 24 hour sustained loading was very small (Fig. 8.21a). It seems that developing ageing effects suppressed the creep strain rate (as illustrated in Fig. 7.11a).

Relation 6→7 presented in **Fig. 8.22** illustrates the 1D stress - strain behaviour during an extremely slow sedimentation process with continuing ageing effect in a natural clay deposit inferred based on the test result presented in Fig. 8.21. Point 8 represents the current state at a certain depth in a homogeneous natural clay deposit that has been reached after some long secondary consolidation 7→8. A trace of such points as point 8

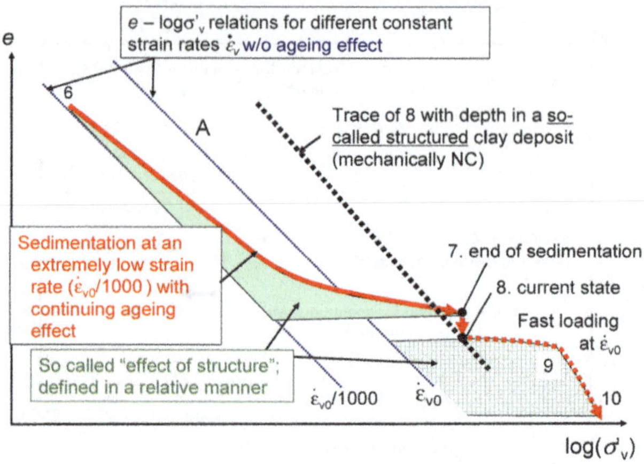

Fig. 8.22 Inferred simplified behaviour during extremely slow sedimentation of clay with continuing ageing effect.

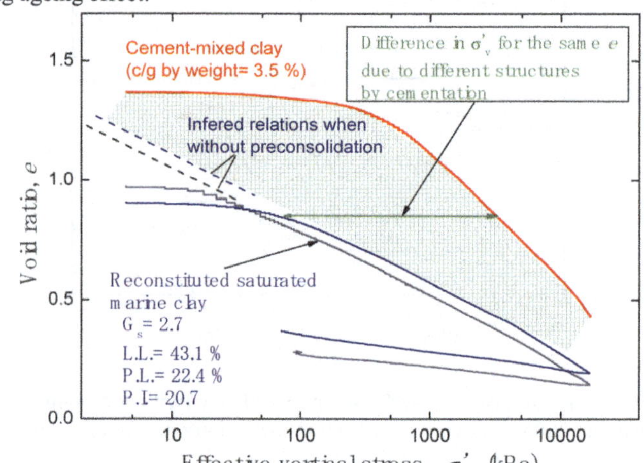

Fig. 8.23 1D compression behaviour of reconstituted clay and cement-mixed clay (Sugai et al., 2000; Sugai & Tatsuoka, 2003).

along the depth in the clay deposit is represented by a thick broken line. Such an e - $\log \sigma'_v$ relation as above is usually located at higher stresses than the relation of remoulded clay (such as relation A) (Burland, 1990). For this reason, natural clay in-situ is often called "structural clay".

When ML starts at a relatively large strain rate from point 8, the stress–strain relation becomes like the one of mechanically over-consolidated clay, exhibiting high-stiffness behaviour until yield point 9. This large high-stiffness stress zone (8→9) has developed by both: a) ageing effect that developed at the fixed effective stress state kept for a long duration (7→8); and b) an increase in the viscous stress associated with a sudden increase in the strain rate when ML is restarted from point 8. It should be noted that the so-called structural effect can be defined and evaluated only by comparing different stress - strain relations obtained at the same irreversible strain rate and for the same recent loading history to exclude the viscous effect, as illustrated in Fig. 8.22. For example, relation 9→10 should be compared with the corresponding relation *A* for remoulded clay obtained for the same or similar strain rate, $\dot{\varepsilon}_{v0}$. **Fig. 8.23** shows such a comparison as above that was obtained by performing CRS 1D compression tests on remould marine clay and its cement-mixture under otherwise the similar test conditions. To ensure drained conditions of the specimens (a diameter of 6 cm and a height of 2 cm), low axial strain rates (0.0055 %/min and 0.015 %/min for cement-mixed and untreated clays) were used. The difference between the two relations would have been slightly larger than the one seen in Fig. 8.23 if the tests had been performed at the same strain rate.

8.5 *Summary*

It is necessary to take into account the viscous property to properly analyze and correctly predict the stress - strain - time behaviour of a natural clay deposit during both so-called primary and secondary consolidation stages. To understand different behaviours of natural clay and remoulded clay, the ageing process and its effects on the stress - strain behaviour during subsequent loading stages should also be taken into account. The "so-called structure" produced by ageing effects can be properly defined only when comparing stress-strain relations at the same irreversible strain rate and for the same recent loading history. The "so-called structure" produced by ageing effects at a certain stress state may be damaged by subsequent irreversible straining. According to the three-component model (Fig. 3.7a), this damaging process is taken into account through its effects on the inviscid strain-hardening property of component P. Therefore, it is not possible to describe and analyze the secondary consolidation process and subsequent ML stress - strain behaviour only by taking into account changes in the so-called structure by irreversible straining while ignoring the viscous effect as well as the ageing effect.

9. CONCLUSIONS

In this paper, three major factors related to the development of inelastic strain increments; *plastic yielding*, *viscous effect* and *inviscid cyclic loading effect*, and their interactions are discussed. It is shown that the simplified non-linear three-component model described in Fig. 3.7a is relevant to describe these three major factors, although the detailed structure of the respective components of the model should be defined specifically and appropriately based on experimental results.

The followings can also be derived from the test data and analysis presented in this paper:

1) Any strain-additive model in which three strain increment components representing these three factors are connected in series is not relevant. Any isochronous model, which describes the viscous effect in terms of general time, is not relevant either.
2) The inviscid yielding consists of shear and volumetric yielding mechanisms (i.e., the double hardening model). Their relative importance depends on the soil type.
3) With respect to the viscous effect when subjected to shearing, at least three basic types of viscosity, *Isotach*, *TESRA* and *Positive & Negative,* have been observed. A

general expression to describe these and others as well as the transition from one to another is suggested.

4) Inviscid cyclic loading effect becomes more significant with an increase in the cyclic stress amplitude and the number of loading cycles for a given duration of loading.
5) Particle shape has systematic effects on the viscosity type and the relative importance between the viscous effect and the inviscid cyclic loading effect. In particular, round granular materials tend to exhibit the *Positive and Negative* type viscosity while exhibiting relatively smaller creep strains and larger inviscid cyclic loading effects.
6) Both elastic and inelastic deformation characteristics are affected by ageing effects.
7) A method to incorporate ageing effect into the three-component model is suggested.
8) To properly understand the stress - strain - time relation during both primary and secondary 1D consolidation processes of clay, it is necessary to take into account both viscous and ageing effects in addition to delayed excess pore water dissipation.

ACKNOWLEDGEMENTS

The author would like to express his sincere thanks to all his previous and present colleagues of Geotechnical Engineering Laboratories of the Institute of Industrial Science and the Department of Civil Engineering, the University of Tokyo, and the Department of Civil Engineering, Tokyo University of Science. Without their help and cooperation, it was not possible for him to write this paper. A long-term cooperative research program with Prof. Di Benedetto, H. and his colleagues at Département Génie Civil et Bâtiment, Ecole Nationale des Travaux Publics de l'Etat (ENTPE), Lyon, France, was another major essence for the content of this paper. Corundum (Al_2O_3) refereed to in this paper was provided by Prof. Gudehus, G., University of Karlsruhe, Germany.

REFERENCES

1) Acosta-Martínez, H., Tatsuoka, F. and Li, Jiangh-Zhong (2005): "Viscous property of clay in 1-D compression: evaluation and modelling", *Proc. 16th ICSMGE*, Osaka.
2) Anh Dan, L. Q., Tatsuoka, F., and Koseki, J. (2006): "Viscous shear stress-strain characteristics of dense gravel in triaxial compression," *Geotechnical Testing Journal*, ASTM, Vol.29, No.4, pp.330-340.
3) Anderson, D. G. and Woods, R. D. (1975): "Time-dependent increase in shear modulus of clay", *Jour, GE Div., Proc. ASCE*, No.102-GT5, pp.525-537.
4) Aqil, U., Tatsuoka, F., Uchimura, T., Lohani, T.N., Tomita, Y. and Matsushima, K. (2005): "Strength and deformation characteristics of recycled concrete aggregate as a backfill material", *Soils and Foundations*, Vol.45, No.4, pp.53-72.
5) Burland, J. B. (1990): "On the compressibility and shear strength of natural clays", Rankine Lecture, *Géotechnique,* Vol.40, No.3, pp.329-378.
6) Chambon, G., Schmittbuhl, J. and Corfdir, A. (2002): "Laboratory gouge friction: seismic-like slip weakening and secondary rate- and state-effects", *Geophysical Research Letters*, Vol.29, No.10, 10.10.1029/2001GL014467, pp.4-1 - 4-4.
7) Deng, J. and Tatsuoka, F. (2004): "Ageing and viscous effects on the deformation of clay in 1D compression" , *Proc. of GeoFrontier 2005 Congress, GeoInstitute, ASCE*, Austin, Texas, GSP 138, Site characterization and modeling (Mayne et al. eds).

8) Deng, J.-L. and Tatsuoka, F. (2006): "Viscous property of kaolin clay with and without ageing effects by cement-mixing in drained triaxial compression", *Soil Stress-Strain Behavior: Measurement, Modeling and Analysis, Proc. of Geotechnical Symposium in Roma, March 16 & 17, 2006* (Ling et al., eds.) (this volume).
9) Di Benedetto, H. and Hameury, O. (1991): "Constitutive law for granular skeleton materials: description of the anisotropic and viscous effects", *Comp. Met. and Ad. In Geomechanics* (Beer et al. eds.), Rotterdam, Balkema, pp.599-603.
10) Di Benedetto, H. and Tatsuoka, F. (1997): "Small strain behaviour of geomaterials: modelling of strain effects", *Soils and Foundations,* Vol.37, No.2, pp.127-138.
11) Di Benedetto, H., Tatsuoka, F. and Ishihara, M. (2002): "Time-dependent deformation characteristics of sand and their constitutive modeling", *Soils and Foundations*, Vol. 42, No.2, pp.1-22.
12) Di Benedetto, H., Tatsuoka, F., Lo Presti, D., Sauzéat, C. and Geoffroy H. (2004): "Time effects on the behaviour of geomaterials", Keynote Lecture, , *Proc. 3rd Int. Sym. on Deformation Characteristics of Geomaterials,* IS Lyon 03 (Di Benedetto et al. eds.), Balkema, September, 2003, Vol.2, pp.59-123.
13) Duttine, A., Kongkitkul, W., Hirakawa, D. and Tatsuoka, F. (2006): "Effects of particle properties on the viscous behaviour in direct shear of unbound granular materials", *Proc. 41st Japanese National Conference on Geotechnical Engineering, the Japanese Geotechnical Society (JGS),* Kagoshima.
14) Enomoto, T., Tatsuoka, F., Shishime, M., Kawabe, S. and Di Benedetto. H. (2006): "Viscous property of granular material in drained triaxial compression", *Soil Stress-Strain Behavior: Measurement, Modeling and Analysis, Proc. of Geotechnical Symposium in Roma, March 16 & 17, 2006* (Ling et al., eds.) (this volume)..
15) Gens, A. (1986): "A state boundary surface for soils not obeying Rendulic's principle", *Proc. 11th IC on SMFE, San Francisco,* Vol.2, pp.473-476.
16) Hayano, K., Matsumoto, M., Tatsuoka, F. and Koseki, J. (2001): "Evaluation of time-dependent deformation property of sedimentary soft rock and its constitutive modelling", *Soils and Foundations*, Vol.41, No.2, pp. 21-38.
17) Hayashi, T., Moriyama, M., Tatsuoka, F. and Hirakawa, D. (2005): "Residual deformations by cyclic and sustained loading of sand and their relation", *Proc. 40th Japanese National Conference on Geotechnical Engineering, JGS,* Hakodate (in Japanese).
18) Hayashi, T., Sakurano, H, Tatuoka, F and Hirakawa, D. (2006): "Residual strains by cyclic loading effects and viscous property of various granular materials and their relation", *Proc. 41st Japanese National Conference on Geotechnical Engineering, JGS*, Kagoshima (in Japanese).
19) Henkel, D. J. (1960): "The relationships between the effective stresses and water content in saturated clays", *Géotechnique*, Vol. X, pp.41-54.
20) Henkel, D. J. and Sowa, V. A. (1963): "The influence of stress history in undrained triaxial tests on clays", *ASTM, STP361*. pp.280-291.
21) Hirakawa, D., Kongkitkul, W., Tatsuoka, F. and Uchimura, T. (2003): "Time-dependent stress-strain behaviour due to viscous property of geosynthetic reinforcement", *Geosynthetics International*, IGS, Vo.10, No.6, pp.176-199.
22) Hoque, E. and Tatsuoka, F. (1998): "Anisotropy in the elastic deformation of materials", *Soils and Foundations*, Vol.38, No.1, pp.163-179.

23) Howie, J.A., Shozen, T. and Vaid, Y. P. (2001): "Effect of ageing on stiffness on loose Fraser Rover sand", *Advanced laboratory stress-strain testing of geomaterials* (Tatsuoka et al. eds.), Balkema, pp.235-243.

24) Imai, G. (1981): "Experimental studies on sedimentation mechanism and sediment formation of clay materials", *Soils and Foundations*, Vol.21, No.1, pp.7-20.

25) Imai, G. and Tang, X.-Y. (1992): "A constitutive equation of one-dimensional consolidation derived from interconnected tests, *Soils and Foundations*, Vol.32, No.2, pp.82-96.

26) Imai, G. (1995): "Analytical examination of the foundations to formulate consolidation phenomena with inherent time-dependence", Keynote Lecture, *Proc. Int. Symp. On Compression and Consolidation of Clayey Soils*, IS Hiroshima '95, Rotterdam: Balkema, Vol.2, pp.891-935.

27) Imai, G. (2006): "Objectives, roles and perspectives of standard consolidation tests of clay in practice", *Tsuchi-to-Kiso, Monthly Journal of Japanese Geotechnical Society*, Vol. 54, No. 2, pp.18-21 (in Japanese).

28) Ishihara, K. and Okada, S. (1978): "Effects of Stress History on Cyclic Behavior of Sand," *Soils and Foundations,* Vol.18, No.4, pp.31-45.

29) Jardine, R., Standing, J. R. and Kovacevic, N. (2005): "Lessons learned from full scale observations and the practical application of advanced testing and modelling", *Keynote Lecture, Deformation Characterisation of Geomaterials, Proc. IS Lyon 2003* (Di Benedetto et al., eds.), Vol. 2, pp.201-245

30) Kawabe, S., Enomoto, T. and Tatsuoka, F. (2006): "Viscous properties of round granular material in drained triaxial compression test", *Proc. 41st Japanese National Conference on Geotechnical Engineering, JGS*, Kagoshima (in Japanese).

31) Kiyota, T., Tatsuoka, F. and Yamamuro, J. (2005): "Drained and undrained creep characteristics of loose saturated sand and their relation", *Proc. of GeoFrontier 2005 Congress, GeoInstitute, ASCE*, Austin, Texas, GSP 138, Site characterization and modeling (Mayne et al. eds).

32) Kiyota, T. and Tatsuoka, F. (2006), "Viscous property of loose sand in triaxial compression, extension and cyclic loading", *Soils and Foundations*, Vol.46 (to appear).

33) Ko, D.-H., Ito. H., Tatsuoka. F. and Nishi. T. (2003): "Significance of viscous effects in the development of residual strain incyclic triaxial tests on sand", *Proc. 3rd Int. Sym. on Deformation Characteristics of Geomaterials*, IS Lyon 03 (Di Benedetto et al. eds.), Balkema, September, 2003, pp.559-568.

34) Komoto, N., Nishi, T., Li, J.-Z. and Tatsuoka, F. (2003): "Viscous stress-strain properties of undisturbed Pleistocene clay and its constitutive modelling", *Proc. 3rd Int. Sym. on Deformation Characteristics of Geomaterials*, IS Lyon 03 (Di Benedetto et al. eds.), Balkema, September, 2003, pp.579-587.

35) Kongkitkul, W., Hirakawa, D., Tatsuoka, F. and Uchimura, T. (2004): "Viscous deformation of geogrid reinforcement under cyclic loading conditions and its model simulation" , *Geosynthetics International*, Vol.GS11, No.2, pp.73-99.

36) Kongsukprasert, L., Kuwano, R. and Tatsuoka, F. (2001): "Effects of ageing with shear stress on the stress-strain behavior of cement-mixed sand", *Advanced Laboratory Stress-Strain Testing of Geomaterials* (Tatsuoka et al. eds.), Balkema, pp.251-258.

37) Kongsukprasert, L., Tatsuoka, F. and Tateyama, M. (2004): "Several factors affecting the strength and deformation characteristics of cement-mixed gravel", *Soils and Foundations*, Vol. 45, No. 3, pp.107-124.
38) Kongsukprasert, L. and Tatsuoka, F. (2005): "Ageing and viscous effects on the deformation and strength characteristics of cement-mixed gravely soil in triaxial compression", *Soils and Foundations,* Vo.45, No. 6, pp.55-74.
39) Kuwano, R. and Jardine, R. J. (2002): "On measuring creep behaviour in granular materials through triaxial testing", *Canadian Geotechnical Journal,* Voi.39, No.5, pp.1061-1074.
40) Lade, P. V. and Duncan, J. M. (1975): "Elasto-plastic stress-strain theory for cohesionless soil", *ASCE, Journal of Geotechnical Division,* Vol.101, GT.100, 1037-1053.
41) Lade, P. V. (1976): "Stress-path dependent behavior of cohesionless soil", *Jour. of the Geotechnical Engineering Div., ASCE,* Vol.102, No.GT1, pp.51-68.
42) Lade, P. V., and Liu, C. T. (1998): "Experimental study of drained creep behavior of sand", *Journal of Engineering Mechanics,* ASCE124(8): 912-920.
43) Lade, P. V. and Liu, C.-T. (2001): "Modeling creep behaviour of granular materials", *Computer Methods and Advances in Geomechanics* (Desai et al. eds.), Balkema, pp.277-284.
44) Leroueil, S. and Marques, M. E. S. (1996) : "Importance of strain rate and temperature effects in geotechnical engineering", S-O-A Report, *Measuring and Modeling Time Dependent Soil Behavior, ASCE Geotech. Special Publication 61*: pp.1-60.
45) Li, Jiangh-Zhong, Acosta-Martínez, H., Tatsuoka, F. and Deng, J.-L. (2004): "Viscous property of soft clay and its modelling", *Engineering Practice and Performance of Soft Deposits,* Proc. of IS Osaka 2004, pp.1-6.
46) Mair, K. and Marone, C. (1999): "Friction of simulated fault gouge for a wide range of velocities and normal stresses", *Journal of Geophysical Research,* Vol. 104, No.B12, pp.28,8999-28,914, December 10.
47) Matsushita, M., Tatsuoka, F., Koseki, J., Cazacliu, B., Di Benedetto, H. and Yasin, S. J. M. (1999): "Time effects on the pre-peak deformation properties of sands", *Proc. Second Int. Conf. on Pre-Failure Deformation Characteristics of Geomaterials, IS Torino '99* (Jamiolkowski et al., eds.), Balkema, Vol.1, pp.681-689.
48) Mejia, C. A., Vaid, Y. P. and Negussey, D. (1988): "Time-dependent behaviour of sand", *Proc. Int. Conf. On Rheology and Soil Mechanics* (Keedwell eds.), Elsevier Applied Science, pp.312-326.
49) Molenkamp, F. (1980): "Elasto-plastic double hardening model MONOT", *Delft Soil Mechanics Laboratory,* Report No. Co.218595.
50) Momoya, M. (1998): "Time effect and consolidation stress path on the deformation characteristics of clay", *Master of Engineering thesis,* Department of Civil Engineering, University of Tokyo (in Japanese).
51) Muir-Wood, D. (1990): "Soil behaviour and critical state soil mechanics", *Cambridge University Press.*
52) Murayama, S., Michiro, K., and Sakagami, T. (1984): "Creep characteristics of sands*", Soils and Foundations*, Vol.24, No.2, pp.1–15.
53) Nakai, T. (1989): "An isotropic hardening elasto-plastic model for sand considering

the stress path dependency in three dimensional stresses", *Soils and Foundations* Vol.29, No.1, pp.119-137.

54) Nakamura, Y., Kuwano, J. and Hashimoto, S. (1999): "Small strain stiffness and creep of Toyoura sand measured by a hollow cylinder apparatus", *Proc. of the Second International Conference on Pre-failure Deformation Characteristics of Geomaterials, Torino, 1999,* Balkema (Jamiolkowski et al., eds.), Vol.1, pp.141-148.

55) Nawir, H., Tatsuoka, F. and Kuwano, R. (2003a): "Experimental evaluation of the viscous properties of sand in shear", *Soils and Foundations,* Vol.43, No.6, pp.13-31.

56) Nawir, H., Tatsuoka, F. and Kuwano, R. (2003b): "Viscous effects on the shear yielding characteristics of sand", *Soils and Foundations*, Vol.43, No.6, pp.33-50.

57) Nirmalan, S. and Uchimura, T. (2006): "Viscous properties and strength of scrapped tire chips", *Proc. 41st Japanese National Conference on Geotechnical Engineering, JGS*, Kagoshima.

58) Oie, M, Sato, N. Okuyama. Y., Yoshida, Teru, Yoshida, Tetuya, Yamada, S., Tatsuoka, F. (2003): " Shear banding characteristics in plane strain compression of granular materials", *Proc. 3rd Int. Symp. on Deformation Characteristics of Geomaterials*, IS Lyon 03 (Di Benedetto et al. eds.), Balkema, September, 2003, pp.597-606.

59) Park, C.-S. and Tatsuoka, F. (1994): "Anisotropic strength and deformations of sands in plane strain compression", *Proc. of the 13th Int. Conf. on Soil Mechanics and Foundation Engineering, New Delhi,* Vol.13, No.1, pp.1-4.

60) Perzyna, P. (1963): "The constitutive equations for work-hardening and rate-sensitive plastic materials", *Proc. of Vibrational Problems,* Warsaw, 4(3), pp.281-290.

61) Poorooshasb, H. B., Holubec, I. and Sherbourne, A. N. (1967): "Yielding and flow of sand in triaxial compression: Parts II and III", *Canadian Geotechnical Journal*, Vol.IV, No.4, pp.376-397.

62) Poorooshasb,H.B. (1971): "Deformation of sand in triaxial compression", *Proc., 4th Asian Regional Conf. on SMFE*, Bangkok, Vol.1. pp.63-66.

63) Rendulic, L. (1936): "Relation between void ratio and effective principal stresses for a remouldedsilty clay", *Proc. 1st International Conference on Soil Mechanics*, Vol.3, pp.48-51.

64) Schanz, T., Vermeer, P. A. and Bonnier, P. G. (1999): "The hardening soil model: Formulation and verification", *Beyond 2000 in Computational Geotechnics* (Brinkgreve eds.), Balkema, pp.281-296.

65) Schofield, A. N. and Wroth, C. P. (1968): "Critical State Soil Mechanics", *McGraw Hill.*

66) Shibuya, S., Mitachi, T., Tanaka, H., Kawaguchi, T. And Lee, I.-M. (2001): "Measurement and application of quasi-elastic properties in geotechnical site characterization", Keynote Lecture, *Prof. 11th Asian Regional Conference on SMGE,* Seoul (Hong et al., eds.), Vol. 2, pp.639-710.

67) Siddiquee, M. S. A., Tatsuoka, F. and Tanaka, T. (2006), "FEM simulation of the viscous effects on the stress-strain behaviour of sand in plane strain compression", *Soils and Foundations*, Vol.46, No.1, pp.99-108.

68) Sorensen, Kenny K., Baudet, Beatrice A. and Tatsuoka, F. (2006): "Coupling of ageing and viscous effects in an artificially structured clay", *Soil Stress-Strain Behavior: Measurement, Modeling and Analysis, Proc. of Geotechnical Symposium*

in Roma, March 16 & 17, 2006 (Ling et al., eds.).
69) Stroud, M. A. (1971): "The behaviour of sand at low stress levels in the simple shear apparatus", *Ph.D Dissertation*, University of Cambridge.
70) Sugai, M., Tatsuoka, F., Kuwabara, M. and Sugo, K. (2000): "Strength and deformation characteristics of cement-Mixed soft clay", *Coastal Geotechnical Engineering in Practice, Proc. IS Yokohama* (Nakase & Tsuchida eds.), Balkema, Vol.1, pp. 521-52.
71) Sugai, M. and Tatsuoka, F. (2003): "Ageing and loading rate effects on the stress-strain behaviour of a cement-mixed soft clay", *Proc. 3rd Int. Sym. on Deformation Characteristics of Geomaterials*, IS Lyon 03 (Di Benedetto et al. eds.), Balkema, September, 2003, pp.627-635.
72) Suklje, L. (1969): "Rheological aspects of soil mechanics", *Wiley-Interscience,* London.
73) Tanaka, H. (2005a): "Consolidation behaviour of natural soils around p_c value – Long-term consolidation test", *Soils and Foundations*, Vol.45, No 3, pp.83-96.
74) Tanaka, H. (2005b): "Consolidation behaviour of natural soils around p_c value – Inter-connected oedometer test", *Soils and Foundations*, Vol.45, No 3, pp.97-106.
75) Tatsuoka, F. (1973): "Fundamental study on the deformation characteristics of sand by triaxial tests", *Dr of Engineering thesis, University of Tokyo* (in Japanese).
76) Tatsuoka, F. and Ishihara, K. (1974): "Yielding of sand in triaxial compression", *Soils and Foundations*, 14(2), 51-65.
77) Tatsuoka, F. (1980): "Stress-strain behaviour of an idealized anisotropic granular material", *Soils and Foundations*, Vo.20, No.3, pp.75-90.
78) Tatsuoka, F., and Molenkamp, F. (1983): "Discussion on yield loci for sands", *Mechanics of Granular Materials: New Models and Constitutive Relations*, Elsevier Science Publisher B.V., pp.75-87.
79) Tatsuoka, F. and Shibuya, S. (1991): "Deformation characteristics of soils and rocks from field and laboratory tests", Keynote Lecture for Session No.1, *Proc. of the 9th Asian Regional Conf. on SMFE,* Bangkok, Vol.II, pp.101-170.
80) Tatsuoka, F. and Kohata, Y. (1995): "Stiffness of hard soils and soft rocks in engineering applications", Keynote Lecture, *Proc. of Int. Symposium Pre-Failure Deformation of Geomaterials* (Shibuya et al., eds.), Balkema, Vol. 2, pp.947-1063.
81) Tatsuoka, F., Lo Presti, D. C. F. and Kohata, Y. (1995): "Deformation characteristics of soils and soft rocks under monotonic and cyclic loads and their relationships", SOA Report, *Proc. of the Third Int. Conf. on Recent Advances in Geotechnical Earthquake Engineering and Soil Dynamics, St Louis* (Prakash eds.), Vol.2, pp.851-879.
82) Tatsuoka, F., Jardine, R. J., Lo Presti, D. C. F., Di Benedetto, H. and Kodaka, T. (1999a): "Characterising the Pre-Failure Deformation Properties of Geomaterials", Theme Lecture for the Plenary Session No.1, *Proc. of XIV IC on SMFE, Hamburg,* September 1997, Volume 4, pp.2129-2164.
83) Tatsuoka, F., Modoni, G., Jiang, G.-L., Anh Dan, L. Q., Flora, A., Matsushita, M., and Koseki, J. (1999b): "Stress-Strain Behaviour at Small Strains of Unbound Granular Materials and its Laboratory Tests, Keynote Lecture", *Proc. of Workshop on Modelling and Advanced testing for Unbound Granular Materials, January 21 and 22, 1999, Lisboa* (Correia eds.), Balkema, pp.17-61.

84) Tatsuoka, F., Santucci de Magistris, F. and Momoya, M. and Maruyama, N. (1999c): "Isotach behaviour of geomaterials and its modelling", *Proc. Second Int. Conf. on Pre-Failure Deformation Characteristics of Geomaterials, IS Torino '99* (Jamiolkowski et al., eds.), Balkema, Vol.1, pp.491-499.
85) Tatsuoka, F., Santucci de Magistris, F., Hayano, K., Momoya, Y. and Koseki, J. (2000): "Some new aspects of time effects on the stress-strain behaviour of stiff geomaterials", *Keynote Lecture, The Geotechnics of Hard Soils – Soft Rocks, Proc. of Second Int. Conf. on Hard Soils and Soft Rocks, Napoli,* 1998 (Evamgelista and Picarelli eds.), Balkema, Vol.2, pp.1285-1371.
86) Tatsuoka, F., Uchimura, T., Hayano, K., Di Benedetto, H., Koseki, J. and Siddiquee, M. S. A. (2001): "Time-dependent deformation characteristics of stiff geomaterials in engineering practice", the Theme Lecture, *Proc. of the Second International Conference on Pre-failure Deformation Characteristics of Geomaterials, Torino, 1999, Balkema (Jamiolkowski et al., eds.),* Vol.2, pp.1161-1262.
87) Tatsuoka, F., Ishihara, M., Di Benedetto, H. and Kuwano, R. (2002): "Time-dependent deformation characteristics of geomaterials and their simulation", *Soils and Foundations*, Vol.42, No.2, pp.103-129.
88) Tatsuoka, F., Di Benedetto, H. and Nishi, T. (2003a): "A framework for modelling of the time effects on the stress-strain behaviour of geomaterials", *Proc. 3rd Int. Sym. on Deformation Characteristics of Geomaterials*, IS Lyon 03 (Di Benedetto et al. eds.), Balkema, September, 2003, pp.1135-1143.
89) Tatsuoka, F., Acosta-Martinez, H. E. and Li, J.-Z. (2003b): "Viscosity in one-dimensional deformation of clay and its modelling and simulation", *Proc. 38th Japan National Conf. on Geotechnical Eingieering, JGS,* Akita.
90) Tatsuoka, F. Nawir, H., and Kuwano, R. (2004a): "A modelling procedure of shear yielding characteristics affected by viscous properties of sand in triaxial compression", *Soils and Foundations*, Vol.44, No.6, pp.83-99.
91) Tatsuoka, F. (2004): "Effects of viscous properties and ageing on the stress-strain behaviour of geomaterials." *Geomechanics- Testing, Modeling and Simulation, Proceedings of the GI-JGS workshop, Boston, ASCE Geotechnical Special Publication GSP No. 143 (Yamamuro & Koseki eds.),* pp.1-60.
92) Tatsuoka, F., Hirakawa, D., Shinoda, M., Kongkitkul, W. and Uchimura,T. (2004b): "An old but new issue; viscous properties of polymer geosynthetic reinforcement and geosynthetic-reinforced soil structures," Keynote lecture, Proc. GeoAsia04, Seoul, pp.29-77.
93) Tatsuoka, F. and Tani, K. (2006): Fundamental issues in clay consolidation, *Monthly Journal Kiso-Ko (the Foundation Engineering and Equipment),* Vol.34, No. 396, June, pp. 12 - 22 (in Japanese).
94) Tatsuoka, F., Enomoto, T. and Kiyota, T. (2006): "Viscous properties of geomaterials in drained shear", *Geomechanics- Testing, Modeling and Simulation, Proceedings of the Second GI-JGS workshop, Osaka, September 2005, ASCE Geotechnical Special Publication GSP (Lade et al. eds.)* (to appear).
95) Vermeer, P. A. (1978): "A double hardening model for sand", *Géotechnique* Vol.28, No.4, pp.413-433.
96) Vermeer, P. A. and Neher, H. P. (1999): "A soft soil model that accounts for creep", *Beyond 2000 in Computational Geotechnics* (Brinkgreve eds.), Balkema, pp. 249-261.

97) Yamamuro, J. A. and Lade, P. V. (1993): "Effects of strain rate on instability of granular soils", *Geotechnical Testing Journal*, Vol.16, No.3, pp.304-313.
98) Yasin, S. J. M. and Tatsuoka, F. (2000): "Stress history-dependent deformation characteristics of dense sand in plane strain", *Soils and Foundations*, Vo.40, No.2, pp.77-98.
99) Yasin, S. J. M. and Tatsuoka, F. (2003): "New strain energy hardening functions for sand based on the double yielding concept", *Proc. 3rd Int. Symp. on Deformation Characteristics of Geomaterials*, IS Lyon 03 (Di Benedetto et al. eds.), Balkema, Sept. 2003, pp.1127-1134.

APPENDIX A: Hypo-elastic model for cross-anisotropic deformation

The vertical, intermediate and horizontal irreversible strain increments, $d\varepsilon_v^{ir}$, $d\varepsilon_m^{ir}$ and $d\varepsilon_h^{ir}$, in plane strain compression (PSC) tests as well as $d\varepsilon_v^{ir}$ and $d\varepsilon_h^{ir}$ $(=d\varepsilon_m^{ir})$ in triaxial compression (TC) tests reported in this paper were obtained by following the hypo-elastic model (e.g., Hoque & Tatsuoka, 1998, 2001; Tatsuoka et al., 1999a, b).

$$d\varepsilon_v^{ir} = d\varepsilon_v^{total} - d\varepsilon_v^e;\ d\varepsilon_h^{ir} = d\varepsilon_h^{total} - d\varepsilon_h^e;\ \&\ d\varepsilon_m^{ir} = d\varepsilon_m^{total} - d\varepsilon_m^e \tag{A1a}$$

$$d\varepsilon_v^e = \frac{d\sigma_v'}{E_v} - \upsilon_{hmv}\frac{d\sigma_m'}{E_{hm}} - \upsilon_{hv}\frac{d\sigma_h'}{E_h};$$
$$d\varepsilon_h^e = \frac{d\sigma_h'}{E_h} - \upsilon_{hh}\frac{d\sigma_m'}{E_{hm}} - \upsilon_{vh}\frac{d\sigma_v'}{E_v};\ \&$$
$$d\varepsilon_m^e = \frac{d\sigma_m'}{E_{hm}} - \upsilon_{hm}\frac{d\sigma_h'}{E_h} - \upsilon_{vm}\frac{d\sigma_v'}{E_v} \tag{A1b}$$

$$E_v = E_{v0}\left(\frac{\sigma_v'}{\sigma_0}\right)^m;\ E_{h0} = (1-I_0)E_{v0};$$
$$E_h = E_{h0}\left(\frac{\sigma_h'}{\sigma_0}\right)^m;\ \&\ E_{hm} = E_{h0}\left(\frac{\sigma_m'}{\sigma_0}\right)^m \tag{A1c}$$

$$\upsilon_{hh} = \upsilon_0;\ \upsilon_{vh} = \upsilon_0\left(\frac{E_{v0}}{E_{h0}}\right)^{1/2}\left(\frac{\sigma_v'}{\sigma_h'}\right)^{m/2};$$
$$\upsilon_{hv} = \upsilon_0\left(\frac{E_{h0}}{E_{v0}}\right)^{1/2}\left(\frac{\sigma_h'}{\sigma_v'}\right)^{m/2};\ \&\ \upsilon_{hm} = \upsilon_0\left(\frac{E_{h0}}{E_{hm0}}\right)^{1/2}\left(\frac{\sigma_h'}{\sigma_{hm}'}\right)^{m/2} \tag{A1d}$$

where E_v, E_h and E_{hm} are the elastic Young's moduli in the vertical and horizontal directions at a given stress state; E_{v0}, E_{h0} and E_{hm0} are the values of E_v, E_h and E_{hm} when the respective normal stress is equal to the reference isotropic stress, σ_0; υ_{hh}, υ_{vh}, υ_{hv} and so on are the elastic Poisson's ratios; and υ_0 $(=\upsilon_{hv}=\upsilon_{vh})$ is the basic Poisson's ratio when the elastic property becomes isotropic; and power m is the material constant. For example, the parameters for Toyoura sand, v_0= 0.3 kgf/cm^2 (= 29.5 kPa), m= 0.494, υ_0= 0.17; I_0= 0.1 and E_0 = 880 kgf/cm^2 (= 86.6 MPa).

APPENDIX B: Simplified non-linear three-component model to describe Isotach viscosity and TESRA viscosity

B1: *General framework of the non-linear three-component modelling (Fig. 3.7a)*

1. A given strain increment, $d\varepsilon$, consists of an elastic (i.e., rate-independent and reversible) component, $d\varepsilon^e$, and a rate-dependent and irreversible (i.e., inelastic or visco-plastic) component, $d\varepsilon^{ir}$, as:

$$d\varepsilon = d\varepsilon^e + d\varepsilon^{ir} \qquad \text{(B1)}$$

   $d\varepsilon^e$ takes place only in component *E*, and is obtained by a hypo-elastic model (Appendix A), which has a set of elastic moduli that are all a function of instantaneous stress state (and also strain history when relevant).

2. A given effective stress, σ, consists of an inviscid (i.e., rate-independent) component, σ^f, and a viscous (i.e., rate-dependent) component, σ^v, as:

$$\sigma = \sigma^f + \sigma^v \qquad \text{(B2)}$$

3. σ^f is a unique function of irreversible strain, ε^{ir}, in the monotonic loading (ML) case along a fixed stress path in which the irreversible strain rate, $\dot{\varepsilon}^{ir} = \partial\varepsilon^{ir}/\partial t$, is always positive irrespective of the sign of stress rate, $\dot{\sigma}$. The σ^f - ε^{ir} relation becomes hysteretic under cyclic loading conditions. The related flow rule is modelled in terms of σ^f similarly as the conventional elasto-plastic theories. So, any elasto-plastic model can be extended to a non-linear three-component model by adding the σ^v component appropriately.
4. The viscous stress increment, $d\sigma^v$, develops by either $d\varepsilon^{ir}$ or its rate, $d\dot{\varepsilon}^{ir}$, or both. In the case of Isotach viscosity, the current σ^v is always proportional to the instantaneous σ^f. Then, "the increment $d\sigma^v$ when $\varepsilon^{ir} = \tau$" is given as:

$$[d\sigma^v]_{(\tau)} = [d\{\sigma^f(\varepsilon^{ir}) \cdot g_v(\dot{\varepsilon}^{ir})\}]_{(\tau)} \qquad \text{(B3)}$$

   It is assumed that Eq. B3 can be applied to all the types of viscosity described in this paper. $g_v(\dot{\varepsilon}^{ir})$ is the viscosity function, which is always zero or positive, given as follows for any strain (ε^{ir}) or stress (σ^f) path (with or without cyclic loading; i.e., irrespective of the sign of $\dot{\varepsilon}^{ir}$):

$$g_v(\dot{\varepsilon}^{ir}) = \alpha \cdot [1 - \exp\{1 - (\frac{|\dot{\varepsilon}^{ir}|}{\dot{\varepsilon}_r^{ir}} + 1)^m\}] \quad (\geq 0) \qquad \text{(B4)}$$

where $\left|\dot{\varepsilon}^{ir}\right|$ is the absolute value of $\dot{\varepsilon}^{ir}$; and α, $\dot{\varepsilon}_r^{ir}$ and m are positive material constants. As explained in Di Benedetto et al. (2002), Tatsuoka et al. (2002, 2006) and Tatsuoka (2004), these constants for a given type of geomaterial are determined based on the rate-sensitivity coefficient, β.

The relations lettered *E* and *P* in Fig. 4.5 represent schematically the stress-strain properties of components *E* and *P*, while the one lettered *E+P* represents the stress-strain relation of an elasto-plastic material having components *E* and *P* connected in series. The relation lettered *E+P+V* represents the behaviour of the three-component model (in the case of Isotach viscosity), while the one when $\dot{\varepsilon}^{ir} = 0$ denotes the reference stress-strain relation for loading, which is the same with the one lettered *E+P*.

B2: *Isotach viscosity*
Di Benedetto et al. (2002, 2005), Tatsuoka et al. (2002) and Tatsuoka (2005) showed that different formulations of σ^v are necessary for different geomaterial types that exhibit different effects of recent history of $\dot{\varepsilon}^{ir}$ on the current σ^v value. Firstly, a stress - strain model called the "new isotach" was proposed to describe the viscous property of sedimentary soft rock (Hayano et al., 2001) and some clay types (Tatsuoka et al., 1999c, 2001). In this case, the current σ^v during primary ML is obtained by directly integrating Eq. B3 with respect to ε^{ir} without referring to the intermediate strain history as:

$$\sigma^v\left(\varepsilon^{ir},\dot{\varepsilon}^{ir}\right) = \sigma^f\left(\varepsilon^{ir}\right)\cdot g_v(\dot{\varepsilon}^{ir}) \tag{B5}$$

From Eqs. B2 and B5, we have:

$$\sigma = \sigma^f\left(\varepsilon^{ir}\right)\cdot\{1 + g_v(\dot{\varepsilon}^{ir})\} \tag{B6}$$

The unique dependency of the current stress, σ, on the instantaneous strain, ε, and its rate, $\dot{\varepsilon}$, was originally called the isotach property (Suklje, 1969). It is to be recalled that, in Eq. B6 (for the ML case), the current stress, σ, is a unique function of instantaneous irreversible strain rate, $\dot{\varepsilon}^{ir}$, and irreversible strain, ε^{ir}. The use of ε^{ir} is necessary to describe realistically the stress-strain behaviour, in particular those during stress relaxation and when the strain changes at a fast rate (Tatsuoka et al., 1999c, 2000 & 2001).

B3: *TESRA viscosity*
Matsushita et al. (1999), Tatsuoka et al. (2000, 2001, 2002) and Di Benedetto et al. (2002) reported that, with two fine uniform sands having sub-angular particles, Hostun and Toyoura sands, the viscous stress increment according to Eq. B3, $[d\sigma^v]_{(\tau)}$, decays with an increase in the irreversible strain, ε^{ir}, during the subsequent ML. When $[d\sigma^v]_{(\tau)}$ decays with an increase in ε^{ir}, the current viscous stress (when $\varepsilon^{ir} = \varepsilon^{ir}$), $[\sigma^v]_{(\varepsilon^{ir})}$, and the current stress, $[\sigma]_{(\varepsilon^{ir})}$, become no longer a unique function of instantaneous ε^{ir} and $\dot{\varepsilon}^{ir}$. Di Benedetto et al. (2002) and Tatsuoka et al. (2002) modified the new isotach model by introducing the decay function, $g_{decay}(\varepsilon^{ir} - \tau)$, as follows:

$$[\sigma^v]_{(\varepsilon^{ir})} = \int_{\tau=\varepsilon_1^{ir}}^{\varepsilon^{ir}} \left[d\sigma^v\right]_{(\tau,\varepsilon^{ir})} = \int_{\tau=\varepsilon_1^{ir}}^{\varepsilon^{ir}} \left[d\{\sigma^f \cdot g_v(\dot{\varepsilon}^{ir})\}\right]_{(\tau)} \cdot g_{decay}(\varepsilon^{ir}-\tau) \tag{B7}$$

where $\left[d(\sigma^f \cdot g_v(\dot{\varepsilon}^{ir})\right]_{(\tau)}$ is the viscous stress increment that developed when $\varepsilon^{ir}=\tau$ (Eq. B3) ; $\left[d\sigma^v\right]_{(\tau,\varepsilon^{ir})}$ is the viscous stress increment that developed in the past (when $\varepsilon^{ir}=\tau$) and then has decayed until the present (when $\varepsilon^{ir}=\varepsilon^{ir}$); and ε_1^{ir} is the irreversible strain at the start of integration, where σ^v= 0. Tatsuoka et al., (2001, 2002) and Di Benedetto et al. (2002) proposed the following power function based on experimental data:

$$g_{decay}(\varepsilon^{ir}-\tau) = r_1^{(\varepsilon^{ir}-\tau)} \tag{B8}$$

where r_1 is a positive constant smaller than unity. The power form has a fundamental advantage in the integration of Eq. B7 (Tatsuoka et al., 2002). That is, we obtain:

$$r_1^{(\varepsilon^{ir}-\tau)} = r_1^{(\varepsilon^{ir}-\Delta\varepsilon^{ir}-\tau)} \cdot r_1^{\Delta\varepsilon^{ir}} \tag{B9}$$

Then, Eq. B7 becomes

$$\begin{aligned}[\sigma^v]_{(\varepsilon^{ir})} &= \left[\int_{\tau=\varepsilon_1^{ir}}^{\varepsilon^{ir}-\Delta\varepsilon^{ir}} \left[d\{\sigma^f \cdot g_v(\dot{\varepsilon}^{ir})\}\right]_{(\tau)} \cdot r_1^{(\varepsilon^{ir}-\Delta\varepsilon^{ir}-\tau)}\right] \cdot r_1^{\Delta\varepsilon^{ir}} + \Delta\{\sigma^f \cdot g_v(\dot{\varepsilon}^{ir})\} \cdot r_1^{\Delta\varepsilon^{ir}/2} \\ &= [\sigma^v]_{(\varepsilon^{ir}-\Delta\varepsilon^{ir})} \cdot r_1^{\Delta\varepsilon^{ir}} + \Delta\{\sigma^f \cdot g_v(\dot{\varepsilon}^{ir})\} \cdot r_1^{\Delta\varepsilon^{ir}/2}\end{aligned} \tag{B10}$$

The current viscous stress, $[\sigma^v]_{(\varepsilon^{ir})}$, can be obtained from the known value at one step before, $[\sigma^v]_{(\varepsilon^{ir}-\Delta\varepsilon^{ir})}$, with no need to repeat the integration Eq. B7 at every incremental step. Siddiquee et al. (2006) showed that FEM analysis incorporating the TESRA model becomes feasible by representing the viscous stress in the incremental form, Eq. B10.

The physical meaning of the decay parameter, r_1, can be readily seen by rewriting Eq. B8 to:

$$g_{decay}(\varepsilon^{ir}-\tau) = (0.5)^{\frac{\varepsilon^{ir}-\tau}{H}} \tag{B11}$$

where H is the irreversible strain difference, $\varepsilon^{ir}-\tau$, by which the viscous stress increment $[d\sigma^v]_{(\tau)}$ has decayed to a half of the initial value during ML at a constant $\dot{\varepsilon}^{ir}$. The relationship between the parameters r_1 and H is given as:

$$r_1 = \left(\frac{1}{2}\right)^{\frac{1}{H}}; \text{ or } H = \frac{\log(1/2)}{\log(r_1)} \tag{B12}$$

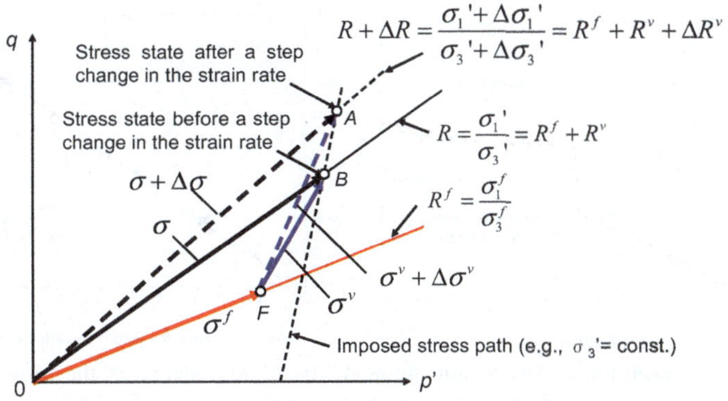

Fig. B1 Viscosity function defined in terms of effective principal stress ratio, *R*.

When r_1 = 1.0, *H* becomes infinitive, and Eq. B7 becomes totally differential, returning to Eq. B5 (the Isotach model). *H* decreases with a decrease in r_1. When r_1 = 0, *H* becomes zero.

Due to such a decay feature as expressed by Eqs. B7 and B8, the effects of irreversible strain rate, $\dot{\varepsilon}^{ir}$, and its rate (i.e., irreversible strain acceleration), $\ddot{\varepsilon}^{ir}=\partial^2\varepsilon^{ir}/\partial t^2$, on the σ^v value during the subsequent loading become transient, or temporary. The new model is therefore called the TESRA model (i.e., temporary or transient effects of irreversible strain rate and irreversible strain acceleration on the viscous stress component). Then, the value of σ^v could become either positive or zero or negative depending on recent loading history even when $\dot{\varepsilon}^{ir}$ has always been kept positive.

B4: *Viscosity function*

Specific form for geomaterials: Di Benedetto et al. (2002) and Tatsuoka et al. (2002) defined the viscosity function for unbound geomaterial (i.e., clay, sand and gravel) using the effective principal stress ratio, $R=\sigma_1'/\sigma_3'$, as the stress parameter (i.e., R for σ and $R^f=\sigma_1^f{}'/\sigma_3^f{}'$ for σ') of the three-component model (Fig. 3.7a) and expressed Eq. B5 (for the Isotach viscosity) as summarised below.

$$R^v\left(\gamma^{ir},\dot{\gamma}^{ir}\right)=R^f\left(\gamma^{ir}\right)\cdot g_v(\dot{\gamma}^{ir}) \tag{B13a}$$

where γ is the shear strain ($=\varepsilon_1-\varepsilon_3$). The incremental form is:

$$d\{R^v\left(\gamma^{ir},\dot{\gamma}^{ir}\right)\}=d\{R^f\left(\gamma^{ir}\right)\cdot g_v(\dot{\gamma}^{ir})\} \tag{B13b}$$

Referring to **Fig. B1**, the viscous stress ratio , R^v, is obtained as:

$$R^v=R-R^f \tag{B14}$$

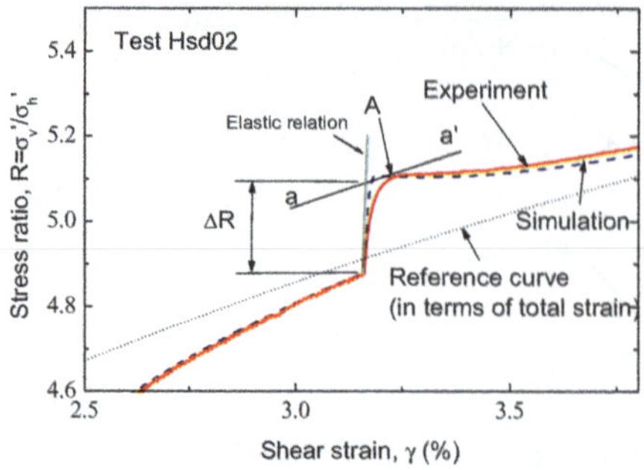

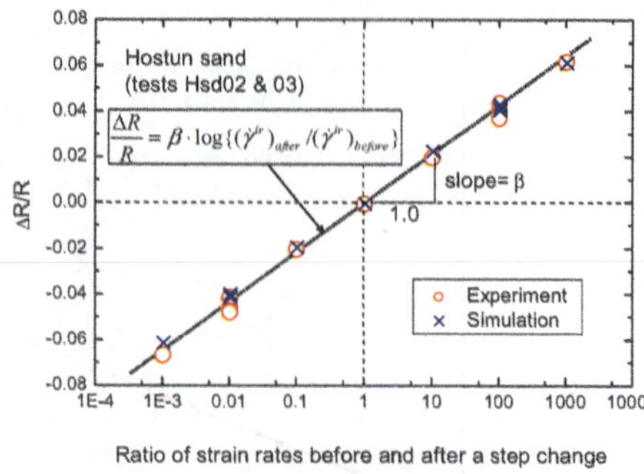

Fig. B2 (left) Definition of stress ratio jump ΔR by a step change in the irreversible shear strain rate in a drained PSC test on Hosun sand (Di Benedetto et al., 2002).

Fig. B3 (right) Definition of rate-sensitivity coefficient β in drained PSC tests on Hostun sand (Di Benedetto et al., 2002).

where R is the measured values of σ_1'/σ_3', which is equal to $(\sigma_1^f + \sigma_1^v)/(\sigma_3^f + \sigma_3^v)$. The current stress state (σ_1', σ_3') (before a step change in the strain rate) is represented by point B in Fig. B1. R^f is the inviscid principal stress ratio, equal to σ_1^f / σ_3^f. The current inviscid stress state (σ_1^f, σ_3^f) is represented by point F. Note that R^v is not equal to σ_1^v / σ_3^v, but equal to $(\sigma_1^f + \sigma_1^v)/(\sigma_3^f + \sigma_3^v)$ - σ_1^f / σ_3^f. For example, in TC at a constant σ_3', R^v is equal to σ_1^v / σ_3^f if $\sigma_3^v = 0$. Kiyota and Tatsuoka (2006) showed that Eq. B13b together with Eq. B14 are relevant also to describe the viscous property of sand in the triaxial extension tests at a fixed confining pressure, σ_1'.

Derivation of the viscosity function from experimental results: The viscosity function, $g_v(\dot{\gamma}^{ir})$, of Eq. B13, which is relevant to the TC, TE and PSC test conditions, is obtained from Eq. 4 as:

$$g_v(\dot{\gamma}^{ir}) = \alpha \cdot [1 - \exp\{1 - (\frac{|\dot{\gamma}^{ir}|}{\dot{\gamma}_r^{ir}} + 1)^m\}] \quad (\geq 0) \qquad \text{(B15)}$$

The parameters of Eq. B15 are determined from experimental data as follows. Points B and A in Fig. B1 represent the stress states, respectively, before and after a step change in the irreversible shear strain rate, $\dot{\gamma}^{ir}$. The associated jump in R, ΔR, is due solely to a change in $\dot{\gamma}^{ir}$ made at a fixed irreversible shear strain keeping R^f constant. **Fig. B2** shows a typical test result showing the definition of ΔR. **Fig. B3** shows typical data showing the relationships between the ratio of ΔR to the instantaneous value of R when the strain rate is stepwise changed and the logarithm of the ratio of the axial strain rates before and after a step change, which is essentially the same as the ratio of the irreversible shear strain rates, $(\dot{\gamma}^{ir})_{after}/(\dot{\gamma}^{ir})_{before}$. The results from the simulation by the three-component model (Fig. 3.7a) of these data are also presented in this figure. It may be seen from this figure that the following linear relation, which is independent of R, fits the data:

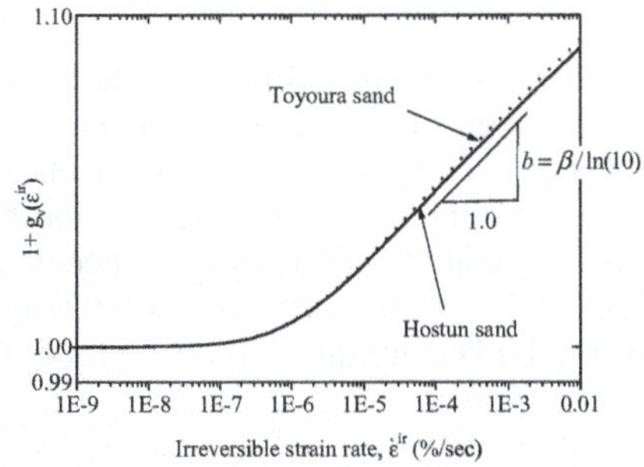

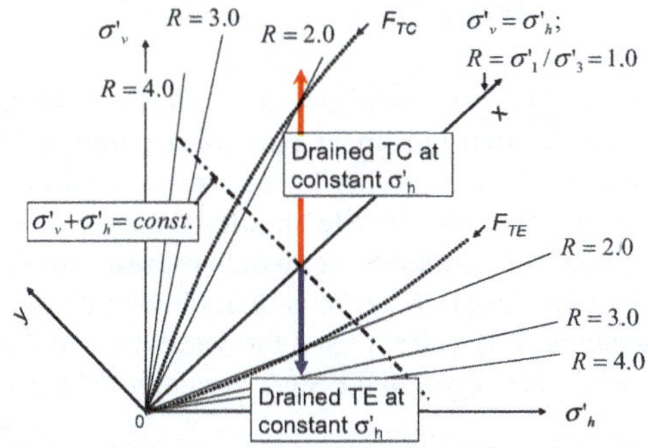

Fig. B4 (left) Viscosity functions of Toyoura and Hostun sands determined based on the values of β measured by drained PSC tests (Di Benedetto et al., 2002).

Fig. B5 (right) Another possible stress parameter to define the viscosity function showing non-linear curves in the stress space (Kiyota & Tatsuoka, 2006).

$$\frac{\Delta R}{R} = \beta \cdot \log_{10}\left(\frac{\dot{\gamma}^{ir}_{after}}{\dot{\gamma}^{ir}_{before}}\right) = b \cdot \ln\left(\frac{\dot{\gamma}^{ir}_{after}}{\dot{\gamma}^{ir}_{before}}\right) \quad \text{(B16)}$$

where β is the rate-sensitivity coefficient; and $b=\beta/\ln 10$. The value of β of sand is rather insensitive to changes in the void ratio, the effective confining pressure and the wet condition (Nawir et al., 2003a; Tatsuoka et al., 2006). Eq. B16 can be rewritten to the incremental form:

$$\frac{dR}{R} = b \cdot d\left(\ln \dot{\gamma}^{ir}\right) \quad \text{(B17)}$$

dR in Eq. B17 is defined for a fixed value of γ^{ir} and so for a fixed value of R^f (i.e., the value when the stress jump starts). Therefore, we obtain $dR = dR^v = d\{R^f \cdot g_v(\dot{\gamma}^{ir})\} = R^f \cdot d\{g_v(\dot{\gamma}^{ir})\}$ referring to Eq. B13b. Then, referring to Eqs. 13a and B14, we obtain:

$$\frac{R^f \cdot d\{g_v(\dot{\gamma}^{ir})\}}{R^f + R^v} = \frac{R^f \cdot d\{g_v(\dot{\gamma}^{ir})\}}{R^f \cdot \{1+g_v(\dot{\gamma}^{ir})\}} = \frac{d\{g_v(\dot{\gamma}^{ir})\}}{1+g_v(\dot{\gamma}^{ir})} = b \cdot d\left(\ln \dot{\gamma}^{ir}\right) \quad \text{(B18)}$$

This equation is assumed to be valid to any imposed stress paths satisfying the loading conditions, $\dot{\gamma}^{ir} > 0$, changing σ_1' or σ_3' or both. To obtain the viscosity function, $g_v(\dot{\gamma}^{ir})$, we do not need to obtain the location of point F (i.e., we do not need to obtain the values of σ_1^f and σ_3^f as well as σ_1^v and σ_3^v), and actually we cannot obtain these values only from such experimental data as shown in Figs. B2 and B3. Then, we obtain:

$$d\{\ln(1+g_v(\dot{\gamma}^{ir})\} = b \cdot d\left(\ln \dot{\gamma}^{ir}\right) \quad \text{(B19a)}$$

By integrating Eq. B19a with respect to $\dot{\gamma}^{ir}$, we obtain:

$$1+g_v(\dot{\gamma}^{ir})=c_v\cdot(\dot{\gamma}^{ir})^b \qquad \text{(B19b)}$$

where c_v is a constant. As shown in **Fig. B4**, the viscosity function (Eq. B15) should be defined so that the linear part for a range of $\dot{\gamma}^{ir}$ for which Eq. B19b was derived has a slope equal to $b=\beta/\ln 10$. That is, Eq. B19b is valid only for a range of $\dot{\gamma}^{ir}$ larger than a certain lower limit while smaller than a certain upper limit. A relevant value should be assumed for parameter α, which represents the upper bound of $g_v(\dot{\gamma}^{ir})$ when $\dot{\gamma}^{ir}$ becomes infinitive. A parameter m is then obtained by try and error. Di Benedetto et al. (2002) and Tatsuoka et al. (2002) used the viscosity function (Eq. B15) determined as above for both Isotach viscosity and TESRA viscosity for consistency.

To apply to cyclic triaxial loading conditions, Kiyota and Tatsuoka (2006) modified Eq. B13 by replacing the inviscid hardening function $R^f(\gamma^{ir})$ with another relevant one that takes into account the hysteretic nature of stress-strain relation under cyclic loading conditions, without changing the other parts of the equation.

A more relevant stress parameter to define the viscosity function: On the stress plane showing the TC and TE stress conditions presented in **Fig. B5**, a constant effective principal stress ratio, $R=\sigma'_1/\sigma'_3$, means two straight lines radiating from the origin that are symmetric about the hydrostatic axis for the same mean stress, $\sigma'_m=(\sigma'_v+\sigma'_h)/2$. On the other hand, as discussed related to Fig. 2.6, the shear yield loci of granular material are slightly curved as represented by the broken curves, denoted by F_{TC} and F_{TE}, in this figure. It is also the case with the failure envelop of granular material. It is likely therefore that $R=\sigma'_1/\sigma'_3$ may not be fully relevant as the stress parameter for the viscosity function when dealing with a wide range of σ'_m.

Curves F_{TC} and F_{TE} can be represented as follows using a stress parameter, r:

$$y=\pm A\cdot(r-1)\cdot x^n \quad \text{(for TC and TE respectively)} \qquad \text{(B20)}$$

where n is a positive constant lower than unity and A is the parameter, which can be selected so that the value of r becomes similar to the value of R for a wide range of R. As Eq. B20 is parabolic, it does not intersect with the hydrostatic axis even when σ'_m becomes infinitive as when using R. As the x and y axes are inclined at an angle equal to 45 degrees relative to the hydrostatic axis, we obtain:

$$y=\frac{1}{\sqrt{2}p_a}(\sigma'_v-\sigma'_h) \qquad \text{(B21a)}$$

$$x=\frac{1}{\sqrt{2}p_a}(\sigma'_v+\sigma'_h) \qquad \text{(B21b)}$$

where p_a is equal to 98 kPa. Then, the relation between r and R becomes:

$$r = \frac{1}{A \cdot (\sqrt{2} p_a)^{1-n}} \cdot \frac{\sigma'_1 - \sigma'_3}{(\sigma'_1 + \sigma'_3)^n} + 1 = \frac{1}{A} \cdot \left(\frac{\sigma'_3}{\sqrt{2} p_a} \right)^{1-n} \cdot \frac{R-1}{(R+1)^n} + 1 \quad \text{(B22)}$$

where $\sigma_3' = \sigma_h'$ in TC and $\sigma_3' = \sigma_v'$ in TE. By replacing R with r in Eqs. B11, B12 and B14, we obtain:

$$r^v \left(\gamma^{ir}, \dot{\gamma}^{ir}\right) = r^f \left(\gamma^{ir}\right) \cdot g_v(\dot{\gamma}^{ir}) \quad \text{(B23)}$$

$$r^v = r - r^f \quad \text{(B24)}$$

$$\frac{\Delta r}{r} = \beta_r \cdot \log_{10} \left(\frac{\dot{\gamma}^{ir}_{after}}{\dot{\gamma}^{ir}_{before}} \right) \quad \text{(B25)}$$

where β_r is the rate-sensitivity coefficient when the stress parameter is r; r is obtained by Eq. B22 from the measured effective stress ratio, $R = \sigma_1' / \sigma_3' = (\sigma_1^f + \sigma_1^v) / (\sigma_3^f + \sigma_3^v)$; and r^f is also obtained by Eq. 22 from the inviscid stress ratio, $R^f = \sigma_1^f / \sigma_3^f$. Despite that it is known that Eqs. B23 through 25 can be applied to the TC and TE stress conditions (Kiyota & Tatsuoka, 2006), it is not known whether they can also be applied to general 3D stress conditions.

B5: *Viscosity function for bound geomaterials*

For bound geomaterials, the stress parameter R, which is used in the formulation of the viscous property for unbound geomaterials as shown above, should be replaced with a more relevant stress parameter. The results from drained TC tests at constant σ_3' on compacted cement-mixed well-graded gravel (Kongsukprasert et al., 2004) showed that the following relation is relevant in place of Eq. B16:

$$\frac{(\Delta q / p_a)_{d\varepsilon_a^{ir}=0}}{(q + q_c) / p_a} = \beta \cdot \log[(\dot{\varepsilon}_a)_{after} / (\dot{\varepsilon}_a)_{before}] \approx \beta \cdot \log[(\dot{\gamma}^{ir})_{after} / (\dot{\gamma}^{ir})_{before}] \quad \text{(B26)}$$

where Δq is the jump of the deviator stress, q, upon a step change in the strain rate; q_c is a constant, independent of q at which Δq is obtained; and p_a= 98 kPa. As σ_3' is kept constant, the left side term of Eq. B26 becomes:

$$\frac{\Delta \sigma_1'}{(\sigma_1' + c)} = \frac{\Delta(\sigma_1' + c)}{(\sigma_1' + c)} = \frac{\Delta(\sigma_1' + c) / \{(\sigma_3' + c)}{(\sigma_1' + c) / \{(\sigma_3' + c)} \quad \text{(B27)}$$

Eq. B27 indicates that, for bound geomaterials, it is relevant to redefined R as $(\sigma_1' + c) / (\sigma_3' + c)$, where c is a constant equal to $q_c - \sigma_3'$, in Eqs. 13 and 14.

B6: *Direction of* $\dot{\sigma}^f$

Under the multiple stress conditions, the direction of $\dot{\sigma}^f$ should be determined to obtain the elasto-plastic solution for component P (Fig. 3.7a). With the Isotach viscosity, stress vector FB in **Fig. B6a**, which represents σ^v, gradually disappears with time during

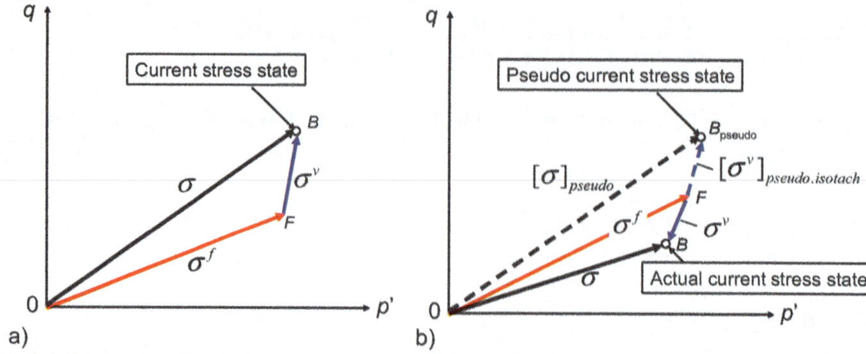

Fig. B6 Direction of inviscid stress increment vector: a) Isotach type; and b) TESRA type.

sustained loading at a fixed stress condition (i.e., $\sigma \equiv$ constant with fixed vector OB). So it is natural to assume that the direction of instantaneous $\dot{\sigma}^f$ is parallel to the instantaneous direction of σ^v (i.e., $\dot{\sigma}^f // \sigma^v$).

The situation in the case of TESRA viscosity is much more complicated, as the current viscous stress, σ^v , can become negative even under the loading or neutral stress conditions, where $d\varepsilon^{ir} \geq 0$ and therefore $\dot{\sigma}^f$ is positive or zero. One possible methodology is to assume that the direction of instantaneous $\dot{\sigma}^f$ is parallel to the direction of the instantaneous pseudo Isotach-type viscous stress, $[\sigma^v]_{pseudo.isotach}$, which is obtained as a unique function of the instantaneous ε^{ir} and $\dot{\varepsilon}^{ir}$ like the ordinary Isotach type (i.e., $\dot{\sigma}^f // [\sigma^v]_{pseudo.isotach}$: **Fig. B6b**). According to this assumption, in the 1D case, $[\sigma^v]_{pseudo.isotach}$ is always positive or zero even when the current viscous stress, σ^v, is negative, therefore $\dot{\sigma}^f$ becomes positive or zero, which is consistent with the loading or neutral conditions, $d\varepsilon^{ir} \geq 0$. The rationale for this assumption is that the instantaneous incremental viscous property is be of Isotach type even when the viscous stress decays with an increase in ε^{ir} according to the three-component model (Fig. 3.7a). It is not known whether the assumption described above is relevant to the other types of viscosity, including the P & N viscosity type.

Soil Stress-Strain Behavior: Measurement, Modeling and Analysis
Geotechnical Symposium in Roma, March 16 & 17, 2006

CHARACTERIZATION OF SOIL DEPOSITS FOR SEISMIC RESPONSE ANALYSIS

Diego Lo Presti, Oronzo Pallara[1] and Elena Mensi[1]
Department of Civil Engineering
University of Pisa, via Diotisalvi 2, 56126 Pisa, Italy
e-mail: d.lopresti@ing.unipi.it

Abstract:

The paper critically reviews in situ and laboratory testing methods used to characterize soil deposits for seismic response analyses. Cyclic loading triaxial tests (CLTX), Cyclic loading torsional shear tests (CLTST) and Resonant column tests (RCT) are considered. As for the in situ testing, geophysical seismic tests and dynamic penetration tests are discussed. Influence of ground conditions on seismic response analyses in a number of real cases is shown. The database made available by the Regional Government of Tuscany (RT) has been used.

1 INTRODUCTION

Ground motion characteristics at a site are strongly influenced by the so – called "*local site conditions*" (local geology and geomorphology, subsurface stratigraphy and geotechnical conditions at the site, the vicinity to seismogenic active faults, etc.). The influence of the local site conditions in modifying the characteristics of ground motion at a site is commonly referred to as "*local site effects*". Surface geology, topography, subsurface stratigraphy and geotechnical characteristics of the upper 50 m of a soil deposit are generally considered the most important factors contributing to local site effects (Aki, 1988; Faccioli, 1991). The evaluation of local site effects is accomplished through ground response analyses. Several national and international building codes (ICBO, 2000; EBC, 1998; OPCM 3274, 2003; Norme Tecniche, 2005) provide simplified criteria and prescribe specific procedures to account for local site effects. In some building codes the latter include also clauses for the assessment of the topographic effects (EBC, 1998; OPCM 3274, 2003). The simplified criteria mainly consist in the prescription of different elastic response spectra for different types of soils. Differences concern both the shape of the spectra and the values of the spectral ordinates. On the other hand, the whole procedure provides both the elastic response spectrum and

[1] Department of Structural and Geotechnical Engineering, Politecnico di Torino, C.so Duca degli Abruzzi 24, 10129 Torino, Italy; e-mail: oronzo.pallara@polito.it

Hoe I. Ling et al. (eds.), Soil Stress-Strain Behavior: Measurement, Modeling and Analysis, 109–157.

accelerograms at the top of the soil deposit. Soil motion is assessed in the free – field condition.

Therefore, ground response analysis is a multi-disciplinary task involving various types of professional competencies including engineering seismology, structural geology, geophysics, geotechnical earthquake engineering. In fact, the whole process requires: i) definition of the expected ground motion at the outcropping rock, ii) assessment of the geo-morphological features of the area under study and iii) determination of the mechanical properties of the soil deposit and of the underlying bedrock by means of geophysical and geotechnical investigation campaigns. Planning of such investigations depends primarily on the definition of the following aspects of the problem:

- Geology and geo-morphology of the area under study;
- Kinematics of the wave-field;
- Constitutive modelling of the subsurface;
- Methods of analysis used to solve the equations of motion

Ground response analyses based on one-dimensional geometry and kinematics are the most commonly used not only because of their simplicity but also because horizontal layering of soil deposits and wave field governed by SH-waves are often reasonable assumptions.

With regards to constitutive modelling of the subsurface soil, two frequently used models are linear and equivalent linear viscoelasticity, mainly for their computational convenience. However equivalent linear analyses in the frequency domain are unable to correctly reproduce the behaviour of a non-linear system (Constantopoulos et al., 1973). Therefore, non-linear analysis by direct numerical integration of the equations of motion is preferable, even though it may be computationally more expensive.

The main focus of this paper is on the soil investigations that are necessary for seismic response analyses. More specifically the following topics are discussed:

- capability and limitations of laboratory testing (resonant column test- RCT, cyclic loading torsional shear test - CLTS, cyclic loading triaxial test - CLTX);
- capability and limitations of in situ testing (Down hole – DH, Seismic refraction – SH, Standard Penetration Test – SPT, Dynamic Penetration Test - DP);
- influence of soil parameters on the elastic response spectra and peak ground acceleration at the top of a soil deposit.

The paper takes advantage of a large data – base provided by the Regional Government of Tuscany (RT) and consisting of laboratory and in situ test results performed in some areas of Tuscany.

2 LABORATORY TESTING

2.1 Some remarks on equipments, testing procedures, data processing and test result interpretation

The key role of laboratory testing, when dealing with soil characterization for seismic response analyses, is the assessment of stress-strain characteristics (or soil stiffness) and material damping ratio of soils. For the type of problem under consideration it is necessary to determine the stress-strain and damping ratio characteristics from very small strain up to peak. Traditionally this has been accomplished by means of Resonant Column Tests (RCT). More recently, it has been demonstrated that Cyclic Loading Triaxial Tests (CLTX) and Cyclic Loading Torsional Shear Tests (CLTST) can also provide the stress-strain characteristics and damping ratio of soils from very small strains to peak (Tatsuoka, 1988).

In order to succeed in determining stresses and strains over a wide strain range (from very small to high) in CLTX, the following are necessary: i) improvement of sensor accuracy, ii) reduction of measuring errors of strains, which are due to the system compliance and the irregular contact between specimen end-faces and top cap and base pedestal (bedding and seating errors), iii) reduction of measuring errors of stresses due to friction on the loading ram and iv) improvement of the resolution of the loading system.

More specifically, the following items are strongly recommended:

- Local axial strain measurements are always preferable and are strongly recommended for any kind of soil. In particular, local strain measurements are imperative when testing hard soils or soft rocks that usually exhibit very small strains during the reconsolidation to the in situ geostatic stress. Global measurement, performed outside the cell (external measurement) which also includes the system compliance should be avoided. On the other hand, global measurement, performed inside the cell (internal measurement) monitoring the relative displacement between top cap and base pedestal may be as accurate as local measurement, especially in the case of cyclic tests (Pallara, 1995).
- LDTs Goto et al. (1991) have proved to be very effective in the measurement of local axial strains of hard soils (gravels, sands) and soft rocks. In the case of samples that experience large consolidation strains, the use of submersible LVDTs or proximity transducers seems to be preferable. The ability to re-setting the sensor position from outside the cell (Fioravante et al., 1994) can be very useful.
- Different techniques have been proposed for local radial or lateral strain measurements (Fioravante et al., 1994; Tatsuoka et al., 1994a; 1994b; Gomes Correia & Gillett, 1996). This type of measurement is important because volumetric strain measurements, during drained stages, are rather inaccurate. Unfortunately, the correct measurement of lateral strain has not been satisfactorily solved yet. Probably the limited accuracy in the lateral strain measurement, especially relevant at small strains, is due to the membrane compliance.
- A cell structure with very low compliance and a loading ram virtually frictionless are also critical (Tatsuoka, 1988).

- Load cell should be located inside the cell pressure. Appropriate amplification should be used when measuring small forces.
- A minimum of 20 measurements per cycle should be done in order to have an accurate measurement of the loop area and accurate assessment of the so-called hysteretic damping ratio.
- The actuator resolution and its ability to apply a given constant strain rate is another essential feature of laboratory testing that should be carefully considered. The actuator should also be able to apply small and large cyclic loading under displacement control without backlash. Two examples of system having these characteristics can be found in the literature. Tatsuoka et al. (1994a) used an analogue motor with electro-magnetic clutches to change the direction of loading ram motion without backlash. Shibuya & Mitachi (1997) used a digital servomotor to control a minimum axial displacement of 0.00015 micrometer spanning over several orders of the rate of axial straining.
- Particularly for cyclic or fast loading tests, it is necessary to have a simultaneous data acquisition for stress and strain measurements or, at least, the time lag between measured stress and strain should be enough small to accurately determine damping ratios of less than 1% (Tatsuoka et al., 1994b).
- Membrane penetration effects are important in the case of granular soils. Such effects reduce as the specimen diameter and the membrane thickness increase. From this point of view the membrane penetration effects should be less important in the case of gravely samples. Unfortunately, the peripheral voids of specimens increase with the soil grain size and therefore, the membrane penetration effects become extremely important in the case of reconstituted gravel samples (Evans, 1987; Hynes, 1988). Several methods have been proposed in order to mitigate the membrane compliance effects (Nicholson et al. 1993a; 1993b, Tanaka et al. 1991), even though, the problem cannot be considered yet solved. Anyway, in the case of undisturbed samples, retrieved by means of in situ freezing, the lateral and end surfaces of the specimens are very smooth and consequently it is possible to assume that both bedding error and membrane compliance effects are small.

It is important to stress that, in the case of CLTST or RCT, negligible differences have been found between local and global shear strain measurements (Drnevich, 1978; Porovic & Jardine, 1994; Ionescu, 1999). RCT is one of the most accurate and repeatable way of determining the small strain shear modulus. Unfortunately, such a test have some disadvantages. The following considerations apply to hybrid equipments capable of performing both RCT and CLTST on small size specimens (Diameter of less than 70 mm and H/D =2) fixed at the base and free of rotating at the top:

- RCT applies very high frequencies 30 Hz $< f <$ 200 Hz. On the contrary, soils act as low-pass filters. As a consequence, soil vibrations have a frequency content mainly in the interval between 0.1 Hz and 10 Hz, with the only exception of near-source motion. It is also important to remember that most of the existing constructions has natural frequency falling in such an interval (0.1 -10 Hz).
- Many researchers have adapted the RC apparatus to perform CLTST (Isenhower et al., 1987; Alarcon-Guzman et al., 1986; Lo Presti et al., 1993; Kim & Stokoe, 1994; d'Onofrio et al., 1999). The advantage of this hybrid apparatus is that it is possible

to determine G_o from RCT and perform cyclic loading torsional shear tests at various frequencies.

- CLTST and RCT performed by means of the same hybrid apparatus are performed under stress control. Usually, a sinusoidal wave is used, which implies a non – uniform strain rate during each cycle. Strain rate is infinitive at the beginning of the loading cycle and approach zero at loading inversion. Anyway, it is possible to define an equivalent strain rate for each cycle as $\dot{\gamma} = 4 \cdot \gamma_{SA} \cdot f$ [%/s] ($\dot{\gamma}$ = equivalent shear strain rate; γ_{SA} = single amplitude shear strain [%]; f = frequency [Hz]). In the case of RCT, equivalent strain rates increase from several %/min to several thousands %/min as the strain increases from 0.001 % to 0.1 % (Lo Presti et al., 1996). In the case of CLTST, performed under stress – control at a given loading frequency (0.1 to 1 Hz), equivalent strain rates also increase 10000 times when the strain increases from 0.001 % to 0.1 % (Lo Presti et al., 1996). Anyway, equivalent strain rates, experienced in CLTST, are usually from one to three order of magnitude smaller than those applied in RCT.
- The influence of strain rate on the small strain shear modulus is quite negligible for a great variety of geomaterials (Tatsuoka et al., 1997) but it becomes increasingly important with an increase of the strain level. More specifically, an increase in strain rate produces an increase of stiffness. As a consequence, both RCT and CLTST could enlarge the linear (elastic) range, arbitrarily modifying the shape of the G-γ curves. The above – described effects are more pronounced in the case of RCT because of the very high strain rates used in such a test.
- Material damping ratio (D) is much more dependent on frequency or rate than G and even at very small strains the frequency dependency of D has been observed (Papa et al., 1988; Shibuya et al., 1995; Tatsuoka & Kohata, 1995; Stokoe et al., 1995; Lo Presti et al., 1997; Cavallaro et al., 1998; d'Onofrio et al., 1999). In particular, the damping ratio values obtained from RCT are markedly greater than that inferred from cyclic tests at frequency from 0.1 to 1.0 Hz as shown in Figure 1. The frequency dependency of D, as that depicted in Figure 1, has been firstly shown by Shibuya et al. (1995) who explained the result by considering that at very low frequencies, creep effects are predominant, while at very high frequencies the effects of viscosity become more relevant. For intermediate values of frequencies, D seems to be rather independent by such a parameter.
- Another reason why very large values of D are obtained in RCT is that electro-magnetic forces (EMF), which drive the RC motor, generate the so-called "back-EMF" which act opposite to the motion and generate an apparent damping ratio (Stokoe et al., 1995; d'Onofrio et al., 1999; Cascante et al., 2003; Wang et al. 2003; Meng & Rix, 2003; 2004). Such an apparent value does not represent a soil property. According to Stokoe et al. (1995) the D values obtained with a RC apparatus should be corrected by subtracting the equipment-generated damping (D_{app}) which is a frequency dependent parameter. The D_{app}-f calibration curve is, of course, different for each apparatus and require an appropriate calibration (Stokoe et al., 1995). Anyway, D_{app} in the already mentioned studies can be as large as 0.04 and reduces to about 0.002 as the resonance frequency increases.

Therefore, the above described inconvenient can be very relevant at small strains where the D_{app} can be much larger than the real material damping ratio.

- In order to minimize the above described inconvenient when determining D, several approaches have been proposed. Cascante et al. (2003) have proposed to replace the usual voltage measurements, which are accomplished in a RCT, with current measurements which implicitly take into account the "back-EMF". As it is easier to measure voltage, the same authors propose a transfer function to convert voltage into current. Meng & Rix (2003; 2004) have proposed to perform RCT at "ideally"-constant current by means of a specially-devised operational amplifier circuit (i.e. a voltage to current converter).
- The above described inconvenient does not occur in the case of CLTST if the applied torque is measured at the top of the specimen or if a calibration curve (voltage – torque) is available for the typically used frequencies. In this case the electro-magnetic forces which act opposite to the motion can be regarded as the friction acting on the loading ram in a triaxial test. The effective torque measurement in a CLTST is equivalent to current measurement in RCT.

It is worthwhile to remark that Lai & Rix (1998), Lai et al. (2001) and Rix & Meng (2005) have successfully used a typical Resonant Column/Torsional Shear apparatus to study the frequency dependency of stiffness and material damping ratio. For such a purpose a non-resonance method have been implemented which mainly consists of the simultaneous measurement of shear wave velocity and material damping ratio. Experimentally, the method is based on the measurement of the response function between the applied torque and resulting angular displacement in the frequency interval of interest. The method assumes that the solution of a harmonic boundary value problem in linear visco-elasticity can be obtained from the solution of the corresponding elastic problem by extending the validity of the elastic solution to complex values of the field variables. Rix & Meng (2005) have experimentally found with the above described method a frequency dependency of D similar to hat depicted in Figure 1 in the frequency interval of 0.01-30 Hz (Figure 2).

On the other hand, several hollow cylinder torsional shear apparatuses have been developed at various research laboratories (Hight et al., 1983; Miura et al., 1986; Pradhan et al., 1988; Techavorasinskun, 1989; Alarcon-Guzman et al., 1986; Vaid et al., 1990; Yasuda & Matsumoto, 1993; Ampadu & Tatsuoka 1993; Cazacliu, 1996; Di Benedetto et al., 1997; Ionescu, 1999; Yamashita & Suzuki, 1999). The main advantage of these devices is that they can operate at constant strain rate and over a very wide strain interval. Some of these devices have been developed to study the stress-strain relationship of geomaterials under a more general stress state and stress-path. Other devices have been developed as an alternative to triaxial tests. An effort to standardize such a test has been undertaken in Japan (Toki et al., 1995).

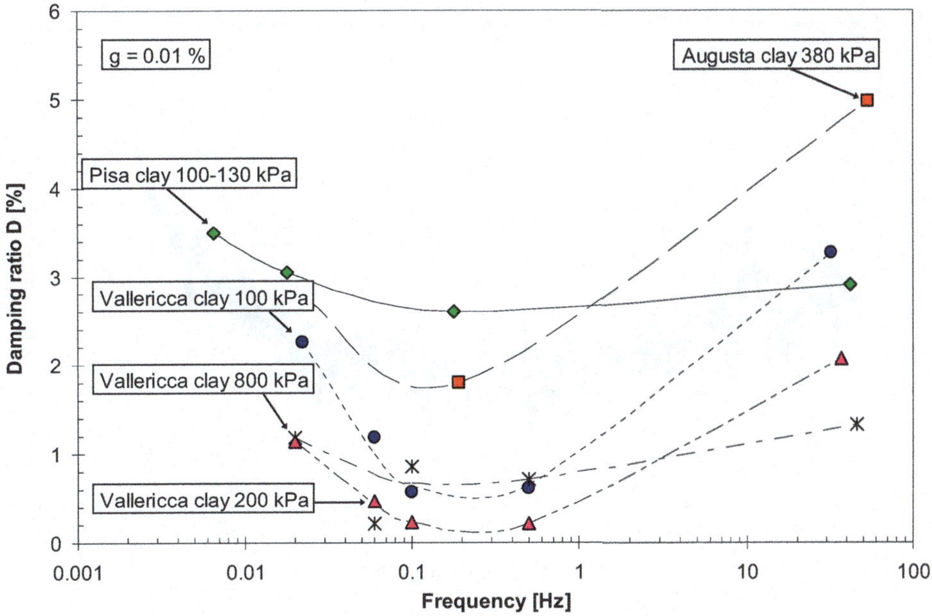

Figure 1 Damping ratio vs. frequency. (Lo Presti et al. 1997, Cavallaro et al. 1998, d'Onofrio et al.1999)

2.2 Relevant results from laboratory testing

Figure 3a, 3b and 3c show respectively the normalized stiffness decay curves (G/G_o – γ) as obtained from RCT, CLTST and CLTX. Each RCT and CLTST has been performed on the same specimen according to the following procedure:

- firstly RCT was performed in undrained conditions;
- at the end of RCT, drainage was opened and the specimen experienced a 24 hrs rest period;
- G_o was measured after the drained rest period, verifying that no relevant change has occurred in the small strain shear modulus before and after the RCT plus the rest period;
- CLTST was then performed in undrained conditions at a frequency of 0.1 Hz, using a triangle wave.

CLTX have been performed on other specimens obtained from the same samples. Cyclic triaxial tests have been performed in undrained conditions, following a cyclic compression loading stress-path at constant total horizontal stress, and at constant strain rate. More precisely a strain rate of 0.1 to 0.3%/min was applied.

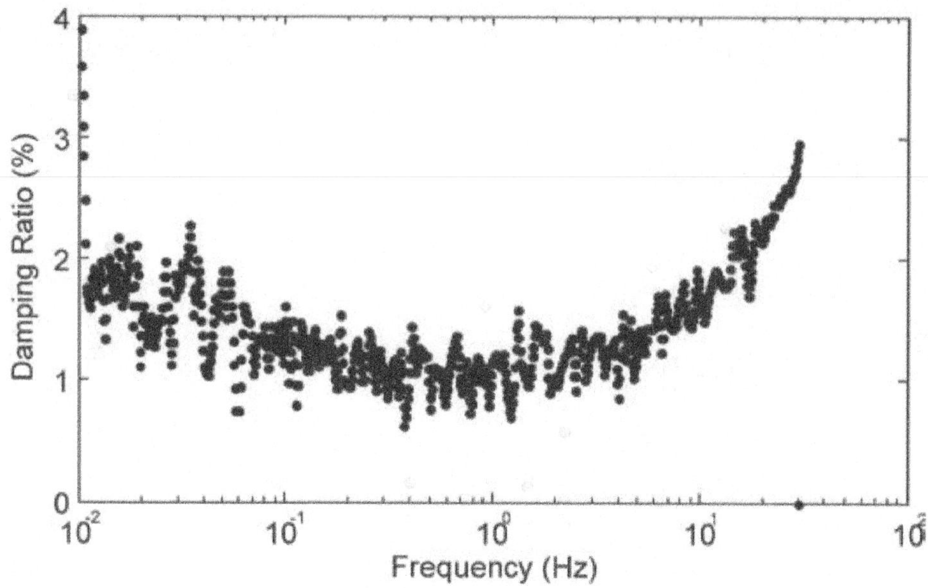

Figure 2 Preliminary experiments with the NR method on natural sandy silty clay specimen over the frequency range of 0.01-30 Hz (Rix & Meng, 2005)

Therefore the following differences mainly exist among different types of tests:

- strain rate
    - very high in RCT and increasing with strain level;
    - high in CLTST and increasing with strain level;
    - equal to 0.1 to 0.3%/min in CLTX;

- stress path
    - in the case of CLTST and RCT, the specimen experiences the same stress-path which involves application of cyclic shear stresses on the horizontal plane. Consequently the directions of principal stresses continuously change during the application of cyclic loading while the total and horizontal normal stresses remain constant;
    - in the case of CLTX, the imposed stress-path involves the cyclic (two-way) variation of the total vertical normal stress, while the total horizontal stress remain constant. The directions of the principal stresses do not change during cyclic loading.

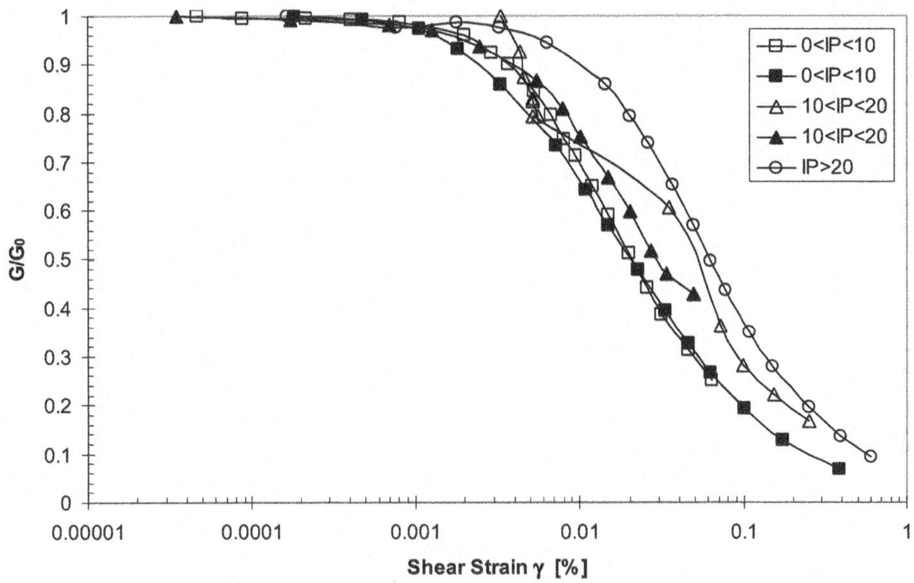

3a) Resonant column tests

The following comments can be done by comparing the data reported in Figure 3a, 3b and 3c:

- RCT results clearly show an increase of the elastic domain with an increase of the plasticity index. Moreover, the stiffness decay is more pronounced for specimen characterized by lower values of the plasticity index. It is believed that what observed in the case of RCT is mainly a consequence of the very high strain rate, increasing with the strain level, which arbitrarily modify the shape of the G-γ curves.
- the above – reported comments do not apply to the CLTST, even though the strain rate increases with strain level also for this type of test. Anyway, strain rates in CLTST are from one to three order of magnitude smaller. Probably, the same phenomenon previously indicated is less pronounced in the case of CLTST;
- CLTX exhibit a more pronounced stiffness decay with strain level which is probably a consequence of different factors (use of a lower and constant strain rate, type of stress path).

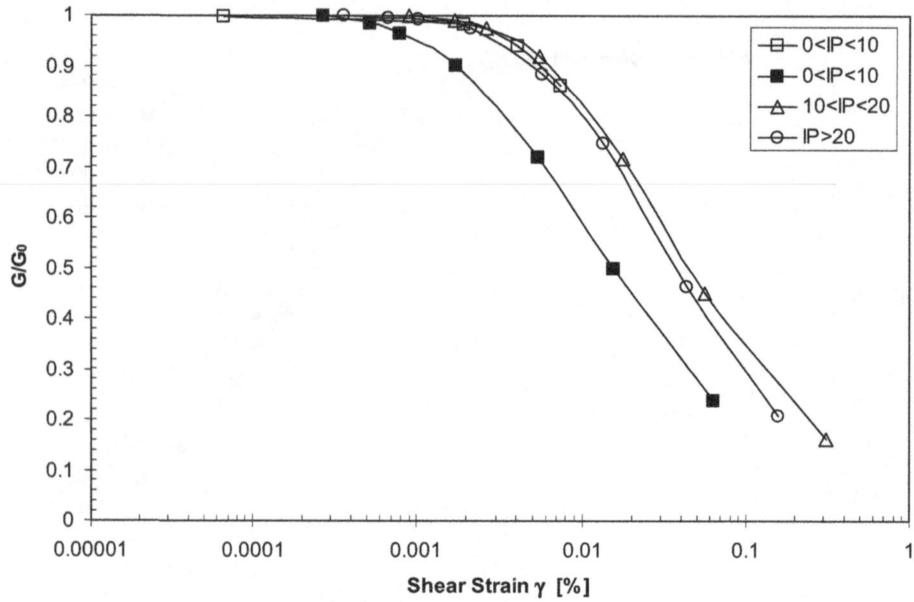

3b) Cyclic loading torsional shear tests

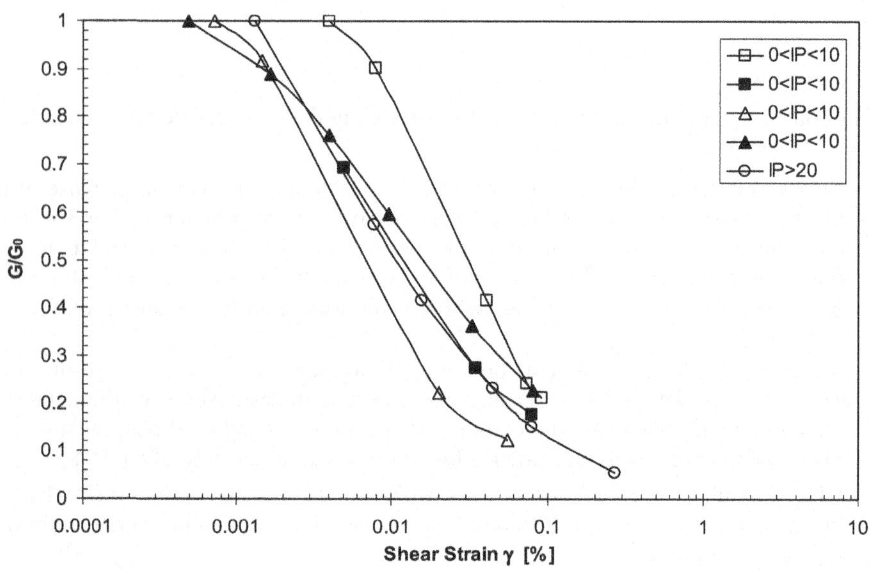

3c) Cyclic loading triaxial tests

Figure 3 Normalized stiffness decay curves obtained from: *3a)* Resonant column tests; *3b)* Cyclic loading torsional shear tests; *3c)* Cyclic loading triaxial tests

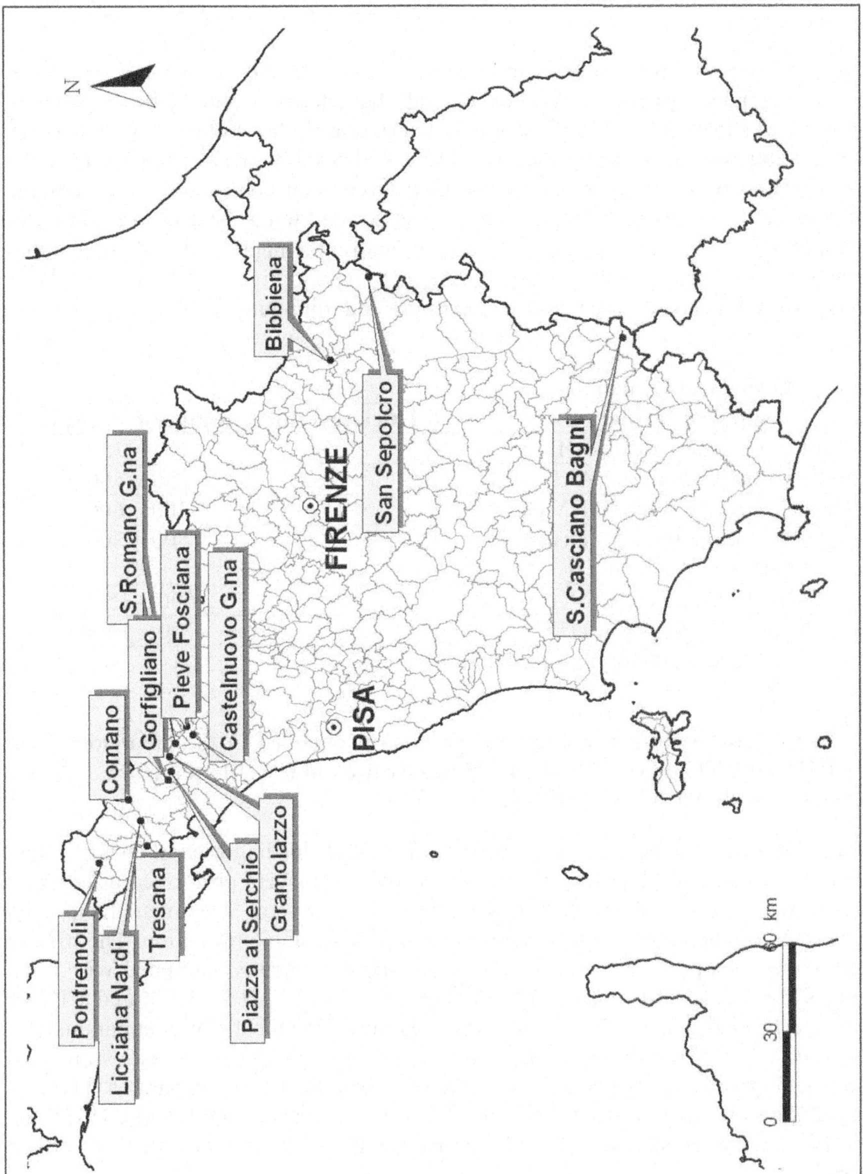

Figure 4 The study areas

Nonetheless CLTX are capable of giving a more complete soil characterization, including strength parameters, cyclic degradation of stiffness, damping and strength parameters, pore pressure build-up with strain level and number of loading cycles, the stress-path imposed in CLTST and RCT seem more appropriate when performing seismic response analyses.

Usually, seismic response analyses involve the characterization of a large portion of soil. As a consequence, spatial variability of soil characteristics should be considered. Extension of investigation is limited by budget restriction. In the authors experience, for seismic microzonation of Tuscany (see in Figure 4 the study areas) samples from 26 boreholes (located in 13 different municipalities) have been considered. The geologic formations of Table 1 exist in the study area. It was decided to consider the variability of G/G_o-γ and D-γ curves in the analyses. More specifically, the upper and lower envelopes of experimental data was established (see Figure 5a to 5g and Figure 6a to 6g). Analyses have been performed considering both the maximum and minimum envelope.

Table 1 Geological formations

Geological formation	**Thickness (m)**	**Range of V_s (m/s)**
Holocene debris deposits (DT)	< 15	100 – 350
Holocene alluvial deposits (ALL – CT)	< 20	250 – 500
Pleistocene alluvial terrace (CT/MG – AT)	< 15	300 – 600
Plio – Pleistocene fluvial – lacustrine deposits (ARG – CG)	< 72	500 – 800
Oligocene sandstone (MG)	-	600 – 1500
Argille – Calcari (AC) (*)	-	700 – 1400

(*) Paleocene – Eocene Claystone-Limestone

Data reported in Figure 5 and Figure 6 have been obtained in the laboratory from RCT, CLTST and CLTX. Triaxial tests were mainly used to test stiff soil (rocks, gravels and Pliocene formations). More specifically:

- two undrained compression loading (two way) cyclic loading triaxial tests (CLTX) were performed on ARG specimens. The triaxial tests were performed at constant strain rate of 0.3 %/min). RCT and CLTST (on the same specimen) have been performed on the other available specimens of this geologic formation. The triaxial tests have been performed by means of a triaxial apparatus, equipped with local gauges for local strain measurements (Lo Presti et al., 1995). RCT and CLTST were performed using a hybrid Resonant Column/Torsional Shear apparatus (Lo Presti et al., 1993). All the specimens were reconsolidated to the best estimate of the in situ effective geostatic stresses, when tested in the triaxial apparatus (i.e. K_o – consolidated by imposing zero radial strain). In the case of RCT and CLTST the specimen were reconsolidated to the best estimate of the in situ effective vertical stress;

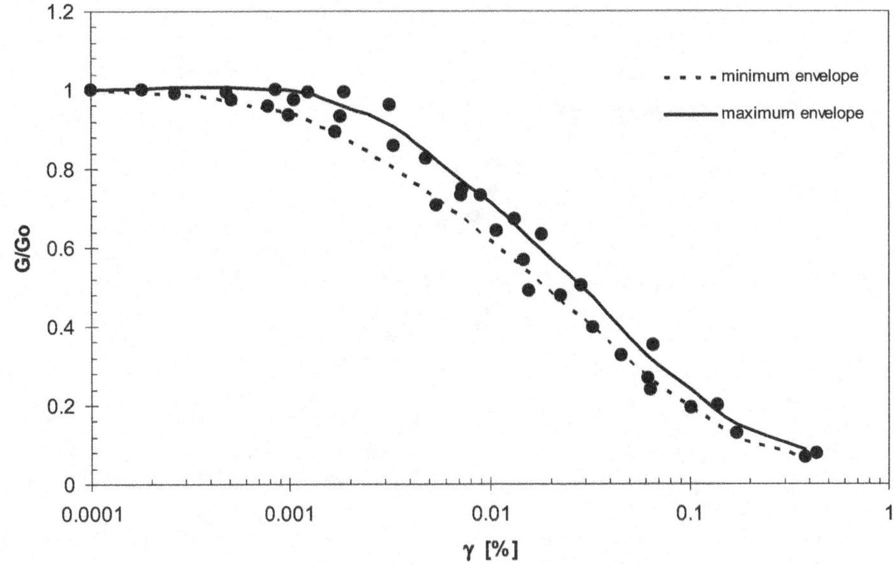

5a) Holocene Debris deposits (DT)

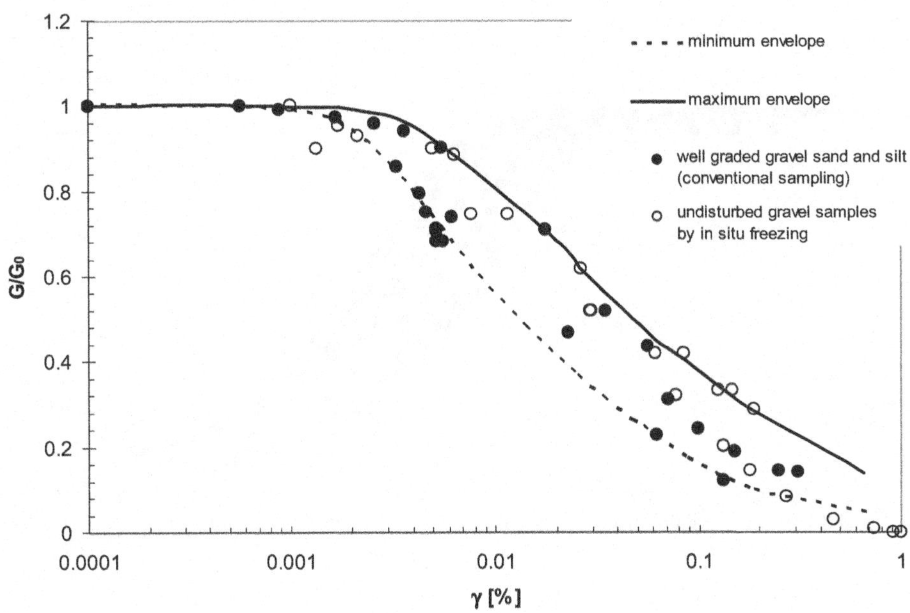

5b) Holocene Alluvial deposits (ALL-CT)

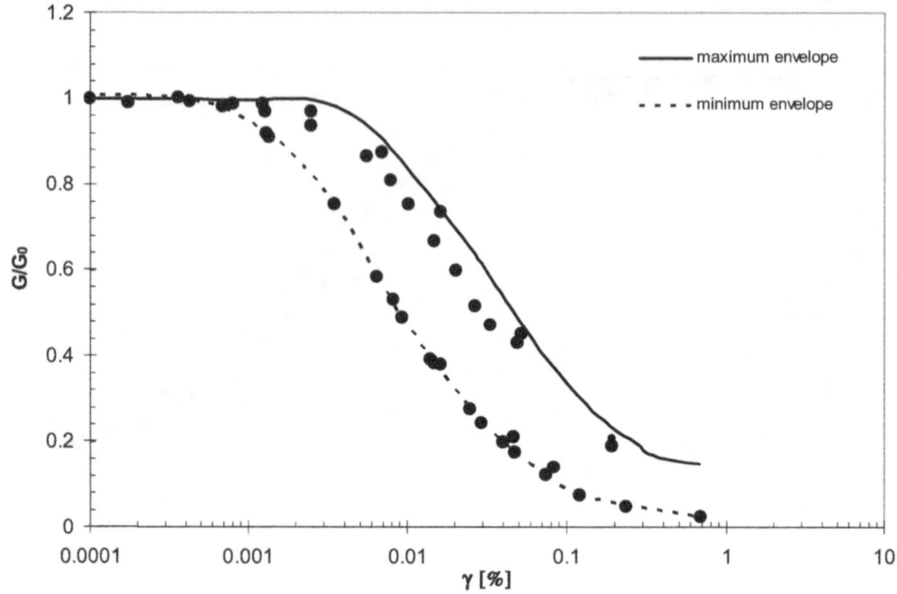

5c) Pleistocene Alluvial terrace (CT/MG-AT)

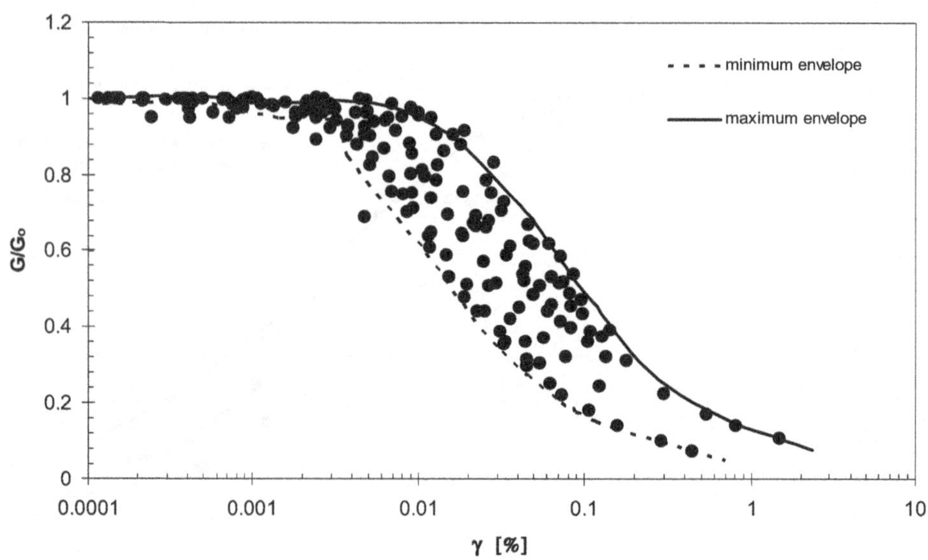

5d) Fluvial Lacustrine formations from Plio-Pleistocene (ARG-CG)

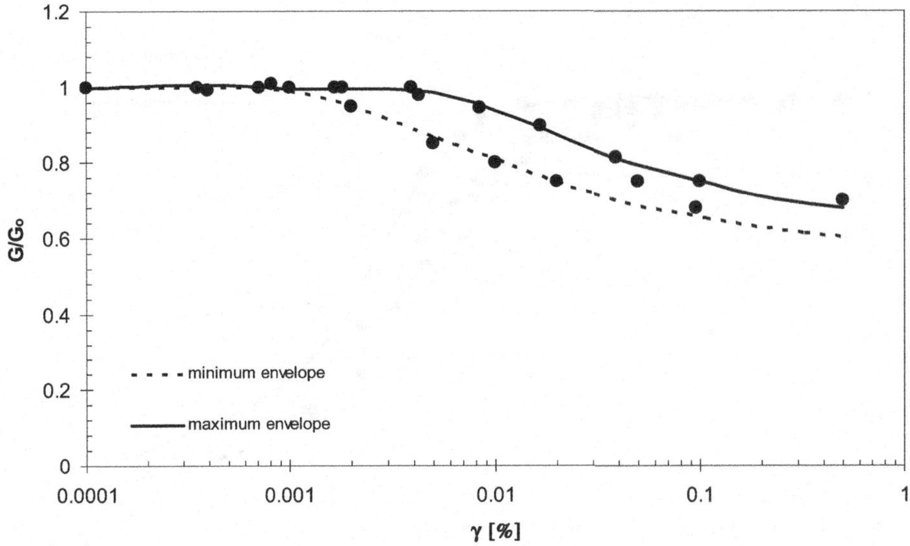

5e) Oligocene Sandstone (MG)

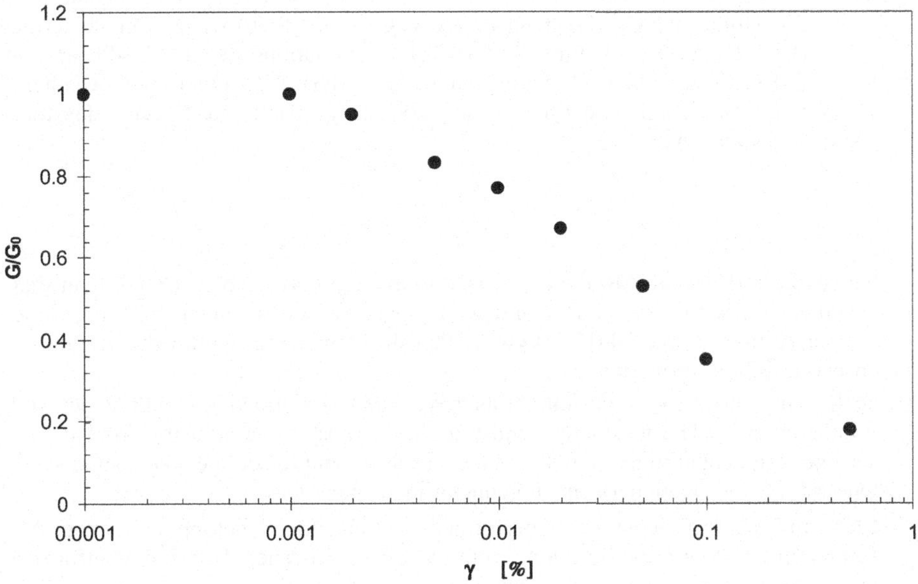

5f) Oligocene weathered Sandstone (MG) (Lo Presti et al., 2002)

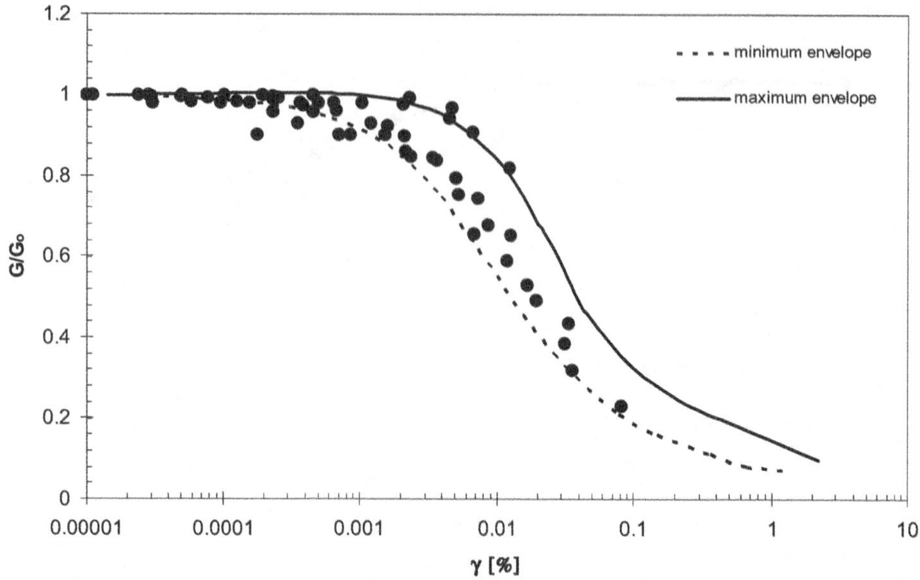

5g) Paleocene-Eocene claystone-limestone (AC)

Figure 5 Normalized stiffness vs. shear strain for: *5a)* Holocene Debris deposits (DT); *5b)* Holocene Alluvial deposits (ALL-CT); *5c)* Pleistocene Alluvial terrace (CT/MG-AT); *5d)* Fluvial Lacustrine formations from Plio-Pleistocene (ARG-CG); *5e)* Oligocene Sandstone (MG); *5f)* Oligocene weathered Sandstone (MG) (Lo Presti et al., 2002); *5g)* Paleocene-Eocene claystone-limestone (AC)

- the stiffness decay of ARG specimens was more pronounced in CLTX than that observed in the case of CLTST and RCT performed on specimen from the same sample (Calosi et al., 2001). Anyway, this difference was within the variability observed for that formation;
- as for the Oligocene sandstone tests have been performed on both intact and weathered rock. The weathered sandstone has a shear wave velocity (from in situ seismic tests) equal to about 600 m/s (see Table 1) and looks like a cemented sand. One way, CLTX were performed on unconfined specimens;
- RCT and CLTST have been performed on specimens belonging to the AC formation (claystone and limestone from Paleocene–Eocene). Tests were performed on mostly weathered claystone specimens;

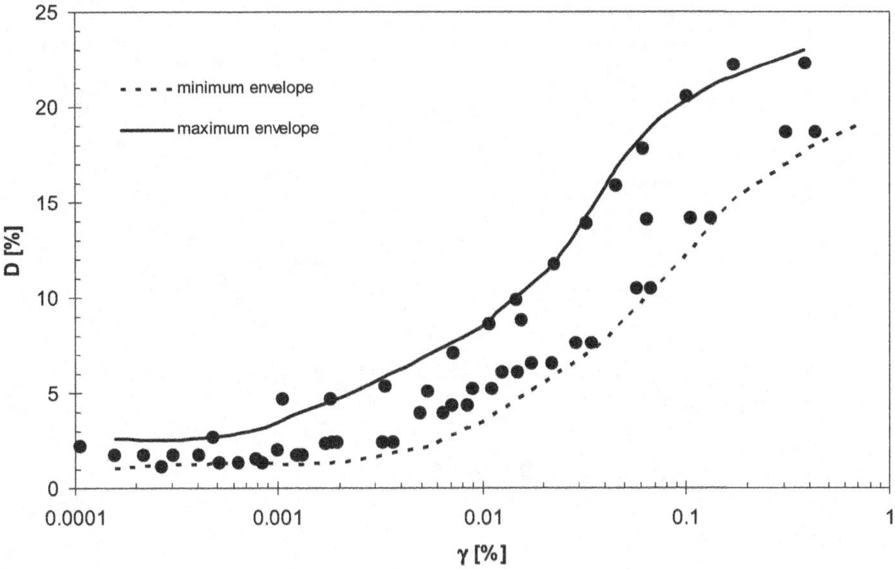

6a) Holocene Debris deposits (DT)

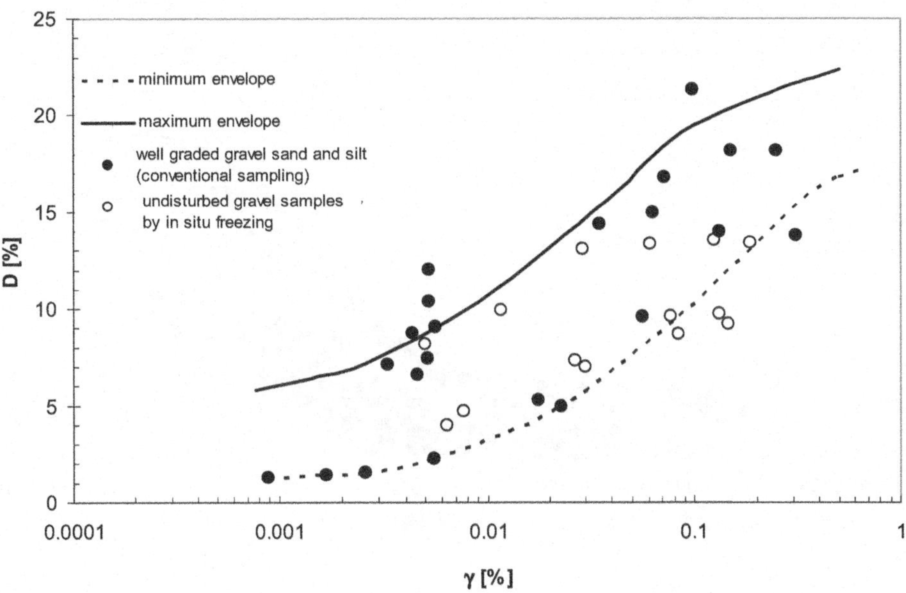

6b) Holocene Alluvial deposits (ALL-CT)

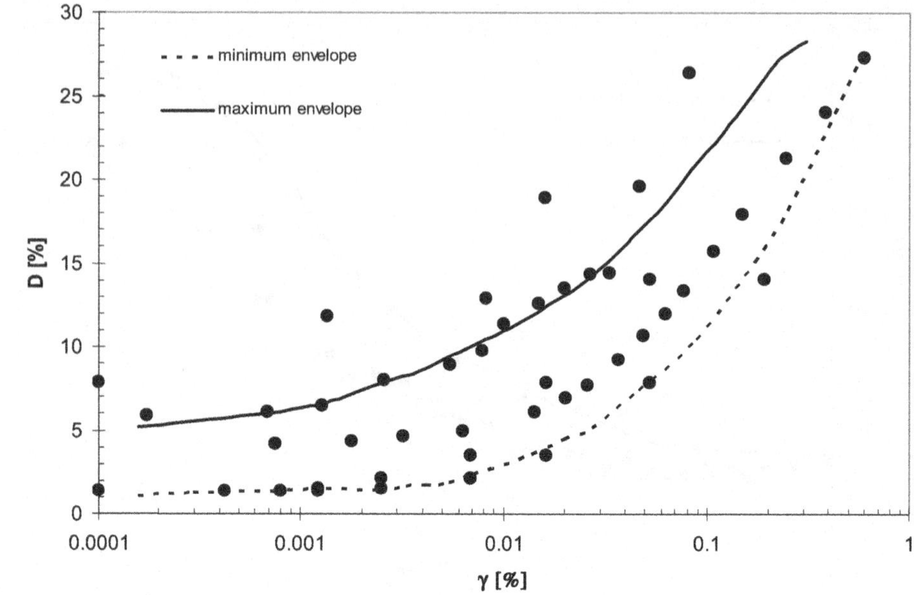

6c) Pleistocene Alluvial terrace (CT/MG-AT)

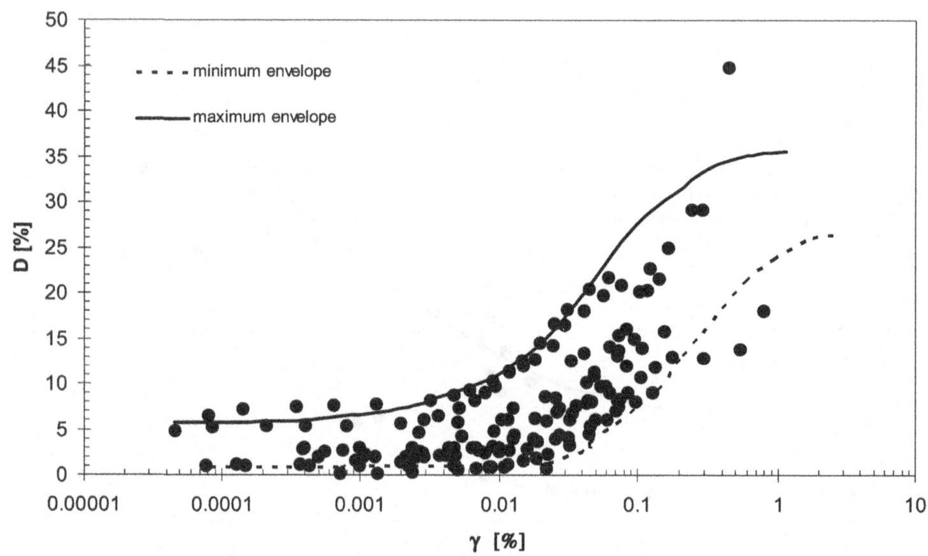

6d) Fluvial Lacustrine formations from Plio-Pleistocene (ARG-CG)

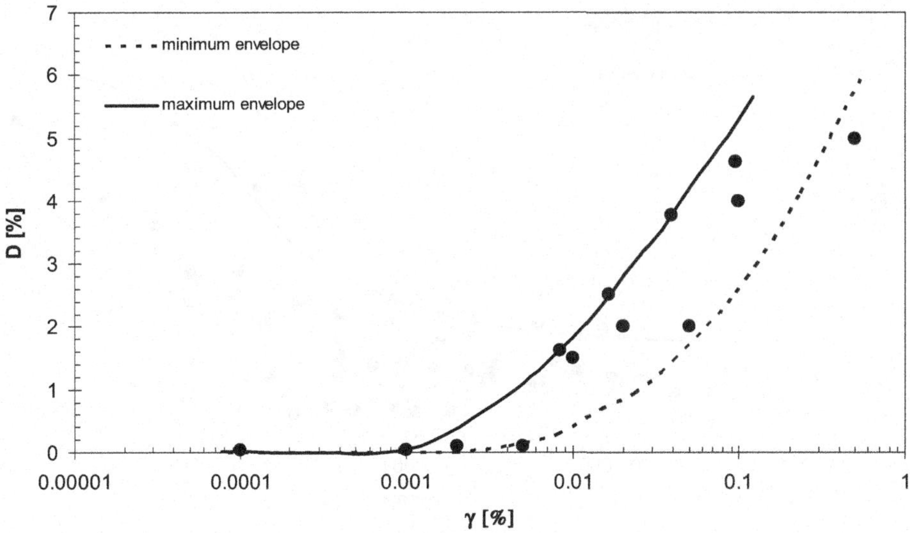

6e) Oligocene Sandstone (MG)

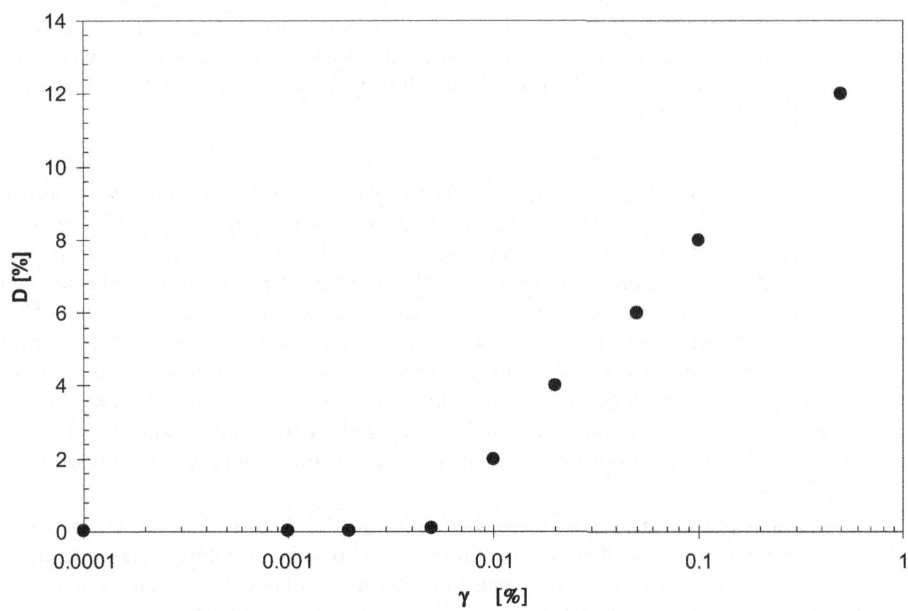

6f) Oligocene weathered Sandstone (MG) (Lo Presti et al., 2002)

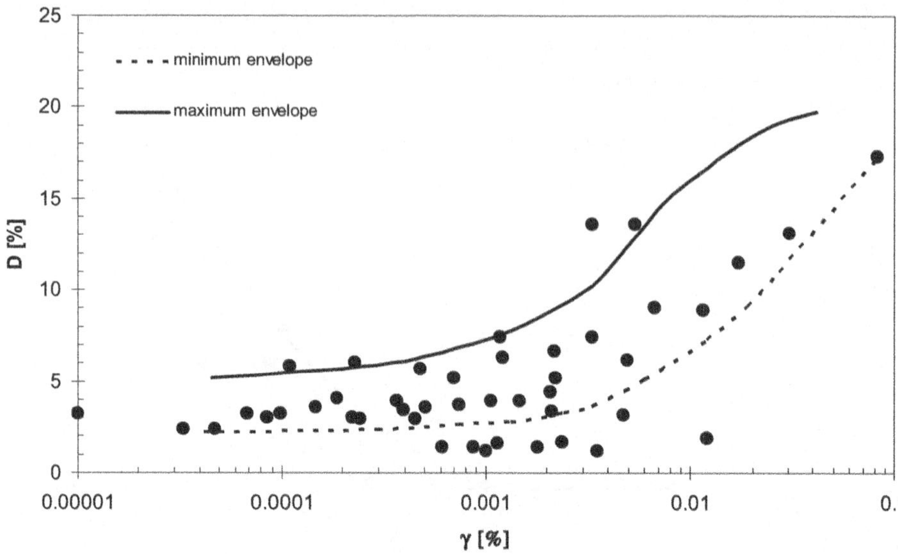

6g) Paleocene-Eocene claystone-limestone (AC)

Figure 6 Damping vs. shear strain for: *6a)* Holocene Debris deposits (DT); *6b)* Holocene Alluvial deposits (ALL-CT); *6c)* Pleistocene Alluvial terrace (CT/MG-AT); *6d)* Fluvial Lacustrine formations from Plio-Pleistocene (ARG-CG); *6e)* Oligocene Sandstone (MG); *6f)* Oligocene weathered Sandstone (MG) (Lo Presti et al., 2002); *6g)* Paleocene-Eocene claystone-limestone (AC)

- triaxial tests have been performed on large size undisturbed samples (300 mm in diameter and 600 mm height) of gravels and sands belonging to the ALL formation (Holocene alluvial deposits). Undisturbed samples were retrieved by in situ freezing method (Figure 7). One-way cyclic compression loading triaxial tests were performed in undrained conditions on isotropically consolidated specimens. Tests were performed under strain control using a triangular waveform and strain rates ranging from 0.2 to 0.5 %/min . Triaxial tests were repeated on samples reconstituted in the laboratory using the same material previously tested in the intact condition. It was found that the normalized stiffness decay curve (G/G_o-γ) is almost the same for both undisturbed and reconstituted samples (Lo Presti et al., 2005);
- it was possible to retrieve samples, belonging to the ALL formation, by means of conventional method (Shelby tube samples). This was possible because strata of silty sands and sandy silts with small percentages of clay exist in that formation. It is interesting to observe that the normalized stiffness decay data for these samples are very similar to that obtained for undisturbed gravel samples retrieved by in situ freezing;

Figure 7 Undisturbed gravel samples by in situ freezing

- the very limited variability observed for the Holocene debris deposits is a consequence of the very small number of samples belonging to this formation, which is a consequence of the difficulties in retrieving coarse-grained soils. However, the data range of DT formation is very close to that of ALL formation.

Additional considerations are necessary when dealing with granular soils. In fact costs for retrieving undisturbed samples by in situ freezing are very high and therefore it is difficult to use this type of sampling technique in the current practice.

Fortunately, the coincidence of the normalized stiffness decay curve of undisturbed and reconstituted samples of granular soils is confirmed in many cases (Goto et al., 1992; 1994; Yasuda et al., 1994; Hatanaka & Uchida, 1995). The curves representing damping ratio D vs. γ are also the same for undisturbed and reconstituted specimens (Goto et al., 1992; 1994; Hatanaka & Uchida, 1995). There are few cases where researcher have pointed out differences between undisturbed and reconstituted specimen of granular soils. As an example, Kokusho & Tanaka (1994) showed that the shear modulus decay is more pronounced in the case of undisturbed specimens that for reconstituted samples, while the D - γ curve of intact samples plots above that of reconstituted. As for the D - γ curves of intact and reconstituted samples Yasuda et al. (1994) have opposite indication. Anyway, from a practical point of view, it is possible to conclude that for the problem under consideration the same curves of stiffness decay and material damping ratio are obtained from undisturbed and reconstituted specimens.

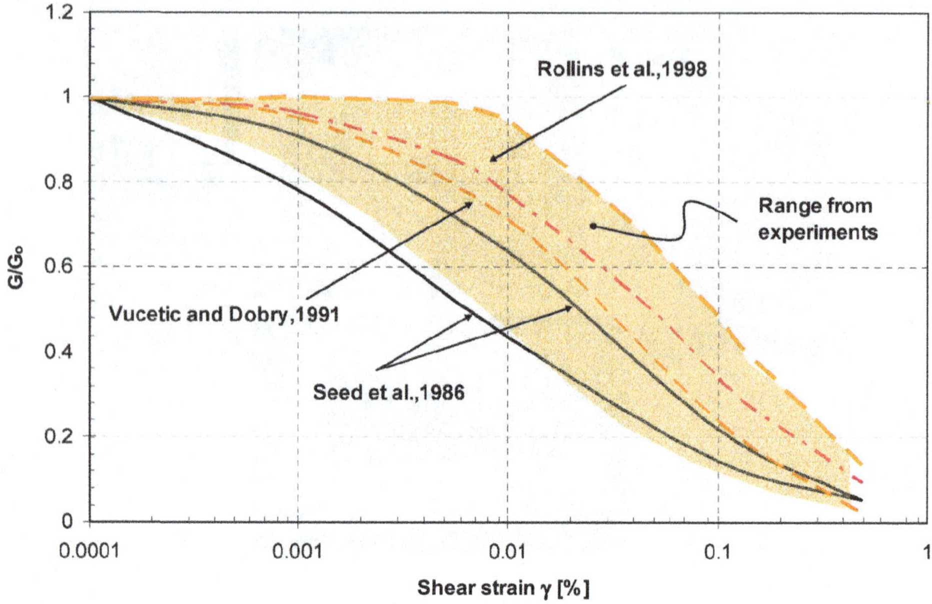

Figure 8 G-γ curves: literature data

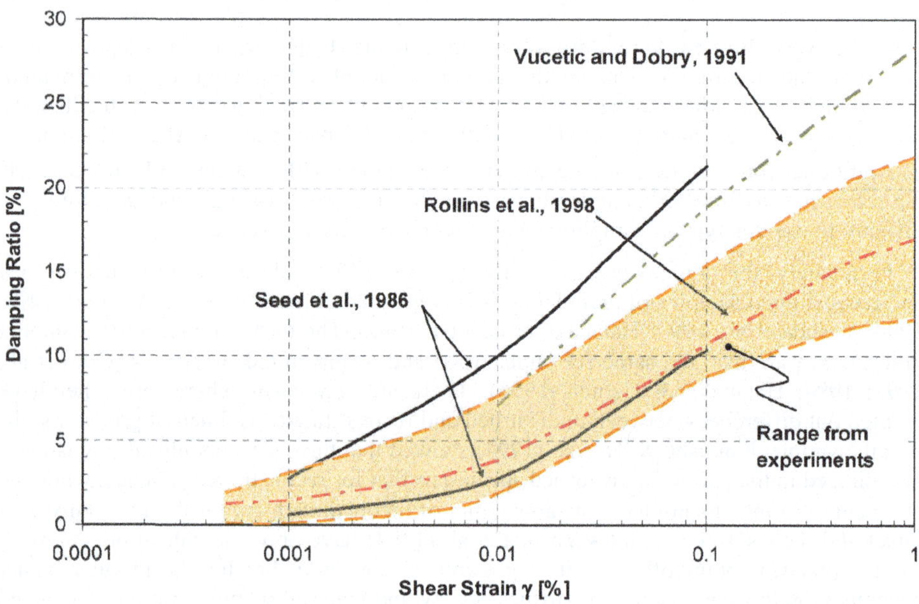

Figure 9 D-γ curves: literature data

As an alternative, it is possible to use literature data.

Figure 8 shows the range of $G/G_0 - \gamma$ values measured in cyclic triaxial tests on undisturbed gravel samples reconsolidated to the in situ vertical geostatic stress (Goto et al., 1987; 1992; 1994; Hatanaka et al., 1988; Hatanaka & Uchida, 1995; Yasuda et al., 1994; Kokusho & Tanaka, 1994; Tanaka et al., 2000). On the same Figure, for comparison, the following curves are also shown:

- the stiffness decay curve proposed by Rollins et al. (1998) based on undisturbed and reconstituted sample test results on granular soils

  $$\left\{\frac{G}{G_o} = \frac{1}{[1+16\gamma\cdot(1.2+10^{-20\gamma})]}\right\};$$

- the limits for granular soils indicated by Seed et al. (1986);
- the curve proposed by Vucetic & Dobry (1991).

Figure 9 shows the range of $D - \gamma$ values measured in cyclic triaxial tests on undisturbed gravel samples reconsolidated to the in situ vertical geostatic stress (Goto et al., 1987; 1992; 1994; Hatanaka et al. 1988; Hatanaka & Uchida, 1995; Yasuda et al., 1994; Kokusho & Tanaka, 1994; Tanaka et al., 2000). On the same Figure, for comparison, the following curves are also shown:

- the stiffness decay curve proposed by Rollins et al (1998) based on undisturbed and reconstituted sample test results $[D = 0.8 + 18\cdot(1+0.15\gamma^{-0.9})^{-0.75}]$;
- the limits for granular soils indicated by Seed et al. (1986);
- the curve proposed by Vucetic & Dobry (1991).

The following comments can be made on Figure 8 and Figure 9:

- The shear modulus decay, according to Seed & Idriss (1986), seems to be too much large, while the curve proposed by Rollins et al. (1998) and especially that by Vucetic & Dobry (1991) fit much better the experimental data.
- As for the damping ratio, the data scatter is large, especially at large strains. Anyway, the curves proposed by Vucetic & Dobry (1991) and Rollins et al. (1998) roughly represent upper an lower limits of the experimental data which plot within the limits proposed by Seed et al. (1986).

Therefore, the available literature indications, concerning the $G - \gamma$ and $D - \gamma$ curves, give a preliminary assessment of the parameters to be used in linear equivalent seismic response analyses (see as an example the indications reported in EBC, 1998).

For more accurate analyses it is better to experimentally determine the $G - \gamma$ and $D - \gamma$ in the laboratory on intact or reconstituted specimens. In this case, special attention should be paid to the stress and strain measurements, using local gauges, internal load cell, virtually frictionless loading ram and smooth lateral and end surfaces of the specimens. In the case of reconstituted samples a crucial problem is the knowledge of in situ density and effective geostatic stresses that should be reproduced in the laboratory.

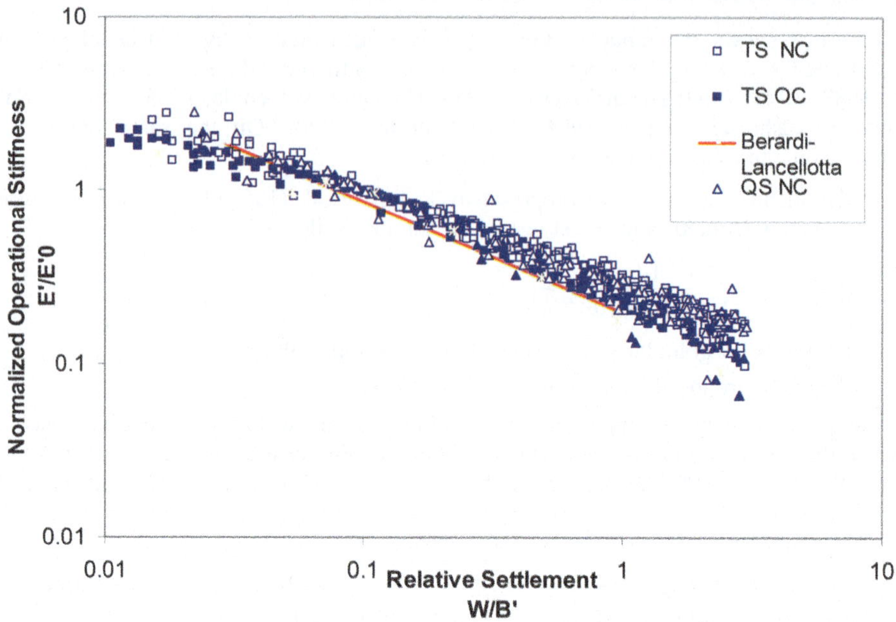

Figure 10 Stiffness non linearity of granular soils (Mussa, 2004)

An additional confirmation of the advantages obtained by normalization of the stiffness with reference to the initial value (G_o) is given in Figure 10 (Mussa, 2004). In Figure 10, the empirical correlation, proposed by Berardi & Lancellotta (1991) to account for the stiffness non-linearity in granular soils, fits quite fairly the experimental results obtained from Plate Load Tests on freshly deposited sands. The following comments explain the comparison of Figure 10:

- Berardi & Lancellotta (1991) used the database of Burland & Burbridge (1985) to back-calculate and operational stiffness. The database consists of more than 200 cases of shallow foundations on granular soils with width spanning from 0.8 m to 135 m. For each case record, the observed settlement and the N_{spt} profile is available;
- Berardi & Lancellotta (1991) back-calculated the operational stiffness (E') through the theory of elasticity. The computed values have been divided by the operational stiffness observed for a relative settlement (w/B) of 0.1 % ($E'_{0.1}$). The best fit of $E'/E'_{0.1}$ vs. w/B is represented by the following equation (also shown in Figure 10): $E'/E'_{0.1} = 0.008\left(\frac{w}{B}\right)^{-0.7}$;

- experimental data shown in Figure 10 have been obtained, according to the above described procedure from deep PLTs (Ghionna et al., 1994; Fioravante et al., 1998) performed in CC (Bellotti et al., 1982) on Ticino (TS) and Quiou (QS) freshly deposited samples and shallow PLTs performed on large size model of TS (Gabrielaitis et al., 2000).

As already stated, Figure 10 confirms the importance of normalizing the stiffness, even though the small strain value has not been used in this case, and that practically the same normalized stiffness decay is observed for both undisturbed and reconstituted granular soils.

3 IN SITU TESTING

3.1 Some remarks on geophysical seismic tests and dynamic penetration tests

In this paper geophysical seismic tests and dynamic penetration tests (DP) are considered. Penetration tests mainly provide a penetration resistance parameter, which enable one to obtain relevant soil parameters through empirical correlations. Usually the penetration resistance leads to the assessment of strength parameters (peak friction angle for granular soils and undrained shear strength for fine – grained soils). On the other hand, geophysical seismic tests enable one to directly measure the surface and body wave velocities into the soil which are related to the small strain stiffness. In particular, assuming that at small strain soil behaves as a continuum elastic isotropic medium, the following well – known relationship exists between the small strain shear stiffness (G_o) and the shear wave velocity (V_s), (where $\rho =$ mass density):

$$G_o = \rho \cdot V_s^2 \qquad [1.]$$

Several empirical correlations have been published in the literature that correlate shear wave velocity and penetration resistance. The effectiveness of these correlations is discussed later on.

Several state of the art reports, concerning the available equipments, testing procedures, data processing and test result interpretation have been published (see just as an example Lo Presti et al., 2001; Lo Presti et al., 2004 and Lo Presti & Puci, 2001).

In the following only few comments concerning the most recent and relevant innovations in the field of DP tests and geophysical seismic tests are reported.

The following comments concern the DP tests:

- Cubrinowski & Ishihara (1999) have definitively shown that the penetration resistance N_{spt} from Standard Penetration Tests (SPT), performed at the bottom of a borehole using a standard Raymond sampler, is strongly affected not only by the soil density and in situ effective geostatic stresses, but also by the grain size;
- There are evidences showing that DP tests, using a closed end bit, are more repetitive and (may be) less influenced by the grain size (Figure 11);

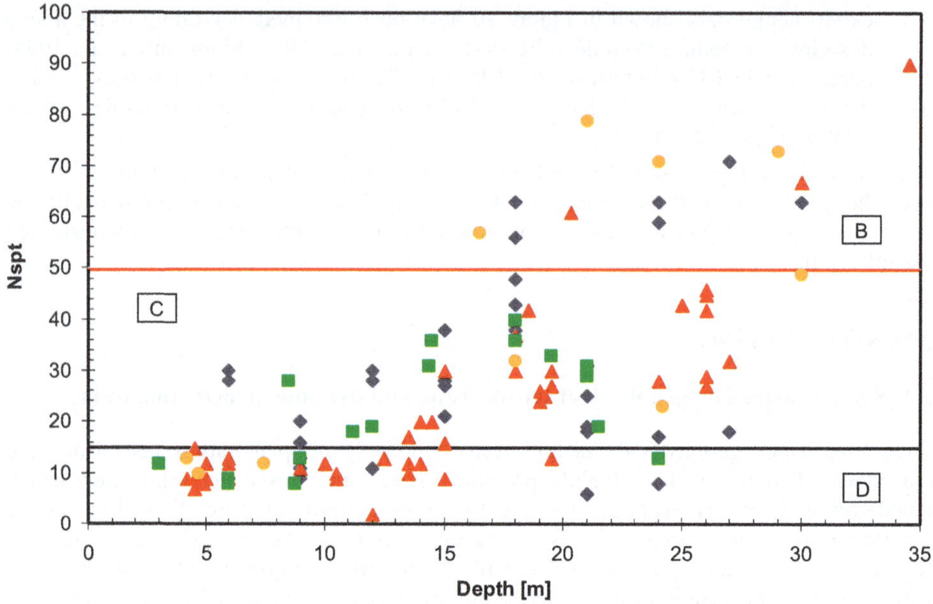

11a) Standard Penetration Test (SPT) performed using a standard sampler

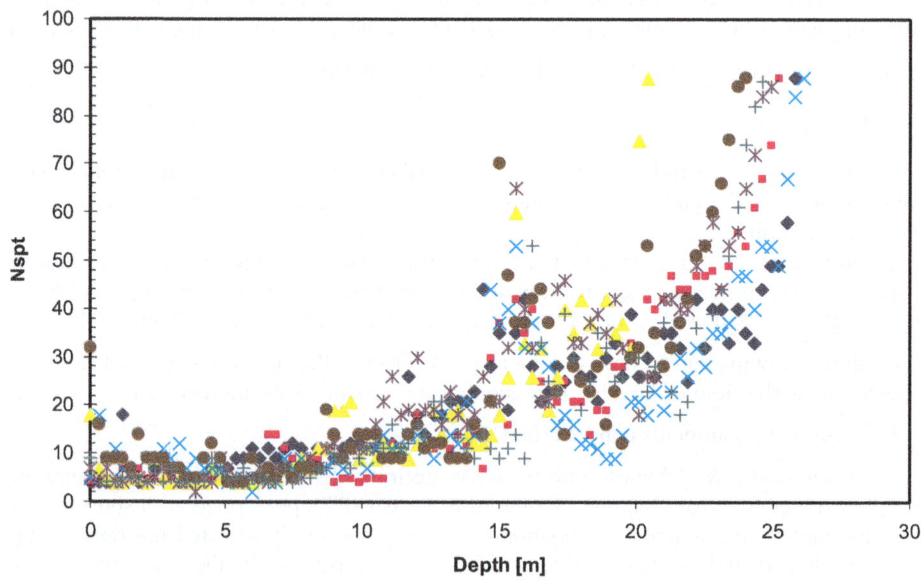

11b) Standard Penetration Test (DP) performed using a closed end bit

Figure 11 Penetration test results at a site along the Arno river (silty sands and silty gravels) performed using: *11a)* a standard sampler (SPT); *11b)* a closed end bit (DP)

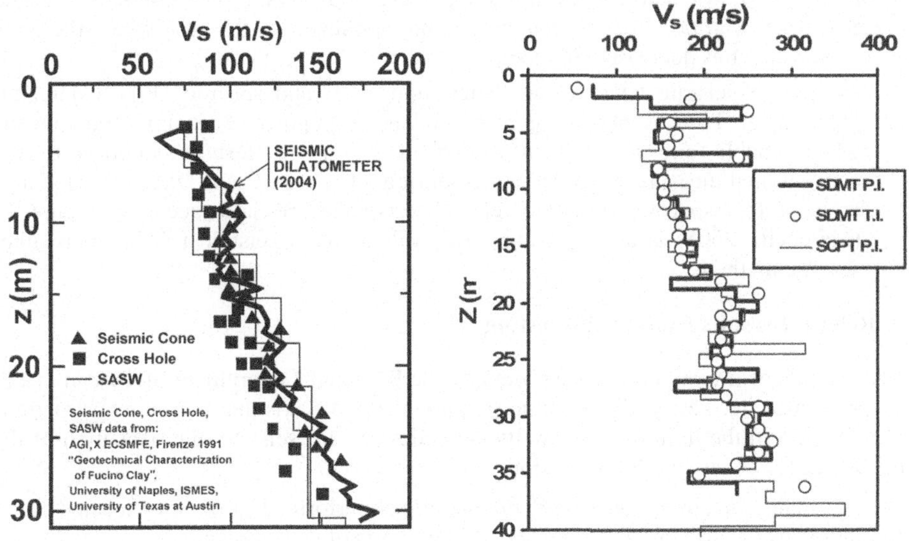

Figure 12 Comparison between seismic DMT (Marchetti, 2005) and other seismic methods at Fucino and Venezia Treporti sites

- The effective Energy Ratio (ER) of both SPT and DP should be known through appropriate equipment calibrations. As an example, the DPSH penetrometer of Pagani Geotechnical Equipment has an ER=73% (Pagani, 2005) quite different than the typical ER=60% of the Italian standard for SPT.

The following comments concern the seismic geophysical tests:

- surface (non – invasive) seismic methods have several advantages with respect to the borehole seismic tests. More specifically, surface tests are economic, mainly not affected by problems of site accessibility and require short time. On the contrary, surface tests are affected by several restrictive hypotheses which limit the accuracy and reliability of test results;
- it is well known that the use of Rayleigh waves involves the assumption that soil strata are plane and parallel. Moreover, the inversion of the dispersion curve does not lead to a unique solution;

- it is also known that seismic refraction using body waves (Compression – P and Shear – S) can be rather inaccurate and not applicable to the case where the body wave velocities decrease with depth;
- the use of seismic CPT (Cone Penetration Test) and seismic DMT (Marchetti Dilatometer Test) is becoming more and more popular. The above equipments which enable one to perform a sort of down-hole test in addition to the conventional measurements performed during a typical CPT or DMT offer reduced costs (with respect to borehole seismic tests) and are quite accurate. Figure 12 (Marchetti, 2005) is a clear example of validation of seismic DMT versus other seismic methods.

3.2 Relevant results from in situ testing

The results of investigations, supported by the Regional Government of Tuscany (RT) and performed to verify the seismic requirements of existing public constructions (schools, etc.) in the light of the new Italian codes and for seismic microzonation of the study areas, are discussed in this section.

The available database consists of investigations performed in 62 sites located in 34 different towns of Tuscany. For each site, data consists of the following:

- stratigraphic profile from boreholes, with SPT measurements;
- shear wave velocity (V_S) profile from Down – Hole (DH) tests performed in the boreholes;
- shear wave velocity (V_S) profile from seismic refraction tests performed generating both compression (P) and horizontally polarized shear waves (SH);
- resonant column tests (RCT), cyclic loading torsional shear tests (CLTST), cyclic loading triaxial tests (CLTX) performed on undisturbed samples (not available for all boreholes and any type of soil);
- dynamic penetration tests (DP), performed by using a cone 51 mm in diameter and with an apex angle of 60° (hammer weight and falling height are the same as for the SPT, blowcounts every 20 cm penetration, ER = 73%). This type of tests has been performed in few sites.

The results of laboratory tests have already been commented. In the following the discussion is focused on in situ test results. More specifically, seismic tests and penetration tests have been used and compared in order to define a shear wave velocity profile, which is necessary to perform seismic response analyses.

EBC (1998) part 1, prescribes two sets (Type 1 & 2) of elastic response spectra, depending on the expected Magnitude of the design earthquake. Each set consists of five different spectra which depend on the type of soil. Therefore EBC (1998) prescribe a simplified way to account for local site effects.

While Italian Codes prescribe slightly different spectra (a sort of simplification of EBC, 1998), soil types are defined in the same way by EBC (1998), OPCM – 3274 (2003) and Norme Tecniche per le Costruzioni (2005). Such a definition is based on stratigraphic profile, average shear wave velocity of the first 30 m (V_{s30}), number of blow/(30 cm) from SPT (N_{spt}), and undrained shear strength (C_u).

Table 2 summarizes the criteria, prescribed to identify the soil type, in terms of V_{s30} and N_{spt}. The above mentioned codes also consider very loose or soft soils (S1 and S2) which are not discussed in this paper.

Table 2 Ground conditions

Soil type	V_{s30} (m/s)	N_{spt} (blow/30cm)
A	> 800	-
B	360 - 800	> 50
C	180 - 360	15 -50
D	< 180	< 15
E	(*)	-

(*) Soil type E consists of 5 to 20 m of soil like C or D underlain by soil type A. This definition is expressed in EC8 not very clearly because refer to V_{s30}.

In the framework of the above Table 2 described simplified approach, soil types have been identified on the basis of the V_{s30} parameter and N_{spt}.

The V_{s30} parameter has been determined from both DH and SH tests at a number of sites. Figure 13 clearly shows the good correlation existing between V_{s30} obtained from DH tests and that inferred from SH tests. Use of surface test is therefore recommended to save costs. In any case, intrinsic limitations of each seismic method should be considered (see as an example Lo Presti et al., 2004). It is worthwhile to remember that velocity profiles, obtained from refraction tests using both P and SH waves, in some circumstances are not consistent each other. The presence of water table in soil, at a given depth, make the P-refraction test unreliable. On the other hand, the measurements, obtained using SH waves, are generally, but not always, consistent with DH test results and stratigraphic profile.

Identification of soil types (according to EBC, 1998) have been done, by means of both N_{spt} and V_{s30},at 18 sites. Only in three cases different indications have been obtained. SPT results classify the soils in the category immediately above or below that selected on the basis of V_{s30}. Consequently, penetration test results can be used for soil-type identification, even though the use of surface test appears more accurate and less expensive.

3.2.1 Assessment of Vs from SPT (DP)

Shear wave velocity has been inferred from penetration test results. For this purpose the Otha & Goto (1978) equation and the Schnaid (1997) approach have been used.

More precisely, Ohta & Goto (1978) equation can be written in the following way:

$$V_s = 69 \cdot N_{60}^{0.17} \cdot Z^{0.2} \cdot F_A \cdot F_S \qquad [2.]$$

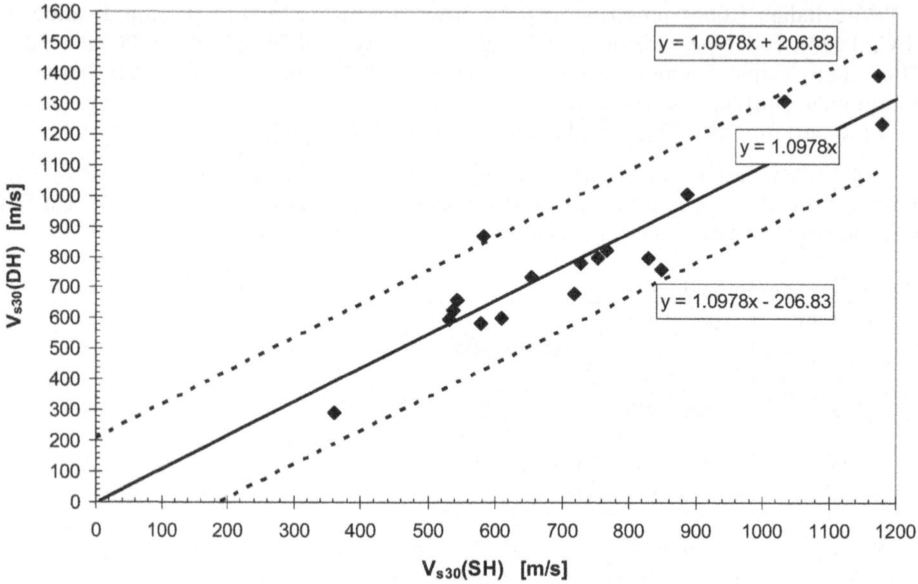

Figure 13 Average shear wave velocity from down-hole tests [V_{S30} (DH)] vs. that obtained from seismic refraction [V_{S30} (SH)]

where:

V_s = Shear wave velocity in m/s

N_{60} = blowcounts/30 cm normalized to an Energy Ratio ER = 60 %

Z = depth in m

F_A, F_S = Correction factors which accounts for deposit age and soil composition.

Penetration test results, terminated at 50 blowcounts, have been linearly extrapolated by considering the effective penetration after 50 blows.

DP test results have been converted into N_{60} by multiplying the data by 1.83 to account for different penetration and ER (usually 60 % in SPT and 73% in DP).

Comparison of shear wave velocities inferred from penetration tests, using Otha & Goto (1978) equation, and those directly measured in DH tests is shown in Figure 14. The Figure clearly shows that eq. [2.], on average, underestimate of about 38 % the measured values with a SD of about 205 m/s. In any case, the correlation is quite poor.

More recently, Schnaid (1997) proposed empirical correlations between $\mathrm{Log}V_s/(N_1)_{60}$ & $\mathrm{Log}(N_1)_{60}$. This correlation has the advantage that, unlike Otha e Goto (1978) eq., the knowledge of soil composition is not necessary. Figure 15 shows the experimental values

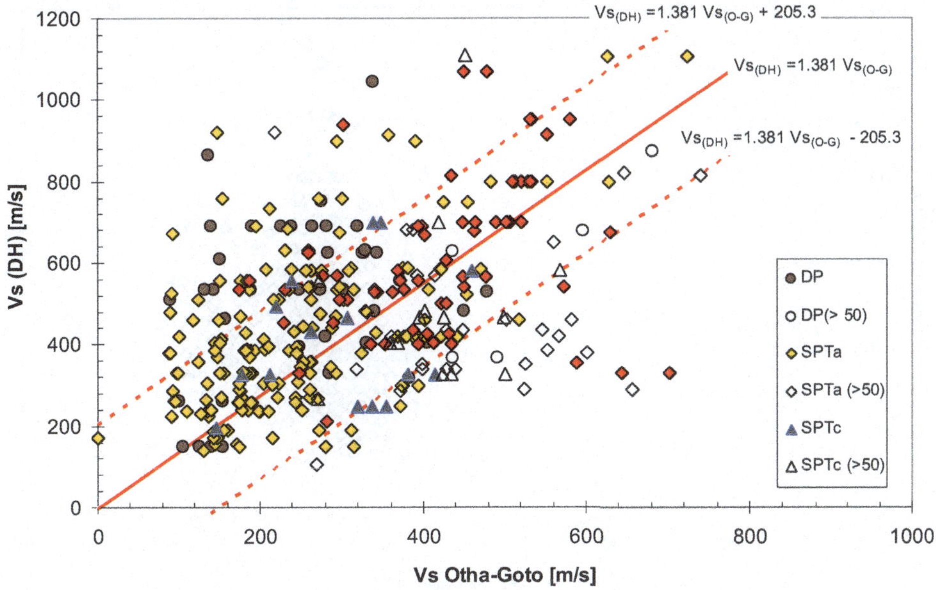

Figure 14 Measured shear wave velocity in down-hole tests vs. that inferred from N_{SPT} by means of Otha and Goto equation

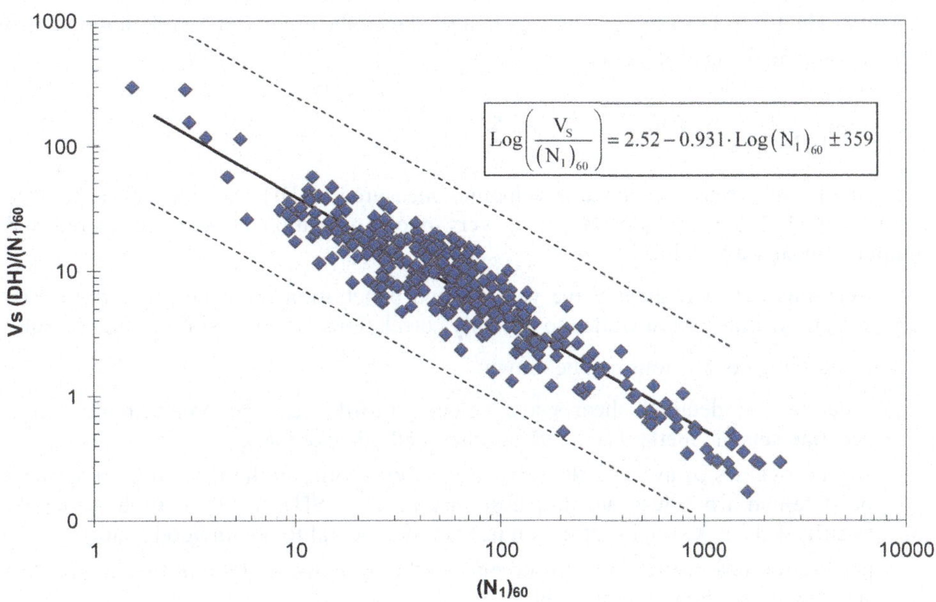

Figure 15 Normalized shear wave velocity vs. $(N_1)_{60}$ according to Schnaid (1997) approach

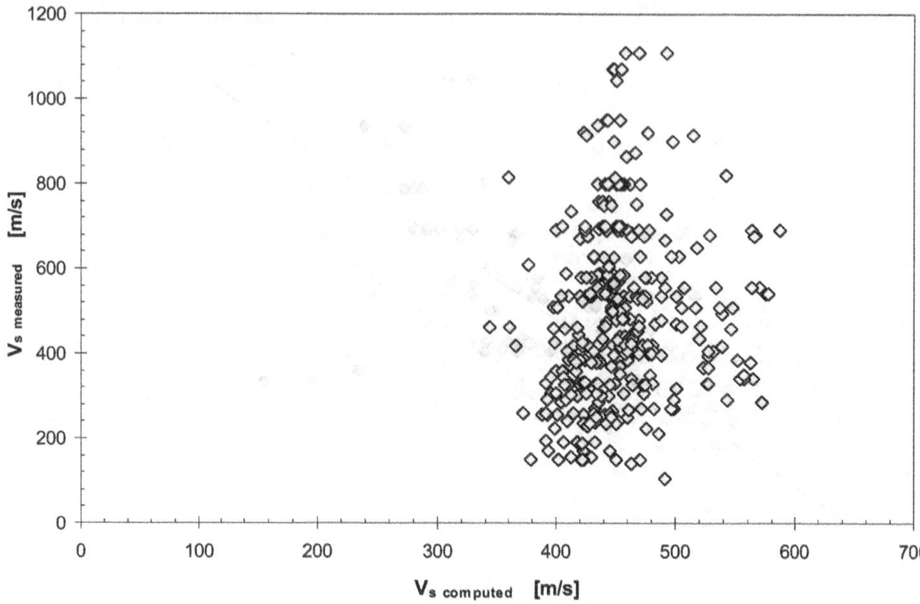

Figure 16 Measured (down-hole tests) vs. computed (according to Schnaid, 1997) shear wave velocity

of $\mathrm{Log}V_s/(N_1)_{60}$ & $\mathrm{Log}(N_1)_{60}$. The Figure also shows the empirical correlation obtained by regression analysis of the data.

$$\mathrm{Log}\left(\frac{V_s}{(N_1)_{60}}\right) = -0.927 \cdot (N_1)_{60} + 2.52 \qquad [3.]$$

Figure 16 shows the shear wave velocities measured in DH tests and those inferred using eq. [3.]. Figure 16 clearly shows very poor correlation between measured and computed shear wave velocity.

Correlations between shear wave velocity and penetration resistance should be used very cautiously. It is recommended to use only correlations established for a specific site.

The following conclusions can be drawn:

- accurate and detailed shear wave velocity profiles can be obtained only from accurate seismic method (i.e. DH or other methods like that);
- accurate values of average shear wave velocity profile in the first 30 m of soil can be obtained from less accurate test method (i.e. SH). The use of horizontally polarized shear waves in refraction test is fundamental to obtain good results;
- penetration test results can be used only in the framework of simplified approaches (i.e. that prescribed by EBC, 1998);

- accurate prediction of shear wave velocity from penetration test results could be possible only using specific correlation for a given site.

4. SEISMIC RESPONSE ANALYSES

In order to evaluate the influence of soil parameters on the seismic response, the elastic response spectra ($\xi = 5\%$) have been computed at 19 different sites by means of 1D equivalent – linear analyses run by means of EERA (Bardet et al., 2000). Analyses have been repeated to accounts for local variability of shear wave velocity of strata.

The local variability refers to that observed at a given site by comparing stratigraphic profile, DH and SH data.

As already mentioned, laboratory tests (Resonant Column Test – RCT, Cyclic Loading Torsional Shear Test – CLTST, Cyclic Loading Triaxial Test – CLTX) were used to determine the variation of stiffness (G) and damping ratio (D) with shear strain (γ). For a given geologic formation the variability of stiffness and damping ratio has been considered by repeating the computations with the upper and lower envelopes of the G/Go-γ D-γ curves. Because of the limited number of laboratory tests, in comparison to the extension of the investigated area, the whole variability of a given geologic formation has been considered.

Input motion (accelerograms at the rock outcrop) has been obtained by means of the following procedure (Mensi et al., 2004; Lai et al., 2005):

- definition of PGA at each site by means of standard probabilistic hazard analysis;
- de-aggregation of the hazard analysis (T_R = 475 years) to obtain the Magnitude-distance couple which mostly contribute to determine the hazard. The following couples have been obtained: M=5.3 d=10.2 km (Mensi et al., 2004), M=5.4 d= 13 km (Lai et al., 2005);
- selection of free-field natural accelerograms seismically compatible with the M-d couples, given a certain tolerance. The largest tolerance was adopted by Lai et al. (2005) in order to guarantee the compatibility with the elastic response spectra prescribed by the Italian code. More specifically a tolerance of ΔM=0.2-0.5 and Δd=9-17 km and 5-20 km were adopted by Lai et al. (2005).
- On the basis of the above described procedure, three accelerograms selected by Mensi et al. (2004) (A1, A2, A3) and another three selected by Lai et al. (2005) (A6.1, A6.2, A6.3) have been used.

As an alternative, an artificial accelerogram, obtained from the probabilistic elastic spectrum on rock, was used (Ferrini et al., 2000).

For each of the seven selected accelerograms the analyses summarized in Table 3 have been done.

Table 3 Analyses performed

Case Number	Vs (bedrock)	Vs (surface layers)	G-γ D-γ envelopes
1	Max	Min	Max
2	Max	Min	Min
3	Min	Min	Max
4	Min	Min	Min
5	Max	Max	Max
6	Max	Max	Min

To evaluate analysis results, PGA, normalized elastic response spectra and a synthetic amplification parameter have been considered.

The amplification parameter is defined in the following way:

$$F_a = \frac{\int_{0.1}^{0.5} \mathrm{PSA(deposit)}}{\int_{0.1}^{0.5} \mathrm{PSA(outcrop)}} \qquad [4.]$$

where: PSA = spectral acceleration computed at the top of the deposit or at outcrop, integrated between 0.1 and 0.5 s (corresponding to the most recurrent periods of the constructions existing in the study area).

The above definition of a synthetic amplification parameter is similar to that proposed by Pergalani et al. (1999).

Figure 17a to 17c show the PGA obtained at the top of soil deposits, by means of linear equivalent analyses at three sites, classified respectively as Type A, B and C soils.

Figure 17a refers to type A soil. Obviously, for this case no amplification exists and effects of different ground conditions and input accelerograms seems negligible. The level of a_{gR} (OPCM-3274, 2003), prescribed by the Italian code for that area and the PGA obtained from hazard study, are also reported in the Figure 17. The degree of conservativisme introduced by the Italian code is evident.

Figure 17b and 17c refer to type B and C soil respectively. In these Figures the effect of both input accelerograms and ground conditions on PGA is seen. It appears that the variability of PGA because of differences in ground conditions is comparable to that due to differences in input motion. As for the ground conditions, variability in the shear wave velocity profile seems to have the prominent effect.

Design accelerations have been plotted in Figure 17b and 17c. The design accelerations have been obtained multiplying a_{gR}, prescribed by the Italian Code, by the soil factor S prescribed for type B and C soils by OPCM 3274 (2003) and those prescribed by EBC (1998) for the same type of soils in the case of type 1 (EC8-1) and 2 (EC8-2) spectra. It is evident that in some cases the prescribed design accelerations are lower than that obtained from linear – equivalent seismic response analyses.

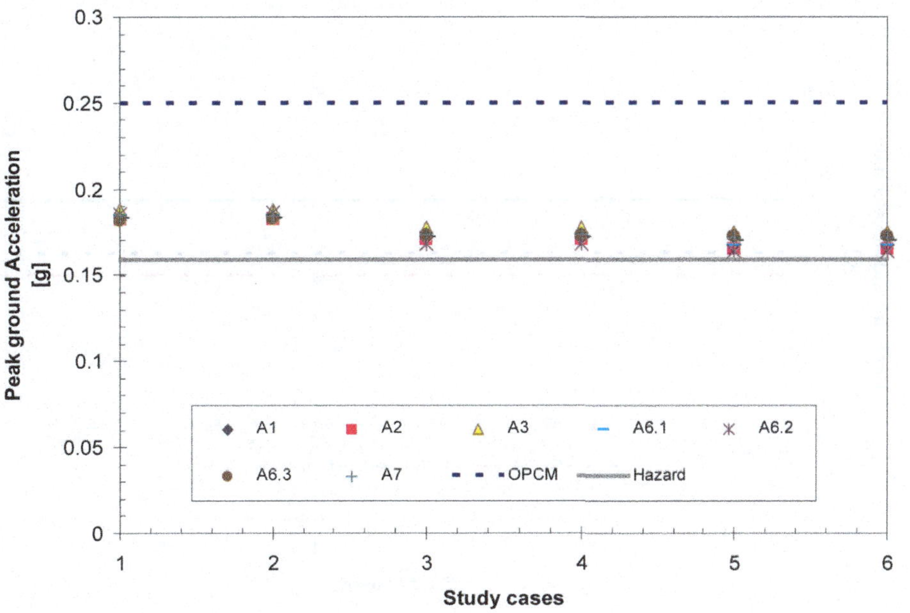

17a) PGA - Type A soil

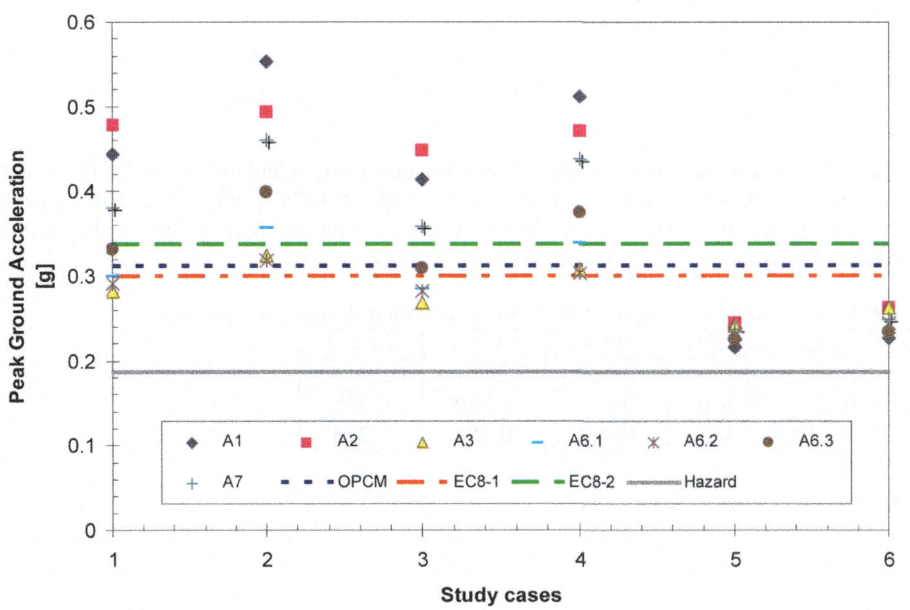

17b) PGA - Type B soil

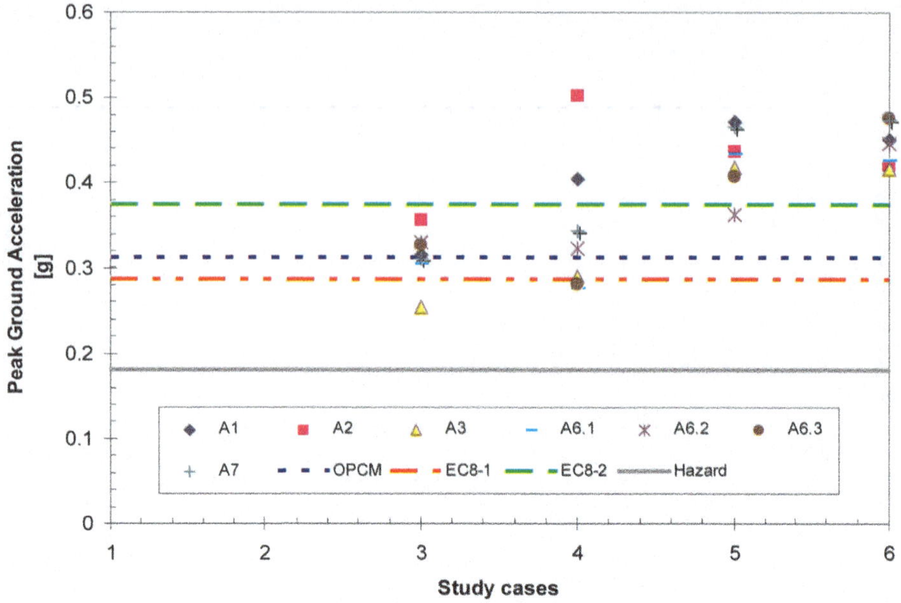

17c) PGA - Type C soil

Figure 17 Computed vs. prescribed PGA: *17a)* type A soil; *17b)* type B soil; *17c)* type C soil;

Table 4 summarizes the average PGA, obtained from calculations for the three sites, and the design accelerations prescribed by OPCM-3274 (2003) and EBC (1998). Anyway, the discussion on the effectiveness of European and Italian Codes is beyond the scope of this paper.

Table 4 Table 8 Computed PGA and prescribed design accelerations

Soil type	PGA(*)	OPCM	EC8-1	EC8-2
A	0.176	0.250	0.250 (-)	0.250 (-)
B	0.330	0.312	0.300	0.338
C	0.385	0.312	0.288	0.375

(*) Average; (-) value from Italian code

Figure 18a to 18c show the amplification parameter F_a, computed according to eq. [4.] at three sites, classified respectively as Type A, B and C soils.

It is evident that the amplification parameter is mainly affected by the type of accelerogram, the shear wave velocity profile and to a less extent the stiffness and material damping ratio parameters.

A comparison between the computed response spectra and the prescribed ones is shown in Figure 19a-19b-19c and Figure 20a-20b-20c. The computed spectra have been normalized with respect to the PGA on rock, while the prescribed ones have been divided by the relevant value of a_{gR}.

Figures clearly show that both input motion and ground conditions influence the response spectra. More specifically:

- natural accelerograms, (A1, A2, A3) selected to represent earthquakes with M=5.4 and d=13 km, lead to high spectral accelerations (PSA) in the high frequency range (T=0.1-0.2s) while PSA are much smaller than that prescribed by EC8 and OPCM 3274 (2003) at T>0.2 s;
- the second set of natural accelerograms (A6.1, A6.2, A6.3), selected considering a range of Magnitude and distances (ΔM=0.2-0.5; Δd=9-17 km; 5-20 km), give the same picture, but the PSA spikes in the high frequency range are less pronounced and, at larger periods (T>0.2s) there is a better agreement between prescribed and computed spectra;
- the artificial accelerogram leads, on the whole, to a better agreement between computed and prescribed spectra;
- computed PSA may be greater than prescribed PSA for the most unfavourable ground conditions;
- use of minimum envelope for the stiffness and material damping ratio has mainly the effect of increasing the response and moving the maximum value of the spectral acceleration towards higher periods, which is consistent with reduced values of material damping and increased non-linearity;
- maximum amplification of the seismic response is observed when combining the highest velocity of the bedrock with the lowest velocity of the soil layers. Minimum amplification of the seismic response is observed when combining the highest velocity of the bedrock with the highest velocity of the soil layers. The third combination usually leads to intermediate values of the seismic response.

It is also interesting to compare (Figure 21) the average response spectrum obtained for a type B soil to those prescribed by the European and Italian codes.

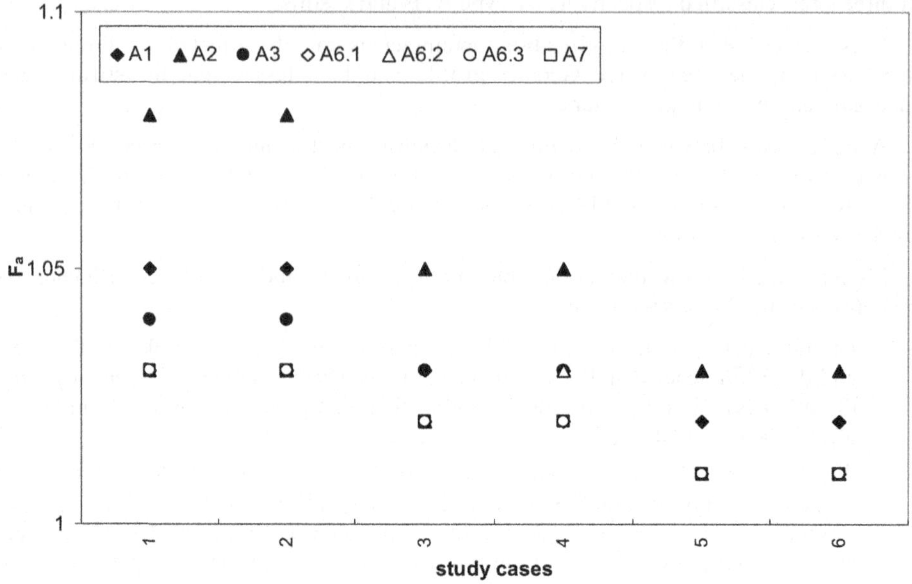

18a) Amplification parameter F_a: type A soil

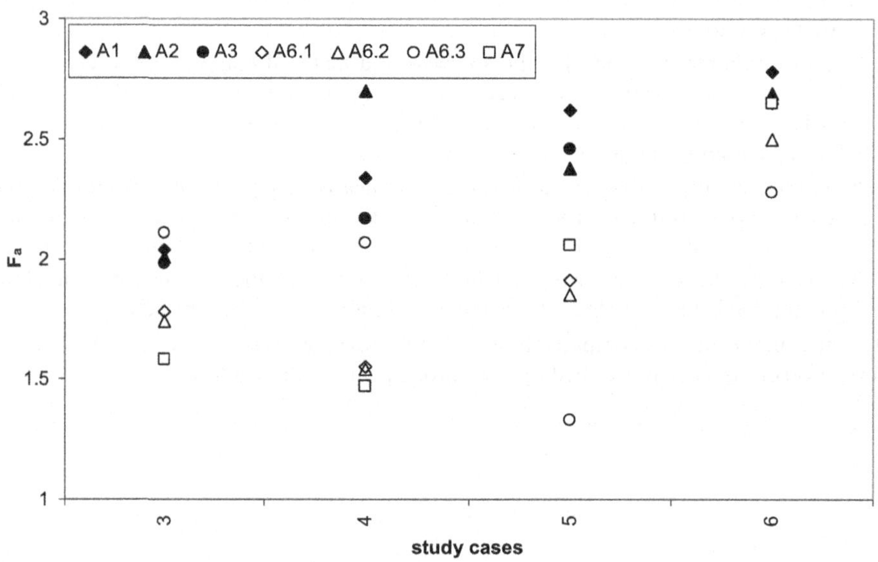

18b) Amplification parameter F_a: type B soil

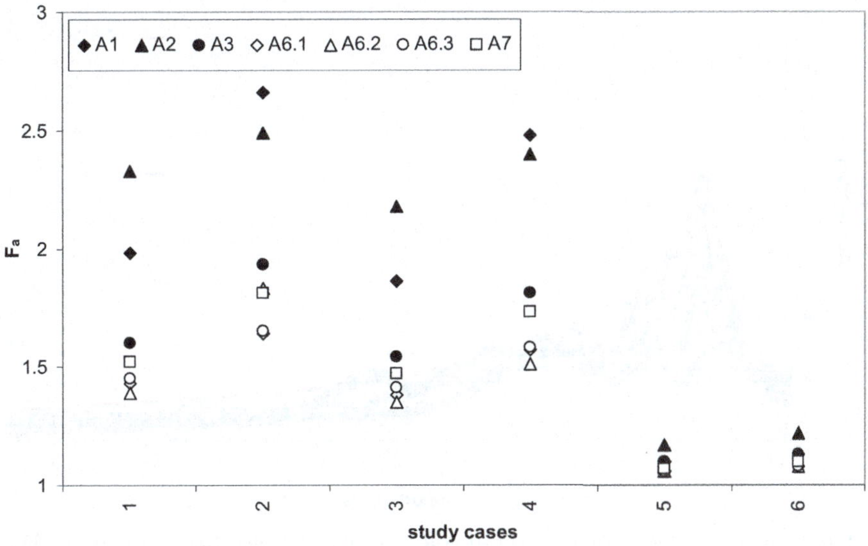

18c) Amplification parameter F_a: type C soil

Figure 18 Amplification parameter F_a: *18a)* type A soil; *18b)* type B soil; *18c)* type C soil

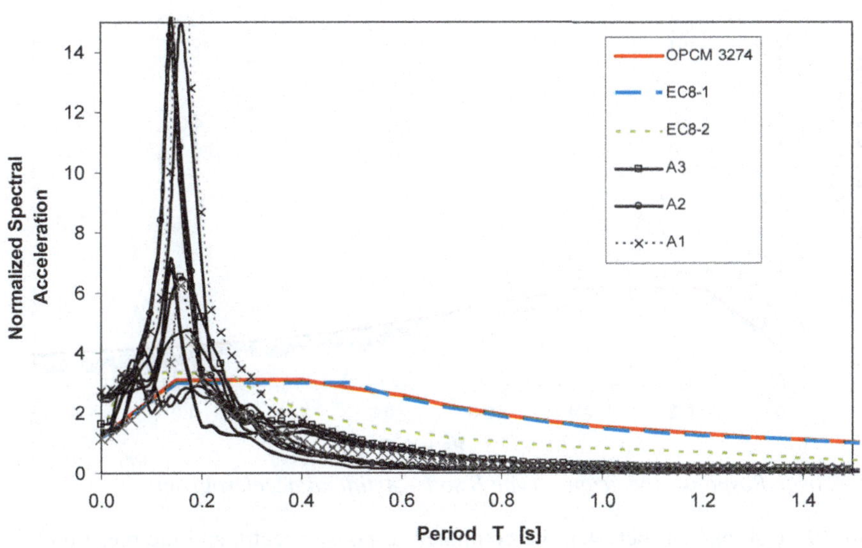

19a) Elastic Response spectrum – Type B soil – Accelerograms A1-A2-A3: M = 5.3 d = 10.2 km

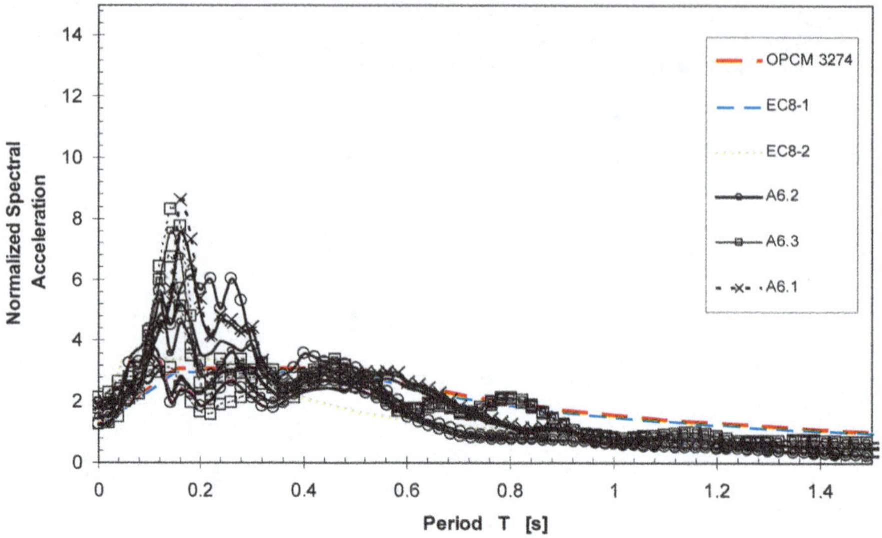

19b) Elastic Response Spectrum – Type B soil – Accelerograms A6.1-A6.2-A6.3: ΔM = 0.2-0.5 Δd = 9-17; 5-20 km

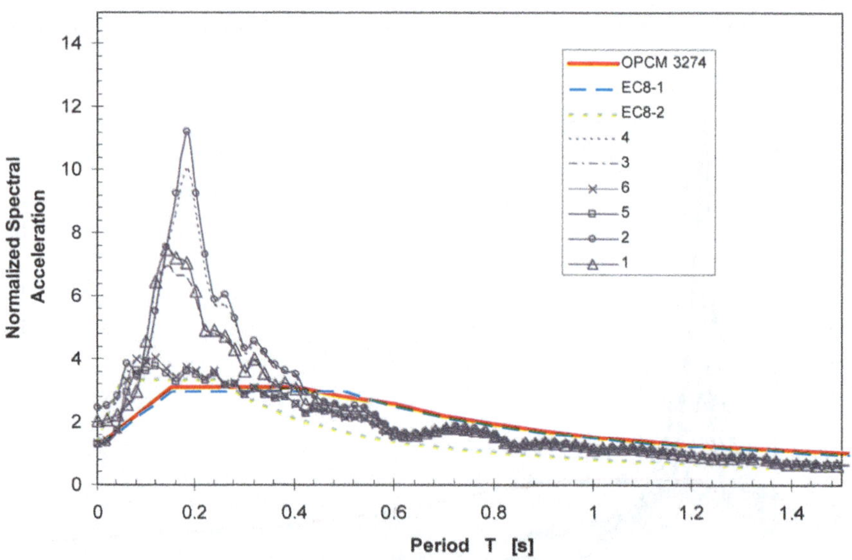

19c) Elastic Response spectrum – Type B soil – Artificial accelerogram

Figure 19 Comparison between the computed response spectra and the prescribed ones for type B soil, using: *19a)* accelerograms A1-A2-A3: M = 5.3 d = 10.2 km; *19b)* accelerograms A6.1-A6.2-A6.3: ΔM = 0.2-0.5, Δd = 9-17; 5-20 km; *19c)* artificial accelerogram

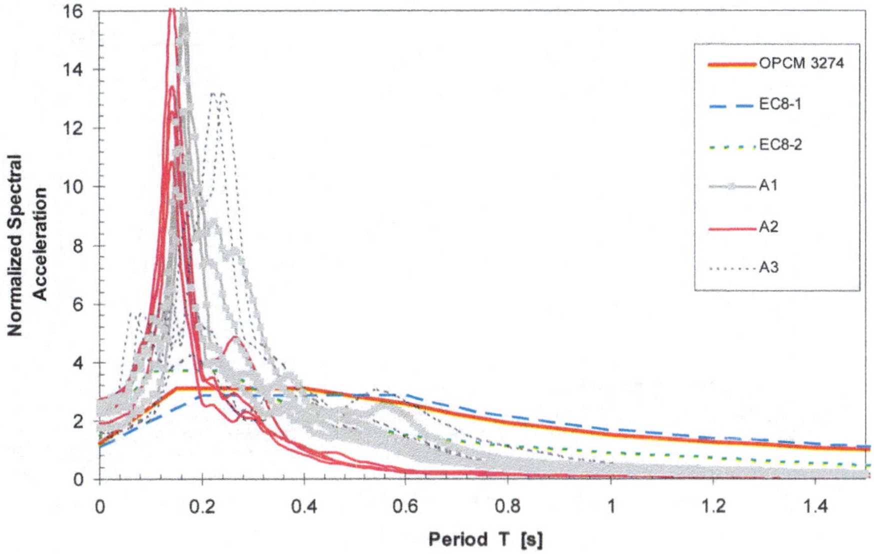

20a) Elastic Response Spectrum – Type C soil – Accelerograms A1-A2-A3: M = 5.3 d = 10.2 km

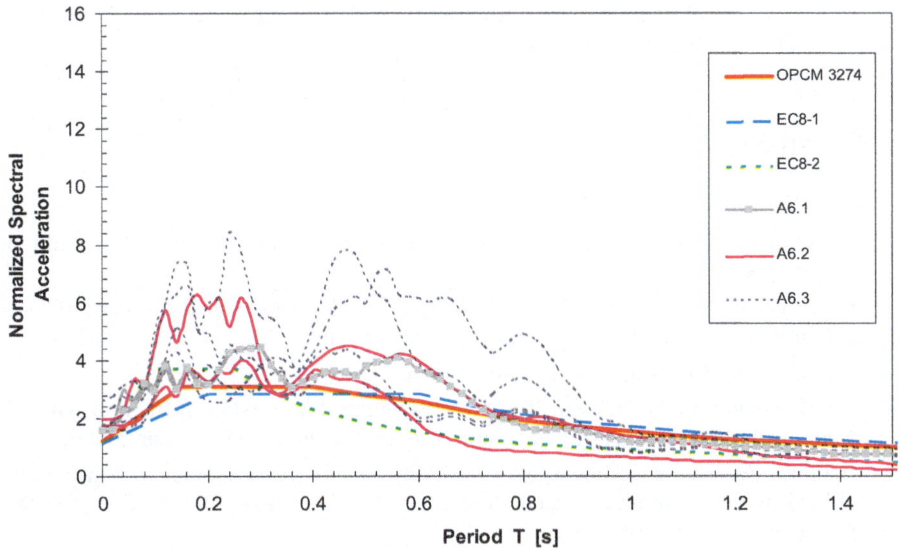

20b) Elastic Response Spectrum – Type C soil – Accelerograms A6.1-A6.2-A6.3: ΔM = 0.2-0.5 Δd = 9-17; 5-20 km

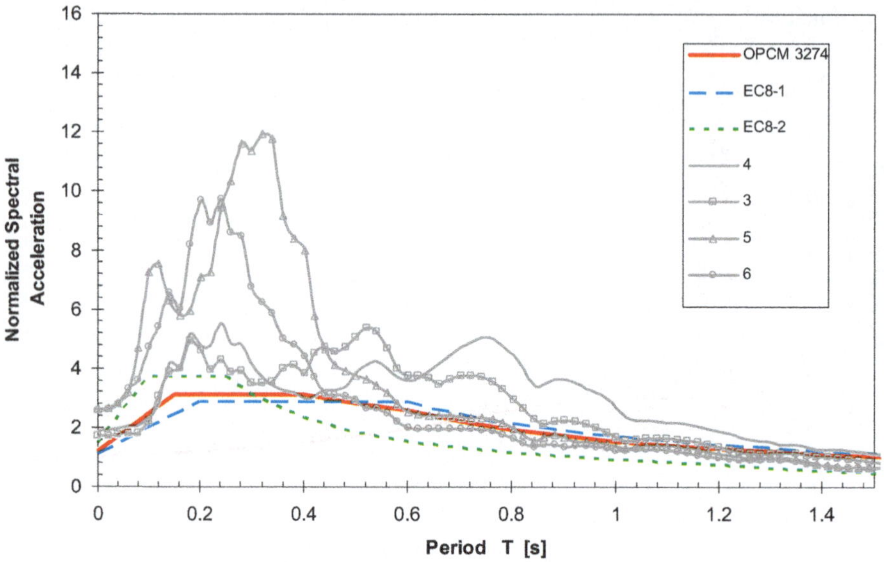

20c) Elastic Response Spectrum – Type C soil – Artificial Accelerogram

Figure 20 Comparison between the computed response spectra and the prescribed ones for type C soil, using: *20a)* accelerograms A1-A2-A3: M = 5.3 d = 10.2 km; *20b)* accelerograms A6.1-A6.2-A6.3: ΔM = 0.2-0.5, Δd = 9-17; 5-20 km; *20c)* artificial accelerogram

CONCLUSIONS

The following conclusions can be drawn:

- use of penetration test results to infer shear wave velocity profiles should be discouraged;
- it is recommended to use only correlations ($N_{spt} - V_s$) established for a specific site and only in the case of spatially uniform soil deposits;
- N_{spt} and V_s lead to consistent definition of soil type;
- use of seismic methods, even if not very accurate like seismic refraction with SH waves, is recommended as an economic way to determine the V_{s30} parameter;
- accurate assessment of shear wave velocity profile from top layer down to the bedrock is very important. Accurate evaluation of the spatial variability of such a parameter is also important;
- accurate assessment of stiffness and material damping ratio and of their spatial variability is also important.

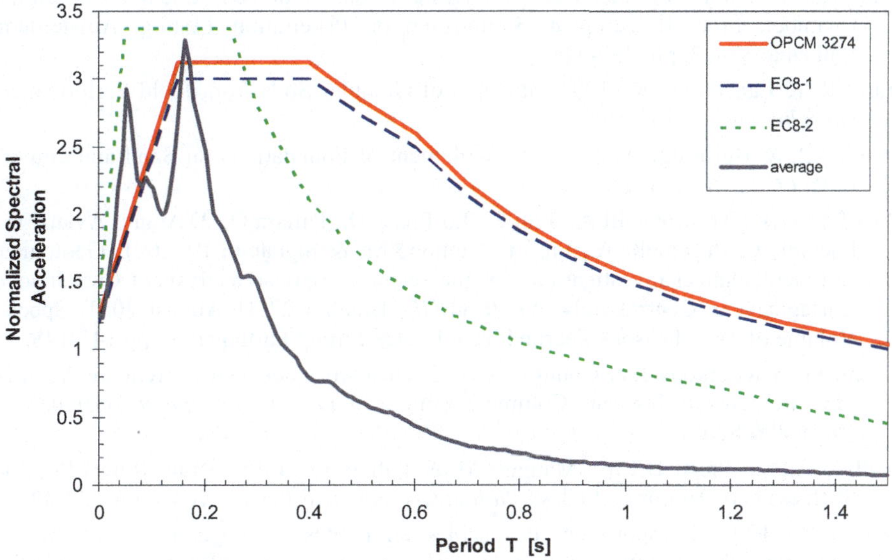

Figure 21 Computed average elastic response spectrum (Type B soil)

6. Acknowledgements

The authors would like to thank Mr. Ferrini (Head of the Seismic Survey of RT) for making available the RT database.

7. References

Aki, K., 1988. Local Site Effects on Strong Ground Motion. J. Lawrence Van Thun (Ed.), Earthquake Engineering and Soil Dynamic II – Recent Advances in Ground-Motion Evaluation, *ASCE Geotechnical Special Publication* No 20, pp. 103-155, Park City, Utah, USA.

Alarcon-Guzman A., Chameau J.L. & Leonards G.A., 1986. A New Apparatus for Investigating the Stress-Strain Characteristics of Sands. *Geotechnical Testing Journal,* 9(4): 204-212

Ampadu S.K. & Tatsuoka F., 1993. A Hollow Cylinder Torsional Simple Shear Apparatus Capable of a Wide Range of Shear Strain Measurement. *Geotechnical Testing Journal*, 16(1): 3-17

Bardet, J.P., Ichii, K. & Lin C.H., 2000. EERA – A Computer Program for Equivalent-Linear Earthquake Site Response Analyses of Layered Soil Deposits. Department of Civil Engineering, *University of Southern California*, http://geoinfo.usc.edu/gees.

Bellotti R., Bizzi G. & Ghionna V.N. 1982 Design, Construction and Use of a Calibration Chamber. Proc. II European Symposium on Penetration Testing. Amsterdam, Balkema, Vol. 2, pp: 439-446

Berardi R. & Lancellotta R., 1991. Stiffness of Granular Soils from Field Performance. *Gétechnique* , 1: 149-157

Burland J.B. & Burbridge M.C., 1985. Settlement of Foundations on Sand and Gravel. Proc. *I.C.E.,* 78, 1: 1325

Calosi E., Ferrini M., Cancelli A., Foti S., Lo Presti D., Pallara O., D'Amato Avanzi G., Pochini A., Puccinelli A., Luzi L. Rainone M. & Signanini P., 2001. Geological and Geotechnical investigations for the seismic response analysis at Castelnuovo Garfagnana in Central Italy. *XV ICSMGE*, Istanbul 27-31 August 2001, Special Volume of TC4, Lessons Learned from Recent Strong Earthquakes. pp 141-148.

Cascante G., Vanderkooy J. & Chung W., 2003. Difference between Current and Voltage Measurements in Resonant Column Testing. *Canadian Geotechnical Journal*, Vol. 40(4): 806-820.

Cavallaro A., Lo Presti D.C.F., Maugeri M. & Pallara O., 1998. Strain Rate Effect on Stiffness and Damping of Clays. *Italian Geotechnical Review*, XXXII(4): 30-49

Cazacliu B., 1996. Comportment des sables en petites et moyennes déformations; réalisation d'un prototype d'essai de torsion compression confinement sur cylindre creux. *Doctorat*, ECP-ENTPE, Paris

Constantopoulos I.V., Roësset J.M. & Christian J.T., 1973. A Comparison of Linear and Exact Nonlinear Analysis of Soil Amplification. *5th WCEE*, Roma, pp: 1806-1815

Cubrinovski M. & Ishihara K., 1999. Empirical Correlation between SPT N-value and Relative Density of Sandy Soils. *Soils and Foundations*. 5: 61-71

Di Benedetto H., CazacliuB., Boutin C., Donah T & Touret J.P., 1997. Comportement des sables avec rotation d'axes. *XIV ICSMFE,* Hamburg, Balkema, 1: 279-282

d'Onofrio A. Silvestri F. & Vinale F., 1999. Strain rate dependent behaviour of a natural stiff clay. *Soils and Foundations*. 39(2):69-82.

Drnevich V.P., 1978. Resonant Column Testing-Problems and Solutions. *Dynamic Geotechnical Testing ASTM STP 654*. pp: 384-398

EBC, 1998. Eurocode 8. Design Provisions for Earthquake Resistance of Structures - Part 1-1: General Rules for the Representation of Seismic Actions. Part 5: Foundations, Retaining Structures and Geotechnical Aspects.

Evans, M.D., 1987. Undrained cyclic triaxial testing of gravels: the effect of membrane compliance. *Ph.D Dissertation*, University of California at Berkeley

Faccioli, E., 1991. Seismic Amplification in the Presence of Geological and Topographic Irregularities. Proceedings *International Conference on Recent Advances in Geotechnical Earthquake Engineering and Soil Dynamics*, St. Louis, Missouri, pp. 1779-1797.

Ferrini M. (coord.), Foti S., Lo Presti D., Luzi L., Pergalani F. Petrini V. Pochini A., Puccinelli A., Signanini P. & Socco V., 2000. La riduzione del rischio sismico nella pianificazione del territorio: le indagini geologico tecniche e geofisiche per la valutazione degli effetti locali, *CISM*, Lucca 3-6 Maggio 2000.

Fioravante V., Jamiolkowski M. & Lo Presti D.C.F., 1994. Stiffness of Carbonatic Quiou Sand. *Proceedings of the XIII ICSMFE,* New Delhi, India,1: 163-167

Fioravante V., Ghionna V.N., Jamiolkowski M. & Pedroni S., 1998. Stiffness of Carbonatic Quiou Sand from CPT. *ISC 98,* Atlanta GA. Balkema Vol. 2, pp:1039-1049.

Gabrielaitis L., Jamiolkowski M. Lo Presti D. & Puci I., 2000. Operational sand stiffness from large scale load tests on reconstituted sand beds. *Baltic Geotechnics IX 2000,* Pärnu, Estonia 9-12 May 2000. Estonian Geotechnical Society. Published by Akadeemia Trükk, Niine, 10414 Tallinn, Estonia. pp: 84-91, paper 1.14.

Ghionna, V.N. Jamiolkowski, M. Pedroni, S. & Salgado, R., 1994. The tip displacement of drilled shafts in sands. *Settlement 94, ASCE Specialty Conference.*

Gomes Correja A. & Gillet S., 1996. A large triaxial apparatus for the study of granular materials under repeated loading used at LNEC. *Proc. of the European Symposium Eurofles 1993,* Lisbon, Balkema. 45-52

Goto, S., Shamoto, Y. & Tamaoki, K., 1987. Dynamic properties of undisturbed gravel sample by the in situ frozen. *Proc.8th ARCSMFE,* No.1, pp 233-236

Goto, S., F. Tatsuoka, S. Shibuya, Y.S. Kim & Sato T., 1991. A simple gauge for local small strain measurements in the laboratory. *Soils and Foundations*: 31(1): 169-180

Goto, S., Suzuki, Y., Nishio, S. & Oh Oka, H., 1992. Mechanical properties of undisturbed Tone-river gravel obtained by in situ freezing method. *Soils and Foundations*, No.3, pp.15-25

Goto S., Nishio, S. & Yoshimi, Y., 1994. Dynamic properties of gravels sampled by ground freezing. Ground failures under seismic conditions, *GSP* No. 44, *ASCE*, pp.141-157

Hatanaka, M., Suzuki, Y., Kawasaki, T. & Endo, M., 1988. Cyclic undrained shear properties of high quality undisturbed Tokyo gravel. *Soil and Foundations*, No.4, pp.57-68

Hatanaka, M. & Uchida, A., 1995. Effects of test methods on the cyclic deformation characteristics of high quality undisturbed gravel samples. *GSP* No.56, *ASCE*, pp.136-151

Hight D.W., Gens A. & Symes M.S., 1983. The Development of a New Hollow Cylinder Apparatus for Investigating the Effects of Principal Stress Rotation in Soils. *Gétechnique* , 33(4): 355-383

Hynes, M.L., 1988. Pore pressure generation characteristics of gravel under undrained cyclic loading. *Ph.D. Dissertation*, University of California, Berkeley

ICBO, 2000. International Building Code, International Code Council (ICC), WEB site address: http://www.icbo.org/

Ionescu F., 1999. Comportamento sforzi-deformazioni della sabbia di Toyoura da prove di taglio torsionale in condizioni di carico monotono e ciclico. *Ph.D. Thesis,* Politecnico di Torino Department of Structural and Geotechnical Engineering

Isenhower W. M., Stokoe K. H. II & Allen J. C., 1987. Instrumentation for Torsional Shear Resonant Column Measurements Under Anisotropic Stresses. *Geotechnical Testing Journal*, 10(4): 183-191

Kim D-S. & Stokoe K.H. II, 1994. Torsional Motion Monitoring System for Small Strain (10^{-5} to 10^{-3}) Soil Testing. *Geotechnical Testing Journal*. 17(1): 17-26.

Kokusho, T. & Tanaka, Y. 1994. Dynamic properties of gravel layers investigated by In-situ freezing sampling. Ground failures under seismic conditions, *GSP* No.44, *ASCE*, pp.121-140

Lai C.G. & Rix G.J., 1998. Simultaneous Inversion of Rayleigh Phase Velocity and Attenuation for Near Surface Site Characterization. *GIT-CEE/GEO-98-2*, School of Civil and Environmental Engineering, Georgia Institute of Technology.

Lai C.G., Pallara O., Lo Presti D. C. F. & Turco E., 2001. Low-strain stiffness and material damping ratio coupling in soils. *XV ICSMGE,* Istanbul 27-31 August 2001, Special Volume of TC29, Advanced Laboratory stress-strain testing of geomaterials. F. Tatsuoka, S. Shibuya and R. Kuwano Eds. Balkema pp 265-274.

Lai C., Strobbia C., Dall'ara, 2005. Elaborazione di raccomandazioni e di linee guida per la definizione dell'input sismico e delle modellazioni da adottare nei territori della Regione Toscana nell'ambito dei progetti VEL – Parte 1. *Eucentre Pavia Report*

Lo Presti D.C.F., Pallara O, Lancellotta R., Armandi M. & Maniscalco R., 1993. Monotonic and Cyclic Loading Behaviour of Two Sands at Small Strains. *Geotechnical Testing Journal*, 16(4): 409-424.

Lo Presti D.C.F., Jamiolkowski M., Pallara O. & Cavallaro A., 1996. Rate and Creep Effect on the Stiffness of Soils. *GSP* No. 61, *ASCE*: 166-180

Lo Presti D.C.F., Pallara O. & Cavallaro A., 1997. Damping Ratio of Soils from Laboratory and In-Situ Tests *Proc. XIV ICSMFE, Seismic Behaviour of Ground and Geotechnical Structures*, Balkema, Rotterdam: 391-400

Lo Presti D.C.F. & Puci I., 2001. Impiego delle prove penetrometriche dinamiche per la caratterizzazione meccanica dei terreni. *XVIII Ciclo delle Conferenze di Geotecnica di Torino*

Lo Presti D.C.F., Shibuya S. & Rix G., 2001. Innovation in Soil Testing. *Proc. of 2nd Int. Symposium on Pre-Failure Deformation Characteristics of Geomaterials*, Torino 27-30 September 1999, Balkema, Vol. 2, 1027-1076.

Lo Presti D., Luzi L., Pergalani F., Petrini V., Puci I & Signanini P. 2002 Determinazione della risposta sismica dei terreni a Castelnuovo Garfagnana (Lucca), Rivista Italiana di Geotecnica. XXXVI, n. 3, pp. 61 – 74.

Lo Presti D., Lai C. & Foti S., 2004. Geophysical and Geotechnical Investigations for ground response analyses. *Recent Advances in Earthquake Geotechnical Engineering and Microzonation*, Kluwer Academic Publisher, Editor A. Ansal, Capitolo di un libro. pp: 101-138.

Lo Presti D., Pallara O., Froio F., Rinolfi A. & Jamiolkowski M., 2005. Stress-Strain-Strength Behaviour of Undisturbed and Reconstituted Gravely Soil Samples. Submitted to *Italian Geotechnical Review* for possible publication, September 2005

Marchetti S., 2005. Shear Wave velocity (Vs) measurement, using a seismic DMT. *Workshop 12 April 2005, University of Pisa.*

Meng J. & Rix G.J., 2003. Reduction of Equipment –Generated Damping in Resonant Column. *Géotechnique*, Vol LIII(5):503-512.

Meng J. & Rix G.J., 2004. Reduction of Equipment –Generated Damping in the Low Frequency Range. *Géotechnique*, Vol LIV(3):219-222.

Mensi E., Lai C., Spallarossa D. Pallara O. & Lo Presti D., 2004. Risposta sismica in alcune aree della Toscana : un confronto con le indicazioni dell'Ordinanza dell'OPCM 3274. *Politecnico di Torino, Department of Structural and Geotechnical Engineering Report.*

Miura K., Miura S & Toki S., 1986. Deformation Behaviour of Anisotropic Dense Sand Under Principal Stress Axes Rotation. *Soils and Foundations*. 26(1) 36-52.

Mussa M., 2004. Calcolo dei cedimenti delle fondazioni superficiali nei terreni granulari mediante la prova CPT. *B. Sc. Thesis.* Politecnico di Torino, II Faculty of Engineering in Vercelli.

Nicholson, P.G., Seed, R.B. & Anwar, H.A., 1993a. Elimination of membrane compliance in undrained triaxial testing. I. Measurement and evaluation. *Canadian Geotechnical Journal*, pp.727-738

Nicholson, P.G., Seed, R.B. & Anwar, H.A., 1993b. Elimination of membrane compliance in undrained triaxial testing. II. Mitigation by injection compensation. *Canadian Geotechnical Journal*, pp.739-746

Norme Tecniche per le Costruzioni, 2005. Ministero delle Infrastrutture e dei Trasporti. *Decreto 14 Settembre 2005.*

OPCM 3274, 2003. Primi elementi in materia di criteri generali per la classificazione sismica del territorio nazionale e di normative tecniche per le costruzioni in zona sismica. *Gazzetta Ufficiale della Repubblica Italiana 8 maggio 2003, n. 108*

Ohta Y. & Goto N., 1978. Empirical Shear Wave Velocity Equations in Terms of Characteristic Soil Indexes. *Earthquake Engineering and Structural Dynamics*. 6

Pagani, 2005. www.pagani-geotechnical.com/italiano/RendimentoDP/Renddp.htm#INI

Pallara O., 1995. Comportamento sforzi-deformazioni di due sabbie soggette a sollecitazioni monotone e cicliche. *Ph. D. Thesis,* Department of Structural Engineering, Politecnico di Torino

Papa V., Silvestri F. & Vinale F., 1988. Analisi delle Proprietà di un Tipico Terreno Piroclastico mediante Prove di taglio Semplice. *Proc. Gruppo Nazionale di Coordinamento per gli Studi di Ingegneria Geotecnica.* Monselice, Italy, 1: 265-286

Pergalani F., Romeo R., Luzi L., Petrini V., Pugliese A. & Sano T., 1999. Seismic microzoning of the area struck by Umbria-Marche (Central Italy) Ms 5.9 earthquake of 26 september 1997. *Soil Dyn. and Earth. Eng.* 18, pp. 279-296.

Porovic E. & Jardine R., 1994. Some observations on the static and dynamic shear stiffness of Ham River sand *IS Hokkaido 1994,* Balkema,1: 25-30

Pradhan T.B.S., Tatsuoka F. & Horii, N., 1988. Strength and Deformation Characteristics of Sand in Torsional Simple Shear *Soils and Foundations*, 28(3): 131-148

Rix G.J. & Meng J., 2005. A Non-Resonance Method for Measuring Dynamic Soil Properties. *Geotechnical Testing Journal*, Vol 28, No. 1

Rollins, K., Evans, M.D., Dhiel, N.B. & Daily, W.D., 1998a. Shear modulus and damping relationships for gravels. *JGGE, ASCE*, No.5, pp.396-405

Schnaid F., 1997. Panel Discussion. Evaluation of in situ tests in cohesive frictional materials. *14th Int. Conf. on SMGE.* Hamburg. 4: 2189-2190

Seed, H.B., Wong, R.T., Idriss, I.M. & Tokimatsu, K., 1986. Moduli and damping factors for dynamic analyses of cohesionless soils. *JGE, ASCE*, 11, pp.1016-1032.

Shibuya S., Mitachi T., Fukuda F. & Degoshi T., 1995. Strain Rate Effect on Shear Modulus and Damping of Normally Consolidated Clay. *Geotechnical Testing Journal.*18(3): 365-375

Shibuya S. & Mitachi T., 1997. Development of a fully digitized triaxial apparatus for testing soils and soft rocks. *Geotechnical Engineering.* 28(2): 183-207

Stokoe, K.H. II, Hwang S.K., Lee, J.N.K. & Andrus R.D. 1995. Effects of Various Parameters on the Stiffness and Damping of Soils at Small to Medium Strains. *Keynote Lecture 2, IS Hokkaido*. 2: 785-816. Balkema

Tanaka Y., Kokusho T., Yoshida Y & Kudo K., 1991. A method for evaluating membrane compliance and system compliance in undrained cyclic shear tests. *Soils and Foundations*. 31(3): 30-42

Tanaka, Y., Kudo, K., Nishi, K., Okamoto, T., Kataoka, T. & Ueshima, T., 2000. Small Strain Characteristics of Soils in Hualien, Taiwan. *Soils and Foundations*, No.3, pp. 111-126

Tatsuoka F., 1988. Some recent developments in triaxial testing systems for cohesionless soils. *ASTM STP 977, Advanced Triaxial Testing of Soil and Rock.* pp: 7-67

Tatsuoka, F., Sato, T., Park, C.S., Kim, Y.S, Mukabi, J.N. & Koha-ta, Y., 1994a. Measurements of Elastic Properties of Geomaterials in Laboratory Compression Tests. *Geotechnical Testing Journal*, N°1, pp.80-94

Tatsuoka, F., Teachavorasinskun, S., Dong, J., Kohata, Y. & Sato, T., 1994b. Importance of measuring local strains in cyclic triaxial tests on granular materials. *Proc. ASTM Symposium on Dynamic Geotechnical Testing II, STP 1213*, ASTM, pp.288-302

Tatsuoka F. & Kohata Y., 1995. Stiffness of Hard Soils and Soft Rocks in Engineering Applications. *Keynote Lecture 8, IS Hokkaido 1994.* 2: 947-1066. Balkema

Tatsuoka F., Jardine R.J., Lo Presti D.C.F., Di Benedetto H. & Kodaka T., 1997. Characterising the Pre-Failure Deformation Properties of Geomaterials. *Theme Lecture, Plenary Session 1, XIV ICSMFE,* Hamburg, in print. Balkema

Teachavorasinskun S., 1989. Deformation Characteristics of Sands at Small Strains. *M.Sc. Thesis,* University of Tokyo, Japan

Toki S., Shibuya S. & Yamashita S., 1995. Standardization of laboratory test methods to determine the cyclic deformation properties of geomaterials in Japan. *Keynote Lecture 1, IS Hokkaido:* 2: 741-784, Balkema

Vaid Y.P., Sayao A., Hou E. & Negussey D., 1990. Generalized stress-path-dependent soil behaviour with a new hollow cylinder torsional apparatus. *Canadian Geotechnical Journal*, 27: 601-616.

Vucetic, M. & Dobry, R. 1991. Effect of soil plasticity on cyclic response. *JGE, ASCE*, No.1, pp. 89-107

Wang Y.-H., Cascante G. & Santamarina J.C., 2003. Resonant Column Testing: the Inherent Counter-EMF Effect. *Geotechnical Testing Journal* Vol. 26(3):

Yamashita S. & Suzuki T., 1999. Young's and shear moduli under different principal stress directions of sand, *Proc. of IS Torino 99*, Balkema, 1: 149-158

Yasuda N. & Matsumoto N., 1993. Dynamic deformation characteristics of sands and rockfill materials. *Canadian Geotechnical Journal*, 30: 747-757

Yasuda, N., Otha, N. & Nakamura, A., 1994. Deformation Characteristics of Undisturbed Riverbed Gravel by In-Situ Freezin Sampling Method. *Proc. Int. Symp. on Pre-Failure Deformation Characteristic of Geomaterials, IS Hokkaido 94* (Shibuya et al. eds.), Rotterdam: Balkema, pp.41-46.

Soil Stress-Strain Behavior: Measurement, Modeling and Analysis
Geotechnical Symposium in Roma, March 16 & 17, 2006

SMALL STRAIN BEHAVIOUR AND VISCOUS EFFECTS ON SANDS AND SAND-CLAY MIXTURES

Di Benedetto H.
DGCB – ENTPE Rue Maurice Audin, 69518 Vaulx-en-Velin Cedex, France ;
e-mail: herve.dibenedetto@entpe.fr

ABSTRACT
This lecture paper focuses on sands and sand-clay mixtures behaviour in the small strain domain. Non viscous and viscous components are measured, identified and modelled within the framework of a 3 component model. Two precision prototype devices (triaxial and hollow cylinder) both equipped with piezoelectric sensors are used. Non viscous measured behaviour considering small quasi-static cycles and wave properties are compared with simulations obtained from 2 recently formulated anisotropic hypoelastic models (DBGS and DBGSP). Then, viscous experimental part is compared with the proposed model prediction. This model is an asymptotic expression, for the small strain domain, of a viscous evanescent formalism proposed by the author. It takes into account very peculiar behaviour observed on sands. Simulation for loadings with and without rotation of axes and for different rate histories, are quite satisfactory.

KEYWORDS : Sands, Sand/clay mixtures, Experiments, Triaxial tests, Hollow Cylinder tests, Modelling, Small strain, Waves propagation, Viscous effects, Anisotropy.

1. INTRODUCTION

This paper focuses on sands and sand-clay mixtures but most of the concerns can be extended to other geomaterials. The development concerns the analysis of small strain behaviour (for strain up to some 10^{-5} m/m) and viscous effects. Both experimental observation and modelling are treated. Due to the lack of space, only a rapid presentation of the results obtained these last ten years at the laboratory DGCB of ENTPE is given. More details can be obtained in the following references (Cazacliu 1996, Sauzéat 2003, Pham Van Bang 2004, Duttine 2005, Di Benedetto et al. 1999, 2001, 2005).

The time effects on the skeleton (which exclude pore water and inertia effects) concerns "**ageing effects**" and "**viscous or loading rate effects**". It is necessary to properly define these two different aspects of the time effects, which are linked to different basic mechanisms behind, such as "changes with time in the intrinsic material properties due to changes in interface and/or internal particle properties caused by a physico-chemical process" (ageing effects), "viscous sliding at inter-particle contact points" (viscous effects), "time dependent property of the particles", among others. By definition time effects exist if a change in the stress-strain curve(s) is observed when changing the rate of loading

Non-ageing geomaterials exhibit the property that *gives the same response for two identical loading histories started at different time values (t1 and t2) on the material*

Hoe I. Ling et al. (eds.), Soil Stress-Strain Behavior: Measurement, Modeling and Analysis, 159–190.

previously at rest (stress and strain rate that are equal to zero). A definition taking into account previous loadings, which is more convenient for experimental use, is given in Di Benedetto et al. (2005a).

This definition has a very important consequence on the relevant selection of time variable for constitutive modelling. For non-ageing materials (which can be non-viscous or viscous), the time origin can be chosen arbitrarily and is not linked to the material evolution (ageing). In any case, the time "t" should not appear in the constitutive equation.

For ageing materials, a specific material time "t_c", which characterizes the ageing of the material, has to be added to the common time "t".

For the domain of loading considered in this paper, sands and sand-clay mixtures exhibit a non ageing behaviour (ie: experimental observations show that the previous non ageing definition is respected). Then only viscous effects are considered in the following.

Investigation of small strain behaviour including viscous and non viscous properties of geomaterials such as sands or mixtures of sand/clay requires some special laboratory equipments with a high degree of accuracy as revealed by previous studies (Tatsuoka et al. 1994, Jardine et al. 1999, Kim and Stokoe 1994, Chaudhary et al. 2004, among others). In fact, amplitudes of viscous phenomena, which remain low, can be easily hidden by inaccurate measurements. However, these phenomena may not be ignored and may exhibit non negligible effects at an engineering scale (Tatsuoka et al. 1997, Jardine et al. 2005, Di Benedetto et al. 2005a).

The two devices designed and used at ENTPE to measure the properties of soils at small strain are rapidly presented in section 2. One called "Triaxial StaDy" allows performing classical triaxial compression and extension (TC and TE) tests. Hollow cylinder samples are tested with the second one "T4C StaDy". Both are equipped with piezoelectric sensors and comparisons are possible between quasi-static and dynamic measurements.

The observed stress-strain and time evolution behaviour is given in section 3. A very peculiar behaviour of sands and mixtures called "Viscous evanescent" or "Temporary effect of strain rate acceleration" (TESRA) could be shown for large straining. But, as indicated before the concern of this paper is focused on small strain behaviour, where viscous and non viscous effects can be isolated and identified.

The Framework of the three component model, which is used as basis for all our simulations, is presented. The proposed non viscous part of this model, which is hypoelastic and anisotropic is introduced in section 5. Comparison between modelling, quasi-static measurement and dynamic wave propagation back analysis are presented. At least modelling and simulation of the viscous behaviour are proposed in section 6.

2. TESTING APPARATUSES AND PROCEDURE

Testing apparatuses

A hollow cylinder experimental prototype has been designed and developed at ENTPE since 1991. It is a precision device of Torsion, Compression, Confinement on Hollow Cylindrical sample for Static and Dynamic loading ("T4C StaDy"). A precision triaxial prototype ("Triaxial StaDy") has also been developed following the same requirements in terms of local strain measurements and dynamic loadings.

The "T4C StaDy" (figure 1) sample has a 12 cm height, an outer diameter of 20 cm and an inner diameter of 16 cm. Two Neoprene membranes (0.5 mm thickness) constitute the lateral side while two rigid platens close the sample at the top and at the bottom. The top cap, connected to the press piston, is mobile in rotation and translation. The "Triaxial StaDy" (figure 2) specimen has a 14 cm height and a 7 cm diameter. Axial loading is applied by an electromechanical testing machine. Investigation of soil response from very small to large strain domains is possible with these devices thanks to local strain measurement systems designed in the same way. Vertical (and angular) displacements are measured on two levels thanks to two light rings (made of duralumin) hanged by 3 points on the outer membrane and carrying targets (made of aluminium) aimed by non contact transducers. Radial (outer and/or inner) displacements are also measured by non contact transducers pointing towards sheets of aluminium paper placed on the inner side of the membrane(s). All the transducers are fixed on (electronically or manually) movable supports. The displacements of the transducers, monitored from outside the cell, during a test allow covering a large range of deformation, while keeping a good accuracy. The number of non contact transducers used for each device is respectively 14 and 6.

Moreover, both devices are equipped with piezoelectric sensors (compression elements and bender elements) located in each platen (figure 3). Concerning the "T4C StaDy" apparatus (figure 3a), two pairs of bender elements give characteristics of waves propagating in the axial direction. The first pair concerns waves polarized in radial (S_r) direction and the other in orthoradial (S_θ) direction. The two pairs of compression elements are identical. They are noted P_r and P_θ and are close to the respective sensors S. As regards the "Triaxial StaDy" is concerned (figure 3b), one single pair of compression element and one single pair of bender elements are used in the platens. Recently (Ezaoui et al. 2006), three couples of piezoelectric transducers developed at ISMES (Fioravante & Capoferri 2001) were added to measure wave propagation in horizontal direction (figure 3c). These transducers of bender type are all identical and placed in different ways along the membrane in order to generate shear or compressive radial waves. These arrangements are described in figure 3c.

Back analysis of the waves travel times provide "dynamic" elastic parameters.

For more details, the "T4C StaDy" device has been more extensively presented for example in Cazacliu (1996), Di Benedetto et al. (2001), Sauzeat (2003), Duttine (2005), the "triaxial StaDy" is described in Pham van Bang (2004), Ezaoui et al. (2006).

Considered materials and test conditions

Tested materials include air-dried poor graded sands (Hostun and Toyoura sands) and two moist mixtures of Hostun sand and Kaolin clay (M15 and M30). Hostun and Toyoura sands are quartz dominated angular shaped sands. The M15 sand/clay mixture is composed by 15% of Kaolin clay (w_l=35%, PI=14%) and 85% of Hostun sand (by dry weight) and by an initial global water content of 4.5%. The M30 sand/clay mixture is composed by 30% of Kaolin clay (w_l=35%, PI=14%) and 70% of Hostun sand (by dry weight) and by an initial global water content of 9.0%. More details on the materials characteristics an on the samples preparation are given in Sauzéat (2003), Duttine (2005), Pham van Bang (2004).

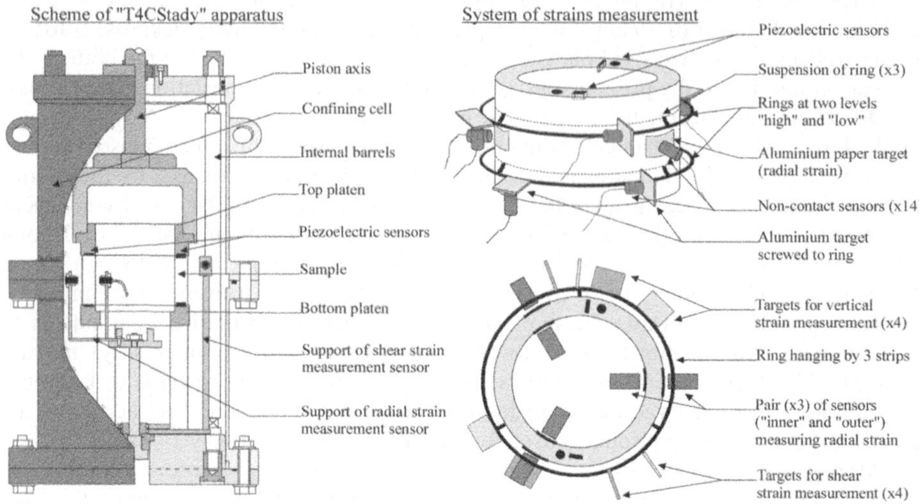

Figure 1. Schematic view of the “T4C StaDy” hollow cylinder apparatus and of its system of strains measurement.

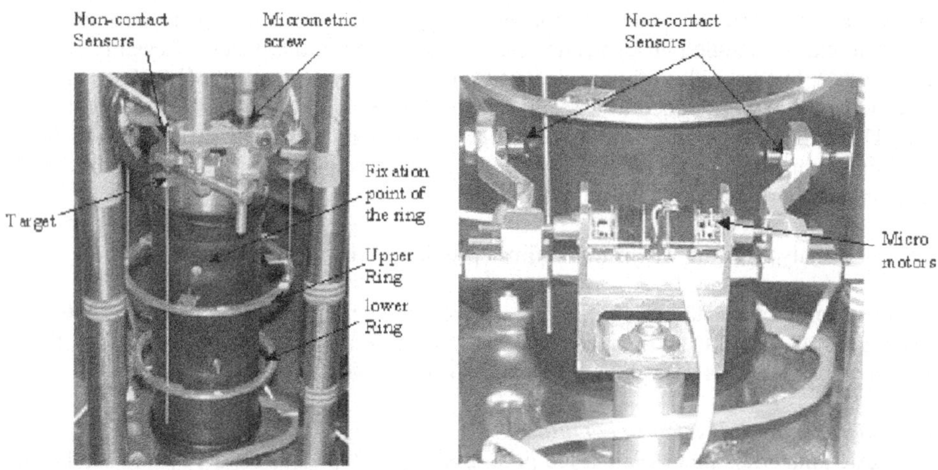

Figure 2. “Triaxial StaDy” apparatus: sample and axial strains measurement (a), radial strain measurement using micro motors (b).

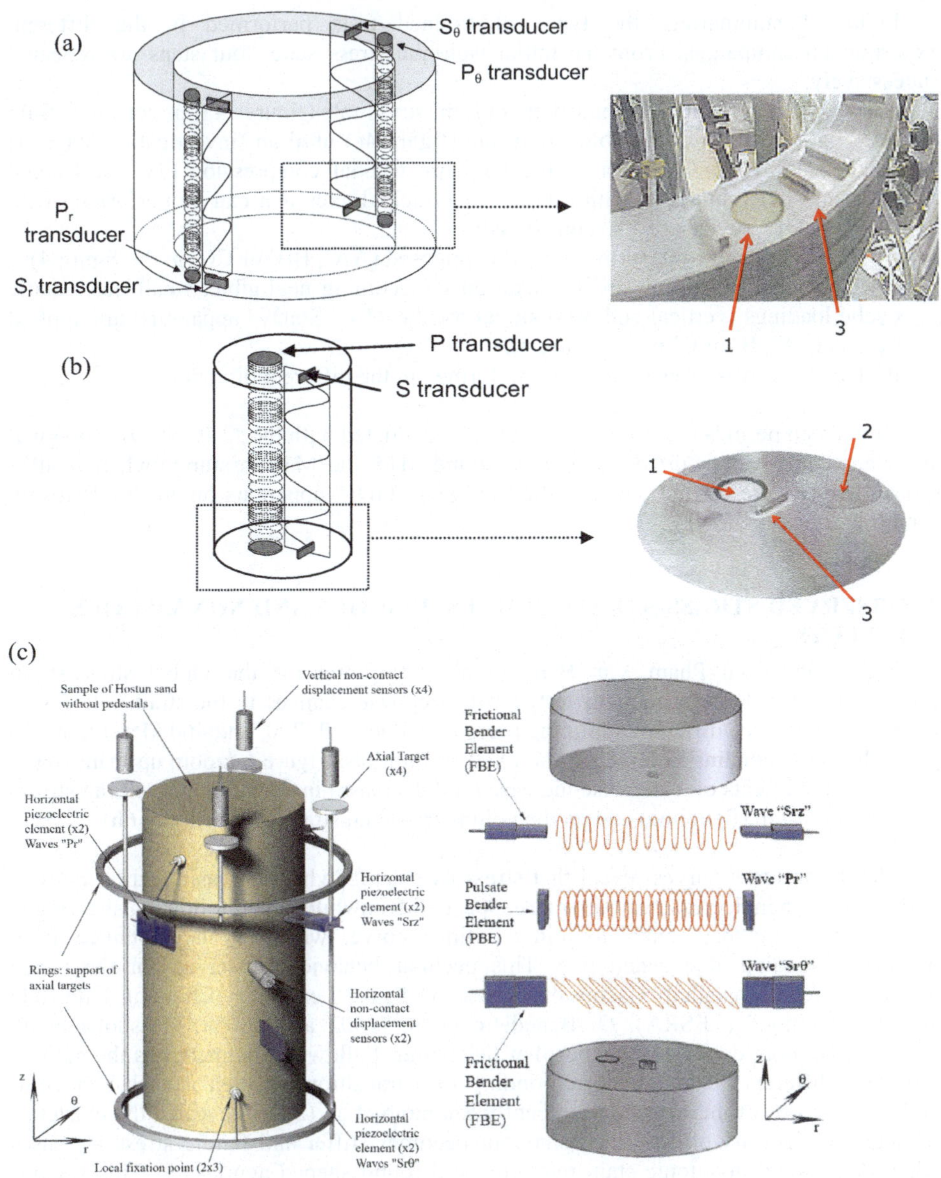

Figure 3. Location of piezoelectric transducers in the hollow cylinder device "T4C StaDy" (a), and platen of the triaxial device "Triaxial StaDy" (1. compression element, 2. porous stone, 3. bender element) (b), and along the lateral surface of the sample (3 couples of bender) (c).

Figure 4 summarizes the types of drained tests performed in the different experimental campaigns. From the initial isotropic stress state, four steps are repeated successively:

i) The sample is vertically loaded at constant stress rate (figures 4a) or constant strain rate (figures 4b) or at variable strain rate (figure 4c) until an 'investigation stage' is reached (A, B or C). Most of the tests are triaxial compression (TC) or triaxial extension (TE), but tests with rotation of principal axes that can be performed with "T4C StaDy" device are also considered;

ii) Then, a creep (or relaxation) period is imposed (AA', BB' or CC' in the figure 4);

iii) Once creep strain or stress relaxation is becoming negligible, small quasi static cyclic loadings (vertical and/or torsional for the "T4C StaDy" apparatus) are applied (at points A', B' or C' in the figure 4);

iv) P and S waves propagations are performed in the different directions.

Only creep periods (i.e. figure 4a) could be conducted with the "T4C StaDy" torsional shear apparatus (on air-dried Toyoura sand and M15 and M30 mixtures) whereas all 3 types of tests were carried out with the "Triaxial StaDy" apparatus on air-dried Hostun sand (i.e. figure 4a,b&c).

3. OBSERVED STRESS-STRAIN CURVES: VISCOUS AND NON VISCOUS EFFECTS

Figure 5a (from Pham Van Bang et al. 2006) presents the global stress-strain relationships for dense specimens tested with stepwise changes in the strain rate ("sra" type of test) at three different confining pressure values: 80, 200, and 400 kPa. Figure 5b gives the corresponding volumetric and axial strain plots. Figure 6 (zoom up from figure 5) reveals the effects of a stepwise increase (and decrease) in the strain rate on the stress-strain relationship (figure 6a) and on the volumetric-axial strain curve (figure 6b).

The stress-strain curves reveal that stress overshoot (when increasing strain rate) or undershoot (when decreasing strain rate) appears. Meanwhile, when following shearing at constant rate, the data tends to join a unique curve, which is also obtained from monotonic loading at constant rate. This peculiar behaviour observed for the tested materials have been called "viscous evanescent"(VE) or "Temporary Effect of Stain Rate and Acceleration" (TESRA). Di Benedetto et al. (2002 and 2005a), Tatsuoka et al. (2002), give more details on this peculiar behaviour. Following the materials the authors introduce different behaviour corresponding to a transition between isotach behaviour and pure viscous evanescent. More recently Enomoto et al. (2006) revealed the existence of negative viscosity effects (the behaviour becomes stiffer and failure stress increases when decreasing monotonic stain rate) for hard round shaped aggregates. This viscous aspect of the behaviour is not detailed in this presentation focused on the small strain domain because it concerns average and large strain domains. It is more extensively treated by Tatsuoka (2006). Some synthetic information for modelling these different types of behaviour are also given in section 6.

When the strain rate is increased at point A (resp. decreased at point B) by a factor 100, stress overshoot (resp. undershoot) is noted in figure 6a. The specimen behaviour seems to be more dilative (resp. compressive) as is illustrated in figure 6b. This more dilative (resp. more contractive) behaviour is observed once the elastic [or quasi-elastic] contractive (resp. dilative) response due to the fast stress overshoot (resp. undershoot) occurred. These phenomena are specifically studied and modelled in the following.

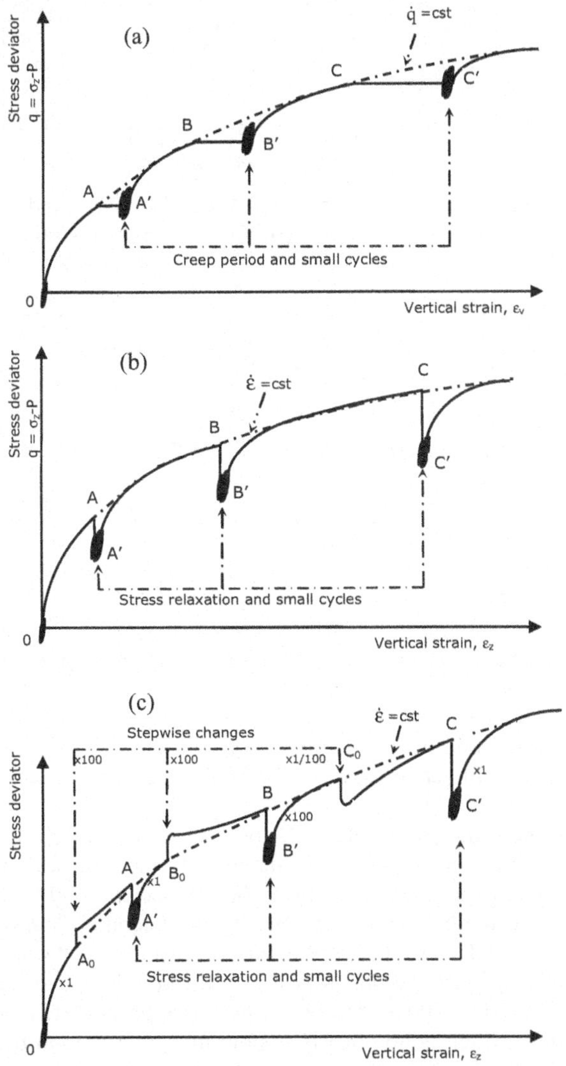

Figure 4. Schematic global loading curves of drained TC or TE tests performed in this study: a. Test with creep periods, b. test with stress relaxation periods, c. tests with stepwise changes in the strain rate.

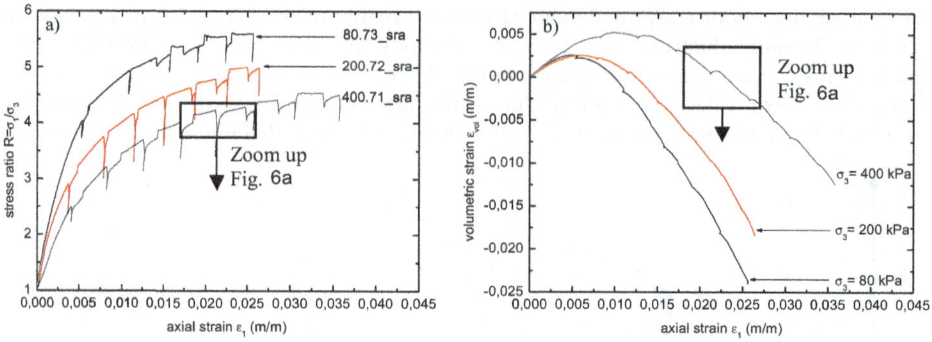

Figure 5. Different tests with stepwise changes in train rate ("sra" type) for dense sand (three values of confining pressure): (a) stress-strain curves; (b) volumetric-axial strain curves.

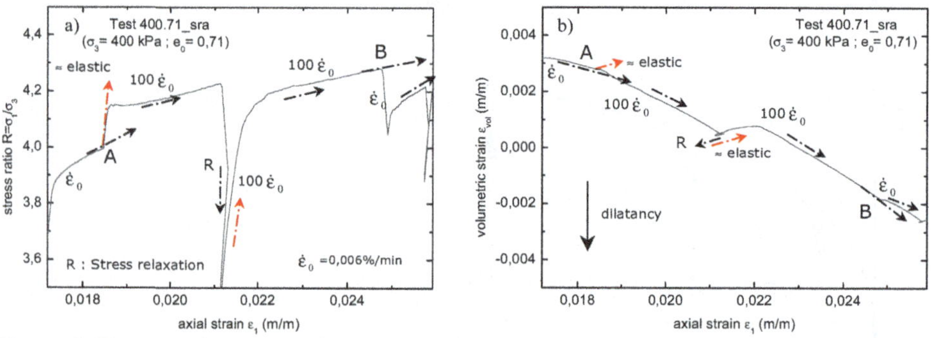

Figure 6. Focus on the test results presented in figure 5: (a) stress-strain relationship ; (b) volumetric-axial strain curve.

Example of creep periods during drained triaxial compression (TC) tests performed on dry Hostun sand with the hollow cylinder apparatus (T4CstaDy) are presented in figure 7, from Di Benedetto (1997).

During loading after a creep at the same strain rate as before the start of creep, the stiffness is very high (E_0 in figure 7). If the creep period is long enough (the time scale is about one hour for dry sand), it was experimentally shown (Di Benedetto 1997, Di Benedetto et al. 1999a and 1999b) that the behaviour for small cycles becomes "quasi" elastic. Then the behaviour can be modelled by a hypoelastic law inside this "quasi" elastic domain. The specific hypoelastic formulation proposed at ENTPE is called the DBGS model (Di Benedetto et al., 2001a and 2001b; Sauzéat, 2003). It appears that creep (as well as the "small" cyclic loading, which is not clearly visible in figure 7) modifies only very locally the response, while general monotonic loading curve is closely rejoined after reloading. The previous tendencies are confirmed by the results on air-dried Hostun and Toyoura sand and sand-clay mixtures.

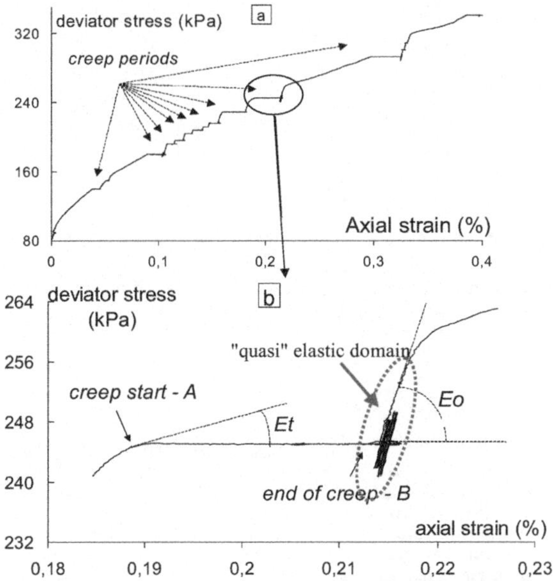

Figure 7 Drained hollow cylinder (T4C Stady) triaxial test on dry medium density RF Hostun sand with creep periods (σ_3 =80 kPa) from Di Benedetto (1997).

From experimental observation on (dry) sands and sand-clay mixtures it can be concluded that:

i) These materials show viscous behaviour,

ii) The loading stress-strain curve is not a stable state for sand, in the sense that the stress and strain values on that curve cannot be both kept constant: i.e., the stress-strain values do not stay on that curve if loading is stopped or if loading rate is stepwise changed. One could imagine that a monotonic loading at an infinitively slow strain rate could give a stable state. This question remains open, since even at the slowest strain rates allowed by the actual devices, creep (or relaxation) has always been observed.

iii) After a rather "long" period of creep, which depends on the type of materials, an elastic domain seems to exist around the actual stress-strain state. In fact, the size of this domain, which remains very small, depends on the previous history and is rapidly covered and forgotten after the restart of monotonic loading. This conclusion is confirmed by the results presented in figure 8 from Di Benedetto et al. (2005a) and Duttine (2005).

In figure 8, a TC test on dry Toyoura sand with creep periods of different times is considered. The general axial strain versus axial stress curve is presented in part a). The secant moduli during the reloading steps (just after the creep periods) are plotted, in function of the logarithm of the axial strain, in parts b) and c). As can be seen the linear domain is larger after the creep periods of 45 minutes than the one after the creep periods of 5 or 10 minutes. This result seems also verified for the creep periods applied during the cyclic loading of large amplitude, but this large cycle and the distance from the stress reversal also seem to affect the size of the linear domain.

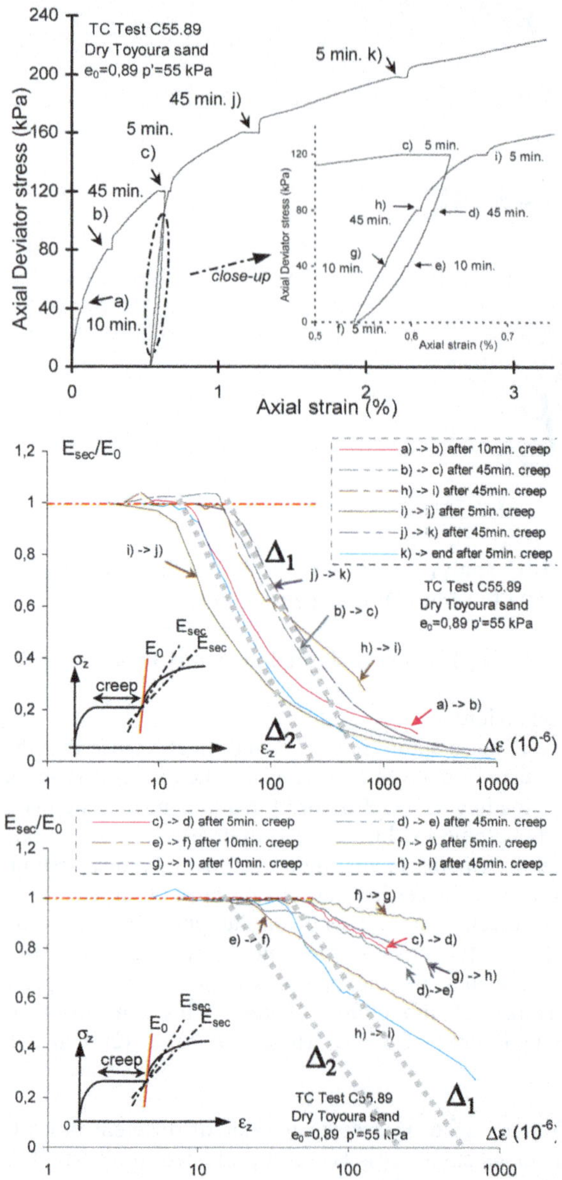

Figure 8. Triaxial compression (TC) test on dry Toyoura sand with creep periods of different durations. The influence of the creep period duration on the size of the « quasi » elastic domain is visible in the middle and lower plots (from Di Benedetto et al. 2005a).

Considering the topic of this paper, which is focused on small strain behaviour including both viscous and non viscous responses, only some parts of the global stress-strain curve are analysed in the next sections. The viscous properties are studied considering the creep and relaxation periods (figures 4a & b) and stepwise changes in the strain rate (figure 4c). The non viscous behaviour concerns the small cycles applied in the quasi static domain (step iii) and inverse analysis of waves propagations (step iv).

4. FRAMEWORK FOR MODELLING: THE 3 COMPONENT MODEL

Based on the experimental observations, it is assumed that the strain is the sum of a non-viscous (or instantaneous) part $\dot{\varepsilon}^{nv}$ and a viscous (or deferred) part $\dot{\varepsilon}^{v}$ (Di Benedetto 1987; Di Benedetto and Tatsuoka 1997, Di Benedetto et al. 2001b and 2005a):

$$\dot{\varepsilon} = \dot{\varepsilon}^{nv} + \dot{\varepsilon}^{v} \qquad (4.1)$$

In addition, the framework of the general three-component model (figure 9) (Di Benedetto, 1987; Di Benedetto and Tatsuoka, 1997; Di Benedetto *et al.*, 1999a, 1999b) is assumed. The decomposition is given in equation 4.1.

The bodies EP1 and EP2 of figure 9 have a non-viscous (sometimes called elastoplastic) behaviour. A great number of constitutive laws have been proposed to describe a non viscous (or elastoplastic) behaviour, such as: elasticity, plasticity, elastoplasticity, hypoplasticity, interpolation type, among many others (cf. figure 9). It can be shown (Darve, 1978) that the general form of the strain increment given by the body of EP type, is:

$$d\varepsilon = M\{h, dir(d\sigma)\}\ \mathrm{d}\sigma \qquad (4.2)$$

Or, if expressed with objective rates, as $\dot{\varepsilon} = \mathrm{d}\varepsilon / \mathrm{dt}$ and $\dot{\sigma} = \mathrm{d}\sigma / \mathrm{dt}$:

$$\dot{\varepsilon} = M(h, dir\dot{\sigma})\ \dot{\sigma} \qquad (4.3)$$

Where these quantities are tensors and $dir(d\sigma) = d\sigma / \|d\sigma\|$ $(= dir\dot{\sigma} = \dot{\sigma} / \|\dot{\sigma}\|)$ is the direction of the stress increment (or objective stress rate) whose norm is 1 ($\|dir(d\sigma)\| = \|dir\dot{\sigma}\| = 1$).The parameter h represents the whole history parameters, also called memory, hardening, state… parameters. M is the constitutive (or compliance) tensor, which depends on h and dir(dσ).

The introduction of $dir\,\dot{\sigma}$ expresses the irreversibility and the parameters h, which may be scalars, vectors or tensors, describes the stress history dependence.

Body V of figure 9 creates the viscous property dependency of the material. It represents a specific time-dependent behaviour, which is expressed by the following equation 4.4:

$$\sigma\ = F(h\ , \dot{\varepsilon}\ ) \qquad (4.4)$$

Where F is the viscous tensor, which depends on h and $\dot{\varepsilon}$ (objective strain rate) applied to the V body.

During small quasi static cycles (cf figure 4) or wave propagation tests, experiments reveal that no viscous deformation appears ($\dot{\varepsilon}^{v} = 0$). Then only body EP1 is activated for the considered granular materials (the body V locks the viscous deformation in the considered domain of loading). As the experimental behaviour is (quasi) elastic (as shown before) a hypoelastic formulation, the DBGS (Di Benedetto, Geoffroy, Sauzéat) model has been developed. This model and its extension (DBGSP), which includes strain anisotropic effects, are presented in the next section.

During creep or relaxation periods or stepwise changes, bodies EP2 and V play a key role and have to be considered. As only small strain domain is treated in this paper, only the asymptotic behaviour of these bodies is introduced. This consideration allows simplifying the equations as presented in section 6.

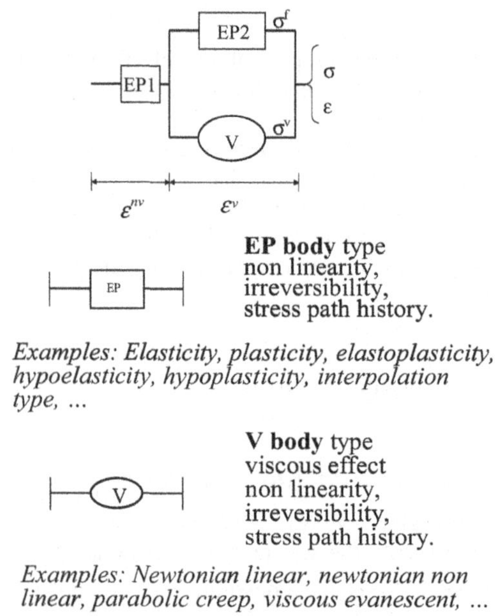

Figure 9. Framework of the non-linear three-component model (Di Benedettto 1987, Di Benedettto and Tatsuoka 1997, Di Benedetto et al. 2002, Tatsuoka et al. 2002) (schematic). For the considered materials EP1 is the DBGS hypoelastic model (cf. section 5).

5. MODELLING AND SIMULATIONS OF NON VISCOUS BEHAVIOUR (EP1 BODY IN FIGURE 9)

The DBGS (Di Benedetto, Geoffroy, Sauzéat) and the DBGSP(Di Benedetto, Geoffroy, Sauzéat, Pham) models

Many experimental results Hardin et al. (1989), Tatsuoka et al. (1999), Jardine et al. (1999), Hoque et al. (1996), Hameury (1995), among others, confirm that a linear domain can be found for very different geomaterials when loading in one direction. This result was also verified in the different experimental campaigns performed in the ENTPE laboratory on sands and sand-clay mixtures. The linearity domains could be estimated below strain amplitudes of some 10^{-5}. In addition Cazacliu (1996), Di Benedetto (1997), Di Benedetto et al. (2001), Sauzeat (2003), showed experimentally for air-dried sands that the compliance tensor M (equation 4.2 or 4.3) can be assumed as independent of stress increment direction in the small strain domain. The result was extended to sand clay mixture by Duttine (2005). This important result could be obtained when comparing the responses of cycles performed in different "directions" (only $\sigma_{\theta z}$ varies or only σ_{zz} varies or both $\sigma_{\theta z}$ and σ_{zz} vary).

Therefore the hypoelastic formulation (relation 5.1) can be assumed for body EP1 (figure 9).

$$\delta\varepsilon^{nv} = M(h).\delta\sigma \qquad (5.1)$$

Moreover, additional experimental results obtained at ENTPE by the previous authors reveals that air-dried sands exhibit a symmetrical compliance tensor M. This interesting result could be obtained when comparing the terms $M_{\gamma z}$ obtained when performing small axial cycles (only σ_{zz} varies) and the term $M_{z\gamma}$ resulting from pure torsion small cycles (only $\sigma_{\theta z}$ varies). These terms appear as very close, as shown in figure 10 for Toyoura sand, and could be considered as equal for the different tested geomaterials.

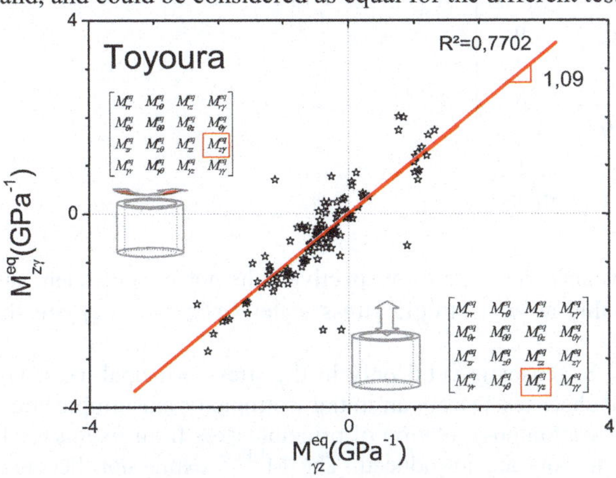

Figure 10. Experimental values for terms $M_{\gamma z}$ obtained from small axial cycles (only σ_{zz} varies) and term $M_{z\gamma}$ resulting from pure torsion small cycles (only $\sigma_{\theta z}$ varies) on Toyoura sand (from Duttine 2005)

As a consequence, Di Benedetto et al. (2001) suggest the following anisotropic (ortho-tropic) and symmetrical expression of tensor M :

$$M^{DBGS} = \frac{1}{f(e)} \frac{S_v.\Sigma_p + {}^t\Sigma_p.{}^tS_v}{2} \quad (5.2)$$

with f(e) the function of void ratio (e) approximating void ratio effects. " t " denotes the function transpose. $\{S_v ; \Sigma_p\}$ are the tensors originally defined by Hardin & Blandford (1989):

$$S_v = \begin{bmatrix} 1 & -v_0 & -v_0 & 0 & 0 & 0 \\ -v_0 & 1 & -v_0 & 0 & 0 & 0 \\ -v_0 & -v_0 & 1 & 0 & 0 & 0 \\ 0 & 0 & 0 & 1+v_0 & 0 & 0 \\ 0 & 0 & 0 & 0 & 1+v_0 & 0 \\ 0 & 0 & 0 & 0 & 0 & 1+v_0 \end{bmatrix} \quad (5.3)$$

$$\Sigma_p = \begin{bmatrix} \frac{1}{\sigma_1^m} & 0 & 0 & 0 & 0 & 0 \\ 0 & \frac{1}{\sigma_2^m} & 0 & 0 & 0 & 0 \\ 0 & 0 & \frac{1}{\sigma_3^m} & 0 & 0 & 0 \\ 0 & 0 & 0 & \frac{1}{\sigma_2^{m/2}\sigma_3^{m/2}} & 0 & 0 \\ 0 & 0 & 0 & 0 & \frac{1}{\sigma_1^{m/2}\sigma_3^{m/2}} & 0 \\ 0 & 0 & 0 & 0 & 0 & \frac{1}{\sigma_1^{m/2}\sigma_2^{m/2}} \end{bmatrix} \quad (5.4)$$

Where $\{m ; v_0\}$ are two constants and stand for respectively the power coefficient and the isotropic Poisson's ratio value at an isotropic stress state. $\{\sigma_1 ; \sigma_2 ; \sigma_3\}$ are the principal stress values.

Note that expression 5.4 of Σ_p tensor is valid only in the stress principal axes. For instance, as far as pure torsional shear tests (from an initial isotropic or anisotropic stress state) are concerned, (sudden or continuous) rotation of principle axes from fixed sample axes are involved and rotation tensors are introduced. The M^{DBGS} expression becomes therefore more complicated (Cazacliu 2003, Duttine, 2005).

When considering for example the triaxial test conditions Expression 5.2 can be rewritten in the cylindrical coordinates (r, θ, z) by relation 5.5.

The considered vertical, horizontal Young's moduli, and shear modulus in the (θ ; z) plane are thus given by relations 5.6 to 5.8. The concerned Poisson's ratios $\{v_{rz};v_{\theta z};v_{r\theta}\}$ are expressed by relations 5.9 and 5.10.

$$M^{DBGS}=\frac{1}{f(e)}.\begin{bmatrix} \frac{1}{\sigma_r^m} & \frac{-\nu_0}{2}\left(\frac{1}{\sigma_\theta^m}+\frac{1}{\sigma_r^m}\right) & \frac{-\nu_0}{2}\left(\frac{1}{\sigma_z^m}+\frac{1}{\sigma_r^m}\right) & 0 \\ \frac{-\nu_0}{2}\left(\frac{1}{\sigma_r^m}+\frac{1}{\sigma_\theta^m}\right) & \frac{1}{\sigma_\theta^m} & \frac{-\nu_0}{2}\left(\frac{1}{\sigma_z^m}+\frac{1}{\sigma_\theta^m}\right) & 0 \\ \frac{-\nu_0}{2}\left(\frac{1}{\sigma_r^m}+\frac{1}{\sigma_z^m}\right) & \frac{-\nu_0}{2}\left(\frac{1}{\sigma_\theta^m}+\frac{1}{\sigma_z^m}\right) & \frac{1}{\sigma_z^m} & 0 \\ 0 & 0 & 0 & \frac{1+\nu_0}{(\sigma_\theta.\sigma_z)^{m/2}} \end{bmatrix} \quad (5.5)$$

Keeping in mind that $\sigma_r=\sigma_\theta=P$, thus:

$$E_z=\frac{\partial\sigma_z}{\partial\varepsilon_z}=f(e).\sigma_z^m \;\; ; \;\; E_r=E_\theta=f(e).\sigma_r^m \quad (5.6\&5.7)$$

$$G_{\theta z}=\frac{\partial\tau_{\theta z}}{\partial\gamma_{\theta z}}=\frac{f(e)}{2(1+\nu_0)}.\sigma_r^{m/2}.\sigma_z^{m/2} \quad (5.8)$$

$$\nu_{rz}=\frac{\partial\varepsilon_r}{\partial\varepsilon_z}=\nu_{\theta z}=\frac{\partial\varepsilon_\theta}{\partial\varepsilon_z}=\frac{\nu_0}{2}.\left(1+\frac{\sigma_z^m}{\sigma_r^m}\right) \quad (5.9)$$

$$\nu_{r\theta}=\nu_{\theta r}=\nu_0 \quad (5.10)$$

Equation 5.5 for general triaxial condition is also valid when the tensor M^{DBGS} is expressed in the principal axes of stress, which are the orthotropic axes of the model (in that case, σ_r, σ_θ, σ_z, should be replaced by the principal stress σ_1, σ_2, σ_3).

The DBGS model does not introduce eventual anisotropy due to strain (Ibraim 1998, Pham Van Bang 2004) and considers that at any isotropic stress state the behaviour is isotropic. This drawback as been recently tackled and a modified version of the model called "DBGSP" (Di Benedetto, Geoffroy, Sauzéat, Pham) has been proposed (Pham Van Bang (2004) and Ezaoui et al. (2006)). This new version introduces fabric and strain induced anisotropy.

The modification of the *DBGS* model into *DBGSP* model consists in adding a new history parameter, the total irreversible deviatoric strain ε^{Tot}. The introduction of such a parameter is realised by the mean of the new tensor Γ, which is diagonal when expressed in the principal axes of total irreversible strain and depends only of ε^{Tot}. Relation (5.2) is then replaced by expression 5.11.

$$M^{DBGSP}=\frac{1}{f(e)}\frac{S_v.\Gamma.\Sigma_p+{}^t(S_v.\Gamma.\Sigma_p)}{2} \quad (5.11)$$

More details on the DBGSP model can be obtained in Pham Van Bang (2004) and Ezaoui et al. (2006).

Simulations are presented in the next paragraphs

Simulation with the DBGS and DBGSP models

5.2.1) Small cycles data

Figure 11 presents the experimental data and DBGS simulations for Young's and shear moduli and for Poisson's ratios at different stress states divided by their reference values at the initial K stress state. The "K type" tests are torsional shear tests from an anisotropic stress state defined by $K=\sigma'_r/\sigma'_z$ equal to 0.5 for these tests. After being isotropically consolidated up to 80kPa, the sample is axially (vertically) loaded to reach the initial given K stress-state. Then, the specimen is subjected to large cyclic torsional shearing given by $-0.6 \leq \tau_{\theta z}/p \leq 0.6$ (p is the constant mean stress). "Quasi-elastic" properties are obtained at different steps of this large shearing, by very small cyclic loading applied after several hours creep periods to remove viscous or time dependent properties as explained before.

The decreasing tendency of these rheological parameters while increasing the shearing is well reproduced by the model. More generally, very different simulations performed for different loading conditions on sands and sand-clay mixtures reveals the good ability of the model to translate the materials behaviour as far as, i) the initial conditioning state is close to isotropy and, ii) total strain amplitude is not too big (see for example Di Benedetto et al. 2001 & 2005, Sauzéat 2003, Geoffroy et al. 2003, Pham Van Bang 2004, Duttine 2005).

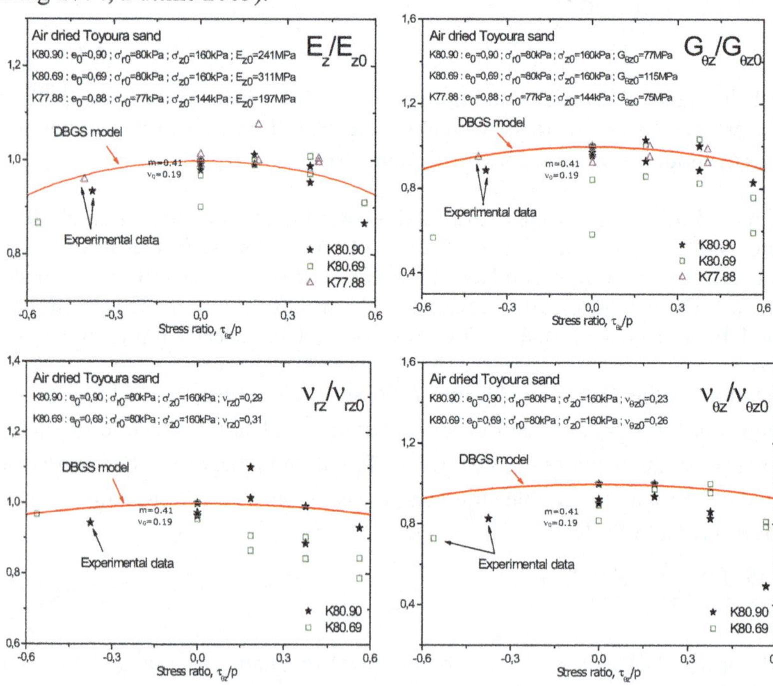

Figure 11 : DBGS simulations and experimental data of Young's and shear moduli and of Poisson's ratios for air-dried Toyoura sand, obtained for K type tests (indice 0 stands for values before shearing, changes of void ratio are neglected during tests, volume change reaching less than 1.5%). σ_r, σ_θ and σ_z are constant. (from Di Benedetto et al. 2005b).

5.2.2) Wave propagation data

The analysis of the data provided by the piezoelectric transducers placed in our 2 devices (cf figure 3) gives complementary information on the rheological hypoelastic tensor. Following the initial configuration of the sample, the inverse analysis is made considering either isotropic or transverse isotropic or even orthotropic hypothesis. Comparison between quasi static results (from small quasi static cycles) and dynamic analysis (from wave propagation test) shows that the parameters obtained by the 2 methods are very close (less than about 10% on the moduli) if the inverse analysis is made considering correct hypotheses. In particular, transverse isotropic (or orthotropic) behaviour should be considered in most of the laboratory or in situ cases. figures 12 to 14 from Duttine (2004) confirm the previous statement. Comparison of figures 13 and 14 shows the interest to consider transverse isotropic assumption.

The relations between the moduli, the Poisson's ratios and the waves rates are given in equations 5.12 and 5.13, in the case of transverse isotropic behaviour.

$$\rho.(V_z^P)^2 = \frac{E_z^2(\nu_{rr}-1)}{(\nu_{rr}-1)E_z + 2\nu_{rz}^2 E_r} \qquad \rho.(V_{zr}^S)^2 = G_{rz} \qquad (5.12)$$

$$\rho.(V_r^P)^2 = E_r \frac{\nu_{rz}^2 E_r - E_z}{(\nu_{rr}^2-1)E_z + 2\nu_{rz}^2(1+\nu_{rr})E_r} \qquad \rho.(V_{r\theta}^S)^2 = \frac{E_r}{2(1+\nu_{rr})} \qquad (5.13)$$

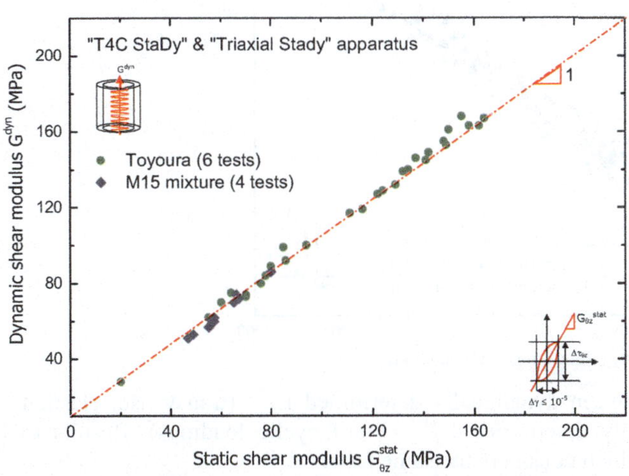

Figure 12. Comparison between statically determined shear modulus $G_{\theta z}^{stat}$ and dynamically determined shear modulus G^{dyn} (isotropic elastic assumption, which gives same results as transverse isotropic assumption) for TC tests performed on "T4C StaDy" apparatus.

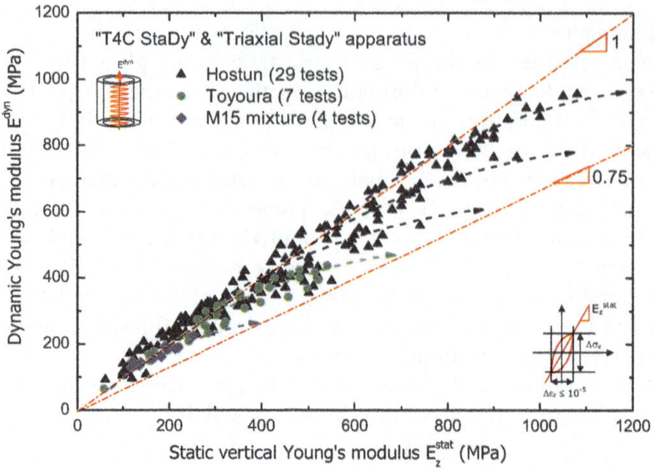

Figure 13. Comparison between statically determined vertical Young's modulus E_z^{stat} and dynamically determined Young's modulus E^{dyn} (isotropic elastic assumption) for TC tests.

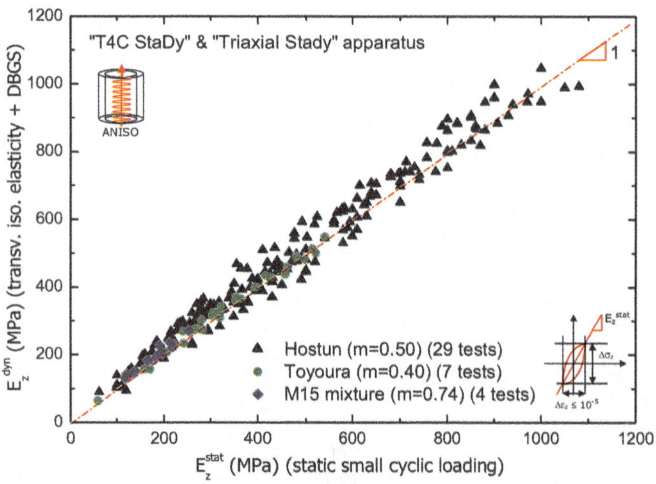

Figure 14. Comparison between dynamically determined E_z^{dyn} (transv. iso. elasticity + DBGS model) and statically determined E_z^{stat} (small cyclic loadings). Results to be compared with figure 13, which considers the same tests.

Recently Ezaoui et al. (2006) obtained at any investigation point of the precision triaxial device ("Triaxial StaDy") the 5 rheological parameters of the transverse isotropic sample. The analysis is made combining the information from small axial quasi static cycles and dynamic measurements (cf. figures 3b and 3c). Figure 15, which is only interpretation of experimental data without any postulated rheological elastic model, gives comparison between dynamic and static measurement. It confirms the previous statement.

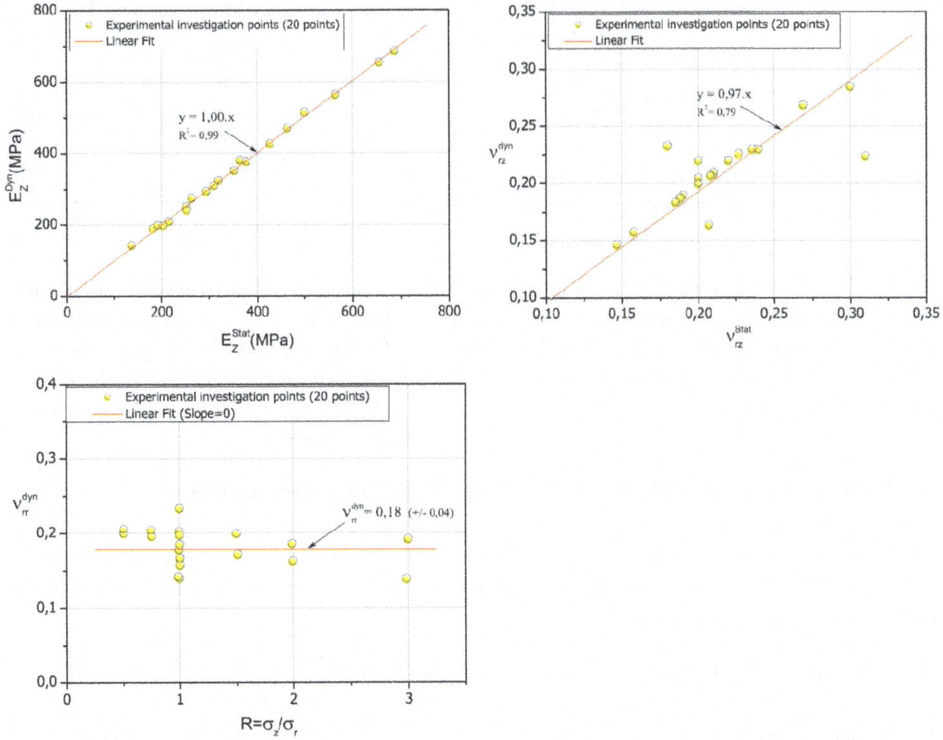

Figure 15. Dynamically (called "dynamic") and statically (called "static") determined elastic parameters and optimized values of ν_{rr} plotted versus stress ratio R, obtained from one triaxial extension (TE) and one triaxial compression (TC) tests on dry Hostun sand (from Ezaoui et al. 2006).

Figure 16 from Ezaoui et al. (2006) presents the 3 moduli (E_z, E_r and G) experimentally obtained during isotropic consolidation of dry Hostun sand samples prepared by pluviation in air. The difference of about 10% between the axial (E_z) and radial (E_r) moduli indicates that the sample is not isotropic. The modulus in the axial direction is lower than in radial direction. Simulations with the DBGS and the DBGSP model are also plotted in this figure. The DBGSP model, which introduces initial anisotropy, gives good fitting.

The inverse analysis can also be done for in situ tests. For example Di Benedetto et al. 2005b present inverse analysis of dynamic cross-hole tests in saturated soils and soft rocks using transverse isotropic hypothesis and the DBGS model.

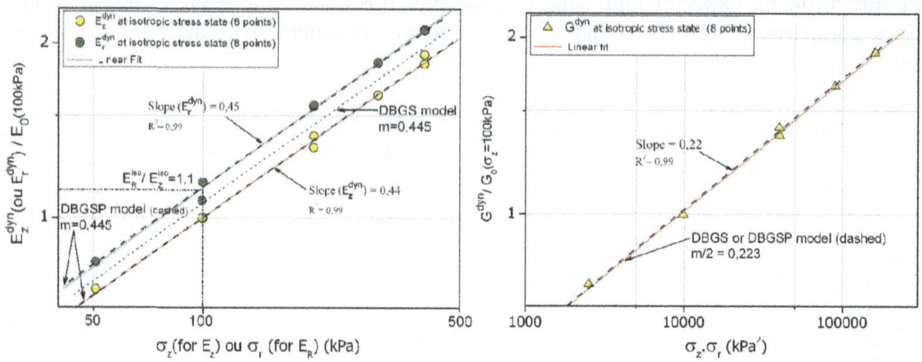

Figure 16. Dynamic Young Moduli (E_z and E_r) (left) normalized by vertical Young modulus at 100kPa and shear modulus (G) (right) normalized by the shear modulus at 100kPa. Experimental results obtained from investigation points during isotropic consolidation loading. Lines represent simulation with DBGS and DBGSP model (from Ezaoui et al. 2006).

6. MODELLING AND SIMULATIONS OF VISCOUS BEHAVIOUR

General consideration

The viscous strain increment $d\varepsilon^v$ is obtained from the EP^f and V bodies (EP2 and V bodies of figure 9) whose general tensorial rheological expressions are given respectively in equations 4.2 and 4.4. It comes:

$$d\varepsilon_{ij}^{v} = M_{ijkl}^{f}\left(\mathrm{dir}\left\{\dot{\sigma}^{f}\right\},h_3\right).d\sigma_{kl}^{f} \qquad \text{or} \qquad \dot{\varepsilon}^{v} = M^{f}\left(\mathrm{dir}\left\{\dot{\sigma}^{f}\right\},h_3\right).\dot{\sigma}^{f} \tag{6.1}$$

$$\sigma^{v} = F(h_2,\dot{\varepsilon}^{v}) \tag{6.2}$$

Where M^f is the inviscid tensor, h_3 is a set of history or memory parameters, $d\sigma^f$ and $\dot{\sigma}^f$ are an objective inviscid stress increment and an objective inviscid stress rate ($d\sigma^f = \dot{\sigma}^f.dt$). $\mathrm{dir}\left\{\dot{\sigma}^f\right\} = \dot{\sigma}^f / \left\|\dot{\sigma}^f\right\|$ is the direction of the inviscid stress rate (or stress increment).

F is the viscous stress tensor function, σ^v is the viscous stress, $\dot{\varepsilon}^v$ is the objective viscous strain rate, h_2 a set of history parameters which can differ from the set h_3.

To describe the peculiar viscous behaviour observed for sands, the following general expression of the viscous stress is considered in the viscous evanescent model (VE) model considering otherwise a 1D case (Di Benedetto et al. 1999b) :

$$\sigma_{(t)}^{v} = \int_{\chi=0}^{t}\left[d\left\{f(\dot{\varepsilon}_{(\chi)}^{v})\right\}\right].\, g_{decay}(\varepsilon_{(t)}^{v} - \varepsilon_{(\chi)}^{v}) \tag{6.3}$$

This can be rewritten:

$$\sigma^v_{(t)} = f(\dot{\varepsilon}^v_{(t)}) + \int_{\chi=0}^{t} f(\dot{\varepsilon}^v_{(\chi)})\left(g'_{decay}(\varepsilon^v_{(t)} - \varepsilon^v_{(\chi)})\right) d(\varepsilon^v_{(\chi)}) \quad (6.4)$$

where the viscosity function $f(\dot{\varepsilon}^v)$ represent the viscous stress jump just after imposing a constant strain rate $\dot{\varepsilon}^v$ if the material is previously at rest ($\dot{\varepsilon}^v = 0$). Then, $d\{f(\dot{\varepsilon}^v_{(\chi)})\}$ corresponds to the viscous stress increment at time χ (that can also be noted $[d\sigma^v](\chi)$) and, $d\{f(\dot{\varepsilon}^v_{(\chi)})\}\ g_{decay}(\varepsilon^v_{(t)} - \varepsilon^v_{(\chi)})$ corresponds to the value (increment of viscous stress) at current time *t* created by the event produced at time χ. The weighting function, g_{decay} (whose derivative is g'_{decay}) expresses the decrease with straining upon the influence of any strain rate changes. This function tends monotonously towards zero for a viscous evanescent (VE) behaviour and is equal to unity to describe classical isotach behaviour. Examples of classical isotach behaviour (Clays, bituminous materials (see for example Di benedetto & De La Roche 1998),..) intermediate behaviour (some gravels,..) and evanescent behaviour (clean sands,..) are given in Di Benedetto et al. (2002, 2005a). Recently Enomoto et al (2006) experimentally obtained that the stress, for the same level of strain during monotonic loading at a constant strain rate, decreases with an increase in the strain rate for poorly graded unbound round granular materials. This peculiar behaviour called "negative viscosity" has also been observed on fault gouge by Mair et al (1999). The figure 17 indicates the evolution of the g_{decay} function for viscous evanescent and isotach behaviour. A schematisation of the different types of behaviour observed on geomaterials is given in figure 18, which takes some elements of figure 2-22 of Di Benedetto et al. (2005a) and completes it. The analogical forms of the viscous V body (cf; figure 9), which can be used to simulate these different types of behaviour are also indicated in figure 18.

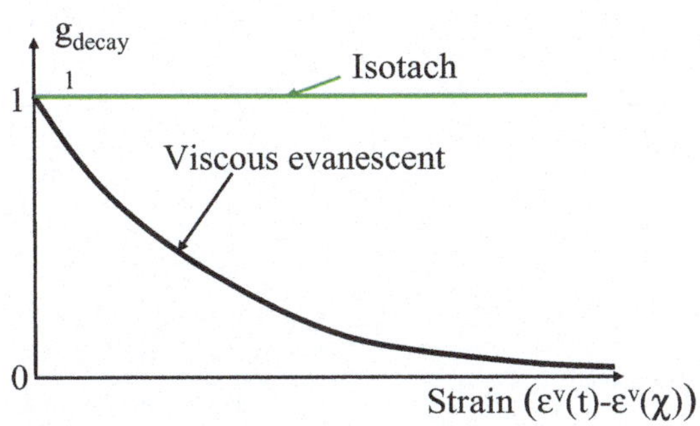

Figure 17. Shape of the g_{decay} function for isotach and viscous evanescent (VE) behaviour.

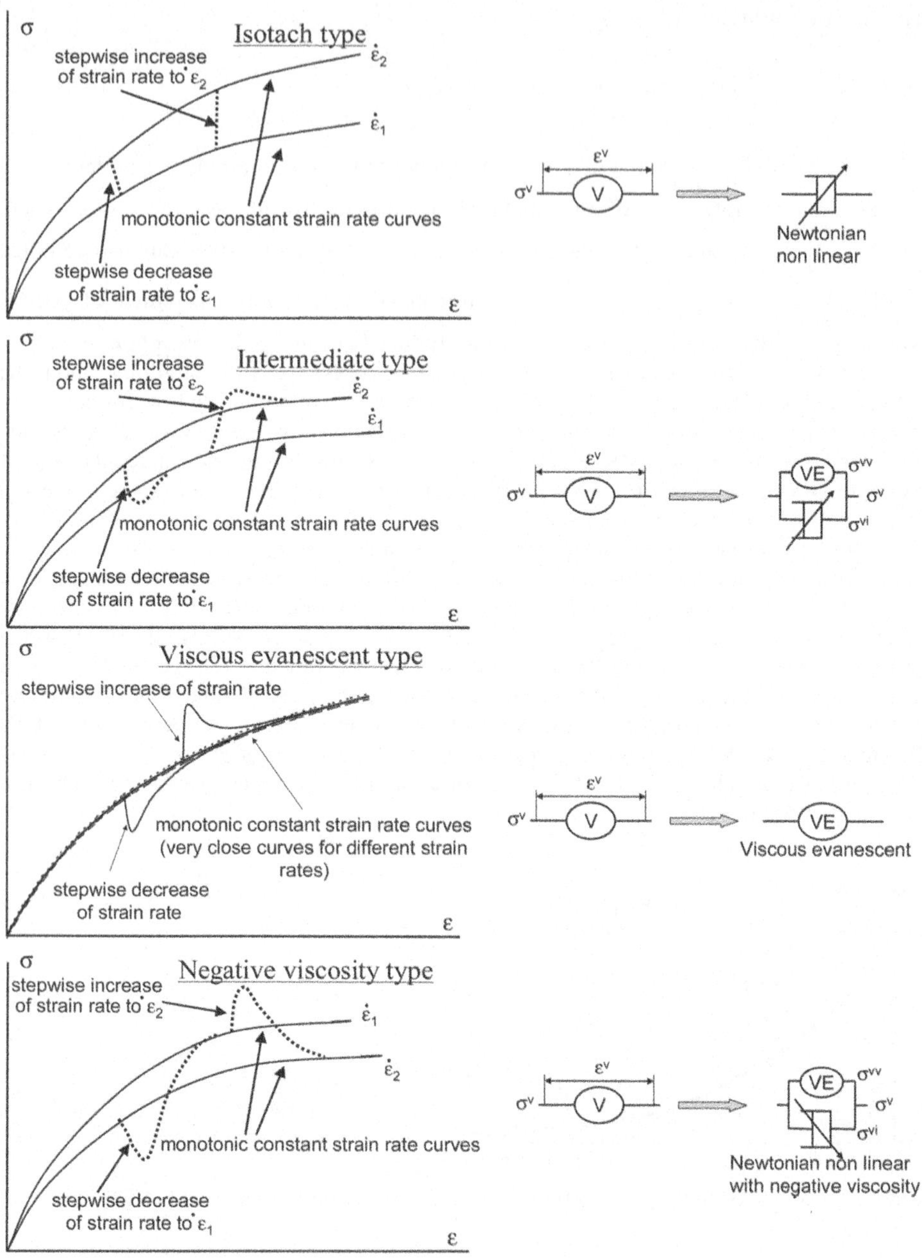

Figure 18. Types of stress-strain behaviour (schematized) observed on geomaterials (left) and analogical form of the viscous V body of the three component model (Cf. figure 9) (right).

Previous studies (Di Benedetto et al. 2001, Sauzeat et al. 2003, Pham Van Bang & Di Benedetto 2003, Sauzeat 2003, Pham Van Bang 2004, Duttine 2005) showed that the following expressions of the viscosity function f and of the decay function g_{decay} are relevant for tests performed on air-dried sands and the considered sand-clay mixtures, which showed a viscous evanescent type behaviour (Figure 18) :

$$\begin{cases} \sigma^{v}_{isotach} = f(\dot{\varepsilon}^{v}) = \eta_0 \cdot \left(\dfrac{\dot{\varepsilon}^{v}}{\dot{\varepsilon}^{v}_0} \right)^{1+b} \quad \text{for } \dot{\varepsilon}^{v} \geq 0 \\ g_{decay}\left(\varepsilon^{v}_{(t)} - \varepsilon^{v}_{(\chi)}\right) = \exp\left(- \dfrac{\varepsilon^{v}_{(t)} - \varepsilon^{v}_{(\chi)}}{\varepsilon_{ref}} \right) \end{cases} \tag{6.5}$$

Where f is the viscosity function and $\{\eta_0 \,;\, b \,;\, \varepsilon^{ref}\}$ are 3 model parameters to be determined. There value may vary following the previous loading path. $\dot{\varepsilon}^{v}_0$ is the reference strain rate ($=10^{-6}$/s). Note that equation 6.5 can be extended for negative value of $\dot{\varepsilon}^{v}$: $\sigma^{v} = sg(\dot{\varepsilon}^{v}).f(|\dot{\varepsilon}^{v}|)$ where $sg(\dot{\varepsilon}^{v})$ is the sign of $\dot{\varepsilon}^{v}$ and $|\dot{\varepsilon}^{v}|$ the norm of $\dot{\varepsilon}^{v}$. In addition, $\varepsilon^{v}_{(t)} - \varepsilon^{v}_{(\chi)}$ becomes the stain path length between time χ and t.
b must be chosen in the range from –1 (no viscous effects) to 0 (Newtonian viscosity).

A possible extension of the 1D VE model (equations 6.3 to 6.5) to a more general 3D model is presented in Sauzeat et al. (2003) and Di Benedetto et al. (2005a). This generalization is based on the co-linearity of the tensors $\underline{\underline{\sigma}}^{v}$ and $\underline{\underline{\dot{\sigma}}}^{f}$ and on similar equations as relations 6.3 to 6.5 but linking the norm of the different tensors ($\|\underline{\underline{\sigma}}^{v}\|$; $\|\underline{\underline{\dot{\varepsilon}}}^{v}\|$ and so on).

Simplification in the case of small and average strain

When considering a small range of strain evolution the g_{decay} function can be considered as a constant. This means that the decay effect can be neglected within a range of strain variation. The range of validity of this assumption can be evaluated to some 10^{-4} m/m (Di Benedetto 1999b, Sauzéat 2003). This assumption can be adopted to model the creep and relaxation periods as well as the stress jump when changing stepwise the strain rate (see figures 4 and 23), which correspond to the points on focus in this paper. Within these considerations, the three component model can be represented by its asymptotic form as represented in figure 19. The EP2 body can be assimilated to a hypoelastic body whose compliance tensor M^{f} is obtained by the tangent behaviour in the considered direction of loading (note that the direction of the stress σ^{f}, which depends on the previous loading history can be considered as constant). The V body reduces to a non linear Newtonian material having a viscosity η for triaxial compression (CT) and triaxial extension (TE) tests.

$$\eta = \frac{\sigma^v}{\dot{\varepsilon}^v} = \frac{f(\dot{\varepsilon}^v)}{\dot{\varepsilon}^v} = \eta_0 \cdot \left(\frac{\dot{\varepsilon}^v}{\left(\dot{\varepsilon}_0^v\right)^{(b+1)/b}} \right)^b \quad \text{for } \dot{\varepsilon}^v \geq 0 \tag{6.6}$$

Where ε is the axial stain, which is close to the norm of the strain. In the case of pure torsion (ie. only the shear stress is not constant) applying from any isotropic or anisotropic initial stress state, the 3D expression of the model gives:

$$\eta_{PT} = \frac{\tau^v}{\dot{\gamma}^v} = \frac{\|\sigma^v\|}{\sqrt{2}\|\dot{\varepsilon}^v\|} = \frac{f(\|\dot{\varepsilon}^v\|)}{\sqrt{2}\dot{\gamma}^v} = \eta_{0PT} \left(\frac{\dot{\gamma}^v}{\left(\dot{\varepsilon}_0^v\right)^{(b+1)/b}} \right)^b \quad \text{for } \dot{\gamma}^v \geq 0 \tag{6.7}$$

With $\eta_{0PT} = \frac{\eta_0}{2^{(1+b/2)}}$

Equations 6.6 and 6.7 deduced from the 3D expression of the model allow treating respectively the results obtained with the Triaxial StaDy and T4C StaDy results.

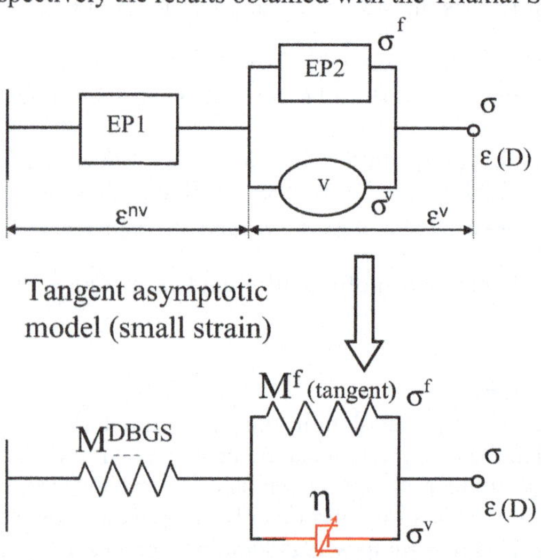

Figure 19. Tangent asymptotic expression of the 3 component model. Considered form when the strain evolution is limited (up to some 10^{-4} m/m).

Application to creep tests and calibration of the model

The procedure to obtain from experimental data the viscous stress (σ^v) for triaxial or pure torsion loading modes during creep periods, is explained in previous publications (Di Benedetto 1999b, Sauzéat 2003). Figure 20 gives evolution of the viscosities η and η_{PS} (equations 6.6 and 6.7) during pure torsion and triaxial creep periods (Duttine 2005) for the different considered materials. It could be verified from our experimental campaigns that the parameter b (equations 6.5 to 6.7) is a constant. This parameter is found to be equal to –0.95, respectively –0.90, for all the creep periods performed on air dried sands, respectively on sand/clay mixtures.

The last parameter η_0 (equations 6.5 to 6.7) has a different value for each creep period. It is determined by the best fitting between experimental and calculated viscous strains. Figure 21, from Pham et al. (2006), gives examples of simulation and experimental creep for test on Hostun sand. The obtained values of the viscosity parameter η_0 for all the creep periods performed are gathered and discussed in section 6 (cf. figure 24).

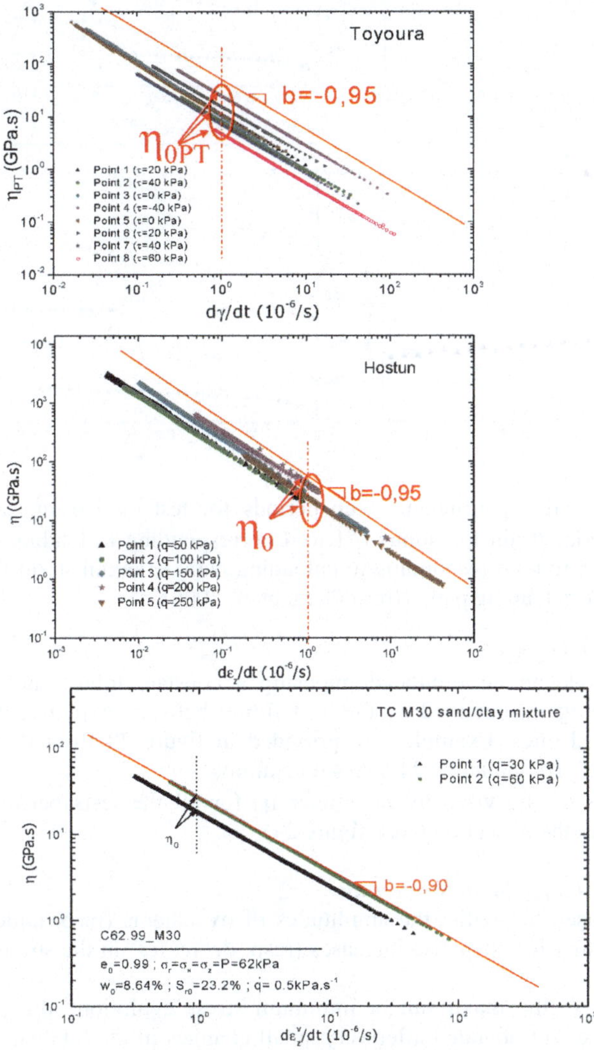

Figure 20. Viscosity of body V in function of the strain rate during creep periods : pure torsion for Toyoura sand (up), triaxial for Hostun sand (middle) and triaxial for sand clay mixture (down) (from Dutine 2005).

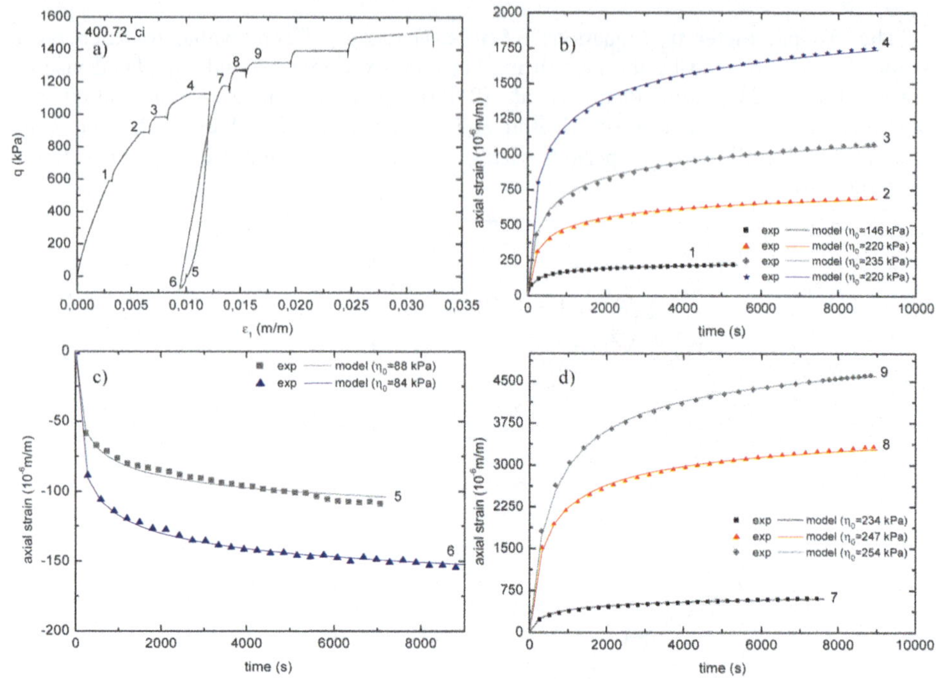

Figure 21. Simulations and experimental creep periods for test on Hostun sand : (a) global q-ε_1 curve, (b) axial strain for points n°1 to 4 corresponding to loading path; (c) axial strain for points n°5 to 6 corresponding to unloading path; (d) axial strain for points n°7 to 9 corresponding to reloading path. (from Pham et al. 2006).

Application to relaxation periods

The relaxation period can be simulated imposing a constant total strain. The η_0 parameter is then, once again, obtained by the best fitting between experimental stress evolutions and calculated ones. Examples are provided in figure 22 from Pham et al. (2006) for the test 200.95_10i including 11 stress relaxations.

The obtained values of the viscosity parameter η_0 for all the tests performed are gathered and discussed in the in section 6 (cf. figure 24).

Application to stepwise change in the strain rate

The model can be used to predict the amplitudes of overshoots (resp. undershoots) that systematically appear when stepwise increases (resp. decreases) in the strain rate are applied.

As shown in figure 6, the maximum or minimum stress evolution, after stepwise changes in the strain rate, are obtained after very small changes in the total strain. Thus evanescent properties can again be neglected. And, referring to equations 6.4 and 6.5, the variation of stress $\Delta\sigma$ (figure 23) during stepwise changes in the strain rate may be expressed as :

$$\Delta\sigma = \left[d\sigma^v\right] = \sigma^v_{after} - \sigma^v_{before} = \eta_0 \cdot \left(\frac{\dot{\varepsilon}^v_{before}}{\dot{\varepsilon}_0}\right)^{1+b} \cdot \left(\left(\frac{\dot{\varepsilon}^v_{after}}{\dot{\varepsilon}^v_{before}}\right)^{1+b} - 1\right) \qquad 6.8$$

Where $[d\sigma^v]$ is the viscous stress variation, the subscripts "after" and "before" stand respectively for the value just after or before the stepwise change in the strain rate. $\dot{\varepsilon}_0$ is the reference strain rate ($=10^{-6}$ /s).

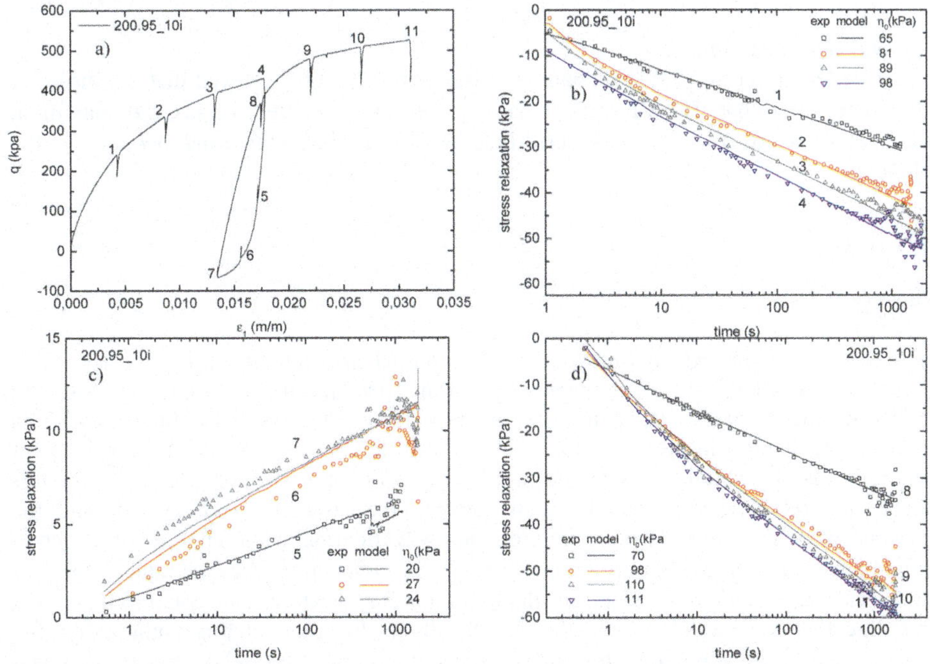

Figure 22. Comparison between simulations and experimental data for test on Hostun sand: (a) global stress-strain relationship; experimental and simulation of stress evolutions with time logarithm : (b) points n°1 to 4 corresponding to loading path, (c) points n°5 to 7 corresponding to unloading path, (d) points n°8 to 11 corresponding to reloading path (from Pham et al. 2006).

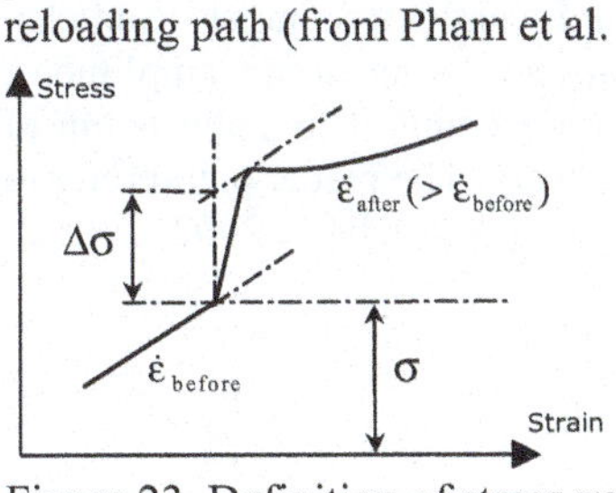

Figure 23. Definition of stress variations or stress jumps $\Delta\sigma$ (case of stepwise increase in the strain rate).

The viscous strain rates $\dot{\varepsilon}^{v}$ in **equation 6.8** can be reasonably assumed as very close to the total strain rate $\dot{\varepsilon}$, because the non viscous part $\dot{\varepsilon}^{nv}$ is small in the considered cases.

η_0 values can then be obtained in the case of stepwise changes in the strain rate by the means of equation 6.8 from experimental stress variations and the considered value of the b parameter. The obtained η_0 values are reported hereafter in figure 24.

Viscosity parameter η_0

Di Benedetto et al. (2005a), Pham Van Bang et al. (2006), showed that the viscosity parameter η_0 of the 1D VE model can be relevantly plotted versus the maximum principal stress σ_1 for TC tests performed on air-dried Hostun sand, leading to the relation:

$$\eta_0 = \alpha\, \sigma_1 \tag{6.9}$$

where α=0.15.

In figure 24 are plotted, in function of the maximal principal stress σ_1, the η_0 values obtained from all the different types of performed triaxial and pure torsion tests. The results includes a total of more than 100 creep periods, 200 stress relaxation periods and 40 stepwise changes in the strain rate realised at different stress levels and for different loading paths on triaxial StaDy device and hollow cylinder T4C StaDy device. From this figure, a good correlation can be seen between the viscosity parameter η_0 and the maximal principal stress, which therefore allow validating equation 6.9 for different stress paths (with and without rotation of axes). The constant α takes a unique value. For the 2 sands, it is equal to 0.15 and for the two sand-clay mixtures α is equal to 0.20. Lets underline that the model is validated for the three different loading paths exhibiting viscous effects considered in this paper: creep periods, stress relaxation periods and stepwise changes in the strain rate.

Based on these results, the three-component formalism associated to the viscosity function (equations 6.6 and 6.7) (whose 3D expression is given) appears to be relevant to describe correctly very different loading paths involving the viscous behaviour of the considered materials (two air-dried poorly-graded sands and two moist sand/clay mixtures). Nevertheless, it should be kept in mind that, for larger strain amplitudes, evanescent properties (equations 6.5) have to be taken into account. This point is out of the scope of this paper and is not treated as indicated previously. The reader could refer to the following publication for more details: Di Benedetto et al. (2001b&c, 2002, 2005a), Sauzéat (2003).

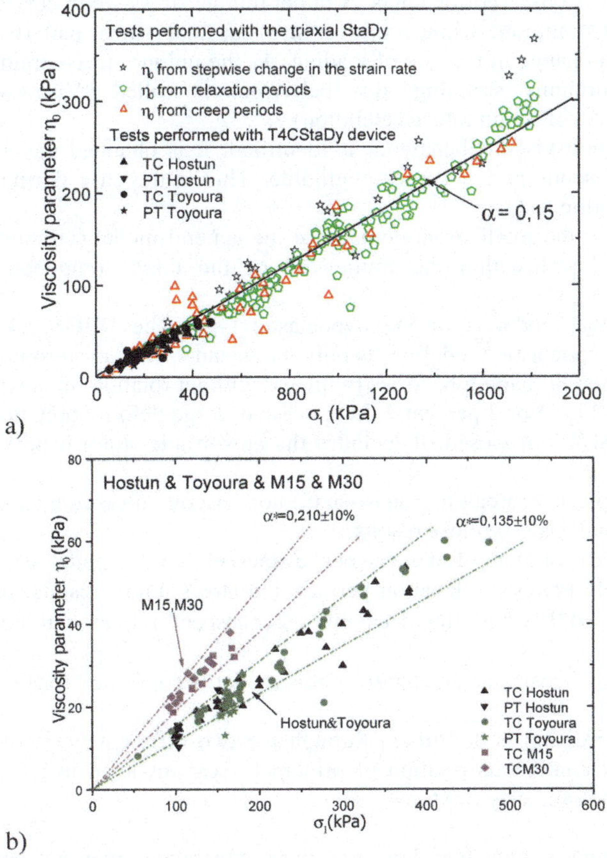

Figure 24. Viscosity parameter η_0 (equations 6.5 to 6.7) of the VE model in function of maximum principal stress σ_1 : tests with triaxial StaDy device (Pham Van Bang et al., 2006) (a) and tests performed on hollow cylinder device T4C StaDy (b) (Duttine et al. 2006). More than 350 data from the two dry sands and the two sand-clay mixtures are considered.

7. CONCLUSIONS

This paper gives a rapid overview of the results obtained on dry sand and sand/clay mixtures in the small strain domain, these last years at DGCB of ENTPE. From the results presented in this paper may be derived the following conclusions:

i) Experimental studies in that range of deformation (from some 10^{-6}m/m) need specifically designed devices with local strain measurements. The 2 designed prototypes (triaxial StaDy and hollow cylinder T4C StaDy), which are used for the experimental investigations, are rapidly presented. They allow accurate analyses and are both equipped with piezoelectric sensors.

ii) A viscous peculiar behaviour is observed for sands. A rather unique stress-strain curve is obtained at different constant strain rate, while very over-stress or under-stress patterns are observed just after a sudden change in the rate of loading. As the unique stress-strain curve is re-joined when continuing straining, this behaviour is called "Viscous Evanescent" (or Temporary Effect of Strain rate acceleration).
iii) In the small stain domain non viscous behaviour is identified. It is obtained under certain conditions, when the viscous part becomes negligible. This is the case during small cycles and waves propagation tests.
iv) The asymptotic expression in the small strain domain, of the general model (viscous evanescent (VE) model) developed within the framework of the three component approach is presented.
v) The non viscous part of the model is of the hypoelastic type. The DBGS (Di Benedetto, Geoffroy Sauzéat) hypoelastic model needs only 3 constants. It is anisotropic and fits well with data from different loading histories (with and without rotation of axes) if the strain path remains not to big (1 or 2 percents). In the case of large deformation, an evolution of the model (DBGSP) is proposed. It includes the anisotropic strain history effects.
vi) When considering anisotropic interpretation, quasi-static non viscous measurements and waves propagations back analysis give close results.
vii) The asymptotic (simplified) version of the viscous evanescent (VE) model can simulate in a relevant way creep periods, relaxation periods and stress jumps results. It needs only two constants $\{b;\ \alpha^*\}$ when the decaying ("evanescent") term can be neglected.
viii) A simple expression of the viscosity parameter η_0 of this model is confirmed for the two kinds of tested materials.
ix) A 3D extension of this expression is confirmed through the two different types of loadings (i.e. with and without continuous rotation of principal axes) involved in tests performed on air-dried sands and sand clay mixtures.

Amplitudes of viscous phenomena remain low, however, these phenomena may not be ignored and may exhibit non negligible effects at an engineering scale (Tatsuoka et al. 1999, Jardine et al. 2005, Di Benedetto et al. 2005a).

ACKNOWLEDGMENT

I wish to thank all my previous PhD students on the topic: Ollivier, Bogdan, Erdin, Cédric, Antoine, Damien and Alan. They allow obtaining all the nice experimental results following long and painful hours in the lab. Without these data our scientific exchanges and production would have been much less fruitful and rich.
I also deeply thank Professor Fumio Tatsuoka from the Tokyo University of Sciences for the long, animated and interesting discussions we had and still have on this long term cooperation research topic.

REFERENCES

Cazacliu, B. (1996). Comportement des sables en petites et moyennes déformations - réalisation d'un prototype d'essai de torsion compression confinement sur cylindre creux. PhD thesis , ECP/ENTPE, Paris. [in French]

Chaudhari S.K., Kuwano J., Hayano Y. (2004). Measurement of quasi-elastic stiffness parameters of dense Toyoura sand in hollow cylinder apparatus and triaxial apparatus with bender elements. Geotechnical Testing jl, Vol. 27, N. 1, p. 13.

Darve F. (1974). Contribution à la détermination de la loi incrémentale des sols. Thèse de Doctorat, Université Scientifique et Médicale de Grenoble, USMG Grenoble, 1974 ; 176. [in French]

Di Benedetto H. (1987). Modélisation du comportement des géomatériaux : application aux enrobés bitumineux et aux bitumes. Thèse de docteur d'Etat. Grenoble : USTMG, 1987. [in French]

Di Benedetto,H. (1997): "Viscous effect and anisotropy for sand". Pannel discussion. *I.1 of XIVth Int. Conf. ISSMFE (1997),* Vol. 4 (1998), Ed. Balkema, Hamburg, pp 2177-79, [in French].

Di Benedetto,H. and Tatsuoka,F. (1997): "Small strain behaviour of geomaterials: Modelling of strain rate effects", *Soils and Foundations*, Vol.37, No.2, pp.127-138.

Di Benedetto,H., De la Roche,C., (1998): "State of the art on stiffness modulus and fatigue of bituminous mixtures", *Bituminous binders and mixtures: state of the art and interlaboratory tests on mechanical behavior and mix design*, E&FN Spon, Ed. L. Francken, pp 137-180.

Di Benedetto,H., Ibraim,E. and Cazacliu,C. (1999a): "Time dependent behavior of sand" ", *Proc. 2nd Int. Symp. on Pre-failure Deformation Characteristics of Geomaterials, IS Torino* (Jamiolkowski et al. eds.), Vol.1, Balkema, pp. 459-466.

Di Benedetto,H., Sauzeat,C. and Geoffroy,H. (1999b): "Modelling viscous effects for sand and behaviour in the small strain domain", *Proc. 2nd Int. Symp. on Pre-failure Deformation Characteristics of Geomaterials, IS Torino* (Jamiolkowski *et al. eds.), panel pres*entation, Vol.2 (published 2001), Balkema, pp. 1357-1367.

Di Benedetto, H., Cazacliu, B., Geoffroy, H. & Sauzéat, C. (1999c). Sand behavior in the very small to medium strain domains. In Jamiolkowski & al (ed.), Proc. of IS on Pre-failure Def. Char. of Geomat., Torino. Balkema, 89-96.

Di Benedetto,H., Geoffroy,H. and Sauzeat,C. (2001a): "Hollow cylinder test and modelling of pre-failure behavior of sands", *Int. Conf. Albert Caquot*, Paris, p. 8.

Di Benedetto,H., Geoffroy,H. and Sauzeat,C. (2001b): "Viscous and non viscous behaviour of sand obtained from hollow cylinder tests", *Advanced Laboratory Stress-Strain Testing of Geomaterials* (Tatsuoka et al. eds.), Balkema, pp.217-226.

Di Benedetto,H., Tatsuoka,F. and Ishihara,M. (2002): Time-dependent shear deformation characteristics of sand and their constitutive modelling, Soils and Foundations, 42-2, pp.1-22.

Di Benedetto, H., Tatsuoka, F., Lo Presti, D., Sauzeat C., Geoffroy H., (2005a). Time effects on the behaviour of geomaterials. Keynote lecture in : Di Benedetto H. et al. Eds. Proc. of the 3rd Int. symp. on Deformation Characteristics of Geomaterials, sept. 2003, Lyon. Rotterdam, Pays-Bas : Balkema, vol.2, pp59-124

Di Benedetto, H., Geoffroy H., Duttine, A., Sauzéat, C., Chau B. (2005b). Anisotropic behaviour of soils and site investigation based on wave propagation tests. Proc. 16th Int. Conf. on Soils Mechanics and Geotechnical Engineering., Osaka. [in French]

Duttine, A. (2005). Comportement des sables et des mélanges sable/argile sous sollicitations statiques et dynamiques avec et sans « rotation d'axes ». Ph.D thesis, ENTPE, Lyon, France. [in French]

Duttine, A., Di Benedetto, H., Pham Van Bang, D. (2006). Viscous properties of dry sands and mixtures of sand/clay from hollow cylinder tests. Proc. *Geotechnical Symposium in Roma.* Roma 2006, March 15-16th

Enomoto T., Tatsuoka F., Shishimine M., Kawabe S., Di Benedetto H. Viscous properties of granular material in drained triaxial compression, *Geotechnical Symposium in Roma.* Roma 2006.

Ezaoui, A., Di Benedetto, H., Pham Van Bang, D. (2006) Anisotropic behaviour of sand in the small strain domain. Experimental measurements and modelling., *Geotechnical Symposium in Roma.* Roma

Fioravante, V., Capoferri, R. (2001). "On the use of multi-directional piezoelectric transducers in triaxial testing." Geotechnical Testing Journal 24(3): 243-255

Geoffroy, H., Di Benedetto,H., Duttine, A. and Sauzeat,C. (2003): "Dynamic and cyclic loadings on sands: results and modeling for general stress strain conditions", *Proc. 3rd Int. Symp. on Deformation Characteristics of Geomaterials, IS Lyon 03* (Di Benedetto et al. eds.), Balkema, Sept. 2003, pp. 353-363.

Hameuy O. (1995): “Quelques aspects du comportement des sables avec ou sans rotation des axes principaux”, thèse de Doctorat, ENTPE-ECP [in French]

Hardin, B.O. and Blandford, G. E. (1989). Elasticity of particulate materials. Journal of Geotechnical Engineering, ASCE, 115(6), 788–805.

Hoque,E. (1996): "Elastic deformation of sands in triaxial tests", Doctor Engineering thesis, University of Tokyo.

Ibraim, E., (1998): “Différents aspects du comportement à partir d’essais triaxiaux : des petites déformations à la liquéfaction statique”, Thèse de Doctorat ENTPE-INSA Lyon, [in French].

Jardine R.J. Kuwano R., Zdravkovic L. & Thornton C. (1999). Some fundamental aspects of the pre-failure behaviour of granular soils, Keynote lecture in IS on Pre-failure Deformation Characteristics of Geomaterials, Torino, vol. II : 1077-1111

Jardine R.J., Standing J.R., Kovacenic N. (2005) : “Lessons learned from full scale observations and the practical application of advanced testing and modelling”, In : Di Benedetto H. et al. Eds. Proc. of the 3rd Int. symp. on Deformation Characteristics of Geomaterials, sept. 2003, Lyon. Rotterdam, Pays-Bas : Balkema, vol.2, pp201-246

Kim, D.S. and Stokoe, K.H. (1994). Torsional motion monitoring system for small strain (10^{-5}% to 10^{-3}%) soil testing. *Geotechnical Testing Journal* **17**(1) : 17–26.

Mair, K. and Marone, C. (1999): Friction of simulated fault gouge for a wide range of velocities and normal stresses, Journal of Geophysical Research, Vol. 104, No.B12, pp.28,899-28,914.

Pham Van Bang D, Di Benedetto H. (2003) Effect of strain rate on the behaviour of dry sand. Proceedings of the third international symposium on deformation characteristics of geomaterials, IS Lyon03, Eds Di Benedetto et al, 2003; Vol. I : 365-373.

Pham Van Bang, D. (2004). Comportement instantané et différé des sables des petites aux moyennes déformations : expérimentation et modélisation, Ph.D Thesis, ENTPE, Lyon. [in French]

Pham Van Bang, D., Di Benedetto, H., Duttine, A., Ezaoui, A. (2006). Viscous behaviour of sands : air dried and triaxial conditions. International Journal for Numerical and Analytical Methods in Geomechanics (2006, to be published)

Sauzeat, C. (2003). Comportement du sable dans le domaine des petites et moyennes deformations : rotations “d’axes” et effets visqueux, Phd thesis, ENTPE, Lyon, France. [in French]

Sauzéat C, Di Benedetto H, Chau B, Pham Van Bang D. (2003) A rheological model for viscous behaviour of sand. Proc. of Proceedings of the third international symposium on deformation characteristics of geomaterials, IS Lyon03, Eds Di Benedetto et al, 2003; 1201-1209.

Tatsuoka F., Teachavorasinsun S., Dong J., Kohata Y. & Sato T. (1994). Importance of measuring local strains in cyclic triaxial tests on granular materials, *Dynamic Geotechnical Testing II, ASTM STP 1213, R. Ebelhar, V. Drnevich, B. Kutter Eds., American Society for Testing and Materials,* 288-302, Philadelphia.

Tatsuoka,F., Jardine,R.J., Lo Presti,D., Di Benedetto,H. and Kodaka,T. (1997) : “Characterising the Pre-Failure Deformation Properties of Geomaterials”, *Theme Lecture for the Plenary Session No.1, Proc. of XIV IC on SMFE*, Hamburg, September 1997, Volume 4, pp.2129-2164.

Tatsuoka F, Uchimura T, Hayano K, Di Benedetto H, Koseki J, Siddiquee M.S.A. (1999) Time dependent deformation characteristics of stiff geomaterials in engineering practice. *Theme Lecture in Proc. of 2nd IS on Pre Failure Deformation Characteristics of Geomaterials*, IS Torino, 1999; Vol. II : 1161-1262

Tatsuoka F, Ishihara M, Di Benedetto H, Kuwano R.(2002). Time dependent shear deformation characteristics of geomaterials and their simulation. Soils and Foundations, vol.42, n°2, pp.103-129.

Tatsuoka F. (2006), Inelastic deformation characteristics of geomaterials, Special lecture, *Geotechnical Symposium in Roma*. Roma, March 2006.

Advanced Laboratory Stress-Strain and Strength Testing of Geomaterials in Geotechnical Engineering Practice

Satoru Shibuya
Department of Architecture and Civil Engineering, Kobe University, Kobe, 657-8501, Japan
e-mail: sshibuya@kobe-u.ac.jp

Takayuki Kawaguchi
Hakodate National College of Technology, Hakodate, 042-8501, Japan
e-mail:kawa@hakodate-ct.ac.jp

ABSTRACT: Advanced laboratory stress-strain and strength testing of geomaterials is a need for geotechnical engineering practice. In this keynote paper, recent developments in advanced stress-strain and strength testing of geomaterials, together with engineering application of the test results are reviewed by showing two case histories with reference to excavation work in urban area, and the failure of reinforced wall due to heavy rainfalls.

1 INTRODUCTION

Development and engineering application of advanced laboratory stress-strain and strength testing of geomaterials was a centralized subject at the 3rd International Symposium on Deformation and Strength Behaviour of Geomaterials (IS-Lyon'03) sponsored by TC29 of ISSMGE. In the keynote paper by Shibuya, Koseki and Kawaguchi (2005), it is clearly pointed out that in the current world-class standard of geotechnical laboratories, the deformation behaviour over the whole pre-failure regime, including the elastic stiffness at very small strains can successfully be observed in triaxial test using a cylindrical specimen. Conversely, direct shear test or direct shear box (DSB) test suits well for providing soil strength for stability analysis of slopes.

The objective of this keynote paper is to demonstrate importance of developments in equipments and techniques of laboratory tests on geomaterials. A couple of case histories, each in which advanced triaxial test or DSB test was employed in order to tackle cited problem are in detail described.

2 CASE HISTORY USING ADVANCED TRIAXIAL TEST

The direct-drive motor coupled with nearly zero-backlash reduction unit has been developed at Hokkaido University (Shibuya and Mitachi, 1997, Kawaguchi et al., 2002, Kawaguchi, 2002). In this system, 'nearly-zero' backlash on reversal of loading direction can be achieved with the combination of the servo-motor, reduction unit and ballspline screw. The control of axial loading can be fully automated by using a personal computer. Furthermore, the loading system shows a distinct capability of achieving a minimum control for the axial displacement of 0.00015 micrometer spanning over several orders of the rate of axial straining. Note that the resolution of axial displacement is equivalent to 1.5×10^{-9} of strain for a sample with 10 cm high (see also Shibuya, Koseki and Kawaguchi, 2005).

A case history in Thailand on deep excavation in soft clay with concrete diaphragm wall is described with attention paid to the ground deformation behind the diaphragm wall. As summarized in Table 1, there have been several empirical proposals how we should select the soil stiffness employed for FE analysis. For excavation works, the use of the equivalent Young's modulus that is several hundreds of undrained shear strength s_u has been proposed. In contrast, the use of half of the small strain stiffness E_{max} obtained by in-situ seismic survey has been proposed by Simpson (2000).

Figure 1 shows the variations of secant Young's modulus normalized using s_u and E_{max} with axial strain, that were obtained in undrained triaxial compression test on normally consolidated clays collected from worldwide (Temma et al., 2000, 2001). In the present case history, the ground strains behind the diaphragm wall was on the order of 0.1 %, which is consistent with the proposals by Simpson and Hock. It should be pointed out that the axial strain associated with $E = E_{max}/2$ corresponds to a narrow range from 0.03 % to 0.3 %. The use of $E = E_{max}/2$, therefore, matches well the ground strains of soft clay behind diaphragm wall

Hoe I. Ling et al. (eds.), Soil Stress-Strain Behavior: Measurement, Modeling and Analysis, 191–199.

Table 1 Elastic deformation modulus employed for FE analysis in the past.

$E = \alpha S_u$	Author	
Embankment loading analysis		
E_u = 200–500S_u	Bjerrum (1964)	
E' = 70–250$S_{u\,FVS}$	Balasubramaniam et al. (1981)	
E_u = 15–40$S_{u\,FVS}$	Bergado et al. (1990)	(Back-calculation)
Excavation works analysis		
E_u = 200–500S_u	Bowels (1988)	
E_u = 280–500S_u	Hock (1997)	(soft clay)
E_u = 1200–1600S_u	Hock (1997)	(stiff clay)
$E' = E'_{max}/2$	Simpson (2000)	

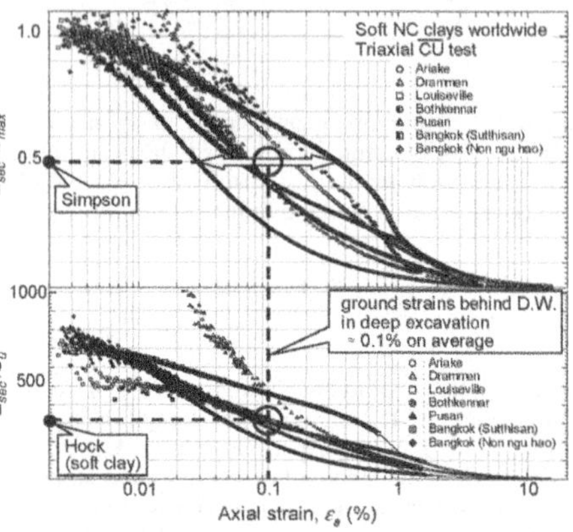

Figure 1 Normalized secant Young's modulus of normally consolidated clays (Temma et al., 2000, 2001)

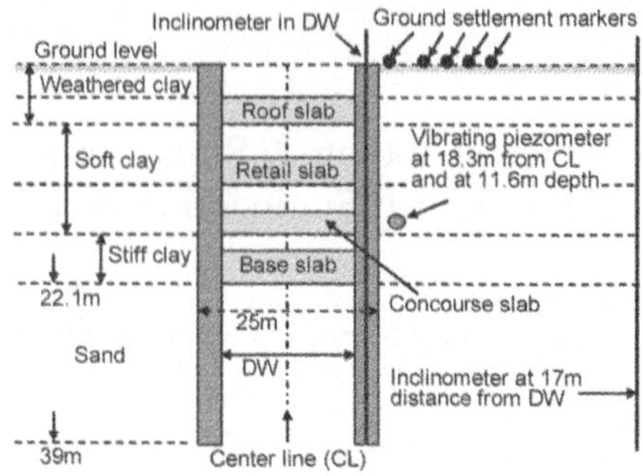

Figure 2 Cross-section of deep excavation with diaphragm wall in Bangkok (Tamrakar et al., 2001).

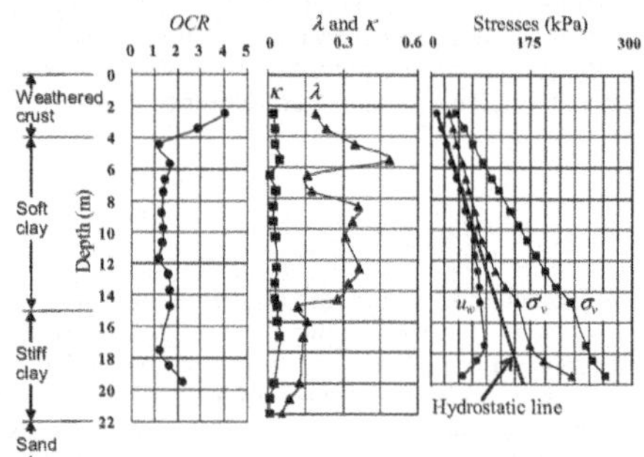

Figure 3 Soil profiles for deep excavation work in Bangkok (Tamrakar et al., 2001).

induced by deep excavation. Note also that the E_{sec}/s_u value corresponding to the axial strain of 0.1% ranges from 200 to 400.

Figure 2 shows cross-section of deep excavation with diaphragm wall for the construction of subway tunnels in Bangkok by open-cut method. The concrete diaphragm walls were pre-installed to a depth of 39 m, and deep excavation was afterwards carried out in soft and stiff clays down to a depth of 22 m (Tamrakar et al., 2001). Instrumentations were a piezometer, two series of inclinometers set along the wall axis and the borehole located at a horizontal distance of 17 m from the diaphragm wall, and a series of markers for monitoring ground settlement.

Figure 3 shows the stratigraphy with representative profiles of *OCR*, compression and swell indices, λ and κ; and geostatic stresses and pore pressures. Figure 4 shows the results of a series of undrained triaxial compression tests on undisturbed clay specimens retrieved from several depths. Although the small strain stiffness values vary with depth, the degradation curves of the normalized stiffness were almost similar to each other.

By considering the proposal by Simpson, half of the small strain stiffness corresponding to the strain level of about 0.1 % was employed in the numerical simulation of the full scale behavior using an equivalent-linear elastic approach. In addition, Kovacevic et al. (2003) has performed a non-linear elastic analysis using a model called the small strain stiffness or SSS model. The results of boundary values predicted using these two kinds of analysis are compared in Figures 5 and 6. The horizontal deformation of the diaphragm wall was compared at several excavation stages. From a practical point of view, the result from equivalent-linear analysis presented using dash lines was not bad, while the deformation profile could be better captured by the non-linear analysis shown in solid lines assuming full moment connection between the roof slab and the diaphragm wall. Similarly, the non-linear analysis using a small-strain stiffness (SSS) model could better capture the ground settlement profile than the linear analysis, suggesting an importance of using proper soil stiffness considering its strain level and stress state dependencies.

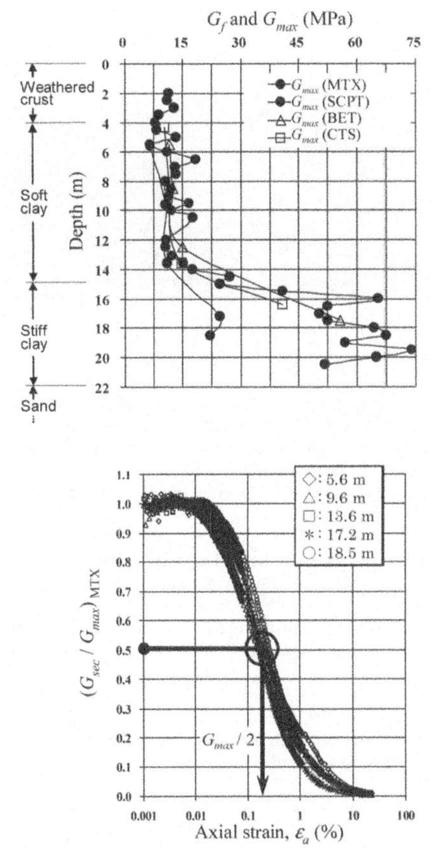

Figure 4 Undrained triaxial compression test results on undisturbed clay specimens (Shibuya et al., 2001).

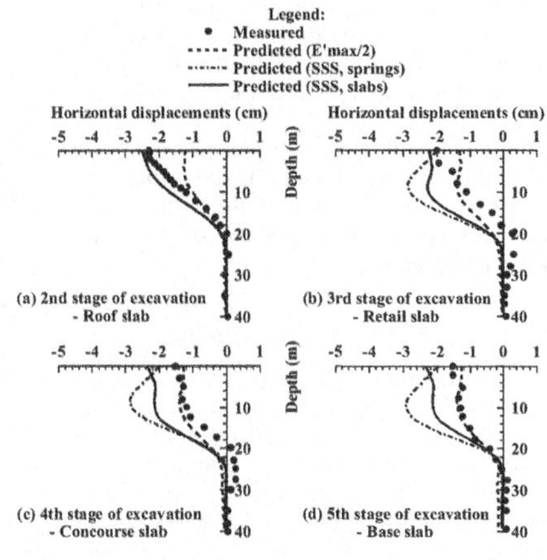

Figure 5 Horizontal deformation of diaphragm wall at four excavation stages (Kovacevic et al., 2003).

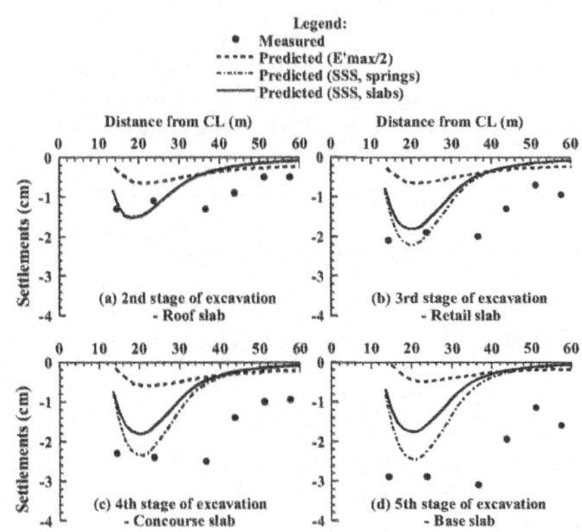

Figure 6 Ground surface settlement at four excavation stages (Kovacevic et al., 2003).

3 CASE HISTORY USING ADVANCED DIRECT SHEAR BOX (DSB) TEST

3.1 *Introduction*

In general, the design and construction of Terr Armée wall does not consider any water in and adjacent to the reinforced soil. Similarly, little attention is paid to infiltration of rainfall water from the surrounding area. Nevertheless, the Terr Armée wall has been popular in use for constructing local roads in mountainous area in Japan, where the attack of seasonal heavy rainfalls is commonly encountered.

At around 1:40 am on 21st of October 2004, a huge landslide involved with a catastrophic failure of Terr Armée wall took place in mountainous area of Yabu city in northern part of Hyogo Prefecture in west Japan. The occurrence of the incident was quite exact by a record of emergency phone call from a local resident to Yabu local government office. It should be mentioned that the data was recorded at an observatory very close to the site. Heavy rain with the intensity in excess of 10 mm per hour continued over eight hours (1pm – 8pm) on 20th. Note that the amount of rainfall a day of 226 mm was certainly the record over the past eight years in this region. It may well be postulated that this heavy rainfall was the trigger of the disaster.

Figure 7 shows a picture of the Terr Armée wall failure that was taken a few months after the incident. More details of the failure have been reported in the paper by Shibuya and Kawaguchi (2006).

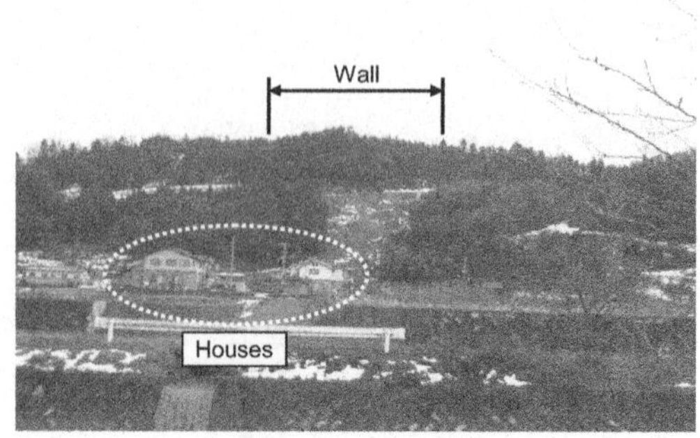

Figure 7 Site after the collapse of Terr Armée wall

3.2 *Outline of Terr Armée wall failure*

The construction of the reinforced wall was completed in the year of 2000, i.e., four years before the incident. The landslide took place over the length of about 150m along the slope involved with the collapse of Terr Armée wall over 80m long along the road. It was obvious in the map that the slope over which the wall failure occurred consisted of a couple of small valleys. Moreover, the road was inclined about 7% from West (right hand side in Fig.2) to East. These topographical surroundings would have brought about some concentration of surface/ground water into the collapsed portion of the road during the heavy rainfall.

Several key observations from comprehensive survey performed immediately after the incident (Shibuya and Kawaguchi, 2006) are;

i) parts of Terr Armée wall (i.e., concrete skins and metal strips) reached to the end of debris flow,
ii) neither damage of metal skins nor breakage of metal strip/concrete skin joints was found,
iii) the remainder of the Terr Armée wall stands in good shape,
iv) the metal strips remained on the side wall of Terr Armée wall were all inclined with an angle of 24-26 degrees from the horizontal that was equal to the supposed slip surface angle on the foundation (Fig.3), and
v) rainfall water pours into the collapsed area when raining.

It should be mentioned that the collapsed Terr Armée wall was as high as 23m. To the authors' best knowledge, no such a huge collapse of Terr Armée wall has been reported in the literature.

3.3 *Terr Armée wall collapsed*

The wall was constructed in the year of 2000. Some features regarding the design and construction details are summarized in the following (Shibuya and Kawaguchi, 2006);

i) local geomaterials, i.e., weathered silty soil originated from yellow tuff, with the fines content well in excess of 25% was employed for constructing the wall (Fig.6),
ii) drain pipe with a small diameter of 200mm was employed at the bottom of the wall, but it did not extend to cover an area behind the wall (see Fig.5),
iii) the surface of the road had been unpaved over the past four years, and
iv) SPT N-value in the foundation ranged between fifteen and twenty, which was surprisingly low to support the soil wall with the maximum height of 23m.

It should be mentioned that in the construction of the wall, the local soil was mixed with cement-based stabilizer in order to secure prescribed friction between metal strips and the soil. As stated later on, the soil stabilization treatment have reduced the permeability of in-wall soil to a considerable extent.

3.4 *Laboratory DSB test*

Two series of constant-volume direct shear box test was performed using reconstituted samples retrieved from the slip surface. A block sample was first prepared from the slurry with the initial water content approximately equal to liquid limit (i.e., 50-60%). Figure 8 shows the direct shear box apparatus developed at Hok-

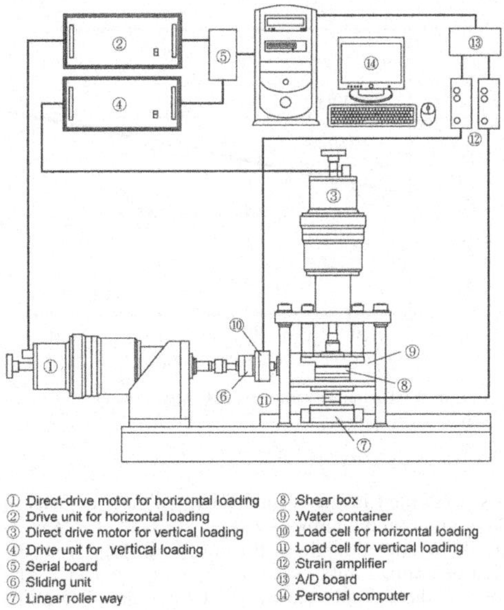

Figure 8 Direct shear box apparatus

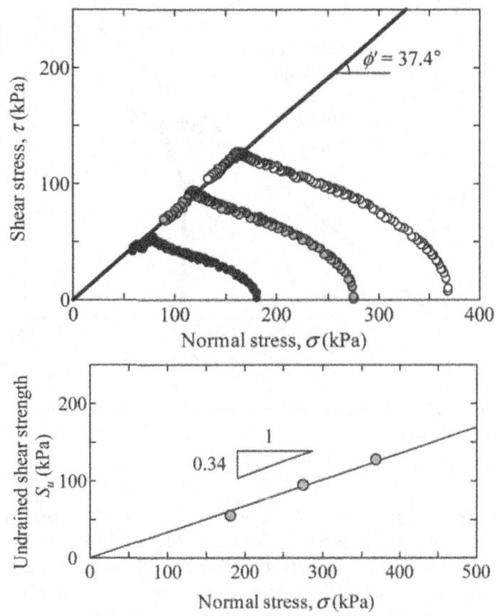

Figure 10 Results of conventional direct shear test ; a) effective stress paths and b) undrained shear strength.

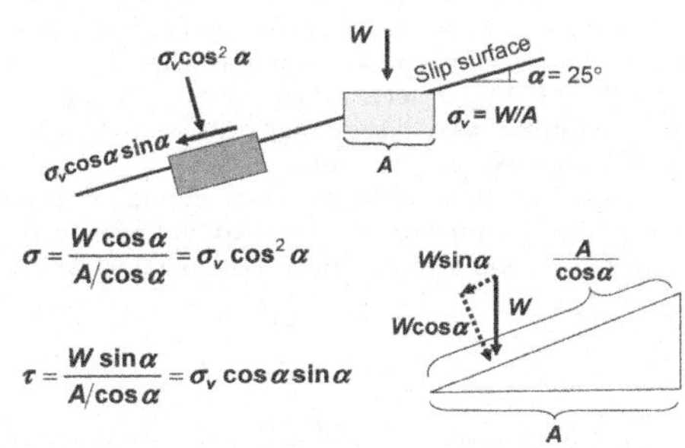

Figure 9 Stress conditions along the slip surface

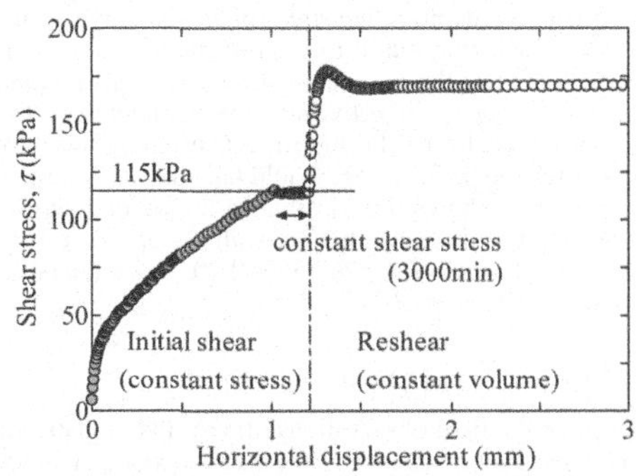

Figure 11 Relationship between horizontal shear stress and horizontal displacement in a test with initial shear.

kaido University (Shibuya et al, 2001). The load control as well as data acquisition is fully automated with an aid of PC.

Regarding a sketch shown in Fig.9, three samples in the first series were each consolidated to the vertical consolidation stress, σ_{vc} of about 200, 300 and 400kPa, respectively. Afterwards, the samples were sheared under constant-volume conditions by using a constant rate of horizontal displacement of 2mm/min. The undrained effective stress paths of this conventional consolidated-undrained (CU) test are shown in Fig.10, in which the variation of normal (=vertical) effective stress is plotted against the horizontal shear stress, τ. The effective strength parameters of (c', ϕ') =(0, 37.4°), together with the total stress parameters, S_u/σ_{vc}=0.34 were obtained for the long-term and short-term stability analysis, respectively.

In the second series, in-situ stress conditions on the collapsed slope were more closely simulated in each sample by applying initial shear stress under drained conditions. In general, in-situ soil element on a slope with angle α from the horizontal is subjected to the initial stress conditions as $\sigma_i = \sigma_{vc}\cos^2\alpha$ and $\tau_i = \sigma_{vc}\cos\alpha$

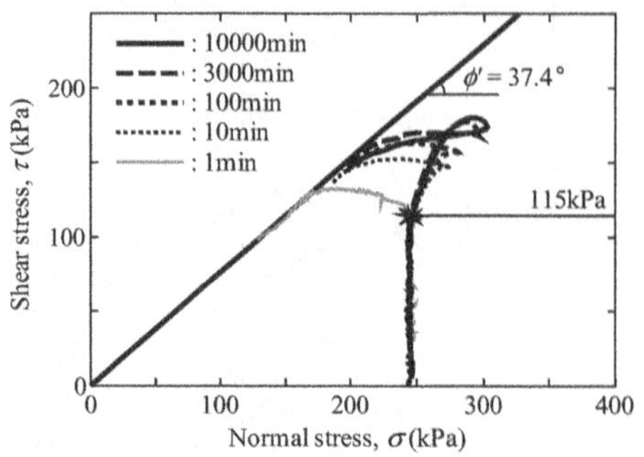

Figure 12 Undrained effective stress paths in direct shear box test with initial shear.

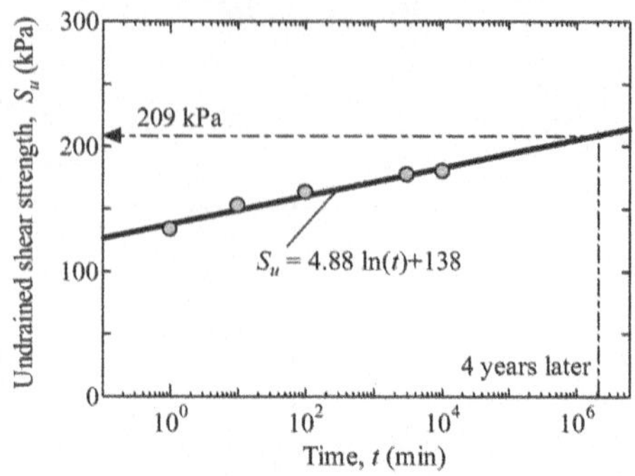

Figure 13 Undrained shear strength versus duration of initial shear stress applied

singα .Figures 11 and 12 show the results of five tests performed in the second series. As can be seen in Fig.11, each sample was first consolidated to a prescribed initial (and common) vertical stress of σ_i =246 kPa (=300× cos^{2}25°), where the angle of 25°corresponds to the averaged angle of the collapsed slope from the horizontal. The sample was then subjected to application of drained initial shear to the value of τ_i=115 kPa (=300×cos25° sing25°) by using a slower rate of horizontal shear displacement of 0.02mm/min. The initial shear stress was then maintained constant over a prescribed period in each test. The sample was finally sheared to failure under the conditions of constant volume by using a faster rate of 2mm/min. The initial shear stress was maintained over different, but fixed, periods of 1, 10, 100, 3,000 and 10,000 min, respectively. Surprisingly, the effective stress path was very much influenced by the duration of the initial shear stress application, showing that it rose more sharply as the duration increased. As a consequence, the undrained shear strength S_u (i.e., the maximum shear stress) also increased with the sustained period of initial shear stress. On the other hand, the effective strength parameters of (c', ϕ') =(0, 37.4°) were not affected at all by the application of initial shear. The significant softening behavior observed in the second series suggests strongly that a catastrophic type of failure would take place once the foundation soil exceeds the peak.

Figure 13 shows the S_u value in the second series when examined against the sustained period of initial shear. Bearing in mind that the wall collapsed at 4 years after the construction, the value of 209 kPa for σ_{vc} =300 kPa, hence S_u/σ_{vc}=209/300=0.70, was attained by extrapolating the S_u values from the laboratory test to an instant after 4 years,

3.5 *Senerio of Terr Armée wall failur*

Based on the field observations, the results of geotechnical investigation and comprehensive review of design and construction scheme of the wall, it was concluded that this particular Terr Armée wall was properly designed and constructed in the light of "design and construction manual" available in Japan. However, it is the fact that the wall failed due to the heavy rainfall.

The wall failure may be attributed to simultaneous occurrence of several, not a single, causes as listed below;

i) concentration of in-soil seepage and surface water flow into the collapsed area,
ii) relatively low permeability of embankment as compared to the weathered rock foundation,
iii) poor drain system behind the wall in particular,
iv) low bearing capacity of the foundation, and
v) softening characteristic of the foundation soil as sheared undrained.

It should be stressed that the first three possible causes have something to do with problems of in-soil seepage flow. As stead earlier, according to "design and construction manual" in Japan like in other countries perhaps, any existence of water/water flow is not considered at all in the design of Terr Armée wall. In fact, however, rainfall water invaded into the reinforced embankment in this case record.

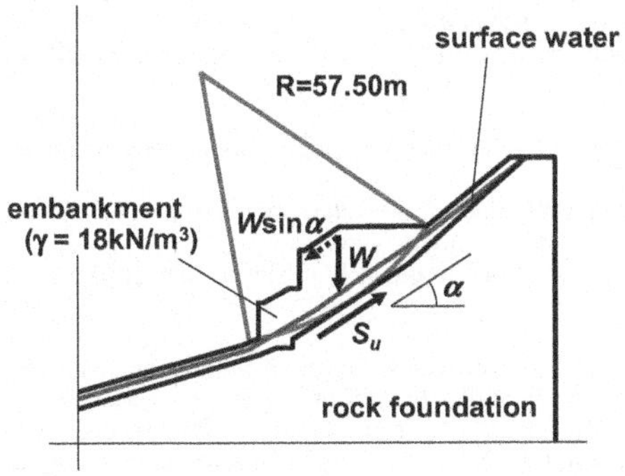

Figure 14 Short-term stability analysis

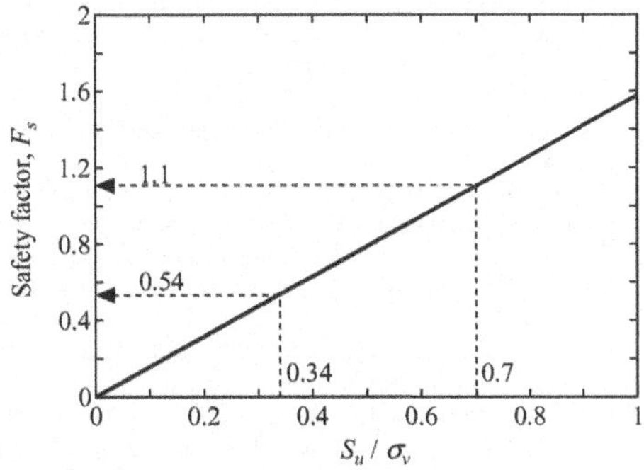

Figure 15 Back calculated factor of safety with S_u/σ_{vc}

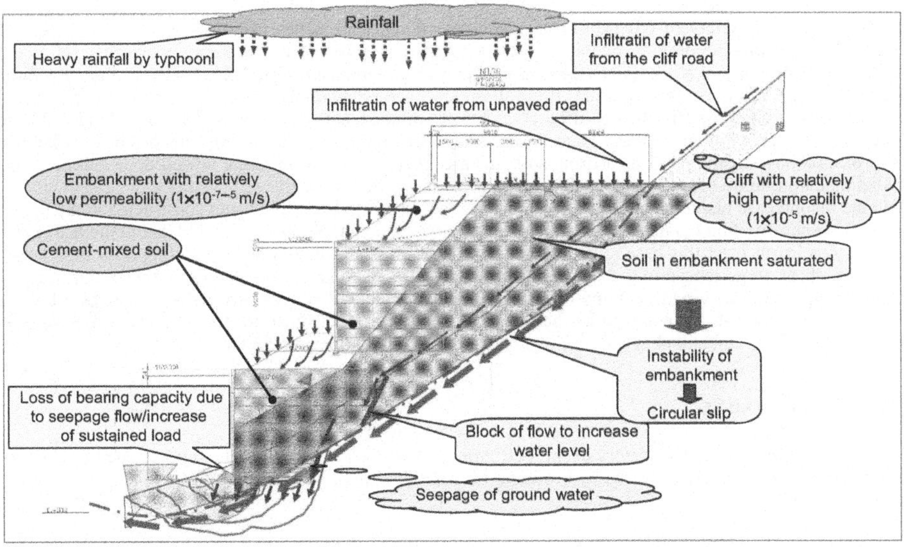

Figure 16 Scenario of the Terr Armée wall failure

The short-term stability analysis by assuming circular slip surface was carried out by using the strength parameters from triaxial UU test and constant-volume direct shear test. Figure 14 shows the short-tem stability analysis performed at the central portion of the collapsed wall. The surface water level was postulated based on in-situ measurement that was made immediately after the incident. The results of back-analysis are shown in Fig.15, in which the back-calculated factor of safety, F_s, is plotted against the total stress strength parameter, S_u/σ_{vc}. Note that F_s is close to unity (i.e., F_s =1.1) when the S_u/σ_{vc} value of 0.70 is employed, whereas F_s was far less than unity when S_u/σ_{vc} of 0.34 from conventional direct shear test is used for the calculation. The result strongly suggests that we should use soil strength from direct shear box test in which the magnitude as well as history of initial stress conditions on the supposed slip surface are closely simulated in the laboratory.

Regarding the sketch shown in Fig.16, the scenario of the wall failure could be described as;

i) a great deal of in-soil and surface water poured into the collapsed area over a short period of time,
ii) the water infiltrated into the embankment and gradually stored behind the wall since the permeability of the wall was relatively small,
iii) total load increased due to the water infiltration,
iv) seepage water pressure in the weathered rock foundation having relatively high permeability increased gradually,
v) the overall stability of the embankment reduced substantially due to the water storage in and behind the wall, together with the development of seepage pressure in the foundation soil,
vi) the localized failure on the foot of the wall may have occurred due to the lack of bearing capacity, and
vii) a type of circular slip occurred across the foundation.

The wall failure as such could be described simply as "the undrained shear failure of foundation since Terr Armée wall acted as if a reservoir dam in the event of heavy rainfall". Finally, it is strongly recommended that the design and construction manual of Terr Armée wall should be revised properly so as to prevent any occurrence of this type of failure. The case record described in this paper also suggests the strong need for co-operation between wall engineers and geotechnical engineers.

4 CONCLUDING REMARKS

In order to meet a variety of engineering needs, conventional approaches based on stability analysis using conservative soil strength or deformation analysis using conservative soil stiffness may not work.

The use of proper soil stiffness considering its strain level dependency and stress state dependency in the deformation analysis is vitally important. If the strain level can be predicted in advance, a linear elastic analysis using a well-chosen stiffness would be of practical value. If not, however, non-linear elastic or elasto-plastic approaches considering these factors would be more appropriate. In the case history in Bangkok, the value of triaxial test capable of measuring soil stiffness over a wide strain range is well demonstrated when predicting deformation of soft ground subjected to deep excavation work.

In the second case history, Terr Armée wall collapsed was properly designed and constructed in the light of "design and construction manual" in Japan. However, it was the fact that the wall failed due to the heavy rainfall. Lessons learnt from this case record is that the use of proper soil strength considering both time history and initial shear experienced in the field may be required for establishing rational design procedures against the short-term (i.e., undrained) stability of reinforced embankment on slopes. Advanced direct shear box test played an important role for manifesting the effects of time and initial shear on strength, resulting in satisfactory result when the embankment stability was back-analyzed with the strength data.

ACKNOWLEDGEMENT

Field data reported in the case history of the Terr Armée failure is referenced to the report issued by the technical committee of Yabu city chaired by Professor Makoto Nishigaki, Okayama University.

REFERENCES

Balasubramaniam, A.S. and Brenner, R.P. 1981. Consolidation and settlement of soft clay, *Soft clay engineering*, Brand, E.W. and Brenner, R.P., eds.: Elsevier Scientific Publishing Co.

Bergado, D.T., Ahmeed, S., Sampaco, C.L. and Balasubramaniam, A.S. 1990. Settlement of Bangna-Bangapakong Highway on soft Bangkok clay, *Geotechnical engineering* 116(1): 136-155.

Bjerrum, L. 1964 Observed versus computed settlements of structures on clay and sand, *Lecture presented at Massachusetts Institute of Technology*, Cambridge, Massachusetts.

Bowels, J.E. 1988. Foundation analysis and Design, *4th Edition, McGraw-Hill.*

Hock, G.C. 1997. Review and analysis of ground movements of braced excavation in Bangkok subsoil using diaphragm walls, *M.Eng.Thesis*, Bangkok Thailand.

Kawaguchi, T., Mitachi, T., Shibuya, S. and Sano, Y. 2002. Development of an elaborate triaxial testing system for deformation of clay, *Journal of Japan Society of Civil Engineering* 708(III-59): 175-186. (in Japanese)

Kawaguchi, T. 2002. Study on measurement and evaluation of elastic modulus in clays, *PhD thesis*, Hokkaido University. (in Japanese)

Kovacevic, N., Hight, D.W. and Potts, D.M. 2003. A comparison between observed and predicted behaviour of a deep excavation in soft Bangkok clay, *Deformation Characteristics of Geomaterials* 1, IS-Lyon, Balkema: 983-989.

Shibuya, S. and Mitachi, T. 1997. Development of a fully digitized triaxial apparatus for testing soils and soft rocks, *Geotechnical Engineering Journal* 28(2): 183-207.

Shibuya, S., Mitachi, T., Tanaka, H., Kawaguchi, T. & Lee, I-M. 2001. Measurement and application of quasi-elastic properties in geotechnical site characterization, Theme Lecture for Plenary Session 1, *Proc. of 11th Asian Regional Conference on Soil Mechanics and Geotechnical Engineering.* 2, Balkema: 639-710.

Shibuya, S., Koseki, J. and Kawaguchi, T. 2005. Recent developments in deformation and strength testing of geomaterials, Keynote Lecture, *Deformation Characteristics of Geomaterials -Recent Investigations and Prospects-*, IS-Lyon, Balkema: 3-26.

Shibuya, S. and Kawaguchi, T. 2006. Lessons Learnt from a Catastrophic Failure of Terr Armée Wall due to Heavy Rainfall, *Proc. of Second GI-JGS workshop*, ASCE, Kyoto (in print).

Simpson, B. 2000. Engineering needs. *Prefailure Deformatioms Characteristics of Geomaterials.* 2, Theme and Keynote Lectures, IS-Torino, Balkema: 1011-1026.

Tamrakar, S.B., Shibuya, S. and Mitachi, T. 2001. A practical FE analysis for predicting deformation of soft clay subjected to deep excavation, *Proc. of 3rd International Conference on Soft Soil Engineering* 1, Hong Kong: 377-382.

Temma, M., Shibuya, S. and Mitachi, T. 2000. Evaluating ageing effects of undrained shear strength of soft clays, *Proc. of International Symposium on Coastal Geotechnical Engineering in Practice* 1, Yokohama: 173-179.

Temma, M., Shibuya, S., Mitachi, T. and Yamamoto, N. 2001. Interlink between metastability index, MI(G) and undrained shear strength in aged holocene clay deposit, *Soils and Foundations* 41(2): 133-142. (in Japanese)

ASSESSMENT OF STRENGTH AND DEFORMATION OF COARSE GRAINED SOILS BY MEANS OF PENETRATION TESTS AND LABORATORY TESTS ON UNDISTURBED SAMPLES

O. Pallara, F. Froio, A. Rinolfi and D. Lo Presti[1]
Department of Structural and Geotechnical Engineering
Politecnico di Torino, C.so Duca degli Abruzzi 24, 10129 Torino, Italy
e-mail: oronzo.pallara@polito.it

ABSTRACT

The paper summarizes the activities undertaken to retrieve large size undisturbed sample of gravel and sand by in situ freezing at Licciana Nardi (Lucca, Italy). The following laboratory tests have been performed on the retrieved sampled: grain size distribution, soil density and relative density, undrained cyclic compression loading triaxial tests under strain control, undrained monotonic compression loading cyclic tests under strain control, triaxial liquefaction tests under stress control. In situ LPT and other dynamic penetration tests have been performed at the investigation site.

1. INTRODUCTION

The scope of this paper is to show some experimental results obtained in the framework of a more comprehensive research project aimed at the evaluation of the "Seismic Response of granular deposits in some areas of Tuscany". Such a project (PRIN 2002 – coordinated by M. Jamiolkowski of the Politecnico di Torino) was supported by the Italian Ministry of University and Research and co-financed by the Regional Government of Tuscany (RT).

More specifically, the paper refers on the geotechnical characterization of Holocene granular deposits by means of in situ and laboratory tests, performed on undisturbed and reconstituted samples. Undisturbed samples were retrieved by means of the in situ freezing method.

As already mentioned, the experimental activities have been done in the framework of a Research Contract between RT and the Politecnico di Torino aimed at the characterization of some areas of recent industrial development [LO PRESTI ET AL. 2004].

---

[1] Associate Professor, Department of Civil Engineering, University of Pisa, via Diotisalvi 2, 56126 Pisa, Italy; e-mail: d.lopresti@ing.unipi.it

Hoe I. Ling et al. (eds.), Soil Stress-Strain Behavior: Measurement, Modeling and Analysis, 201–213.

2. TEST SITE SELECTION

Selection of the test site has been accomplished according to the following criteria: a) geological evidences; b) water table depth; c) site accessibility and logistic aspects; d) interests of RT.

Several sites have been considered. Eventually, after several surveys conducted in conjunction with geologists of the RT Seismic Survey and of the Universities of Pisa and Chieti, the selected site was Terrarossa at Licciana Nardi (Massa Carrara), along the banks of the Magra river (Figure 1).

In particular the geological maps show the existence of recent alluvial deposits with a thickness of about 10 to 15 m overlying the bedrock, which mainly consists of grey-blackish claystone.

In order to check the lithological conditions, a boreholes has been performed. More specifically a rotary borehole (158 mm in diameter) with Large Penetration Test (LPT) [JAMIOLKOWSKI AND LO PRESTI (2003)] measurements every 1.5 m has been carried out.

An open tube piezometer was installed in each borehole for the purpose of periodically checking the water table.

BOZZOLA (2003) performed grain size analyses of the retrieved soil and analyzed the LPT measurements to assess the in situ relative density (Dr). The method proposed by CUBRINOWSKI AND ISHIHARA (1999) was adopted for such a purpose. It was found that Dr was mainly in between 30 and 55 %.

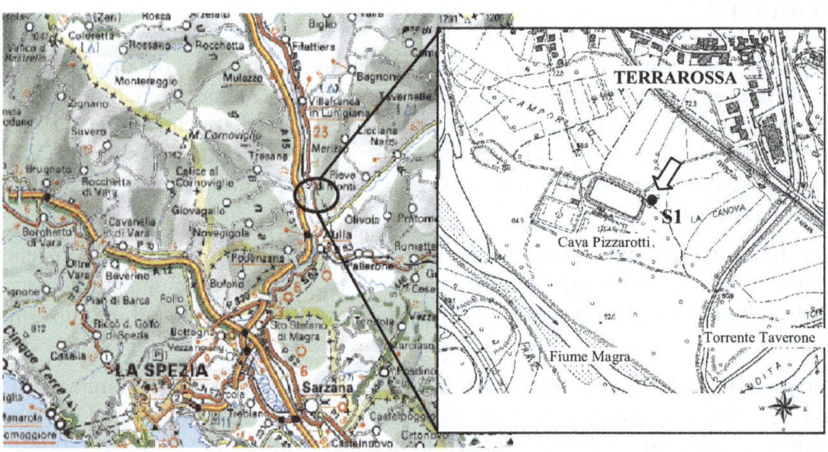

Figure 1 Site location

3. SAMPLING AT TERRAROSSA, LICCIANA NARDI (MASSA CARRARA)

Due to budget restrain, it was decided to retrieve frozen samples from a depth of 2.0 m below the ground level down to 6.0 m. For this purpose guide pipes have been preliminarily installed in agreement with the usual practice [HATANAKA ET AL. 1988, GOTO ET AL. 1992]. More specifically five steel pipes about 2.0 m long connected by two flanges at the ends have been installed in a pre-drilled hole of about 1.5 m in diameter (Figure 2). The central pipe was used to install the double - tube freezing - pipe, necessary for the circulation of the LN_2. The freezing pipe extended down to 8.0 m below the ground level. The shallower portion of the freezing pipe (about 2.0 m) was thermally isolated. Three pipes were necessary to guide the double core barrel; the fourth pipe was used to install a number of thermocouples. This pipe was inclined of about 4° to locate the thermocouples at increasing radial distances with depth.

Unlike the usual method [GOTO ET AL 1992] only three frozen samples were retrieved at each depth, because of budget restrain. On the other end, instead of retrieving a fourth sample, the temperature into the soil was monitored at different depths and radial distances. More specifically, 7 thermocouples were installed. The shallower thermocouple was located at 2.0 m depth with a radial distance of 0.35 m. The other thermocouples were installed at increasing depth and radial distances as precisely shown in Figure 3a.

Figure 3b and 3c show respectively, the LN_2 consumption monitoring and temperature with time. Thermocouples were fixed onto a PVC tube, which was inserted into the inclined guide-pipe and inclined hole. The space between the PVC tube and the hole was filled with mortar pumped from the bottom. Initial high temperatures are due to the chemical reactions of mortar. As for the temperature, it is clear that a target value of about –10°C was reached in most of the target zone after about 5 days (120 hours). On the other end the deeper and more external thermocouples never reached temperature below zero °C. It is supposed that the frozen zone had a radius of no more than 0.70 m. Moreover, temperature into the soil rapidly increases above zero °C at short distance from the frozen front, showing a very high gradient.

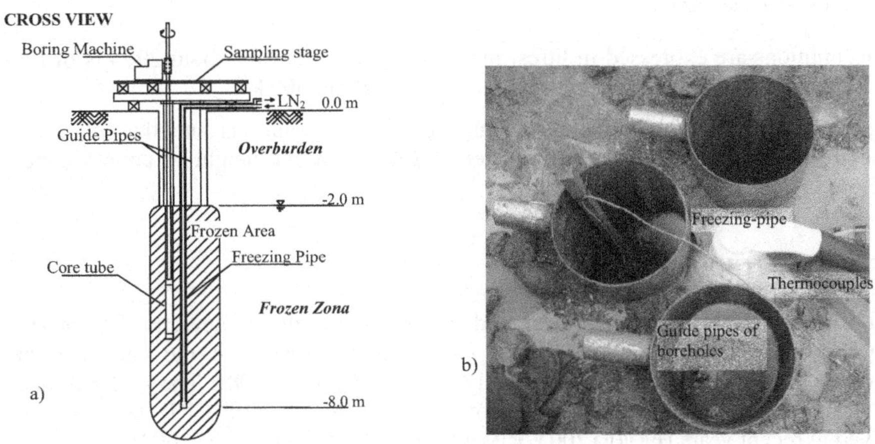

Figure 2 Sampling procedure by in situ freezing [after GOTO ET AL., 1992]: a) cross view; b) picture of plane view in the site test.

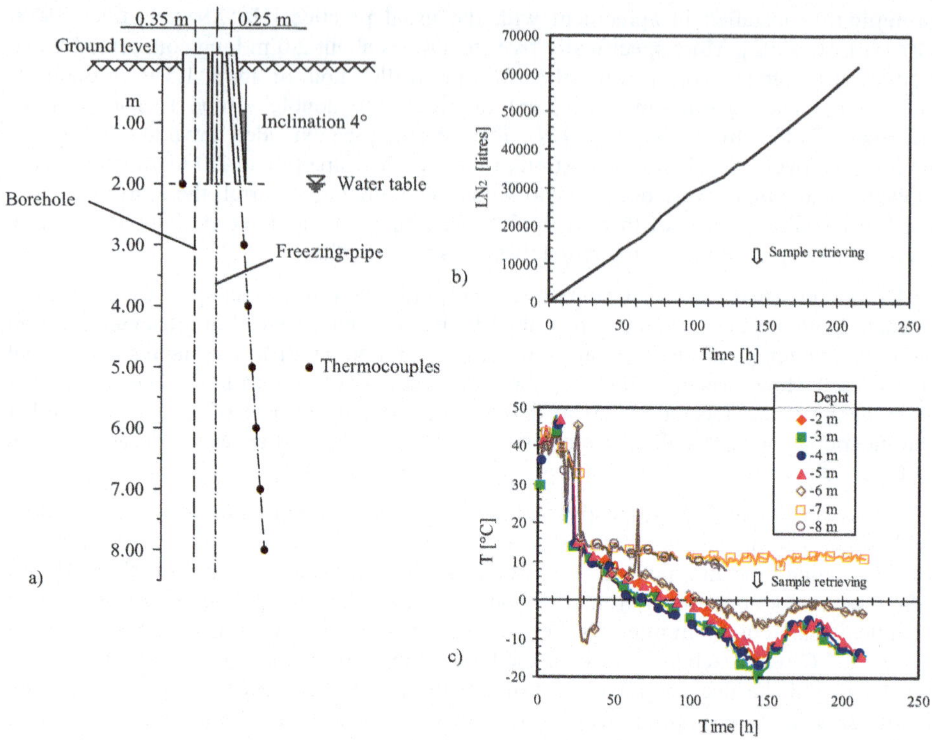

Figure 3 Freezing at Licciana Nardi: a) Thermocouples location; b) LN_2 consumption monitoring; c) temperatures monitoring

Consumptions are expressed in litres. In order to freeze the soil about 290 kN of LN_2 were necessary. The total consumption of LN_2 was of about 500 kN.

Figure 4 shows the individual geotechnical drilling operations and the full set of samples retrieved. Samples have a diameter of 298.0 mm and height of approximately 600.0 mm.

4. TRIAXIAL CELL

The triaxial cell (Figure 5) was designed and realized in the framework of a research contract between ENEL-CRIS of Milan (which is the owner of the cell) and the Politecnico di Torino [Lo Presti et al. 1997, Pietrobono 1998]. The cell has been given in use to the Politecnico di Torino by the courtesy of ENEL-CRIS and has been improved in recent years [Fiorio 2003, Rinolfi 2005, Lo Presti et al. 2004].

The main characteristics of the triaxial cell structure are similar to those accepted for advanced equipments [Tatsuoka 1988, Lo Presti et al. 1994] and are summarized in Lo Presti et al. 2005

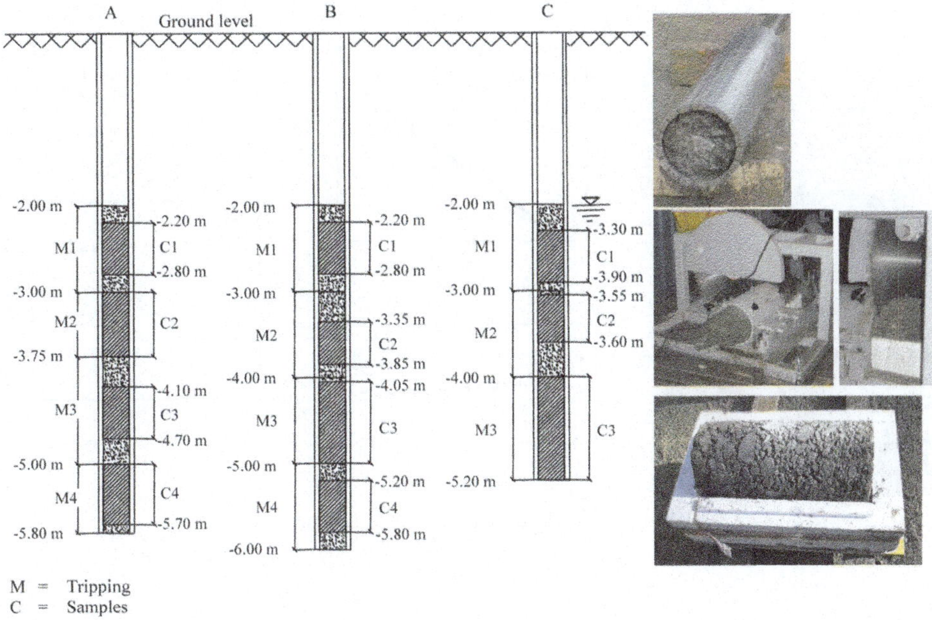

Figure 4 Retrieved sample

The following stresses and strains are measured during a test:

- External axial strain by means of an LVDT (stroke ±100 mm, resolution < 0.01 mm);
- Local axial strain by means of three LDT [GOTO ET AL. 1991], located along the lateral surface of the specimen at 120°. (stroke 20 mm, resolution < 0.002 mm)
- Axial load by means of an internal load cell (stroke = ±200 kN, resolution < 0.02 kN);
- Cell and pore pressure by means of pressure transducers (maximum capacity 10 bar);
- Measurements of volume change have been done using a low-compliance burette system equipped with a differential pressure transducer in agreement with the suggestion by PRADHAN ET AL. (1986). The smallest accurate measurement of volumetric strain is 0.0018%.

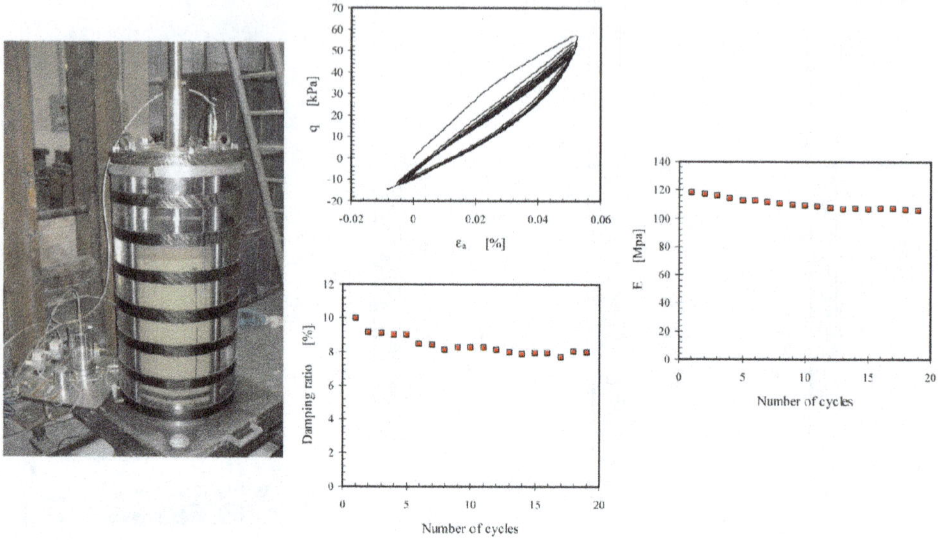

Figure 5 Triaxial cell with a typical results from (strain controlled) cyclic tests

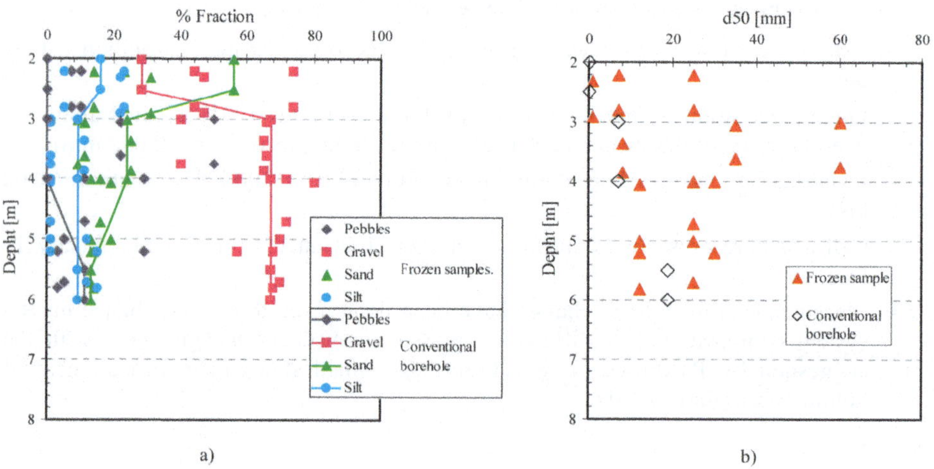

Figure 6 Grading at Licciana Nardi a) grain size distribution, b) mean grain size

5. EXPERIMENTAL RESULTS: GRAIN SIZE DISTRIBUTION AND RELATIVE DENSITY

Grading of the soil retrieved from the conventional borehole and of frozen samples was determined by means of conventional equipments and referring to the usually accepted standards.

It is worthwhile to compare the percentages of different soil types as obtained from conventional borehole [BOZZOLA 2003] and in the case of frozen samples [SICILIANO 2005]. It is clear (see Figure 6a) that the percentage of very coarse soil (pebbles) is small in the conventional borehole while is quite large in some frozen sample. This fact could be a consequence of the different diameter between conventional borehole (158 mm) and frozen samples (298 mm). Anyway, relevant differences in grading are also noticed between adjacent frozen samples (Figure 6a). This fact can be explained by considering a relevant non - uniformity of soil deposit and non - planarity of strata. Such a non – uniformity is also confirmed in terms of mean grain size (Figure 6b).

The relative density of frozen samples was determined. The maximum and minimum dry unit weights were determined according to the ASTM 4253-00 and ASTM 4254-00 [SICILIANO 2005].

6. EXPERIMENTAL RESULTS: SOIL STIFFNESS AND DAMPING RATIO

Triaxial tests were performed on isotropically consolidated specimens. Table 1a summarizes the test conditions at the end of consolidation.

One-way cyclic compression loading triaxial tests were performed in undrained condition under strain control using a triangular waveform with strain rate ranging from $\dot{\varepsilon}_a = 0.2$ %/min to $\dot{\varepsilon}_a = 0.5$ %/min.

The single amplitude cyclic strain levels applied during each test were tentatively: 0.001, 0.002, 0.005, 0.01, 0.02, 0.05, 0.1 %. For each step, 19 loading cycles were applied. Figure 5 show typical test results for a given (0.05%) loading step.

(a) Frozen samples			
	C2	*A3*	*B4*
z (m)	3.05-3.60	4.10-4.70	5.20-5.80
e (-)	0.271	0.246	0.302
σ'_c (kPa)	60.5	69.0	82.7
Reconstituted samples			
z (m)	3.05-3.60	4.10-4.70	5.20-5.80
e (-)	0.373	0.302	0.488
σ'_c (kPa)	46.2	47.6	47.2

(b) Frozen samples			
	A1	*C1*	*B3*
z (m)	2.20-2.80	2.30-2.90	4.05-4.70
e (-)	0.290	0.333	0.332
σ'_c (kPa)	47.6	45.8	54.1
σ_d *(CSR) (kPa)*	95 (0.98)	60 (0.66)	80 (0.74)
Reconstituted samples			
z (m)		2.30-2.90	4.05-4.70
e (-)		0.388	0.373
σ'_c (kPa)		42.6	46.2

Table 1 Test condition: a) Undrained Triaxial Monotonic and Cyclic Compression Loading Tests; b) Liquefaction tests

In each test, after the largest strain level has been imposed to the specimen, the drainage was opened and the specimen was let in drained conditions for, at least, 24 hours. After such a rest period, the specimen was subjected to monotonic loading triaxial compression in undrained conditions under constant rate of strain $\dot{\varepsilon}_a = 0.5$ %/min.

The same type of tests (cyclic and monotonic) were repeated on reconstituted samples. Samples were reconstituted by compacting the soil in 10 strata. In order to guarantee a high saturation degree, the water level inside the forming sample was progressively raised during the reconstitution process so that the water height was coincident with the current specimen height.

Table 1a summarizes test conditions also for the reconstituted samples. Table 1a shows differences in terms of e and σ'_c between undisturbed and reconstituted samples. Small differences of void ratio are due to intrinsic and unavoidable limitations of the adopted reconstitution method.

Figure 7 compares the secant stiffness obtained from undrained monotonic loading triaxial compression tests and that inferred from undrained cyclic loading triaxial compression tests. The results shown in Figure 7a can be commented in the following way:

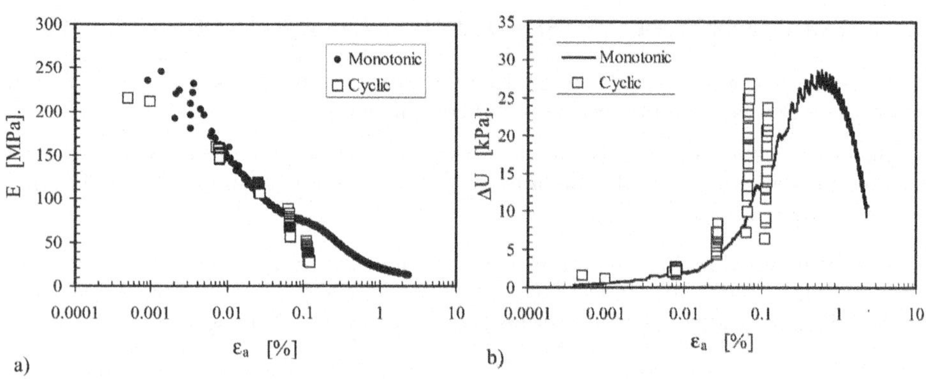

Figure 7 Comparison between undrained monotonic and cyclic loading triaxial compression test: a) Young's modulus; b) pore pressure.

- stiffness from undrained cyclic and monotonic loading tests is almost the same, up to strain values of 0.01%. For larger strains the stiffness from cyclic loading tests is smaller than that observed in monotonic loading tests, especially after the application of a certain number of loading cycles (N). This is probably due to the different mechanism of pore pressure accumulation. In cyclic tests the pore pressure continuously increases, whilst in monotonic tests a dilatant behaviour occurs at large strains. Consequently the pore pressure accumulated in cyclic tests at large strains is higher than that observed in monotonic loading tests (see Figure 7b);

- external axial strain measurements largely underestimate the soil stiffness in comparison to local measurements up to a strain level of about 0.1%.

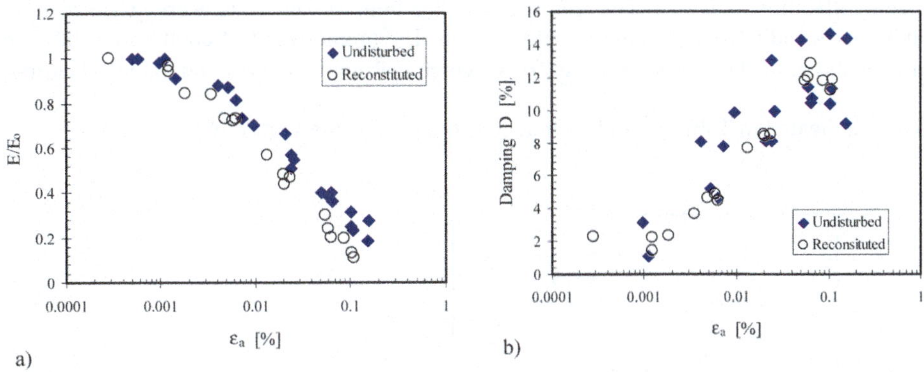

Figure 8 Results of cyclic loading triaxial compression test performed on undisturbed and reconstituted gravel samples: a) Normalized stiffness E/E_o; b) Damping ratio.

Figure 8a and 8b compare the results obtained on undisturbed and reconstituted samples. More specifically Figure 8a compares the stiffness E, as obtained from cyclic loading tests, of undisturbed and reconstituted samples. Soil stiffness has been divided by the initial or small strain modulus E_o.. Figure 8b compares the damping ratio. Both Figures indicate that the normalized stiffness and damping ratio of undisturbed and reconstituted samples are very similar. As for the stiffness, this result is in agreement with to data shown by several researchers [GOTO ET AL. 1987, 1992, 1994, HATANAKA ET AL. 1988, HATANAKA AND UCHIDA 1995, YASUDA ET AL. 1994]. Even data by KOKUSHO AND TANAKA (1994) do not show very relevant differences between the normalized stiffness of undisturbed and reconstituted samples. As for the damping ratio, the already mentioned works show contradictory indications.

The small strain stiffness of undisturbed samples resulted to be up to 20% greater than that of reconstituted samples, but in most cases E_o. of undisturbed samples was very close to that of reconstituted samples tested under same conditions (void ratio - e, consolidation pressure - σ'_c).

To account for differences in terms of e and σ'_c , data have been normalized according to the following dimensionless equation [LO PRESTI 1989]:

$$E_o = S \cdot e^{-1,1} \cdot \sigma'^{0.5}_c \cdot p_a^{0.5}$$

where: p_a = atmospheric pressure expressed in the same unit as stiffness and consolidation pressure and S = soil constant, ranging between 400 and 800. Data available in literature indicate that the small strain stiffness of undisturbed samples from recent Holocene deposits is on average 30 % greater than that measured in the case of reconstituted samples are summarized in JAMIOLKOWSKI ET AL. 2004.

7. EXPERIMENTAL RESULTS: CYCLIC UNDRAINED STRENGTH

Liquefaction tests were performed on isotropically consolidated specimens. More specifically, one – way cyclic compression loading triaxial tests were performed in undrained condition under stress control. Table 1b summarizes test condition at the end of consolidation. The deviator stress (σ_d), which was applied during undrained shearing, is also indicated in Table 1b and, in brackets, the cyclic stress ratio $\left(CSR = \frac{\sigma_d}{2 \cdot \sigma'_c} \right)$.

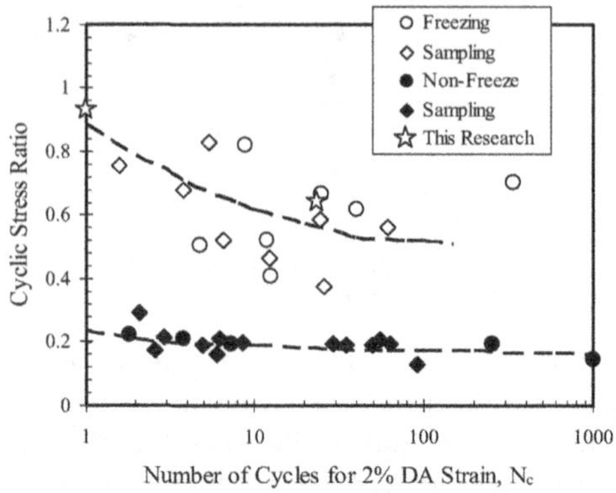

Figure 9 Comparison of CSR obtained in this research on undisturbed samples, to those obtained by KOKUSHO AND TANAKA (1994)

Figure 9 compares the CSR, obtained in this research on undisturbed samples, to those obtained by KOKUSHO AND TANAKA (1994) on gravely undisturbed samples. The CSR was defined as the normalized deviator stress, which causes 2% single amplitude axial strain in a given number of cycles. This condition (for the tests performed in this research) coincides with a pore pressure ratio equal to 1.0 (zero –effective consolidation pressure). It is worthwhile to remark that the agreement with the data by KOKUSHO AND TANAKA (1994) is a coincidence and the only possible conclusion is that the liquefaction strength of undisturbed samples is much larger than that usually obtained from reconstituted samples. It is also important to notice that only two of the three tested samples liquefied.

It is worthwhile to remark that the liquefaction condition appears to be more controlled by compositional factors than global relative density. It should be stressed that the samples that liquefied exhibit a quite important percentage of medium to fine sand, whilst the sample that did not liquefy consists of gravel and coarse to medium sand.

8. CONCLUSIONS

The results shown in this paper allow the following remarks:

- undisturbed sampling by in-situ freezing is technically affordable but extremely expensive and therefore not possible in ordinary projects;
- the small strain stiffness of undisturbed samples was, as a maximum, 20% greater than that observed in the case of reconstituted samples, tested under the same test conditions;
- the normalized stiffness-decay $(E/E_o - \varepsilon_a)$ and damping ratio curves $(D - \varepsilon_a)$ of undisturbed and reconstituted samples are almost the same;
- the undrained cyclic strength of undisturbed samples is much greater than that usually observed in the case of reconstituted samples;

9. ACKNOWLEDGMENTS

Dr. Ferrini Head of the Seismic Survey of the RT, who enthusiastically supported the research project.
Prof. Ilatanaka, Dr. Abe and Mr. Ohara (Tokyo Soil Research Company) for their precious advises about the in situ freezing method and their kind hospitality in Japan where we learnt to use such technique.
Dr. Redepaolini (RCT – Trevi Group) for the accurate sampling of frozen soil.
Dr. Cogliati (Linde group) for providing Liquid Nitrogen and monitoring soil freezing.
Pagani group performed in situ SCPT, free of costs.
Dr. Manenti (Major of Licciana Nardi) and *Mr. Ferdani* (technical staff of the town of Licciana Nardi) for the assistance during all the in situ activities.
Drs. Fiorio, Bozzola, Mensi, Sebastiani, who prepared their M. Sc. Thesis under the guidance of first Author, on topics related to the present study, for their enthusiasm and their presence on site during soil freezing.

REFERENCES

Bozzola F. 2003. *La tecnica del congelamento per il prelievo di campioni di terreno a grana grossa: vantaggi e benefici.* M. Sc. Thesis, Faculty of Engineering at Vercelli, Politecnico di Torino.

Cubrinovski M. and Ishihara K. 1999. *Empirical correlation between SPT-N value and relative density for sandy soils.* Soils and Foundations, No. 5, pp. 61-71.

Fiorio S. 2003. *Rigidezza di terreni a grana grossa in prove triassiali.* M. Sc. Thesis, Faculty of Engineering at Vercelli, Politecnico di Torino.

Goto S., Shamoto Y. and Tamaoki K. 1987. *Dynamic properties of undisturbed gravel sample by the in situ frozen.* Proc. 8th ARCSMFE. 1: 233-236.

Goto, S., F. Tatsuoka, S. Shibuya, Y.S. Kim and Sato T. 1991. *A simple gauge for local small strain measurements in the laboratory.* Soils and Foundations:31(1): 169-180.

Goto, S., Suzuki, Y., Nishio, S. and Oh Oka, H., 1992. *Mechanical Properties of Undisturbed Tone-River Gravel obtained by In-Situ Freezing Method.* Soils and Foundations No. 3, pp. 15-25.

Goto S., Nishio S. and Yoshimi Y. 1994. *Dynamic properties of gravels sampled by ground freezing*. ASCE GSP No. 44, Ground failures under seismic conditions, pp: 141-157.

Hatanaka, M., Suzuki, Y., Kawasaki, T. and Endo, M. 1988. *Cyclic Undrained Shear Properties of High Quality Undisturbed Tokyo Gravel*. Soil and Foundations, Vol. 28, No. 4, pp. 57-68.

Hatanaka M. and Uchida A. 1995. *Effects of test methods on the cyclic deformation characteristics of high quality undisturbed gravel samples*. ASCE GSP No. 56, pp. 136-151.

Kokusho T. and Tanaka Y. 1994. *Dynamic properties of gravel layers investigated by In-situ freezing sampling*. ASCE GSP No. 44, Ground failures under seismic conditions, pp: 121-140.

Jamiolkowski M. and Lo Presti D. 2003. *Geotechnical Characterization of Holocene and Pleistocene Messina sand and gravel deposits*. Invited Lecture International Workshop on Characterization and Engineering Properties of Natural Soils (National University of Singapore, Dec. 2002), Balkema, Vol. 2, 1087-1120.

Jamiolkowski M., Kongsukprasert L. and Lo Presti D.C.F. 2004. *Characterization of gravelly geomaterials*. Proceedings of the Fifteenth Southeast Asian Geotechnical Conference, Bangkok, Keynote lecture, Vol. 2, pp. 29-56.

Lo Presti D.C.F. 1989. *Proprietà dinamiche dei terreni*, XIV CGT, Politecnico di Torino, Department of Structural Engineering.

Lo Presti D., Pallara O., Rainò M. and Maniscalco R., 1994. *A Computer Controlled Triaxial Apparatus; Preliminary Results*, Rivista Italiana di Geotecnica, vol. XXVIII, n. 1, pp 43-60.

Lo Presti D., Pallara O. and Froio F. 1997. *Progetto e realizzazione di una cella triassiale per materiale di grossa pezzatura*. Convenzione tra Politecnico di Torino ed ENEL CRIS, Contratto No. 951/96, Volume 6.1 Relazione Finale.

Lo Presti D., Froio F., Pallara O., Rinolfi A. and Kongsukprasert L. 2004. *Caratterizzazione meccanica di terreni a grana grossa mediante prove di laboratorio su campioni indisturbati e rimaneggiati nei depositi alluvionali delle aree produttive dei Comuni della Garfagnana e Media Valle del Serchio, Lunigiana nell'ambito del programma DOCUP 2000-2006*. Relazione Finale

Lo Presti D., Pallara O., Froio F., Rinolfi A., and Jamiolkowski M., 2005. *Stress-strain-strength behaviour of undisturbed and reconstituted gravely soil samples*, Submitted to RIG for possible publication, October 2005

Pietrobono S. 1998. *Rigidezza dei terreni granulari ricostituiti da prove in una cella triassiale di grandi dimensioni*. M. Sc. Thesis, Department of Structural and Geotechnical Engineering, Politecnico di Torino.

Pradhan B.S., Tatsuoka F. and Molenkamp F. 1986. *Accuracy of automated volume change measurement by means of a differential pressure transducer*, Soils and Foundation, 26(4): 150-158.

Rinolfi A. 2005. *Sul prelievo di campioni indisturbati di ghiaia mediante il congelamento. Determinazione della rigidezza, del rapporto di smorzamento e della resistenza ciclica non drenata*. M. Sc. Thesis, Faculty of Engineering at Vercelli, Politecnico di Torino.

Siciliano D. 2005. *Verifica di approcci empirici per la determinazione della densità relativa dei terreni a grana grossa mediante prove penetrometriche*. Bachelor Thesis, Faculty of Engineering at Vercelli, Politecnico di Torino.

Tatsuoka F. 1988. *Some recent developments in triaxial testing systems for cohesionless soils*. ASTM STP 977, Advanced Triaxial Testing of Soil and Rock. pp: 7-67.

Yasuda N., Otha N. and Nakamura A. 1994. *Deformation Characteristics of Undisturbed Riverbed Gravel by In-Situ Freezin Sampling Method*. Proc. 1st Int. Sympopsium on Pre-Failure Deformation of Geomaterials. Sapporo 1994. Balkema, 1:41-46.

Soil Stress-Strain Behavior: Measurement, Modeling and Analysis
Geotechnical Symposium in Roma, March 16 & 17, 2006

STRENGTH PROPERTIES OF SAND BY TILTING TEST, BOX SHEAR TEST AND PLANE STRAIN COMPRESSION TEST

Kimio Umetsu
Department of Civil Engineering
College of Science and Technology , Nihon University
Surugadai1-8, Kanda, Chiyoda-ku, Tokyo, 〒101-8308, Japan
e-mail: umetsu@civil.cst.nihon-u.ac.jp

ABSTRACT
This report describes comparative examinations of several sands' strength properties using tilting test, box shear test and plane strain compression test. The soil behaviors in these tests are under plane strain condition in common. In tilting test a handy tilting box was contrived to research strength anisotropy with facility. In box shear test a new apparatus which applies stress uniformly was devised to make comparative discussion theoretically with the results of plane strain compression test.

1. INTRODUCTION
The collapse angle of dry sand in a tilted box is also called the angle of repose. Although this angle is said to be equivalent to the angle of internal friction at the loosest state of the material (Miura et al. 1997), the angle is not adopted as a mechanical indicator at present. However, a tilting test performed in University College London(UCL) can research strength anisotropy of sand. In the tilting box the sand surface and bottom are dampened. For that reason, shear failure occurs in the dry middle part. The author contrived a simple tilt box(handy tilt box) to apply this tilting test, and made a comparative examination between the internal friction angle (ϕ) by the tilting test (TT) and the ϕ by the plane strain compression test (PSCT).

The relationship between ϕ by the box shear test (BST) and that by the plane strain compression test has been studied for a long time. However, the box shear test is generally performed such that the upper box is forcibly moved from the lower box to shear the soil specimen within it. In so doing, the stress condition of the specimen becomes very complicated and progressive failure might occur in the specimen. Therefore, results of such box shear tests(traditional box shear test) are unsuitable for a comparative theoretical discussion with the other element test result because the box shear test lacks both "uniformity of applied stress" and "uniformity of deformation", both of which are necessary for element tests. For that reason, the author devised an obliquely divided box shear test (ODBST) apparatus, which uniformly applies stress, and made a comparative examination between the ϕ by BST and ϕ by PSCT. With this comparative examination, we tried to carry out a PSCT in which the axial stress does not influenced by the friction between the specimen and the confining plate.

Hoe I. Ling et al. (eds.), Soil Stress-Strain Behavior: Measurement, Modeling and Analysis, 215–224.

Table-1 Specific gravity, mean grain size and uniformity coefficient of materials

Materials		Specific gravity Gs	Mean grain size D_{50}	Uniformity coefficient Uc
Toyoura sand	Batch-3	2.64	—	—
	Batch-N	2.63	0.24	2.93
SLB sand	Batch-U	2.66	0.62	—
Gifu sand (No.6)		2.64	0.32	1.50
Gifu sand (No.4)	Batch-A	2.64	0.98	1.69
	Batch-B	2.64	0.96	1.69
Oiso sand (ES)		2.75	0.76	2.00
Glass Beads		2.48	2.00	1.00

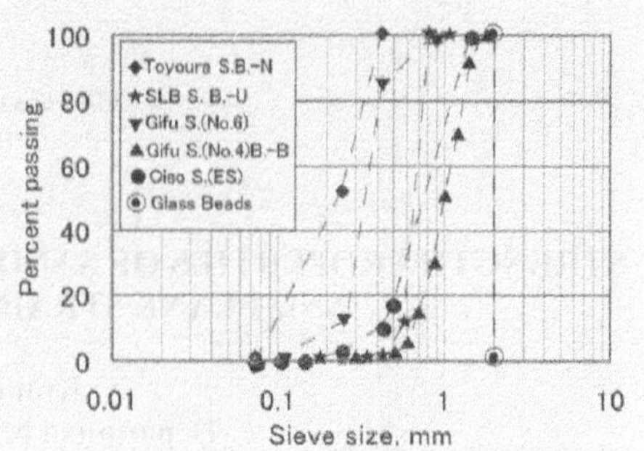

Fig. 1 Grading curves

2. SAND MATERIALS

Eight kinds of materials were used in this study. The specific gravity, mean grain size and uniformity coefficient of each material are shown in **Table 1**. The grading curves are shown in **Fig. 1**. And these grain shapes are in **Photo 1**. Furthermore, **Table 2** shows the correspondence between each material used and the test apparatus, although there is a test apparatus by another researcher among them (as indicated by the Δ mark).

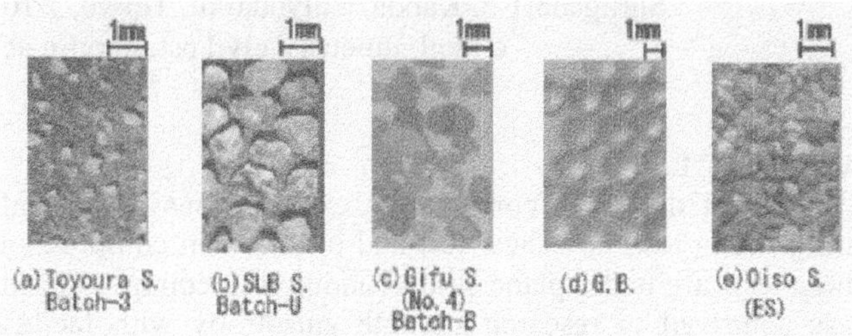

Photo 1 Grain Shape

Table-2 Materials and tests

Materials		Tilting test		Box shear test	Plane strain test
		UCL type	Handy type		
Toyoura sand	Batch-3	○			Δ
	Batch-N			○	○
SLB sand	Batch-U	○			Δ
Gifu sand (No.6)		○	○	○	
Gifu sand (No.4)	Batch-A				○
	Batch-B			○	○
Oiso sand (ES)				○	○
Glass Beads				○	○

○: test of this report Δ: another test

3. TILTING TEST (TT)

3-1 Test apparatus and procedures

(1) Tilt box of the UCL type

A schematic diagram of the tilt box of the UCL type is shown in **Fig. 2**. The test apparatus is a tilt box, which was used in the soil mechanics laboratory of University College London. The box has a 30-degree angle in its top, as shown in **Fig. 2,** and the materials used are glued on the bottom of the box. The test was performed as follows: ① the bottom of this box is dampened by a sprayer, ② dry sands are pluviated into the box uniformly, and the surfaces are prepared in a set thickness, ③ before tilting, the surface is dampened by the sprayer and wet paper is placed on the surface, and ④ shear failure occurs in the middle of the dry part by tilting and the tilt angle is measured.

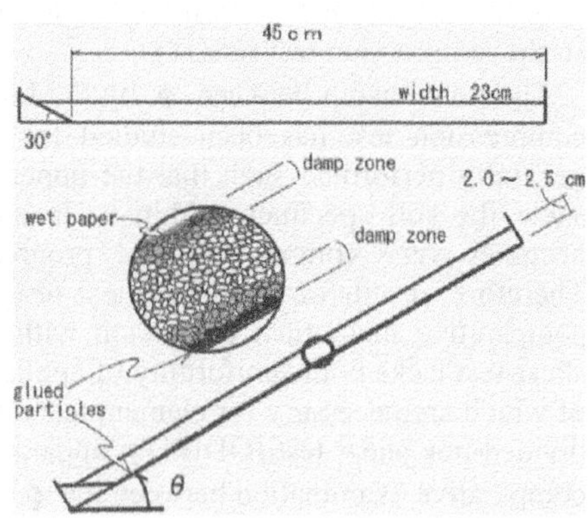

Fig. 2 Tilt Box of UCL type

Fig. 3 shows the method for the **δ** sand layer. The value of **δ** is the angle of the direction of σ_1 relative to the bedding plane and is calculated as shown in **Fig. 3**. The void ratio (**e**) of this sand layer was estimated by measuring the **e** of the side mold.

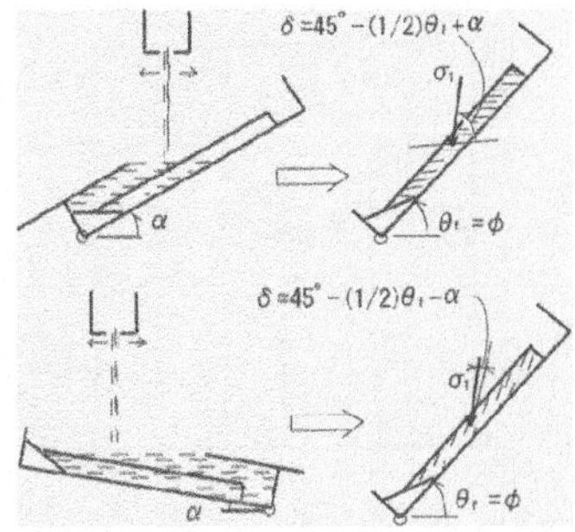

Fig. 3 Way of δ sand layer

(2) Handy tilt box

Fig. 4 shows a schematic diagram of the handy tilt box and the method of testing. This tilt box is an obliquely divided box, as shown in **Fig. 4 (a)**.

Between the upper box and lower box, there is a styrene paper frame (t=5 mm; **Fig. 4 (b)**) and the top of the upper box is open. There are four kinds of obliqueness ($\alpha°$) in these divided boxes, as shown in **Fig. 4 (a)**. The size of the styrene paper frame is the same in all four kinds. The test was performed as follows: ①the sands are pluviated into this obliquely divided box, ② after that, the upper box is removed, ③ the sands on the styrene paper frame are removed and laid flat in order not to disturb them, ④ before box tilting, the surface is dampened by a sprayer, ⑤the failure angle and void ratio are measured in the same way as in the UCL type.

By using the styrene paper frame, the influence of the top in the UCL type can be clarified and the size of the tilt box can be small. And, by preparing four kinds of tilt boxes, as mentioned above, the test of strength anisotropy can easily be carried out.

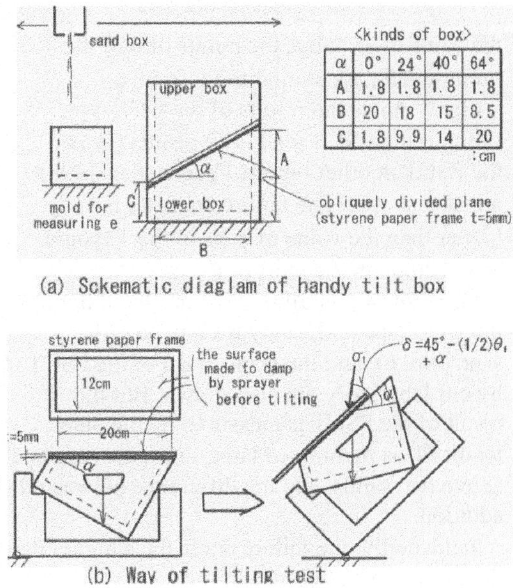

Fig. 4 Handy tilt box

3-2 Test results

Fig. 5 shows the results of the UCL type tilting test on Toyoura sand and the relationship between the failure angle (θ_f) and void ratio (**e**). As shown in the figure, the value of θ_f corresponds to the value of **e**.

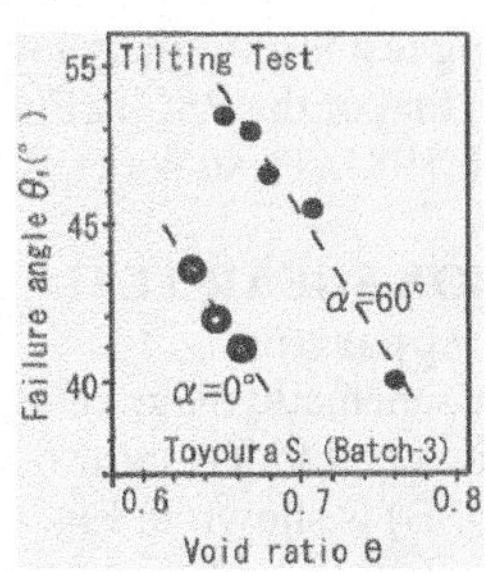

Fig. 5 Failure angle θ_f versus intial void ratio e of Toyoura S. by UCLtype

Fig. 6 shows θ_f plotted against **δ** at **e**=0.66, according to the relationship of **Fig. 5**. The results (ϕ) of the PSCT in other reports (Park 1993, Yasin et al. 1998) are also shown.Although the failure angle θ_f is higher than the value of ϕ by about 5 degrees, the angle θ_f corresponds to

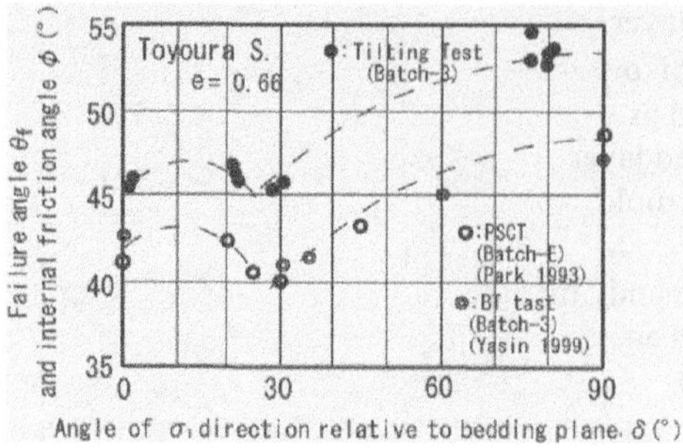

Fig. 6 Strength anisotoropy of Toyoura S. by tilting test (UCL type)

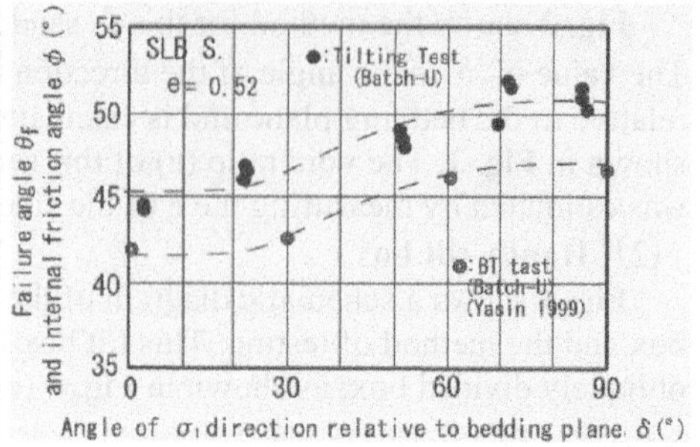

Fig. 7 Strength anisotoropy of SLB S. by tilting test (UCL type)

the value of ϕ ; also, the points of δ of the minimum θ_f are similar to those in ϕ.

Fig. 7 shows the results of the UCL type tilting test on SLB sand. The results (ϕ) of the PSCT in other reports (Yasin et al. 1998) are also shown. The failure angle θ_f is higher than the value of ϕ ,as in the Toyoura sand, although the plots of θ_f vary.

Fig. 8 shows the tilting test results of both the UCL type and handy tilt box on Gifu sand (No. 6), and the results (ϕ) of the PSCT by our laboratory are also shown. But this result of the PSCT is measured by the outer load cell, as mentioned later. From the results, these two kinds of θ_f are similar to the above two sands, and the difference between the two tilting tests is not recognized in addition.

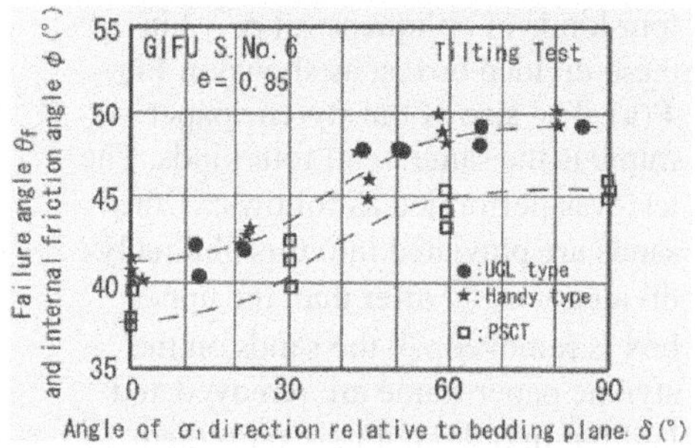

Fig. 8 Strength anisotoropy of Gifu S. No. 6 by tilting test (UCL type and handy tilt box)

Incidentally, the failure angle θ_f is higher than the internal angle ϕ of the PSCT, and this seems to be because of the confining stress dependency, that is, the confining stress in the tilting test is extremely lower than that of the PSCT (Fukusima et al. 1984, Tatsuoka et al. 1986).

4. BOX SHEAR TEST

4-1 Apparatus

A schematic diagram of the obliquely divided shear box (ODSB) is shown in **Fig. 9**.The shear box is made up of upper and lower obliquely divided boxes of 60° obliqueness. A

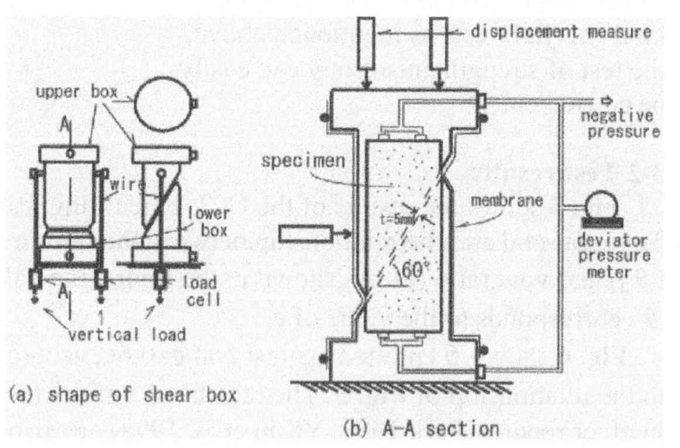

Fig. 9 Obliquely divided box shear test appratus

rectangular specimen, which is described later, is placed in the box and covered with a membrane. Negative pressure is applied in the box and a vertical load is applied to the upper box, as shown in **Fig. 9**.

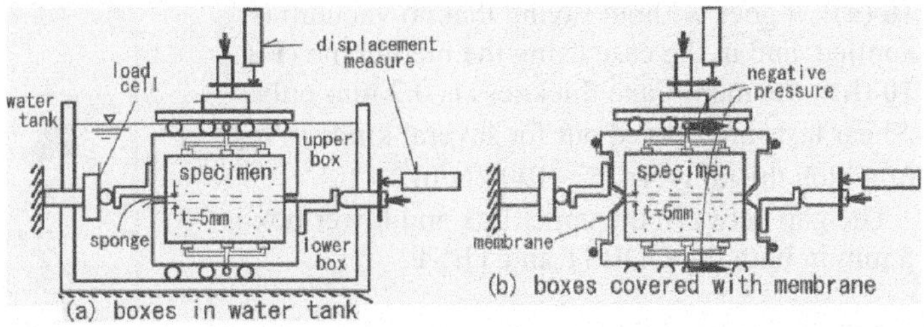

Fig. 10 Traditional box shear test ppratus

Fig. 10 shows a traditional shear box (TSB), which is compared with the result of the ODSB. There are two kinds of TSBs. One is shown in **Fig. 10 (a)** and the other is shown in **Fig. 10 (b)**. **Fig. 10 (a)** shows a shear box, which is lowered into the water to saturate the specimen. **Fig. 10 (b)** shows a shear box covered with a membrane, as in the ODSB.

4-2 Specimens and test procedures

Fig. 11 shows the specimen shapes and sizes of the ODSB and TSB. In both tests the specimens are prepared by dry sand pluviated into a mold and subsequently frozen. The frozen specimen is set in the box. However, in setting the specimen, the direction of bedding plane must be as shown in **Fig. 11** in order that the influence of the bedding plane direction does not appear. In these tests, the void ratio (e) is 0.72～0.74.

Test procedures in the obliquely divided box shear test (ODBST) are as follows: ① the box holding the frozen specimen is covered with a membrane, and then is thawed under a vacuum of 49 kN/m^2. ② the thawed specimen is saturated, ③ the vacuum is applied (the stress condition on the divided plane in the box is labeled A in **Fig. 12**), ④ each displacement measure is set as shown in **Fig. 9**, ⑤ a vertical load is applied to the upper box (the stress path on divided plane is labeled A→B' in **Fig. 12**), ⑥ the amount of vacuum is reduced and the specimen is sheared at the divided part (the stress path is B'→B in **Fig. 12**), ⑦ as a result, after the process of ⑤ and ⑥ are carried out gradually and together, the test of σ_n=constant can be carried out as in the traditional box shear test (TBST).

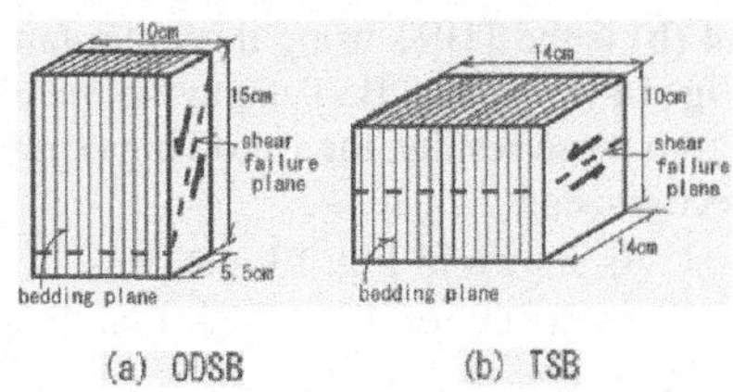

Fig. 11 Specimen (ODSB and TSB)

All of the following results are under the stress path mentioned above. The thicknesses of the membrane are 0.2, 0.3 and 0.4 mm. The traditional box shear test (TBST) is performed the same as the ODBST and σ_n is constant during the shear test. But, in the case using the water tank (**Fig.**

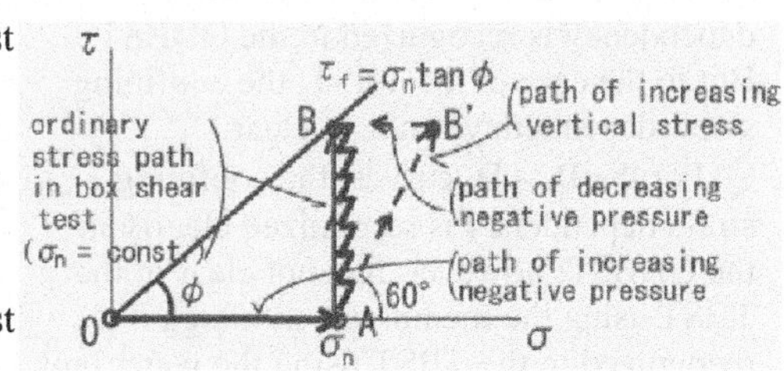

Fig. 12 Stress path (ODSB)

10 (a)), it goes without saying that no vacuum is applied, and in the case using the membrane (**Fig. 10 (b)**), the membrane thickness is 0.2 mm only. Shear tests are carried out for several kinds of σ_n, of which the range is 15～100 kN/m^2.

The gap between the upper box and lower box is 5 mm in both the ODBST and TBST.

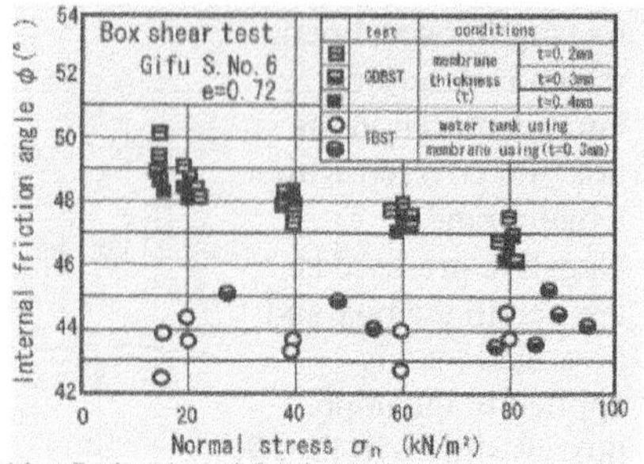

Fig. 13 Relationship between ϕ and σ_n

4-3 Test results

Fig. 13 shows the internal friction angles (ϕ) of the above three box shear tests (one ODBST and two TBSTs) on Gifu sand No. 6 against σ_n at e=0.72. All ϕ values by the ODBST are higher than those of the TSB-T and the confining stress dependency is recognized in the ODBST. The influences of membrane thickness are not recognized in the ODBST. In the results of TBST, the values of ϕ when using the membrane are a little bit higher than those when using the water tank.

Fig. 14 shows representative stress ratio (τ/σ_n) - shear displacement (**D$_s$**) curves and normal displacement (**D$_n$**) - shear displacement (**D$_s$**) curves of three test apparatuses: **Fig. 14 (a)** is the ODBST, **Fig. 14 (b)** is the TBST using the water tank and **Fig. 14 (c)** is the TBST using the membrane. From these results, the following points are recognized.

① As shown in **Fig. 13**, the peak (τ/σ_n) of the ODBST is higher than those of the others.

② The value of **D$_s$** at the peak (τ/σ_n) is small in the ODBST and the confining stress dependency is recognized in the ODBST. But in the case of the TBST, the confining stress dependency is not so clear.

③ For the **D$_n$** - **D$_s$** curves, the confining stress dependency is recognized clearly in the ODBST; however, it is not clear in the TBST using the membrane, although it is recognized in the TBST using the water tank.

In the TBST, progressive failure may occur in the specimen because the upper box is forcibly made to move from the lower box to shear the soil specimen

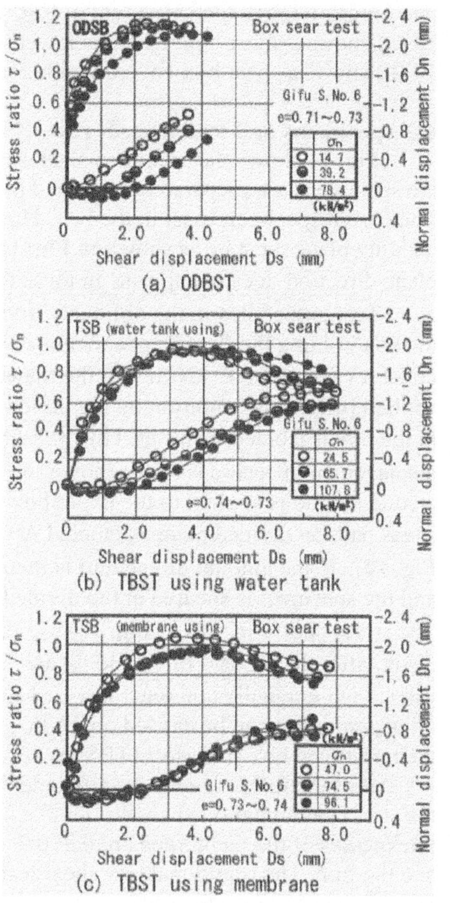

Fig. 14 Stress ratio τ/σ_n - shear displacementand Ds curves and normal displacement Dn - shear displacementand Ds curves

that is in it. On the other hand, in the ODBST, both the upper and lower obliquely divided boxes are covered with membrane and shear failure occurs between those two boxes by way of both the vertical stress and the pore negative pressure. The above results seem to describe those two different shear methods.

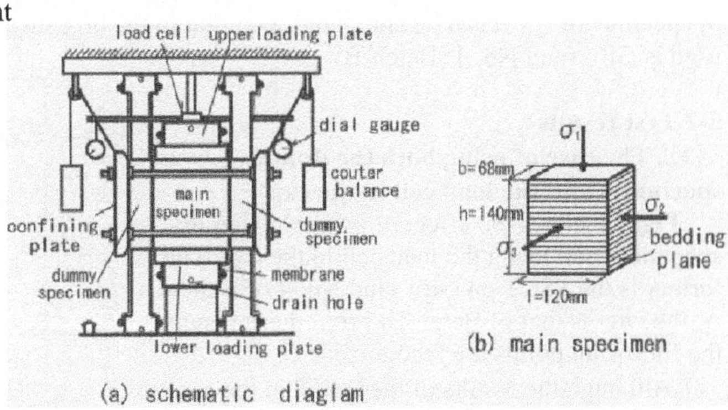

Fig. 15 PSCT used with dummy specimen and main specimen

5. PlANE STRAIN COMPRESSION TEST (PSCT)

5-1 Test apparatus and procedures

In the PSCT, the friction on the confining plane is caused by the confined side deformation of the specimen. So, the PSCT, as shown in **Fig. 15(a),** which has dummy specimens (**l**:50×**b**:68×**h**:140 mm) on both sides of the main specimen (**l**:120×**b**:68× **h**:140 mm), was performed. These specimens are prepared similarly: that is, dry Gifu sand No. 4 (Batch-A) is pluviated into a mold, moistened and subsequently frozen. After freezing, the specimen which is set between the cap (upper loading plate) and pedestal (lower loading plate) is covered with a membrane (**t**=0.3 mm) and then is thawed under a vacuum of 19.6 kN/m^2. Each surface of the cap and pedestal is lubricated with a 0.3-mm thick single rubber sheet smeared with grease. The σ_2 surfaces of the main specimen are also smeared with grease; however, the test, in which the surface is smeared with bond instead of grease, is also performed. After the specimen is saturated, a vacuum (σ_3) is applied and then the PSCT is performed under a drained condition.

The bedding plane of this specimen is normal to the direction of σ_1 (δ=90°), as shown in **Fig. 15 (b).**

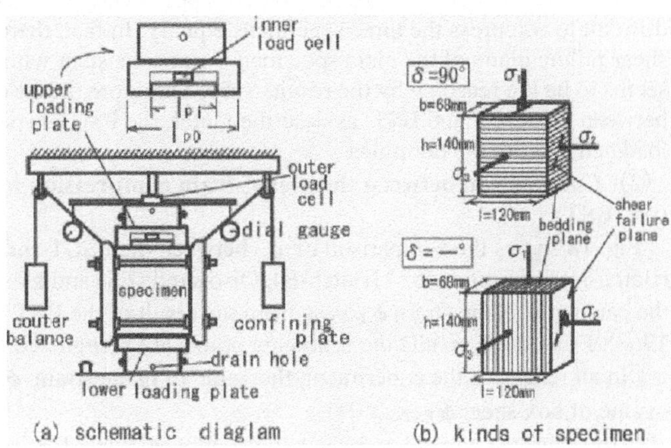

Fig. 16 PSCT with load cell in upper loading plate and kinds of specimen

Fig. 16 (a) shows the PSCT, in which a load cell is set in the cap in order to measure σ_1 without the influence of the friction on the confining plate. The specimen is the same as above (**l**:120×**b**:68× **h**:140 mm). The size of the cap (upper loading plate) is $\mathbf{l_{po}}\times\mathbf{b_p}$=118×75 mm. In the center part of this upper plate, there is a plate connected to an inner load cell. There two sizes of this plate. One is $l_{pi} \times b_p$=100×75 mm and the other is $l_{pi} \times b_p$=60×75 mm. Although the method of preparing the specimen is the same as that above, two kinds of specimens are prepared, as shown in **Fig. 16 (b):** that is, one is **δ**=90°, the other is **δ**=−°. This type of specimen(**δ**=−°) can be compared with that in the box shear test(BST). The material used is Gifu sand No. 4 (Batch-B).

5-2 Test results

(1) The case of using both the dummy specimens and the load cell in the cap

Fig. 17 shows the effect of using the dummy specimens and using the load cell in the cap. The former is the PSCT on Gifu sand No. 4 of Batch-A, the later is that of Batch-B. From these results, the following points are recognized.

①Although the results of the PSCT in the case using the dummies vary, they are distributed under the results of the PSCT without using the dummies, and the effect of the dummies appears. In the result of the PSCT, in which the σ_2 surfaces are smeared with bond, although there is only one result, the effect of the dummies does not appear.

② The results of the PSCT in the case using the inner load cell are lower by a mere 1° than those measured with the outer load cell. The length of the plate connected to the inner load cell does not affect the result.

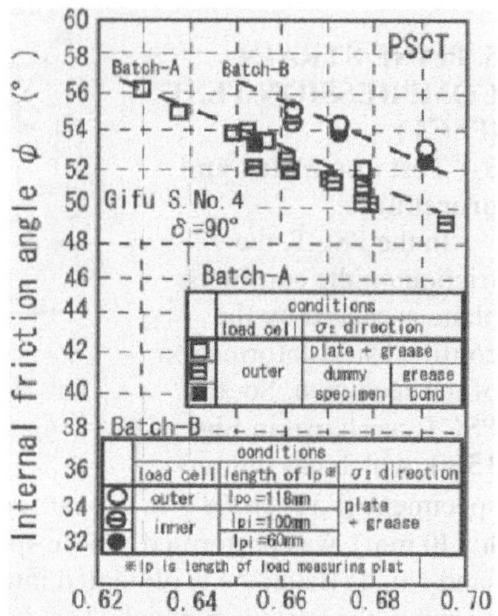

Fig. 17 The value of φ versus initial void ratio e by PSCT used dummy specimens and an inner load cell (Gifu sand No. 4, Batch-A, B)

In the PSCT using the dummies and the bond on the σ_2 surfaces, the effect of the dummies does not appear contrary to our expectations. This seems to mean that it is difficult to compress the three specimens equally. In fact, there are many cases where the shear failure plane of the main specimen is not consistent with that of the dummy. This seems to be the reason why the results vary. Therefore, in the case of the comparison between the PSCT and BST, as described later, the PSCT is performed by using inner load cell and not the dummies.

(2) Comparison between the plane strain compression test (PSCT) and box shear test (BST)

Fig. 18 shows the comparison of ϕ between the PSCT and BST on Toyoura sand (Batch-N),Gifu sand No. 4 (Batch-B), Oiso sand (ES) and glass beads. In the same figure, the calculated plane strain $\phi_{C\text{-}PSCT}$ from the result of the BST by Davis is plotted (Davis 1968). From these results the following points are recognized.

①In all results of the experiment, the value of plane strain ϕ_{PSCT} is higher than the value of box shear ϕ_{BST}.

②However, the correspondence between the calculated $\phi_{C\text{-}PSCT}$ and the experimental ϕ_{PSCT} differs according to the kinds of sand.

③ On Toyoura sand and glass beads, the value of $\phi_{\text{C-PSCT}}$ approximates the value of ϕ_{PSCT}, but on Oiso sand the calculated $\phi_{\text{C-PSCT}}$ is smaller than theexperimental ϕ_{PSCT}, on the other hand, on Gifu sand (No. 6) the calculated $\phi_{\text{C-PSCT}}$ is higher than the experimental ϕ_{PSCT} by about 10 degrees.

The equation by Davis is made using the dilatancy property in BST. The dilatancy is affected by the stress path and rotation of the principal stress. However, the stress path in the PSCT is different from that of the BST; besides, the PSCT has no rotation of the principal stress, whereas BST does have principal stress rotation. Moreover it seems that the dilatancy property is influenced by the grain shape considerably. These differences between $\phi_{\text{C-PSCT}}$ and ϕ_{PSCT} by different kinds of sand were expected to be depend on the angularity of grain shape first, that is, the less angular the grain shape is, the more correspondent the calculated $\phi_{\text{C-PSCT}}$ to the experimental ϕ_{PSCT} . However, on Toyoura sand, although the grain shape is angular(**Photo 1(a)**), the calculations ($\phi_{\text{C-PSCT}}$) is similar to the experimental result (ϕ_{PSCT}), but then on Oiso sand, although the grain shape is less angular(**Photo 1(e)**), the calculations ($\phi_{\text{C-PSCT}}$) is smaller than the experimental result (ϕ_{PSCT}), After all above results cannot be interpreted clearly.

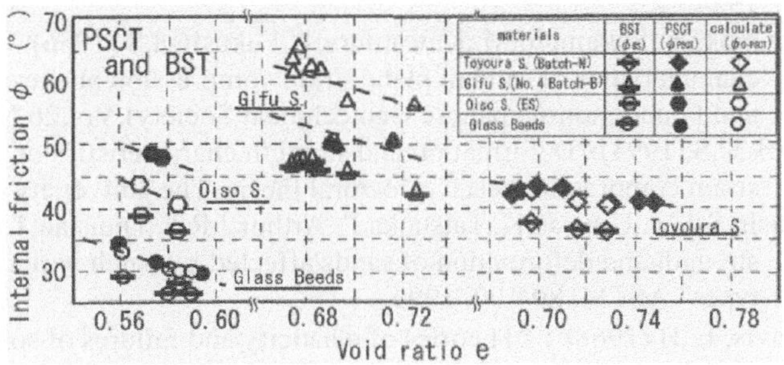

Fig. 18 The value of φ versus initial void ratio e by PSCT(Toyoura S., Gifu S., Oiso S. and Glass beeds)

6. CONCLUSION

On the basis of a limited number of tilting tests (TT), box shear tests (BST) and plane strain compression tests (PSCT) on several kinds of sand, as reported in this paper, the following was found.

(1) The failure angle θ_f in the TT is higher than the internal friction angle ϕ in the PSCT.

(2) The value of ϕ in the PSCT is higher than that in the BST.

(3) The correspondence between the calculated ϕ of the PSCT by Davis and the measured ϕ in the PSCT is depend on the kinds of grain materials.

Table-3 Tests and δ of sample

Materials		Tilting test		Box shear test	Plane strain test
		UCL type	Handy type		
Toyoura sand	Batch-3	δ=0~90°			δ=0~90°
	Batch-N			δ=-°	δ=-°
SLB sand	Batch-U	δ=0~90°			δ=0~90°
Gifu sand (No.6)		δ=0~90°	δ=0~90	δ=-°	δ=0~90°
Gifu sand (No.4)	Batch-A				δ= 90°
	Batch-B			δ=-°	δ=-° , 90°
Oiso sand (ES)				δ=-°	δ=-°
Glass Beads				δ=-°	δ=-°

δ : angle of σ_1 direction relative to bedding plane (σ_2 direction is normal to bedding plane)
δ=-° : σ_2 direction is parallel to bedding plane

Above these apparatuses are under the plane strain condition, therefore the relationship among the TT, BST and PSCT can be examined theoretically. However, as shown in **Table 3**, in order to examine these relationships, the results of **δ**=−° in the TT are necessary.

ACKNOWLEDGMENT

The beginning of this study was included in the research activities at the University College in London (UCL), where the author was in 1993, thanks to Prof. Tatsuoka. The author expresses gratitude to Prof. J.R.F. Authur, Dr. T. Dunstan and other staff of UCL, as well as Prof. Tatsuoka. The many experiments in this report were carried out mainly by A. Ishigami, Y. Ogiwara, J. Hamayose, S. Konno, and T. Mori who were graduate students at that time. The author gratefully acknowledges them.

REFERENCE

Miura,K.,Maeda,K.,Toki,S. (1997): "Method of measurement for the angle of repose of sands" Soils and Foundations(Japanese Geotechnical Society),Vol.37,No.2,89-96, June 1997

Fukusima,S., Tatsuoka,F. (1984): "Strength and deformation characteristics of saturated sand at extremely low pressures" Soils and Foundations, Vol.24,No.4,30-48, Dec. 1984

Tatsuoka,F.,Sakamoto,M.,Kawamura,T.,Fukusima,S.(1986): "Strength and deformation characteristics of sand in plane strain compression at extremely low pressures" Soils and Foundations(Japanese Geotechnical Society),Vol.26,No.1,65-84, Mar. 1986

Park,C.S.(1993):"Deformation and strength characteristics of a variety of sands by plane strain compression tests", Doctoral thesis ,The University of Tokyo

Yasin,S.J.M.,Umetsu,K.,Tatsuoka,F.,Arthur,J.R.F.,Dunstan,T.(1999): "Plane strain strength and deformation of sands affected by batch variations and different apparatus types" ASTM, 80-100, 1999

Davis, E.,H.(1968) : "Theories of plasticity and failures of soil masses" Soil Mechanics, selected topics(ed. I. K. Lee), London : Butterworth.

Soil Stress-Strain Behavior: Measurement, Modeling and Analysis
Geotechnical Symposium in Roma, March 16 & 17, 2006

MICRO X-RAY CT AT SPRING-8 FOR GRANULAR MECHANICS

Takashi Matsushima [1], Jun Katagiri [2], Kentaro Uesugi [3], Tsukasa Nakano [4] and Akira Tsuchiyama [5]

[1] Department of Engineering Mechanics and Energy,
University of Tsukuba, 1-1-1, Tennodai, Tsukuba, Ibaraki 305-8573, JAPAN.
E-mail: tmatsu@kz.tsukuba.ac.jp
[2] Undergraduate student, College of Engineering Systems,
University of Tsukuba
[3] Japan Synchrotron Radiation Research Institute (JASRI),
SPring-8, Hyogo, JAPAN
[4] Institute of Geology and Geo-information,
National Institute of Advance Industrial Science and Technology (AIST),
Tsukuba, JAPAN.
[5] Department of Earth and Space Science, Graduate School of Science,
Osaka University, Osaka, JAPAN.

ABSTRACT
This paper deals with the microscopic visualization technique of sand specimen using a micro x-ray CT at SPring-8. SPring-8 is the world's largest third-generation synchrotron radiation facility and it has laboratories, BL47XU and BL20B2, at which a high-resolution x-ray CT system is available. A series of experiments performed in 2003 and 2004 are overviewed to demonstrate the performance of the system. Then the paper focuses on a newly-developed image technique to identify each grain in the specimen. The obtained grain information is further processed to some micromechanical properties such as grain size distribution and contact point statistics.

1. INTRODUCTION
The mechanical properties of sand are essentially governed by the constituent grains' properties and their structure. Grain properties include its size, shape, crushability and so on, while the granular structure can be represented by void ratio, fabric tensor, the orientation of grain's axes, and so on. Recent development of computer ability has made it possible to handle such micro properties in particle-based simulations like Discrete Element Method (DEM)[1], and accordingly a lot of important numerical data have been piled up [2][3]. On the other hand, experimental data to directly enhance such numerical results are still lacking. In other words, particle-level observation in physical experiment is definitely needed to prove the numerical findings correct.
Experimental researches on this matter so far always have encountered the difficulty how to measure such 3-D microstructural information. X-ray CT is a powerful tool to do that [4], but standard sands frequently used in geotechnical laboratory tests have their mean diameter of about 0.1mm to 0.5mm, which is less than the resolution of usual medical or industrial x-ray CT devices. Micro x-ray CT using a synchrotron radiation facility is one

Hoe I. Ling et al. (eds.), Soil Stress-Strain Behavior: Measurement, Modeling and Analysis, 225–234.

of the very few exceptions [5]. In particular, SPring-8, the world's largest third-generation synchrotron radiation facility, has laboratories, BL20B2 and BL47XU, at which a well designed micro x-ray system (SP-μCT) is available. Its high flux density x-ray beam is naturally well collimated and monochromatic, which enables us a very high spatial resolution of x-ray CT (around 10μm and 1 μm at BL20B2 and BL47XU, respectively)[6][7].

This paper describes a series of micro x-ray CT experiments on sands conducted at BL20B2 at SPring-8 during 2003 and 2004 [8][9][10]. The overview of the experiments is presented at first to demonstrate the performance of the system. Then a newly-established image analysis procedure to identify the constituent grains in the specimens is described in details, and its accuracy is examined. Finally some micromechanical properties such as grain size distribution and contact point statistics are presented and discussed.

2. MICRO X-RAY CT EXPERIMENTS ON SANDS AT SPring-8

X-ray CT system at BL20B2 at SPring-8

The x-ray CT system (SP-μCT) at BL20B2 consists of an x-ray light source, double crystal monochromator, high precision stages and high resolution x-ray image detector (**Figure 1**) [6][11]. The intense x-ray from the storage ring is first monochromatized with a Si (311) double crystal monochromator to make the subsequent CT process simple and accurate. Since the original energy is immense, the available monochromatized x-ray energy at the sample stage is still as much as 9-72 keV. Passing through the specimen the x-ray arrives at an x-ray image detector, which consists of thin scintillator, optic system and CCD camera. The effective pixel size depends more or less on the choice of the optics and the CCD camera. In the following experiments the format of the CCD camera was 1000 by 1018 and the effective pixel size of the detector was 5.8μm by 5.8μm.

The x-ray energy used in the following experiments was 25 keV. The image was taken every 0.5 degree for 180 degrees rotation, that means 360 images are used for a single CT reconstruction.

Visualization of grain shape and packing structure of sands and glass beads

We conducted a series of experiment with different standard sands including Toyoura sand, Ottawa sand and others together with artificial glass beads (nearly spherical grains).

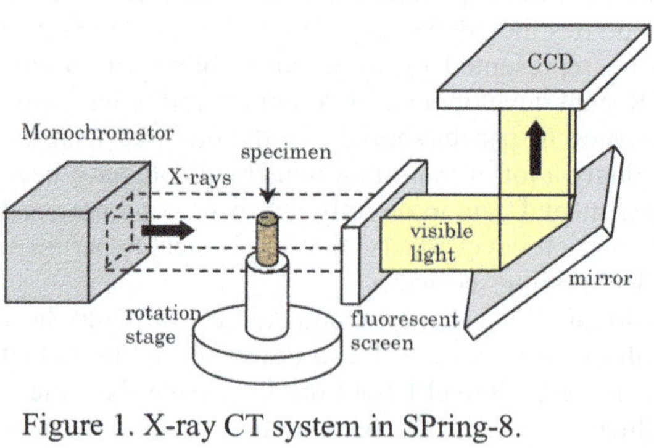

Figure 1. X-ray CT system in SPring-8.

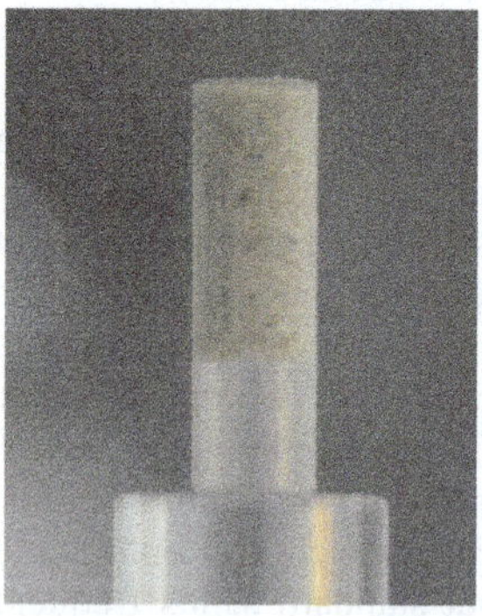

Figure 2. An example of specimen

Each material was prepared in a cylindrical vessel whose inner and outer diameters are 3.5mm and 4.5mm, respectively (**Figure 2**). A loose specimen was prepared with a special funnel that can be inserted in the vessel. Firstly we put the funnel in the vessel and placed the sand in the funnel. Then the funnel was slowly pulled out so that the loosest condition corresponding to the maximum void ratio test [12] is attained. A dense specimen was made by tamping with a rod.

After the x-ray shooting, we measured the weight of the specimens to calculate the void ratio.

Figure 3 shows examples of the CT images (a horizontal cross section) converted to 8-bit images. The reconstruction of CT image was conducted with a program based on the convolution back projection (CBP) method [13]. The inclination of rotation axis was considered to obtain clear images. The pixel intensity 0 and 100 in the images correspond to the observed LAC (the value of linear attenuation coefficient of x-ray) of 0 and 3.663 cm^{-1}, respectively. The latter value is the theoretical LAC of quartz (SiO_2) at 25 keV. Since Most of sand grains and glass beads consist of quartz, the grains in **Figure 3** are mainly in monotonic gray. Slightly brighter grains than quartz are feldspars. Very few grains have rather white, which indicates that they consist of more heavy minerals.

The further image analysis to identify the constituent grains is described in the following sections.

Micro triaxial tests

In order to observe microstructural change inside a triaxial specimen due to deviatoric compression, we developed a micro triaxial test apparatus suitable for SP-μCT at BL20B2 (**Figure 4**). It is so compact and light that we can easily deriver to SPring-8 and set on the rotational stage. The severe requirement for SP-μCT is that the apparatus width in the scanning area (see **Figure 4**) must not exceed the threshold (10 mm) during 180 degree rotation. This threshold is determined by the image detector size. Therefore, in this apparatus, the reaction force for the loading is sustained by the specimen cell. It is an acrylic cylinder of 1mm thick, and its outer and inner diameters are 10 mm and 8 mm, respectively. The specimen cell is connected with the lower pressure tank so that the confining pressure can be kept constant. The precise loading motor is inside the pressure tank and the axial strain is applied from the bottom of the specimen with a constant speed.

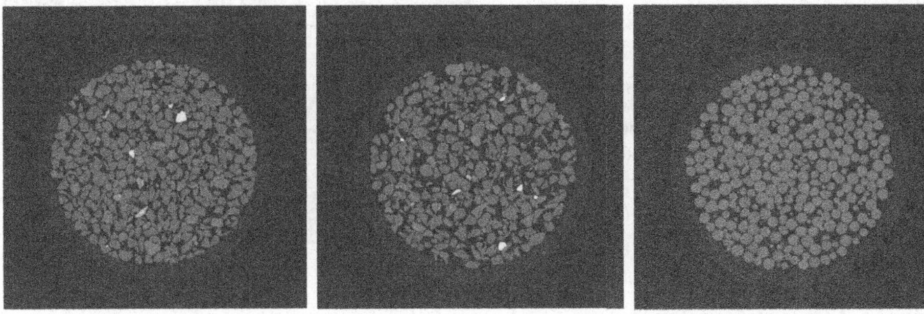

(a) Toyoura sand (dense) (b) Toyoura sand (loose) (c) Glass beads (dense)

Figure 3. Examples of reconstructed CT image (horizontal cross section)

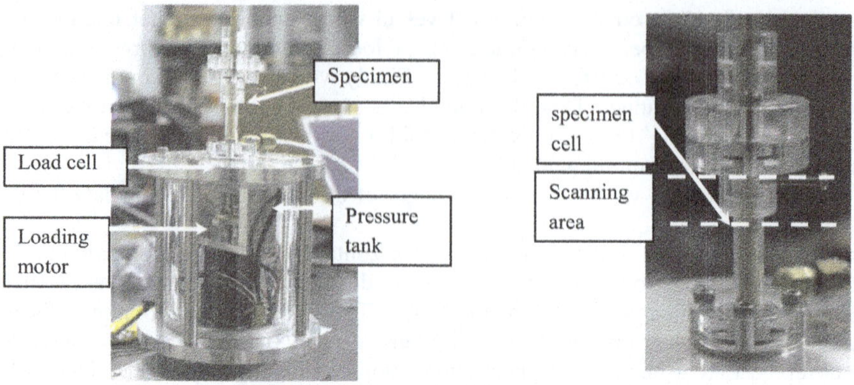

Figure 4. Micro triaxial test apparatus

Specimen size is 3.4 mm in diameter and 10 mm in height. **Figure 5** shows stress-strain curves for dense and loose Toyoura sand specimens. The confining pressure is 100 kPa and the axial strain rate is about 2.5%/min. The resultant peak axial stresses for both specimens are rather higher than those with a conventional triaxial test whose specimen size is over 50 mm in diameter and 100mm in height. It is mainly due to the effect of relatively thick (0.15mm) membrane wrapping the specimens.

Loading was stopped several times during the test, as it is observed in **Figure 5** by the vertical stress drop, to take a series of X-ray images. **Figure 6** shows the reconstructed CT images of a central vertical cross section of the dense specimen before and after loading. It can be recognized that the shear deformation is concentrated on the upper part of the specimen, accompanying a considerable dilation. The images are clear enough to quantify each grain motion including its rotation. However, since the number of grains is huge and their motion is 3-D, it is necessary to develop an appropriate grain identification technique.

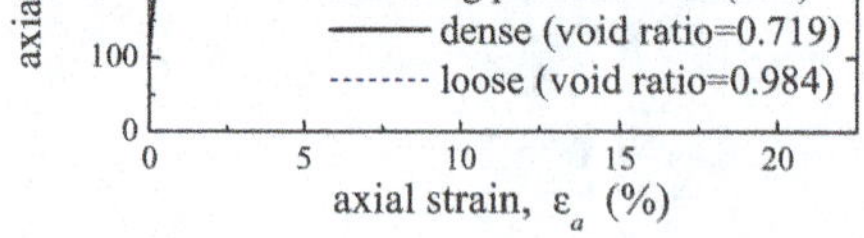

Figure 5. Stress-strain relation in triaxial compression

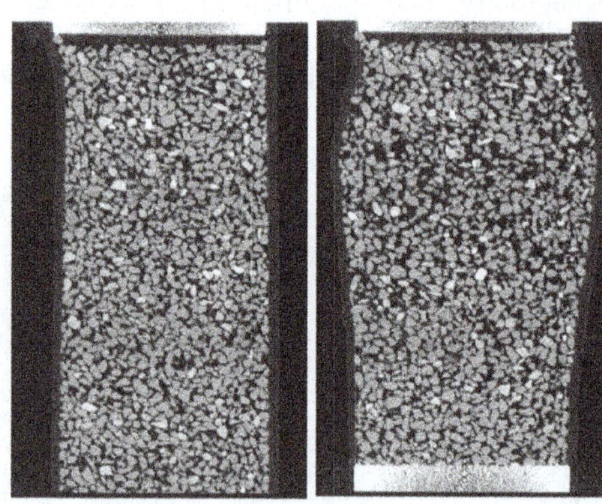

(a) $\varepsilon_a = 0\%$ (b) $\varepsilon_a = 15\%$

Figure 6. CT images in triaxial compression.

3. IMAGE PROCESSING FOR GRAIN IDENTIFICATION

As described in the previous sections, it is essential to establish an accurate grain identification method to make the CT images more valuable in granular mechanics researches. This section describes our efficient image processing technique for grain identification. It is composed of several processes. All the processes are done automatically and no visual judgment by an operator is necessary. This is very important because the amount of data is extremely huge.

Binarization

The first step is to distinguish solid part from void by the binarization of the CT images. **Figure 7** shows the frequency of the observed LAC for various specimens. The curves show that the materials in the measured region can be classified into two; the material of the lower LAC corresponds to air (void) or acrylic cell, while the material of the higher LAC corresponds to solid grains. The peak LAC for solid grains is about 3.2 (cm^{-1}) for the sands (Toyoura sand, Ottawa sand and Hostun sand), and around 3.7 (cm^{-1}) for the glass beads. This difference is due to the difference of quartz contents between the sands (over 90%) and the glass beads (72%). It should be noted that the peak observed LAC of the sands are rather smaller than the theoretical LAC of quartz (SiO_2) as shown in the figure, because of the system absorption. Tsuchiyama [14] studied the relation between the theoretical LAC and the observed LAC in SP-μCT at SPring-8, and obtained the following relation:

$$\text{observed LAC} = (0.8870 \pm 0.0039) \times \text{theoretical LAC}$$

for LAC<20 (cm^{-1}). **Figure 7** is consistent with their result.

In order to distinguish sand grains from void, the threshold LAC (or pixel intensity) between them should be determined in a rational manner. **Figure 8** shows the averaged area of solid grains in a horizontal cross section with different threshold value for various specimens. The vertical dashed lines in the figure are the best threshold values with which the solid area from the CT image coincides with those calculated from the measurement of specimen weight. They range mainly from 1.8 to 2.0 (cm^{-1}) except the

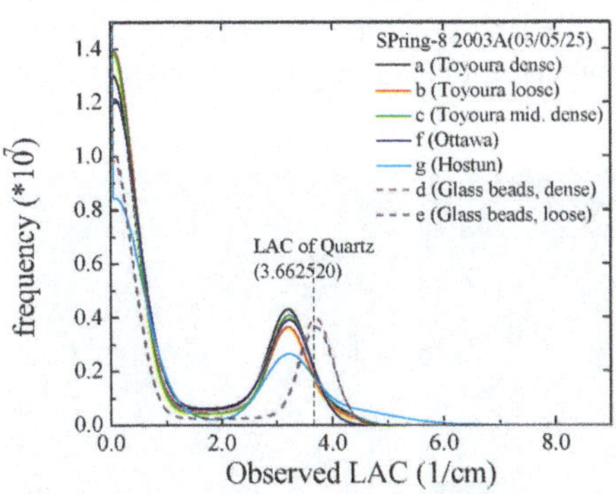

Figure 7. Frequency of observed LAC for varous specimens

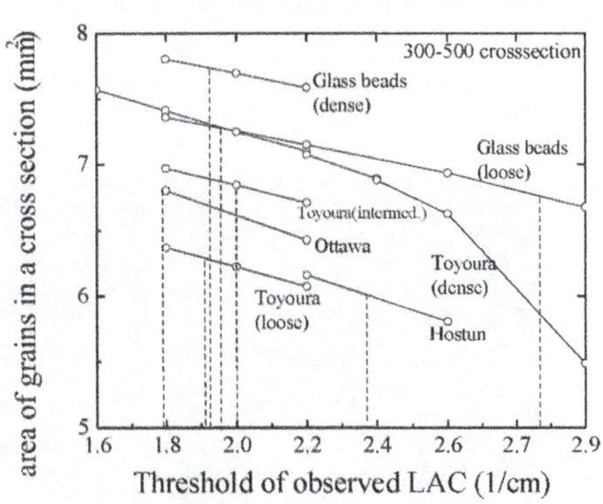

Figure 8. Area of grains in a horizontal cross section with respect to threshold of LAC

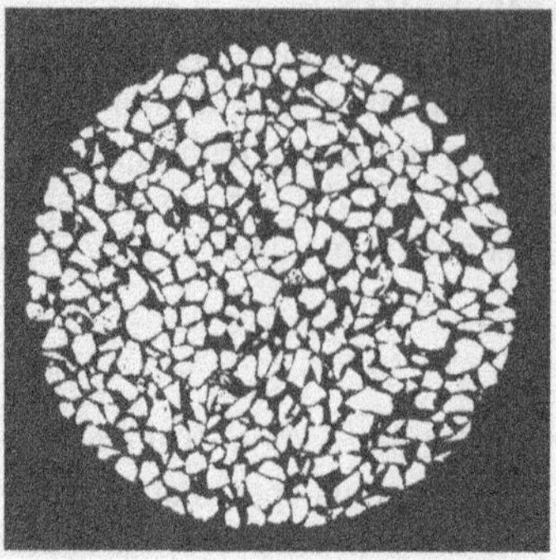

Figure 9. Result of binarization threshold LAC=2.0 (cm^{-1}) (dense Toyoura sand)

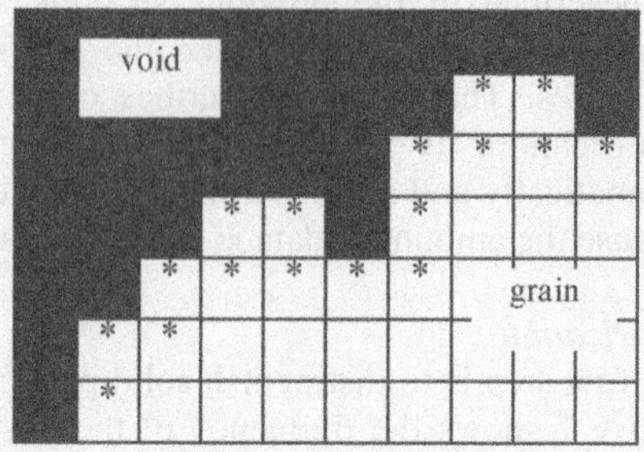

Figure 10. Erosion process

case of Hostun sand and the loose glass beads specimen. Hostun sand consists of rather bigger grains and the effect of the top and bottom boundaries may not be negligible, which is a possible reason for this discrepancy. In the case of loose glass beads, the error on the weight measurement or inhomogeneity of the specimen could be the reason.
Based on the above consideration, we used the threshold value of 2.0 (cm^{-1}) in the following procedure. **Figure 9** shows the result of binarization for dense Toyoura sand specimen (horizontal cross section z=400). (The original image before binarization is shown in **Figure 3**(a).)

Erosion
Since each grain has contact points with the neighboring grains in a specimen, the solid part of the binarized image is necessarily connected with each other. In order to identify individual grains, we need an image processing to separate each grain from the others. Here we adopt so called 'erosion' process, that is schematically illustrated in **Figure 10**. Note that the figure is drawn in 2-D, but the actual process is done in 3-D. White pixels represent the solid grains and black pixels signify the voids. In the adopted erosion process any white pixel having at least one neighboring pixel in black (adding '*' in the figure) is changed into black. Therefore the pixels composing the grain edge will be

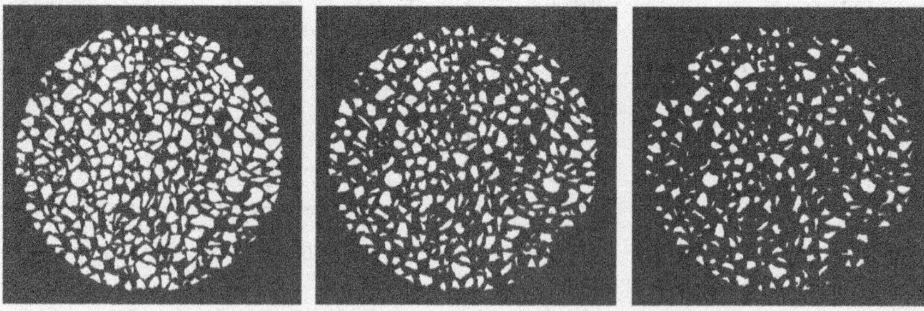

(a) after the first erosion (b) after the second erosion (c) after the third erosion
Figure 11. Cross sectional image after erosion process

eliminated and the grain becomes smaller (eroded).

Figure 11 shows its application into a cross section (z=400) of dense Toyoura specimen. The number of erosion processes to be applied for the best result is determined by the following cluster labeling result and the final grain size distribution.

Cluster labeling

After a certain number of erosion processes, the remaining 'core' pixel clusters are considered as individual grains. Cluster labeling is a process to numbering those individual grains for identification. A well established efficient cluster labeling algorithm [15] enables us to handle our huge data with an allocable memory size.

It is important to compare the resulting cluster size distribution with the grain size distribution by sieving for the determination of the most suitable number of erosion process application. According to the figure on Toyoura sand (**Figure 12(a)**), whose grain diameter ranges from 0.1 to 0.3 mm, the result after the second erosion still contains rather bigger clusters. This implies that the separation of grains by erosion process does not work enough. The result after the third erosion seems to be better.

The other thing to be considered is the number of small-sized clusters, as shown in **Figure 12(b)**. It is noticeable from the figure that the result after the third erosion contains a considerable number of small-sized clusters. It is easily found from the image that most of such small clusters are generated by the erosion process; one single grain is recognized as several clusters. The number of such small clusters increases with increasing the number of applied erosion process. So we adopt the result after third erosion process in the following process.

Attribution of eroded pixels

The final step is to attribute the eroded pixels to the identified grains. In order to do that, we take the image difference before and after erosion, and the recognized eroded pixels are attributed to the adjacent core pixel cluster. If some pixels are adjacent to plural core clusters, we search the number of neighboring pixels for each candidate clusters to find the most adequate core clusters. Since some eroded pixels are not directly adjacent to the core clusters, this attribution process is applied repeatedly until the number of unattributed pixels becomes sufficiently small.

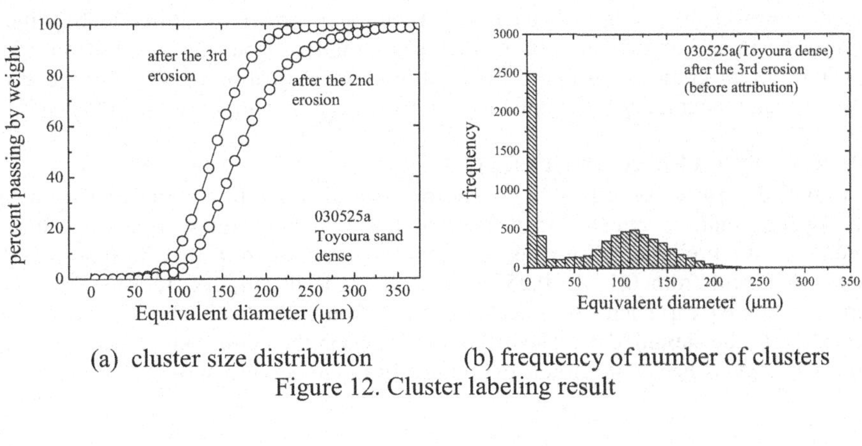

(a) cluster size distribution (b) frequency of number of clusters

Figure 12. Cluster labeling result

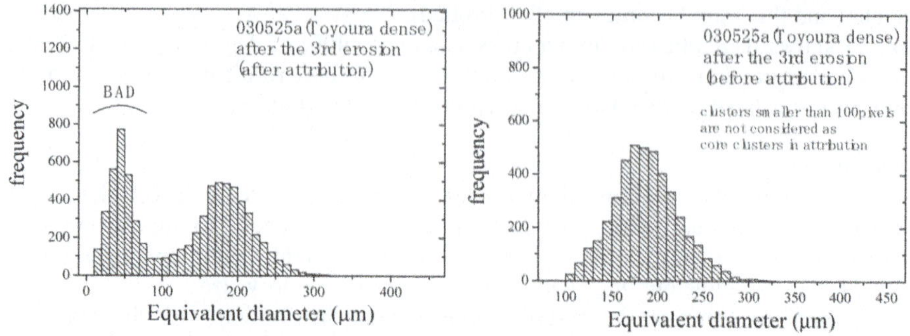

(a) without unlabeling process (b) with unlabeling process

Figure 13 Frequency of the finally-obtained clusters in size

(a) whole image (with unlabeling) (b) a part (with unlabeling) (c) a part (without unlabeling)

Figure 14. Cross-sectional image after attribution process

When we directly apply this process to the cluster labeling result shown in **Figure 12(b)**, the frequency of the finally-obtained clusters in size contains a considerable number of misrecognized small clusters as shown in **Figure 13(a)**. To avoid this, such small clusters are not considered as the core clusters in the initial attribution process. This 'unlabeling' treatment improves the resulting frequency distribution as shown in **Figure 13(b)**. **Figure 14** shows the image of the same cross section as **Figure 9** but the color of the grain differs in different grain according to the final grain identification result. Looking at the details, one can recognize that the unlabeling treatment gives a more reasonable result.

4. MICROSTRUCTURAL PROPERTIES

This section describes some results on microstructural properties for three specimens; dense Toyoura sand specimen and dense and loose spherical glass beads specimens. According to the preliminary sieving, the grain size ranges from 0.10 to 0.30 mm for Toyoura sand, and from 0.18 to 0.25 mm for glass beads. **Figure 15** shows the size distribution curves for such specimens obtained from CT data, which have a good agreement with the sieving data. The difference between the dense glass beads specimen and the loose glass beads specimen for relatively bigger grains may be caused by the error at the grain identification; the grains in the dense specimen look to be connected each other more tightly and are more difficult to be separated by the erosion process.

Once grain identification is complete, contact surface can be defined by the pair of adjacent pixels that are attributed to different grains. **Figure 16** shows the distribution of coordination number (average number of contact points per grain), N_C, for the three specimens. In order to exclude the boundary effect, only the central portion of the specimens is used for counting contacts. The figure includes the result of the early work by Smith et al. [16] for the experiment with loose spherical grains of lead. The distributions are quite Gaussian, and their range is quite reasonable.

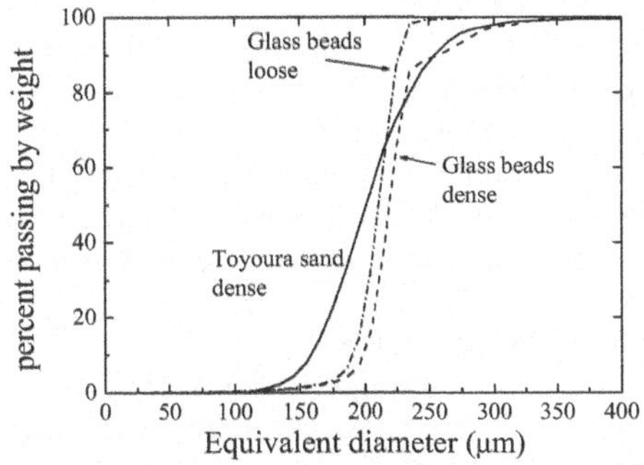

Figure 15. Grain size distribution from CT data

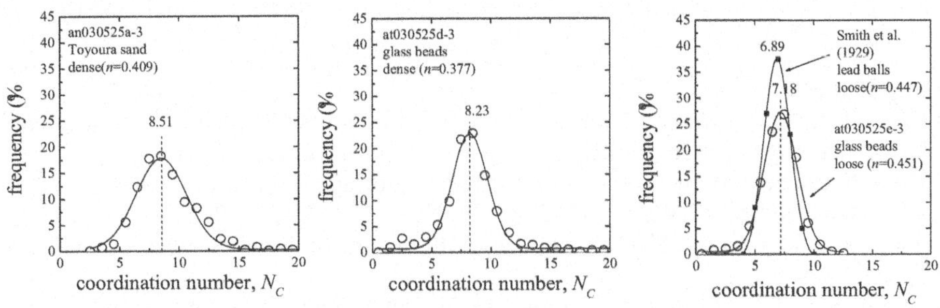

(a) Toyoura sand (dense) (b) Glass beads (dense) (c) Glass beads (loose)

Figure 16. Distribution of coordination number

Figure 17 shows the coordination number with respect to the porosity. The figure includes the theoretical relation by Smith et al. [16] for random packing of equal-sized spheres. It was found that the CT results are in good accordance with the theory even in Toyoura sand specimen, the packing of irregularly-shaped grains.

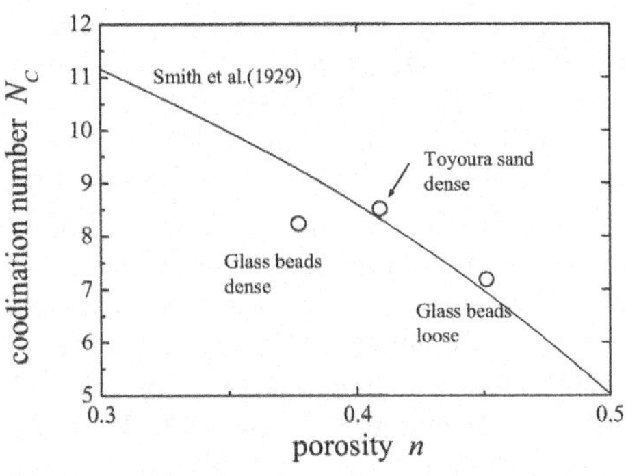

Figure 17. $N_C - n$ relationship

5. CONCLUSION

Micro X-ray CT experiments provide us with a lot of useful knowledge on micromechanics of granular assembly. This paper studied in details a newly-developed image procedure, which is quite important to make the CT results more fruitful to micromechanics of granular materials. The obtained contact information was in good agreement with the result in the literature, which implies that the proposed image analysis has a sufficient accuracy. Further analysis on contact points will be presented elsewhere.

ACKNOWLEGMENT

The authors would like to thank Mr. M. Iidaka, technician of University of Tsukuba for making a special vessel, funnel and micro triaxial apparatus. Mr. K. Toda, Mr. Y. Kawamura and Mr. K. Sato, the former graduate students of University of Tsukuba, are also acknowledged for their help in conducting the experiment at SPring-8. The experiment was carried out at the Spring-8 with the approval of the Japan Synchrotron Radiation Research Institute (Proposal Numbers 2003A0127-NDL2-np and 2003B0479-ND2b-np).

REFERENCES

[1] Cundall, P. A., A computer model for simulating progressive, large-scale movements in blocky rock systems. Symp. ISRM, Nancy, France. Proc., 2: 129-136, 1971.

[2] Discrete Element Methods –Numerical Modeling of discontinua, B.K. Cook and R.P. Jensen editors, Geotechnical special Publication No. 117, ASCE, 2002.

[3] Powders and Grains 2005, Garcia-Rojo, Herrmann, McNamara editors, Balkema, 2005.

[4] X-ray CT for geomaterials: Proc. International workshop on X-ray CT for geomaterials, J. Otani and Y. Obara editors, Balkema, 2004.

[5] Bonse, U. and Busch , F., X-ray computed microtomography (μCT) using synchrotron radiation (SR), Prog. Bio-phys. molec., 65, 133-169, 1996.

[6] Uesugi, K., Tsuchiyama, A., Nakano, T., Suzuki, Y., Yagi, N., Umetani, K. and Kohmura, Y., Development of micro-tomography imaging system for rocks and mineral samples, Proc. SPIE, Developments in X-ray Tomography II, 3772, 214-221, 1999.

[7] JSSMFE(Japanese Society of soil Mechanics and Foundation engineering), Method of Soil Testing, JSF T161-1990, JSSMFE 1990, (in Japanese).

[8] Nakano, T. and Fujii, N., The multiphase grain control percolation: its implication for a partially molten rock, J. Geophysics. Res., 94, 15653-61, 1989.

[9] Tsuchiyama, A., Uesugi, K., Nakano, T., Ikeda, S., Quantitative evaluation of attenuation contrast of X-ray computed tomography images using monochromatized beams, American Mineralogist, 90, 132-142, 2005.

[10] J. Hoshen and R. Kopelman, Percolation and cluster distribution. I. Cluster multiple labeling technique and critical concentration algorithm, Physical Review B, 14, 8, 3238-3445, 1976.

[11] Smith, W.O., Foote, P.D., Busang, P.F., Packing of homogeneous spheres, Physical Review, 34, 1271-1274, 1929.

Soil Stress-Strain Behavior: Measurement, Modeling and Analysis
Geotechnical Symposium in Roma, March 16 & 17, 2006

VISUALIZATION OF PARTICLE-FLUID SYSTEM BY LASER-AIDED TOMOGRAPHY

Hidetaka Saomoto[1], Takashi Matsushima[2] and Yasuo Yamada[2]
[1]Active Fault Research Center, Geological Survey of Japan, National Institute of Advanced Industrial Science and Technology, 1-1-1 Higashi ,Tsukuba, JAPAN
e-mail: h-saomoto@aist.go.jp
[2]Department of Engineering Mechanics and Energy, University of Tsukuba, 1-1-1 Tennodai, Tsukuba, JAPAN

ABSTRACT
Mechanical behavior of particle-fluid system is widely related to various engineering problems such as boiling, piping and liquefaction in the field of geotechnical engineering. This study employed LAT (Laser-Aided Tomography) technique in permeability and boiling tests and proved it's usefulness for visualizing grain-pore fluid interaction. During these tests, the local velocity profile was also obtained by PIV (Particle Image Velocimetry) analysis. As a result, the permeability coefficient measured in the tests was in good agreement with that estimated by Ergun's equation. The interaction between grains and liquid monitored during boiling condition showed a high degree of locality and nonstationality.

1. INTRODUCTION
The Mechanical behavior of particle-fluid system plays important role in various engineering problems; in the field of geotechnical engineering, such phenomena as boiling, piping and liquefaction are of this type. It is, however, substantially difficult to observe the internal behavior of such phenomena, and few visualization techniques are available.

One of well known methods is that using X-ray technique [1] which envisages the inside of particle-fluid system by detecting the differences in material density. The X-ray technique has the advantage of being applicable to actual materials, i.e. sand particles and pore water, while it can not visualize the flow of pore water because of the homogeneity of its density.

A new technique, LAT-PIV, combining LAT (Laser-Aided Tomography) [2][3][4] and PIV (Particle Image Velocimetry) [5], was developed for visualizing both movement of particles and fluid flow: LAT for 3-D movement of each particle, and PIV for velocity field of fluid. A drawback of this method is that crushed glass is to be used as particulate media instead of actual geomaterials. The method, however, can treat digitalized images directly, which enables us to obtain higher resolution in the future with the progress in digital technology.

2. VISUALIZATION TECHNIQUE AND MEASUREING SYSTEM
Figure 1 shows a schematic illustration of the observation system used in this study. LAT, a kind of laser-slicing method, visualizes each particle motion including rotation inside a

Hoe I. Ling et al. (eds.), Soil Stress-Strain Behavior: Measurement, Modeling and Analysis, 235–242.

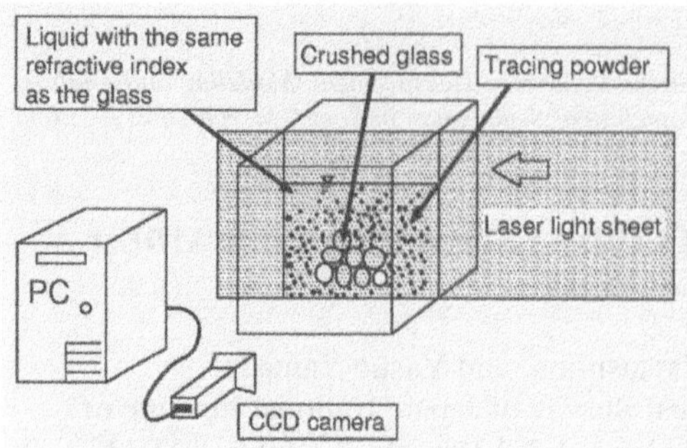

Fig.1. Overview of observation system

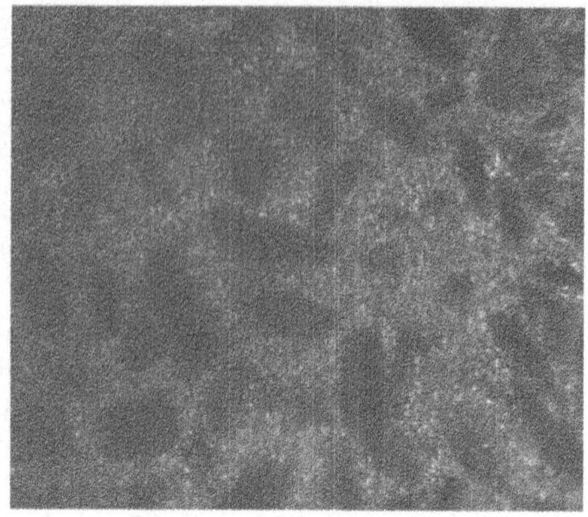

Fig.2. Snap shot by LAT-PIV

3-D specimen composed of crushed glass and pore fluid which are of the same refractive index. PIV, also a kind of laser-slicing method, visualizes fluid motion by pursuing a pattern created by tracing powder mixed into the fluid.

Simultaneous application of LAT to particle part and PIV to fluid part enables us to observe the interaction of the particle-fluid system. A newly developed measuring system, consisting of a CCD monochrome camera (1 million pixels, 500 frames/sec) and a PC, is able to take photographs in digitized form for almost 10 seconds with 500 fps as the marginal performance.

Figure 2 shows a photograph taken by the LAT-PIV technique with numerous tracing powder (flecks) and cross sections of the glass grains (no pattern regions) on the laser light sheet. To separate the pore fluid area from the LAT-PIV image and improve the accuracy of image processing, fluorescent powder that gleams pale orange on the green laser sheet and green-cut filter were employed here.

3. VISUALIZATION OF PERMEABILITY TEST AND SEEPAGE FAILURE WITH LAT-PIV

In order to investigate the applicability of the LAT-PIV, a series of permeability tests including seepage failure was conducted. Crushed glass and silicone oil were adopted to compose specimens. The LAT-PIV requires the same refractive index between particles and pore fluid, which is attained by interblending two kinds of silicone oil with different refractive indexes.

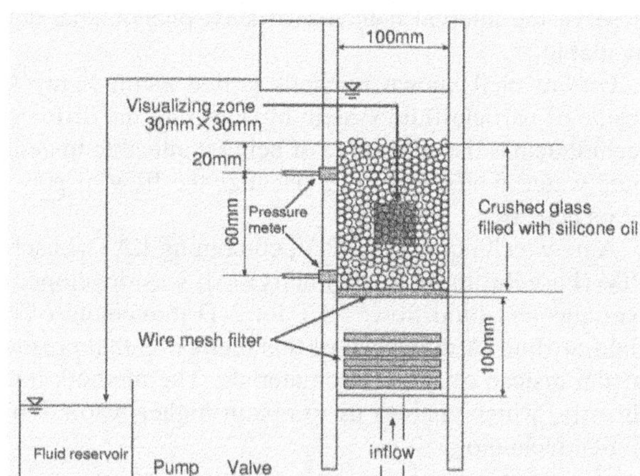

Fig.3. Schematic diagram of experimental apparatus

Figure 3 indicates a schematic diagram of the experimental

apparatus and Table 1 lists the material properties. The apparatus has a circulation loop for the silicone oil which flows upward from the bottom of the loosely packed specimen made by a pluviation method. Wire mesh filters (1mm grid) placed close to the inlet zone are designed to avoid a local seepage failure caused by non- rectified inflow.

Table 1. Material properties

	Silicone oil	Glass grain	Tracing powder
Density	1.02 g/cm³	2.52 g/cm³	1.02 g/cm³
Diameter	---	2 ~ 5mm	40 μm
Viscosity	20.6 mm²/s	---	---
Refractive index	1.514	1.514	---

The experimental procedures are as follows:

(1) Pump the silicone oil into the specimen by opening the valve, then keep the valve open to stabilize the outflow with a constant rate.

(2) After the stabilization, measure the pressure gradient and the mass of outflow.

(3) Visualize the interior of the specimen with the LAT-PIV during the steady state.

(4) Increase the flow rate slightly, and repeat the procedures (1) to (3) step by step.

The experiment was comprised of 20 stages, varying from the seepage state to the boiling state with increasing flow rate. During the experiment, the laser light sheet always passed through 3cm inside the specimen with the same intensity and hundreds of cross-sectional digital images were taken at each stage.

4. RESULTS OF EXPERIMENT

Figure 4 plots the relationship between the mean fluid velocity and the hydraulic gradient obtained by the experiment. The data points are clearly divided into two regions: one corresponds to the seepage state (stage No. 0 - No. 14) and the other to the boiling state (stage No. 15 - No. 20).

The value of critical hydraulic gradient is 0.71 at stage No. 14, which is comparative but a little smaller compared with the theoretical range of 0.78 to 0.92 corresponding to the maximum and the minimum void ratios, i.e. 0.90 and 0.59, respectively. In contrast to this result, the critical hydraulic gradient measured for another specimen made with bigger glass grains from 5 to 10 mm in diameter was definitely smaller than the theoretical range because of its high perviousness.

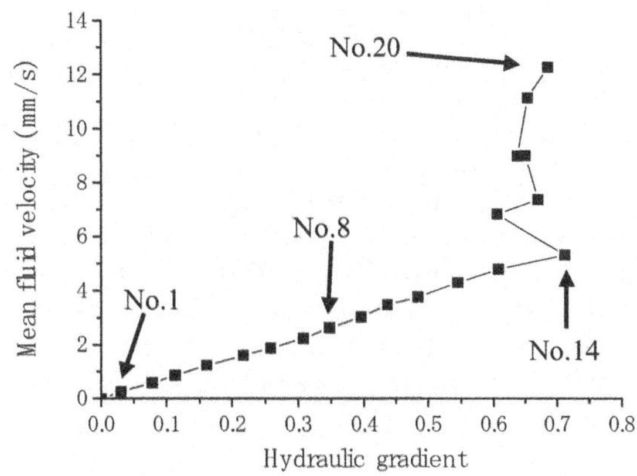

Fig.4. Plot of mean fluid velocity vs. hydraulic gradient

Within the stage from No. 1 to No. 14, a linear relationship exists between the mean fluid velocity and the hydraulic gradient, which represents the Darcy's law. The coefficient of permeability obtained from the slope of the line is 7.68×10-1 cm/s.

In the field of geotechnical engineering, Although Hazen's equation or Taylor's equation are well known and often used for the estimation of permeability, in this study the measured permeability was evaluated using Ergun's equation [6]:

$$i = \frac{150\mu(1-\varepsilon)^2}{\rho g(\phi d_m)^2\varepsilon^3}v + \frac{1.75(1-\varepsilon)}{g(\phi d_m)\varepsilon^3}v^2$$

where the notations i and v indicate the hydraulic gradient and the mean fluid velocity; m and r correspond to the viscosity and the density of the fluid; e is the porosity and f and dm are spericity and diameter of the grains, respectively; g is the acceleration of gravity. The Ergun's equation, obtained on the assumption that the viscous losses and the kinetic energy losses are additive, covers the entire range of flow rates and is widely referred in chemical engineering. Since the Ergun's equation takes account of viscosity, density and porosity as well, it can be easily applicable to a special pore fluid like silicone oil.

By substituting the values listed in Table 1 and the maximum porosity into this equation, then the sphericity only remains uncertain for evaluating the hydraulic losses. According to the references [7] [8], the sphericity of grains often encountered in engineering fields is almost within the range from 0.6 to 0.8: e.g. 0.65 for crushed glass, 0.83 for rounded sand.

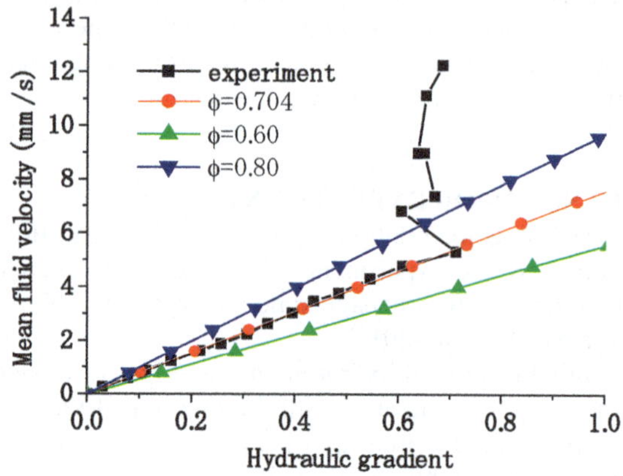

Fig. 5. Mean fluid velocity vs. hydraulic gradient with the fitting results of the Ergun's equation

Figure 5 shows the relationship between the mean fluid velocity and the hydraulic gradient obtained from the Ergun's equation for different values of sphericity. As can be seen in Fig. 5, when the sphericity equals to 0.704, the Ergun's equation yields a good agreement with the experimental results within the seepage stages. This value is, however, a little greater than the reference value of 0.65 for glass grains subjected to abrasion by means of ball milling. The Ergun' s equation gives the possible values of permeability bounded by the upper and the lower lines with triangle marks, which corresponds to the range from 5.53×10-1 cm/s to 9.76×10-1 cm/s. However, it is to be noted that the Ergun's equation can not predict the drastic change in the permeability due to boiling.

Figures 6, 7 and 8 represent the distributions of fluid velocity vectors inside the specimen at stages No.1, No.8 and No.14 indicated in Fig. 4, respectively. To obtain these figures, a PIV image analysis with a sub-pixel accuracy was employed.

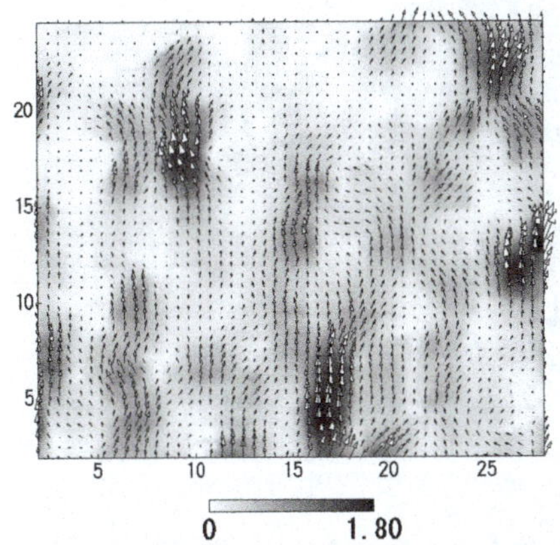

Fig.6. Distribution of fluid velocity at stage No. 1, 60fps

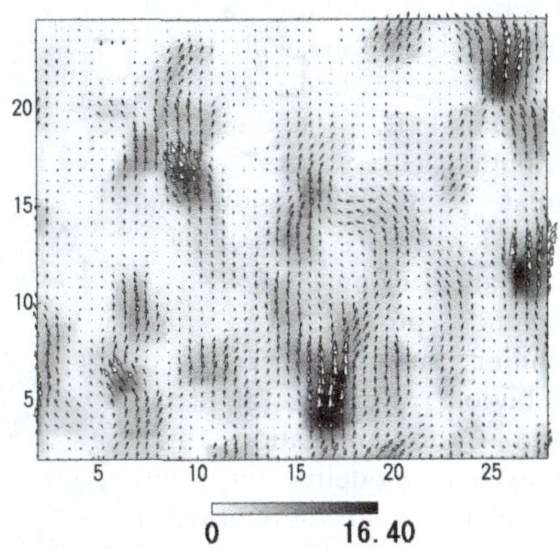

Fig.7. Distribution of fluid velocity at stage No. 8, 250fps

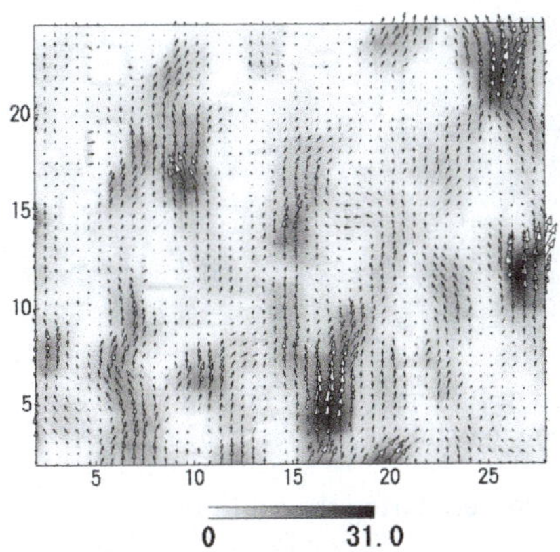

Fig.8. Distribution of fluid velocity at stage No. 14, 500fps

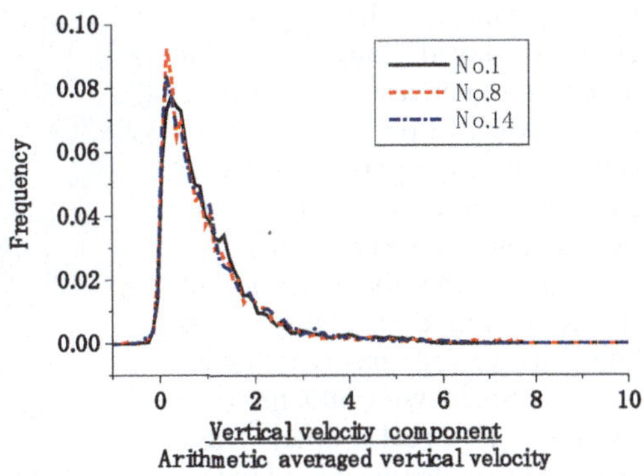

Fig.9. Histograms of vertical velocity component

In the image analysis the magnifying power was set at 5.53×10-2 mm/pixel, and then a square template of 33 pixels (1.82 mm) was used as a pattern matching area. The fluid velocity vectors were measured at 7885 points for each image with a lattice-like arrangement; the values of grayscale indexes correspond to the norms of the velocity vectors in terms of millimeter per second.
As can be seen in these figures the velocity field is not homogeneous, and the region of faster flow always appears at almost similar spot at each stage within seepage state, which suggests an immutability of the pore structure.

Figure 9 shows the histograms of the vertical velocity components, normalized by the arithmetic averaged vertical velocity, for stage No. 1, No. 8 and No. 14. The three lines for different stages coincide well and obviously indicate a non-Gaussian distribution with a long tail reaching to almost 8 on the horizontal axis. The measured distribution curves seem to be simulated by a log-normal distribution. Since the velocity measurement using the LAT-PIV technique provides a high resolution, such a non-Gaussian distribution is very useful in the Monte Carlo simulation often employed in modeling the long-term behavior of groundwater pollution.

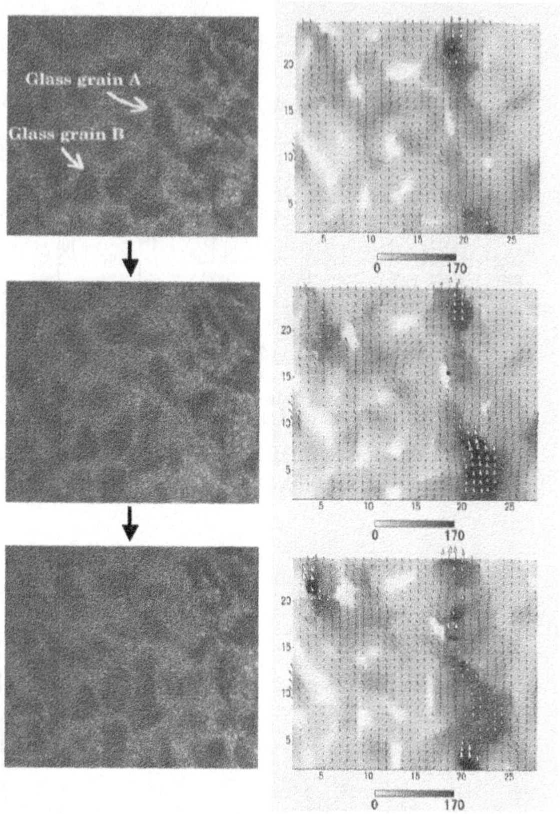

Fig.10. Snapshots of boiling
(from top to bottom, t=0.198, 0.252, 0.306 s)

Once exceeding the critical stage represented by No. 14, a quick condition or boiling occurred in the specimen, the hydraulic gradient stayed fairly constant despite the increase in the inflow rate. Meanwhile the layer consisting of the glass grains was entirely destroyed by the hydrodynamic force; both the glass grain and the silicone oil started to move randomly. Two hundreds photographs were taken at stage No.20 with 500 fps (0.4 seconds in real time) and analyzed to construct complicate velocity fields observed during the boiling.

Figure 10 shows snapshots taken during the boiling together with the results of image analysis. As shown in those figures, the LAT-PIV successfully captured a time series of localized upward flow in the region of $20 < x < 25$. The change in the maximum vertical velocity component was traced from these images so as to characterize the flow state during boiling in terms of the Reynolds number.

In consequence, the Reynolds number calculated by the mean grain diameter, the fluid viscosity and the maximum vertical velocity component was as low as 35, which suggests that the intense upward flow during boiling still remains in a laminar flow state. The LAT-PIV can also quantify the configurations of glass grains by tracing the boundaries between the fluorescent powder regions and the no pattern regions.

Figure 11 illustrates the trajectories of two glass grains A and B which are indicated in Fig. 10. There is an obvious difference in the behaviors of these two grains.

The grain A was acutely under the influence of the strong upward flow, causing a heavy anticlockwise rotation reaching to almost 70 degrees and a considerable translation within 0.4 seconds. On the other hand, the glass grain B, being positioned out of the strong upward flow, exhibited much less rotation and translation compared with grain A in spite of the small distance between them.

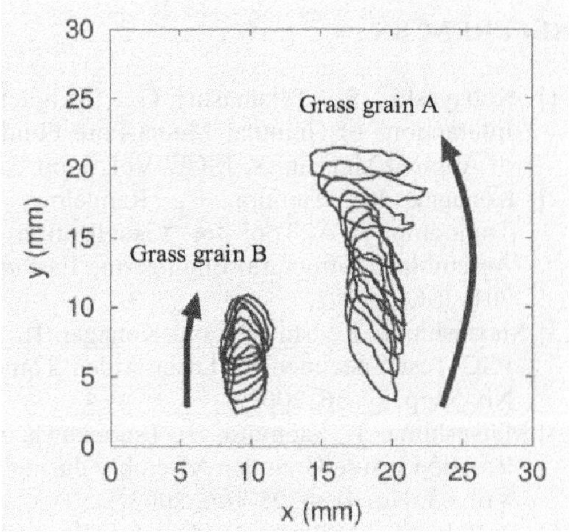

Fig.11. Trajectories of glass grains

As abovementioned the LAT-PIV technique provides us with attractive information on the behavior of the particle-fluid system. It also has an advantage in terms of scalability, because the resolution shall easily be improved with rapid innovation of digital technology such as those employed in CCD cameras.

5. CONCLUSIONS

A visualization technique for the particle-fluid system, the LAT-PIV, was developed and applied to permeability tests including seepage failure.

The test results are as follows:

(1) The critical hydraulic gradient given by a classical theory and that observed in the experimental were almost the same.

(2) The measured data proved that the Ergun's equation is able to represent the relationship between the mean fluid velocity and the hydraulic gradient with a good accuracy during the seepage state. However it was not applicable to the state of flow whose hydraulic gradient exceeds the critical value.

(3) The distribution of the fluid velocity during seepage state was quantified by the LAT-PIV, and consequently it is clearly recognized as a non-Gaussian distribution.

(4) The distribution of the fluid velocity during boiling was also quantified, then the Reynolds number was calculated on the basis of the maximum vertical component. The Reynolds number was found to be almost 35, which indicated a laminar flow state.

Since the LAT-PIV is essentially a two-dimensional visualization technique, its development for the three-dimensional problems remains as a future prospect.

REFERENCES

[1] Kobayashi, S., Takahashi, G., Sekiguchi, H.: Visualization of the Dynamic Interactions of Granular Media-Pore Fluid Systems by X-ray TV Imaging, Journal of Applied Mechanics, JSCE, Vol. 3, pp. 521-531, 2000 (in Japanese).

[2] Konagai, K., Tamura, C., Rangelow, P. and Matsushima, T.: Laser-Aided Tomography: A Tool for Visualization of Changes in the Fabric of Granular Assemblage, Structural Engineering Earthquake Engineering, Vol. 9, No. 3, pp. 193-201, JSCE, 1992.

[3] Matsushima, T., Ishii, T. and Konagai, K.: Observation of Grain Dislocation inside PSC Test Specimen by Laser-Aided Tomography,Soils and Foundations, Vol. 42, No. 5, pp. 27-36, 2002.

[4] Matsushima, T., Saomoto, H., Tsubokawa, Y. and Yamada, Y.: Observation of Grain Rotation inside Granular Assembly during Shear Deformation, Soils & Foundations, Vol. 43, No. 4, pp. 95-106, 2003.

[5] Raffel, M., Willert, C. E. and Kompenhans, J.: Particle Image Velocimetry (Translation from the English language edition), Springer-Verlag, Berlin Heidelberg, 1998.

[6] Ergun, S. : Fluid Flow Through Packed Columns, Chemical Engineering Progress, Vol. 48, No. 2, pp. 89-94, 1952

[7] McCabe, Smith and Harriott: Unit Operations of Chemical Engineering, Mcgraw-Hill, 2000.

[8] Perry R. H., Green, D. W. and Maloney, J. O.: Perry's Chemical Engineers' Handbook, Mcgraw-Hill, 1997.

[9] Saomoto, H., Matsushima, T. and Yamada, Y.: Experimental Observation and Direct Simulation of Particle-Fluid System, 16th Engineering Mechanics Conference, CD-ROM, ASCE, 2003.

[10] Saomoto, H., Matsushima, T. and Yamada, Y.: Development of LAT - PIV Visualization Technique for Particle-Fluid System, Journal of Applied Mechanics, JSCE, Vol. 8, pp. 601-608, 2005 (in Japanese).

Soil Stress-Strain Behavior: Measurement, Modeling and Analysis
Geotechnical Symposium in Roma, March 16 & 17, 2006

STRENGTH AND STIFFNESS OF COARSE GRANULAR SOILS

Ramon Verdugo and Karem de la Hoz
Department of Civil Engineering
University of Chile, Santiago, Chile
e-mail: rverdugo@ing.uchile.cl kdelahoz@ing.uchile.cl

ABSTRACT

A methodology for estimating the shear strength parameters of coarse granular material by means of conventional testing equipment, avoiding the oversized particles and using the finer part of the original soil, is investigated. The procedure uses soil samples with a parallel gradation and a maximum particle size compatible with the available testing facilities. Experimental results obtained for five different coarse granular materials are presented, showing the feasibility of the application of this method.

1. INTRODUCTION

Coarse granular soils are extensively used in the construction of large earth dams and massive fills due to the well recognized mechanical properties of these soils (Leps, 1970; Marachi et al, 1972; Varadajan et al, 2003). On the other hand, the existence of soil deposits constituted by coarse granular materials is very common in cities that are located close to high mountains as the case of many Latin American cities established along the proximity of Los Andes Range, where it is necessary to deal with these soils in a wide variety of constructions. In the mining sector, the operation of open pit mines generate huge deposits of waste material constituted by coarse particles that have to be managed and stored in economical and stable conditions. Therefore, it is apparent the importance of evaluating the geotechnical properties of coarse granular soils. However, there is, in general, a lack of suitable equipment for testing soils constituted by large particles due to the high cost and effort required to work with large samples. Additionally, conventional procedures to recover "undisturbed" samples, frequently used in sandy and fines soils, cannot be used with coarse materials, unless innovative techniques are developed and implemented, which are seldom applied in ordinary projects. On top of this, given the empirical fact that coarse soils have excellent geomechanical properties, there is usually pressure from the construction sector to minimize cost, requesting designs as close as possible to the maximum soil capacity.

Consequently, it is imperative to study alternative methods to evaluate the geomechanical properties of coarse soils using ordinary equipment, where only the finer part of the original soil is tested. In this context, the experimental procedure generally referred to as the parallel gradation method is used and analyzed.

Hoe I. Ling et al. (eds.), Soil Stress-Strain Behavior: Measurement, Modeling and Analysis, 243–252.

2. PROCEDURES TO EVALUATE STRENGTH OF COARSE SOILS

To estimate the strength of coarse soils by means of laboratory tests, different methods have been proposed, which involve testing of "equivalent" soil samples, free of oversized particles. The more widely used procedures are the matrix model method, the parallel gradation method and the scalping and replacement method. In the matrix model, the original coarse soil is divided in two parts: oversized particles and matrix material. The definition of oversize is arbitrary, and it is related to the maximum particle size that can be tested in the available equipment. It is assumed in this method that the oversized particles are in a "floating" state, meaning that these particles have little or no contact between them. The matrix material to be tested is compacted to a density that has to be estimated, corresponding to the actual density of the soil matrix in the field (Siddiqi et al, 1987; 1991; Fragaszy et al, 1990; 1992). Therefore, the use if this procedure is limited by the validity of the assumption that the oversized particles are "floating" and the accuracy of the procedure to estimate the actual density of the soil matrix in the field.

In the scalping-replacement method, all those particles that are considered oversized with respect to the available testing equipment are scalped and replaced with an equal weight of a smaller particle range (Donaghe and Torrey, 1979). This procedure changes drastically the original soil gradation and, although some experimental data have shown promising results, there is no real evidence to support the equivalence between the original soil and the artificially created batch of soil scalped and replaced.

In the parallel gradation method, the oversized particles are scalped and a new batch of soil is prepared using the original material, which has a grain size distribution curve parallel (in the common semi log scale) to that of the original sample (Lowe, 1964; Marachi et al, 1972; Verdugo et al, 2003; Varadajan et al, 2003). The main advantage of this procedure is that the soil gradation is maintained. However, depending upon the particular characteristics of each soil, the mineralogy and hardness of grains, particle shape, and particle roughness, may be different and function of the particle size (Al-Hussaini, 1983; Cho et al, 2005; Lee et al, 1967; Santamarina et al, 2003 & 2004). In granular soils where these factors are similar for all particle sizes, the parallel gradation method can be seen as an attractive alternative. One additional limitation arises in those cases where the original coarse material has a considerable content of fines, such that the finer parallel gradation results in sample with an important fines content, for example greater than 10%. The geomechanical response of this new batch is strongly controlled by the fines and therefore, it cannot reproduce the behavior of the original coarse material.

Considering the advantages and shortcomings of these procedures, the parallel gradation method was selected for this study and its capability to estimate the geomechanical properties of coarse soils was investigated.

3. MATERIAL TESTED

Five different samples of gravelly soils identified as A-1, M-1, M-2, M-3 and P-1, were tested. These granular materials were retrieved from the bed of three different Chilean rivers. Therefore, they and can be considered as fluvial soils. The sample M-1 also comes from a river bed, but it was obtained from a plant that produces concrete aggregate, so this sample contains crushed particles, which are angular. Each of these five samples was used as a prototype soil, from where new batches were artificially created having parallel grain size distribution curves of finer particles. The procedure required that the original samples were sieved and separated into their particles according to their sizes. Then, new batches were made by weighting and selecting the exact amount of soil particles required for each size. The fines content was limited to a maximum value of 10%. Consequently, in the case of samples P-1 and M-2, the gradation of the finer batches was not completely parallel to that of the prototype soil.

The maximum and minimum densities of each batch were evaluated using the ASTM procedures. The grain size distribution curves and the maximum and minimum densities of the material tested are presented in Figs 1 to 5. It is interesting to observe that in those batches that have parallel gradations (A-1, M-1, M-2 and M-3), the maximum and minimum densities are rather constant, regardless of the mean grain size, D_{50}. On the other hand, when the grain sizes are not totally parallel to each other (P-1), there is a clear effect of the mean grain size on the maximum and minimum densities.

4. TEST PROCEDURES

Series of CID triaxial tests were carried out at confining pressures in the range of 20 to 600 kPa. Depending upon the maximum particle size of the soil batch, triaxial samples with diameters of 5, 10 and 15 cm were used. All the samples had height/diameter ratio of 2. End plates without lubrication were used. All the tests were conducted under strain-controlled with a deformation rate of 0.1 %/min. The saturation of the sample was considered sufficient when the B-value was greater than 0.95.

The samples were prepared by the method of wet tamping using distilled water in a proportion of 5% in weight. Samples A-1, M-1 and P-1 were initially compacted to a relative density of 80%, while samples M-2 and M-3 were initially compacted to a relative density of 70 %. The soil was compacted inside a split mold in six layers of equal height and amount of wet soil.

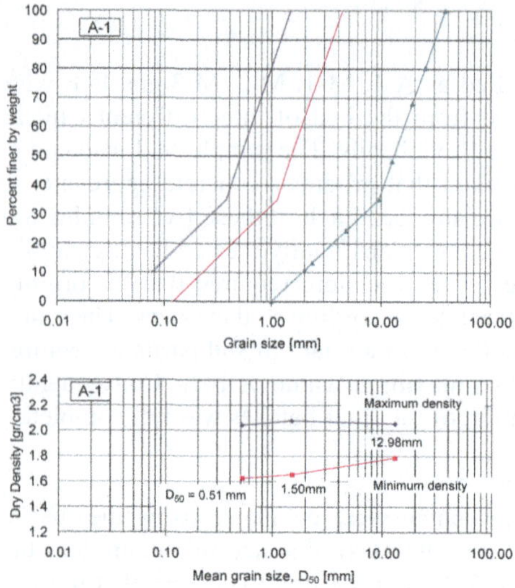

Fig.1. Grain size distribution, maximum and minimum densities. A-1

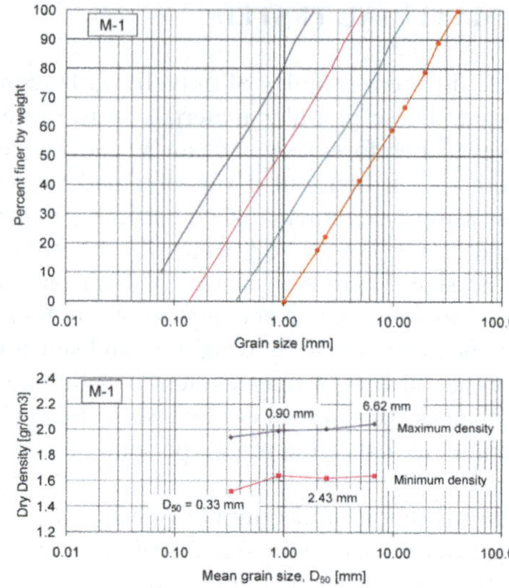

Fig.2. Grain size distribution, maximum and minimum densities. M-1

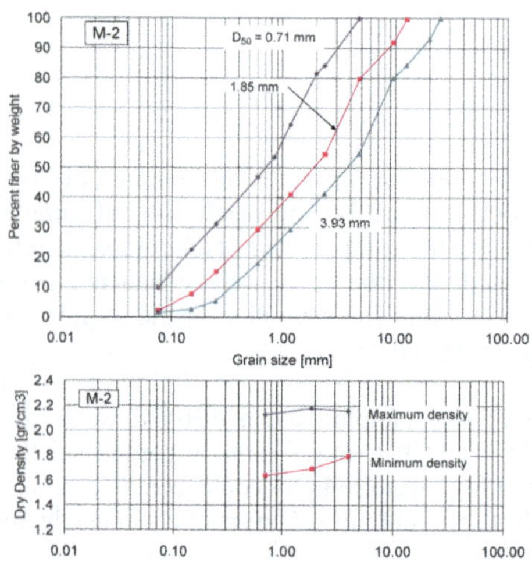

Fig.3. Grain size distribution, maximum and minimum densities. M-2

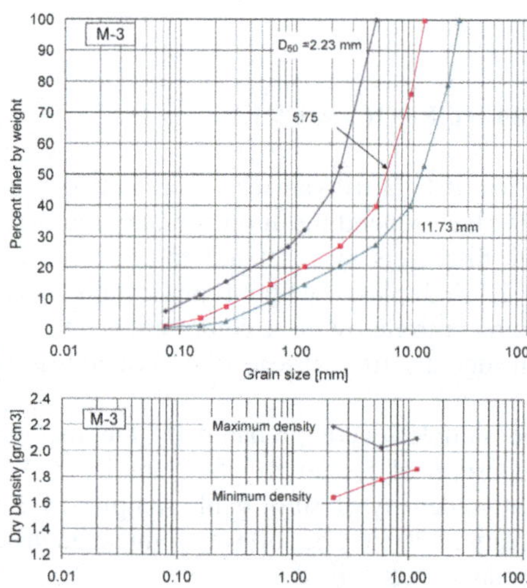

Fig.4 Grain size distribution, maximum and minimum densities. M-3

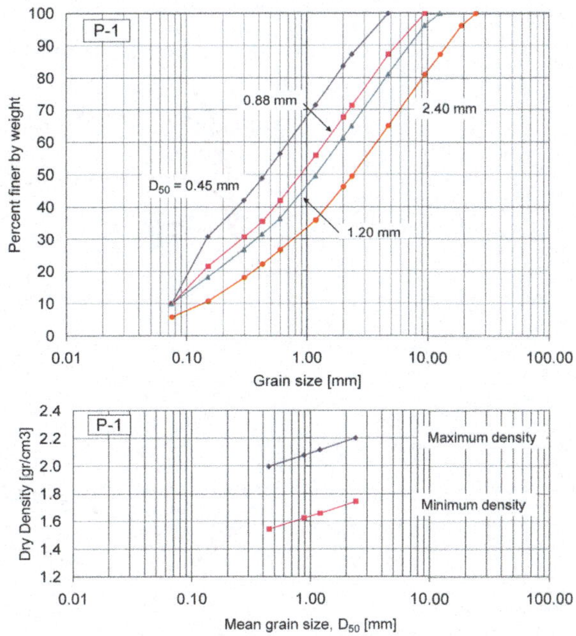

Fig.5. Grain size distribution, maximum and minimum densities. P-1.

5. STRESS-STRAIN AND VOLUMETRIC STRAIN CURVES

The stress-strain curves and the volumetric strain responses of the five materials that have been tested at different confining pressure are presented in Figs. 6 to 10. It can be observed that both peak strength and stiffness are similar in those series with samples of parallel gradation, especially in samples M-2 and M-3. The larger difference in behavior is presented by soil P-1. This can be explained by the restriction adopted in the fines content, which generated soil batches with gradations not perfectly parallel to the original soil.

Considering that all the tests have been performed on samples compacted to an initial relative density of 70 - 80% and in a range of confining pressure between 20 and 600 kPa, the stress-strain curves present an initial linear portion that can be represented by the deformation modulus, E_{50} (stiffness associated with a stress level equal to half of the peak strength). The obtained results are presented in Figs. 11 and 12, the large scatter is observed in samples M-1 and A-1. However, even in these samples is possible to draw a unique curve for the E_{50}. In general, these results indicate that the parallel gradations are able to capture the essential mechanical response of the soils.

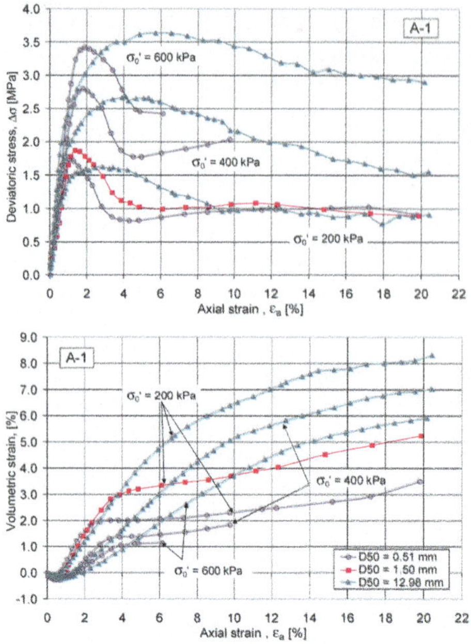

Fig.6. Stress-strain curves and volumetric strain response. A-1

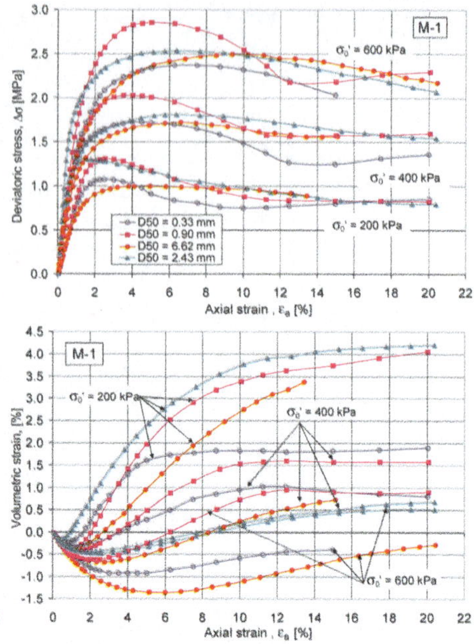

Fig.7. Stress-strain curves and volumetric strain response. M-1.

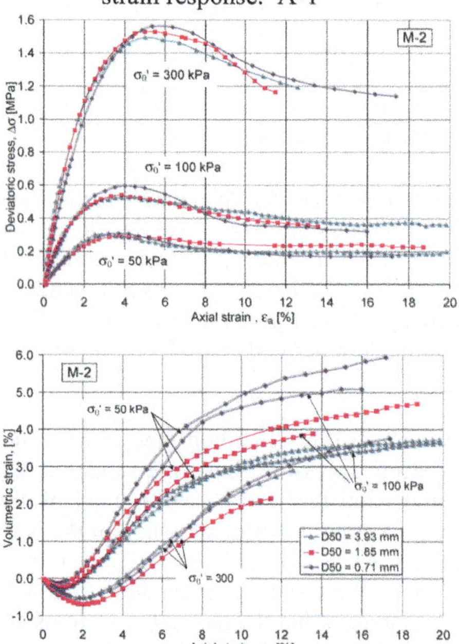

Fig.8. Stress-strain curves and volumetric strain response. M-2.

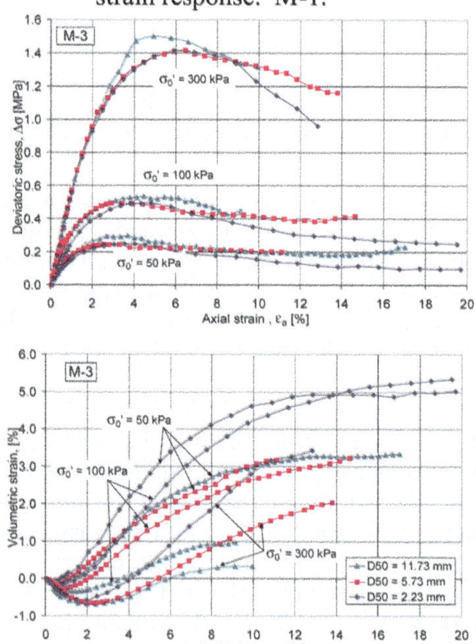

Fig.9. Stress-strain curves and volumetric strain response. M-3.

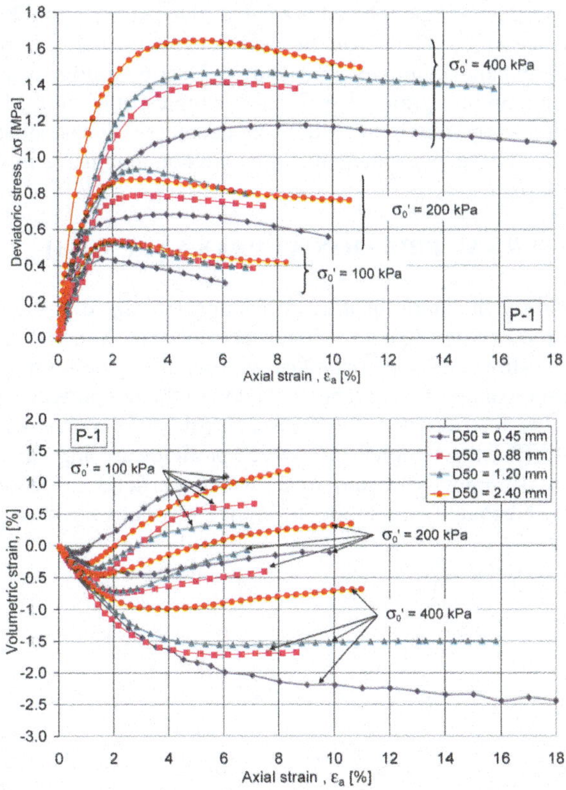

Fig. 10. Stress-strain curves and volumetric strain response. P-1.

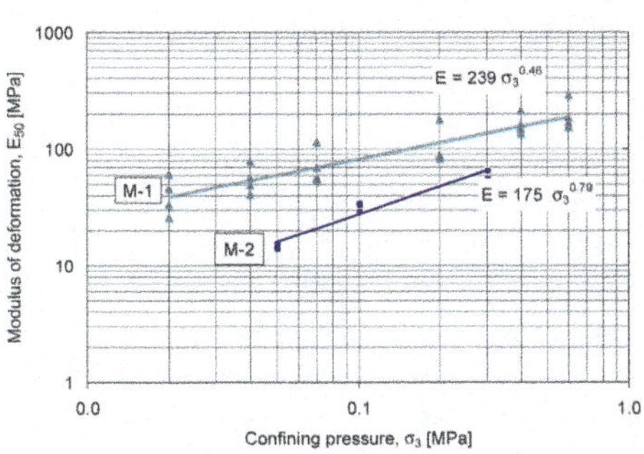

Fig.11. Modulus of deformation E_{50} as function of confining pressure. Samples M-1 and M-2.

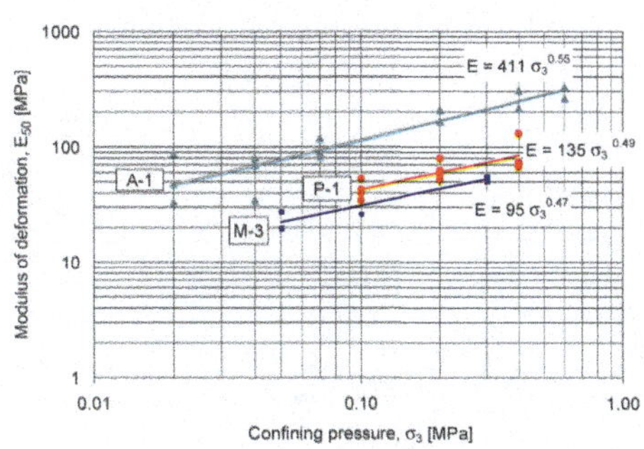

Fig.12. Modulus of deformation E_{50} as function of confining pressure. Samples A-1 , P-1 and M-3.

Another interesting aspect to be analyzed is related to the stress-strain behaviour after the peak strength has occurred, which can be associated with the soil response after the occurrence of a shear band. Samples M-2 and M-3 show similar post-peak behaviour, regardless of the maximum grain size. While samples A-1 and M-1 show a more pronounced softening as the maximum grain size decreases. In case of sample P-1 a clear trend is not observed. These experimental results are not conclusive in the effect of grain size in the post-peak response.

6. ANGLE OF INTERNAL FRICTION AT PEAK STRENGTH

The obtained results for the angle of internal friction mobilized at the condition of peak strength are presented in Figs. 13 to 17. It can be seen that, in general, the peak friction angle is adequately estimated by the samples with parallel gradation. Again, the sample P-1 shows some discrepancies that can be attributed to the not perfectly parallel gradation that resulted from the restriction in the fines content. It is interesting to notice that at pressure below 0.3 MPa, the samples M-1 which particles are angulars, mobilize greater friction angles than samples M-2 and M-3, which particles are more rounded. This difference cannot be attributed only to the difference in relative density.

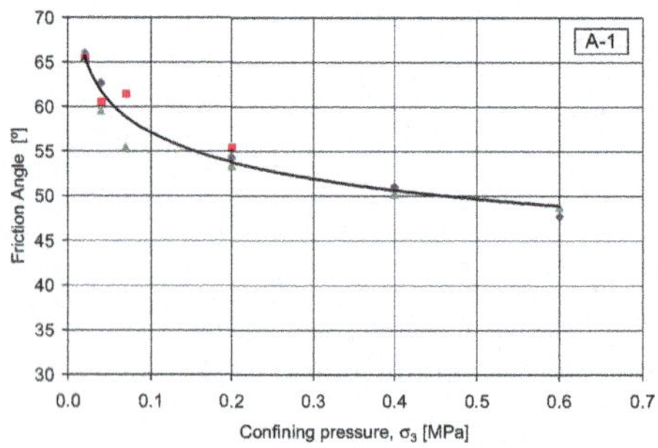

Fig.13. Angle of internal friction at peak strength. A-1

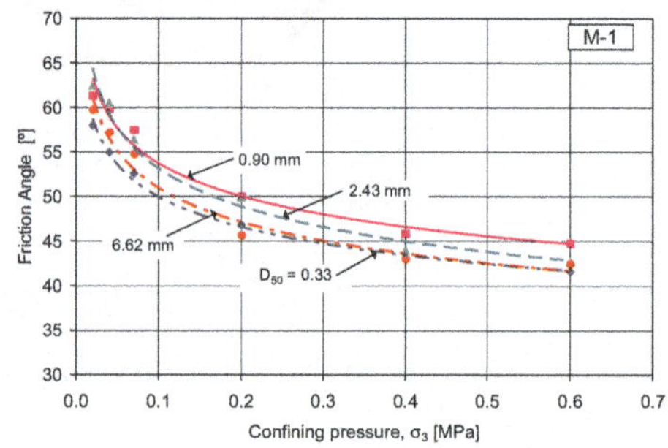

Fig.14. Angle of internal friction at peak strength. M-1

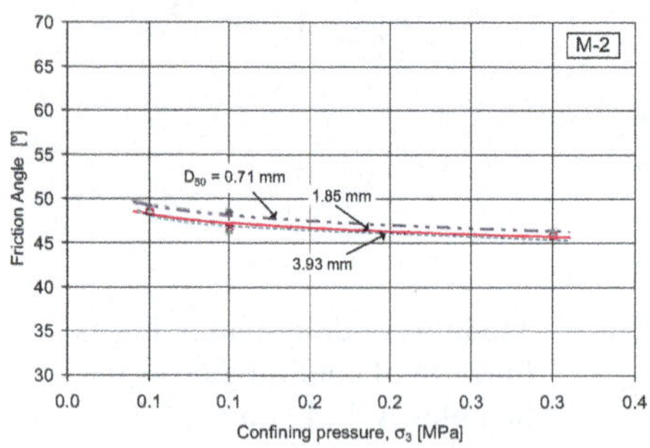

Fig.15. Angle of internal friction at peakstrength. M-2

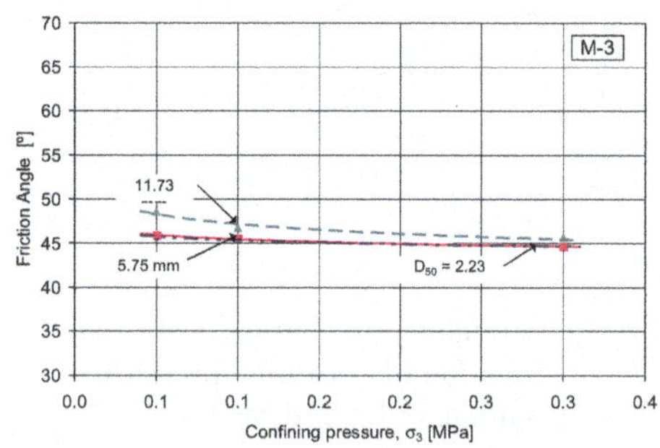

Fig.16. Angle of internal friction at peak strength. M-3

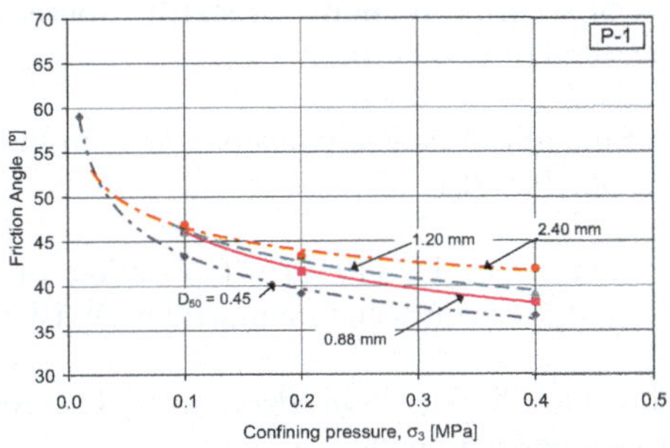

Fig. 17. Angle of internal friction at peak strength. P-1

7. CONCLUSIONS

Under a dense state of a granular material and for the range of grain sizes and confining pressures utilized in this study, the experimental results indicate that the parallel gradation method provides a quite reasonable procedure to evaluate the geomechanical response of coarse granular materials. However, a limitation in the results was observed when the parallel gradation cannot be completely satisfied due to the restriction in the content of fines of the finer batches.

8. REFERENCES

Al-Hussaini, M. (1983): "Effect of particle size and strain conditions on the strength of crushed basalt," Canadian Geotechnical Journal, 20(4): 706-717.

Cho, G., Dodds, J. and Santamarina, J. (2005): "Particle Shape effects on parking density, stiffness and strength: Natural and Crushed sands," Internal report – Georgia Institute of Technology.
http://www.ce.gatech.edu/~carlos/laboratory/tool/Particleshape/Particleshape.htm

Donaghe, R. and Torrey, V. (1979): "Scalping and replacement effects on strength parameters of earth-rock mixtures,". Proc. Conf. on Design Parameters in Geotechnical Engineering, London, Vol.2, pp.29-34.

Fragaszy, R., Su, W. and Siddiqi, F. (1990): "Effects of oversize particles on the density of clean granular soils," Geotechnical Testing Journal, 13(2):106-114.

Fragaszy, R., Su, W., Siddiqi, F. and Ho, C. (1992): "Modeling strength sandy gravel," Journal of Geotechnical Engineering, 118(6):920-935.

Lee, K. and Seed, H. (1967): "Drained strength characteristics of sands," Journal of the Soil Mechanics and Foundations Division, 93(6):117-141.

Leps, T. (1970): "Review of shearing strength of rockfill," Journal of the Soil Mechanics and Foundations Division, 96(4):1159-1170.

Lowe, J. (1964): "Shear Strength of coarse embankment dam material," Procedings, 8th Congress on Large Dams, pp. 745-761.

Marachi, D., Chan, C. and Seed, H. (1972): "Evaluation of properties of rockfill materials," Journal of the Soil Mechanics and Foundations Division, 98(1):95-114.

Santamarina, J. and Cho, G. (2004): "Soil Behaviour: The role of particle shape," Proceedings Skempton Conference. London.
http://www.ce.gatech.edu/~carlos/laboratory/tool/Particleshape/Particleshape.htm

Santamarina, J. and Díaz-Rodríguez, J. (2003): "Friction in Soils: Micro and Macroscale Observations," Pan-American Conference, Boston, 2003.

Siddiqi, F. and Fragaszy, R. (1991): "Strength evaluation of coarse grain dam material," IX Pan-American Conference, Viña del Mar, Chile, pp 1293-1302.

Siddiqi, F., Seed, R., Chan, C., Seed, H. and Pyke, R. (1987): "Strength evaluation of coarse-grained soils," Earthquake Engineering Research Center, University of California, Berkeley, California. Report Nº UCB/EERC-87/22.

Varadajan, A., Sharma, K., Venkatachalam, K. and Gupta, K. (2003): "Testing and Modeling two rockfill materials," Journal of Geotechnical and Geoenvironmental Engineering, 129(3):206-218.

Verdugo, R., Gesche, R. and De La Hoz, K. (2003): "Metodología de evaluación de parámetros de resistencia al corte de suelos granulares gruesos," 12th Pan American Conference on Soil Mechanics & Geotechnical Engineering, Cambridge, MA, pp. 691-696.

DEVIATORIC STRESS RESPONSE ENVELOPES FROM MULTIAXIAL TESTS ON SAND

D Muir Wood*, T Sadek, L Dihoru, ML Lings
Department of Civil Engineering, University of Bristol, UK
H Javaheri
Department of Civil Engineering, Sharif University of Technology, Iran
*e-mail: d.muir-wood@bristol.ac.uk

ABSTRACT
Sets of rosettes of stress probes have been performed on Hostun sand in a flexible boundary true triaxial apparatus. These have shown the effect of stress history on the local distortional stiffness of the sand. Histories have included isotropic compression, and deviatoric paths with increasingly complex shapes.

These tests were performed and interpreted against a background of kinematic hardening and bounding surface plasticity, using stress response envelopes to reveal evolving distortional stiffness. Use of response envelopes to study volumetric stiffness appears to be less helpful but the phase transformation surfaces that can be extracted seem somewhat independent of recent stress history.

INTRODUCTION
Conventional laboratory element testing of soils concentrates on the axisymmetric triaxial apparatus which has two degrees of freedom and provides very limited possibilities for exploration of stress space. In fact, in much testing it is used simply to perform confined uniaxial tests and even the second degree of stress freedom is not exploited. In practice, elements of soil in the ground or around any geotechnical system will experience variations of six independent stress variables (Fig 1). Any constitutive model that is used for numerical analysis – finite element or finite difference – will find itself being expected to make predictions for the way in which the soil will behave under these completely general changes in stress. Laboratory testing is used to calibrate constitutive models (and also to inspire their development) and there is obvious advantage in trying to extend as far as possible the regions of the 6 dimensional stress space that can be explored.

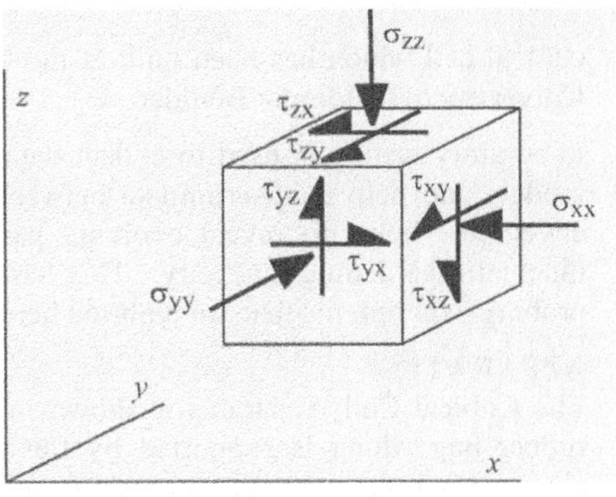

Figure 1: General stress state

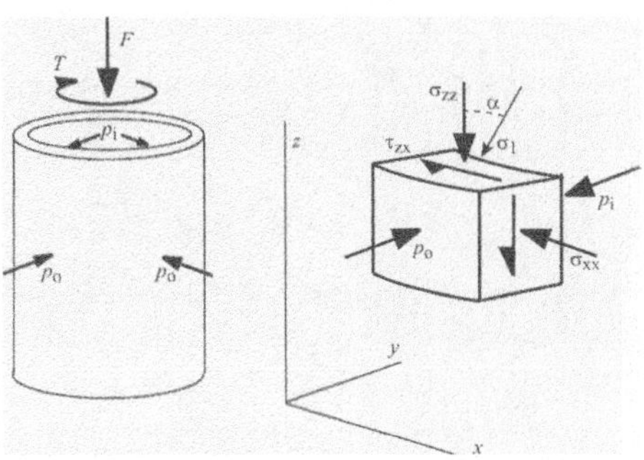

Figure 2: Torsional hollow cylinder apparatus

The torsional hollow cylinder apparatus (Fig 2) provides the possibility of controlling four of the six degrees of freedom but at the expense of radial variation of stress and strain components

Hoe I. Ling et al. (eds.), Soil Stress-Strain Behavior: Measurement, Modeling and Analysis, 253–262.

through the thickness of the wall of the apparatus. The device is thus a *system* from which the constitutive response has to be deduced by back analysis or by assumption of average expressions for stress and strain quantities.

True triaxial apparatus (Fig 3) provide the possibility of controlling the three principal stresses and strains without allowing the directions of the principal axes to change. In principle it is possible to maintain uniformity of stresses and strains within the cuboidal sample. Two routes to effecting these three degrees of freedom have been followed: use of three pairs of rigid boundaries, mechanically linked to ensure that the sample is fully contained at all times (True Triaxial Apparatus: various devices inspired by Hambly (1969) and also described by Airey & Muir Wood (1988)); and use of an arrangement of six pressurised bags to contain the sample within a loading frame (Cubical Cell: the original cubical cell was described by Ko & Scott (1967) and there have been a number of variants of the basic design since that time). Other techniques for applying and controlling three principal stresses lack the essential symmetry of these rigid boundary or flexible boundary devices. The results reported here have come from experiments conducted in a flexible boundary cubical cell which has been built at the University of Bristol following designs provided by University of Colorado, Boulder.

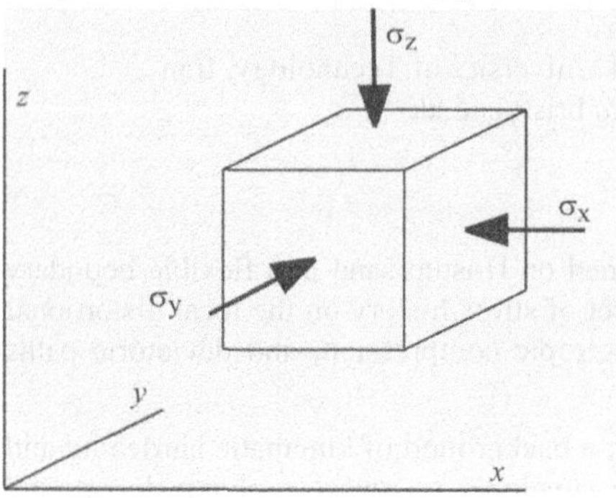

Figure 3: True triaxial apparatus: independent control of three principal stresses

Laboratory testing is used to collect data which can inform the development of constitutive models and help to discriminate between competing models. The use of 'stress response envelopes' helps to reveal evolving patterns of response in the hypothesised context of kinematic hardening plasticity. This has been used as an underlying philosophy behind the probing experiments that are reported here.

APPARATUS

The Cubical Cell Apparatus is shown in Figs 4 and 5. The sample is contained in a latex rubber bag which is supported by flexible 'top-hat' shaped air-filled cushions which are clamped into the sides of a cubical frame. Opposite pairs of cushions are connected so that changes in the applied stresses are synchronised. Figure 4 shows an isometric view of the

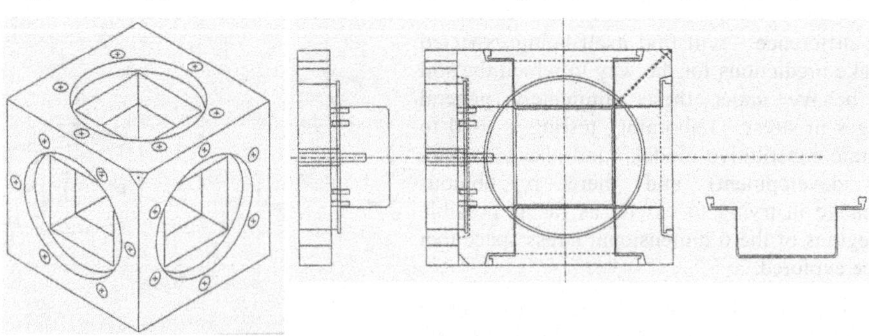

Figure 4: Cubical Cell Apparatus: isometric view of containment frame; sections through face plate, cell, and pressure cushion

cubical frame together with sections through the assembled frame with the clamping plate and a section through one of the cushions. Figure 5 shows a photograph of the assembled cubical cell with the LVDTs used for measurement of the displacements on each of the six faces clearly visible. Pressures in the cushions are measured with pressure transducers located immediately behind the cushions. Data from all transducers are collected by a computer running LabVIEW™ software. This software provides both a data logging and a control service. Stress controlled tests of the type to be described here have been performed by direct control of the cushion pressures using voltage-pressure converters.

Figure 5: Cubical cell apparatus

All tests have been performed on dry Hostun RF sand (Figs 6 and 7, taken from Benahmed, 2001). Samples of sand were prepared by pluviation through a set of sieves (following the general procedure developed by Miura & Toki (1982)) into a membrane held by vacuum against the sides of a cubical mould. Once full, the top of the sample was carefully sealed and a thin tube used to establish a vacuum of about 30kPa in the sand. This provided sufficient effective stress within the sand for it to be stiff and strong enough to be lifted in its containing membrane from the mould and located in the cubical cell. Once all the six pressure cushions and their retaining plates had been reassembled the cushion pressures were increased in stages and the internal vacuum released at the same time so that the effective mean stress remained rather constant. Once the vacuum had been completely released then the tests proceeded as drained stress-controlled tests, with no internal pore (air) pressures.

Figure 6: Hostun RF sand (Benahmed, 2001)

STRESS RESPONSE ENVELOPES

Stress response envelopes were introduced by Gudehus (1979) as a way of illustrating the contrasting nature of the character of stiffness response predicted by different classes of constitutive model (see also Muir Wood, 2004a, b).

From a given initial stress state, a series of *strain* probes is applied to the model soil. For each strain increment the amplitude is the same but the ratio of volumetric to distortional strain increment is different: the strain probes have identical normalised magnitude (Fig 8a). A solid curve joins all the resulting stress increments from the common initial stress. The resulting envelope of *stress* responses provides a visual indication of the stiffness character of the model (Fig 8b). At each point on this *stress response envelope* a short line may be drawn to indicate the direction of the corresponding strain probe.

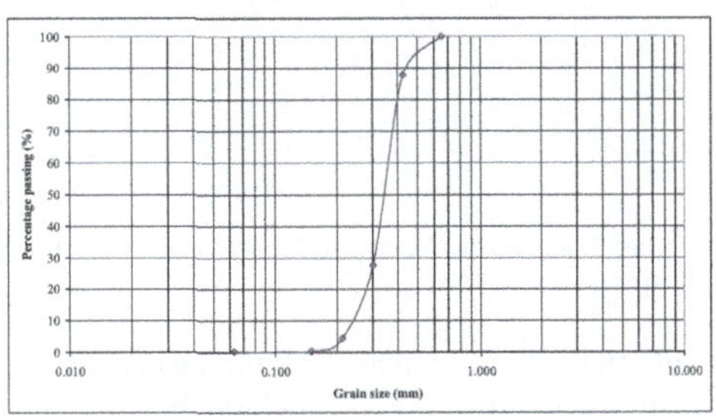

Figure 7: Hostun RF sand: particle size distribution (Benahmed, 2001)

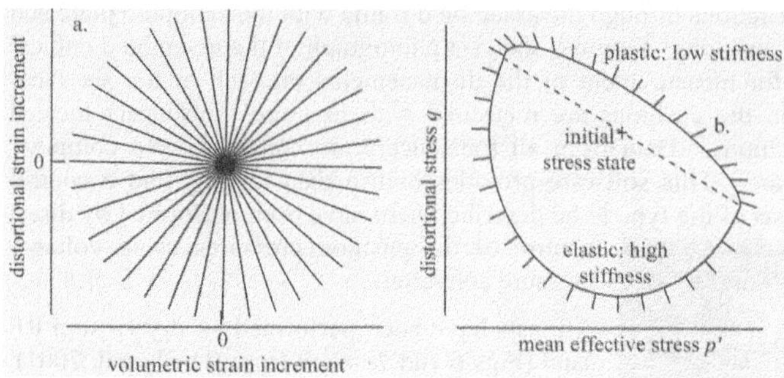

Figure 8: (a) Rosette of strain increments; and (b) envelope of resulting stress responses for elastic-plastic material. (Short lines in (b) indicate directions of strain increments.)

Thus, for example, for an elastic material (or an elastic-plastic material with initial stress state inside the current yield surface) the stress response envelope will be an ellipse centred on the initial stress. For an elastic-plastic material with a starting stress on the yield surface, the stress response envelope typically consists of sections of two ellipses (Fig 8b). For strain probes backing away from the yield boundary, into the elastic region, the stiffness is high and the distance travelled in stress space for a given magnitude of strain is also high. For strain probes which immediately mobilise plastic strains, the overall stiffness is lower and the distance travelled in stress space is also low.

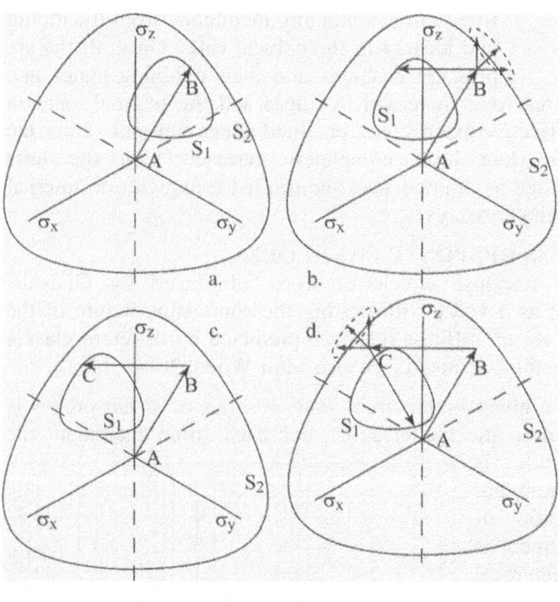

Figure 9: Kinematic hardening framework for deviatoric stress histories

Inevitably, such response envelopes will usually be presented as two dimensional curves such as Fig 8b (for ease of presentation and viewing) but these curves are sections through stress response hypersurfaces – in principle, six dimensional hypersurfaces. If we restrict ourselves to more limited spaces then this will usually be because of the limitations of the testing apparatus which are available to us. Thus, for axially symmetric states of stress attainable in the conventional triaxial apparatus, envelopes can be shown in terms of volumetric strain and distortional strain. For the true triaxial tests to be shown here the envelopes will be presented primarily in terms of deviatoric stress and strain components. Some data for the third – volumetric – dimension of the response to the deviatoric probes will be shown but the response to probes which include the volumetric (or isotropic) stress will also be important.

KINEMATIC HARDENING: MODELLING FRAMEWORK

Stress response envelopes provide a discipline for gathering data of evolving stiffness. It is helpful also to have some model in mind which is being tested by the collection of such envelopes. All the evidence from laboratory element tests suggests that some sort of kinematic hardening plasticity model is likely to be appropriate (Gajo & Muir Wood, 1999). Figure 9 shows an outline of the operation of such a model in the deviatoric plane. The shapes of the curves shown are somewhat arbitrary.

We imagine that a small region of high stiffness (S_1 in Fig 9) is carried round with the current stress state. It is pragmatically convenient to suppose that such a region represents the elastically attainable states of stress for a given stress history and thus that its boundary is the current yield locus for the sand. Thus after the deviatoric stress path AB, the elastic region has the position shown in Fig 9a. The stiffness along this and other paths is assumed to vary according to the current separation of surfaces S_1 and S_2 – the latter acting as a bounding surface which changes in size as the density of the sand changes.

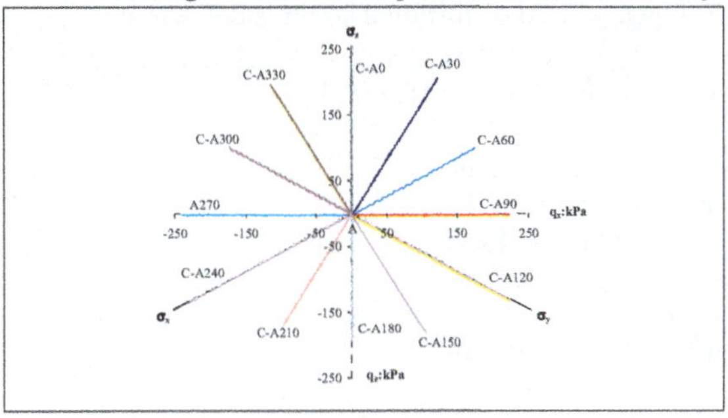

Figure 10: Deviatoric stress probes forming rosette A

For stress probes from B (Fig 9b) a distinction then needs to be drawn between probes such as those linked by the dotted ellipse which are essentially continuing the loading AB, and which will therefore be expected to show low plastic incremental stiffness, and probes such as BC which are backward pointing and tending initially to pick up the elastic response (compare Fig 8b). Some generalised mobilised friction – perhaps described by a criterion such as that proposed by Lade & Duncan (1975) – might be defined for the stress state reached at B, and this would then logically divide loading and unloading.

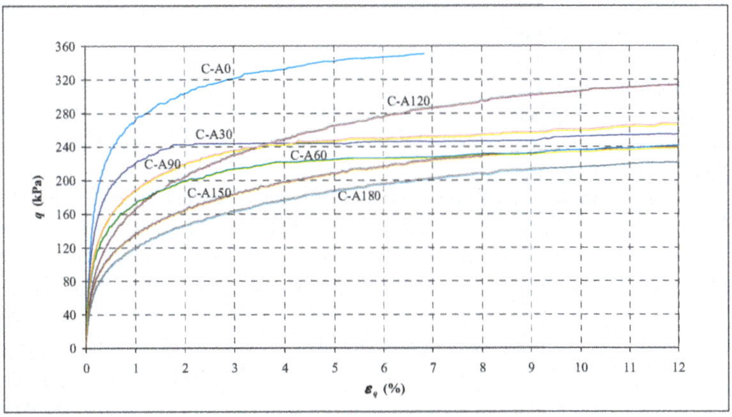

Figure 11: Deviatoric stress:strain response for half the probes of rosette A

On the way from B to C (Fig 9c) the incremental stiffness is again expected to fall according to the current separation of the kinematic yield locus S_1 and the bounding surface S_2.

Once the stress path has reached C then the elastic region S_1 will have been carried to C. Probing from C will once again divide into those probes which are essentially continuing the stress path from B (linked by the dotted ellipse in Fig 9d) and those which are reversing the direction of the stress path. It is hypothesised that there are two factors at work: one linked with the recent stress path and one linked with the current stress, in order to deduce whether the soil is being 'loaded' or not and therefore whether the stiffness should be high or low – whether the soil is engaging the plastic or the elastic sector of the stress response envelope. The tests reported here have explored the consequences for incremental stiffness of these three histories: A, AB and ABC.

DEVIATORIC STRESS PROBES

A first group of twelve samples was compressed isotropically to a mean stress of 200kPa and then subjected to a set of deviatoric stress paths at constant mean stress. These tests form rosette A (Fig 10). The deviatoric stress:strain responses for the paths which form the right-hand half of the rosette in Fig 10 are shown in Fig 11. The other half is essentially similar. The generalised distortional strain is defined as:

$$\varepsilon_q = \frac{1}{3}\sqrt{2[(\varepsilon_y - \varepsilon_z)^2 + (\varepsilon_z - \varepsilon_x)^2 + (\varepsilon_x - \varepsilon_y)^2]}$$

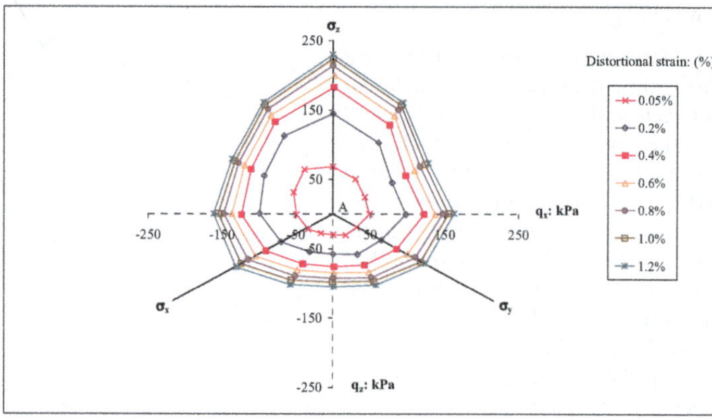

Figure 12: Stress response envelopes for rosette A

The same information can be presented in the form of stress response envelopes for different levels of deviatoric strain up to 1.2% (Fig 12). The set of nested envelopes represents a generalisation of the secant shear stiffness of the sand for this given history. While secant stiffness is not of itself particularly useful as an indicator of constitutive response – because it tells you about the past average behaviour (where the soil has been) rather than about the incremental present stiffness (where the soil might be about to go), for which a tangent quantity is needed – these collected envelopes do help to reveal the evolving stiffness of the soil from a common starting point.

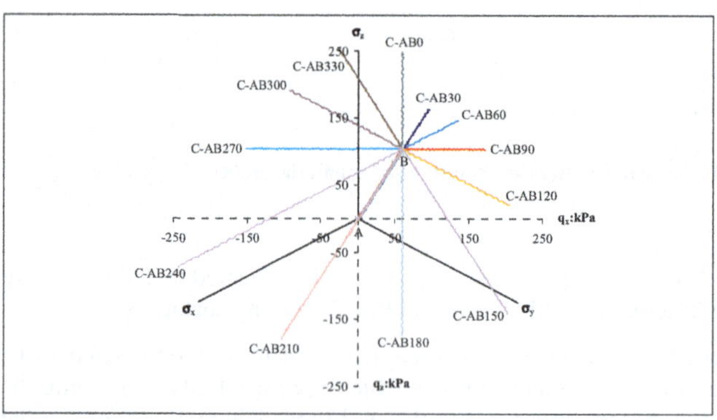

Figure 13: Deviatoric stress paths for rosette B with common history AB

If the sample were initially isotropic as a result of the sample preparation process then the response envelopes (Fig 12) would have a three-fold symmetry about the centre of the deviatoric plane. It is clear that the envelopes are pushed upwards along the σ_z axis which corresponds to the direction of deposition during sand pluviation. This is an indication that

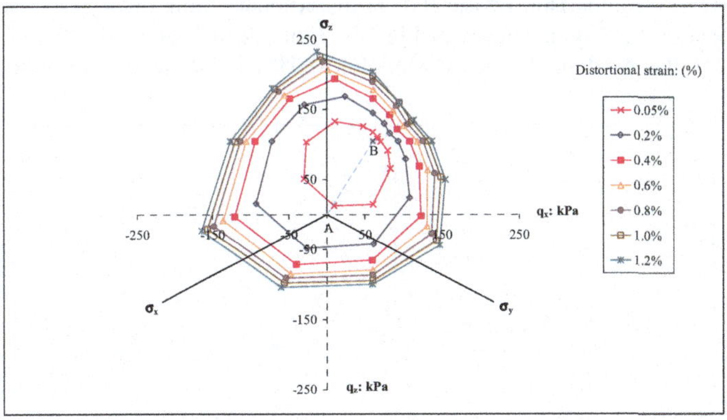

Figure 14: Stress response envelopes for history AB

the stiffness of the sand is higher for loading in the vertical direction than it is for loading in either of the horizontal directions: the stress probes reach a given strain level more rapidly for these latter probes. The symmetry of these response envelopes about the vertical axis in Fig 12 is an indication that the anisotropy is itself more or less symmetric about the axis of deposition: implying transverse isotropy/cross-anisotropy. This is reassuring: there should be nothing about the preparation process that leads to a distinction between the horizontal directions.

A second group of samples was subjected to rosette B with a common stress history. Isotropic compression to A (mean stress 200kPa) was followed by a radial stress probe with constant mean stress to point B with stresses (σ_x, σ_y, σ_z = 115, 200, 285kPa). From that point a rosette of deviatoric stress paths at constant mean stress was applied (Fig 13). As before, the deviatoric response is presented in the form of stress response envelopes for samples with this common history (Fig 14). The deviatoric strain is zeroed at B so that these curves all relate to secant stiffnesses from B. Comparison with Fig 12 shows that the size of the 0.05% envelope (the smallest that can be reliably resolved) has changed a little but its location has moved with the stress history AB. The envelopes for the larger strains – shown here only up to 1.2% – have not changed very much: there is a small dragging of the envelopes towards the upper right of the deviatoric plane with the recent history.

A third set of deviatoric stress probes followed the slightly more elaborate deviatoric history ABC (Fig 15): isotropic compression to A followed by radial shearing to B and then shearing again to C which is the mirror image of B in the deviatoric stress plane: (σ_x, σ_y, σ_z = 285, 200, 115kPa). Again the stress response envelopes have been deduced: they are shown in Fig 16.

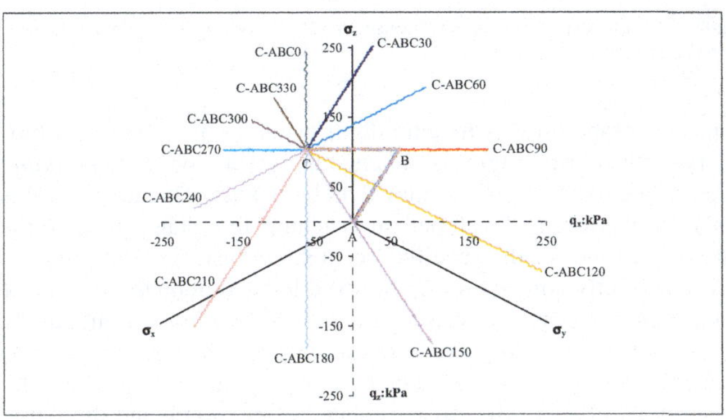

Figure 15: Stress probes for rosette C with common history ABC

It is of interest now to compare the response envelopes for rosettes B and C. The difference in stress states is one of reflection in the σ_z axis but there is of course a significant difference in the directions of the stress path taken to reach B (purely radial from A)

and C (non-radial from B). A comparison of the 0.05% distortional strain stress response envelopes for the three rosettes A, B, C is shown in Fig 17. From A to B and from B to C there is some hardening of the sand in the sense that this 0.05% response envelope has become slightly larger in each case – so that the sand has become stiffer.

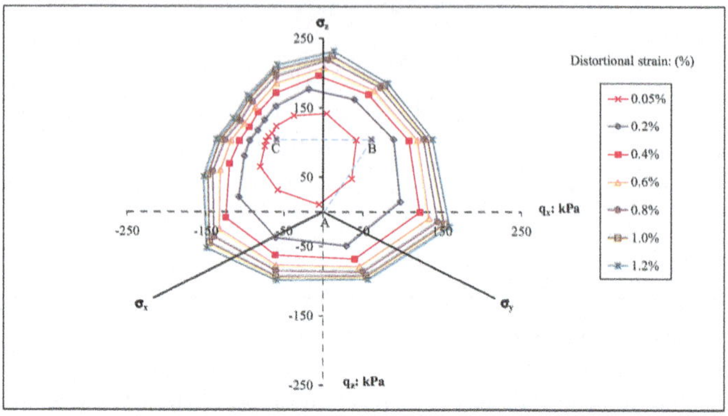

Figure 16: Stress response envelopes for rosette C with common history ABC

But if one tries to establish the 'centres' of these roughly circular envelopes – points a, b, c shown in Fig 17 – then it is evident that the movement of the 'centre' does reflect the recent history: the centre c for history ABC is further from the centre of the deviatoric plane (A) than the centre b for history AB. This is consistent with earlier studies of evolving deviatoric anisotropy reported by Alawaji *et al.* (1990).

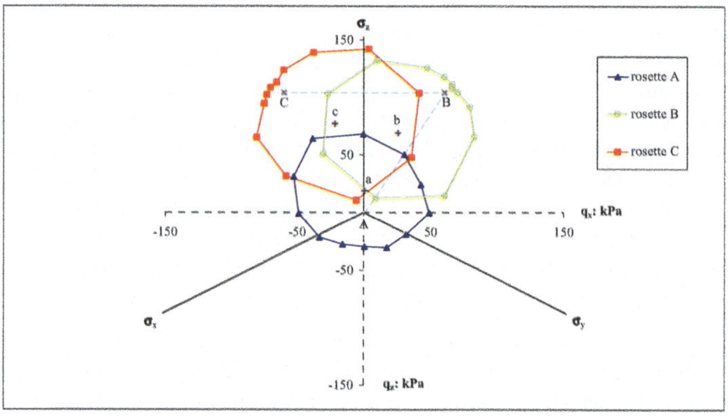

Figure 17: Comparison of stress response envelopes for rosettes A, B and C for 0.05% deviatoric strain

The truly elastic region for sands extends probably no further than a shear strain of the order of 0.001% (see, for example, LoPresti *et al.*, 1997), which is well below the strain that can be reliably resolved in the Cubical Cell. A stress response envelope for this magnitude of distortional strain could thus be associated with the boundary of the elastic region and hence the yield locus *for unloading stress paths*. Loading stress paths are expected to be immediately plastic. The centre of this yield locus (the 'back stress'), or the strain developed in reaching this state, gives some indication of the current fabric of the sand for the specific recent stress history. The response envelope for 0.05% strain will not strictly correspond to the yield locus (although it might be pragmatic to give the yield locus such a size in developing a constitutive model) but can be inspected to deduce the link between stress history and evolving fabric. The conclusion to be drawn links back to a conclusion from analysis of behaviour of particulate assemblies by Chen *et al.* (1988) and Ishibashi *et al.* (1988): 'the stress is ahead of the fabric and the fabric is ahead of the strain'. In other words, after a corner in the stress history the stress marches ahead on a new imposed path but the fabric lags behind – influenced by the strain. From a plasticity point of view, one would expect the immediate response on reversal – after a sharp corner – to be essentially elastic with high stiffness and strain increments linked with stress increments. As plasticity takes over then one would expect plastic strains to be controlled much more by stresses than by stress increments.

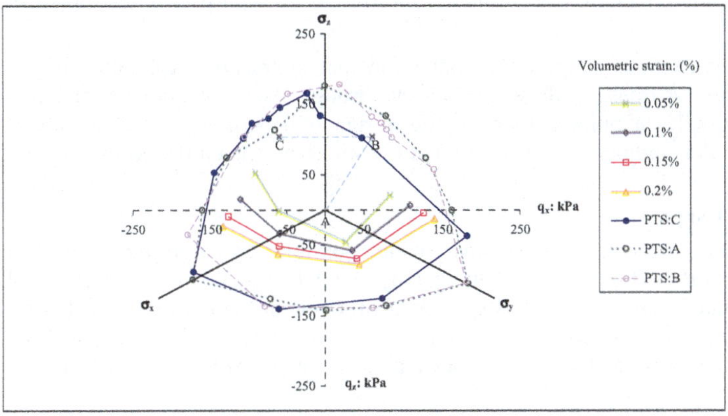

Figure 18: Stress response envelopes for volumetric strain for rosette C, with phase transformation surfaces (PTS) for rosettes A, B, C

The distortional strains give only one part of the picture of the response to the stress probes: there are also volumetric strains. Stress response envelopes for volumetric strains for rosette C with common history ABC are shown in Fig 18. Volumetric strain typically shows compression followed by expansion – especially for reversal stress probes within the rosette. This has the implication that on any such probe a given magnitude of *compression* strain may be reached either twice or not at all.

In the context of elastic-plastic modelling, where as well as a criterion of yielding we require some definition of the plastic deformation mechanism accompanying yield, it is helpful to look at incremental rates of change of volumetric strain with distortional strain. One particular rate is 'zero': and we can identify the point at which the volumetric response changes from compressive to dilatant. This has been described as the characteristic state (Luong, 1979) or phase transformation state (Tatsuoka & Ishihara, 1974). Phase transformation points have been deduced from the probes of each of the three rosettes: these are all shown in Fig 18 – marked PTS. It is remarkable that, whereas the stress response envelopes for different values of volumetric strain and distortional strain are clearly influenced by the recent stress history, the phase transformation points appear, within experimental uncertainty, to be probably somewhat insensitive to history. Such a finding, if supported by further experimental study, begins to simplify to some degree the development of the appropriate accompanying constitutive model, indicating the strong role that some generalisation of mobilised friction plays in controlling dilatancy. In fact the shape of the phase transformation surface shown in Fig 18 appears similar to that described by the deviatoric function proposed by Lade & Duncan (1975).

CONCLUSION

Some results from stress probing tests on Hostun sand conducted in a Cubical Cell Apparatus have been presented here. The associated constitutive picture is starting to emerge. The kinematic nature of the deviatoric response of the sand seems to be well revealed at least qualitatively. Work is continuing in order to convert the stress response envelope information into clear statements concerning the deviatoric kinematic hardening of the material.

The stress-dilatancy relationship describing the development of volumetric strains has also been investigated. The kinematic hardening structure links current volumetric stiffness strongly with recent stress history. Exploration of volumetric strains shows that, when the rate at which volumetric strain develops with distortional strain is evaluated, then, at least to first order, the locus of points in the deviatoric plane at which shearing occurs without volume change (characteristic state or phase transformation) seems to be somewhat independent of the recent history.

These experimental observations support the contention that systematic exploration of the evolution of incremental stiffness with stress history through the establishment of stress response envelopes provides a rational route to the design of laboratory test programmes intended to support the development (or refutation) of constitutive conjectures (Muir Wood *et al.*, 1997).

ACKNOWLEDGEMENTS

The experimental work described here was made possible by a research project GR/R05260/01 *Multiaxial stiffness of sand* funded by the UK Engineering and Physical Sciences Research Council. Hosein Javaheri spent part of a year at the University of Bristol with a scholarship from the Ministry of Science, Research and Technology of Iran. Laboratory work cannot proceed without technician support for which we are very grateful.

REFERENCES

Airey, DW & Muir Wood, D (1988) Cambridge True Triaxial Apparatus. *Advanced triaxial testing of soil and rock* (eds RT Donaghe, RC Chaney and ML Silver) ASTM:STP 977, 796-805.

Alawaji, H, Alawi, M, Ko, H-Y, Sture, S Peters, JF & Muir Wood, D (1990): Experimental observations of anisotropy in some stress-controlled tests on dry sand. In: *Yielding, damage, and failure of anisotropic solids (EGF5)*. (Ed: Boehler,JP) Mechanical Engineering Publications, London, 251-264.

Benahmed, N (2001) *Comportement mécanique d'un sable sous cisaillement monotone et cyclique: application aux phénomènes de liquéfaction et de mobilité cyclique.* Thèse de doctorat, École Nationale des Ponts et Chaussées, Paris.

Chen, Y-C, Ishibashi, I & Jenkins, JT (1988) Dynamic shear modulus and fabric: Part I: Depositional and induced anisotropy. *Géotechnique* **38** (1) 25-32.

Gajo, A & Muir Wood, D (1999) A kinematic hardening constitutive model for sands: the multiaxial formulation. *International Journal for Numerical and Analytical Methods in Geomechanics* **23** 925-965.

Gudehus, G (1979) A comparison of some constitutive laws for soils under radially symmetric loading and unloading. *Proc. 3rd Conf. Numerical methods in geomechanics*, Aachen (ed W Wittke) AA Balkema, Rotterdam 1309-1323.

Hambly, EC (1969) A new true triaxial apparatus. *Géotechnique* **19** (2) 307-309.

Ishibashi, I, Chen, Y-C & Jenkins, JT (1988) Dynamic shear modulus and fabric: Part II: Stress reversal. *Géotechnique* **38** (1) 33-37.

Ko, H-Y & Scott, RF (1967) A new soil testing apparatus. *Géotechnique* **17** (1) 40-57.

Lade, PV & Duncan, JM (1975) Elasto-plastic stress-strain theory for cohesionless soil. *Proc ASCE, Journal of the Geotechnical Engineering Division* **101** (GT10) 1037-1053.

LoPresti, DCF, Jamiolkowski, M, Pallara, O, Cavallaro, A & Pedroni, S (1997) Shear modulus and damping of soils. *Géotechnique* **47** (3) 603-617.

Luong, MP (1979) Les phénomènes cycliques dans les sables. *Journée de Rhéologie: Cycles dans les sols – ruptures – instabilités.* (Vaulx-en-Velin: École Nationale des Travaux Publics d'État), publication 2.

Miura, S & Toki, S (1982): A sample preparation methods and its effect on static and cyclic deformation-strength properties of sand. *Soils and Foundations* **22** (1), 61-77.

Muir Wood, D (2004a) *Geotechnical modelling.* E. & F.N. Spon (488pp)

Muir Wood, D (2004b) Experimental inspiration for kinematic hardening soil models. *Journal of Engineering Mechanics, ASCE* **130** (6), 656-664

Muir Wood, D, Alsayed, MI & Stewart, WM (1997) Constitutive conjectures for response of sand under multiaxial loading. *Numerical models in geomechanics (NUMOG VI: Montreal)* (eds GN Pande and S Pietruszczak) Balkema 51-56.

Tatsuoka, F & Ishihara, K (1974) Drained deformation of sand under cyclic stresses reversing direction. *Soils and Foundations* **14** (3) 51-65

STRESS-STRAIN BEHAVIOUR OF A MICACIOUS SAND IN PLANE STRAIN CONDITION

S.J.M.Yasin
Department of Civil Engineering
Bangladesh University of Engineering and Technology (BUET), Dhaka-1000, Bangladesh
e-mail:sjmyasin@ce.buet.ac.bd

F.Tatsuoka
Department of Civil Engineering
Tokyo University of Science, 2641, Yamazaki, Noda, Chiba 278-8510, Japan
e-mail: tatsuoka@rs.noda.tus.ac.jp

ABSTRACT

Unusual failures of river banks and river training structures have been reported during construction and shortly after commissioning of several structures along Jamuna river in Bangladesh that raised widespread questions regarding the design principles and parameters used. The natural sand deposit along the Jamuna river contain relatively larger amount of mica than most other natural soils. Jamuna sand needs to be studied under wide range of loading conditions (such as triaxial, plane strain, simple shear etc.), drainage and density conditions (i.e. drained / undrained, dry / saturated, dense/loose state etc.) to reveal the extent of variation of its strength and deformation characteristics in order to facilitate understanding of the mechanism of past failures of structures and suggest rational design parameters. A series of plane strain compression tests were performed on Jamuna sand. It is observed that Jamuna sand is highly contractive under shear and more anisotropic than other non-mica sands.

INTRODUCTION

Bangladesh, the largest delta in the world, is criss-crossed with many rivers. The Jamuna, one of the major rivers flowing throw the country (Fig.1) originates in the Himalaya mountains then flows over China as Tsang-po, then over India as Brahmaputra and finally over Bangladesh as Jamuna before falling into the Bay of Bengal. It carries huge amount of sediment from the upstream hills and a large part of the country is formed by deposition of such sediments. The Jamuna is a relatively unstable (i.e. shifting its course), braided river with high fluctuation in water level between winter (dry season) and the monsoon (rainy season). The river is noted for high bank erosion. In the last decade several bank protection and river control structures have been made along this river. Unusual failures of river banks and river training structures have

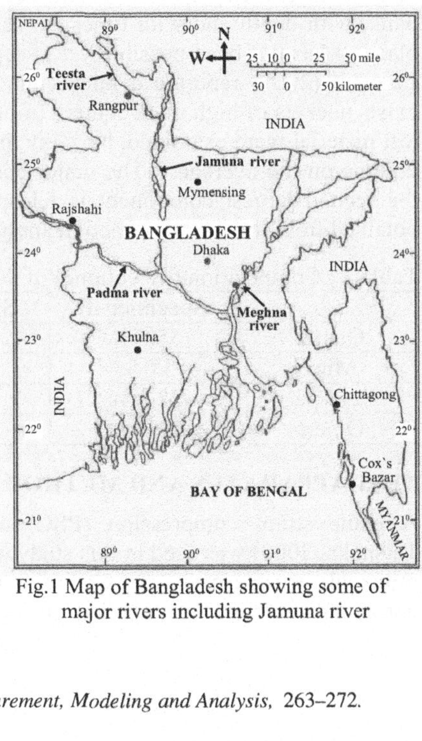

Fig.1 Map of Bangladesh showing some of major rivers including Jamuna river

Hoe I. Ling et al. (eds.), Soil Stress-Strain Behavior: Measurement, Modeling and Analysis, 263–272.

been reported during construction and shortly after commissioning of several structures that raised widespread questions regarding the design principles and parameters used. Failure of groyne and hard points occurred in the monsoon when the water level rised near to the top of the bank and flow volume increased. Slope failure also occurred in both natural deposit and dredge-fill at low water level during construction of the guide bund for the Jamuna bridge. Thus the soil along Jamuna river drew attention for comprehensive study. To this end, this paper presents the results from a series of plane strain compression tests performed on Jamuna sand and compares them with the behaviour of Toyoura (which is a widely studied sand) and other sands.

TEST MATERIAL

Fig.2 compares the grain size distribution of Jamuna sand with that of Toyoura sand (Japan), Hostun sand (France) and SLB sand (UK). The physical and mechanical characteristics of the Jamuna sand is: G_s= 2.7, D_{50}= 0.16, e_{max}= 1.173 and e_{min}= 0.690 and FC= 7 %. Apart from difference in grain size distribution and possible variation in the shape of particles, Jamuna sand contains relatively larger amount of mica compared to many other natural sands. The quantity of mica, its distribution and orientation in the natural deposit varies from place to place. The reason for such variation is that during high current and high flow the material gets lifted as sediment and later gets deposited at places of low current. Since the river current varies from place to place along and across the bank, with depth and with time, the nature of deposit varies significantly from place to place. It has not been possible yet to separate the mica particles mechanically. Hight and Leroueil (2003) reported a mica content of 5-10 % by grain counting, however there exists pockets of high mica content (more than 30%). In the present study, 3 specimens of test material were examined, by x-ray diffraction method which showed 4.7 % of mica by volume on the average. The major component is quartz (around 55 % by volume) and the second largest component is feldspar (around 25 % by volume). The composition obtained from the x-ray diffraction analysis are presented in Table 1.

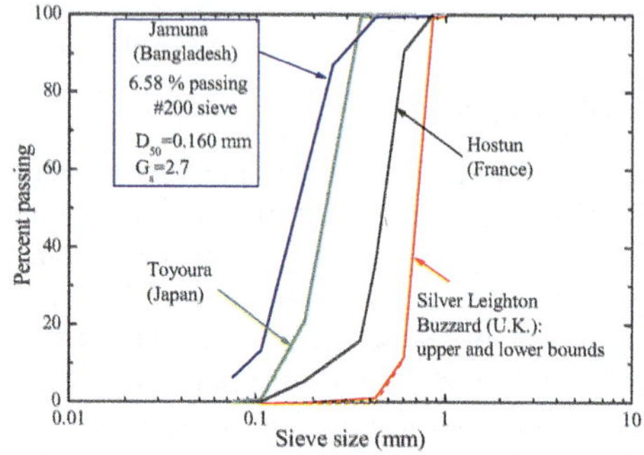

Fig.2 Particle size distribution of Jamuna sand

Table 1 : Composition (by volume) of the test material by x-ray diffraction method.

	Specimen-1	Specimen-2	Specimen-3	Average
Quartz	55.8 %	56.6 %	53.7 %	55.4 %
Mica	5.0 %	4.5 %	4.5 %	4.7 %
Feldsper	22.5 %	27.4 %	27.8 %	25.9 %
Others	16.7 %	11.5 %	14.0 %	14.0 %

TEST APPARATUS AND METHOD

A plane strain compression (PSC) apparatus (Park and Tatsuoka 1994;Yasin and Tatsuoka, 2000) was used in this study which is a strain controlled one in which the major principal stress (in the vertical direction) and the intermediate principal stress (in horizontal direction) are applied via rigid (steel platen) and lubricated boundaries. The direction of loading on the major principal stress boundary can be reversed (i.e. up or down) without backlash error and the strain rate can be changed instantaneously by use of servo motor, Electro Magnetic Clutch, gear and Timer. The minor principal stress (in the horizontal direction) boundary is flexible (membrane) and controlled by cell pressure. The specimen is prismatic; 8 cm. (in the direction of σ_3') by 16cm. in cross-section and 20 cm in height. The load on the major and intermediate principal stress boundaries are measured by load cells and the cell pressure is measured by a High Capacity Differential Pressure Transducer (HCDPT) and regulated by an Electro-pneumatic Pressure (EP) Transducer. The friction forces on the confining platens are also measured by load cells attached to them and the vertical stresses (σ_1') presented are those calculated at the mid height of the specimen and corrected for the platen friction. The vertical compression of specimen is recorded locally (i.e. on the specimen surface) by a pair of LDTs (up to around 4% axial strain which was the maximum limit for the LDTs) and the externally by a dial gauge (up to the end of the test). The bedding error, deduced from local measurements, is found to be very small for Jamuna sand. The deformation in the intermediate principal stress direction is recorded by four pairs of proximity transducers. In each test, the control of loading conditions (i.e. strain rate, stress path, start and duration of creep, relaxation, stress state for unload-reload cycle including amplitude, number of cycles etc.) and data logging are done automatically by an in-house developed computer program.

Hight et al. (1999) showed that the effect of specimen preparation on the strength and deformation characteristics of micacious sand is significant. From undrained triaxial tests they obtained that a specimen formed by dry spooning and at a relative density of 40% dilated significantly while another specimen of similar relative density but formed by moist temping at a similar void ratio collapsed . Vaid et al. (1995) performed undrained simple shear tests on loose specimens of Syncrude sand prepared by moist tamping, air-pluviation and water pluviation. They observed that the moist tamped specimen was highly contractive, the air-pluviated specimen was also contractive but to a lesser degree. In contrast, the water pluviated specimen was dilative. They also reported considerable difference in the shear stress ~ shear strain response of these specimens. Therefore, a consistent method of specimen preparation that requires relatively less human involvement was adopted (for example tamping of different specimen may produce different fabric). To control density and to obtain a consistent method of preparing successive specimens, a sieve stack was used. Dry sand was poured from a box through the sieve stack into the specimen mold. Several trials showed that the method was effective in preparing successive specimens with reasonably close density.

STRESS PATHS FOLLOWED

The tests performed by the PSC apparatus can be broadly divided into two types depending on the stress path followed. For one group of tests (group-1) the specimens were sheared along 'σ_h = minor principal stress = constant' path starting from the initial isotropic stress state of $(\sigma_h)_c$ and no change in strain rate was made (Fig.3a). For the other group of tests (group-2) the specimens were initially sheared along the 'σ_h = minor principal stress = constant' path to bring the stress ratio $\sigma_{major}/\sigma_{minor}$ to 3.0 which was then maintained while increasing σ_{minor} to $\sigma_{minor} = (\sigma_h)_{cd}$. The specimens were finally sheared along the '$(\sigma_h)_{cd}$ =constant' path (Fig.3b). In the latter group of tests there were changes in strain rate or creep at different stress levels along path cd. $\sigma_{major}/\sigma_{minor}$ =3.0 was employed so that no separation occurs along the intermediate principal stress plane while the minor principal stress is increased by increasing the cell pressure.

Tests were performed on both air dry and saturated state at different initial density and at different $(\sigma_h)_c$ (for group 1) and $(\sigma_h)_{cd}$ (for group 2).

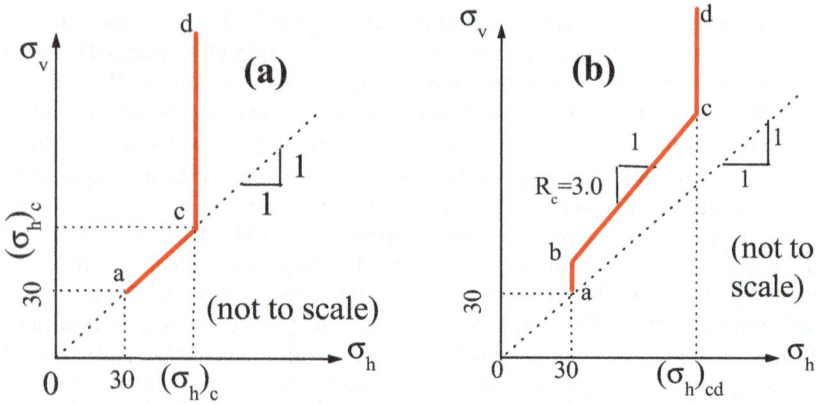

Fig.3 Stress paths followed in the tests (a) `$\sigma_h$=constant=$(\sigma_h)_c$` shearing from initial isotropic stress state (b) initial `$\sigma_h$=constant` shearing from isotropic state to reach $\sigma_v/\sigma_h=R_c=3.0$ which is maintained up to $\sigma_v = (\sigma_h)_{cd}$ and finally shearing along $(\sigma_h)_{cd}$ path.

TEST RESULTS

Stress - Strain Behaviour

Fig.4 and Fig.5 show the effect of initial density and confining stress on the Axial stress-Axial strain-volume change behaviour from 'σ_h = minor principal stress = constant' tests. The axial strain rate in these tests was 0.0125% per minute throughout. The specimens were air dry and drained to air. The volumetric strains were calculated as $\varepsilon_{vol}=\varepsilon_v+\varepsilon_h$ where ε_v = Vertical strain (which is in the σ_1 direction), ε_h =Lateral strain (which is in the σ_3 direction; strain in the other lateral direction is zero due to plane strain condition). The starting point of each curve correspond to isotropic stress states of either $(\sigma_h)_c$ = 100 kPa or $(\sigma_h)_c$ = 400 kPa. The lateral strain measurements (which are by gap sensors) became ineffective at post peak region due to formation of the shear band. So, the volumetric strains plotted in Fig.4 are plotted up to peak stress ratio (i.e. the end points of the curves correspond to maximum σ_v/σ_h level). Test J319C with dense initial

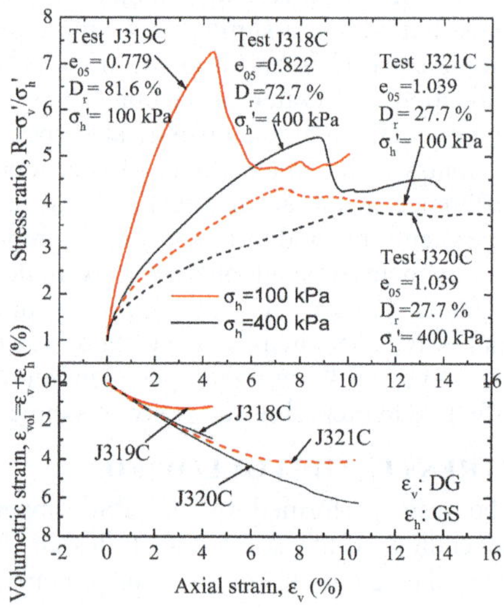

Fig.4 `Stress ratio~axial strain` and `Volumetric strain~axial strain` relationship from σ_h=constant tests starting from isotropic stress state (sheared air-dried; $\dot{\varepsilon}_0$ =0.125 % per minute)

state and low confining stress showed sharp peak in the 'axial stress~axial strain' relation and the specimen also started dilating only near peak stress level and still ε_{vol} is positive (i.e. the volume is lower than the initial volume) at the peak. On the other hand test J318C with dense initial state and high confining pressure showed a more ductile behaviour without any sign of dilation even at the peak state that resemble the volume change behaviour of loose specimens (Tests J320C and J321C). This clearly demonstrates that the so called "dense" and "loose" behaviour which usually state that "dense sand dilates and loose sand undergoes volume compression during shear" do not represent the proper picture without reference to confining pressure level. Rather a dense sand at higher confining stress may behave like a loose sand at a relatively low confining stress. Such behaviour is also observed from triaxial tests on Sacramento river sand by Lee (1965) as reported by Leroueil & Hight (2003). The effect of confining pressure on stress~strain response appeared to be relatively high for the specimens with dense initial state compared to that at loose initial state. Fig.5 shows that for high initial density and low confining stress the specimen contract at low stress level but starts dilating at stress levels near peak. However, for specimens with low initial density and low confining stress or specimens with high confining stress (either low or high initial density) only contraction occurred. Fig.6 compares the volumetric strain at failure of Jamuna sand for different relative density with those of Toyoura sand. It may be noted that the effect of confining pressure on volumetric strain is much more significant for Jamuna sand for either dense or loose initial state than that for Toyoura sand at dense state and within the range of σ_h=100~400 kPa.

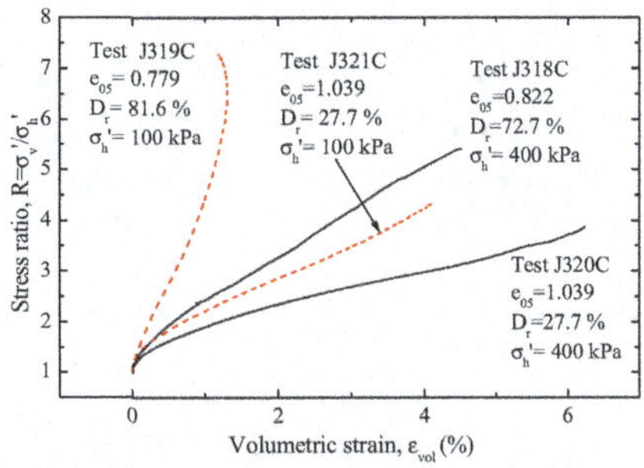

Fig.5 Stress ratio vs volumetric strain from σ_h=constant tests starting from isotropic stress states

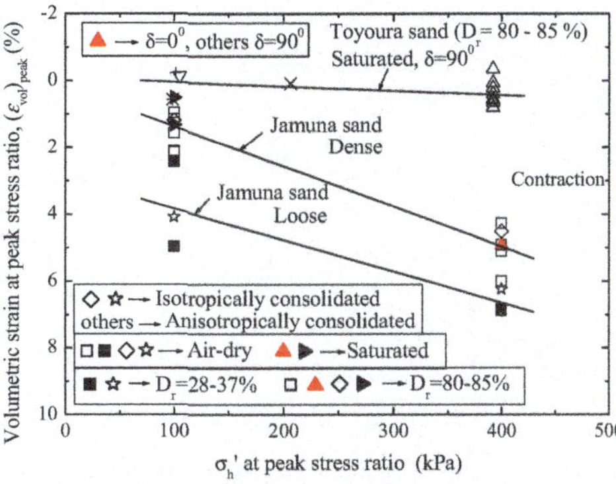

Fig.6 Variation of volumetric strain at failure with minor principal stress (data of Toyoura sand from Yasin and Tatsuoka, 2000)

Angle Of Internal Friction

Fig.7 and Fig.8 show the variation of angle of internal friction at peak stress ratio and at residual stress state respectively. It can be seen that for a change in relative density from 30% to 80% the peak angle of internal friction increases in the range of 10 to 15 degrees whereas the residual angle of internal friction increases about 5 degrees. The effect of density on peak friction angle appear to be slightly higher at high confining stress than at relatively low confining stress. Fig.9 shows the variation of friction angle at peak and residual states with minor principal stress at failure. There is significant effect of confining stress on ϕ_{peak} in the range of σ_3=100 kPa to 400 kPa (1 kgf/cm^2 to 4 kgf/cm^2). At dense state ϕ_{peak} increases in the order of 5 degree for a change in confining stress from

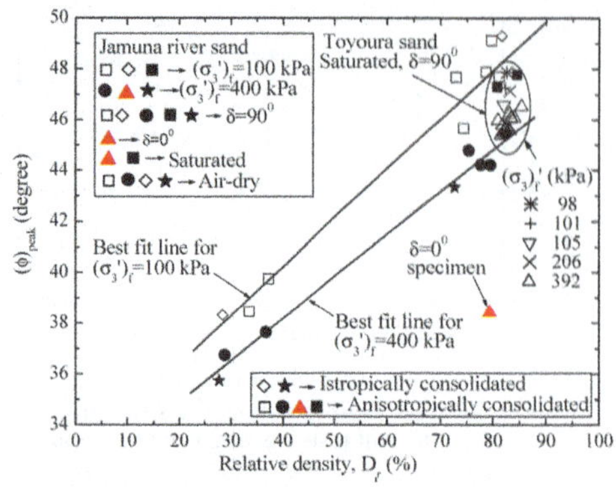

Fig.7 Variation of friction angle at peak with relative density

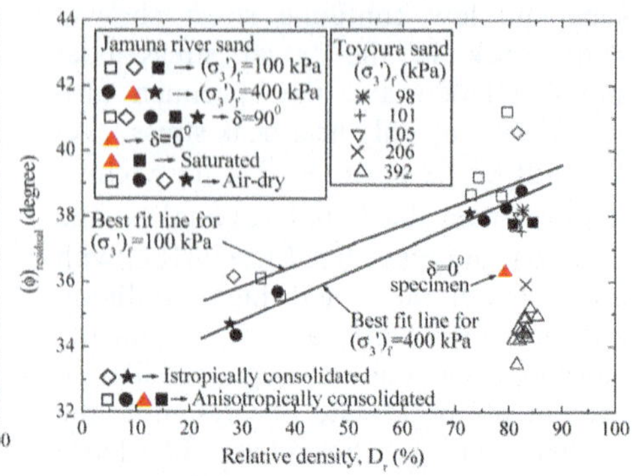

Fig.8 Variation of residual friction angle with relative density

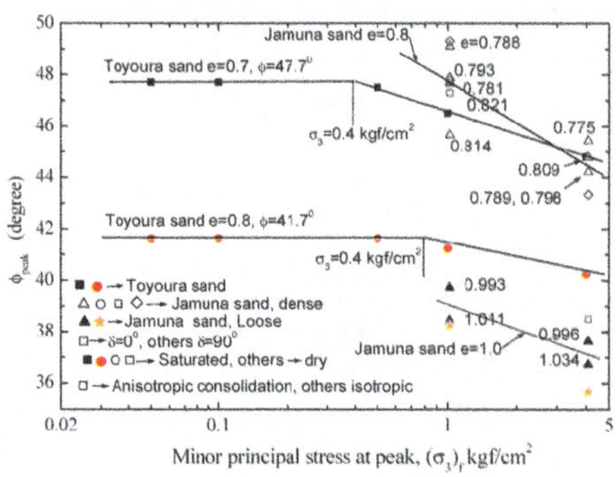

Fig.9 Variation of friction angle at peak and residual state with minor principal stress at peak (Toyoura sand data from Tatsuoka et al.1986)

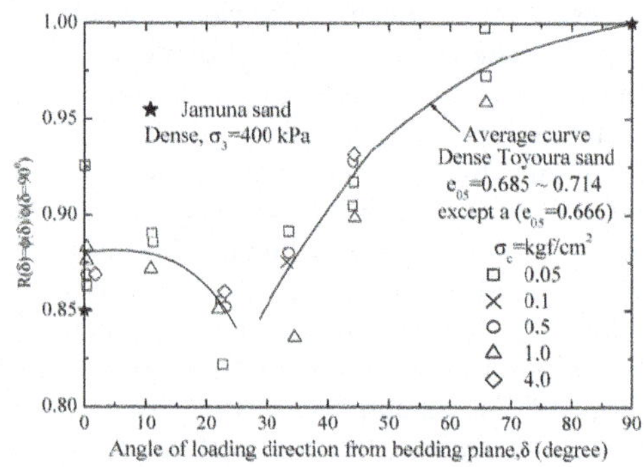

Fig.11 Effect of loading direction with respect to bedding plane on ϕ_{peak}

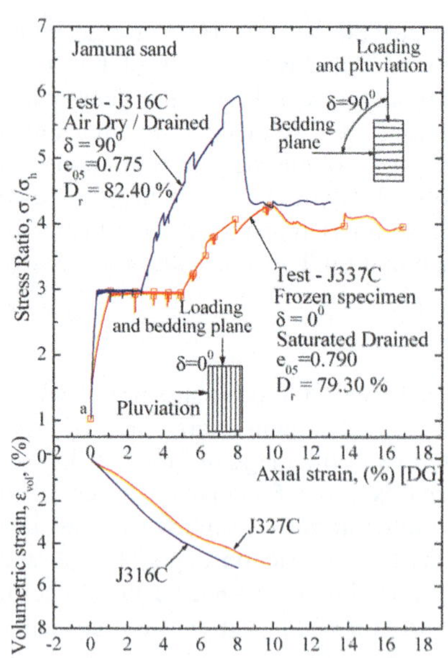

Fig.10 Stress-strain relationship from δ=0° and δ=90° specimens of Jamuna sand

100 kPa to 400 kPa. There is similar effect of confining stress on residual angle of internal friction. Figs.7, 8 and 9 also shows data for Toyoura sand. Despite the difference in gradation (Fig.2) and large difference in volume change behaviour (Fig.6) between Jamuna sand and Toyoura sand, it is observed, the peak and residual angle of internal friction of these sands are not much different at dense state (Fig.7 and 8). However, the effect of minor principal stress on ϕ_{peak} is slightly greater for Jamuna sand compared to that of Toyoura sand (Fig.9).

Anisotropy

Fig.10 compares the stress-strain relation for δ=0° and δ=90° of Jamuna sand for PSC Tests. δ is the angle between bedding plane and loading direction. It may be observed that the peak stress ratio for δ=0° specimen is as low as the stress ratio at residual stress state of $\delta=90^0$ specimen. Compared to Toyoura sand the degree of anisotropy of Jamuna sand is slightly large. For Toyoura sand $\phi_{peak}(\delta=0^0)/\ \phi_{peak}(\delta=90^0)=0.87$ where as for Jamuna sand $\phi_{peak}(\delta=0^0)/\ \phi_{peak}(\delta=90^0)=0.85$ (Fig11).

TIME DEPENDENT CHARACTERISTICS

Shearing Under Monotonic Strain Rates

The rate of loading or speed of construction may be a key concern in certain soils such as saturated clay which may give rise to pore-water pressure and consequent loss of effective stress and failure. For pervious material such as sand, the rate of loading may be important for prediction of deformation, particularly by stress~strain models used in FEM. Different geomaterials are reported to show different characteristics depending on monotonic strain rate during shearing. Matsushita (1998a,b) have shown that stress-strain relationships of Toyoura and Hostun sands, which are clean sands, are virtually same when sheared under different constant axial strain rates. Clay on the other hand shows different stress-strain relations for different strain rates (Shibuya et al. 1996; Mukabi, 1995; Vaid and Campanella, 1977; Tatsuoka et al. 1999; Tatsuoka et al. 2002) which was first recognized by Suklje (1969) as isotach behaviour. To investigate the effect of rate of loading on the stress-deformation characteristics of Jamuna sand, different specimens were tested in PSC with 100 times difference in monotonic axial strain rate during the final $(\sigma_h)_{cd}$=100 or 400 kPa shearing phase. Only dense specimens were tested. Fig.12 and Fig.13 presents the stress-strain-volume change behaviour from these tests. These tests do not show any significant effect of change in strain rate of this magnitude on the stress-strain-volume change behaviour i.e. Jamuna sand, a much finer sand than Toyoura and Houstun sand has little strain rate dependency in stress~deformation behaviour under constant monotonic strain rate.

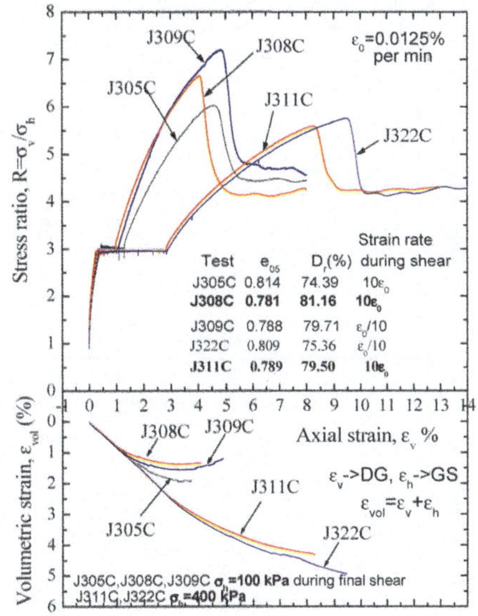

Fig.12 Effect of different magnitude of constant strain rate along the shearing path σ_h=100 kPa and σ_h=400 kPa

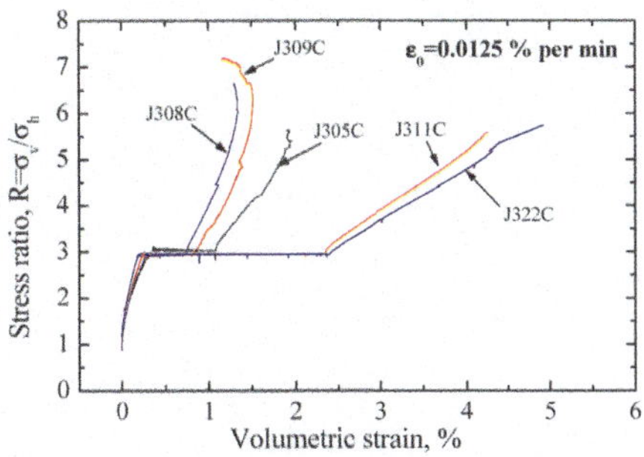

Fig.13 Stress ratio vs volumetric strain from tests with constant axial strain rate during shearing

Shearing with sudden change in monotonic strain rate

Most geomaterials show sudden drop or rise in the stress~strain response corresponding to sudden reduction or increase in monotonic strain rate (Hayano et al., 2001; Matsushita et al,1999, Tatsuoka et al.,2002). Figs. 14 and 15 show typical stress ratio and volumetric strain response from tests on both loose specimens where the specimens were subjected to

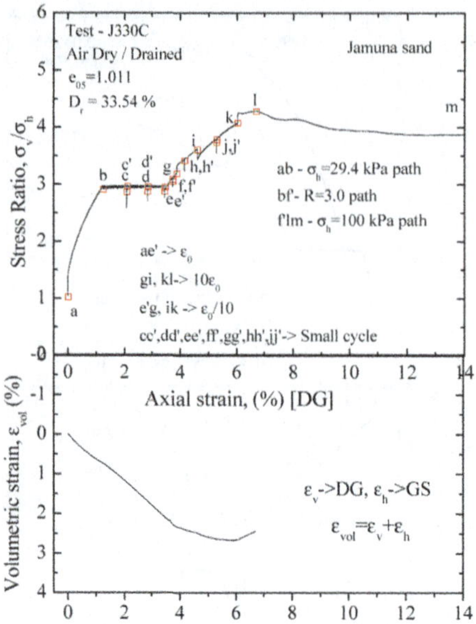

Fig.14 Stress~strain curve with 100 times change in strain rate during σ_h=constant=100 kPa shearing (loose specimen)

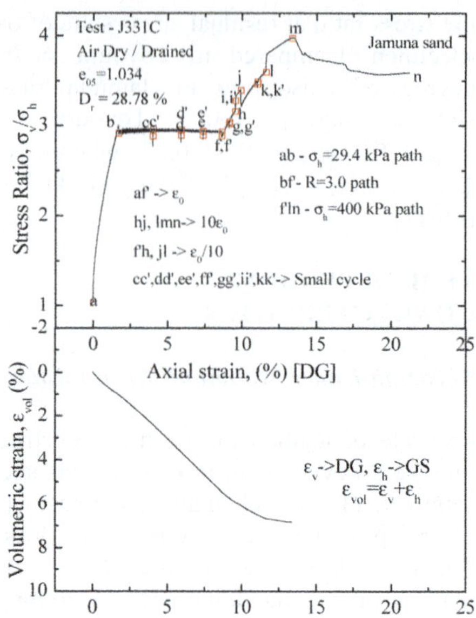

Fig.15 Stress~strain curve with 100 times change in strain rate during σ_h=constant= 400 kPa shearing (loose specimen)

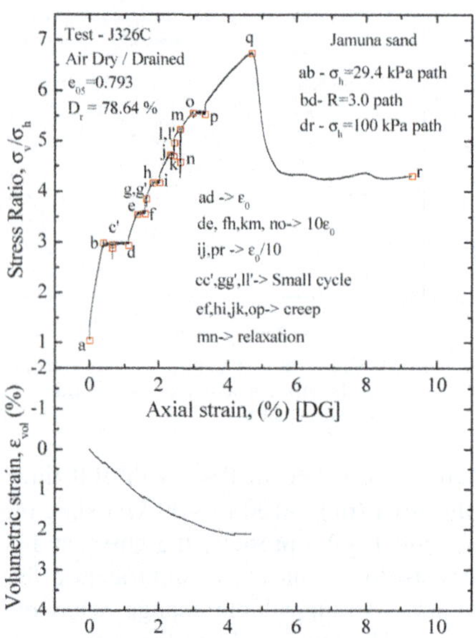

Fig.16 Stress strain curves from test with creep, relaxation and change in axial strain rate during shearing along σ_h=constant=100 kPa (dense specimen)

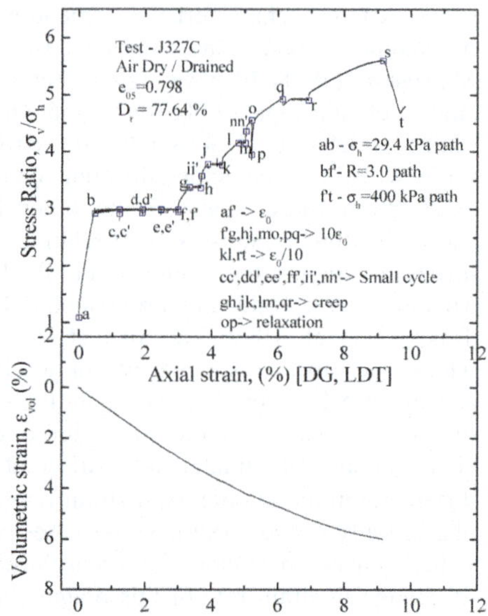

Fig.17 Stress strain curves from test with creep, relaxation and change in axial strain rate during shearing along σ_h=constant=400 kPa (dense specimen)

100 times change (reduction or increase) in axial strain rate at specified stress levels. At confining pressures of both σ_h=100 and 400 kPa it is observed that there was sudden drop in the stress~strain response when the strain rate was reduced. This was also observed for dense specimens. Such effects appear to be larger at higher stress levels. Also the stress~strain response became stiffer when the strain rate was increased. Although ageing

can also cause such stiffer response, but in these test there was no time delay in changing the strain rate, so this phenomenon can be attributed entirely to change in strain rate.

Creep and Relaxation during shear

In almost all civil engineering constructions, the phenomenon of Creep and Relaxation are almost inevitable in the field and study about the nature and extent of their influence is important for both design and prediction of performance of the soil mass under the field loading conditions. Figs.16 and 17 presents typical relationships between stress ratio and axial strain observed from tests on dense specimens in which creep, relaxation and change in axial strain rate were applied at different stress levels along the stress path $(\sigma_h)_{cd}$=const. (Fig.3b). A stiffer zone of response is observed when loading is started just after a creep or relaxation stage. Also the subsequent stress~strain response is found to depend on the prevailing strain rate when loading starts after creep or relaxation. No abrupt change could be seen in the strain paths. Creep deformation appears to be relatively larger at higher stress ratio compared to that at low stress ratio. Similar response was observed from tests on loose specimens. A possible consequence of the creep phenomenon is that field loading tests (e.g. plate load test) may show high initial stiffness and clear yielding in the load~settlement relationship which is often interpreted as failure stage but actually be away from the ultimate strength (i.e. peak strength). Thus designs based on field loading tests may be overly conservative. On the other hand, deformations predicted from conventional tests (such as monotonic loading laboratory tests) may be underestimated.

CONCLUSIONS

Stress deformation characteristics of Jamuna sand, a micacious sand from Banglaesh, as found from plane strain compression tests, are presented. The following conclusions may be drawn from the test results presented.

(1) Jamuna sand is very contractive i.e. there is virtually no volume increase during shearing even at dense state. This behaviour is significantly different from non-mica sands such as Toyoura sand. The magnitude of volumetric strain is significantly dependent on minor principal stress.

(2) Jamuna sand appear to be highly anisotropic i.e. stress~strain characteristics and angle of internal friction at peak is very much dependent on the direction of loading.

(3) There is almost no effect of strain rate on stress~strain behaviour for shearing at monotonic strain rates. However, the material exhibits considerable viscous effects for change in the monotonic strain rate during shearing including creep and relaxation.

ACKNOWLEDGEMENT

Dr. S.J.M.Yasin was invited at the University of Tokyo by the Japan Society for Promotion of Science under "The JSPS Postdoctoral Fellowship Program for Foreign Researchers". The support received is gratefully acknowledged. The laboratory tests on Jamuna sand was carried out at the Geotechnical Laboratory of the University of Tokyo.

REFERENCES

Hayano, K., Matsumoto, M, Tatsuoka, F. and Koseki, J. (2001) Evaluation of Time-dependent deformation properties of sedimentary soft rock and their constitutive modeling, Soils and Foundatons, vol.41, no.2, pp.21-38.

Hight, D.W., Georgiannou, V. N., Martin, P.L. & Mundegar, A. K. (1999) Flow slides in micacious sands, Proc. Of the International symposium on Problematic Soils, IS Tohoku, Sendai, Japan, Balkema, Rotterdam, Vol.2:945-958

Hight, D.W. and Leroueil, S. (2003) "Characterization of soils for engineering purpose", Characterisation and Engineering Properties of Natural Soils, Tan et al.(eds.), A.A.Balkema Publishers, ISBN 90 5809 537 1.

Lee, K.L. (1965) Triaxial compressive strength of saturated sand under seismic loading conditions, Ph.D. thesis, University of California, Berkely (reported by Leroueil & Hight, 2003).

Leroueil S. and Hight D.W. (2003) "Behaviour and properties of natural soils and soft rocks", Characterisation and Engineering Properties of Natural Soils, Tan et al.(eds.), A.A.Balkema Publishers, ISBN 90 5809 537 1.

Matsushita, M., Yasin,S.J.M., Cazacliu, B., Tatsuoka, F. and Koseki, J. (1998a) Strain acceleration dependency of sand deformation in plane strain compression. Proc. 33rd Japan National Conference on Geotechnical Engineering, JGS, Yamaguchi, 515-516 (in Japanese).

Matsushita, M., Yasin,S.J.M., Cazacliu, B., Tatsuoka, F. and Koseki, J. (1998b) Effect of strain acceleration on deformation strength characteristics of sand, Proc.53rd Annual Conf. of Japanese Society of Civil Engs., III-A, Kobe, 62-63 (in Japanese).

Matsushita, M., Tatsuoka, F., Koseki, J., Cazacliu, B., Di Benedetto, H. and Yasin, S. J. M. (1999) Time effects on the pre peak deformation properties of sands, Proc. Second Int. Conf. on Pre-Failure deformation characteristics of geomaterials, IS Torino, Jamiolkowski et al., Balkema, 1, 681-689.

Mukabi, J.N. (1995) Deformation characteristics of small strains of clays in triaxial tests, Dr. Engng. Theis, University of Tokyo.

Park,C.S and Tatsuoka, F. (1994) Anisotropic strength and deformation of sands in plane strain compression, Proc. Of the 13th Int. Conf. on SMFE, New Delhi, 1, pp.1-6.

Shibuya,S, Mitachi, T, Hosomi,A. and Whang,S.C. (1996) Strain rate effects on stress strain behaviour of clay as observed in monotonic and cyclic triaxial test, Measuring and Modeling Time Dependent Soil Behaviour, ASCE, Geotechnical Special Publication 61, 214-227.

Suklje, L. (1969) Rheological aspects of soil mechanics, Wiley Interscience, London.

Tatsuoka, F., Sakamoto, M., Kawamura, T. and Fukushima, S. (1986) Strength and deformation characteristics of sand in plane strain compression at extremely low pressures, Soils and Foundations, Vol.26, No.1, pp.65-84.

Tatsuoka, F., Murayama, N., Santucci de Magistris, F. and Momoya, M. (1999) Isotach behaviour of geomaterials, Jamiolkowski, Lancellotta & Lo Presti (eds), Balkema.

Tatsuoka,F., Ishihara, M., Di Benedetto, H., Kuwano, R. (2002) Time-dependent shear deformation characteristics of geomaterials and their simulation, Soils and Foundations, Vol.42, no.2, pp.103-129.

Vaid, Y.P. and Campanella, R.G. (1977) Time-dependent behaviour of undisturbed clay, Journal of Geotechnical Engineering Division, ASCE, 103(7), 693-709.

Vaid, Y.P., Uthayakumar, M., Sivathayalan, S., Robertson, P.K. & Hofmann, B. (1995) Laboratory testing of Syncrude sand, Proc. 48th Canadian Geotech Conference, Vancouver, vol.1, 223-232.

Yasin, S.J.M. and Tatsuoka, F. (2000) "Stress history-dependent deformation characteristics of dense sand in plane strain, Soils & Foundations, Vol.40, No.2, pp.77-98, April 2000.

Soil Stress-Strain Behavior: Measurement, Modeling and Analysis
Geotechnical Symposium in Roma, March 16 & 17, 2006

UNDRAINED SHEAR BEHAVIORS OF HIGH PLASTIC NORMALLY K_0-CONSOLIDATED MARINE CLAYS

S.Nishie, L.Wang and I.Seko
Chuo Kaihatsu Co.Ltd.
Nishi-Waseda 3-13-5, Shinjuku-ku , Tokyo 169-8612, Japan
e-mail: wang@ckcnet.co.jp

ABSTRACT

Isotropic normally consolidated-undrained and K_0-consolidated-undrained triaxial compression tests were carried out in order to investigate the effects of K_0-consolidation and secondary consolidation on the undrained shear behavior of undisturbed marine clay with varying plasticity index in this study. The deviator stress ratio M^* mobilized at the maximum deviator stress q_{max} of normally K_0-consolidated young clays at the end of primary consolidation became smaller than the critical stress ratio M due to stress induced anisotropy caused by the K_0-consolidation, this ratio M/M^* was 0.85 regardless of plasticity index. On the other hands, the effective stress path for normally K_0-consolidated young clays was parabolic curves with plastic softening beyond the maximum deviator stress, which became more marked as the plasticity index increased. It was clearly showed that the shape of the curve will strongly depend on the plasticity index. Furthermore, it was found that experimental effective stress path of high plastic normally K_0-consolidated young clays showed remarkable plastic softening beyond the maximum deviator stress, and this phenomenon could be modeled using a quadratic equation and the parameters M^* and initial deviator stress ratio after K_0 consolidation, η_0, modified by an anisotropic kinematical hardening rule.

The normalized maximum deviator stress increased linearly with the log of the normalized consolidation time and high plasticity K_0 consolidated aged clays subjected to secondary consolidation showed more brittle behavior. The effective stress paths for high plasticity K_0 consolidated aged clays initially maintained a constant mean normal stress p' until they reached a deviator stress ratio η_1 which itself increased with consolidation time.It was found that the experimental effective stress path in the region with a deviator stress ratio η greater than η_1 could be modeled using M^* and an extended quadratic equation in which the anisotropic hardening rule modified η_1 instead of η_0. This model was applicable across the wide range of consolidation time histories from young clay to aged clay subjected to secondary consolidation.

KEY WORDS:

Undisturbed sample; Marine clay; K0-consolidation; Undrained triaxial compression test; Effective stress path; Stress induced anisotropy; Secondary consolidation.

Hoe I. Ling et al. (eds.), Soil Stress-Strain Behavior: Measurement, Modeling and Analysis, 273–286.

1. INTRODUCTION

Soft alluvial marine clays are widely deposited around urban coastal areas in Japan. These soft alluvial marine clays are highly plastic and compressible such that large and long-term consolidation settlements occur under reclaimed landfill and artificial islands such as those constructed for the Kansai and Tokyo Haneda International Airports.

It seems reasonable to suppose that such reclamation landfill over a wide area combined with a previous sedimentary consolidation history results in one dimensional K_0 consolidation settlement and that in-situ stress conditions of the marine clays could not be regarded as isotropic. It is already well established that the undrained shear characteristics of normally consolidated clays are strongly affected by stress induced anisotropy due to K_0-consolidation. Ladd et al. (1977) summarized the effects of this stress induced anisotropy on the stress-strain relationship and the effective stress paths of K_0 consolidated clays. Although there was evidence of the influence of stress induced anisotropy on the undrained shear characteristics of normally consolidated clays in data presented by Diaz-Rodriguez et al. (1992) and Watanabe et al. (2001) neither this nor the effect of the plasticity index was discussed in detail. Not only static undrained shear behaviour but also undrained cyclic shear behaviour is affected by initial stress anisotropy. Based on the results of cyclic triaxial tests, Hyodo et al. (1999) pointed out that the cyclic undrained shear strength of anisotropically consolidated high plasticity clays subjected to a high initial drained shear stresses decreased as shown in Figure 1. Accordingly, it is important to investigate the effects of plasticity index and stress induced anisotropy due to K_0-consolidation on the undrained shear characteristics of normally consolidated clays for future constitutive modeling of these materials.

It is also well known that undrained shear characteristics of clays are influenced by aging under long term consolidation as has been pointed out by many authors such as Bjerrum et al. (1963). Yasuhara et al. (1983), Mitachi et al. (1987), Burland (1990) and Nishie et al. (2000, 2002) whose experimental results indicated that the undrained shear strength of K_0 consolidated aged clays subjected to secondary consolidation increased in proportion to the logarithm of consolidation time. However as the purposes of this particular research was mainly focused on the undrained shear strength and modulus of deformation, the effective stress paths for K_0 consolidated aged clay consolidation were not quantitatively evaluated. Although Shen et al. (1973) quantitatively evaluated the effective stress paths for clays consolidated for one day and one week, the discussion was limited to isotropically consolidated clays. Thus little quantitative research has been published to date on the undrained shear behaviour of highly plastic marine clays subjected to secondary consolidation in addition to stress induced anisotropy due to K_0 consolidation.

The effects of secondary consolidation on one dimensional normally consolidated clays of varying plasticity subjected to triaxial compression tests are presented in this study.

Table.1 Material properties of undisturbed marine clays

Clay Name	Symbol	I_P	OCR	C_C
Ariake	AA1	54.3	1.3	1.16
Kobe Port	AK2	61.9	1.2	0.90
Mikawa Bay	AM2	59.2	1.3	1.47
Osaka Bay	AO1	74.3	1.4	1.24

2.CLAY SAMPLES AND TEST PROCEDURES

Clay Samples

Table.2 Axial stress used in triaxial compression tests (TRX)

TRX	AA1	AK2	AM2	AO1
CIUC	-	-	-	102kPa
CK_0UC-A	200kPa	120kPa	150/200kPa	120kPa
CK_0UC-B	150/200kPa	120kPa	100/150/200kP	120kPa

Four kinds of undisturbed marine clay were sampled at Ariake Sea, Kobe Port, Mikawa Bay and Osaka Bay by the thin-wall sampling method. The material properties are shown in Table 1. All marine clay used in this study are high plastic clay which the plasticity index I_P is in the range of 54.3 to 74.3. Also, theses clays are quasi-over-consolidated with OCR being in the range of 1.2 to 1.4. Furthermore, the compressibility of clay is very high as the compression index C_C is in the range of 0.90 to 1.47.

Test Procedures

Isotropic consolidated-undrained triaxial compression test CIUC and two kinds of K_0-consolidated-undrained triaxial compression tests CK_0UC-A, CK_0UC-B were carried out. All specimens were tested in the normally consolidation region on the condition of the axial stress being more than yielding consolidation pressure p_c by factor of 1.4 to 2.7. Axial stress used in triaxial compression test was shown in table 2.

In this study, K_0-consolidated specimens used in CK_0UC-A that the a normalized consolidation time T in the range of 1.0 to 1.5 were called "normally K_0-consolidated young clay", and secondary K_0-consolidated specimens used in CK_0UC-B that T was greater than 1.5 were called "normally K_0-consolidated aged clay". Here, T was defined as the ratio of the actual consolidation time t_s divided by the primary consolidation time t_p obtained by the Casagrande's method.

The double cell type triaxial apparatus that could be controlled automatically with high accuracy during consolidation process to shear process by the personal

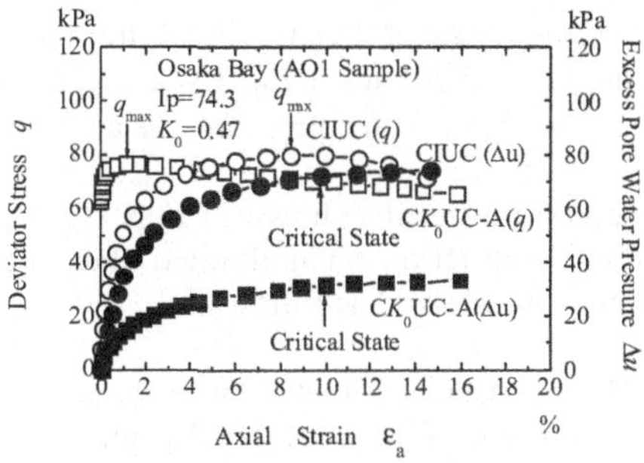

Fig.1 q-ε_a and Δu- ε_a relation of normally consolidated young clay

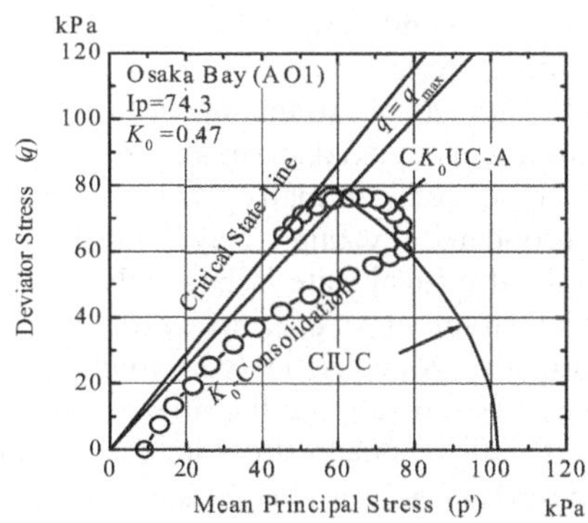

Fig.2 Effective stress path of normally consolidated young clay

computer system was used for CK_0UC-A and CK_0UC-B. In this study, moreover, the lateral strain was limited to ±0.02% in a whole process of primary consolidation with incremental axial loading to secondary consolidation for CK_0UC-A and CK_0UC-B.

As a result, the coefficient of earth pressure at rest K_0^{NC} for normally K_0-consolidated young clays was in the range of 0.44 to 0.47, and the K_0-value K_0^{QOC} of normally K_0-consolidated aged clays remains unchanged K_0^{NC} in this study.

The stress control method was used in consolidation process and the strain control method was used in undrained shear process. The axial loading rate during incremental axial loading process of CK_0UC-A and CK_0UC-B was employed in 0.147kPa/min, and the axial strain rate of CIUC, CK_0UC-A and CK_0UC-B was employed in 0.01%/min. at the undrained shear process, respectively.

3.NORMALLY CONSOLIDATED YOUNG CLAY

Effects of stress induced anisotropy on undrained shear behaviors

In order to investigate effects of stress induced anisotropy caused by K_0-consolidation on undrained shear behaviors of normally K_0-consolidated young clay using results obtained by CIUC and CK_0UC-A, Fig.1 shows deviator stress q-axial strain ε_a relation and excess pore water pressure Δu axial strain ε_a relation, Fig.2 shows effective stress path.

Based on the results shown in Figs.1-2, the effects of stress induced anisotropy caused by K_0-consolidation on undrained shear behaviors of normally K_0-consolidated young clay are summarized as follows

(1) The maximum deviator stress q_{max} of normally K_0-consolidated young clay occurs at small axial strain given by about ε_a =1%, and the stress-strain curve shows strain-softening in the region of axial strain beyond q=q_{max} such as a brittle material.

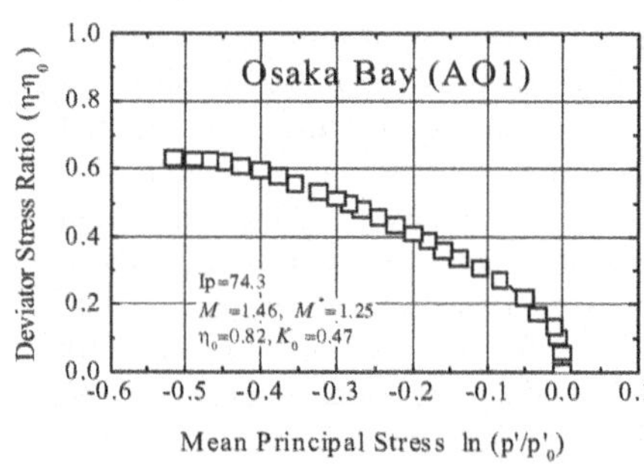

Fig.3 relation between $(\eta-\eta_0)$ and $ln(p'/p_0')$ of K_0-young clay (AO1)

(2) Although axial strain at critical state equals to axial strain at q=q_{max} in case of normally isotropic consolidated young clay, axial strain at critical state is larger than axial strain at q=q_{max} in case of normally K_0-consolidated young clay. As a result, the mobilized stress ratio M^* at q=q_{max} is smaller than the critical stress ratio M due to stress induced anisotropy.

(3) The excess pore water pressure Δu of normally K_0-consolidated young clay built up during shear test is larger than that of normally isotropic consolidated young clay. Accordingly, the coefficient of pore water pressure A_f decreases by stress induced anisotropy.

(4) The effective stress path of normally isotropic consolidated young clay reaches the

critical state line CSL drawing the smooth curve which is similar to the Cam-Clay model by the Cambridge theory (Roscoe et al., 1963; Roscoe and Schofield, 1963). In the case of normally K_0-consolidated young clay, maximum deviator stress occurs before critical state, and the effective stress path shows a shell-typed unique curve with plastic-softening beyond $q=q_{max}$. Thus a unique curve of effective stress path can be observed in results of natural sedimentary clay subjected to inherent anisotropy demonstrated by Adachi et al. 1991.

(5) The CSL of normally K_0-consolidated young clay is same as that of normally isotropic consolidated young clay. It can be said that the CSL is an inherent characteristic of clay which is not affected by stress induced anisotropy.

From above results, it can be concluded that undrained shear characteristics of normally consolidated young clay, except for the CSL, are strongly affected by stress induced anisotropy due to K_0-consolidation.

Experimental equation of effective stress path for NC K_0-young clay

Figs.3-4 show the result of relationship between shear stress ratio and mean principal stress ratio of normally K_0-consolidated young clay obtained by CK_0UC-A for AO1 sample. While the vertical axis is the deviator stress ratio of $(\eta-\eta_0)$ in shown Fig.3, the vertical-axis is the square deviator stress ratio of $(\eta-\eta_0)^2$ in shown Fig.4. Where, η_0 is the initial deviator stress ratio caused by K_0-consolidation. The horizontal axis is the normalized mean principal stress ratio of $ln(p'/p_0')$ on scale of natural logarithm in both Fig.3 and Fig.4.

As shown in Figs.4, it was found that between $(\eta-\eta_0)^2$ and $ln(p'/p_0')$ is a liner relationship in case of normally K_0-consolidated young clay. This relationship can be written as the follows using experimental soil parameter N^*.

$$(\eta-\eta_0)^2+N^*\times ln(p'/p_0')=0 \quad (1)$$

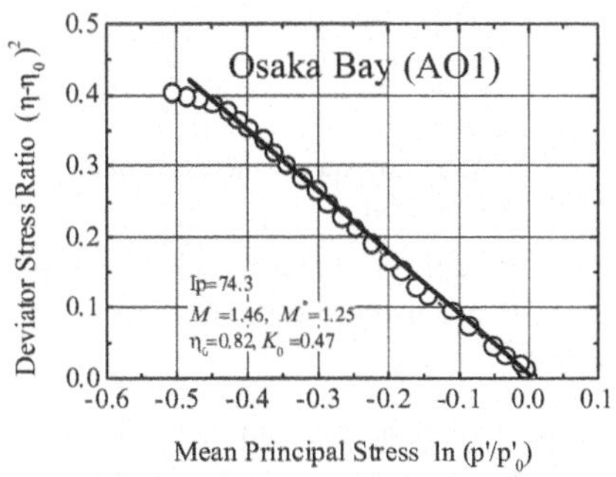

Fig.4 relation between $(\eta-\eta_0)^2$ and $ln(p'/p_0')$ of K_0-young clay (AO1)

Where, η_0, η, p_0' and p' are initial deviator stress ratio $\eta_0=q_0/p_0'$, deviator stress ratio $\eta=q/p'$, initial mean principal stress and mean principal stress, respectively. N^* is experimental parameters for K_0 consolidated young clays depending on the plasticity index. The relationship given by Eq.1 is same equation which was derived by Yasufuku et al. (1991) for anisotropic consolidated dense sand.

The parameter N^* is determined in Eq.2 by substituting Eq.1 for the condition of $\partial\eta/\partial p'=0$ at $\eta = M^*$. Where, M^* is the mobilized stress ratio at the maximum deviator stress as mentioned above.

$$N^* = 2M^*\times\left(M^*-\eta_0\right) \quad (2)$$

Consequently, the equation of effective stress path for normally K_0-consolidated young clay can be written as Eq.3.

$$(\eta-\eta_0)^2+2M^*\times(M^*-\eta_0)\times ln(p'/p_0')=0 \tag{3}$$

Comparison between experimental data and computed result

Figs.5-7 show the comparison of the effective stress path normalized by the initial mean principal stress p_0' between the computed results by using Eq.3 and the experimental results of AA1, AK2 and AM2 samples as shown in Table 1, respectively. In Fig.5-7, η_0 and M^* are used in computation as shown in Table 3.

Table.3 Soil parameter for normally K_0-consolidated young clays

Clay Sample	K_0^{NC}	η_0	M^*	M
AA1	0.45	0.868	1.16	1.39
AK2	0.44	0.894	1.22	1.46
AM2	0.47	0.820	1.21	1.41
AO1	0.47	0.820	1.25	1.46

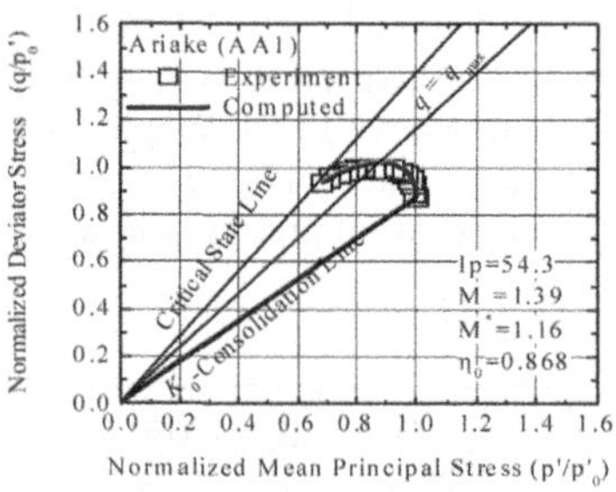

Fig.5 Comparison experiment and computed (AA1 sample)

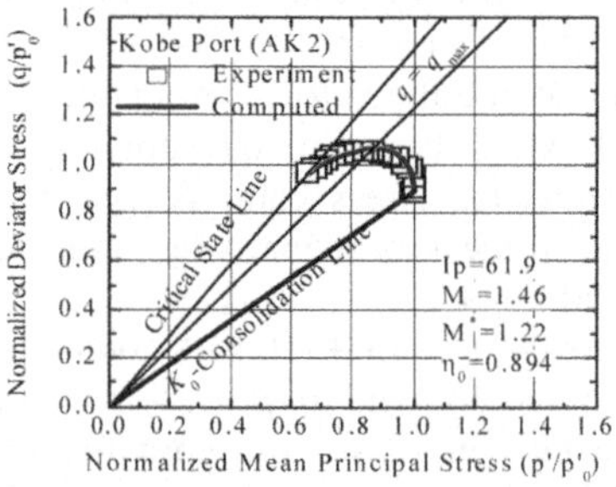

Fig.6 Comparison experiment and computed (AK2 sample)

It is clear that the computed results are in a good agreement with experimental results. Therefore, it was found that effective stress path of normally K_0-consolidated young clays could be given by Eq.3 using a parabolic function and kinematical hardening rule.

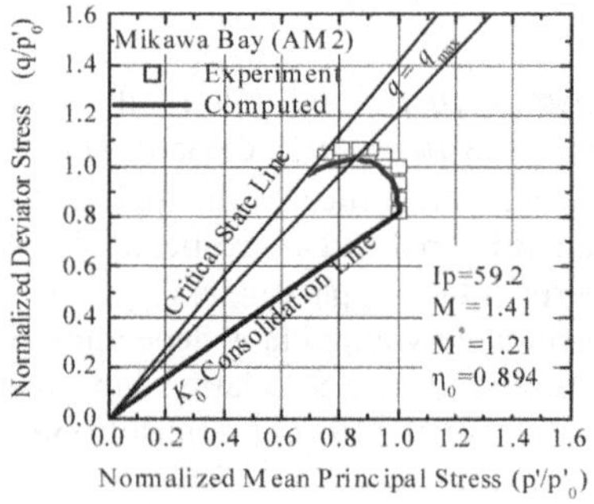

Fig.7 Comparison experiment and computed (AM2 sample)

4. NORMALLY K_0-CONSOLIDATED AGED CLAY

Change in undrained shear behaviors due to secondary consolidation

Figs.8-9 show a comparison q-ε_a and Δu-ε_a relation of the representative specimens (AK2, AO1) between normally K_0-consolidated young clays and normally K_0-consolidated aged clays subjected to secondary consolidation in addition to the K_0-consolidation. The effects of secondary consolidation on q-ε_a and Δu -ε_a relation as shown in Figs.8-9 are as followings;

(1) The q-ε_a curves of normally K_0-consolidated aged clays are stiffer stress-strain response just after shear start, and the secondary consolidation cause the increase of the maximum deviator stress.
(2) The strain-softening phenomena were showed by the secondary consolidation.
(3) On the contrary, the excess pore water pressures are deceased by the secondary consolidation.

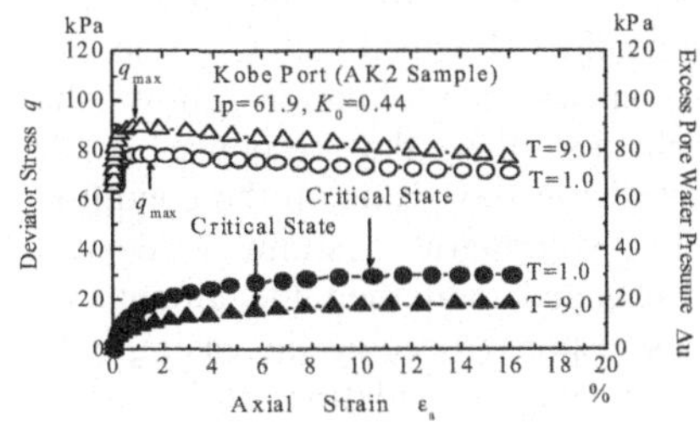

Fig.8 Change in q-ε_a, Δu-ε_a due to secondary consolidation (AK2)

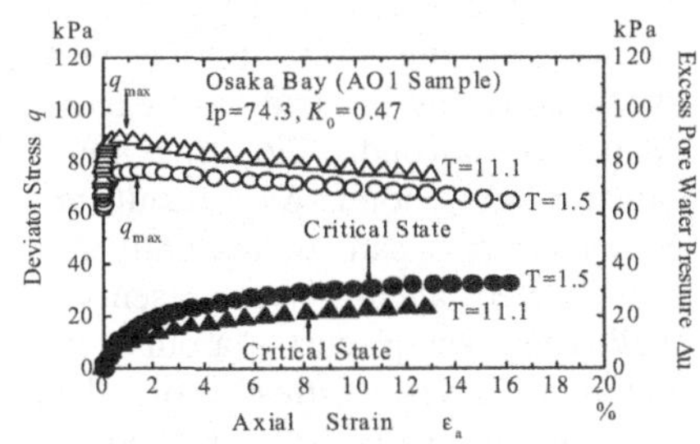

Fig.9 Change in q-ε_a, Δu-ε_a due to secondary consolidation (AO1)

Fig.10 shows a relationship between the maximum deviator stress ratio q_{max}/σ_a and the normalized consolidation time T. In Fig.10, the vertical axis is the maximum stress ratio normalized by the axial consolidation pressure σ_a; the horizontal axis is the consolidation time normalized by the primary consolidation time t_p. As shown in Fig.10, it was found that the maximum stress ratio is increasing in proportional to the normalized consolidation time on the scale of logarithm approximated by Eq.4 in case of this experimental results.

$$(q_{max} / \sigma_a) = 0.65 + 0.10 \times log(T) \tag{4}$$

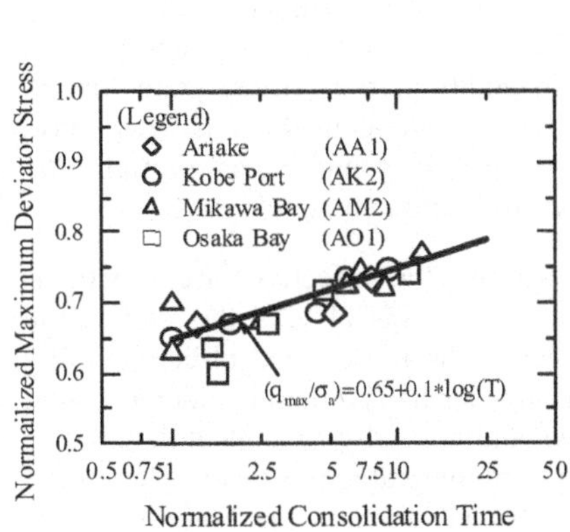

Fig.10 The relation of q_{max}/σ_a and $log\ (T)$

Fig.11 and Fig.12 show a change in the effective stress path of the representative specimens (AK2, AM2) during secondary consolidation. The effects of secondary consolidation on the effective stress path shown in Figs.11-12 are summarized as followings;

(4) Just after the shear start, the effective stress path of normally K_0-consolidated aged clays due to secondary consolidation show the behavior that the response occur parallel to the q-axis on the condition of volumetric strain remains unchanged, and then the amount of this response become increasing with secondary consolidation progress.

(5) The mean principal stress becomes decreasing beyond a singular point that is a limitation point of the response paralleled to the q-axis, and the effective stress path approaches the critical state line (CSL) with plastic-softening and negative dilatancy beyond the maximum deviator stress point. As a result, the effective stress path of normally K_0-consolidated aged clays presents a shell-typed sharp curve around the maximum deviator stress more than that of normally K_0-consolidated young clays.

(6) The mobilized stress ratio M^* at $q=q_{max}$ and the critical stress ratio M remains unchanged that are independent of the consolidation time. In addition, the critical state line of normally K_0-consolidated aged clays has also the same line as normally K_0-consolidated young clays.

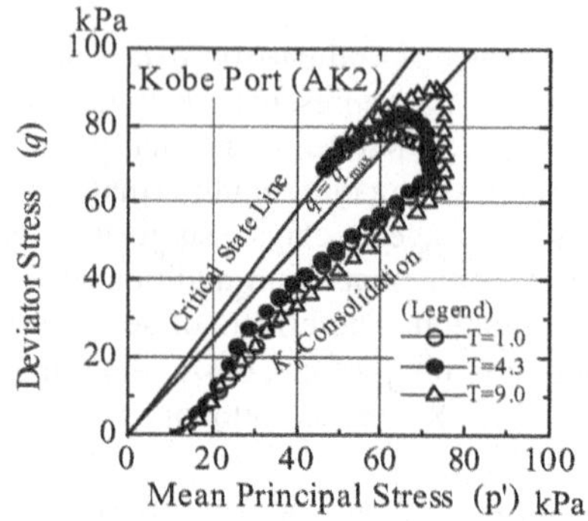

Fig.11 Effective stress path of K_0-young and K_0-aged clays (AK2)

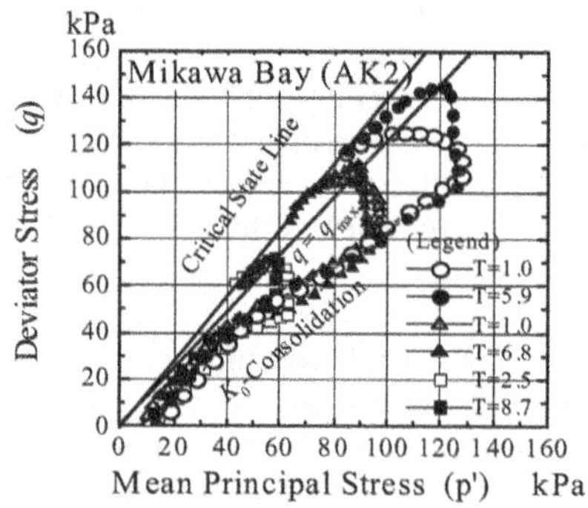

Fig.12 Effective stress path of K_0-young and K_0- aged clays (AM2)

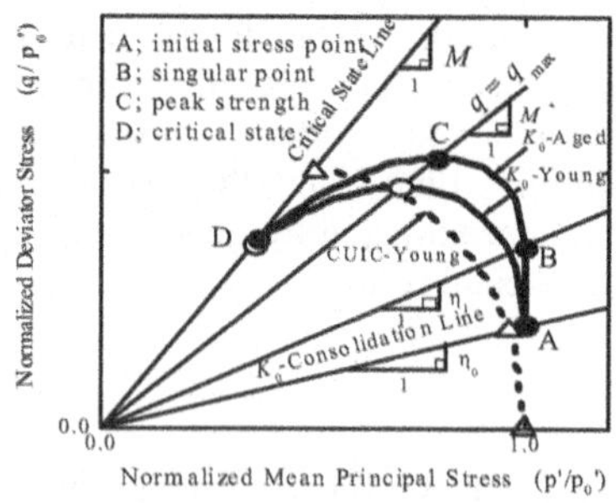

Fig.13 Schematic model of the effective stress path for K_0-clay

Based on the effective stress path as shown in Figs.11-12, the schematic model of the normalized effective stress path for normally K_0-consolidated young clays and normally K_0-consolidated aged clays are illustrated in Fig.13. Here, both the vertical axis and the horizontal axis are displayed the normalized stress divided by the initial mean principal stress. That is, the secondary K_0-

consolidation causes the effective stress path of clays to change a unique curve as follows;

(a)The response paralleled to the q-axis occurs from point A just after the shear start to a singular point B on the condition of volumetric strain remains unchanged.

(b)The effective stress path moves to a point C at the maximum deviator stress with decreasing mean principal stress beyond point B at a singular point.

(c)The effective stress path approaches a point D on the CSL with remarkable plastic softening in a region of decreasing mean principal stress beyond point C.

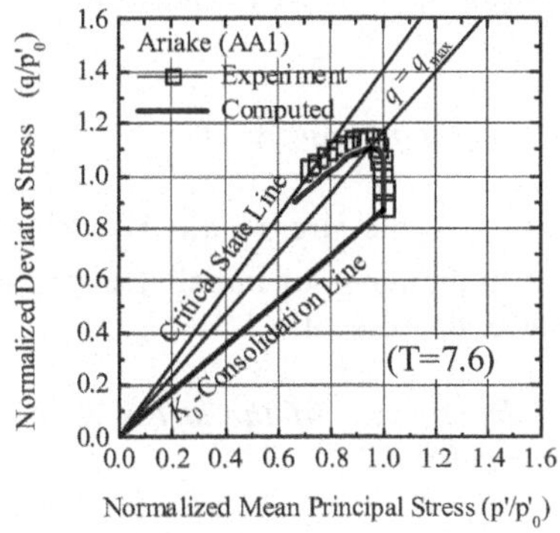

Fig.14. Comparison experiment and computed (AA1 sample, T=7.6)

5.Experimental equation of effective stress path for K_0-aged clay

Assumption of two regions for effective stress path

Based on the schematic model as shown in Fig.13, the effective stress path presents a different behavior on boundary of a singular point B. That is, the effective stress path behaves as a material typed of von-Mises in a region-A when $\eta_0 \le \eta < \eta_1$ because no plastic volumetric strain occurs. On the other hand, the effective stress path shows a unique shell-typed curve with plastic softening which is similar to the effective stress path of normally K_0-consolidated young clay in a region-B when $\eta_1 \le \eta < M$.

Accordingly, it seems to be reasonable that each equation of the effective stress path for normally K_0-consolidated aged clay in two regions of region-A and region-B will be independently derived.

Region-A ($\eta_0 \le \eta < \eta_1$)

In the region-A, equation of the effective stress path is defined as a simple equation given by Eq.5 because the mean principal stress remains unchanged.

$$p' = p_0' \text{ or } p'/p_0' = 1 \tag{5}$$

Region-B ($\eta_1 \le \eta < M$)

In a region-B, equation of the effective stress path is assumed on the parabolic function given by Eq.6 which the stress ratio η_1 at a singular point instead of the

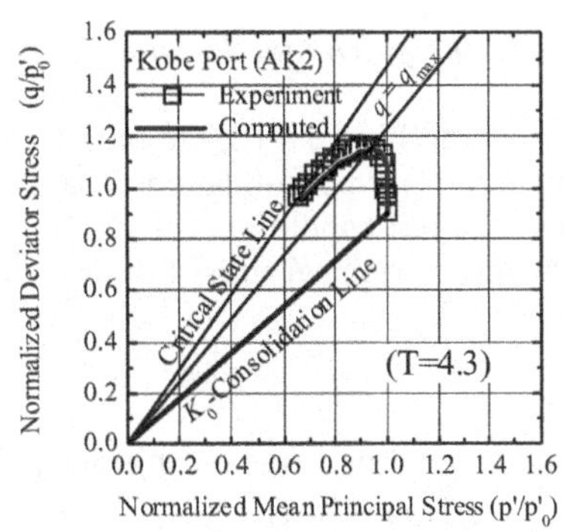

Fig.15. Comparison experiment and computed (AK2 sample, T=4.3)

initial stress η_0 is used in the former Eq.3. That is,

$$(\eta-\eta_1)^2+2M^*\times(\eta-\eta_1)\times ln(p'/p_0')=0 \quad (6)$$

Where, η_1 is the stress ratio at a singular point B in the range of initial stress ratio η_0 to mobilized stress ratio M^*.

Time dependence of the stress ratio at a singular point B

The stress ratio η_1 at a singular point B as shown Fig.13 can be derived by next procedures.

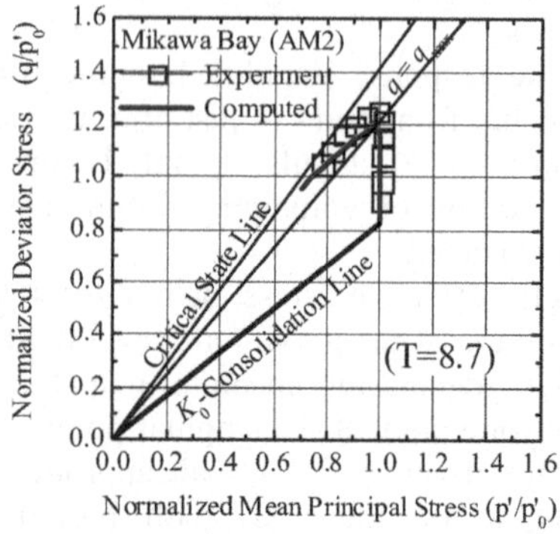

Fig.16. Comparison experiment and computed (AM2 sample, T=8.7)

Predicting equation of q_{max} during secondary consolidation

As shown in Fig.10, the maximum deviator stress is increasing in proportional to the normalized consolidation time T on the scale of logarithm. According to the former reports (Yasuhara and Ue 1983; Nishie et al. 2000; 2002), it was known that the change in the maximum deviator stress ratio of normally K_0-consolidated aged clays at $t=t_s$ during secondary consolidation can be predicted by Eq.7 using the maximum stress ratio at the just finished primary consolidation and the non-dimensional soil parameter R which is the coefficient of secondary consolidation C_α divided by the compression index C_c.

$$\left(\frac{q_{max}}{\sigma_a}\right)_{t=t_s}=\left(\frac{q_{max}}{\sigma_a}\right)_{t=t_p}\times T^R \quad (7)$$

Relation the maximum deviator stress and the mean principal stress

As shown in Figs. 11-12, the mobilized stress ratio M^* at the maximum deviator stress during secondary consolidation is same as that of the just finished primary consolidated clays on the test condition of this study. A relation between the maximum deviator stress and the mean principal stress for normally K_0-consolidated clays can be given as Eq.8.

$$q_{max}(t_s)/q_{max}(t_p)=p_2'/p_1' \quad (8)$$

Where, p_2' and p_1' is the mean principal stress at $q=q_{max}(t_s)$ and the mean principal stress at $q=q_{max}(t_p)$, respectively.

Furthermore, the relationship between p_0' and p_1' is rewritten as Eq.9 using Eq.7.

$$(p_2'/p_1')=T^R \quad (9)$$

The stress ratio η_1 at a singular point B

Instituting the mean principal stress p_1' at $\eta = M^*$ of normally K_0-consolidated young clay for Eq.3, Eq.10 can be obtained.

$$\left(M^* - \eta_0\right)^2 + 2M^* \times \left(M^* - \eta_0\right) \times ln\left(p_1'/p_0'\right) = 0 \tag{10}$$

Similarly, instituting the mean principal stress p_2' at $\eta = M^*$ of normally K_0-consolidated aged clay for Eq.6, Eq.11 can be obtained.

$$\left(M^* - \eta_1\right)^2 + 2M^* \times \left(M^* - \eta_1\right) \times ln\left(p_2'/p_0'\right) = 0 \tag{11}$$

Accordingly, the relationship between the stress ratio η_0 and the stress ratio η_1 is derived as Eq.12 using both Eq.10 and Eq.11.

$$\eta_1 = \eta_0 + 2M^* \times ln\left(p_2'/p_1'\right) \tag{12}$$

Finally, the stress ratio (η_1) at a singular point B can be given by Eq.13 using both Eq.9 and Eq.12.

$$\eta_1 = \eta_0 + 2M^* \times R \times ln(T) \tag{13}$$

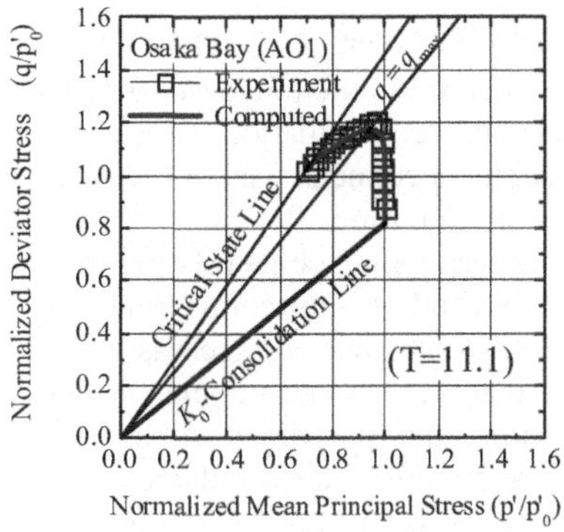

Fig.17. Comparison experiment and computed (AO1 sample, T=11.1)

Here, as η_1 equals to η_0 in case of T=1, it is clear that Eq.6 and Eq.13 includes the effective stress path of normally K_0-consolidated young clays as shown in Eq.3. And, as each soil parameter η_0, M^* and R can be obtained by the practical triaxial compression test, η_1 depended on the normalized consolidation time T can be easily determined.

6.Comparison experimental results and computed results

Figs.14-17 shows the comparison between experimental effective stress path and computed effective stress path of the representative normally K_0-consolidated aged clay specimens used in this study. In these figures, the vertical axis and the horizontal axis is employed the normalized value divided by the initial mean principal stress, respectively. The soil parameter η_0 and M^* used in computation are shown in the former Table 3, and R is used the value shown in Table 4.

Table 4. Soil parameter R used in computation

Sample	AA1	AK2	AM2	AO1
$R = C_\alpha / C_c$	0.042	0.053	0.046	0.065

As shown in Figs.14-17, the computed results are in a good agreement with the experimental results for the practical use. Accordingly, it was found that the quantitative evaluation of the effective stress path for normally K_0-consolidated aged clays could be done by the proposed equations in this study.

7. Applicable limitation of the normalized consolidation time T

On the condition of $\eta_0 \leq \eta_1 < M^*$ as mentioned before, the normalized consolidation time (T) used in Eq.13 has a limitation time ($T_{lim.}$) as Eq.14.

$$T_{lim.} = exp\left(\frac{M^* - \eta_0}{2M^* \times R}\right) \qquad (14)$$

A limitation time $T_{lim.}$ obtained by Eq.14 was about in the range of 10 to 30, and actual consolidation time of experimental specimens was in the range of this limitation time.

If an experimental consolidation time will be beyond a limitation time, the mobilized stress ratio M^* seems to be greater than the experimental value. Furthermore, the mobilized stress ratio M^* of over-consolidated clays (AA1 sample) as shown in Fig.18 and/or aged clays subjected to chemical bonding will be larger than the critical stress ratio M.

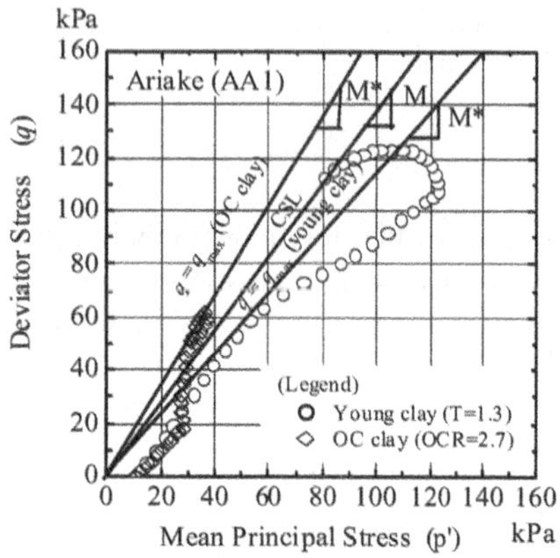

Fig.18 The effective stress path of over-consolidated clay (AA1)

The further study is required because it is beyond the scope of this study to discus the effective stress path of clays at such consolidation state.

8. CONCLUSIONS

Undrained shear behavior of normally K_0-consolidated young clays is strongly affected by the stress induced anisotropy caused by K_0-consolidation. Particularly, the effective stress path draws a shell-typed unique curve because the mobilized stress ratio M^* is smaller than the critical stress ratio M.

Based on the experimental results, the experimental equation of the effective stress path for normally K_0-consolidated young clays was derived as next equation using a parabolic function and the kinematical hardening rule. The computed results were in a good agreement with the experimental results.

$$(\eta - \eta_0)^2 + 2M^* \times (M^* - \eta_0) \times ln(p'/p_0') = 0 \qquad (15)$$

Undrained shear behaviors of normally K_0-consolidated aged clays during secondary consolidation are strongly affected by not only the stress induced anisotropy cause by K_0-consolidation but also the secondary consolidation. The effective stress path of normally K_0-consolidated aged clays shows a unique shape which is divided into two regions on the boundary of a singular point depended on the consolidation time as shown Eq.16.

$$\eta_1 = \eta_0 + 2M^* \times R \times ln(T) \tag{16}$$

The Eq.17 extended Eq.15 in a region with decrease mean principal stress can be used for the effective stress path of normally K_0-consolidated aged clays.

$$(\eta - \eta_1)^2 + 2M^* \times (\eta - \eta_1) \times ln(p'/p_0') = 0 \tag{17}$$

As Eq.16 and Eq.17 include the effective stress path of normally K_0-consolidated young clays as shown in Eq.15, the quantitative evaluation for the effective stress path of normally K_0-consolidated young and aged clays can be done.

REFERENCES

Adachi,T., Oka,F., Hirata,T., Hashimoto,T., Pradhan,T.B.S., Nagaya,J. and Miura,M. (1991): "Triaxial and torsional hollow cylinder tests of sensitive natural clay and viscoplastic constitutive model", *Proc. 10th European Conf. on SMFE*, Florence, **1**, 3-6.

Bjerrum,L. and Lo,K.Y. (1963): "Effect of aging on the shear strength properties of a normally consolidated clay", *Geotechnique*, **13**(2), 147-157.

Diaz-Rodriguez,J.A., Leroueil,S. and Aleman,J.D. (1992): "Yielding of Mexico City Clay and other natural clays", *J.Geotech.Engrg.Division, ASCE*, **118**(GT7), 981-995.

Hyodo,M., Yamamoto,Y. and Sugiyama,M. (1994): "Undrained cyclic shear behaviour of normally consolidated clay subjected to initial static shear stress", *Soils and Foundations*, **34**(4),1-11.

Hyodo,M., Hyde,A.F.L., Yamamoto,Y. and Fujii,T. (1999): "Cyclic shear strength of undisturbed and remoulded marine clays", *Soils and Foundations*, **39**(2),45-58.

Ladd,C.C., Foott,R., Ishihara,K., Schlosser,F. and Poulos,H.G. (1977): "Stress-deformation and strength characteristics", S.o.A. Report, *Proc. 9th Int. Conf. on SMFE*, Tokyo, **2**, 421-494.

Mitachi,T. and Fujiwara,Y. (1987): "Undrained shear behavior of clays undergoing long-term anisotropic consolidation", *Soils and Foundations*, **27**(4),45-61.

Nishie,S. and Wang,L. (2000): "K_0-consolidation behavior and dependence of undrained shear strength on consolidation time for marine clay", *Coastal Getechnical Engineering in Practice, Balkema*, **1**, 89-94.

Nishie,S., Wang,L., Kutsuzawa,S. and Hyodo.M. (2002): "The K0 consolidation behavior and undrained shear characteristics of undisturbed marine clays during the secondary consolidation", *Journal of Geotechnical Eng., JSCE,* No.708/III-59, pp.53-68, (in Japanese).

Roscoe,K.H., Schofield,A.N. and Thrairajah,A. (1963): "Yielding of clays in state wetter than critical", *Geotechnique*, **13**(3), 211-240.

Roscoe,K.H. and Schofield,A.N. (1963): "Behavior of an idealized 'wet clay' ", *Proc. 2nd European Conf. on Soil Mech.*, Wiebaden, **1**, 47-54.

Shen,C.-K., Arulanandan,K. and Smith,W.S. (1973): "Secondary consolidation and strength of a clay", *Proc. Of ASCE*, **99**(SM1), pp.95-110.

Yasufuku,N., Murata,H., Hyodo,M. and Hyde,A.F.L. (1991): "A stress-strain relationship for anisotropically consolidated sand over a wide stress region", *Soils and Foundations*, **31**(4),75-92.

Yasuhara,K. and Ue, S. (1983): "Increase in undrained shear strength due to secondary compression", *Soils and Foundations*, **23**(3), 50-64.

Watanabe,Y. and Tsuchida,T. (2001): "Comparative study on undrained shear strength of Osaka Bay Pleistocene clay determined by several kinds of laboratory test", *Soils and Foundations*, **41**(5),47-59

Soil Stress-Strain Behavior: Measurement, Modeling and Analysis
Geotechnical Symposium in Roma, March 16 & 17, 2006

THE EFFECTS OF CONTROLLED DESTRUCTURING ON THE SMALL STRAIN SHEAR STIFFNESS G_0 OF BOTHKENNAR CLAY

David Nash, Martin Lings, Nadia Benahmed, Jiraroth Sukolrat, and David Muir Wood
Department of Civil Engineering,
University of Bristol, Bristol BS8 1TR, UK
email: David.Nash@bristol.ac.uk

ABSTRACT

The important differences between natural and reconstituted soils are well recognised. In a natural structured clay, the microstructure enables it to exist at states outside the state boundary surface for the reconstituted soil, resulting in greater peak undrained strength and yield stress at a given void ratio. The structure of soft clays is gradually destroyed by strain; understanding the process of destructuring is important both in developing constitutive models and in understanding the differences between field and laboratory behaviour.

A systematic investigation of the effects of destructuring on properties of the Bothkennar clay has been carried out, using changes of small strain shear stiffness G_0 as an indicator of damage. Tests were conducted on both natural and reconstituted material, so that the effects of microstructure could be isolated. After initial reconsolidation under in-situ stresses to establish a baseline condition, samples were subjected to controlled cycles of undrained compression/extension strain. These have shown that such strains result in significant temporary reduction of G_0, but with time, the clay regains much of its original small strain stiffness on reconsolidation to the initial stress state. While small changes of G_0 after reconsolidation appear to be consistent with small changes to the peak strength, they do not reflect the damage that affects the medium-strain stiffness. To correctly identify effects of microstructure, it proved important to normalise the data to a common void ratio. Drained compression tests at constant stress ratio (approx 1-D) revealed that the normalised stiffness of the reconstituted clay does not form a lower bound to that of the natural clay. These findings appear to have important implications for the formulation of constitutive models of structured clays.

1. INTRODUCTION: STRUCTURE IN SOFT CLAYS

The important differences between natural and reconstituted soils are well recognised (e.g. Burland, 1990; Leroueil & Vaughan, 1990), where the term 'structure' has been used to refer to the distinctive fabric and bonding that characterise intact natural soils. More recently, Burland *et al.* (1996) have used the term 'microstructure' to refer to both the arrangement of the soil particles (microfabric) and the bonding between them. It is well established that natural soils may exist at states well outside the bounding surface for the same soil when reconstituted. In a natural structured clay, the microstructure results in greater peak undrained strength and yield stress at a given void ratio than for the same clay when reconstituted.

Hoe I. Ling et al. (eds.), Soil Stress-Strain Behavior: Measurement, Modeling and Analysis, 287–298.

The microstructure is gradually destroyed by strain (Clayton *et al.*, 1992; Hight *et al.*, 1992) which results in shrinking of the bounding surface towards that of the reconstituted clay as indicated in Figure 1. Here the stresses are normalised by p_e', the value of mean effective stress at the appropriate void ratio on the normal consolidation line for the reconstituted material. Destructuration of the natural clay may result from compression, shearing or swelling (Leroueil and Vaughan, 1990) – all three have been systematically explored in the research discussed here – and may lead to (Hight and Leroueil 2003):

- decrease of stiffness inside the bounding surface;
- decrease of peak strength, of yield stress and shrinkage of the bounding surface;
- decrease of compression index.

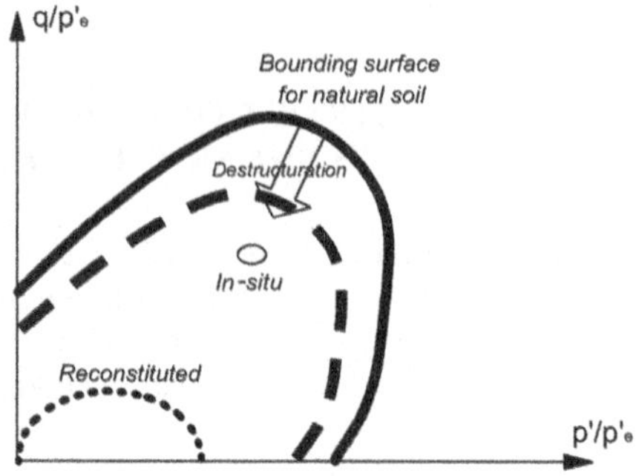

Figure 1: Bounding surface of soft clay showing effect of destructuration

Destructuration occurs progressively both during construction at full-scale, and during sampling and preparation of laboratory samples. Understanding the process of destructuration is important both in developing constitutive models (Muir Wood, 1995; Rouainia and Muir Wood, 2000) and in understanding the differences between field and laboratory behaviour. Nevertheless, in modelling construction on soft clays it is often necessary to resort to constitutive models such as Modified Cam Clay which do not include the effects of structure.

Sampling disturbance causes changes to the microstructure of the soil in ways which cannot be reversed by reinstating the *in-situ* effective stresses (Clayton *et al.*, 1992): a prime argument for higher quality sampling is the ability to obtain more appropriate laboratory measurements of field stiffness and strength. Too often, poor quality samples are used to obtain compressibility and strength data which then lead to conservative design.

The main indicators that might be used to assess destructuration are change of state (reduction of void ratio at a given stress state), increase of creep rate, changes of small strain shear stiffness G_0, changes of small and intermediate strain axial stiffness E_v and volumetric stiffness K, reduction of yield stress, and reduction of undrained strength. Of these observations the first three may be made at a given state and are thus non-destructive, while the others necessitate changing the stresses. In current practice it is common to examine volumetric strains that occur as a laboratory sample is reconsolidated under in-situ stress (Lunne *et al.*, 1997) and to use this as an indication of sample quality, but this can only be done after testing.

In tests on Bothkennar clay to explore the effect of sampling disturbance, Clayton *et al.* (1992) showed that axial secant and tangent stiffness falls during monotonic undrained loading at strains greater than 0.002%. As part of a collaborative research project on Bothkennar clay in 1997, undrained triaxial tests at Bristol found a similar loss of axial stiffness with strain, but also showed that the elastic stiffness G_0 measured by bender elements was not affected by the occurrence of plastic strains below about 1% axial strain.

Similar work was reported from Japan and Italy by Lo Presti *et al.* (1999). It was concluded that sampling (or construction) effects producing this level of inelastic strain may not affect the small strain, truly elastic properties. Subsequently Hight (1998) proposed that measurement of shear wave velocity V_S combined with suction measurements provides an *a priori* non-destructive means of detecting differences in sample quality.

2. OVERVIEW OF TEST PROGRAMME

The research described in this paper has explored the changes of G_0 resulting from controlled destructuration of Bothkennar clay. Bothkennar clay was chosen because of the availability of some high quality natural soil samples, obtained using the Sherbrooke sampler during the 1997 collaborative research project. These samples were found to be in a good condition despite having been stored for more than 5 years. At Bothkennar there are different facies with distinct properties (Paul *et al.*, 1992, Clayton *et al.*, 1992), but all the samples tested were from the mottled facies which is significantly structured with a void index of 1.0 or more, with the majority of samples taken from around 8.0m depth. Each Sherbrooke sample was carefully divided and trimmed using a wire saw, into several specimens of natural clay 75 or 100 mm diameter. Index tests were carried out and reconstituted samples prepared from the trimmings by mixing with de-ionised water and subsequent 1-D consolidation from slurry.

Testing was carried out in 75mm and 100mm Bishop-Wesley stress path apparatus on cylindrical samples fitted with mid-height pore pressure transducers and bender elements enabling measurements to be made of the three shear wave velocities V_{Svh} V_{Shv} and V_{Shh} similar to that described by Pennington *et al.* (1997). In all tests the same sine-pulse was used to energise the benders, with the travel times determined from its first arrival. Software and hardware have been developed to enable automated bender element readings to be obtained. The corresponding shear moduli G_{0vh} G_{0hv} and G_{0hh} have then been calculated from:

$$G_0 = \rho . V_s^2 \tag{1}$$

where ρ is the bulk density of the clay.

In tests on natural clay, samples were first reconsolidated under in-situ stresses to establish a baseline condition. The tests then explored destructuration resulting from three types of loading: isotropic swelling to low stresses, drained compression at constant stress ratio (approx 1-D) and undrained shearing. Typically tests lasted 2 to 3 months with lengthy hold stages to allow excess pore pressures to dissipate and creep rates to reduce to 0.01%/day or less.

During preparation of reconstituted clay specimens these were first consolidated from slurry in tubes under axial effective stresses of about 100kPa so that they were sufficiently robust for transfer to the triaxial cell. They were then reconsolidated under the same stresses as had been used for the natural clay samples (resulting in an OCR of about 1.6) before they too were subjected to drained compression at constant stress ratio or undrained shearing.

3. EFFECT OF STRUCTURE ON SMALL STRAIN STIFFNESS OF CLAYS

Research on the small strain stiffness of soils in the laboratory (e.g. Jamiolkowski *et al.*, 1995) has shown that G_0 depends only on the following factors:

- the soil structure and fabric resulting from deposition, ageing processes and strain history;
- the soil state described by a combination of stresses and void ratio.

Bristol work on Cambridge Gault clay (Pennington, 1999) confirmed that small strain shear stiffness G_0 is dependent on effective stress state and void ratio, and suggested that at a given state the natural soil has a higher stiffness than the reconstituted soil. An empirical expression for the void ratio and stress-dependency of the shear moduli can be written:

$$G_{0ij} = \frac{S_{ij}.f(e).(\sigma'_i.\sigma'_j)^{nij}}{(\sigma'_{ref})^{2nij}} \tag{2}$$

where S_{ij} is a structure term, $f(e)$ is a void ratio function (taken here as e^{-m}, where e is the void ratio), σ'_{ref} is a reference stress (taken here as 1kPa), and i and j are the direction of wave propagation and polarisation respectively.

Figure 2 indicates the expected relationships that stem from equation (2) in which it is assumed that the effects of structure can be expressed purely by the structure terms S_{ij}. In this study measurements of G_0 have first been made on reconstituted (fully destructured) Bothkennar clay from which the void ratio function and stress indices were determined. In comparing natural and reconstituted soils to correctly identify effects of microstructure, it has proved important to normalise the data to a common void ratio and to a common effective stress state. For this purpose we have chosen to apply the void ratio function from the tests on reconstituted clay when interpreting the data for the natural clay.

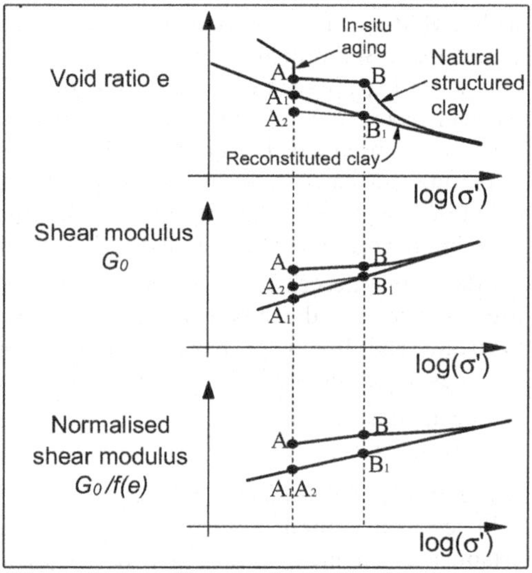

Figure 2: Expected relationships between void ratio, G_0 and effective stress

4. BEHAVIOUR OF RECONSTITUTED BOTHKENNAR CLAY

The small strain stiffness G_0 of reconstituted Bothkennar clay was examined assuming that at all soil states the structure terms S and the indices m and n in equation (2) remain constant. One way of obtaining the indices is to undertake quasi-1D consolidation tests, in which the sample is periodically returned to a common stress state (Nash *et al.*, 1999). Starting from an initial isotropic stress of 20 kPa, the clay was consolidated incrementally under successively greater stresses at a stress ratio $K = 0.65$. After each consolidation stage the stresses were held constant for several days until the creep rate reduced to 0.01%/day before bender readings were taken. The specimens were periodically allowed to swell back to an isotropic stress state of 20 kPa so that the void ratio functions m_{ij} could be obtained. G_{0ij} normalised by the void ratio function could then be plotted against $(\sigma'_i.\sigma'_j)$ on a double log plot resulting in reasonably unique straight lines enabling S_{ij} and

n_{ij} to be determined. Several tests were carried out in which the moduli obtained show a consistent trend of $G_{0hh} > G_{0hv} > G_{0vh}$. The tests resulted in similar indices with indices n_{ij} ranging from 0.11 to 0.15. One set of values, obtained from test BR8, are given in Table 1; the data were interpreted assuming the indices m_{ij} and n_{ij} are the same in each direction - a deliberate choice, rather than a definite trend evident in the data.

Table 1: Indices for reconstituted clay (test BR8)

	vh	hv	hh
Structure term S_{ij}	42	52	72
Void ratio index m_{ij}	-3.3	-3.3	-3.3
Stress index n_{ij}	0.13	0.13	0.13

5. BEHAVIOUR OF NATURAL BOTHKENNAR CLAY

Initial stiffness of natural samples after reconsolidation to in-situ stresses

All specimens were mounted on dry porous stones and initially consolidated under isotropic stresses equal to the measured or estimated suction in the block samples. After saturation they were consolidated to in-situ stresses ($K_0 \approx 0.65$) along a linear stress path and the stresses held for several days for the creep rate to reduce. Overall volume changes during specimen saturation and reconsolidation ranged from 0.3% to 1% confirming the excellent quality of the samples.

At *in-situ* stress, no direct comparison of G_0 values is possible between natural (N) and reconstituted (R) samples (since the N samples lie outside the state boundary surface for the R material). An indirect comparison may be made using the values from Table 1 in equation (2) to extrapolate the stiffnesses of the R material to the state of the N samples. Comparing data from eight tests on natural samples with the R data showed average ratios G_{0N}/G_{0R} of 1.38 (vh), 1.23 (hv) and 0.91 (hh). This surprising finding is consistent with the different anisotropy ratios G_{0hh}/G_{0hv} of the natural and reconstituted soils; while the natural material is almost isotropic (average G_{0hh}/G_{0hv} = 0.97) the reconstituted material has a ratio G_{0hh}/G_{0hv} = 1.39. The marked difference may arise from the dissimilarity of the depositional environments, the natural material is believed to have been bioturbated after deposition whereas the reconstituted clay was 1-D consolidated in tubes.

Effect of swelling and reconsolidation

The possibility that swelling to low effective stresses might itself cause destructuration was examined in several tests. Figure 3 shows data from a test in which a sample of natural clay was initially consolidated under 20 kPa isotropic stress, then swelled to 2.5 kPa and then reconsolidated to 20 kPa followed by further consolidation to the in-situ stress condition. Very small

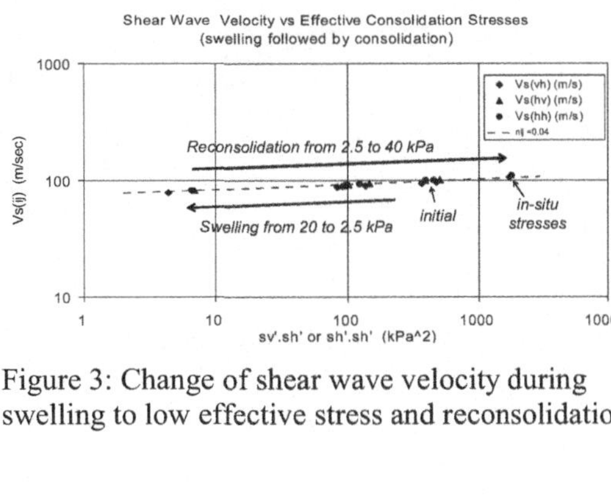

Figure 3: Change of shear wave velocity during swelling to low effective stress and reconsolidation.

recoverable volume changes occurred (corresponding to κ=0.01) and it may be seen from the linear relationship between log(V_s) and log($\sigma_i'.\sigma_j'$), that swelling has had little effect.

Effect of 1-D volumetric straining

Changes of G_0 during destructuration by compression have been explored in some detail. It is generally assumed that a reconstituted soil is fully destructured and forms a lower limit to the behaviour of a structured natural soil. In one dimensional compression the volumetric effects of structure may be expressed by the void index (Burland, 1990, Sheahan, 2005) or metastability index (Lo Presti, Shibuya and Rix, 1999); Shibuya has also examined the volumetric effects of structure on G_0 in a similar way (Shibuya, 2000). In this research we have adopted the framework of equation (2) as the starting point.

As noted above, at *in-situ* stress and void ratio, no direct comparison of G_0 values is possible between natural (N) and reconstituted (R) samples. Under increasing K_0 stresses, it is possible to bring normally consolidated N and R samples closer together on an *e*-log*p'* plot (Figure 4) as the natural material is progressively destructured. However, the only place where the two can be directly compared at the same state is in the over-consolidated region to the left of the NCL. During K_0 consolidation of N samples, the structure, void ratio and stresses are all changing, so the individual terms cannot be isolated in the manner used for the R tests. Nevertheless it has been assumed here that the void ratio power index m_{ij} found from R samples is also valid for N samples.

If G_{0ij} is plotted against ($\sigma'_i.\sigma'_j$), there is a general trend for G_{0ij} to increase throughout the test for both R and N samples – a consequence of the decreasing void ratio. But whereas when G_{0ij} for R samples is normalised by the void ratio function and plotted

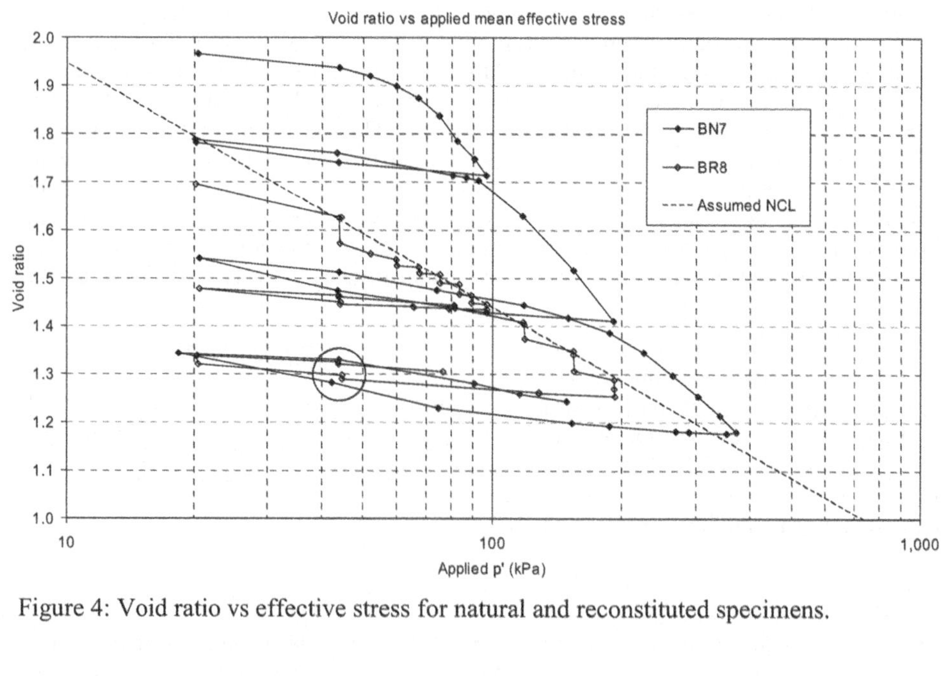

Figure 4: Void ratio vs effective stress for natural and reconstituted specimens.

using double-logarithmic axes, a reasonably unique straight line is found (whose slope defines the value of n), a more complex picture is found with N samples. An example of this is given in Figure 5 for the test BN7 for the G_{0hv} data. The initial pre-yield part of the test gives $n = 0.08$, similar to the value found during swelling from *in-situ* to very low stresses. Subsequent parts of the test show increased G_{0hv} as the void ratio is reduced, but once the decrease in void ratio is taken into account, it can be seen that normalised G_{0hv} reduces. With the index m taken as -3.3 the plot suggests there is a modest loss of structure; using a larger value of m would result in greater reduction of normalised G_0 but the same trends obtain. It may be seen that at *in-situ* stress level, the swelling and reconsolidation stages post-yield show a clear reduction of normalised G_{0hv} for the natural material.

Surprisingly when the data are compared with trend-line of normalised G_{0ij} data for R samples, it is evident that, not only is the slope different, but the R data intersect the N data (dotted trendline in Figure 5). While at high stresses there is a clear convergence of the natural and reconstituted data, at higher OCR the stiffness of the N material falls well below the trend for the R data. It should be emphasised that this finding does not rest on any normalisation assumptions: N and R samples brought to the same in-situ stress state at low void ratio (circled in Figure 4) show that measured G_0 values for N samples are lower than for R samples.

The normalised data for swelling and reconsolidation stages lie on well-defined straight lines, indicating that these stages do not cause significant destructuration. However, it is clear that these swelling stages have progressively steeper slopes, with n values rising to more than 0.2, a value higher than obtained in R tests (typically 0.11 to 0.15). Although only G_{0hv} data are shown, similar data were obtained for G_{0vh} and G_{0hh}, albeit with

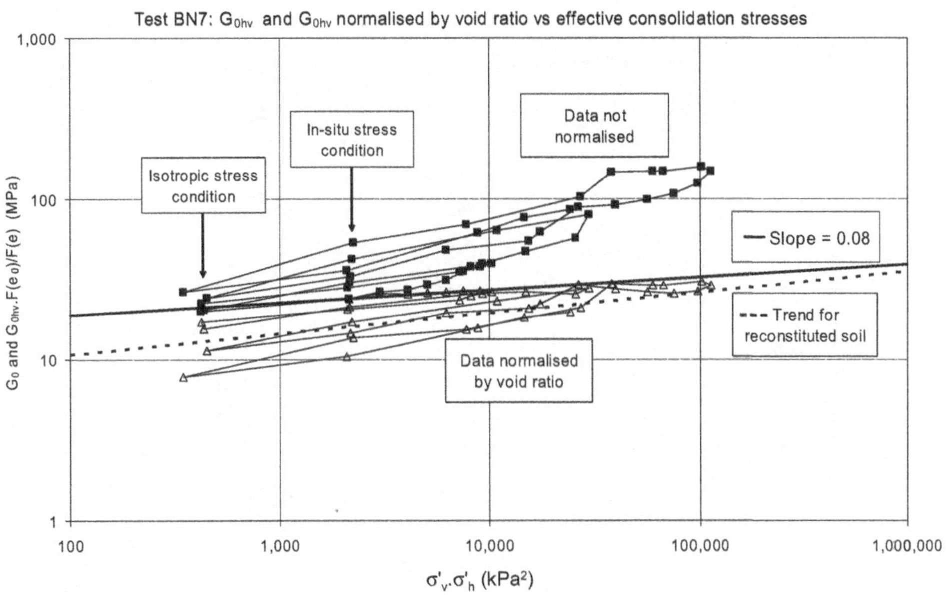

Figure 5: G_{0hv} and normalised G_{0hv} for natural specimen BN7 *vs* effective stress[2].

differing degrees of anisotropy for R and N samples. Thus the picture that emerges for N tests is more complicated than suggested by equation (2): At the same time as the structure term S is decreasing, the stress power index n is increasing. The value of n is initially smaller than for R tests, but eventually becomes larger. Major loss of structure is primarily driven by plastic volumetric straining, and initial modelling suggests that the functions describing it will be non-linear.

It is interesting to observe a parallel phenomenon in the compressibility data (see Figure 4). The values of swell index κ (slope of the swelling lines) are initially smaller for N samples, but then become larger than equivalent R samples. Thus the κ and n values show the same trends when comparing N and R samples. This finding, together with the fact that normalised stiffness for N and R tests do not converge to a common value, highlights a fundamental problem in creating R samples that can successfully mimic fully-destructured N samples. This was an unexpected and important finding, because destructuration models implicitly assume that R samples define the properties of a fully-destructured N sample.

Effect of undrained shearing

Tests on high quality samples of Bothkennar clay have shown that peak undrained strength is reached after an axial strain of less than 1%. It is generally accepted that tube sampling in clays is an undrained process during which the soil is subjected to a strain path involving compression and extension. Clayton *et al.* (1992) simulated the effects of tube sampling by subjecting three undisturbed samples reconsolidated under in-situ stresses to a cycle of ±0.5, ±1 and ±2% strain respectively. They showed that such strains reduce the subsequent peak strength and medium strain stiffness, behaviour which has been interpreted as a destructuration and shrinkage of the bounding surface.

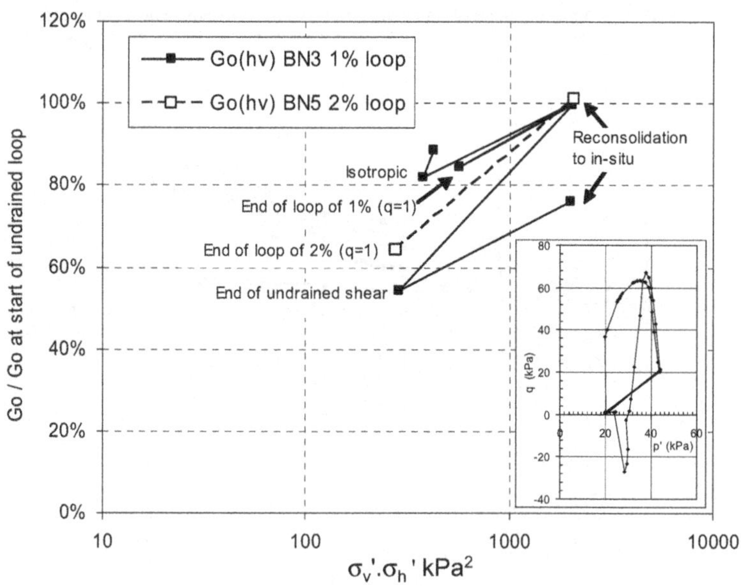

Figure 6: Changes in G_0 due to undrained cycles of strain

Similar tests in the present research have confirmed this behaviour, and have shown that destructuration by undrained shearing results in a reduction of G_0. Figure 6 illustrates the changes in G_0 as samples are first reconsolidated to in-situ stresses, then subjected to undrained loops of ±1% or ±2% strain, then reconsolidated and held under in-situ stresses, then subject to undrained shear in compression and finally reconsolidated and again held under in-situ stresses. The data are normalised by the initial value before the undrained loop. The inset box shows a typical stress path.

One test (BN11) has examined destructuration resulting from several successive undrained cycles of strain (±2, ±4 and ±8%); each cycle was followed by reconsolidation back to the in-situ stress state accompanied by reduction of volume. The stress-strain data

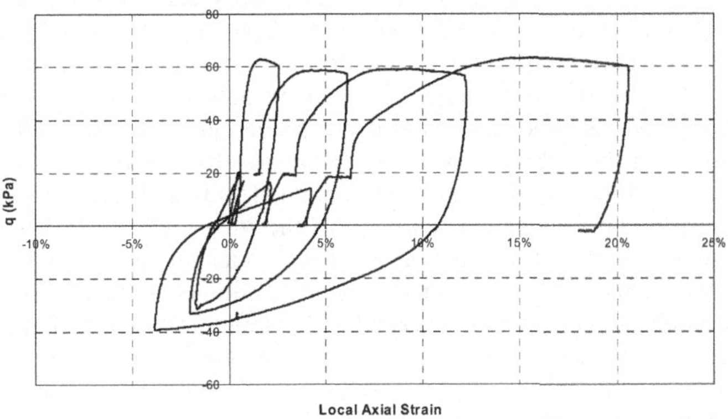

Figure 7: Stress strain behaviour of specimen BN11 subject to undrained cycles of ±2, ±4 and ±8% strain followed by undrained shear.

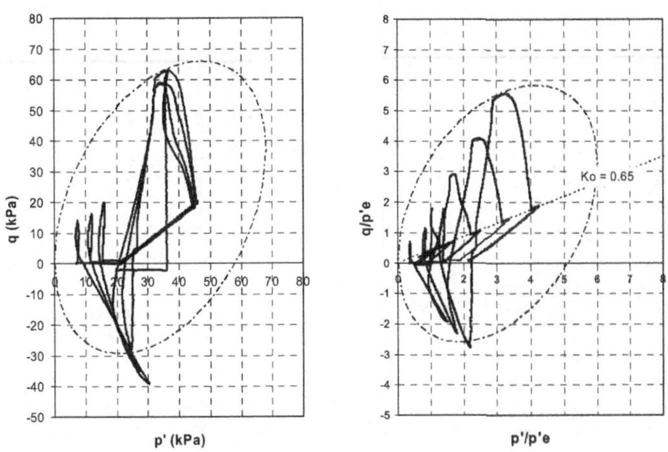

Figure 8: Stress paths in test BN11 during ±2, ±4 and ±8% undrained cycles of strain followed by undrained shear, a) without normalisation and b) with normalisation by p_e'.

are shown in Figure 7 which illustrates clearly the effect of the undrained shear cycles on the medium and large strain stiffness. At first sight the peak undrained strength is apparently not affected by the shearing, but here the effects of destructuration are offset by volumetric hardening. This can be illustrated by examination of the stress paths. Figure 8a shows that on a plot of q vs p' the successive stress paths are similar; in Figure 8b the same data are normalised by p_e' (the value of mean effective stress at the appropriate void ratio on the normal consolidation line for R material shown in Figure 4), and this clearly reveals the successive destructuration. Figure 8 also indicates a possible initial bounding surface of the natural clay before destructuration.

Figure 9 compares the stiffness data from all such tests after undrained loops with that when the samples were first consolidated under in-situ stresses; data are shown plotted against the size of the strain loop both at the isotropic stress state immediately after each undrained loop (before any reconsolidation) and after reconsolidation under in-situ stresses. Data are presented both without normalisation for void ratio and normalised to take account of changes of void ratio, and indicate several important trends. Such undrained strain excursions produce a significant reduction of G_0, to well below that which would arise from the reduction of effective stresses by swelling alone; the degree of reduction is linked to the magnitude of the undrained strains. Then, for modest strain loops, when specimens are reconsolidated back to the in-situ stress state, G_0 increases with hold time such that the clay apparently regains much of its original small strain stiffness. Figure 9a shows that ±2% strains result in temporary reduction of G_0 of 30%, which is recovered on reconsolidation. Indeed ±8% strains in test BN11 resulted in an increase of measured stiffness of up to 25% after reconsolidation. This recovery or increase of stiffness contrasts with the effects on medium-strain stiffness where significant reductions are observed. The trend is clearer once G_0 is normalised to take

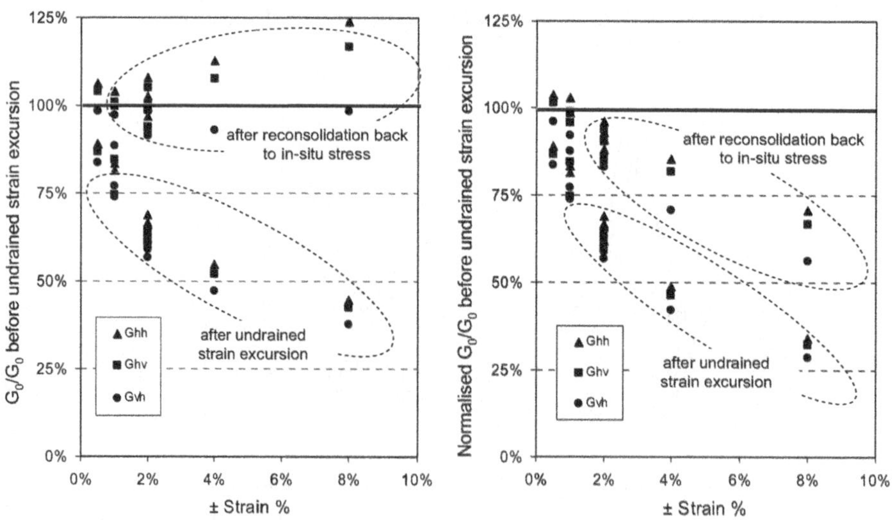

Figure 9. Change of stiffness due to undrained loops, a) without and b) with normalisation for void ratio changes.

account of changes on void ratio. Figure 9b shows the normalised data which show the permanent effects of destructuration clearly. Now it may be seen that although the ±8% strains in test BN11 resulted in an increase of measured stiffness of up to 25%, once the change in void ratio is taken into account (volume reduction of about 9%), there was actually a 30-40% reduction of normalised stiffness which is clearly associated with the destructuration. These results again demonstrate the importance of taking volume changes into account when interpreting the stiffness data.

6. CONCLUSIONS

A systematic investigation of the effects of destructuring on properties of the Bothkennar clay has been carried out, using changes of small strain shear stiffness G_0 as an indicator of damage. Tests were conducted on both natural and reconstituted material, so that the effects of microstructure could be isolated. After initial reconsolidation under in-situ stresses to establish a baseline condition, samples were subjected to controlled cycles of undrained compression/extension strain. These have shown that such strains result in significant temporary reduction of G_0, but with time, the clay regains much of its original small strain stiffness on reconsolidation to the initial stress state. While small changes of G_0 after reconsolidation appear to be consistent with small changes to the peak strength, they do not reflect the damage that affects the medium-strain stiffness. To correctly identify effects of damage to microstructure, it proved important to normalise the data to a common void ratio.

The research shows that comparisons of V_s between laboratory and field should be made after samples are reconsolidated to in-situ stress conditions and the same data are normalised by p_e' (the value of mean effective stress at the appropriate void ratio on the normal consolidation line for R material take account of void ratio changes. Although others have reported fairly good agreement between lab and field (for example Lo Presti *et al.*, 1999), there has previously been uncertainty whether the damaging effects of sample disturbance might have been offset by the volumetric strains during reconsolidation, particularly in less structured clays. It has thrown light on our previous observations of reduction of shear wave velocity V_s in various cores from tube samples of Bothkennar clay (Hight, 1998).

The research has revealed features of soft clay behaviour which are not captured by current constitutive models. For example degradation of stiffness during undrained shear was expected but not its recovery on reconsolidation. While G_0 normalised for void ratio clearly indicates permanent damage resulting from compression and shear, the subsequent changes with time conceal reduction in the medium strain stiffness, which is a more useful indication of practical damage. This seems to imply that G_0 cannot so easily be used as an anchor for evolution of medium strain stiffness - the shape of the stiffness degradation curve with strain is changing more significantly than the initial zero strain value. The tests reveal the difficulty of defining a reference structureless material, since G_0 for the natural material was not found to be asymptotic to G_0 of the reconstituted material. Just as damage in natural material increases its swell index κ, so it also increases the stress indices $n_{(ij)}$ – in contrast to the reconstituted material which shows none of these progressive changes when subjected to similar cycles of compression stress.

ACKNOWLEDGEMENTS

The authors gratefully acknowledge the support of the laboratory staff Mike Pope, Steve Iles and Mark Fitzgerald. The research was partly funded by the UK Engineering and Physical Sciences Research Council.

REFERENCES

Burland, JB (1990) On the compressibility and shear strength of natural clays. *Géotechnique* 40:3:327-378.

Burland, JB, Rampello, S, Georgiannou, VN and Calabresi, G (1996) A laboratory study of the strength of four stiff clays. *Géotechnique* 46:3:491-514.

Callisto, L and Calabresi, G (1998) Mechanical behaviour of a natural soft clay. *Géotechnique* 48:4:495-513.

Clayton, CRI, Hight DW and Hopper RJ (1992) Progressive destructuring of Bothkennar clay: implications for sampling and reconsolidation procedures. *Géotechnique* 42:2:219-240.

Hight, DW (1998) Soil characterisation: the importance of structure, anisotropy and natural variability. 38th Rankine Lecture. *Géotechnique* (to appear)

Hight, DW, Böese, R, Butcher, AP, Clayton, CRI and Smith, PR (1992) Disturbance of the Bothkennar clay prior to laboratory testing. *Géotechnique* 42:2:199-217.

Hight, DW and Leroueil, S (2003). Characterisation of soils for engineering purposes. *Characterisation and Engineering Properties of Natural Soils* – Tan *et al.* (eds.) Swets & Zeitlinger, Lisse. 1:255-360.

Jamiolkowski, M, Lancellotta, R & Lo Presti, DCF (1995) Remarks of the stiffness at small strains of six Italian clays. *Developments in deep foundations and ground improvement schemes.* (Ed: Balasubramaniam *et al*) Balkema, Rotterdam, 197-216.

Leroueil, S and Vaughan, PR (1990) The general and congruent effects of structure in natural soils and weak rocks. *Géotechnique* 40:3:467-488.

Lo Presti, DCF, Shibuya, S and Rix, GJ (1999). Innovation in soil testing. Theme lecture to *2nd International Symposium on Pre-Failure Deformation Characteristics of Geomaterials (IS TORINO 99).*

Lunne, T., Berre, T., and Strandvik, S. (1997). Sample Disturbance Effects in Soft Low Plastic Norwegian Clay. In *Recent Developments in Soil and Pavement Mechanics (pp. 81-102) (Ed:Almeida).* Balkema: Rotterdam. reprinted in *Norwegian Geotechnical Institute report no 204.*

Muir Wood, D (1995) Kinematic hardening model for structured soil. *Numerical Models in Geomechanics (*NUMOG V*)* (eds GN Pande and S Pietruszczak), Balkema, Rotterdam 83-88.

Nash DFT, Pennington DS and Lings, ML (1999). The dependence of anisotropic G_0 shear moduli on void ratio and stress level for reconstituted Gault Clay. *Pre-Failure Deformation Characteristics of Geomaterials (IS TORINO 99)* Balkema, Rotterdam 1:229-238.

Paul, MA, Peacock, JD and Wood, BF (1992) The engineering geology of the Carse clay of the National Soft Clay Research Site, Bothkennar. *Géotechnique* 42:2:183-198.

Pennington, DS (1999) The anisotropic small strain stiffness of Cambridge Gault clay. PhD thesis, Univ. of Bristol.

Pennington, DS, Nash, DFT and Lings, ML (1997) Anisotropy of G_0 shear stiffness in Gault clay. *Géotechnique* 47:3:391-398.

Rouainia, M and Muir Wood, D (2000) A kinematic hardening constitutive model for natural clays with loss of structure. *Géotechnique* 60:2:153-164.

Sheahan, T.C. (2005). "A Soil Structure Index to Predict Rate Dependence of Stress-Strain Behavior," in *Site Characterisation and Modelling*, ASCE GSP No. 38 (to be published)

Shibuya, S (2000). Assessing structure of aged natural sedimentary clays. *Soils and Foundations* 40:3:1-16.

Soil Stress-Strain Behavior: Measurement, Modeling and Analysis
Geotechnical Symposium in Roma, March 16 & 17, 2006

SMALL STRAIN STIFFNESS OF A SOFT CLAY ALONG STRESS PATHS TYPICAL OF EXCAVATIONS

S. Fortuna, L. Callisto & S. Rampello
Department of Structural and Geotechnical Engineering
University of Rome 'La Sapienza', 00186, IT
e-mail: sonia.fortuna@uniroma1.it

Abstract
In this paper, an experimental work carried out on Pisa clay is described. Undisturbed samples are subjected to stess paths analogous to those followed by soil elements adjacent to excavations. The stress paths adopted in the experimental programme derive from a simplified analysis of a propped retaining wall, using limit equilibrium and allowing, through an approximate procedure, for wall flexibility and for seepage around the wall. Such stress paths are corroborated by in situ measurements and numerical results taken from the literature. It comes out that the range of stress path directions for soil elements close to an excavation can be quite wide. The experimental results are analysed in terms of secant shear stiffness and compared to results obtained from bender element measurements. Also, a generalised definition of stiffness is used, which accounts for both spherical and deviatoric strain components. Results obtained at small strains are compared with predictions of cross-anisotropic elasticity.

Introduction

The stress-strain behaviour of a clayey soil is known to depend on the stress path direction, the initial stress state and the recent stress history. Figure 1(a) shows two soil elements located behind (A) and in front (B) of a retaining structure: it can be anticipated that they will experience stress paths completely different from each other. On a first approximation, it can be assumed that element A will undergo a compressive stress path with constant vertical total stress σ_v and decreasing horizontal total stress σ_h, while element B will undergo an extension stress path with decreasing σ_v and approximately constant values of σ_h (Muir Wood 1984). The corresponding effective stress paths are shown qualitatively in Figure 1(b) for both drained and undrained conditions; all of them are characterised by a decrease in mean effective stress p'. However, in situ measurements (Tedd *et al* 1984, Ng 1999), numerical analyses (Potts & Fourie 1984, Hashash & Whittle 2002), and theoretical considerations lead to the conclusion that the soil can experience a range of stress paths much wider than that shown in Figure 1, depending on: drainage conditions; depth of the soil element and distance from the retaining wall; soil-wall relative stiffness; soil-wall friction.

Back analyses of full-scale prototypes show that the deviatoric strains ε_s experienced by most of the soil interacting with the excavation are smaller than about 0.1% (Burland 1989). On the basis of experimental evidence (e.g. Jardine 1992, Smith *et al.* 1992, Callisto & Calabresi 1998), pre-failure soil behaviour can be described using concepts derived from elasto-plastic constitutive models with kinematic hardening as shown, for instance, by Callisto *et al.* (2002). In the conceptual model proposed by Jardine (1992), two kinematic yield surfaces exist in stress space, within a gross yield surface. The position of the kinematic surfaces with respect to the gross yield surface illustrates the effect of overconsolidation, while the location of the current stress state with respect to the kinematic surfaces describes the effect of the recent stress history, that affects soil behaviour for small to medium strains. In fact, the effect of recent stress history can be described as the effect of the direction of the stress path previously followed by the soil on subsequent stress-strain behaviour. It has been observed (e.g. Atkinson *et al.* 1990) that soil stiffness increases as the angle between the

Hoe I. Ling et al. (eds.), Soil Stress-Strain Behavior: Measurement, Modeling and Analysis, 299–310.

previous and the new stress path direction increases. Therefore, stiffness for soil elements behind a retaining wall can be either larger or smaller than that observed in soil elements in front of a retaining wall, depending on the recent stress history and the overconsolidatio ratio (Amorosi *et al.* 1999).

The mobilised strength in the soil involved in an excavation can vary significantly, from the initial stress state to limit active or passive conditions. Depending on the initial stress state, strain values can range eventually over orders of magnitude, stability being guaranteed by the contemporary presence of soil elements in plastic conditions and soil elements far from failure, interacting with each other. Therefore, it is important to investigate accurately the complete stress-strain behaviour of the material from small strains up to failure in order to allow for a satisfactory analytical description of the observed behaviour.

Stress paths associated to excavations

Stress paths followed by soil elements behind and in front of an excavation can be drawn by evaluating the vertical and horizontal effective stresses acting on an embedded retaining wall through a simple limit equilibrium analysis.

A 6 m deep excavation supported by a 20 m long diaphragm wall braced at the top by a rigid constraint is sketched in Figures 2 and 3. Excavation is assumed to be carried out under drained conditions in a soft clay deposit with an angle of shearing resistance $\varphi' = 22°$, typical for Pisa clay (Rampello & Callisto 1998) and zero cohesion. It is assumed that ground water level is initially located at the ground surface and it is maintained at dredge level inside the excavation. Seepage effect is accounted for by the simplifying assumption of one-dimensional flow around the retaining wall.

It is assumed that horizontal effective stresses acting on the wall have a linear (Fig. 2) or a bi-linear (Fig. 3) distribution. The first hypothesis is relevant for a stiff wall, for which the degree of mobilisation of the available strength can be assumed constant along its length. The second hypothesis can be associated to a relatively flexible structure, which is long and slender enough to avoid movements at its base. Therefore, in determining the distribution of horizontal stresses on the wall, it is assumed that the horizontal effective stresses at the toe are equal to $K_0\,\sigma'_v$, where K_0 is the earth coefficient at rest. For each excavation depth, the length of the wall along which active and passive states are attained is that necessary to ensure equilibrium.

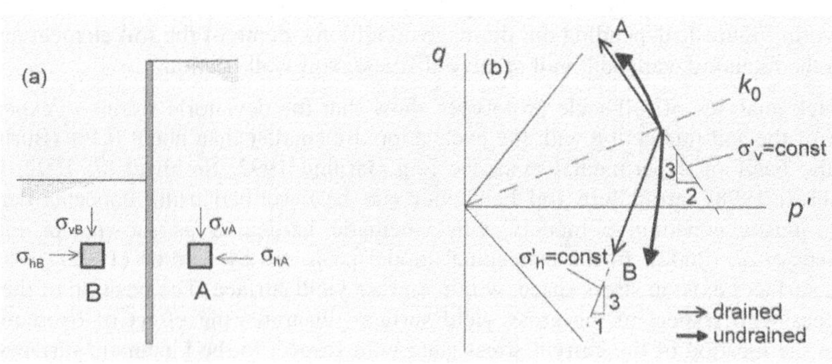

Fig. 1 – Stress paths associated to two soil elements located behind and in front of a retaining wall.

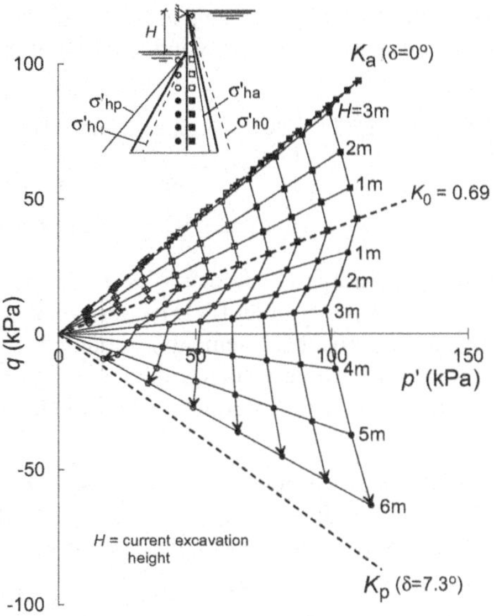

Fig. 2 – Stress paths for soil elements located at different depths, assuming a linear distribution of horizontal effective stress.

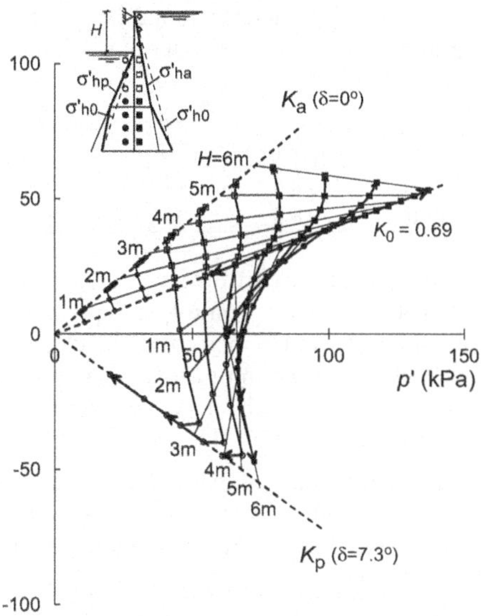

Fig. 3 – Stress paths for soil elements located at different depths, assuming a bi-linear distribution of horizontal effective stress.

Figure 2 shows the resulting stress paths experienced by soil elements located at various depths adjacent to the stiffer retaining wall. It is assumed that principal stresses act in the vertical and horizontal directions, therefore stress invariants are defined as $q = (\sigma'_v - \sigma'_h)$; $p' = (\sigma'_v + 2\,\sigma'_h)/3$. 'Active' stress paths are linear, parallel to each other, and characterised by some increase in σ'_v as an effect of seepage forces. 'Passive' stress paths are initially similar to the drained path B in Fig. 1, in that σ'_v decreases at constant σ'_h. As limit state is reached at the back of the wall, horizontal stresses needed for equilibrium increase, and 'passive' stress paths veer to the right.

Figure 3 shows the stress paths experienced by soil elements located at various depths, resulting from the second scheme. At shallow depths, 'active' stress paths are not dissimilar from the ones shown in Fig. 1(b) for soil element A. However, as depth increases, 'active' paths rotate clockwise, ad this effect is due to both seepage forces and the assumed bilinear distribution of horizontal stresses. 'Passive' stress paths are characterised by a substantial decrease in σ'_v, together with some increase in σ'_h.

A range of possible stress paths followed by soil elements close to an excavation, derived from the above simplified analysis, is plotted in Figure 4. Similar stress paths have been derived from field measurements by Tedd *et al.* (1984) and by Ng (1999), and by numerical analyses of idealised (Potts & Fourie 1984) and real (Calabresi *et al.* 2002) excavations.

Experimental programme

Triaxial stress path tests were carried out on 38.1 mm diameter specimens obtained from undisturbed Laval samples (La Rochelle *et al.*, 1981), retrieved from a depth of about 18 m in the upper clayey deposit found below the Tower of Pisa. Physical and index properties are reported in Table 1. The material is a high plasticity clay with a liquidity index of about 0.6.

Table 1 – Physical and index properties of samples

Sample	depth (m)	γ (kN/m^3)	W_0 (%)	W_L (%)	I_P (%)	CF (%)	I_L
30 A	18.6	16.7	56.0	71.5	40	66.0	0.61
30 B	18.8	16.7	56.7	73.2	43	68.5	0.62

All tests were carried out in a stress-path triaxial apparatus (Bishop & Wesley, 1975) with automatic feedback control, equipped with internal submergible LVDTs for axial strain measurement and a pair of piezoelectric transducers (bender elements) mounted in the pedestal and in the top platen, for measurement of shear wave velocity. A single sinusoidal input wave was used for bender element measurements, with a frequency of 10 kHz and a repeat frequency of 50 Hz.

In the following, the stress and strain state is described in terms of invariants (q, p') and (ε_s, ε_v), defined as $q = (\sigma'_a - \sigma'_r)$; $p' = (\sigma'_a + 2\,\sigma'_r)/3$; $\varepsilon_s = 2(\varepsilon_a - \varepsilon_r)/3$; $\varepsilon_v = \varepsilon_a + 2\,\varepsilon_r$, where indexes *a* e *r* refer to the axial and radial directions in a cylindrical specimen respectively.

All specimens were first consolidated to the in situ stress state O along path ABO, as shown in Fig. 5. Before initiating the probing paths, a standard waiting period of 60 hours was applied corresponding to a rate of volume strains less than about 0.02 % per day, and a bender element measurement was carried out. Subsequently, different tests were carried out as shown in Figures 5 and 6.

A first series of tests, shown in Fig. 5, consisted of a rosette of drained rectilinear stress paths starting from the in situ stress state (O) and pointing along the different directions expected in the vicinity of an excavation (Fig. 4). They are denoted by the prefix D, followed by the angle formed by the stress path direction with the horizontal. Undrained compression (UC) and extension (UE) tests were also performed, starting from the in situ stress state.

A second series of tests, shown in Fig. 6, was carried out, in which point P was reached from the initial stress state (O). From point P, a new rosette of drained stress paths was performed. These tests are denoted by a prefix P, followed by the angle formed by the stress path direction with the horizontal.

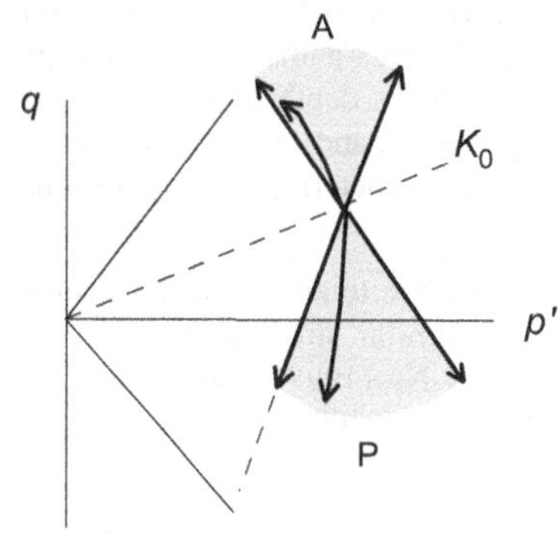

Fig. 4 –Range of stress paths associated to two soil elements located behind and in front of a retaining wall, deriving from schemes shown in figures 2-3.

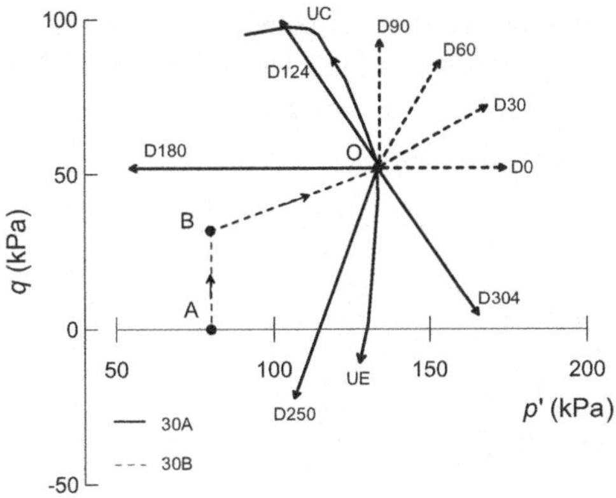

Fig. 5 – Triaxial stress paths performed starting from in situ stress state (O).

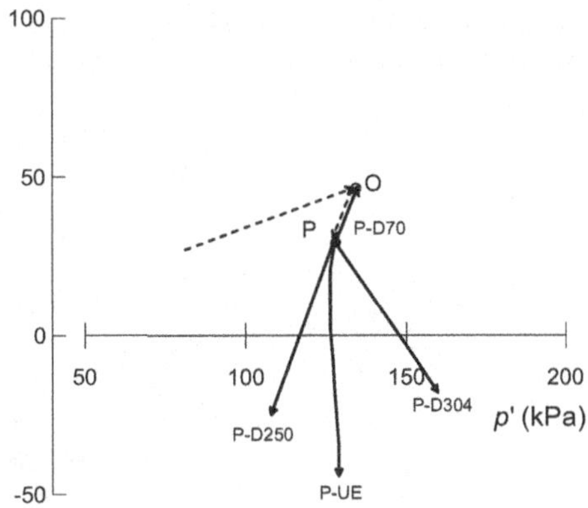

Fig. 6 – Stress paths performed starting from perturbed stress state (P).

Stress-strain behaviour

Figure 7(a) shows the stress-strain curves in the q-ε_s plane, relative to tests starting from the in situ state (O). The deviatoric strains shown in Fig. 7 were derived from both local and external measurements: strains computed from external transducers were significantly higher, up to twice for $\varepsilon_s < 0.1\%$.

For tests carried out under strain controlled conditions (UC, D90, D124, D250) the curves show a ductile soil behaviour. For test UE, carried out under stress controlled conditions, observation of any post-peak softening was prevented and failure was assumed to occur when the maximum value of q was reached and a sudden increase in the measured rate of deformation was observed. In the triaxial compression tests, failure was reached at deviatoric strains smaller than 3 %, whereas extension tests failed at much larger strain ($\varepsilon_s \approx 8$ %). It was not possible to observe failure conditions in test D60 since it ended prematurely, while tests D30 and D304 showed increasing values of q in the strain range investigated. Extension tests D250, D304 and UE show a much stiffer behaviour than the compression tests for the entire range of strains, and a more gradual decrease of stiffness.

Figure 7(b) shows the volumetric strains ε_v observed in drained tests starting from in situ state (O), plotted versus the corresponding deviatoric strains ε_s. Tests D180 and D250, for which p' decreases, show a dilatant behaviour, while tests D0, D30, D60, D90 and D304, characterized by increases in p', show a decrease in volume. The ratio $\varepsilon_s/\varepsilon_v$ is somewhat variable along each single test. However, it can be seen that the average values of $\varepsilon_s/\varepsilon_v$ is proportional to the corresponding q/p' ratio, that is, the relative amount of deviatoric strain is proportional to the relative amount of deviatoric stress.

The volumetric strains developed in drained tests D124 and D250 are relatively small, consistently with the similarity observed between drained stress paths and the stress paths obtained in undrained compression (UC) and extension (UE) tests, respectively (Fig. 5). Specifically, for test D124 the maximum volumetric strain is $\varepsilon_{vmax} = 2.9$ %, while for test D250 it is $\varepsilon_{vmax} = 1.2$ %.

Figure 7(c) shows the excess pore water pressure Δu_q produced, in the undrained tests UC and UE, by the increments in the deviator stress q, plotted as a function of ε_s. This is computed as $\Delta u_q = p'_0 - p'$ where p'_0 is the mean effective stress at point (O). Values of Δu_q increase continuously in each test and tend to become stationary for ε_s larger than about 12 %.

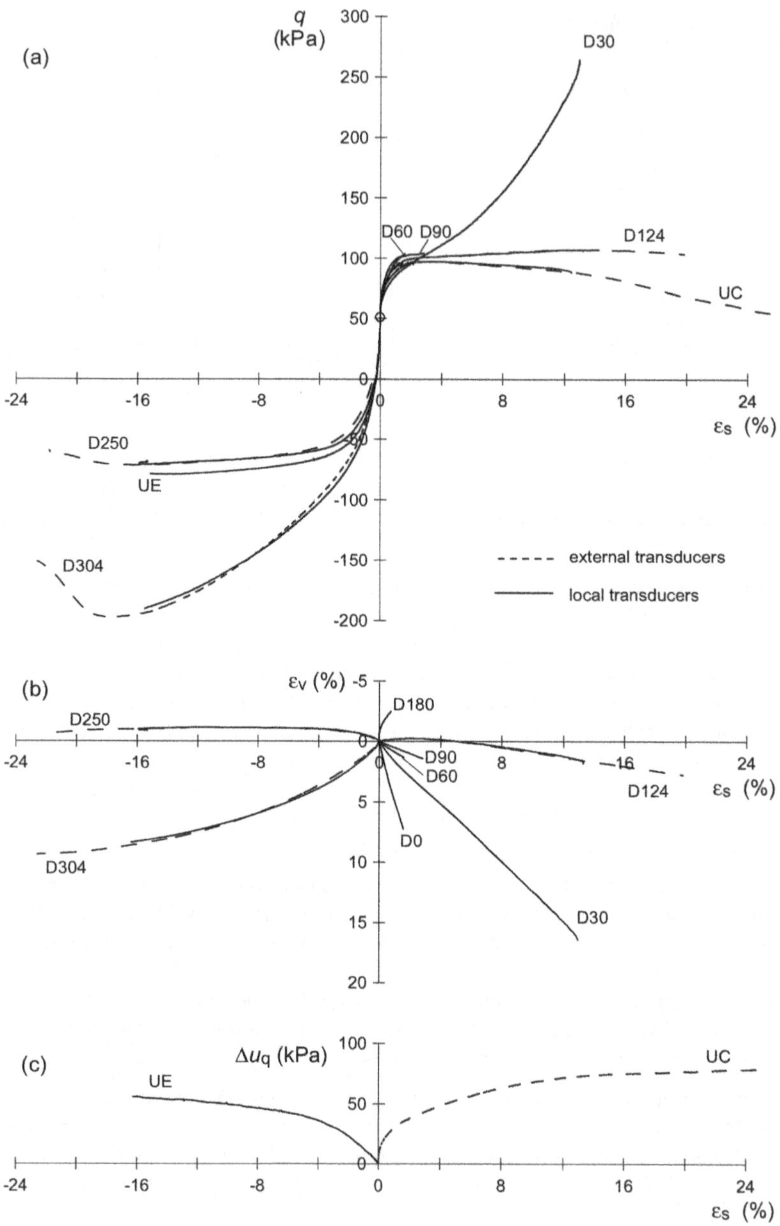

Fig.7 – Stress-strain behaviour for tests starting from (O).

Small strain stiffness

Figure 8 shows the secant shear stiffness, conventionally defined as $G_{sec}=\Delta q/3\Delta\varepsilon_s$, observed along stress paths starting from the in situ state (O), plotted as a function of deviatoric strains ε_s. In the same plot the range of values of the small-strain shear modulus G_0 measured using the bender elements is represented: these values can be assumed to correspond to deviatoric strains in the range of 0.0001 to 0.001% (Dyvik & Madshus 1985). Most of the tests show an initial value of G_{sec}, measured at $\varepsilon_s = 0.001$ %, ranging between 25.5 and 28 MPa, with the exception of test D30, that shows a lower initial stiffness. In the average, initial secant stiffness measured in the triaxial tests seems to be in a good agreement with the small strain shear stiffness measured using the bender elements.

As the strain level increases, the stiffness decays with a rate that depends significantly on the stress path direction. Extension tests UE, D250 and D304 show the lowest rate of decay: for $0.002 < \varepsilon_s < 0.1$ % the G_{sec}-log ε_s curves show a nearly constant gradient. More specifically, test D304 shows a slightly stiffer soil behaviour than test UE, which in turn exhibits a higher stiffness than test D250. Curves relative to the compression tests UC and D124 are very similar, consistently with their similar stress-paths. Compression test D90 also shows a similar stiffness, though for this test only data for $\varepsilon_s > 0.02$ % are available. Secant shear stiffness observed in these compression tests is smaller than that measured in the extension tests, by about 20 % at $\varepsilon_s = 0.01$ % and by about 45 % at $\varepsilon_s = 0.1$ %. Secant shear stiffness from tests D30 and D60 show a much higher rate of decay with deviatoric strains.

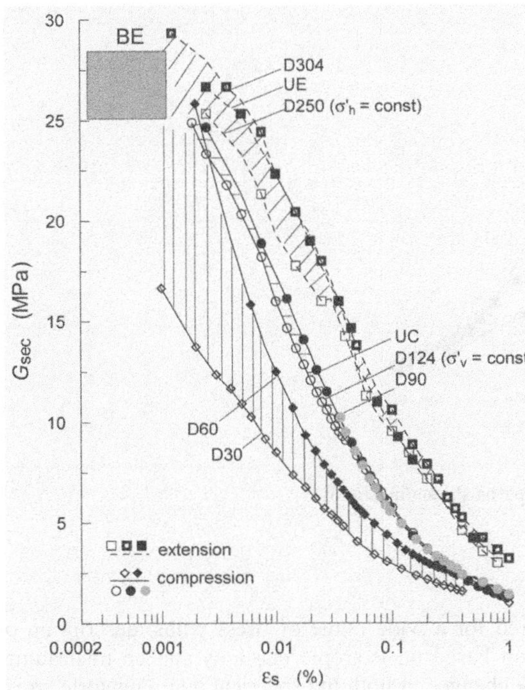

Fig. 8 – Secant shear stiffness along stress paths shown in Fig. 5.

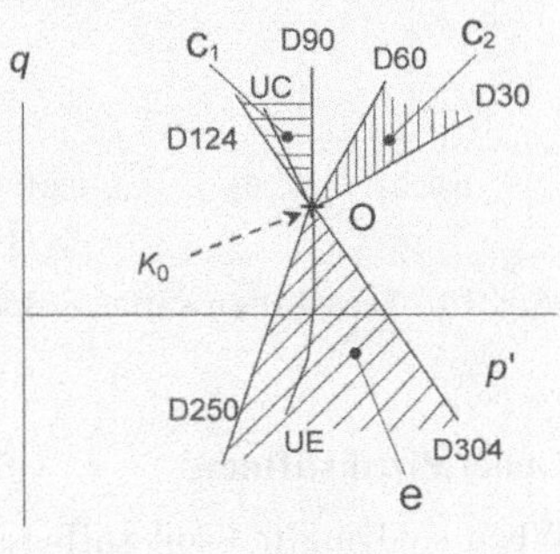

Fig. 9 – Stress paths with common behaviour in terms of secant shear stiffness.

In summary, along extension stress paths, similar to the 'passive' ones followed by soil elements below the dredge line of a supported excavation (D250, UE, D304), soil behaviour is stiffer than along compressive, 'active' stress paths. This result is in agreement with experimental findings by Amorosi *et al.* (1998) on reconstituted specimens of Vallericca clay subjected to similar paths starting from an initial anisotropic, normally consolidated state.

The previous observations permit to identify three different zones in the q-p' plane, as depicted in Figure 9. Stress paths in zone (e) show a high secant shear stiffness and a slow stiffness decay with ε_s, which depend only slightly on the specific direction of the stress path; stress paths in zone (c_1) are characterised by an intermediate shear stiffness; secant shear stiffness observed along stress paths belonging to zone (c_2) is quite smaller than that associated to the remaining zones, decays steeply with ε_s, and shows a significant dependence on stress path direction.

Figure 10 shows the secant shear stiffness observed along stress paths starting from the stress state perturbed towards the passive direction (P), plotted as a function of deviatoric strain ε_s. The maximum secant shear stiffness is again observed in extension, along path P-D304, followed by the undrained tests P-UE, and in turn by test P-D70 and P-D250. At deviatoric strains larger than 0.1 % the curves relative to the extension tests tend to converge.

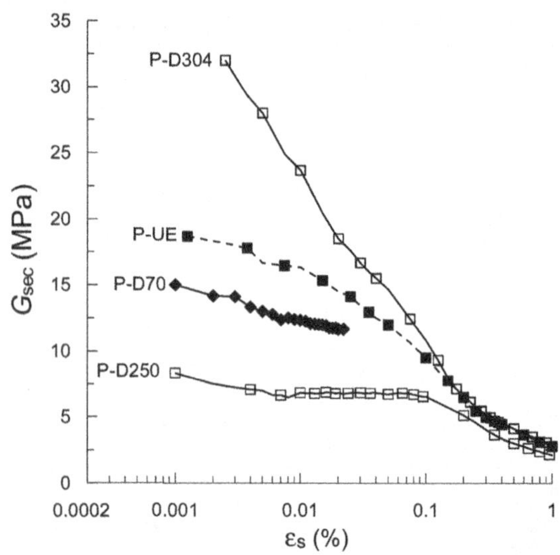

Fig. 10 – Secant shear stiffness along stress paths shown in Fig. 6.

Generalized stiffness

When studying the soil stiffness measured for a wide range of stress paths, description of stiffness using the conventional definition based on isotropic elasticity can be misleading, since each stress path is characterised by changes in both the spherical and deviatoric stress component, and each component can be thought to produce both volumetric and deviatoric strains.

Muir Wood (2004) proposed, in order to represent more effectively experimental results and allow for an easier comparison with the predictions of constitutive models, to plot results in terms of a generalised strain:

$$\varepsilon = \sqrt{\varepsilon_v^2 + \varepsilon_s^2}$$

and defined a generalised stiffness as:

$$S = \frac{\sqrt{\Delta p'^2 + \Delta q^2}}{\sqrt{\Delta\varepsilon_v^2 + \Delta\varepsilon_s^2}}$$

Figures 11 and 12 show, in the q-p' plane, contours of equal generalised strain ε for tests starting from in situ state (O): in Fig. 11 ε ranges between 0.01 and 0.1 %, while Fig. 12 is relative to $0.1 < \varepsilon < 1$ %.

The distance, gauged along a stress path, from the stress path origin to a specific contour is a measure of the generalised secant stiffness S. In Fig. 11 it can be seen that, for $0.01 < \varepsilon < 0.1$ %, the q-p' plane can be sub-divided into two parts by a straight line *a-a*: response to stress paths pointing to the left of line *a-a* is much stiffer than that observed along stress paths pointing to the right of line *a-a*. At a given ε, the contour associated to an elastic response is an ellipse, whose axes are vertical and horizontal for isotropic elasticity and are inclined for a cross-anisotropic elastic behaviour. The dashed ellipse in Fig. 11 was plotted for $\varepsilon = 0.01$ % using the elastic cross-anisotropic model proposed by Graham & Houslby (1983) with the following properties: Young's modulus in the vertical direction $E'_v = 45$ MPa; $\nu'_{vh} = 0.25$; anisotropy ratio $E'_h/E'_v = 1.5$. It can be seen that, for the paths on the left of *a-a* such ellipse is close to the observed $\varepsilon = 0.01$ % contour. The difference between the predicted elastic contour and the experimental one along a given stress path can then be regarded as a global measure of the amount of plastic strain developed along that path. Therefore, it can be inferred that, even for deviatoric strains as small as 0.01 %, plastic strains are significant for tests D0, D30 and D60. This is consistent with the rapid decay in secant shear stiffness shown by tests D30 and D60 in Fig. 8. At larger strains (Fig. 12), the shape of the contours changes, the decay of generalised stiffness becoming similar for all stress path directions.

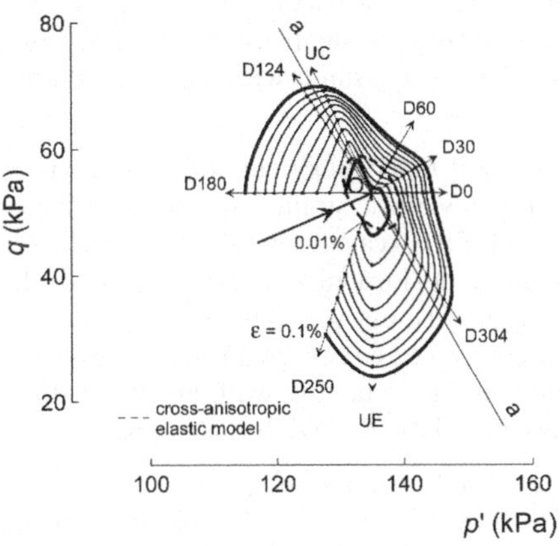

Fig.11 – Contour lines of $\varepsilon = \sqrt{(\Delta\varepsilon_v^2 + \Delta\varepsilon_s^2)}$ from initial stress state (O), for $0.01 < \varepsilon < 0.1$%.

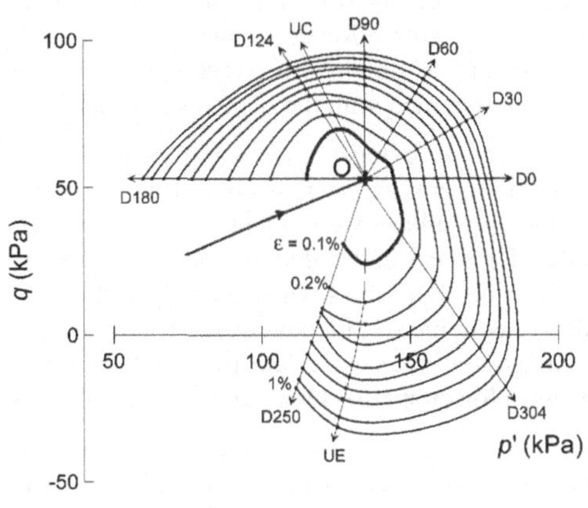

Fig.12 – Contour lines of $\varepsilon = \sqrt{(\Delta\varepsilon_v^2 + \Delta\varepsilon_s^2)}$ from initial stress state (O), for $0.1 < \varepsilon < 1$%.

For the stress paths starting from point P, the shape of the ε contours, shown in Figure 13, is consistent with a different elliptic contour, characterised by a different anisotropy ratio. The one shown with a dashed line in Fig. 13 has been plotted for $\varepsilon = 0.01$ % using $E'_v = 40$ MPa, $\nu'_{vh} = 0.25$ and $E'_h/E'_v = 0.7$. This suggests that a stress perturbation as small as path OP can produce substantial changes in the degree of elastic anisotropy. In this case, the maximum difference from the assumed elastic behaviour is observed along stress path D250, coinciding with the recent reconsolidation path to point P, while along the remaining paths, characterised by a change in the direction of the stress path from the reconsolidation one, the measured $\varepsilon = 0.01$ % contour is similar to elastic one.

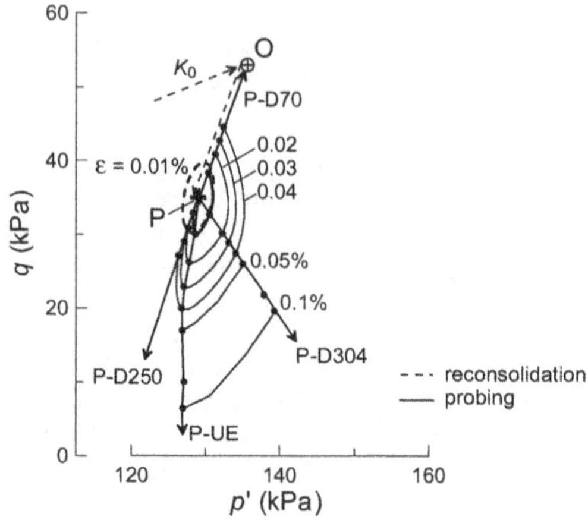

Fig.13 – Contour lines of $\varepsilon = \sqrt{(\Delta\varepsilon^2_v + \Delta\varepsilon^2_s)}$ from perturbed state (P), for $0.01 < \varepsilon < 0.1$%.

Concluding remarks

Undisturbed samples of soft Pisa clay were tested along a range of stress paths resembling those followed by soil elements close to excavations. Local displacement transducers and bender elements allowed to measure soil stiffness at strains as small as 0.001 %. A good agreement was obtained between small-strain shear stiffness measured with local transducers and that obtained with the bender elements.

In general, extension tests exhibited a much stiffer soil behaviour than compression tests and a more gradual decrease of stiffness with increasing deviatoric strain. Two different zones in the q-p' plane were identified, which are relevant for excavations. Zone (e) includes extension stress paths, that are likely to occur in soil elements located below the bottom of the excavation. For stress paths located in this zone, a high stiffness was observed, that decays slowly with increasing deviatoric strains. Zone (c_l) incorporates compression stress paths with decreasing p', that can occur to soil elements located behind a retaining wall. For these stress paths, a lower stiffness was found, which decays more rapidly with deviatoric strains.

Test results at small to medium strains were interpreted in terms of generalised strain ε. This allowed to estimate the generalised stiffness S from the contours of equal generalised strain, and to interpret test results assuming cross-anisotropic soil behaviour at small strains (0.01 %). The major differences between test results and predictions of cross-anisotropic elasticity were observed for compression tests carried out from the in situ stress state and characterised by increasing mean effective stress. A small perturbation from the in situ state was seen to produce significant changes in the degree of elastic anisotropy. Behaviour observed along stress paths starting from the perturbed state are characterised by different generalised stiffness as an effect of their recent stress history.

References

Amorosi A., Callisto L. & Rampello S. (1999). Observed behaviour of reconstituted clay under stress paths typical of excavations. In: M. Jamiolkowski, R. Lancellotta & D. Lo Presti (eds.) *2nd Int. Symp. on pre-failure Deformation Characteristics of Geomaterials.* Torino, Balkema, **1,** 35-42.

Atkinson J.H., Richardson D. & Stallebrass S.E. (1990). Effect of recent stress history on the stiffness of overconsolidated soil. *Géotechnique* **40**, No. 3, 531-540.

Bishop A. W. & Wesley L. D. (1975). A hydraulic triaxial apparatus for controlled stress path testing. *Géotechnique* **25**, No 4, 657-670.

Burland, J.B. (1989). Ninth Bjerrum Memorial Lecture: "Small is beautiful" – the stiffness of soils at small strains. *Canadian Geotechnical Journal* **26**, No. 4, 499-516.

Callisto L. & Calabresi G. (1998). Mechanical behaviour of a natural soft clay. *Gèotechnique* **48**, No.4, 495-513.

Callisto L., Gajo, A. & Muir Wood, D. (2002). Simulation of triaxial and true triaxial tests on natural and reconstituted Pisa clay. *Gèotechnique* **52**, No. 9, 649-666.

Calabresi G., Callisto L. & Rampello, S. (2002). Il ruolo delle pressioni interstiziali nella previsione del comportamento di uno scavo profondo. *Atti XXI Convegno Nazionale di Geotecnica* – L'Aquila, 273-282.

Dyvik, R. & Madshus, C. (1985). Laboratory measurements of G_{max} using bender elements. *Proc. of the ASCE Annual Convention: advances in the art of testing soils under cyclic conditions*. Detroit:186-196.

Hashash Youssef M. A. & Whittle A. J., (2002). Mechanisms of load transfer and arching for braced excavations in clay. *Journal of Géotechnical Engineering and Geoenvironmental Engineering*, **128**, No. 3, 187-197.

Graham J. & Houlsby G. T. (1983). Anisotropic elasticity of a natural clay. *Géotechnique* **33**, No 2, 165-180.

Jardine R.J. (1992). Some observations on the kinematic nature of soil stiffness. *Soils and Foundation* **32**, No. 2, 111-124.

La Rochelle P., Sarrailh J., Tavenas F., Roy M. & Leroueil S. (1981). Causes of sampling disturbance and design of a new sampler for sensitive soils. *Canadian Geotechical Journal* **18**, 52-66.

Muir Wood D. (1984). Choice of models for Geotechnical Predictions. In: *Mechanics of Engineering materials*, edited by C. S. Desai and R. H. Gallagher, John Wiley & sons Ltd, Chapter 12, 633-654.

Muir Wood D. (2004). *Geotechnical modelling*. Spon Press.

Ng C.W.W. (1999). Stress paths in relation to deep excavations, *Journal of Geotechnical Engineering*, **125**, No. 5, 357-363.

Potts D.M. & Fourie A. B., (1984). The behaviour of a propped retaining wall: results of a numerical experiment. *Géotechnique* **34**, No.3, 383-404.

Rampello S. & Callisto L. (1998). A study on the subsoil of the Tower of Pisa based on results from standard and high-quality samples. *Canadian Geotechnical Journal* **35**, No. 6, 1074-1092.

Smith P.R., Jardine R.J. & Hight D.W. (1992). The yielding of Bothkennar clay. *Géotechnique* **42**, N.2, 257-274.

Tedd P., Chard B. M., Charles J. A., Symons I. F. (1984). Behaviour of a propped embedded retaining wall in stiff clay at Bell Common Tunnel, *Géotechnique* **34**, No. 4, 513-532.

Soil Stress-Strain Behavior: Measurement, Modeling and Analysis
Geotechnical Symposium in Roma, March 16 & 17, 2006

MECHANICAL BEHAVIOR OF FLORENCE CLAY AT THE HIGH-SPEED TRAIN STATION

Angelina Parlato*, Anna d'Onofrio*, Augusto Penna*, Filippo Santucci de Magistris**
* Department of Geotechnical Engineering - University of Naples Federico II, Naples, Italy
email: aparlato@unina.it
** S.A.V.A. Department – University of Molise, Campobasso, Italy

ABSTRACT

To develop the design of the Florence high-speed train station, Arup's office in Milan committed to the geotechnical laboratory of the University of Naples the execution of an extensive experimental program.

Special triaxial and RCTS (resonant column-torsional shear) tests were performed using deep undisturbed samples of Florence clay. To reduce initial damage to specimens, the coil wire suspension method, proposed by the Imperial College, was adopted. Moreover, two tests were performed on specimens set up with the more traditional wet setting and dry setting method. The best results were obtained through the dry setting method.

Using both RCTS and TXJ it was possible to analyze the small and medium strain behavior of specimens, while the strength was studied under triaxial condition only. Some triaxial extension tests were also performed. Data obtained from THOR and TXJ at small and at medium strain were compared in terms of undrained shear modulus. A fairly good agreement has been found between the experimental results.

1. INTRODUCTION

The solution of boundary value problems often requires the employment of advanced laboratory tests, in order to properly characterize the soil formations and to carefully define the soil parameters for the constitutive models. This was the case of the design of some part of the TAV station in Florence, Italy. TAV stands for Treni ad Alta Velocità (High-Speed Trains in English) that is a new railway system that will connect, with fast trains, some parts of Italy. Arup's office in Milan, working in joint venture with Foster & Partners who will be acting as architect on the project, has been awarded the design contract for the new railway station in Florence. The total area of the site is approximately 45000 sq m. The platform level in the new station is located 25 meters below the ground. The station chamber consists of a single volume, 454 meters long and 52 meters wide, built using cut-and-cover techniques. The composition is capped by an arching glazed roof, which evokes the great railway structures of the nineteenth century.

To develop this project, the Department of Geotechnical Engineering, University of Naples was entrusted to carry on extensive experimental program using deep undisturbed samples of Florence clay, following some indication and procedures specified by the contractor, as described below. Apart from some oedometer tests, a set of special triaxial and resonant column-torsional shear (RCTS) tests were performed. With both the experimental apparatuses, it was possible to analyze the small and the medium strain behavior of specimens, while the strength was studied under triaxial condition only. Some triaxial extension tests were also performed.

Hoe I. Ling et al. (eds.), Soil Stress-Strain Behavior: Measurement, Modeling and Analysis, 311–321.

In the first part of the paper, the attention is focused on the description of the experimental set-up procedures that were adopted to reduce the damage of the specimens at the beginning of the test. Laboratory specimens might, in fact, swell and destructurate when put in contact with the pore water system of the testing apparatus. In an attempt to reduce this problem, the coil wire suspension method, proposed by the Imperial College, is adopted and compared with the more traditional wet and dry setting methods.

All the experimental results are shown in the second part of the paper, where the data obtained with the torsional shear and the triaxial devices are compared.

2. PHYSICAL PROPERTIES OF FLORENCE CLAY

The employed material is a homogenous clay in which some weathered limestone fragments are present. The material was extracted from a single vertical using a piston sampler at depths spanning from 19 to 37 m below the ground level. Seven samples were sent to the laboratory, six of them were used for physical and mechanical tests. The ground water level was at approximately 7 m below the ground level.

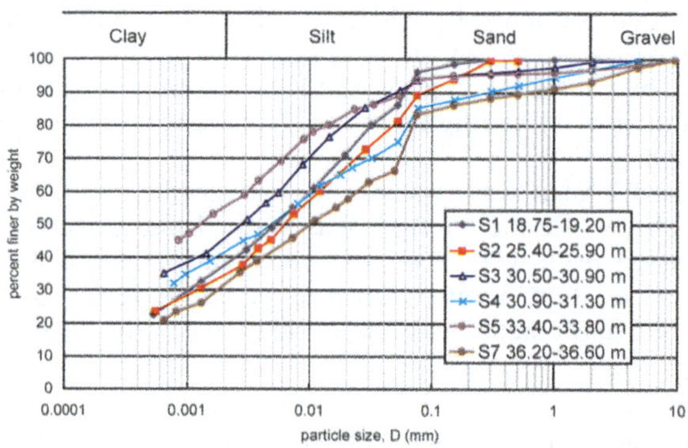

Figure1. Grading curves of the tested materials.

Figure1 shows the grading curves of the material, detected by sieving and sedimentation. The material can be classified as clayey silt. Results are relatively dispersed, with the maximum particle diameter spanning from 0.3 to 10 mm, due to the possible presence of lithoid fragment.

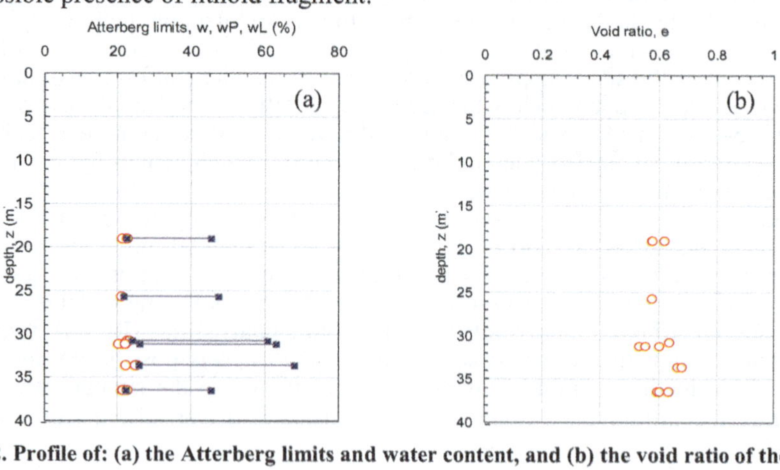

Figure2. Profile of: (a) the Atterberg limits and water content, and (b) the void ratio of the tested samples.

In Figure2 (a) the profile of Atterberg limits is reported, together with measurements of water content obtained on and near the tested specimen. While plastic limit shows a limited variation around the average value of 23.6%, liquid limit seems to increase between 30 and 35 m below the ground level, with an average value on the whole vertical of 54.8%. Soil water content is similar or a little below the plastic limit, and therefore the consistency index is slightly above the unit. All specimens have a saturation degree close to the unit and, being the soil particle density Gs = 2.72, Figure2 (b) shows the void ratio profile. It can be seen that void ratio is almost constant with depth and it averages at around 0.60.

3. EMPLOYED APPARATUSES

Three apparatuses were employed to detect the mechanical properties of Florence clay: a high-pressure oedometer, a triaxial apparatus and a resonant column-torsional shear device (Figure3).

Oedometer tests were executed using a high-pressure device designed at the University of Naples. The machine can accommodate specimens having diameter equal to 56 mm and height equal to 20 mm. Drainage is allowed on both the top and bottom base of the specimen. Vertical loads are applied by dead weights that are transmitted to the specimen by means of a loading frame. The loading frame was designed to multiply dead loads by a factor of 41. The maximum allowable σ'_v on the soil specimen is equal to 20 MPa. Vertical deformation of the specimen is manually monitored by a dial gauge, which measures the vertical movement of the loading frame.

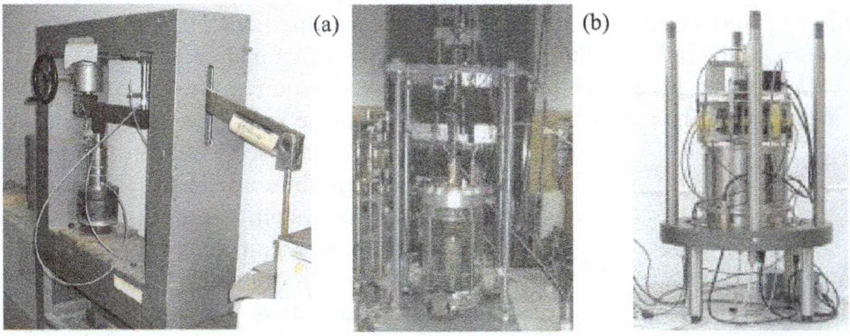

Figure3. (a) The dead loads multiplier device designed to perform high-pressure oedometer tests, (b) the Triaxial cell and (c) the RCTS machine.

The apparatus employed for the triaxial tests has been originally designed and developed at the University of Tokyo, Japan, and is described in detail by Tatsuoka (1988), Tatsuoka et al. (1994) and Santucci de Magistris et al. (1999). The triaxial chamber can accommodate 50 millimeters diameter cylindrical specimens with top and bottom drainage lines. The axial strains are measured locally by a pair of LDTs set at the opposite ends of the lateral surface of the specimen (Goto et al., 1991). A high capacity load cell set inside the triaxial chamber between the loading cap and the loading ram measures the axial load. Vertical loads are applied on specimens through an AC motor equipped with a digital control system that is connected to the axial piston.

Some triaxial tests were also performed using a conventional hydraulic stress-path apparatus (Bishop and Wensly, 1975) modified at the University of Naples (Aversa and Vinale, 1996).

The torsional shear tests carried out under monotonic and dynamic conditions were performed using a single apparatus developed at the University of Naples and described in d'Onofrio et al. (1999).

The torque is applied by a powerful electro-magnetic motor (drive plate) at the free top of the specimen, and is directly proportional to the applied current intensity. The drive plate supply system was designed controlling electric current rather than voltage, thus avoiding errors on damping measurements, which may be induced by internal equipment compliance related to the "back electro-magnetic force".

The rotation θ is measured by two couples of gap sensors, allowing accurate readings of shear strains spanning from γ=5x10^{-5}% to 0.5 % on a specimen of 36 mm in diameter. A LVDT and a double-bellofram volume gauge respectively measure axial and volumetric deformations, while a miniaturized pressure transducer allows reliable pore pressure measurements.

4. TESTING PROCEDURES

Conventional ASTM D 2435-03 standard test methods were adopted to execute the oedometer tests. However, the first incremental load was set such as to approximately reproduce the in situ overburden pressure and then, following a variable increment ratio, the vertical stress was brought up to a value of 15,36 MPa.

Particular attention should be paid to the experimental procedures adopted in the triaxial and torsional tests. Most of the triaxial tests, and all the torsional shear tests, were executed with the experimental procedures that are described hereafter. Two specimens were tested with more conventional methods: the wet setting method (Cavallaro et al., 2002) and the dry setting method (Ampadu e Tatsuoka, 1993).

The experimental procedure prevalently adopted requires the use of dry filter paper (the lateral paper has to be 2-3 mm shorter than the specimen height), and a kind of spacer, which is interposed between the porous stone and the specimen bottom base. The specimen is suspended over the system in order to avoid the contact with water before the confining stress application: in fact an early contact could make the clay swelling, thus leading to destructuration. This procedure was chosen by the Arup company on the basis of the suggestions reported by Burland and Maswoswe (1982). The latter Authors use two semi-circles of fuse wire between specimen and porous stones, and they claim that the air trapped in the drainage line would be dissolved by backpressurizzation. They also said that the fuse wire penetrate into the specimen during the application of the confining stress, when the soil entered in contact with the drainage water. Figure4 shows a picture of the system adopted in Naples that was made by three small pieces of lead wire originally used for welding, placed at 120° and joined in the centre. The wires had a 0.97 mm diameter, which was slightly larger than the diameter of 0.864 mm of the coil wires originally adopted by Burland and Maswoswe.1982).

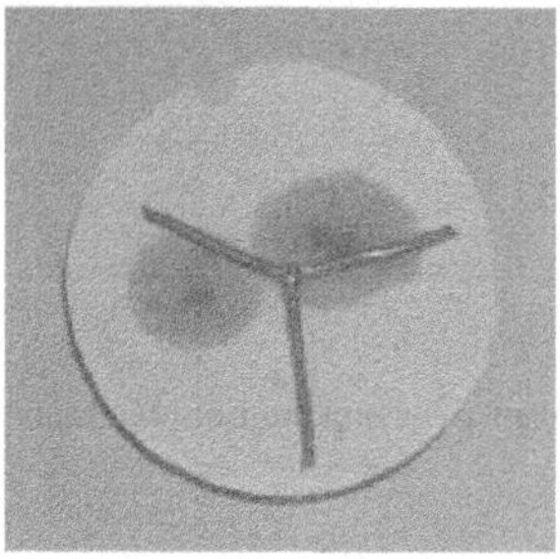

Figure4. Spacer used in the experiments on Florence clay.

Then, after that all the transducers have been put in place, and the cell has been closed, a single cell pressure increment (Δp) equivalent to the estimated in-situ confining stress p_0' is isotropically applied with the drainage closed. When the cell pressure is applied, the suspension system is pressed into the specimen and the soil enters in contact with the water of the drainage system only under the assigned pressure. At this stage, measuring the pore pressure, the mean effective stress (p_1') acting over the specimen is determined; it was always checked that $p_1' < p_0'$. Then cell pressure and backpressure are gradually stepped up together to provide a backpressure of +200 kPa, while keeping $p' = p_1'$. The Skempton B-value measured at this stage turns out to be always lesser than 1, probably because of the air trapped between the specimen and the porous stone during the assembling. This point will be discussed later in the paper.
At the end of B test the isotropic compression is carried out, increasing the effective confining stress from the initial p_1' up to the p_0', that is the value of the mean effective stress estimated in situ, assuming K_0=0.8. A stress rate equal to 10 kPa/hr was used for the triaxial while a stress rate of 5 kPa/hr was adopted for the torsional tests. At the end of the compression stage the mean effective stress is kept constant until the measured axial strain rate is less than 1% of the subsequent shearing., adopting again a procedure that is currently employed at the Imperial College.
During the isotropic compression stage in the RCTS tests measurements of small strain stiffness, G_0, are carried out by performing RC tests at very small strain level. E_{max}, is also measured during the isotropic compression stage in the triaxial device, through small strain cyclic tests. Then, if a triaxial test has to be executed, the drainage circuit is kept closed, and compression and extension tests are carried out, keeping the cell pressure constant. At first (ε_a<1%), to allow for a homogeneous distribution of pore pressure, the axial strain rate is kept equal to ε_a=0.1%/hr; then (ε_a>1%), to speed up the test, the axial strain rate is increased up to ε_a=1%/hr.
Instead, if a torsional shear test has to be performed, with sample drainage closed a steady continuous monotonic path of torsional shear is applied to the sample at constant cell pressure. The torsional system at the University of Naples allows applying torque load under stress controlled condition only. To make the torsional and the triaxial tests comparable, in the former the shear strain rate applied is such that the rate matched the rate given for the triaxial tests over the equivalent range of strain.

5. EXPERIMENTAL RESULTS

Oedometer test results

The data of all the oedometer tests are reported in Figure5. Three tests were executed on intact specimens (sample 01, 04, 07) and one test was carried out on reconstituted soil. The latter was obtained pouring a slurry of disaggregated soil and water directly into the oedometer device. All the tests carried out on intact soil approximately start from the same void ratio and have a similar behavior up to the yielding point, which can be easily seen from data and which is highlighted in Figure5 by an arrow. The yielding stress ranges between 0.6 and 1.2 MPa . and the maximum value has been measured on the shallowest sample. The clay is then rather overconsolidated, with an overconsolidation ratio OCR varying approximately from 1.5 to 4.5.
After yielding, specimens taken from sample 04 and 07 show a similar behavior in terms of compressibility and porosity, while the specimen taken from sample 01 is more compressible than the others. The different compressibility might be attributed to the absence of any sand fraction in the grading curve of the shallow sample.

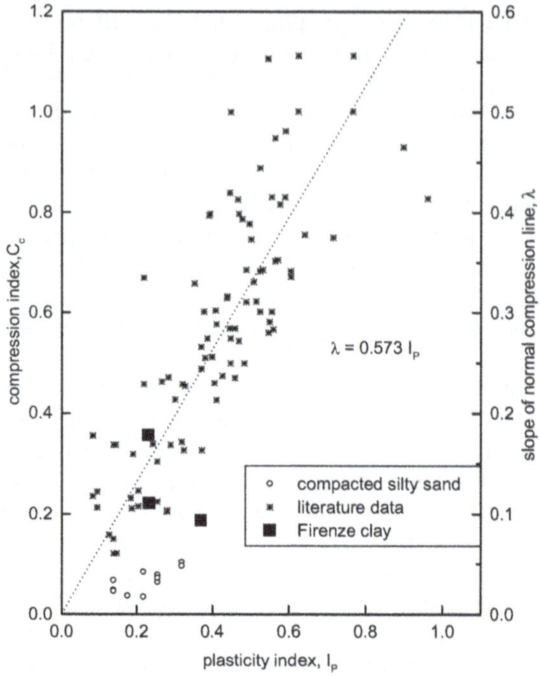

Figure6. Compression indexes of Florence clay versus the plasticity index (modified after Wood, 1990)

No yielding point is shown for the reconstituted specimen. However, its compressibility is comparable to that of the specimen 04 and 07, and its unloading behavior is similar to that of specimen 01.

From an analytic viewpoint, the compressibility index Cc and the swelling index Cs calculated over sample 04, sample 07 and over the reconstituted sample fall in a narrow range, whose mean value are Cc=0.2189 and Cs=0.0236.

Overall it can be noticed that the compression index is about ten times larger than the swelling index, and this is in agreement with most of literature data on clays. For the natural samples of Florence clay the compression index versus the plasticity index is plotted in Figure6, together with literature data on clays and on compacted silty sand.

It might be noticed that the compressibility of Florence clay is in agreement with the literature information, but the scatter in plasticity index between the different specimens cannot justify the scatter in the compression index.

As underlined before, the compressibility of natural specimens are close to that of the remoulded one, while the yielding point of the intact samples always lies on the right of the reconstituted one-dimensional compression line thus indicating a non negligible soil structure effect . Expressing the yield stress as a ratio of the corresponding stress on the reconstituted compression line (i.e., at the same void ratio) it could be possible to measure the strength of the soil structure in excess of that possessed by

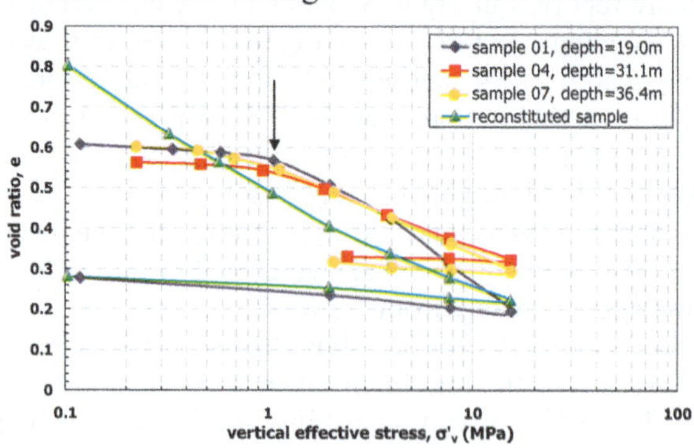

	Cc	Cs
sample 01	0.3569	0.0403
sampe 04	0.1867	0.0177
sample 07	0.2214	0.0255
rec sample	0.2485	0.0277
mean value	0.2189	0.0236

Figure5. Oedometer tests on intact and reconstituted samples.

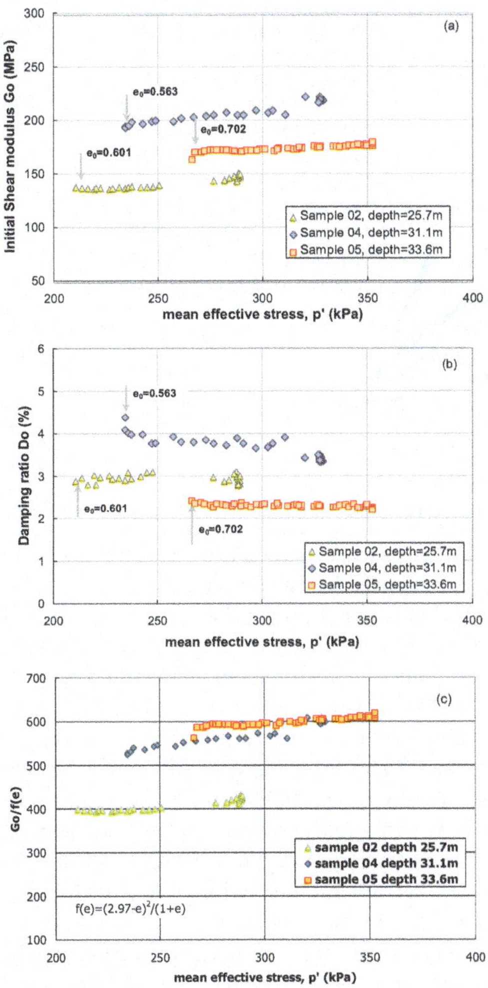

Figure7. Resonant column results: a) initial shear modulus vs; b) initial damping ratio; c) normalised shear modulus plotted against the mean effective stress

reconstituted soil. The ratio computed for the three intact samples ranges between 1.2 and 1.5, showing a quite significant structure effect.

Small strain data

As indicated before, all the specimens undergo an isotropic compression stage, in which the effective pressure p' is brought from its initial value p'_I up to the estimated in situ value p'_0. The values of the initial shear modulus and damping ratio measured during the compression stage of the RCTS tests are plotted in Figure7. The G_0 and D_0 values are hardly affected by the confining stress, even though there is a little increase in G_0 with p', probably because of the small variation in volume during compression and the overconsolidation ratio of the soil tested. The three samples show quite different initial stiffness values, and this could be partly due to the different initial void ratio. The shear modulus data have been then normalized by adopting a proper void ratio function and the resulting values are plotted in Figure7c. Again the shallow sample shows a more deformable behavior if compared to that exhibited by the sample 4 and 5, which appear quite similar. As for D_0, whose trend is almost horizontal, it comes out that the densest specimen is also the one with the highest damping ratio.

Overall, the difference in the initial stiffness and damping ratio between the three specimens cannot be justified looking at the differences between specimens in physical properties only.

Pre-peak behavior

The undrained Young's modulus measured during the triaxial compression and extension tests are shown in Figure 8 together with the stiffness decay curves obtained during the monotonic torsional shear tests. To convert the shear stiffness from monotonic torsional tests into Young's modulus, the hypothesis of an isotropic homogeneous medium, for which E_u=3G is made. Also, shear strain was converted

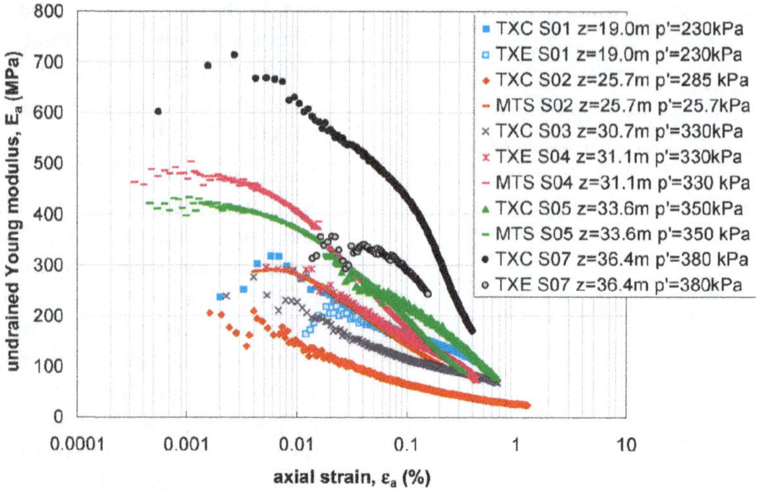

Figure8. Undrained Young modulus measured by triaxial and monotonic torsional shear tests.

into axial strain using $\varepsilon = 2\gamma/3$. The raw data from triaxial tests are scattered up to an axial strain of 0.01% even though they were measured by LDT transducers.
All the decay curves obtained from triaxial tests seems to have the same trend except for the two measured on the deepest specimen, which shown a quit unusual shape both in compression and extension. The experimental results from MTS and TX tests, compared starting from an axial strain of 0.01%, follow the same trend even though the MTS stiffness decay curves always lie above that measured by triaxial test on the same sample, at the same confining pressure. The considerable difference in the stiffness measured by TX and MTS tests could be partially ascribed to the non-isotropic behavior of the tested soil. It is worth highlight that the triaxial test on sample S02 and S05 have been carried out by adopting a different setting up procedure: S02 has been saturated by a wet setting method while for S05 the dry setting method was chosen. MTS tests instead were always performed by using the spacer described in the previous paragraph. The different setting up procedure could induce a various degree of sample disturbance that in turn largely affects the initial stiffness value.

Behavior after peak stress
In Figure9 is represented a compound of the results obtained in the triaxial apparatus. The drawn data have been captured through the external axial strain transducer, in order to show also the results obtained in the Bishop and Wesley cell that does not have any internal transducer.
As mentioned before, following the indication from Arup, the shear test is performed at two different strain rates: when the axial strain reached the value ε_a=1%, the axial strain rate is increased by a factor of 10. When strain rate suddenly changes, there is a clear stress jump, but later the stress strain curve seems to rejoin the original one: this behavior could remind the TESRA model (Tatsuoka et al. 2002), originally conceived to describe the behavior of sands. However, it is significant that the peak stress has already been reached: in this field, data can be affected by the presence of shear bands. Instead, the behavior of sample 02, which was prepared with the wet setting method, looks somehow different. When strain rate is changed at ε_a>1%, stress-strain

curve has not reached its peak yet, and then it seems that the stress jump is permanent, like in the ISOTACH model (Tatsuoka et al. 2002).

Analyzing the mechanical parameters of the material, it was found that cohesion is equal to c'=140 kPa both in compression and in extension, while friction angle was ϕ'=27° in compression, and ϕ'=14° in extension. These values were estimated by the method of least squares, by interpolating the coordinates representative of the critical state in terms of Mohr circles (centre and radius).

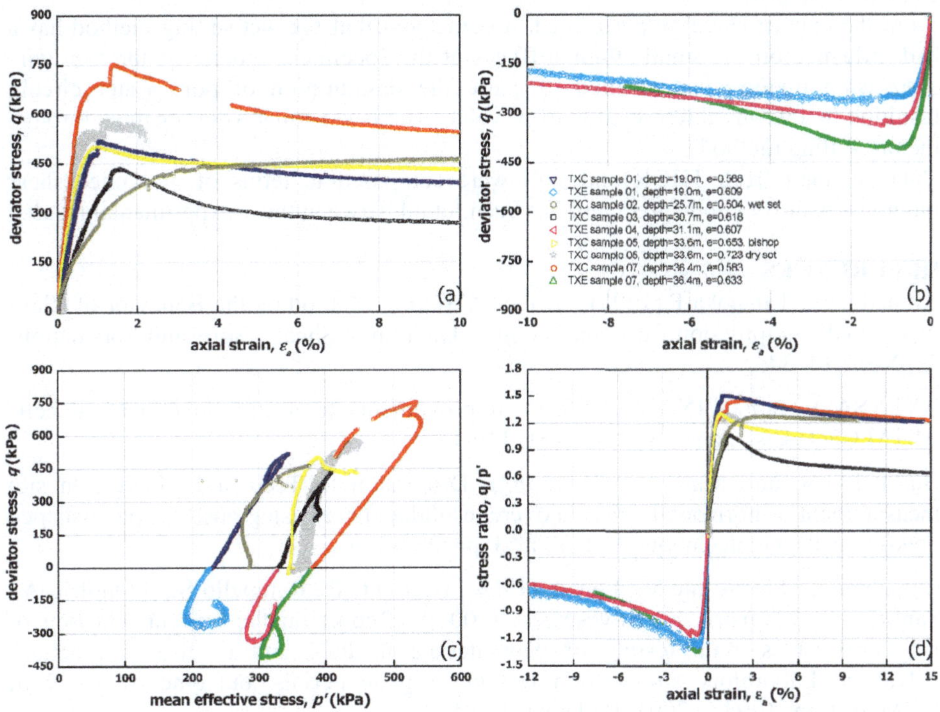

Figure9. Stress-strain behavior of the material: (a) TX compression tests, (b) TX extension tests, (c) q-p' behavior and (d) q/p' versus ε_a.

Comparison between different testing procedures

In Figure9 the stress-strain curves obtained by adopting different set-up procedures are reported. The specimen set up with the wet setting method (dark yellow curve) has a behavior less stiffer than the specimen set up with the dry setting and the coil wire method; this behavior is still clearer in Figure8 (red diamonds), where the specimen stiffness is directly displayed. Moreover, the specimen set up with the dry setting shows stiffness comparable with that of the specimens set up with the coil wire method. According to the results obtained through the tests performed, the usefulness of adopting an experimental procedure that reduces the risk of destructuration sounds clear.

However, also if in the evaluation of stiffness the thin wire procedure and the dry setting procedure are similar, in terms of q-p' behavior the coil wire method shows its limits. In fact, when using the coil wire method, it was not possible to grant the perfect saturation of the pore water line, thus affecting the reading of the effective pressure acting on the specimen. The dry setting method, instead, gives no problem with desaturation.

6. CONCLUSIONS

This experimental note is a further confirmation that the stress-strain behavior of a specimen is strongly influenced by the experimental procedures adopted during the re-consolidation phase.

From the experimentation performed, it comes out that the wet setting method has a bad influence on the small strain stiffness of the specimen. Moreover, the thin wire suspension method has resulted to cause the desaturation of pore water circuit, resulting in a wrong interpretation of p' value. The best results were obtained through the dry setting method.

Data obtained from THOR and TXJ were compared in terms of undrained shear modulus. A fairly good agreement has been found between these experimental results.

REFERENCES

Ampadu SK, Tatsuoka F.(1993): "Effect of Setting Method on the Behavior of Clays in Triaxial Compression from Saturation to Undrained Shear," Soils and Foundations 33, No. 2,14–34.

AVERSA S. e F. VINALE (1995): "Improvements to a stress-path triaxial cell" ASTM Geot. Test. Journ. Vol. 18 No1, 116-120

Burland, J.B. and Maswoswe J. (1982): Discussion to Tedd and Charles, In situ measurement of horizontal stress in overconsolidated clay using push-in spade-shaped pressure cells, Geotechnique Vol. 32 n° 3 pp. 285-286.

Cavallaro A., Fioravante V., Lanzo, G. Lo Presti D.C.F., Rampello S., d'Onofrio A., Santucci de Magistris F., Silvestri, F. (2002): "Report on the current situation of laboratori stress-strain testing of geomaterials in Italy and its use in practice" Advanced Laboratory stress-strain testing of geomaterials, outcome of TC29 of ISSMGE from 1994 to 2001, Balkema, 15-45.

D'Onofrio A., Silvestri F., Vinale F.(2000): "A new torsional shear device" ASTM Geotechnical Testing Journal 22, No.2, 107-117

Goto, S., Tatsuoka, F., Shibuya, S., Kim, Y.S. and Sato T. (1991): A simple gauge for local small strain measurement in the laboratory, Soils and Foundations, 31(1), 169-180.

Parlato, A. e Santucci de Magistris, F. (2003): "Confronto tra una cella triassiale a stress path controllato ed una cella "TXJ"", IARG 2003.

Burland, J. B. (1990):"On the compressibility and shear strength of natural clays". Géotechnique 40, No. 3, 329-378.

Hight, D. W. & Georgiannou, V. N. (1995): "Effects of sampling on the undrained behavior of clayey sands" Géotechnique 45, No. 2, 237-247.

Santucci de Magistris, F., Koseki, J., Amaya, M., Hamaya, S., Sato, T. and Tatsuoka F. (1999): A triaxial testing system to evaluate stress-strain behavior of soils for wide range of strain and strain rate, Geotechnical Testing Journal, ASTM, 22(1), 44-60.

Skempton, A. W. e V. A. Sowa, V. A (1963): "The behavior of saturated clays during sampling and testing". Géotechnique 13, No 4,.269-290.

Tatsuoka, F. (1988): "Some Recent Developments in Triaxial Testing System for Cohesionless Soils," Advanced Triaxial Testing of Soil and Rock, STP 977, R.T. Donaghe, R.C. Chaney and M.L., Silver Eds., American Society for Testing and Materials, West Conshohocken, PA, pp. 7-67.

Tatsuoka, F., Sato, T., Park,C.-S., Kim, Y.-S., Mukabi, J.N., and Kohata, Y., (1994): "Measurements of Elastic Properties of Geomaterials in Laboratory Compression Tests," Geotechnical Testing Journal, Vol. 17, No. 1, pp. 80-94.

Tatsuoka,F., Ishihara,M., Di Benedetto,H. and Kuwano,R. (2002): Time-dependent shear deformation characteristics of geomaterials and their simulation, Soils and Foundations, 42-2, pp.103-129.

Soil Stress-Strain Behavior: Measurement, Modeling and Analysis
Geotechnical Symposium in Roma, March 16 & 17, 2006

STIFFNESS OF NATURAL AND RECONSTITUTED AUGUSTA CLAY AT SMALL TO MEDIUM STRAINS

Giuseppe Lanzo, Alessandro Pagliaroli
Dipartimento di Ingegneria Strutturale e Geotecnica
Università di Roma "La Sapienza", Rome, Italy
e-mail: giuseppe.lanzo@uniroma1.it, alessandro.pagliaroli@uniroma1.it

ABSTRACT

Cyclic simple shear tests were conducted on undisturbed and reconstituted specimens of high-plasticity Augusta clay. Test results were examined in terms of stiffness characteristics either in the small and medium strain range. It has been found that the small-strain shear modulus of natural soil is always higher than that of the reconstituted material. A normalization procedure was proved useful in interpreting the small-strain test results. At larger strain levels, the normalized shear modulus of natural soil showed a steeper decrease with shear strain amplitude with respect to that detected on the reconstituted material, thus evidencing a more marked non-linear behavior. The influence of soil structure was invoked to explain the observed differences.

1. INTRODUCTION

In the last two decades a wealth of experimental works have compared the mechanical properties of several natural clays with those pertaining to the reconstituted material (e.g. Leroueil & Vaughan 1990, Burland 1990, Burland et al. 1996). In these studies the behavior of the reconstituted material, taken as reference, has proved useful in interpreting the results for natural soil. In fact, the observed differences in strength and compressibility characteristics allowed a better understanding of the effect of microstructure, which implies the combined contribution of fabric and inter-particles bonding (Mitchell 1976).

The soil structure also significantly affects the stiffness properties at different shear strain levels. For instance, Shibuya (2000) and Li et al. (2003) recently attempted to quantitatively assess the influence of soil structure by taking into account the characteristic behavior of small-strain shear modulus using various natural and reconstituted soft clays. Variations of stiffness characteristics either in the small and in the medium strain range were investigated for natural and reconstituted stiff clays by Rampello & Silvestri (1993), Rampello et al. (1994), d'Onofrio et al. (1998) and Rampello et al. (2003), among others.

In this paper the influence of soil structure is investigated by comparing the stiffness characteristics of natural stiff clay with those of the reconstituted soil as observed in cyclic simple shear tests under small and medium shear strain levels.

2. SOIL TESTED, EXPERIMENTAL APPARATUS AND TEST PROGRAMME

The clay used for the experimental investigation is named Augusta clay. This is a

Hoe I. Ling et al. (eds.), Soil Stress-Strain Behavior: Measurement, Modeling and Analysis, 323–331.

medium stiff, overconsolidated high plasticity clay of marine origin and Pleistocene age, coming from a quite homogeneous soil deposit located close to the town of Augusta in south-eastern Sicily, Italy (Maugeri & Frenna 1995, Lo Presti et al. 1998).

Because of its homogeneity, Augusta clay has been selected in 1998 as test material for a round robin investigation involving different universities and research teams worldwide to evaluate the scatter in deformation characteristics obtained by different types of laboratory tests (Cavallaro et al. 2001).

The undisturbed samples used in this study were retrieved at the depth of about 25 m. In Table 1, relevant physical properties of Augusta clay are listed. The preconsolidation pressure, σ'_p, determined from oedometer test using Casagrande's method, is around 1100 kPa. The overconsolidation ratio $OCR=\sigma'_p/\sigma'_{v0}=4.8$, where σ'_{v0} is the vertical effective overburden pressure.

Table 1. Basic properties of natural Augusta clay.

G_s	w_o (%)	e_0	S_{r0} (%)	w_L	PI	LI	CF (%)
2.74	34.1	0.977	97	74.5	43.7	0.08	60

The reconstituted samples were prepared from a slurry obtained by mixing oven-dried soil, powdered to pass a 0.42 mm sieve, with distilled water at a water content of about 1.5 w_L. The slurry was then subjected to one-dimensional compression at a vertical effective stress $\sigma'_v = 300$ kPa.

The experimental investigation was carried out by the double specimen direct simple shear (DSDSS) device. A complete description of this apparatus along with its capabilities and limitations can be found in Doroudian & Vucetic (1995). It is worth mentioning that the device allows to investigate, in a single test, the cyclic properties of soils in a very wide range of strains, from about 0.0004% to in excess of 1% (Lanzo et al. 1997, Lanzo et al. 1999). A DSDSS device was recently constructed at the University of Rome *La Sapienza*, with minor modifications with respect to the prototype (D'Elia et al. 2003).

Tests were conducted on saturated specimens, 66 mm in diameter and 20 mm high, following the constant-volume equivalent-undrained simple shear testing procedure (Bjerrum & Landva 1966, Dyvik et al. 1987).

The testing program is summarized for the natural and reconstituted samples in Table 2. Tests on natural samples were carried out at the University of California at Los Angeles as part of the round robin test project (Cavallaro et al. 2001). Tests on reconstituted samples were conducted at the University of Rome *La Sapienza*. Approximately the same vertical loading sequence, composed of loading-unloading-reloading stages, was followed for both natural and reconstituted samples. The vertical effective stress σ'_v was varied between 250 and 1670 kPa, that is larger than the maximum stress experienced by the natural clay during its geological history. At the completion of primary consolidation under constant σ'_v, the specimens were subjected to several steps of cyclic strain-controlled tests gradually increasing the magnitude of the cyclic shear strain amplitude, γ_c. A wide range of shear strains was covered, between 0.0004% and 0.1% approximately. In Table 2 the void ratio at the end of primary consolidation (*e*), the frequency of cyclic loading (f) and the OCR values are also reported.

Table 2. Summary of cyclic testing conditions on natural (N) and reconstituted (R) Augusta clay. (LD = Loading, ULD = Unloading, RLD = Reloading).

Test #	N1	N2	N3	N4	N5	N6	N7	N8	N9	N10	N11
Load sequence	LD	LD	LD	LD	LD	ULD	ULD	ULD	ULD	RLD	RLD
σ'_v (kPa)	249	418	857	1254	1672	1254	857	418	249	857	1672
e	0.929	0.906	0.85	0.813	0.772	0.776	0.785	0.809	0.835	0.797	0.758
OCR	4.4	2.6	1.3	1	1	1.3	1.95	4	6.7	1.95	1
γ_c (%)	0.0004-0.034	0.0004-0.012	0.0004-0.037	0.0004-0.12	0.0004-0.039	0.0004-0.029	0.0004-0.012	0.0004-0.12	0.0004-0.051	0.0004-0.053	0.0004-0.013
f (Hz)	0.03-0.05	0.01-0.038	0.014-0.034	0.01-0.02	0.02-0.035	0.013-0.023	0.02-0.035	0.023-0.05	0.008-0.83	0.005-1.28	0.02-1.0
Test #	R1	R2	R3	R4	R5	R6	R7	R8	R9	R10	R11
Load sequence	LD	LD	LD	LD	LD	ULD	ULD	ULD	ULD	RLD	RLD
σ'_v (kPa)	250	400	800	1250	1670	1250	800	400	250	800	1670
e	0.870	0.813	0.678	0.618	0.551	0.556	0.571	0.600	0.625	0.593	0.557
OCR	1.2	1	1	1	1	1.3	2.1	4.2	6.7	2.1	1
γ_c (%)	0.0003-0.082	0.0003-0.097	0.0004-0.011	0.0003-0.11	0.0003-0.10	0.0004-0.10	0.0004-0.12	0.0004-0.12	0.0004-0.11	0.0004-0.12	0.0004-0.10
f (Hz)	0.09-0.23	0.10-0.23	0.12-0.26	0.08-0.19	0.08-0.23	0.11-0.26	0.10-0.27	0.10-0.26	0.09-030	0.10-0.24	0.08-0.21

In Fig. 1 the *e*-log σ'_v data referred to the consolidation stages in the DSDSS device for the natural and reconstituted clay are plotted by closed and open symbols, respectively. At a given stress level σ'_v, the curve for the natural clay plots on the right with respect to that of the reconstituted clay. The intrinsic compression (C_c^*) and swelling (C_s^*) indices, which describe the compressibility of the reconstituted clay for the normally consolidated and the unloading-reloading lines (drawn as dashed line in Fig. 1), are respectively equal to 0.42 and 0.10. For the natural clay the corresponding C_c and C_s values were determined from both the compressibility curves referred to the DSDSS consolidation stages and to an oedometer test (not displayed in Fig. 1); it results C_c =0.36 and C_s =0.09.

The void ratio data in Fig. 1 have been normalized in terms of void index I_v (Burland 1990) and the normalized compression curves I_v-σ'_v are shown in Fig. 2. In the same plot the normalized oedometer compression curve on natural soil is also reported for comparison. Further, in Fig. 2 the relationships I_v-σ'_v obtained by Burland (1990) for natural and reconstituted soils, defined as the sedimentation compression line (SCL) and the intrinsic compression line (ICL) respectively, were also plotted. It can be seen that the SCL and ICL lines of Augusta clay plot very close to the corresponding lines obtained by Burland. The relative position of these two curves indicates therefore that the effect of the clay structure is very similar to that displayed by many others natural clays.

3. RESULTS AT SMALL STRAINS

The small-strain stiffness G_0 is known to be mainly affected by mean effective consolidation stress σ'_m, overconsolidation ratio OCR and void ratio *e*. The dependency of G_0 on the above mentioned parameters have been usually expressed by relationships in the form:

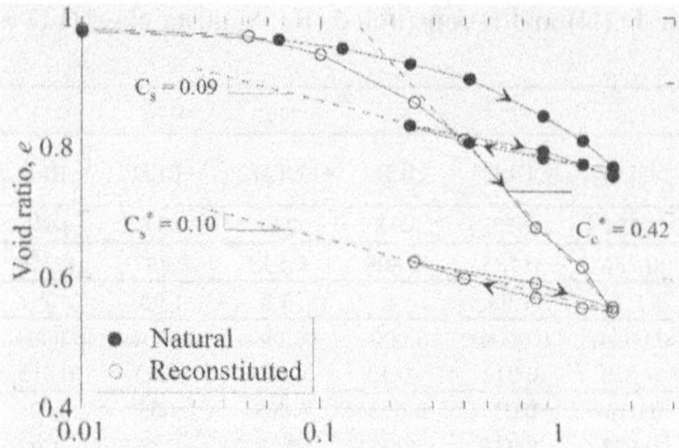

Figure 1. One-dimensional compression curves in DSDSS tests.

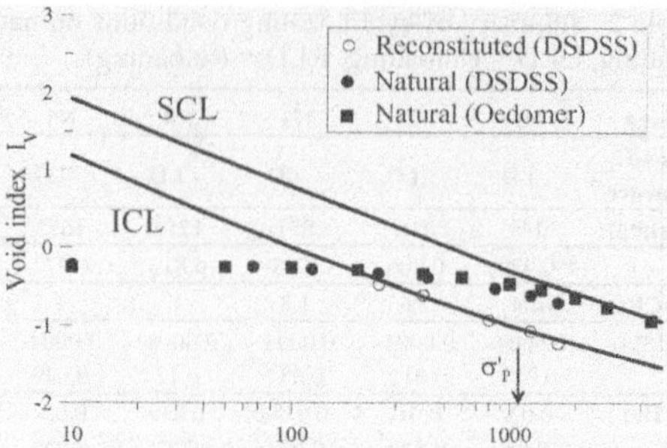

Figure 2. Normalized one-dimensional compression curves.

$$G_0 = S\,F(e)\,\sigma'^{\,n}_m\,\sigma_r^{1-n}\,OCR^k \tag{1}$$

where F(*e*) is a void ratio function, σ_r is a reference stress and S, n, and k are adimensional stiffness parameters accounting for the nature of soil. In particular S represents the maximum shear modulus for normally consolidated state at the reference stress σ_r; the exponents n and k express the dependence of G_0 on stress level, for normally and overconsolidated conditions respectively. The main difference among the various relationships depends on the choice of the void ratio function (e.g. Hardin, 1978).

Equation (1) does not take into account the compressibility relationship existing between void ratio and effective stress state, which implies that one of the three variables *e*, OCR and σ'_m is unnecessary.

Based on this consideration, Viggiani (1992) and Rampello et al. (1994) proposed an alternative procedure in order to consider the dependency of G_0 on state and stress history. The Authors, for isotropic condition of confinement, expressed G_0 only as a function of the mean effective stress p' and the isotropic overconsolidation ratio R, according to the following equation:

$$G_0 = S'\,p'^{\,n'}\,p_r^{1-n'}\,R^m \tag{2}$$

where the constants S', n' and m have similar physical meanings of S, n and k but different values. The contribution of void ratio in Equation (2) is indirectly considered since *e* is uniquely linked to p' for normally consolidated states and accounted for by p' and R for overconsolidated states. Equation (2) can also be written:

$$G_0/p_r = S'\,(p'/p_r)^{n'}(p'_e/p')^c \tag{3}$$

where p'_e is the equivalent pressure, $c=m/\Lambda$ where $\Lambda=(\lambda - k)/\lambda$, λ and k being the slopes of the normal compression and swelling lines in the ln *p'*-v plane, v being the specific volume.

A similar approach may be applied to describe the effect of stress level and history on the small-strain stiffness of a soil compressed in oedometer conditions by using the following expression:

$$G_0 = S'\,\sigma'^{\,n'}_m\,\sigma_r^{1-n'}\,OCR^m\ . \tag{4}$$

It must be considered that:

$$OCR = \sigma'_p/\sigma'_v = (\sigma'_{ve}/\sigma'_v)^{\Lambda^{-1}} \tag{5}$$

where σ'_{ve} is the equivalent one-dimensional consolidation pressure (Fig. 3) defined as:

$$\sigma'_{ve} = \sigma_r\, 10^{(e_r - e)/C_c} \tag{6}$$

e_r being the reference void ratio at $\sigma'_v = \sigma_r = 1$ kPa (Fig. 3). In Equation (5) the quantity Λ, previously introduced for isotropic consolidation, can be used to a good approximation also for K_0 consolidation. So that, because $\lambda = 0.434C_c$ and $k \sim 0.434C_s$, it results:

$$\Lambda = (\lambda - k)/\lambda \cong (C_c - C_s)/C_c\,. \tag{7}$$

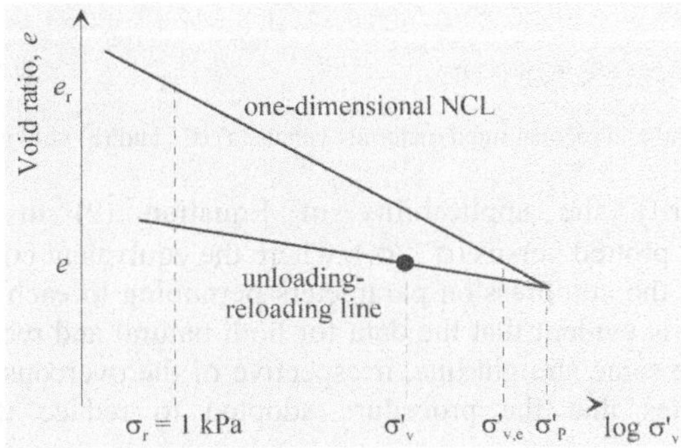

Figure 3. Equivalent one-dimensional consolidation pressure σ'_{ve}.

Taking into account Equation (5), it is possible to write Equation (4) as follows:

$$G_0 = S'\, \sigma'^{\,n'}_m\, \sigma_r^{1-n'}\, (\sigma'_{ve}/\sigma'_v)^c \tag{8}$$

or

$$\frac{G_0/\sigma_r}{(\sigma'_{ve}/\sigma'_v)^c} = S'\, (\sigma'_m/\sigma_r)^{n'} \tag{9}$$

where $c = m/\Lambda$. If the reconstituted soil is considered, the equivalent consolidation pressure, the reference void ratio and the stiffness parameter will be denoted by the notations σ'^*_{ve}, e_r^* and S'^* respectively.

The above relationship (9) was used to reduce the G_0 values estimated from the equivalent shear modulus, G_{eq} versus γ_c curves, by extrapolating at $\gamma_c = 0.0001\%$ the measured results (Lanzo et al. 1999). In Fig. 4 the G_0 values for both natural and reconstituted soils are plotted versus σ'_m and e in a log-log scale. The mean effective stress σ'_m was calculated as $(\sigma'_v + 2\sigma'_h)/3$, where $\sigma'_h = K_0\sigma'_v$ is the horizontal effective stress and K_0 is the coefficient of earth pressure at rest. The K_0 values for the various stages of the vertical loading-unloading-reloading sequence were obtained by means of the empirical relationships suggested by Mayne & Kulhavy (1982), assuming the effective stress friction angle φ' from empirical correlation. In Fig. 4a the G_0 data of both natural and reconstituted soils plot very close, especially at higher confining stresses.

Conversely, in Fig. 4b it is apparent that, at a given void ratio, G_0 data of natural soil are higher than those pertaining to the reconstituted one.

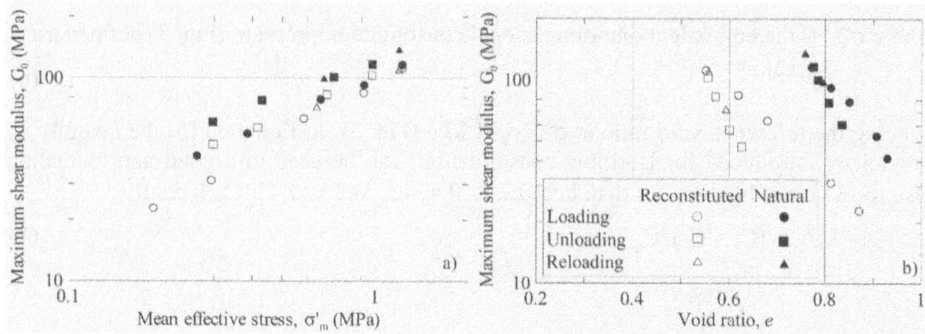

Figure 4. G_0 data of natural and reconstituted materials versus (a) σ'_m and (b) void ratio e.

In order to verify the applicability of Equation (9) in Fig. 5 the ratio $(G_0/\sigma_r)/(\sigma'_{ve}/(\sigma'_v)^c$ is plotted versus (σ'_m/σ_r), where the equivalent consolidation pressure was computed using the compression parameters pertaining to each set of data (natural and reconstituted). It is evident that the data for both natural and reconstituted soils plot approximately on the same straight line, irrespective of the overconsolidation ratio. This circumstance indicates that the procedure adopted to reduce the G_0 data yields satisfactory results.

In order to highlight the effect of structure, the reconstituted material was assumed as reference and the small-strain stiffness of natural soil was compared with that of the reconstituted one. Following a normalization procedure similar to that employed by Rampello & Viggiani (2001), both sides of Equation (9) were divided by the equivalent consolidation pressure of the reconstituted material σ'^*_{ve} raised to the power of c. After normalization the following equation was obtained:

$$\frac{(G_0/\sigma_r)}{(\sigma'^*_{ve}/\sigma'_v)^c} = S'_e\,(\sigma'_m/\sigma_r)^{n'} \qquad (10)$$

with the stiffness parameter S'_e being equal to:

$$S'_e = S'\left(\frac{\sigma'_{ve}}{\sigma'^*_{ve}}\right)^c = S'\,10^{\frac{c}{C_c^*}(e_r - e_r^*)} \qquad (11)$$

where was assumed $C_c = C_c^*$.

According to Equation (10), the normalized G_0 data are plotted versus the ratio σ'_m/σ_r in Fig. 6. A constant value of n' was assumed in elaborating the data, determined from multiple regression analysis of G_0 reconstituted soil data; therefore, for each set of data, two parallel lines were used for fitting, drawn as dashed lines in Fig. 6. After normalization, the data pertaining to the natural soil plot above those corresponding to the reconstituted one. The distance between these two lines highlights the greater stiffness of the natural soil due to its structure. The ratio S'_e/S'^*_e between the stiffness parameter of the natural soil (S'_e) and that of the reconstituted one ($S_e'^{*} = S'^{*}$), is equal to 1.74 and represents a measure of the natural soil structure.

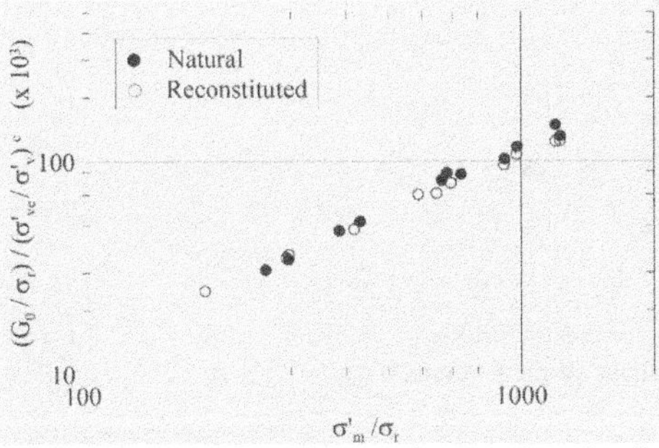

Figure 5. G_0 data of natural and reconstituted soils, normalized by the corresponding equivalent pressure, versus σ'_m/σ_r.

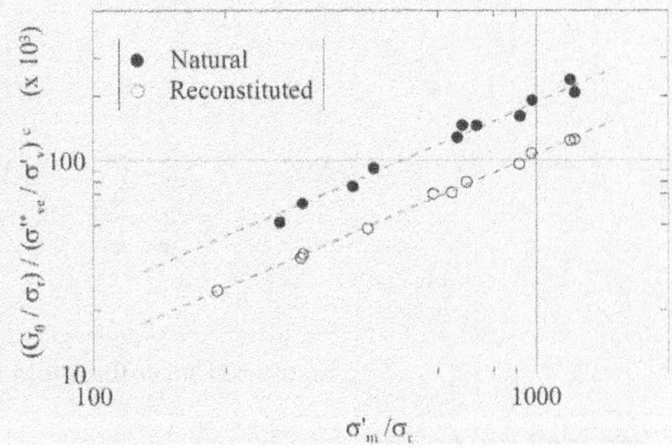

Figure 6. G_0 data, normalized by the intrinsic equivalent consolidation pressure, versus σ'_m/σ_r.

4. RESULTS AT MEDIUM STRAINS

In Fig. 7a and 7b the G_{eq}/G_0-γ_c curves corresponding to the loading and unloading stages, respectively, are plotted for natural and reconstituted materials. It can be seen that, for each set of data, the G_{eq}/G_0-γ_c curves are not much influenced by σ'_v as all the curves describe a narrow band. It can also be noted that the reduction of the normalized shear modulus is more pronounced for the natural soil than the reconstituted one. In fact, the G_s/G_0-γ_c data points of the reconstituted soil plot above those corresponding to the natural soil, thus indicating a more linear behavior.

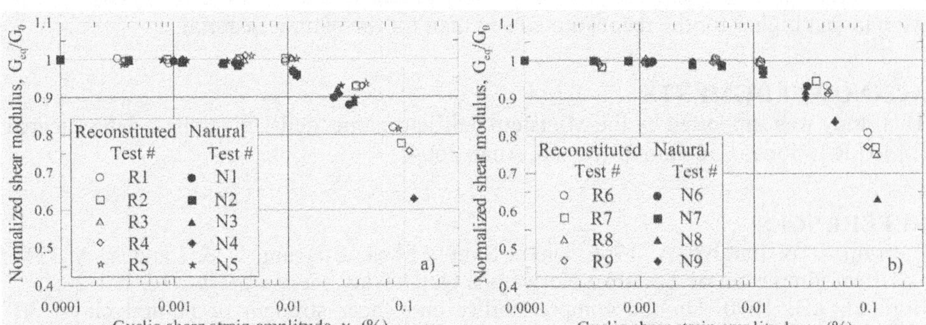

Figure. 7. G_{eq}/G_0 data versus γ_c for (a) loading and (b) unloading stages.

The departure of the above mentioned curves from the linear behavior can be described by the linear threshold shear strain γ_l, defined as the value of γ_c corresponding to G_{eq}/G_0 equals to 0.95 (Vucetic 1994). The γ_l values determined from the G_{eq}/G_0 - γ_c curves in are plotted versus σ'_m for both natural and reconstituted soils in Fig. 8. It is apparent that the values of γ_l for the reconstituted soil plot always higher than those pertaining to the natural one. These differences in γ_l values are presumably related to the different micro-structure.

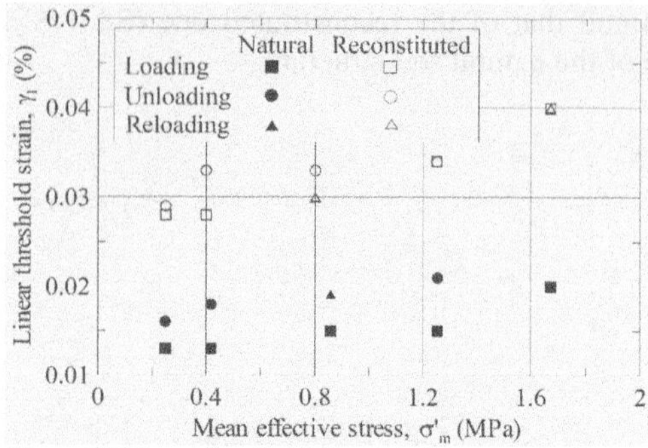

Figure 8. Linear threshold shear strain γ_l versus σ'_m.

5. CONCLUSIONS

The influence of structure on stiffness characteristics of natural and reconstituted Augusta clay was investigated by means of cyclic simple shear tests in the range of small and medium strains.

Tests results were expressed in terms of maximum and normalized equivalent shear modulus. It has been shown that the maximum shear modulus is greatly influenced by the arrangement and inter-particle bondings of particles considering that the small-strain shear modulus of natural soil is always higher than that measured on the reconstituted soil, at a given state and stress history. The effect of structure was examined and quantitatively taken into account by means of a normalization procedure, considering the reconstituted soil as reference.

The normalized stiffness curves also showed a dependency from structure. The linear threshold shear strains, calculated from the normalized shear modulus versus γ_c curves, were in fact higher for the reconstituted soil than for the natural material.

ACKNOWLEDGMENTS

This study was supported by the Ministero dell'Istruzione, dell'Università e della Ricerca (M.I.U.R.) - Finanziamento di Ateneo, Anno 2004.

REFERENCES

Bjerrum, L. & Landva, A. 1966. Direct Simple-Shear Bjerrum, L. & Landva, A. 1966. Direct Simple-Shear Test on a Norwegian Quick Clay. *Géotechnique*, 16, 1, 1-20.

Burland, J.B. 1990. On the compressibility and shear strength of natural clays, 30th Rankine Lecture, *Géotechnique*, 40, 3, 329-378.

Burland, J.B., Rampello, S., Georgiannou, V.N. & Calabresi, G. 1996. A laboratory study of the strength of four stiff clays, *Géotechnique*, 46, 3, 491-514

Cavallaro, A., Fioravante, V., Lanzo, G., Lo Presti, D.C.F., Pallara, O., Rampello, S., d'Onofrio, A., de Magistris, F.S. & Silvestri, F. 2001. Report on the current situation of laboratory stress-strain testing of geomaterials in Italy and its use in practice. In Tatsuoka et al. (eds), *Advanced Laboratory Stress-Strain Testing of Geomaterials*, Swets & Zeitlinger, Lisse, 15-44.

D'Elia, B., Lanzo, G. & Pagliaroli, A. 2003. Small-strain stiffness and damping of soils in a direct simple shear device, *Pacific Conf. on Earth. Eng.*, Christchurch, New

Zealand.

d'Onofrio, A, Vinale, F. & Silvestri, F. 1998. Effects of micro-structure on the stress-strain behaviour of two natural clays. In Jamiolkowski et al. (eds), *Pre-Failure Deformation Characteristics of Geomaterials,* Balkema, Rotterdam, 257-264.

Doroudian, M. & Vucetic, M. 1995. A Direct Simple Shear Device for Measuring Small-Strain Behavior. *Geotechnical Testing Journal*, 18, 1, 69-85.

Dyvik, R., Berre, T., Lacasse, S., & Raadim, S. 1987. Comparison of truly undrained and constant volume direct simple shear tests. *Géotechnique*, 37, 1, 3-10.

Hardin, B.O. 1978. The nature of stress-strain behaviour for soils. State-of-the-art-report. *Proc. Spec. Conf. On Earthquake Eng. and Soil Dynamics*, 3-90, Pasadena, ASCE.

Lanzo, G., Vucetic, M, & Doroudian, M. 1997. Reduction of shear modulus at small strains in simple shear, *Journal of Geotech. and Geoenv. Eng.*, 123, 11, 1035-1042.

Lanzo, G., Doroudian, M. & Vucetic, M. 1999. Small-strain cyclic behavior of Augusta clay in simple shear. In Jamiolkowski et al. (eds), *Pre-Failure Deformation Characteristics of Geomaterials,* Balkema, Rotterdam, 213-220.

Leroueil, S. & Vaughan P.R. 1990. The general and congruent effects of the structure in natural soils and weak rocks, *Géotechnique*, 40, 3, 467-488.

Li, D.J., Shibuya, S., Mitachi, T. & Kawaguchi, T. 2003. Judging fabric bonding of natural sedimentary clay. In Di Benedetto et al. (eds), *Deformation Characteristics of Geomaterials*, Swets & Zeitlinger, Lisse, 203-209.

Lo Presti, D.C.F., Pallara, O., Maugeri, M., & Cavallaro, A. 1998. Shear modulus and damping of stiff marine clay from in situ and laboratory tests, *Proc. First Int. Conf. on Site Characterization*, Atlanta, Georgia, 2, 1293-1300.

Maugeri, M. & Frenna, S.M. 1995. Soil-response analyses for the 1990 South-East Sicily earthquake, *Proc. Third Int. Conf. on Recent Adv. in Geotech. Earth. Eng. Soil Dynam.*, 2, 653-658.

Mayne, P.W. & Kulhavy, F.H. 1982.K_0-OCR relationships in soils, *Journal of Geotech. Eng.*, ASCE, vol. 108, no. GT6, pp. 851-872.

Mitchell, J.K. 1976. *Fundamentals of soil behaviour*, New York, John Wiley & Sons.

Rampello, S. & Silvestri, F. 1993. The stress-strain behaviour of natural and reconstituted samples of two overconsolidated clays. *1st Int. Symp. on Geotechnical Engineering of Hard Soils-Soft Rocks*, Athens, Balkema, 1, 769-778.

Rampello, S., Silvestri, F. & Viggiani, G. 1994. The dependance of small strain stiffness on stress and history for fined grained soils: the example of Vallericca clay. *1st Int. Symp. on Pre-Failure Deformation Characteristics of Geomaterials,* IS-Hokkaido,1, Sapporo, Balkema, 273-278.

Rampello, S. & Viggiani, G.M.B. 2001. Pre-failure Deformation Characteristics of Geomaterials. Discussion Leader Report on Session 1a: Laboratory test. In M. Jamiolkowski, R. Lancellotta & D. Lo Presti (eds.) *2nd Int. Symp. on Pre-Failure Deformation Characteristics of Geomaterials, 2*, Torino, Balkema, 1279-1289.

Rampello, S., Calabresi, G. & Callisto, L. 2003. Characterisation and engineering properties of a stiff clay deposit. In Tan et al. (eds), *Characterisation and Engineering Properties of Natural Soils,* Swets & Zeitlinger, Lisse, 1021-1045.

Shibuya, S. 2000. Assessing structure of aged natural sedimentary clays, *Soils and Foundations*, 40, 3, 1-16.

Viggiani, G. 1992. *Small strain stiffness of fine grained soils*, PhD thesis, City Univ.,U.K.

Vucetic, M. 1994. Cyclic threshold shear strains in soils. *Journal of Geotechnical Engineering*, ASCE, 120, 12, 2208-2229.

Soil Stress-Strain Behavior: Measurement, Modeling and Analysis
Geotechnical Symposium in Roma, March 16 & 17, 2006

THE INFLUENCE OF MESO-STRUCTURE ON THE MECHANICAL BEHAVIOUR OF A MARLY CLAY FROM LOW TO HIGH STRAINS

Francesco Silvestri[1], Claudia Vitone[2], Anna d'Onofrio[3], Federica Cotecchia[2]
Rodolfo Puglia[1], Filippo Santucci de Magistris[4]

[1] Department of Soil Defense, University of Calabria, Italy
[2] Department of Civil and Environmental Engineering, Technical University of Bari, Italy
[3] Department of Geotechnical Engineering, University of Napoli Federico II, Italy
[4] Department of Animal, Plant and Environmental Sciences, University of Molise, Italy

ABSTRACT

On October 31, 2002, a M_L=5.5 earthquake struck the Molise region in Southern Italy. The strongly non-uniform damage distribution observed in the town of San Giuliano di Puglia suggested that site amplification significantly affected the seismic response of the Toppo Capuana marly clay formation.

The geotechnical laboratories of both the University of Napoli and the Technical University of Bari were entrusted by the Department of Civil Protection of the Italian Government with the experimental investigation of the geotechnical properties of the marly clay, in order to develop seismic microzonation studies. Laboratory tests were carried out on undisturbed borehole samples taken at depth along two orthogonal sections crossing the town center. The testing programme consisted of standard classification tests, one-dimensional and isotropic compression tests at medium-high pressures, cyclic and monotonic triaxial tests and both cyclic and dynamic torsional shear tests carried out at variable frequencies. The geotechnical investigation identified the presence of three main geotechnical units within the Toppo Capuana formation: grey marly clays at depth, overlayed by a few metre thick stratum of intensely fissured tawny clay, and a thin cover of softer soils disturbed by repeated sliding; the laboratory tests were developed on samples from each of these units.

The paper discusses the main experimental results and the consequent interpretation of the stress-strain behaviour of the tested soils, characterizing in particular the dependency of the soil pre-failure behaviour and strength on its stress state and history, strain level and strain rate.

The influence of the different meso-strucures was not recognisable on the physical properties, but clearly affected the mechanical behaviour of tawny and grey clays, which exhibited more pronounced scale effects. The experimental data permitted to evaluate the effects of the degree of fissuring on compressibility, strength and small strain behaviour; in particular, the different meso-structures of the tawny and grey clays were reflected by the comparison between yielding stresses in one-dimensional compression, the peak strength values, as well as by the small strain stiffness and damping ratio.

All such effects are conceptually consistent, and significantly reflect on the subsoil modelling for seismic response analyses of the subsoil to the seismic sequence.

Hoe I. Ling et al. (eds.), Soil Stress-Strain Behavior: Measurement, Modeling and Analysis, 333–350.

1. INTRODUCTION

On the 31st of October 2002, a M_L=5.5 earthquake struck the Molise region in Southern Italy. The earthquake was followed by a comparable aftershock (M_L=5.4) nearby on the day after, and by a series of seismic events of lower energy in the following month (Fig. 1).

In the town of San Giuliano di Puglia, the distribution of the damage consequent to the mainshocks was strongly non-uniform: the old part, which was less damaged, lies on outcropping rock, while the most severe damages were surprisingly concentrated in the new part of the town, lying on fine-grained soils. This evidence suggested that differential site amplification significantly affected the seismic response in the town. This hypothesis has been confirmed by aftershock records logged by two temporary accelerometric stations, located both in the ancient and in the new portions of the town (Puglia, 2005).

In order to develop a seismic microzonation of the area (Baranello et al., 2003), the *Department of Civil Protection* of the Italian Government committed a comprehensive investigation on the subsoil properties. The experimental programme included: boreholes, in situ piezometric measurements, CPT, cross-hole and down-hole tests and laboratory investigations on undisturbed samples, namely oedometer, triaxial and cyclic/dynamic torsional shear tests.

The geotechnical laboratories of both the *University of Naples* and the *Technical University of Bari* were entrusted by the *Department of Civil Protection* with the laboratory investigations, aiming at the characterisation of the mechanical behaviour of the fine-grained soils.

The present paper discusses the main experimental results of such laboratory investigations, as well as a comparison with seismic in-situ tests, giving evidence to the dependency of both the soil strength properties and stress-strain behaviour on the soil structural features.

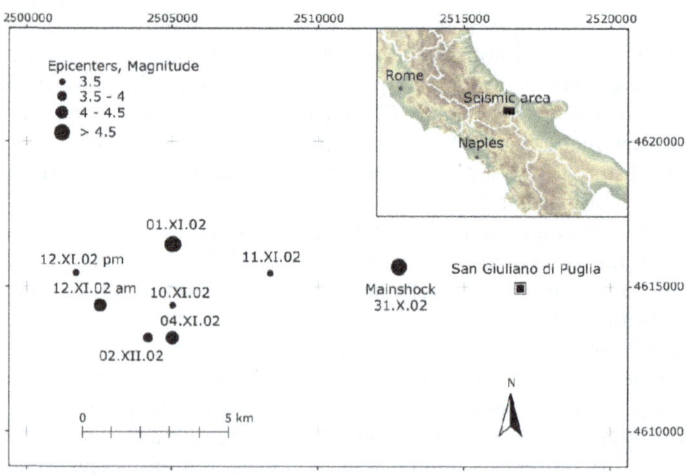

Figure 1. Location of the epicenters of the 2002 Molise earthquake and of the town of S. Giuliano di Puglia.

2. GEOLOGY AND LITHOLOGY

2.1 Geological setting of San Giuliano di Puglia

On the basis of surface surveys and of the analysis of 35 borehole corings (15 of which cored on-purpose after the earthquake), Melidoro (2004) and Guerricchio (2005) have proposed an interpretation of the town geological setting, according to which the main formations present at San Giuliano di Puglia are (Fig. 2):

- the Faeto flysch (F), that is a sedimentary succession of mainly calcareous soils, either coarse or fine-grained; in particular these can be calcirudites, limestones, calcareous marls, white marls and green clays, differently fractured and fissured;
- a deep layer of Toppo Capuana marly clays (MC), whose maximum thickness has not been assessed yet; at the top, these clays are weathered down to few metres; they are overtopped by a shallow cover of disturbed soil and landslide debris;
- a chaotic complex (C) formed by Varicoloured Scaly clays, limestones, calcareous marls, calcarenites and fragments of the Faeto flysch.

As shown in Fig. 2, the MC formation is in lateral contact with formation F, that emerges in the Southern part of the ridge, where the flysch appears to be less fractured and constitutes the foundation soil of the historical part of San Giuliano di Puglia. In the Northern part of the town, the Faeto flysch is heavily tectonised and broken up.

Two different hypotheses have been formulated about the origin of the current geological setting at San Giuliano di Puglia. According to the first one (Baranello et al., 2003), the marly clay deposit lies within the 'bowl shaped' top surface of the flysch, resulting form a sinclinal deformation of its structure. However, this interpretation is in contrast with the results of micro-palaeonthological analyses reported by Guerricchio (2005), which show that the age of the marly clays increases with the distance from the flysch outcrop.

A second hypothesis (Giaccio et al., 2004) assumes that the contact between the marly clays and the flysch is sub-vertical, and results from the thrusting and oversliding of the Faeto flysch against and above the Toppo Capuana formation. This latter hypothesis is in agreement with the dating results; more objective indications are expected from deep geophysical investigations in the next future (Puglia, 2005).

2.2 Lithological properties of the marly clays

According to Melidoro (2004) and Guerricchio (2005), the Toppo Capuana marly clay formation at San Giuliano di Puglia consists of three principal units:

- a 'debris cover', of less than five metres thickness, including black organic carbonaceous elements, lumps and lenses of white powdery calcite and small calcareous litho-clasts;
- a layer, of two to ten metres thickness, of 'weathered tawny clays', characterised by medium to intense fissuring, resulting from the weathering and disturbance of the uppermost part of Toppo Capuana marly clays;
- a deep layer of Toppo Capuana marly clays, called 'grey clays' hereafter. The thickness of this layer seems to be around three hundred metres, as back-figured from the natural frequencies of the aftershocks recorded in the new part of the town (Puglia, 2005).

The grey Toppo Capuana marly clays are less intensely fissured than the weathered tawny clays; in some cases the fissure surfaces are either ochraceous or covered by a black oxidation patina. The polyhedral clay elements, of 2-4 cm maximum size, are sharply edged and well embedded within the fissure network. Crystals of selenitic gypsum are detectable in rare lenses or in thin layers of fine sand.

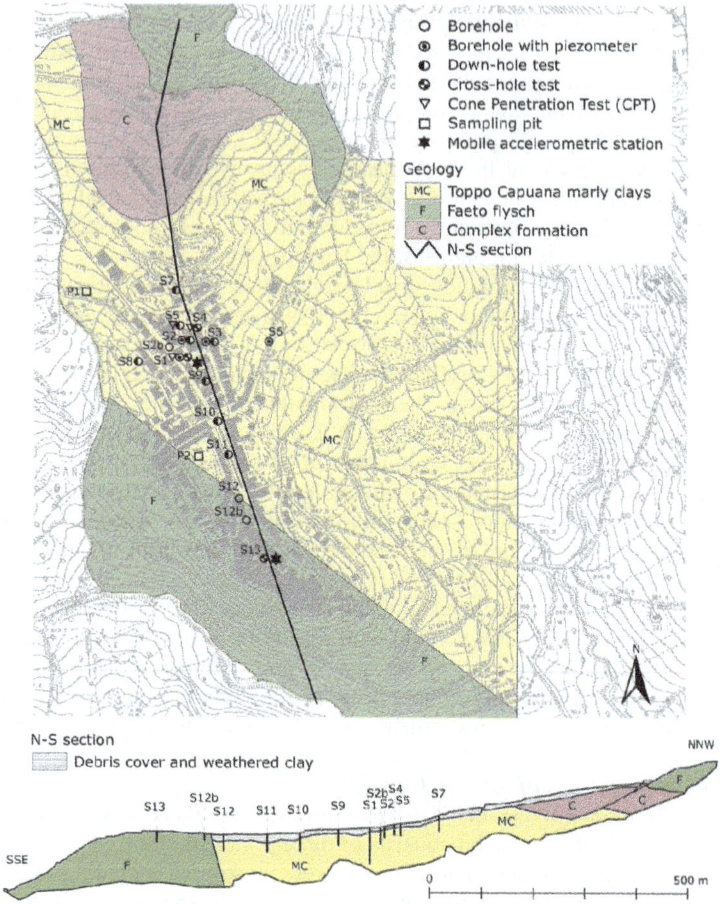

Figure 2. Geological map (a) and N-S section (b) of San Giuliano di Puglia.

2.3 Micro-structure and meso-structure of the marly clays

Mineralogical analyses (Melidoro 2004; Guerricchio, 2005) were carried out by X-ray diffraction on samples of both the shallow (Fig. 3a) and the deep marly clays (Fig. 3b); they showed that the mineral compositions of the two clays are similar, except for the presence of gypsum in the shallowest clays. The principal minerals are phyllosilicates (ΣPhyl = 57-68%), followed by quartz (Qtz = 12-14%) and calcite (Cal = 10-22%). Within the clay fraction, the swelling minerals (Esp = 69-77%) are more abundant than illite (I = 14-16%), chlorite (Chl = 5-8%) and kaolinite (Kao = 4-6%). These results are consistent with the similarity in index properties for the tawny and the grey clay samples (see § 3.2), confirming that the tawny clay results from the weathering of the grey one. The uniformity in composition between weathered and unweathered clays, reflected in the similarity of the soil index properties, has been often reported in the literature (Chandler & Apted, 1988; Cafaro & Cotecchia, 2001).

Vitone (2005) and Vitone et al. (2005) have proposed a classification criterion for the fissuring of homogeneous fine-grained soils, that allows to characterise clay meso-structures according to the lithology and consistency of the soil matrix and to the nature, geometry and orientation of the discontinuities (Table I). This classification has been proposed as a tool to compare the meso-structures of different fissured clays and to identify how given fissuring features affect the clay mechanical behaviour.

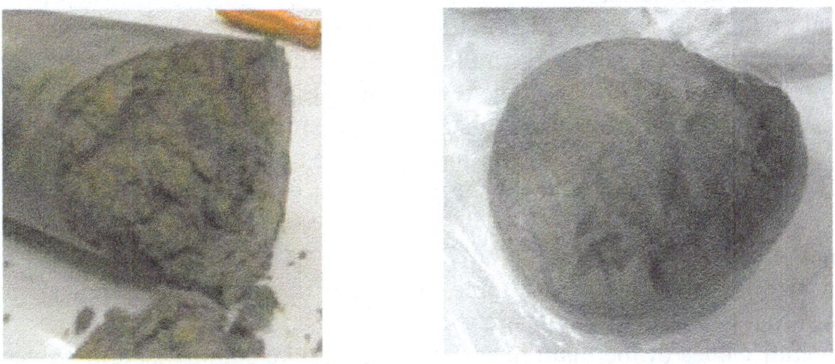

Figure 3. Pictures taken on ∅80 mm samples of tawny (a) and grey clay (b).

	INTACT SOIL		DISCONTINUITY NATURE			DISCONTINUITY ORIENTATION	DISCONTINUITY GEOMETRY		
	Main sediment fraction	*Undrained strength°*	*Type*	*Roughness**	*State*		*Shape˙*	*Continuity˙*	*Intensity⁺*
	A	B	C	D	E	F	G	H	I
1	Clay	Mudstone	Depositional	Very rough	Fresh	Single	**Planar**	Continuous	Very low: $< 3\ m^2/m^3 - > 1\ m^3$
2	Silt	**Stiff clay**	**Stress relief**	Rough	**Slightly weathered**	Few	Curved	Many intersections	Low: $3 \div 10\ m^2/m^3 - 0.027 \div 1\ m^3$
3	Silty clay	Firm clay	**Shear induced**	Slightly rough	Highly weathered	**Random**	Hinged	**Some intersections**	Moderate: $10 \div 30\ m^2/m^3\ 0.001 \div 0.027\ m^3$
4	Clayey silt	Soft clay		**Smooth**	**Iron stained**		Folded	Very few intersections	**High: $30 \div 100\ m^2/m^3 - 27 \div 1000\ cm^3$**
5	**Marly clay**			Very smooth	Coated		Conchoidal		**Very high: $100 \div 300\ m^2/m^3 - 1 \div 27\ cm^3$**
6				Slickensided					Excessive: $> 300\ m^2/m^3 - < 1\ cm^3$

° After Morgenstern and Eigenbrod (1974) and BS 8004 (1986)
* Sandpaper grade (Fookes and Denness, 1969); roughness classification (ISRM, 1993)
⁺ Area of discontinuities per unit volume (m^2/m^3) and average size (m^3, cm^3) of intact block (Fookes and Denness, 1969)
˙ After Coffey & Partners in Walker et al. (1987)

Table I. Fissuring classification chart for homogeneous fine-grained soils (Vitone, 2005). Bold characters correspond to the fissuring features of the Toppo Capuana marly clays.

As shown in Table I, the fissured marly clays of San Giuliano di Puglia, either tawny or grey, are made up of stiff (*B2*) marly clay elements or peds (*A5* in Table I), separated by fissures induced by either stress relief or shearing *(C2÷C3)*. The element surfaces are smooth (*D4*) and slightly weathered or stained (*E2* or *E4*). The planar (*G1*) discontinuities seem to be randomly oriented (*F3*) and intersect each other (*H3*). On average, the polyhedral elements of the grey clays (Fig. 3b) have a maximum dimension of 2-4 cm, with a volume of about 27-30 cm^3 (*I4*). The samples of weathered tawny clay (Fig. 3a), instead, are characterised by a more intense degree of fissuring and can be ascribed to level *I5*.

The meso-structure of the debris cover is similar to that of the tawny clay samples, although the weathering level in the cover is higher.

3. LABORATORY TESTS

3.1 Experimental programme

An unusual quantity of advanced laboratory tests were made possible by the availability of a large number of undisturbed samples of marly clays, taken at depths down to 21 metres from the boreholes drilled in the town centre (Fig. 2).

The testing programme consisted of standard classification tests, one-dimensional and isotropic compression tests at medium-high pressures, undrained triaxial tests and both cyclic and dynamic torsional shear tests, carried out at variable frequencies.

3.2 Physical properties

The grain size distributions of the three units part of the Toppo Capuana marly clay formation at San Giuliano di Puglia are shown in Fig. 4.

The overall values of the main physical properties are summarised in Table II, which reports the number of measurements, the average value and the standard deviation of each data set. The statistics for the grading data are referred to interpolation curves, drawn through the original grain size data points by means of a logistic dose response function in the form p=a+b/(1+(d/c)^k).

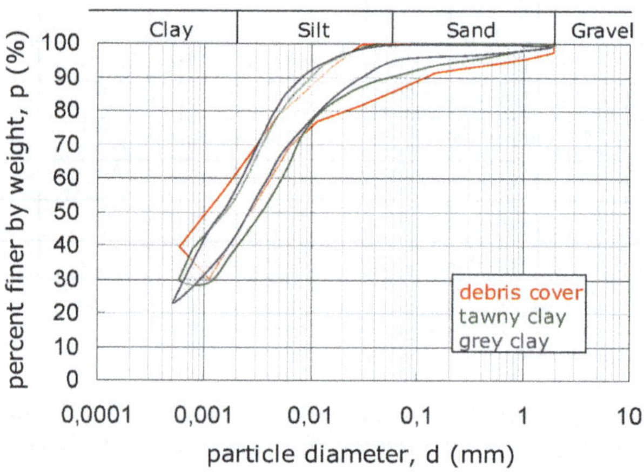

Figure 4. Particle size distributions of the three units.

	debris cover			tawny clay			grey clay		
	n.	average	standard dev.	n.	average	standard dev.	n.	average	standard dev.
Particle specific gravity, Gs	5	2.71	0.02	13	2.71	0.04	16	2.73	0.04
Unit volume weight, γ (kN/m^3)	5	19.65	0.90	13	21.09	0.37	16	21.23	0.85
Void ratio, e	5	0.72	0.18	13	0.54	0.03	16	0.49	0.08
Water content, w (%)	5	22.4	2.6	18	19.5	1.2	21	17.4	2.9
Plasticity limit, w_P (%)	5	23.8	4.0	12	23.2	3.5	16	23.2	2.0
Liquidity limit, w_L (%)	5	63.4	16.9	12	53.8	6.3	16	53.2	6.0
Sand fraction (%)	5	5.8	5.0	11	3.1	2.7	18	2.4	1.3
Silt fraction (%)	5	42.0	8.2	11	48.2	3.8	18	51.0	1.8
Clay fraction (%)	5	51.7	6.5	11	48.6	4.9	18	46.5	2.5

Table II. Physical properties of the three units.

According to Fig. 4 and Table II, very slight differences exist among the particle size distributions of the three units, which show a degree of variability decreasing from the shallowest (debris cover) to the deepest formation (grey clay). Consistently with the increase of disturbance and heterogeneity which can be associated to weathering, the uppermost samples exhibit higher sand and clay fractions, where the deepest ones appear richer in silt.

Figure 5a shows the variations of the Atterberg limits with depth, together with the average water content measured on each undisturbed sample. Below the debris cover, the values of plastic and liquid limits do not vary significantly with depth, being similar for both the tawny and the grey clay (Table II); the average liquid limit w_L is 53.5% and the average plasticity index, I_P, is 30.3%.

In both the tawny and the grey clay layers, the natural water content is always just a little below the plastic limit, the consistency index being always above unity. A slight increase of consistency index with depth can be observed. On the other hand, the water content of the shallow debris is higher, resulting in a significant increase of void ratio close to ground surface (Fig. 5b).

3.3 Compressibility

In order to investigate the response to one-dimensional compression of the Toppo Capuana fissured marly clays, nine restrained-swelling oedometer tests were carried out. A vertical effective stress $\sigma'_{v,max} \approx$ 15 MPa was reached in tests on the deepest samples. The resulting one-dimensional compression curves are reported in Fig. 6. The initial void ratio of the samples decreases with increasing samples depth, the void ratios of both the debris cover and the tawny clay samples being more scattered, in agreement with their more opened and disturbed meso-structure (Vitone, 2005).

The compression curves in the figure show that, while for both the debris cover and the tawny clay samples the vertical effective stress at gross yield, σ'_Y, falls between 1.0 and 1.3 MPa, for the grey clays σ'_Y is higher, i.e. around 3 MPa. As observed for other fissured clays (Vitone et al., 2005), also for the Toppo Capuana fissured marly clays at gross yield there is no evidence of a significant increase of curvature in the one-dimensional compression curve.

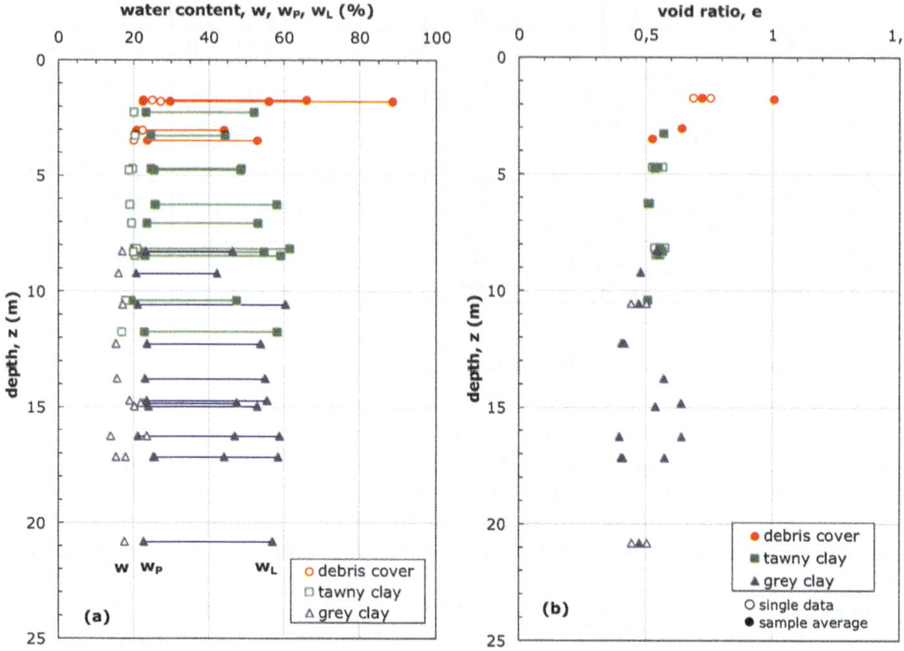

Figure 5. Vertical profiles of Atterberg limits (a) and void ratio (b) of the marly clays.

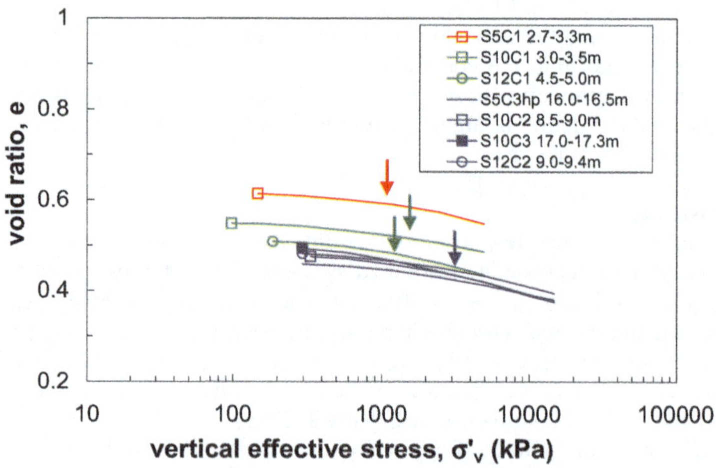

Figure 6. One-dimensional compression curves of the debris cover (red curve), tawny clays (green curves) and grey clays (blue curves). The arrows correspond to the gross yield pressures (data after Vitone, 2005).

The maximum compression indexes measured in the oedometer tests are in general very low, being C_c equal to 0.07÷0.10 for σ'_v=2÷15 MPa.

The mechanical behaviour of the same clay samples in isotropic compression (Vitone, 2005) is consistent with what observed in one-dimensional compression; the gross yield states identified along the compression paths are found to correspond to consolidation states at which the clay exhibits a *wet* behaviour during shearing, as would be expected according to critical state soil mechanics (Schofield & Wroth, 1968) and discussed later.

3.4 Shear strength

Figures 7a,b show the stress-strain curves resulting from triaxial undrained shear tests carried out at Technical University of Bari on representative samples of the Toppo Capuana marly clays; the corresponding stress paths are shown in Fig. 7c. The tests on the most intensely fissured clays (debris and tawny clays) were carried out on 38mm diameter samples, whereas for the grey clays 50mm diameter samples were used.

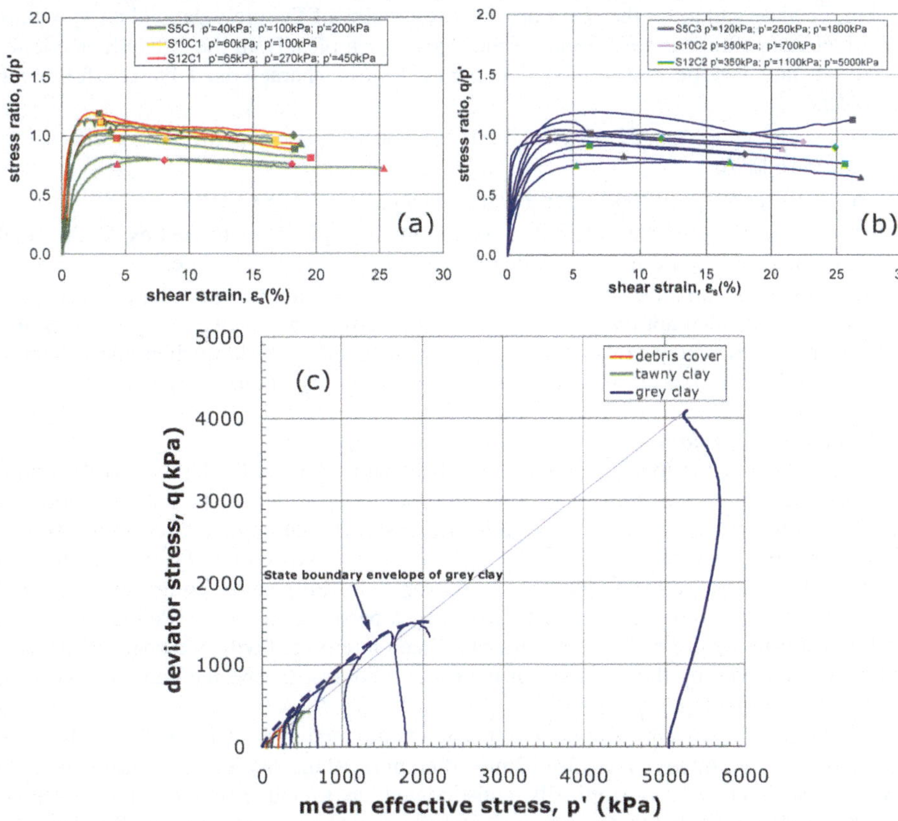

Figure 7. Stress-strain behaviour of (a) the shallow samples (debris cover and tawny clays) and of (b) the grey clay samples; (c) stress-paths and strength envelopes corresponding to the tests in both (a) and (b). In the above plots, the rhombus, the square and the triangle correspond to the lowest, the medium and the highest mean effective consolidation stress respectively (data after Vitone, 2005).

Fig. 7a reports the stress ratio, *q/p'*, plotted against the shear strain, ε_s, for samples of both the debris cover (red lines) and the tawny clay (green lines), whereas the data for the grey clays are shown in Fig.7b. In one of the tests in Fig. 7 (S12C2 sample) the sample has been consolidated to very high pressure (p' = 5 MPa) and it exhibits *wet* behaviour, whereas in all the other tests the samples exhibit dilation along with softening at large strains, characteristic of a *dry* behaviour.

The large dilation with shearing exhibited by the dry samples is indicative of a significant overconsolidation of the samples, i.e. of a significant distance of the sample consolidation state from the gross yield state in compression.

In all the tests either a single or more failure surfaces formed across the sample. In both Figs. 7a and 7b, the stress-strain states corresponding to the onset of sliding along the failure surface have been indicated by means of a symbol along the curve. Sliding seems to take place for shear strains about $\varepsilon_s \approx$ 5% and 7-10%, for the shallow samples (debris and tawny clay) and the grey clay samples respectively (Vitone, 2005).

Fig. 7c shows that, on the *dry* side, the stress-paths of the grey clays identify a peak strength envelope different from that of the shallow samples. Both the strength envelopes are curved and can be approximated to straight lines characterised by the following parameters:

- ϕ'_p = 18° and c'_p = 32 kPa for both the debris cover and the tawny clay (p' = 182÷516 kPa);
- ϕ'_p = 20° and c'_p = 126 kPa for the grey clay (p' = 358÷1882 kPa).

On the *wet* side, the grey clay exhibits a maximum strength characterised by ϕ'=20°, that is close to the post-rupture friction angle of the same clay on the *dry* side.

Summarising, the triaxial tests data show that the state boundary envelope of the grey clays is larger than that applying to either the debris cover or the tawny clays. The poorer strength properties of the tawny clays is likely to be effect of both their more intense fissuring and higher void ratios with respect to the stiffer less fissured grey clays.

3.5 Pre-failure behaviour

Cyclic and dynamic torsional shear tests have been carried out at the University of Napoli on undisturbed samples of the marly clays, in order to characterise the soil behaviour for numerical simulations of the local seismic response. The equipment used is a resonant column/torsional shear device (THOR) developed by d'Onofrio et al. (1999). It has been recently updated to perform continuous isotropic loading paths controlling the cell pressure via an E/P converter, remotely driven by a personal computer (Penna, 2001). As usual, the non-linear pre-failure behaviour has been interpreted with the linear equivalent model, characterised by the variation of the shear modulus, *G*, and the damping ratio, *D*, with the shear strain level, γ.

Sequences of resonant column (RC) and cyclic torsional shear (CTS) tests were carried out on specimens sized 36x72mm after multi-stage isotropic consolidation. At first, the specimen was isotropically consolidated up to the estimated in situ stress, thereafter it was subjected to continuous isotropic compression up to the final stress state, at a constant rate of 5 kPa/h. During all loading stages, the soil small strain response was investigated by means of non-destructive low-amplitude RC tests. At the end of the isotropic loading path, an undrained sequence of CTS and RC tests was performed with increasing strain levels, in order to investigate the behaviour of the clay from small to medium strains. In all the tests, a back-pressure typically around 200 kPa was adopted.

The dependency of small strain stiffness and damping on stress state has been investigated through the non-destructive RC test results pertaining to 11 different samples.

The initial shear modulus, G_0, measured at the end of the consolidation stages, is plotted in Fig. 8a against mean effective stress, p'. The data measured on specimens of the same unit have been grouped together (same colour) and different trends are recognised for the different groups.

The results obtained for the debris cover are not enough to assess its behaviour, while the data collected for the tawny clay samples and the grey ones clearly identify two different trends in the $G_0{:}p'$ plane. Both data sets were fitted by the power function:

$$G_0 = A \cdot (p')^b \tag{1}$$

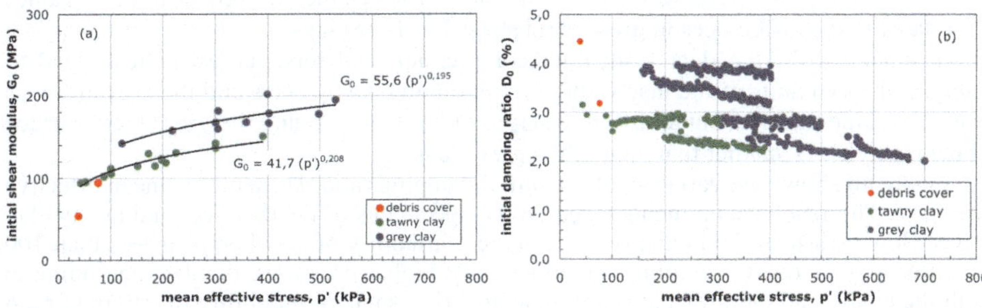

Figure 8. Dependency of the initial shear modulus (a) and damping ratio (b) on the stress state.

The values of the parameters A (in MPa) and b in eqn. (1) are reported on the same plots. Note that the exponent value, b, of p' is practically the same for the two sets of samples (around 0.2), whereas the coefficient A for the tawny clay is about 20% lower than that for the grey clay.

As discussed in § 3.3, the three clay units are strongly overconsolidated and characterised by different gross yield stresses. Therefore, the isotropic compression states of each set of samples plot along different unloading-reloading curves; in particular, the compression curve applying to the tawny clay lays above that pertaining to the grey clays in the *e-p'* plane.

Following the approach introduced by Rampello et al. (1994), the relationship between G_0, the current stress state and the stress history of the soil can be expressed as follows:

$$\frac{G_0}{p_r} = S \cdot \left(\frac{p'}{p_r}\right)^n \cdot \left(\frac{p'_y}{p'}\right)^m \tag{2}$$

In eqn. (2), the coefficient S represents the stiffness of the clay when normally consolidated at a reference stress state $p'=p_r$ (typically taken equal to 1 kPa or to the atmospheric pressure), the exponent n depends on the rate of variation of G_0 with the normalconsolidation stress p', and m accounts for the dependency of G_0 on the overconsolidation ratio. These parameters are usually affected by micro-structural features for natural non-fissured clays (Rampello et al., 1995; d'Onofrio & Silvestri, 2001).

With reference to a given unloading-reloading compression path characterised by a given gross yield stress p'_y, eqn. (2) can be re-arranged as follows:

$$G_0 = S(p_r)^{1-n}(p'_y)^m(p')^{n-m} \tag{3}$$

By comparing the power functions (1) and (3), it can be observed that the exponent b is equal to $(n\text{-}m)$, while the coefficient A corresponds to $S(p_r)^{1-n}(p'_y)^m$, and, therefore, it is proportional to the gross yield stress. Thus, it can be argued that the equal values of b found for both clays are consistent with the similarities in compression indexes (both before and post-gross yield) of the different clays, whereas the differences in the A values are related to the differences in gross yield stress for these clays.

In other words, the data from both compression and torsional shear tests lead to interpret the two units of the marly clay formation, the tawny clays and the grey clays, as being the same material but at different stages of hardening, being the gross yield surface of the tawny clays smaller than that of the grey clays.

Figure 8b shows the variation of the initial damping ratio, D_0, with the mean effective stress, p'. The few measurements taken on the specimens of debris cover and tawny clay at very low pressures are indicative of a higher variability of D_0 when p' is less than 100 kPa; this may be due to sampling disturbance. At higher pressures, the data pertaining to both the tawny and the grey clays are indicative of a smaller variability of damping ratio with stress level. The D_0 values for the grey clays are higher than for the tawny clays at p' less than 300-400 kPa, while at higher stresses the trends are about the same for both clays.

This apparent discrepancy might be ascribed to the difference in intensity of the discontinuities for the two clays, which is likely to affect the damping ratio at low pressures, when the fissures are still rather open and unlocked, and to give rise to dissipation of energy in shearing. With increasing confining stress, the discontinuities close progressively, and the soil dissipative behaviour becomes the same irrespective of the level of fissuring, becoming closer to that of the intact soil.

The non-linear pre-failure behaviour of the three units of the marly clay formation was analysed by means of resonant column test data at medium strain levels. Figure 9 shows the experimental results obtained for the different sets of samples in terms of normalised shear modulus, G/G_0 (plots a,b,c), and damping ratio, D (plots d,e,f), versus shear strain, γ.

Since each set of experimental results defines quite homogeneous trends, they were interpreted using the Ramberg-Osgood model, obtaining the analytical curves drawn in the plots. In Table III the R-O coefficients, C and R, obtained for the normalised stiffness-strain curves of the three units, are reported, together with the values of the linear threshold strain, γ_l, and the reference strain, γ_r, corresponding to $G/G_0 = 0.95$ and 0.5, respectively. The difference in the decay curves of the three units are not very significant: with the increase of depth of the layers, the linear threshold strain γ_l increases slightly and a sharper decay of stiffness is identified by the decrease of γ_r. Table III also reports the coefficients C' and R', obtained from the interpretation of the damping curves;

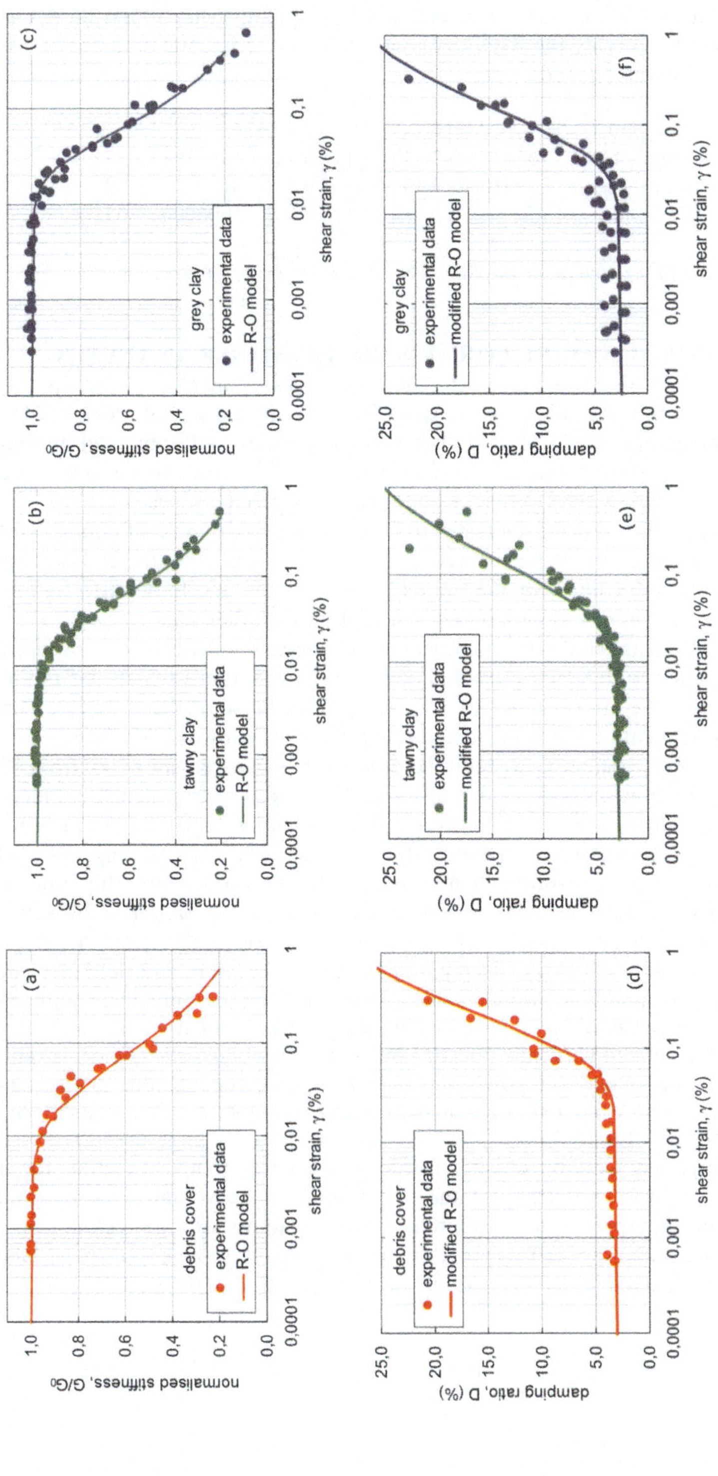

Figure 9. Variation of normalised shear modulus (a,b,c) and damping ratio (d,e,f) with shear strain for the three sub-units after RC tests.

these latter values were not obtained referring to the Masing criteria (Hardin & Drnevich, 1972), but directly applying the R-O regression to the experimental data points, assuming a non-zero initial damping ratio.

Sub-unit	C	R	γ_l (%)	γ_r (%)	C'	R'
Debris cover	365627	2.71	0.011	0.113	1950798	3.29
Tawny clay	14903068	3.17	0.014	0.106	3910	2.28
Grey clay	$1.31*10^9$	3.75	0.017	0.096	1000000	3.05

Table III. Analytical parameters for the Ramberg-Osgood curves in Fig. 9.

3. GEOTECHNICAL MODEL FOR SEISMIC RESPONSE ANALYSES

Laboratory test data were used as a reference for the assessment of shear wave velocity profiles measured in the field by the cross-hole (CH) and down-hole tests (DH) carried out in the verticals indicated in Fig. 2. The vertical profiles of V_S obtained by these tests are shown in Fig. 10, where the data have been separated with reference to the three units (plots a,b,c) and drawn with different graphics according to the source.

The V_S values from down-hole tests are constant for each range of depths, since they were obtained after an inversion procedure of the test data (Petillo, 2004). They have been plotted with solid lines for the verticals close to the accelerometric station in the northern part of the town, and with dotted lines for all the other sites.

The cross-hole data tend to overestimate the DH measurements of V_S, because the CH tests were executed with a non-polarised source type ('sparker') which did not allow to clearly distinguish the arrival times of SV waves from those of the P waves. Therefore, only the down-hole data can be considered reliable for the subsoil modelling.

The plots in Fig. 10a,b,c also show the laboratory measurement of V_S from RC tests driven at consolidation stresses comparable with the in-situ overburden stress (see § 3.5); it can be noted that for the tawny and grey clay, the laboratory data points tend to fall below the average V_S values from field DH measurements. This finding, for stiff clays typically due to sampling disturbance and re-consolidation procedure, this time could be also conditioned by the variable degree of fissuring of the samples. In fact, as the fissuring spacing increases from tawny to grey clays, the scale effect for these latter seems to more sensibly affect the discrepancy between laboratory and field measurements.

In Fig. 10d the laboratory estimates of V_S, deduced from the same test results plotted in Fig. 8a, are compared to the down-hole data ranges, as obtained for the three clay units along all the investigated verticals. For the tawny and grey clays, the law of variation of V_S with depth was assumed as a power function of z, as for eqn. (1); due to the quadratic dependency of G_0 on V_S, the exponent of such power functions resulted one half of those of eqn. (1) itself, i.e. around 0.1.

Summarising, the subsoil velocity profile at the accelerometric station for the site response analyses was modelled, for the tawny and grey clay units, merging the field profiles with the interpretation of laboratory tests in terms of $V_S(z)$ relationship. As a result, the dashed line represents a power function with the same exponent of laboratory data, but which fits the average of field DH ranges measured in the station area.

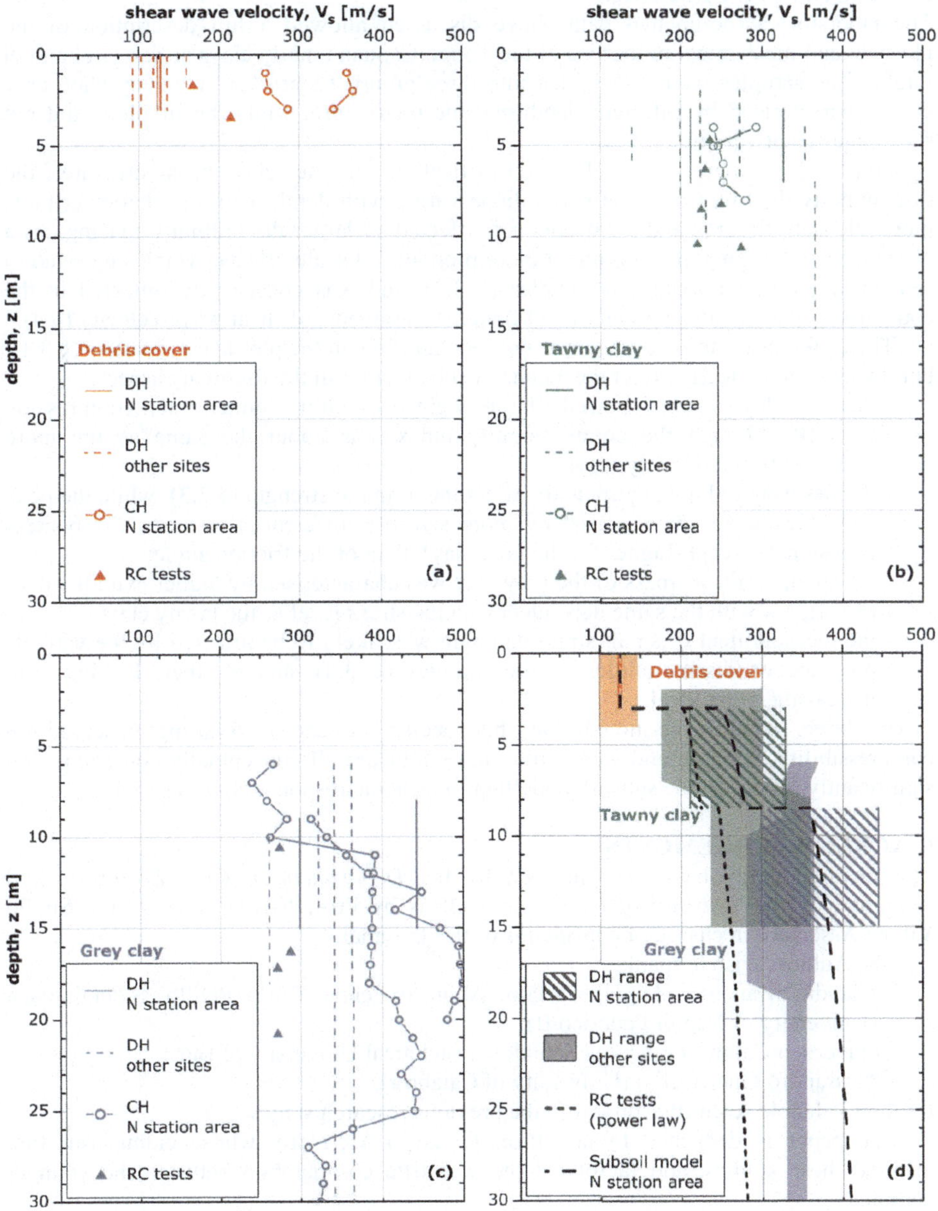

Figure 10. Shear wave velocity profiles in the debris cover (a), tawny clay (b), grey clay (c), and comparison between the different trends (d).

5. CONCLUSIVE REMARKS

The laboratory experimental data above discussed allowed a full description of the physical and mechanical properties of the Toppo Capuana marly clays at San Giuliano di Puglia. The samples were subdivided into three groups (debris cover, tawny clay, grey clay), characterised by different depth and meso-structure, while the lithology did not show significant variations.

Following a new proposal of classification of the clay meso-structure, the discontinuity distribution in the soil volume varied with depth, with the shallower units more intensely fissured and the deeper soil affected by larger discontinuity spacing. As a consequence, the grey clay exhibited more pronounced scale effects, which suggested to test on larger size triaxial test samples (§ 3.3), and was consistently reflected in the comparison between the laboratory and field measurements of shear wave velocity (§ 4).

The different fissuring degree was not recognisable on the physical properties (§ 3.1), but it was seen to clearly affect the mechanical behaviour in the following aspects:

- the grey clay samples showed a larger yield stress in one-dimensional compression (§ 3.2), although the compressibility index was about the same as the more intensely fissured tawny clay;
- the fissuring did not significantly affect the ultimate strength (§ 3.3), while the peak envelopes were characterised by more sensible differences in terms of apparent cohesion intercept (higher for the grey clay), than of the friction angle;
- the small strain stiffness of the grey clay was characterised by higher overall values (§ 3.4), but with the same dependency on the stress level as the tawny clay;
- the measurement of small strain damping was likely to be affected by the variable presence of discontinuities in the samples (§ 3.4), and by their locking with increasing stress level.

From above, it must be noted that the specific effects of fissuring observed on compressibility, strength and small strain behaviour are all conceptually consistent, and significantly reflect on the subsoil modelling for seismic response analyses (§ 4).

6. ACKNOWLEDGEMENTS

The research study has been financed by the *Department of Civil Protection* and supported by the *Tribunal of Larino*, as well as by Prof. Nicola Augenti and Studio Vitone Associati, in charge of consultants of the Tribunal.

The Authors wish to thank:

- Claudio Mancuso, Angela Parlato, Augusto Penna, Carlo Petillo, Stefania Sica (University of Napoli Federico II),
- Francesco Cafaro, Francesca Santaloia (Technical University of Bari),
- Alessandro Guerricchio (University of Calabria),

for their valuable scientific support in the previous research stages.

The paper is dedicated to late Prof. Gregorio Melidoro, whose enthusiasm first inspired these studies, and facilitated the scientific co-operation between the younger authors.

7. REFERENCES

ASTM (1976). *Annual Book of ASTM standard, Part 19*. Philadelphia.

Baranello S., Bernabini M., Dolce M., Pappone G., Rosskopf C., Sanò T., Cara P.L., De Nardis R., Di Pasquale G., Goretti A., Gorini A., Lembo P., Marcucci S., Marsan P., Martini M.G., Naso G. (2003). *Rapporto finale sulla Microzonazione Sismica del centro abitato di San Giuliano di Puglia*. Department of Civil Protection, Rome, Italy.

BS 8004 (1986). *Code of practice for foundations*. British Standard Institutions, London.

Cafaro F.& Cotecchia F. (2001). Structure degradation and changes in the mechanical behaviour of a stiff clay due to weathering. *Géotechnique* 51:441-453.

Chandler R.J. & Apted J.P. (1988). The effect of weathering on the strength of London clay. *Quarterly Journal of Engineering Geology*, 21:59-68.

d'Onofrio A., Silvestri F., & Vinale F. (1999). A new torsional shear device. *ASTM Geotechnical Testing Journal* 22(2):107-117

d'Onofrio A. & Silvestri F. (2001). Influence of micro-structure on small-strain stiffness and damping of fine grained soils and effects on local site response. *Proc. IV international Conference on Recent Advances in Geotechnical Earthquake Engineering and Soil Dynamics, San Diego*. CD-ROM, University of Missouri, Rolla.

Fookes P.G. & Denness B. (1969). Observational studies on fissure patterns in cretaceous sediments of South-East England. *Géotechnique*, 19 (4): 453-477.

Giaccio B., Ciancia S., Messina P., Pizzi A., Saroli M., Sposato A., Cittadini A., Di Donato V., Esposito P. & Galadini F. (2004). *Caratteristiche geologico-geomorfologiche ed effetti di sito a San Giuliano di Puglia (CB) e in altri abitati colpiti dalla sequenza sismica dell'ottobre-novembre 2002*. Il Quaternario (Italian Journal of Quaternary Sciences), 17(1):83-99.

Guerricchio A. (2005). *Private communication*. University of Calabria.

Hardin B.O.& Drnevich V.P. (1972). Shear modulus and damping in soils: design equations and curves. *Journal of the Soil Mechanics and Foundations Division, ASCE*, 98(SM7):667-692.

ISRM (1993). Metodologie per la descrizione quantitativa delle discontinuità nelle masse rocciose. *Rivista Italiana di Geotecnica*, 2:151-197.

Melidoro G. (2004). *Private communication*. Technical University of Bari.

Morgenstern N.R. & Eigenbrod K.D. (1974). Classification of argillaceous soils and rocks. *Journal of the Geotechnical Enginering Division, ASCE*, 100(GT10):1137-1156.

Penna A. (2001). *Effetti delle tecniche di preparazione sul comportamento meccanico di un limo argilloso costipato*. Master thesis (in italian), University of Napoli Federico II.

Petillo C. (2004). *Risposta sismica del centro abitato di San Giuliano di Puglia*. Master thesis (in italian), University of Napoli Federico II.

Puglia R. (2005). *Analisi della risposta sismica locale di San Giuliano di Puglia*. Research report (in italian), University of Calabria.

Rampello S., Silvestri F. & Viggiani G. (1994). The dependence of small strain stiffness on stress state and history of fine grained soils: the example of Vallericca clay. *Proc. I Intern. Symp. on 'Pre-failure Deformation Characteristics of Geomaterials', Sapporo*, 1:273-279. Balkema, Rotterdam.

Rampello S., Silvestri F. & Viggiani G. (1995). The dependence of G_0 on stress state and history in cohesive soils. Panel discussion. *Proc. I Intern. Symp. on 'Pre-Failure Deformation Characteristics of Geomaterials', Sapporo*, 2:1155-1160. Balkema, Rotterdam.

Schofield A.N. & Wroth C.P. (1968). *Critical state soil mechanics*. McGraw-Hill, London.

Vitone C. (2005). *Comportamento meccanico di argille da intensamente a mediamente fessurate.* Ph.D. Thesis (in italian), Technical University of Bari.

Vitone C., Cotecchia F., Santaloia F. & Cafaro F. (2005). Preliminary results of a comparative study of the compression behaviour of clays of different fissuring. *Proc. Intern. Conference on Problematic Soils,* Cyprus, 1173-1181.

Walker B.F., Blong R.J. & McGregor J. P. (1987). Landslide classification, geomorphology and site investigation. *Soil Slope Instability and Stabilisation*, 1-52. Balkema, Rotterdam.

Soil Stress-Strain Behavior: Measurement, Modeling and Analysis
Geotechnical Symposium in Roma, March 16 & 17, 2006

STRESS STATE AND STRESS RATE DEPENDENCIES OF STIFFNESS OF SOFT CLAYS

Supot Teachavorasinskun
Department of Civil Engineering, Chulalongkorn University, Bangkok, Thailand
tsupot@chula.ac.th

ABSTRACT

The influence of the stress anisotropy imposed during consolidation on the stiffness of soft Bangkok clays was explored using the triaxial equipment. Several testing conditions were imposed on the samples to examine the effects of stress state as well as the rate of loading. It was found the stiffness at moderate strain levels was almost independent to the stress state; i.e., the deviator stress level. On the contrary, the rate of stress application played a very important role. The faster the rate of stress application, the higher the values of the stiffness at moderate strains. Nevertheless, a simple empirical equation can be given based on the test results to represent the influence of rate of application on the stiffness of soft clay.

INTRODUCTION

Stiffness of soil; i.e., Young's modulus, at small to moderate strains is one of the most important parameters in the area of soil dynamics. Although abundant information and general correlation for soil stiffness does exist in the literature (e.g., Hardin and Drnevich, 1972, Tatsuoka and Shibuya, 1992), their highly site-specific nature requires an individual study for a specific soil. For Bangkok clays, a few laboratory investigations on its stiffness characteristics have been conducted (Teachavorasinskun et al. 2001 and 2002 and Teachavorasinksun and Amornwithayalax, 2002). As the government has prepared to enforce a seismic resistant design code in Bangkok, information concerned the dynamic characteristics of the Bangkok subsoil; especially the topmost soft clay deposit, must be extensively explored in various aspects. The paper aims to extend the information obtained from the previous studies to specifically describe the influences of the consolidation state and rate of stress application on the stress-strain relationship of Bangkok soft clays.

PHYSICAL PROPERTIES OF TESTED SPECIMENS

Test results obtained from triaxial tests conducted on specimens collected from 3 locations in Bangkok were reviewed and reproduced. The general sub-soil profiles of those three sites are schematically depicted in Fig.1. Their general descriptions are given below.

Hoe I. Ling et al. (eds.), Soil Stress-Strain Behavior: Measurement, Modeling and Analysis, 351–356.

(1) *Chulalongkorn University site (Chula site) and Mahidol University site (MU site).* These two sites are located in the center of Bangkok. Soft clay samples at depths of about 5.0 – 7.0 m from ground surface were used.

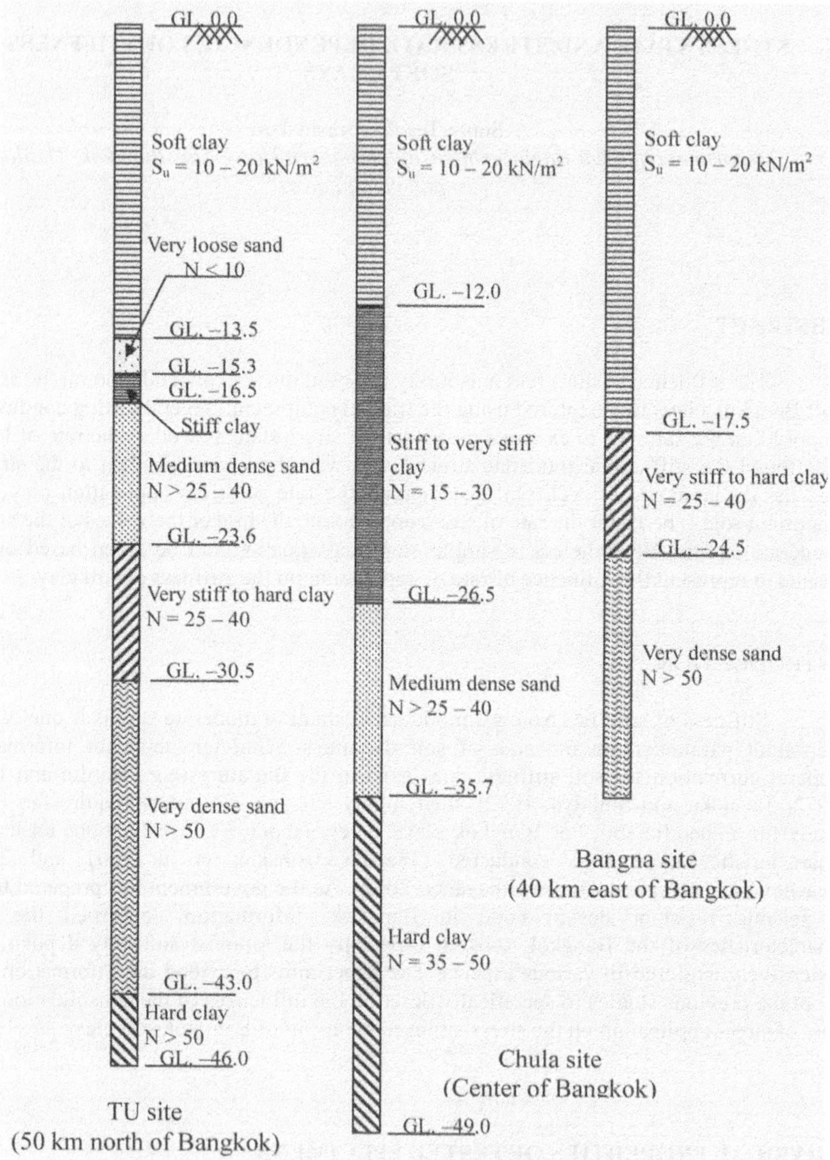

Fig.1 Typical soil profiles at tested sites

Table 1 Summary of the Young's modulus obtained at various total stress paths

Total stress path (α)	**E_{sec} (MPa) at $\varepsilon_1 \approx 0.02\%$ Chula site : PI = 30-40%***		**E_{sec} (MPa) at $\varepsilon_1 \approx 0.02\%$ Bangna site : PI = 55-70%***	
	OCR = 1.8 p' = 28 kPa	OCR =1.0 P' = 200 kPa	OCR = 1.8 p' = 40 kPa	OCR =1.0 p' = 170 kPa
45°	16	60	7	37
90°	20	-	9	-
135°	25	50	9	40

* General description of sites is summarized in Table 2

Table 2 Results of triaxial tests used in the present study

Loading Condition	**Drainage condition**	**Initial stress**	**Sampling location**	**Rate of loading (kPa/min.)**	**Reference**
Cyclic triaxial	Undrained	Isotropic	Chula site [1)] MU site [2)]	1300	Teachavorasinskun et al. (2002(b))
Triaxial compression	Undrained	k_0 consolidation	Chula site [1)] Bangna site [3)]	1.0 – 6.0	Yuttana (2002)
Triaxial compression	Undrained	Isotropic	Chula [1)]	0.05 – 5.0	Teachavorasinskun et al. (2002(a))
Triaxial compression	Undrained bender element	Isotropic	Chula [1)] TU [4)]	–	Teachavorasinskun and Amornwithayalax (2002)

1) Chulalongkorn University, Central of Bangkok, 2) Mahidol University, Central of Bangkok
3) Bangna, 40 km East of Bangkok, 4) Thammasat University, 50 km North of Bangkok

(2) *Thammsat University site (TU site)*. The site is located about 50 km north of Bangkok. Soft clay samples at depths of about 7.0 m from ground surface were used.
(3) *Bangna site.* The site is located about 40 km east of Bangkok. Soft clay layer found at this site is the thickest among the three sampling sites (Fig.1). The samples collected at depth of about 8.0 m were used.

EFFECT OF THE TOTAL STRESS PATH ON YOUNG'S MODULUS

Kurojjanawong (2002) had carried out a series of undrained triaxial compressions on k_0-consolidated samples collected from *CU* and *Bangna* sites. Undrained compressions were carried out under three different total stress paths, namely α = 45°, 90° and 135° (see Fig.2 for definition). For conventional triaxial compression, α is equal to 45°.

The secant Young's modulus, E_{sec}, used in this study is defined as;

$$E_{\text{sec}} = \frac{(\sigma_1' - \sigma_3')}{\varepsilon_1}$$

Where σ_1' and σ_3' are the effective vertical and confining pressures and ε_1 is the vertical strain. Fig.3 shows the typical degradation characteristics of E_{sec} obtained from the normally consolidated samples. The data, though scattering at small strain levels, imply minor effect of total stress paths on the Young's modulus. Table 1 summarizes the values of the secant Young's modulus obtained at strain level of about 0.02%. The samples tested under higher values of α exhibit only slightly higher values of Young's modulus at moderate strains ($\varepsilon_1 \approx$ 0.02%). This is because the observed effective stress paths obtained from samples sheared under different total stress paths are very similar (Kurojjanawong, 2002). Since stiffness is highly dependent of effective stress states, the total stress paths used in their test does not much affect the value of Young's modulus.

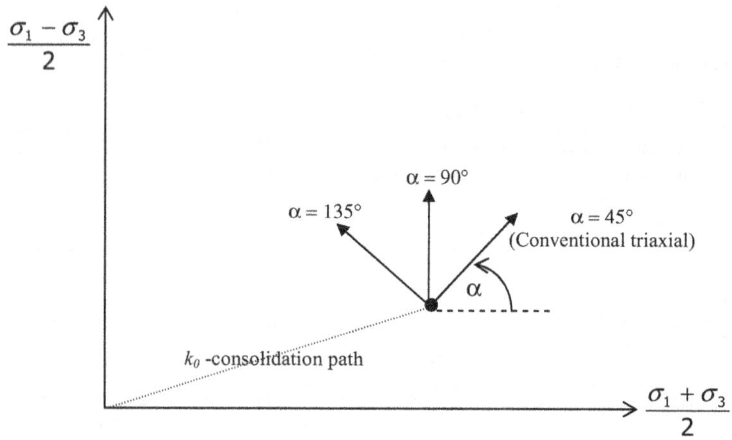

Fig. 2 Definition of total stress path adopted in Kurojjanawong (2002).

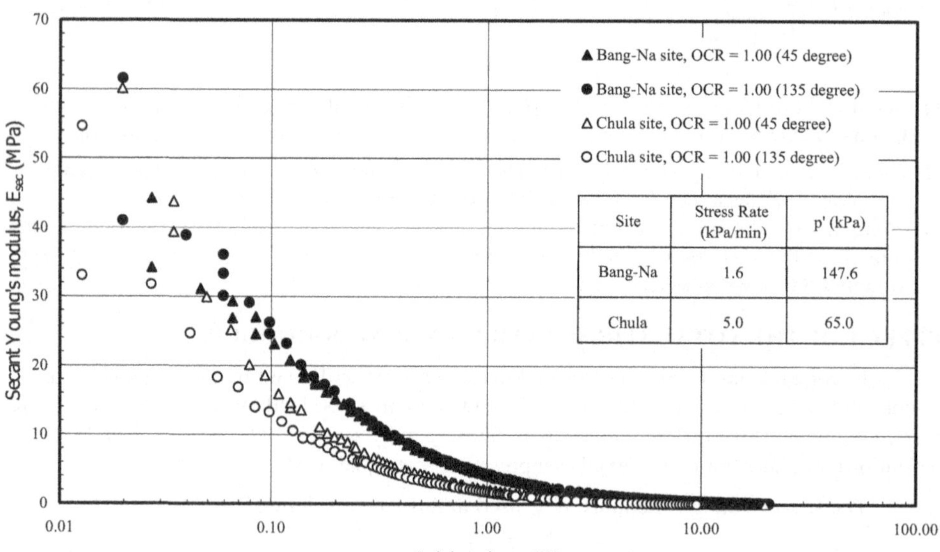

Fig. 3 Effects of total stress paths on secant Young's modulus of soft clays

INDEPENDENT OF YOUNG'S MODULUS ON THE INITIAL SHEAR STRESS

Descriptions of tests used to interpret the effects of initial shear stress and rate of loading on the secant Young's modulus, E_{sec}, are provided in Table 2. They are briefly described herein;

(1) *Cyclic triaxial loading test on isotropically consolidated samples (Teachavorasinskun et al. 2002(a))*: At a single amplitude axial strain of 0.02% and load frequency of 0.1 Hz, the rate of load application used in their tests was as fast as 800 kPa/min.
(2) *Triaxial compression test on isotropically consolidated samples (Teachavorasinskun et al. 2002(b))*: Undrained compression tests were conducted at rates of loading between 0.05 – 50 kPa/min.
(3) *Triaxial compression test on k_0-consolidated samples (Kurojjanawong, 2002)*: Tests were conducted at rates of loading between 1 – 6 kPa/min.
(4) *Measurement of shear wave velocity using bender element. (Teachavorasinksun and Amornwithayalax, 2002)*: The bender element installed in triaxial equipment directly detected the variation of shear wave velocity during isotropic consolidation and undrained triaxial compression. The pulse generated by bender element is considered to be dynamic in nature.

The relation between the secant Young's modulus, E_{sec}, determined at moderate strain level ($\varepsilon_l \cong 0.02\%$) and the initial effective mean stress, $p'_{ini} = (\sigma_1' + 2\sigma_3')/3$, has been prepared from the above mentioned literature and plotted in Fig. 4. The corresponding rates of stress application are also shown. With regarding of the effective mean stress and rate of loading, the initial deviator stress plays minor influence on E_{sec} of Bangkok clays. Namely, E_{sec} obtained from the k_0-consolidated samples are similar to those obtained from the isotropically consolidated ones. This is in well corresponding to Teachavorasinskun and Amornwithayalax (2002) who reported that the shear wave velocities – p' relation obtained during undrained triaxial compression, which incorporated high deviator stress, is similar to that measured during isotropic consolidation.

Fig. 4 further indicates that the $E_{sec} - p'$ relationship is strongly dependent on the rate of stress application. Samples tested with faster rate of loading exhibit higher values of E_{sec}. As a consequence, the line obtained from bender element forms the upper boundary due to its dynamic nature. In order to quantitatively indicate the effect of rate of loading, the plot between the normalized Young's modulus, $(E_{sec}/p_a)\big/(p'_{ini}/p_a)^{0.5}$ and rate of stress application ($\dot{q}$) is prepared in Fig.5 (where p_a is the atmospheric pressure). The normalized Young's modulus is proposed in order to eliminate the effect of the mean effective stress. The power constant of 0.5 is adopted following Teachavorasinskun et al. (2002a). The semi-logarithmic plot provides a linear relationship between normalized Young's modulus and rate of stress application as;

$$\frac{(E_{sec}/p_a)}{(p'_{ini}/p_a)^{0.5}} = A + B\log\left(\dot{q}\right)$$

Where $\boldsymbol{A}$ = 40 (dimensionless) and $\boldsymbol{B}$ = 116 for results obtained from the present study. It should be emphasized herein that the data points shown in Fig.5 are obtained from samples tested under various conditions as indicated previously. The rate of stress application is seen to solely dominate the stiffness at moderate strains of soft clays. In summary, it is suggested that, in exploring, comparing and studying of the stiffness characteristics of clays, the rate of stress application should be seriously taken into consideration.

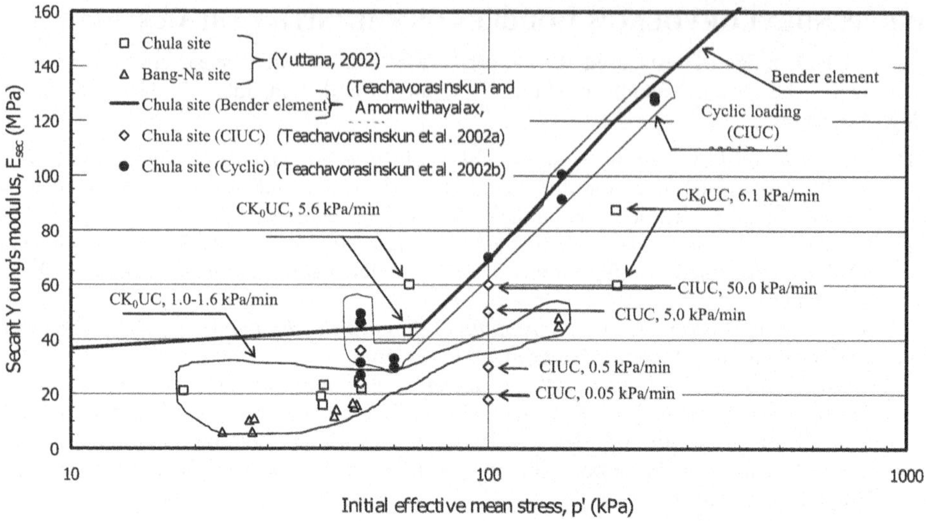

Fig. 4 Relations between secant Young's modulus, mean effective stress and rate of stress application

CONCLUSIONS

The rate of stress application was the most influential factor affecting the secant Young's modulus at moderate strains of soft clays. A sample sheared with faster rate of loading generally exhibited larger value of secant Young's modulus. The initial deviator stress, type of loadings and total stress path during undrained shear had minor effect on the of stiffness of Bangkok clays.

REFERENCES

Teachavorasinskun, S., Thongchim, P. and Lukkunaprasit, P. 2002(b). Stress rate effect on the stiffness of a soft clay from cyclic, compression and extension triaxial tests. Geotechnique, 52(1), pp.51-54.

Tatsuoka, F. and Shibuya, S. 1992. Deformation characteristics of soils and rocks from field and laboratory tests. Keynote Lecture. In Proceeding of the 9th Asian Regional Conference on Soil Mechanics and Foundation Engineering, Bangkok, December 1991, Asian Institute of Technology, Vol.2, pp.101-170.

Teachavorasinskun, S., Thongchim, P. and Lukkunaprasit, P. 2002(a). Shear modulus and damping of soft Bangkok clays. Canadian Geotechnical Journal, 39(5), pp. 1201-1208.

Teachavorasinskun, S. and Amornwithayalax, T. 2002. Elastic shear modulus of Bangkok clay during undrained triaxial compression. Geotechnique, 52(7), pp.537-540.

Kurojjanawong, Y. 2002. Effects of total stress path's directions on undrained stress-strain-strength characteristics of aging marine Bangkok clay. Master Thesis, Faculty of Engineering, Chulalongkorn University.

Hardin, B.O. and Drenvich, V.P. 1972. Shear modulus and damping in soil: Measurement and parameter effects. Journal of the Soil Mechanics and Foundation Engineering Division, ASCE, 98(SM6), pp. 353-369.

Soil Stress-Strain Behavior: Measurement, Modeling and Analysis
Geotechnical Symposium in Roma, March 16 & 17, 2006

COUPLING OF AGEING AND VISCOUS EFFECTS IN AN ARTIFICIALLY STRUCTURED CLAY

Kenny K. Sorensen
Department of Civil and Environmental Engineering
University College London, United Kingdom
e-mail: k.sorensen@ucl.ac.uk

Beatrice A. Baudet
Department of Civil and Environmental Engineering
University College London, United Kingdom
e-mail: b.baudet@ucl.ac.uk

Fumio Tatsuoka
Department of Civil Engineering
Tokyo University of Science, Japan
email: tatsuoka@rs.noda.tus.ac.jp

ABSTRACT

A series of short term isotropically consolidated drained triaxial compression tests was conducted to investigate the influence of cementation, total curing time and strain rate on the stress-strain behaviour of cement-mixed kaolin. The research suggests that the behaviour of cement-mixed kaolin can be described by a unique stress-plastic strain-time relationship independently of strain (curing) history. Both the peak strength and the small strain stiffness were observed to be dependent on the total curing time. The small strain stiffness normalised for stress level showed a continuous linear increase with logarithm of total curing time, while the tested samples of cement-mixed kaolin reached an apparent plateau for peak strength after about one day of curing. The post-peak critical state strength was in contrast seen to be constant with curing time. In relation to findings in the literature and in this study, the coupling of ageing and viscous effects is discussed. It is suggested that there must be a point (characteristic strain rate) at which the behaviour of both artificially and naturally structured clays changes from being dominated by ageing effect to being predominantly viscous.

1. INTRODUCTION

Time effects in clays result from the combination of ageing effects and strain rate effects. Ageing effects are usually associated with the formation of new bonds or cementation between particles due to physico-chemical processes and depend on time. Strain rate effects refer to the response of a soil subjected to different strain rates due to its viscosity. Ageing and strain rate effects occur simultaneously, and it is likely that they influence each other. The coupling of ageing and viscous effects has been investigated by Kongsukprasert & Tatsuoka (2003) using test results on cement-mixed gravel. It was found that at early curing times both ageing and viscous effects were significant in

Hoe I. Ling et al. (eds.), Soil Stress-Strain Behavior: Measurement, Modeling and Analysis, 357–366.

cement-mixed gravel. Due to ageing effects both the elastic stiffness of the soil mixture and its peak strength increased with total curing time. Significant viscous effects were evident from creep at fixed stress and from the response of the soil mixture after step changes in strain rate. The influence of curing time on the stress-strain relation was found to be dependent on curing stress state. Similar coupling of ageing and viscous effects has also been observed in some natural clays. In these cases however the effect of ageing was much less pronounced (Tatsuoka et al., 2000). In this paper, the interaction between ageing and viscous effects is examined for cement-mixed kaolin. The curing of the cement in the soil mixture represents the ageing that would occur in natural clay during its geological history, but at a much faster rate so that it can be more easily observed in the laboratory. Different strain rates were applied to examine the viscous response of the soil. The observed behaviour was compared to available data from the literature for natural clays.

2. SAMPLE PREPARATION AND TESTING PROCEDURE

Samples 50mm in diameter and approximately 100mm in height were prepared by compacting a dry mixture of TA kaolin clay powder mixed with 3% (by weight) Rapid Hardening Portland (RHP) cement into a tall split mould. After compaction the samples were set up in the triaxial cell and generally saturated under an effective isotropic confining pressure p′=100kPa unless otherwise stated. Approximately two hours after saturation the initial shearing stage was commenced with radial, top and bottom drainage permitted. A triaxial testing system allowing automated step-wise change in axial strain rate was employed with external strain measurements and internal load cell (Komoto, 2004). The commercially available kaolin clay is an inorganic plastic clay, which is characterised by its relatively high degree of permeability. Upon contact with water RHP cement experiences rapid hydration with the majority of the strength increase expected within the first 2 to 3 days. The index properties of the tested materials are given in Table 1 below.

Table 1 Properties of tested materials after Komoto (2004)

	w_p [%]	w_L [%]	PI [%]	G_s
TA kaolin	21	46	25	2.68
RHP cement	-	-	-	3.13

To investigate the influence of total curing time six samples (no. 2-6 and 8) were prepared identically. Following saturation under isotropic stress conditions (p'_0=100kPa) the samples were sheared drained to a stress state of q=230kPa, at which point each sample experienced a prolonged constant effective stress creep period between 0 and 9 days. After the creep stage the samples were sheared to failure. Two supplementary tests (no. 1 and 7) were carried out to investigate the influence of stress state during curing, but the limited results did not indicate any clear effect of curing state.

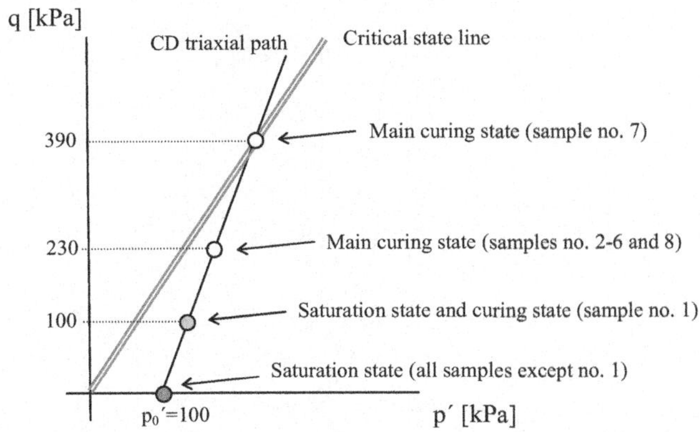

Figure 1 Illustration of testing procedures followed for the investigation

To investigate the influence of strain rate and indirectly the accumulated total curing time in monotonic constant rate of strain shearing, three additional tests (no. 9-11) were carried out at different fixed strain rates. The samples were sheared from the initial isotropic stress state of p'_0=100kPa at an initial strain rate of 0.6%/hr until ε_a= 0.2%. The nominal axial strain rates used thereafter were; 0.08%/hr ($2.2\times10^{-7}s^{-1}$), 0.6%/hr ($1.7\times10^{-6}s^{-1}$) and 2.6%/hr ($7.2\times10^{-6}s^{-1}$). In the majority of the tests short unload-reload cycles with an amplitude of about $\Delta\varepsilon_a\sim0.02\%$ were performed during shearing to determine the pre-peak small strain stiffness.

3. EFFECT OF STRAIN RATE ON SOIL BEHAVIOUR

Laboratory data reported in the literature show that most natural clays have different compression curves when subjected to compression at different loading rates. This type of behaviour is usually referred to as Isotach behaviour. For example the soft normally consolidated Canadian Batiscan clay (Leroueil et al., 1985) shows parallel compression curves for different strain rates, the yield stresses increasing for faster rates of straining (Fig.2). This is typical of the behaviour observed in soft clays (Mitchell et al., 1997). However when the compression has been performed at a very slow strain rate ($1.69\times10^{-8}s^{-1}$ or 6.08×10^{-3}%/hr), the pattern of behaviour changed and the compression curve for that strain rate shifted to follow a higher yield stress locus. In that test the slow straining allowed new bonds to develop in excess of the bonds destroyed by compression, resulting in extra yield strength. In this case it can be simplified that ageing effects have become prominent and overshadowed effects of strain rate.

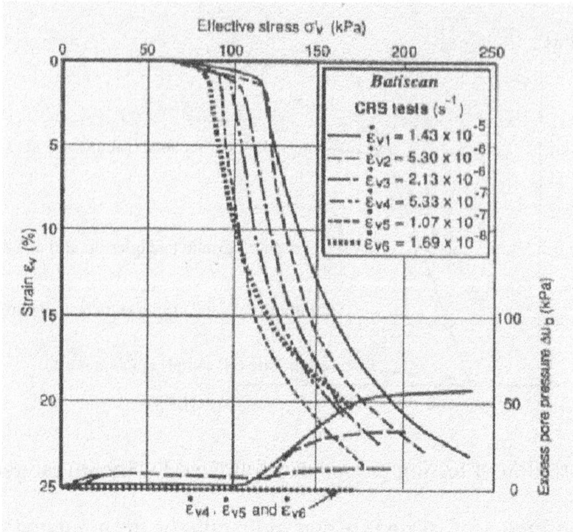

Figure 2 Response of Batiscan clay to Constant Strain Rate oedometer tests (Leroueil et al., 1985)

The Isotach response of clays to shearing manifests itself by different stress-strain curves for different strain rates similarly to the response observed in compression (Tatsuoka et al., 2000). As in compression, extra strength is gained when the clay is sheared at faster strain rates. Figure 3 shows the response of cement-mixed kaolin specimens (no. 9-11) to triaxial drained shearing from isotropically consolidated states (p'_0=100kPa) at strain rates varying between 0.08%/hr ($2.2\times10^{-7}s^{-1}$) and 2.6%/hr ($7.2\times10^{-6}s^{-1}$). Unlike typically seen behaviour the soil mixture reached higher strengths when tested at slower strain rates. It can also be observed that the behaviour gradually changes from being ductile to being brittle. This can be attributed to the behaviour being driven by the bonds (here the cement) rather than the soil matrix. In this example ageing effects overshadow viscous effects even at relatively high applied strain rates. It is to be noted that the net ageing effect on the stress value for the same strain in the pre-peak regime is slightly larger than the one seen among the data presented in Figure 3 due to the opposing effect of viscosity. The net ageing effect can be obtained by correcting the measured stress values to those for the same strain rate.

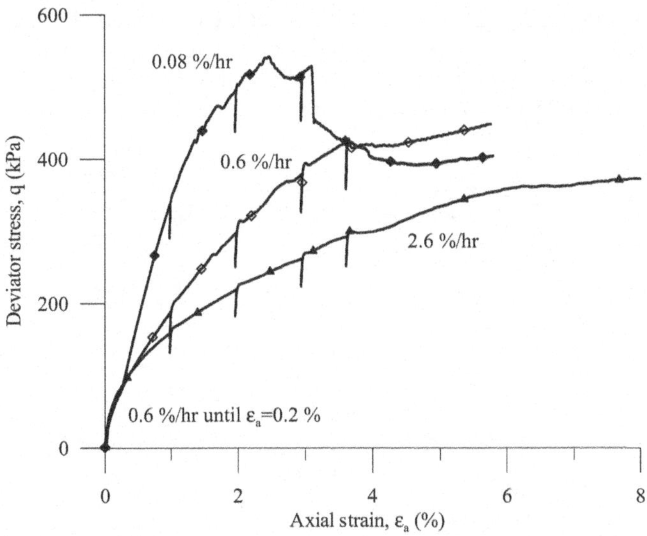

Figure 3 Response of cement-mixed kaolin to drained shearing at different strain rates, from an isotropically consolidated state (p_0'=100kPa)

It is suggested that for a given clay there is a characteristic strain rate that determines whether the response of that clay is dominated by ageing or viscous effects. This characteristic strain rate represents an upper bound to the ageing rate, and is therefore very slow in natural clays. It is reached during creep tests when the creep strain rate becomes sufficiently slow. In standard laboratory tests the behaviour is investigated until large strains. For a test to reach 20% strain at a characteristic strain rate of e.g. 6.08×10^{-3}%/hr (Leroueil et al., 1985) the test would take 137 days. In practice this is almost never achieved and faster strain rates are applied to characterise clay behaviour. To carry out the tests on natural clay at standard rates (e.g. 0.1%/hr) would thus only highlight the viscous effects and hide the ageing effects that may occur during a period of rest when creep rates become lower than the characteristic strain rate.

4. EFFECT OF TOTAL CURING TIME ON BEHAVIOUR

It has been seen above that the behaviour of clay observed during laboratory testing is dominated by either viscous or ageing effects, depending on the strain rate applied. Very slow strain rates allow bonds to develop faster than they are being destroyed by compression or shearing. But slow strain rates also mean longer curing times to reach a specific strain. In the following the effect of curing time on strength, both peak and critical state, and on stiffness is investigated. The total curing time, that is including the time taken during testing, was calculated for different stages of different tests.

Figure 4 shows the influence of total curing time on the peak strength and critical state strength of cement-mixed kaolin. The post-peak critical state strength does not seem to be affected by ageing, and is unique for the soil mixture. The peak strength however initially increases over the first day, to reach an apparent plateau of maximum strength. Curing occurring after that initial day does not seem to affect the peak strength noticeably. It can also been seen from Figure 5 that not only the peak strength but the whole stress-strain curve evolves with total curing time during the first day of curing. This suggests that there is a unique stress-plastic strain-time relationship for a given clay.

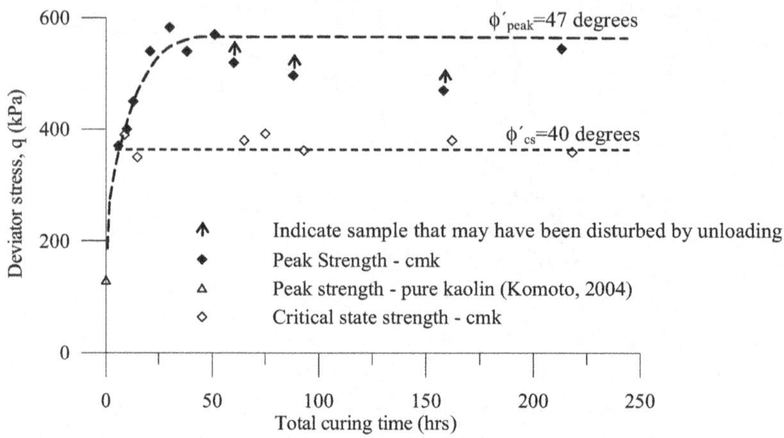

Figure 4 Effect of curing time on peak and critical state strengths in cement-mixed kaolin (cmk)

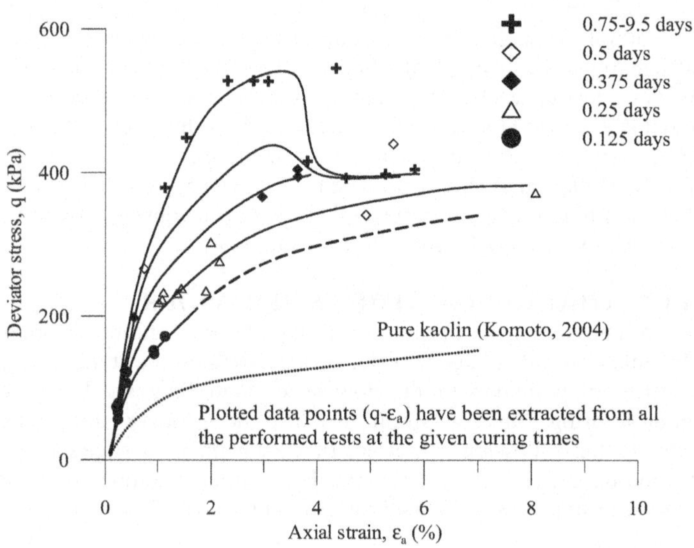

Figure 5 Effect of curing time on pre-peak stress-strain response in cement-mixed kaolin

For the range of tests investigated the small strain secant Young's modulus, E′ was derived from small unload-reload cycles of about $\Delta\varepsilon_a = 0.02\%$ at different stress states and different total curing times. Viggiani & Atkinson (1995) showed that the very small strain shear modulus of normally consolidated clays is dependent on the mean effective stress. The elastic Young's modulus can be related to the elastic shear modulus using Poisson's ratio. In Figure 6 it was assumed as a rough approximation that small strain secant Young's modulus of cement-mixed kaolin would be related to mean effective stress in a similar way, and the relationship proposed by Viggiani & Atkinson (1995) was plotted, using a value of 0.2 for Poisson's ratio and coefficients A=1964 and n=0.653 for pure kaolin (reference pressure p'_r=1kPa). This first approximation, which corresponds to the relationship for a total curing time of zero, plots as a straight line in lnE′-lnp′ plot. Values obtained for the tests performed on cement-mixed kaolin are also plotted on the graph, with the values for the total curing time for each point. The data points for short periods of curing (0.1 to 0.8 hour) plot close to the reference line based on Viggiani & Atkinson (1995). This shows that the first approximation is not as unreasonable as might have been expected. Data points for longer total curing times plot above the reference line, the further away from that line the longer the total curing time. Figure 7 shows the same data points normalised for stress with respect to the reference line. There is a clear linear increase in Young's modulus with the logarithm of total curing time. This suggests that there is a unique relationship between mean effective stress, total curing time and very small strain stiffness for normally consolidated clays.

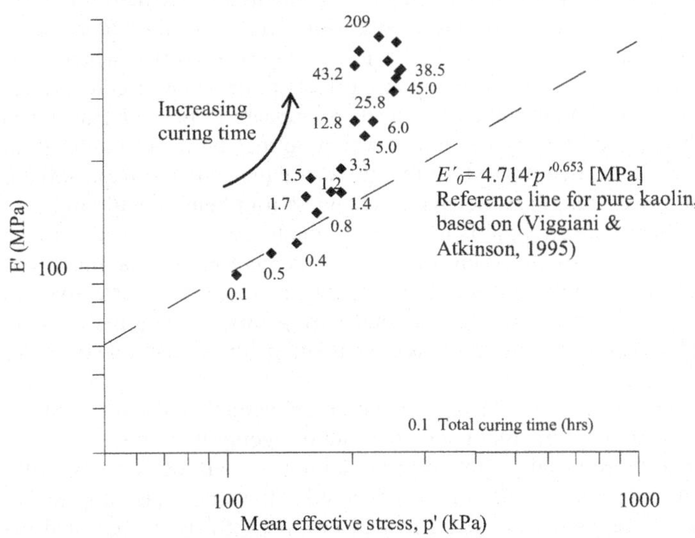

Figure 6 Effect of mean effective stress and curing time on Young's modulus in cement-mixed kaolin

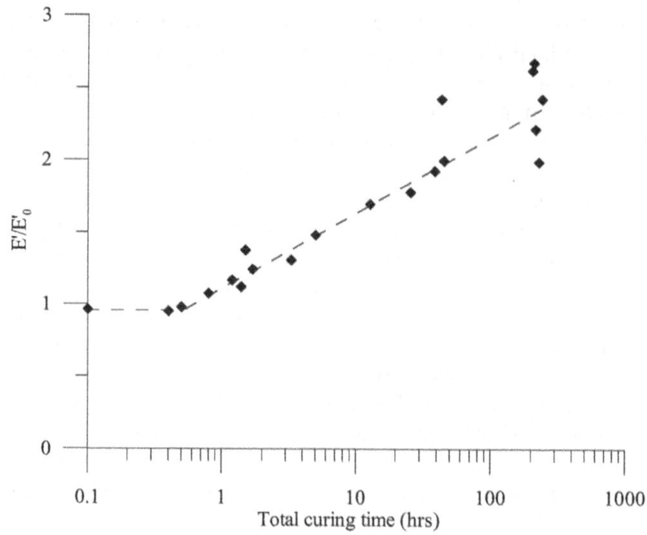

Figure 7 Effect of curing time on Young's modulus in cement-mixed kaolin

5. CONCLUSION AND DISCUSSION

The results from this study suggest that the behaviour of cement-mixed kaolin can be described by a unique stress-plastic strain-time relationship independently of strain (curing) history. It has been seen that in initial stages of curing, when the curing rate is fast the behaviour is dominated by ageing effects rather than viscous effects. In contrast to expectations, in cement-mixed kaolin there appears to be a plateau for peak strength achieved after about one day of curing, which defines an upper bound strength. Further tests are however needed to confirm this trend. As the cement-mixed soil approaches the upper bound peak strength its behaviour changes from being ductile to being brittle. The post-peak critical state strength is in contrast seen to be constant with curing. Curing also seems to affect the small strain stiffness of cement-mixed kaolin. The small strain stiffness normalised with respect to the assumed influence of effective stress shows a linear increase with the logarithm of total curing time. Hence the results suggest that there might be a unique stiffness-stress-time relationship for natural structured clays.

The cement-mixed kaolin is a representation of natural soils but with an accelerated ageing rate. In natural structured soils the rate of ageing is very slow, and at strain rates of the order of those usually applied in the laboratory, viscous effects will dominate the behaviour. Ageing effects will only be observed if the strain rates imposed are very slow (e.g. $1.7\times10^{-8}s^{-1}$ for Batiscan clay, Leroueil et al., 1985). It is suggested that at a given time after hydration there must be a point (characteristic strain rate) at which the behaviour of cement-mixed kaolin changes from being dominated by cementation (curing) effects to being predominantly viscous, as illustrated in Figure 8a. The net cementation effect can be defined as the strength increase per unit time due to development of cement bonds, subtracting any destruction caused by straining, while the viscous effect can be defined as the strength increase per unit change in strain rate. At strain rates lower than the characteristic value, bonds are allowed to develop faster than they are being destroyed by the compression or shearing action and hence the cementation effects will dominate the observed stress-strain behaviour (Figure 8b). While for applied strain rates above the characteristic rate, new bonds are continuously destroyed by the straining, resulting in the net cementation effect being zero and the stress-strain relationship being dominated by the viscous effects. It should be noted that the characteristic strain rate is highly dependent on the magnitude of the cementation effects, i.e. the rate at which the strength increase per unit time due to curing. Since the cementation (curing) effect is seen to reduce rapidly with time after initial hydration in the artificially cemented kaolin, a comparable reduction in the characteristic strain rate can therefore be expected with time. Similar coupling between ageing effects and viscous effects may be expected in natural clays, which have been subjected to recent disturbance. Undisturbed natural clays on the other hand, which have been aged over a geological time scale are likely to have a characteristic strain rate, which is extremely low and primarily affected by changes in the surrounding environment rather than anything else.

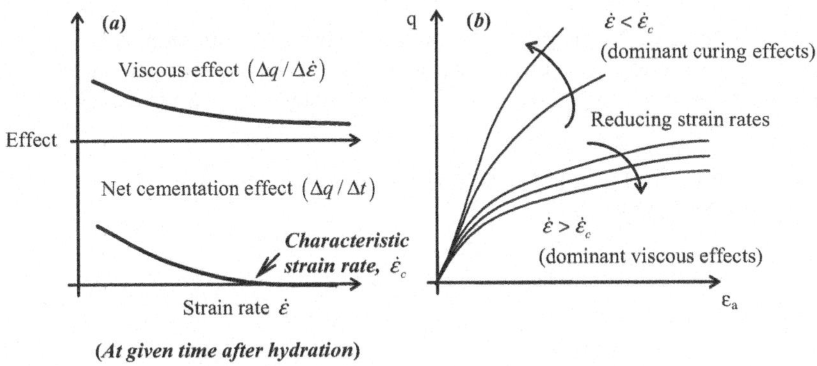

Figure 8 Illustration of interaction between cementation and viscous effects and definition of characteristic strain rate

ACKNOWLEDGEMENTS

All laboratory tests in this study were carried out at the Institute of Industrial Science, University of Tokyo, Japan as part of a short term study visit. A travel grant was co-sponsored by the Royal Academy of Engineering, the University College London Graduate School and the Department of Civil Engineering, University of Tokyo. The research was also made possible through funding from EPSRC's Cooperation Awards in Science and Engineering (CASE) in collaboration with Ove Arup and Partners.

REFERENCES

Komoto, N. (2004). *Experimental study on ageing effect using cement-mixed clay.* MSc thesis, University of Tokyo (in Japanese).

Kongsukprasert, L. & Tatsuoka, F. (2003). Viscous effects coupled with ageing effects on the stress-strain behaviour of cement-mixed gravel. *Proc. 3rd Int. Symp. Deformation Characteristics of Geomaterials, IS Lyon*, 569-577.

Leroueil, S., Kabbaj, M., Tavenas, F., & Bouchard, R. (1985). Stress-strain-strain rate relationship for the compressibility of sensitive natural clays. *Géotechnique* 35, No. 2, 159-180.

Mitchell, J. K., Baxter, C. D. P., & Soga, K. (1997). Time effects on the stress-deformation behaviour of soils. *Proc. of Professor Sakuro Murayama Memorial Symp., Kyoto University*, 1-64.

Tatsuoka, F., Santucci de Magistris, F., Hayano, K., Koseki, J., & Momoya, Y. (2000). Some new aspects of time effects on the stress-strain behaviour of stiff geomaterials. *The Geotechnics of Hard Soil - Soft Rocks, Proc. 2nd Int. Symp. Hard Soils and Soft Rocks, Napoli*, 1285-1371.

Viggiani, G. & Atkinson, J. H. (1995). Stiffness of fine-grained soil at very small strains. *Géotechnique* 45, No. 2, 249-265.

Soil Stress-Strain Behavior: Measurement, Modeling and Analysis
Geotechnical Symposium in Roma, March 16 & 17, 2006

VISCOUS PROPERTIES OF SANDS AND MIXTURES OF SAND/CLAY FROM HOLLOW CYLINDER TESTS

Antoine Duttine
Department of Civil Engineering, Tokyo University of Science
2641, Yamazaki Noda-City, Chiba-pref, 278-8510, JAPAN (formerly DGCB-ENTPE)
e-mail: antoine@rs.noda.tus.ac.jp
Herve Di Benedetto
Department of Civil Engineering DGCB
ENTPE Rue Maurice Audin, 69518 Vaulx-en-Velin Cedex, France
e-mail: herve.dibenedetto@entpe.fr
Damien Pham Van Bang
Electricité De France (EDF), Laboratoire National d'Hydraulique et d'Environnement,
Equipe de recherche EDF-CETMEF, 6 quai Watier, 78401 Chatou Cedex, France
e-mail: damien.pham-van-bang@edf.fr

ABSTRACT

Tests on air dried Hostun and Toyoura sands and on two moist mixtures of mainly Hostun sand with Kaolin clay were performed with a precision hollow cylinder device "T4C StaDy". Viscous properties are investigated through creep tests with and without rotation of stress principle axes (i.e. during triaxial compression - TC - and torsional shear - TS - tests) from small strain domain (some 10^{-5} m/m) up to large strain (some 10^{-2} m/m). A simplified version of the viscous evanescent model (VE), developed specifically at DGCB/ENTPE to model the peculiar viscous behaviour of sand, can be considered for creep tests. Good correlations are obtained between simulated and experimental creep strains for all the materials tested. A simple relation is confirmed for the viscous parameter of the VE model for sands (considering tests with and without rotation of stress principle axes) as well as for sand/clay mixtures (considering tests without rotation of stress principle axes).

1. INTRODUCTION

Investigation of viscous properties of geomaterials such as sands or mixtures of sand/clay requires some special laboratory equipment with a high degree of accuracy, mainly because amplitudes of viscous phenomena remain much lower than for clays and can be easily hidden by innacurate measurements. However, these phenomena may not be ignored and may exhibit non negligible effects at an engineerical scale (Tatsuoka et al., 1999, 2001, Jardine et al., 2005, Di Benedetto et al., 2005).

A number of previous studies with the use of relevant advanced testing apparatuses showed that sand exhibits noticeable creep deformation in drained TC, plane strain compression (PSC), and TS tests (Matsushita et al., 1999, Sauzeat et al., 2003, Di

Hoe I. Ling et al. (eds.), Soil Stress-Strain Behavior: Measurement, Modeling and Analysis, 367–382.

Benedetto et al., 2005, among others). Dual phenomenon i.e. stress relaxations have also been exhibited in drained TC (Matsushita et al., 1999, Pham Van Bang, 2004, Pham Van Bang et al., 2006, among others). Moreover, viscous behaviour of sand has been enlightened through stress jumps (resp. overshoots or undershoots) taking place upon step changes (resp. increases or decreases) in the strain rate during otherwise monotonic loadings (ML) at a constant strain rate. These stress jumps decay then with straining (while ML). Meanwhile, a rather unique stress-strain relationship is exhibited for several ML performed at constant but different strain rates under otherwise the same conditions (Matsushita et al., 1999, Di Benedetto et al., 2005, Pham Van Bang, 2004, Pham Van Bang et al., 2006). In this regard, the peculiar viscous behaviour observed for sand may be characterised as "viscous evanescent " (Di Benedetto et al., 2001, 2005) and has also been observed for saturated as well as air dried specimens so must be considered as free from effects of pore water (delayed dissipation of excess pore water pressure and so on) (Matsushita et al., 1999, Nawir et al., 2003, Pham Van Bang et al., 2003, Di Benedetto et al., 2005).

To describe this behaviour, a viscous evanescent (VE) model has been developped at ENTPE within the frawemork of a 3 component formalism. This model has been found to be relevant to simulate air dried Hostun sand viscous behaviour (Di Benedetto et al., 2001, Pham Van Bang et al., 2003, 2006, Sauzeat et al., 2003, Di Benedetto et al., 2005). From this model, a viscous parameter (η_0) can be exhibited and may be seen as a quantification of the magnitude of viscous properties for a given geomaterial.

In this paper, are reported additional experimental results on the viscous properties of airdried Hostun and Toyoura sands and also on two moist mixtures of Hostun sand/Kaolin clay (respectively including 85% of Hostun sand/15% of Kaolin clay – percentage by dry weight - with an initial water content of 4.5% and 70/30% of Hostun sand/Kaolin clay with an initial water content of 9.0%). These results have been obtained with the advanced prototype of hollow cylinder test "T4C StaDy" developped at the Civil Engineering Departement of ENTPE (Cazacliu, 1996, Cazacliu&Di Benedetto, 1998, Sauzeat, 2003, Duttine, 2005). This apparatus and the testing procedure are described in paragraph 2. In paragraph 3, the 3 component formalism and the VE model are presented. Application to our test conditions and simulations of experimental data are then reported and discussed.

2. TESTING APPARATUS AND PROCEDURE

2.1 Testing apparatus

The "T4C StaDy" (Figure 1) sample has a 12 cm height, an outer diameter of 20 cm and an inner diameter of 16 cm. These dimensions allow to assume reasonable stress and strain homogeneity as pointed out by Hight et al. (1983) or Sayao&Vaid (1991) among others. Two Neoprene membranes (0.5 mm thickness) constitute the lateral side while two rigid platens close the sample at the top and at the bottom. The top cap, connected to the press piston, is mobile in rotation and translation. The quasi-static loading of compression/ extension and torsion are ensured by a servo controlled hydraulic Instron press. Confinement is applied by depression inside the sample and by pressure in the confining cell.

Investigation of soil response from very small to large strain domains is possible with this device thanks to local strain measurement systems. Vertical (and/or angular) displacements are measured on two levels by two light rings (duralumin) glued on the outer membrane and carrying targets (aluminium) for non contact transducers. Radial (outer and inner) displacements are also measured by non contact transducers pointing towards sheets of aluminium paper placed on the inner side of the membranes. All the 14 non contact transducers are fixed on movable supports.
Moreover, the prototype is equipped with piezoelectric sensors (compression elements and bender elements) located in each platen closing the sample. Two bender elements are aligned following two different directions. One emits shear waves polarized in radial (S_r) direction and the other in orthoradial (S_θ) direction. The two compression elements (emitting compression waves) are identical. They are noted P_r and P_θ and are close to the respective sensors S. By back analysis of the waves travel times, dynamic elastic parameters of the specimen may be infered.

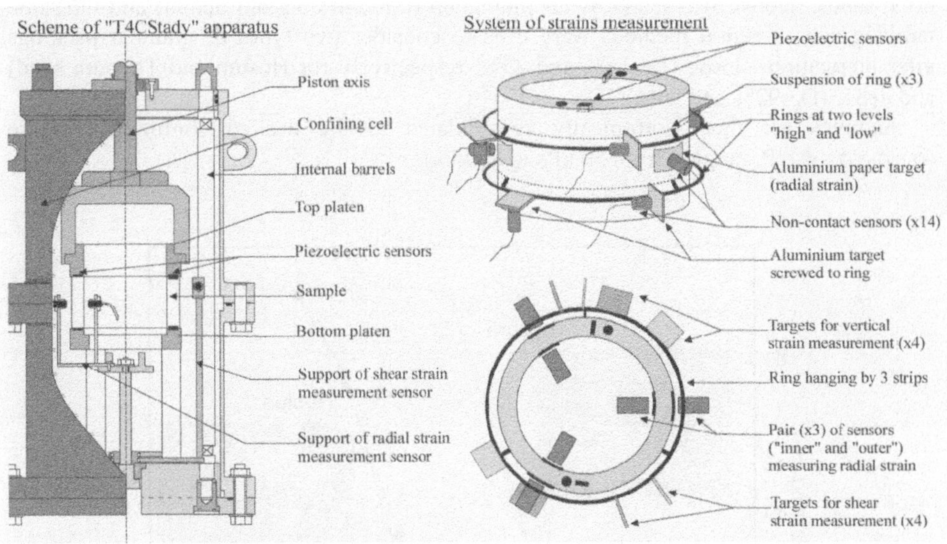

Figure1. Schematic view of the "T4C Stady" apparatus and of its system of strain measurement

For more details, the "T4C StaDy" device has been more extensively presented for example in Cazacliu (1996), Di Benedetto et al. (2001), Sauzeat (2003), Duttine (2005).

2.2 Tested materials and sample preparation

Tested materials include air-dried poor graded sands (Hostun and Toyoura sands) and two moist mixtures of Hostun sand and Kaolin clay (M15 and M30). Hostun and Toyoura sands are quartz dominated angular shaped sands whose grading curves are reported on figure 2 (characteristics on table 1). Particle shapes can also be seen on this figure.

The M15 sand/clay mixture is composed by 15% of Kaolin clay (w_l=35%, PI=14%) and 85% of Hostun sand (by dry weight) and by an initial global water content of 4.5%. The M30 sand/clay mixture is composed by 30% of Kaolin clay (w_l=35%, PI=14%) and 70% of Hostun sand (by dry weight) and by an initial global water content of 9.0%. For an initial global void ratio of 0.98 (defined as the ratio of the volume of clay and sand solid grains by the volume of void – water+air), preliminary conducted TC tests have shown that these materials exhibits respectively an apparent cohesion of 15.2 kPa and 25.8 kPa and a maximum friction angle of 30.5° and 25.2°, which appear to be lower than for air-dried Hostun sand only – 33.5° - during approximately the same conditions (Duttine, 2005). On a second hand, these tests showed that can be assumed an unique stress-strain relationship when ML at constant but different strain rates under otherwise the same conditions.

"T4C StaDy" samples of M15 and M30 mixtures were filled following 6 sub-layers with height and mass controlled and deposit using spoon. Tamping and vibration methods were used to reach the lowest possible fabric void ratio (e_0=0.98~0.99). Concerning air-dried sands, deposit was made by air pluviation (through constant height) and the same tamping and vibration methods were used to consider two types of granular packings after fabrication : loose (D_r≈35% and 25%, respectively for Hostun and Toyoura sand) and dense (D_r≈92% and 90%).

Samples are then isotropically consolidated to the desired confining pressure (σ_r=σ_θ=σ_z=σ_0=P, ranging from 50 kPa to 80 kPa).

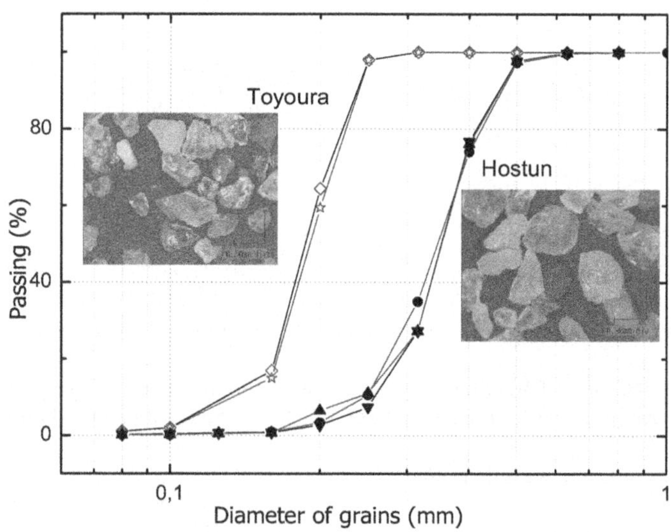

Figure2. Examples of the gradings of Hostun and Toyoura sands batches used in the present study and view of particle shapes

Passing	Diameter (mm)			Coefficients		Void ratios	
	D_{10}*	D_{30}*	D_{60}*	Cu**	Cc**	e_{min}***	e_{max}***
Hostun	0.26	0.32	0.37	1.42	1.06	0.648	1.041
Toyoura	0.13	0.17	0.20	1.33	1.02	0.605	0.977

* D_x defined by x% passing particle size
** Coefficient of uniformity: $Cu=D_{60}/D_{10}$ and coefficient of curvature : $Cc=(D_{30})^2/(D_{10}D_{60})$
*** Hostun : after Flavigny et al. (1990), Toyoura : data provided by Kitami Institute of Technology, Japan

Table 1. Grading characteristics of Hostun and Toyoura sands used in the present study

2.3 Experimental campaign

Figure 3 summarizes the different types of drained TC or TS tests performed in the different experimental campaigns. From the initial isotropic stress state, four steps are repeated successively :

i) the sample is vertically or torsionally loaded at constant stress rate ($\dot{\sigma}_z$ or $\dot{\tau}_{\theta z}$ constant) until an 'investigation point' is reached (A, B or C in figure 3);
ii) then, a creep period is imposed (AA', BB' or CC' in the figure 3);
iii) P and S waves propagations are performed;
iv) small quasi static cyclic loading (vertical and/or torsional) are applied after the creep period (at points A',B' or C' in the figure 3).

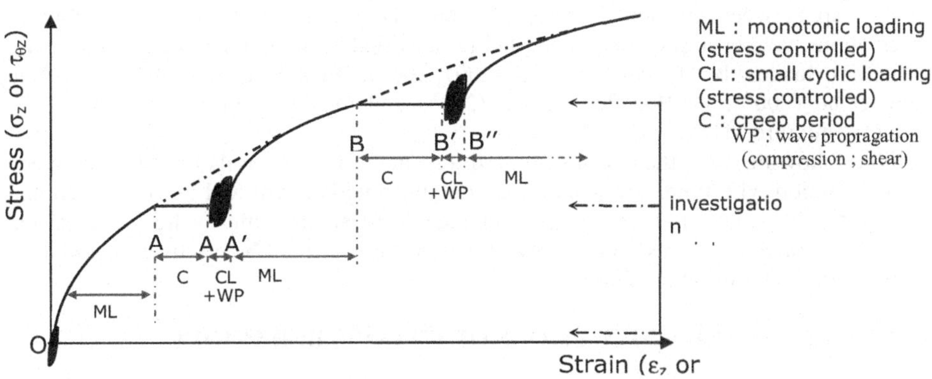

Figure2. Typical stress-strain relationships for TC and TS tests performed in the present study

A total of 29 tests has been performed on the different materials : 9 on airdried Hostun sand (4 on loose specimens , 5 on dense ones), 11 on airdried Toyoura sand (6 on loose, 5 on dense) and 9 on sand/c lay mixtures (4 on M15 and 5 on M30). Tests include TC tests from an isotropic stress state and TS tests from an anisotropic isotropic state (preceded by a monotonic TC to reach the corresponding stress ratio : σ_z / P =K=0.5). TC and TS tests were conducted on airdried sands whereas only TC tests were preliminary carried out on sand/clay mixtures. Moreover, considering these latter tests on sand/clay mixtures, the number of investigation stages was intentionnally limited in order to avoid the influence

of important ageing effect like water content evolution and induced cementation (degree of saturation was roughly constant during the tests, $S_r \approx 12.0\%$ and 23.0% respectively for M15 and M30 mixtures).
Test names follow the convention : Xyy.zz_M where X stand for the type of tests : C (for TC test) or K (for TS tests from anisotropic K stress state), yy, the confining pressure (P in kPa), zz the initial void ratio (percent, after fabrication) and M the type of material (H for Hostun, T for Toyoura; M15 or M30).
Figure 4 gives an example of typical results from tests K80.90_T (TS test from initial K stress state on air dried Toyoura sand ; e_0=0.90, P=80kPa) and C65.99_M15 (TC tests on M15 mixture ; e_0 =0.99, P=65kPa). On figures 4a)b)c) are respectively reported typical shear stress-shear strain relationship, the corresponding volumetric strain – shear strain relationship and the evolution with time of shear strain increment during creep at each investigation stage for tests K80.90_T. On figures 4d)e)f) are shown the similar plots for test C65.99_M15.
Attention needs to be attracted on figures 4b)&e) and on figures 4c)&f). From figures 4b)&e) one may notice that specimen behaviour becomes more contractive, resp. less dilative during creep straining when behaviour is initially (i.e. before creep) compressive, resp. dilative. These observations are consistent with TC tests results reported by AhnDan et al. (2001) on Chiba gravel and by Pham Van Bang (2004) and Pham Van Bang et al. (2006) on air dried Hostun sand considering otherwise creep periods but also upon step changes in the strain rate. From figures 4c)&f) may be noted that creep strain amplitudes and signs depend globally on actual stress states and stress history. More precisely, creep strain sign and amplitude may depend on last loading phase (loading or unloading) and on the gap between the actual stress state and the last reversal stress state. Similar observations have been previously reported for air dried Hostun or Toyoura sands during drained TC loading by DiBenedetto et al. (2001, 2002), Tatsuoka et al. (2002), Pham Van Bang et al. (2003), Di Benedetto et al. (2005) and may be consistenly reduced for sand/clay mixture M15 (figure 4f).

In the following, only the viscous properties of the tested materials linked to the creep strain evolution with time, stress and so on (as illustrated above) will be considered and simulated. The small strain properties obtained from small static cyclic loadings and dynamic loadings performed at each investigation stage will not be discussed herein (see Duttine, 2005, Duttine et al., 2006).

3. MODELLING AND SIMULATIONS OF VISCOUS BEHAVIOUR

3.1 General 3-component formalism and 1D VE model

The general three-component formalism (Di Benedetto, 1987) has been found to be relevant to describe correctly the viscous behaviour of many geomaterials (Di Benedetto et al., 2001, 2002, 2005, Tatsuoka et al., 2002, Tatsuoka, 2005). Its analogical representation is shown in figure 5, assuming the decomposition of strain increment into the sum of a non viscous (or instantaneous or inviscid) part and a viscous (or deferred or delayed or viscid) part (equation 1).

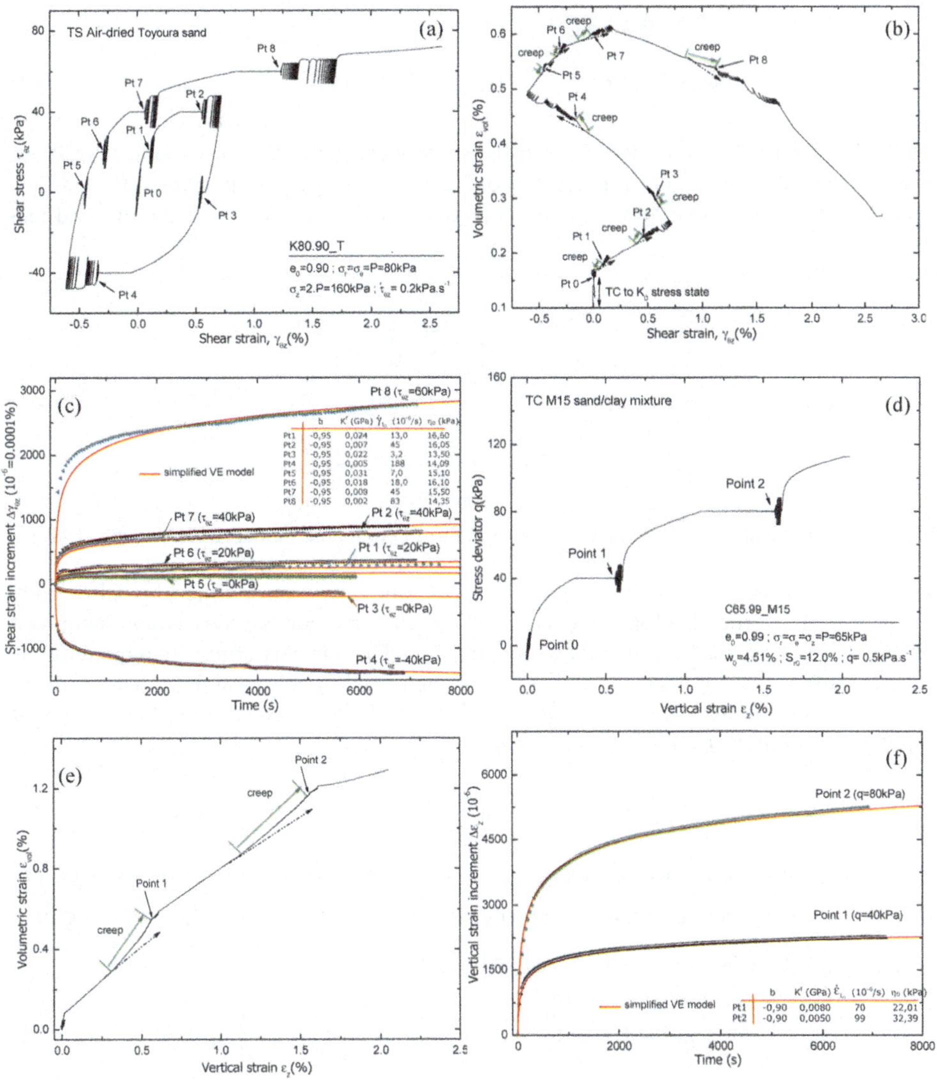

Figure 4. Typical experimental results respectively from TS test K80.90_T on air-dried Toyoura sand and TC test C65.99_M15 on moist M15 sand/clay mixture : a&c.shear (resp. deviator) stress- shear (resp. vertical) strain relationship, b&d. evolution of the volumetric strain with shear (resp. vertical) strain, c&f. experimental creep shear (resp. vertical) strain evolution with time and simulations by simplified VE model (see §3.3)

In addition, the stress increment acting on the viscous part of the strain increment is also divided into an inviscid part and a viscous part (equation 2).

$$\delta\varepsilon_{ij}=\delta\varepsilon_{ij}^{nv}+\delta\varepsilon_{ij}^{v} \text{ or } \underline{\underline{\dot{\varepsilon}}}=\underline{\underline{\dot{\varepsilon}}}^{nv}+\underline{\underline{\dot{\varepsilon}}}^{v} \quad (1)$$

$$\delta\sigma_{ij}=\delta\sigma_{ij}^{f}+\delta\sigma_{ij}^{v} \text{ or } \underline{\underline{\dot{\sigma}}}=\underline{\underline{\dot{\sigma}}}^{f}+\underline{\underline{\dot{\sigma}}}^{v} \quad (2)$$

where the exponents, "nv" and "v", over the strain increment "$\delta\varepsilon_{ij}$" or the objective strain rate "$\underline{\underline{\dot{\varepsilon}}}$" stand respectively for non viscous and viscous. The exponents, "f" and "v", over the stress increment "$\delta\sigma_{ij}$" or the objective stress rate "$\underline{\underline{\dot{\sigma}}}$" stand for inviscid and viscous.

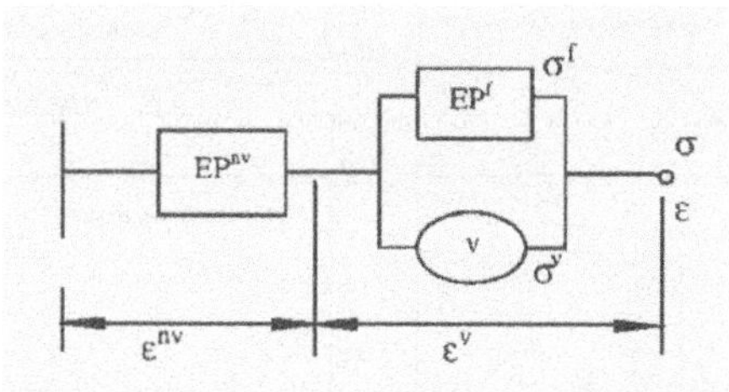

Figure 5. Analogical representation of the general three component model (Di Benedetto, 1987)

The non viscous part of the behaviour is obtained by considering the EP^{nv} body characterized by the rheological tensor M^{nv} linking the non viscous strain increment $\delta\varepsilon^{nv}$ to the total stress increment $\delta\sigma$ (equation 3). The viscous strain increment $\delta\varepsilon^{v}$ is obtained from the EP^f and V bodies whose tensorial rheological expressions differ (equation 4 and 5), the rheological tensor M^f (of the EP^f body, equation 4) having similar expression with the tensor M^{nv} (equation 3).

$$\delta\varepsilon_{ij}^{nv}=M_{ijkl}^{nv}\left(\text{dir}\{\underline{\underline{\dot{\sigma}}}\},h_1\right).\delta\sigma_{kl} \text{ or } \underline{\underline{\dot{\varepsilon}}}^{nv}=M^{nv}\left(\text{dir}\{\underline{\underline{\dot{\sigma}}}\},h_1\right).\underline{\underline{\dot{\sigma}}} \quad (3)$$

where M^{nv} is the non viscous tensor, h_1 is a set of history or memory parameters, $\delta\underline{\underline{\sigma}}$ and $\underline{\underline{\dot{\sigma}}}$ are an objective stress increment and an objective stress rate ($\delta\underline{\underline{\sigma}}=\underline{\underline{\dot{\sigma}}}.\delta t$). $\text{dir}\{\underline{\underline{\dot{\sigma}}}\}=\underline{\underline{\dot{\sigma}}}/\|\underline{\underline{\dot{\sigma}}}\|$ is the direction of the objective stress rate.

$$\delta\varepsilon_{ij}^{v}=M_{ijkl}^{f}\left(\text{dir}\{\underline{\underline{\dot{\sigma}}}^{f}\},h_3\right).\delta\sigma_{kl}^{f} \text{ or } \underline{\underline{\dot{\varepsilon}}}^{v}=M^{f}\left(\text{dir}\{\underline{\underline{\dot{\sigma}}}^{f}\},h_3\right).\underline{\underline{\dot{\sigma}}}^{f} \quad (4)$$

where M^f is the inviscid tensor, h_3 is a set of history or memory parameters, $\delta\underline{\underline{\sigma}}^f$ and $\underline{\underline{\dot{\sigma}}}^f$ are an objective inviscid stress increment and an objective inviscid stress rate ($\delta\underline{\underline{\sigma}}^f=\underline{\underline{\dot{\sigma}}}^f.\delta t$). $\text{dir}\{\underline{\underline{\dot{\sigma}}}^f\}=\underline{\underline{\dot{\sigma}}}^f/\|\underline{\underline{\dot{\sigma}}}^f\|$ is the direction of the inviscid stress rate.

$$\delta\varepsilon^{v}_{ij} = N_{ijkl}(h_2).\sigma^{v}_{kl}.\delta t \quad \text{or} \quad \underline{\underline{\dot{\varepsilon}}}^{v} = N(h_2)\underline{\underline{\sigma}}^{v} \quad \text{or} \quad \underline{\underline{\sigma}}^{v} = F(h_2, \underline{\underline{\dot{\varepsilon}}}^{v}) \tag{5}$$

where N is the viscous tensor and F the viscous stress tensor function, $\underline{\underline{\sigma}}^{v}$ is the viscous stress, $\underline{\underline{\dot{\varepsilon}}}^{v}$ is the objective viscous strain rate, h_2 a set of history parameters which can differ from the sets h_1 and h_3.

To describe the peculiar viscous behaviour observed for sands, the following general expression of the viscous stress is considered in the VE model considering otherwise a 1D case (Di Benedetto et al., 2001) :

$$\sigma^{v}_{(t)} = \int_{\chi=0}^{t} \left[d\{f(\dot{\varepsilon}^{v}_{(\chi)})\} \right] . g_{decay}(\varepsilon^{v}_{(t)} - \varepsilon^{v}_{(\chi)}) \tag{6}$$

where $d\{f(\dot{\varepsilon}^{v}_{(\chi)})\}$ corresponds to the viscous stress increment at time χ, or $[d\sigma^{v}]_{(\chi)}$. $g_{decay}(\varepsilon^{v}_{(t)} - \varepsilon^{v}_{(\chi)})$ corresponds to the value at current time t of a unity event produced at time χ. The weighting function, g_{decay} expresses the decrease with straining upon the influence of any strain rate changes. This function tends monotonously towards zero for a pure evanescent behaviour and can be chosen equal to unity to describe a classical isotach behaviour (Di Benedetto et al., 2002, 2005). Previous studies (Di Benedetto et al., 2001, Sauzeat et al., 2003, Pham Van Bang&Di Benedetto, 2003, Sauzeat, 2003, Pham Van Bang, 2004) showed that the following expressions of the viscosity function f and of the decay function g_{decay} are relevant for TC and TS tests performed on air-dried Hostun sand :

$$\begin{cases} \sigma^{v}_{isotach} = f(\dot{\varepsilon}^{v}) = \eta_0 . \left(\dfrac{\dot{\varepsilon}^{v}}{\dot{\varepsilon}^{v}_{0}} \right)^{1+b} & \text{for } \dot{\varepsilon}^{v} \geq 0 \\ g_{decay}\left(\varepsilon^{v}_{(t)} - \varepsilon^{v}_{(\chi)}\right) = \exp\left(-\dfrac{\varepsilon^{v}_{(t)} - \varepsilon^{v}_{(\chi)}}{\varepsilon_{ref}} \right) \end{cases} \tag{7}$$

where f is the viscosity function and $\{\eta_0 \,;\, b \,;\, \varepsilon^{ref}\}$ 3 model parameters to be determined. $\dot{\varepsilon}^{v}_{0}$ is the reference strain rate ($=10^{-6}$/s). Note that equation 7 can be extended for negative value of $\dot{\varepsilon}^{v}$: $\sigma^{v} = sg(\dot{\varepsilon}^{v}).f(|\dot{\varepsilon}^{v}|)$ where $sg(\dot{\varepsilon}^{v})$ is the sign of $\dot{\varepsilon}^{v}$ and $|\dot{\varepsilon}^{v}|$ the norm of $\dot{\varepsilon}^{v}$. b must be chosen in the range from –1 (no viscous effects) to 0 (Newtonian viscosity).

A possible extension of the 1D VE model (equations 6&7) to a more general 3D model is presented in Sauzeat et al. (2003) and Di Benedetto et al. (2005). This generalization is based on the colinearity of the tensors $\underline{\underline{\sigma}}^{v}$ and $\underline{\underline{\dot{\sigma}}}^{f}$ and on similar equations as relations 6 and 7 but linking the norm of the differents tensors ($\|\underline{\underline{\sigma}}^{v}\|$; $\|\underline{\underline{\dot{\varepsilon}}}^{v}\|$ and so on).

3.2 Application to creep tests and calibration of the model

As shown typically in figure 4, experimental creep results reveal that creep strains are restricted to some 10^{-3} m/m for the materials tested. In that case and considering otherwise that typical values of ε^{ref} acting on the weighting function g_{decay} (equation 7) are found to be of the order of 10^{-2} m/m, Sauzeat (2003), Sauzeat et al. (2003) showed that the VE model can be relevantly simplified by neglecting the evanescent property, i.e. by considering equation 7 only. Thus, from the condition $\dot{\sigma} = \dot{\sigma}^f + \dot{\sigma}^v = 0$ (*) (for a creep period starting at a time $t{=}t_0$) may be derived the following differential equation ruling the creep strain evolution with time :

$$0 = K^f \dot{\varepsilon}^v (\dot{\varepsilon}_0)^{1+b} + (1+b)\ddot{\varepsilon}^v \eta_0 (\dot{\varepsilon}^v)^b \tag{8}$$

where K^f is the tangent modulus of the EP^f body, η_0 and b are the viscous model parameters to be determined (equation 7). $\dot{\varepsilon}_0$ is a reference strain rate ($=10^{-6}$/s).
Due to small strain evolution, parameters $\{K^f\ ;\ \eta_0\ ;\ b\}$ may be considered as constants. Double integration of equation 8 leads to :

$$\varepsilon^{creep}_{(t)} = \varepsilon^v_{(t)} - \varepsilon^v_{(t_0)} = \frac{\eta_0}{K^f} \cdot \left[\left(\frac{\dot{\varepsilon}^v_{(t_0)}}{\dot{\varepsilon}_0} \right)^{1+b} - \left[\left(\frac{\dot{\varepsilon}^v_{(t_0)}}{\dot{\varepsilon}_0} \right)^b - \frac{b.K^f.\dot{\varepsilon}_0.(t-t_0)}{\eta_0.(1+b)} \right]^{\frac{b+1}{b}} \right] \tag{9}$$

K^f value can be determined by equation 10, referring to figure 6 :

$$\frac{1}{E_{tan}} = \frac{1}{K^{nv}} + \frac{1}{K^f} \tag{10}$$

where E_{tan} stands for the tangent Young modulus just before point A (start of the creep period, figure 6), K^{nv} is the small strain quasi elastic Young modulus, corresponding to the tangent modulus of the EP^{nv} body and which can be obtained experimentally by the small cycling loadings performed in our tests. Figure 6 shows the different stress-strain curves and moduli involved for a loading including a creep period.
By plotting experimentally the viscosity η versus the creep strain rate can be verified the constance of the parameter b (equation 11 and figure 7). This parameter is found to be constant and equal to –0.95, respectively –0.90, for all the creep periods performed on air dried sands, respectively on sand/clay mixtures.
The last parameter η_0 may then be the most accurately determined by the best fitting between experimental and calculated viscous strains (equation 9), as shown in the next paragraph.

(*) As 1D case is considered, the equations established in this section are relevant for TC as well as for TS tests by simply replacing ε by ε_z or $\gamma_{\theta z}$ and σ by σ_z or $\tau_{\theta z}$

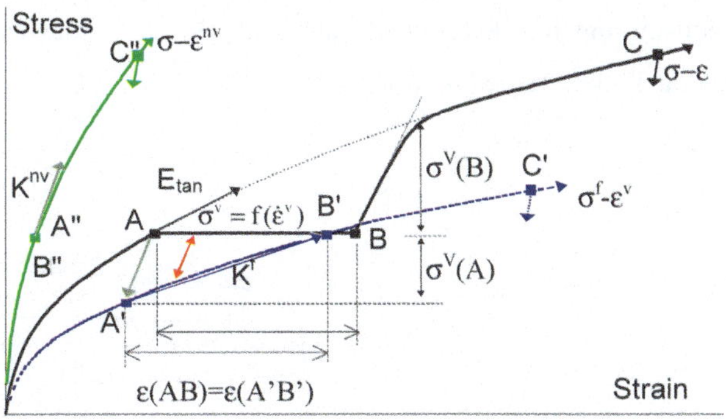

Figure 6. Viscous and non viscous stress-strain curves during a loading with a creep period.

$$\eta = \frac{\sigma^v(t)}{\dot{\varepsilon}_1^v(t)} = \frac{K^f(\varepsilon^{fin} - \varepsilon(t))}{\dot{\varepsilon}_1^v(t)} \tag{11}$$

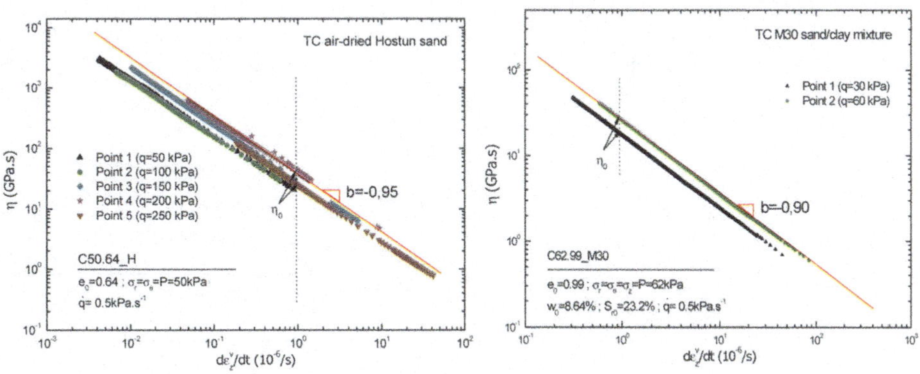

Figure 7. Viscosity η (equation 11) and viscosity parameter η_0 (equation 7) obtained for creep periods during tests C50.64_H and C62.99_M30

The last parameter η_0 may then be the most accurately determined by the best fitting between experimental and calculated viscous strains (equation 9), as shown in the next paragraph.

3.3 Simulations of creep strain evolution with time

Simulations of creep strain evolution with time can be seen on figure 4c) resp. 4f) for tests K80.90_T on air-dried Toyoura sand, resp. C65.99_M15 on sand/clay mixture M15. They are obtained by a least square optimization between experimental and calculated viscous strains considering the determined values of $\{K^f\ ;\ \dot{\varepsilon}^v_{(t_0)}\}$ at each investigation stage and the previously determined b-value.

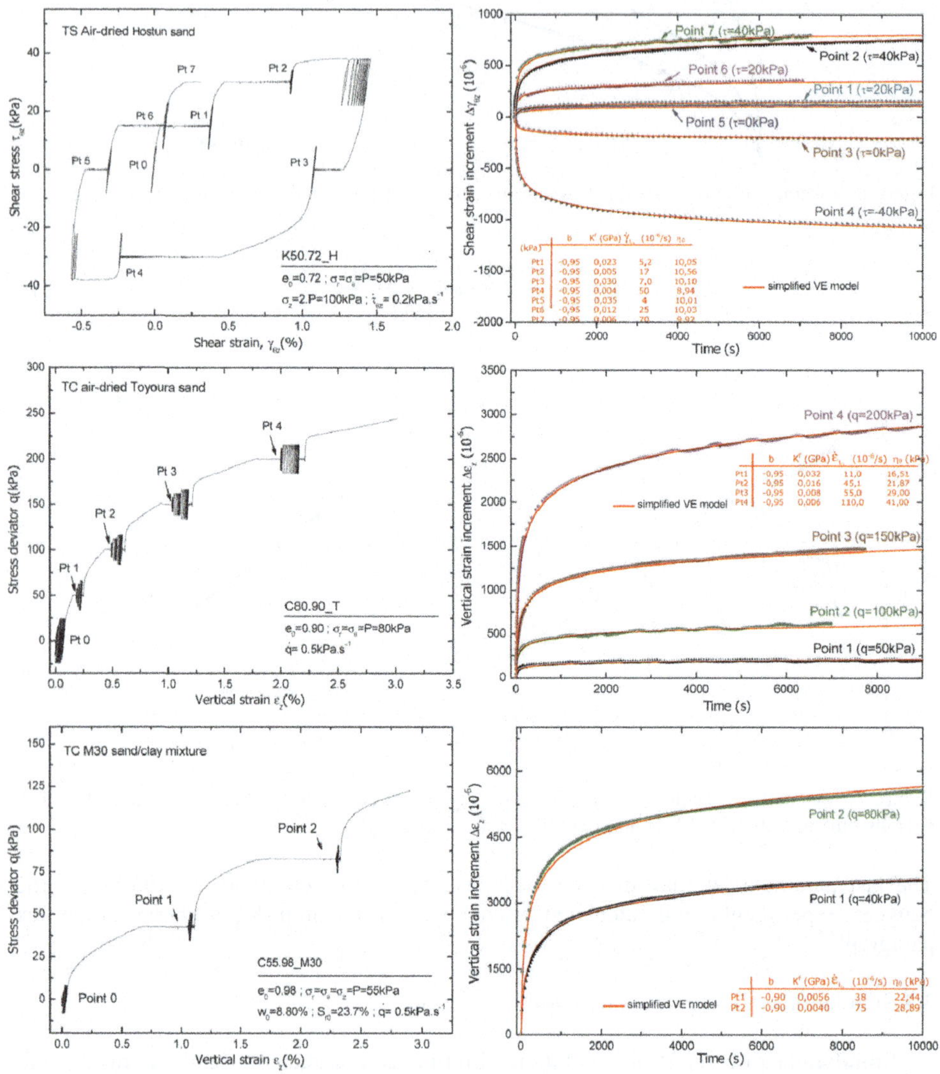

Figure 8. Stress-strain relationships and creep strain evolution with time with simplified VE model simulations and the corresponding parameters values for tests K50.72_H (air-dried Hostun sand), C80.90_T (air-dried Toyoura sand) and C55.98_M30 (M30 san/clay mixture)

On these figures are precisely reported the corresponding values of $\{K^f ; \eta_0 ; \dot{\varepsilon}^v_{(t_0)} ; b\}$ for each simulation. Figure 8 shows another examples of simulations for tests K50.72_H, C80.90_T, and C55.98_M30. From these figures may be seen that the creep strain evolution with time are correctly described by the simplified VE model, i.e by equation 9, for air-dried Hostun and Toyoura sands under different experimental conditions (TC or TS on loose and dense specimens) as well as for moist sand/clay mixtures (M15 or M30) under TC.
The obtained values of viscosity parameter η_0 for all the tests performed are gathered and discussed on the next paragraph.

3.4 Viscosity parameter η_0

Pham Van Bang et al.(2006), Di Benedetto et al.(2005) showed that the viscosity parameter η_0 of the 1D VE model can be relevantly plotted versus the maximum principal stress σ_1 for TC tests performed on air-dried Hostun sand, leading to the relation :

$$\eta_0 = \alpha^* . \sigma_1 \quad (12)$$

where $\alpha^*=0.15$.
Figure 9a follows this consideration for all the tests performed in this study. It has to be noted that for TS test data, the following 3D extension of equation 7 has been considered (Sauzeat et al., 2003, Di Benedetto et al., 2005):

$$\left\| \underline{\underline{\sigma}}^v \right\| = \eta_0 . \left(\frac{\left\| \dot{\underline{\underline{\varepsilon}}}^v \right\|}{\dot{\varepsilon}^v_0} \right)^{1+b} \quad (13)$$

therefore leading to the generalized expression of the viscosity parameter η_0 for TS test :

$$\begin{cases} \left\| \underline{\underline{\sigma}}^v \right\| = \sqrt{2} . \tau_{\theta z}{}^v \\ \left\| \dot{\underline{\underline{\varepsilon}}}^v \right\| \approx \sqrt{(\dot{\gamma}_{\theta z}{}^v)^2 / 2} \end{cases} \Rightarrow \quad \eta_0 \approx 2^{1+\frac{b}{2}} (\eta_0)_{TS} \quad (14)$$

assuming otherwise that the other terms of $\dot{\underline{\underline{\varepsilon}}}^v$ are negligible in front of $\dot{\gamma}_{\theta z}{}^v$.
On figure 9b are reported the data from TC and TS tests performed in this study on air-dried sands. They are associated with the η_0 values obtained from TC tests performed on air-dried Hostun sands thanks to a precision device by Pham Van Bang et al. (2006) through not only creep tests but also relaxation tests and stepwise changes in the strain rate.
From figure 9a may be seen that equation 12 is relevant for the two air-dried poorly-graded sands and also for the two moist sand/clay mixtures tested in this study. The 3D equation 13 (linked with relation 12) may also be validated through the consistency between the results obtained from TC and TS tests.

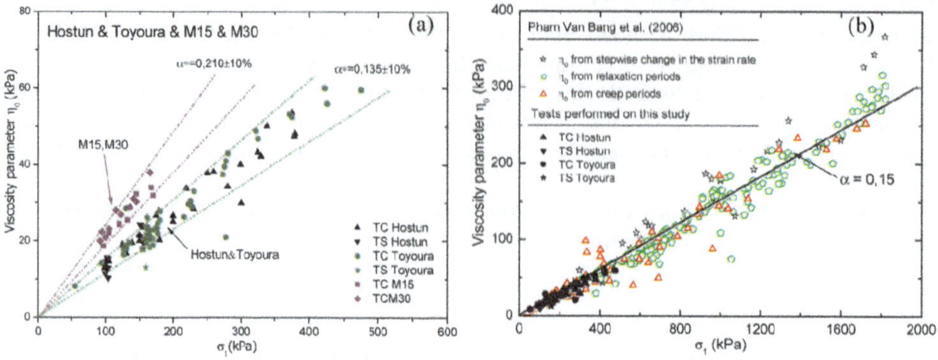

Figure 9. Viscosity parameter η_0 (equation 11) of the VE model function of maximum principal stress σ_1 for all tests performed in this study (a.) and for tests performed on air-dried sands only with data from TC tests performed on air-dried Hostun sand with a precision triaxial device and reported by Pham Van Bang et al. (2006) (b.)

In addition, a common value of α^* can be exhibited at $\pm 10\%$ separately for the two kinds of materials. For sand/clay mixtures, further studies are required to precise the complex influence of each component and to explain these results. For the poorly-graded angular sands, these results are somehow consistent with remarks reported in Tatsuoka (2005) on the viscous properties of these two sands, the author using otherwise a similar non-linear 3 component formalism. Moreover, TC tests performed on air-dried Hostun sand with a triaxial precision device covering otherwise a larger range of σ_1 and different types of loadings involving viscous properties exhibit a similar α^* value, as can be seen in figure 9b.

4. CONCLUSION

TC and TS tests have been performed on air dried Hostun and Toyoura sands thanks to a hollow cylinder precision device ("T4C StaDy"). Additionnaly, TC tests have been carried out with the same apparatus on two moist mixtures of mainly Hostun sand with Kaolin clay. Viscous properties are investigated through creep tests from small strain domain (some 10^{-5} m/m) up to large strain (some 10^{-2} m/m) and simulated by a simplified version of the viscous evanescent model (VE), developed specifically at DGCB/ENTPE to model the peculiar viscous behaviour of sand.
From the tests results presented in this paper may be derived the following conclusions :
i) the simplified version of the VE model can simulate in a relevant way the creep straining for air-dried Hostun and Toyoura sands under TC and TS loadings, and for two moist sand/clay mixtures under TC loading,
ii) a simple expression of the viscosity parameter η_0 of this model is confirmed for the two kinds of materials tested in this study

iii) a 3D extension of this expression is confirmed through the two different types of loadings (TC and TS, i.e. with and without continuous rotation of principal axes) involved in tests performed on air-dried sands
iv) the magnitude of purely viscous properties are found to be equivalent between the two moist sand/clay mixtures, and between the two poorly graded angular sands, in a consistent manner with previous studies.

ACKNOWLEDGMENT

The authors wish to thank Electricité de France (EDF) for their collaboration and financial support during this study.

REFERENCES

1) AhnDan L., Koseki J., Tatsuoka F. (2001). Viscous deformation in triaxial compression of dense well-graded gravels and its model simulation. In : Tatsuoka F., Shibuya S., Kuwano R. Eds. Advanced laboratory stress-strain testing of geomaterials. Rotterdam, Pays-Bas : Balkema, 2001, pp.187-194
2) Cazacliu, B. (1996). Comportement des sables en petites et moyennes déformations - réalisation d'un prototype d'essai de torsion compression confinement sur cylindre creux. PhD thesis , ECP/ENTPE, Paris.
3) Cazacliu, B. and Di Benedetto, H. (1998). Nouvel essai sur cylindre creux de sable. Revue Française de Génie Civil, 2, No. 27
4) Di Benedetto H. (1987). Modélisation du comportement des géomatériaux : application aux enrobés bitumineux et aux bitumes. Thèse de docteur d'Etat. Grenoble : USTMG, 1987.
5) Di Benedetto H., Sauzeat C., Geoffroy H. (2001). Time dependent behaviour of sand. In : Jamiolkowski et al. Eds. Proc. of the 2nd Int. Symp. on Deformation Characteristics of Geomaterials, sept. 1999, Torino, Italie. Rotterdam, Pays-Bas : Balkema, 2001, vol. 2, pp. 1357-1367
6) Di Benedetto, H., Tatsuoka, F., Ishihara M. (2002) Time dependent shear deformation characteristics of sand and their constitutive modelling. Soils and Foundations; vol.42, n°2,pp.1-22.
7) Di Benedetto, H., Tatsuoka, F., Lo Presti, D., Sauzeat C., Geoffroy H., (2005). Time effects on the behaviour of geomaterials. In : Di Benedetto H. et al. Eds. Proc. of the 3rd Int. symp. on Deformation Characteristics of Geomaterials, sept. 2003, Lyon. Rotterdam, Pays-Bas : Balkema, vol.2, pp59-124
8) Duttine, A. (2005). Comportement des sables et des mélanges sable/argile sous sollicitations statiques et dynamiques avec et sans « rotation d'axes ». Ph.D thesis, ENTPE, Lyon, France
9) Duttine A., Di Benedetto H.,Pham Van Bang D.,Ezaoui A. (2006). Anisotropic small strain elastic properties of sands and mixture of sand/clay measured by dynamic and static methods. Soils and Foundations (submitted)
10) Flavigny E., Desrues J., Palayer B. (1990). Note technique : le sable d'Hostun RF. Revue Française de Géotechnique, 1990, vol. 53, pp. 67-70
11) Hight, D. W., Gens, A. and Symes, M. J. (1983). The development of a new hollow cylinder apparatus for investigating the effects of principle stress rotation in soils. Géotechnique vol.33, n°4, pp. 355–384.
12) Jardine R.J., Standing J.R., Kovacenic N. (2005) : "Lessons learned from full scale observations and the practical application of advanced testing and modelling", In : Di Benedetto

H. et al. Eds. Proc. of the 3rd Int. symp. on Deformation Characteristics of Geomaterials, sept. 2003, Lyon. Rotterdam, Pays-Bas : Balkema, vol.2, pp201-246

13) Matsushita M., Tatsuoka F., Koseki J., Cazacliu C., Di Benedetto H., Yasin S.J.M (1999). Time effects on the pre-preak deformations properties of sand. In : Jamiolkowski et al. Eds. Proc. of the 2nd Int. Conf. on Deformation Characteristics of Geomaterials, sept. 1999, Torino, Italie. Rotterdam, Pays Bas : Balkema, 1999, vol.1, pp.681-689

14) Nawir, H., Tatsuoka, F., Kuwano, R. (2003). Experimental evaluation of the viscous properties of sand in shear. Soils and Foundation, 2003, vol. 43, n°6, pp. 13-31

15) Pham Van Bang, D., Di Benedetto, H. (2003). Effects of strain rate on the behaviour of dry sand. In : Di Benedetto H. et al. Eds. Proc. of the 3rd Int. symp. on Deformation Characteristics of Geomaterials, sept. 2003, Lyon. Rotterdam, Pays-Bas : Balkema, 2003, vol.1, pp. 365-374

16) Pham Van Bang, D. (2004). Comportement instantané et différé des sables des petites aux moyennes déformations : expérimentation et modélisation. Ph.D thesis. Lyon : ENTPE/INSA Lyon, 2004

17) Pham Van Bang, D., Di Benedetto, H., Duttine, A., Ezaoui, A. (2006). Viscous behaviour of sands : airdried and triaxial conditions. International Journal for Numerical and Analytical Methods in Geomechanics (accepted)

18) Sauzeat, C. (2003). Comportement du sable dans le domaine des petites et moyennes deformations : rotations "d'axes" et effets visqueux, Phd thesis, ENTPE, Lyon, France.

19) Sauzeat, C., Di Benedetto, H., Chau, B., Pham Van Bang, D. (2003). A rheological model for the viscous behaviour of sand. Di Benedetto H. et al. Eds. Proc. of the 3rd Int. symp. on Deformation Characteristics of Geomaterials, sept. 2003, Lyon. Rotterdam, Pays-Bas : Balkema, 2003, vol.1, pp. 1201-1209

20) Sayao A., Vaid Y.P. (1991). A critical assessment of stress nonuniformities in hollow cylinder tests specimens. Soils and Foundations, vol. 31, n°1, pp.60-72

21) Tatsuoka F., Jardine R.J., Lo Presti D., Di Bendetto H., Kodaka T. (1999) : "Characterising the pre-failure deformation properties of geomaterials", Theme Lecture, Proc. of 14th Int. Conf. on Soil Mechanics and Foundation Engineering, Hamburg, vol. 4, pp. 2129-2164

22) Tatsuoka F., Shibuya S., Kuwano R. (2001) : "Recent advances in stress-strain testing of geomaterials in laboratory", Advanced laboratory stress-strain testing of geomaterials, In (Tatsuoka et al. eds), Balkema, pp. 1-12

23) Tatsuoka F, Ishihara M, Di Benedetto H, Kuwano R.(2002). Time dependent shear deformation characteristics of geomaterials and their simulation. Soils and Foundations, vol.42, n°2, pp.103-129.

24) Tatsuoka F..(2005) : "Effects of viscous properties and ageing on the stress-strain behaviour of geomaterials" Proc. Of the GI-JGS workshop, Boston, ASCE SPT 143 (Yamamuro&Koseki eds),pp.1-60

VISCOUS PROPERTY OF GRANULAR MATERIAL IN DRAINED TRIAXIAL COMPRESSION

Enomoto, T.
Department of Civil Engineering, The University of Tokyo, 4-6-1 Komaba, Meguro-ku, Tokyo 153-8505, Japan, e-mail: tenomoto@iis.u-tokyo.ac.jp

Tatsuoka, F.
Department of Civil Engineering, Tokyo University of Science, 2641 Yamazaki, Noda City, Chiba 278-8510, Japan, e-mail: tatsuoka@rs.noda.tus.ac.jp

Shishime, M. and Kawabe, S.
ditto

Di Benedetto, H.
Département Génie Civil et Bâtiment, Ecole Nationale des Travaux Publics de l'Etat, France, e-mail: herve.dibenedetto@entpe.fr

ABSTRACT

The viscous properties of a wide variety of reconstituted loose and dense unbound granular materials, including natural sands and gravels, were evaluated by a series of drained triaxial compression (TC) tests at fixed confining pressure. In total sixteen granular materials having different mean particle diameters, D_{50}, coefficients of uniformities, U_c, fines contents, *FC*, degrees of crushability and particle shapes were newly tested. The viscous properties were quantified basically by many times changing stepwise the axial strain rate and partially by performing drained sustained loading during otherwise drained monotonic loading (ML) at a constant strain rate. It is shown that the viscous properties can be represented by the rate-sensitivity of the stress upon a step change in the strain rate, the decay rate of a viscous stress increment during the subsequent ML at a constant strain rate and the dependency of the residual stress during ML at a constant strain rate. The effects of the particle characteristics on the viscous properties were evaluated by summarising the results from the present and previous studies. As a new and surprising fact, with poorly graded unbound round granular materials, the stress for the same strain during ML at a constant strain rate decreases with an increase in the strain rate.

1. INTRODUCTION

It is often required to evaluate the residual deformation of ground and residual structural displacements for serviceability design of civil engineering structures. To this end, it is necessary to understand correctly the viscous properties of soil. Despite that the long-term compression of sand and gravel sometime becomes an important engineering

Hoe I. Ling et al. (eds.), Soil Stress-Strain Behavior: Measurement, Modeling and Analysis, 383–397.

issue (e.g., Tatsuoka & Kohata, 1995; Tatsuoka et al., 1999; Jardine et al., 2004; Di Benedetto et al., 2004; Day, 2005), the study on the viscous properties of unbound granular materials has been rather limited when compared with that of clay (e.g., secondary consolidation). Yet, it has been shown by a number of previous studies that sand exhibits significant creep deformation in drained TC tests and plane strain compression (PSC) tests (e.g., Matsushita et al., 1999) and torsional shear tests (e.g., Benedetto et al., 2004). Namely, unbound granular materials have significantly viscous properties. Tatsuoka et al. (2001, 2006), Tatsuoka (2004) and Di Benedetto et al. (2004) showed that the stress-strain behaviour of unbound and bound geomaterials (e.g., natural and reconstituted clay, sand, gravel, sedimentary soft rock and cement-mixed soil) is all rate-dependent (or more specifically, elasto-viscoplastic) even when free from the effects of pore water as well as delayed dissipation of excess pore water pressure; i.e., unbound granular material is not an exception in this regard.

The recent findings with respect to the viscous properties of granular material can be summarized as follows:

1) The magnitude of the viscous property of a given geomaterial can be adequately represented by the rate-sensitivity coefficient, β (explained later).
   a) The β value is basically independent of particle size (Tatsuoka, 2004).
   b) With a poorly graded fine sand (i.e., Toyoura sand) in drained TC tests (Nawir et al., 2003a & b) and drained PSC tests (Kongkitkul et al., 2005), the effects of confining pressure and dry density on the β value are insignificant, if any.
   c) Kiyota and Tatsuoka (2006) showed that, with three types of poorly graded relatively angular sand (Toyoura, Hostun and silica No. 8 sands), the same definition for the rate-sensitivity coefficients, β, is relevant to the TC and triaxial extension (TE) tests. They also showed that the effects of over-consolidation on the β value are insignificant.
2) The most striking trend of the viscous behaviour of poorly graded granular materials has been that a stress jump that takes place upon a step change in the strain rate during otherwise ML at a constant strain rate decays with an increase in the strain during the subsequent ML at a constant strain rate. By this feature, the stress-strain relations from ML tests at constant but different strain rates performed under otherwise the same conditions tend to collapse into a single and unique one. At the same time, significant creep deformation and stress relaxation takes place. This type of viscous property is called the TESRA viscosity (i.e., temporary effects of strain rate and strain acceleration on the viscous property) and has been formulated in the framework of a non-linear three-component model (explained later).

Despite these findings, there are a number of other factors that may control the viscous properties of unbound granular materials that are not well understood, such as grading characteristics, particle shape and so on. In particular, the range of particle size that has been examined is not wide, while most of the granular materials examined are poorly graded, relatively angular and basically uncrushable. Furthermore, several test results with respect to the rate-dependency of the stress-strain relation that cannot be explained or simulated by the TESRA viscosity have been found. For example, **Fig. 1a** shows the results from a series of CD TC tests at different constant strain rates on a crushed concrete aggregate consisting of relatively round, strong and stiff core particles covered with a thin weak and soft mortar layer (see Fig. 3 for the grading curve). It may be seen that, with an increase in the axial strain rate, the stress at the same axial strain decreases in the pre-peak regime but the opposite is true in the post-peak regime. On the other hand, this material exhibits a sudden increase and decrease in the stress upon a step increase and decrease in the axial strain rate (**Fig. 1b**). It was observed in the post-peak regime in direct shear tests on Ottawa sand (a round uniform sand) (Mair & Marone, 1999) and simple shear tests on glass beads (Chambon et al., 2002) that, upon a step increase in the shear displacement rate, $\dot{d}$, during otherwise ML at a constant $\dot{d}$, the shear stress exhibits a sudden increase, followed by a rapid decay in the shear stress towards to the residual value that is lower than the one during the precedent ML at a lower $\dot{d}$ (i.e., the residual shear stress during ML decreases with an increase in $\dot{d}$, like the pre-peak behaviour of a crushed concrete aggregate shown in Fig. 1a).

In view of the above, a new series of drained TC tests at fixed effective confining pressure were performed to evaluate the effects of particle characteristics (i.e., particle size, coefficient of uniformity, fines content, particle shape and particle crushability) on the viscous properties of unbound granular materials.

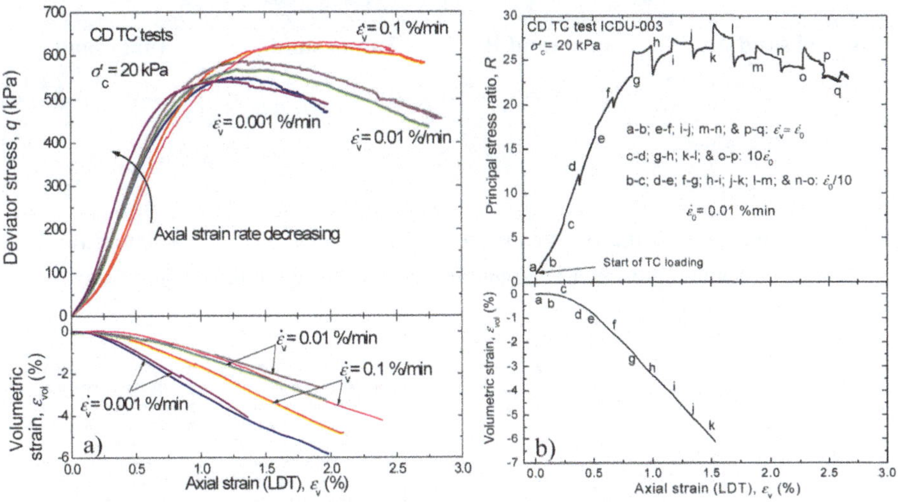

Fig. 1: Stress-strain relations from CD TC tests : a) ML at different constant strain rates; and b) ML with step changes in the strain rate, crushed concrete aggregate (G_s= 2.65; D_{max}= 19 mm; D_{50}= 5.84 mm; U_c= 18.76; FC= 1.32 %) (Aqil et al., 2005).

2. TEST MATERIALS

The particle pictures of some representative materials tested in the present study are shown in **Fig. 2**. **Fig. 3** shows the grading curves of the granular materials newly tested in the present study and those from the previous studies of which the data are referred to in this paper.

Silica Nos. 3, 4, 5, 6 and No. 8 sands, which are all poorly graded, have different D_{50} values with similar relatively angular particles. Mixed silica sand was made by mixing these silica sands to have a larger U_c value. Coral sands A and B have different grading curves while coral sand B includes crushable shell fractions. Ishihama beach sand is poorly graded and sub-angular. Tanno and Inagi sands are inland weathered sands having crushable particles. Corundum A (granular aluminum oxide), Albany silica sand and Hime gravel have relatively round uncrushable particles.

Fig. 2: Some representative granular materials tested in the present study.

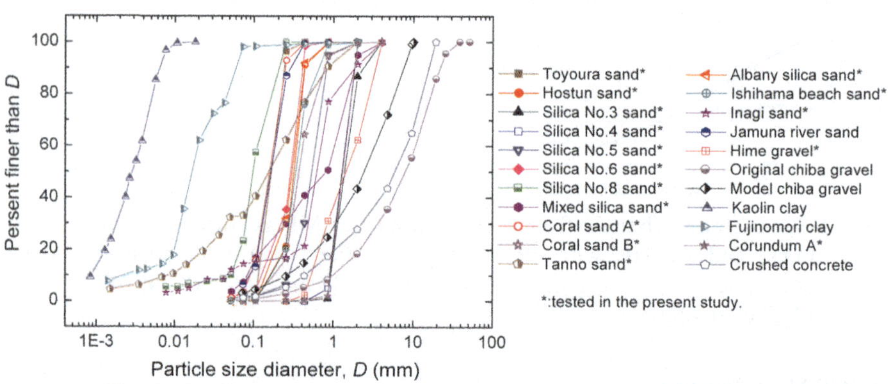

Fig. 3: Grading curves of the granular materials referred to in this paper.

3. TEST PROCEDURES

An automated triaxial apparatus (**Fig. 4**) was used. For each test material, loose (initial relative density, D_r= 20 ~ 50 %) and dense (D_r= 65 ~ 95 %) specimens (d= 70 mm & h= 150 ~ 155 mm) were prepared by the air-pluviation method. The top and bottom ends of each specimen were well-lubricated by using a 0.3 mm-thick latex rubber smeared with a 0.05 mm-thick silicone grease layer (Tatsuoka et al., 1984). The specimens were tested under either saturated or air-dried condition.

The axial deformation was measured by external deformation transducer and local deformation transducer (LDT; Goto et al., 1991). Externally measured axial strains and results of analysis based on them are reported in this paper unless otherwise noted. It was confirmed that, despite that the effects of bedding error on the measured axial strains cannot be ignored, their effects on the parameters describing the viscous properties presented in this paper are negligible. The volume change of a saturated specimen was obtained by measuring the water height in a burette connected to a specimen by using a low-capacity differential pressure transducer (LC-DPT). With air-dried specimens, the volume change of each specimen was estimated based on the Rowe's stress-dilatancy relation obtained from the corresponding drained TC tests on saturated specimens.

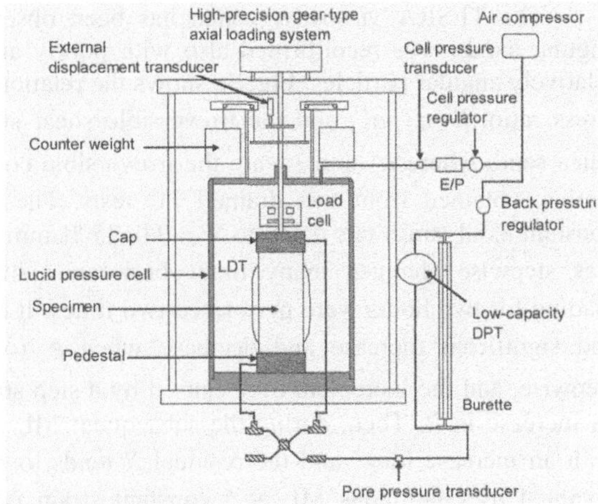

Fig. 4: Automated triaxial apparatus used in the present study.

Axial compressive loading was performed in an automated way using a high-precision gear-type axial loading system driven by a servo-motor, together with an electrical pneumatic pressure transducer (E/P) for the automated cell pressure control. Isotropic compression was performed at an axial strain rate of 0.00625 %/min towards a mean principal effective stress $p'=(\sigma'_v+2\sigma'_h)/3$ equal to 400 kPa, where σ'_v and σ'_h are the effective vertical and horizontal principal stresses. At p'= 50, 100, 200 and 300 kPa during the isotropic compression process, eight cycles of an axial strain (double amplitude) of 0.001 ~ 0.003 % were applied to evaluate the vertical quasi-elastic Young's modulus, E_v. From a full-log plot of the σ'_v (=$q+\sigma'_h$) - E_v relation, the coefficients m and E_{v0} of Eq. 1a (Hoque & Tatsuoka, 1998), which was used to evaluate elastic axial strain increments, were obtained.

$$E_v = E_{vo}\left(\frac{\sigma'_v}{98}\right)^m \quad (1a); \qquad \nu_{vh} = -\frac{d\varepsilon_h^e}{d\varepsilon_v^e} = \sqrt{\frac{E_{v0}}{E_{h0}}}\cdot\nu_0\cdot\left(\frac{\sigma'_v}{\sigma'_h}\right)^{\frac{m}{2}} = \sqrt{a}\cdot\nu_0\cdot\left(\frac{\sigma'_v}{\sigma'_h}\right)^{\frac{m}{2}} \quad (1b)$$

ν_{vh} (Eq. 1b) is the Poisson's ratio, which was used to obtain elastic lateral strain increments. a= 1.0 and ν_0 = 0.168 were used for all the test materials. After drained sustained loading for thirty minutes at p' = 400 kPa, drained TC was started.

4. TEST RESULTS AND DISCUSSIONS

4.1 Reconfirmation of TESRA viscosity

The TESRA viscosity, which has been observed with poorly graded relatively angular sands, was reconfirmed also with poorly and well graded silica sands, having relatively angular particles. **Fig. 5a** shows the relationship between the effective principal stress ratio, $R=\sigma'_v/\sigma'_h$, and the irreversible shear strain, $\gamma^{ir}=\varepsilon_v^{ir}-\varepsilon_h^{ir}$, on dense mixed silica sand, where ε_v^{ir} and ε_h^{ir} are the irreversible components of vertical and horizontal strains, obtained from two drained TC tests. The first is a continuous ML test at a constant axial strain rate equal to $\dot{\varepsilon}_0=0.0625$ %/min and, in the other test, the strain rate was stepwise changed many times between $\dot{\varepsilon}_0/10$ and $20\dot{\varepsilon}_0$ and drained sustained loading for two hours were performed two times. It can be seen that *R* exhibits a sudden and significant increase and decrease when $\dot{\varepsilon}_v$ (or $\dot{\gamma}^{ir}$) is increased and decreased stepwise, and the increment of *R* caused by a step strain rate change, *ΔR*, increases with an increase in *R*. Then, during the subsequent ML at a constant strain rate, *ΔR* decays with an increase in γ^{ir} and the residual *R* tends to become the same as the one that is obtained by continuous ML at a constant strain rate (i.e., the TESRA viscosity). As shown later, the decay rate is affected by particle characteristics. Furthermore, this sand exhibits noticeable creep deformation. **Fig. 5b** shows the relationship between the irreversible volumetric strain, $\varepsilon_{vol}^{ir}=\varepsilon_v^{ir}+2\varepsilon_h^{ir}$, and γ^{ir}.

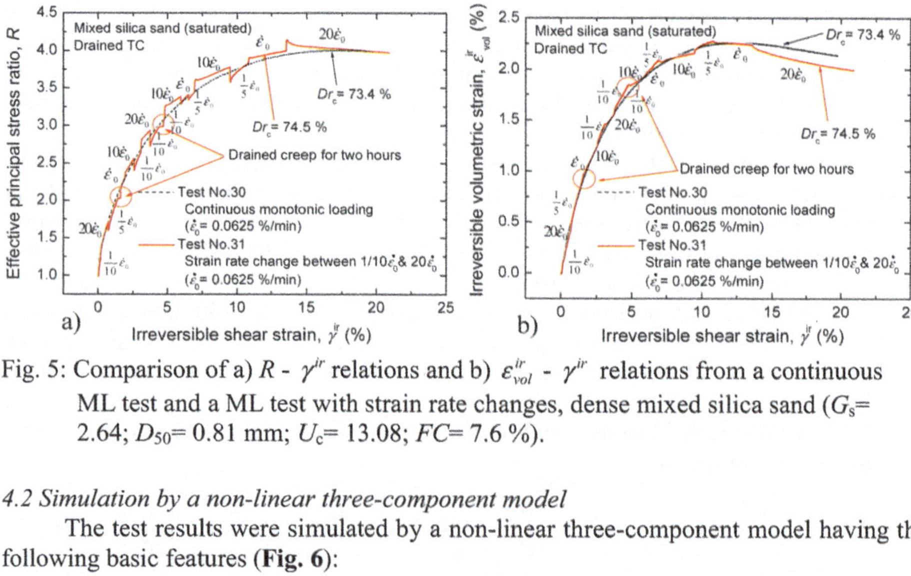

Fig. 5: Comparison of a) R - γ^{ir} relations and b) ε_{vol}^{ir} - γ^{ir} relations from a continuous ML test and a ML test with strain rate changes, dense mixed silica sand (G_s= 2.64; D_{50}= 0.81 mm; U_c= 13.08; FC= 7.6 %).

4.2 Simulation by a non-linear three-component model

The test results were simulated by a non-linear three-component model having the following basic features (**Fig. 6**):

1) A given strain increment, $d\varepsilon$, consists of elastic and rate-dependent irreversible components, $d\varepsilon^e$ and $d\varepsilon^{ir}$. A given effective stress, σ, consists of an inviscid stress, σ^f, which is a unique function of ε^{ir} in the case of ML, and a rate-dependent viscous stress, σ^v. That is:

$$d\varepsilon=d\varepsilon^e+d\varepsilon^{ir}\ ,\quad \sigma=\sigma^f+\sigma^v \tag{2}$$

2) When σ^v is a unique function of ε^{ir} and its rate, $\dot{\varepsilon}^{ir}$, and always proportional to the instantaneous σ^f (i.e., the isotach type viscosity), we obtain:

$$\sigma^v\left(\dot{\varepsilon}^{ir},\varepsilon^{ir}\right)=g_v\left(\dot{\varepsilon}^{ir}\right)\cdot\sigma^f\left(\varepsilon^{ir}\right);\quad \text{where } g_v\left(\dot{\varepsilon}^{ir}\right)=\alpha\cdot\{1-\exp\{1-\left(\left|\dot{\varepsilon}^{ir}\right|/\dot{\varepsilon}_r^{ir}+1\right)^m\}\tag{3}$$

where $g_v(\dot{\varepsilon}^{ir})$ is the viscous function; $|\dot{\varepsilon}^{ir}|$ is the absolute value of $\dot{\varepsilon}^{ir}$; and α, $\dot{\varepsilon}_r^{ir}$ and m are the positive material constants.

3) When the viscosity is of TESRA type, the current viscous stress when $\varepsilon^{ir}=\varepsilon^{ir}$, σ^v, is no longer a unique function of the instantaneous values of ε^{ir} and $\dot{\varepsilon}^{ir}$, but it is obtained as:

$$\sigma^v=\int^{\varepsilon^{ir}}\left[d\sigma^v\right]_{(\tau,\varepsilon^{ir})}=\int^{\varepsilon^{ir}}\left[d\{\sigma^f\cdot g_v\left(\dot{\varepsilon}^{ir}\right)\}\right]_{(\tau)}\cdot g_{decay}\left(\varepsilon^{ir}-\tau\right)\tag{4}$$

where $[d\sigma^v]_{(\tau,\varepsilon^{ir})}$ is the viscous stress increment that developed when $\varepsilon^{ir}=\tau$ and then has decayed until the present ($\varepsilon^{ir}=\varepsilon^{ir}$); and $[d\{\sigma^f\cdot g_v(\dot{\varepsilon}^{ir})\}]_{(\tau)}$ is the viscous stress increment that developed by a change in either ε^{ir} or $\dot{\varepsilon}^{ir}$, or both when $\varepsilon^{ir}=\tau$; and $g_{decay}(\varepsilon^{ir}-\tau)$ is the decay function (Tatsuoka et al., 2001, 2002) defined as:

$$g_{decay}\left(\varepsilon^{ir}-\tau\right)=r_1^{\varepsilon^{ir}-\tau}=(0.5)^{\frac{\varepsilon^{ir}-\tau}{H}};\quad H=\log(1/2)/\log(r_1)\tag{5}$$

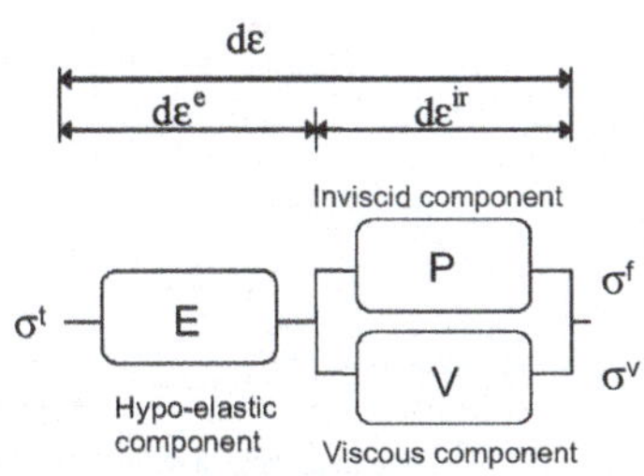

Fig. 6: Non-linear three-component model (Di Benedetto et al., 2002; Tatsuoka et al., 2002).

where r_1 is the decay parameter, which is a positive constant smaller than unity, affected by particle characteristics; and H is the half-strain defined as the irreversible strain difference, $\varepsilon^{ir}-\tau$, until $[d\sigma^v]_{(\tau,\varepsilon^{ir})}$ decays to a half of the initial value during ML at a constant strain rate (**Fig. 7**). **Δ** and Δ' are differences between "the components of σ^v due to an increase in ε^{ir}" for the strain rates after and before a step change, which continuously decay with ε^{ir}. The value of Δ is usually negligible unless r_1 is close to 1.0.

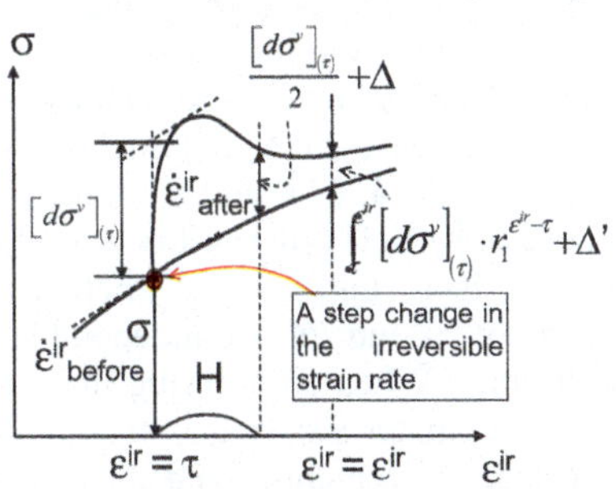

Fig. 7: Definition of the half-strain and the current viscous stress.

Figs. 8a and **8b** show the behaviour of loose silica No. 5 sand, which typically exhibits the TESRA viscosity. In these figures, the simulation by the TESRA model is also presented. It may be seen that the model can simulate the stress-strain relations including viscous effects very well.

Figs. 9a, b and **c** summarise the decay parameters, r_1, defined in terms of the irreversible axial strain (%), obtained from the simulation by the TESRA model of the R - ε_v relation from the drained TC tests, plotted in the logarithmic scale against the mean particle diameter, D_{50}, the coefficient of

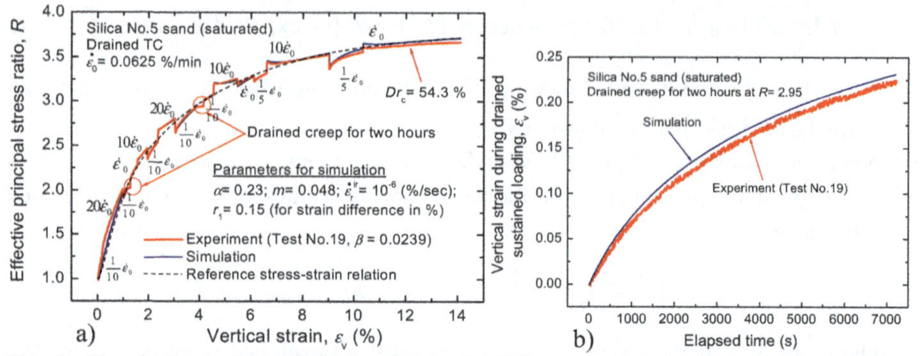

Fig. 8: Simulation by the TESRA model of a TC test on loose silica No. 5 sand (G_s= 2.65; D_{50}= 0.554 mm; U_c= 2.24; FC= 1.8 %); a) overall R - ε_v relation; and b) creep strain.

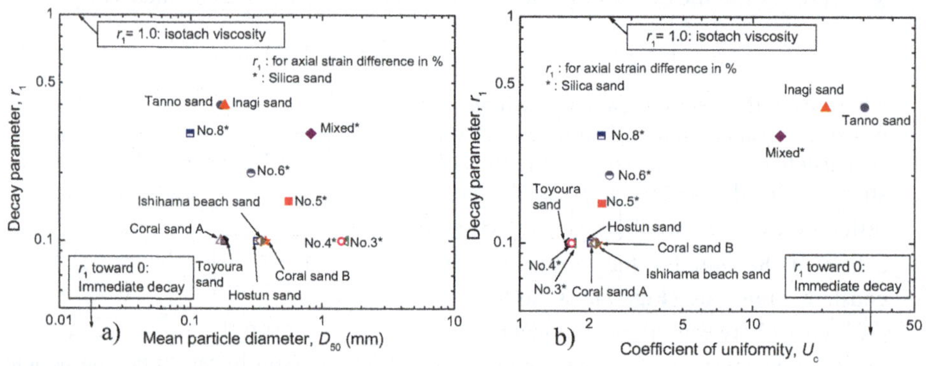

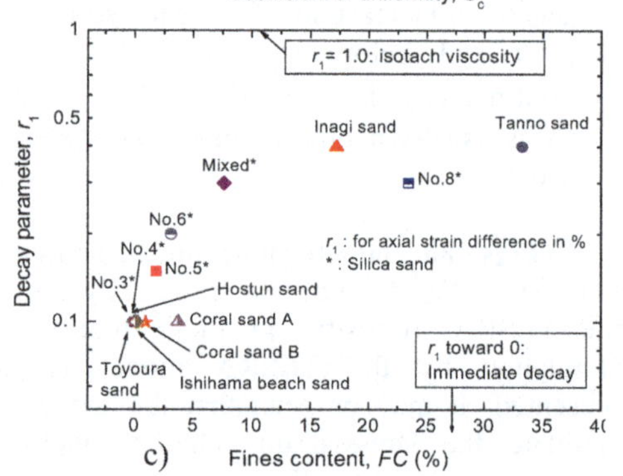

Fig. 9: Decay parameters r_1 plotted against; a) D_{50}; b) U_c; and c) FC.

uniformity, U_c, and the fines content, FC. It may be seen that the r_1 value noticeably increases with an increase in U_c and FC. Despite that it is to a lesser extent, the r_1 value tends to increase also with a decrease in D_{50} (typically with silica Nos. 3, 4, 5, 6 and 8 sands). The effects of particle crushability are subtle, if any. It is not known whether the effects of U_c and FC are independent of each other. When considering that more coherent materials (e.g., sedimentary soft rock and cement-mixed soil) exhibit the isotach type viscosity (i.e., r_1= 1.0), the results presented in Fig. 9 can be interpreted in such that the r_1 value decreases as the micro-structure becomes less stable with a decrease in the coordination number (i.e., the average number of contact points between particles) associated with a decrease in U_c or FC or both.

4.3 Rate-sensitivity coefficient

The change in R upon a stepwise change in the irreversible shear strain rate from $\dot{\gamma}^{ir}_{before}$ to $\dot{\gamma}^{ir}_{after}$ at the fixed irreversible shear strain, denoted as $\Delta R = \Delta(\sigma'_v/\sigma'_h) = (\partial R/\partial \dot{\gamma}^{ir}) \cdot \Delta \dot{\gamma}^{ir}$, is always proportional to the effective principal stress ratio, R, at which $\dot{\gamma}^{ir}$ (or $\dot{\varepsilon}_v$) was stepwise changed (**Fig. 10**). The ratios, $\Delta R / R$, from the drained TC test described in Fig. 5a are plotted against $\log\{(\dot{\varepsilon}_v)_{after}/(\dot{\varepsilon}_v)_{before}\}$ or $\log(\dot{\gamma}^{ir}_{after}/\dot{\gamma}^{ir}_{before})$ in **Fig. 11**. The relations in terms of $\dot{\varepsilon}_v$ (measured externally) and $\dot{\gamma}^{ir}$ (based on locally measured axial strain) are nearly the same. The slope of the relation, which is essentially linear, is defined as the rate-sensitivity coefficient β:

$$\frac{\Delta R}{R} = \beta \cdot \log\left(\frac{\dot{\gamma}^{ir}_{after}}{\dot{\gamma}^{ir}_{before}}\right) \approx \beta \cdot \log\left[\frac{(\dot{\varepsilon}_v)_{after}}{(\dot{\varepsilon}_v)_{before}}\right] \quad (6)$$

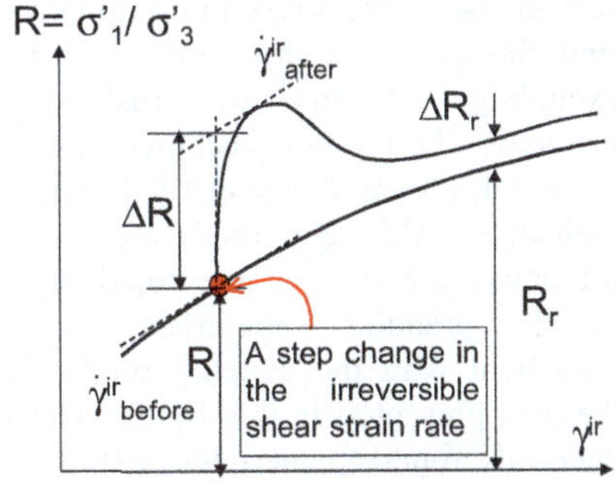

Fig. 10: Definition of ΔR and ΔR_r.

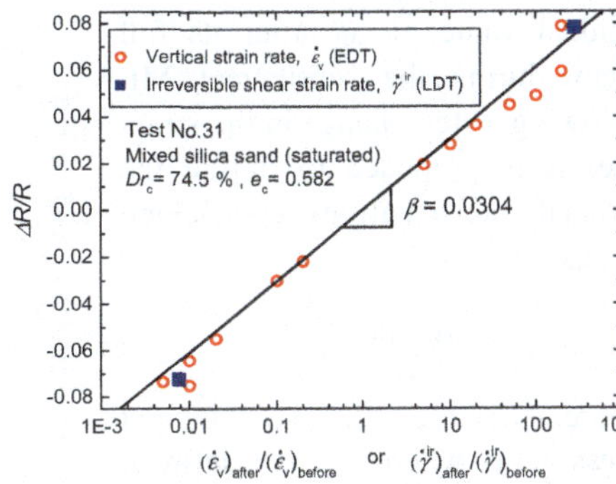

Fig. 11: β value obtained from a drained TC test on dense mixed silica sand (EDT; external displacement transducer, Fig. 4).

The β values evaluated by the drained TC tests performed in the present study and those by the previous studies (Tatsuoka, 2004), which are all free from the effects of pore water, were plotted against D_{50} (**Fig. 12a**), U_c (**Fig. 12b**) and FC (**Fig. 12c**). The following trends of behaviour may be noted:

1. The data points in a broken curve in Fig. 12a are for the unbound granular materials that are relatively angular while not crushable. The effects of D_{50} on these β values are insignificant. The major reason for some variation in these β values is variations in U_c or FC or both.
2. The β value tends to increase with an increase in U_c (Fig. 12b) and FC (Fig. 12c). It is not known whether both parameters U_c and FC are necessary, or either is enough, to explain the effects of grading characteristics. When excluding the data of crushable and round materials, the β - U_c and β - FC relations have respectively a very small range indicated by a pair of broken curves (with a few exceptions in Fig. 12c).
3. The β values of the relatively round granular materials, which were not crushable in the TC tests, are noticeably smaller than the relatively angular and uncrushable materials having similar values of U_c or FC. This trend is more obvious in Fig. 12b.

Fig. 13 shows the relationships between the logarithm of the decay parameter, r_1, and the rate-sensitivity coefficient, β. It may be seen that r_1 tends to increase with an increase in β, showing that the viscous property becomes closer to the isotach type with an increase in the stress jump upon a step change in the strain rate. When excluding the data of crushable materials, the relation becomes quite linear with a small scatter of data as indicated. Although they are not presented, the data of the relatively round granular materials are consistent with the general trend of the data presented in this figure. The data of round materials will be reported in the near future.

ΔR_r in Fig. 10 denotes the residual value of ΔR after its full decay during the subsequent ML following a step change in the strain rate. It is proposed to define the residual rate-sensitivity coefficient, β_r, as:

$$\frac{\Delta R_r}{R_r} = \beta_r \cdot \log\left(\frac{\dot{\gamma}^{ir}_{after}}{\dot{\gamma}^{ir}_{before}}\right) \qquad (7)$$

where R_r is the effective principal stress ratio where ΔR_r is defined. Then, different viscosity types can be classified based on the ratio, β_r/β. That is, $\beta_r/\beta = 1.0$ and 0.0 for the isotach and TESRA viscosity types. It is shown below that the relatively round granular materials exhibit negative values of β_r compared to positive β values.

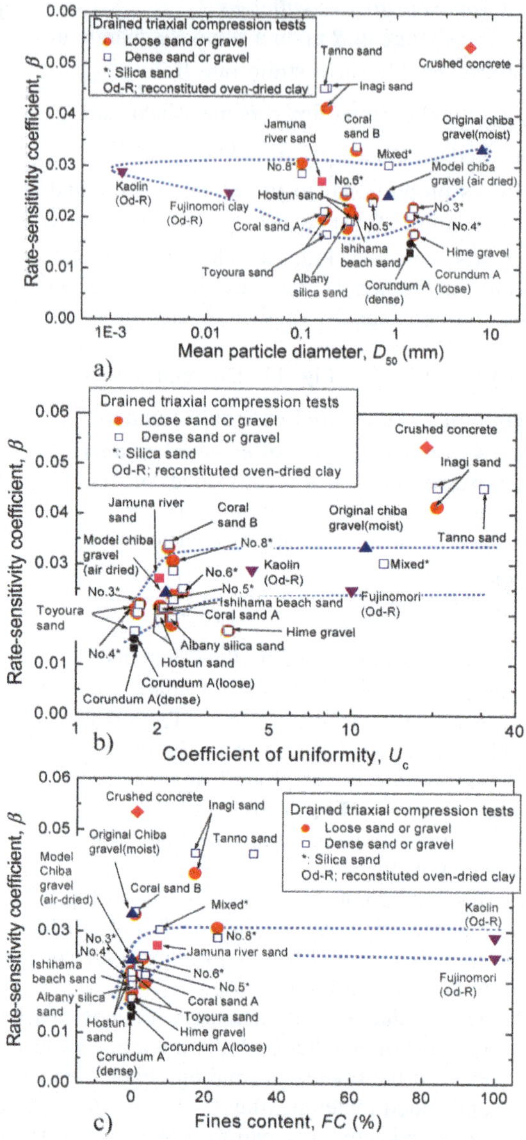

Fig. 12: Rate-sensitivity coefficients β plotted against; a) D_{50}; b) U_c; and c) FC.

4.4 Positive and negative viscosity

In the present study, the relatively round granular materials, Corundum A, Albany silica sand and Hime gravel, were tested in addition to the relatively angular ones as used in the previous study. Figs. 14a and 14b show the R - γ^{ir} relations from, respectively, four drained TC tests at constant but different axial strain rates, $\dot{\varepsilon}_v$, on dense specimens of corundum A and Albany silica sand. It may be seen that, unlike the relatively angular granular materials, the stress at the same axial strain decreases with an increase in the axial strain rate. Hime gravel also exhibited a similar trend of behaviour, although it is to a lesser extent. These test results are consistent with those by the direct and simple shear tests on round materials (Mair & Marone, 1999; Chambon et al., 2002). This surprising trend of behaviour is considered due to the *negative* isotach viscosity, compared to the *positive* isotach viscosity that has been observed with coherent geomaterials (e.g., sedimentary soft rock and plastic clay; Tatsuoka et al., 2001, 2002; Tatsuoka, 2004).

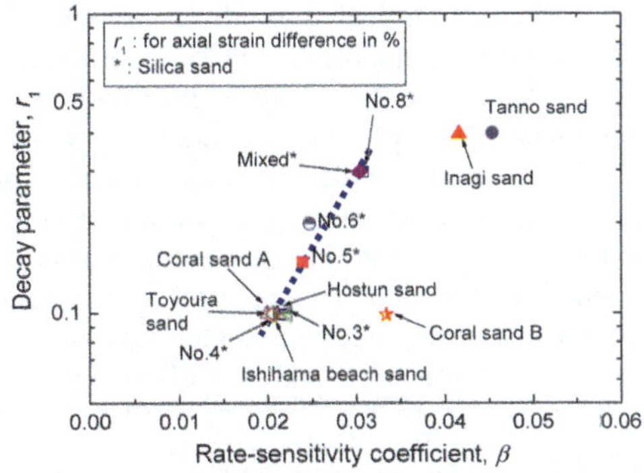

Fig. 13: Relationship between r_1 and β.

This trend of behaviour, described above, may be characteristic with the relatively round and poorly graded unbound granular materials. On the other hand, a crushed concrete aggregate consisting of relatively round coarse particles covered with a thin mortar layer exhibits a similar trend of behaviour only in the pre-peak regime (Fig. 1a). Further study is necessary to find the mechanism of the *negative* isotach viscosity.

The TESRA type viscous property can also be called the *positive* viscosity in the sense that the stress increases suddenly upon a step increase in the strain rate, as typically seen from Figs. 5a and 8a. The relatively round and poorly graded granular materials also exhibit this type of *positive* viscosity as seen from **Figs. 15** and **16**. It is to be noted that, the residual stress after the viscous stress has fully decayed tends to become <u>smaller</u> than the stress before a step <u>increase</u> in the strain rate and vice versa, showing negative values of β_r due to the *negative* isotach viscosity.

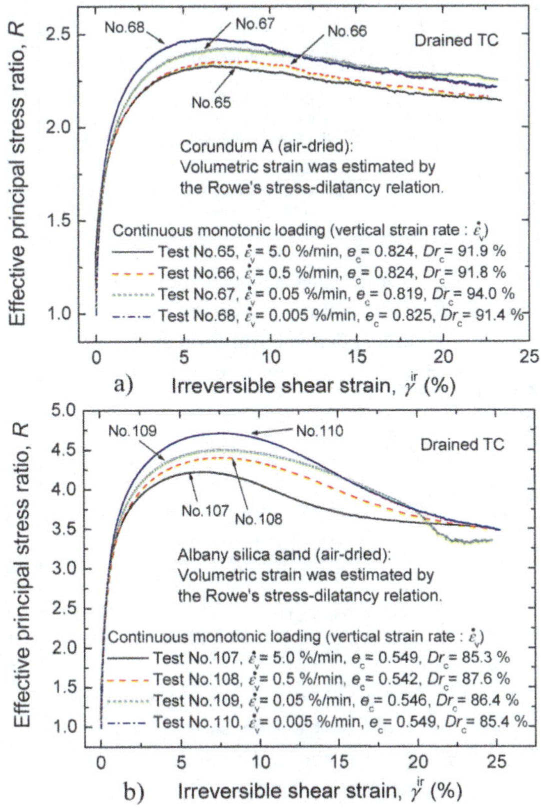

Fig. 14: R - γ^{ir} relations from, respectively, four drained TC tests at constant but different axial strain rates, $\dot{\varepsilon}_v$, dense specimens of; a) corundum A (G_s= 3.90; D_{50}= 1.42 mm; U_c= 1.62; FC= 0 %); and b) Albany silica sand (G_s= 2.67; D_{50}= 0.30 mm; U_c= 2.22; FC= 0.1 %).

These complicated trends of viscous behaviour of these relatively round and poorly graded granular materials can be interpreted in such that the viscous stress increment, $\Delta\sigma^{v}$, consists of the TESRA (*positive*) viscosity component and the *negative* isotach component (**Fig. 17**). The fact that the rate-sensitivity coefficients, β, of these relatively round and poorly graded granular materials are smaller than those of relatively angular ones under otherwise the same conditions (Figs. 12a, b and c) may be due to this structure of the viscous stress: i.e., the actual stress jump observed upon a step *increase* in the strain rate is a sum of a *positive* component by the TESRA viscosity and a *negative* component by the *negative* isotach viscosity. The correlations of the β_r values with other viscosity parameters, β and r_1, will be reported in the near future.

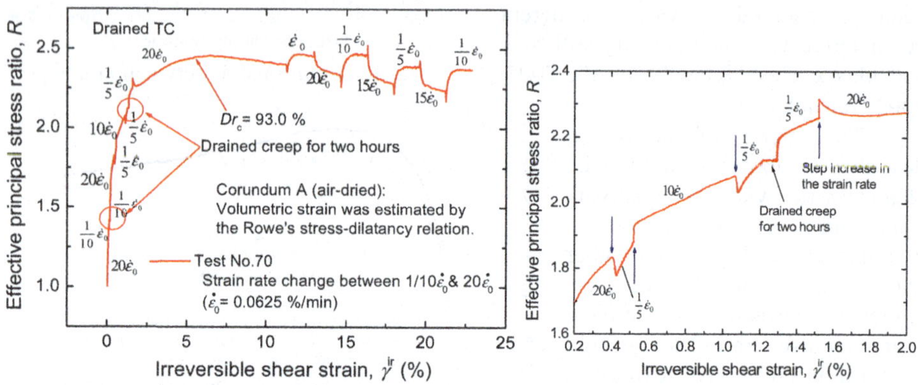

Fig. 15: R - γ^{ir} relation (left) and a close-up (right) from a ML test with strain rate changes, dense corundum A.

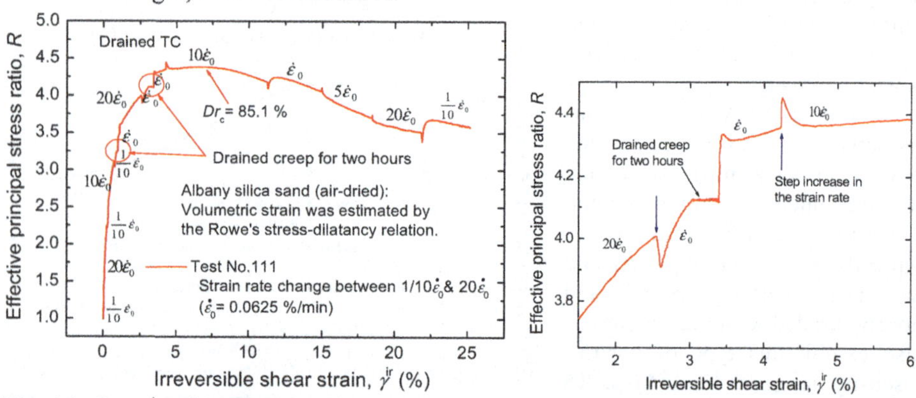

Fig. 16: R - γ^{ir} relation and a close-up from a ML test with strain rate changes, dense Albany silica sand.

Finally, it was found that the creep strain of the relatively round and poorly graded granular materials was smaller than that of the relatively angular ones under otherwise the same conditions, as typically seen from Figs. 5a and 8a. **Fig. 18** compares the creep axial strains by drained sustained loading for ten hours and the stress state (i.e., the ratio of R to its maximum value, R_{max}) where the respective sustained loading was performed of two relatively round granular materials (corundum A and Albany silica sand) and two relatively angular ones (silica No. 4 sand and coral sand A) having similar poor grading characteristics. These drained sustained loading tests were performed during otherwise drained ML at a constant axial strain rate, $\dot{\varepsilon}_v = 0.0625$ %/min. It may be seen that the creep deformation of the relatively round granular materials are noticeably smaller.

Further study is necessary to explain the effects of particle characteristics on the viscous behaviour of granular materials shown in this paper.

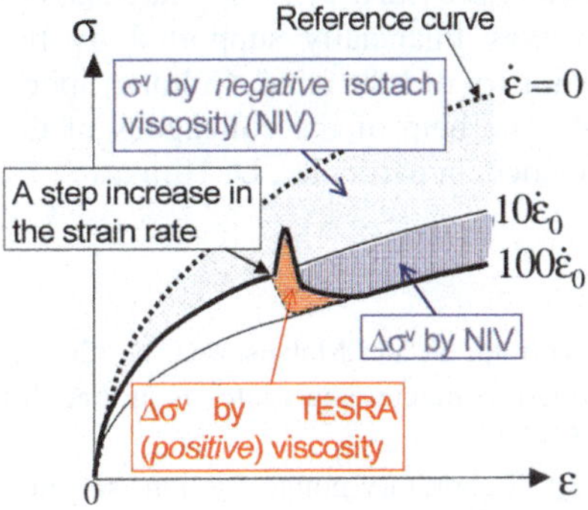

Fig. 17: *Positive* and *negative* viscosity.

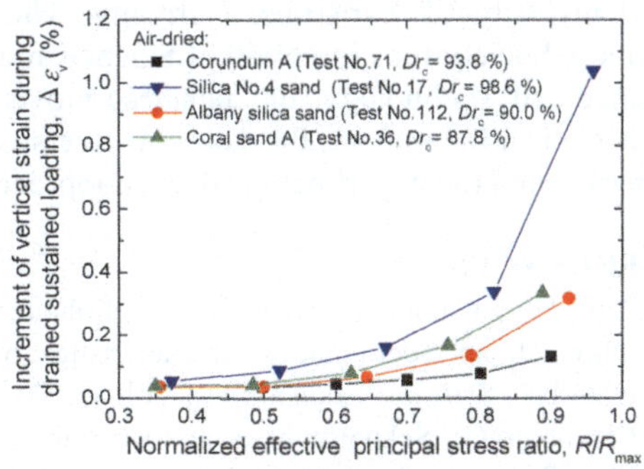

Fig. 18: Comparison of creep deformation between relatively angular and round granular materials.

5. CONCLUSIONS

The following conclusions can be derived from the drained TC test data presented in this paper.

1) The viscous properties of geomaterials that have been observed in the previous and present studies can be represented by: a) the rate-sensitivity coefficient, β, b) the decay parameter, r_1, and c) the residual rate-sensitivity coefficient, β_r, described and defined in this paper.
2) With respect to the rate-sensitivity coefficient, β; a) the effect of relative density, D_r, on the β value is insignificant; b) the overall effects of D_{50} on the β value are insignificant; and c) the β value tends to increase with an increase in the uniformity coefficient, U_c, the fines content, FC, and the particle crushability and decrease with an increase in the particle roundness.
3) Compared with the isotach viscosity, for which $r_1 = 1.0$ and $\beta_r = \beta$ (i.e., no decay of the viscous stress with an increase in the strain), relatively angular unbound granular materials have the TESRA viscosity, for which r_1 is a positive value lower than 1.0 and $\beta_r = 0$ (i.e., eventually full decay of the viscous stress with an increase in the strain). The decay parameter, r_1, tends to decrease with a decrease in U_c and FC (i.e., as U_c and FC decrease, the viscous stress increment decays at a higher rate during ML at a constant strain rate).
4) The viscosity of the unbound relatively round and poorly graded granular materials consists of the TESRA viscosity (for which r_1 is a positive value lower than 1.0) and

the *negative* isotach viscosity (for which β_r is negative). Due to this peculiar property, the stress at the same strain during ML at a constant strain rate decreases with an increase in the strain rate.

5) The creep strain rate of the unbound relatively round and poorly graded materials is smaller than that of the unbound relatively angular granular ones under otherwise the same conditions.

ACKNOWLEDGEMENTS

The corundum used in the present study was kindly provided by Prof. Gudehus, G., the University of Karlsruhe, Germany. The study was financially supported by the Japanese Society for Promotion of Science and the Ministry of Education, Culture, Sports, Science and Technology, the Japanese Government. The help of the colleagues of the Geotechnical Laboratory, the Tokyo University of Science, in particular, Dr Hirakawa, D., in performing the experiment is deeply appreciated.

REFERENCES

1) Aqil, U., Tatsuoka, F., Uchimura, T., Lohani, T.N., Tomita, Y. and Matsushima, K. (2005); "Strength and deformation characteristics of recycled concrete aggregate as a backfill material", *Soils and Foundations*, Vol. 45, No. 4, pp.53-72.
2) Chambon, G., Schmittbuhl, J. and Corfdir, A. (2002): "Laboratory gouge friction: seismic-like slip weakening and secondary rate- and state-effects", Geophysical Research Letters, Vol.29, No.10, 10.10.1029/2001GL014467, pp.4-1 – 4-4.
3) Day, P. (2005), "Long term settlement of granular fills", Summary of Practioner/Academic Forum, Preprint for 16th ICSMGE, Osaka.
4) Di Benedetto, H., Tatsuoka, F. and Ishihara, M. (2002): "Time-dependent shear deformation characteristics of sand and their constitutive modeling", *Soils and Foundations*, Vol.42, No.2, pp.1-22.
5) Di Benedetto, H., Tatsuoka, F., Lo Presti, D., Sauzéat, C. and Geoffroy, H. (2004): "Time effects on the behaviour of geomaterials", Keynote Lecture, *Proc. 3rd Int. Sym. on Deformation Characteristics of Geomaterials*, IS Lyon 03 (Di Benedetto et al. eds.), Balkema, September, 2003, Vol.2, pp.59-123.
6) Goto, S., Tatsuoka, F., Shibuya, S., Kim, Y.-S., and Sato, T. (1991), "A simple gauge for local small strain measurements in the laboratory", *Soils and Foundations*, Vol.31, No.1, pp.169-180.
7) Hoque, E. and Tatsuoka, F. (1998): "Anisotropy in the elastic deformation of granular materials", *Soils and Foundations*, Vol.38, No.1, pp.163-179.
8) Jardine, R.J., Standing, J.R. and Kovacevic, N. (2004), "Lessons learned from sull scale observations and the practical application of advanced testing and modelling", Keynote Lecture, *Proc. 3rd Int. Sym. on Deformation Characteristics of Geomaterials*, IS Lyon 03 (Di Benedetto et al. eds.), Balkema, September, 2003, Vol.2, pp.201-245.
9) Kiyota, T. and Tatsuoka, F. (2006): "Viscous property of loose sand in triaxial compression, extension and cyclic loading", *Soils and Foundations* (accepted for publication).
10) Kongkitkul, W., Tatsuoka, F. and Hirakawa, D. (2005); "Behaviour of geogrid-reinforced sand subjected to sustained loading in plane strain compression", *Geosynthetics and Geosynthetic-Engineered Soil Structures, Symposium Honoring Prof. Robert M. Koener* (Ling et al. eds.), pp.251-280.

11) Leroueil, S. and Hight, D.W. (2003): "Behaviour and properties of natural soils and soft rocks", *Characterisation and engineering properties of natural soil* (Tan et al. eds.), Balkema, Vol.1, pp.29-254.
12) Mair, K. and Marone, C. (1999): "Friction of simulated fault gouge for a wide range of velocities and normal stresses", *Journal of Geophysical Research*, Vol. 104, No.B12, pp.28,899-28,914, December 10.
13) Matsushita, M., Tatsuoka, F., Koseki, J., Cazacliu, B., Di Benedetto, H. and Yasin, S.J.M. (1999): "Time effects on the pre-peak deformation properties of sands", *Proc. Second Int. Conf. on Pre-Failure Deformation Characteristics of Geomaterials, IS Torino '99* (Jamiolkowski et al., eds.), Balkema, Vol.1, pp.681-689.
14) Nawir, H., Tatsuoka, F. and Kuwano, R. (2003a). "Experimental evaluation of the viscous properties of sand in shear" *Soils and Foundations*, Vol.43, No.6, pp.13-31.
15) Nawir, H., Tatsuoka, F. and Kuwano, R. (2003b): "Viscous effects on the shear yielding characteristics of sand", *Soils and Foundations*, Vol.43, No.6, pp.33-50.
16) Tatsuoka, F., Molenkamp, F., Torii, T. and Hino, T. (1984), "Behavior of lubrication layers of platens in element tests", *Soils and Foundations,* Vol.24, No.1, pp.113-128.
17) Tatsuoka, F. and Kohata, Y. (1995), "Stiffness of hard soils and soft rocks in engineering applications", Keynote Lecture, *Proc. of Int. Symposium Pre-Failure eformation of Geomaterials* (Shibuya et al., eds.), Balkema, Vol. 2, pp.947-1063.
18) Tatsuoka, F., Jardine, R.J., Lo Presti, D., Di Benedetto, H. and Kodaka, T. (1999), "Characterising the Pre-Failure Deformation Properties of Geomaterials", Theme Lecture for the Plenary Session No.1, *Proc. of XIV IC on SMFE, Hamburg,* September 1997, Volume 4, pp.2129-2164.
19) Tatsuoka, F., Uchimura, T., Hayano, K., Di Benedetto, H., Koseki, J. and Siddiquee, M.S.A. (2001); "Time-dependent deformation characteristics of stiff geomaterials in engineering practice", the Theme Lecture, *Proc. of the Second International Conference on Pre-failure Deformation Characteristics of Geomaterials, Torino, 1999, Balkema* (Jamiolkowski et al., eds.), Vol. 2, pp.1161-1262.
20) Tatsuoka, F., Ishihara, M., Di Benedetto, H. and Kuwano, R. (2002): Time-dependent compression deformation characteristics of geomaterials and their simulation, *Soil and foundation,* Vol.42, No.2, pp.103-138.
21) Tatsuoka, F. (2004). "Effects of viscous properties and ageing on the stress-strain behaviour of geomaterials", Geomechanics- Testing, Modeling and Simulation, *Proceedings of the GI-JGS workshop, Boston, ASCE Special Geotechnical SPT No. 143* (Yamamuro & Koseki eds.), pp.1-60.
22) Tatsuoka, F., Enomoto, T. and Kiyota, T. (2006): " Viscous properties of geomaterials in drained shear", Geomechanics- Testing, Modeling and Simulation, *Proceedings of the Second GI-JGS workshop, Osaka, ASCE Geotechnical Special Publication GSP* (Lade et al., eds.)

Soil Stress-Strain Behavior: Measurement, Modeling and Analysis
Geotechnical Symposium in Roma, March 16 & 17, 2006

VISCOUS PROPERTY OF KAOLIN CLAY WITH AND WITHOUT AGEING EFFECTS BY CEMENT-MIXING IN DRAINED TRIAXIAL COMPRESSION

J.-L. Deng[1] and F. Tatsuoka[2]

[1]Department of Civil Engineering, University of Tokyo, 4-6-1, Komaba, Meguro-ku, Tokyo, Japan; e-mail: deng@iis.u-tokyo.ac.jp (on leave from Department of Architecture Engineering, Fundamental Mechanics and Material Engineering Institute, Xiangtan University, Hunan, China, 411105)

[2]Department of Civil Engineering, Tokyo University of Science, 2641 Yamazaki, Noda-City, Chiba-Pref, Japan 278-8510; PH +81 (04) 7122-9819; FAX +81 (04) 7123-9766; e-mail: tatsuoka@rs.noda.tus.ac.jp

ABSTRACT
The viscous property of kaolin and the ageing effect on its stress-strain behaviour as well as their interactions were evaluated by performing a series of drained triaxial compression (TC) tests on air-dried and saturated specimen with and without cement-mixing. The axial strain rate was changed stepwise many times as well as drained creep and relaxation tests were performed during otherwise monotonic loading (ML). Different types of viscous properties (called Isotach type as well as weak and strong TESRA types) were observed depending on test conditions. The ageing effects by hydration of cement, interacted with the viscous property, were observed with saturated cement-mixed kaolin. It is shown that the effects of viscosity property and ageing, which interact with each other, can be simulated by a non-linear three-component rheology model modified from the original one to account for ageing effects.

Key word: triaxial compression, cement-mixed kaolin, viscous properties, rate-sensitivity coefficient

1. INTRODUCTION
A number of researchers (e.g., Matsui & Abe 1985; Yin & Graham 1999; Tatsuoka et al. 2000, 2003; Imai et al. 2003; Oka et al. 2003) investigated the viscous property of geomaterial. Di Benedetto et al. (2002) and Tatsuoka et al. (2002) proposed a non-linear three-component rheology model (**Fig. 1**) that can simulate several different types of the viscous property that are observed with different types of geomaterial. On the other hand, a limited number of researchers (e.g., Leroueil & Marques 1996; Vaughn, 1997; Tatsuoka et al. 2003, 2004; Kongsukprasert & Tatsuoka 2005) studied the ageing effect on the stress-strain behaviour of geomaterial. In relatively long terms in the field and in relatively short terms in the laboratory, ageing effects take place usually concurrently with viscous

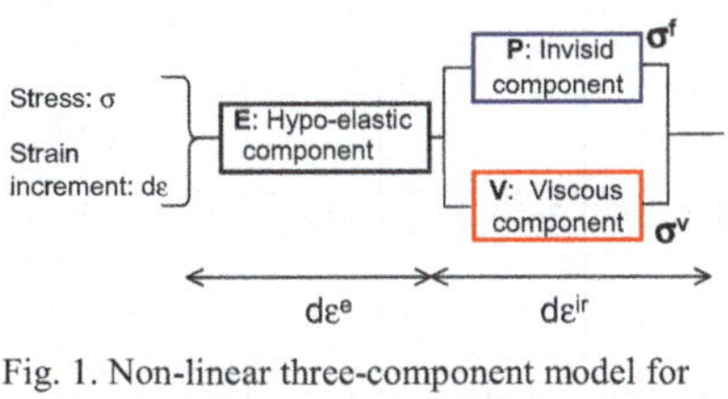

Fig. 1. Non-linear three-component model for geomaterial (Di Benedetto et al., 2002; Tatsuoka et al., 2002)

Hoe I. Ling et al. (eds.), Soil Stress-Strain Behavior: Measurement, Modeling and Analysis, 399–412.

effects. Tatsuoka et al. (2003) proposed to modify the three-component model (Fig. 1) to account for the ageing effects on the stress-strain-time behaviour of geomaterial. Deng et al. (2005a) and Kongsukprasert & Tatsuoka (2005) investigated experimentally interactions between the effects of viscosity and ageing by, respectively, one-dimensional (1D) compression tests on kaolin and drained triaxial compression (TC) tests on compacted cement-mixed well-graded gravelly soil.

In the present study, to study the interactions between these two factors on the stress-strain-time behaviour of clay, a series of drained TC tests were performed on four specimen types of kaolin: a) air-dried pure kaolin; b) saturated pure kaolin (prepared air-dried); c) air-dried cement-mixed kaolin; and d) saturated cement-mixed kaolin (prepared air-dried). Different types of viscous property were observed depending on these different test conditions. All the test results were simulated by the three-component model.

2. TEST PROCEDURES

The following two types of specimens (50 mm in diameter and 100 mm high) were prepared in ten sub-layers by vertically compressing each sub-layer at a vertical stress σ'_{v0}= 340 kPa for a period of 2 - 5 minutes in a split model with an internal diameter of 50 mm: a) air-dried powder of kaolin (D_{50}= 0.0013 mm; PI= 41.6; & LL= 79.6 %) ; and b) air-dried kaolin powder mixed with 3 % (by weight) of high-early-strength Portland cement (a specific gravity G_{cement}= 3.13; Sumitomo Osaka Cement Co. Ltd.). The specimens that were prepared as above, which could self-stand, were then set in a triaxial cell. The specimens that would be made saturated at the later stage were provided with side drain of filter paper as drainage in addition to drainage from the top and bottom ends of specimen.

The specimens were isotropically consolidated toward σ'_c= 100 kPa. Subsequently, with or without applying an initial deviator stress, the specimens were made saturated by percolating

Table 1. Kaolin specimens without cement-mixing

Test number	Water content (%)[a)]	Initial void ratio[b)]	B value	Saturation conditions and σ_3' during TC loading	TC loading conditions ($\dot{\varepsilon}_1$: the axial strain rate in second^{-1})	β value
40622 (***P***)	35.61	1.05	0.24	Saturated taking 1 day when q= 0 kPa & σ_3'= 100 kPa; σ_3' (TC)=100 kPa	Step changes in $\dot{\varepsilon}_1$ between 1.64x10^{-5} and 1.35x10^{-6}; and creep for 11 hours at q= 165 kPa	0.029
40625(***P***)	36.96	1.01	0.25	Saturated taking 1 day when q= 0 kPa & σ_3'= 100 kPa; σ_3' (TC)=100 kPa	Step changes in $\dot{\varepsilon}_1$ between 2.47x10^{-5} and 1.49*10^{-6}	0.034
40720(***P***)	36.09	1.08	0.20	Saturated taking 1 day when q= 0 kPa & σ_3'= 100 kPa; σ_3' (TC)=100 kPa	Step changes in $\dot{\varepsilon}_1$ between 2.62x10^{-5} and 1.49x10^{-6}; and creep for 31 hours at q= 183 kPa	0.031
40810(***P***)	Air-dried	1.38	-	Not saturated; σ_3' (TC)=100 kPa	Step changes in $\dot{\varepsilon}_1$ between 2.22x10^{-5} and 1.27x10^{-6}; and creep for 12 hours at q= 158 kPa; and relaxation test for 0.6 hours starting from R=3.31	0.029

a) measured after the end of test; b) measured before the start of either saturation or TC loading for the unsaturated specimen.; ***P***: made by compacting air-dried kaolin powder;

a sufficient amount of tap water and then supplying a sufficient back pressure (equal to 100 kPa or more). The specimens were then cured for a respectively prescribed period (one day or two days) before the start of drained TC. The axial strains of specimen were measured externally with a LVDT. It was confirmed by performing several tests also locally measuring axial strains with a pair of LDTs that the effects of bedding error are insignificant in the present case. The volume changes of a saturated specimen were obtained by measuring the amount of water that was expelled from or sucked into the specimen, which was actually by automatically measuring the height of the water in a burette with a low capacity differential pressure transducer.

The other specimens were kept air-dried throughout the drained TC loading scheme without changing the water content from the one when prepared. As the volume changes of a air-dried specimen were not measured, the volumetric strain increment for a given axial strain increment at a given stress state was estimated by substituting the respective measured axial strain increment and the instantaneous effective principal stress ratio into the Rowe's stress-dilatancy relation obtained from CD TC tests on saturated specimens.

The drained TC tests were performed on the following specimens with and without cement-mixing:

Table 2. Cement-mixed kaolin specimens

Test number	Water content[a)] (%)	Initial void ratio[b)]	B value	Saturation conditions and σ_3' during TC loading	TC loading conditions ($\dot{\varepsilon}_1$: the axial strain rate in second^{-1})	β value
40618(***P***)	41.05	1.16	0.51	Saturated taking 2 day when q= 0 kPa & σ_3'= 100 kPa; σ_3' (TC)=100 kPa	Step changes in $\dot{\varepsilon}_1$ between 1.91x10^{-5} and 8.69x10^{-7}	0.0595
40630(***P***)	40.94	1.13	0.26	Saturated taking 2 day when q= 0 kPa & σ_3'= 100 kPa; σ_3' (TC)=100 kPa	Step changes in $\dot{\varepsilon}_1$ between 6.42x10^{-5} and 1.32x10^{-6}	0.062
40712(***P***)	39.77	1.11	0.63	Saturated taking 2 day when q= 50 kPa & σ_3'= 100 kPa; σ_3' (TC)=100 kPa	Step changes in $\dot{\varepsilon}_1$ between 5.96x10^{-5} and 1.27x10^{-6} and drained creep for 24 hours at q= 420 kPa	0.054
40716(***P***)	38.91	1.07	0.46	Saturated taking 2 day when q= 0 kPa & σ_3'= 100 kPa; σ_3' (TC)=200 kPa	Step changes in $\dot{\varepsilon}_1$ between 6.01x10^{-5} and 1.21x10^{-6} and drained creep for 24 hours at q= 430 kPa	0.067
40807(***P***)	41.88	1.10	0.47	Saturated taking 2 day when q= -10 kPa & σ_3'= 100 kPa; σ_3' (TC)=200 kPa	Step changes in $\dot{\varepsilon}_1$ between 6.48x10^{-5} and 1.33x10^{-6}	0.056
40811(***P***)	Air-dried	1.41	-	Not saturated; σ_3' (TC)=100 kPa	Step changes in $\dot{\varepsilon}_1$ between 2.17x10^{-5} and 1.29x10^{-6}; and creep for 16 hours at q= 158 kPa; and relaxation test for 46 second and stopped	0.030

a) measured after the end of test; b) measured before the start of either saturation or loading;
P: made by compacting air-dried cement-mixed kaolin powder

a) Uncemented pure kaolin (**Table 1**): Three saturated specimens (40622, 40625 & 49720) and a single air-dried specimen (40810) were tested. The saturation was made taking one day under the stress conditions of q= 0 kPa and σ_3'= 100 kPa. In all the tests, the confining stress, σ'_3, during drained TC loading was equal to 100 kPa.

b) Cement-mixed kaolin (**Table 2**): Specimen (40811) was kept air-dried throughout drained TC loading at σ_3'= 100 kPa. The other five specimens were made saturated when σ'_3=100 kPa before the start of drained TC loading and cured for two days at either q= 0 kPa (40618, 40630 & 40716); or q= 50 kPa (40712); or q= -10 kPa (40807). The σ'_3 value during drained TC was σ'_3= 100kPa except for specimen 40716, for which σ'_3= 200 kPa.

During otherwise drained monotonic TC loading, the axial strain rate was changed stepwise many times and drained creep tests for 12 – 31 hours were performed. Stress relaxation tests were also performed on specimens 40810 & 40811 (Tables 1 & 2).

3. TEST RESULTS & DISCUSSIONS

<u>*General trends of stress-strain behaviour during drained TC:*</u> **Fig. 2** shows the measured relationships between the effective principal stress ratio, $R = \sigma'_v/\sigma'_h = \sigma'_1/\sigma'_3$, and the vertical (axial) strain, $\varepsilon_v = \varepsilon_1$, from two drained TC tests on air-dried kaolin without (40810) and with cement-mixing (40811). In this figure, the axial strains in the two tests are plotted in the shifted axes. The simulated relations are explained later. The TC test on specimen 40811 was ended at ε_v = 5.19 %, immediately after the start of a stress relaxation test, due to malfunction of the test system. It may be seen that the R - ε_v relations of the two specimens are similar, showing that the mixing of a small amount of cement does not alter the stress-strain behaviour of kaolin when air-dried. It is important to note that, even without pore water, the specimens exhibited a significant trend of viscous behaviour. A similar trend of viscous behaviour was also observed in 1D compression tests on air-dried and oven-dried kaolin (Deng & Tatsuoka, 2005b).

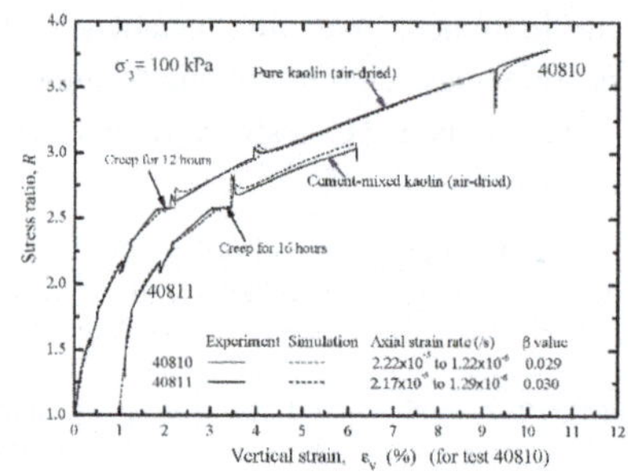

Fig. 3 shows two R - γ relations of saturated kaolin without cement-mixing, where γ is the shear strain, $\varepsilon_v - \varepsilon_h$. **Fig. 4** shows similar three relations of saturated cement-mixed kaolin. It may be seen by comparing Figs. 2 and 3 that, despite significantly lower initial void ratios (Tables 1 & 2), the saturated specimens without

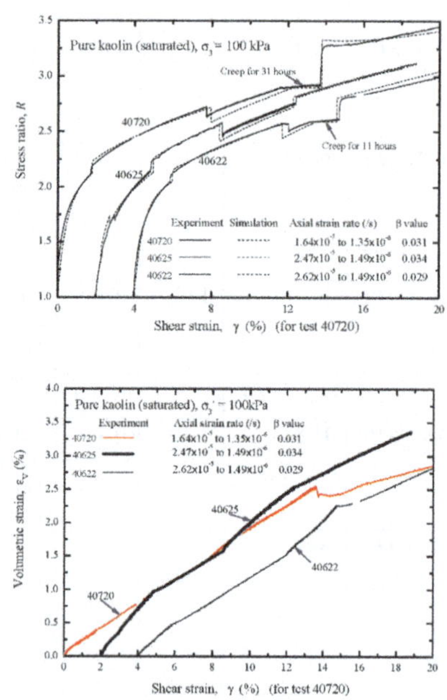

Fig. 3. R - relationship of saturated uncemented kaolin (σ'_3= 100 kPa).

cement-mixing are generally weaker than the air-dried ones. This result indicates that the particle surface conditions are significantly different between the saturated and air-dried

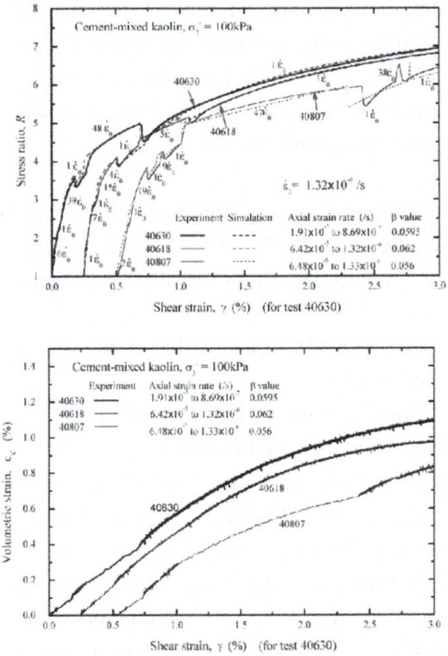

Fig. 4a. *R* - relationship of saturated cement-mixed kaolin (specimens 40618, 40630 & 40807, without drained creep; σ'_3= 100 kPa).

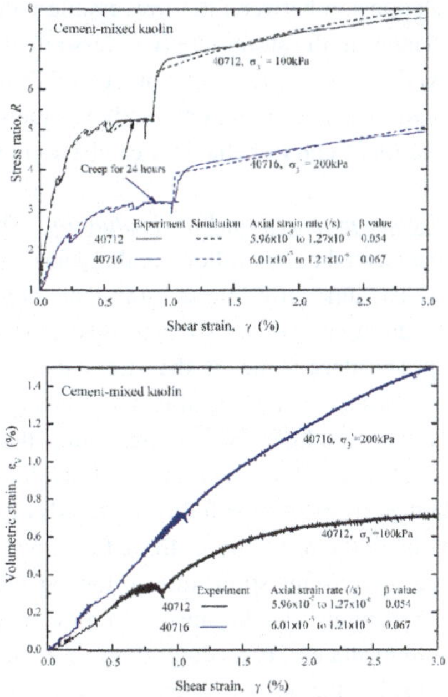

Fig. 4b. *R* - relationship of saturated cement-mixed kaolin (with drained creep for 24 hours; σ'_3= 100 kPa and 200 kPa).

specimens. As seen from Fig. 3, the three specimens without cement-mixing exhibited significant viscous effects, despite that the behaviour is free from the effects of delayed dissipation of excess pore water pressure.

It may be seen by comparing Figs. 3 and 4 that, when saturated, the specimens became significantly stiffer and stronger by cement-mixing. In Fig. 4a, the general features of the *R* - γ relations, including the viscous response, of specimens 40618, 40630 and 40807 tested under nearly the same conditions are rather similar. In Fig. 4b, the peak *R* value of specimen 40716 (σ'_3= 200 kPa) is significantly lower than the other tests (σ'_3= 100 kPa). This result indicates that the normalization, $R=\sigma'_1/\sigma'_3$, is not relevant to the stress normalization in this case.

It may be seen by comparing Figs. 4a and 4b that the strength of specimen 40712 (Fig. 4b) is significantly higher than the other three specimens tested at σ'_3= 100 kPa (Fig. 4a). This trend with specimen 40716 is due only partly to the fact that the initial drained sustained loading for two days was performed at an anisotropic stress state at *q*= 50 kPa (i.e., *R*= 1.5), but it is due much more to additional drained sustained loading for 24 hours at a higher stress ratio, *R*= 5.2. A similar test result was obtained by Komoto et al. (2004). The effects of drained sustained loading at an anisotropric stress state during otherwise ML on the subsequent stress-strain behaviour can also be seen in the test result of specimen 40716 (Fig. 4b).

The saturated pure kaolin exhibits highly contractive behaviour in drained TC (Fig. 3). This trend became smaller by cement-mixing (Fig. 4). The flow characteristics (i.e., the relationship between the irreversible volumetric and shear strains) are much less sensitive to changes in the strain rate (i.e., less affected by the viscous property) than the deviator stress – axial or shear strain relation. Despite the above, it may be seen that uncemented saturated kaolin becomes more contractive as the strain rate decreases (Fig. 3). Despite that it is subtle, a similar trend of behaviour can be seen with saturated cement-mixed clay (Fig. 4).

Viscous property and its evaluation: The viscous property of a given geomaterial can be represented by the following three factors (Di Benedetto et al. 2002; Tatsuoka et al., 2002):

1) The amount of stress jump that takes place upon a step change in the strain rate applied during otherwise ML at a constant strain rate.
2) The decay rate of the stress jump during subsequent ML scheme that continues at a constant strain rate after a step change in the strain rate.
3) The residual viscous stress after the viscous stress increment that developed by a step change in the strain rate has fully decayed during ML at a constant strain rate.

In these respects, the following trends of viscous behaviour may be seen from Figs. 2, 3 and 4:

1) In all the tests, the stiffness for some large stress range immediately after the restart of ML at a constant strain rate following drained sustained loading is very high. This trend of behaviour can be attributed first to the creep deformation that takes place during the immediately preceding drained sustained loading stage. With saturated cement-mixed kaolin, the positive ageing effect (i.e., cement hydration in the present case) during drained sustained loading is another factor, which makes the high stiffness zone larger.
2) Whether cement-mixed or not and whether saturated or air-dried, all the kaolin specimens exhibited a sudden increase or decrease in the R value when the strain rate is stepwise increased and decreased, showing that all the specimens had noticeable viscous property.

Table 3. Viscosity type of kaolin in 1D compression and drained TC

	Drained TC	1D compression
Uncemented (made by compressing air-dried); and saturated during TC	Weak TESRA	Isotach
Uncemented and cement-mixed; and both air-dried during TC	Very strong TESRA	Weak TESRA
Cement-mixed (made by compressing air-dried); and saturated during TC	Strong TESRA	Strong TESRA

3) The trend of stress-strain behaviour during the subsequent ML at a constant strain rate after a sudden stress change upon a step change in the strain rate is different depending on the test conditions as summarized below (see **Table 3**):
   a) With air-dried kaolin (Fig. 2), whether cement-mixed or not, the jump of deviator stress upon a step change in the strain rate decays at a very high rate with an increase in the strain during the subsequent ML at a constant strain rate. Di Benedetto et al. (2002) and Tatsuoka et al. (2002) called this type of viscous behaviour *the TESRA viscosity* (n.b.,

TESRA stands for "temporary effects of strain rate and strain acceleration"). In this case, the R - ε_v or R - γ relations from continuous ML tests at constant but different strain rates become essentially independent of strain rate. On the other hand, when the strain rate changes during ML, the current stress depends on not only the instantaneous irreversible strain and its rate but also strain history. The test results from the present study show that the inclusion of a small amount of cement has no significant effects on this trend of viscous behaviour when air-dried.

b) Without cement-mixing (Fig. 3), the viscous property of saturated kaolin is also of the TESRA type, but the decay rate of the viscous stress is significantly smaller than when air-dried.

c) The viscous property of saturated cement-mixed kaolin is also of strong TESRA type (Fig. 4), although it is not as strong as when air-dried.

Table 3 summarises these different types of viscous property described above as well as the corresponding observations in the 1D compression tests on kaolin (Deng & Tatsuoka, 2005b). It may be seen that, although the general features are similar in the 1D and drained TC tests, the trend of TESRA viscosity is generally stronger in the drained TC tests than in the 1D compression tests.

Stress jump upon a step change in the strain rate: In all the tests, for a given ratio of the axial strain rates before and after a step change, $(\dot{\varepsilon}_v)_{after}/(\dot{\varepsilon}_v)_{before}$, the stress jump denoted as $\Delta R = \Delta(\sigma_v'/\sigma_h') = \Delta(\sigma_v')/\sigma_h'$ was always proportional to the stress ratio, R, at which the strain rate was changed. Based on this fact, the measured values of $\Delta R/R$ (= $\Delta\sigma_v'/\sigma_v'$) were plotted against the respective value of $\log[(\dot{\varepsilon}_v)_{after}/(\dot{\varepsilon}_v)_{before}]$. **Figs. 5a** and **5b** show results from two typical tests. The fact that the scatter in the data is rather small shows that the stress normalization, $\Delta R/R$, is relevant. It seems that the degree of cementation with the saturated cement-mixed kaolin tested in the present study is not as strong as the one with the compacted cement-mixed gravelly soil tested by Kongsukprasert and Tatsuoka (2003), in which the normalization $\Delta\sigma_v'/(\sigma_v'+c)$, where c is a positive constant, was relevant. The relations presented in Figs. 5a and 5b (and others) are rather linear. The slope of the respective fitted linear relation is defined as the rate-sensitivity coefficient, β (Tatsuoka, 2004).

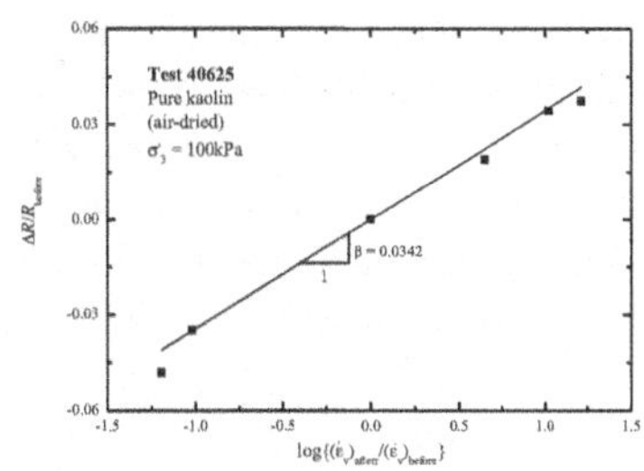

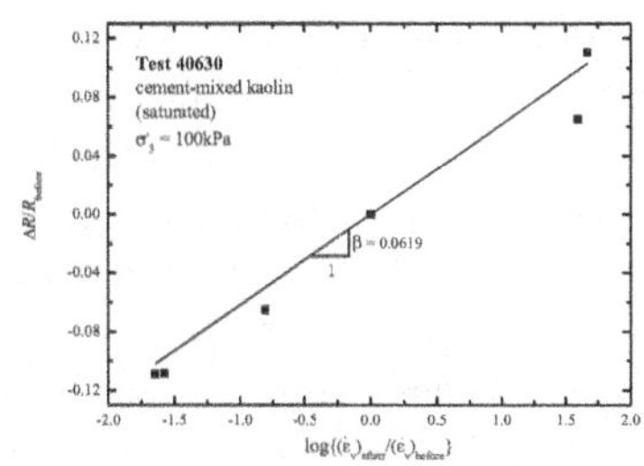

Fig. 5. Evaluation of β (compacted air-dried kaolin power, made saturated at q= 0 kPa & =100 kPa): a) without; and b) with cement-mixing.

Fig. 6 summarized the β values as a function of the degree of saturation (measured after the respective test) obtained from the drained TC tests on air-dried and saturated kaolin specimens (with and without cement-mixing) performed in the present study as well as those from the 1D compression tests on oven-dried, air-dried and saturated specimens of different types of clay (Deng & Tatsuoka, 2005b). **Fig. 7** compares the β values when

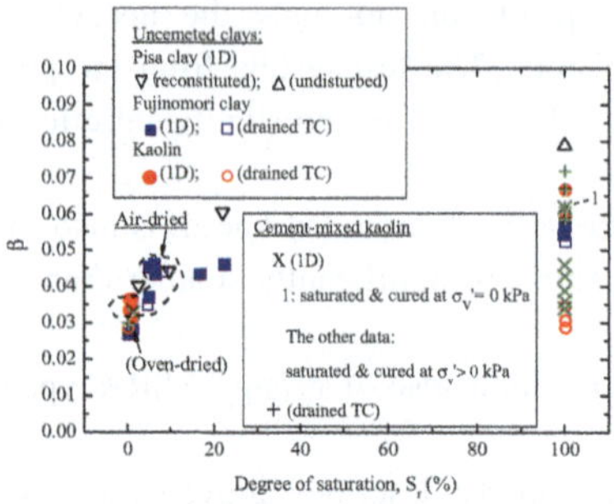

Fig. 6. Relationship between β values and the degree of saturation from 1D and drained triaxial compression tests on clay.

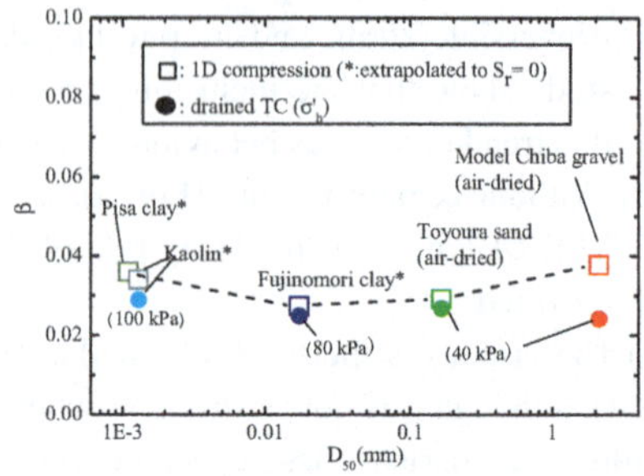

Fig. 7. Effects of particle size on β value free from effects of power water

$S_r = 0$ of kaolin, Fujinomori clay and Pisa clay (all without cement-mixing) obtained from the 1D compression tests (Deng & Tatsuoka, 2005b). The values β when $S_r = 0$ of kaolin and Fujinomori clay from drained TC tests are also plotted in Fig. 7. The β values of kaolin (1D compression and drained TC) and Fujinomori clay (drained TC) were obtained by extrapolating the β - S_r relations to $S_r = 0$ by referring to the 1D compression test data of Fujinomori clay (see Fig. 6). The data of the following other unbound geomaterials (without cement-mixing) obtained from 1D compression and drained TC tests are also presented in Fig. 7: air-dried specimens of a quartz-rich sub-angular fine sand, Toyoura sand (D_{50}= 0.18 mm, U_c= 1.64, G_s= 2.65, e_{max}= 0.99 & e_{min}= 0.62) and a well-graded gravel of crushed sandstone, called model Chiba gravel (D_{50}= 0.8 mm, U_c= 2.1, D_{max}= 5.0 mm, G_s= 2.74, e_{max}= 0.727 & e_{min}= 0.363) (Tatsuoka, 2004; Hirakawa et al., 2003). The S_r values of these air-dried specimens were of the order of 1 % while the β values of saturated and air-dried Toyoura sand specimens are essentially the same (Nawir et al. 2003). It seems therefore that the effects of S_r on the β values of Toyoura sand and model Chiba gravel are insignificant, and all the β values plotted in Fig. 7 can be compared on the same basis. The following trends of behaviour may be seen from Fig. 7:

1) The β values from drained TC tests are generally slightly smaller than those from 1D compression tests.
2) The β values of the three types of clay are similar to those of air-dried Toyoura sand and well-graded gravel. This fact indicates surprising small effects of particle size when the viscous property is free from the effects of pore pressure. Tatsuoka et al. (2006) and Enomoto et al. (2006) report a more detailed analysis of the effects of particle property on the β value.

On the other hand, the effects of cement-mixing on the β value of saturated kaolin are complicated (Fig. 6); that is, by cement-mixing, the β value decreases in the 1D compression test, while it increases in the drained TC test. Furthermore, by comparing the β values of specimens 40712 and 40716 (Table 2), it can be seen that the β value increases slightly when σ_3' increases double. Further study will be necessary to find causes for these trends described above.

4. THREE-COMPONENT MODEL:

According to the three-component model (Fig. 1), a given strain rate, $\dot{\varepsilon}$, consists of the elastic component, $\dot{\varepsilon}^e = \dot{\sigma}/E^e(\sigma)$, where $E^e(\sigma)$ is the hypo-elastic stiffness, and the irreversible component, $\dot{\varepsilon}^{ir}$, while a given stress, σ, consists of the inviscid component, σ^f, and the viscous component, σ^v:

$$\dot{\varepsilon} = \dot{\varepsilon}^e + \dot{\varepsilon}^{ir}; \qquad \text{and} \qquad \sigma = \sigma^f(\varepsilon^{ir}, h) + \sigma^v \tag{1}$$

where σ^f is a unique function of ε^{ir} for primary ML along a fixed stress path. The σ^f - ε^{ir} relation is called the reference stress-strain relation. For general loading histories including cyclic loading, the loading history parameter, *h*, becomes necessary. The results from this and previous studies by the authors and their colleagues showed that, when the Isotach viscosity is relevant, σ^v is linked to σ^f as:

$$\sigma^v(\varepsilon^{ir}, \dot{\varepsilon}^{ir}) = \sigma^f(\varepsilon^{ir}) \cdot g_v(\dot{\varepsilon}^{ir}) \qquad \text{(for primary ML)} \tag{2}$$

where $g_v(\dot{\varepsilon}^{ir})$ is the viscosity function, given as:

$$g_v(\dot{\varepsilon}^{ir}) = \alpha \cdot [1 - \exp\{1 - (|\dot{\varepsilon}^{ir}| / \dot{\varepsilon}_r^{ir} + 1)^m\}] \quad (\geq 0) \tag{3}$$

where $|\dot{\varepsilon}^{ir}|$ is the absolute value of $\dot{\varepsilon}^{ir}$; α, m and $\dot{\varepsilon}_r^{ir}$ are the positive constants, which should be determined based on the respective measured β value (Di Benedetto et al. 2002). The value of $g_v(\dot{\varepsilon}^{ir})$ becomes zero and a positive value, α, when $|\dot{\varepsilon}^{ir}|$ becomes respectively zero and infinitive, while the increasing rate, $\mathrm{d}[\log\{1 + g_v(\dot{\varepsilon}^{ir})\}] / \mathrm{d}\{\log|\dot{\varepsilon}^{ir}|\}$, when $|\dot{\varepsilon}^{ir}|$ is within a range between $\dot{\varepsilon}_r^{ir}$ and a certain larger value is equal to $\beta / \log 10$.

Decay of viscous stress: Matsushita et al. (1999), Di Benedetto et al. (2002) and Tatsuoka et al. (2002) showed that, with poorly-graded sands, the stress jump, $\Delta\sigma$, that has developed upon a step change in the strain rate is essentially the same with a viscous stress increment, $\Delta\sigma^v$, which decays with an increase in $\dot{\varepsilon}^{ir}$ during the subsequent ML at a constant strain rate. Then, the current viscous stress, σ^v, when the irreversible strain is ε^{ir} (denoted as $[\sigma^v_{TESRA}]_{(\varepsilon^{ir})}$) can be obtained by integrating the increment $\Delta\sigma^v$ (i.e., $d\sigma^v$) with respect to irreversible strain (not with respect to time) as:

$$[\sigma^v_{TESRA}]_{(\varepsilon^{ir})} = \int_{\varepsilon_1^{ir}}^{\varepsilon^{ir}} [d\sigma^v]_{(\tau, \varepsilon^{ir})} = \int_{\varepsilon_1^{ir}}^{\varepsilon^{ir}} [d\sigma^v_{iso}]_{(\tau)} \cdot g_{decay}(\varepsilon^{ir} - \tau) \tag{4}$$

where $[d\sigma^v_{iso}]_{(\tau)}$ is the viscous stress increment that develops by $d\varepsilon^{ir}$ or $d\dot{\varepsilon}^{ir}$ or both when ε^{ir} is equal to τ if the viscous property were of the Isotach type (Eq. 2); and $g_{decay}(\varepsilon^{ir} - \tau)$ is the decay function, which is a function of the difference between the current irreversible strain, ε^{ir}, and τ, given as:

$$g_{decay}\left(\varepsilon^{ir}-\tau\right)=r_1^{(\varepsilon^{ir}-\tau)} \quad (5)$$

where r_1 is a positive constant, which is equal to unity when the viscous property is of the Isotach type and is for less than unity when $[d\sigma^{v}_{iso}]_{(\tau)}$ in Eq. 3 decays with ε^{ir}. From Figs. 2, 3 and 4, it may be seen that the r_1 value of air-dried pure kaolin (with and without cement-mixing) is smallest among the three types of kaolin specimen, indicating the strongest TESRA property.

5. AGEING EFFECT

Ageing effect is defined as changes with time in the material property (i.e., the strength and deformation characteristics in the present case). An increase in the strength with time, which takes place typically during drained sustained loading at a fixed effective stress state with saturated cement-mixed kaolin, is defined as positive ageing effect. Ageing effect and loading rate effect by the viscous property are caused by different mechanisms as summarized in **Table 4**. In the illustrations presented in **Fig. 8**, where the Isotach viscosity is assumed for simplicity, a soil specimen is supposed to be subjected to different loading histories (1)-(5) in drained TC (**Fig. 8a**). Then, a unique stress-strain curve (i.e., curve (1) in **Fig. 8b**) is predicted by an elasto-plastic model not accounting for ageing effects. For an elasto-viscoplastic model not accounting for ageing effects, different stress-strain curves due to the viscous effect are obtained (**Fig. 8b**). For loading history (3), apparent ageing effect due to the viscous effects is observed when ML at the original constant strain rate is restarted following creep deformation *a-b* (Fig. 8b). The same stress-strain curve is obtained for loading histories (1) & (2). When ageing effect becomes active, different stress-strain curves due to different effects of ageing and viscosity are obtained (**Fig. 8c**). For loading history (3), when ML at a constant strain rate is restarted following creep deformation *a-b*, the stress-strain behaviour becomes very stiff for a large stress range. Without an interaction between the ageing and viscosity effects, the same stress state is ultimately obtained for different loading histories when the instantaneous irreversible strain rate and ageing period become the same. With a positive interaction between the ageing and viscosity effects (**Fig. 8d**), the ultimate strength for the same irreversible strain rate and ageing period becomes larger as aged longer at higher deviator stresses. This is the case with highly-compacted cement-mixed well-graded gravelly soil (Kongsukprasert & Tatsuoka, 2005).

Table 4. Definitions of ageing and loading rate effects (Tatsuoka et al., 2003; Kongsukprasert & Tatsuoka, 2005).

	Phenomenon	Mechanism or Material Property	Parameter	
e.g., cement-mixed soil & soil for a geological period	Ageing effect	*Time-dependent:* Change of material property with time, e.g., cementation, weathering, etc.	Time with the fixed origin, t_c	* excluding geological effects
	Apparent Ageing Loading rate effect (creep, stress relaxation, etc.)	*Rate-dependent:* Response of material due to viscous property	Irreversible strain rate, $\dot{\varepsilon}^{ir}$	e.g., air-dried sand*

In Fig. 2 (for air-dried kaolin with and without cement-mixing), the stress-strain behaviour after the restart of ML following a sustained loading stage tends to rejoin the relation that would have been obtained when ML had continued at a constant strain rate without an intermission of sustained loading. This is the case illustrated in Fig. 8b when the viscosity is

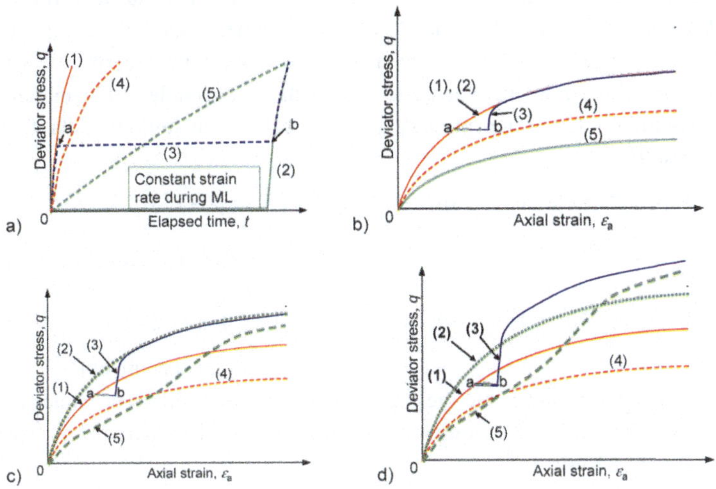

Fig. 8. a) Various loading histories in drained TC; and stress-strain curves for elasto-viscoplastic models; b) without ageing; c) with ageing (no coupling); and d) with ageing (positive coupling) (Tatsuoka et al., 2006).

of TESRA type. In this case, the ageing effect is insignificant, if any. In Fig. 4b, on the other hand (for saturated cement-mixed kaolin), significant ageing effects can be observed in the stress-strain relation immediately after the restart of ML at a constant strain rate following a sustained loading stage. That is, the broken curves presented in **Fig. 9** are the reference relations inferred by removing viscous effects from the segmental stress-strain relations measured for different strain rates. In so doing, the respective reference curve was obtained by scaling the segmental R - γ relations when the axial strain rate was $\dot{\varepsilon}_0 = 1.21 \times 10^{-6} / s$ before the start of drained sustained loading, It may be seen that the measured stress-strain relation after the restart of ML following a sustained loading stage largely overshoots the reference relation that is extrapolated to the regime after the start of sustained loading. The trend of over-shooting is due partly to the viscous effect of the TESRA type, but largely to the ageing effect that developed during the sustained loading stage. This is the case illustrated in Fig. 8c when the viscosity is of TESRA type (Komoto et al., 2004).

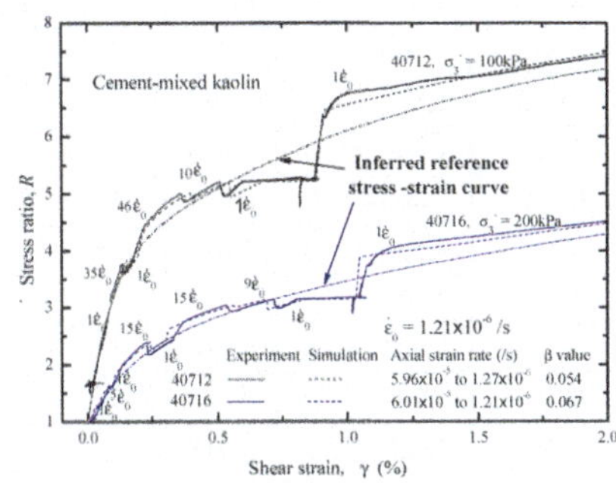

Fig. 9. Ageing effects on saturated cement-mixed kaolin

6. SIMULATION

Reference relation: In the simulation explained below, the stress parameter, σ, is the stress ratio, R, while the strain parameter, ε, is the shear strain, $\gamma = \varepsilon_v - \varepsilon_h$, for saturated specimens and the vertical strain, ε_v, for air-dried specimens. The elastic stiffness obtained from unload/reload cycles with a small strain amplitude, which is a function of the instantaneous stress state, was used to evaluate the elastic strain increments, which were removed from the total strain increments to obtain the irreversible strain increments. The reference stress-strain relation for primary loading defined as follows was fitted to the respective test result:

$$R^f = A_1[1-\exp(-\gamma^{ir}/t_1)] + A_2[1-\exp(-\gamma^{ir}/t_2)] + A_3[1-\exp(-\gamma^{ir}/t_3)]$$

(for saturated specimens) (6a)

$$R^f = A_1[1-\exp(-\varepsilon_v^{ir}/t_1)] + A_2[1-\exp(-\varepsilon_v^{ir}/t_2)] + A_3[1-\exp(-\varepsilon_v^{ir}/t_3)]$$

(for air-dried specimens) (6b)

where A_1, A_2, A_3, t_1, t_2 and t_3 are the parameters, for which different values are defined for primary, unloading and reloading stress-strain curves. The parameters used in the simulation are listed in **Table 5**.

Table 5. Model parameters for simulation (E^* is the elastic modulus defined for the R and γ_v or ε_v relation)

Specimen	E^*	α	m	$\dot{\varepsilon}_r^{ir}$	r_1	A_1	A_2	A_3	t_1	t_2	t_3
40810	3,600	1	0.015	3×10^{-9}	0.001	0.40	0.96	5.17	0.75	0.75	32.32
40811	3,600	1	0.015	3×10^{-9}	0.001	0.80	1.5	0.7	0.2	3.2	42.32
40622	1,800	1	0.023	10^{-9}	0.8	0.70	0.33	2.95	0.4	5.8	40.32
40625	1,800	1	0.023	10^{-9}	0.8	0.50	0.53	2.75	0.4	3.5	36.32
40720	1,800	1	0.023	10^{-9}	0.8	0.70	0.33	2.95	0.4	2.8	36.32
40618	3,600	1.5	0.018	10^{-9}	0.1	1.8	4.2	0	0.04	1.1	-
40630	3,600	1.5	0.018	10^{-9}	0.1	1.8	4.4	0	0.06	1.261	-
40712	3,600	1.5	0.018	10^{-9}	0.1	2.4	4.1	0	0.082	1.2	-
40716	3,600	1.5	0.018	10^{-9}	0.1	1.2	3.5	0	0.185	2.62	-
40807	3,600	1.5	0.018	10^{-9}	0.1	2.5	4.2	0	0.145	2.22	-

Simulation of ageing effect: Yielding is defined as the development of $d\varepsilon^{ir} > 0$ taking place when:

$$\left[\sigma^f\right]_{(\tau,t)} = \left[\sigma_y^f\right]_{(\tau,t)} \text{ and } \left[d\sigma^f\right]_{(\tau,t)} = \left[d\sigma_y^f\right]_{(\tau,t)} \tag{7}$$

where σ_y^f (= R_y^f) is the yield inviscid stress that is subjected to ageing effect. When σ_y^f is assumed to be independent of loading history, we have:

$$\left[\sigma_y^f\right]_{(\varepsilon^{ir},t_c)} = \sigma_y^f(\varepsilon^{ir},t_c) = \sigma_0^f(\varepsilon^{ir}) \cdot A^f(t_c) \quad (8)$$

where $\sigma_0^f(\varepsilon^{\mathrm{ir}})$ is the inviscid yield stress when ageing effect is not active; and $A^f(t_c)$ is the ageing function, for which $log_{10}(10\times(t_c + t_{ageing})/t_{ageing})$ (t_{ageing} = 302,000 sec) was used in the present simulation.

Figs. 2, 3 & 4 compare the measured and simulated stress-strain relations. It may be seen that the observed trends of viscous behaviour with and without ageing affects are all simulated very well.

7. CONCLUSIONS

The following conclusions can be derived from the results of drained triaxial compression tests on reconstituted specimens of kaolin under different conditions presented in this paper:

1) Air-dried kaolin, free from the effects of pore water, exhibits a significant trend of viscous behaviour, while drained saturated kaolin, free from the effects of delayed dissipation of excess pore water pressure, also exhibits a significant trend of viscous behaviour.
2) In all the tests, the magnitude of viscous stress could be quantified by the rate-sensitivity coefficient, β, defined in the paper. When free from the effects of pore water, the β values of clays were surprising similar to those of sands and gravels, showing that, when free from the effects of pore water, the β value is essentially independent of particle size. When air-dried, the β value does not change by cement-mixing. When saturated, on the other hand, the β value tended to increase by cement-mixing.
3) The manner how the viscous stress decays with an increase in the strain during monotonic loading at a constant strain rate depends on the test condition (air-dried or saturated; and with or without cement-mixing). The mechanism for this variation is not known.
4) Saturated cement-mixed kaolin exhibited very stiff behaviour for a large stress range immediately after the restart of ML at a constant strain rate following a sustained loading stage, which was due to coupled effects of creep strain and ageing that developed during the sustained loading.
5) The observed effects of viscosity and ageing on the stress-strain behaviour of kaolin were successfully simulated by a non-linear three-component rheology model that was modified from the original one to account for ageing effects.

ACKNOWLEDGEMENT

The writers would like to thank Prof. Koseki, J. of Institute of Industrial Science, the University of Tokyo for helpful suggestions on the present study.

REFERENCE

Deng, J. and Tatsuoka,F. (2005a): Comprehensive effects of ageing and viscosity on the deformation of clay in 1D compression, *Natural Science Journal of Xiangtan University*, Vol.27, No.1, pp.102-105.

Deng, J. and Tatsuoka,F. (2005b) : Ageing and viscous effects on the deformation of clay in 1D compression, *Geotechnical Special Publication, No.130-142, Proc. Geo-Frontiers 2005*, p 2311-2322

Di Benedetto,H., Tatsuoka,F. and Ishihara,M. (2002): Time-dependent shear deformation characteristics of sand and their constitutive modelling, *Soils and Foundations*, Vol.42, No.2, pp.1-22.

Enomoto. T., Tatsuoka. F., Kawabe, S. and Di Benedetto, H. (2006), "Viscous property of granular material in drained triaxial compression", this symposium.

Hirakawa.D., Tatsuoka.F. and Siddiquee,M.S.A. (2003): Viscous effects on bearing capacity characteristic of shallow foundation on sand, *Proc. 38th Japan National Conf. on Geotechnical Engineering,* JGS, Akita, June (in Japanese).

Imai,G., Tanaka,Y. and Saegusa,H. (2003): One-dimensional consolidation modeling based on the Isotach law for normally consolidated clays, *Soils and Foundation,* Vol.43, No.4, pp.173-188.

Komoto, N., Tatsuoka, F., Koseki, J., Sato, T. and Oka, H. (2004): Deformation and strength characteristics of cement-mixed clay, *Proc. 39th Japanese Geotechnical Symposium*, pp.239-240(in Japanese).

Kongsukprasert, L. and Tatsuoka, F. (2003): Viscous effects coupled with ageing effects on the stress-strain behaviour of cement-mixed granular materials and a model simulation, *Proc. 3rd Int. Symp. on Deformation Characteristics of Geomaterials*, IS Lyon 03 (Di Benedetto et al. eds.), Balkema, Sept. 2003, pp.569-577.

Kongsukprasert, L. and Tatsuoka, F. (2005): "Ageing and viscous effects on the deformation and strength characteristics of cement-mixed gravely soil in triaxial compression", *Soils and Foundations* (accepted)

Leroueil, S. and Marques, M. E. (1996). "Importance of strain rate and temperature effects in geotechncial engineering." *S-O-A Report for Session on Measuring and Modeling Time Dependent Soil Behaviour, ASCE Convention, Washington*, Geot. Special Publication 61: 1-60.

Li,J.-Z., Acosta-Martínez,H., Tatsuoka,F. And Deng,J.-L. (2004): Viscous property of soft clay and its modeling, *Engineering Practice and Performance of Soft Deposits*, Proc. of IS Osaka 2004.

Matsui,T. and Abe,N. (1985): Elasto/viscoplastic constitutive equation of normally consolidated clays based on flow surface theory, *International Conference on Numerical Methods in Geomechanics*, Vol.5, No.1, pp.407-413.

Matsushita,M., Tatsuoka,F., Koseki,J., Cazacliu,B., DiBenedetto,H. and Yasin,S.J.M. 1999. Time effects onthe pre-peak deformation properties of sands, *Proc.Second Int. Conf. on Pre-Failure Deformation Characteristicsof Geomaterials* (Jamiolkowski et al. eds.),Balkema, 1: 681-689.

Nawir,H., Tatsuoka,F. and Kuwano,R. (2003): Experimental evaluation of the viscous properties of sand in shear, *Soils and Foundations,* Vol.43, No.6, pp.13-31.

Oka,F., Kodaka,T., Kimoto,S., Ishigaki,S. and Tsuji,C. (2003): Step-changed strain rate effect on the stress-strain relations of clay and a constitutive modeling, *Soils and Foundation,* Vol.43, No.4, pp.189-201.

Tatsuoka,F., Santucci de Magistris,F., Hayano,K., Momoya,Y. and Koseki,J. (2000): "Some new aspects of time effects on the stress-strain behaviour of stiff geomaterials", Keynote Lecture, *The Geotechnics of Hard Soils – Soft Rocks, Proc. of Second Int. Conf. on Hard Soils and Soft Rocks, Napoli,* 1998 (Evamgelista & Picarelli eds.), Balkema, Vol.2, pp1285-1371.

Tatsuoka,F., Ishihara,M., Di Benedetto,H. and Kuwano,R. (2002): Time-dependent compression deformation characteristics of geomaterials and their simulation, *Soil and foundation,* Vol.42, No.2, pp.103-138.

Tatsuoka,F., Di Benedetto,H. and Nishi,T. (2003): A framework for modelling of the time effects on the stress-strain behaviour of geomaterials, *Proc. 3rd Int. Sym. on Deformation Characteristics of Geomaterials*, IS Lyon 03 (Di Benedetto et al. eds.), Balkema, September, 2003, pp.1135-1143.

Tatsuoka,F. (2004). "Effects of viscous properties and ageing on the stress-strain behaviour of geomaterials." Geomechanics-Testing, Modeling and Simulation, Proceedings of the GI-JGS workshop, Boston, ASCE Special Geotechnical SPT No. 143 (Yamamuro & Koseki eds.), pp.1-60.

Tatsuoka,F., Kiyota,T. and Enomoto,T. (2005). "Viscous properties of geomaterials in drained shear" *Geomechanics-Testing, Modeling and Simulation, Proceedings of the Second GI-JGS workshop, Osaka, ASCE Geotechnical Special Publication GSP No. ???* (Lade et al. eds.).

Vaughn, P. R. (1997). "Engineering behaviour of weak rock: Some answers and some questions." *Geotechnical Engineering of Hard Soils and Soft Rocks,* Balkema, 3, 1741-1765.

Yin,J.-H and Graham,J. (1999): Elastic viscoplastic modelling of the time-dependent stress-strain behaviour of soils, *Canadian geotechnical journal*, Vol.36, No.4, pp.736-745.

Soil Stress-Strain Behavior: Measurement, Modeling and Analysis
Geotechnical Symposium in Roma, March 16 & 17, 2006

EFFECTS OF CURING TIME AND STRESS ON THE SHEAR STRENGTH AND DEFORMATION CHARACTERISTICS OF CEMENT-MIXED SAND

Jiro Kuwano
Geosphere Research Institute
Saitama University, Japan
e-mail: jkuwano@post.saitama-u.ac.jp

Tay Wee Boon
Singapore Government, Singapore
(Formerly Tokyo Institute of Technology)

ABSTRACT

This study investigates the coupled effect of curing time and stress on the strength and deformation characteristics of cement-mixed sand over a long period of time, e.g. 180 days, as compared to 7 to 60 days in past studies.

1. INTRODUCTION

Similar to concrete materials, hydration of cement in cement-mixed soil continues with time, thereby improving the strength and deformation characteristics of cement-mixed soil over time (e.g. Taguchi et al 2002; Kongsukprasert et al 2003). On the other hand, it has been found that there are differences between measured results in the laboratory and those deduced from field behavior. In fact, the cementation bonds found in *in situ* soil were formed under stress. However, in the usual testing techniques, cementing under stress has not been considered. This leads to an underestimation of the stress-strain behavior of cement-mixed soil (Consoli et al 2000).

The objective of this study is to investigate the coupled effect of curing time and stress on the strength and deformation characteristics of cement-mixed sand over a long period of time, e.g. 180 days, as compared to 7 to 60 days in past studies.

2. EXPERIMENTAL CONDITIONS

The amount of high-early-strength Portland cement used was 60kg per $1m^3$ of sand, to

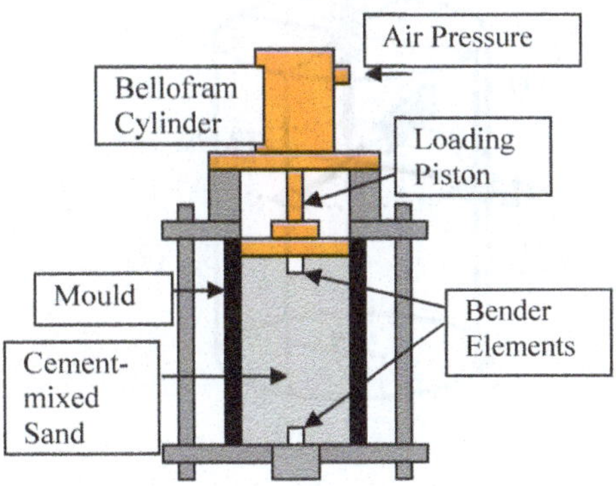

Fig.1. Apparatus for measuring G_{vh} during curing with stress

Hoe I. Ling et al. (eds.), Soil Stress-Strain Behavior: Measurement, Modeling and Analysis, 413–418.

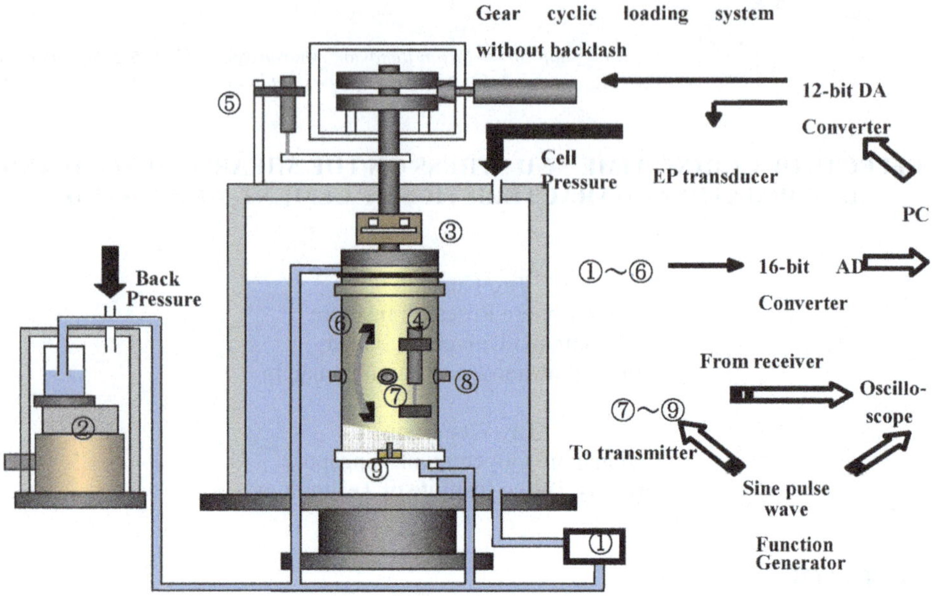

Fig. 2. Automated triaxial system

achieve cement-mixed sand with a dry density of γ_t =1.6g/cm^3 and an unconfined compressive strength of about 500 kPa after 7 curing days. The composition of cement-mixed sand used in this study is the same as that used in the centrifuge model tests on reinforced embankments using cement-mixed sand and geogrids (Ito et al. 2002).

In this study, specimens with 2 different curing stresses, i.e. without stress at 0 kPa

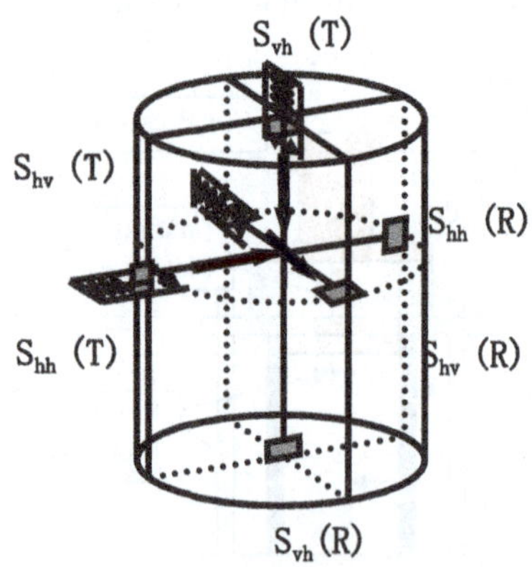

Fig. 3. Arrangement of bender elements

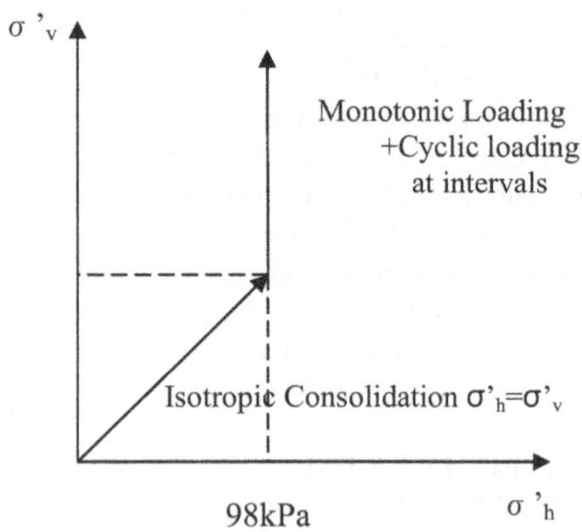

Fig. 4. Stress path for all the test cases

and under stress at 98 kPa, and curing times of 7, 28, 90 and 180 days for each curing stress were considered. The change in shear modulus G_{vh} during curing with stress was measured using a mould with a pair of bender elements fixed at the vertical ends as shown in Figure 1. After their respective curing days, specimens were set in a triaxial apparatus shown in Figures 2 and 3 (Chaudhary et al. 2004) and isotropically consolidated to an effective confining stress of 98kPa, followed by drained monotonic loading as shown by the stress path in Figure 4. Elastic modulus E_v and shear moduli G_{hh}, G_{hv}, G_{vh} were measured at intervals.

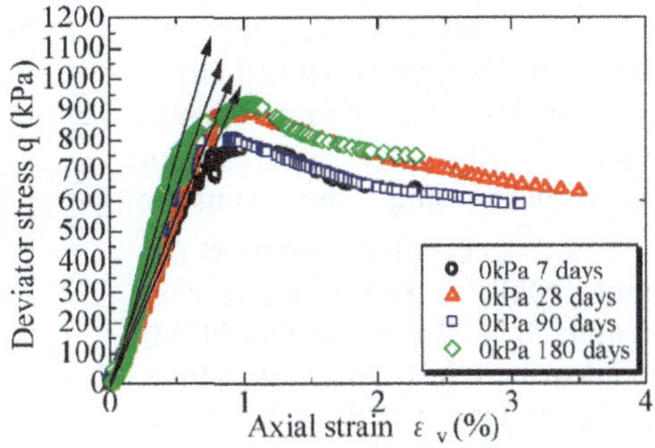

Fig. 5. Cured without stress

3. CONSOLIDATION AND DRAINED MONOTONIC LOADING

Figures 5 and 6 show the stress-strain relations during monotonic loading for specimens cured without and under stress respectively. It can be observed that deviator stress q increases with axial strain ε_v, reaches a peak before decreasing. Cement-mixed sand is stiff and brittle ascompared to pure sand. Stiffness increases with curing time, and specimens cured under stress are noted to be stiffer. Peak strength q_{max} also increases with curing time, and specimens cured under stress have higher peak strength during monotonic loading, as shown in Figure 7. Figure 8 shows that specimens become less compressive and more dilatant with increase in curing time. Specimens cured under stress are more dilatant, but the reverse is seen in specimens cured for 90 days and above. Concentration of strain at the slip surface may have caused less dilatant character in specimens.

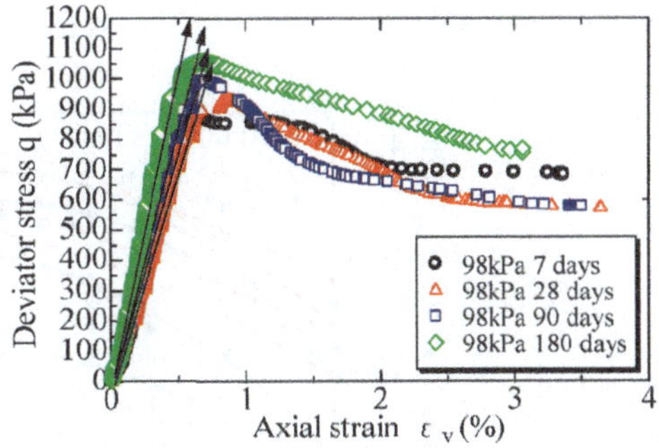

Fig. 6. Cured under stress

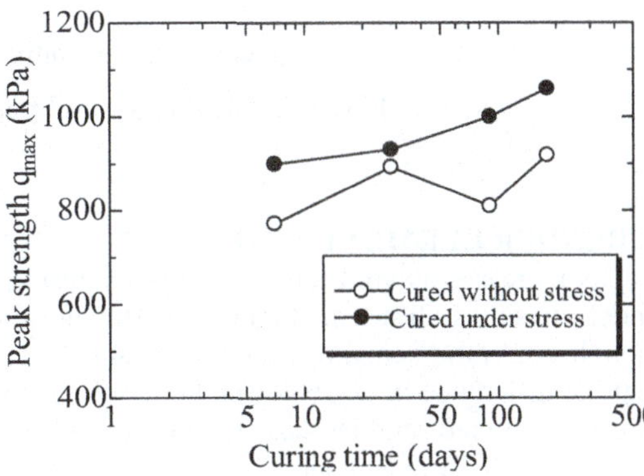

Fig. 7. Peak strength

4. CYCLIC LOADING TEST

E_v was measured by cyclic loading of small amplitude during drained monotonic loading test with a constant effective horizontal stress. It can be observed in Figure 9 that E_v increases with σ'_v, and can be represented using the equation $E_v = A(\sigma'_v)^{n_v}$. E_v drops when σ'_v is about 400kPa, which is before peak strength. It is also noted that elastic modulus increases with curing time, and specimens cured under stress have a higher elastic modulus than their correspondents. On the other hand, rate of increase in E_v with σ'_v is higher for specimens cured without stress.

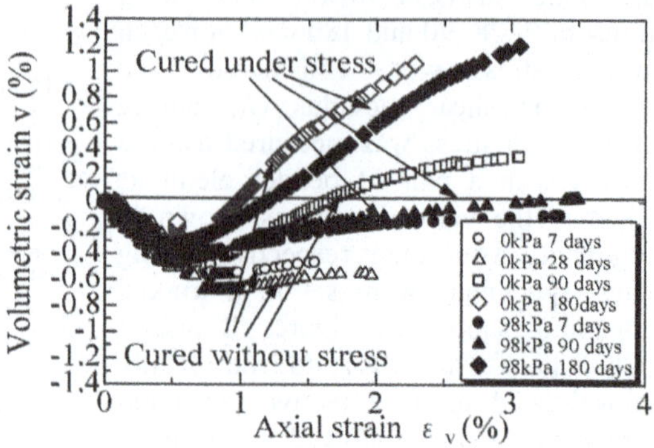

Fig. 8. Volumetric strain

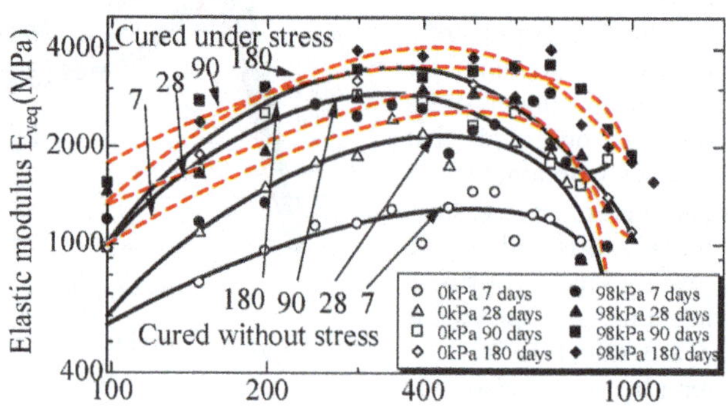

Fig. 9. Change in G_{vh} and void ratio during

5. BENDER ELEMENT TEST

G_{vh} increases as void ratio e decreases during loading of overburden stress for curing under stress, as shown in Figure 10. But during curing, G_{vh} is noted to increase even though void ratio remains almost constant for both curing stresses. It can also be noted that the difference in G_{vh} between both curing stresses decreases with time. Shear moduli of cement-mixed sand increase with time, and specimens cured under stress have a higher G_{vh} than their correspondents. It is also observed that the rate of increase in G_{vh} with σ'_v is higher for specimens cured without stress.

6. CONCLUSIONS

1) Stiffness, peak strength q_{max}, elastic modulus E_v and shear moduli G_{hh}, G_{hv}, G_{vh} of all specimens increase with curing time regardless of the availability of acting stresses during curing.
2) Specimens become less compressive and more dilatant with increase in curing time.
3) Specimens cured under stress are noted to have higher stiffness, peak strength, elastic modulus and G_{vh}, regardless of the number of curing days.
4) Difference in G_{vh} between specimens cured without and under stress decreases with time.
5) Specimens cured without stress have a higher rate of increase in both E_v and G_{vh} with σ'_v.

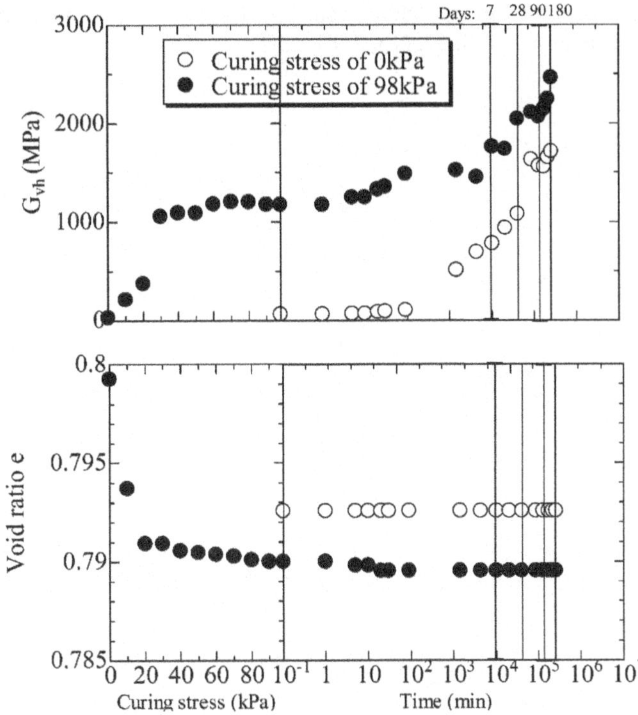

Fig. 10 Change in G_{vh} and void ratio during

REFERENCES

1) Chaudhary, S.K., Kuwano, J. & Hayano, Y. 2004 Measurement of quasi-elastic stiffness parameters of dense Toyoura sand in hollow cylinder apparatus and triaxial apparatus with bender elements, Geotechnical Testing Journal, ASTM, 27(1), 23-35.
2) Consoli, N.C., Rotta, G.V. & Prietto, P.D.M. 2000. Influence of curing under stress on the triaxial response of cemented sands. Geotechnique, 50(1), 99-105.
3) Itoh, H., Saitoh, T., Kuwano, J. & Izawa, J. 2003. Development of reinforcement wall using cement-mixed soil and geogrids. Geosynthetics Technical Information, JCIGS, 19(3), 42-49.
4) Kongsukprasert, L. 2003. Time effects on the strength and deformation characteristics of cement-mixed gravel. Dr Thesis. University of Tokyo.

5) Taguchi, T., Suzuki, M., Yamamoto, T., Fujino, H., Okabayashi, S. & Fujimoto, T. 2002. Influence of consolidation stress history on unconfined compressive strength of cement-stabilized soil. Technical Report, Department of Engineering, Yamaguchi University, Vol. 52, No. 2: 87-92

Soil Stress-Strain Behavior: Measurement, Modeling and Analysis
Geotechnical Symposium in Roma, March 16 & 17, 2006

EFFECTS OF SOME FACTORS ON THE STRENGTH AND STIFFNESS OF CRUSHED CONCRETE AGGREGATE

L. Lovati
Politecnico di Torino, Torino, Italy

F. Tatsuoka & Y. Tomita
Department of Civil Engineering,
Tokyo University of Science, Chiba, Japan

Abstract: A series of consolidated drained triaxial compression (TC) tests were performed on a crushed concrete aggregate (CCA) compacted using three different levels of energy. A wide range of moulding water content, *w*, and two different confining pressures were employed. The compressive strength and stiffness of the tested CCA when highly compacted at water content close or slightly higher than the optimum value, w_{opt}, were very high, higher than those of a typical natural well-graded gravelly soil having similar grading characteristics used as the backfill material of highest quality. The compressive strength and stiffness of the tested CCA was not highly sensitive to changes in *w*, in particular when $w \geq w_{opt}$, but it decreased sharply when *w* became lower than w_{opt}. The strength and stiffness was very sensitive to compaction energy, therefore the degree of compaction. All the test results show that highly compacted CCA can be used as the backfill material for important civil engineering soil structures, such as retaining walls and bridge abutments, that need a high stability while allowing limited deformation.

1. INTRODUCTION

Efficient recycling of concrete scrap, for example by the reuse in construction projects, is becoming more important for both environmental and economic reasons in a number of developed countries. Changes in functional requirements have often reduced the effective life time of civil engineering steel-reinforced concrete (RC) structures (e.g., buildings and bridges), which has resulted in their relatively early demolishment. A great amount of concrete waste produced from such events has resulted and will result in a shortage of dumping area, while a high cost for the scrap transport and disposal is becoming more serious. Moreover, natural aggregate needed for new RC structures is consistently becoming more in short.

One of the realistic solutions to these problems is the reuse of crushed concrete as the aggregate of concrete to be newly produced after necessary crushing and other treatments. Actually a number of researchers in different parts of the world studied on this issue (e.g., Wainwright et al., 1993). It has been revealed however that the cost to remove thin mortar layers from the surface of coarse gravel particles to the extent that is sufficient to recover the original strength of concrete produced by using fresh aggregate is prohibitively high.

On the other hand, a limited number of research were performed on the strength and deformation characteristics of crushed concrete aggregate (CCA) as the backfill material for geotechnical engineering soil structures, such as railway and highway embankments and soil retaining walls. For example, based on the results of a series of drained triaxial compression

Hoe I. Ling et al. (eds.), Soil Stress-Strain Behavior: Measurement, Modeling and Analysis, 419–427.

(TC) tests, Aqil et al. (2005) and Tatsuoka et al. (2006) showed that CCA can become a very good backfill material in terms of strength and stiffness when highly compacted, preferably at water content around the optimum. They also showed that the strength and stiffness of CCA thus obtained is similar to, or ever better than, that of similarly compacted well-graded gravelly soil used as the backfill of highest quality. At the same time, they are rather insensitive to changes in the water content and confined saturation. Despite several important findings by their studies, the comparison of strength and deformation characteristics between CCA and natural well-graded gravelly soil having similar grading characteristics has not been made in a systematic way in terms of the range of water content during compaction, compaction energy and compacted dry density. It is the objective of the present study to study on the effects of these factors on the strength and deformation characteristics of CCA in comparison of those of a typical natural well-graded gravelly soil.

2. TESTING PROCEDURE

2.1 Materials and specimen preparation

Two materials were used. The first was a well-graded CCA, called REPA, which was obtained from an industrial recycling process of the waste of electricity supply poles. Therefore, the quality of the original concrete was very high. The other material is a well-graded gravelly soil consisting of crushed sandstone from a quarry (named model Chiba gravel), which has nearly the same grading characteristics as REPA. This gravelly soil is considered as the backfill of highest-quality. The physical properties of these materials are listed in **Table 1**.

Table 1 Physical properties of the test materials (*: fines content)

Material type	G_s	D_{max} (mm)	D_{50} (mm)	U_c	F_c*	Energy level	w_{opt} (%)	$\rho_{d,max}$ (g/cm^3)
REPA	2.654	19	6.72	13.99	0.80	**E0**	6.6	1.778
						E1	8.9	1.868
						E2	11	1.930
Chiba gravel	2.737	19	3.83	15.77	2.40	**E0**	5.2	2.137
						E1	5.8	2.300
						E2	4.9	2.337

Both materials were sieved to remove particles with a diameter larger than 19 mm in order to fit the maximum specimen dimensions allowed for the available triaxial apparatus (i.e., 10 cm in diameter and 20 cm in height). The specimens were produced by manually tamping the material in five sub-layers in a cylindrical split mould at different water contents using three different levels of compaction energy. The water content was between 4 % and 12 % for REPA and between 4 %

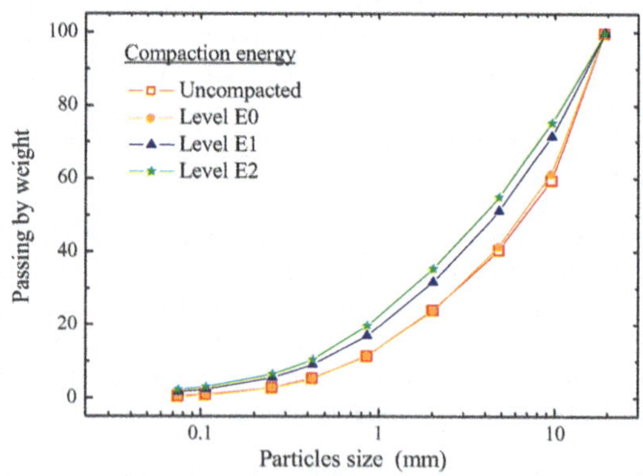

Fig. 1 Gradation curves of REPA before and after compaction by using different energy levels.

and 6 % for model Chiba gravel. The three levels of compaction energy were: 560 kN·m/m^3 (E0), 2530 kN·m/m^3 (E1) and 5060 kN·m/m^3 (E2) (n.b., the compaction curves are presented in Fig. 2).

Fig. 1 shows the gradation curves of CCA (REPA) before and after compaction while before a TC test. The grading curve of model Chiba gravel is similar to that of REPA before compaction. The curve was shifted left by particle breakage, of which the amount increased with an increase in the compaction energy. It was found by sieving REPA after a TC test that the gradation curve did not change noticeably by TC shearing, showing that the particle breakage during TC was insignificant.

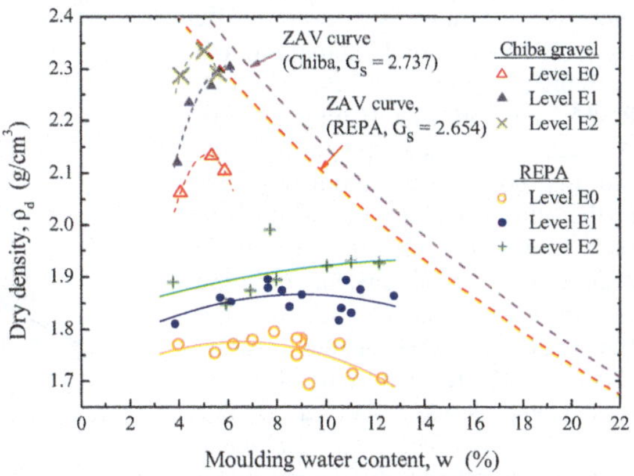

Fig. 2 Compaction curves of the test materials (CCA, REPA and a natural gravelly soil, model Chiba gravel).

2.2 Triaxial apparatus and loading method

A state-of-the-art automated strain-controlled triaxial testing apparatus was used (e.g., Santucci de Magistris et al., 1999). The axial load was measured with a load cell located inside the triaxial cell to eliminate the effects of piston friction (Tatsuoka 1988). Both axial and lateral strains were measured locally by using, respectively, a pair of Local Deformation Transducer (LDTs) (Goto et al., 1991; Hoque et al., 1997) and three clip gauges set at 5/6, 1/2 and 1/6 of the specimen height (Lohani et al., 2004). Local axial and radial strains obtained by averaging these locally measured quantities are reported in this paper. During the respective test, axial strains were monitored by axial displacements of the loading piston measured with a Linear Variable Displacement Transducer (LVDT) set outside the triaxial cell. All the acquired data were recorded automatically.

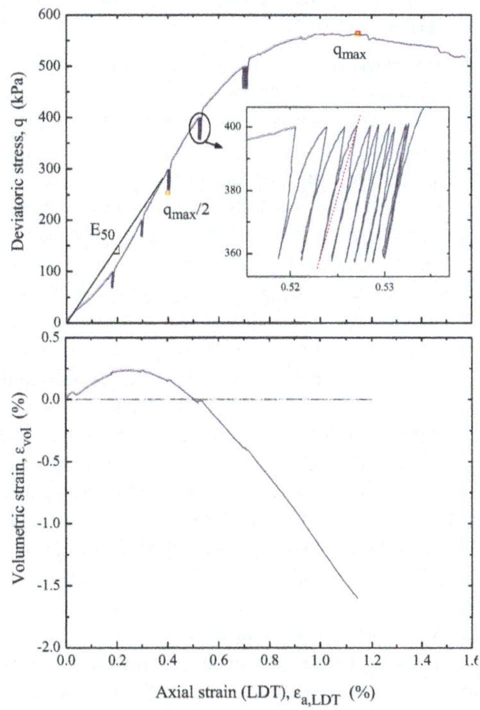

Fig. 3 Typical test result from a TC test (σ_3'= 30 kPa) of REPA (test 36, compacted at w= 10 % using energy level E0).

Without made saturated, the respective specimen was isotropically consolidated by partial vacuuming to a confining pressure of 30 kPa or 90 kPa (REPA) and 30 kPa (Chiba gravel). After leaving the specimen at the final confining pressure for about one hour for the micro-structure of specimen to reach equilibrium, drained monotonic loading (ML) TC at constant confining pressure was started at constant axial strain rate of 0.003 %/min. During otherwise ML, ten load-unload cycles of deviator stress with a double amplitude of 40 kPa were applied at every deviator stress increment equal to 100 kPa (as shown in Fig. 3).

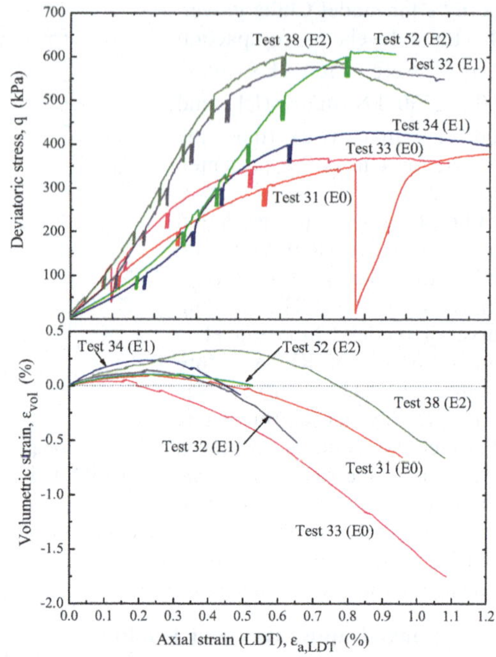

Fig. 4 Effects of compaction energy and water content during compaction, REPA at σ_3'= 30 kPa.

3 TEST RESULTS AND DISCUSSIONS

3.1 Gradation curves and compaction characteristic

Fig. 2 shows the compaction curves of the two test materials. The following trends of behaviour may be seen:

1) For both materials, the compaction curve is shifted upward with an increase in the compaction energy. Changes in the optimum water content by the change in the compaction energy are not obvious. The maximum dry density, $\rho_{d,max}$, and optimum moisture content, w_{opt}, in the respective case are listed in Table 1.
2) The effects of moulding water content on the compacted dry density, ρ_d, for the same compaction energy are less significant with the CCA, REPA, than the natural gravelly soil, model Chiba gravel. This result indicates that the water content control could be less strict with CCA than with natural gravelly soil in obtaining as possible as a high compacted dry density for a given compaction energy level.
3) For the same compaction energy, the ρ_d values of the CCA are much smaller than those of the natural gravelly soil. This trend is due only partly to a low specific gravity of the CCA, but mostly to higher compacted void ratios. Such a result as above has led to a notion that CCA is generally significantly inferior in the strength and stiffness than natural well-graded gravelly soil. It is shown below that this notion is not correct.

3.2 Stress-strain relations

Fig. 3 shows a typical test result from a TC test (σ'_c = 30 kPa) of REPA. It may be seen that the compressive strength, q_{max}, of well compacted REPA is very large compared with the confining pressure, σ_3'. The q_{max} value and the secant Young's modulus at $q = q_{max}/2$, E_{50}, were obtained from this and other test results for the analysis shown below. **Fig. 4** shows other similar results of REPA compacted at different w values by using the three different levels of compaction energy. In test 31, the TC loading was terminated at ε_a= about 0.85 % and the specimen was brought to unconfined conditions due to malfunction of the test system. TC loading was restarted. The difference seen in the results for the same test conditions is due mostly to effects of specimen density and water content. It may be seen that the CCA generally becomes stronger, stiffer and more dilative with an increase in the compaction energy. More detailed analysis is given below.

3.3 Effects of moulding water content

Figs. 5a and **b** show the values of q_{max} and E_{50} when σ'_c= 30 kPa of REPA plotted against the respective moulding water content for three different levels of compaction energy. Despite a relatively large scatter in the data (in particular the data points with an arrow), the following trends of behaviour may be seen:

1) The values of q_{max} and E_{50} significantly increase with an increase in the compaction energy.
2) For the same compaction energy level, the values of q_{max} and E_{50} tend to become the respective maximum value when the water content is around the optimum water content, w_{opt}.
3) The decrease in the values of q_{max} and E_{50} with a decrease in the water content from w_{opt} is significantly larger than with an increase from w_{opt}. This trend of behaviour is opposite to the one usually observed with compacted finer geomaterials (e.g., Santucci de Magistris & Tatsuoka, 2004).

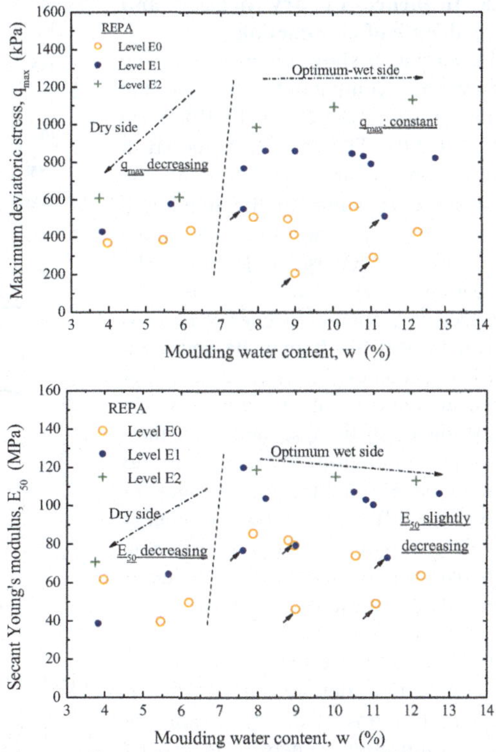

Fig. 5 Effects of *w*, on: a) (top) q_{max}; and b) (bottom) E_{50} of REPA (σ_3'= 30 kPa)

3.4 Effects of compaction energy

Figs. 6a and **b** show the effects of compaction energy on the stress-strain behaviour when σ_3'= 30 kPa and 90 kPa of REPA specimens compacted around w_{opt}. It may also be seen that, when compacted denser using higher compaction energy, REPA becomes stronger, stiffer and more dilative. **Figs. 7a** and **b** summarise the effects of compaction energy on the q_{max} and E_{50} values of REPA when σ'_3= 30 kPa and 90 kPa (the data of model Chiba gravel are discussed later in this paper). It may be seen that the effects of compaction energy on the peak strength, q_{max}, is significant, while the effects on the E_{50} value are less significant, in particular between the compaction energy levels, *E1* and *E2*. This trend of behaviour can also be noted in the pre-peak stress-strain relations at smaller strains (Fig. 6). The effects of moulding water content discussed in Section 3.3 can also be seen from Fig. 7.

3.5 Influence of dry density and degree of compaction

Fig. 8a and **b** show the effects of the degree of compaction, D_c, on the values of q_{max} and E_{50} of REPA when σ_3'= 30 kPa. The D_c values shown in these figures are the ratio of the respective ρ_d value to the maximum value, $\rho_{d.max}$, for the respective compaction energy level. The maximum value of D_c for the respective compaction energy level is equal to 100 %. It may be seen that REPA becomes stronger and stiffer with an increase in D_c. However, the dependency of the q_{max} and E_{50} values on D_c is not obvious in Fig.8. This was due to the fact that the $\rho_{d.max}$ values for the three different compaction energy levels are naturally largely different, whereas the respective q_{max} value is not linked to these $\rho_{d.max}$ values.

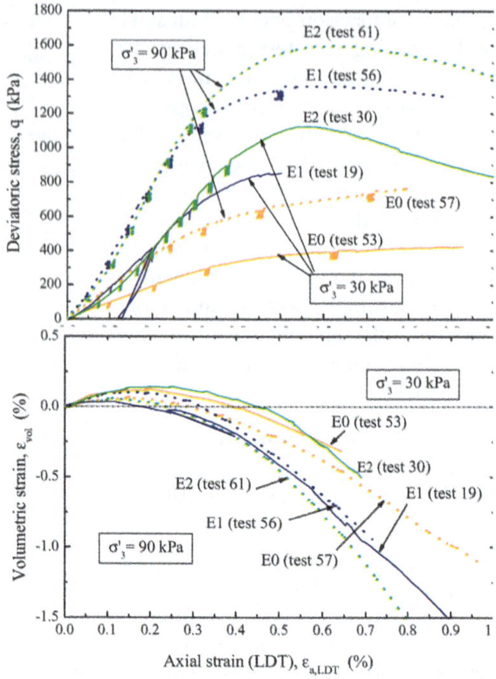

Fig. 6 Effects of compaction energy and confining pressure, REPA (compacted around wopt), σ_3'= 30 kPa and 90 kPa.

Fig. 9 shows the effects of the degree of compaction, denoted as D_c*, that is defined in terms of the ρ_{dmax} value for the compaction energy level E2 commonly for the three different compaction energy levels. Then, the q_{max} value of REPA becomes a rather unique function of D_c*. It may be seen that the q_{max} value is very sensitive to the degree of compaction, D_c*. Often in engineering practice, the allowable minimum degree of compaction is equal to 90 %. It is seen from Fig. 9 however that this value is too low to ensure sufficiently large q_{max} and E_{50} values of CCA. The test results indicate that the allowable minimum degree of compaction for CCA should be, say, 95 %.

Fig. 10 shows the relationship between q_{max} and the compacted density, ρ_d, which is equivalent to Fig. 9a. The data of the CCA when σ_3'= 90 kPa and the gravelly soil when σ_3'= 30 kPa are also plotted in this figure. The following important trends of behaviour of REPA may be seen:

1) For both σ_3' values, the q_{max} - ρ_d relation is rather independent of the compaction energy level. The effects of water content during compaction on the respective relation are small.
2) For both σ_3' values, the q_{max} value is highly sensitive to the ρ_d value. However, the sensitivity becomes smaller with an increase in σ_3'.

3.6 Effects of confining pressure

Significant effects of confining pressure on the strength and deformation characteristics of the CCA may be noted from Figs 6, 7 and 10. This result is consistent with the conclusion obtained by Aqil et al. (2005) from theresults of another series of CD TC on another, but similar type of CCA. The results from this and previous studies show that the peak strength and stiffness of CCA increases with an increase in the confining pressure in a similar way as ordinary unbound sand and gravel.

3.7 Comparison of stress-strain behaviour with a natural gravelly soil

It may be seen from Fig. 2 that, when compacted using the same energy, the dry density, ρ_d, of the CCA is much lower than a natural gravelly soil having similar grading characteristics. Despite the above, as seen from Figs. 7 and 10, when highly compacted, for the same compaction energy, the CCA, REPA, is much stronger and stiffer than the gravelly soil, model Chiba gravel. It may also be seen from Fig. 7 that the gravelly soil becomes stronger and stiffer when compacted drier than when compacted wetter under otherwise the same test conditions. This trend of behaviour is the same as compacted finer geomaterials (e.g., Santucci de Magistris & Tatsuoka, 2004), but it is opposite to the trend of the CCA.

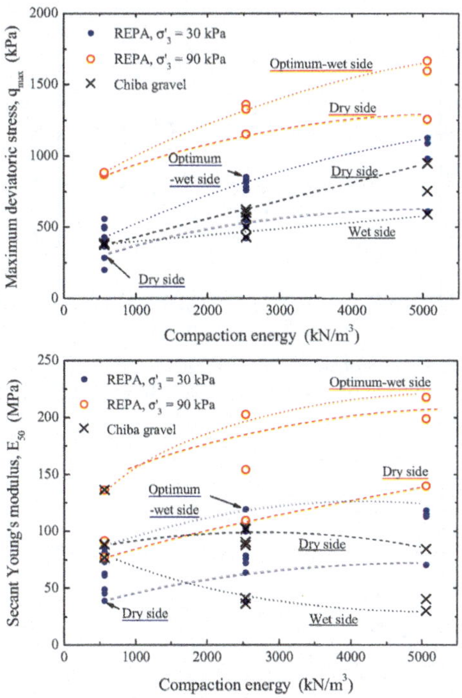

Figure 7 Effects of compaction energy on; a) (top) q_{max}; and b) (bottom) E_{50}, σ'_3 = 30 kPa & 90 kPa, REPA and model Chiba gravel.

To evaluate differences in the effects of compacted dry density on the compressive strength, q_{max}, between REPA and model Chiba gravel, the respective q_{max} value was divided by its maximum value (for the same σ_3' and the same material) and plotted against D_c*=ρ_d/"ρ_{dmax} for E2 (for the same σ_3' and the same material)" (**Fig. 11**). The following trends of behaviour may be seen:

1) The dependency of the q_{max} value of REPA on the compacted dry density, ρ_d, or the degree of compaction, D_c*, when σ_3'= 30 kPa is much larger than when σ_3'= 90 kPa. This means that, at lower confining pressure, loosely compacted CCA is particularly weak when compared with highly compacted ones. However, this defect becomes lighter with an increase in the confining pressure.
2) For the same σ_3' (= 30 kPa), the dependency of the q_{max} value of REPA on ρ_d, or D_c*, is much larger than Chiba gravel. This indicates that high compaction is very efficient in obtaining higher strength and stiffness with CCA than with natural gravelly soil.

Summarising the above, we conclude that we can use crushed concrete aggregate (CCA) as the backfill material for important civil engineering structures if highly compacted, preferably at the optimum water content or water content somehow higher than the optimum.

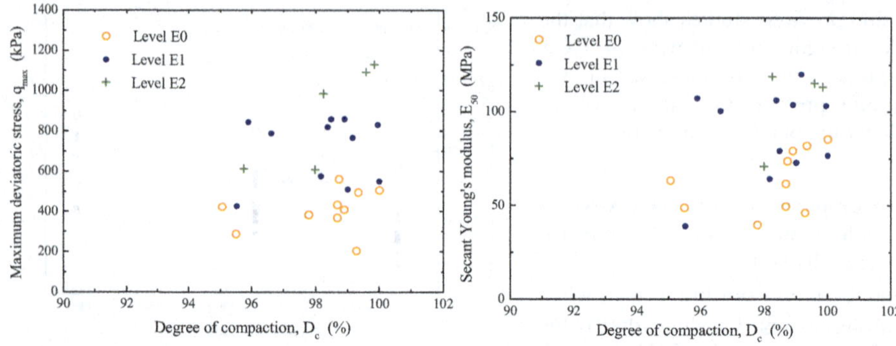

Fig. 8. Effects of the degree of compaction defined in terms of ρ_{dmax} for the respective compaction energy, D_c, on; a) (left) q_{max}; and b) (right) E_{50} at σ_3'= 30 kPa, REPA.

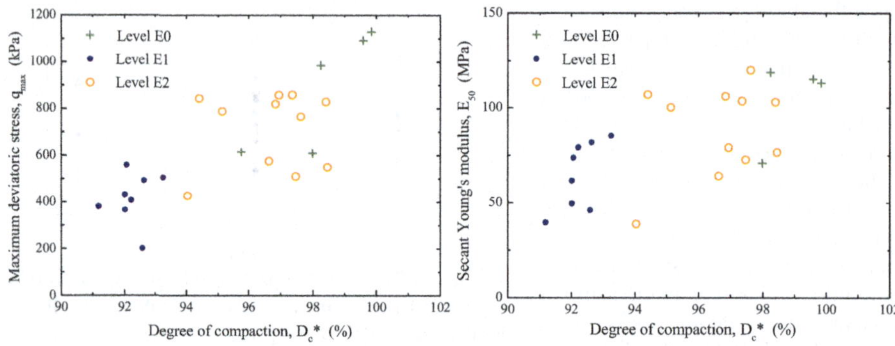

Fig. 9. Effects of the degree of compaction defined in terms of ρ_{dmax} for E2, D_c*, on; a) (left) q_{max}; and b) (right) E_{50} at σ_3'= 30 kPa, REPA.

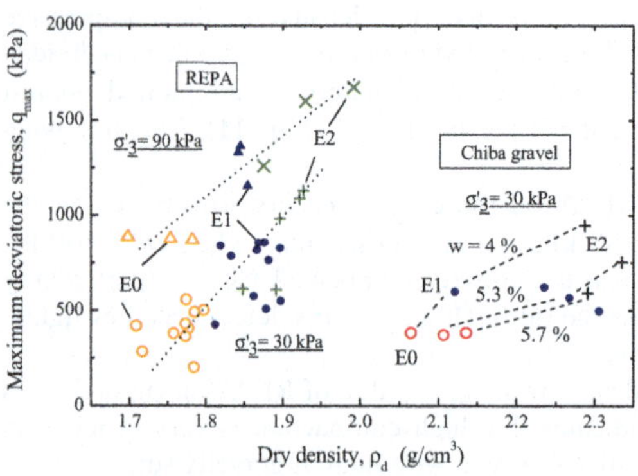

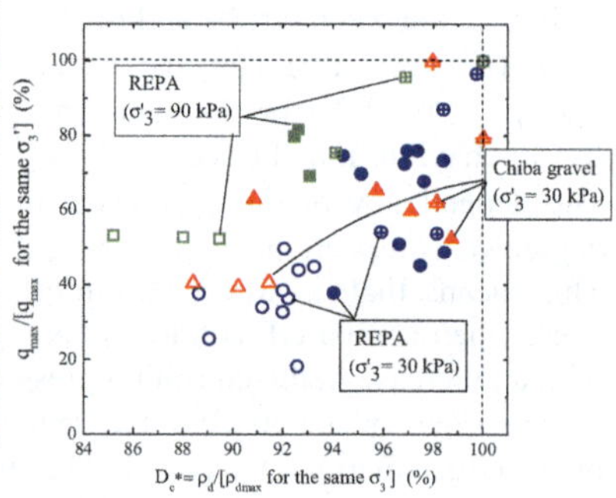

Fig. 10 (left). Effects of compacted dry density rd on q_{max} at s3'= 30 kPa and 90 kPa, REPA and model Chiba gravel.

Fig. 11 (right) Relationship between normalised compressive strength and normalised dry density, REPA and model Chiba gravel.

4 CONCLUSIONS

The following conclusions can be derived from the test results presented in this paper:

1) The particles of crushed concrete aggregate (CCA) were slightly crushed during compaction, of which the amount increased with an increase in the compaction energy level. It seems that crushing took place in thin layers of mortar covering the surface of stiff and strong core gravel particles.
2) The strength and stiffness of the tested CCA was not very sensitive to changes in the water content, w, during compaction in particular when $w \geq$ the optimum (w_{opt}). The sensitivity was smaller than a typical well-graded gravelly soil having similar grading characteristics tested in the present study which is considered as the backfill of highest-quality. The compressive strength and stiffness of the CCA decreased sharply when w became much lower than w_{opt}. Therefore, the use of too low water content, lower than about 7 % with the CCA, during compaction is not recommended to ensure a sufficiently high effectiveness of compaction in obtain high strength and stiffness.
3) The strength and stiffness, in particular the former when the confining pressure was 30 kPa, of the CCA was very sensitive to the degree of compaction. The sensitivity was more than the tested gravelly soil. Therefore, higher compaction is highly effective to obtain higher strength and stiffness of CCA.
4) Except when poorly-compacted, the CCA exhibited higher strength and stiffness than the tested gravelly gravel.
5) The effects of confining pressure on the strength and stiffness of the tested CCA were large, similar to other natural granular materials.

REFERENCES

Aqil, U., Tatsuoka, F., Uchimura, T., Lohani,T.N., Tomita,Y. and Matsushima, K. (2005); "Strength and deformation characteristics of recycled concrete aggregate as a backfill material", *Soils and Foundations*, Vol. 45, No. 4, pp.53-72.

Goto, S., Tatsuoka, F., Shibuya, S., Kim, Y.S. & Sato, T. (1991), "A simple gauge for local strain measurements in the laboratory", Soils and Foundations 31(1), pp.169-180.

Hoque,E., Sato,T. and Tatsuoka,F. (1997), "Performance evaluation of LDTs for the use in triaxial tests", *Geotechnical Testing Journal, ASTM,* Vol.20, No.2, pp.149-167.

Lohani, T.N., Kongsukprasert, L., Watanabe, K. and Tatsuoka, F. (2004) "Strength and deformation properties of compacted cement-mixed gravel evaluated by triaxial compression tests, *Soils and Foundations*, Vol.44, No.5, pp.95-108.

Santucci de Magistris, F., Koseki, J., Amaya, M., Hamaya, S., Sato, T. and Tatsuoka,F. (1999), "A triaxial testing system to evaluate stress-strain behavior of soils for wide range of strain and strain rate", *Geotechnical Testing Journal, ASTM,* 22(1): 44-60.

Santucci de Magistris,F. and Tatusoka,F. (2004): Effects of moulding water content on the stress-strain behaviour of a compacted silty sand, *Soils and Foundations,* Vol.44, No.2, pp.85-102.

Tatsuoka,F. (1988), "Some recent developments in triaxial testing system for cohesionless soils", *ASTM STP No.977*, pp.7-67.

Tatsuoka, F., Tomita, Y., Lovati, L. and Aqil, U. (2006), "Crushed concrete aggregate as a backfill material for civil engineering soil structures", *Proc. of Workshop of TC3 of the ISSMGE, 16th ICSMGE*, Osaka (eds. Correia).

Wainwright P. J., Yu J. and Wang Y. (1993): Modifying the performance of concrete made with coarse and fine recycled concrete aggregates, EIK Lauritezen (Ek.), *Demolition and Reuse of Concrete, Guidelines for Demolition and Reuse of Concrete and Masonry*, pp 319-330.

Soil Stress-Strain Behavior: Measurement, Modeling and Analysis
Geotechnical Symposium in Roma, March 16 & 17, 2006

FREEZING AND ICE GROWTH IN FROST-SUSCEPTIBLE SOILS

Radoslaw L. Michalowski and Ming Zhu
Department of Civil and Environmental Engineering
University of Michigan, Ann Arbor, U.S.A
e-mail: rlmich@umich.edu

ABSTRACT
A model for energy transfer, and freezing and thawing of soils is described first. This model is then incorporated into a description of heaving in frost-susceptible soils. Frost heaving is caused by formation of ice lenses, a result of transfer of unfrozen water and freezing at the cold side of the *frozen fringe*. The description of frost heave presented here is based on a porosity rate function. This description does not model formation of individual ice lenses; rather, it yields the average growth in porosity due to growth of ice. Application of the model is illustrated in examples of a chilled gas pipeline and a retaining wall with frost-susceptible backfill.

1. INTRODUCTION
Frost-susceptible soils are characterized by their sensitivity to freezing that is manifested in heaving of the ground surface. While significant contributions to explaining the nature of frost heave in soils were published as early as 1920's, modeling efforts did not start till three decades later. Several models have been introduced in the past to describe this process, but none of them has been generally accepted as a reliable tool in engineering applications.

Among the proposals are the capillary theory of frost heaving and the *secondary frost heaving* theory, which led to the development of the *rigid ice model* in the 1980's. Although very appealing from the physics standpoint, the rigid ice model is limited to one-dimensional simulations. The approach explored in this presentation will be based on the concept of porosity growth function dependent on two primary material parameters: maximum rate, and the temperature at which the maximum rate occurs. The advantage of this approach stems from a formulation consistent with continuum mechanics that makes it possible to generalize the model to arbitrary three-dimensional processes.

The porosity rate function concept will be presented. The physical premise for the model will be discussed first, and the development of the constitutive model will be outlined. The model will be implemented in a finite element code, and boundary value problems will be simulated to indicate its effectiveness.

Hoe I. Ling et al. (eds.), Soil Stress-Strain Behavior: Measurement, Modeling and Analysis, 429–441.

Freezing and frost heaving is part of the seasonal freeze-thaw cycle, and the model presented will constitute a component of a more comprehensive model of freezing and thaw-softening of frost-susceptible soils.

2. THE PHYSICS OF FROST HEAVING

Frost heaving is a process caused by transfer of moisture and freezing. Ice lenses nucleate behind the freezing front, on the cold side of a region called the *frozen fringe*, Fig. 1. The ice lenses grow, fed by the moisture moving into the frozen fringe driven from the unfrozen region of the soil by cryogenic suction. A common misconception is attributing frost heave to expansion of water upon freezing. Early tests by Taber (1929) showed clearly that frost heave occurs in susceptible soils even if the water is replaced with a liquid that contracts upon freezing.

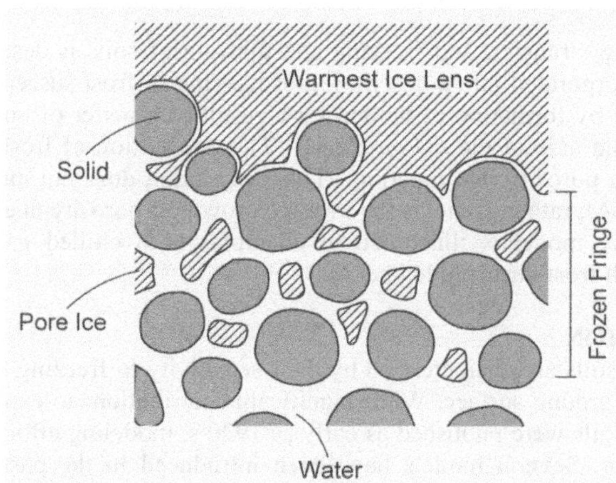

Figure 1. Freezing region in a frost-susceptible soil.

A plausible explanation of the frost heave mechanics is that suggested by Miller (1978), which later gave rise to the *rigid ice* model.

If a wire is draped over a block of ice, with both ends of the wire loaded with weights, the wire will gradually cut into the block and move through the block. The ice in direct contact beneath the wire gradually melts as the melting point of water is depressed by the contact stress. The melted water travels around the wire and refreezes above it, allowing the wire to travel through the ice. This mechanism of *regelation* was central to Miller's concept of the *secondary frost heaving* that led to the *rigid ice model.* If a small mineral particle is embedded in a block of ice subjected to a temperature gradient, the particle will travel toward the warmer side of the block (up the temperature gradient). This is caused by the very same mechanism of regelation, where the ice melts at the warm side of the particle, melted water travels around the particle, and refreezes at the cold side. The key experiment for the particle migration was presented by Römkens and Miller (1973). A frost-susceptible soil subjected to freezing is now viewed as an assembly of particles, with in-situ frozen pore water, but connected, forming one ice body. Hence, the particles are embedded in what can be considered a block of ice, and they attempt to move up the temperature gradient (downward). However, they are kinematically constrained by other particles beneath; therefore, it is the ice that moves upward, the relative motion being consistent with the migration of particle embedded in ice. An ice lens is initiated when the pore pressure (combined suction in unfrozen water and pressure in the ice frozen in the soil pores) becomes equal to the overburden. The model of frost heave based on the description above is called the *rigid ice model.* While this is a reasonable, physically-based explanation of the frost heave process, efforts toward producing a computational model ended with a one-dimensional numerical scheme, the most recent one described in Sheng et al. (1995).

While the rigid ice concept based on the regelation phenomenon is a reasonable explanation of the physical process behind the frost heaving, an engineering model predicting frost heave as an integral of the ice lenses growth does not appear feasible from the computational standpoint. Therefore, a model suggested here will be based on introducing a phenomenological ice growth function.

3. CONSTITUTIVE FUNCTIONS FOR FROST HEAVE DESCRIPTION

Mathematical description of frost heave requires modeling of heat flow, moisture transfer, and the ice growth in freezing soil. The components of that description are presented in the following subsections.

3.1 Heat flow

As the freezing process requires that a temperature gradient be maintained in the soil, modeling of frost heave must include a heat transfer law. Here, the Fourier law of heat conduction is used with one thermal conductivity $\lambda(T)$ for the mixture

$$\mathbf{Q} = -\lambda(T)\nabla\mathbf{T} \tag{1}$$

The heat conductivity is a function of the composition of the soil, and therefore, a function of temperature.

The soil is assumed to be saturated, with the volumetric fractions θ_s, θ_i, and θ_w describing the fraction of mineral (solid particles), ice, and unfrozen water, respectively. The thermal conductivity is bound by the conductivity of a parallel (λ_{par}) and series (λ_{ser}) configurations of the components. The former is a weighted arithmetic mean

$$\lambda_{par} = \theta_s\lambda_s + \theta_w\lambda_w + \theta_i\lambda_i \tag{2}$$

and the weighted harmonic mean describes the conductivity for the serial arrangement of the components

$$\frac{1}{\lambda_{ser}} = \frac{\theta_s}{\lambda_s} + \frac{\theta_w}{\lambda_w} + \frac{\theta_i}{\lambda_i} \tag{3}$$

The true (effective) thermal conductivity of soil is contained in the range between the two bounds, and it was selected to be governed by a logarithmic law

$$\log \lambda_e = \theta_s \log \lambda_s + \theta_w \log \lambda_w + \theta_i \log \lambda_i \tag{4}$$

or

$$\lambda_e = \lambda_s^{\theta_s} \lambda_w^{\theta_w} \lambda_i^{\theta_i} \tag{5}$$

which does fall in the bounded range described by eqs. (2) and (3). The typical values of thermal conductivity for the constituents of the frozen soil are given in Table 1. These values were used later to simulate the boundary value problems.

Table 1. Thermal properties of soil constituents (after Williams and Smith, 1989).

	Density ρ (kg/m^3)	Mass heat capacity c (J/kg·°C)	Volumetric heat capacity C (J/m^3·°C)	Thermal conductivity λ (W/m·°C)
Soil particles (clay mineral)	2620	900	2.36×10^6	2.92
Water	1000	4180	4.18×10^6	0.56
ice	917	2100	1.93×10^6	2.24

3.2 Unfrozen water in frozen soil

Frost-susceptible soils are characterized by large specific surface (combined particle surface per unit mass) and a substantial portion of the water in the soil is adsorbed to the particles. This water does not freeze at the freezing point of free water, leading to the presence of unfrozen water in the "frozen soil." The water content at freezing point T_0 drops down to some amount $\overline{w}$, and it then decays to a small content w^* at some reference low temperature. This moisture content can be described analytically with the following equation (Michalowski 1993)

$$w = w^* + (\overline{w} - w^*)e^{a(T-T_0)} \tag{6}$$

Parameter a describes the rate of unfrozen moisture decay. This relation is illustrated in Fig. 2.

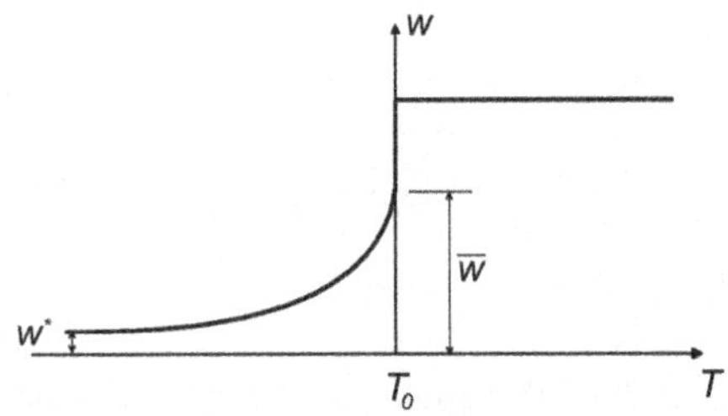

Figure 2. Unfrozen moisture content in frozen soil.

Unfrozen moisture content during freezing ($\partial T/\partial t < 0$, $T < 0$) and thawing ($\partial T/\partial t > 0$, $T < 0$) does not fall along the same curve, and the process is hysteretic.

3.3 Energy balance

The volumetric heat capacity of the mixture is equal to the sum of the heat capacities of the three phases multiplied by their volumetric fractions

$$C = \rho_w c_w \theta_w + \rho_i c_i \theta_i + \rho_s c_s \theta_s \tag{7}$$

where c is the mass heat capacity (J/kg·°C), and subscripts w, i and s denote water, ice and soil particles, respectively. The product of the density and the mass heat capacity of a constituent is equal to the volumetric heat capacity of that constituent, for example, $C_i = \rho_i c_i$.

Considering the heat conduction as the only form of energy exchange in the soil, the energy balance takes the form

$$C\frac{\partial T}{\partial t} - L\rho_i \frac{\partial \theta_i}{\partial t} - \nabla(\lambda \nabla T) = 0 \tag{8}$$

where L is the latent heat of fusion of water. For numerical reasons (convergence) the first two terms in eq. (8) were combined, with an apparent volumetric heat capacity defined as $C - L\rho_i\, \partial\theta_i / \partial T$, and adjusted at every step of the freezing process.

3.4 Porosity growth tensor

What makes the model presented here different from most models proposed in the last 30 years is the way the growth of ice in freezing soil is simulated. The formation of individual ice lenses is not modeled here, instead, an average increase in porosity is simulated that is equivalent to the global increase in volume produced by ice lens growth. While this concept was introduced earlier (Frémond 1987, Michalowski 1993), only recently was the fundamental function of *porosity growth* calibrated and validated

(Michalowski and Zhu, 2006a, 2006b). The *porosity rate function* $\dot{n}$ is proposed in the following form

$$\dot{n} = \dot{n}_m \left(\frac{T - T_0}{T_m} \right)^2 \cdot e^{1-\left(\frac{T-T_0}{T_m}\right)^2} \cdot \frac{\left|\frac{\partial T}{\partial l}\right|}{g_T} \cdot e^{-\frac{|\sigma_{kk}|}{\zeta}}; \qquad T < T_0, \quad \frac{\partial T}{\partial t} < 0 \tag{9}$$

where the first portion represents the growth of ice as a function of temperature T, and with two last factors including the influence of the temperature gradient and the stress state. The maximum porosity rate $\dot{n}_m$ is a material property, and it reflects the maximum rate determined at one well-defined temperature gradient g_T. The quotient $\dot{n}_m / g_T$ (with $\dot{n}_m$ determined at temperature gradient g_T) is a material constant for a given soil. T_m is the temperature at which maximum of porosity rate occurs, and it is a material property, as well as the stress-related parameter ζ. The porosity rate function is illustrated in Fig. 3.

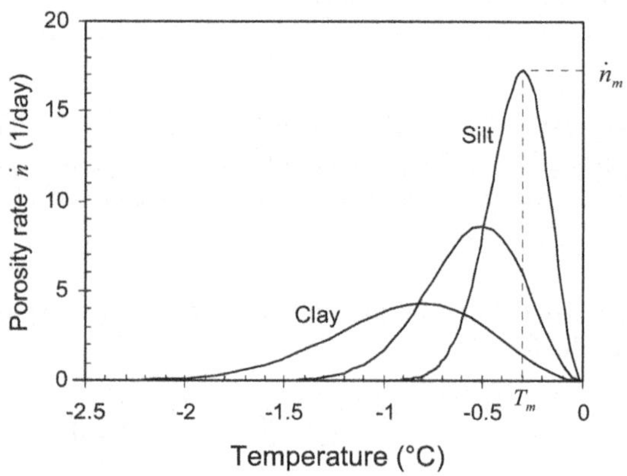

Figure 3. Porosity rate function.

The scalar function in eq. (9) cannot account for anisotropic growth of ice lenses; therefore, a unit growth tensor α_{ij} (with the trace equal to 1) is introduced to describe anisotropic growth of ice in freezing soil

$$\alpha_{ij} = \begin{vmatrix} \alpha_{11} & \alpha_{12} & \alpha_{13} \\ \alpha_{21} & \alpha_{22} & \alpha_{23} \\ \alpha_{31} & \alpha_{32} & \alpha_{33} \end{vmatrix} = \begin{vmatrix} \xi & 0 & 0 \\ 0 & (1-\xi)/2 & 0 \\ 0 & 0 & (1-\xi)/2 \end{vmatrix} \tag{10}$$

Coefficient ξ can range from 0.33 (isotropic growth) to 1.0 (1-D growth). The co-ordinate system for the representation of the unit growth tensor in eq. (10) was chosen such so that the direction x_1 coincides with the heat flow direction (direction of the maximum temperature gradient). The porosity growth tensor, $\dot{n}_{ij}$, is then obtained as a product of the porosity rate (scalar) and the unit growth tensor α_{ij} (Michalowski, 1993)

$$\dot{n}_{ij} = \dot{n}\alpha_{ij} \tag{11}$$

The porosity rate tensor in eq. (11) and the strain rate tensor of the soil due to loading are additive.

3.5 Deformation of freezing soil

It is assumed here that the frozen soil behaves as an elastic solid. The total deformation is then taken as a sum of both the elastic strain increment and the increment induced by the porosity growth ($d\varepsilon_{ij}^{p} = dn_{ij}$)

$$d\varepsilon_{ij} = d\varepsilon_{ij}^{e} + d\varepsilon_{ij}^{p} \tag{12}$$

The elastic increment is defined by the elastic constitutive law

$$d\varepsilon_{ij}^{e} = B_{ijkl}\, d\sigma_{kl} \tag{13}$$

with $d\sigma_{kl}$ being the Cauchy total stress tensor increment and the elastic compliance tensor B_{ijkl} being dependent on the temperature, and the increment induced by the porosity growth governed by eq. (11) ($d\varepsilon_{ij}^{p} = dn_{ij}$).

4. COMPUTATIONAL EXAMPLES

Two examples are presented to illustrate the application of the model. The first one includes only the thermal component of the model where the propagation of freezing and thawing zones is shown around a pipeline carrying chilled gas. The second example includes freezing of a frost-susceptible backfill behind a retaining wall.

4.1 Chilled gas pipeline

The problem described first is the freezing and thawing of soil around a chilled gas pipeline. The purpose of transporting natural gas at temperature below freezing is to preserve permafrost in cold regions. However, when the pipeline travels through a seasonally freezing and thawing area, the soil around the pipeline will freeze, and it may cause damage to the pipeline due to uneven heaving. Here, however, the soil is not frost-susceptible, and we are interested only in the extent of freezing in the ground.

The bottom of the pipeline is placed 1.72 m below the ground surface (Fig. 4(a)). The pipeline is made of steel with an external diameter of 0.90 m. The wall thickness of the pipe is 8.5 mm. The pipe is not insulated. The width of the model is 6.0 m and the height is 8.0 m. The 4-node linear heat transfer quadrilateral element (ABAQUS type DC2D4) is used to discretize the model. The total number of elements is 920 and the total number of nodes is 987. The finite element mesh is shown in Fig. 4(b). The simulation is performed using a reasonable set of material properties estimated for a silty sand, and readily available properties of steel.

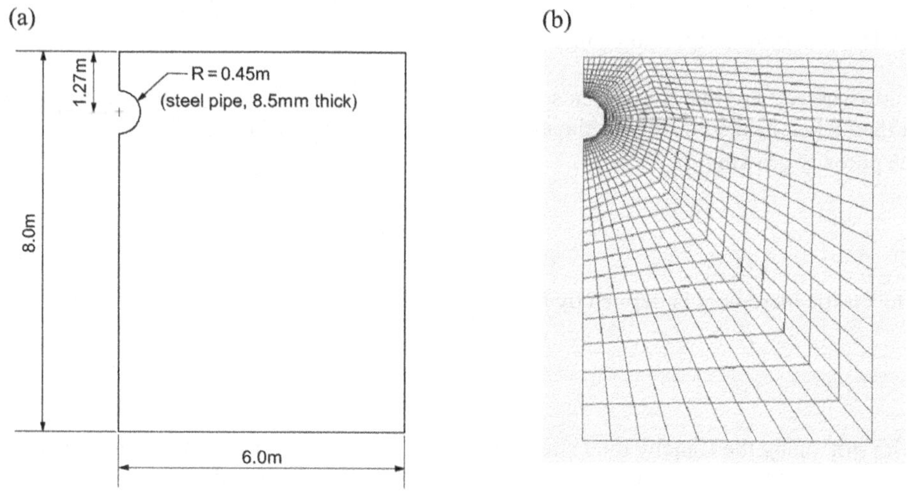

Figure 4. (a) Pipeline problem definition, and (b) the finite element discretization.

It is assumed that the initial temperatures at the inner surface of the pipe, the top surface, and the bottom surface are 1°C, 3°C, and 6°C, respectively. A steady-state distribution of temperature associated with these boundary temperatures is shown in Fig. 5(a). The thermal boundary conditions are illustrated in Fig. 5(b).

At time t = 0, the temperature along the inner surface of the pipe is suddenly dropped from 1°C down to -10°C and is kept at this level for 60 days, simulating the transport of the chilled natural gas. The temperature along the ground surface decreases linearly from 3°C at time t = 0 to -15°C at time t = 30 days; it then increases linearly to 5°C at time t = 60 days. The temperature at the base is kept constant at 6°C over the entire process. The two vertical boundaries of the model are adiabatic.

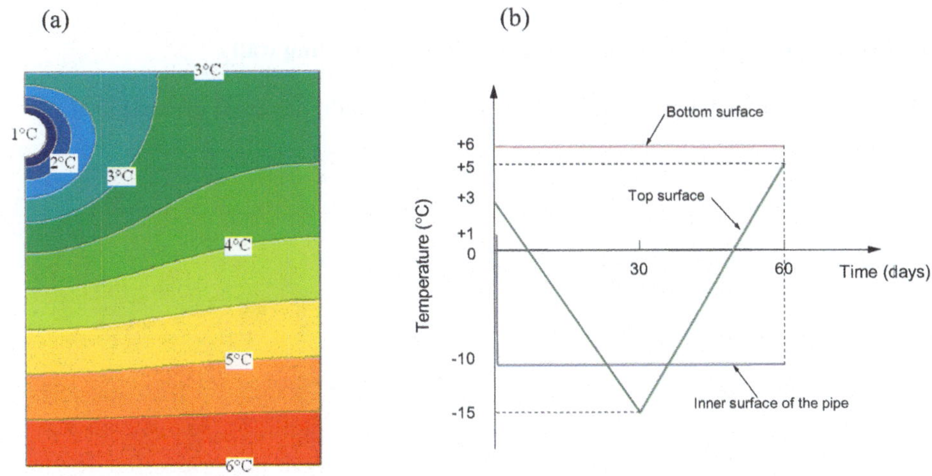

Figure 5. (a) Initial temperature distribution, and (b) thermal boundary conditions.

The temperature profile at the end of 60 days is shown in Fig. 6, along with the distribution of ice content in the soil. Between two 0°C isotherms is the "frozen belt". The volumetric ice content in the soil near the pipe is now about 0.3.

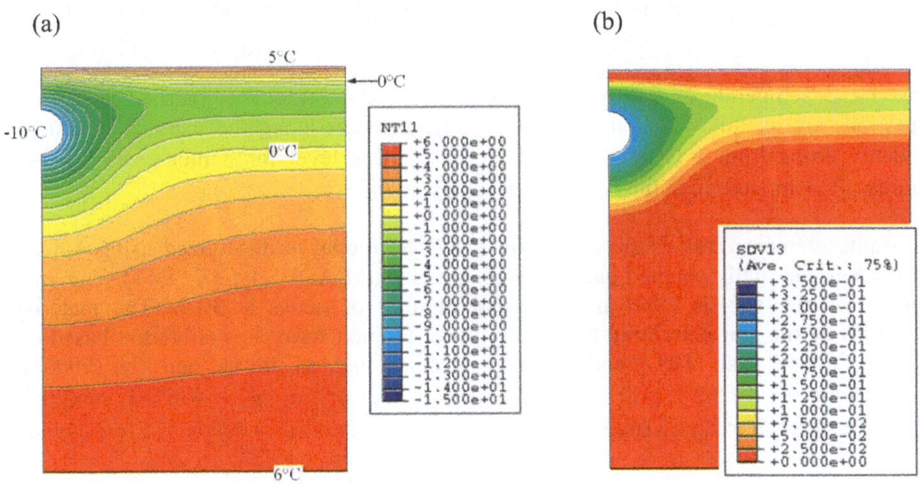

Figure 6. (a) Temperature distribution after 60 days, and (b) the distribution of the ice content.

The pipeline clearly causes a disturbance in otherwise a one-dimensional heat transfer regime.

4.2 Freezing of frost-susceptible backfill behind a retaining wall

The geometry of the retaining wall and the boundary conditions are shown in Fig. 7.

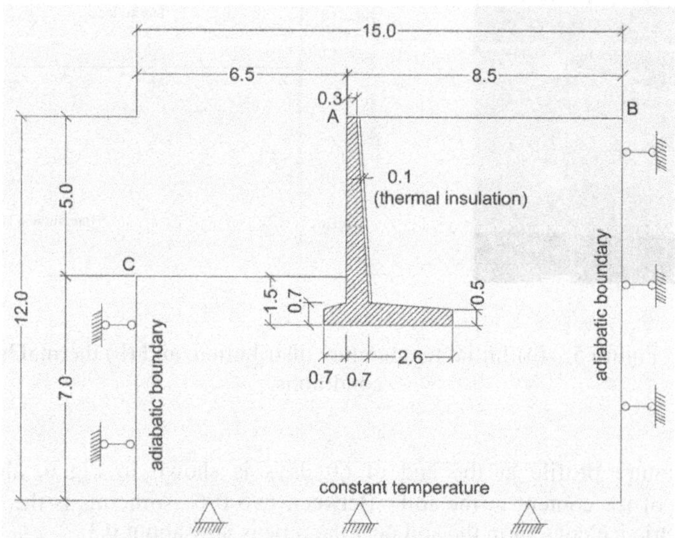

Figure 7. Geometry of the retaining wall system (dimensions in meters).

The initial temperature distribution is shown in Fig. 8(a). The thermal boundary conditions are as follows: at time $t = 0$, the temperature along the external boundary suddenly drops from 1°C to -5°C, and is kept at this level for 3 months (92 days), whereas the temperature at the base is kept steady at 3°C.

The finite element mesh is shown in Fig 8(b). The model is discretized using 4-node plane strain thermally coupled quadrilateral elements (ABAQUS type CPE4T). The total number of elements is 2950 and the total number of nodes is 3076. The material properties of the soil pertaining to heaving were adopted from the calibration based on tests by Fukuda et al. (1997) and presented in detail in Michalowski and Zhu (2006a): $\dot{n}_m = 6.02 \cdot 10^{-5}\ s^{-1}$ (or 5.2·1/24h) at g_T = 100 °C/m, or $\dot{n}_m / g_T = 6.02 \cdot 10^{-7}\ m°C^{-1}s^{-1}$, T_m = - 0.87°C, and ζ = 0.6 MPa.

The elasticity parameters for the soil were taken as follows: Young's modulus equal to 11.2 MPa for unfrozen soil, temperature-dependent $E = 13.75\ |T|^{1.18}$ MPa (T in °C) for frozen soil below -1°C (after Ladanyi and Shen, 1993), and linear interpolation in the range 0 to -1°C; Poisson's ratio was taken as $\mu = 0.3$ for both the frozen and unfrozen soil. The remaining thermal parameters were: thermal conductivities: 1.95, 0.56, and 2.24 $Wm^{-1}K^{-1}$ for solid skeleton, water, and ice, respectively; heat capacities: 900, 4180, and 2100 $Jkg^{-1}K^{-1}$ for solid skeleton, water, and ice, respectively; latent heat of fusion of water: $3.33 \cdot 10^5$ $Jkg^{-1}K^{-1}$. These were extracted from Williams and Smith (1989) and Selvadurai et al. (1999). Parameter ξ that governs the anisotropy of the ice growth in eq. (10) is difficult to assess, since no laboratory measurements are available for its evaluation. It is known, however, that the ice lenses grow predominantly in the direction of heat flow, and the value $\xi = 0.9$ was adopted.

The material properties of the concrete wall and the thermal insulation are listed in Table 2. Three cases are considered to study the effect of the thermal insulation: (a) no thermal insulation behind the wall, (b) a 0.1m thick layer of insulation on the backfill side of the wall, with heat conductivity of $\lambda = 0.2$ W/m·°C, and (c) a 0.1m thick layer of insulation with a heat conductivity of $\lambda = 0.03$ W/m·°C.

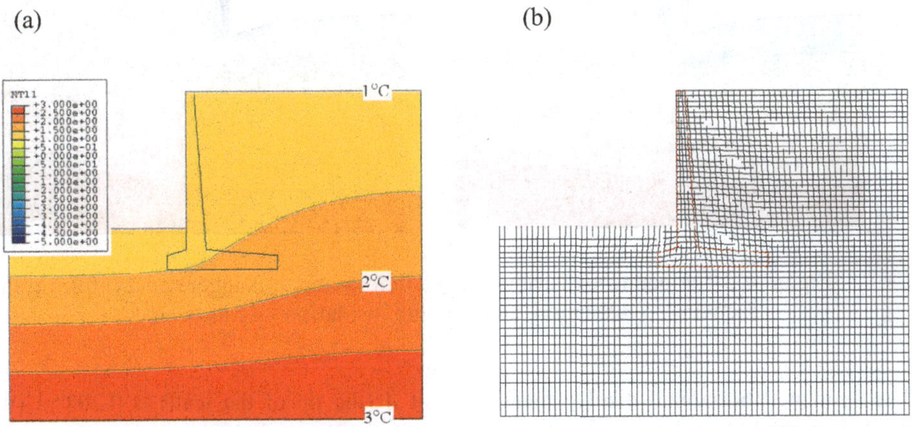

Figure 8. (a) Initial temperature distribution, and (b) the finite element mesh.

The temperature distribution in the uninsulated wall and the soil at the end of the third month is shown in Fig. 9(a) for the case of no thermal insulation behind the retaining wall. Also shown in the figure is the deformation. The contours relate to the temperature distribution in the soil. The displacements are exaggerated by a factor of 2. As the frost heave occurs in the direction of the heat flow, vertical displacements are considerable along the horizontal surfaces subjected to the freezing temperature. Along the wall,

Table 2. Material properties of concrete and thermal insulation

	Density ρ (kg/m³)	Mass heat capacity *C* (J/kg·°C)	Thermal conductivity λ (W/m·°C)	Young's modulus *E* (Pa)	Poisson's ratio μ
Concrete	2242	970	1.2	2×10^{10}	0.38
Thermal insulation	50	2000	0.03 or 0.2	1×10^{7}	0.3

however, the isotherms are oriented vertically, thus the heaving occurs horizontally. The retaining structure is an obstacle to heaving, and the wall tilts.

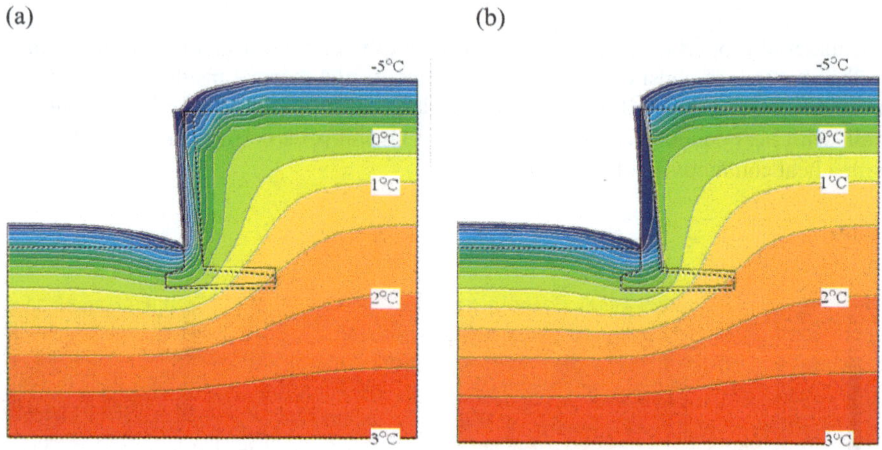

Figure 9. Temperature distribution and deformation (exaggerated 2x) for: (a) the case with no insulation, and (b) with insulation λ = 0.03 W/m·°C.

The simulated horizontal displacement at point *A* (the tip of the wall) is 0.208 m (the displacements in Fig. 9 are exaggerated by a factor of 2). With the insulation of λ = 0.20 W/m·°C the horizontal displacement at point *A* was reduced by almost 20%, and when a more effective insulation was used λ = 0.03 W/m·°C, the horizontal displacement at the wall crown was reduced by roughly 42%.

5. FINAL REMARKS

A constitutive model for simulations of energy transfer and heave in frost-susceptible soils was presented, and implemented in the finite element method. Calibration of the model shown earlier (Michalowski and Zhu 2006a) indicated its good predictive capabilities, and the boundary value problems presented here appear to yield very reasonable results.

Acknowledgement

The research presented in this paper was supported by the U.S. Army Research Office, Grant No. DAAD19-03-1-0063. This support is greatly appreciated.

REFERENCES

Frémond, M., (1987). Personal communication.

Fukuda, M., Kim H. and Kim Y. (1997). "Preliminary results of frost heave experiments using standard test sample provided by TC8." In: Proceedings of the International Symposium on Ground Freezing and Frost Action in Soils, Luleå, Sweden, 25-30.

Ladanyi, B. and Shen, M. (1993). "Freezing pressure development on a buried chilled pipeline." In: *Proc. 2nd Int. Symp. Frost in Geotechnical Engineering*, Anchorage, AK, 23-33

Michalowski, R. L. (1993). "A constitutive model of saturated soils for frost heave simulations." *Cold. Reg. Sci. Technol.*, 22(1), 47-63.

Michalowski, R.L. and Zhu, M. (2006a). "Frost heave modeling using porosity rate function." *Int. J. Num. Analyt. Meth. Geomech.*, in print.

Michalowski, R.L. and Zhu, M. (2006b). "Modeling and simulation of frost heave using porosity rate function." *2nd Japan-U.S. Workshop on Testing, Modeling and Simulation in Geomechanics*, Kyoto, Japan, Sept. 8 – 10, 2005; ASCE Geotechnical Special Publication, in print.

Miller, R. D. (1978). "Frost heaving in non-colloidal soils." *Third Int. Conf. Permafrost*. Edmonton, 707-713.

Römkens, M. J. M. and Miller, R. D. (1973). "Migration of mineral particles in ice with a temperature gradient." *J. Colloid Interf. Sci.*, 42, 103-111.

Selvadurai A.P.S., Hu J., and Konuk I. (1999). "Computational modeling of frost heave induced soil-pipeline interaction: I. Modeling of frost heave." *Cold Regions Science and Technology*, 29, 215-228.

Sheng, D., Axelsson, K. and Knuttson, S. (1995). "Frost heave due to ice lens formation in freezing soils. 1. Theory and verification." *Nordic Hydrology*, 26, 125-146.

Taber, S. (1929). "Frost heaving." *Journal of Geology*, 37, 428-461.

Williams, P. J. and Smith, M. W. (1989). *The Frozen Earth, Fundamentals of Geocryology*. Cambridge University Press.

[illegible] (unpublished).
[illegible] Acoustical [illegible]
[illegible] soils [illegible]
[illegible] (2004) [illegible]
[illegible]
[illegible] of [illegible] geophysical [illegible]
[illegible] Acoustical [illegible]
[illegible]
[illegible] Physical Review [illegible]
[illegible]
[illegible] Canada [illegible]
[illegible] Society [illegible]
[illegible]
[illegible] between [illegible]
[illegible] 40 [illegible]
[illegible] 15 [illegible]
[illegible] University Press.

Soil Stress-Strain Behavior: Measurements, Modeling and Analysis
Geotechnical Symposium in Roma, March 16 & 17, 2006

EFFECT OF FLY ASH STABILIZATION ON GEOTECHNICAL PROPERTIES OF CHITTAGONG COASTAL SOIL

M. A. Ansary, M. A. Noor, M. Islam
Department of Civil Engineering
Bangladesh University of Engineering and Technology, Dhaka 1000, Bangladesh
e-mail: mnoor@ce.buet.ac.bd

ABSTRACT

The use of fly ash has been studied to investigate the strength properties of stabilized soils collected from two sites of Chittagong coastal region namely, Anwara and Banshkhali. In the present study unconfined compressive strength (q_u), compaction properties and flexural properties has been studied. Compaction apparatus was employed to determine the strength of the stabilized soils. Strength tests were carried out on the specimens up to 28 days curing period. The investigated admixture was fly ash with lime; the amount of lime was fixed at 3 percent with the amount of fly ash 0, 6, 12 and 18 percent. The results from the experimental investigation shows that by increasing the amount of fly ash the strength properties of lime-fly ash stabilized soils improve. The use of fly ash with lime gave better strength and it may be more economical. For samples of both the coastal soils, compared with the untreated samples, unconfined compressive strength of fly ash and lime treated increased significantly, depending on the additive content and curing age. Compared with the untreated samples, flexural strength and modulus increased considerably, depending on the additive content. Compared with the untreated sample, the flexural strength and flexural modulus of fly ash treated samples increased up to about 4.6 and 4.7 times and 3 and 4.3 times respectively for both the soils. It could be concluded that fly ash stabilization of the coastal soils studied would be suitable for use in road construction.

1. INTRODUCTION

In early days, engineers could avoid unsuitable sites or unsuitable construction material sources whenever the required conditions for the construction were not fulfilled. It was ease of construction and ease in obtaining material, which governed the choice of site rather than economic factors. As time passed people became more cautious about the economy or, for different reasons, it has been difficult to find suitable sites for construction or suitable material sites for earth structures, such as highways, dams, or runways, within an economic range. It is evident that earth structures, such as embankments, highways, airport runways, dams, or reclamation appurtenance require soils with sufficiently good engineering properties: low plasticity, high bearing capacity, low settlement, etc. Since unsuitable materials, which have low bearing capacity, coupled with low stability and high settlement or excessive swelling or squeezing properties, are frequently encountered, it has been necessary to improve unsuitable materials to make them acceptable for construction. Improvement of soil by altering its properties is known as soil stabilization. An increment in strength, a reduction in compressibility, improvement of the swelling or squeezing characteristics, and increasing the durability of soil are the main aims of soil stabilization. The aim of this research work is to develop a

Hoe I. Ling et al. (eds.), Soil Stress-Strain Behavior: Measurement, Modeling and Analysis, 443–454.

general approach to soil improvement in coastal area by using fly ash for road construction. The study was limited to chemical stabilization by fly ash. A detailed study of the improvement of the strength properties of coastal soil of Bangladesh has been made using fly ash as additives. Investigation was made using commercial fly ash from brickfield, foundry shop and restaurant etc.

2. METHODOLOGY

Index property tests of the two coastal soils without any treatment were carried out to characterize the soils. Index tests include Atterberg limit tests, specific gravity and grain size analysis. Index property tests of the two soils stabilized with different fly ash having 3% lime as constant were also performed.

Modified compaction test, unconfined compressive strength test on molded cylindrical samples of 2.8 inch (71 mm) diameter by 5.6 inch (142 mm) high and flexural strength test using simple beam with third point loading system were carried out on the two coastal soils without any treatment and stabilized with three different fly ash contents (6%, 12% and 18%); having every time 3% lime and both the soil stabilized with 3% lime.

Unconfined compressive strength tests and flexural strength test using simple beam with third point loading were carried out on fly ash stabilized samples cured at 7,14 and 28 days in order to investigate the effect of curing age on the measured compressive strength and flexural strength and stiffness.

3. REVIEW

Fly ash is regularly used as a partial replacement for cement in concrete because of its pozzolanic properties; it is also the form of ash, which has the greatest potential for use in ground modification. Increased use as a partial cement or lime replacement would also represent a savings in energy (fly ash has been called a high-energy waste material).

Besides using fly ash alone as a structural fill material scope exists for employing techniques of ground modification to find more medium-to high-volume applications in the following ways: add cement or lime to stabilize the fly ash, stabilize soil with cement-lime-fly-ash mixes and use fly ash in the containment of toxic wastes.

3.1 ENGINEERING PROPERTIES

The specific gravity of the ash particle ranges from 1.9 to 2.5, which is below that normally measured for soil solids. The average grain size D_{50} of fly ash is likely to be in the range of 0.02 to 0.06mm. The friction angle as measured in consolidated drained triaxial tests is typically on the order of 30^0, but values as low as 20^0 and as high as 40^0 have been reported. The permeability of a fly ash compacted to standard maximum dry density depends on the coal type it is derived from [EPRI (1986)]. Considerable capillary rise of water in fly ash fills can occur on the order of 2 m and possibly more. Negative environmental impacts from a fly ash fill are unlikely, but a study has to be made of the chemical composition of its leachate; its corrosivity on buried pipes, culverts, or other structural elements; and its radioactivity (Radium-226).

3.2 FLY ASH STABILIZED WITH LIME, CEMENT, AND/OR AGGREGATE

The use of mixtures of lime (L) or cement (C) and fly ash (F) with aggregate (A) giving LFA, CFA, or LCFA bases or subbases for pavements is relatively well established in most countries. Guidelines for design and construction were given by Barenberg (1974) and other. Many local authorities have published criteria for the incorporation of pozzolanic materials with cement or lime in aggregate layers, either rated as bound or unbound layers, depending, e.g., on whether their indirect tensile strength is above or below 80 kPa (NAASRA, 1986). To build a sub base or base course with lime-or cement-stabilized ash alone is not yet common, but this is one high-volume ash applications being promoted by ash producers (British and American experience, EPRI (1986)).

3.3 SOIL MODIFIED WITH FLY ASH AND CEMENT

For cohesion less soils or soils with very low plasticity (plasticity index <10), cement will be more effective than lime, either alone or when combined with fly ash. For more plastic soils, either cement or lime may be added with fly ash. Only a soils testing program can indicate optimal mixes and relative economies. Fly ash could also serve as filler in the bituminous stabilization of coarse-grained materials. Stabilization of a sandy road base with a fly-ash-cement mix, rather than cement alone, creates a less-permeable stiffer layer. This may result in reduced long-term maintenance. Initial financial benefits depend on local material and transport costs.

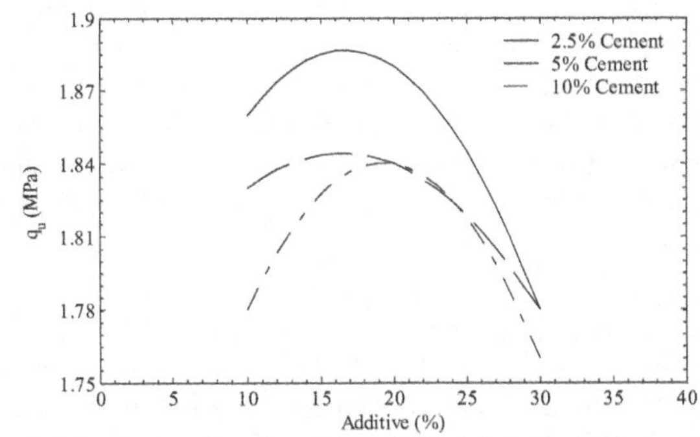

Figure 1: Maximum dry density and fly ash (after hausmann 1990)

Figure 1 demonstrates the effect of fly ash on the strength of cement-stabilized sand. The sand in question is of medium gain size (D50 = 0.3 mm), is fairly uniform (USCS classification SP), and is from the Woy Woy area, New South Wales. Stabilization of a sandy road base with a fly-ash-cement mix, rather than cement alone, creates a less-permeable stiffer layer. This may result in reduced long-term maintenance. Initial financial benefits depend on local material and transport costs. It has also been demonstrated that cement-fly-ash-sand or cement-fly-ash-gravel mixtures shrink less than soil-cement mixtures. Greater shrinkage is observed in these combinations if the cement is replaced by lime.

3.4 LIME FLY ASH-SOIL MIXTURE

Improvement of the strength properties of residual clay using fly ash as sole additive was shown by Nettleton (1962). However, he further observed that the additional of a small amount of lime in fly ash-soil mixtures further improved the strength of the stabilized soil. Subsequently, many authors have indicated the use of fly ash as an additive in lime-soil mixtures to achieve better strength of the mixture. However, there exist about little literature describing the use of fly ash a lesson soil stabilization.

In mechanism of lime, fly ash, and soil in stabilization as describe by Chu et al (1955) as follows: Fly ash is a gray, dust like ash which results from burning powered coal. The coal is burned while in suspension in air, and the resulting ash consists largely of tiny spheres of silica and alumina glass. The ash is similar to volcanic ash used in early Roma construction. It is a pozzolanic material; at is, it is not itself a cement, but it reacts with lime and water to form cementitious material. However, it is the reaction of lime and fly ash which is utilized to stabilize soils. After mixing the proper proportion the mixtures in a moist, non plastic state, but it can be readily compacted to form a dense mass.

Leonard and Davidson (1959) reported that because of the slow reaction of lime absorption, the development of compressive strength of a soil, lime and fly ash mixture is slow. Therefore, the rate of development of compressive strength of lime-fly ash reaction is directly related to the rate of lime absorption by the fly ash. The rate of lime absorption is limited by the rate of diffusion of the calcium through the reaction product. Minnick and Miller (1950) found that the coarser the material to the stabilized with lime and fly ash, the higher the volume of fly ash that is required.

One of the main questions in soil, lime, and fly ash stabilization is how much lime and fly ash are needed. The amount and proportions of the lime and fly ash admixtures are governed by the desired strength in the stabilized soil and by economy. Mateos and Davidson (1962) stated that there is no optimum amount, nor optimum ratio, of lime and fly ash for stabilizing all soils. The amounts of lime and fly ash to be used depend greatly on the kinds of fly ash and soil, and some what on the kind of lime. The authors found that the amount of hydrated lime for granular soils should be from 3 to 6 percent, and the amount of fly ash between 10 and 25 percent. For clay soils, the amount of lime should be between 5 and 9 percent, and the amount of fly ash between 10 and 25 percent.

Viscochil et al (1958) have shown that the density of soil, lime, and fly ash mixtures is dependent on the compactive effort applied, but the density also depends on the lime to fly ash ratio. The density is decreased by higher contents of lime because of two factors: Lime itself is less dense than soil or fly ash, and lime cause aggregation of clay. The authors further stated that the unconfined compressive strength is primarily influenced by cementation and does not give a true measure of the frictional strength developed in a confined state. Therefore, a stabilized granular material with relatively low unconfined compressive strength may show satisfactory stability.

The stability of lime-fly ash-soil mixtures is affected by many variables. According to Chu et al (1955), the following are main factors which affect the stability of a stabilized soil: Properties of Soil, amount and ratio of lime and fly ash in the mixture, properties of Lime and Fly ash, aging of lime, moisture contents of mixture, method and degree of compaction, length of curing period, and condition during curing.

4. LABORATORY INVESTIGATION AND TESTING PROGRAM

The laboratory investigations carried out on the untreated and stabilized samples of the two soil samples collected from coastal region of Chittagong have been described in details in the following sections.

Disturbed soils from two selected sites, namely Anwara and Banshkhali of Chittagong coastal region were collected for the present investigation. Soil sampling was carried out according to the procedure outlined in ASTM D420-87. Proper care was taken to remove any loose material, debris, coarse aggregates and vegetation from the bottom of the excavated pit. Disturbed samples were collected from the bottom of the borrow pit through excavation by hand shovels. The soil samples were designated as follows: Soil-A: collected from Anwara and Soil-B: collected from Banshkhali.

A comprehensive laboratory investigation program was undertaken in order to examine the physical, index and engineering characteristics of base soils (i.e., untreated soils) and soils stabilized with fly ash and lime. Fly ash and air-slaked lime were used as additives for stabilization. Both Soil-A and Soil-B were stabilized with Portland fly ash in percentages of 6, 12 and 18 keeping 3% lime constant. The whole laboratory-testing program consisted of carrying out the following tests on samples of the two coastal soils:

Index property tests on samples of the two coastal soils without any treatment with 3% lime and with different fly ash (6, 12, 18). Index tests included specific gravity test, Atterberg limit tests, and linear shrinkage test and grain size analysis.

Modified compaction test, unconfined compressive strength test on molded cylindrical samples of 2.8-inch diameter by 5.6 inch (142 mm) high, California Bearing Ratio (CBR) tests, and flexural strength test using simple beam with third point loading system were carried out.

Unconfined compressive strength test and flexural strength tests using simple beam with third point loading were carried out on fly ash and lime stabilized samples cured at three different ages (7 days, 14 days and 28 days). CBR tests were carried out on the untreated samples and samples treated with different fly ash with 3% lime contents using three levels of compaction. Table 1 Details of laboratory tests performed on samples of the two coastal soils.

4.1 PHYSICAL AND INDEX PROPERTIES OF UNTREATED SOILS

The samples collected from the field were disturbed samples. These samples were then air-dried and the soil lumps were broken carefully with a wooden hammer so as to avoid breakage of soil particle. The required quantities of soil were then sieved through sieve No.40. (0.435 mm). The following Standard test procedure were followed in determining the physical and index properties of the untreated soils: Specific gravity (ASTM D854), liquid limit (Cone penetrometer Method) (BS 1377), plastic limit and plasticity index (BS 1377), shrinkage limit (ASTM D427), linear shrinkage (BS 1377), percent of material in soils finer than No. 200 sieve (ASTM D1140), grain size distribution (ASTM D422) were carried out both on treated and untreated soil.

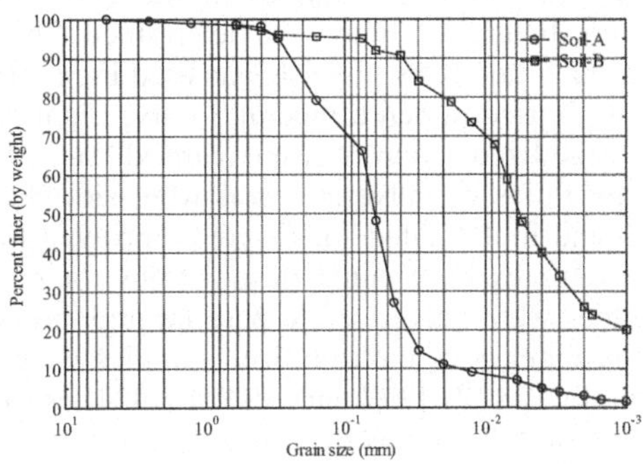

Figure 2: Grain size distribution curve of soil A and soil B.

The grain size distribution curves of samples of the two coastal are presented in Figure 2. The different fractions of sand, silt and clay of samples of Soil-A and Soil- B were found from the grain size distribution curves following the MIT Textural Classification System (1931). The soils were classified according to Unified Soil Classification System (ASTM

D2487) and the positions of the two soils (Soil-A and Soil-B) on the plasticity. The soils were also classified according to ASSET Soil Classification System (AASHTO M145-49). Table 1 presents are values of index and shrinkage properties, grain size distribution and classifications of Soil-A and Soil-B.

Table 1 Index properties and Classification of the coastal soils used

Index Properties and Classification	**Soil-A**	**Soil-B**
Specific Gravity	2.70	2.80
Liquid Limit	30	44
Plastic Limit	23	25
Plasticity Index	7	19
Shrinkage Limit	20	23
Linear Shrinkage	7	8
% Sand (0.60 mm to 2 mm)	34	6
% Silt (0.002 mm to 0.06 mm)	62	68
% Clay (< 0.002 mm)	4	26
% of Material Finer than No. 200 Sieve (0.074mm)	68	94
Unified Soil Classification	ML	CL
AASHTO Soil Classification	A-4	A-7-6

4.2 PROPERTIES OF FLY ASH USED FOR SOIL STABILISATION

The fly ash was obtained from different source. The chemical analysis was made by the Department of Chemistry, DU, and Bangladesh. Presented in bellow Table 2 the chemical composition of fly ash:

Table. 2 Chemical Analysis of Fly ash

Sample	SiO_2	Al_2O_3	Fe_2O_3	CaO	MgO	SiO_3	Total
Fly ash	10.7	5.00	16.0	40.5	24.7	3.4	100.3

Data based on chemical analysis by department of chemistry, Dhaka University

4.3 INDEX PROPERTIES OF STABILISED SOIL SAMPLES

Liquid limit, plastic limit, plasticity index and shrinkage characteristics including shrinkage limit and linear shrinkage of samples of the two soils (from Anwara and Banshkhali) stabilized with fly ash and lime were determined. Fly ash and hydrated lime (i.e., slaked lime) were used as additives. Fly ash was used in percentages of 6, 12 and 18 while the lime contents were used in percentage of 3 only; Liquid limit, plastic limit and plasticity index of the stabilized samples were carried out on air-dried pulverized samples. The required quantities of pulverized soil were sieved through sieve. No. 40 (0.425 mm). The fly ash and lime treated soils were compacted following ASTM D558 method. The compacted samples were cured in moist environment for 7 days and air-dried. The air-dried samples were pulverized to pass through no. 40 sieve. Liquid limit, plastic limit and plasticity indexes of the stabilized samples were determined following the standard procedure outline in BS 1377 and ASTM D424 respectively. The shrinkage factor comprising the shrinkage limit was determined in accordance with the procedure specified in ASTM D427. Linear Shrinkage of the fly ash and lime treated samples were determined following the procedure outlined in BS 1377.

The samples were never cured with direct water spray or under submerged condition. The samples were always protected from free water for the specified most curing periods of 7, 14 and 28 days. It may be mentioned that the soil samples that were prepared without adding fly ash or lime, i.e., the untreated samples were not cured.

5. RESULTS AND DISCUSSIONS

The findings of the laboratory investigations on the characteristics of untreated and stabilized samples of the two coastal soils are presented and discussed in the following sections. In the following sections the physical and engineering characteristics comprising plasticity and shrinkage properties, moisture-density relations, unconfined compressive strength, California Bearing Ratio (CBR), flexural properties, of untreated and fly ash-treated samples of the two coastal soils are presented and discussed.

5.1 PLASTICITY AND SHRINGKAGE CHARACTERSTICS

The values of plasticity and shrinkage properties of the untreated and fly ash-treated soil samples are presented Table 3 for Soil-A and Soil-B respectively. Compared with the untreated sample, the value of liquid limit of the treated sample increased in Soil A while it is reduced Soil-B. It can be seen from Table 3 that for Soil-A (LL=30, PI=7), both liquid limit and plastic limit increased while for Soil-Between (LL=44, PI=19) liquid limit reduced and plastic limit Increased with increasing fly ash content. Table 3 also shows the changes in shrinkage limit due to increase in fly ash content and the variation of linear shrinkage with the increase in fly ash content. It can be seen that for both the soils shrinkage limit and linear shrinkage reduced slightly with the increase in fly ash content.

5.2 MOISTURE DENSITY RELATIONS

The moisture-density relations of untreated and fly ash treated samples of Soil-A and Soil-B have been determined for both the soils, with the increase in fly ash content with 3% lime, values of γ_{max} increased while the values of ω_{opt} reduced. The increase in ϒmax with the increase in fly ash content for the two soils is shown in Figure 3. Compared with the untreated sample, the values of ϒmax increased up to 8% and 7% for Soil-A and Soil-B respectively. The values of ω_{opt} reduced up to 9% and 10% respectively for Soil-A and Soil-B.

Table 3: Atterberg limits of two coastal soils.

Properties	0% Lime		3% Lime		3% Lime with					
					6% Fly ash		12% Fly ash		18% Fly ash	
	A	B	A	B	A	B	A	B	A	B
LL	43.7	25.5	42.8	29.7	42.5	31.2	41.0	32.3	41.5	33.7
PL	30.4	23.5	29.8	24.8	32.9	28.9	37.1	33.5	38.1	35.8
SL	20.0	23.5	19.8	23.8	21.5	18.3	20.2	18.0	19.0	6.0
LS	6.8	8.0	6.0	6.9	5.5	6.5	5.0	6.0	4.5	5.5

5.3 UNCONFINED COMPRESSIVE STRENGTH

Figures 4 to 7 show the unconfined compression test results for Soil-A and Soil-B. It can be seen from Figures 4 to 7 that for both the soils, compared with the untreated samples, the values of q_u of the treated samples increased significantly, depending on the fly ash content and curing age. Leonard and Davidson (1959) reported that because of the slow reaction of lime absorption, the development of compressive strength of soil directly related with lime absorption by fly ash. It can be seen from Figure 4 to 7 that the values of q_u of samples of Soil-A and Soil-B treated with 6% fly ash and cured at 28 days were found to be about 4 times higher than the strength of the untreated samples and with 18% fly ash it about 5 times higher than untreated sample. It is also evident that the gain in strength with increasing fly ash content and curing age is higher in less plastic Soil-A (Pl=7) than in more plastic Soil-B (Pl=19). Figures also show that for all fly ash contents and all curing ages, the values of q_u of treated samples fulfilled the requirements of soil-fly ash road sub-base for light traffic as proposed by Ingles and Metcalf (1972) with cement. It can also be seen from Figures that compared with the untreated samples, the values of ε_f of the stabilized samples reduced and those values of ε_f of the treated samples reduced with the increase in fly ash content. It can't be seen from Figures 4 to 7 that the values of q_u of treated samples increased with increasing fly ash content and curing age.

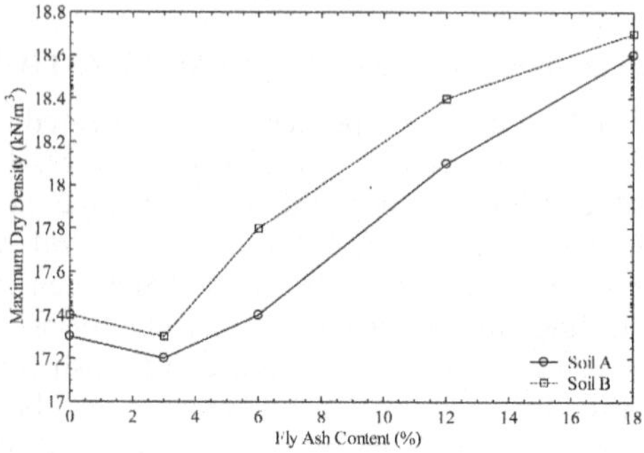

Figure 3: Effect of fly ash content on maximum dry density of fly ash treated soil A and soil B.

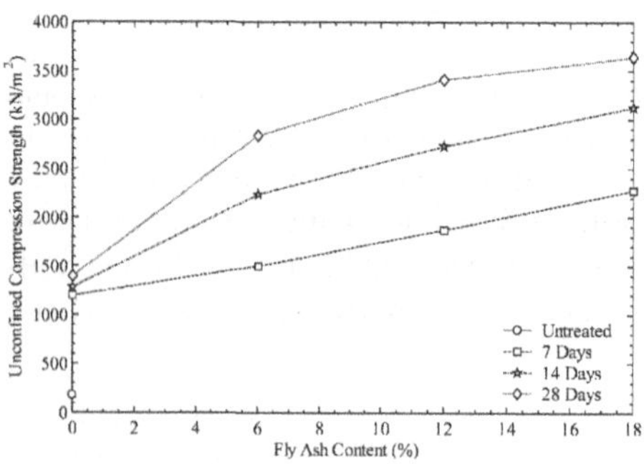

Figure 4: Unconfined compressive strength of lime, fly ash treated soil A.

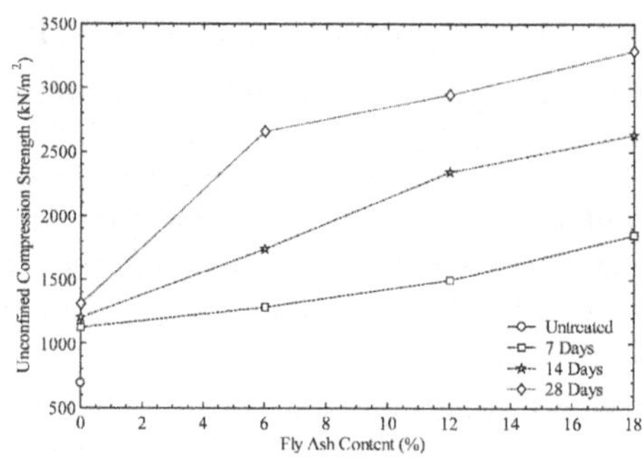

Figure 5: Unconfined compressive strength of lime, fly ash treated soil B.

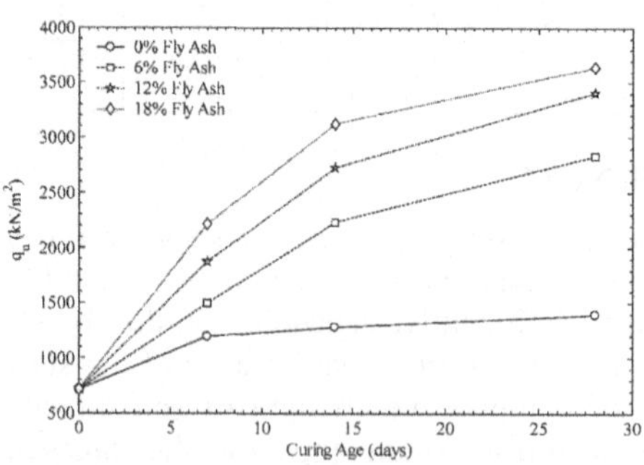

Figure 6: Unconfined compressive strength of lime, fly ash treated soil A.

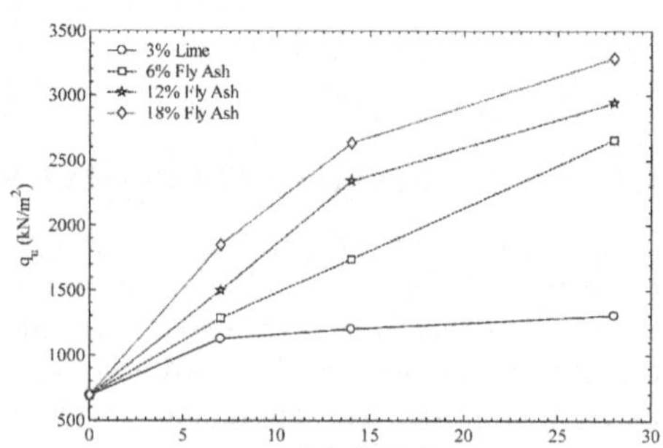

Figure 7: Unconfined compressive strength of lime, fly ash treated soil B.

5.4 FLEXURAL STRENGTH AND MODULUS

The flexural properties of untreated and stabilized samples of the two soils have been investigated by carrying out flexural strength test using simple beam test with third point loading. Typical flexural stress versus defection curves for two stabilized samples of Soil-A and Soil-B are presented in Figure 8 respectively. It can be seen from Figure 8 that flexural stress-deflection curves are approximately linear. From the flexural stress and deflection data flexural strength and modulus were determined. It can be seen from Figures 9 to 12 that for both Soil-A and Soil-B, compared with the untreated sample, flexural strength and modulus of the treated samples cured at 7, 14 and 28 days increased significantly. It can be seen from Figures that compared with the untreated sample, the flexural strength and modulus of Soil-A treated with 6%, 12% and 18% fly ash and cured at 28 days are respectively about 1.5, 3, 4.6 times and 1.8, 2 and 3 times higher respectively. Figures 9 to 12 show that the flexural strength and modulus of Soil-B treated with 6%, 12% and 18% fly ash and cured at 28 days are respectively about 2, 1.5, 6.7 times and 2.6, 3, 4.4 times higher respectively than those of the untreated samples. The maximum deflection and of untreated and stabilized soil-fly ash beams were in the range of 0.15 mm to 0.35 mm respectively. Comparing the flexural strength and modulus of Soil-A with those of Soil-B, it is evident that the values of flexural strength and modulus of samples of more plastic Soil-B (PI=19) is higher than the less plastic Soil-A (PI=7).

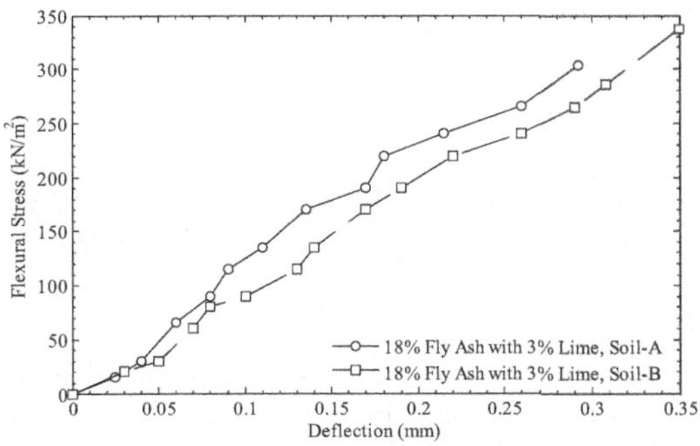

Figure 8: Flexural stress with deflection of fly ash treated soil A and B.

The effect of fly ash content on flexural strength for Soil-A and Soil-B are shown in Figures 24 and 25 respectively while Figures 11 and 12 present the effect of fly ash content on flexural modulus of Soil-A and Soil-B respectively. Figures 9 to 12 show that flexural strength and modulus increases with increasing fly ash content. It is evident from Figures 9 to 12 that curing age has got insignificant effect on increase in flexural strength and modulus.

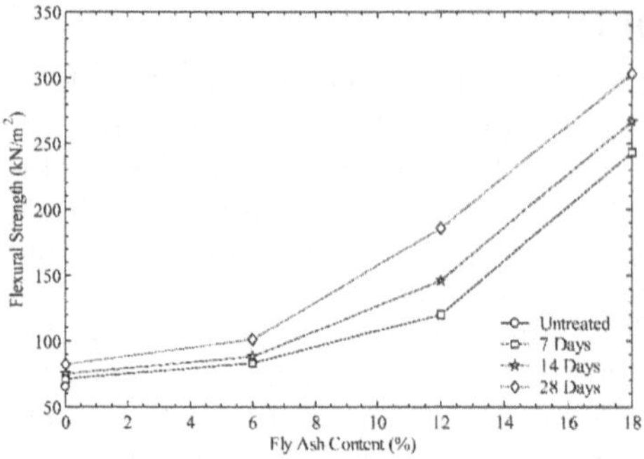

Figure 9: Effect of fly ash content on flexural strength of soil A.

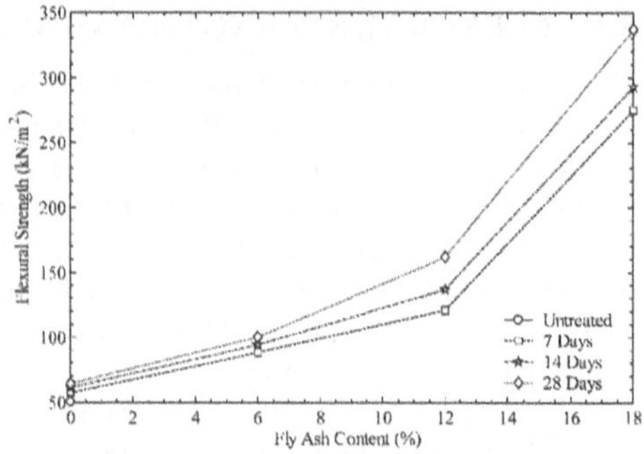

Figure 10: Effect of fly ash content on flexural strength of soil B.

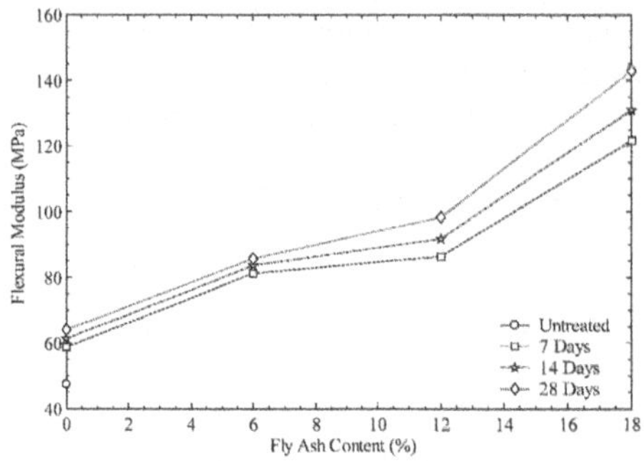

Figure 11: Effect of fly ash content on flexural modulus of soil A.

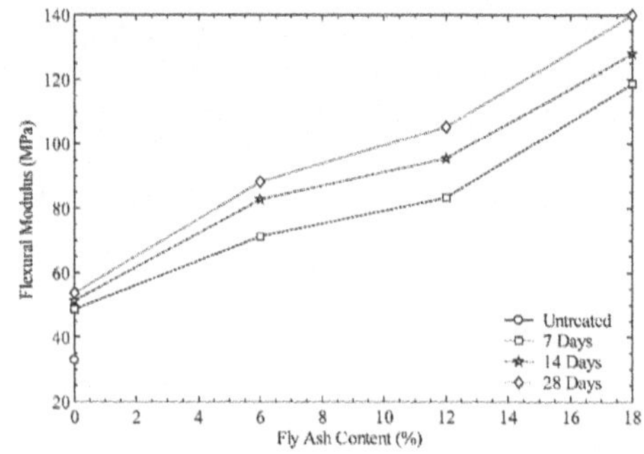

Figure 12: Effect of fly ash content on flexural modulus of soil B.

6. CONCLUSIONS

In this research work, fly ash stabilization of two selected soil (collected from Anwara and Banshkhali) of Chittagong coastal region have been carried out. Fly ash has been used in percentage of 6, 12 and 18 while lime has been added in percentages of 3 as additives with fly ash. The physical and engineering properties of fly ash and lime stabilized soil have been determined in order to asses the suitability of fly ash and lime stabilization further use in road construction. The major findings and conclusions have been separated into three sections relating to the following areas:

Compared with the untreated samples of Soil-A and Soil-B, plastic limit of the stabilized samples increased while plasticity index, shrinkage limit and linear shrinkage reduced. Compared with the untreated sample, the value of liquid limit of the treated sample increased in Soil-A while it is reduced in case of Soil-B. For Soil-A (LL= 30 PI =7) both liquid and plastic limit increased while for Soil-B (LL=44, PI=19) liquid limit reduced and plastic limit increased with increased fly ash content.

For samples of both the soils, compared with the untreated samples, the values of maximum dry density (γ_{max}) increased with fly ash content. Compared with the untreated sample, the values of γ_{max} increased up to 7.5% for both the soil with 18% fly ash. The values of ω_{opt} reduced upto 15% and 10% respectively for samples of Soil-A and Soil-B.

For samples of both the coastal soils, compared with the untreated samples, the values unconfined compressive strength (q_u) of the treated samples increases significantly, depending on the fly ash content and curing age. The values of q_u of samples of Soil-A and Soil-B treated with 6% and 18% fly ash and cured at 28 days were found to be about 4 and 5 times higher than the strength of he untreated samples respectively. It has also been fount that the gain in strength with increasing fly ash content and curing age is higher in the less plastic Soil-A (PI= 7) than in the more plastic Soil-B (PI=19). Compared with the untreated samples, the values of axial strain at failure (ε_f) of the stabilized samples reduced with the increase in fly ash content which evidently indicated that the treated samples became more brittle as fly ash content increased. The rate of strength gain with curing time (determined in terms of strength Development Index, SDI) for samples of Soil-A and Soil-B treated with 6% fly ash are relatively much slower than those of samples treated with 12% and 18% fly ash.

It was found that values of q_u of samples of Soil-A (belonging to A-4 group) and Soil-B (belonging to A-7 group) treated with 6%, 12% and 18% fly ash with 3% lime and cured for 7, 14 and 28 days satisfied the requirements of PCA (1956) for the unconfined compressive strength of soil-fly ash mix.

The flexural stress versus deflection curves have been found to be approximately linear for both Soil-A and Soil-B, compared with the untreated sample, flexural strength and flexural modulus of the treated samples increased significantly, depending on the fly ash content. For comparison, the flexural strength and flexural modulus of Soil-A treated with 18% fly ash and cured at 28 days are respectively about 4 times and 2.7 times higher than those for the untreated sample. The flexural strength and modulus of Soil-B treated with 18% fly ash and cured at 28 days are respectively about 6 times and 4.3 times higher than those of the untreated samples. The curing age, however, has got insignificant effect on increase in flexural strength and modulus. It was also found that the values of flexural strength and modulus of samples of more plastic Soil-B (PI=19) is higher than the less plastic Soil-A (PI=7). The maximum deflection and failure strain of untreated and stabilized soil-fly ash beams were very small and have been found in the range of 0.15 mm to 0.35 mm and 0.11% to 0.24% respectively.

From the aforementioned findings, it is evident that for both samples of the two coastal soils studied, fly ash stabilization provided a substantial improvement in the engineering properties as compared with the samples of the untreated soils. It has also been found that, in general, samples of both the soils stabilized with 12% and 18% fly ash with 3% Limes satisfied the requirements of compressive strength, and durability for their use as base or sub-base materials in roads subjected to light traffic.

REFERENCES

ASTM (1989), "Annual Book of ASTM Standards" Volume 04.08 Soil and Rock, Building Stones, Geotextiles.

Barenberg, E.J. (1974). "Lime Fly ash Aggregate Mixtures in Pavement Construction", Process and Technical Data Publication, National Ash Association.

BS 1377 (1975), "Method of Tests for Soils Civil Engineering Purposes", British Standards Institution, Gaylard and Son Ltd., London.

Chu, T.Y. Davidson, D.T, Goecker, W.L. and Moh, z.c. (1975), Soil stabilization with Lime-Fly ash Mixtures: Preliminary Studies with Silty and Clayey Soils, Hwy. Res. Bd. Bull 108, Washington, D.C, pp- 102-112

EPRI (Electrical Power Research Institute), (1986). Fly Ash Design Manual for Road and Site Application", Vol. 1: Dry or Conditioned Placement, CS-4419, Research Project 2422-2, Interim Report.

Hausmann, M.R. (1990), "Engineering Principles of Ground Modification," McGraw-Hill Publishing Company, Singapore.

Ingles, O.G. and Metcalf, J.B (1972), "Soil Stabilization –Principles and Practice," Butterworths, Pty. Limited, Australia.

Leonard, B. J and Davidson, D.T (1959), Pozzolanic Reactivity Study of Fly ash, Hwy. Res. Bd. Bull. 231, Washington, D.C, pp. 1-15.

Mateos, M. and Davidson, D.T. (1961), Further Evaluation of promising Chemical Additives for Accelerating Hardening of Soil-Lime-Fly ash Mixtures, Hwy. Res. Bd. Bull. 304. Washington, D.C, PP –32-50.

Mateos, M. and Davidson, D.T. (1962) Lime and Fly ash Proportions in Soil, Lime and Fly ash Mixtures, and Some Aspects of Soil Lime Stabilization, Hwy. Res Bd Bull – 335, Washington, D.C, pp-4064.

NAASRA (1986), "Guide to Stabilization in Roadorks," Sydney, Australia.

Nettleton, A.F.S. (1962), The Stabilization of Sydney Basin Wianamatta Derived Residual Clay With Fly ash and Chemical Control of Environment, Australian Road Research, Vol-1. No. 4 pp-29-40.

PCA (Portland Cement Association) (1956), "Soil-Cement Construction Handbook," Chicago, Illinois, U .S .A.

Serajuddin, M. (1993), "Studies on Fine-Grained Soils for Road subgrade," Proc. of the First Bangladesh-Japan Joint Geotechnical Seminar on Ground Improvement, Dhaka.

Viskochil, R. K, Handy, R. L, and Davidson, D.T. (1958), Effect of Density on Strength of Lime-Fly ash Stabilized Soil, Hwa. Res. Bd. Bull-183. Washington, D.C, pp. 5-15.

Soil Stress-Strain Behavior: Measurement, Modeling and Analysis
Geotechnical Symposium in Roma, March 16 & 17, 2006

STRENGTHENING OF WEAKLY-CEMENTED GRAVELLY SOIL WITH CURING PERIOD

Tara Nidhi Lohani
JSPS Post-Doctoral Fellow, Kobe University, Japan
email: tnlohani@gmail.com

Fumio Tatsuoka
Professor,Tokyo University of Science, Japan
email: tatsuoka@rs.noda.tus.ac.jp

Masaru Tateyama
Research engineer, Railway Technical Research Institute, Japan
email: tate@rtri.or.jp

Satoru Shibuya
Professor, Kobe University, Japan
email: sshibuya@kobe-u.ac.jp

ABSTRACT: A series of drained triaxial compression (TC) tests were conducted on a lightly cement-mixed gravelly soil (CMG) compacted at three different water contents (w), on a dry side, at the optimum (w_{opt}) and at a wet side. The specimens were cured at constant water content under atmospheric pressure for periods of 7, 28 and 180 days. The TC tests performed under 20 kPa confining pressure by measuring local strains showed that the compressive strength, q_{max}, and the initial Young's modulus, E_0, in particular the former, became the respective maximum value when compacted at $w=w_{opt}$ regardless of curing period (t_c). Each of the q_{max} and E_0 value increased at a rather constant rate with respect to log (t_c) even after t_c= 28 days, at which the design strength is determined in the usual design practice. Such an increasing trend for CMG specimens is close enough to the conclusion drawn for cement-mixed sand specimens but differs largely from that of ordinary cement concrete, found in the existing literatures. The stress-strain behavior became more linear with t_c irrespective to their molding water content.

1. INTRODUCTION

Natural soils are often mixed with cement to enhance their strength and stiffness. When compared with the deep mixing technology for clay, the use of cement-mixed gravelly soil is relatively new (e.g., Watanabe et al., 2002, 2003; Tatsuoka, 2004). Despite the target strength and therefore the amount of cement content are largely different, the effects of cement-mixing on the strength and deformation characteristics of cement-mixed soil are basically the same

Hoe I. Ling et al. (eds.), Soil Stress-Strain Behavior: Measurement, Modeling and Analysis, 455–462.

with those of ordinary concrete and the design method of CMG in engineering practice follows that of concrete engineering. That is, the design compressive strength of cement-mixed soil, including CMG, is usually determined by unconfined compression tests on specimens that have been cured at constant water content under the atmospheric pressure for 28 days. However, there are a number of aspects specific to cement-mixed soil including the followings.

1) Soils to produce cement-mixed soil have various grading and proportions. CMG uses usually gravelly soils with some fines. The inclusion of some fines content, which is not preferred when producing concrete, contributes to an increase in the compacted density and ultimately enhances the compressive strength of CMG (Lohani et al., 2003b).
2) CMG is usually compacted in the field. It has been found that compaction at the optimum water content results into the maximum compressive strength (Lohani & Tatsuoka, 2003; Lohani et al., 2003b & 2004), which is not only by the maximum compacted dry density but also by the optimum water content condition for cement slurry (Kongsukprasert et al., 2004).
3) With ordinary concrete, the compressive strength increase by curing becomes almost saturated, reaching 90 to 95 % of the maximum, in about 28 days. An additional increase in the strength by longer curing is ignored in usual design practice. On the other hand, cemented-mixed sand exhibits a continuous increase in the strength even until three years (e.g., Shibuya et al. 2001; Barbosa-Cruz & Tatsuoka 1999 & 2000).
4) Despite that the effects of confining pressure are usually ignored in concrete engineering, the effects are significant with cement-mixed soil.

As the conventional use of CMG has been limited to secondary structures, the effects of these factors listed above are usually ignored for the cement-mixed soil as well. On the other hand, when used as the backfill of critical structures that need high stability and stiffness (e.g., bridge abutments), the strength and deformation characteristics of CMG should be evaluated more reliably and realistically considering aforementioned factors and others. In the previous studies (Lohani et al., 2003; Kongsukprasert et al., 2004), the effects of factor 2 was examined on the CMG at a curing period, t_c = 7 days. In the

Table 1. Properties of test specimens

Specimen name	Molding water content, w (%)	Dry density, ρ_d (g/cm^3)	Curing period, t_c (days)
A1-7	3.28	2.150	7
A1-28	4.02	2.178	28
A1-180	3.82	2.185	180
A2-7	5.64	2.203	7
A2-28	5.74	2.222	28
A2-180	5.85	2.214	180
A3-7	7.56	2.200	7
A3-28	7.82	2.200	28
A3-180	7.56	2.204	180

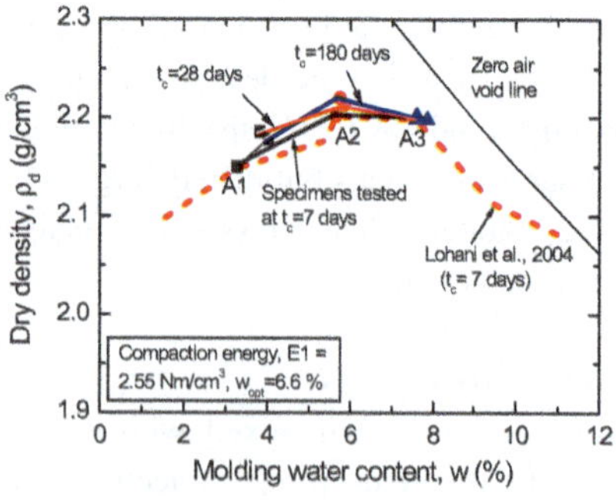

Fig. 1 Compaction curves of all three series of specimens, A1, A2 & A3 prepared for CD TC tests

present study, by using the same type of well-graded gravelly soil including fines and a constant confining pressure (factor 1 & 4), the effects of factor 2 as well as factor 3 were investigated for a wide range of t_c from 7 to 180 days by means of drained triaxial compression (TC) tests.

2. TEST METHOD

Test material: A crushed sandstone gravel, called model Chiba gravel (Lohani et al., 2004), with D_{max}=10 mm, U_c=3.2 and G_s=2.74 was used. First, air-dried material was mixed with the ordinary Portland cement (2.5 % by gravel weight) and then with tap water. Standard compaction tests performed on this mixture showed the optimum water content of 6.6 % at compaction energy, E1 (=2.55 Nm/cm^3).

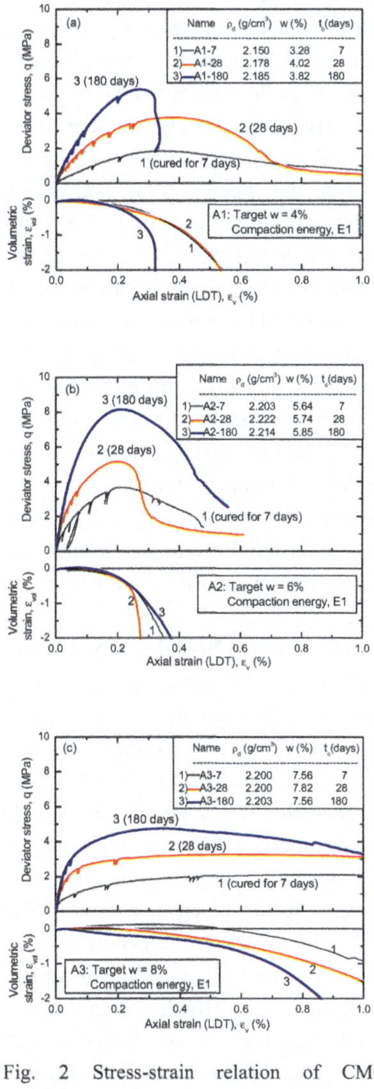

Fig. 2 Stress-strain relation of CMG compacted at water content of (a) 4 %, (b) 6 % and (c) 8 %, cured for 7, 28 and 180 days.

Specimens: Cylindrical specimens (75 mm ϕ x 150 mm high) were prepared by compaction in a split mold at three target molding water contents of 4 %, 6 % and 8 % (series A1, A2 & A3) by applying compaction energy E1 (**Table 1**). **Figure 1** compares the relationships between the compacted dry density, ρ_d, and the molding water content, w, of the specimens that were prepared for drained TC tests at t_c = 7, 28 & 180 days. Figure 1 also includes the curve at t_c = 7 for the same test conditions by Lohani et al. (2004).

Curing: The compacted specimens were cured inside the mold as compacted for the initial five days and were then taken out and cured outside for the remaining days. The curing was carried out at constant water content in a temperature-controlled room (at 25° C). To maintain the water content as constant as possible during the total span of curing period, the specimens, both while inside and outside of the compaction mold, were sealed with polypropylene kitchen wrapping sheet and then stored inside a thick glass desiccator.

Drained triaxial compression (TC) tests: After the respective specified curing period, both top and bottom ends of the specimen were capped with gypsum to ensure a uniform stress distribution and was set in a triaxial cell. The capping was essential for a high repeatability of the test results as well as for a high accuracy of local axial and lateral strain measurements. Drainage path was prepared by using filter paper strips around the capped specimen. A relatively low confining pressure of 20 kPa was applied by partial vacuuming in all the tests to simulate the typical stress conditions in the shallow backfills, such as bridge abutments. After having been left at isotropic conditions of 20 kPa for about 2 hours, which was sufficient to reach essentially zero creep strain rate, the specimen (without saturation) was subjected to drained TC at an axial strain rate of 0.03 %/min. Axial strains, ε_v, were measured locally by using a pair of local deformation transducers (LDT) (Goto et al. 1991). The use of LDTs was discontinued at ε_v= about 1 %, whereas the external measurement by means of a LVDT was continued until the end of the respective test. Lateral strains, ε_h, were measured with three clip gauges attached at 1/6th, 1/2 and 5/6th of the specimen height. The average local axial and lateral strains were used to obtain volumetric strain, $\varepsilon_{vol} = \varepsilon_v + 2\varepsilon_h$. A special axial loading system with almost zero backlash, which makes it possible to reverse the loading direction without any noticeable delay (Tatsuoka et al., 1994), was used. During otherwise monotonic drained TC loading at a constant ε_v, two unload-reload cycles with a very small axial strain amplitude were applied to evaluate instantaneous elastic Young's modulus values at different loading stages.

3. RESULTS AND DISCUSSIONS

Table 2 lists the maximum deviator stress, q_{max}, the axial strain at q_{max} condition, $\varepsilon_{v,peak}$, and the initial Young's modulus, E_0 (n.b., the E_0 value was obtained from the initial part of the respective q - ε_v relation at axial strains less than about 0.001 %: i.e., the tangent Young's modulus, E_{tan}, at about q= 0, presented in Fig. 6b). **Figures 2a, b** and **c** show the stress-strain relations at t_c= 7, 28 and 180 days of the specimens compacted, respectively; a) at a dry side; b) at the optimum water content; and c) at a wet side. **Figure 3** illustrates the relationship between the compressive strength and the axial strain at the peak stress state. The following trends of behavior may be noted:

Table 2. Summary of the test results

Specimen name	Compressive strength, q_{max} (MPa)	ε_v at peak, $\varepsilon_{v,peak}$ (%)	Initial Young's modulus, E_0 (GPa)
A1-7	1.84	0.34	3.3
A1-28	3.84	0.38	6.4
A1-180	5.44	0.28	6.8
A2-7	3.66	0.21	8.1
A2-28	5.23	0.20	10.0
A2-180	8.23	0.21	12.8
A3-7	2.10	0.85	6.5
A3-28	3.33	0.53	8.6
A3-180	4.80	0.35	12.3

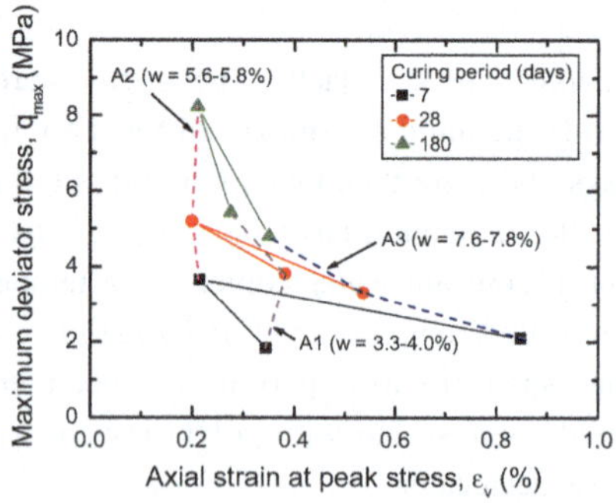

Fig. 3 Relationships between q_{max} and axial strain at peak stress at t_c= 7, 28 and 180 days.

1) With an increase in the curing period until 180 days, the initial stiffness, E_0, peak strength, q_{max} and dilatancy rate all increase significantly.

2) Regardless of curing period, the specimens compacted at the optimum water content (A2 series) exhibit the highest values of peak strength, pre-peak stiffness and dilatancy rate.
3) When compacted at a wet side (A3 series), the axial strain at the peak stress becomes relatively large, particularly at younger curing periods (Fig. 3). In other two series, where the strength is generally larger, the axial strain at the peak is rather independent of curing period.
4) The post-peak stress-strain behavior becomes more ductile with an increase in the compaction water content. Based on this fact, when a structure of CMG is designed ensuring ductile post-peak behavior, it is suggested to compact CMG somehow at a wet side.

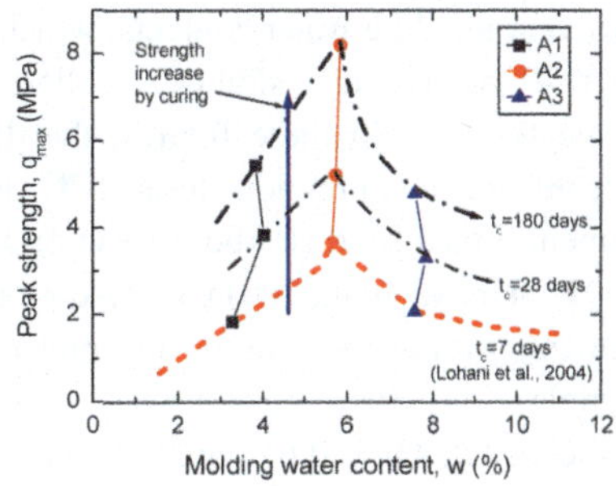

Fig. 4 Relationships between q_{max} and molding water content at t_c= 7, 28 and 180 days.

Figure 4 shows the relationships between the peak strength, q_{max}, and the molding water content, w, for the three different curing periods (t_c), corresponding to Fig. 1. The curve for t_c= 7 days from Lohani et al. (2004) is also presented. It is observed that the increasing rate of q_{max} with t_c is similar among the three compaction water contents and, therefore, the pattern of q_{max} - w curves look similar for all three t_c values. That is, the trend that q_{max} becomes the maximum when compacted at w around w_{opt} does not change with an increase in t_c.

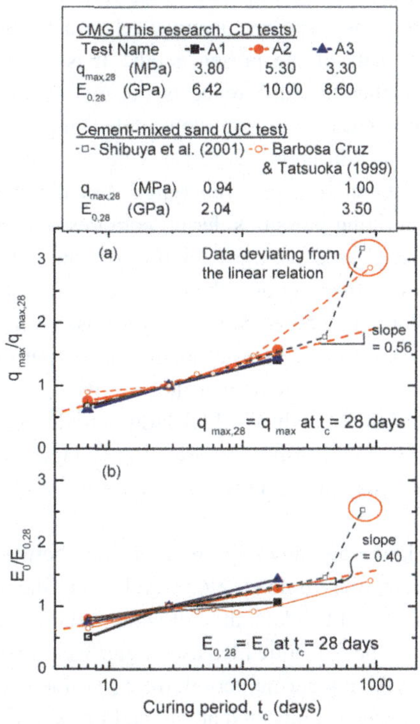

Fig. 5 Effects of curing period on q_{max} and E0 for different molding water contents from CD TC tests of CMG (present study), compared with those from UC tests on cement-mixed sand.

Figure 5a shows the relationships between the normalized compressive strength, $q_{max}/q_{max,28}$, and the logarithm of curing period, t_c, where $q_{max,28}$ is q_{max} at t_c= 28 days, obtained from the data presented in Fig. 2. It is distinguishable that the increasing rate, $d(q_{max}/q_{max.28})/d(\log t_c)$, for a range of t_c from 7 days to 180 days is nearly the same among the specimens prepared at different molding water contents, which is equal to 0.56 on average. **Figure 5b** shows the relationship between the normalized initial Young's modulus, $E_0/E_{0,28}$, and t_c, corresponding to Fig. 5a. Here, $E_{0,28}$ is E_0 at t_c = 28 days. It is perceivable that the increasing rate, $d(E_0/E_{0.28})/d(\log t_c)$, for a range of t_c from 7 days to 180 days is nearly the same among the different molding water contents, which is equal to 0.40 on average, slightly smaller than the value for q_{max}. In these figures, the data from unconfined compression tests (UC tests) on cement-mixed sand (Barbosa-Cruz & Tatsuoka, 1999; Shibuya et al., 2001) are also plotted. The cement proportions used in preparing the samples in these references, which are respectively, 4% and 6%, are not so far from the 2.5% proportion adopted here. It may be seen that the values of $q_{max}/q_{max,28}$ and $E_0/E_{0.28}$ at t_c = 800 - 900 days (i.e., about 2 - 3 years) from their UC tests are quite high, largely deviating from the respective average relation for the CD TC data of the present study. It is not known whether this difference is due to the different soil types (sand and well-graded gravelly soil) or the different cement-mixing conditions (so different compressive strengths) or the different confining pressures during compression tests or else. Based on the deviation of this particular result from among their own data in semi-log plot (Figs. 5a & 5b), such increases may even be the true features at longer curing periods. In any case, it seems that the use of $q_{max,28}$ as the design strength of CMG largely underestimates the compressive strength at longer curing periods. The same is also true for the stiffness.

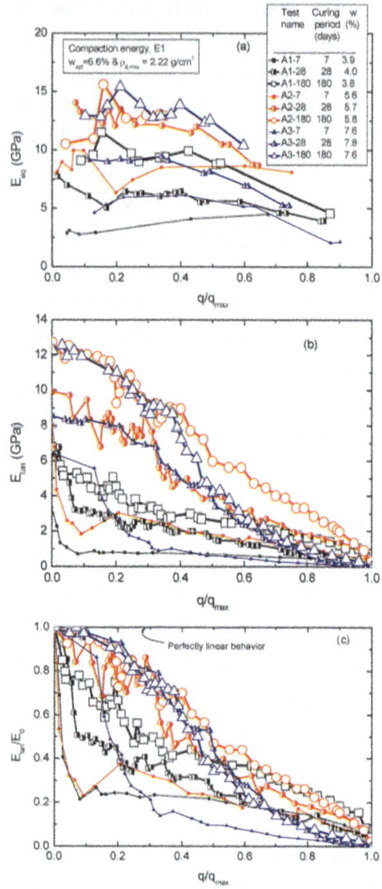

Fig. 6 Relationships between; a) E_{eq}; b) E_{tan}; an c) E_{tan}/E_0, and q/q_{max} from CD TC tests on CMG.

Figure 6a shows the relationships between the equivalent Young's modulus, E_{eq}, obtained from small unload-reload cycles and the shear stress level, q/q_{max}. Despite a relatively large scatter of the data, the following trend of behavior is witnessed:

1) The E_{eq} value increases significantly with an increase in the curing period, t_c= 7 - 180 days. At any curing period, the E_{eq} value is consistently larger when compacted at $w = w_{opt}$ than when compacted at wet and dry sides. That is, a large difference due to different molding water contents do not disappear even at t_c = 180 days.
2) With an increase in q/q_{max}, the E_{eq} value first tends to increase slightly, followed by a marked decrease as approaching the peak stress state. The initial increase is consistent with the trend of unbound granular material, while the subsequent decrease is likely due to

damage to bonding at inter-particle contact points associated with shearing (Tatsuoka et al., 1997).

Figure 6b shows the relationships between the tangent Young's modulus, E_{tan}, and q/q_{max}, corresponding to Fig. 6a. **Figure 6c** shows the relationships between E_{tan}/E_0 and q/q_{max}, corresponding to Fig. 6b. It may be noticed that, with an increase in q_{max} associated with an increase in t_c, not only the E_{tan} value at a given condition, but also the E_{tan}/E_0 value, therefore, the linearity of stress-strain curve, increase.

4. CONCLUSIONS

The following conclusions with respect to the strength and deformation characteristics of cement-mixed gravelly soil, which are noticeably different from those of ordinary concrete, can be derived from the test results discussed so far:

1) The initial stiffness, linearity of pre-peak stress-strain relation and peak strength become the respective maximum value when compacted at water content close to the optimum value. This trend does not change with curing period until 180 days.
2) The post-peak strain-softening becomes smaller with an increase in the molding water content. When compacted at a dry side of w_{opt}, the material becomes highly brittle.
3) The initial stiffness, linearity of pre-peak stress-strain relation and peak strength increases by curing even after the curing period becomes longer than 28 days. In the present study, the increase was observed until t_c= 180 days. The increasing rate is rather similar among the CMG samples compacted at a dry side, at the optimum and at a wet side. On looking the absolute values of the tested specimens, wetter samples maintained relatively higher stiffness magnitudes as compared to the drier ones.
4) Based on the data within t_c= 180 days, the trend of E_0 and q_{max} increase against curing period is rather similar for sands and gravels.

ACKNOWLEDGEMENT

This research was performed at the Department of Civil Engineering, the University of Tokyo. The authors appreciate a great help of their colleagues in the geotechnical laboratory. Financial support from the Ministry of Education, Culture and Sport, Government of Japan and the Program for Promoting Fundamental Transport Technology Research from Corporation for Advanced Transport & Technology is greatly acknowledged.

REFERENCES

Barbosa-Cruz, E.R. and Tatsuoka, F. (1999): Effects of stress state during curing on stress-strain behavior of cement-mixed sand. *Proc. 2nd Int. Conf. on Pre-Failure Deformation Characteristics of Geomaterials* (Jamiolkowski et al. eds.), Balkema, I, pp.509-516.

Barbosa-Cruz, E.R. and Tatsuoka, F. (2000): Stress-strain properties from elastic behavior to peak strength of compacted cement-mixed sand, *Grouting, Soil Improvement, Geosystems including Reinforcement*, pp.3-10.

Goto, S., Tatsuoka, F., Shibuya, S., Kim, Y. S. and Sato, T. (1991): A simple gauge for local small strain measurement in the laboratory, *Soils and Foundations*, 31 (1), 169-180.

Kongsukprasert, L., Tatsuoka, F. and Tateyama, M. (2004): “Several factors affecting the strength and deformation characteristics of cement-mixed gravel”, *Soils and Foundations*, Vol. 45, No. 3, pp.107-124.

Lohani, T. N. and Tatsuoka, F., (2003). Effects of molding water content on the strength of cement-mixed gravel. *Proc. of 12th Asian Regional Conference on SMGE, Singapore (Leung et al. eds.)*, Vol. 1, pp.497-500.

Lohani, T. N., Kongsukprasent, L., Watanabe, K., and Tatsuoka, F. (2003b). Strength and deformation characteristics of cement-mixed gravel for engineering use, *Proc. 3rd Int. Symp. on Deformation Characteristics of Geomaterials*, IS Lyon 03 (Di Benedetto et al. eds.), pp.637-643.

Lohani, T.N., Kongsukprasert, L., Watanabe, K., and Tatsuoka, F., 2004. Strength and deformation properties of a compacted cement-mixed gravel evaluated by triaxial compression tests. *Soils and Foundations*, Vol. 44, No. 5, pp.95-108.

Shibuya, S., Mitachi, T., Tanaka, H., Kawaguchi, T. and Lee, I.M. (2001). Measurement and application of quasi-elastic properties in geotechnical site characterization. *Proc. of 11th Asian Regional Conference on SMGE, 1999* (Sun-Wang et al. eds.), Vol. 2, pp.639-710.

Tatsuoka, F., Jardine, R.J., Lo Presti, D., Di Benedetto, H. and Kodaka, T. (1997). Characterizing the pre-failure deformation properties of geomaterials, Keynote Lecture for session No. 1. *Proc. of 14th ICSMFE*, Hamburg, Vol.4, pp.2129-2164.

Tatsuoka, F., Teachavorasinskun, S., Dong, J., Kohata, Y. and Sato, T. (1994), “Importance of measuring local strains in cyclic triaxial tests on granular materials”, *Proc. of ASTM Symposium Dynamic Geotechnical Testing II, ASTM, STP 1213*, pp.288-302.

Tatsuoka,F. (2004): Cement-mixed soil for Trans-Tokyo Bay Highway and railway bridge abutments, Geotechnical Engineering for Transportation Projects, *Proc. of GeoTrans 04, GI, ASCE, Los Angels, ASCE Geotechnical Special Publication GSP No. 126 (Yegian and Kavazanjian eds.),* pp.18-76.

Watanabe, K., Tateyama, M., Yonezawa, T., Aoki, H., Tatsuoka, F., and Koseki, J., 2002. Shaking table tests on a new type bridge abutment with geogrid-reincorced cement treated backfill. *Proc. of the 7th international Conf. on geosynthetics*, France, pp.119-122.

Watanabe K., Tateyama M., Jiang G., Tatsuoka F. and Lohani T.N., 2003. Strength characteristics of cement-mixed gravel evaluated by large triaxial compression tests. *Proc. 3rd Int. Symp. on Deformation Characteristics of Geomaterials*, IS Lyon 03 (Di Benedetto et al. eds.), pp.683-693.

Soil Stress-Strain Behavior: Measurement, Modeling and Analysis
Geotechnical Symposium in Roma, March 16 & 17, 2006

THE LOSS OF STRENGTH OF AN UNSATURATED LOCAL SOIL ON SOAKING

Samuel I.K. Ampadu
Department of Civil Engineering
Kwame Nkrumah University of Science and Technology, Kumasi GHANA
e-mail: sikampadu@yahoo.co.uk

ABSTRACT

Residual soils are known to undergo strength loss when unsaturated soils are soaked. This study looks at the nature of such strength loss for both disturbed and undisturbed samples of a local soil as determined in the direct shear apparatus. Remoulded and undisturbed samples of a local soil were tested both soaked and unsoaked in the direct shear apparatus under vertical consolidation pressures ranging from 20kPa to 320 kPa. The effect of the soaking on the strength as defined by the maximum shear stress and the parameters c and ϕ are compared and discussed. The results are discussed in terms of unsaturated soil mechanics principles.

KEYWORDS: Direct Shear, remoulded samples, undisturbed samples, unsaturated soils, shear strength, soaking.

INTRODUCTION

One of the most common residual soils in the sub-region is decomposed phyllites. These soils occur at various degrees of laterization and are used extensively as subgrade and sub-base layers for many road and highway pavements in the sub-region as well as construction material for engineered fills. Many cuts are also made in these materials during road and other construction projects. However, one of the difficulties encountered with this material during construction and also when in service is that they undergo appreciable strength reduction when they are soaked in water. In recognition of this fact, attempts have been made to take this phenomenon into account in design. For example, the California Bearing Ratio (CBR) Pavement Design Methodology (TRRL 1999) specifies the use of the four-day soaked CBR value (ASTM D1883-87). On the other hand unsaturated soil mechanics principles have not been applied to local soils. This paper seeks to initiate an investigation into local unsaturated soils.

This paper describes the effect of soaking on the shear strength of undisturbed and remoulded samples of a local soil formation in direct shear tests under consolidation pressures ranging from 20 to 320kN/m^2. The test results of the saturated and the unsaturated samples are discussed and compared with unsaturated soil strength model at high degree of saturation.

METHODOLOGY

Sample Preparation

Both undisturbed and disturbed samples of the decomposed phyllites were obtained from a depth of 3.0m in a trial pit manually sunk, near the central business district of the city of

Hoe I. Ling et al. (eds.), Soil Stress-Strain Behavior: Measurement, Modeling and Analysis, 463–472.

Kumasi. For undisturbed samples (i.e. PU-samples), the CBR mould of dimensions 150mm in diameter and 178mm long, was hammered into the bottom of the trial pit and retrieved by excavating around it. The sample was covered with plastic to prevent moisture loss and then sent to the laboratory where the 10cm x10cm x 2cm specimen was cut for the direct shear test.

For the remoulded samples designated PR-samples in this study, disturbed samples retrieved from the pit were air dried and sieved through ¾" BS-sieve to remove particles larger than 19 mm. Samples were then prepared by compaction at the optimum moisture content using ASTM Test Method for Laboratory Compaction Characteristics of Soil Using Modified Effort (ASTM D 1557-91) except that 25 blows of the rammer per layer were used instead. The 10cm x10cm x 2cm samples were then cut from the compacted sample for the direct shear test.

Direct Shear Test Procedure

A commercially available direct shear apparatus was used for this study. For each type of sample, two test series, unsoaked (-U) and soaked (-S) were conducted for consolidation pressures ranging from 20 kN/m^2 to 320 kN/m^2. For the unsoaked test series, after setting the sample in the apparatus, the specified consolidation pressure, σ_v' was applied and consolidation was allowed to proceed for about 30 minutes. After that, the locking screws were disengaged to free the upper and lower shearbox halves and the specimen was subjected to the drained direct shear test at a constant rate of 0.048mm/min under the prevailing consolidation pressure. The shearing force was measured by a proving ring while the vertical and horizontal displacements were measured using dial gauges.

For the soaked test series, after assembling the specimen in the apparatus, a nominal consolidation pressure of 5kN/m^2 was applied and the carriage was filled with water to flood the sample under this pressure for 24 hours. After soaking for 24 hours, the rebound on saturation was noted and the specified consolidation pressure was applied. The specimen was consolidated under this consolidation pressure for about 1 hour after which the locking screws were disengaged for the drained shear test to proceed.

DISCUSSION OF RESULTS

Properties of Samples Tested

Geologically, phyllites are metamorphosed sediments of the Precambrian age belonging to the lower Birrimian geological system of Ghana. The physical characteristics of the samples tested which are residual products of these phyllites are summarized in Table 1. The sample may be described as a clayey silt with some sand.

Table 1 Physical Characteristics of the Test Material

Grading (%)				Atterberg Limits		Compaction		Gs
Gravel	Sand	Silt	Clay	LL (%)	PI (%)	OMC (%)	MDD (Mg/m^3)	
6.1	23.8	44.0	26.2	51	19	18.0	1.72	2.70

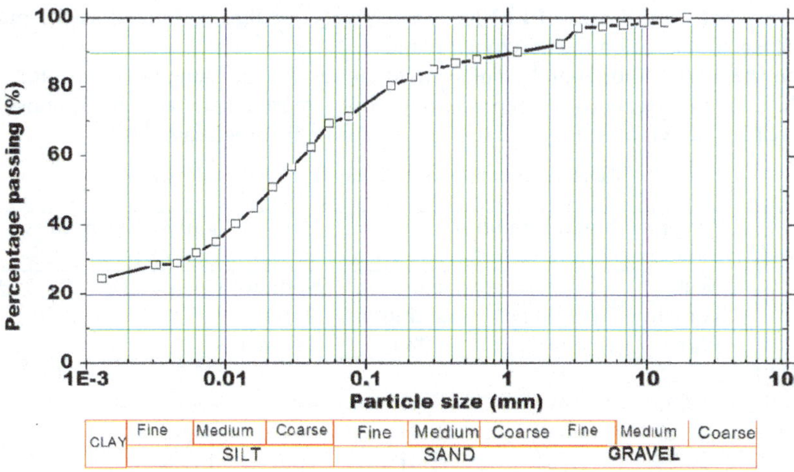

Figure 1 Grading Characteristics of P-Samples Tested

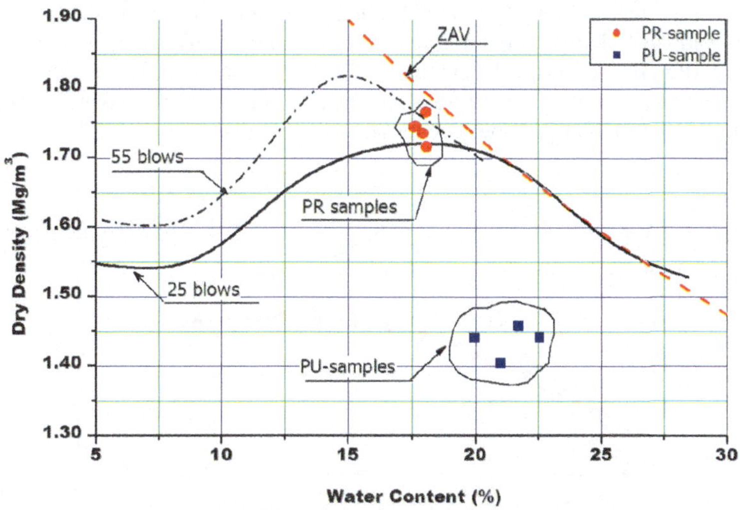

Figure 2 Compaction Characteristics and Initial Condition of P-samples

Undisturbed Samples (PU-samples)

In this testing programme, it was observed in many cases that primary consolidation was completed in less than 1 minute. The rebound on saturation was very small, typically less than 0.01 mm corresponding to a volumetric strain of less than 0.05%.

Table 2 shows the PU-sample initial condition and failure stresses. The initial sample condition is the average for the particular batch of specimens tested. The maximum individual deviation of PU-samples from the average initial water content and the average dry density was 0.9% and 0.10 kN/m³ respectively. Failure is defined as the maximum shear stress measured on the horizontal plane.

The shear stress-horizontal displacement curves for the samples tested under 20, 80 and 320 kN/m² respectively representing low, intermediate and high consolidation pressures are shown from Figure 3 to Figure 5 for undisturbed samples.

Table 2 Summary of test results for PU- samples

Test Name	Average Initial Condition				Failure Shear Stress (kN/m²)	
	Water content (%)	Dry Density (Mg/m³)	Degree of Saturation, Sr (%)	Normal Stress (kN/m²)	Unsoaked τ_u	Soaked τ_s
PU-X[1]20	22.55	1.442	70.22	20	25.2	15.8
PU-X40	19.97	1.441	63.41	40	46.1	36.4
PU-X80	21.69	1.459	69.41	80	83.9	59.8
PU-X160	20.99	1.405	62.01	160	141.8	104.4
PU-X320	21.72	1.458	69.45	320	264.2	198.0

The results show that not only does soaking lead to a reduction in the failure shear stress, but it also leads to a change in the nature of the shear stress-displacement curves. Soaking appears to flatten the stress-displacement curves. Even though the test results reported did not appear to have attained the residual state, they suggest that soaking of undisturbed samples leads to a smaller post-peak reduction in strength.

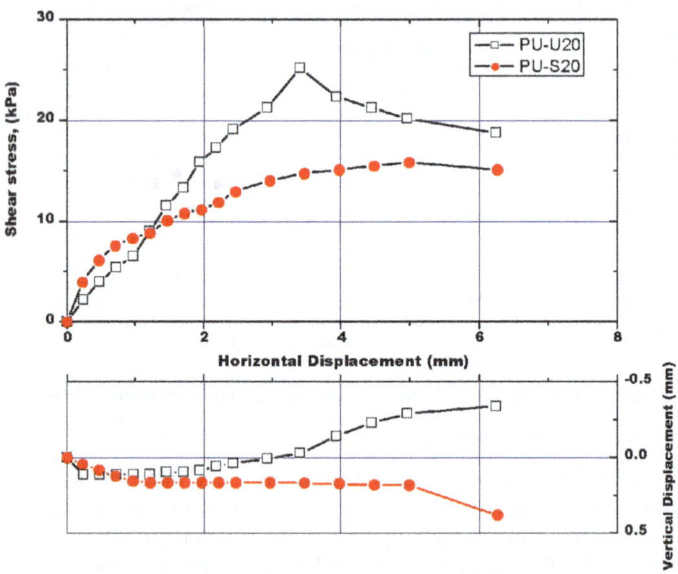

Figure 3 Stress-displacement curves for PU-samples at 20kN/m².

[1] X takes values of U for unsoaked and S for soaked test conditions respectively

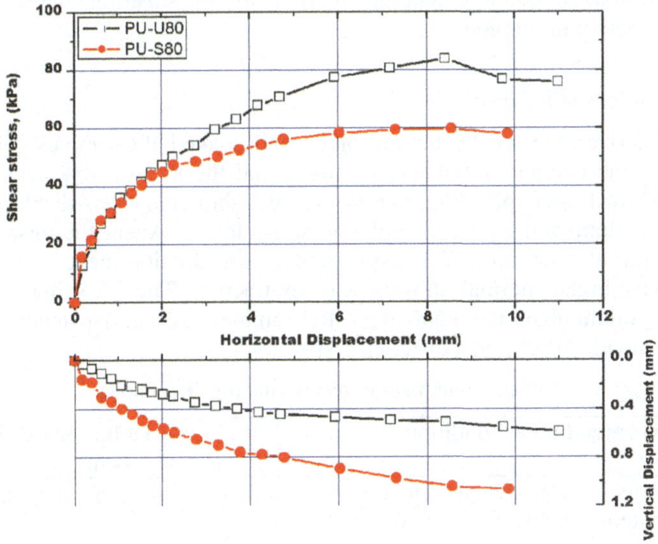

Figure **4** Stress-displacement curves for PU-samples at 80kN/m^2

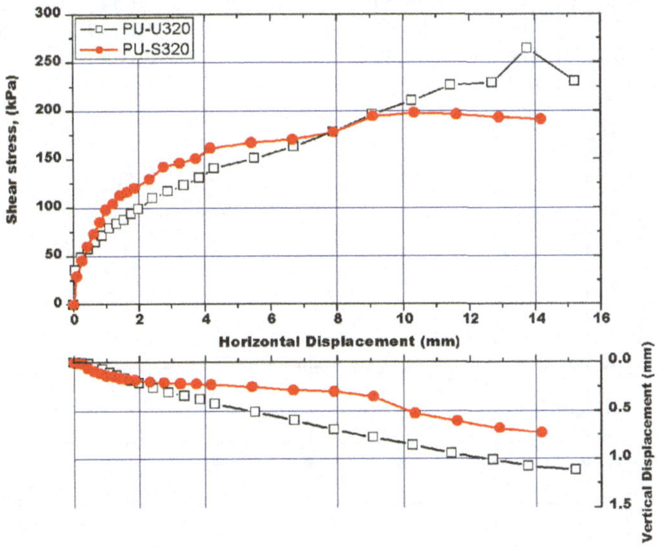

Figure 5 Stress-displacement curves for PU-samples at 320kN/m^2

The plots also show the progress of the vertical displacement, v, during shearing. Assuming the horizontal direction to be the direction of zero linear incremental strain then the ratio of the incremental boundary displacements gives the current angle of dilation in the deforming material. The results show that at low normal stress levels whereas the unsoaked samples undergo dilation, during shearing, the soaked samples remained contractive throughout

Remoulded Samples (PR-Samples)

Table 3 summarizes the results for unsoaked and soaked PR-samples. The maximum deviation from the average initial water content and the average dry density was 0.4% and 0.6 kN/m^3 for PR-samples. Figure 6 to Figure 8 show the stress-displacement curves for low, intermediate and high consolidation stress levels. Again the unsoaked samples during shearing at low normal stresses underwent dilation near failure while the behaviour under higher normal stresses was contractive. The PR-samples showed less tendencies to dilate than the equivalent PU-samples. A similar tendency has been reported by Ampadu (1998) on a different test series.

Table 3 Summary of test results for PR-samples

Test Name	Average Initial Condition				Failure Shear Stress (kN/m^2)	
	Water content (%)	Dry Density (Mg/m^3)	Degree of Saturation, Sr (%)	Normal Stress (kN/m^2)	Unsoaked τ_u	Soaked τ_s
PR-X20	18.08	1.717	86.36	20	35.12	10.83
PR-X40	17.95	1.736	88.54	40	70.54	29.61
PR-X80	17.65	1.746	88.81	80	101.49	52.78
PR-X160	17.59	1.745	88.59	160	203.87	105.28
PR-X320	18.05	1.767	93.55	320	265.77	125.00

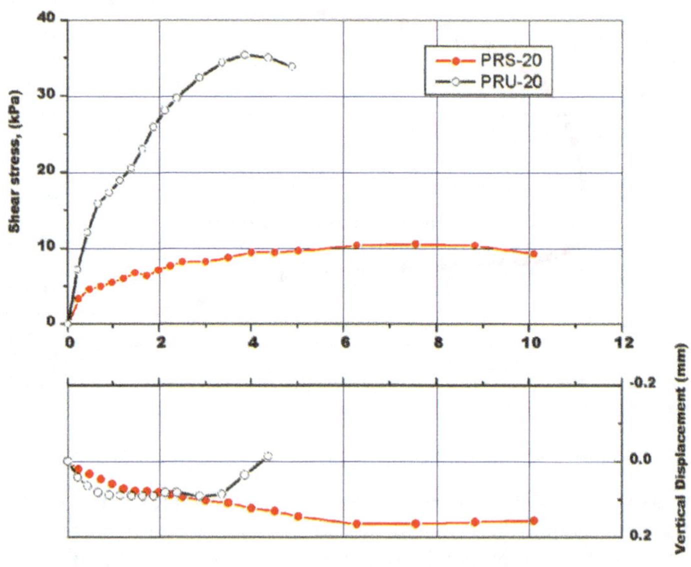

Figure 6 Stress-displacement curves for PR-samples at 20kN/m^2

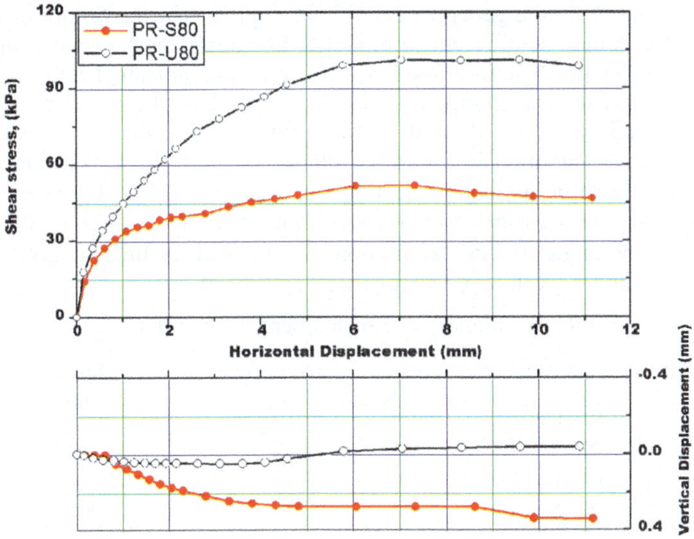

Figure 7 Stress-displacement curves for PR-samples at $80kN/m^2$

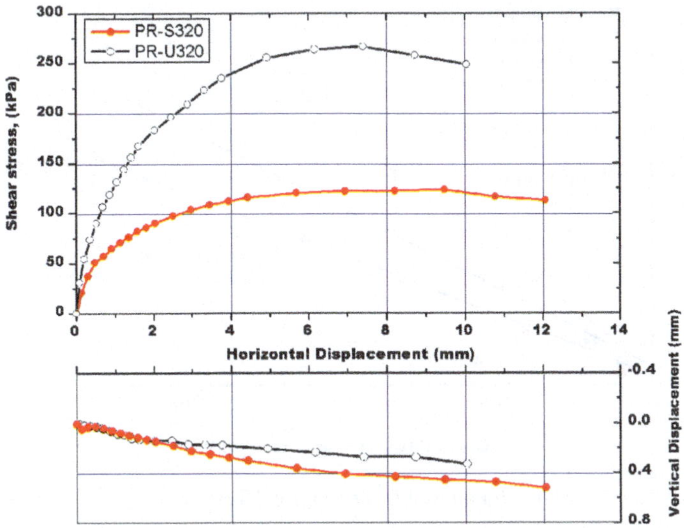

Figure 8 Stress-displacement curves for PR-samples at $320kN/m^2$

Failure Envelope

The drained direct shear failure envelopes for the PR-samples and for the PU-samples are shown in Figure 9. The PU results showed well defined failure envelopes over the whole range of stresses studied. For both unsoaked and soaked PR-samples, however, it was observed that the failure shear stress values for σ_v' of $320kN/m^2$ were lower than the expected values based on the general trend. This implied that the failure envelope for high normal stresses may be different from that at low normal stress levels suggesting a bi-linear failure envelope. A similar trend was observed in Ampadu (1998). Even though bi-segment shear strength envelopes for saprolitic soils have been reported elsewhere (Irfan 1998) the results reported in this study may be considered inconclusive and requires further investigation. The failure envelopes for both the soaked and unsoaked PR-samples were therefore defined neglecting the values for σ_v' of 320kN/m^2 and were based on consolidation pressures of 160kN/m^2 and less. The values of the cohesion intercept, c_d and the angle of internal friction, ϕ_d obtained by linear regression of the failure shear stress and σ_v' values are summarized in Table 4.

Table 4 Summary of Failure Strength Parameters

Sample Condition	Sample No.	c_d (kN/m^2	ϕ_d (degrees)	Degree of Saturation S_r (%)
Remoulded	PR-U	15	49.4	89.2
	PR-S	0	33.4	100
Undisturbed	PU-U	18	37.8	67.0
	PU-S	10	30.7	100

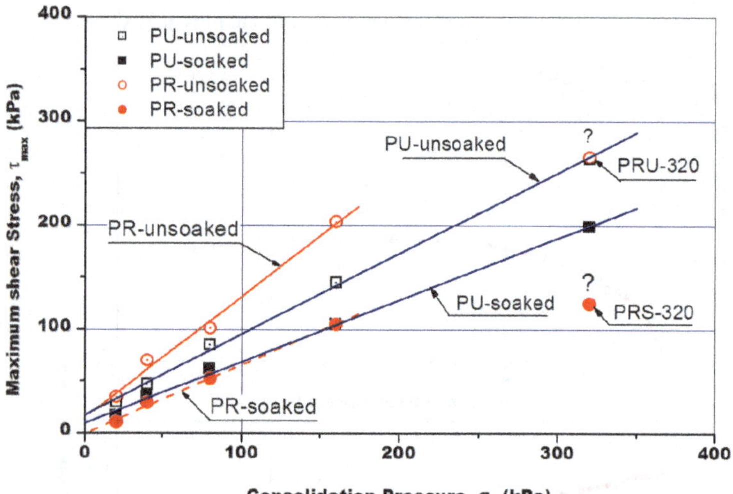

Figure 9 Failure Envelope for PU Test Series

Reduction in Shear Strength

For this study, the reduction in shear strength on soaking may be measured by the shear strength reduction ratio (SRR) which is defined as the reduction in shear strength of an unsaturated soil on soaking divided by the unsoaked shear strength. Figure 10 shows the variation of SRR with the effective normal stress or the consolidation pressure as measured in this study. The figure shows large SRR values especially at low confining pressures and this initially reduces with increasing pressure and attains an almost constant value at confining pressures exceeding about 80kN/m^2. This trend is similar for both PR and PU-samples, but the magnitudes are different. For PR-samples there is an almost 70% reduction in strength at a normal stress of 20 kN/m^2. The difference in SRR values of about 20-30% between the PR and PU samples appears to be true for all normal stress levels studied. The results of Yoshitake and Onitsuka (1990) on decomposed granitic soils from Japan, on the other hand suggest that at large normal stresses the disturbed and undisturbed samples tended towards similar magnitudes of strength reduction. It must be noted that the PR and PU-samples have different matrix structure and different degrees of saturation. Remoulding and subsequent re-compaction therefore lead to a substantial loss of strength.

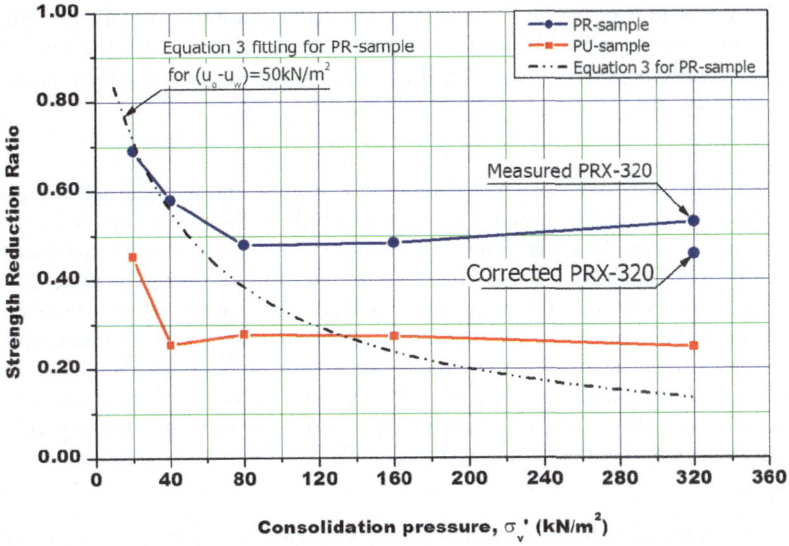

Figure 10 Variation of strength reduction ratio with pressure

Unsaturated Soil Mechanics Principles

The mechanical response of unsaturated soils is controlled by the net stress and the matrix suction. In this testing programme, suction was not measured. Thus the unsoaked test results are tests for which suction is uncontrolled and unmeasured, while those for soaked samples represent zero suction. The shear strength equation for saturated samples, τ_s, based on Mohr-Coulomb failure criterion is given by Equation 1 for a total normal stress, σ_n, pore water pressure, u_w, effective angle of internal friction, ϕ' and effective cohesion c'.

$$\tau_s = c' + (\sigma_n - u_w)\tan\phi' \qquad \text{Equation 1}$$

Based on the effective stress concept (Bishop 1959) Equation 2 has been proposed for unsaturated soils where χ is the effective stress parameter, Lu and Likos (2004).

$$\tau_s = c' + (\sigma_n - u_a)\tan\phi' + \chi(u_a - u_w)\tan\phi' \qquad \text{Equation 2}$$

The results analyzed by Khalili and Khabbaz (1998) show that for suction values less than the air entry pressure, χ is constant at 1.0. On the soil-water characteristic curve, air-entry pressures correspond to high degrees of saturation. For PR-samples in this study at 89.2% degree of saturation it may be assumed that the suction value is below the air-entry pressure and that χ=1. The shear strength reduction ratio for PR-samples is therefore given by Equation 3.

$$\frac{\tau_u - \tau_s}{\tau_u} = \frac{(u_a - u_w)\tan\phi'}{c' + (\sigma_n - u_a)\tan\phi' + (u_a - u_w)\tan\phi} \qquad \text{Equation 3}$$

The variation of SSR with (σ_n-u_a) as modeled by Equation 3 for PR-samples using the strength values in Table 3 and an assumed u_a-u_w=50 kN/m^2 is compared with the laboratory PR curve in Figure 10. The results show that Equation 3 models the trends at low stress levels rather well. However at higher stress levels, whereas the test results indicate almost constant SSR values with increasing σ_n-u_a, the model shows continued reduction in SSR values.

CONCLUSIONS

Drained direct shear tests were conducted on both disturbed and undisturbed samples of completely decomposed phyllites under both soaked and unsoaked conditions for normal stresses varying between 20 kN/m^2 and 320 kN/m^2. The results suggest that there is a large reduction in direct shear strength for decomposed phyllites due to soaking and that this reduction is most severe at low normal stress levels, but appears almost independent of normal stresses for stresses larger than about 80kN/m^2. Remoulding of decomposed phyllites increases the propensity for strength reduction on soaking by about 20-30% for normal stresses in the range of 20-320kN/m^2. The unsaturated soil mechanics model based on the effective stress approach at high degrees of saturation appears to predict the strength loss trends very well only at low stress levels.

REFERENCES

Ampadu SIK 1998 A laboratory investigation of the effect of soaking on strength characteristics of local soils. *International Symposium on Problematic Soils, IS-TOHOKU '98, Sendai Japan 28-30 October*. A.A. Balkema, Rotterdam/Brookfield, 665-668 pp.

ASTM D 1557-91 Test Method for Laboratory Compaction Characteristics of Soil Using Modified Effort. *Annual Book of Standards*.

ASTM D1883-87 Standard Test Method for CBR (California Bearing Ratio) of Laboratory-Compacted Soils. *Annual Book of Standards*

Bishop AW 1959 The principle of effective stress. *Teknisk Ukeblad I Samarbeide Med Teknikk*, Oslow, Norway, 106(39), 859-863

Irfan TY 1988 Fabric variability and index testing of a granitic saprolite. *Proceedings of the Second International Conf. on Geomechanics in Tropical Soils*, Singapore, Vol. 1, pp. 25-35

Khalili N, Khabbaz, MH 1998 A unique relationship for χ for the determination of the shear strength of unsaturated soils. *Geotechnique* 48, No. 5, 681-687.

Lu N. and Likos WJ 2004, "Unsaturated soil mechanics", *John Wiley & Sons, Inc*.

TRRL 1999 Overseas Road Note 31 A guide to the pavement evaluation and maintenance of bitumen-surfaced roads in tropical and sub-tropical countries" *Transport Research Laboratory*, 1999,

Yoshitake A, Onitsuka K 1990 Factors Influencing Strength Parameters (c' and ϕ) Behaviour of Decomposed Granite Soil. *Residual Soils in Japan*, JSSFME Sugayama Bldg. 4F, Kanda Awaji-cho 2-23, Chiyoda-ku

Soil Stress-Strain Behavior: Measurement, Modeling and Analysis
Geotechnical Symposium in Roma, March 16 & 17, 2006

MATERIAL PROPERIES OF INTERMEDIATE MATERIALS BETWEEN CONCRETE AND GRAVELLY SOIL

Taro Uchimura, Yuko Kuramochi and Bach Thuan Thai
Department of Civil Engineering
University of Tokyo, 7-3-1 Hongo, Bunkyo-ku Tokyo 113-8656 Japan
e-mail: uchimura@civil.t.u-tokyo.ac.jp

ABSTRACT

Compaction and strength properties of cement-mixed well-graded gravel are studied. Such materials can also be considered as a kind of concrete materials which has much lower cement contents than usual. New concepts on material properties related to their mixture ratio of cement, gravel (aggregate) and water, as well as their compaction density, are proposed, unifying the concepts of geotechnical engineering and concrete engineering. For materials with higher cement contents, the compaction curve becomes flat, with lower maximum compaction density, and higher optimum water contents. The triaxial compressive strength are clearly affected by the dry density, as well as the cement contents.

1. INTRODUCTION

Concrete engineers have developed RCC (roller compacted concrete) and CSG (cemented sand and gravel) methods, which use mixture materials of aggregate, cement and water with lower cement contents than usual concrete. Such materials can be filled and mechanically compacted like soil structures. Figure 1 shows an example of dam constructed using CSG materials (Taiho Side Dam in Okinawa, Japan). Cheap and relatively low quality aggregates are used with lower cement contents (60 kg/m^3), without reinforcement. The stress in the dam body is dispersed to the wide base area of the trapezoidal cross-section.

On the other hand, geotechnical engineers have developed cemente-mixed well-graded gravels as a new backfill material, in order to construct more stable soil structures. Figure 2 shows an example of a bridge abutment for an express railway in Kyushu, Japan, constructed by using cement-mixed well-graded gravel with geogrid-reinforcement. As the cement-treated backfill can stand by itself without the support by RC wall even in a case of strong earthquake, the RC wall can be much simpler than conventional soil retaining structures without cement-mixing. Usually, cement-treatment technique is used for low quality soils such as high water content clay to improve their properties. Cement-mixing for high quality soils like well-graded gravel is a relatively new technique.

These materials are quite similar to each other in their quality of aggregate (or well-graded gravel), mixture ratio of cement and water. They could be classified in the same category of "intermediate between concrete and soils", while their concepts of mixture design, quality control and construction methods have different origin in concrete and geotechnical engineering respectively. This paper compares these materials, and reports some experimental results on their compaction and strength characteristics.

Hoe I. Ling et al. (eds.), Soil Stress-Strain Behavior: Measurement, Modeling and Analysis, 473–478.

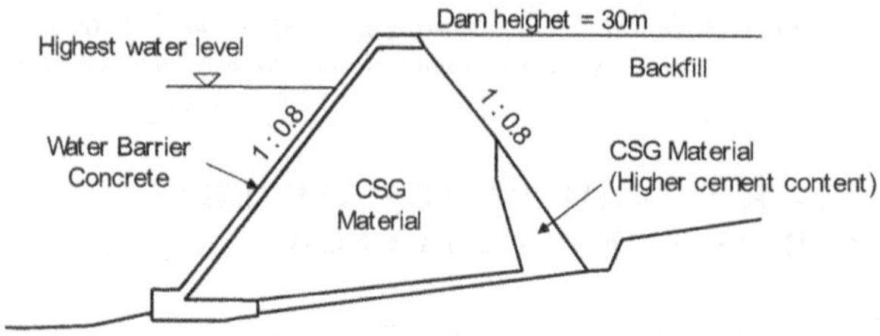

Figure 1. Example of CSG dam.

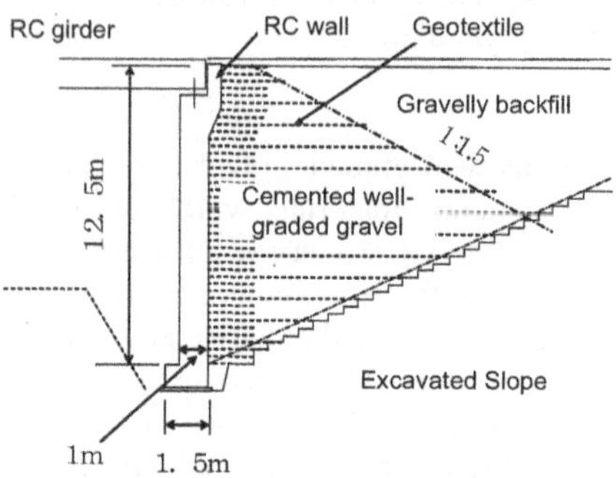

Figure 2. Bridge abutment using cement-mixed well-graded gravel.

2. COMPARISON WITH SOILS AND CONCRETE

Figure 3 compares the typical particle size distributions of concrete aggregate (Figure 3a, referring to JIS (Japanese Industrial Standards); coarse aggregate (JIS-A-5005) and fine sand (JIS-A-5001) are mixed with weight ratio of 1:1) and well-graded gravel (Figure 3b, so called M40 type in railway engineering). The particle size distribution of the backfill gravel used in the abutment shown in Figure 2 is also plotted in the both figures. Their particle size distributions are quite similar to each other. In addition, other indexes such as specific gravity, water absorption ratio, abrasion losses of the particles, etc., are also defined in the regulations for concrete aggregates and well-graded gravels respectively. Thus, it can be said that the aggregates for concrete and well-graded gravel for soil structures are the same materials.

Moreover, the cement contents of the CSG used in the dam in Figure 1 is 60 kg/m^3, the water contents is 100 to 120 kg/m^3, and the density is 2.3 to 2.4 ton/m^3 including the weight of the aggregate, water and cement. This corresponds to c/g = 2.7 % of cement ratio (weight of cement per weight of aggregate), w/g = 5 % of water contents (weight of water per weight of aggregate), ρ_d = 2.3 g/cm^3 of dry density (excluding weight of water) in geotechnical terminology. On the other hand, the indexes of the cement-mixed well-graded gravel used in the abutment in Figure 2 are c/g = 4 %, w/g = 4 % and ρ_d = 2.7 g/cm^3 (this high density is due to high specific gravity of the gravel of 3.03). The mixture ratio and density of these materials are similar to the CSG.

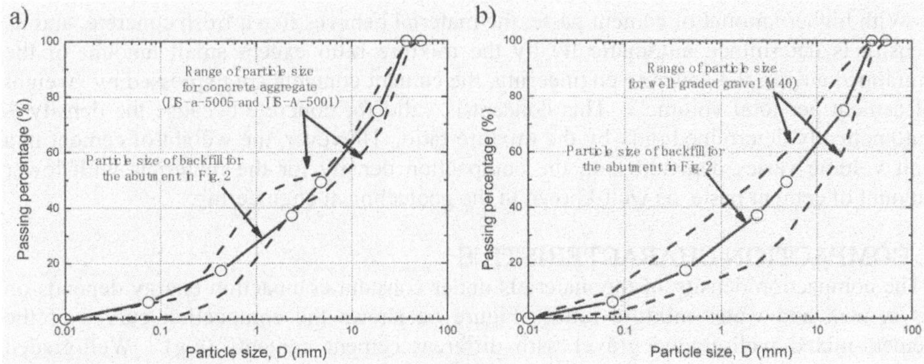

Figure 3. Particle size distributions of typical materials:
a) Concrete aggregate; b) Well-graded gravel

Thus, CSG (or RCC) and cement-mixed well-graded gravel can be considered to be the same materials intermediate between concrete and soils. Such materials should be evaluated using unified concepts of concrete engineering and geotechnical engineering. For example, the mixture ratio can be summarized as Figure 4. The lateral axis w/g (weight of cement per weight of aggregate) and the vertical axis c/g (weight of water per weight of aggregate) are in geotechnical terms. Then, each line passing the origin corresponds to a water-cement ratio (w/c), which is an important index of concrete engineering. Examples of several materials between soil and concrete are indicated in the figure. We can compare the materials in a unified way by this chart, while their particle size of the gravel (or aggregate) may be different.

Further, the density of the material can be summarized as Figure 5. This figure is an example showing a case along the line of w/c = 50 % in Figure 4. The vertical axis is the dry density, exclude the cement weight. With lower amount of cement paste (i.e. c/g and w/g), the material behaves like a soil, that is, there are maximum possible density (zero void) curve and minimum possible density curve. The compaction curve is between them.

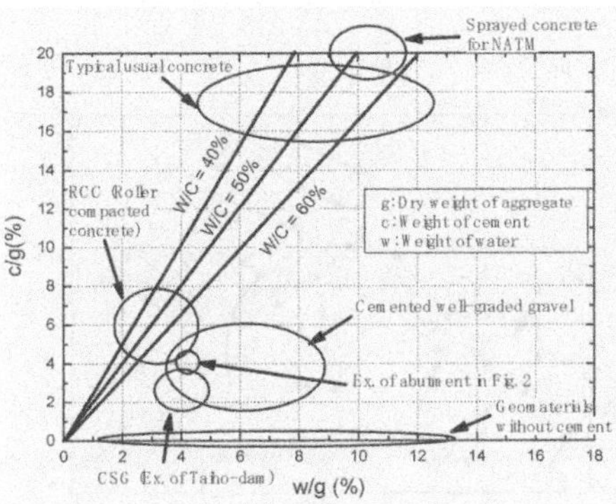

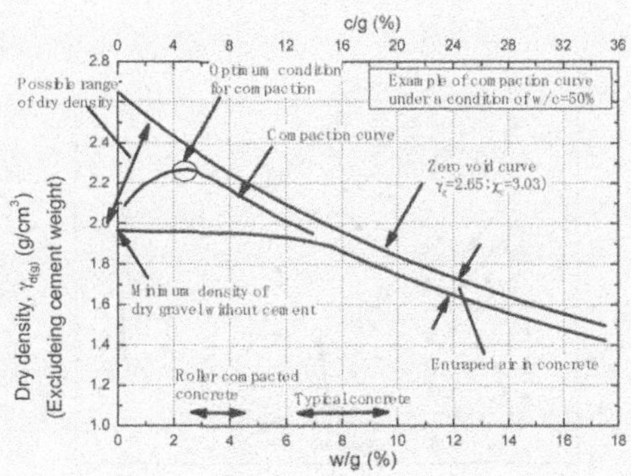

Figure 4. Chart of mixture ratio of various materials with cement.

Figure 5. Schematic diagram of density of materials containing cement.

With higher amount of cement paste, the material behaves like a fresh concrete, and its density is determined automatically by the mixture ratio except small amount of the entrapped air. In the concrete engineering, the cement contents are expressed by "weight of cement per total volume". This concept is valid for concrete because the density is automatically determined only by the mixture ratio. However, the weight of cement in a unit volume varies depending on the compaction density for the materials with lower amount of cement paste, as well-known in the geotechnical engineering.

3. COMPACTION CHARACTERISTICS

The compaction density of the materials under constant compaction energy depends on its cement and water mixture ratio. Figure 6a shows the compaction curves of the cement-mixed well-graded gravel with different cement contents (c/g). Well-graded gravel ($Gs = 2.71$, $D_{max} = 10mm$, $D_{50} = 2.03mm$, $Uc = 15.8$, fine contents = 4.3 %) and standard portland cement were used. The compaction energy is 4.56 MN/m^3 (testing method type A in Japanese standard). The vertical axis of Figure 6a is dry density excluding the cement weight. For materials with higher cement contents, the compaction curve becomes flat, with lower maximum compaction density, and higher optimum water contents. It may be natural that the dry density in Figure 6a is lower with higher cement contents, because some part of volume is occupied by cement in the material. So, Figure 6b compares the compaction curves with the dry density including the cement weight for the vertical axis. The compaction density is similar for the cement contents lower than 4 %, but it dropped with higher cement contents. In addition, the zero void curve moves upward with higher cement contents because the specific gravity of cement is higher than that of gravel. Thus we can conclude that higher cement contents results in lower compaction efficiency.

In order to find out the reason for this result, a material containing 12 % by weight of kaolinite instead of cement was tested. As seen in the Figure 6a and b, the compaction curve with kaolinite is similar to that with 12 % of cement. This means that the lower compaction efficiency with high cement contents is not due to specific properties of cement, but due to the large change in the particle size distribution of the material with high fine contents of cement powder.

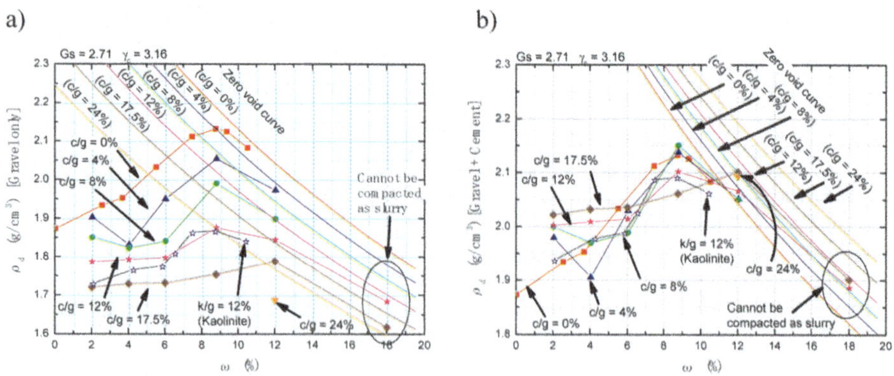

Figure 6. Compaction curves of well-graded gravel with various cement contents. Cement weight is a) excluded; and b) included in the dry density in vertical axis.

4. STRENGTH CHARACTERISTICS:

Triaxial specimens of cement-mixed well-graded gravel were tested to measure their compressive strength with various cement and water contents. The density of each specimen was determined based on the compaction test result mentioned above under constant compaction energy. The specimen was 50 mm in diameter and 100 mm in height, and cured for 7 days in the mold. The effective confining pressure was constant of 20 kPa, and the loading strain rate was 0.03 %/min.

Figure 7 shows the compressive strength of the specimens with each mixture ratio and their contours. It is clear that higher cement contents results in higher strength. Comparing the specimens with the same cement contents, there is optimum water content which gives the highest strength, and this water contents becomes higher with higher cement contents. It is a common knowledge in concrete engineering that lower water-cement ratio (w/c) results in higher the strength of concrete. But, water-cement ratio is not a dominant factor to determine the strength of concrete with lower cement contents.

Figure 8 shows the contours of the dry density obtained from compaction tests with various cement contents and water contents already shown in Figure 6a. The optimum water contents for density for every cement contents have similar trend as the optimum water contents for strength shown in Figure 7. Thus, we conclude that the compaction density of the cement-mixed well-graded gravel, as well as the cement content, is important factor to determine the compressive strength of the specimen.

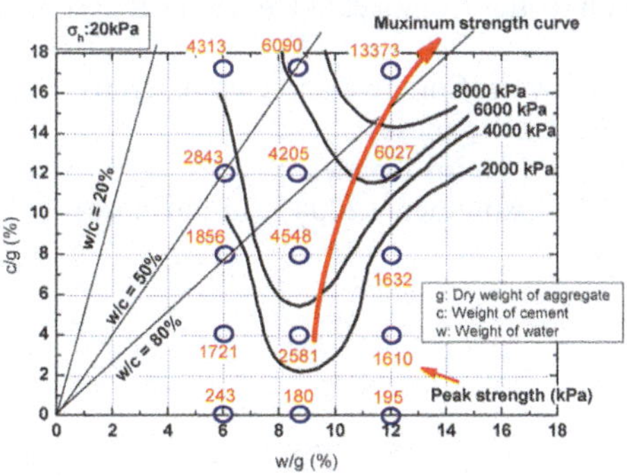

Figure 7 Contour of triaxial strength of cement-mixed well-graded gravel.

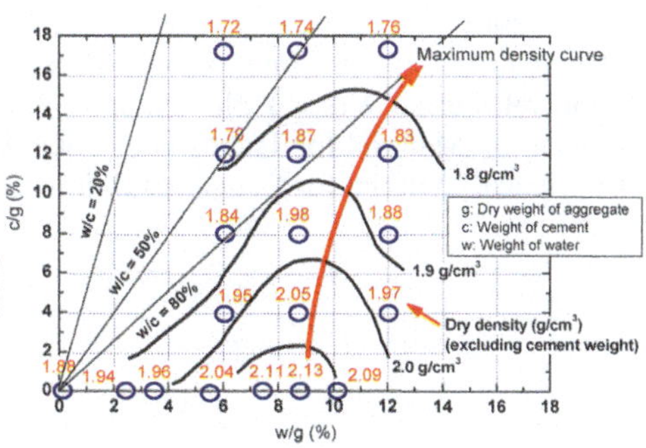

Figure 8 Contour of compaction density of cement-mixed well-graded gravel.

5. CONCLUSIONS

The basic properties of cement-mixed well-graded gravel and low cement content concrete are discussed. Each material has been developed by geotechnical engineers and concrete engineers respectively, but they are quite similar to each other. New concepts were proposed to evaluate the basic indexes of such materials, such as water contents, cement contents, water cement ratio, dry density etc., unifying the concepts of geotechnical and concrete engineering.

The density of compacted cement-mixed gravel under constant compaction energy largely depends on the water contents and cement contents. For materials with higher cement contents, the compaction curve becomes flat, with lower maximum compaction density, and higher optimum water contents. This is probably due to the large change in the particle size distribution of the material with high fine contents of cement powder.

The strength of the cement-mixed well-graded gravel was examined by triaxial tests. The results showed a common-sense fact that higher cement content results in higher strength. However, it was also found that, if the cement mixture ratio to gravel by weight is the same, the highest strength is obtained at the optimum water contents which give the maximum compaction density. This seems to be reasonable from geotechnical viewpoints, but it also means that the "water-cement ratio" which is the most important parameter of concrete engineering controlling the strength and quality of concrete is not dominant for such materials.

ACKNOWLEDGEMENT

The authors acknowledge the cooperation by "Railway Technical Research Institute" for providing the materials, "Japan Railway Construction, Transport and Technology Agency" for providing information on the abutment in Kyushu, "Maeda-MitsuiSumitomo-Ooshiro JV" for providing information on the CSG dam in Okinarwa, and Prof. Maekawa, University of Tokyo, for advisory discussion.

REFERENCES

1) Nagayama, I. and Jikan, S. (2003) : 30 Years History of Roller-compacted Concrete Dam in Japan, 4th International symposium on Roller Compacted Concrete (RCC) Dam, 2003.11.
2) Uchimura, T. (2005): Intermediate Materials between Concrete and Geomaterials, Concret Journal, Vol. 43, No. 10, pp. 3-8, 2005. (in Japanese)
3) Technical Material for Trapezoidal CSG Dam (2003). (in Japanese)
4) Technical Material for Cemented Reinforced Soil Abutment (2004). (in Japanese)

Soil Stress-Strain Behavior: Measurement, Modeling and Analysis
Geotechnical Symposium in Roma, March 16 & 17, 2006

COMPACTION-INDUCED ANISOTROPY IN THE STRENGTH AND DEFORMATION CHARACTERISTICS OF CEMENT-MIXED GRAVELLY SOILS

Lalana Kongsukprasert[1], Yusuke Sano[2] and Fumio Tatsuoka[3]
[1]Post Doctoral Fellow, Department of Civil Engineering, Tokyo University of Science
2641 Yamazaki, Noda, Chiba prefecture 278-8510, Japan
e-mail: lalana@rs.noda.tus.ac.jp
[2]Graduate student, Department of Civil Engineering, Tokyo University of Science
2641 Yamazaki, Noda, Chiba prefecture 278-8510, Japan
e-mail: j7605614@ed.noda.tus.ac.jp
[3]Professor, Department of Civil Engineering, Tokyo University of Science
2641 Yamazaki, Noda, Chiba prefecture 278-8510, Japan
e-mail: tatsuoka@rs.noda.tus.ac.jp

ABSTRACT
The effects of inherent anisotropy produced by compaction on the strength and deformation characteristics of unsaturated specimens of compacted cement-mixed gravelly soil were evaluated by performing a series of consolidated drained (CD) triaxial compression (TC) test. Specimens were prepared by compacting in the direction of compression in the TC tests and its perpendicular direction. It is shown that these two types of specimen exhibited nearly the same stiffness at small strain and viscous property. The effects of compaction in the two orthogonal directions increase with an increase in the strain with noticeable effects on the peak strength and more in the post-peak regime. The effects are however generally small, much less significant than those observed with air-pluviated unbound sand. The effects of total compacted dry density of solid and curing period on the stress-strain behaviour were found nearly the same between the specimens compacted in the two orthogonal directions.

1. INTRODUCTION
The deformation and strength characteristics of compacted cement-mixed gravelly soil are evaluated usually by unconfined compression tests and occasionally by drained triaxial compression tests in which the specimens are compressed in the direction of compaction. However, the deformation and strength characteristics could be somehow anisotropic due to a compaction-induced anisotropic structure. This issue has not been systematically investigated so far mainly because cement-mixed soil has been used mostly for secondary structures, such as the road base. On the other hand, when used for more critical structures that require a high stability while allowing a limited amount of instant and residual deformation, such as bridge abutments (Tatsuoka, 2004), a more detailed evaluation of the strength and deformation characteristics becomes necessary to perform a realistic analysis of the field full-scale structural behaviour when subjected to

Hoe I. Ling et al. (eds.), Soil Stress-Strain Behavior: Measurement, Modeling and Analysis, 479–490.

multi-directional loadings at different times. In this respect, possible effect of compaction-induced anisotropy is one of the important issues.

In view of the above, a series of consolidated drained (CD) triaxial compression (TC) tests were performed on unsaturated specimens of cement-mixed gravelly soil compacted in two orthogonal directions, the direction of compression loading and its perpendicular direction.

2. TESTING METHODS

Materials and specimen preparation

A well-graded quarry gravelly soil of crushed sandstone (a so-called 'Chiba gravel') was sieved to remove particles larger than 10 mm. The obtained test material is named model Chiba gravel (G_s= 2.74; Kongsukprasert et al., 2005). During a long period of the study (about 6 years), several different batches of this gravelly soil having slightly different grain size distributions and index parameters (Fig. 1a) were obtained at different times. The name of the batch used in the respective test is listed in Table 1 (and Tables 2 and 3 shown later). The same batch of Ordinary Portland cement with G_s = 3.16 was used for all the tests.

Table 1 Specimen specifications and part of the results of CD TC tests to evaluate effects of compaction-induced anisotropy.

Com. 1)	Batch No	Test name	w_i (%)	c/g (%)	ρ_d (g/cm^3) 2)	t_{ini} (day) 3)	σ_h' (kPa)	Loading condition	q_{max} (kPa)	ε_a at q_{max} (%)	Moduli (GPa) E_{50} 4)	E_0 5)	E_0/q_{max}
Horizontal	5	IS009	8.752	2.499	2.000	6.97	19.8	ML	1651	0.2052	3.09	5.37	3.25E+03
	3	IS001	8.747	2.500	2.000	7.13	19.7	ML	1668	0.195	3.35	5.92	3.55E+03
		IS002	8.749	2.500	2.000	7.13	19.7	With creep	2008	0.182		6.05	3.01E+03
		IS003	8.750	2.501	2.000	7.08	19.7	Stepwise 6)	1609	0.225	2.87	5.26	3.27E+03
		IS004	8.756	2.500	2.000	7.12	49.2	ML	1675	0.284	3.04	4.96	2.96E+03
Vertical	4	RD003	8.745	2.500	2.000	6.97	19.7	ML	1917	0.319	2.62	5.81	3.03E+03
	1	J023	8.746	2.500	2.000	7.04	19.8	ML	1988	0.312	3.19	5.33	2.68E+03
	2	JA012	8.750	2.502	2.000	7.04	19.7	ML	1983	0.275	2.55	5.29	2.67E+03
		JA013	8.755	2.502	2.000	7.07	19.8	With creep	2126	0.266		5.70	2.68E+03
	3	SP009	8.749	2.501	2.000	7.03	19.7	ML	1713	0.353	2.24	4.37	2.55E+03
		SP005	8.752	2.501	2.000	6.84	49.2	Stepwise	1966	0.670		4.65	2.37E+03
		SP008	8.748	2.500	2.000	7.09	19.6	Stepwise	1909	0.516		5.71	2.99E+03

1) Direction of compaction
2) Nominal dry density at compacted
3) Initial curing period under unconfined condition
4) Secant modulus defined at $q = q_{max}/2$
5) Initial elastic Young's modulus defined at axial strain less than 0.001%
6) Strain rate changed stepwise during otherwise ML

Fig. 1b shows the compaction curves for batch Nos. 1 and 5 with and without cement-mixing (a cement/gravel ratio by dry weight, c/g = 2.5 and 4.0 %, for batch Nos. 1 and 5, respectively) obtained by using compaction energy $E0$ = 550 kJ/m^3. The compaction curves of the two batched are only slightly different with the optimum water contents, w_{opt}, equal to 8.75 % and 8.85 % for batch Nos. 1 and 5, respectively (Fig. 1b).

The TC specimens were prepared by mixing the gravelly soil with Ordinary Portland cement and water at c/g = 2.5 % and an initial water content at compaction, w_i, equal to the optimum, w_{opt}= 8.75 %. So-called *vertical* specimens were prepared by compacting manually in a rectangular prismatic mould (95 mm x 95 mm x 190 mm) in the direction in parallel to the direction of triaxial compression loading in even five layers, while so-called *horizontal* specimens were by compaction in the direction perpendicular to the direction of compression loading in even four layers (see Fig. 3d). The compaction energy was controlled to attain the respective target nominal dry density of solid, ρ_d. The specimens were cured inside the mould under the atmospheric pressure with constant water content (= w_i) for five days. When the initial curing period (defined zero at the start of compaction), t_{ini}, was greater than five days, the specimen was taken out from the mould, then wrapped with a plastic wrapping sheet to be further cured under the atmospheric pressure and a constant water content.

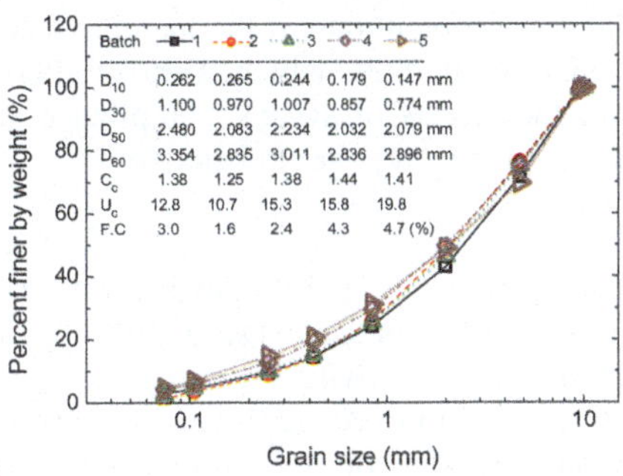

a)

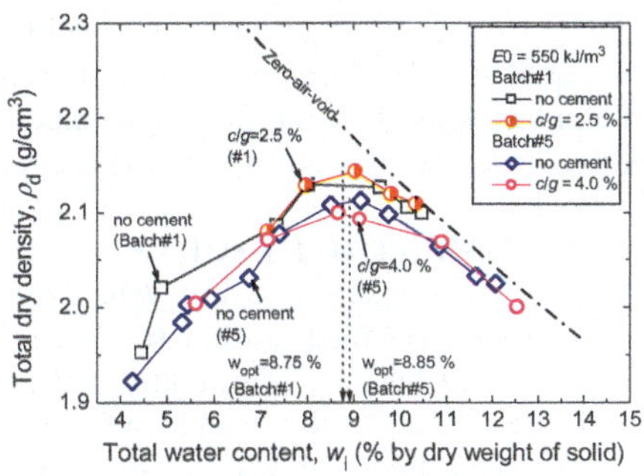

b)

Fig.1 a) Gradation curves of different batches; and b) compaction curves of batch Nos. 1 and 5 with and without cement-mixing.

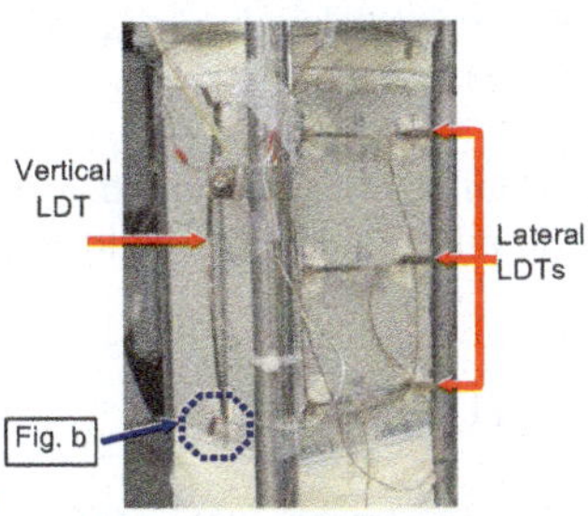

a)

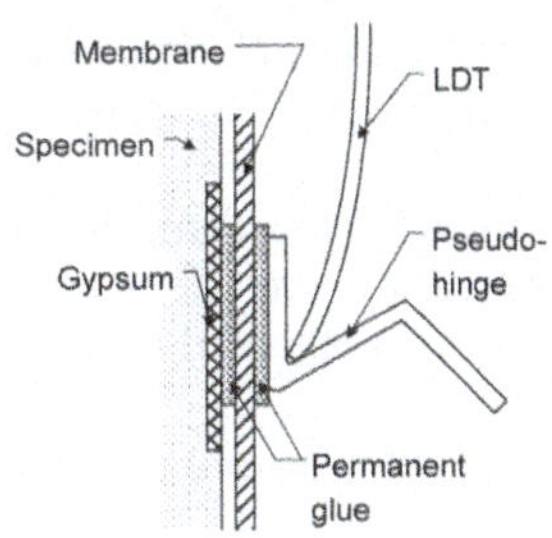

b)

Fig. 2 a) Positioned Local Displacement Transducers (LDTs); and b) details of the hinge to pinch a LDT (Kongsukprasert et al., 2005).

Triaxial compression tests

CD TC tests were performed using a triaxial testing system allowing automated stress path control (Santucci de Magistris et al., 1999). Axial strains were sensitively and accurately measured by using a pair of 160 mm-long longitudinal LDTs (Goto et al., 1991) set on the opposite side faces of the specimen (Fig. 2). Rectangular prismatic specimens were used, instead of cylindrical ones, to alleviate the problem of significant effects of membrane penetration and bedding error on the lateral strains when the effective confining pressure changes (Hayano et al., 1997): i.e., lateral strains were locally measured with three pairs of 70 mm-long lateral LDTs set at 5/6, 3/6 and 1/6 of the specimen height on the opposite specimen side faces (Fig. 2). Moreover, it is much easier to prepare *vertical* and *horizontal* specimens having similar relationships between the bedding planes and the specimen dimensions with a rectangular prismatic shape than with a cylindrical one.

The specimen, which was unsaturated at w_i, was isotropically consolidated to a specified confining pressure (σ_h') (applied by means of partial vacuum) and left for one hour before the start of drained TC at constant σ_h'. At a number of arbitrary stress states during otherwise monotonic loading (ML), sustained loading for 65 hours, step changes in the axial strain rate and one or five small unload/reload cycles with a single axial strain amplitude of about 0.005 % to evaluate elasticity at each stress state were applied. The nominal basic axial strain rate (obtained from the displacement rate of the loading piston) was equal to 0.03 %/min. Locally measured axial strain rates, which were smaller to different extents than the nominal values due to the effects of system compliance and bedding error, were used to evaluate the rate effects on the stress-strain behaviour.

3. EXPERIMENTAL RESULTS

Anisotropy in the strength and deformation characteristics: Fig. 3a shows results from CD TC tests performed on four *vertical* and two *horizontal* specimens (Fig. 3d) prepared using different batches of gravelly soil performed under otherwise the same conditions (see Table 1). Part of the results was reported by Kongsukprasert and Tatsuoka (2005). It may be seen that the effects of using different batches on the test results were insignificant, if any. Fig. 3b shows the equivalent Young's modulus, E_{eq}, from a single unload/reload cycle with a single axial strain amplitude of about 0.005 % applied during otherwise ML, plotted against diaviator stress, q. Fig. 3c shows the relations between the normalized tangent modulus, E_{tan}/E_0, and the normalized shear stress level, q/q_{max}, where E_0 is the initial elastic Young's modulus defined at axial strains less than 0.001 %.

To obtain the test results presented in Fig. 3a and other similar ones presented in this paper, it was assumed that the lateral strain was isotropic in all the lateral directions and the volumetric strain was obtained as follows (method A):

$$\varepsilon_{vol} = \varepsilon_{axial} + 2 \times \varepsilon_{lateral} \text{ (measured with lateral LDTs)} \qquad (1)$$

where $\varepsilon_{lateral}$ (measured with lateral LDTs) is the average local lateral strain from the readings of the six lateral LDTs. This assumption is relevant with the *vertical* specimens when deforming rather uniformly, in particular before the development of shear band(s) in the post-peak regime. With the *horizontal* specimens, this assumption should have become more inadequate as the axial strain became larger, in particular after the specimens started dilating in the post-peak regime. This inference is based on the observation such that, in Fig. 3a, the *horizontal* specimens exhibit dilation rates that are significantly larger than those with the *vertical* specimens. It seems that this trend is due to that soil particles were separated more easily in the lateral direction during TC (i.e., in the direction of compaction) because of the trend of soil particles to be aligned in the direction of the bedding planes, in particularly in the zones closer to the lift joints.

Therefore, it is likely that, with the *horizontal* specimens, after the start of dilation, the absolute values of $\varepsilon_{lateral}$ in the direction of compaction, which were measured with the lateral LDTs, were much larger than those in the direction perpendicular to the direction of compaction (i.e., in parallel to the bedding planes). Then, the true volumetric strain may be in between the two extreme values evaluated by Eq. 1 (method A) and Eq. 2 (assuming that $\varepsilon_{lateral}$= 0 in the direction in parallel to the bedding planes; method B):

$$\varepsilon_{vol} = \varepsilon_{axial} + \varepsilon_{lateral} \text{ (measured with lateral LDTs)} \qquad (2)$$

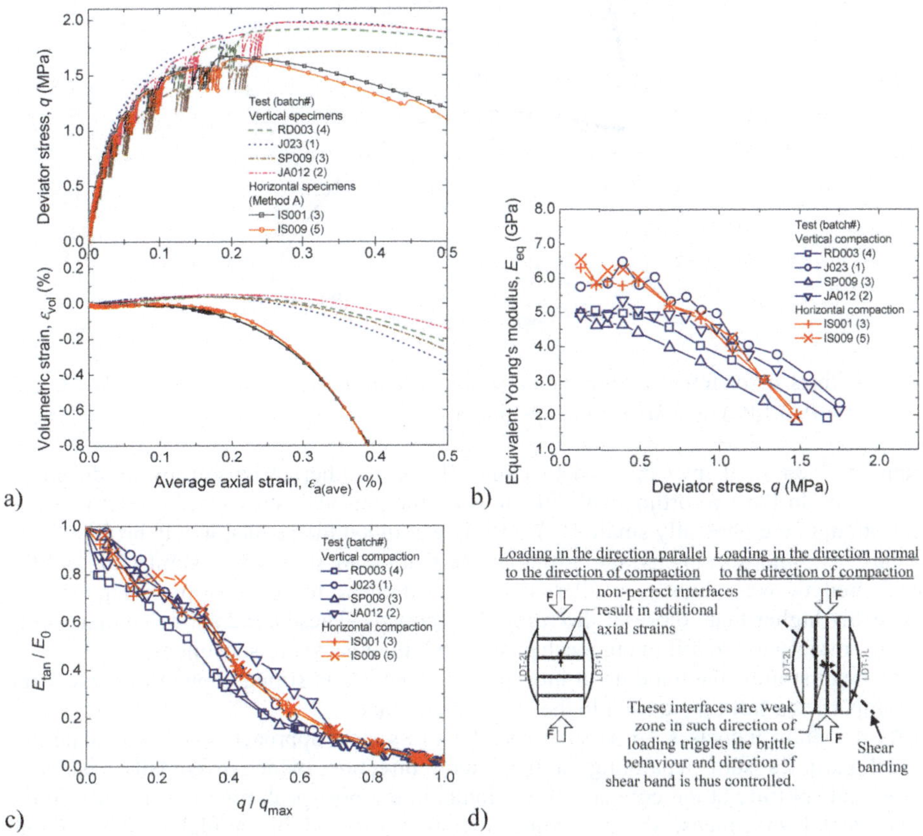

Fig. 3 Effects of compaction-induced anisotropy: a) stress-strain relations from ML TC tests (σ_h' = 19.7 kPa) on *vertical* and *horizontal* specimens; b) E_{eq} - q relations; c) E_{tan}/E_0 – q/q_{max} relations; and d) illustration of *vertical* and *horizontal* specimens (after Kongsukprasert and Tatsuoka, 2005).

Fig. 4 shows the results of a typical *horizontal* specimen in which the ε_{vol} values were obtained based on method A (Eq. 1) and method B (Eq. 2). The differences between the deviator stresses calculated using the cross-sectional areas of the specimen obtained based on these two methods are negligible even at the end of test, which is due to that the largest axial strain was small, only 0.5 %.

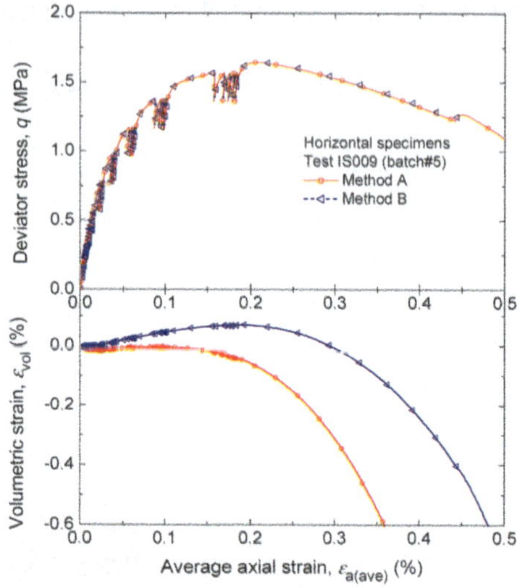

Fig. 4 Effects of different assumptions to obtain lateral strains in a typical CD TC test (σ_h'= 19.7 kPa) on a *horizontal* specimen.

From Figs. 3a~d, it may be seen that the effects of inherent anisotropy produced by compaction in the two orthogonal directions on the pre-peak stress-strain behaviour and peak strength are generally small. By carefully examined the results, it may be noted that, of the *horizontal* specimens, the initial stiffness at small strains was similar to, and the peak strength was slightly smaller and the brittleness in the post-peak regime was noticeably higher than, those of the *vertical* specimens. These trends of behaviour would be due possibly to the following mechanisms with the *horizontal* specimens:

1) At small strains, the bonding at inter-particle contacts is strong enough to prevent the soil particles from separating in the lateral directions.
2) As the strain increases, in particular as the stress state approaches the peak state, the soil particles start separating in the lateral direction, which eventually results in vertical splitting at the compaction lift joints in the post-peak regime (Fig. 3d). In the horizontal specimens, shear banding always developed in parallel to the bedding planes, triggered by splitting along the compaction lift joints. With the *vertical* specimens, on the other hand, shear banding develops in arbitrary directions.

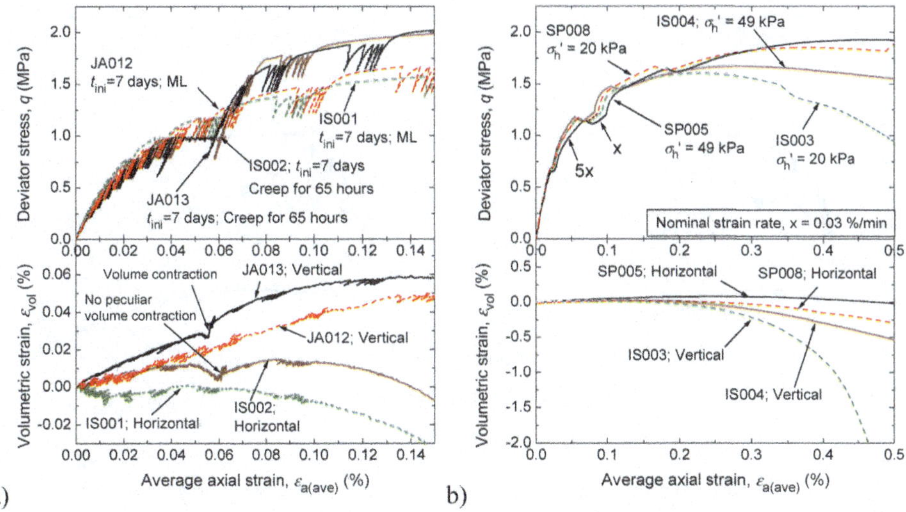

Fig. 5 Effects of compaction-induced anisotropy on; a) sustained loading and subsequent stress-strain behaviour at σ_h' = 19.8 kPa; and b) viscous behaviour at two σ_h' s.

Anisotropy in ageing and viscous effects: Fig. 5a shows results from two pairs of CD TC tests. In the first pair, one *horizontal* specimen (test IS001) and one *vertical* one (test JA012) were subjected to continuous ML. In the second pair, one *horizontal* specimen (test IS002) and one *vertical* one (test JA013) were subjected to drained sustained loading at q = 1 MPa for 65 hours during otherwise ML. It may be seen that the *horizontal* and *vertical* specimens exhibit similar creep deformation and similar effects of sustained loading on the subsequent stress-strain behaviour: i.e., nearly elastic behaviour for a large stress range upon the restart of ML at a constant strain rate, followed by an increased strength compared with the one observed in the first pair of test (without an intermediate sustained loading stage.

Fig. 5b shows results from a set of CD TC tests performed on *horizontal* specimens (tests IS003 and IS004) and *vertical* ones (tests SP005 and SP008) at σ_h' = 19.8 kPa and 49 kPa. The strain rate was stepwise changed at arbitrary stress states during otherwise ML. The relative strength and stiffness between the *horizontal* and *vertical* specimens at these two σ_h' values is nearly the same. The following other trends of behaviour, which are nearly the same with the *horizontal* and *vertical* specimens, may be seen:

1. The initial stiffness is rather insensitive to this change in σ_h', while the effects of σ_h' increase with an increase in the strain. The brittleness in the post-peak regime noticeably decreases with an increase in σ_h'.
2. The volumetric change becomes more contractive with an increase in σ_h'.
3. Due to the viscous property of the material, the deviator stress exhibits a noticeable jump upon a step change in the strain rate.

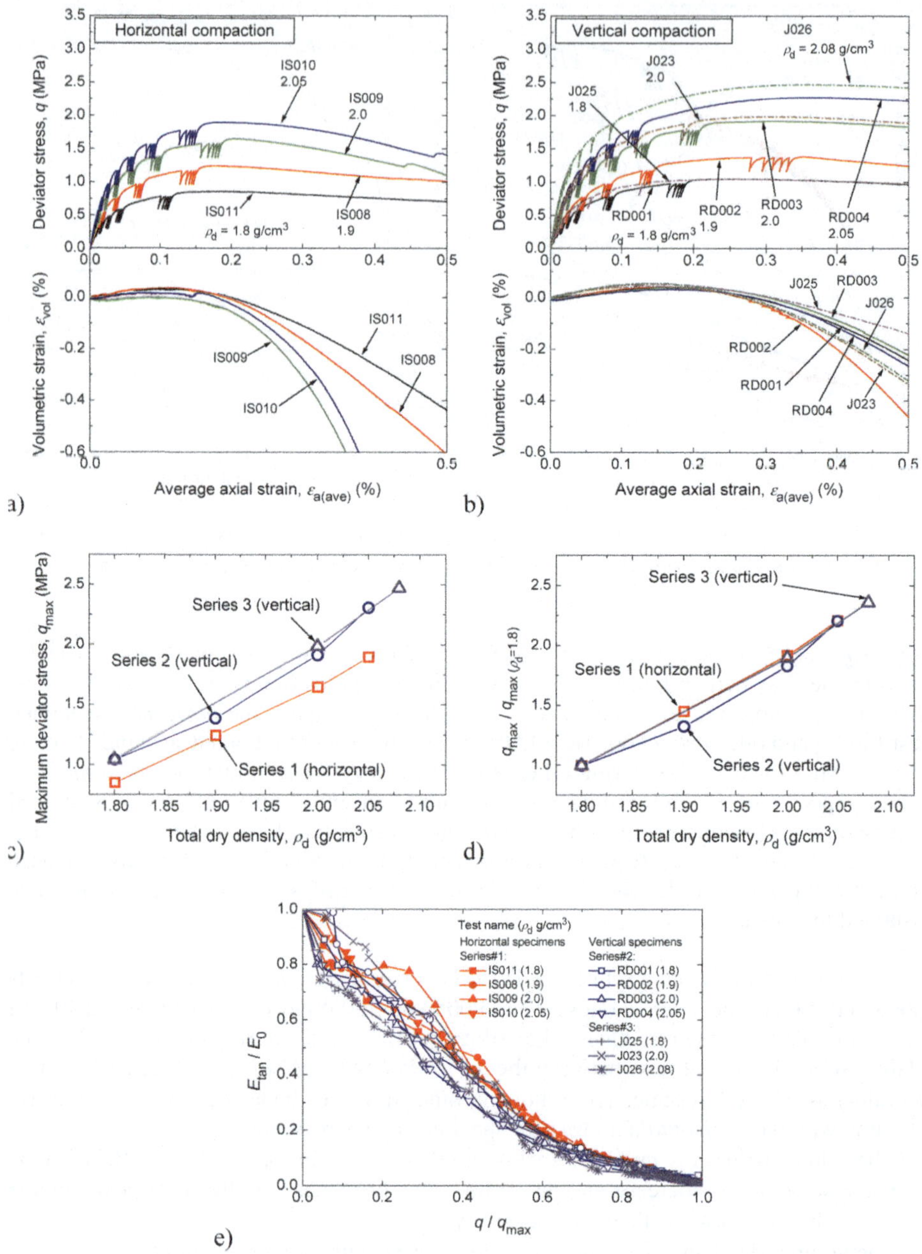

Fig. 6 Effects of dry density: a) *horizontal* specimens (series 1); b) *vertical* specimens (series 2 & 3); c) $q_{max} - \rho_d$ relations; d) $q_{max}/q_{max}(\rho_d=1.8) - \rho_d$ relations; e) and $E_{tan}/E_0 - q/q_{max}$ relations, CD TC tests at σ_h' = 19.8 kPa.

Effects of dry density

The effects of total compacted dry density, ρ_d, on the stress-strain behaviour of the *vertical* and *horizontal* specimens were compared (Table 2). Figs. 6a and 6b show the stress-strain relations from CD TC tests on *horizontal* and *vertical* specimens compacted to different ρ_d values between 1.8 and 2.1 g/cm^3 under otherwise the same conditions. Figs. 6c and 6d summarize the peak strengths, where $q_{max}(\rho_d=1.8)$ is the peak strength when ρ_d = 1.8 g/cm^3. Fig. 6e shows the normalized relationships between E_{tan}/E_0 and $q/q_{max.}$ It may be seen that, despite that the *vertical* specimens are always slightly stiffer and stronger than the *horizontal* ones, the pre-peak stiffness and linearity and the q_{max} value as well as the degree of dilatancy increase with an increase in ρ_d in a very similar way with the *vertical* and *horizontal* specimens.

Table 2 Specimen specifications and part of the results of CD TC tests to evaluate effects of dry density.

Series	Com	Batch No	Test name	w_i (%)	c/g (%)	ρ_d (g/cm^3)	t_{ini} (day)	σ_h' (kPa)	Loading condition	q_{max} (kPa)	ε_a at q_{max} (%)	Moduli (GPa)		E_0/q_{max}	$q_{max}/q_{max,1.8}$ [1)]
												E_{50}	E_0		
1	Horizontal	5	IS011	8.754	2.496	1.800	6.98	19.8	ML	859	0.180	1.91	3.59	4.18E+03	1.000
			IS008	8.756	2.500	1.900	7.14	19.8	ML	1250	0.177	2.68	5.04	4.03E+03	1.454
			IS009	8.752	2.499	2.000	6.97	19.8	ML	1651	0.205	3.09	6.13	3.71E+03	1.921
			IS010	8.751	2.500	2.050	7.02	19.8	ML	1901	0.186	3.99	7.58	3.99E+03	2.212
2	Vertical	4	RD001	8.741	2.500	1.800	7.08	19.8	ML	1048	0.267	1.62	3.18	3.04E+03	1.000
			RD002	8.746	2.501	1.900	6.98	19.7	ML	1391	0.351	2.00	3.81	2.74E+03	1.327
			RD003	8.745	2.500	2.000	6.97	19.7	ML	1917	0.319	2.62	5.81	3.03E+03	1.830
			RD004	8.747	2.501	2.050	7.02	19.7	ML	2312	0.548	3.00	6.66	2.88E+03	2.207
3		1	J025	8.748	2.500	1.800	7.04	19.6	ML	1048	0.257	2.24	3.93	3.75E+03	1.000
			J023	8.746	2.500	2.000	7.04	19.8	ML	1988	0.312	3.19	5.33	2.68E+03	1.896
			J026	8.742	2.505	2.080	7.04	19.8	ML	2475	0.355	4.53	9.41	3.80E+03	2.361

[1)] Normalised by the value obtained from the specimen having ρ_d = 1.8 g/cm^3, under otherwise the same conditions.

Effects of initial curing period

One set of *horizontal* specimens (series 4); and two sets of *vertical* specimens (series 5 and 6) were prepared to evaluate the effects of initial curing period, t_{ini}, on the strength and deformation characteristics. The specimens were compacted to ρ_d = 2.0 g/cm^3 at w_i = 8.75 % and cured under unconfined conditions for various periods: namely, t_{ini} (defined as "the period that has elapsed since the start of compaction until the start of TC loading) = 3, 7 and 15 days (series 4); t_{ini} = 1, 3, 7 and 14 days (series 5); and t_{ini} = 7, 14 and 30 days (series 6; see Table 3). Figs. 7a and 7b show the results from the CD TC tests. Figs. 7c and 7d summarize the peak strengths, where $q_{max}(t_{ini}=7)$ is the peak strength when t_{ini}= 7 days. The additional time that elapsed until the peak stress state during TC loading was on the order of 30 ~ 35 minutes, of which the effects could be considered negligible, if any. Fig. 7e shows the normalized relationships between E_{tan}/E_0 and q/q_{max}. It may be seen from these figures that the *vertical* specimens are always slightly stiffer and stronger than the *horizontal* ones, whereas the pre-peak stiffness and linearity and the q_{max} value as well as the degree of dilatancy increase with an increase in t_{ini} in a very similar way with the *vertical* and *horizontal* specimens.

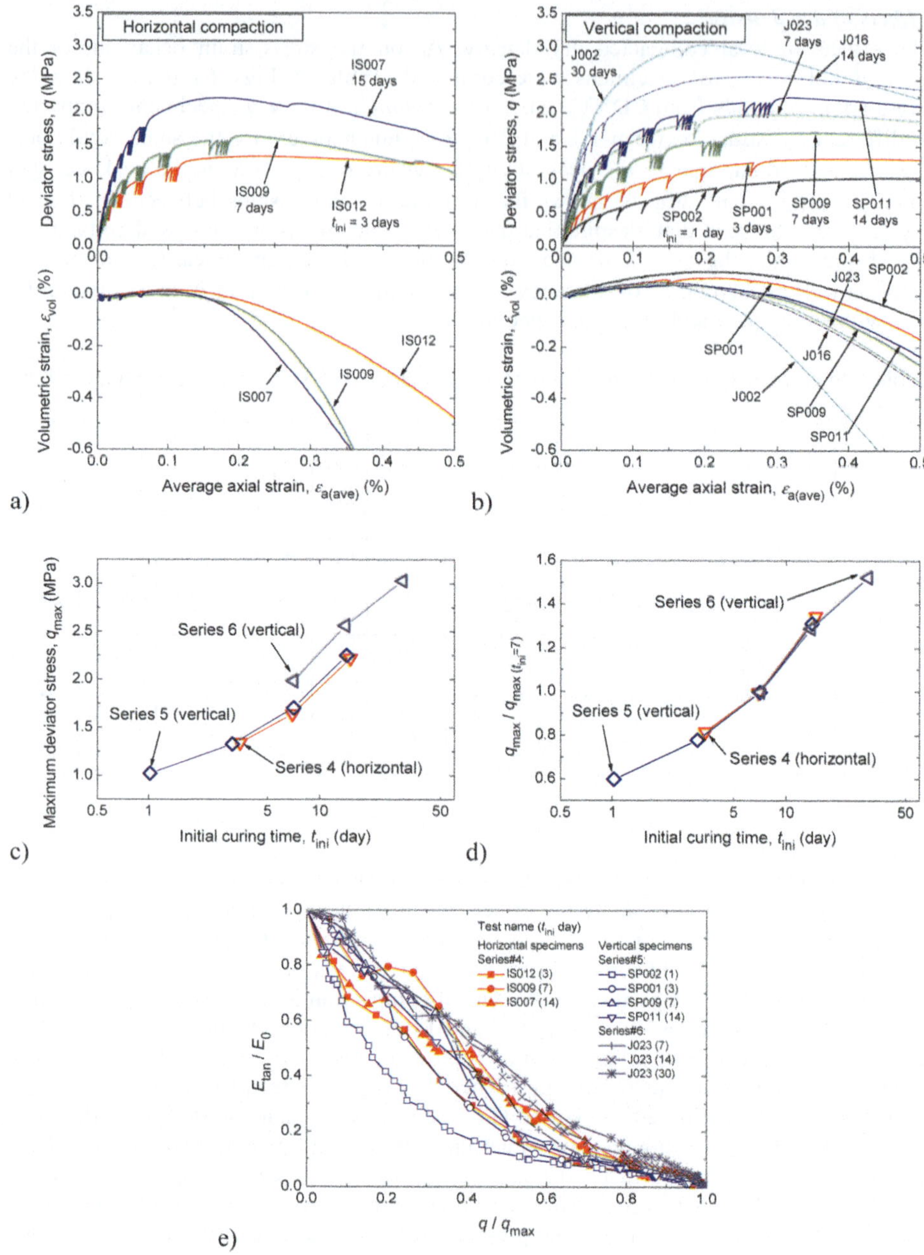

Fig. 7 Effects of curing period: a) *horizontal* specimens (series 4); b) *vertical* specimens (series 5 & 6); c) $q_{max} - \rho_d$ relations; d) $q_{max}/q_{max}(t_{ini}=7) - \rho_d$ relations; e) and $E_{tan}/E_0 - q/q_{max}$ relations, CD TC tests at σ_h'= 19.8 kPa.

Table 3 Specimen specifications and part of the test results from CD TC tests to evaluate effect of initial curing period.

Series	Com	Batch No	Test name	w_i (%)	c/g (%)	ρ_d (g/cm³)	t_{ini} (day)	σ_h' (kPa)	Loading condition	q_{max} (kPa)	ε_a at q_{max} (%)	Moduli (GPa)		E_0/q_{max}	$q_{max}/q_{max,7}$ [1)]
												E_{50}	E_0		
4	Horizontal	5	IS012	8.753	2.498	2.000	3.38	19.8	ML	1354	0.226	3.13	7.22	5.33E+03	0.820
			IS009	8.752	2.499	2.000	6.97	19.8	ML	1651	0.2052	3.09	6.13	3.71E+03	1.000
			IS007	8.757	2.499	2.000	14.94	19.8	ML	2228	0.170	5.21	9.08	4.07E+03	1.350
			IS016	8.759	2.490	2.000	18.87	19.8	ML	2277	0.155		7.36	3.23E+03	1.379
5	Vertical	3	SP002	8.748	2.500	2.000	1.01	19.8	ML	1031	0.489	0.81	2.71	2.63E+03	0.602
			SP001	8.739	2.500	2.000	3.09	19.8	ML	1337	0.393	1.49	3.23	2.42E+03	0.780
			SP009	8.749	2.501	2.000	7.03	19.7	ML	1713	0.353	2.24	4.37	2.55E+03	1.000
			SP011	8.750	2.500	2.000	14.10	19.7	ML	2256	0.319	3.13	6.04	2.68E+03	1.317
6		1	J023	8.746	2.500	2.000	7.04	19.8	ML	1988	0.312	3.19	5.33	2.68E+03	1.000
			J016	8.709	2.491	2.000	13.93	19.9	ML	2565	0.260	5.15	7.23	2.82E+03	1.291
			J002	8.750	2.501	1.999	30.04	19.7	ML	3031	0.199	5.20	7.35	2.43E+03	1.525

1) Normalised by the value obtained from the specimen cured for 7 days, under otherwise the same conditions.

4. CONCLUSIONS

From the experimental results presented above, the following conclusions can be derived:

1) The effects of compaction-induced anisotropy on the stress-strain-time behavior (including viscous property) of the compacted cement-mixed gravelly soil tested in the present study were generally insignificant.
2) The effect was negligible at the small strains but it became more noticeable with an increase in the strain. The specimens compacted in the direction perpendicular to the direction of compression loading (i.e., the *horizontal* specimens) exhibited slightly smaller peak strength, followed by a noticeably larger brittleness in the post-peak regime than those compacted in the direction of compression loading (i.e., the *vertical* specimens).
3) The *vertical* specimens are always slightly stiffer and stronger than the *horizontal* ones for different compacted densities and curing periods. However, the pre-peak stiffness and linearity and the q_{max} value as well as the degree of dilatancy increase with an increase in the compacted density and curing period in a very similar way with the *vertical* and *horizontal* specimens.

References

Barbosa-Cruz, E.R. and Tatsuoka, F. (2000). Stress-strain properties from elastic behaviour to peak strength of compacted cement-mixed sand. *Grouting, Soil Improvement, Goesystems including Reinforcement* (Rathamayer eds.), 3-10.

Goto, S., Tatsuoka, F., Shibuya, S., Kim, Y. S., and Sato, T. (1991). A simple gauge for local small strain measurement in the laboratory. *Soils and Foundations, 31* (1), 169-180.

Hansen, K. D. and Reinhardt W.G. (1990). *Roller-compacted concrete dams*. New York: McGraw-Hill, Inc.

Hayano,K., Sato,T. and Tatsuoka,F. (1997). Deformation characteristics of a sedimentary softrock from triaxial compression tests rectangular prism specimens. *Géotechnique* (Symposium In Print), *47*(3), 439-449.

Kongsukprasert, L. and Tatsuoka, F. (2003). Viscous effects coupled with ageing effects on the stress-strain behavior of cement-mixed granular materials and a model simulation. *Proc. 3rd Int. Conf. on Deformation Characteristics of Geomaterials,* Lyon (Di Benedetto et al. eds.), Balkema, (1), 569-577

Kongsukprasert, L., Tatsuoka, F. and Tateyama, M. (2005). Several factors affecting the strength and deformation characteristics of cement-mixed gravel. *Soils and Foundations*, *45*(3), 107-124

Kongsukprasert. L. (2003). Time effects on the strength and deformation characteristics of cement-mixed gravel. *Doctoral dissertation*, University of Tokyo.

Kongsukprasert. L. and Tatsuoka, F. (2005). effects of compaction-induced anisotropy in the stress-strain-time behaviour of cement-mixed gravelly soil. *Proc. 40th Annual conference,* JGS.

Santucci de Magistris,F., Koseki,J., Amaya,M., Hamaya,S., Sato,T. and Tatsuoka,F. (1999). A triaxial testing system to evaluate stress-strain behaviour of soils for wide range of strain and strain rate. *Geotechnical Testing Journal*, ASTM, *22*(1), 44-60.

Schrader, E.K. (1988). Compaction of roller compacted concrete. *Consolidation of concrete* (SP-96), ACI (S. H. Gebler eds.), 77-101.

Tatsuoka, F., Sato, T., Park, C.-S., Kim, Y.-S., Mukabi, J.N., and Kohata, Y. (1994). Measurements of elastic properties of geomaterials in laboratory compression tests. *Geotechnical Testing Journal* (ASTM), *17*(1), 80-94.

Tatsuoka,F. (2004). Cement-mixed soil for Trans-Tokyo Bay Highway and railway bridge abutments, Geotechnical Engineering for Transportation Projects, *Proc. of GeoTrans 04, L.A., ASCE GSP No. 126 (Yegian & Kavazanjian eds.),* 18-76.

Soil Stress-Strain Behavior: Measurement, Modeling and Analysis
Geotechnical Symposium in Roma, March 16 & 17, 2006

STRESS-STRAIN BEHAVIOR OF A COMPACTED SAND-CLAY MIXTURE

Jui-Pin Wang and Hoe I. Ling
Department of Civil Engineering and Engineering Mechanics
Columbia University, New York
Email : jw2174@columbia.edu; hil9@columbia.edu

Yoshiyuki Mohri
Soil Engineering Laboratory, National Institute of Rural Engineering
The Ministry of Agriculture, Forestry and Fisheries, Japan
Email : accord1@nkk.affrc.go.jp

ABSTRACT

Compacted soils with high fines content are regarded as disqualified backfill materials for retaining earth structure. For economical reasons, their beneficial use warrants to be studied. In this paper, a sand-clay mixture containing 40% fines was prepared and compacted with water content at optimum, and a series of "unconventional" drained compression tests were conducted. The specimens were unsaturated specimens and the confining pressure used for the tests was from 30 kPa to 150 kPa. The volume change of unsaturated specimens was measured from radial displacements using gap sensors. The sand-clay mixtures showed a highly overconsolidated behavior and the Mohr-Coulomb failure envelope was nonlinear. The results also showed that the angle of internal friction decreased with an increase in the confining pressure, whereas the elastic modulus increased with an increase in the confining pressure.

1. INTRODUCTION

In the guidelines of earth retaining structures, soils with fines are disqualified as backfill materials. For example, according to the AASHTO (American Association of State Highway and Transportation Officials) specifications, the content of fines used in a reinforced soil retaining wall must be less than 15%. However, geotechnical engineers usually face practical concerns, like the availability of good quality backfill materials and the construction costs in meeting these criteria.

Many compacted soils with fines are unsaturated materials. The unsaturated behavior of soil has not been studied as widely as saturated soil (eg., Ling and Tatsuoka, 1994, 1997), but it has drawn more attention recently. Fredlund and Rahardjo (1993) and Lu and Likos (2004) provided a comprehensive literature about the engineering aspects of unsaturated soil. One distinguished difference of unsaturated soil from saturated soil is that one stress state variable may no longer be adequate to describe the behavior. Colemen (1962) proposed the usage of both one net normal stress (σ-u_a) and matric suction (u_a-u_w) as stress state variables. Since then, many researchers have followed this work. Using these stress state variables, the shear strength equation for unsaturated soil has been formulated as a linear combination of the stress variables incorporating shear

Hoe I. Ling et al. (eds.), Soil Stress-Strain Behavior: Measurement, Modeling and Analysis, 491–502.

strength parameters: $\tau = c' + (\sigma - \sigma_a)_f \tan\phi' + (u_a - u_w)\tan\phi^b$, where c' is the cohesion at zero matric suction and the subscript f denotes failure. ϕ^b is usually less then ϕ'. Study shows that ϕ^b is a highly nonlinear function of matric suction, and it is nearly equal to ϕ' for near saturated conditions but zero for suction approaching saturation state.

The scope of this paper is to describe the unsaturated behavior of sand-clay mixtures using a series of triaxial compression tests. Two types of soil were blended to produce a mixture having a fines content of 40%. The test procedures were described followed by the observations on its strength-deformation characteristics.

2. SOIL INDEX PROPERTIES

The soil mixture was obtained by blending mechanically cohesive soil and granular soil. The cohesive soil used here was Kanto Loam and the granular soil was a gravel dust that was available commercially. An agricultural tractor was used to mix them from which the sample used in this study was obtained.

The index properties of Kanto loam, gravel gust and soil mixture are given in Table 1. The grain size distribution of three different types of soil, obtained by sieve and hydrometer analyses, are shown in Figure 1. Kanto loam and gravel dust had 92% and 30% of fines respectively, and the soil mixture had 43.3% percent of fines. The results of Atterberg tests showed the liquid limit and plastic limit of the soil mixture was 28.8% and 21.4%, respectively (plastic index is 7.4). According to USNS (Unified Soil Classification System), the soil was classified as SM-SC.

Table 1. Index Properties of Gravel Dust, Kanto Loam and Soil Mixtures

	Gravel Dust	Kanto Loam	Soil Mixture
Specific Gravity, G_s	2.699	2.424	2.66
Natural Water Content, w%	6.2	79.7	17.8
D_{60} (mm)	0.519	0.0462	0.257
D_{50} (mm)	0.2	0.0426	0.105
D_{30} (mm)	0.0693	0.0322	0.0531
D_{10} (mm)	0.0186	-	0.0044
C_u	27.9	-	53.9
C_c	0.498	-	2.7

Based on AASHTO, the amount of fines for a reinforced soil retaining wall must be less than 15% with PI less than 6%, whereas the amount of fines for reinforced slopes must be less than 50% with PI less than 20. Obviously, the blended soil did not meet the requirements of AASHTO in the construction of reinforced retaining walls and reinforced slopes.

Standard Proctor compaction tests were conducted to find out the maximum dry unit weight and optimum water content of the soil mixture. Based on the test results

(Figure 2), the maximum dry unit weight $\gamma_{d, max}$ and the optimum water content w_{opt} were obtained as 16.2 kN/m^3 and 20%, respectively.

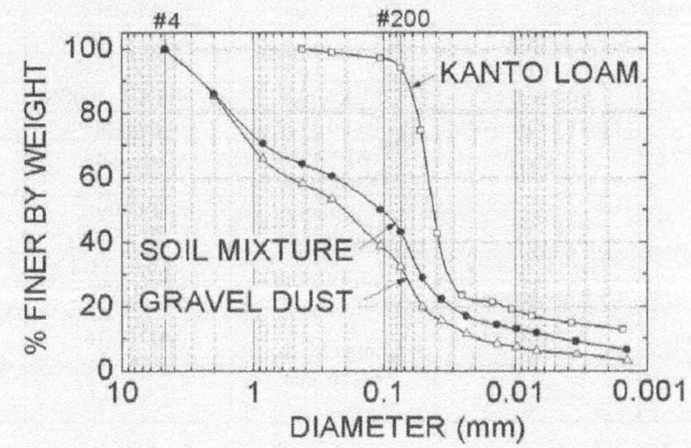

Figure 1. Grain size distribution curve of gravel dust, Kanto Loam and soil mixtures

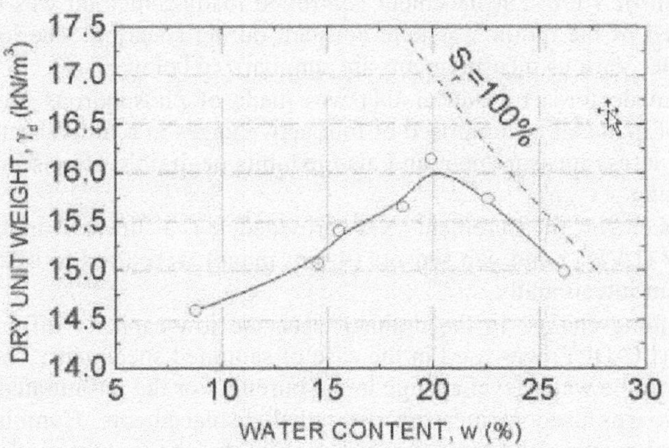

Figure 2. Compaction test result

3. TRIAXIAL TEST DEVICE

The triaxial testing system used in this research is based on the design of Tatsuoka (1988), which is known as Seiken-type triaxial system. The schematic sketch of the system used at Columbia University is shown in the Figure 3. The major components and measuring devices are summarized in the Table 2. The system was automated with the A/D (digital to analog) and D/A (digital to analog) boards. The A/D board was used to convert the analog signals into digital value that was recorded by a computer. The computer was used to control the test such as the start and stop of the motor as well as the direction of loading. The D/A board was also used to control the electro-pneumatic transducer in the stress path testing.

Table 2. Summary of Transducers

Component/Transducer		Manufacture	Model	Range
Loading	Motor	Oriental Motor	USMS40-401W	1500rpm
	Gear Head		5GN10XK	Ratio 1:10
	Screw Jack	Makisinko	NJ40DCL 300SK	300mm
Force	Force Transducer	Tomo Test	Custom made	2 kN
Displacement	LVDT	Tokyo Sokki Kenkyujo	CDP-50	50mm
	Gap Sensor	Applied Electronic	PU-09 (AEC-5509)	4mm
Differential Pressure	High Capacity	Fuji Electirc	FFF35WB2	3200mm H_2O
	Low Capacity		FFC33WB2	1300mm H_2O
Data Acquisition and Control	A/D Board	Contec Micro-Electronic	AD12-16	12 bits, 16 channels
	D/Aboard		DA1-6LC	12 bits, 16 channels

This system provides two sets of caps and pedestals in different sizes. The dimension of the cap and pedestal used in this research was of diameter 3.5 cm, thus the specimen height of 7 cm. Displacement controlled loading method was adopted, which means the speed of the motor was kept constant during shearing. The loading rate was 0.8% per minute. Various measurements are summarized below.

- A custom-made force transducer that was made of phosphorous bronze was used. This type of load cell is comprised of four active gages in a full Wheatstone bridge. It allows low stress measurement and also exhibits negligible effects in hysteresis and force coupling.
- The displacement measurement was provided by a linear variable differential transformer (LVDT) and gap sensors (4-mm range) were used to offer more precise measurement intentionally.
- For the volume change in the drained tests, the low-capacity differential pressure transducer (LC-DPT) was used in the case of saturated specimens. The measurement was based on the water level change in the burette. For the unsaturated tests, a pair of gap sensors was used to measure the radial displacements. Combining the radial displacement provided by gap sensors and axial displacement provided by LVDT or the axial gap sensors, the volume change for unsaturated tests could be determined.
- For undrained tests, the high-capacity differential pressure transducer (HC-DPT) measures the effective stress. During shearing, the confining pressure that was also the chamber pressure was kept constant so that the pressure difference recorded by HC-DPT could be used to determine the excess pore water pressure.

4. TESTING PROCEDURES

A series of triaxial compression tests on unsaturated sand-clay mixture were conducted at a confining pressure of 30, 60, 85 and 150 kPa. Some of the tests were repeated because of fear of nonuniformity of the compacted specimens.

Using the dynamics compaction method (ASTM D 6698), the density of specimens was prepared as that of the maximum density and optimum water content. An aluminum split mold was used to produce cylindrical specimen with diameter and height of 3.5 cm and 7 cm, respectively. The soil was compacted in six layers to simulate the standard Proctor compaction energy. After compaction, the mold was removed and the specimen was installed in the triaxial device. Filter papers were placed on the top and at the bottom of the specimen before the setup was enclosed in a rubber membrane. Aluminum foils were attached on the surface of the membrane to act as the targets for the gap sensors (Figure 4).

When conducting unsaturated tests, confining pressure was provided by suction except the tests of 150 kPa where the confining pressure was provided by the air pressure inside the triaxial chamber. This was because the air pressure applied as suction on the specimens could not go beyond the atmospheric pressure. No water passage was allowed into the specimen during the phase of consolidation and shearing. Due to the small measuring range of the gap sensors, they were in contact with the specimens at a strain level of 10% or less. In order not to interrupt with shearing, they were removed at that point.

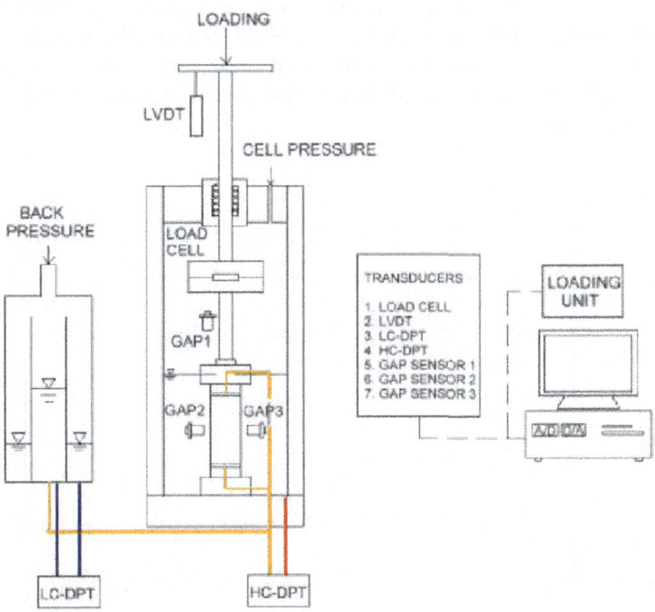

Figure 3. Sketch of triaxial test device

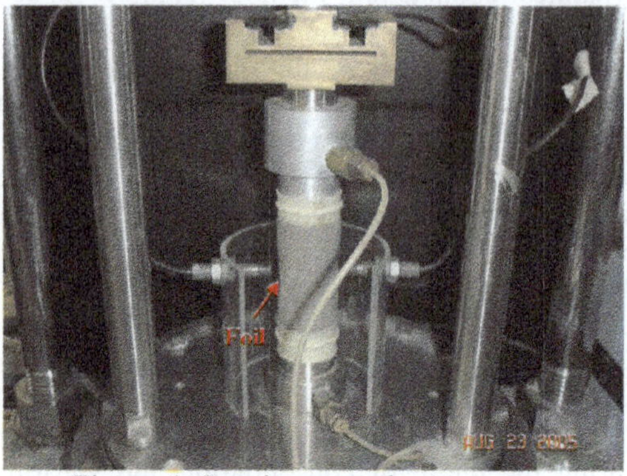

Figure 4. Specimens and gap sensors

5. TEST RESULTS

Figures 5 to 8 show the test results of unsaturated specimens under various confining pressures. The results show a strain-softening behavior in these tests. Although the shear band affected the volume change based on local displacements, the tests results showed that the specimen contracted in the small strain level and then dilated in the large strain level. Clearly, the behavior of compacted soil mixtures resembled that of dense sand or overconsolidated clay.

(a)

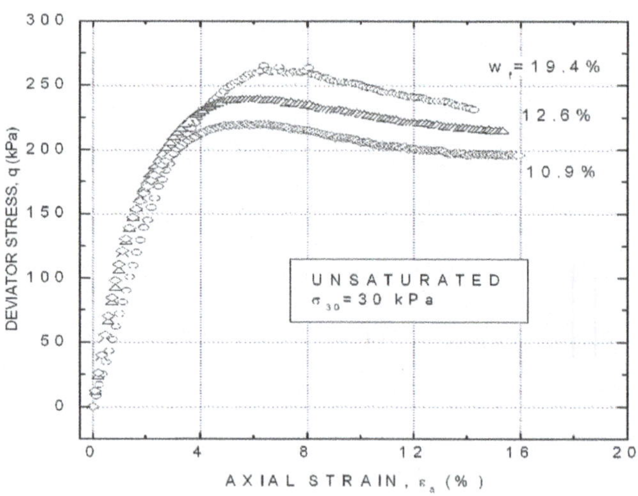

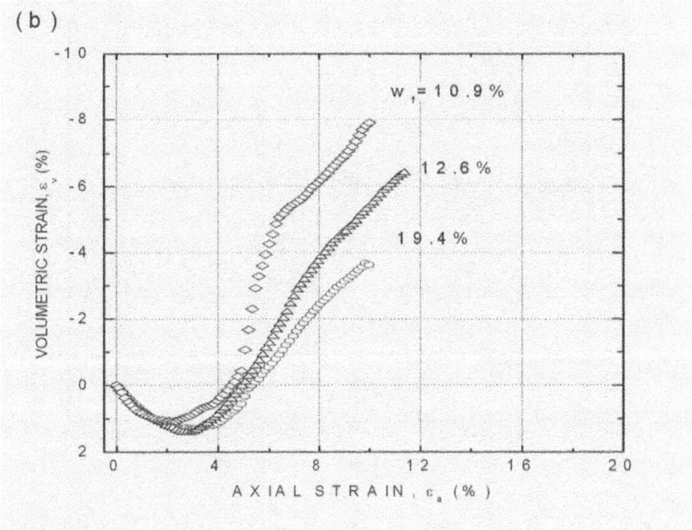

Figure 5. Confining pressure= 30 kPa: (a) Stress strain relationship; (b) volume change

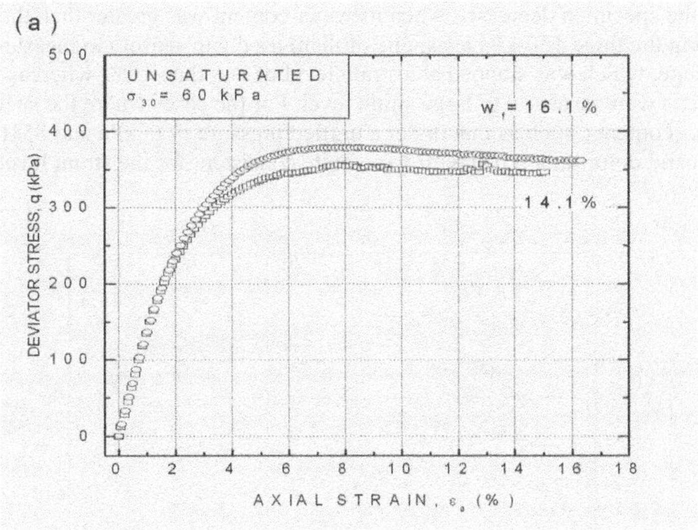

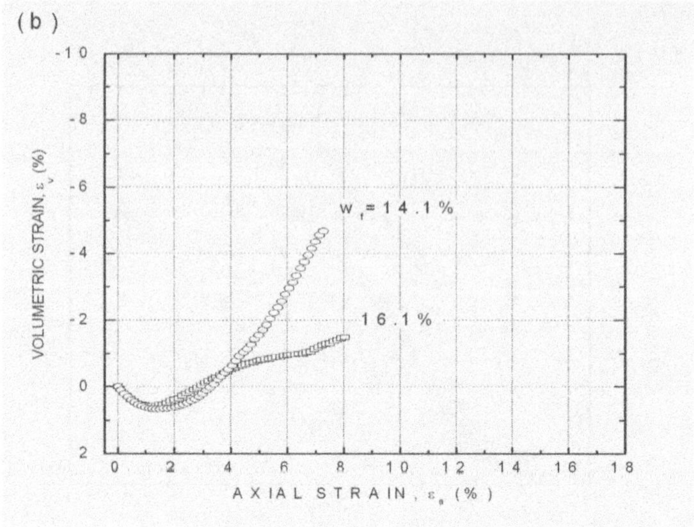

Figure 6. Confining pressure= 60 kPa: (a) Stress strain relationship; (b) volume change

Figure 5 also shows the effects of water content of the specimens on the strength and deformation behavior of compacted soil mixtures. When the water content was close to the optimum value, the strength increased but the dilation reduced. The stiffness did not change significantly with a variation in the water content from 10.9% to 19.4%. The strength of the specimen decreased when its water content was greater than the optimum value. Among the three 150-kPa tests, one of them used gap sensors to measure the volume change, which was stopped at a strain level of less than 10%, whereas the other two specimens were tested up to large strain level. For the cases where the specimens had similar water contents, such as that under a unified pressure of 60kPa and 85kPa, the stress-strain and deformation behavior were quite consistent for the strain level up to 5%.

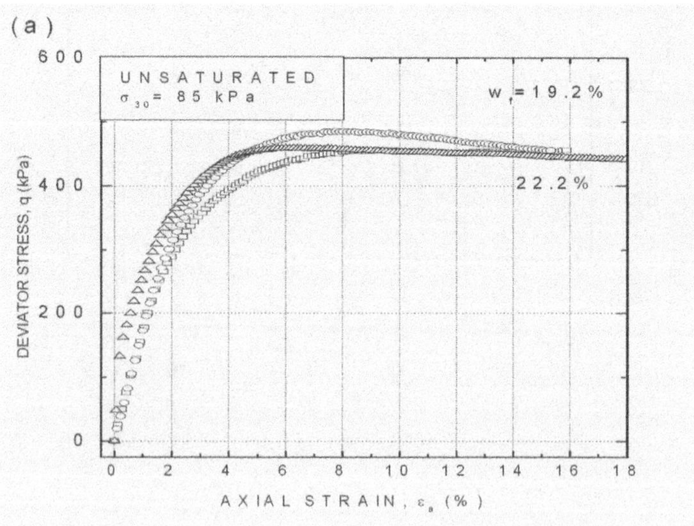

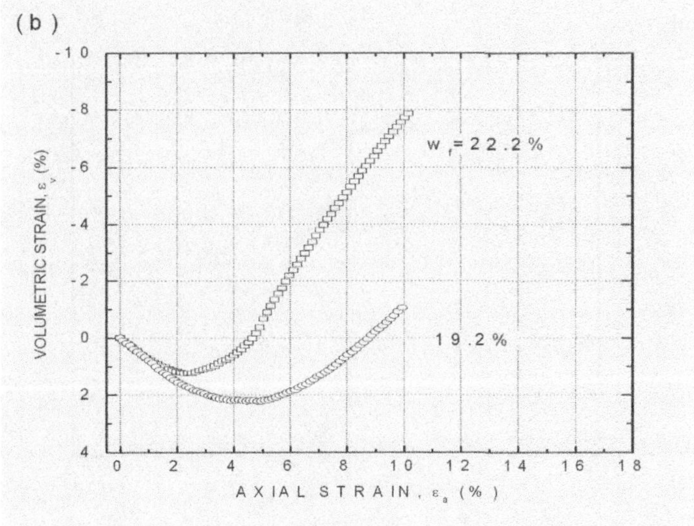

Figure 7. Confining pressure= 85 kPa: (a) Stress strain relationship; (b) volume change

Figure 9 shows that the failure envelope is nonlinear and it is inappropriate to approximate it as a linear curve with an intercept c. The angle of internal friction ϕ, which is expressed as $\phi = \sin^{-1}\dfrac{\sigma_1 - \sigma_3}{\sigma_1 + \sigma_3}$, decreases logarithmically with the confining pressure as shown in Figure 10.

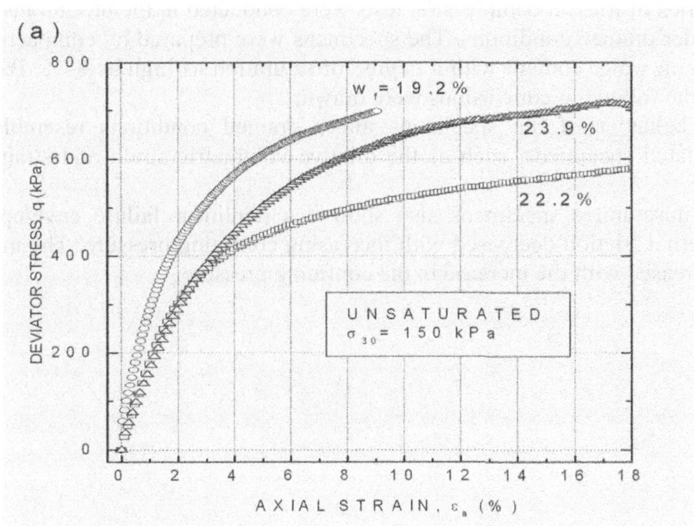

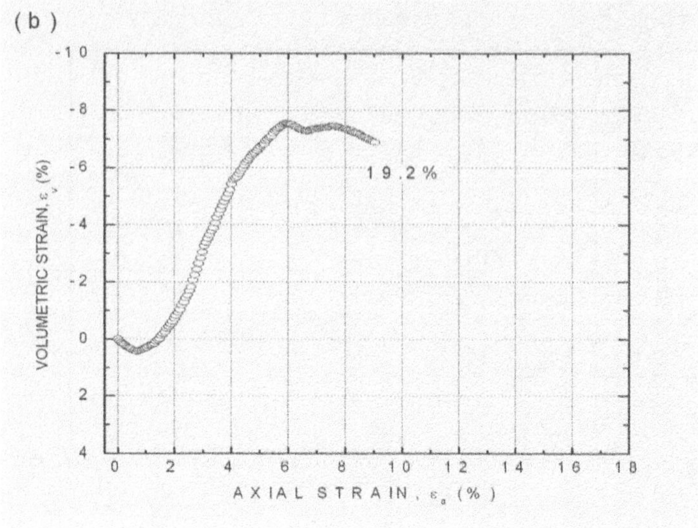

Figure 8. Confining pressure= 150 kPa: (a) Stress strain relationship; (b) volume change

In terms of the initial elastic modulus of unsaturated specimens, Figure 11 shows that the specimen under higher confining pressure featured a higher elastic modulus. Generally speaking, the initial elastic modulus increased logarithmically with the increase of the confining pressure.

6. CONCLUSIONS

A series of triaxial compression tests were conducted in the unsaturated sand-clay mixtures under drained conditions. The specimens were prepared by compaction at close to the optimum water content with a degree of saturation as high as 87%. Based on the test results, the following conclusions were drawn:

The behavior of the specimens under drained conditions resembled that of overconsolidated specimens, such as the dilative volumetric strain and strain softening behavior.

The unsaturated specimens also showed a nonlinear failure envelope, and the angle of internal friction decreased with increasing confining pressure. The initial elastic modulus increased with the increase in the confining pressure.

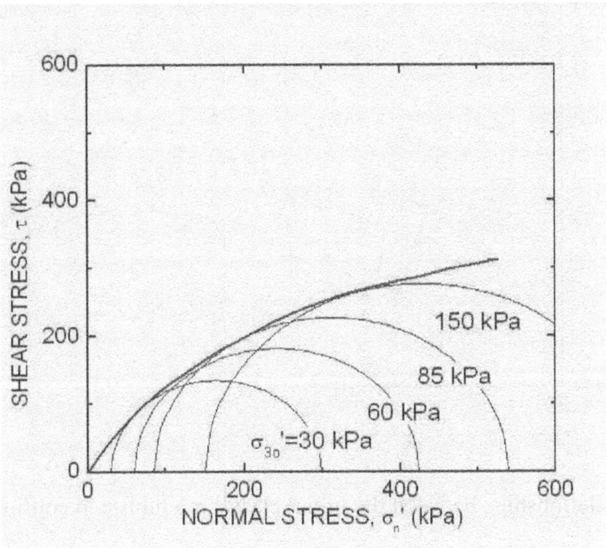

Figure 9. Failure envelope of unsaturated soil

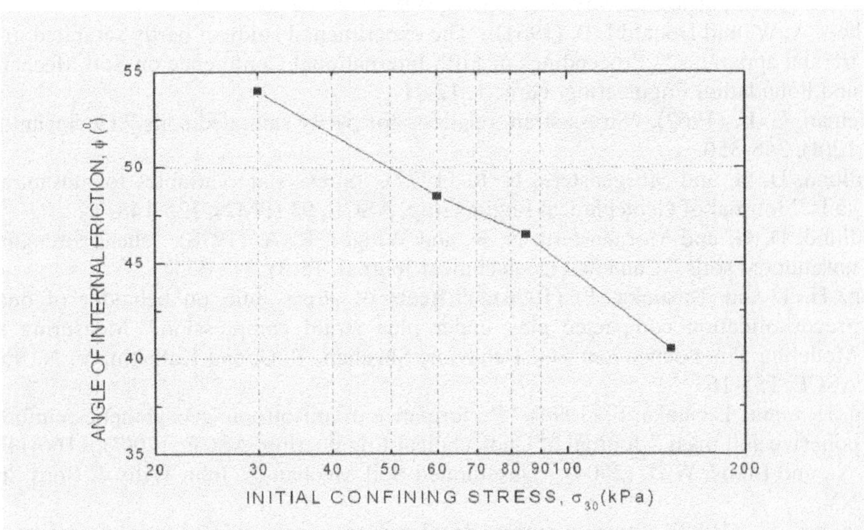

Figure 10. Relationships between the angle of internal friction and initial confining pressure

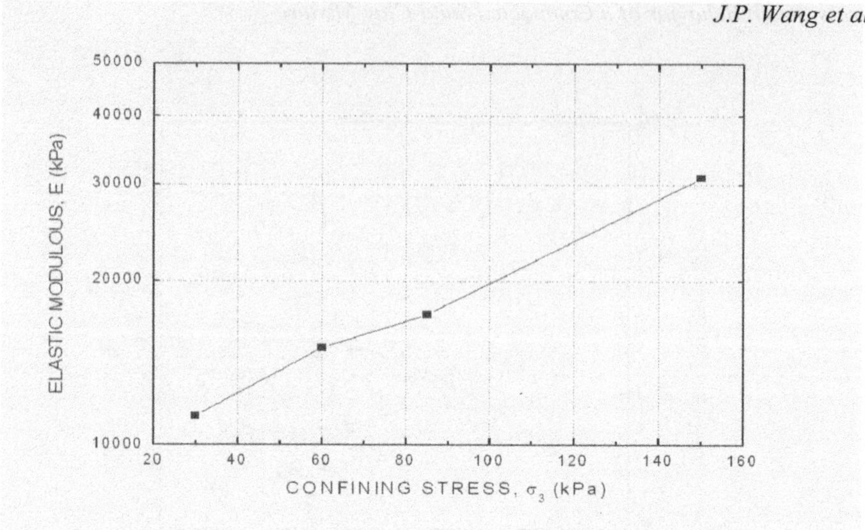

Figure 11. Relationships between the initial elastic modulus and confining pressure

7. REFERENCES

Bishop, A. W. (1959). "The principle of effective stress." Teknisk Ukeblad I Samarbeide Med Teknik, Oslo, Norway, 106 (39), 859-863

Bishop, A. W. and Donald I. B. (1961). "The experimental study of partly saturated in the triaxial apparatus." Proceedings of Fifth International Conference on Soil Mechanics and Foundation Engineering, Paris, 1, 12-21

Coleman, G. E. (1967). "Stress-strain relations for partly saturated soils." Geotechnique, 12(4), 348-350.

Fredlund, D. G. and Morgenstern, N. R. (1977). "Stress state variables for unsaturated soils." Journal of Geotechnical Engineering, ASCE, 93 (SM2), 125-148.

Fredlund, D. G. and Morgenstern, N. R. and Widger, R. A. (1978). "shear Strength of unsaturated soils." Canadian Geotechnical Journal, 15(3), 313-321.

Ling, H. I. and Tatsuoka, F. (1996). "Effects of stress ratio on behavior of quasi-preconsolidation compacted clay under plan strain compression." Measuring and Modeling Time-dependent Soil Behavior, Sheahan, T. C. and Kaliakin, V. N., Eds., ASCE, 151-165

Ling, H. I. and Tatsuoka, F. (1994). "Performance of anisotropic geosynthetic-reinforced cohesive soil mass." Journal of Geotechnical Engineering, ASCE, 120(7), 1166-1184.

Lu, N., and Likos, W. J. (2004). "Unsaturated Soil Mechanics, John Willy & Sons, Inc., NY.

Tatsuoka, F. (1988). "Some recent developments in triaxial testing systems for cohesionless soil." Advanced Triaxial Testing of Soil and Rock, ASTM STP 977, Donaghe, R. T., Chaney, R. C., and Silver, M. L., Editors, American Society for Testing and Material, 7-67.

Soil Stress-Strain Behavior: Measurement, Modeling and Analysis
Geotechnical Symposium in Roma, March 16 & 17, 2006

SMALL STRAIN PROPERTIES AND CYCLIC RESISTANCE OF CLEAN SAND IMPROVED BY SILICATE-BASED PERMEATION GROUTING

Y. Tsukamoto, K. Ishihara, K. Umeda, T. Enomoto, J. Sato, D. Hirakawa and F. Tatsuoka
Department of Civil Engineering
Tokyo University of Science, Chiba, 278-8510, Japan
e-mail: ytsoil@rs.noda.tus.ac.jp

ABSTRACT

Permeation grouting has been developed as one of countermeasures against soil liquefaction during earthquakes. This study aims at examining the small strain properties and cyclic resistance of jellied sand improved by silicate-based permeation grouting. In the first part of the present study, the small strain properties and cyclic resistance of jellied sand were examined based on the measurements of propagation velocities of longitudinal and shear waves travelling through triaxial specimens and the results from subsequently conducted undrained cyclic triaxial tests. In the second part, strain-controlled small-amplitude cyclic triaxial tests are conducted using non-contact gap displacement transducers to examine the initial elastic moduli of silicate gel itself and intact sand as well as jellied sand. The roles of the stiffness at small and large strains of silicate gel in increasing the undrained cyclic resistance of jellied sand are discussed in detail.

1. INTRODUCTION

The technology of permeation grouting has been used for the improvement of subsoils under airport runways and around existing bridge piers against liquefaction during earthquakes. In this method, the specially developed silicate solution is used, which is environmentally harmless and initially permeable enough to travel through soil aggregates and to gradually solidify to become gel-like formation. The steel annular rods with extensible tubes installed inside are then inserted into subsoils. The silicate solution is poured into the extensible tubes, and is then hydraulically pushed to permeate through surrounding soils from several positions along the steel rods. A series of solidified ball-like soil structures are eventually formed within the grounds, which are more resistant against soil liquefaction during earthquakes. In the present study, a series of special triaxial tests were conducted in the laboratory, and the increased undrained cyclic resistance of jellied sand improved by permeation grouting was examined from the standpoint of the moduli at small and large strains of silicate gel compared with those of intact sand and jellied sand. Herein, the silicate gel refers to a solidified form of the silicate solution after a curing period. The jellied sand refers to a sand sample within which part of voids are replaced with silicate gel. The intact sand refers to a sand sample that has not been subjected to permeation grouting.

Hoe I. Ling et al. (eds.), Soil Stress-Strain Behavior: Measurement, Modeling and Analysis, 503–511.

2. UNDRAINED CYCLIC RESISTANCE

Undrained cyclic triaxial tests were conducted to examine if the undrained cyclic resistance can be improved by permeation of the silicate-based solution through sand specimens whereby replacing the pore water with silicate gel. The details of the testing methods and test results are described below.

Toyoura sand was used in all of the tests, which is fine clean sand with no fines less than 0.075 mm, and the mean diameter D_{50} is 0.17 mm. Two methods were used for the preparation of triaxial specimens. In the method of water sedimentation (W.S.), 6 % silicate-based solution was first poured into a 6 cm-diameter mould to a prescribed depth, and sand particles were poured into the solution. There is a certain condition at which a designed soil density is achieved with regard to a depth of sedimentation through the solution and a height of pluviation. The specimens prepared by water sedimentation are thought to be nearly saturated. In the method of wet tamping (W.T.), moist samples were produced by mixing sand with 6 % silicate-based solution. They were uniformly tamped in several layers in the mould. The silicate solution was then circulated through the specimens in the mould. With this method, it was found that approximately 75 % of voids in volume ratio were made saturated with silicate solution and the rest of 25 % were occupied by pore air. All of the W.S and W.T. specimens thus prepared were cured for one month prior to triaxial testing. They were isotropically consolidated to a confining stress of σ'_o = 98 kPa without any back pressure, and subjected to undrained cyclic triaxial loading.

The data of the undrained cyclic triaxial tests are summarized in the form of cyclic stress ratio $\sigma_d/(2\sigma'_o)$ to induce a 4 % double amplitude (DA) axial strain against the number of cycles N_c, as shown in Figs. 1 and 2. Based on the observation of development of axial strain during cyclic loading, it is more convenient to take a DA axial strain of 4 % compared to 5 % adopted in usual practice, because more test data become available, which are comparable to each other among the test series. Such data for the W.S. and W.T. specimens are plotted in Figs. 1 and 2, respectively. The cyclic resistance $\sigma_{d,l}/(2\sigma'_o)$

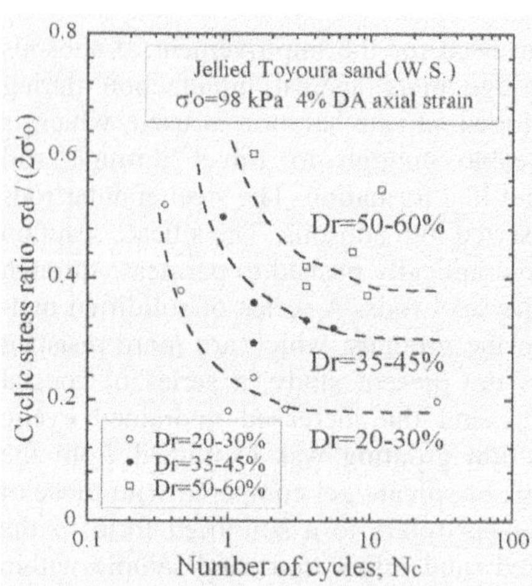

Fig. 1 Plots of cyclic stress ratio against number of cycles, (Jellied sand, W.S.).

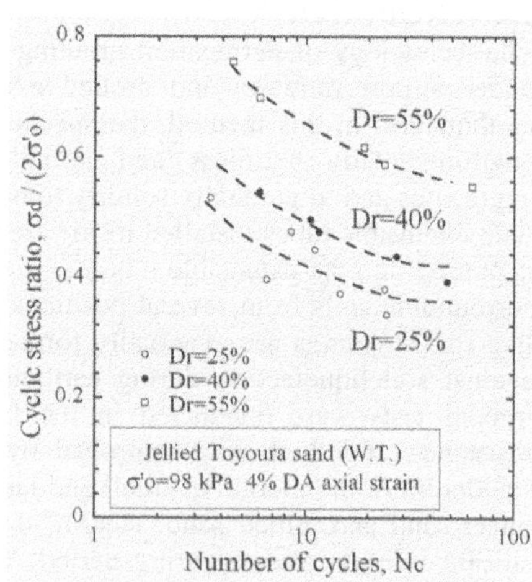

Fig. 2 Plots of cyclic stress ratio against number of cycles, (Jellied sand, W.T.).

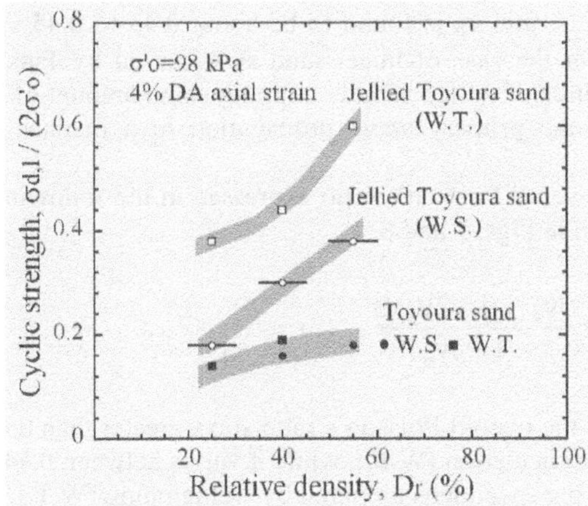

Fig. 3 Plots of cyclic strength against relative density.

was then defined as $\sigma_d/(2\sigma'_o)$ at DA = 4% and N_c = 20. The comparisons of the cyclic resistance $\sigma_{d,l}/(2\sigma'_o)$ plotted against the relative density are shown in Fig. 3. The cyclic resistance of jellied sand is greater than that of intact sand in any respect. It is also seen that the jellied sand prepared by wet tamping shows greater cyclic resistance than that prepared by water sedimentation. This is thought to be due to the fact that wet tamping produced specimens which contain some amount of pore air in the voids, and also due to the difference in the structures of soil fabrics created by these two preparation methods.

3. SMALL STRAIN MODULUS FROM VELOCITY MEASUREMENTS

Prior to large-strain undrained cyclic triaxial tests described above, the non-destructive measurements of propagation velocities of longitudinal and shear waves, V_p and V_s, were conducted. This testing method was previously used by Tsukamoto et al. (2002), in which the influence of imperfect saturation on the liquefaction resistance of sand was evaluated in relation to the V_p and V_s values. The cross-section of a triaxial specimen with the cap and pedestal housing transducers is shown in Fig. 4. The piezo-electric transducer embedded within the cap is used as a source of generating compression waves, while the accelerometer attached to the bottom of a dummy duralumin block embedded within the pedestal is used as a receiver. A pair of bender elements is used to measure the velocity of shear waves. The data of V_p and V_s values are plotted against the B-value, as shown in Figs. 5 and 6. It is seen from Fig. 5 that, for the specimens prepared by water sedimentation, the values of V_p stays greater than the velocity of primary waves travelling through water, V_w = 1492 (m/s), regardless of the B-value. Therefore, the saturated gel-formation seems to prevail within the pores. On the other hand, it is seen from Fig. 6 that, for the specimens prepared by wet tamping, which contain some amount of pore air entrapped inside, the values of V_p change with the B-value varying between 0 and 0.4, following the following theoretical relations assuming a porous elastic medium:

$$(V_p/V_s)^2 = \frac{4}{3} + \frac{2(1+\nu_b)}{3-(1-2\nu_b)(1-B)} \quad . \ (1)$$

The skeleton Poisson's ratio ν_b is found to be about 0.43 to 0.45 and is larger than the value of $\nu_b = 0.35$ for the case of intact sand as observed by Tsukamoto et al. (2002). Therefore, the structures of jellied sand entrapping some amount of pore air in the voids seem to respond to the primary wave propagation in a manner similar to a porous medium.

The values of the overall Poisson's ratio expressed in the following are plotted against the B-value, as shown in Figs. 7 and 8,

$$\nu = \frac{(V_p / V_s)^2 - 2}{2(V_p / V_s)^2 - 2} = \frac{3\nu_b + (1 - 2\nu_b)B}{3 - (1 - 2\nu_b)B} \quad . \quad (2)$$

It is clearly seen that the overall Poisson's ratio stays greater than 0.49 for the specimens prepared by water sedimentation (W.S.), while it varies between 0.44 and 0.47 depending upon the B-value for the specimens prepared by wet tamping (W.T.).

The values of the initial shear modulus calculated from the values of V_s, based on the simple expression for an isotropic medium, $G_o = \rho V_s^2$ are plotted against the void ratio in Fig. 9. For comparison purposes, the general expression proposed by Kokusho (1980) for intact Toyoura sand is also indicated in Fig. 9. It is found that the values of G_o for jellied sand are by far similar to those for

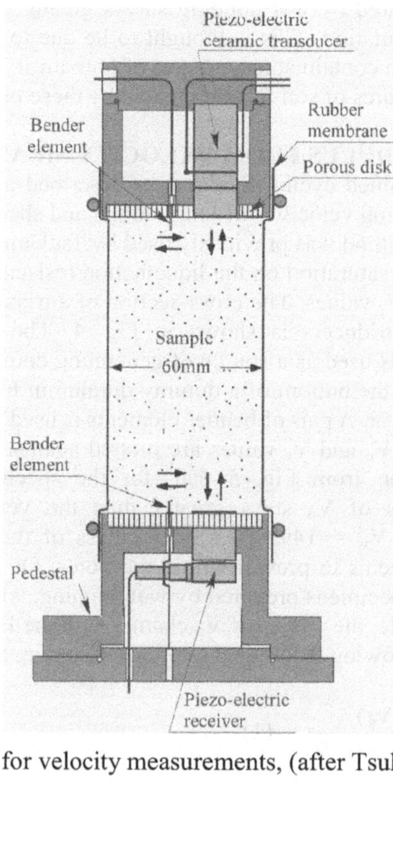

Fig. 4 Transducers for velocity measurements, (after Tsukamoto et al. 2002).

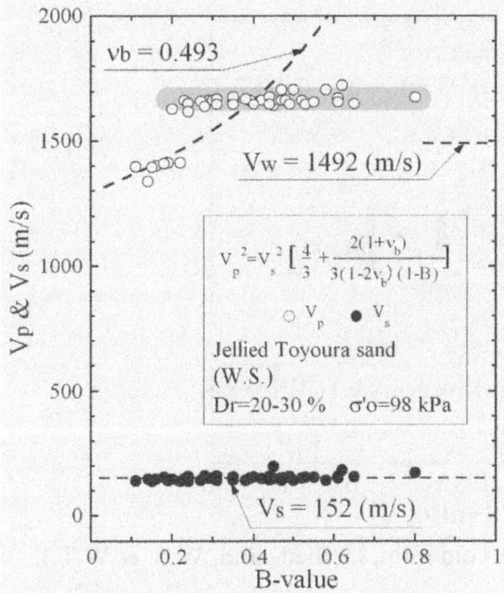

Fig. 5 Plots of V_p & V_s against B-value, (Jellied sand, W.S.).

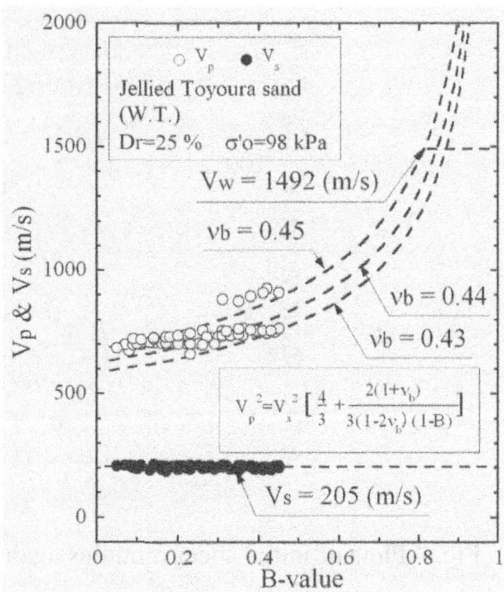

Fig. 6 Plots of V_p & V_s against B-value, (Jellied sand, W.T.).

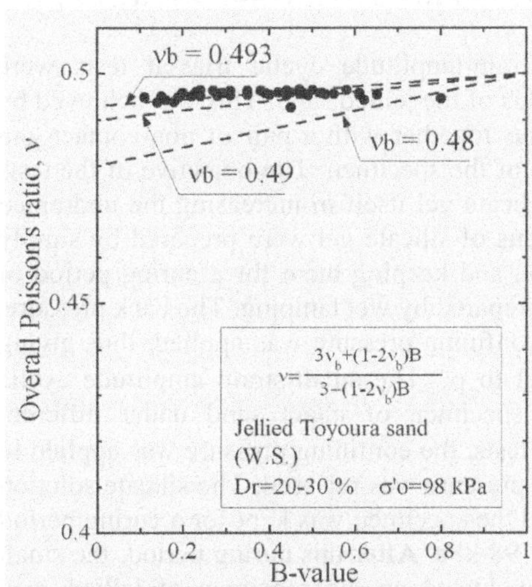

Fig. 7 Plots of overall Poisson's ratio against B-value, (Jellied sand, W.S.).

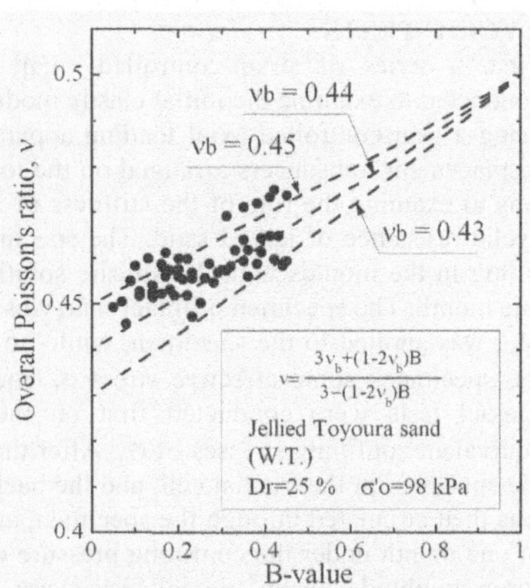

Fig. 8 Plots of overall Poisson's ratio against B-value, (Jellied sand, W.T.).

intact sand. The G_o-values for the specimens prepared by water sedimentation are even lower than those for intact sand. Therefore, the increased large-strain cyclic resistance of jellied sand is not accompanied by an increased small-strain shear modulus. Attempts have been made to monitor the velocity of primary waves travelling through the silicate gel itself. However, the primary waves were not allowed to transmit through the silicate

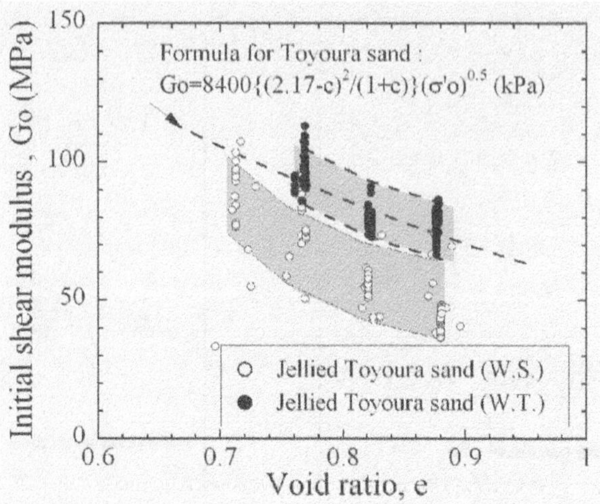

Fig. 9 Plots of initial shear modulus against void ratio, (Jellied sand, W.S. & W.T.).

gel, and such measurements were found to be impossible.

4. SMALL AND LARGE STRAIN STIFFNESS FROM STRAIN-CONTROLLED CYCLIC TESTS

First, a series of strain-controlled small strain-amplitude cyclic triaxial tests were conducted to examine the initial elastic modulus of the jellied sand. This was achieved by using a gear-controlled axial loading apparatus together with a pair of non-contact gap displacement transducers arranged on the top of the specimen. The objective of the tests was to examine the role of the stiffness of silicate gel itself in increasing the undrained cyclic resistance of jellied sand. The specimens of silicate gel were prepared by simply filling in the moulds with the silicate solution and keeping them for a curing period of one month. The specimen of intact sand was prepared by wet tamping. The back pressure, $-p_a$, was applied to the specimen, while no confining pressure was applied, thus giving the specimens some effective stress σ_o equal to p_a. The small-strain amplitude cyclic triaxial tests were conducted first on the specimen of intact sand under different equivalent confining stresses of σ_o. After the tests, the confining pressure was applied to the specimen in the triaxial cell, and the back pressure was released. The silicate solution was then circulated through the specimen, and the specimen was kept for a curing period of one month under the confining pressure of 98 kPa. After this curing period, the small strain-amplitude cyclic triaxial tests were conducted on the specimen of jellied sand under different confining stresses of σ_o at an axial strain rate equal to ± 0.00388 %/min.

The test results on silicate gel are shown in Figs. 10(a) to (d). The continuous increase in the residual axial strain that can be seen even after a couple of loading cycles is due mostly to the viscous property of the test material. The values of the small strain-elastic modulus were inferred by estimating the inclinations of deviator stress against axial strain. The average values of such inclinations observed during cyclic loading were estimated by a simple expression of $E_a = \Delta q/\Delta\varepsilon_a$. The test results on intact sand and jellied sand are shown in Figs. 11 and 12, respectively. The data of the elastic modulus of E_a thus inferred

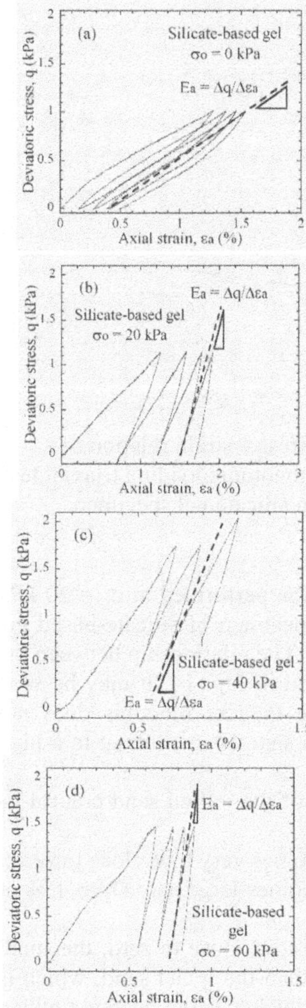

Fig. 10 Cyclic test results, (Silicate gel).

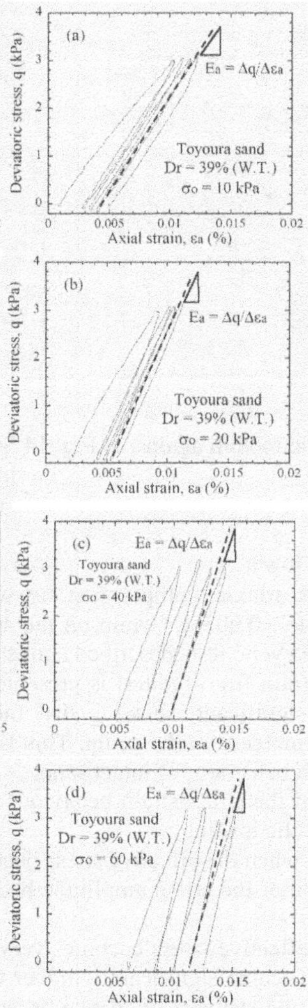

Fig. 11 Cyclic test results, (Intact sand).

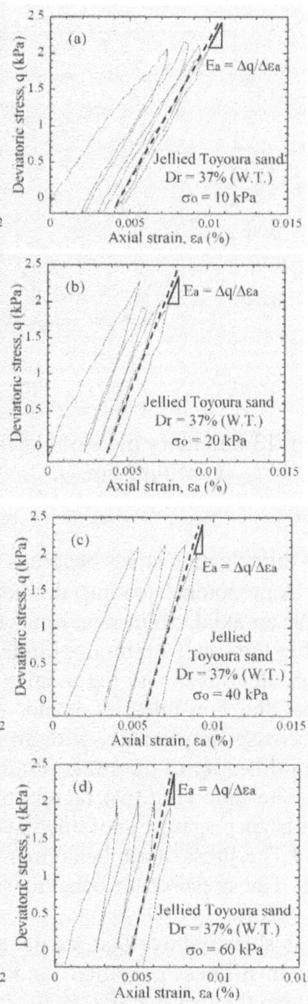

Fig. 12 Cyclic test results, (Jellied sand).

plotted against the confining stress of σ_o are shown in Fig. 13. It may be seen that the values of the elastic modulus for intact sand and jellied sand are similar to each other. It is interesting to notice here that the same results were obtained from the velocity measurements as described above in Fig. 9. It is also to note here that the silicate gel exerts some stiffness during small-strain cyclic loading, thus giving some values of E_a as shown in Fig. 13, while the initial shear modulus were not obtained from the velocity measurements as described above. When subjected to cyclic shear stresses at low effective stresses, the small strain-modulus of the jellied sand became low. However, the elastic modulus of the jellied sand becomes larger than the intact sand to a larger extent

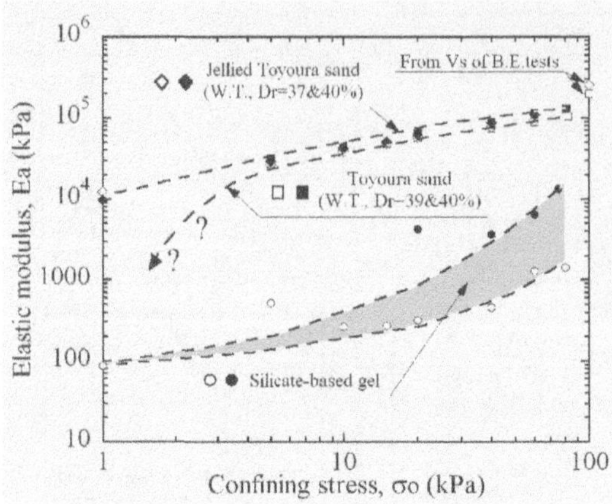

Fig. 13 Summary plots of elastic moduli against confining stress.

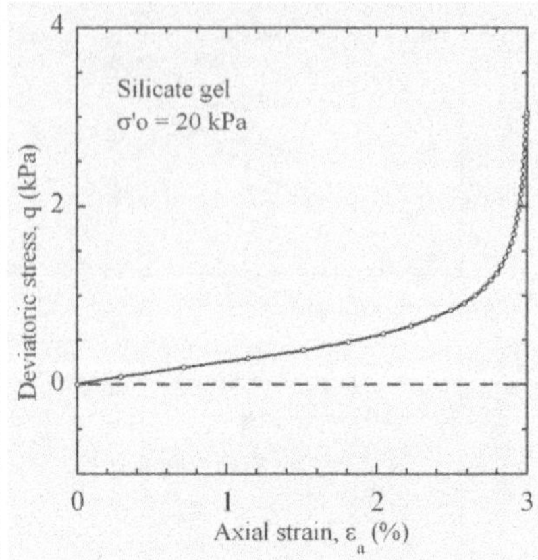

Fig. 14 Stress – strain relation of a monotonic loading triaxial test on silicate gel specimen.

as the effective stress becomes lower.

A monotonic loading drained triaxial compression test was performed at $\sigma_0 = 20$ kPa and an axial strain rate equal to ±0.00388 %/min on the specimen of silicate-based gel after the small strain-amplitude cyclic tests described above. The relationship between the deviator stress and the axial strain from the test is presented in Fig. 14. It may be seen that the non-linearity is not significant, showing that the tangent modulus does not decrease significantly with an increase in the strain. This is sharply in contrast to a high non-linearity of the stress-strain behaviour of intact sand.

The following two factors of the stress-strain behaviour of the jellied sand control its resistance against undrained cyclic loading :

(1) The stress-strain behaviour when the effective stress becomes very low, close to zero.
(2) The stress-strain behavior after the strain amplitude becomes large, say DA= 4 % or more.

It is shown above that, as the effective stress becomes very low, close to zero, the small strain-stiffness of jellied sand became significantly higher than the intact sand, which is due to the fact that the small strain-stiffness of silicate-based gel does not decrease with a decrease in the effective stress at a high rate as intact sand. Moreover, due to a low strain-nonlinearity, the stiffness of silicate-based gel at large strains becomes comparable with that of intact sand under low effective stresses. Then, under such conditions, the sustained modulus of silicate gel under such low effective stresses could have exerted some important role in the increased cyclic resistance of jellied sand.

On the other hand, the small strain-stiffness of jellied sand from wave velocities measured at the initial high effective stress reflects only very part of these two factors. Therefore, it is not possible to evaluate the cyclic undrained strength of jellied sand based on the wave velocities measured at the initial high effective stress.

5. CONCLUSIONS

The small strain properties of jellied sand improved by silicate-based permeation grouting were examined based on the velocity measurements on triaxial specimens and small strain-amplitude cyclic triaxial tests. In both of the tests, the small strain-moduli were found to be by far the same for intact sand and jellied sand. Therefore, there was no increase in the small strain-modulus due to permeation grouting. In other words, the increased undrained cyclic resistance of jellied sand in comparison to intact sand was not accompanied by the increased small strain-modulus, in particular when measured at initial high effective stresses. Therefore, it is not possible to evaluate the cyclic undrained strength of jellied sand based on the wave velocities measured at the initial high effective stress. On the other hand, the sustained modulus of silicate gel under low effective stresses, in particular the one at large strains, could have exerted some important role in the increased cyclic resistance of jellied sand.

ACKNOWLEDGEMENTS

The authors would like to express their sincere appreciation to Mr. A. Yoshida and Mr. T. Hada of Raito Kogyo Co., for their generous support on the present study. This research was funded by Ministry of Education, Science and Technology.

REFERENCES

Kokusho, T. (1980) "Cyclic triaxial test of dynamic soil properties for wide strain range", Soils and Foundations, Vol.20, pp.45 - 60.

Tsukamoto, Y., Ishihara, K., Nakazawa, H., Kamada, K. and Huang, Y. (2002) "Resistance of partly saturated sand to liquefaction with reference to longitudinal and shear wave velocities", Soils and Foundations, Vol.42, No.6, pp.93 - 104.

Soil Stress-Strain Behavior: Measurement, Modeling and Analysis
Geotechnical Symposium in Roma, March 16 & 17, 2006

EFFECTS OF CYCLIC LOADING ON GRAVEL

Giuseppe Modoni * , Le Quang Anh Dan** , Junichi Koseki **, and Sajjad Maqbool**

*University of Cassino (Italy)
**Tokyo University of Science (Japan)

ABSTRACT

The effects of a large number of load repetitions on gravel are discussed based on experimental evidences. To this purpose the results of several triaxial tests carried out on a coarse grained soil, artificially reconstituted at different prescribed levels of density, are reported and compared. The tests reproduce a broad variety of possible field conditions, being performed at different mean effective stresses, by applying sequences of load-unload repetitions, different by number of repetitions and amplitudes of cycles. The analysis of the experimental results has been conducted by focusing the attention on the plastic components of strain, calculated by scaling the non negligible elastic components from the measured ones. A theory for the interpretation of monotonic test results is first briefly presented to form the basis of the analysis of the cyclic test results. This last is referred to the evolution of the stress-strain relationships along with load repetitions, and particularly to the accumulation of distortional and volumetric strains and to the effects of previous cyclic loading on subsequent soil response.

1. THEORETICAL INTRODUCTION

The analysis of the stress-strain response of cyclically loaded gravels has practical consequences in several fields of civil engineering. Earth dams, highway or railway embankments are some of the most highlighting applications where coarse grained soils are massively adopted to take advantage of their excellent mechanical properties. Natural gravels are also frequently found at the foundation of important and heavy structures. In the performance of all these applications a fundamental role is played by the response of soil to the action of repeated loading, which can be caused respectively by water reservoir fluctuations, traffic and wind loading, seismic events etc. It is pointed out that, even though each cycle of loading produces small variations of stress and strains in the soil mass, the cumulative effect of a large number of load applications may turn relevant. In fact, depending on the soil characteristics, cyclic loading may progressively lead to unsuitable serviceability conditions which, in the most favourable cases, require continuous maintenance operations. In case of seismic events, it is of vital importance to evaluate the liquefaction potential of the soil together with post seismic settlements, which may have dramatic consequences on the safety of the structures.

The research shown in this paper is aimed to observe, starting from experimental evidences, the effects of a large number of shearing cycles on the stress-strain response of a gravel from Chiba (Japan), in order to write possible correlations of such effects with the characteristics of load (initial stress level, stress or strain amplitude, number of repetitions) and with the properties of soil.

The analysis hereafter presented is performed in the framework of continuum mechanics. In particular the experimental results are interpreted by observing the variation of the strain variables grouped into a volumetric ε_p and a distortional component ε_q as a function of the deviatoric component of the stress q or the mobilised friction angle ($\sin\phi'_{mob}$). The non linear irreversible

Hoe I. Ling et al. (eds.), Soil Stress-Strain Behavior: Measurement, Modeling and Analysis, 513–522.

mechanical response of gravels is here assumed of elastoplastic nature and any deformation $\delta\varepsilon$ is considered as a combination of an elastic ($\delta\varepsilon^e$) and a plastic ($\delta\varepsilon^p$) component whose relative importance depends on void ratio, stress level, stress path and past stress history.

$$\delta\varepsilon = \delta\varepsilon^e + \delta\varepsilon^p \tag{1}$$

The main discussion is here confined to the evolution of the plastic components of strains which have been calculated as differences between total and elastic strains. These latter have been evaluated with an anisotropic hypoelastic model first introduced by Tatsuoka and Kohata (1994), which takes into account both inherent and stress induced anisotropy and which has proven to satisfactorily simulate the small strain response of the tested soil under a wide range of stress conditions, namely soil void ratios, stress levels and stress paths (see Modoni et al, 1999). Five parameters are required to fully characterize the elastic response of the soil which are, namely, a reference stiffness parameter (E_1), a reference Poisson coefficient (ν_o), an inherent anisotropy factor (I_o) indicating the stiffness anisotropy at isotropic stress state and two exponents (m and n) expressing the dependency of the stiffness on the stress components. Finally, a function f(e) is required to relate the soil stiffness to the soil void ratio. The whole set of parameters calibrated for Chiba gravel is reported in Tab.I.

Tab.I Calibration parameters of the anisotropic hypoelastic model for Chiba gravel (Modoni, 1999)

E_1 (kPa)	$(1-I_o)$	n	m	ν_o	f(e) (Hardin and Richart, 1963)
18480	0.59	0.5	0.252	0.17	$\frac{(2.17-e)^2}{(1+e)}$

It is worth observing that the analysis of the elastic strain components, which is often neglected in the modelling of granular soil response, is not trivial in the present case of cyclically loaded gravels, since, as is shown later, the amount of strains at each cycle is rather small and fully comparable with the elastic component.

In order to characterize the plastic response of soil, reference is made to a critical state model originally introduced by Muir Wood et al. (1994) and then developed by Gajo and Muir Wood (1999). A key factor of such a model is the introduction of a bounding surface through a function of the state parameter ψ defined as difference between the soil void ratio and critical void ratio at the same mean effective stress p' (Jefferies, 1993). By introducing some slight modifications (see Modoni et al, 2003; Modoni et al, 2005) this model has been found suitable to simulate the response of Chiba gravel. In particular the modified version has proven to be capable of catching the effect of soil density, stress level and stress paths on the soil response observed during monotonic and cyclic tests, these latter performed with different loading sequences.

For the purpose of the present paper it is just worth recalling the assumed hardening function for monotonic loading, which is expressed by the following formulation:

$$\frac{\sin\phi'_{mob}}{M_p} = \frac{\varepsilon^p{}_q}{B(\varepsilon^p{}_q)^c + \varepsilon^p{}_q} \tag{2}$$

where $\varepsilon^p{}_q$ represents the plastic distortional strain and M_p is an upper bound limiting value of $\sin\phi'_{mob}$ which is never attained but at infinitive strains where critical state is assumed to occur. The analytical expression of M_p is:

$$M_p = \sin\phi'_{cs} - k\psi = \sin\phi'_{cs} - k(e - \Gamma + \lambda \ln p') \qquad (3)$$

B and c in eq.3 and Γ and λ in eq. 4 are soil parameters. The complete set of soil parameters adopted for Chiba gravel is reported in Tab. II.

Tab.II. Constitutive parameters of Chiba gravel

ϕ'_{cs}	Γ	λ	k	B	c
42.8	2.52	0.17	1	0.07	0.9

For unloading and reloading curves the increments of the modified stress variable are related to the increments of the plastic distortional strains, both computed from the reversal point of loading direction, in the following way:

$$\Delta\left(\frac{\sin\phi'_{mob}}{M_p}\right) = \pm\alpha \frac{\left|\Delta\varepsilon^p{}_q\right|^c}{B + \left|\Delta\varepsilon^p{}_q\right|^c} \qquad (4)$$

where the sign + is referred to reloading the sign – to unloading, α is an amplifying factor simulating the increase of soil stiffness during unloading and reloading compared with primary loading.

2. EXPERIMENTAL SET UP

The tests have been performed with a controlled stress path triaxial cell (Hoque et al., 1996), on large square prismatic samples, 57 cm high and 23 cm large on each side of the base, to account for the large size of particles. Very accurate and stable local axial and lateral displacements could be obtained by averaging the measures taken on the lateral surfaces of the specimen by several LDTs (Local Displacement Transducer, Goto et al. 1991). In particular the vertical displacements have been measured with two LDT's placed on two opposite sides of the specimen, while the horizontal strains have been calculated by means of eight LDT's distributed on the four lateral faces of the specimen. In some cases, where the analysis of soil response is pursued at very large strains (ε>2%), the vertical strains have been calculated from the movements of the loading piston rather than locally on the specimen side.
The tested soil (Chiba gravel) is a crushed sandstone and its grain size characteristics are summarised in Tab. III.

Tab. III Granular characteristics of Chiba gravel.

G_s	D_{max} (mm)	D_{50} (mm)	$U_c=D_{60}/D_{10}$	<0.4mm (%)
2.71	35.8	7.8	10	8

Each samples has been artificially prepared by compacting fourteen layers, each about 4 cm thick, at a water content of 5.5%. Compaction has been achieved by heavily tamping the top of each soil layer with a dead weight. This procedures has proven effective in allowing an accurate control of the initial soil density and in obtaining particularly low values of void ratios (the minimum obtainable void ratio is $e_o \approx 0.2$, with a relative density, defined by ASTM standards D2049-69, higher than 100%). The tests analysed in the present paper have been performed at initial void ratios varying from 0.26 to 0.36.
Each test has been conducted at constant effective confining stress, following the classic triaxial stress path (Δq/Δp'=3). The confining stress has been varied among the different tests between 100 and 490 kPa. In order to explore the role of repeated cyclic loading a large number of unloading-

reloading cycles of different amplitude have been produced on soil samples while shearing them in different tests. Typical stress path and loading sequence followed in any of the performed tests are reported in Fig.1. It is shown in this plot that the initial monotonic and the cyclic loading have been applied at a constant controlled stress rate ($\Delta q/\Delta t=\pm 49$ kPa/min), while the final shearing has been performed at a controlled strain rate ($\Delta \varepsilon_a/\Delta t=0.06$ %/min).

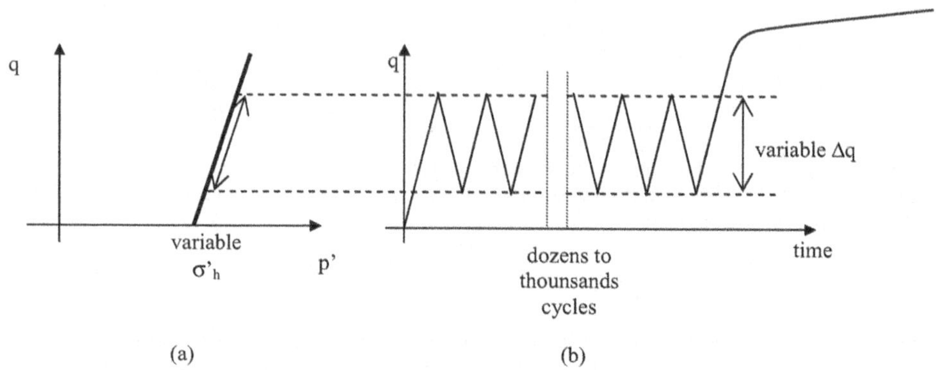

Fig. 1 Typical stress path (a) and loading sequence (b) followed in the performed tests.

3. ANALYSIS OF TEST RESULTS

The first aspect herein tackled is the accumulation of strains produced by load repetitions. This phenomenon, which is often responsible of serviceability loss of the structures, has been investigated by means of different tests performed on samples compacted at different soil densities (Fig.2). Each sample has been subjected to cyclic shearing of different amplitudes, starting from different values of isotropic and deviatoric stresses. In a first case ($\sigma'_h=100$ kPa, $e_o=0.361$) reported in Fig.2.a, one hundred large amplitude cycles have been applied to the soil between q=50 kPa and 350 kPa. The effect of each cycle is an almost close hysteretic loop of the stress-strain curve, with some residual distortional and volumetric strains of a progressively reduced amount as far as the number of cycles increases. From a practical viewpoint it is seen that, even though the amount of strains produced at each cycle is small, the accumulation after many cycles is quite large and comparable with that produced during primary loading. The result of a second test ($\sigma'_h=490$ kPa, $e_o=0.30$) is reported in Fig. 2.b. In this case the soil has been subjected to a sequence of five different cyclic loading, each of them with 5000 repetitions. The deviator stress q fluctuates around an almost constant value of about 670 kPa, but the stress amplitude has been gradually changed, being $\Delta q = 80$ kPa for the first sequence, 160 and 280 for the second and third sequences and 360 kPa for the last two sequences. Large amplitude cycles ($\Delta q = 1350$ kPa) have also been performed between two different sequences. It is clearly seen from this plot that the strain development significantly depends on the amplitude of loading cycles, being the strains accumulation higher for the larger amplitude cycles. It is also seen that the rate of strain accumulation progressively reduces as far as the loading cycles proceeds (see as an example $\Delta\varepsilon_q$ in the fourth and fifth sequences). By comparing the results of this second test with the first one it can be also concluded that a dependency of the strain accumulation with the soil density and the confining stress must be assumed. Similar conclusions can be retrieved from the observation of the third test results (Fig.2.c) where two cyclic loading sequences (Δq ranges between 400 and 940 kPa in the first 5000 cycles, between 350 and 1040 kPa in the second

5000 cycles) have been applied on a soil having a initial void ratio e_o=0.26, subjected to a confining stress σ'_h= 490 kPa.

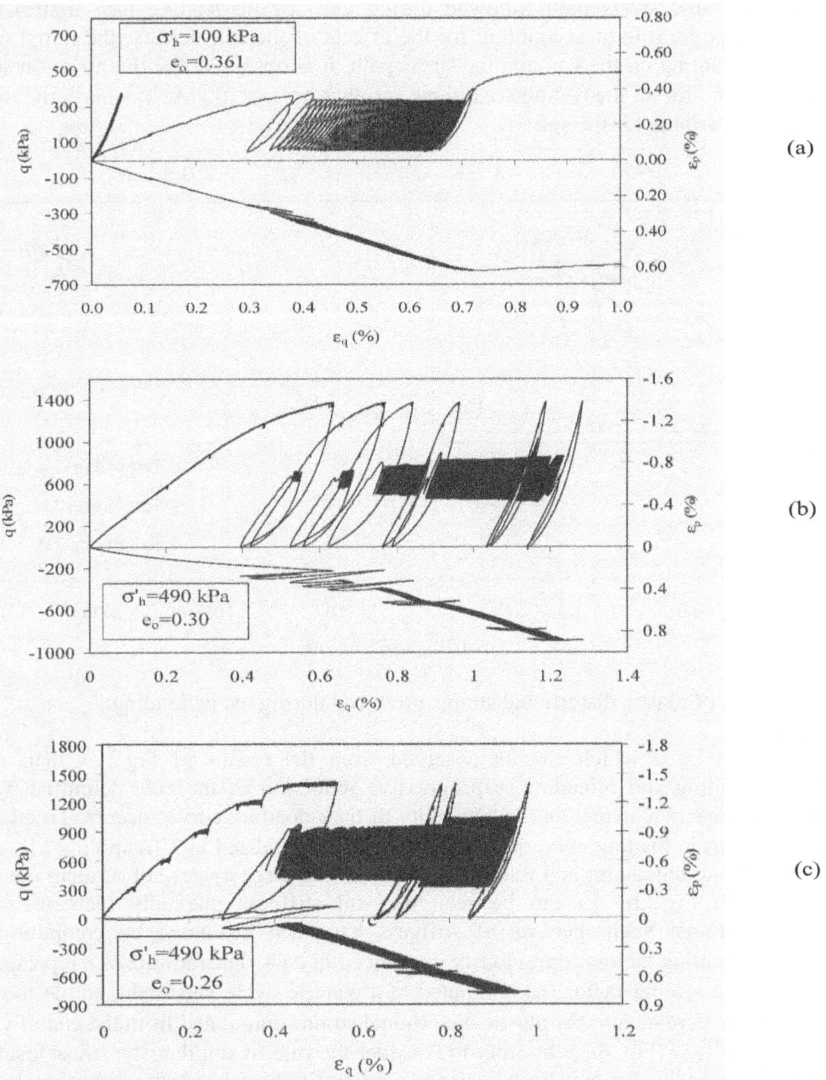

Fig. 2. Results from a triaxial test (σ'_h=const) where a large number of load cycles were repeatedly applied.

Form these tests it can be concluded that the amount of accumulated distortional strains is a function of the number and the amplitude of load repetitions, the average stress level (in terms of spherical and deviatoric component q) acting on the soil and the initial void ratio of the soil. To simply account for all these factors, the experimental results of the previous three tests have been summarised and compared in Fig.3. This figure shows how the plastic distortional strains produced at the end of each cycle ($\Delta\varepsilon^p_{q\ res}$) cumulate with the number and the amplitude cyclic loading, this last quantified by the increments of plastic distortional strains produced in each cycle during unloading ($\Sigma\Delta\varepsilon^{p}_{q\ unl}$). Each increment of $\varepsilon^{p}_{q\ res}$ and of $\varepsilon^{p}_{q\ unl}$ has been scaled with the plastic distortional strains ($\varepsilon^{p}_{q\ p.l}$) produced during primary loading by the same stress path followed during each cyclic loading (see figure). This normalisation has the role of accounting for the effects of the soil density, the initial or the average stresses acting on the soil and the stress path. It is observed that this function seems to be rather unique for all the testing conditions and that $\Delta\varepsilon^{p}_{q\ unloading}/\Delta\varepsilon^{p}_{q\ p.l.}$ plays the role of an endochronic variable for the soil.

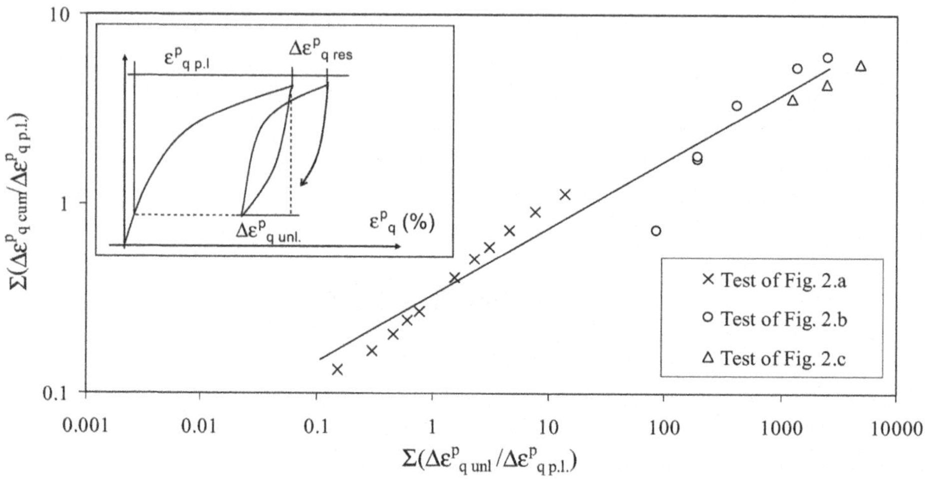

Fig. 3. Evolution of plastic distortional strains produced during cyclic loading.

Another relevant issue which can be observed from the results of Fig.2 is that, while repeatedly unloading and reloading, a progressive reduction of the area delimited by the stress-strain loops and a change in the convexity of the reloading curves occurs. To quantify this aspect the virgin loading curve is compared in the normalised plot ($\Delta\sin\phi'_{mob}/M_p$- $\Delta\varepsilon^{p}_{q}$) of Fig. 4.a with the unloading and reloading curves of the large cycles, all of them obtained from the test of Fig.2.b. As can be seen the soil stiffness markedly increases while accumulating strains. Such increase of stiffness has been evaluated by computing the variation of the scaling factors α previously introduced in eq.4. The ratio (α/α_1), representing the change of the scaling factor (α) computed at a generic cycle versus the one at the first cycle (α_1), has been related to the plastic distortional strains cumulated from the end of virgin loading branch $\varepsilon^{p}_{q\ cum}$ (Fig. 4.b). In order to consider the rule of soil density, stress level and stress path, the cumulated strains have been again normalised by the plastic distortional strain occurred at the same stress level during primary loading $\varepsilon^{p}_{q\ p.l}$. In Fig. 4.b the data from the three different tests of Fig.2 are compared. From this plot it is reasonable to assume that the scaling factor α tends to reach a very high value, possibly infinitive, for a finite level of strain accumulation. This trend, which is consistent with the observation obtained by Anh Dan and Koseki (2004), is quite similar for the three tests and similar to the qualitative curve traced on the same plot.

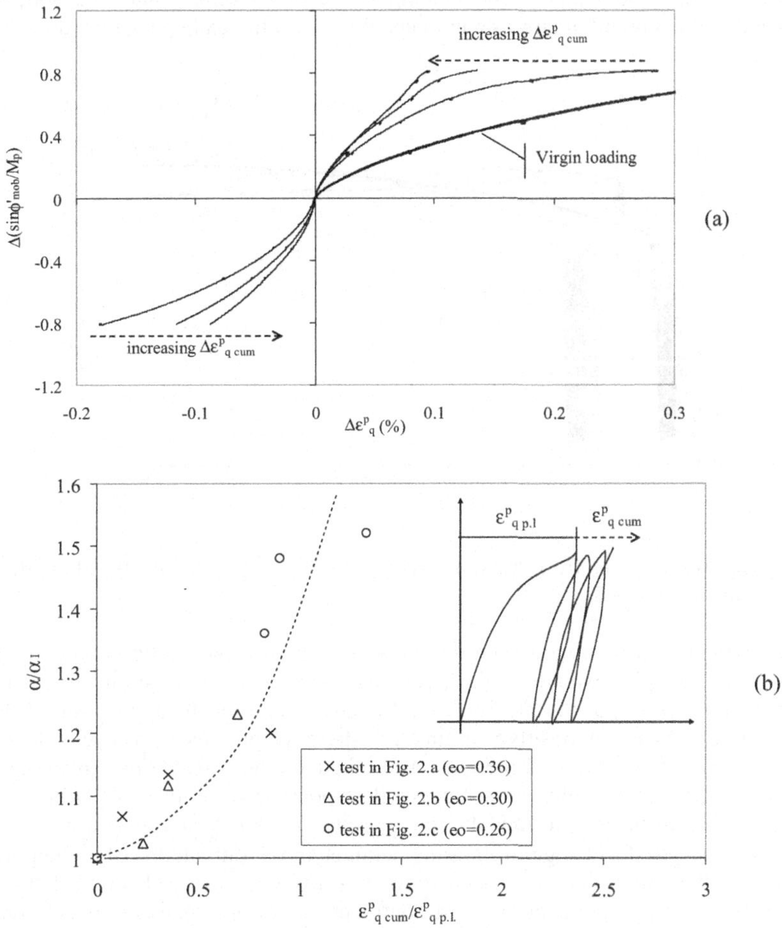

Fig. 4. Increase of soil stiffness induced by cyclic loading.

A more complete picture of the effects induced by cyclic loading on the subsequent stress-strain response of soil can be seen in Fig.5 by observing the results of four different tests performed at the same confining pressure (σ'_h=100 kPa) on samples compacted at similar initial void ratios (e_o=0.340-0.361) by applying different loading sequences. In one case (A) the soil has been monotonically sheared up to failure; in another case (B) 100 unload-reload cycles of 300 kPa amplitude have been applied starting from q=350 kPa; in another case (C) 46 cycles of 200 kPa amplitude have been applied starting from q=450 kPa after 12 hours of creep; in the last case (D) 100 cycles of 400 kPa amplitude have been applied starting from q=450 kPa. In all the cases, after cyclic loading, the soil has been sheared up to failure. In this plot, due to the large amount of strains, the deviator stress is reported as a function of the vertical strain ε_a calculated as a function of the loading piston movements.

The most remarkable effect observed from these four curves is that, after a large number of unload-reload cycles, the soil responses become stiffer and significantly different from the one shown before the cyclic sequence. This stiffer response extends well beyond the past maximum deviator stress, up to a new value where an abrupt yielding of soil occurs. It can be thus generally concluded that cyclic loading produces hardening onto the soil, which is proportional to the amount of strains cumulated during cyclic loading (see the different tests).

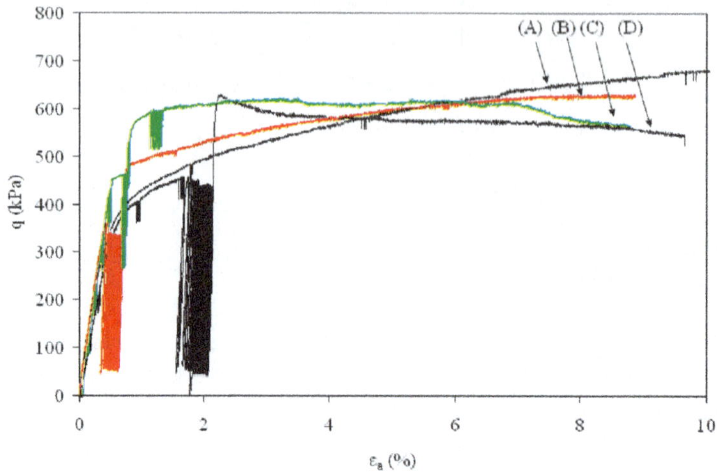

Fig. 5. Stress-strain curves from triaxial tests (σ'_h=100 kPa ; e_o=0.340 -0.361) with different loading sequence.

However it must be also argued from these results that such hardening is much higher than predictable by simply extending the previous monotonic stress-strain rule. In fact a significant overshooting of the previous trend is observed after the application of the cyclic sequence. Even though a positive volumetric strain is produced by cyclic loading, the consequent variation of soil void ratio is very limited and not capable of explaining by itself the observed increase of yield stress. Such soil response can be thus hardly predicted with common continuum theoretical models, (e.g. the one previously introduced where the yield function is related to the change of the state parameters ψ), but additional assumptions must be made. On the other side, it is sometimes postulated in many theoretical models that unloading from a yield point produces a contraction (isotropic hardening) or a downward movement (kinematic hardening) of the yield surface, which are considered responsible of the observed strain accumulations. By looking at the results shown in Fig.5 it is seen, on the contrary, that an increase of the size of the yielding surface is produced by cyclic loading.

Another effect of cyclic loading can be seen when focusing on the response of soil at very large strains. In the reported cases where cyclic loading is applied (B, C and D) the ultimate soil strength becomes always smaller than that observed in the monotonic loading test (A) and, generally, the higher stiffness increase corresponds to the most noticeable strength drop (see for instance test D where the strain accumulation is larger and where a peak is followed by a noticeable softening). Such large observed differences of ultimate strength means that different failure mechanisms must be postulated for each sample and that the hypothesis of continuum which is at the base of Critical State is difficult to keep on at such large strains.

A reasonable physical explanation of all the observed effects is that a change in soil fabric is induced by cyclic loading, and that this fabric modification is responsible for the observed increase of soil stiffness and strength. These effects cannot be simply associated to state variables as usually done in the modelling of the monotonic soil response (e.g. Jefferies, 1993), but a dependency of position and/or shape of the yielding surface and of the hardening function on some internal soil structure variable, possibly sensitive to cyclic loading, must be assumed.

The last aspect herein tackled is the link between volumetric and distortional strain increments, which influence the amount of cumulated settlement. As can be seen from the volumetric-distortional strain plots of Figg.2, the applied loading sequences produced in all the cases a volume contraction. The ratio $\Delta\varepsilon^p{}_p/\Delta\varepsilon^p{}_q$ between plastic volumetric and distortional strains cumulated at the end of cyclic loading is similar in the three examined cases (about 1.1, 1.05 and 1.18 respectively in case a, b and c). Besides this ratio seems to be also independent on the slope of the $\varepsilon^p{}_p$- $\varepsilon^p{}_q$ curve before cyclic loading, which is positive in the case of Fig.2.a and 2.b, negative in the case of Fig.2.c. However it should be considered that the average mobilised friction angle during cyclic loading is quite similar for these three tests while a dependency of the dilatancy on the average stress has been previously observed by Tatsuoka and Ishihara (1973).

For a more complete picture of this phenomenon, the evolution of the stress-dilatancy relation during cyclic loading has been detected. A sample of such analysis is reported in Fig.6 where the results of Fig.2.b have been reanalysed. In particular, in Fig.6.c the ratio $D_p=\delta\varepsilon^p{}_p/\delta\varepsilon^p{}_q$ is plotted for the large unloading-reloading cycles as a function of the ratio $\sin\phi'_{mob}/M_p$ defined by eqs.2 and 3. It is pointed out that volume contraction is associated to positive D_p values during reloading and negative D_p values during unloading, while negative D_p values during reloading and positive D_p values during unloading mean dilation. This figure shows a continuous alternation of contraction and dilation during unloading and reloading, but volume contraction progressively reduces while cyclic loading goes on, as shown by the progressive approach of the two families of stress dilatancy curves. It is worth noting that the last unloading and reloading stress dilatancy curves are almost parallel each other and almost symmetrical around a point marked with a cross in the D_p - $\sin\phi'_{mob}$ plane. Their inclination is similar to that observed in other tests not reported here (see Modoni et al., 2005). Concerning volumetric-distortional strain coupling it can be thus concluded that there is a progressive reduction of the contraction of soil and a tendency of the stress dilatancy curves to a final position.

4. CONCLUSIONS

The results of triaxial tests on gravel subjected to a large number of load repetitions have been analysed with the aim of observing the effects of cyclic loading. The most important observed effects can be summarised in a distortional and volumetric strains accumulation which are responsible of a deep change in soil fabric. Stiffening of the soil stress strain response, hardening of soil and a progressive change of the stress dilatancy relation after cyclic loading are the most remarkable effects of such soil structure modification.

The accumulation of strain has been observed to be continuously increasing while cyclic loading goes on and dependent on the number of loading cycles, their strain amplitudes, the soil void ratio and the effective stresses acting on the soil. A relation accounting for all these variables has been proposed to quantify the plastic distortional strain accumulation.

The volumetric-distortional strain coupling has also been finally detected. An alternation of volume contraction and dilation has been observed in each unloading and reloading step, being the contraction more effective. As a result volume contraction is generally produced at the end of large amplitude repeated loading. A tendency to a common final stress dilatancy relation has been also found.

ACKNOWLEDGEMENTS

The present research has been conducted under the auspices of the Japan Society for the Promotion of Science (JSPS).

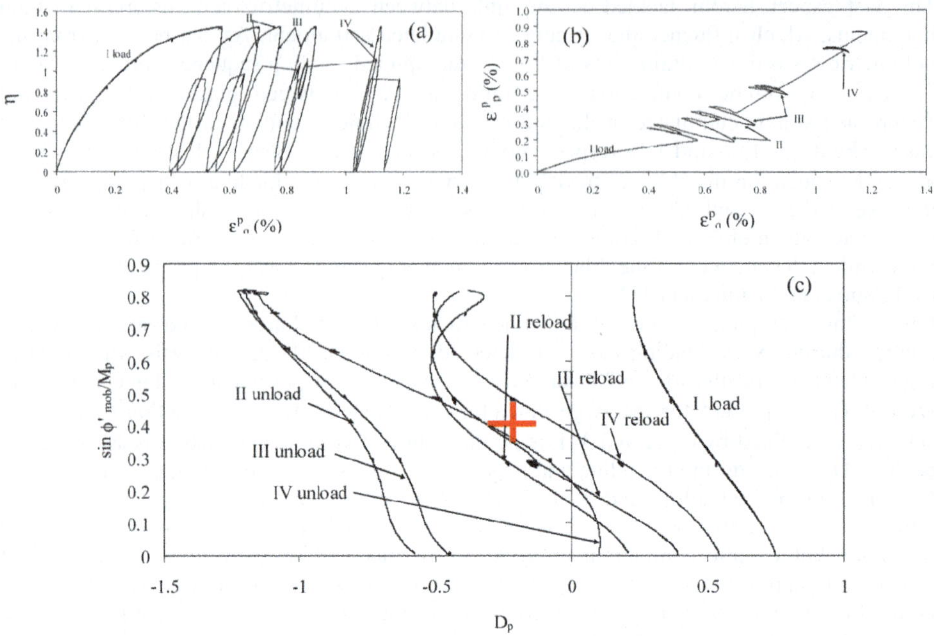

Fig. 6. Stress-dilatancy relation in a cyclic triaxial test (e_o= 0.30 - σ'_h = const = 490 kPa).

REFERENCES

Anh Dan, L.Q. and Koseki, J., 2004. *Effects of large number of cyclic loading on deformation characteristics of dense granular materials*, Soils and Foundations, Vol.44, No.3, pp.115-123.

Gajo, A. and Muir Wood, D., 1999, *Severn-Trent sand: a kinematic hardening constitutive model: the q-p formulation*, Géotechnique 49, No.5, pp.595-614.

Goto, S., Tatsuoka, F., Shibuya, S., Kim, Y.S. and Sato, T., 1991. *A simple gauge for local small strain measurements in the laboratory*, Soils and Foundations 31, No.1, pp.169-180.

Hoque, E., Sato, T. and Tatsuoka, F., 1996. Performance of LDT's for the use in triaxial tests, Geotechnical Testing Journal, ASTM 19, pp.411-420.

Jefferies, M.G., 1993. *Nor-Sand: a simple critical state model for sand*, Géotechnique 43, No.1, pp.91-103.

Modoni, G., 1999. *Analisi sperimentale e modellazione del comportamento meccanico dei terreni a grana grossa*, Ph.D. Thesis, University of Roma "La Sapienza" and Napoli " Federico II" (in italian).

Modoni, G., Flora, A., Mancuso, C., Tatsuoka, F. and Viggiani, C., 2000. *Evaluation of gravel stiffness by pulse wave transmission tests*. Geotechnical Testing Journal, GTJODJ, Vol. 23, N. 4, pp. 506-521, ASTM, West Conshohocken, USA.

Modoni, G. Flora, A., Anh Dan, L.Q. and Tatsuoka, F., 2003. *Experimental investigation and constitutive modelling of pre-failure deformation of a very densely compacted gravel.* III International Symposium on pre-failure deformation characteristics of geomaterials, Lyon.

Modoni, G., Koseki, J. and Anh Dan, L.Q., 2005. *Hysteretic stress-strain response of gravel*, submitted for possible publication, Geotechnique.

Muir Wood, D., Belkheir, K. and Liu, D.F., 1994. *Strain softening and state parameter for sand modelling*, Géotechnique 44, No.2, pp.335-339.

Tatsuoka, F. and Kohata, Y., 1994. *Stiffness of hard soils and soft rocks in engineering applications*, Pre-failure deformations of geomaterials (Shibuya, Mitachi and Miura eds.), A.A. Balkema, Rotterdam, pp.947-1066.

Soil Stress-Strain Behavior: Measurement, Modeling and Analysis
Geotechnical Symposium in Roma, March 16 & 17, 2006

COMPARISON OF SIMULTANEOUS BENDER ELEMENTS AND RESONANT COLUMN TESTS ON PORTO RESIDUAL SOIL

Cristiana Ferreira, António Viana da Fonseca
Department of Civil Engineering, Faculdade de Engenharia da Universidade do Porto
Rua Roberto Frias, s/n, 4200-465 Porto, Portugal
e-mail: cristiana@fe.up.pt, viana@fe.up.pt

Jaime A. Santos
Department of Civil Engineering, Instituto Superior Técnico
Av. Rovisco Pais, 1049-001 Lisboa, Portugal
e-mail: jaime@civil.ist.utl.pt

ABSTRACT
Bender elements are a powerful and increasingly common laboratory tool for determining the shear wave velocity hence the small strain shear stiffness (G_0) in soil samples. There are several advantages of the bender element technique, namely its simplicity and ease of use; however, there is no standard developed for this technique as the interpretation of the results involves some uncertainty and subjectivity. Different approaches have been proposed to deal with these issues, especially in terms of the interpretation techniques, based on the time and on the frequency domain. In the present work, a modified resonant column, equipped with bender elements, has been used, where shear wave velocities can be measured independently and different interpretation methodologies of the bender element results can be applied. For this study, natural samples of Porto granitic residual soil were tested, since this geomaterial has been object of research and interest for many years in the University of Porto. The paper will focus on the comparison of simultaneous results of shear wave velocities by the resonant column and the bender elements. It is intended to provide some contribution to the routine laboratory practice using bender elements, with further insight in the interpretation of the results.

1. INTRODUCTION
Widely recognised as a reference parameter for the definition of constitutive properties of soils, the maximum shear modulus (G_0) or very small strain stiffness can be determined in the laboratory in dynamic tests, by measuring shear wave velocities, namely in the resonant column or in bender element tests. The resonant column (RC) is a classical dynamic laboratory test, in which the vibration frequency can be varied, in order to induce resonance of the system. Its main advantage lies on the reliability of the derived parameters, which are based on one-dimensional wave propagation theory. As an alternative, bender elements (BE) are rapidly becoming common laboratory tools for determining the shear wave velocity in soil samples. The installation of bender elements into a conventional static soil test device, such as the triaxial cell, is relatively easy and the determination of G_0 is simpler and performed more rapidly than in the expensive

Hoe I. Ling et al. (eds.), Soil Stress-Strain Behavior: Measurement, Modeling and Analysis, 523–535.

resonant column apparatus (Gordon & Clayton, 1997). There are indeed several advantages of the bender element technique, namely its simplicity and ease of use; however, there is no standard developed for this technique as the interpretation of the results, under different methodologies, involves some uncertainty and subjectivity.

Considering the current discussion on the interpretation methods for BE tests, it seems interesting to compare results of the application of both tests (RC and BE), not only on the same material under identical conditions, but also if possible in a unique specimen, at the same time. Hence, in an attempt to bring together the benefits of both tests, that is, the reliability of the RC and the usefulness of the BE, simultaneous tests have been carried out, after modification of the resonant column apparatus to accommodate bender elements. Relatively few publications can be found in the literature reporting these experiments. One of the exceptions is the bender elements reference work of Dyvik & Madshus (1985), who found good agreement between G_0 derived from the two techniques. Similar observations have been reported by Fam *et al.* (2002). Before presenting the results of this research, it is worthwhile detailing, even if briefly, the most common techniques for the interpretation of bender element tests, which are the fundamental subject under research.

2. INTERPRETATION METHODS FOR BENDER ELEMENTS TESTS

As previously mentioned, the interpretation of bender elements traces involves some uncertainty and it is widely accepted that whatever simple the transmitted wave is, a far more complex wave will be received (Moncaster, 1997). Arroyo (2001) points out that pulse tests in soils should cope not just with a slow, highly attenuated transmission, but also with an important distortion of the transmitted signal. Different approaches have been proposed to deal with these issues, in terms of the interpretation techniques, usually based on the time or on the frequency domain. The frequency domain method generally produces an estimate of shear wave velocity, which is lower than that from traditional time domain readings (Greening *et al.*, 2003). These discrepancies have not yet been systematized and the reasons are still unclear and diverse. In the present study, first direct arrival (at various input frequencies) and continuous sweep input frequency (using a specific software, ABETS) were applied. A brief description of each of these methodologies is presented as follows.

a) Time domain: First direct arrival method, at various frequencies [TD]

The first direct arrival method is the most simple, common and usual procedure for interpreting BE measurements. It consists on the identification of the first instant of arrival of the wave in the output signal, similarly to the techniques used in geophysical tests (namely Cross-Hole and Down-Hole). While it is sometimes easy to determine first arrival, it is often the cause of much uncertainty. For instance, Arroyo (2001) has estimated uncertainties of up to 100% in estimation of the small strain shear stiffness (G_0). Many authors (Sanchez-Salinero *et al.*, 1986; Viggiani & Atkinson, 1995; Jovicic *et al.*, 1996; Arulnathan *et al.*, 1998; Pennington, 1999) have reported several sources and factors of error and inherent near field effects, which mask and compromise the identification of the arrival point. Several solutions have been proposed in order to minimize the subjectivity of this method. Changing the frequency and shape of the pulse is sometimes advised, either to a square wave (Dyvik and Madshus, 1985; Jamiolkowski *et al*, 1995), a single pulse, a burst, or to a distorted sinusoidal (Jovicic *et al.*, 1996; Pennington, 1999). On the other hand, it is recommended that a number of technical requirements and boundary conditions are accomplished (Jovicic, 2004; Lee and Santamarina, 2005). These requirements comprise good electronics equipment, good shielding and grounding, properly connected and encased transducers, leak free connections, and noise free environment. It is important to be aware that other issues also play a part, especially spatial conditions, such as alignment of the BE, reflections of the wave on the edges and sides of the sample, near field effects, relative distance between transmitter and receiver; contact between the BE and the soil, which might induce poor coupling, especially at low confining pressures; and overshooting, since at high frequencies the bender element changes its mode shape and the response becomes very complex.

Recently, analytical approaches (Fratta and Santamarina,1996; Arroyo, 2001; Lee and Santamarina, 2005) and numerical studies have been reported on the propagation of seismic waves in cylindrical specimens, modelled in finite differences, finite elements, and spectral elements (Arroyo *et al.*, 2002; Hardy *et al.*, 2002; Rio *et al*, 2003). These studies have revealed significant influences of sample geometry and boundary conditions in the shape, frequency, and velocity of the wave travelling through the sample, while highlighting the limitations of time domain interpretations.

b) Frequency domain: Continuous sweep input frequency via ABETS [FD]

The use of continuous signals which require the shear wave velocity to be decoded from measurements of relative phase of transmitted and received signals has been gaining in popularity (e.g. Brocanelli and Rinaldi, 1998; Blewett *et al.*, 1999; 2000; Greening *et al.*, 2003). While this technique has been used across a wide range of fields, Viggiani and Atkinson (1995) were the first to apply phase-delay method to BE testing, using pulse input signals. According to Greening *et al.* (2003), these frequency domain methods have a number of advantages over traditional time-based measurements, namely the possibility of creating an algorithm to determine travel time by establishing the gradient of a graph of phase difference against frequency. Generally, a continuous harmonic sinusoid is used as input signal, though a generic input signal can also be used to evaluate the phase delay by decomposing the signal into its harmonics, using the Fourier transform.

Phase delay methods can be performed reliably using "traditional" equipment i.e. a signal generator and oscilloscope (Kaarsberg, 1975). In this procedure, the frequency of a continuous sine wave is manually changed, and the frequencies at which the transmitted signal and received signal are exactly in and out of phase with one another (so called π-points) are noted. Greening and Nash (2003) showed that this process could be automated and established less onerously using a sweep input signal and a spectrum analyser. The continuous sweep input method enables the acquisition of a continuous phase angle versus frequency relationship.

Greening *et al.* (2003) suggested a setup, consisting of a low-cost spectrum analyser system implemented in *Microsoft Excel*^{TM}, which makes use of a PC on which a specific software, ABETS (Automatic Bender Element Testing System) is loaded to control a high-speed dual channel data acquisition unit. The software details, namely on data processing, can be found in Greening *et al.* (2003). Slight modifications have been introduced to the program, namely for post-processing data and analysis. A screenshot of this program, in its Portuguese version, is presented in Figure 1, with an example of the data acquisition results spreadsheet.

Best results have been obtained with a sweep sine input signal with a 0-20kHz bandwidth. The first graph in Figure 1 shows the input and output signals in the time domain, where it is unfeasible to determine a direct arrival time. Below it, is the graph of the coherence between the two signals (from 0 to 1) against input frequency; by definition, the coherence of two waves indicates how well correlated the waves are, as quantified by the cross-correlation function, which essentially quantifies the ability to predict the value of the second wave by knowing the value of the first. Hence, the higher the coherence, the more correlated the signals will be. The plots at the right show the relative phase against frequency; in the top one, the phase angle is "wrapped", ranging from -π to π, while on the bottom one, it is "unwrapped", starting near zero, and continuously increasing. The travel time is derived directly from the slope of this curve, given from the best fit line, for a selected range of frequencies. Such selection is evidenced by the vertical lines of the bottom right graph.

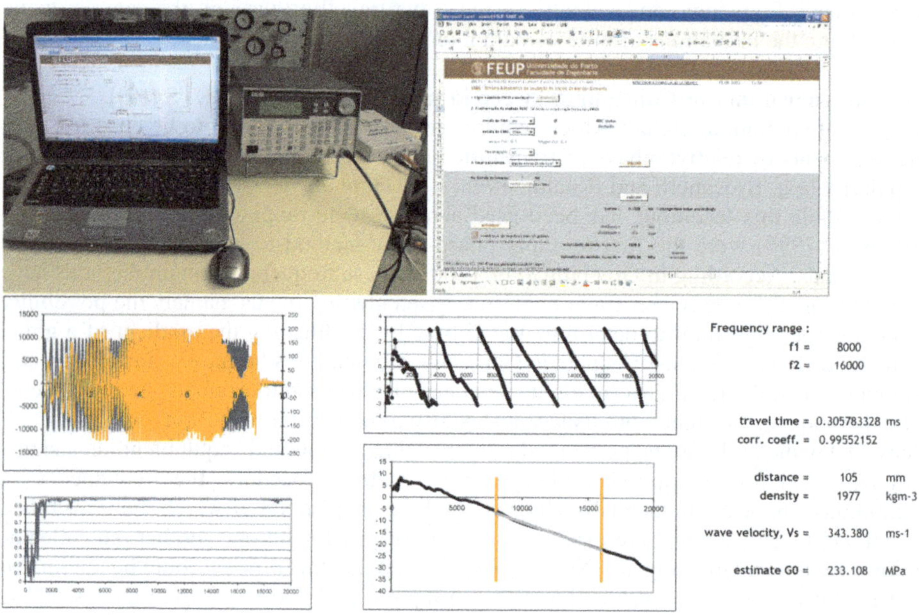

Figure 1 – Data acquisition results using ABETS

A non-linear relationship between the relative phase and the signal frequency can be generally observed, showing that the 0-20kHz range is too broad to provide reasonable results. However, this is useful as an overview of the full coherence function as well as the complete unwrapped phase against frequency relationship, with the purpose of deciding the most adequate ranges to select. A selection of a high coherence range is necessary to obtain low dispersion in the results, thus a high correlation coefficient of the best fit line.

3. BE AND RC TESTS ON RESIDUAL SOIL SAMPLES

3.1 Outline of the tests

In this research, the RC apparatus is a Hardin oscillator, manufactured by Seiken (Japan) and installed at the Geotechnical Laboratory of IST, in Lisbon. The modifications in the platens for the BE were designed by the authors and new platens were made by the manufacturer. The bender elements were manufactured at COFS (in UWA, University of Western Australia) and consist of a T-shaped pair encased in a stainless steel casing. The electronic equipments used for BE measurements comprise a programmable function generator; input and output multiplexers (also specifically developed at COFS); an oscilloscope connected to a PC or, alternatively, a data acquisition with spectral analyser device (*PICOScope*), also connected to a PC, acquiring on ABETS. In each test, simultaneous measurements of the RC and BE were made, using the two interpretation methodologies for the BE results, in order to identify any discrepancies in the measured shear wave velocity, towards determining which method most closely estimates the true travel time. In Figure 2 a photograph of the apparatus is shown, with emphasis to the modified platens equipped with the T-shaped pair of bender elements.

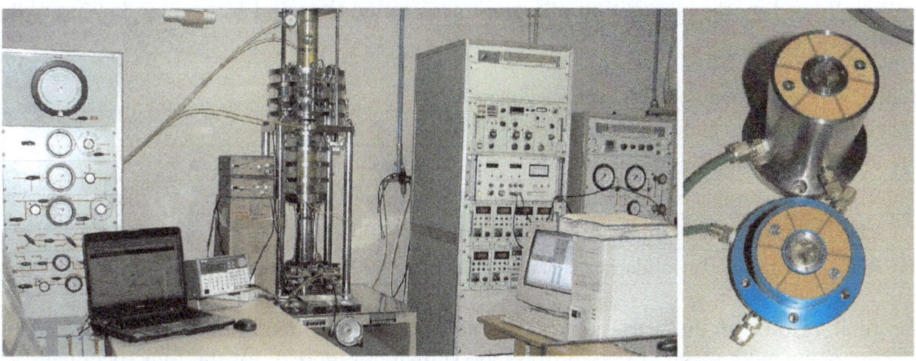

a) b)

Figure 2 – Resonant column device with bender elements: a) test apparatus; b) modified platens

Porto granitic residual soil has been selected in this work as it is regionally dominant and has been thoroughly studied for many years. Residual soils result from physical, chemical and mechanical weathering processes acting on rocks and, in the present case, this soil is a weathered granite. A reference work on its characterization has been presented by Viana da Fonseca (2003). The typical Porto granite is a leucocratic alkaline rock, with two micas and of medium to coarse grain size. The natural variability of the rock generates a fairly heterogeneous mass in macro scale. This soil is characterised by the presence of a bonded structure and fabric that has significant influence on its engineering behaviour. For the present research, it is relevant to mention its wide grain size distribution, usually defined in the Unified Classification of Soils (ASTM D 2487-85), as silty (SM) or well graded (SW) sands, or more rarely as clayey sands (SC). The presence of fines adds higher compressibility than that of sands, though permeability is relatively high ($k=10^{-6}$-10^{-5} m/s).

Four simultaneous RC and BE tests were carried out on natural samples of Porto residual soil, which will be analysed in this work. The main properties of the test specimens are summarized in Table 1; it should be emphasized that each of these were fairly homogeneous in texture and fabric.

Table 1 - Identification and soil index properties of the residual soil samples

Specimen	Borehole	γ kN/m^3	w %	e	Sr %	w_L %	w_P %	<2um %	<#200 %	<#10 %
01CRBE	S2 [6.25-6.50m]	19.5	22.4	0.64	93	n/a	n/a	5.0	42.0	94.5
02CRBE	S2 [2.00-2.30m]	19.3	19.4	0.62	83	30	26	5.5	36.9	89.8
03CRBE	S1 [2.20-2.30m]	19.5	22.9	0.65	94	32	25	7.3	46.4	88.4
04CRBE	S1 [2.10-2.20m]	19.4	23.0	0.66	93					

The tests comprised the application of isotropic and (or) anisotropic confining pressures, at various stages, as outlined in Table 2. All tests were drained, and the samples were almost but not fully saturated during testing, maintaining the natural moist. After a stabilization period of about 30 minutes, RC readings of the resonance frequency were made, followed by the acquisition of wave traces in the time and frequency domain with the BE.

Table 2 - Testing conditions for each specimen

Specimen	Testing conditions
01CRBE	Anisotropic consolidation ($K_c = 0.5$): 6 stages up to $\sigma'_v = 400$kPa
02CRBE	Anisotropic consolidation ($K_c = 0.5$) and shear: 7 stages up to $\sigma'_v = 120$kPa
03CRBE	Isotropic consolidation: 8 stages up to $\sigma'_c = 400$ kPa
04CRBE	Isotropic consolidation: 8 stages up to $\sigma'_c = 400$ kPa, with creep (for 48 days)

The final results of the variation of the shear modulus, determined by RC and BE measurements, with mean effective stress are shown in Figure 3, for the four specimens.

The results exhibit significant similarities in terms of the G_0 versus p' relation and the dispersion between RC and the two BE results are very small. The derived equations for this relation, respective to each method, evidence good convergence. These equations are merely indicative for the dependence of G_0 with p', as it is beyond the scope of this paper to examine this relation. It is assumed that the obtained constants (multiplier and exponent) incorporate very important contributions from void ratio, structure, stress history, among others, which must be addressed separately. It is interesting to note that, despite having being followed diverse stress paths in each test, the overall curve for each method closely matches the individual points.

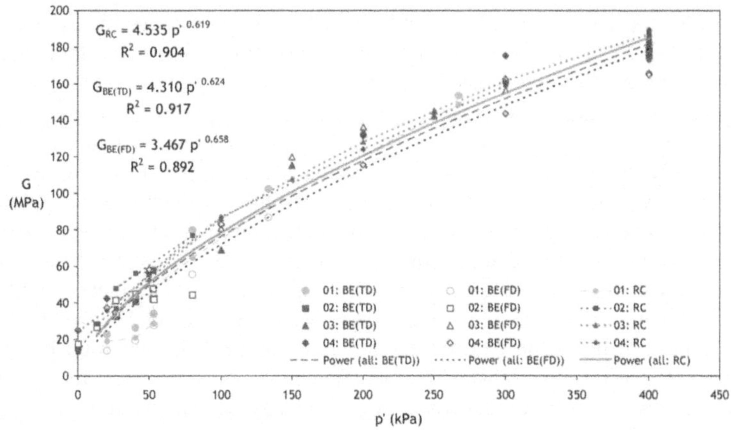

Figure 3 – Variation of G_0 with p' for all samples, from RC and BE readings

This is, however, the final graph obtained after careful analyses of the acquired data. The RC test is known to be very accurate dynamic test and the experience from these (and previous) tests confirm it. The margin of error from one operator to another is generally less than 1Hz for the resonant frequency under standard testing conditions. BE readings are clearly more controversial: time domain results remains subjective and the frequency domain technique, despite being automated and clearly more objective, still requires judgement and an "experienced eye" for post-processing the data. It is therefore worthwhile taking a closer look at one of the tests, to understand the underlying requirements towards good and reliable results.

3.2 Analysis and Discussion of the Results

As an example of the BE test procedure and analysis, test 04CRBE will be examined in detail. In the present tests, the first arrival method was used and several input frequencies

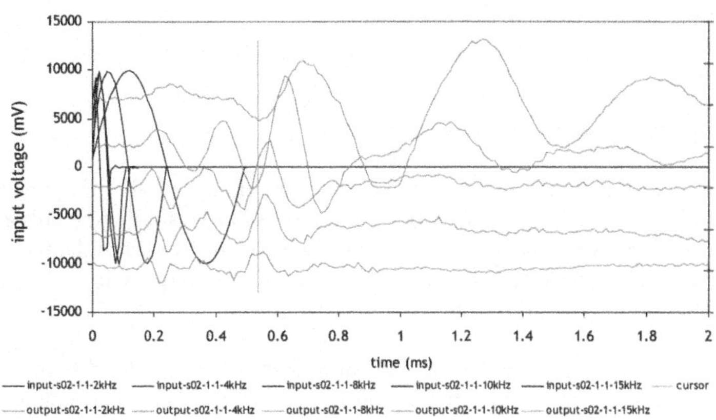

Figure 4 - Time domain results in stage #02 of test 04CRBE

at relevant levels were applied, in order to minimize the sources of errors. Figure 4 illustrates the BE traces for 2, 4, 8, 10 and 15kHz input pulse sine waves, for stage #02, corresponding to an isotropic effective stress of 50kPa.

Despite the diverse configurations of the response waves for different input frequencies, it is possible to identify a "most likely" arrival point, even if questionable. The distortions in the traces evidence effects of different nature as frequency increases. More relevant to the discussion is to observe the aspect of the output waves with the increase of applied confining pressure, as shown in Figure 5. For clarity, the complete set of frequencies acquired is not displayed, and only selected frequencies of the input waves are shown for each test stage. Since the travel time is decreasing as a consequence of higher effective stresses, the most relevant frequencies of the input wave change and need to be readjusted. This comment is based on experience and is theoretically confirmed by the requirement of a minimum normalised distance from the source, defined . The recording of several frequencies, by default, evidences and safeguards this important "rule of thumb".

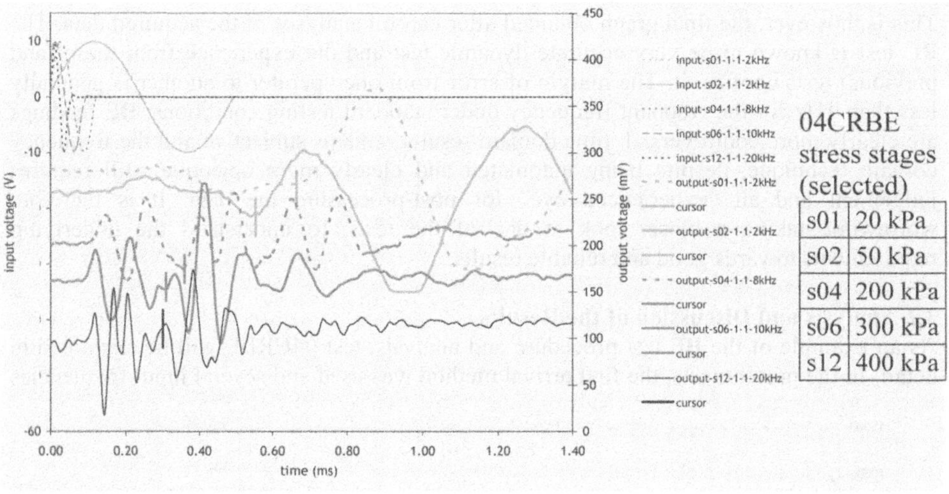

Figure 5 – Time domain results in various stages of test 04CRBE

Similarly, it is interesting to compile the results of the application of the frequency domain technique into a single graph, as presented in Figure 6. As an example, the curves of unwrapped phase angle versus frequency for selected stages of 04CRBE are depicted, where the best fit line is also represented, for the selected range of frequencies of each test stage, from which the travel time of the shear wave is calculated.

It can be seen that in the time domain, changes in confining pressure are noticeable in the shape of the wave, yet more subtle in the distinction of travel time. In the frequency domain, the slope variation is clearer and more illustrative of such stress variations.

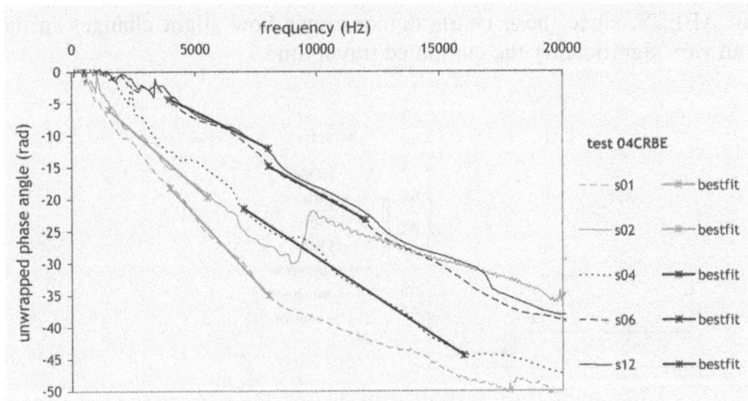

Figure 6 – Travel time determination from the unwrapped phase angle versus frequency

An interesting and useful analysis of the frequency domain results consists on selecting different ranges of frequencies in order to observe the changes in the travel time. With the purpose of further understanding this variation of travel time with this method, another approach has been investigated. This consists on a different method of representing the travel time from the sweep results. A simple Visual Basic program was implemented, to manipulate the sweep data considering "unbiased" and objective pre-established ranges of frequencies and to calculate the respective best fit line. The frequency ranges considered were 0.2, 1, 2, and 4kHz. The generated graph of travel time versus frequency, called "time chart", is presented in Figure 7, where the corresponding first arrival results are included, within the respective frequency range. For the purpose of comparison, the coherence plot, which gives an indication of the dependence of the received signal on the input signal, is combined in the same chart, in a secondary vertical axis. It is only reasonable to compute the travel time for frequencies of highest coherence, near unity.

The time charts show that the lowest range selections are highly variable and very sensitive to noise. Nevertheless, this works well as reference to the original data trends. All charts exhibit significant fluctuation on the travel time calculation, with lower variability for higher frequency ranges. Obviously, these tend to smooth down and average the results. On the other hand, the principle of unity coherence is not sufficient to guarantee a stable and unequivocal determination of the travel time. Correspondence between the variability of the two plots is clear, but only partly does it justify the inconsistencies of the time chart.

It can also be noted that the fluctuations of travel time with frequency for the various ranges decrease substantially with the stress increase, as well as the differences with the travel time measured by the first arrival method. This fact is likely to be associated with a higher coupling between the soil and the BE, aided by the stiffening of the soil.

The time chart, besides enabling an overview of the travel time in a continuous manner, is seen also as a useful and practical way of selecting the most adequate travel time, with less iterations and error. Therefore, it is an improvement and complement of the sweep method, applied in ABETS, since these charts demonstrates how slight changes in the frequency range can vary significantly the computed travel time.

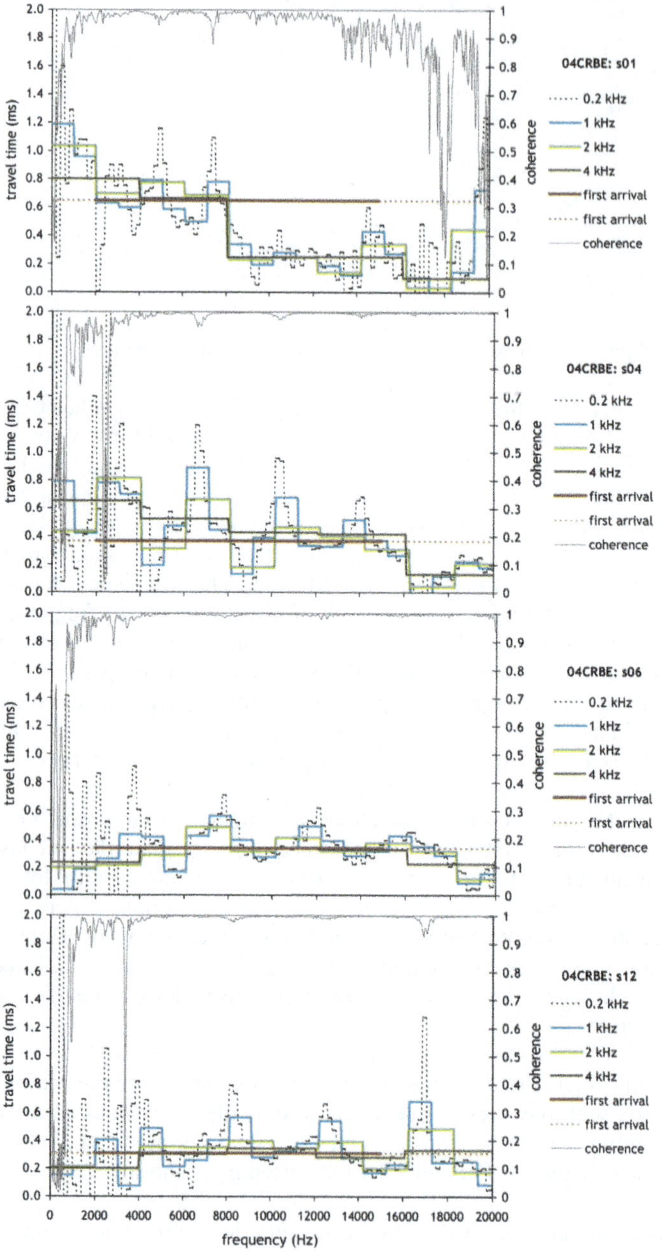

Figure 7 – Time charts for selected frequency ranges (0.5, 1, 2, 4 kHz) for 04CRBE

Having shown the analysis process for the BE methods, it is now important to address the main topic of this work, which resides on the comparison of RC and BE tests. As illustrated in Figure 3, the results from both tests show good agreement and convergence. Taking as reference the RC results for the G_0 values, it is relevant to assess the relative distance between methods. Figure 8 provides such information and enables to conclude that the FD domain technique results on G_0 values that deviate from those measured by the RC by less than 1% in average. The time domain has provided values of G_0 that are within an average of 2% of the RC values. The lower and upper limits of the ratio between G_0 values for the FD compared with the RC are of about 58% and 110%, respectively. These limits are similar to the ones measured in the TD, which are 58% and 125%, respectively.

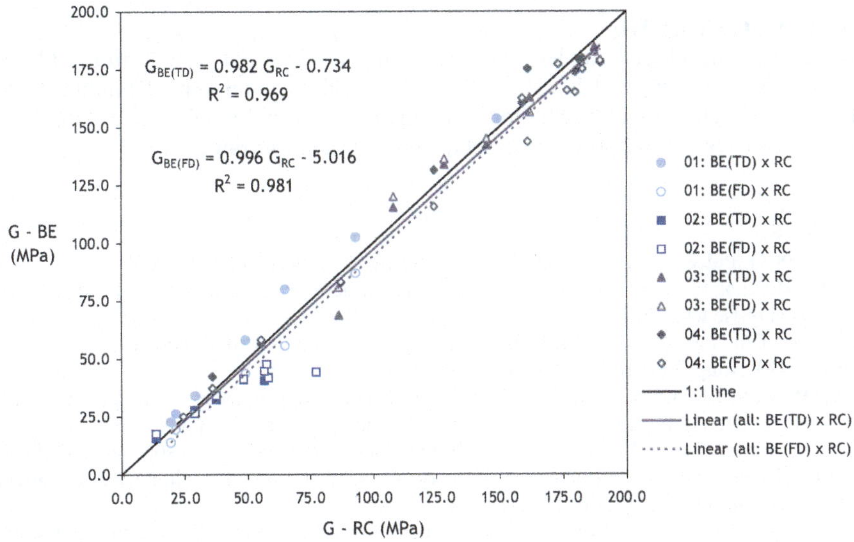

Figure 8 – G_0 values derived from the RC versus from the two BE methods

Based on these results, it can be stated that the bender element tests provide good and valid results, comparable to the resonant column. It is acknowledged that the methods currently available, and herein applied, have some subjectivity and still require some judgement, but at a much lower degree than that associated with a first peak detection on a single pulse reading.

CONCLUSIONS

The use of bender elements in the laboratory as means to determine the shear wave velocity hence the small strain stiffness continues to increase worldwide. Recent and old discussions on the inherent subjectivity of the interpretation of its results continue to leave room for improvement and research. This paper has analysed the performance of the two most common interpretation techniques for the BE test, by comparing the results with the resonant column test, which is renowned for the quality of its measurements of the dynamic parameters. This has been made by modifying a RC apparatus to accommodate BE and by performing simultaneous readings of the RC frequency and the BE waves in the time and the frequency domain.

Results show that the two testing methods compare well. Both BE interpretation techniques can yield very close values of travel time, though its determination cannot be fully automated nor unambiguous, even for the frequency domain. In effect, both techniques require some degree of judgement and experience that has not yet been eliminated. Data acquisition and post-processing procedures are suggested, as these have proven to lead to reliable values of G_0 in the light of the close agreement obtained between RC and BE test results.

ACKNOWLEDGEMENTS
This work was developed under the research activities of CEC from FEUP (POCTI/ECM/55589/2004) and ICIST of IST, supported by multi-annual funding from FCT (Portuguese Science and Technology Foundation). Special thanks are due to Eng. Gouveia, for his technical support during setup and laboratory testing.

REFERENCES

Arroyo, M. (2001). Pulse tests in soil samples. PhD thesis, University of Bristol, UK.

Arroyo, M.; Medina, L. and Muir Wood, D. (2002). Numerical modelling of scale effects in bender-based pulse tests. NUMOG VIII, Pande, G.N. & Pietruszczak, S. (eds), pp. 589-594.

Arulnathan, R.; Boulanger, R.W. and Riemer, M.F. (1998). Analysis of Bender Element Tests. Geotechnical Testing Journal, Vol. 21, No. 2, pp. 120-131.

Blewett, J.; Blewett, I.J. and Woodward, P.K. (1999) Measurement of shear wave velocity using phase sensitive detection techniques. Canadian Geotechnical Journal 36, pp. 934-939.

Blewett, J.; Blewett, I.J. and Woodward, P.K. (2000). Phase and amplitude responses associated with the measurement of shear-wave velocity in sand by bender elements. Canadian Geotechnical Journal, Vol. 37, pp. 1348-1357.

Brocanelli, D. and Rinaldi, V. (1998). Measurement of low strain material damping and wave velocity with bender elements in the frequency domain. Canadian Geotechnical Journal, Vol. 35, pp. 1032–1040.

Dyvik, R. and Madhsus, C. (1985). Lab measurements of Gmax using bender elements. Proceedings ASCE Annual Convention: Advances in the art of testing soils under cyclic conditions, Detroit, Michigan, pp. 186-197.

Fam, M. A.; Cascante, G. and Dusseault, M. B. (2002). Large and Small Strain Properties of Sands Subjected to Local Void Increase. Journal of Geotechnical and Geoenvironmental Engineering, Vol. 128, No.12, pp. 1018-1025.

Fratta, D. and Santamarina, J.C. (1996). Wave propagation in soils: Multi-mode, wide-band testing in a waveguide device. Geotechnical Testing Journal, Vol. 19, No. 2, pp. 130-140.

Gordon, M.A.and Clayton, C.R.I (1997). Measurement of stiffness of soils using small strains triaxial testing and bender elements. Modern Geophysics in Engineering Geology. McCann, D.M.; Eddleston, M.; Fenning, P.J. & Reeves, G.M. (eds.). Geological Society Engineering Geology Special Publication, No. 12, pp. 365-371.

Greening, P.D. and Nash, D.F.T. (2004). Frequency domain determination of G_0 using bender elements. ASTM Geotechnical Testing Journal, Vol. 27, No. 3, pp. 288-294.

Greening, P.D.; Nash, D.F.T.; Benahmed,N.; Viana da Fonseca, A. and Ferreira, C. (2003). Comparison of shear wave velocity measurements in different materials using time and frequency domain techniques. Proceedings of Deformation Characteristics of Geomaterials, Lyon, France, 22-24 September, Lyon, France:Balkema, pp. 381-386.

Hardy, S.; Zdravkovic, L. and Potts, D.M. (2002). Numerical interpretation of continuously cycled bender element tests. Numerical Models in Geomechanics (NUMOG VII). Pande & Pietruszczak (eds.). Swets & Zeitlinger, Lisse, pp. 595-600.

Jamiolkowski, M.; Lancellotta, R. and Lo Presti, D.C.F. (1995). Remarks on the stiffness at small strains of six Italian clays. Pre-failure Deformation of Geomaterials. Shibuya, Mitachi & Miura (eds). Balkema, Rotterdam, pp. 817-836.

Jovicic, V.; Coop, M.R. and Simic, M. (1996). Objective criteria for determining Gmax from bender element tests. Geotéchnique, Vol. 46, No. 2, pp. 357-362.

Jovicic, V. (2004). Rigorous bender element testing. Workshop on Bender Element Testing of Soils, UCL, London.

Kaarsberg, E.A. (1975). Elastic wave velocity measurements in rocks and other materials by phase-delay methods. Geophysics, Vol. 40, pp. 855-901.

Lee, J. S. and Santamarina, C. (2005). Large and Small Strain Properties of Sands Subjected to Local Void Increase. Journal of Geotechnical and Geoenvironmental Engineering, Vol. 131, No. 9, pp. 1063-1070.

Moncaster, A.M. (1997). The shear modulus of sand at very small strains. MSc thesis. Department of Civil Engineering, University of Bristol, UK.

Pennington, D.S. (1999). The anisotropic small strain stiffness of Cambridge Gault clay. PhD thesis, Department of Civil Engineering, Univ. Bristol.

Rio, J.; Greening, P. and Medina, L. (2003). Influence of sample geometry on shear wave propagation using bender elements. Proceedings of Deformation Characteristics of Geomaterials, Lyon, France, 22-24 September, Lyon, France:Balkema, pp. 963-967.

Sánchez-Salinero I, Roesset JM & Stokoe II KH (1987) Analytical studies of wave propagation and attenuation Geotechnical report No GR86-15. Civil Engineering department, University of Texas at Austin.

Viana da Fonseca, A. (2003). Characterising and deriving engineering properties of a saprolitic soil from granite, in Porto. 'Characterisation and Engineering Properties of Natural Soils'. Swets & Zeitlinger, Lisse, pp. 1341-1378.

Viggiani,G.; Atkinson, J.H.(1995). Interpretation of bender element tests. Géotechnique, Vol. 45, No. 1, pp. 149-154.

Soil Stress-Strain Behavior: Measurement, Modeling and Analysis
Geotechnical Symposium in Roma, March 16 & 17, 2006

DYNAMIC MEASUREMENTS AND POROSITY IN SATURATED TRIAXIAL SPECIMENS

Marcos Arroyo*, Cristiana Ferreira** and Jiraroth Sukolrat***
* Department of Geotechnical Engineering and Geosciences, UPC, Barcelona, Spain
**Department of Civil Engineering, Universidade do Porto, Porto, Portugal
***Department of Civil Engineering, University of Bristol, U.K.

ABSTRACT

Recent work has shown how soil porosity may be obtained from elastic wave measurements in the field. The procedure is based on Biot poroelastic theory, and employs both S-wave and P-wave measurements. Only a limited amount of laboratory tests has been employed to date to substantiate this procedure. All of them relate to sand samples, in tests where porosity variation was small. In this work we apply the poroelastic procedure to infer porosity during triaxial tests instrumented with piezoelectric transducers. The materials tested, Bothkennar clay and Porto granite residual soil, have large variations in porosity, variations that are independently measured. The results would offer some light on the applicability of the poroelastic procedure to the frequency testing range common in laboratory applications.

KEYWORDS
LABORATORY, STIFFNESS, DYNAMICS, ELASTICITY, NUMERICAL MODELLING

1. INTRODUCTION

The use of non-destructive techniques to monitor soil tests in the geotechnical laboratory has several advantages. One of them is the useful relation that may be established with field geophysics, allowing, for instance, for techniques established on the field to be employed in the laboratory.

In a recent series of papers, (Foti et al., 2002; Foti & Lancellotta, 2004), Foti and co-workers have described a technique to obtain soil porosity from seismic wave measurements. The technique uses the Biot poroelastic model to invert the wave velocity measurements.

The main motivation of their work was the well-known difficulty of obtaining adequate samples for porosity estimation in many fragile and granular soils. With this purpose in mind, most of the supporting evidence presented by Foti et al. relates to dynamic field measurements. More specifically, they compared porosity deduced from cross-hole measurements with independent on-sample laboratory determinations. The agreement thus obtained between the directly measured and the inferred porosity value was generally very good, thus proving the attractiveness of the proposed technique.

Hoe I. Ling et al. (eds.), Soil Stress-Strain Behavior: Measurement, Modeling and Analysis, 537–546.

Foti et al. do also present some comparisons between porosity deduced from dynamic laboratory data and that independently measured in the laboratory. However, the laboratory data presented (Bates, 1989) was relatively scarce or referred to the more complex, unsaturated case (Tsukamoto et al. 2002). The purpose of this communication is to further explore the potential of the porosity dynamic measurement technique in laboratory saturated triaxial conditions.

While the advantage of obtaining estimates of "in situ" porosity is clear, some benefits may also accrue from the use of this technique at the soils laboratory. For instance, when testing saturated soil samples, the current alternatives to measure volume change rely on gauging the amount of water in or out of the sample, or in accurate multi-dimensional deformation measurements. These two techniques are not without problems and the possibility of having access to the same information by different means will be surely welcome.

On the other hand, laboratory seismic wave measurements in soils are themselves tricky enough. Interpretation difficulties and measurement uncertainty are sometimes important, particularly when shear wave velocity is measured (e.g. Arroyo et al. 2003a). Many difficulties can be traced to the various size effects that appear when probing samples with wavelengths commensurate with their size. Some problems like those derived from operating in the source near field can now be avoided using simple practical rules (Arroyo et al. 2003b) or by inverse analysis (Lee & Santamarina, 2005). No simple rules or inverse model are yet available to avoid the guide effects that may appear when transmitting pulses alongside cylindrical samples (Arroyo et al., 2006).

In this context, the proposals by Foti et al. are here employed in a different way. Instead of using seismic measurements to obtain porosity, the independently measured porosity may be employed to check the coherence of the measured velocities and/or to improve the means of interpreting the seismic test results.

2. EXPERIMENTAL DETAILS

2.1. Bothkennar clay (Bristol University)

The first material employed was Bothkennar clay. Bothkennar is a soft clay from a well known experimental test site whose properties have been intensively investigated (e.g. Hight, Bond & Legge, 1992). The sample was taken from a depth of about 8 m and had an initial water content of circa 73%, typical for the Bothkennar site. At such depth the typical permeability value is 10^{-9} m/s, while the specific gravity is 2.65.

A 150 mm height and 75 mm diameter sample (BN13) was subject to a K_0 consolidation path with stress reversals in a Bishop-Wesley triaxial cell. Details of the apparatus and the static testing program can be found elsewhere (Sukolrat, 2006). Volume change was measured with a bellofram type volumetric transducer of Imperial College design. The reference void ratio thus measured during the dynamic tests period is shown in Figure 1.

Small strain excursions, monitored with static transducers, suggest a Poisson ratio value of 0.11 at a strain level of 0.05%, increasing to 0.21 at higher strains. The Skempton B value after saturation was above 0.95.

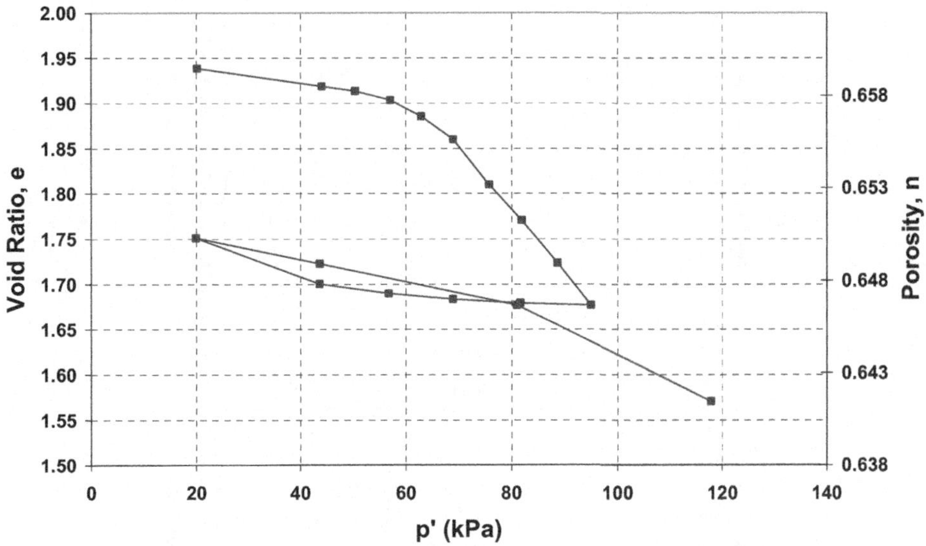

Figure 1 Porosity change in sample BN13 of Bothkennar clay during the dynamic measurement period

P and S wave velocities were measures using bender-extender transducers (Lings & Greening, 2001). The measurement system comprised a TG 1010 function generator, an amplifier for the received signal and a Tektronix TDS3014 oscilloscope. When measuring V_s, a half sine pulse of 5 kHz apparent frequency was used as input to the system. When measuring V_p the apparent frequency was increased up to 30 kHz. Selection of arrival times was made by picking a characteristic point of the time domain trace (Figure 2).

2.2. Granite saprolite (Universidade do Porto)

Granite saprolite is a residual soil common in the North-Western region of Portugal. Viana da Fonseca (2003) gives an overview of its geotechnical characteristics. It is a highly porous soils were quartz particles are bonded by plagioclase bridges.

The permeability values in intact saprolite samples ranges from 10^{-5} to 10^{-6} m/s (Viana da Fonseca, 2003). There is not much direct data about the small strain Poisson ratio, but a value of 0.2 is usually adopted for the soil skeleton (Ferreira, 2003).

A triaxial cell adapted from an ISMES design was employed in a large series of triaxial tests on reconstituted and intact saprolite samples. Details of the apparatus and the static test results can be found elsewhere (Ferreira, 2003). The Skempton B value was always higher than 0.90 and usually higher than 0.93. The specific gravity of each sample was measured, giving values between 2.65 and 2.70.

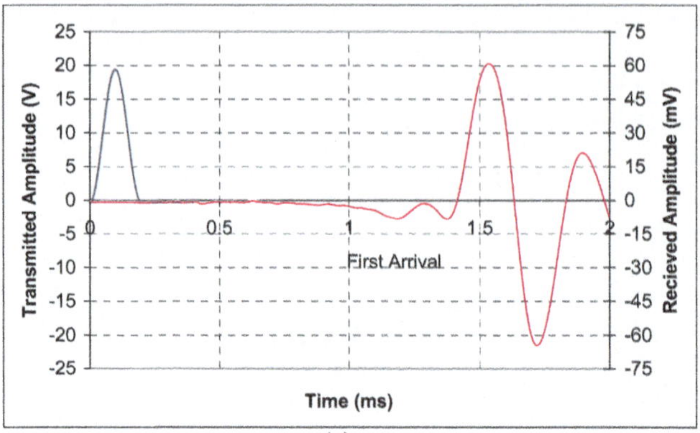

(a)

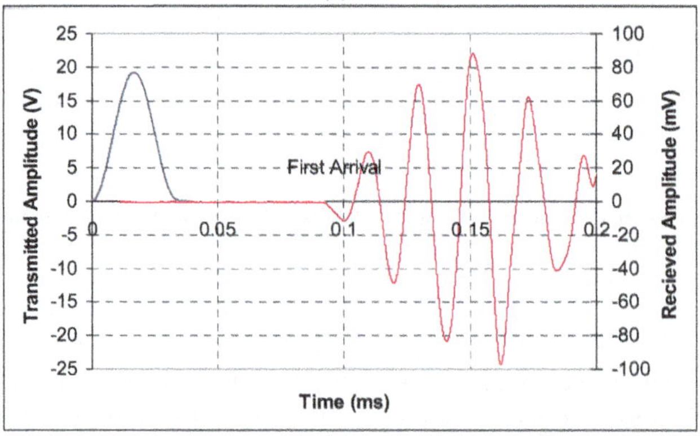

(b)

Figure 2 Example arrival time selection for Bothkennar clay (a) compressive wave (b) shear wave

During the triaxial test program, shear and compressive wave measurements were taken both during consolidation, saturation and shearing. P-wave measurements were taken using compression transducers (Brignoli et al., 1996) using pulses of sinusoidal shape and apparent frequencies of 25, 50 and 75 kHz. S-wave measurements were obtained using bender elements. The electronic equipments used for running the seismic measurements consist of a function generator *(TTi* TG1010); an integrated unit with input and output amplifiers (specifically developed by *ISMES-Enel.Hydro*); and an oscilloscope (*Tektronix* TDS220), connected to a PC.

Again, the arrival time for P-waves was selected in the time domain in a plot representing together the response at different frequencies. For S-waves, the arrival time was obtained both on the time domain by the identification of the first arrival of the output wave and on the frequency domain by the π-point phase comparison method (Greening & Nash, 2004). Figure 3 illustrates the two interpretation methods.

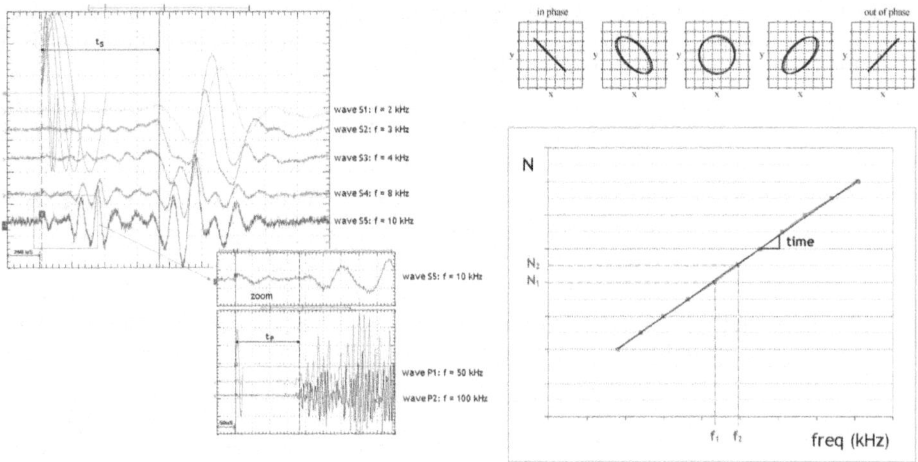

a) time domain b) frequency domain

Figure 3 Determination of shear wave travel time: a) first arrival, using various input frequencies; b) frequency versus number of wavelengths: the slope is the travel time

3. POROELASTIC FORMULAE

The poroelastic Biot model results in dispersive wave propagation, i.e. wave velocities that are generally dependent on frequency (see, for instance, Miura et al., 2001). This dependency is negligible below a certain frequency range. The upper limit of that frequency range is given by a characteristic frequency (Foti et al. 2002), which, in Hertz, is given by

$$f_c = \frac{ng}{2\pi k} \tag{1}$$

Where g is 9.81 m/s^2, n the porosity and k is the Darcy permeability of the soil. For the Bothkennar clay the characteristic frequency is close to 100 MHz. For the undisturbed Porto saprolites the characteristic frequency is slightly above 60 kHz, whereas for the reconstituted samples it goes up to 6 MHz. Some of the P-wave measurements on undisturbed saprolites may have been taken close to this limit.

Under the assumption of a very high bulk modulus of the soil particles, Foti et al.(2002) obtained the following formula

$$n = \frac{\rho^S - \sqrt{\left(\rho^S\right)^2 - \dfrac{4\left(\rho^S - \rho^F\right)K^F}{V_P^2 - 2\left(\dfrac{1-\upsilon^{SK}}{1-2\upsilon^{SK}}\right)V_S^2}}}{2\left(\rho^S - \rho^F\right)} \tag{2}$$

The porosity, n, is here a function of the bulk shear wave velocity, V_S, and the compressive wave velocity, V_P. The other parameters are the mass density of the soil particles, ρ^S, that of the fluid, ρ^F, the fluid bulk modulus, K^F, and the soil skeleton Poisson ratio, υ^{SK}.

The previous equation can be easily inverted to give the shear velocity as

$$V_S = \sqrt{\frac{\left(1-2\upsilon^{SK}\right)}{2\left(1-\upsilon^{SK}\right)}\left\{V_P^2 - \frac{K^F}{n^2\left(\rho^F-\rho^S\right)+n\rho^S}\right\}} \quad (3)$$

It is also interesting to appreciate that a lower bound of the compressive wave velocity, neglecting the skeleton soil stiffness, is given by

$$V_P \geq V_F \sqrt{\frac{1}{n^2+n(1-n)\frac{\rho^S}{\rho^F}}} \quad (4)$$

V_F is the value of sound velocity in the saturating fluid.

4. RESULTS

The previous formulae may be applied to check the velocities measured on the samples. The fluid wave velocity in water, V_F, at 20°C is 1483.74 m/s (Marczak, 1997). This is equivalent to a bulk modulus of K_F = 2.2*10^6 kPa. Using this value and the measured porosity and soil specific gravity, a first check can be made on the compressive wave measurements using equation 4. The results are plotted in Figure 4.

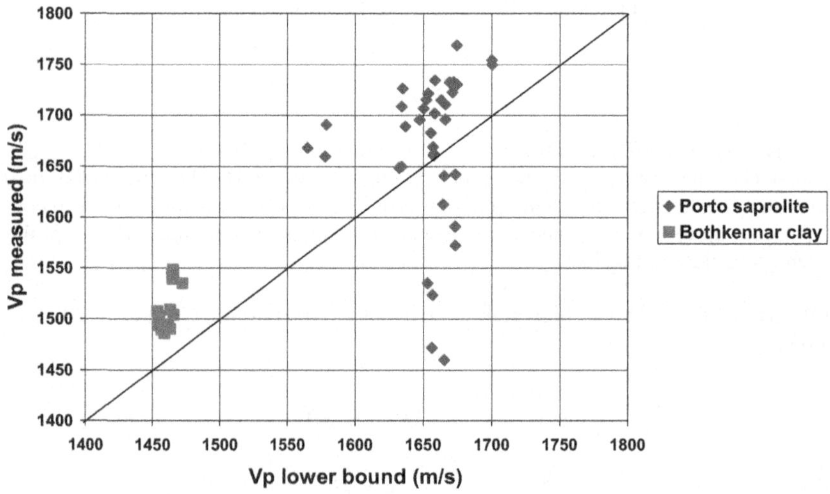

Figure 4 Compressive wave measurements. Estimated value versus lower bound according to eq (4)

It is clear that some of the saprolite measurements are below the theoretical lower bound. The most likely cause for this behaviour is incomplete saturation. This was indeed anticipated in some cases, but not in all, on the basis of a relatively low B value. This illustrates how compressive wave measurements may be more sensitive to saturation than the B value usually measured.

The next step is to estimate the shear wave measurement, according to equation (3), from the measured porosity and P-wave velocity. For the cases of incomplete saturation, evaluation of the fluid bulk modulus would need an estimate of saturation at each measurement point, which is not available. Therefore one would like to exclude the cases which are not saturated. To do this practically we deem not saturated the cases in which the compressive wave measurement was below 1.02 times the lower bound value given by equation (4).

With this criteria and using the Poisson ratio values previously referred, the estimates of shear velocity obtained with equation (3) are compared with the measurements in Figure 5. The measured velocities are well below the values estimated from porosity and compressive measurements. It is also clear that the porosity-based estimates are more sensitive to porosity changes than the measured values.

The average divergence between the porosity based estimate and the measured value is highly dependent on the Poisson ratio value. If a value of 0.4 is assumed the adjustment between both datasets is clearly improved, as shown in Figure 6. With the same criterion the comparison between porosity estimated with equation (2) and measured independently is shown in Figure 7.

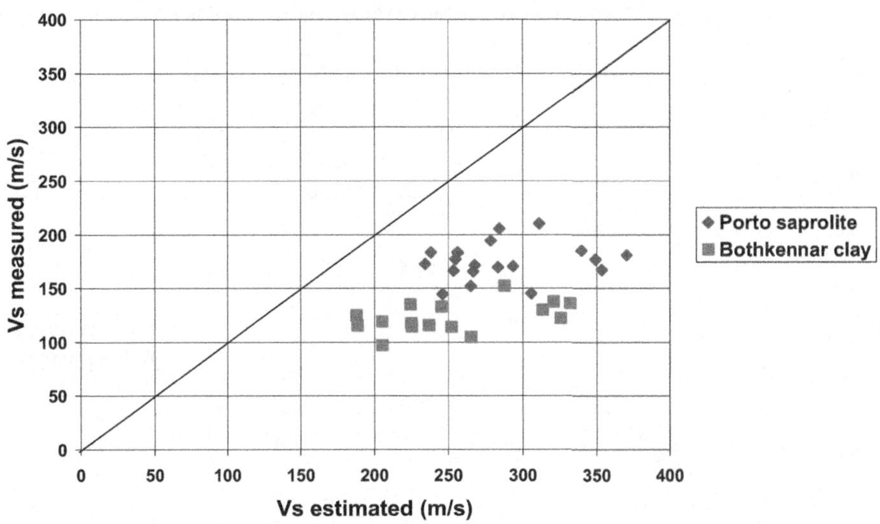

Figure 5 Comparison of estimated and measured shear velocity. Poisson ratio from previous information.

After Poisson ratio optimization, the porosity estimate for the saprolite seems to withstand better the comparison with the measured porosity value than that of Bothkennar clay. This may perhaps be taken as an indication that shear velocity estimates for the saprolite were more precise than those of the clay.

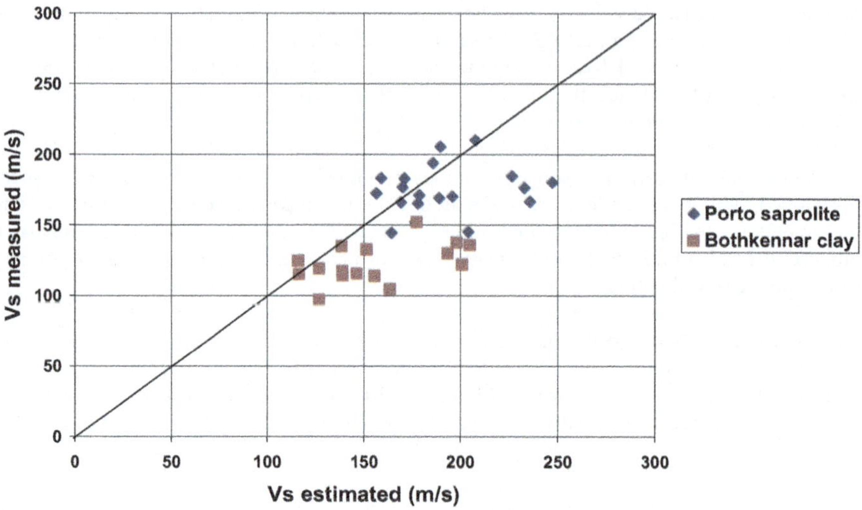

Figure 6 Comparison of estimated and measured shear velocity. Poisson ratio optimized to 0.4.

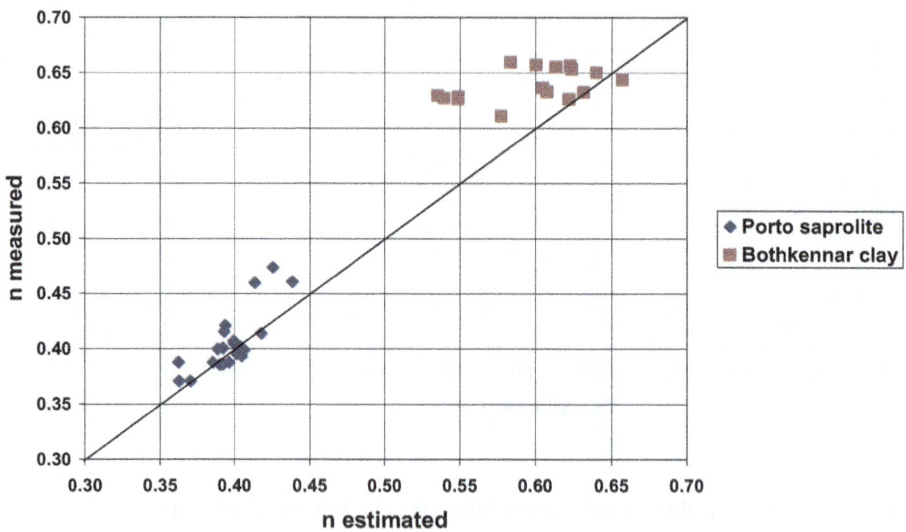

Figure 7 Comparison of estimated and measured porosity. Poisson ratio optimized to 0.4

5. SUMMARY

The use of P-wave measurements alongside the more common S-wave measurements in triaxial samples has some advantages. By means of simple poroelastic formulae it is possible to check the measured wave velocities using the conventionally measured porosity. The effect of incomplete saturation can be easily detected. The estimates of shear wave velocity based on porosity and compressive velocity seem to be too high. The estimates of porosity based on both wave velocities seem somewhat low.

ACKNOWLEDGEMENTS

The work at the University of Porto was developed under the research activities of CEC from FEUP (POCTI/ECM/55589/2004), supported by FCT (Portuguese Science and Technology Foundation).

REFERENCES

Arroyo, M., Muir Wood, D., Greening, P., Rio, J. & Medina, L. (2006) Effects of sample size on bender-based axial G_0 measurements, *Géotechnique*, 56, 1, 39-52

Arroyo, M., Greening, P. & Muir Wood, D. (2003a) An estimate of uncertainty in current pulse test practice. *Rivista Italiana di Geotecnica* 37, 1, 17-35

Arroyo, M., Wood, D.M. & Greening, P. (2003b) Source near-field effects and pulse tests in soil samples, *Géotechnique* 53, 3, 337-345

Bates, C.R. (1989) Dynamic soil property measurements during triaxial testing, *Geotechnique*, 39, 4, 721-726

Brignoli, E.G.M.; Gotti, M.; Stokoe, K.H.II (1996). Measurement of shear waves in laboratory specimens by means of piezoelectric transducers. *Geotechnical Testing Journal*, 19, 4, 384-397

Ferreira, C. (2003). Implementation and application of piezoelectric transducers on the determination of seismic wave velocities in soil specimens. Assessment of sampling quality in residual soils. *MSc Thesis, Universidade do Porto* (in Portuguese)

Foti, S., Lai, C.G. & Lancellotta, R. (2002) Porosity of fluid-saturated porous media from measured seismic wave velocities, *Géotechnique*, 52, 5, 359-373

Foti, S. & Lancellotta, R. (2004) Soil porosity from seismic velocities, *Géotechnique*, 54, 8, 551-554

Greening, P.D. & Nash, D.F.T. (2004) Frequency domain determination of G0 using bender elements, ASTM *Geotechnical Testing Journal*, 27, 3, 1-7

Hight, D.W., Bond, A.J. & Legge, J.D., Characterization of the Bothkennar clay: an overview, *Géotechnique* 42, 2, 303-347

Lee, J.-S. & Santamarina, J.C. (2005) Bender elements: performance and signal interpretation, ASCE *Journal of Geotechnical and Geoenvironmental Engineering*, 131, 9, 1063-1070

Lings, M.L. & Greening, P.D. (2001). A novel bender/extender element for soil testing, *Géotechnique* 51, 8, 713-717

Marczak, W. (1997) Water as a standard in the measurements of speed of sound in liquids, *J. Acoust. Soc. Am.*, 102, 5, 2776-2779

Miura, K., Yoshida, N. & Kim, Y.-S. (2001) Frequency dependent property of waves in saturated soil, *Soils & Foundations*, 41, 2, 1-19

Sukolrat, J. (2006) *Destructuration of Bothkennar Clay* PhD Thesis, University of Bristol. (to be submitted)

Tsukamoto, Y., Ishihara, K., Nakazawa, H., Kamada, K. & Huang, Y. (2002) Resistance of partly saturated sand to liquefaction with reference to longitudinal and shear velocities, *Soils & Foundations*, 42, 6, 93-114

Viana de Fonseca, A. (2003) Characterising and deriving engineering properties of a saprolitic soil from granite, in Porto, *Characterisation and Engineering Properties of Natural Soils – Tan et al. (eds.)* 1341-1378.

Soil Stress-Strain Behavior: Measurement, Modeling and Analysis
Geotechnical Symposium in Roma, March 16 & 17, 2006

CYCLIC PLANE STRAIN COMPRESSION TESTS ON DENSE GRANULAR MATERIALS

Junichi Koseki[1)*], Job Munene Karimi[2)], Yukika Tsutsumi[1)**], Sajjad Maqbool[3)] and Takeshi Sato[1)***]
1) Institute of Industrial Science
the University of Tokyo, Tokyo, 153-8505, JAPAN
e-mail: 1)* koseki@iis.u-tokyo.ac.jp
1)** tsutsumi@iis.u-tokyo.ac.jp
1)*** tsato@iis.u-tokyo.ac.jp
2) Materials Testing and Research Department
Ministry of Roads and Public Works, Nairobi, KENYA
e-mail jmkarimi72@yahoo.com
3) Department of Civil Engineering
University of Engineering and Technology, Lahore, PAKISTAN
e-mail sajjadmaqbool@hotmail.com

ABSTRACT
A series of cyclic plane strain compression tests are performed under drained condition on dense Toyoura sand and compacted Chiba gravel, by using small-scale and large-scale apparatuses, respectively. Comparisons are made with results from monotonic loading tests. Local strain distributions are calculated by conducting image analyses of digital photographs taken at different stages of loading during each test. Based on these results, strain localization properties of dense granular materials are discussed, in particular focusing on possible effects of cyclic loading history.

1. INTRODUCTION
Dense granular materials are known to exhibit strain softening behavior during shearing, which is associated with strain localization or formation of shear band. These properties have been taken into account in evaluating rationally the seismic earth pressure that is exerted to retaining walls from compacted backfill soils (Koseki et al., 1998).

The effects of cyclic loading history on the strain localization behavior of dense granular materials are, however, not well understood. In order to investigate into the strain localization properties with/without cyclic loading history, therefore, a series of plane strain compression tests were performed on dense Toyoura sand (Karimi et al., 2005) and compacted Chiba gravel (Maqbool et al., 2005). In this paper, their results are briefly summarized, while adding the results from image analysis on compacted Chiba gravel with cyclic loading history.

2. MATERIAL, EQUIPMENT AND TEST PROCEDURES
Two kinds of granular materials (uniform sand and well-graded gravel) and two plane strain apparatuses with different sizes were used for this study.

Hoe I. Ling et al. (eds.), Soil Stress-Strain Behavior: Measurement, Modeling and Analysis, 547–555.

For testing on Toyoura sand (Batch G, D_{50}=0.2 mm, e_{max}=0.961, e_{min}=0.601 and Gs=2.635), a small-scale apparatus (Koseki et al., 2005) was employed. A prismatic specimen as shown in Figure 1 (Karimi et al., 2005) was prepared in a mold by pluviating dry sand particles through air at a relative density of 90 %, which was followed by wetting, draining and freezing processes. After rotating the frozen specimen by 90 degrees, it was set to the apparatus and thawed under room temperature. Then, it was saturated and consolidated isotropically to a confining stress of 49 kPa by increasing the partial vacuum that was applied as the back pressure without any cell pressure.

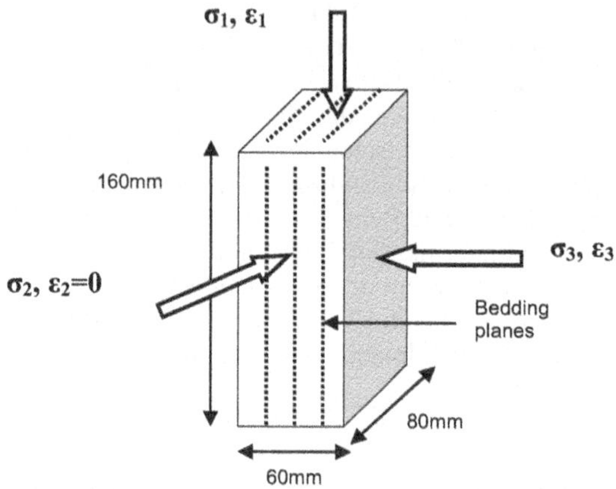

Figure 1. Toyoura sand specimen for small-scale apparatus

For testing on Chiba gravel (D_{max}=38 mm and D_{50}=11 mm), a large-scale apparatus as shown in Figure 2 (Sato et al., 2005) was employed. A prismatic specimen was prepared at an initial water content of 5.5 % by employing heavy dynamic compaction using an automatic compactor with compaction energy of 612 kJ/m3 (Maqbool et al., 2005). The dry densities obtained were within the range of 1.95-1.97 g/cm^3. Under partially saturated condition, the specimen was consolidated isotropically to a confining stress of 78 kPa following the same procedures as for Toyoura sand.

After fixing a pair of confining plates to restrict the specimen deformation in the σ2 direction, vertical loading was applied under drained condition at an axial strain rate of 0.01 %/min and 0.08 %/min, respectively, for Toyoura sand and Chiba gravel. In some of the tests, cyclic loading at constant amplitude of deviator stress was applied in the course of otherwise monotonic loading.

During shearing, digital photographs of the specimen's face were taken through a transparent confining plate. On the membrane, a series of equally spaced points had been imprinted at a spacing of 5 mm, which was used to quantify the deformation in the specimen from the displacements of the grid nodes employing image analysis (Salas-Monge et al., 2003; Tsutsumi et al., 2005).

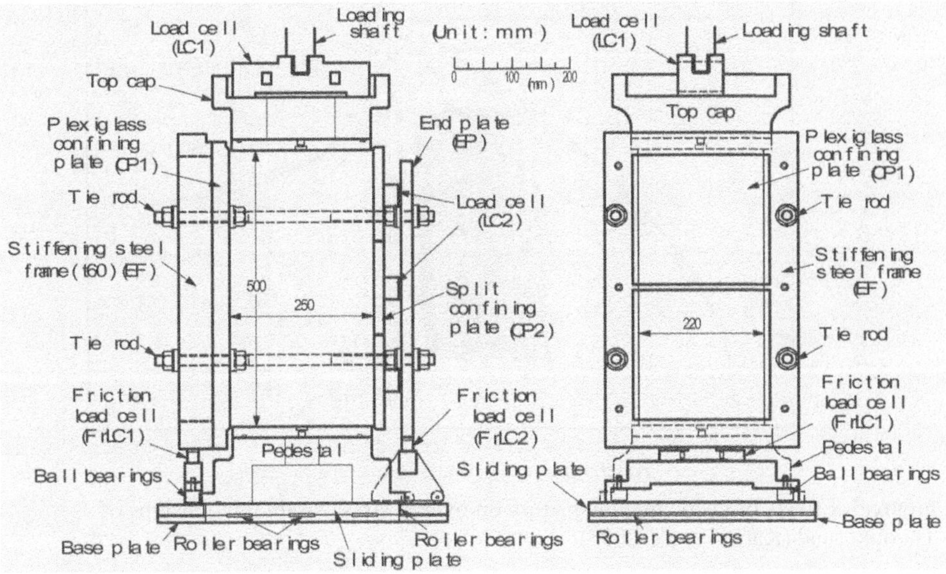

Figure 2. Large-scale plane strain apparatus

3. TEST RESULTS AND DISCUSSIONS

3.1 Toyoura sand

The overall stress-strain relationships of Toyoura sand are compared in Figure 3. Two tests without cyclic loading history (hereafter denoted as "monotonic tests") resulted in the average value of the peak deviator stress, q_{max}, of 266 kPa.

In the tests with cyclic loading history ("cyclic tests"), the maximum and minimum amplitudes of cyclic loading were set at almost 90 % and 10 % of q_{max} in the monotonic tests. In the cyclic tests, with the increase in the number of cycles from 20 to 50, the residual axial strain at the end of the cyclic loading increased accordingly, and the subsequent monotonic loading resulted in higher q_{max} value as well. The peak stress states in the cyclic tests were located along the extension of the average stress-strain curve of monotonic tests, as shown in Figure 3.

Distributions of the maximum shear strain, $\gamma_{max}=\varepsilon_1-\varepsilon_3$, at several loading stages in one of the monotonic tests (test PSCm02) are shown in Figure 4. In this test, the overall stress-strain curve exhibited the first peak at an axial strain measured with external transducer, $\varepsilon_{1,EXT}$, of about 2.1 %, followed by the second peak at $\varepsilon_{1,EXT}$, of about 2.5 % while mobilizing larger q values (Figure 3). In-between these two peaks, strain localization or formation of a shear band progressed rapidly (Figure 4).

Distributions of volumetric strain increments, $\Delta\varepsilon_{vol}=\Delta\varepsilon_1+\Delta\varepsilon_3$, at several loading stages with an increment in the $\varepsilon_{1,EXT}$ values of about 0.1 % are shown in Figure 5 for the same monotonic test. In-between the above-mentioned two peaks in the overall stress-strain curve, dilative regions with $\Delta\varepsilon_{vol}$>0.1 % appeared along the shear band.

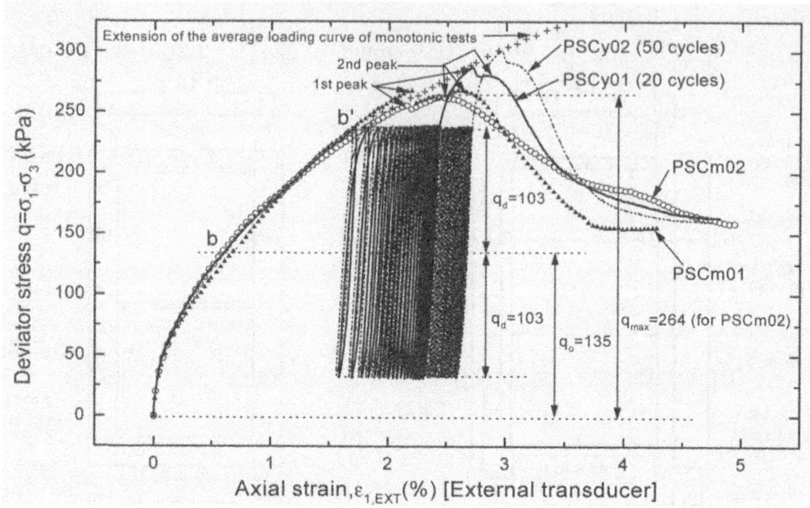

Figure 3. Effects of cyclic loading history on overall stress-strain relationships of Toyoura sand (Karimi et al., 2005)

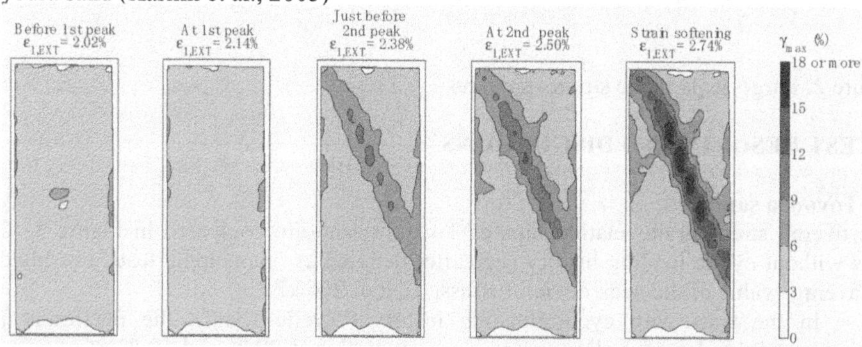

Figure 4. Contours of maximum shear strains in monotonic test (test PSCm02) on Toyoura sand (Karimi et al., 2005)

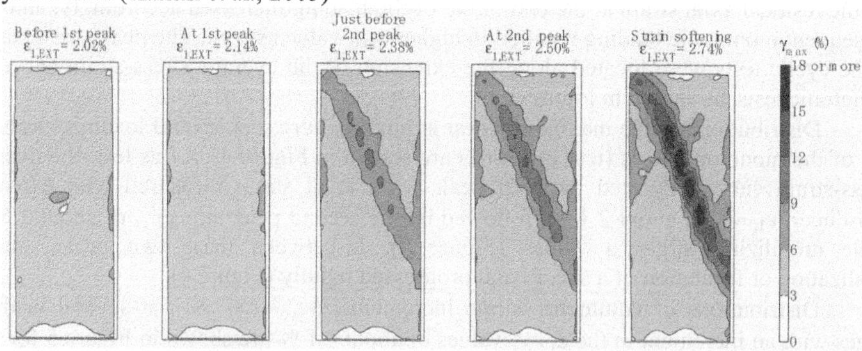

Figure 5. Contours of volumetric strain increments in monotonic test (test PSCm02) on Toyoura sand (Karimi et al., 2005)

For one of the cyclic tests (test PSCy02), distributions of γ_{max} are shown in Figure 6. The extent of strain localization at the end of the cyclic loading stage was not significant. It was the case with the subsequent monotonic loading stage up to $\varepsilon_{1,EXT}$ of about 2.9 %, which is much larger than the $\varepsilon_{1,EXT}$ values at the two peaks during the monotonic test. After reaching the first peak stress state, formation of a shear band progressed rapidly, in the same manner as during the monotonic test.

Distributions of $\Delta\varepsilon_{vol}$ during the cyclic loading stage at an interval of every 10 cycles are shown in Figure 7 for the same cyclic test. With the increase in the number of cycles, dilative regions with $\Delta\varepsilon_{vol}$>0.1 % reduced, in particular up to the 30th cycle, accompanied by slight increase in contractive regions with $\Delta\varepsilon_{vol}$<-0.1 %. Such behavior suggests local densification during cyclic loading. This may have strengthened potentially weak zones in the specimen and caused the increase in the peak strength during the subsequent monotonic loading stage.

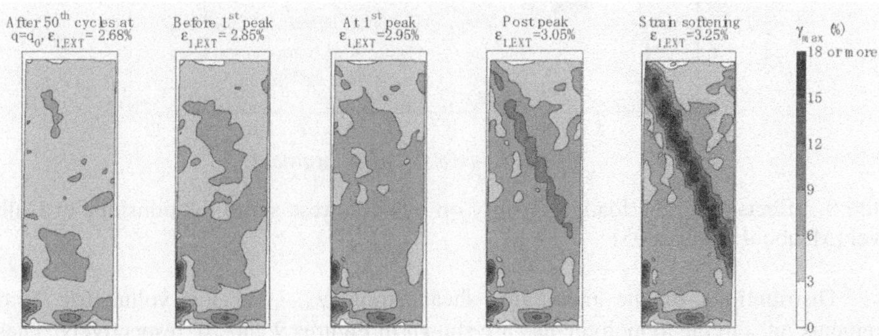

Figure 6. Contours of maximum shear strains in cyclic test (test PSCm02) on Toyoura sand (Karimi et al., 2005)

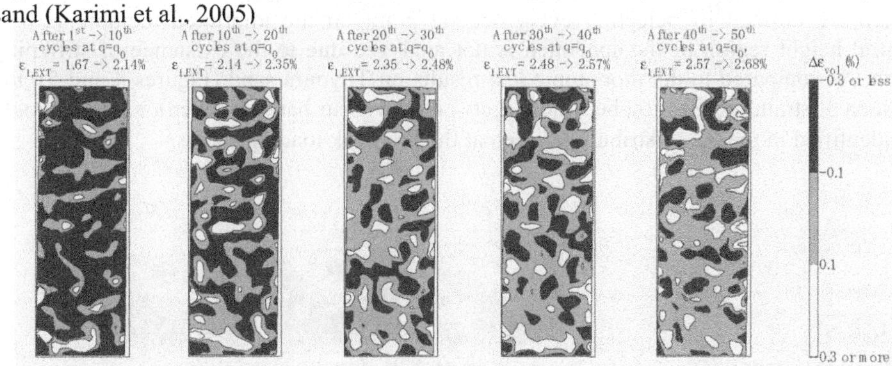

Figure 7. Contours of volumetric strain increments in cyclic test (test PSCm02) on Toyoura sand (Karimi et al., 2005)

3.2 Chiba gravel

One monotonic test and another cyclic test were conducted on Chiba gravel. Their overall stress-strain relationships are compared in Figure 8. The values of peak deviator stress q_{max} in both tests were around 600 kPa and were not increased by the cyclic loading history. However, the post-peak reduction in the q values for the cyclic test was significantly faster than that for the monotonic test.

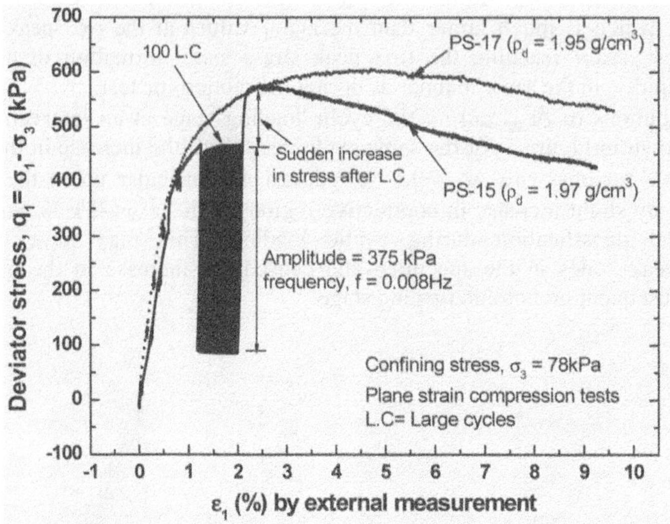

Figure 8. Effects of cyclic loading history on overall stress-strain relationships of Chiba gravel (Maqbool et al., 2005)

Distributions of the maximum shear strain γ_{max} and the volumetric strain increments $\Delta\varepsilon_{vol}$ in the monotonic test are shown in Figures 9 and 10, respectively. These strains are computed from the displacements of the grid nodes at a spacing of 10 mm, which are defined by selecting every second point on the imprinted membrane. The central height region of the specimen is not analyzed due to the existence of stiffening plate. As compared to the monotonic test results on Toyoura sand (Figures 4 and 5), the regions of strain localization became larger (i.e., the shear band widened), and they could be identified in the γ_{max} distributions even at the pre-peak loading stages.

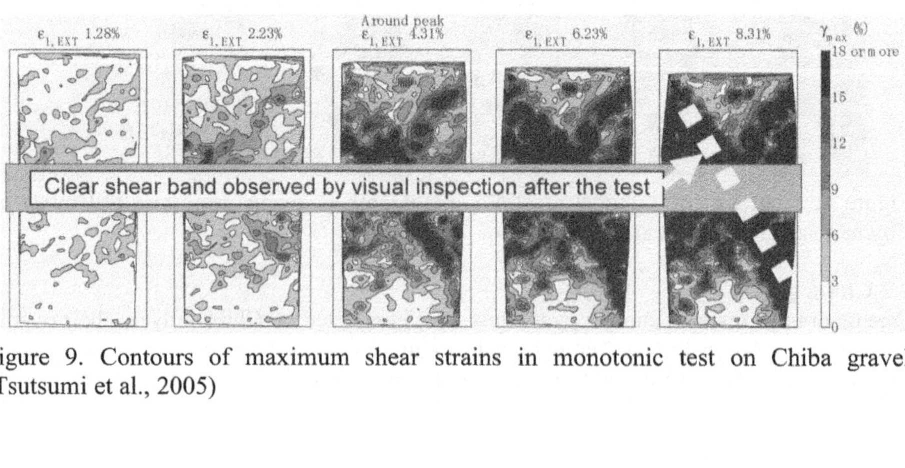

Figure 9. Contours of maximum shear strains in monotonic test on Chiba gravel (Tsutsumi et al., 2005)

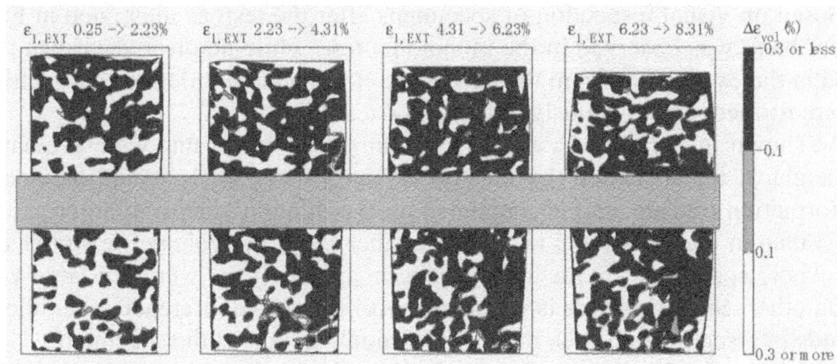

Figure 10. Contours of volumetric strain increments in monotonic test on Chiba gravel (Tsutsumi et al., 2005)

The γ_{max} and $\Delta\varepsilon_{vol}$ distributions in the cyclic test are shown in Figures 11 and 12, respectively. In general, the properties of strain localization were not significantly different from those in the monotonic test, except for less dilative behavior during cyclic loading. Note also that the regions of strain localization could be identified in the γ_{max} distribution at the end of cyclic loading (i.e., pre-peak loading stage).

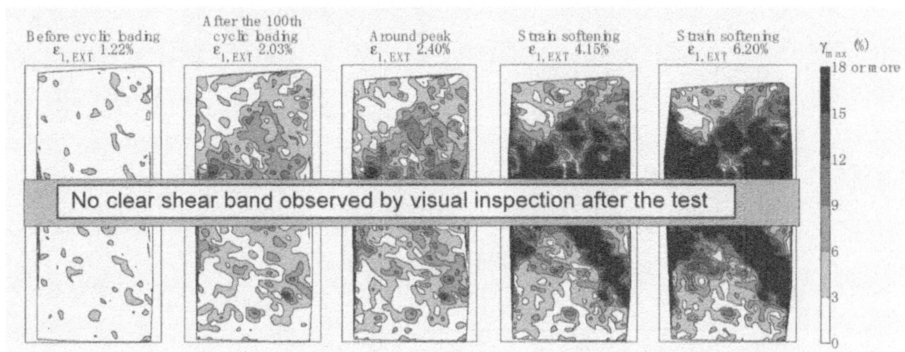

Figure 11. Contours of maximum shear strains in cyclic test on Chiba gravel

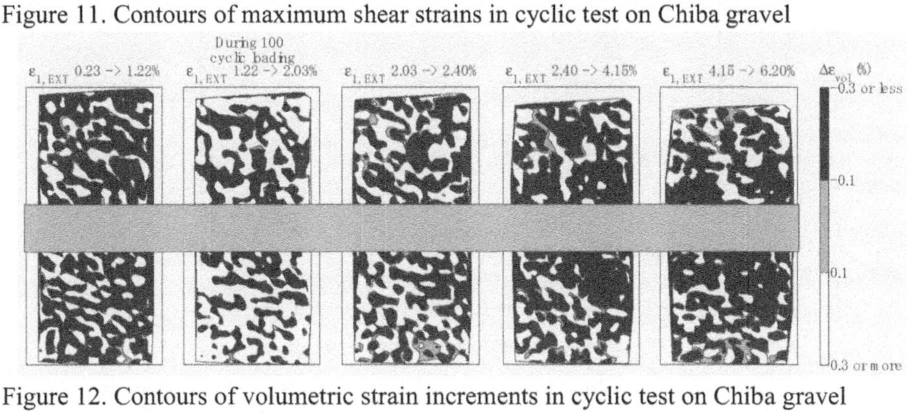

Figure 12. Contours of volumetric strain increments in cyclic test on Chiba gravel

Based on visual inspection of specimens after the test, as illustrated in Figure 9, a clear shear band was observed in the monotonic test, while no clear shear band could be identified in the cyclic test. From the latter behavior, it can be inferred that multiple shear bands were formed simultaneously in the cyclic test.

As shown in Figures 13a and b, the volumetric strains that were evaluated at the middle height of the specimen (by method B using image analysis and method C using local deformation transducers) accumulated on the dilation side to a larger extent in the cyclic test than in the monotonic test. On the other hand, the volumetric strains evaluated over the whole specimen (by method A using image analysis) were not largely different from each other. Such behavior is consistent with the above inference when the multiple shear bands intersected with each other at the middle height of the specimen.

Simultaneous formation of multiple shear bands would result into slower post-peak reduction in the q values on the overall stress-strain behavior, which is opposite to what was observed in the present study (Figure 8). Therefore, further investigations are required, including those on the issue of repeatability of test results.

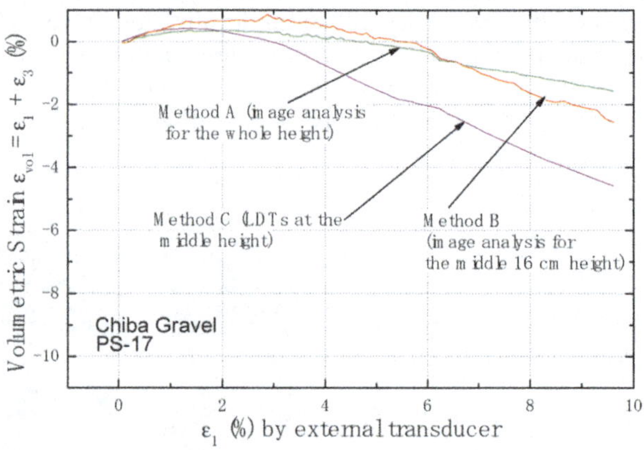

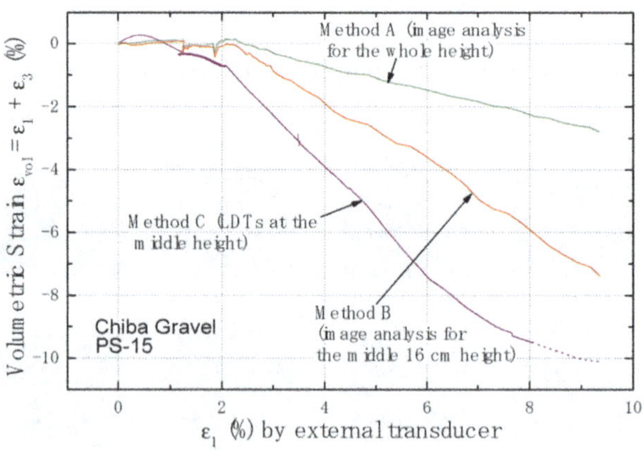

Figure 13. Accumulation of volumetric strains of Chiba gravel; a) in monotonic test and b) in cyclic test

4. CONCLUSIONS

In the drained plane strain compression tests on dense Toyoura sand, strain localization progressed rapidly at peak and post-peak stress states. Cyclic loading history caused an increase in the peak strength during the subsequent monotonic loading stage, due possibly to local densification of potentially weak zones during cyclic loading.

On the other hand, in similar tests on compacted Chiba gravel, strain localization could be identified even at pre-peak stress states. Cyclic loading history did not affect significantly the peak strength, although less dilative behavior was observed during cyclic loading.

5. REFERENCES

1- Karimi, J.M., Sato, T. and Koseki, J.: Plane strain compression tests with image analysis on dense Toyoura sand, *Bulletin of ERS*, No. 38, pp.81-92, 2005.

2- Koseki, J., Tatsuoka, F., Munaf, Y., Tateyama, M. and Kojima, K.: A modified procedure to evaluate active earth pressure at high seismic loads, *Soils and Foundations, Special Issue on Geotechnical Aspects of the January 17 1995 Hyogoken-Nambu Earthquake,* Vol. 2, pp.209-216, 1998.

3- Koseki, J., Salas-Monge, R. and Sato, T.: Plane strain compression tests on cement-treated sands, *Geomechanics: Testing, Modeling and Simulation*, Geotechnical Special Publication No. 143, ASCE, pp.429-443, 2005.

4- Maqbool, S, Koseki, J. and Sato, T.: Effects of large cyclic and creep loading on strength and deformation properties of compacted gravel, *Proc. of 7th International Summer Symposium*, International Activities Committee, JSCE, pp.187-190, 2005.

5- Salas-Monge, R., Koseki, J. and Sato, T.: Cyclic plane strain compression tests on cement treated sand, *Bulletin of ERS*, No. 36, pp.131-141, 2003.

6- Sato, T., Maqbool, S., Tsutsumi, Y. and Koseki, J.: Development of plane strain compression test apparatus for coarse-grained soil with high capacity, *Proc of 40th Japan National Conf. on Geotechnical Engineering*, 2005. (in Japanese)

7- Tsutsumi, Y., Maqbool, S. and Koseki, J.: Plane strain compression tests with image analysis on compacted gravel, *Proc. of 60th Annual Conf. of the Japan Society of Civil Engineers*, 2005. (in Japanese)

Soil Stress-Strain Behavior: Measurement, Modeling and Analysis
Geotechnical Symposium in Roma, March 16 & 17, 2006

SMALL STRAIN DEFORMATION CHARACTERISTICS OF GRANULAR MATERIALS IN TORSIONAL SHEAR AND TRIAXIAL TESTS WITH LOCAL DEFORMATION MEASUREMENTS

T, Kiyota[1)], L.I.N. De Silva[2)], T. Sato[3)] and J. Koseki[4)]
Institute of Industrial Science
The University of Tokyo, 4-6-1 Komaba, Meguro-ku, Tokyo, JAPAN
e-mail: kiyota@iis.u-tokyo.ac.jp[1)]
nalin@iis.u-tokyo.ac.jp[2)]
tsato@iis.u-tokyo.ac.jp[3)]
koseki@iis.u-tokyo.ac.jp[4)]

ABSTRACT

The soil behaviour under normal working loads usually mobilizes strain increments that are less than 0.1 %. Therefore it is very important to measure the small deformation of soil by laboratory experiments accurately. Among others, the local deformation transducer (LDT) has been used to measure locally small deformation of cylindrical or rectangular specimens in triaxial tests. In this study, a hollow cylindrical specimen was used in torsional shear and triaxial tests. By using a modified version of LDT, called PLDT (Pin-typed LDT), to comply with shear deformation of hollow cylindrical specimen under torsional loading, it was confirmed that the modified version could be successfully used to evaluate quasi-elastic deformation properties. The small deformation characteristics of Toyoura sand, Hime gravel and glass beads were investigated using both PLDTs for the local static measurement and gap sensors and a potentiometer for the external static measurement. In addition, a dynamic measurement system by using a set of small triggers and accelerometers was used to evaluate S wave velocity of glass beads. The specimens at different densities were subjected to small vertical and torsional cyclic loadings during otherwise isotropic consolidation from 50 kPa to 400 kPa and isotropic unloading from 400 kPa to 50 or 100 kPa.

It was confirmed that small strain Young's and shear moduli values of the tested granular materials which were measured by PLDTs, external transducers and accelerometers are consistent with each other in terms of their stress state dependencies. The externally measured Young's and shear moduli were by about 5 to 20 % larger than locally measured ones except for Young's moduli of Toyoura sand. The measured Young's and shear moduli showed an increasing trend with the density at the same stress level. Based on these test results, applicability of several void ratio functions was also studied.

Hoe I. Ling et al. (eds.), Soil Stress-Strain Behavior: Measurement, Modeling and Analysis, 557–566.

1. INTRODUCTION

Since the stiffness of geomaterials and its strain-dependency are important, strain measurement systems having a wide range are required. However, it is known that conventional displacement transducers with large capacity cannot measure accurately small strain of less than 0.1 %, which is typical ground deformation under normal working loads. On the other hand, the local deformation transducer (LDT, Goto et al., 1991) has been widely reported as a useful small strain measurement system of specimens in the laboratory. Since the system compliance problems in laboratory testing apparatus, such as end restraint effects and bedding errors, lead to less reliable results with external deformation measurements, the importance of local measurements has been also recognized by many researchers.

In this study, small strain characteristics of some granular materials, i.e. Toyoura sand, Hime gravel and glass beads, under isotropic stress states were investigated using local and external measurements in medium-size hollow cylinder specimens. To evaluate Young's modulus and shear modulus in hollow cylinder specimen, a modified version of LDT, as originally developed by Hong Nam et al (2001), which consists of a set of LDTs in a triangle form was employed. In addition, a set of accelerometers was used to obtain dynamic shear modulus of glass beads.

2. TEST PROCEDURE

An automated hollow cylinder apparatus (Fig.1) with a high loading capacity that was developed at Institute of Industrial Science, the University of Tokyo, was used. This new apparatus can apply very small unload-reload cycles at any stress state in both vertical and torsional directions independently to evaluate quasi-elastic deformation properties of geomaterials. The vertical load and the torque were measured by a two-component load cell placed above the top cap of the specimen (inside the pressure cell), which were used to calculate the deviator stress $q= \sigma'_z - \sigma'_\theta$ and the torsional shear stress $\tau_{z\theta}$, respectively.

Test materials were air-dried Toyoura sand (specific gravity G_s= 2.64, mean particle diameter D_{50}= 0.16 mm), Hime gravel (G_s= 2.65, D_{50}= 1.73 mm) and glass beads (G_s= 2.49, D_{50}= 0.15 mm). The specimens of Hime gravel and glass beads were 20 cm in outer diameter, 12 cm in inner diameter and 30 cm in height, and prepared by air-pluviation method under various densities. Those of Toyoura sand were 15 cm in outer diameter, 9 cm in inner diameter and 30 cm in height (De Silva et al., 2005).

The vertical strain was measured externally by gap sensors (GS2 and GS3), and the shear strain was measured externally by a gap sensor (GS1) and a potentiometer (POT2). To measure local vertical and shear strains, two-sets of PLDTs (Pin-typed LDTs) were attached symmetrically, over the middle-height regions of the specimen in order to avoid possible effects of bedding error and end-restraint. The PLDTs were arranged in a shape of triangle using three LDTs, while they were further modified by using six separated hinges. The details of the further modification and its virtue against the original version of PLDT have been discussed in previous studies (e.g., De Silva et al., 2005). Fig. 2 illustrates the layout of the modified version of PLDT and other transducers used in this study.

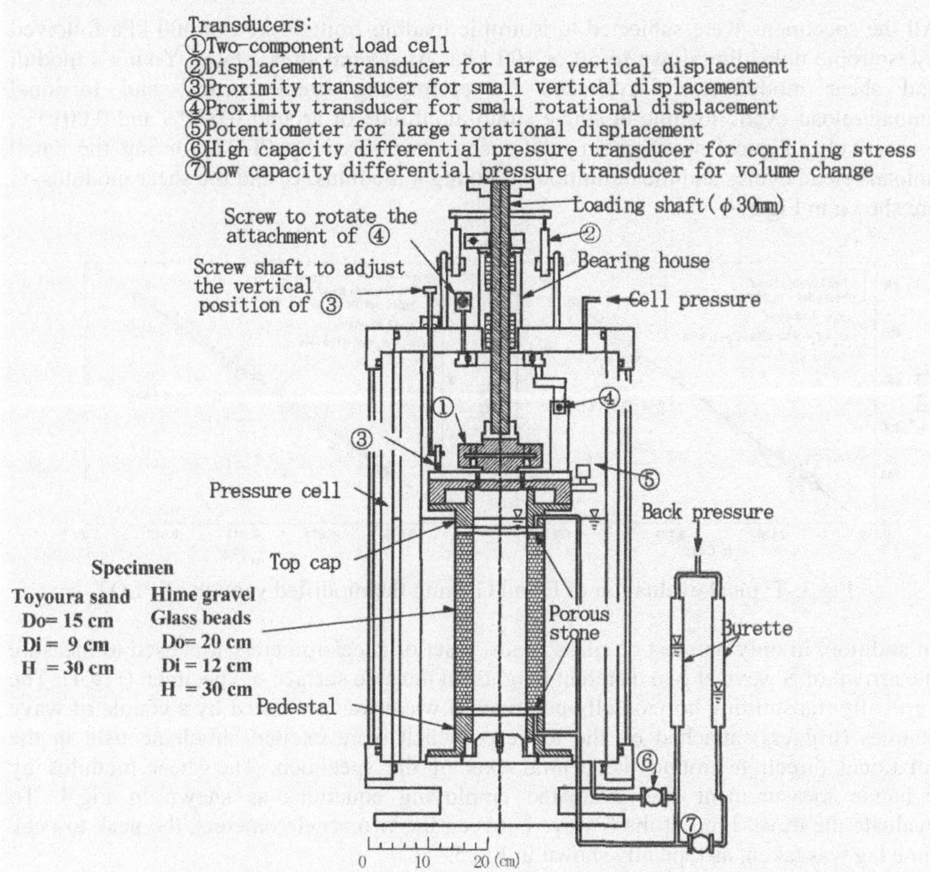

Fig.1. Hollow cylinder apparatus (Hong Nam and Koseki, 2005)

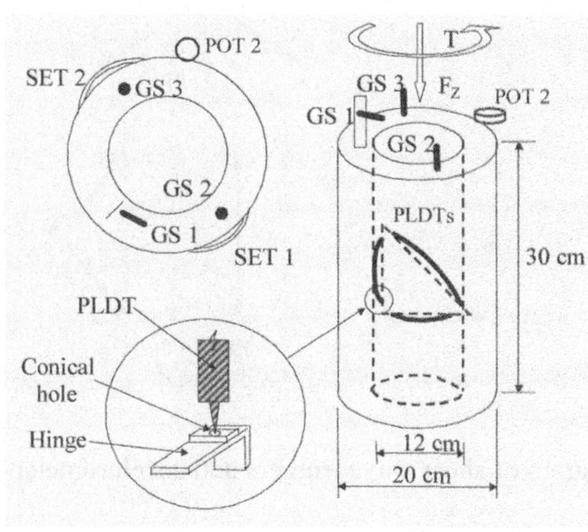

Fig.2. Arrangement of PLDTs and transducers

All the specimens were subjected to isotropic loading from 30 kPa to 400 kPa followed by isotropic unloading down to 50 or 100 kPa. At several stress states, Young's moduli and shear moduli were evaluated by performing small vertical and torsional unload/reload cyclic loading at single strain amplitude of around 0.001% and 0.0015%, respectively. Typical stress-strain relationships measured by PLDTs during the small unload/reload cycles and the definition of Young's modulus, E, and the shear modulus, G, are shown in Fig.3.

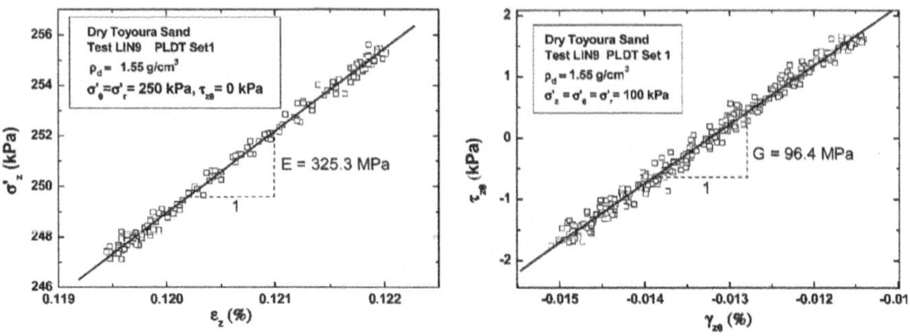

Fig.3. Typical evaluation of E and G using the modified version of PLDT

In addition, in only one test on glass beads, a set of accelerometers was used to measure the arrival of S wave at two different heights on the side surface of specimen (Fig.4). The vertically-transmitting horizontally-polarized S wave was generated by a couple of wave sources (triggers) attached on the top cap, which were excited simultaneously in the rotational direction around the central axis of the specimen. The shear modulus by dynamic measurement was evaluated employing equations as shown in Fig.4. To evaluate the travel time of the S wave between the two accelerometers, the peak to peak time lag was taken, as typically shown in Fig.5.

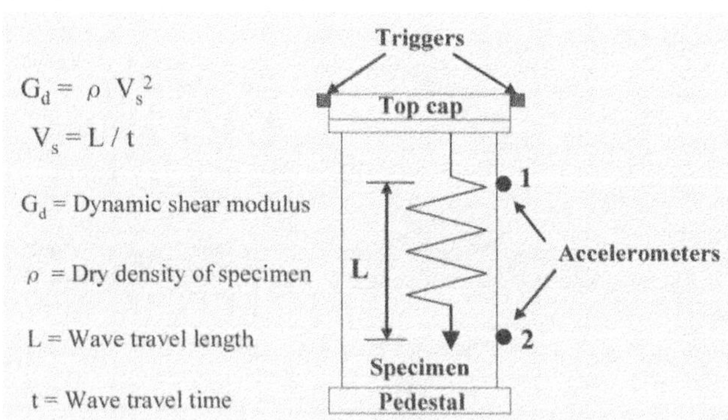

Fig.4. Diagram of shear wave triggers and accelerometers on a specimen

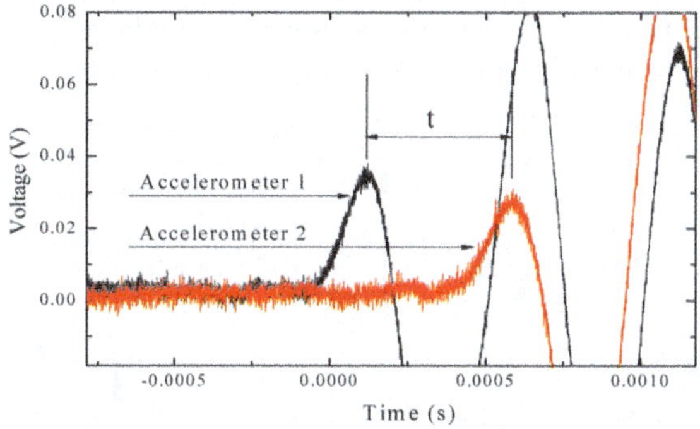

Fig.5. Definition of wave travel time (peak to peak)

3. TEST RESULTS AND DISCUSSION

Typical test results on each material are shown in Fig.6. The vertical Young's modulus, E, and shear modulus, G, were increased with the increase in the isotropic stress level.

E is evaluated from the outputs of four transducers: two gap sensors (GS2 and GS3) as external transducers and two sets of PLDTs as local transducers. It can be seen that E values of Toyoura sand are similar to each other irrespective of the transducers, while the externally measured values of glass beads are by about 5 to 20 % greater than the local ones. In the case of Hime gravel, although the E values by the two sets of PLDTs are different from each other, their mean values are similar to those by external measurements.

G is evaluated from the outputs of five transducers: gap sensor (GS1) and potentiometer (POT2) as external transducers, two sets of PLDTs as local transducers, and accelerometers as dynamic transducers. In the case of static measurement, G values of the three materials by external measurements are by about 5 to 20 % greater than the mean values by local measurements.

The possible influential factors on the quasi-elastic moduli measured by external transducers are the bedding error which is caused by disturbance at the edge of the specimen, and the end restraint which is caused by friction at the interface between the top cap or pedestal and the specimen. It is known that the bedding error and the end restraint lead to underestimate and overestimate the soil stiffness respectively. On the other hand, the locally measured quasi-elastic modulus is much less subjected to these mechanical errors. Therefore, it can be interred that the quasi-elastic moduli by external measurements which are larger than those by the local measurements were possibly affected by the latter effect: the end restraint.

In addition, dynamic G values of glass beads evaluated by accelerometer are equal to or greater than the statically measured ones. Similar trend has been reported by previous studies on many kinds of materials using accelerometers or bender elements (e.g., Maqbool et al, 2004, Yamashita et al, 2003). The G values evaluated by dynamic measurement reflect the response of stiff part of the specimen because the S wave propagates through inter-locking of soil particles, while the G values by static measurement is obtained from global response of the specimen. Thus, it would be reasonable even if the G values by dynamic measurement are larger than those by static measurement, in particular with specimens having a larger degree of heterogeneity in term of local stiffness.

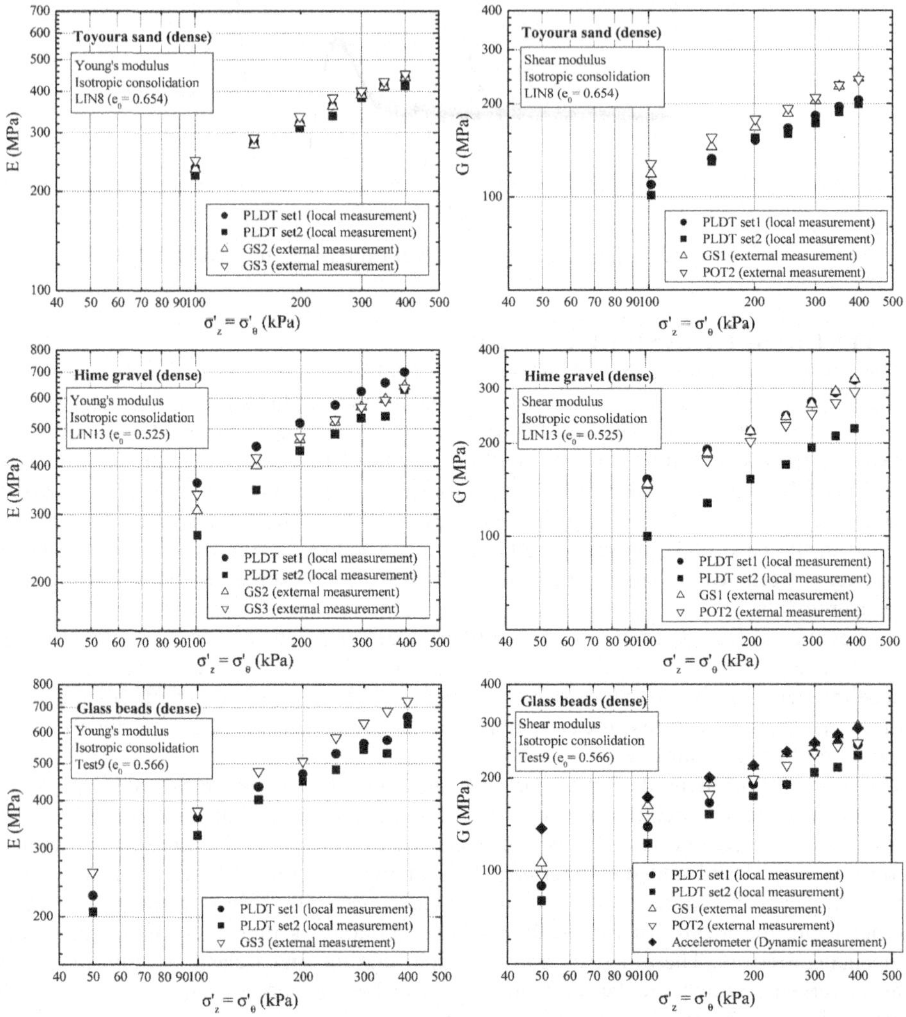

Fig.6. Comparison of externally and locally measured E and G

The values of E and G for all the specimens with different densities measured during isotropic consolidation using the modified version of PLDTs are shown in Fig.7. In order to correct for the effects of different void ratios, the following function proposed by

Hardin and Richart (1963) on round particle soils is applied to normalize the values of E and G.

$$f(e) = (2.17 - e)^2 / (1 + e)$$

After the normalization, the values of E/f(e) and G/f(e) of Toyoura sand become rather unique against the isotropic stress levels. On the other hand, in the case of Hime gravel and glass beads, the E/f(e) and G/f(e) of dense specimens are generally greater than those of loose specimen. This feature indicates that the change of E and G due to the changes of specimen densities is larger than those formulated by the void ratio function proposed by Hardin and Richart (1963). The reason for the above discrepancy may be different soil particle properties, such as particle size, shape and particle breakage among others.

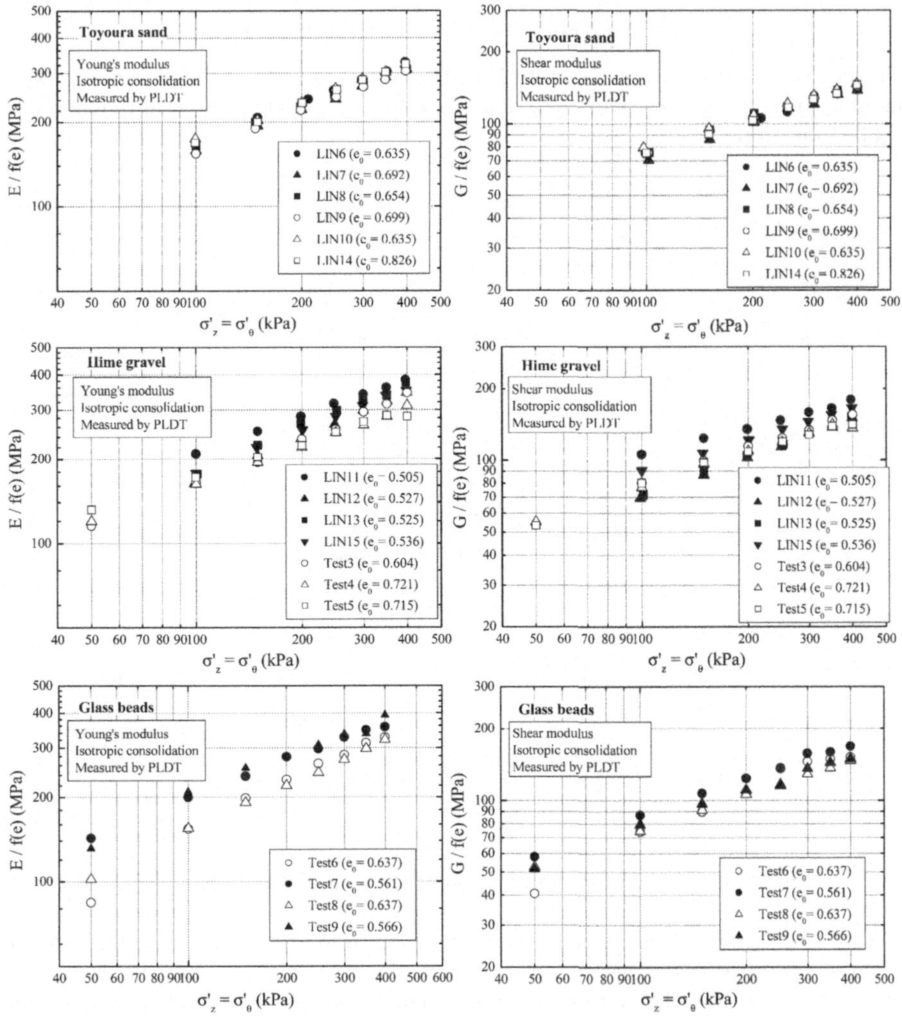

Fig.7. Comparison of E and G at various densities using modified version of PLDTs

In the present study, the particle shape of glass beads and Hime gravel is round or sub-round, while that of Toyoura sand is sub-angular. In addition, the mean particle size of Hime gravel is around 10 times larger than that of Toyoura sand. Therefore, the present test results on three different materials may be affected by the particle shape and possibly the particle size.

The value of E and G of Hime gravel and glass beads at $\sigma_z' = \sigma_\theta' =$ 100, 200 and 400 kPa are plotted versus the void ratio in Fig. 8.

By setting the reference void ratio at 0.536 and 0.561 for Hime gravel and glass beads, respectively, the applicability of the void ratio function by Hardin & Richart (1963) and those by the authors as given below is examined. The latter functions could fit the test results more reasonably.

$$f(e) = e^{-1.8} \text{ (Hime gravel)}, \quad f(e) = e^{-2.2} \text{ (Glass beads)}$$

It can be seen that the dependency of E and G of glass beads on the void ratio is larger than that of Hime gravel. This behaviour may also be affected by the difference in the soil particle properties.

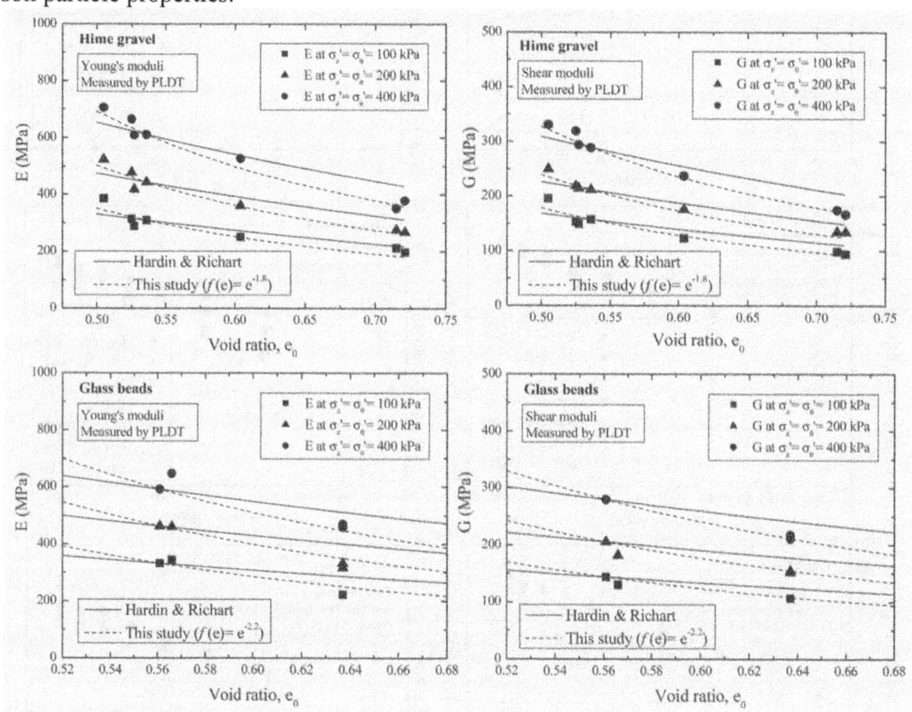

Fig.8. Original void ratio function of Hime gravel and glass beads

In the present study, all the specimens were subjected to very small unloading/reloading cycles at several stress levels during isotropic unloading from p'= 400 kPa to 50 or 100 kPa. Figure 9 shows the ratio of E or G values measured by PLDT during isotropic unloading to those measured during isotropic loading. In the case of Toyoura sand, the E and G values during isotropic unloading were almost the same as those during isotropic loading, and this trend was consistent with the data in the literature (e.g., Hong Nam and Koseki, 2005 among others). On the other hand, Young's moduli of Hime gravel exhibited similar tendency to those of Toyoura sand, while its shear moduli during unloading were occasionally 10 % smaller than those during loading. In addition, E and G values of glass beads were always smaller during unloading than during loading, and the ratio became smaller as the over-consolidation ratio was increased. Although the dynamic shear moduli were obtained on only glass beads specimen, the measured values during unloading were also smaller than those during loading. Since the void ratio at the

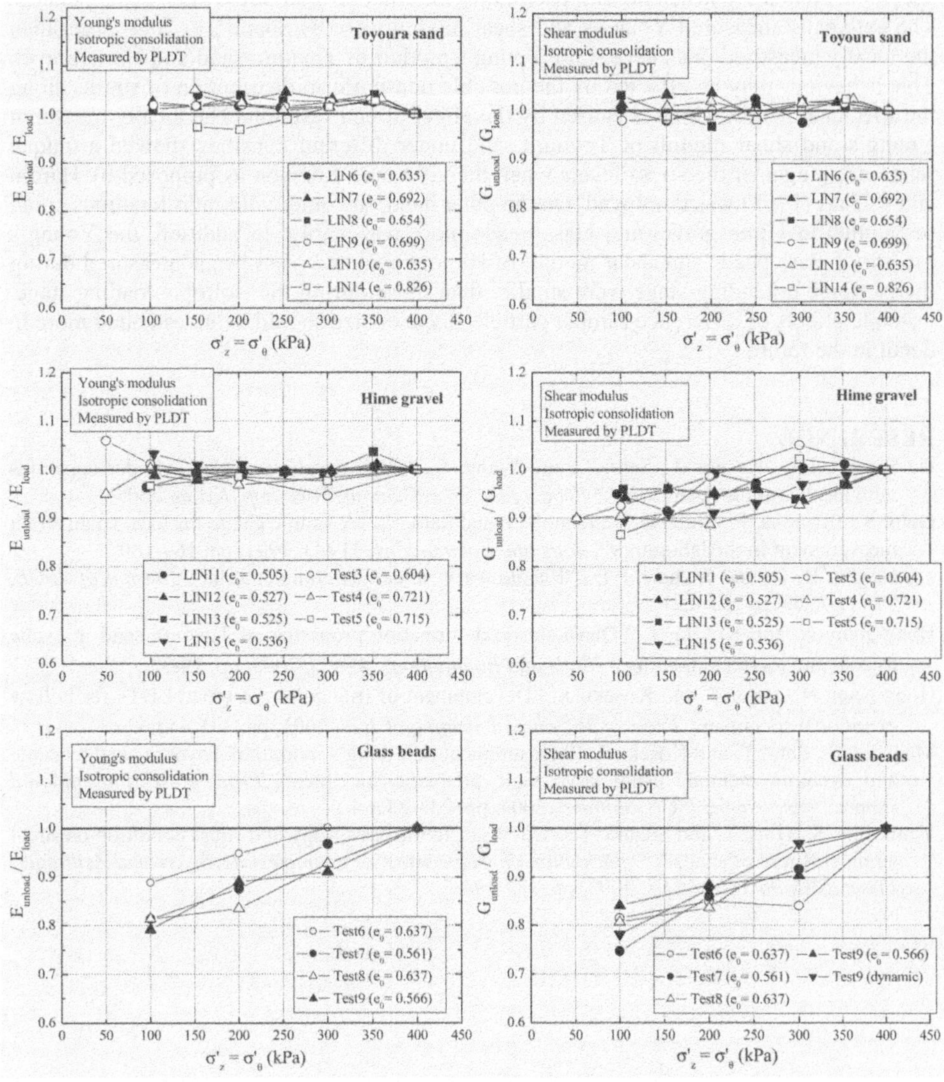

Fig.9. Comparison of E and G during isotropic loading and unloading

same stress level during unloading should be smaller than during loading, larger elastic moduli would be expected when the specimen was unloaded. The present test results on glass beads, however, showed the opposite trend, suggesting a possible change in the soil particle structure. The extent of such change might be affected by the soil particle properties (i.e., shape and size).

4. CONCLUSION

It was confirmed that the values of small strain Young's and shear modulus of the tested granular materials which were measured by PLDTs, external transducers and accelerometers are consistent with each other in terms of their stress state dependencies. The externally measured Young's and shear moduli were by about 5 to 20% larger than the locally measured ones except for Young's moduli of Toyoura sand and Hime gravel. This behaviour may be affected by the possible non-uniform distribution of strains along the specimen height which is caused by the effect of end restraint. The locally measured Young's and shear moduli of Toyoura sand under different densities showed a unique relationship against the stress levels when the void ratio function as proposed by Hardin and Richart (1963) was employed. On the other hand, the other void ratio functions could be applied to Hime gravel and glass beads more reasonably. In addition, the Young's moduli of glass beads and shear moduli of Hime gravel and glass beads measured during the isotropic unloading stage were smaller than those during the isotropic loading stage. Possible effects of difference sample particle shape or size should be investigated more in detail in the future.

REFERENCE

De Silva, L.I.N., Koseki, J., Sato, T., and Wang, L., "High capacity hollow cylinder apparatus with local strain measurements" *Proc. of the 2nd US-Japan Workshop, Kyoto,* 2005 .

Goto, S., Tatsuoka, F., Shibuya, S., Kim, Y.S. and Sato, T., "A simple gauge for local small strain measurement in the laboratory", *Soils and Foundations*, 31-(1), 1991, pp. 169-180.

Hardin, B. O. and Richart, F.E., "Elastic wave velocities granular soils", *Journal of ASCE*, 89-(1), 1963, pp. 33-65.

Hong Nam, N. and Koseki, J., "Quasi-elastic deformation properties of Toyoura sand in cyclic triaxial and torsional loadings", *Soils and Foundations*, 45-(5), 2005, pp. 19-38.

Hong Nam, N., Sato, T. and Koseki, J., "Development of triangular pin-typed LDTs for hollow cylindrical specimen", *Proc. of 36th annual meeting of JGS*, 2001, pp. 441-442.

Maqbool. S., Sato. T. and Koseki. J., "Measurement of Young's moduli of Toyoura sand by static and dynamic methods using large scale prismatic specimen", *Proc. of 6th international summer symposium, JSCE, Saitama*, 2004, pp. 233-236. 4.

Yamashita. S., Hori. T. and Suzuki. T., "Effects of fabric anisotropy and stress condition on small strain stiffness of sands", *Deformation characteristics of geomaterials, Swets and Zeitlinger, Lisse, ISBN 90 5809 604 1,* 2003, pp. 187-194.

Soil Stress-Strain Behavior: Measurement, Modeling and Analysis
Geotechnical Symposium in Roma, March 16 & 17, 2006

DEPENDENCY OF THE MECHANICAL BEHAVIOUR OF GRANULAR SOILS ON LOADING FREQUENCY: EXPERIMENTAL RESULTS AND CONSTITUTIVE MODELLING

Clara Zambelli, Claudio di Prisco
Structural Engineering Department
Politecnico di Milano, Milan, Italy
e-mail: zambelli@stru.polimi.it; cdiprisc@stru.polimi.it

Anna d'Onofrio, Ciro Visone
Department of Geotechnical Engineering
University of Naples Federico II, Naples, Italy
e-mail: donofrio@unina.it; ciro.visone@unina.it

Filippo Santucci de Magistris
S.A.V.A. Department
University of Molise, Campobasso, Italy
e-mail: filsantu@unina.it

ABSTRACT

The paper shows some cyclic torsional shear test results obtained on dense Toyoura sand specimens. The dependency of the mechanical response of this material on both loading amplitude and frequency are discussed, to obtain a framework to analyze the mechanical behaviour of granular soils. The pseudo-elastic shear stiffness and the damping ratio are the variables taken into consideration; their evolution with the number of cycles performed at the different loading amplitudes is analysed to describe the mechanical irreversibility of the material response.

The experimental data are then reproduced by means of a multiple mechanism elasto-viscoplastic constitutive model characterised by one viscoplastic mechanism and by an internal kinematic plastic mechanism capable of capturing the material mechanical response even when small cyclic tests are performed.

1. INTRODUCTION

The analysis and the understanding of the mechanical behaviour of cyclically and/or dynamically loaded soils can be considered one of the most stimulating subjects of soil mechanics, in particular, when large strains as well as a wide range of frequencies are taken into consideration and the problem of the coupling between volumetric and shear irreversible strains is investigated. Usually, this subject is studied by starting from two opposite points of view. The cyclic stress-strain behaviour is highlighted by means of sophisticated constitutive models capable of reproducing the volumetric-deviatoric coupling and the irreversibility of the constitutive relationship, and by disregarding at all

Hoe I. Ling et al. (eds.), Soil Stress-Strain Behavior: Measurement, Modeling and Analysis, 567–582.

the time factor. On the contrary, the dynamic mechanical response is tackled by means of visco-elastic or hysteretic-elastic constitutive models that are linear and allow us to solve boundary value problems in the frequency domain. In such perspective, only two constitutive parameters, the so-called equivalent parameters, are introduced to describe the material mechanical response: the shear modulus G and the damping ratio D.

Several data were published in the last three decades concerning the dependency of the soil stiffness (both shear modulus G and Young's modulus E) and/or the damping ratio at small strains on the loading frequency (see for instance Iwasaki et al, 1978; Isenhower, 1979; Tatsuoka & Shibuya, 1992; Kim et al., 1994; Fioravante et al., 1994; Porovic & Jardine, 1994; Tatsuoka et al., 1994; Tatsuoka et al., 1995; d'Onofrio et al., 1995; Di Benedetto & Tatsuoka, 1997; d'Onofrio et al., 1999b; Vucetic & Tabata, 2003).

Shibuya et al., 1995 state that at low frequency creep phenomena are relevant and therefore the damping ratio decreases as the loading frequency increases; at a medium frequency range damping ratio is hysteretic and then frequency-independent, while at higher frequency soil behaviour is prevalently linear visco-elastic and then damping increases with frequency.

In an attempt to summarize the available experimental results, Tatsuoka et al. (2001) sustain that:

a) the small strain stiffness of hard rocks, clean sands and gravels is very insensitive to the change in the strain rate; and

b) the small strain stiffness of a silty sand, a sedimentary soft rock and stiff and soft clays marginally depends on the strain rate, which means that even at a strain of 0.001 %, the stress and strain relationship is not totally strain-rate independent.

They also observed that the dependency of the Young's modulus E and damping ratio D at small strains from the strain rate becomes smaller at higher strain rate, E and D tending to approach an upper and lower bounds respectively. Therefore, at higher strain rate the material behaviour becomes less rate-dependent, more linear and more reversible.

Few experimental results and constitutive models concern the dependency of the equivalent parameters both on the shear strain amplitude and the loading frequency (Lin et al 1996, Bolton & Wilson 1989). As for the small strain only, here also results and models are somehow contradictory and some time not completely satisfactory.

According to the authors, to define appropriate constitutive models capable of reproducing the material mechanical behaviour, even when large irreversible strains take place, more sophisticated constitutive approaches must be conceived and, consequently, the experimental results must be newly interpreted accordingly to the chosen models. The importance of such a problem is due to the current availability of non-linear numerical codes which allow us to reproduce the seismic response of large domains, to simulate even the occurrence of unstable phenomena like liquefaction of loose deposits and to analyse dynamic soil-structure interaction.

In this paper, some cyclic torsional shear test on dry dense Toyoura sand specimens are discussed by considering a wide range of loading frequencies and different strain levels. The experimental program allowed for analyzing, in a comprehensive way, the dependency of the mechanical response on both the loading frequency and the loading amplitude.

The experimental data are then reproduced by means of a multiple mechanism elasto-visco plastic constitutive model, characterised by one visco-plastic mechanism and by a kinematic plastic mechanism capable of capturing the material mechanical response even when small cyclic tests are performed, as well as at large strains.

2. EXPERIMENTAL DEVICE AND PROCEDURES

A resonant column-torsional shear apparatus developed at the University of Naples was used to obtain the experimental data reported here. Readers can refer to d'Onofrio et al. (1999a) for the description of the apparatus and its performances.

2.1 *Sample preparation: air pluviation method*

Dry sand is filled in a funnel with a metallic tube attached to the end. The terminal cross section of the tube has a width of 2mm and a length of 15mm. A membrane is stretched taut to the inside face of a split mould which is attached to the base pedestal of the test apparatus. The sand is spread in the mould with about 55cm in height of fall at a constant speed until the mould becomes filled with the dry sand. The height of fall is selected to obtain a relative density of specimens about of 80%.

Table 1. State parameters of dense dry Toyoura specimens used in the tests.

Test	γ [g/cm^3]	e	D_r [%]
TOY 0.01 Hz	1.552	0.709	79.7
TOY 0.05 Hz	1.556	0.705	81.0
TOY 0.1 Hz	1.556	0.706	80.8
TOY 0.5 Hz	1.554	0.708	80.2
TOY 1 Hz	1.554	0.708	80.1
TOY 2 Hz	1.554	0.707	80.3

Being for Toyoura sand e_{max}= 0.944 and e_{min}= 0.609, Table 1 shows the physical characteristics of each specimen: the measured relative density on specimens (averaged at D_r = 80.4%) is close to the target value of Dr = 80% and the scatter among data is reasonably low. The top of the sample is levelled with a knife to have an even surface. After the specimen is encased in the membrane with the top cap, a vacuum of 20 kPa is applied to complete the assembling operations of the apparatus.
Notice that in the used experimental device is not possible to saturate the specimens in the cell because there is a drainage line in the lower base only, in order to maintain a fixed-free scheme in performing resonant column tests. For this reason it is possible to test only dry specimens.

2.2 *Experimental procedures*

The experimental results were obtained with the above mentioned torsional device, testing a series of six dry Toyoura dense sand specimens (35 x 72 mm) air pluviated isotropically consolidated under a cell pressure of 200 kPa. Before reaching the consolidation pressure, isotropic compression is performed at a constant stress rate of 10 kPa/hr.

During isotropic compression, resonant column RC tests at a small level of excitation were performed to verify the reproducibility of the data. Figure 1 shows the evolution of initial shear stiffness G_0 and initial damping ratio D_0 with effective confining stress p' for all the tests. From the figures it can be seen that shear modulus and damping ratio respectively increases and decreases with the confining stress accordingly to a non-linear relationship and the experimental data are not much scattered, particularly for the initial stiffness. Initial damping values appear more dispersed because there is more uncertainties in their experimental determination. The numerical values of G_0 and D_0 here are different from those discussed in the following part of the paper due to the interpretation criteria (i.e., the visco-elastic model) and the high strain rate associate to RC tests.

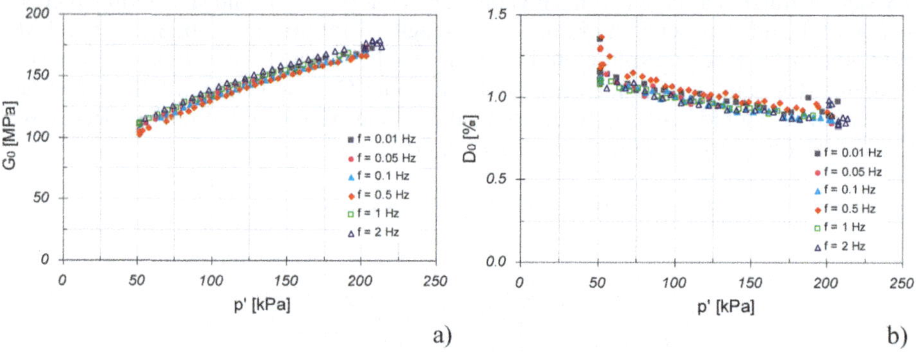

Figure 1. Evolution of initial shear stiffness and initial damping ratio with effective confining stress, measured during resonant column tests at small strain levels on the dense dry Toyoura specimens.

After the compression, a series of cyclic torsional shear tests were performed on each specimen by fixing the loading frequency (varying between 0.01 Hz and 2 Hz) and by increasing progressively the torque amplitude. At each shear stress level, ten cycles were performed.

3. EXPERIMENTAL RESULTS

For the sake of clarity, the experimental results concerning small/medium and large strain ranges are discussed separately.

Figure 2 shows an introducing summary of the experimental results. In this figure, as in all the following ones of the paper, γ_{SA} stands for the semi-amplitude or single-amplitude shear strain. These results are very close to many others already published in literature regarding Toyoura sand in a dense state and illustrate the marked dependency both of shear modulus G and damping ratio D on γ_{SA}.

In Figure 2 the two thresholds of the linear behaviour (γ_L) and of the stable cycles (γ_S), respectively, are put in evidence. When $\gamma_{SA} < \gamma_L$, G and D are approximately constant and respectively equal to the initial shear modulus G_0 and the initial damping ratio D_0; when $\gamma_{SA} < \gamma_S$, both G and D do not depend on the number of cycles.

The evolution of the material mechanical response with the number of cycles, imposed for each shear stress amplitude, was considered to be important for capturing the occurrence of micro-structural rearrangements of the material internal fabric.

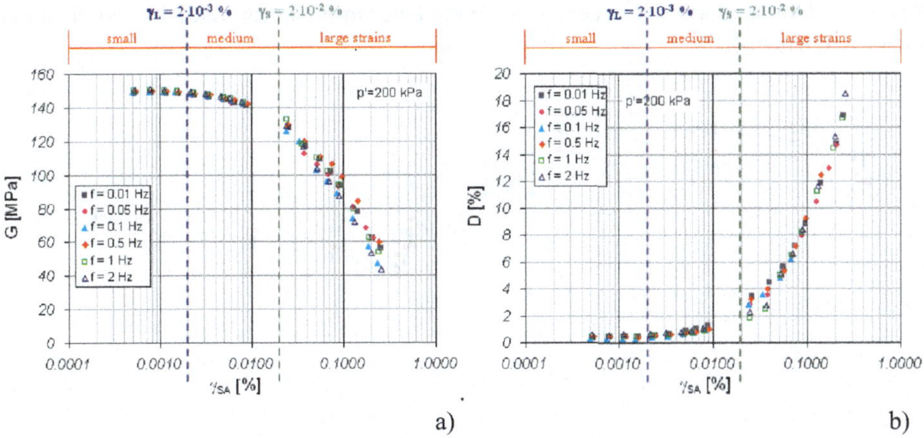

Figure 2. Shear moduli and damping ratio versus single amplitude shear strain, measured during cyclic torsional shear tests on a dense dry Toyoura specimen.

3.1 *Small-medium strains*

In Figure 3 typical stress-strain loops at small and medium strain levels are reported. As is evident from the figure, when small strains are considered, it is difficult to properly individuate the stress-strain cycles because of the scatter in the experimental data. For this reason, to evaluate the damping ratio D the experimental points were fitted by means of sinusoidal curves (Papa et al., 1988). At these strain ranges, the shape of the cycles do not evolve whatever the number of loading cycle are applied, and the damping ratio is very small.

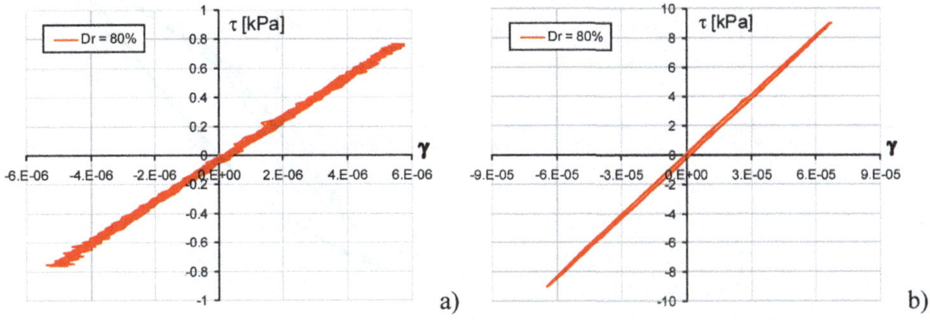

Figure 3. Cyclic torsional shear tests on a dense dry Toyoura specimen: stress-strain cycles at a) small strains, b) medium strains for a loading frequency 0.5 Hz.

Moreover, by plotting G and D with respect to frequency f, it is worth noting that G slightly increases with f, while the dependency of D is described by a non-monotonous trend (Figure 4). Related with the latter, apparently, when the frequency is increased to values greater than 0.5 Hz, the trend seems to become linear accordingly to the already cited test results of Lin (Lin et al 1996). This seems to be due to already highlighted

(Shibuya et al. 1995) two antagonistic effects: (i) the former associated to the delayed plasticity (dominant at low frequencies), (ii) the latter to the linear viscosity (dominant at higher frequencies).

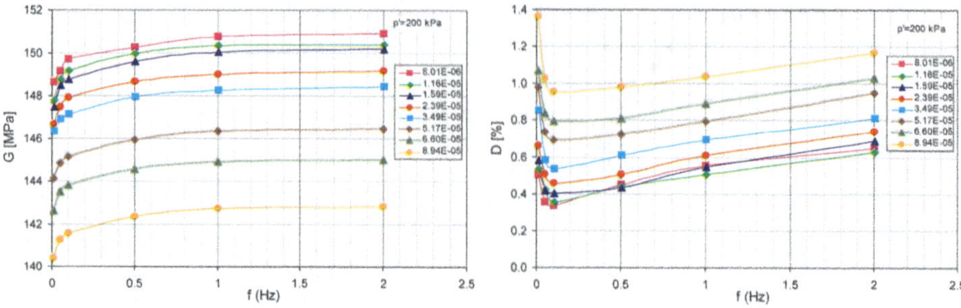

a) b)

Figure 4. Shear moduli and damping ratio versus frequency, measured during cyclic torsional shear tests on a dense dry Toyoura specimen, in the range of the small / medium strains.

3.2 *Large strains*

When large shear strain tests are performed, the mechanical response of the sand specimens changes with time: both the shape and the average shear stiffness evolve progressively with the number of cycles. A typical test result is reported in Figure 5, where the first (Figure 5a) and all the stress-strain cycles (Figure 5b) are drawn with reference with cyclic torsional tests performed at different loading levels applied at a constant frequency of 0.1 Hz.

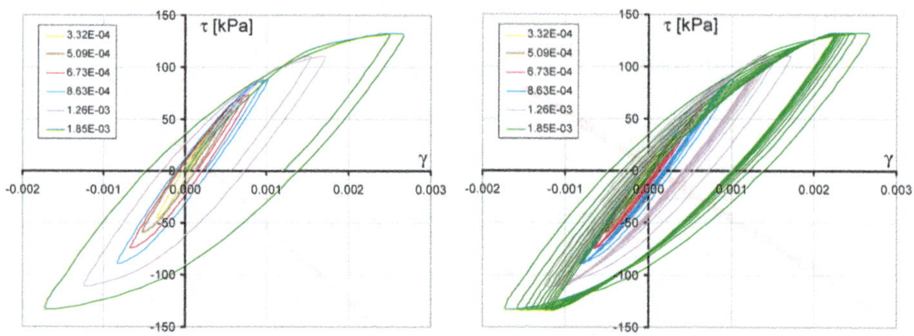

a) b)

Figure 5. Cyclic torsional shear test at medium strains on a dense dry Toyoura specimen: stress-strain cycles a) 1° cycle, b) all nine cycles for a loading frequency 0.1Hz.

The evolution of the shear modulus G and the damping ratio D with the shear strain and the number of cycle is illustrated in Figure 6. Both the damping ratio D and the shear modulus G vary abruptly when the previously defined threshold γ_S is overtaken. Particularly, Figure 6a, b and c show respectively test results performed at a frequency f=

0.01 Hz, f= 0.1 Hz and f=1 Hz, while instead in Figure 6d, the dependency of the damping ratio D from the number of cycles is illustrated in the case of f = 1 Hz. By comparing Figure 6a, b and c, we can derive that the evolution of the mechanical response upon the number of cycles is influenced severely by the frequency imposed. In particular, this dependency arises earlier for lower frequencies (i.e., at lower strain levels) and later (i.e., at larger strain levels) for higher frequencies.

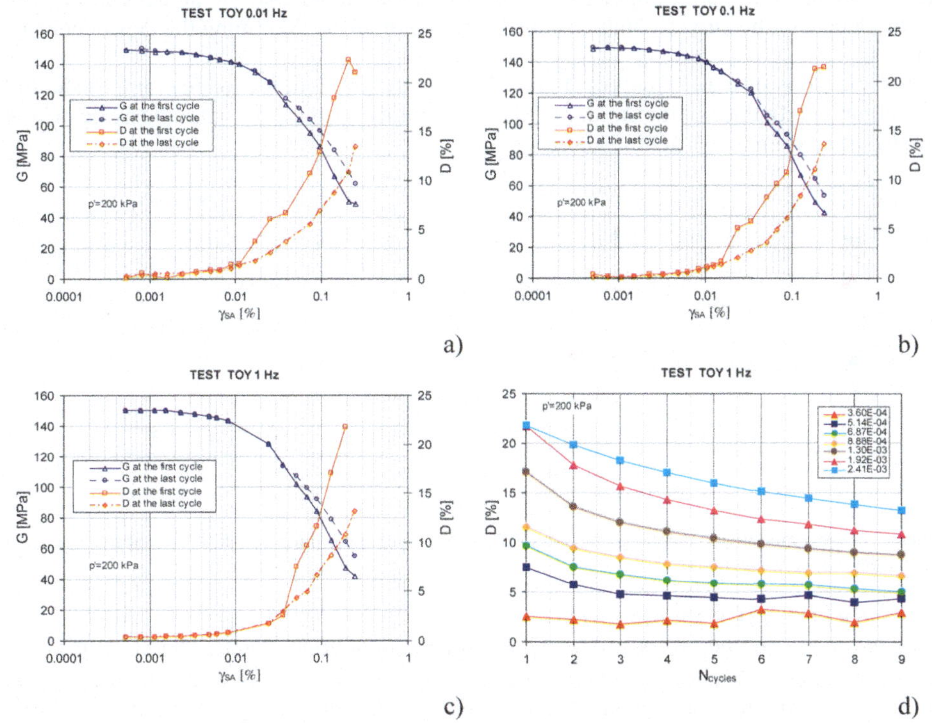

Figure 6. Shear moduli and damping ratio versus single amplitude shear strain, measured during cyclic torsional shear tests on dense dry Toyoura specimens, for the following loading frequencies: 0.01 Hz (a), 0.1 Hz (b), 1 Hz (c) in different loading cycles.

The dependency of the shear modulus G and damping ratio D from the loading frequency can be obtained here combining data from the different tests, in which each single specimen was loaded at a single frequency. Data are summarized in Figure 7, where the shear modulus (Figure 7a) and the damping ratio (Figure 7b) obtained in the last cycle of each series (i.e., the 10th cycle[1]) are plotted versus the loading frequency for the different analyzed loading levels In the investigated range of shear strains, apart from some slight scatter due to the small differences between specimens, we can state that the shear stiffness does not depend on the loading frequency, while a linear increase of D with f can be observed in Figure 7b. Figure 7 apparently contradict Figure 4, particularly when

[1] Please notice that the data pertained to the first cycle were disregarded, since the loop usually does not close and the evaluation of the equivalent parameters is doubtful.

the non linear dependency of D on loading frequency is missing by increasing the shear strains. We should observe, however, that even low frequencies are associated to high average shear stress rates: by increasing the cycle amplitude and by keeping constant the loading frequency, the shear stress rate progressively increases. If we plot the same experimental data, by taking into consideration the imposed average shear stress rate rather then the loading frequency, we can notice that the values of the former are too high to capture the minimum of the curves. For this reason, in our case we do not observe any clear change in the curves obtained at different strain levels describing the dependency of D on f.

Since the slope of the D-f curves (Figure 7b) is constant with the loading amplitude, the quantitative influence of the strain rate on the damping ratio value continuously decreases when shear strain levels are increased.

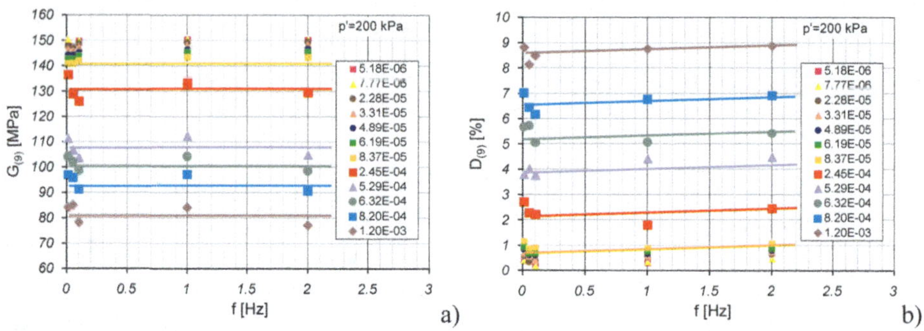

Figure 7. Shear moduli and damping ratio versus frequency, measured during cyclic torsional shear tests on dense dry Toyoura specimens.

4. CONSTITUTIVE MODELLING

The Milan Model 2002 proposed in this paper (Zambelli 2002, Zambelli et al. 2004) is an elastoviscoplastic-cyclic model for sands characterized by two uncoupled plastic mechanisms: the former is associated to a global evolution of the material internal fabric, the latter to small strains and small loops taking into account the fabric rearrangements due to small size cyclic load disturbances. This approach, by separately considering, perhaps artificially, the two distinct contributes, allows the user a simpler calibration of the constitutive parameters.

The strain rate tensor is obtained by the sum of three distinct terms as follow:

$$\dot{\varepsilon}_{ij} = \dot{\varepsilon}^{el}_{ij} + \dot{\varepsilon}^{vp}_{ij} + \dot{\varepsilon}^{cyc}_{ij} \quad (1)$$

where $\dot{\varepsilon}^{el}_{ij}$ is the instantaneous elastic strain rate tensor, $\dot{\varepsilon}^{vp}_{ij}$ is the delayed plastic strain rate tensor, and $\dot{\varepsilon}^{cyc}_{ij}$ is the irreversible strain rate tensor associated to the second cyclic irreversible mechanism. $\dot{\varepsilon}^{el}_{ij}$ is defined by starting from the definition of an elastic potential (Lade & Nelson 1987, Molenkamp 1988) from which the elastic incremental compliance matrix C^{el}_{ijhk} can be derived. The viscoplastic strain rate term is obtained by following the Perzyna's approach (Perzyna 1963, Perzyna 1966), whereas $\dot{\varepsilon}^{cyc}_{ij}$ is evaluated by introducing both an additional yield function and a plastic potential.

A detailed description of the viscoplastic part of the model can be found in di Prisco et al. 1993, 1995, di Prisco & Imposimato 1996, 1997 and di Prisco & Zambelli 2003. Here, for the sake of brevity only some features will be outlined to put in evidence the main differences between the two irreversible mechanisms characterising the constitutive model. On the contrary, the cyclic part of the model will be described in a more detailed manner because this has been considered to be central in the numerical simulation of the experimental results of §.3.

The viscous Perzyna's approach allows evaluating the tensor $\dot{\varepsilon}_{ij}^{vp}$ without imposing the consistency rule and by modifying the flow rule as it follows:

$$\dot{\varepsilon}_{ij}^{vp} = \phi(f_1)\frac{\partial g_1}{\partial \sigma_{ij}'} \tag{2}$$

σ_{ij}' is the effective stress tensor, f_1 is the yield function, g_1 the plastic potential, $\phi(f_1)$ the viscous nucleus. As was described by di Prisco (di Prisco 1993), both f_1 and g_1 are characterized by an anisotropic strain hardening, that is governed by a back stress tensor χ_{ij}. The yield locus f_1 is a function of two additional hidden variables: r_c describing its size, and β_f, its shape. This implies that f_1 is not fixed but widens, rotates and changes shape with viscoplastic straining. For each hidden variable an appropriate evolution rule is defined.

In Eq. (2) the yield function f_1 may be positive or negative, without any constraint, i.e. the stress state may be external or internal to the yield locus. When the time factor is considered, in the material mechanical response a crucial role is thus played by the viscous nucleus definition. According to the authors, in granular continua this aspect can become evident both when creep tests and dynamic loading are taken into consideration (Zambelli et al. 2004) and the viscous nucleus definition must be capable of capturing the mechanical response of the material in both these conditions.

The elastoviscoplastic part of the model is capable of reproducing correctly the behaviour of sand not only along monotonic tests, but even along non-monotonic tests characterised by a continuous rotation of principal stress axes, circular tests in the deviatoric plane and tests in which the loading direction varies. In these latter cases only an anisotropic hardening allows us to capture the mechanical response of the material. When the strain amplitude of cycles is large enough to make active the anisotropic hardening, i.e. the rotation of yield locus f_1, the constitutive model is capable of capturing the main features of the experimental mechanical behaviour of granular soils.

During small size cyclic loading tests the material does not accumulate irreversible strains, because, after the first load, the stress state continues to remain within the yield locus f_1. For this reason, the additional plastic mechanism here called cyclic mechanism is introduced. This requires the definition of a second yield function f_2, an additional plastic potential g_2, a new flow rule and a new hardening rule. As is schematically illustrated in Figure 8a, the cyclic mechanism is characterized by a very elongated cone shaped yield locus f_2, rotating in the effective stress space in accordance with the evolution of its axis χ_{ij2} (that is an additional tensorial hidden variable) and keeping constant its size. The yield locus f_2 is defined as follows:

$$f_2 = a_f I_1^* + \sqrt{J_2^*} = 0 \tag{3}$$

where $I_1^* = \sqrt{3}\sigma'_{ij}\chi_{ij2}$, $J_2^* = s^*_{ij}s^*_{ij}$ and $s^*_{ij} = \sigma'_{ij} - \frac{1}{\sqrt{3}} I_1^* \chi_{ij2}$, while $a_f = -\mathrm{tg}\ \alpha_f$ (Figure 8a) describes the cone opening and is assumed to be so small that it does not influence the numerical simulations. The definition of yield locus derives from the assumption that cyclic irreversible strains take place only when the mobilised frictional angle changes: radial cyclic effective stress paths of small amplitude are assumed to be approximately reversible.

Even this second plastic mechanism has been assumed to be non associated: the shape of the plastic potential g_2 has been taken similar to the yield function, but term a_g replaces term a_f of Eq. (3):

$$g_2 = a_g I_1^* + \sqrt{J_2^*} = a_g \sqrt{3}\sigma'_{ij}\chi_{ij2} + \sqrt{s^*_{ij}s^*_{ij}} \tag{4}$$

with

$$a_g = \frac{a_{g1} + a_{g2}}{2} + \frac{a_{g1} - a_{g2}}{\pi} \mathrm{arctg}\left(-\cdot c_1 \frac{\partial f_2}{\partial \sigma'_{ij}} \delta_{ij} \right) \tag{5}$$

where a_{g1} is the constitutive parameter for the loading, while a_{g2} for the unloading. The scalar product $\partial f_2 / \partial \sigma'_{ij} \delta_{ij}$ defines the criterion for distinguishing between loading and unloading. The term c_1 is a constitutive parameter describing the rapidity of changing of a_g from a_{g1} to a_{g2} and vice versa (Figure 8b). In Figure 9a, with reference to a loose sand specimen, the geometrical meaning of variable a_g is sketched.

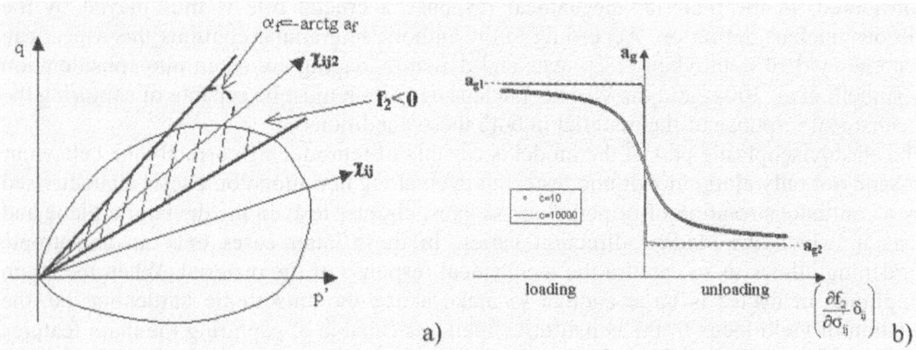

Figure 8. Two mechanisms constitutive model: a) a schematic representation of the two yield loci: the dashed zone represents the quasi-reversible locus; b) dependency of a_g on the current effective stress state.

As far as this cyclic mechanism is concerned, a time independent mechanical relationship has been chosen:

$$\dot{\varepsilon}^{cyc}_{ij} = \dot{\lambda}_{cyc} \frac{\partial g_2}{\partial \sigma'_{ij}} \tag{6}$$

where $\dot{\lambda}_{cyc}$ is the plastic multiplier. Once the definition of an appropriate hardening rule is introduced, this is calculated by imposing the consistency rule. The chosen hardening rule is of Prager's type (Prager & Drucker 1952):

$$\dot{\chi}_{ij_2} = w_0\left(\frac{1}{2}+\frac{1}{\pi}arctg\left(\left(\left\|\hat{\chi}_{ij_2}-\chi_{ij_2}\right\|-\left\|\hat{\chi}_{ij_2}-\delta_{ij}/\sqrt{3}\right\|\right)\cdot c_2\right)\right)\left(\dot{\varepsilon}_{ij}^{cicl}-\dot{v}^{*cicl}\chi_{ij_2}\right) \quad (7)$$

where $\dot{v}^{*cicl}=\dot{\varepsilon}_{ij}^{cicl}\chi_{ij_2}$ and $\hat{\chi}_{ij2}$ is the limit tensor for χ_{ij2} inspired to the definition introduced for the limit locus of χ_{ij} cited above. In this case, the limit locus for $\hat{\chi}_{ij2}$ is a macro-cone with circular section in the deviatoric plane. Eq. (7) allows us to correctly simulate the shape of each cycle in the stress-strain plain. In fact in unloading the value of the term $\left\|\hat{\chi}_{ij_2}-\chi_{ij_2}\right\|$ will be larger than in reloading and this implies a larger stiffness at the beginning of the unloading. On the contrary, w_0 is a constitutive parameter to be calibrated on the energy D dissipated during a single loading-unloading cycle (Figure 9b) and will be a function only of the type of material and of the current relative density.

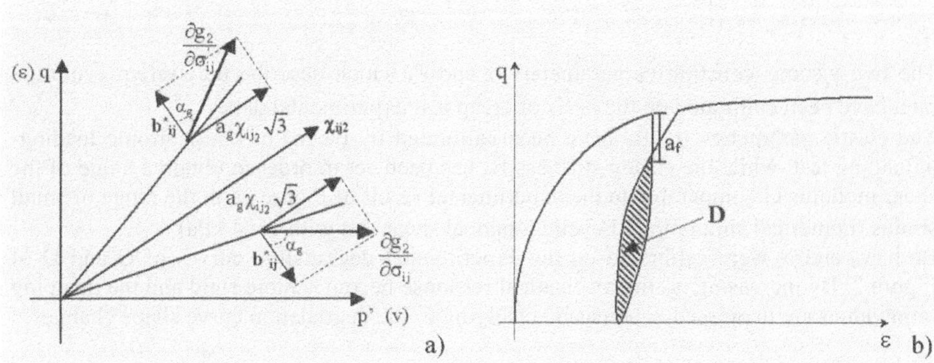

Figure 9. a) Geometrical meaning of term a_g: the loose sand case; b) Dissipated energy during a single loading-unloading cycle.

Summarising the above, the added constitutive parameters are five: a_{g1}, a_{g2}, c_1, c_2 and w_0. The first three parameters are used to capture the volumetric response but, in this case, this was not experimentally available. Their calibration is then missing. As far as the calibration of the last two, an appropriate discussion is introduced in the following section.

5. PARAMETER CALIBRATION

The constitutive model so far presented is characterised by many parameters: three elastic (B_o, α_c, R_F), eleven plastic ($\gamma, r_{c0}, \hat{\theta}_c, \hat{\theta}_e, \xi_c, \xi_e, t_p, c_p, B_p, \beta_{f0}, \hat{\beta}_f$), two visco-plastic ($\alpha, \bar{\gamma}$) and five cyclic. The plastic constitutive parameters have been calibrated on the basis of the results obtained in one isotropic loading-unloading test, two drained triaxial compression/extension tests and two undrained triaxial compression/extension tests all performed in strain-controlled conditions on saturated dense Toyoura specimens at University of Brescia. For satisfactorily capturing the experimental results with respect to the plastic constitutive model already cited, a modification of the plastic potential g function was introduced and this justifies the large value of γ in Table 2. In particular, the parameters $\gamma, \hat{\theta}_c, \hat{\theta}_e, \xi_c, \xi_e$ are linked to the failure behaviour and have been calibrated on the drained triaxial compression and extension tests. B_p, which is the plastic logarithmic volumetric compliance, can be directly established from a loading-unloading isotropic test, while $\hat{\beta}_f$, linked to the yield function shape, can be determined loading the specimen up to failure in triaxial compression and then unloading it. Finally the parameters c_p and t_p control the rate of evolution of the yield locus and therefore the stiffness of the material.

Table 2.: Constitutive model parameters.

D_r [%]	e_{max}			B_0 [kPa]	α_c	R_F			c_2	w_0		
80	0.6403			10600	0.25	0.25			80	32000		
γ	B_p	ξ_c	ξ_e	$\hat{\theta}_c$	$\hat{\theta}_e$	$\hat{\beta}_f$	β_{f0}	t_p	c_p	r_{c0}	$\bar{\gamma}$ [s^{-1}]	α
8.5	0.0004	-0.17	-0.04	0.362	0.136	0.6	0.2	18	85	1	$1 \cdot 10^{-6}$	61

The two viscous constitutive parameters α and $\bar{\gamma}$, which describe the system evolution rate, have been calibrated on the basis of creep test experimental data.

The elastic parameters α_c, R_F have been calibrated by means of one isotropic loading-unloading test, while the elastic stiffness B_0 has been set in order to obtain a value of the shear modulus G comparable to the experimental result of Figure 2a in the range of small strains (numerical simulation of cyclic torsional shear test with $\tau = 4$ kPa).

Both w_0 and c_2 were calibrated on the experimental degradation curves of G and D of Figure 2. By increasing w_0 the mechanical response becomes more rigid and the damping ratio values are depressed, whereas by changing c_2 the degradation curve shape changes.

6. NUMERICAL SIMULATIONS

As was already observed, in the three different ranges of shear strain γ_{SA}, small, medium, large, respectively, the constitutive model activates prevalently three different mechanisms: the elastic one, the cyclic elastoplastic and the anisotropic elasto-viscoplastic.

In Figure 10, the experimental and numerical results are compared. It is evident that the numerical simulations are not capable of reproducing the severe decrease in the shear modulus when $0.01 < \gamma_{SA}$ (%)< 0.1, whereas the damping ratio within the same range is slightly overestimated. As it can be derived by considering Figure 5 and Figure 11, the numerical model does not capture the progressive rotation of cycles when the shear stress amplitude is increased. In this range of γ_{SA}, the numerical curve is characterised by a sharp bend which is due to the activation of the anisotropic viscoplastic mechanism changing abruptly the shape of the cycles.

On the contrary, it is worth noting that for values of γ_{SA} greater than 0.1 %, the experimental points are likely to overestimate G and severely underestimate D.

From Figure 11d we can derive that even the numerical simulations, although in a less marked manner than experiments, show a progressive evolution of the cycle shape. This is due to a partial activation of the viscoplastic mechanism when the shear stress imposed is sufficiently severe.

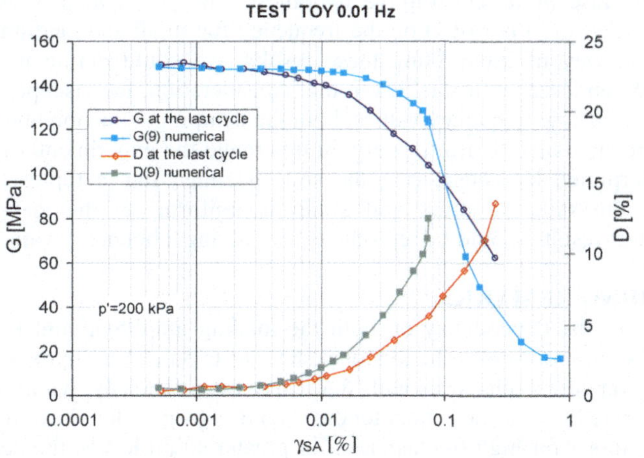

Figure 10. Experimental and numerical results: shear moduli and damping ratio versus single amplitude shear strain, measured during cyclic torsional shear tests on dense dry Toyoura specimens for loading frequencies 0.01 Hz.

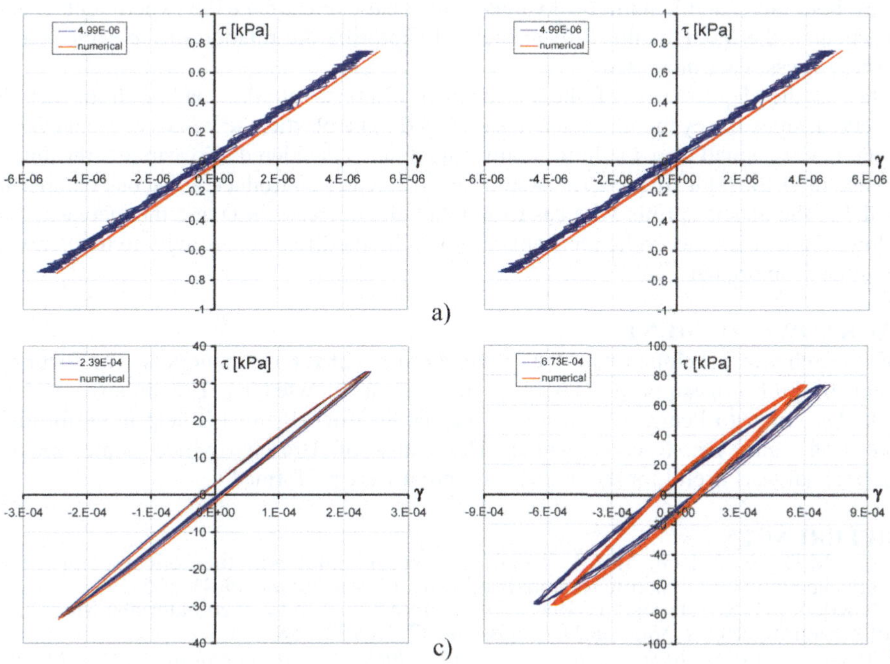

Figure 11. Numerical simulation of cyclic torsional shear tests on dense dry Toyoura specimens for loading frequency 0.01 Hz at different strain levels.

The model, because of its constitutive hypotheses, is not capable of reproducing the dependency neither of G nor of D on the frequency for small and medium strain ranges. In the common way of interpreting these results, we could define it as a hysteretic relationship. When large strains are taken into consideration, the viscoplastic mechanism is activated and for low frequency values, both the damping reduction and the increase in the G value are simulated. Unfortunately, in this range the experimental results obtained cannot be interpreted by considering the soil specimen like a representative volume within which the state of stress and strain is uniform. In this case the torsional experimental test results should be reproduced by solving a boundary value problem.

7. CONCLUDING REMARKS

The paper shows the dependency on both the loading amplitude and frequency of the stress-strain response during cyclic torsional shear tests on dense Toyoura sand. In small-medium strain ranges, G increases with frequency monotonously, while the dependency of D is described by a non-monotonous trend, steeper for low frequencies and approximately linear for high frequencies. For greater strain levels, the dependency of D and G on the loading frequency becomes less clear; moreover, both the shape of cycle and the average shear stiffness evolve progressively with the number of cycles imposed.
By employing a multiple mechanism elastoplastic constitutive model, the authors tried to simulate first the dependency of G and D on the loading amplitude. At medium strain values, both the G and D numerical values somewhat overestimate the experimental data. In particular, the model is not fully capable of capturing the rotation of the cycles when the shear stresses are increased.
As far as the dependency of the mechanical response on the loading frequency is concerned, since the cyclic mechanism is assumed to be plastic, the values of D and G are constant when small and medium strain ranges are considered. To capture the linear increase in D for high frequencies should be sufficient to introduce a viscous damper in parallel to the elastic spring; whereas to simulate the decrease in D and the increase in G for low frequencies, the cyclic mechanism should be modified accordingly to the Perzyna visco-plastic approach.

8. ACKNOWLEDGMENTS

The research was developed with the framework of ALERT Geomaterials. The financial support of the European project DIGA is gratefully acknowledged. The authors whish to thanks Dr. Augusto Penna, University of Naples Federico II, for his help in performing some tests, and Dr. Rocco Lagioia, University of Brescia, where some triaxial experimental tests useful for the model calibration were performed.

9. REFERENCES

Bolton, M.D. & Wilson, J.M.R. 1989. An experimental and theoretical comparison between static and dynamic torsional soil tests. *Géotechnique* 39(4): 585-599.
Di Benedetto, H. & Tatsuoka, F. 1997. Small strain behaviour of geomaterials: Modelling of strain rate effects. *Soils and Foundations* 37(2): 127-138.
di Prisco, C. 1993. *Anisotropia delle sabbie: indagine sperimentale e modellazione matematica*. Milan: PhD Thesis.
di Prisco, C., Nova, R. & Lanier, J. 1993. A mixed isotropic kinematic hardening constitutive law for sand. *Modern approaches to plasticity* (eds D. Kolymbas): 83-124. Rotterdam: Balkema.

di Prisco, C., Matiotti, R. & Nova, R. 1995. Theoretical investigation of the undrained stability of shallow submerged slopes. *Géotechnique* 45(3): 479-496.
di Prisco, C. & Imposimato, S. 1996. Time dependent mechanical behaviour of loose sand. *Mechanics of cohesive-frictional materials* 17(1): 45-73.
di Prisco, C. & Imposimato, S. 1997. Experimental analysis and theoretical interpretation of triaxial load controlled loose sand collapses. *Mechanics of cohesive-frictional materials* 18(2): 93-120.
di Prisco, C. & Zambelli, C. 2003. Cyclic and dynamic mechanical behaviour of granular soils: experimental evidence and constitutive modelling. *Revue francaise de genie civil* 7(7-8): 881-910.
d'Onofrio, A., Santucci de Magistris, F., Silvestri, F., & Vinale, F. 1995. Behaviour of compacted sand-bentonite mixtures from small to medium strains. *First Earthquake Geotech. Eng. Conf.* Tokyo Vol. 1: 133-138, Balkema.
d'Onofrio, A, Silvestri, F, & Vinale, F. 1999a. New torsional shear device. *ASTM, Geotechnical Testing Journal* 22(2): 107-117.
d'Onofrio, A, Silvestri, F, & Vinale, F. 1999b. Strain rate dependent bahaviour of a natural stiff clay. *Soils and Foundations* 39(2): 69-82.
Fioravante, V, Jamiolkowski, M, Lo Presti, DCF. 1994. Stiffness of carbonatic quiou sand. *International conference on soil mechanics and foundation engineering*, 13, New Delhi, Jan. 1994. Proceedings, vol. 1.
Isenhower, W.M. 1979. Torsional simple shear/Resonant column properties of San Francesco Bay mud. *Master of Science thesis*. The University of Texas at Austin.
Iwasaki, T., Tatsouka, T. & Takagi, Y. 1978. Shear modulus of sand under cyclic torsional shear loading. *Soils and Foundations* 18(1): 39-56.
Kim, Y.S., Tatsuoka, F & Ochi, K. 1994. Deformation characteristics at small strains of sedimentary soft rock by triaxial compression tests. *Geotechnique* 44(3): 461-478.
Lade, P.V. & Nelson, R.B. 1987. Modelling the elastic behaviour of granular materials. *Int. J. Num. Anal. Meth. Geomech.* 11: 521-542.
Lin, M.L., Huang, T.H. & You, J.C. 1996. The effects of frequency on damping properties of sand. *Soil dynamics and earthquake engineering* 15: 269-278.
Molenkamp, F. 1988. A simple model for isotropic non-linear elasticity of frictional materials. *Int. J. Num. Anal. Meth. Geomech.* 12: 467-475.
Papa, V., Silvestri, F. & Vinale, F. 1988. Recenti sviluppi e prospettive nelle tecniche di interpretazione di prove dinamiche di taglio semplice. *Atti del Convegno Nazionale del Coordinamento per gli Studi di Ingegneria Geotecnica. Monselice (PD)* (in italian).
Perzyna, P. 1963. The constitutive equations for rate sensitive plastic materials. *Quart. Appl. Math.* 20: 321-332.
Perzyna, P. 1966. Fundamental problems in viscoplasticity. *Advances in applied mechanics, Academic press* 9: 243-377.
Prager, W. & Drucker, D.C. 1952. Soil mechanics and plastic analysis or limit design. *Quart. Appl. Math.* 10(2): 157-165.
Porovic, E. & Jardine, R.W. 1994. Some observations on the static and dynamic shear stiffness of Ham River Sand. *Proc. of Intern. Symp. on Pre-failure deformation of geomaterials. IS-Hokkaido* Vol. 1, 25-30, Balkema.
Shibuya, S., Mitachi, T., Fukuda, F. & Degoshi, T. 1995. Strain rate effects on shear modulus and damping of normally consolidated clay. *Geotechnical testing journal* 18(3): 365-375.
Tatsuoka, F. & Shibuya, S. 1991. Deformation characteristics of soils and rocks from field and laboratory tests, *Keynote Lecture for Session No.1, Proc. of the 9th Asian Regional Conf. on SMFE*, Bangkok, Vol. II, 101-170.
Tatsuoka, F., Sato, T., Park, C-S. Kim, Y-S., Mukabi, J.N. & Kohata, Y. 1994. Measurement of elastic properties of geomaterials in laboratory compression test. *ASTM Geot. Test. Journ.* 17(1): 80-94.
Tatsuoka, F., Lo Presti, D.C.F. & Kohata, Y. 1995. Deformation characteristics of soils and soft rocks under monotonic and cyclic loads and their relationships. *III Int. Conf. on Recent Advances in Geotechnical Earthquake Engineering and Soil Dynamics.* University of Missouri-Rolla. St. Louis MO.

Tatsuoka, F., Santucci de Magistris, F., Hayano, K., Momoya, Y. & Koseki, J. 2001. Some new aspects of time effects on the stress-strain behaviour of stiff geomaterials. *Keynote Lecture, The Geotechnics of Hard Soils – Soft Rocks, Proc. of Second Int. Conf. on Hard Soils and Soft Rocks*, Napoli, 1998 (Evangelista and Picarelli eds.), Balkema, Vol. 2: 1285-1371.

Vucetic, M. & Tabata, K. 2003. Influence of soil type on the effect of strain rate on small-strain cyclic shear modulus. *Soils and Foundations* 43(5): 161-173.

Zambelli, C. 2002. *Modellazione costitutiva del comportamento meccanico in campo dinamico di materiali granulari*. Milan: Tesi di laurea.

Zambelli, C., di Prisco, C. & Imposimato, S. 2004. A cyclic elasto-viscoplastic costitutive model: theoretical discussion and validation. *Cyclic Behaviour of Soils and Liquefaction Phenomena* edited by Triantafyllidis Th., Balkema: 99-106.

Soil Stress-Strain Behavior: Measurement, Modeling and Analysis
Geotechnical Symposium in Roma, March 16 & 17, 2006

DYNAMIC CLAY SOILS BEHAVIOUR BY DIFFERENT LABORATORY AND IN SITU TESTS

A. Cavallaro[(1)], S. Grasso[(2)] and M. Maugeri[(3)]
[(1)]CNR Ibam, 95131 Ct, ITALY
[(2)]University of Catania, 95125 Ct, ITALY
e-mail: acava@dica.unict.it

ABSTRACT

Since the eastern Sicily has been struck by various disastrous earthquakes with a MKS intensity from IX to XI in the last 900 years, an investigation program was performed in different test sites to evaluate the dynamic clayey soils behaviour.

The following dynamic investigations in the laboratory were carried out on undisturbed samples: Resonant Column Tests (RCT), Cyclic Loading Torsional Shear Tests (CLTST) and Cyclic Loading Triaxial Tests (CLTxT). The influence of the type of tests, of the strain rate and of the number of loading cycles on G-γ and D-γ curves, as well as on pore pressure build-up, was evaluated by means of laboratory tests.

Moreover in situ investigations were carried out in order to determine the soil profile and the geotechnical characteristics for the site under consideration. Borings, Standard Penetration Test (SPT), Cone Penetration Test (CPT), Down Hole (DH) Tests and new Seismic Marchetti Dilatometer Tests (SDMT) were performed.

The available data enabled one to obtain a comparison between the small strain shear modulus (DH, SDMT and laboratory tests) for the specific site. It was also possible to compare in situ and laboratory small strain shear and the G-γ curve for soil non-linearity evaluation..

1. INTRODUCTION

The eastern Sicily has been struck by various disastrous earthquakes with a MKS intensity from IX to XI in the last 900 years. It is well known that local geological and geotechnical conditions play a major role on earthquake ground motions and distribution of damage. Thus an investigation program was performed in different areas to evaluate the dynamic clayey soils behaviour. The areas investigated are the city of Catania, the city of Noto and the city of Augusta. All these three cities were destroyed by the 1693 earthquake. Noto and Augusta were also damaged by the recent 1990 eastern Sicily earthquake and the investigation was performed with the aim of repairing the damaged buildings and monuments. The site investigation in the city of Catania was performed in the framework of the Research Project: Detailed Scenarios and Actions for Seismic Prevention of Damage in the Urban Area of Catania, financed by the Department of National Civil Defence of Italy. In the urban area of Catania twelve test site has been established where laboratory and in situ tests were performed.

The following dynamic investigations in the laboratory were carried out on undisturbed samples: Resonant Column Tests (RCT), Cyclic Loading Torsional Shear Tests (CLTST)

Hoe I. Ling et al. (eds.), Soil Stress-Strain Behavior: Measurement, Modeling and Analysis, 583–594.

and Cyclic Loading Triaxial Tests (CLTxT). The influence of the type of tests, of the strain rate and of the number of loading cycles on G-γ and D-γ curves, as well as on pore pressure build up, was evaluated by means of laboratory tests.
Moreover in situ investigations were carried out in order to determine the soil profile and the geotechnical characteristics for the site under consideration. Borings, Standard Penetration Test (SPT), Cone Penetration Test (CPT), Down Hole (DH) Tests and new Seismic Marchetti Dilatometer Test (SDMT) were performed.

2. SHEAR MODULUS AND DAMPING RATIO FROM LABORATORY TESTS

Resonant Column (RCT) and Cyclic Loading Torsional Shear (CLTST) tests have been performed by using the same apparatus (Lo Presti et al. 1993) to evaluate the shear modulus G and damping ratio D for Augusta (Saline site and Hangar site), Noto (Cathedral site) and Catania (Via Stellata and Via Dottor Consoli sites) clay soils. Cyclic Loading Triaxial Tests (CLTxT) has been performed for Noto clay. The laboratory test conditions stress path, strain rate, frequency and type of control, are given in Table 1. The obtained small strain shear modulus G_o are listed in Table 2. The undisturbed specimens were isotropically reconsolidated to the best estimate of the in situ mean effective stress. The same specimens were first subjected to CLTST (Cyclic Loading Torsional Shear Test), then to RCT (Resonant Column Test) after a rest period of 24 hrs with opened drainage. CLTST were performed under stress control condition by applying a torque with triangular time history at a frequency of 0.1 Hz. Only for Augusta Hangar specimen the RCT was performed before the CLTST.

Table 1. Test Condition in Different Tests.

Test	Stress Path	Strain Rate [%/min]	f [Hz]	Type of Control
CLTxT	S_3 = const	0.01/0.1	0.01/0.001	strain
MLTxT	S_3 = const	0.01/0.1	0.01/0.001	strain
CLTST	S = const	0.0025/2.5	0.1	stress
RCT	S = const	1/5000	20/50	stress

Table 2. Test Condition for Eastern Sicily Specimens.

Test Site	H [m]	σ'_{vc} [kPa]	e	PI	CLTST RCT	G_o (1) [MPa]	G_o (2) [MPa]	Speci men	G_o(RCT)/G_o(CLTST)
Augusta Saline (SR)		259	0.889	40	U	70	75	H	1.07
Via Stellata (CT)	39.00	411	0.695	31	U	77	93	S	1.21
Via Dottor Consoli (CT)	10.00	201	0.527	16	U	61	49	S	1.24
Cattedrale di Noto (SR)	15.50	237	0.718	27	U	68	84	S	1.23
Augusta Hangar (SR)	12.25	235	0.805	40	U	69	56	S	0.81

where: U= Undrained. G_o (1) from CLTST, G_o (2) from RCT. H = Hollow cylindrical specimen (R_O = 25 mm; R_i = 15 mm; h = 100 mm). S = Solid cylindrical specimen (R = 25 mm; h = 100 mm).

The G_o values, reported in Table 2, indicate a moderate but measurable influence of strain rate and type of loading even at very small strain where the soil behaviour is supposed to

be elastic (Cavallaro, 1997; Lo Presti et al., 1996; Lo Presti et al., 1997b; Lo Presti et al., 1998). The ratio $G_o(RC)/G_o(CLTST)$ ranges between 1.21 and 1.24 for Noto and Catania clay soils. In order to appreciate the rate effect on G_o, it is worthwhile to remember that the equivalent shear strain rate experienced by the specimens during RCT can be three orders of magnitude greater than those adopted during CLTST. The effects of the rate and loading conditions on the shear modulus are the same up to 0.01 % strain level. This experimental finding is different for that observed for Augusta Saline site where the ratio $G_\gamma(RC)/G_\gamma(CLTST)$ is near one as observed by Cavallaro (1997), Lo Presti et al. (1997b), Lo Presti et al. (1998) and Tatsuoka et al. (1997), who have showed an increasing rate effect as the strain level increase. This different behaviour can be tentatively explained by considering that in this study solid cylindrical specimens with a shear strain variable from zero (at the center of the section) to a maximum value at the edge have been used, while in previous study mainly hollow cylinder specimens were used. In the case of hollow specimens, the shear strain is quite constant along the radius.
Normalized shear modulus G/G_o and damping (D) obtained from RCT and CLTST are shown in Figures 1 and 2.

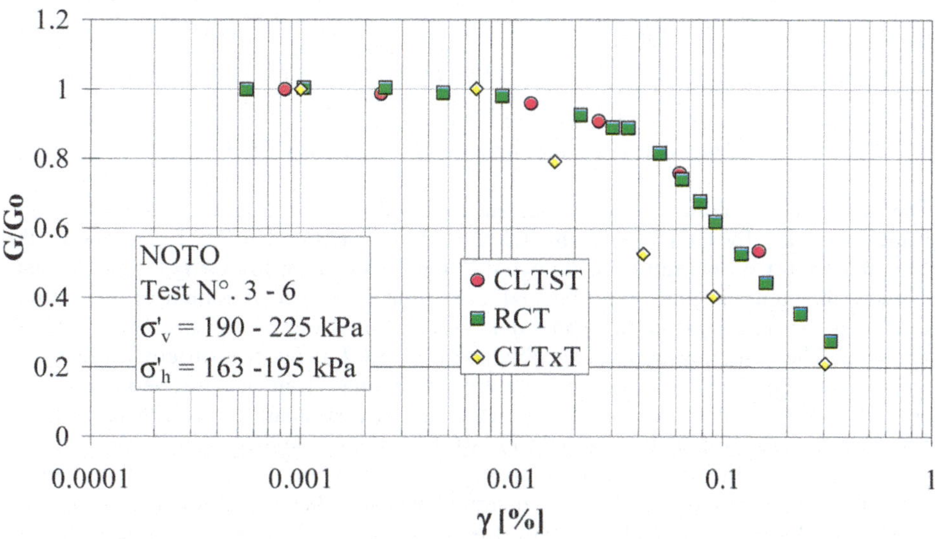

Figure 1. G/G_o-γ curves from CLTST, RCT and CLTxT.

The same shear modulus decay is obtained from both type of tests but the damping ratio values provided by CLTST are smaller than those measured in RCT. For RCTs the damping ratio was determined using two different procedures: following the steady-state method, the damping ratio was obtained during the resonance condition of the sample; following the amplitude decay method it was obtained during the decrement of free vibration. The damping ratio values obtained from RCT using two different procedures are similar even if higher values of D have been obtained from steady-state method. It is possible to see that the damping ratio from CLTST, at very small strains, is equal to about 1 %. Greater values of D are obtained from RCT for the whole investigated strain interval.

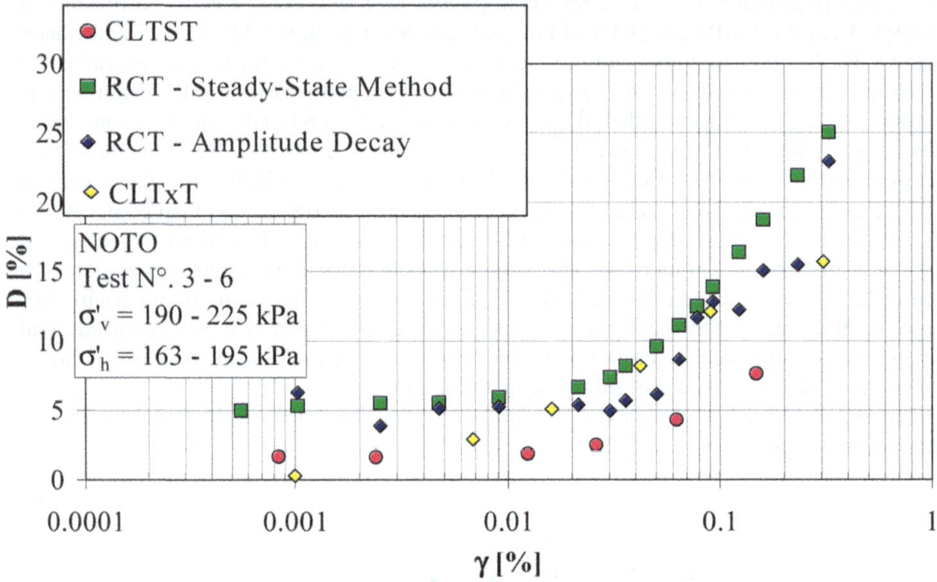

Figure 2. D-γ curves from CLTST, RCT and CLTxT.

The same shear modulus decay is obtained from both type of tests but the damping ratio values provided by CLTST are smaller than those measured in RCT. For RCTs the damping ratio was determined using two different procedures: following the steady-state method, the damping ratio was obtained during the resonance condition of the sample; following the amplitude decay method it was obtained during the decrement of free vibration. The damping ratio values obtained from RCT using two different procedures are similar even if higher values of D have been obtained from steady-state method. It is possible to see that the damping ratio from CLTST, at very small strains, is equal to about 1 %. Greater values of D are obtained from RCT for the whole investigated strain interval. Considering that the influence of number of cycles N on D has been found to be negligible, in the case of clayey soils for strain levels less than 0.1 % (Cavallaro 1997, Lo Presti et al. 1996, 1997a, 1997b, 1998), it is supposed that RCT provide larger values of D than CLTST because of the rate (frequency) effect, in agreement with data shown by Shibuya et al. (1995) and Tatsuoka et al. (1995). According to these researchers the nature of soil damping in soils can be linked to the following phenomena:

- non-linearity which governs the so called hysteretic damping controlled by the current shear strain level. This kind of material damping is absent or negligible at very small strains;
- viscosity of the soil skeleton (creep) which is relevant at very small strain rates;
- viscosity of the pore fluid which is relevant at very high frequencies.

Soil damping, at very small strains, is mainly due to the viscosity of the soil skeleton or of the pore fluid, depending on the strain rates or frequencies. Moreover, according to Tatsuoka et al. (1995) a partial drainage condition can provide very high values of the damping ratio.

One Noto soil specimen was tested in the triaxial apparatus (Cavallaro, 1997; Cavallaro et al., 1998; Cavallaro & Maugeri, 2003). The size of solid cylindrical specimens are: radius = 35 mm and height = 140 mm. This specimen was reconsolidated to the in situ geostatic stresses (K_o condition).

After the consolidation stage, the specimen was subjected to Cyclic Loading Triaxial test (CLTxT), at constant strain rate. Six different strain levels of progressive amplitude were imposed to the specimen. For each strain level 30 cycles were applied. The maximum applied axial strain (single amplitude) was about 0.3 %. The same specimen was subjected, after a rest period of 24 hrs with open drainage, to Monotonic Loading Triaxial test (MLTxT). The results of CLTxT are also compared in Figures 1 and 2 with RCT and CLTST results. It is possible to notice that CLTxT results show a greater non-linearity while the damping ratio values from CLTxT and those from CLTST are comparable for stress level less than 0.01 %. It should be remembered that CLTxT have been performed at constant strain rate equal to 0.01 %/min. Yet the different deformation mechanism (different stress-path) could be responsible for the observed differences.

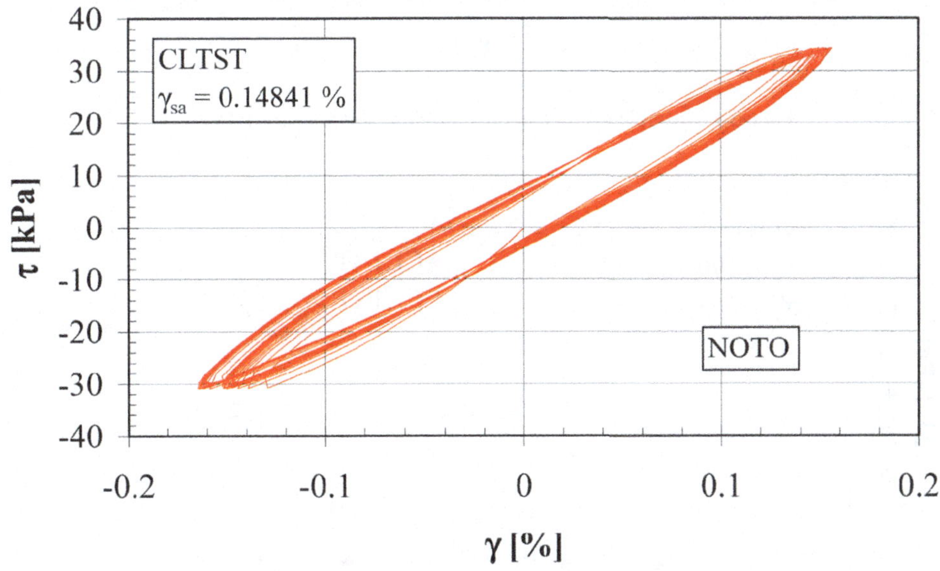

Figure 3. Unstable cyclic loadings for Noto soil during CLTST.

During CLTST and CLTxT unload-reload cycles become unstable and degradation phenomena of material occur when a certain limit strain is exceeded (Figures 3 and 4). This limit strain is called volumetric threshold shear strain (γ_t^v) and is rate dependent. The degradation causes a decrease of stiffness, an increase of D and pore pressure with the increase of N because of cyclic material degradation as obtained from a CLTST and CLTxT on Noto soil.

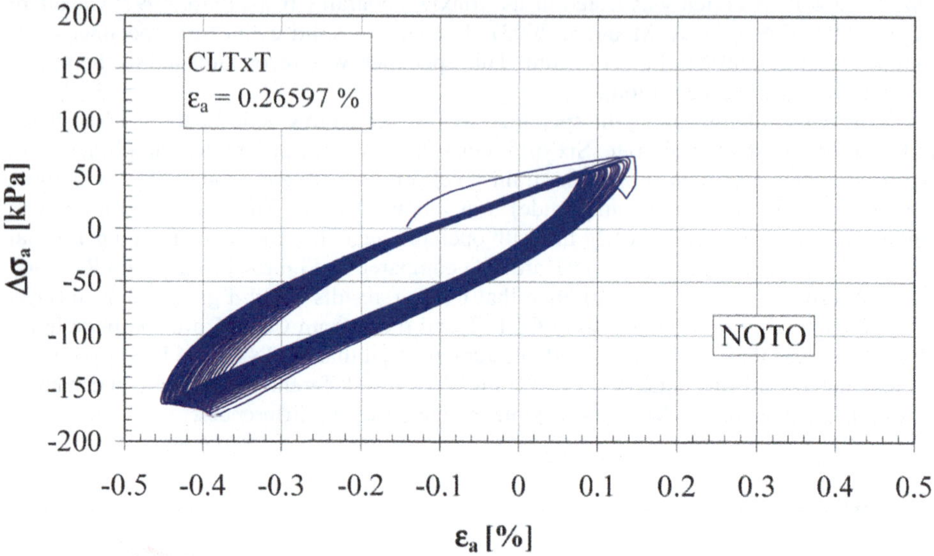

Figure 4. Unstable cyclic loadings for Noto soil during CLTxT.

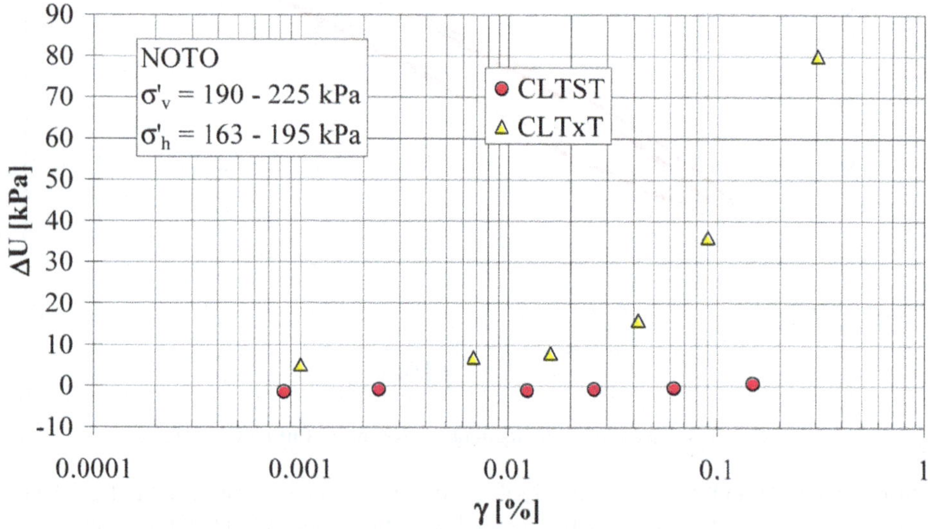

Figure 5. Pore pressure build-up for Noto soil during CLTST and CLTxT.

The Figure 5 showed the pore pressure build up during CLTxT and CLTST. The pore pressure build up during CLTST is so negligible. Only at strain level of about 0.15 % it is possible to observe a moderate increase of pore pressure.
The secant soil stiffness (E_s), which has been obtained from monotonic loading (Figure 6) was compared to that inferred from unloading-reloading (E_{eq}).
The E_o values obtained during CLTxT and MLTxT, indicate a good agreement between cyclic and monotonic tests even at very small strains where the soil behaviour is supposed to be elastic.
The effects of the loading conditions become more and more relevant with an increase of the shear strain level, as can be seen in Figure 6, where the E-ε curves obtained from MLTxT and CLTxT are compared.
It is possible to notice that the lowest decay of E with ε is observed in MLTxT, while the maximum decay occurs during CLTxT. After a strain value of 0.3 % the modulus decay is the same for MLTxT and CLTxT because of the degradation phenomena occurs. It can be seen that the degradation curves coincide in a range between 0.001 - 0.01 %.

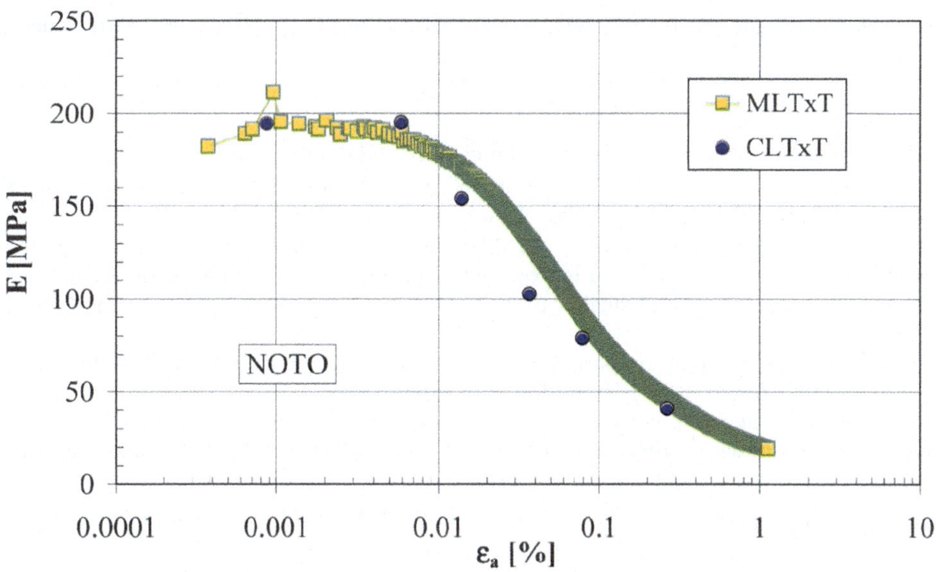

Figure 6. E versus axial strain curves from MLTxT and CLTxT.

After this strain level, the cyclic tests show a progressive decline for the monotonous curve with natural degradation of mechanical properties of materials.
This modulus reduction was probably caused by these factors:
- pore pressure build up with a reduction of effective stress;
- soil degradation caused by the maximum shear strain level investigated during the test.

Moreover, considering that after a period of 24 hrs with opened drainage, the initial Young modulus value at small strain obtained during MLTST is comparable to that obtained at the beginning of the CLTST, it is possible to assume the reduction of shear modulus as the consequence of effective stress reduction for the pore pressure build up. So it is possible to consider that the elastic energy loosed was recovered in a period of 24 hrs with the dissipation of pore pressure.

3. SHEAR MODULUS FROM IN SITU TESTS

SPT, CPT, DH and the new SDMT tests were performed. The seismic flat dilatometer (SDMT) provides a simple means for determining the initial elastic stiffness at very small strain and in situ shear strength parameters at high strain in natural soil deposits. The small strain shear modulus G_o is determined by the theory of elasticity by the well known relationship:

$$G_o = \rho V_s^2 \tag{1}$$

where: ρ = mass density.

A key feature distinguishing SDMT from other seismic tips is that SDMT, besides G_o, should determine a "working strain" shear modulus, G_{WS}. The availability of two datapoints (G_o and G_{WS}) may help in selecting the G-γ decay curve, important in soil dynamics.

G_{WS} can be evaluated by the following equation based on M_{DMT} (Marchetti, 1980) values:

$$G_{WS} = \frac{(1-2\cdot\nu)}{2\cdot(1-\nu)}\cdot M_{DMT} \tag{2}$$

where ν is the Poisson ratio, obtained from in situ or laboratory tests.

As regard the evaluation of "working strain" γ_{WS}, we must distinguish the settlements predicted during the analysis of case histories (γ = 0.05 to 0.1 %) and the real strain investigated by SDMT to measure the dilatometer modulus E_D. In the vicinity of the probe, the flat dilatometer blade is expected to produce shear similar to the cylindrical probes of the piezocone and smaller than the push-in pressuremeter (Lacasse and Lunne, 1988).

In Figure 7 it is tentatively reported the comparison between RCT for different Catania site and SDMT results at large strain for STM M6 test site.

Moreover, for Catania area, it is possible to evaluate the values of G_o versus depth by means of the following empirical correlations based on SPT, SDMT, CPT tests results or laboratory test results available in literature:

a) Hryciw (1990):

$$G_o = \frac{530}{(\sigma'_v/p_a)^{0.25}}\frac{\gamma_D/\gamma_W - 1}{2.7-\gamma_D/\gamma_W}K_o^{0.25}\cdot(\sigma'_v\cdot p_a)^{0.5} \tag{3}$$

where: G_o, σ'_v and p_a are expressed in the same unit; p_a = 1 bar is a reference pressure; γ_D and K_o are respectively the unit weight and the coefficient of earth pressure at rest, as inferred from SDMT results according to Marchetti (1980);

b) Mayne and Rix (1993):

$$G_o = \frac{406\cdot q_c^{0.696}}{e^{1.13}} \tag{4}$$

where: G_o and q_c are both expressed in [kPa] and e is the void ratio. Eq. (4) is applicable to clay deposits only;

c) Jamiolkowski et. al. (1995):

$$G_o = \frac{600\cdot\sigma_m'^{0.5}p_a^{0.5}}{e^{1.3}} \tag{5}$$

where: $\sigma'_m = (\sigma'_v + 2\cdot\sigma'_h)/3$; p_a = 1 bar is a reference pressure; G_o, σ'_m and p_a are expressed in the same unit. The values for parameters which appear in equation (5) are equal to the average values that result from laboratory tests performed on quaternary Italian clays and reconstituted sands.

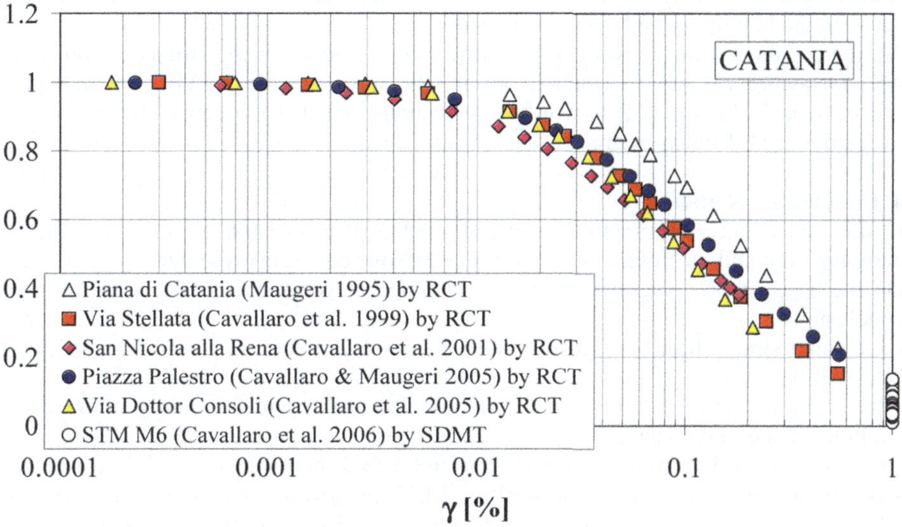

Figure 7. G/G_o vs shear strain for Catania area.

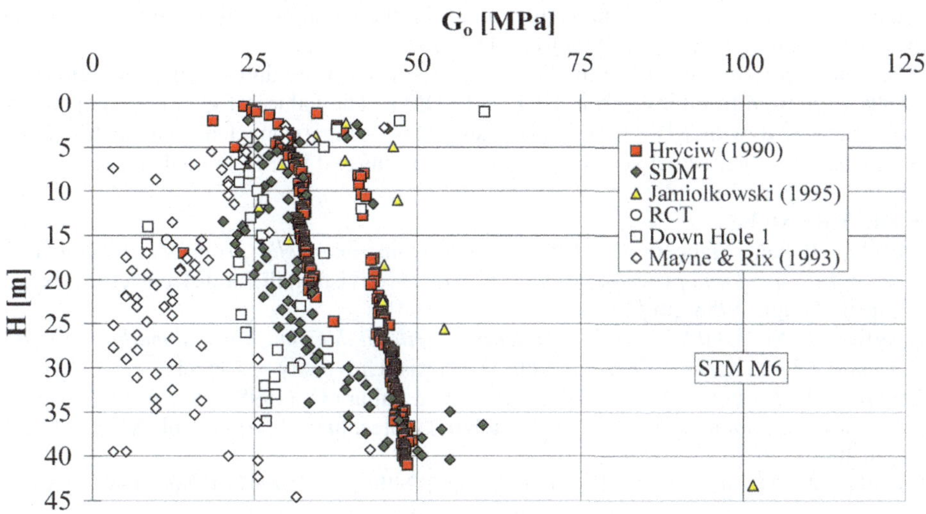

Figure 8. G_o from laboratory tests, in situ tests and different empirical correlations.

The G_o values obtained by SDMT, Down Hole test and with the methods above indicated are plotted against depth in Figure 8. The method by Jamiolkowski et al. (1995) was applied considering a given profile of void ratio. The coefficient of earth pressure at rest was inferred from SDMT. Moreover Figure 8 shows the values of G_o obtained from RCT (Carrubba and Maugeri 1988). It is possible to see that quite a good agreement exists between the laboratory and in situ test results.
All the considered methods show very different G_o values of the Holocene soil. On the whole, equation (3) and (5) seems to provide the most accurate trend of G_o with depth, as can be seen in Figure 8. It is worthwhile to point out that the considered equation (5) overestimate G_o for depths greater than 25 m.

4. Concluding Remarks

In this paper some information on the geotechnical properties, on shear modulus and damping ratio and on Young modulus of clayey soils has been presented.
On the basis of the experimental results obtained, it is possible to draw the following conclusions:
- the small strain shear modulus obtained from CLTST and RCT is influenced by the strain level;
- the damping ratio, measured in the laboratory, resulted to be mainly influenced by strain level and rate effects;
- the normalized shear modulus obtained from CLTxT show a greater non-linearity;
- damping ratio values determined from RCT are greater than those obtained from CLTST while the damping ratio values from CLTxT and those from CLTST are comparable for stress level less than 0.01 %.
- the observed differences between RCT, CLTST and CLTxT results are probably due to rate and/or frequency effects and different deformation mechanism (different stress-path);
- the small strain shear modulus measured in the laboratory is about 90% of that measured in situ by means of SDMT and DH tests;
- empirical correlations between the small strain shear modulus and penetration test results were used to infer G_o from CPT and SDMT. The values of G_o were compared to those measured with SDMT and DH tests. This comparison clearly indicates that a certain agreement exist between empirical correlations and SDMT and DH tests.

5. REFERENCES

Carrubba P. and Maugeri M. (1988) - *Determinazione delle proprietà dinamiche di un'argilla mediante prove di colonna risonante.* Rivista Italiana di Geotecnica, N°. 2, Aprile-Giugno1988: pp.101-113.

Cavallaro A. M. F. (1997) - *Influenza della velocità di deformazione sul modulo di taglio e sullo smorzamento delle argille.* Ph. D. Thesis, University of Catania.

Cavallaro A., Lo Presti D. C. F., Maugeri M. and Pallara O. (1998) - *Strain rate effect on stiffness and damping ratio of clays.* Italian Geotechnical Journal, Vol. XXXII, N°. 4, pp. 30-50.

Cavallaro A., Maugeri M., Lo Presti D.C.F. and Pallara O. (1999) - *Characterising shear modulus and damping from in situ and laboratory tests for the seismic area of Catania.* Proceedings of the 2nd International Symposium on Pre-failure Deformation Characteristics of Geomaterials, Torino, 28 - 30 September 1999: pp. 51-58.

Cavallaro A., Grasso S. and Maugeri M. (2001) - *A dynamic geotechnical characterization of soil at Saint Nicolò alla Rena Church damaged by the south eastern Sicily earthquake of 13 December 1990*. Proceedings of the 15th International Conference on Soil Mechanics and Geotechnical Engineering, Satellite Conference "Lessons Learned from Recent Strong Earthquakes", Istanbul, 25 August 2001:pp. 243-248.

Cavallaro A., Grasso S. and Maugeri M. (2005) - *Site Characterisation and Site Response for a Cohesive Soil in the City of Catania*. Proc. of the Satellite Conference on Recent Developments in Earthquake Geotechnical Engineering. Osaka University Nakanoshima Center, Osaka City, Japan, September 10,2005, pp. 167-174.

Cavallaro A., Grasso S. and Maugeri M. (2006) - *A volcanic soil characterisation and site responseanalysis in the city of Catania*. Proceedings of the 8th National Conference on Earthquake Engineering, San Francisco, 18-22 April.

Cavallaro A. and Maugeri M. (2003) - *Site characterization by in-situ and laboratory tests for the microzonation of Noto*. In Geotechnical analysis of seismic vulnerability of Monuments and Historical Sites. Editors: M. Maugeri & R. Nova, Patron Editore: pp. 237-256.

Cavallaro A. and Maugeri M. (2005) - *Non linear behaviour of sandy soil for the city of Catania*. Seismic Prevention of Damage: A Case Study in a Mediterranean City, Wit Press Publishers, Editor: Maugeri M.: pp. 115-132.

Hryciw R.D. (1990) - *Small strain shear modulus of soil by dilatometer*. Journal of the Geotechnical Engineering Division, ASCE, Vo. 116, No. 11, pp. 1700-1715.

Jamiolkowski M., Lo Presti D. C. F. and Pallara O. (1995) - *Role of in-situ testing in geotechnical earthquake engineering*. Proceedings of the 3rd Int. Conf. on Recent Advances in Geotechnical Earthquake Engineering and Soil Dynamic, State of the Art 7, Vol. 3, pp. 1523-1546.

Lacasse S. and Lunne T. (1988) - *Calibration of dilatometer correlations*. Proceedings of 1st International Symposium on Penetration Testing, ISOPT-1, Orlando: pp. 539-548.

Lo Presti D. C. F., Pallara O., Lancellotta R., Armandi M. and Maniscalco R. (1993) - *Monotonic and cyclic loading behaviour of two sands at small strains*. Geotechnical Testing Journal, December 1993: pp. 409-424.

Lo Presti D. C. F., Jamiolkowski M., Pallara O. and Cavallaro A. (1996) - *Rate and creep effect on the stiffness of soils*. ASCE Convention, Washington, 10-14 Nov. 1996, Geotechnical Special Publication No. 61, pp. 166-180.

Lo Presti D. C. F., Jamiolkowski M., Pallara O., Cavallaro A. and Pedroni S. (1997a) - *Shear modulus and damping of soils*. International Symposium on the Pre-failure Deformation Behaviour of Geomaterials, 50th Geotechnique, London, 4 September 1997, Geotechnique 47(3): pp. 603-617.

Lo Presti D.C.F., Pallara O. and Cavallaro. A. (1997b) - *Damping ratio of soils from laboratory and in situ tests*. Proceedings of the 14th International Conference on Soil Mechanics and Foundations Engineering, Hamburg, Special Volume TC4, 6 - 12 September 1997, pp. 391-400.

Lo Presti D. C. F., Maugeri M., Cavallaro A., and Pallara O. (1998) - *Shear modulus and damping of a stiff clay from in situ and laboratory tests*. Proceedings of the 1st International Conference on Site Characterization, Atlanta, 19 - 22 April 1998, pp. 1293-1300.

Marchetti S. (1980) - *In situ tests by flat dilatometer.* Journal of Geotechnical Engineering Division, ASCE, Vol. 106, No.GT3, pp. 299-321.

Maugeri M. (1995) - *Discussions and replies session IX.* Proceedings of International Conference on Recent Advances in Geotechnical Earthquake Engineering and Soil Dynamics, St. Louis, 2 – 7 April 1995: pp. 1323-1327.

Mayne P.W. and Rix G.J. (1993) - *G_{max}-q_c relationships for clays.* Geotechnical Testing Journal, Vol. 16, N°. 1: pp. 54-60.

Shibuya S., Mitachi T., Fukuda F. and Degoshi T. (1995) - *Strain rate effect on shear modulus and damping of normally consolidated clay.* Geotechnical Testing Journal, 18:3, pp. 365-375.

Tatsuoka F., Lo Presti D. C. F. and Kohata Y. (1995) - *Deformation characteristics of soils and soft rocks under monotonic and cyclic loads and their relations.* Proceedings of the 3rd International Conference on Recent Advances in Geotechnical Earthquake Engineering and Soil Dynamic, State of the Art 1, 2, pp. 851-879.

Tatsuoka F., Jardine R. J., Lo Presti D., Di Benedetto H. and Kodaka T. (1997) - *Characterizing the pre-failure properties of geomaterials.* 14th International Conference on Soil Mechanics and Foundations Engineering, Hamburg, 6 - 12 September 1997, Theme Lecture.

Soil Stress-Strain Behavior: Measurement, Modeling and Analysis
Geotechnical Symposium in Roma, March 16 & 17, 2006

DYNAMICALLY AND STATICALLY MEASURED SMALL STRAIN STIFFNESS OF DENSE TOYOURA SAND

Sajjad Maqbool[1)], Takeshi Sato[2)] and Junichi Koseki[3)]

1) Department of Civil Engineering, University of Engineering and Technology, Lahore, Pakistan
e-mail: sajjadmaqbool@hotmail.com
2) Institute of Industrial Science, The University of Tokyo
Ce201, 4-6-1 Komaba, Meguro-ku, Tokyo 153-8505, Japan
e-mail: tsato@iis.u-tokyo.ac.jp
3) Institute of Industrial Science, The University of Tokyo
Ce201, 4-6-1 Komaba, Meguro-ku, Tokyo 153-8505, Japan
e-mail: koseki@iis.u-tokyo.ac.jp

ABSTRACT

Small strain stiffness of dense Toyoura sand was investigated by performing triaxial compression tests using large scale apparatus. The specimens were rectangular prismatic with dimensions of 50 cm high and 23.5 cm times 23.5 cm in cross-section. To measure the vertical stress σ_1, a load cell is located just above the top cap inside the triaxial cell in order to eliminate the effects of piston friction. The vertical strain ε_1 was measured not only externally but also locally with three pairs of vertical local deformation transducers (LDTs). Three tests were conducted using air-dried Toyoura sand (D_{max} = 0.35 mm, D_{50} = 0.23 mm, U_c = 1.80, e_{max} = 0.966, e_{min} = 0.600 and G_s = 2.635) as the test material. The specimens were prepared by employing air pluviation method and keeping dry densities within the range of 1.62 - 1.63 g/cm^3. Dynamic and static Young's moduli were evaluated by wave velocity measurement and by conducting small unloading / reloading cycles, respectively.

1- INTRODUCTION

Dynamic properties of the ground have been evaluated for decades using in-situ seismic measurements. For the last decade, the dynamic measurements are also becoming popular in the laboratory tests. In the past, "static" and "dynamic" properties were thought to be two different properties, while with the development of precise static small strain measurement devices, it has been recognized that "static" and "dynamic" properties are no more different from each other.

Hoe I. Ling et al. (eds.), Soil Stress-Strain Behavior: Measurement, Modeling and Analysis, 595–604.

In this study, to examine such behavior in Toyoura sand, "static" measurement was made by applying very small unloading/reloading cycles at several stress levels, and strains were measured locally using local deformation transducers (LDTs) within a very small strain range (about 0.001%) at the sides of the specimen. For the "dynamic" measurement, by triggering with single pulse, single sinusoidal or continuous sinusoidal waves at several stress levels, the velocity of vertically transmitting compression waves was evaluated from the arrival time difference between the input and the output wave signals. Finally the comparison between static and dynamic vertical Young's moduli was made.

2- MATERIAL, EQUIPMENT AND TEST PROCEDURES

A large-scale true triaxial apparatus (Sato et al. 2001) was used for this study. By this apparatus, all three-principle stresses (σ_1, σ_2, σ_3) can be controlled independently though in this study it was used just as a triaxial apparatus without confining plates. The specimen, as shown in Fig.1, was rectangular prismatic with dimensions of 50 cm high and 23.5 cm times 23.5 cm in cross-section for all tests. To measure the vertical stress $\sigma 1$, load cell is located just above the top cap inside the triaxial cell. The vertical strain $\varepsilon 1$ was measured not only with external

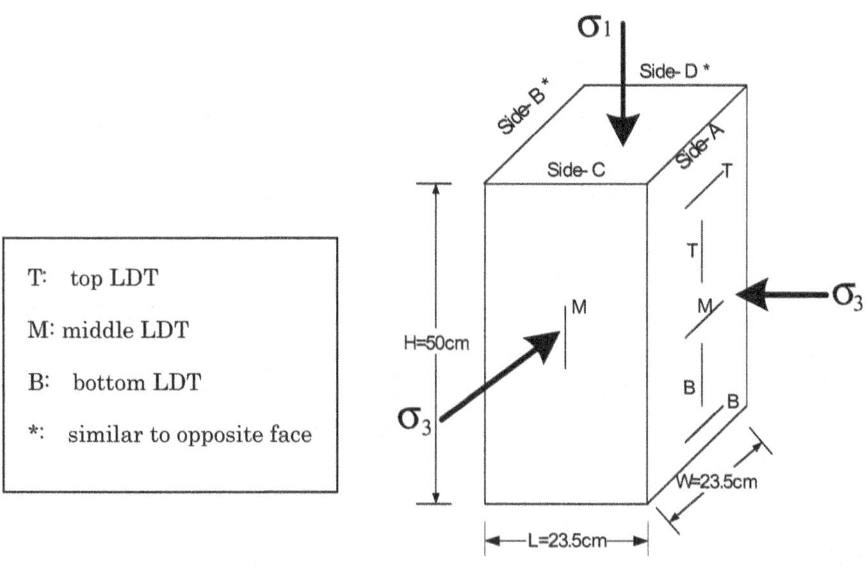

Fig. 1- Positioning of LDTs on the specimen

transducers but also with three pairs of vertical local deformation transducers (LDTs), located at the top and bottom on sides-A & B while at the middle on sides-C & D of the specimen. Minor principle stress σ_3 was applied through the cell pressure, which was measured with a high capacity differential pressure transducer (HCDPT). The corresponding horizontal strain ε_3 was measured with three pairs of horizontal LDTs, located at top, middle and bottom on sides-A & B of the specimen.

Toyoura sand (D_{max}=0.35 mm, D_{50}=0.23 mm, U_c=1.80, e_{max} = 0.966, e_{min} = 0.600 and G_s = 2.635) was employed as the test material. In this study, three dry specimens, named as TC5, TC6 & TC7, were prepared by employing air pluviation method and keeping dry densities within a narrow range of 1.62-1.63 g/cm^3 (relative densities Dr = 92.6-95.4%). The membrane used in these tests was 0.80 mm thick.

2.1 Evaluating static vertical Young's modulus

To evaluate quasi-elastic vertical Young's modulus based on static measurements, at some stress levels, very small unloading/reloading cycles were applied on the specimen in the vertical direction. The stress levels at which small unloading/reloading cycles were applied included 50kPa, 100kPa, 150kPa, 200kPa, 250kPa, 300kPa, 350kPa, 400kPa, 450kPa and 500kPa. Small cyclic loading was applied on the same stress levels during both loading and unloading

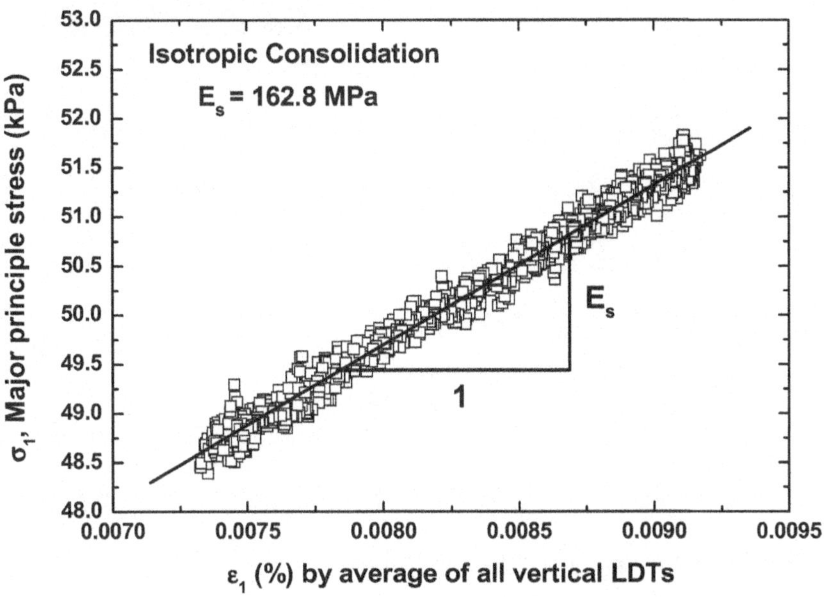

Fig. 2 - Typical stress-strain relationship during a small vertical loading cycle

stages of isotropic consolidation. At each of this stress level, eleven (11) cycles of loading were applied and in computing static vertical Young's modulus, the eleventh cycle was employed.

Existing data on the deformation properties from cyclic loading tests on dense granular soils reported in the literature showed that the behavior at strains lesser than about 0.001% is nearly elastic (Jiang et al. 1997). Typical stress-strain relationship during a small vertical loading cycle is shown in Fig. 2. The increments of the vertical strain and stress were detected with the local deformation transducers (LDTs) and the internal load cell, respectively. Stress-strain relationships were fitted by a straight line, and the quasi-elastic vertical Young's modulus E_s was evaluated from the slope of the line.

2.2 Evaluating dynamic vertical Young's modulus

To evaluate dynamic vertical Young's modulus, compression (or primary) waves were generated and received by a recently developed dynamic measurement system (Maqbool, 2005) as shown in Fig.3. The detailed description of the major components of this system is given in the following paragraphs.

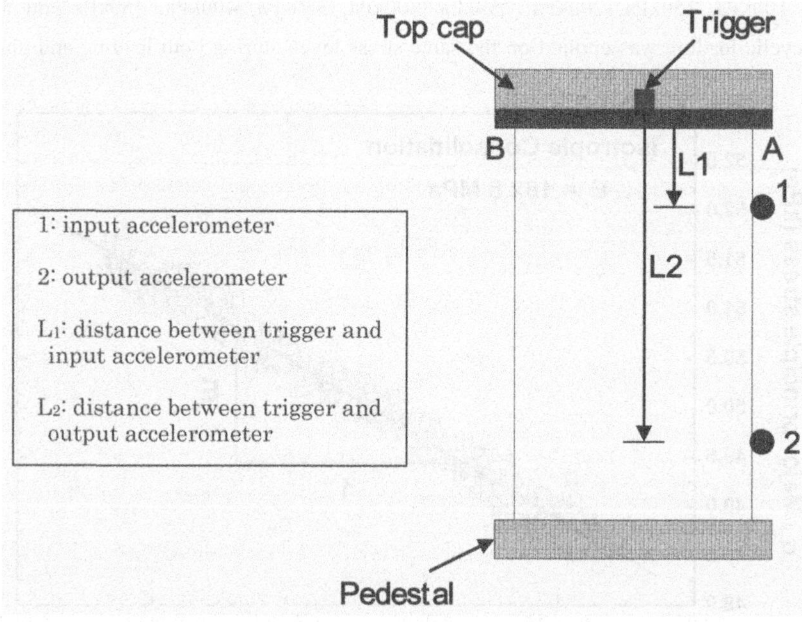

Fig. 3 - Location of trigger and accelerometers

2.3 Generating dynamic waves

In order to generate dynamic waves, a special type of wave source that is denoted as trigger in Fig.3 was employed. The components of trigger are shown in Fig.4. It is a multi-layered piezoelectric actuator made of ceramics (dimensions of 10 mm x 10 mm x 20 mm, mass of 35 gram and natural frequency of 69 kHz) and a thick steel bar (mass of 60 grams) that was bent in a U shape. The actuator was put inside the U-shaped steel bar, and they were covered by another steel plate. To generate the vertical compression wave that transmits from the top to the bottom of the specimen, the trigger was glued inside the top cap.

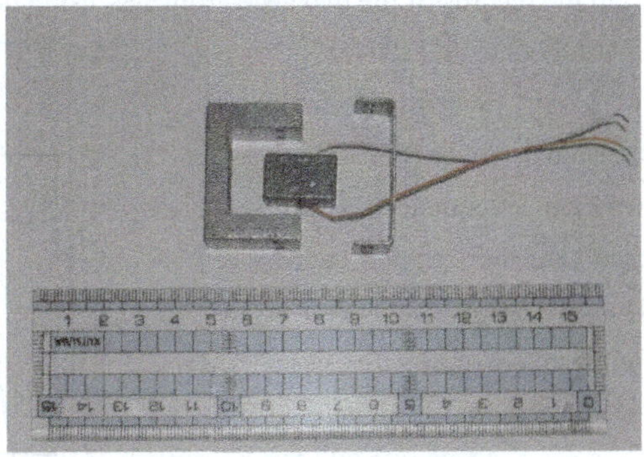

Fig.4 - Components of a trigger

2.4 Receiving dynamic waves

In order to receive dynamic waves at two locations (denoted as input and output) on the side surface of the specimen as shown in Fig.3, piezoelectric accelerometers were employed. As shown in Fig.5, these accelerometers were hexagonal in shape with a diagonal dimension of 6.85 mm, height of 4.50 mm, mass of 0.70 gram and natural frequency of 60 kHz.

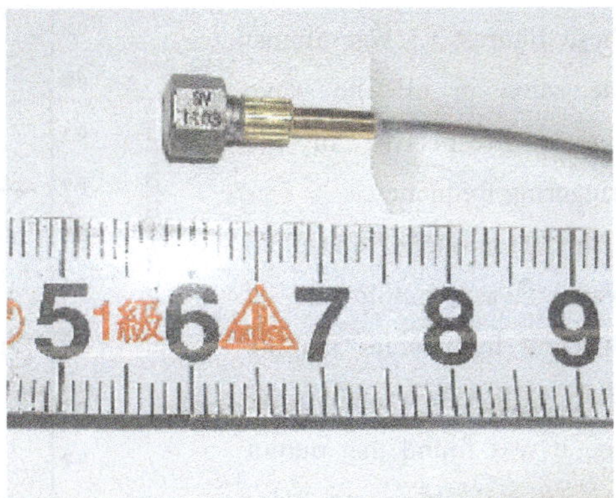

Fig.5 - Accelerometers used to receive waves

3- TEST RESULTS AND DISCUSSIONS

3.1 Wave signals triggered by different waves

The trigger was driven by inputting an electric signal of +25 volt in a form of single pulse, single sinusoidal and continuous sinusoidal waves.

First, while triggering with single pulse wave, to study the effect of triggering frequency, Fast Fourier Transform (FFT) was applied to the first half cycle of the input wave record, and its dominant frequency was computed. In the typical example as given in Fig. 6a, the predominant frequency was 3.90 kHz. The travel time of the generated wave was computed from the records of the two accelerometers using definition of first peak points (t_{pk}) as explained in Maqbool et al. (2004a).

Second, single sinusoidal waves with different frequencies were used as the trigger wave. Typical records are shown in Fig.6b, c, & d. In this case, the output wave signals at frequency of 3.90 $\pm$ 1.00 kHz were with the highest amplitude along with clear wave records. As shown in these figures, t_{pk} was almost the same in all the wave signals irrespective of the triggering frequency.

Third, continuous sinusoidal wave was employed at different frequencies ranging from 1.50 kHz to 5.00 kHz and it was found that output wave signals were very clear only at the frequency of 3.90 kHz as shown in Fig.6e.

While making comparison of wave velocities it was found that the wave velocities obtained by single pulse wave were consistent with those by single sinusoidal waves especially at a frequency that is equal to the predominant one for the first half cycle of the measured wave triggered

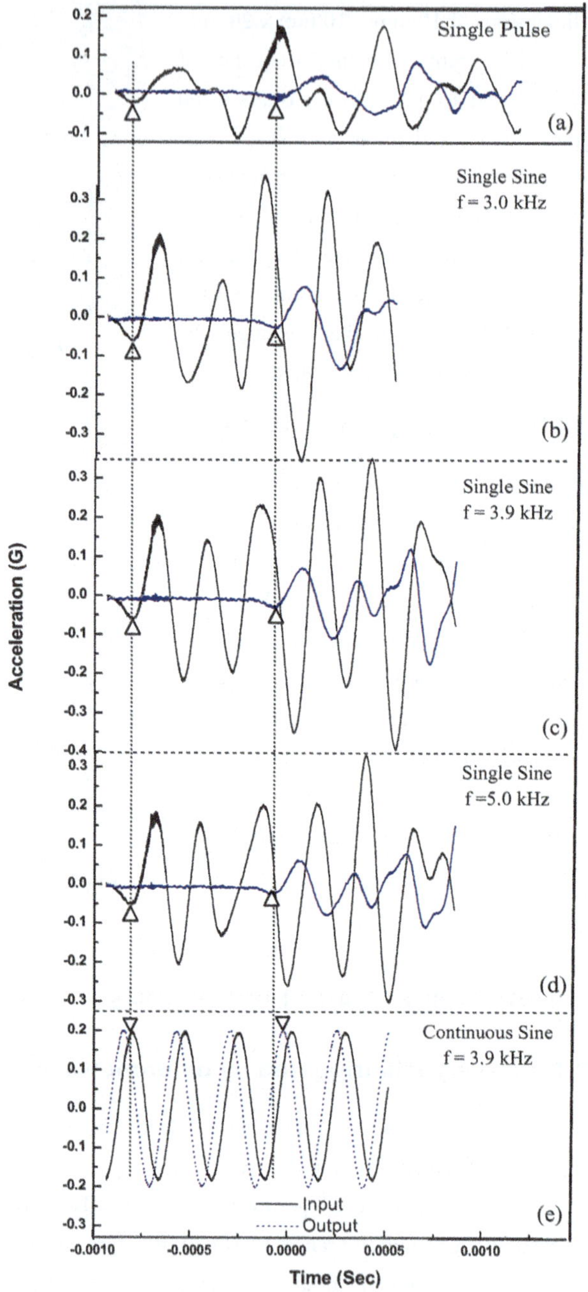

Fig.6 - Effects of triggering conditions on recorded wave signals

by the single pulse. On the other hand wave velocities obtained by continuous sinusoidal triggering were found largely inconsistent with those obtained by single pulse wave except at the predominant frequency as shown in Fig. 7. Moreover while employing continuous sinusoidal triggering, it was found difficult to determine correctly the arrival of the output wave signal that corresponds to the arrival of the respective input wave signal.

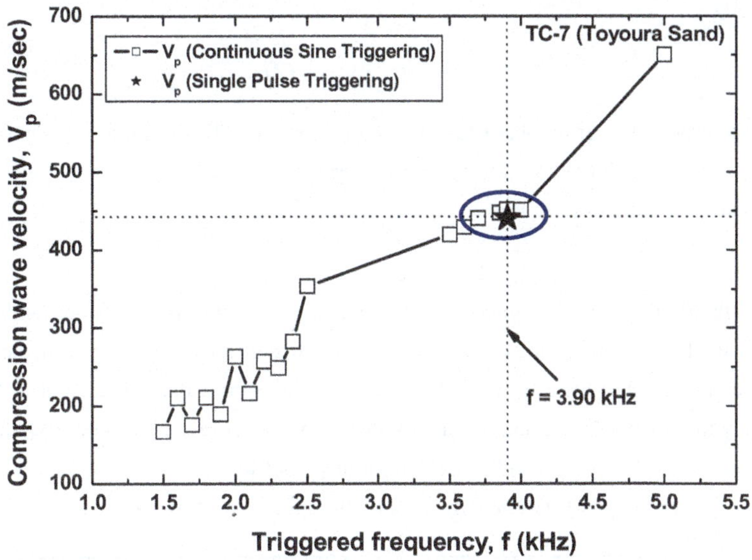

Fig.7 - Wave velocities obtained by different triggered frequencies

Based on these results, in this study, triggering by a single pulse wave with 128 times stacking was mainly employed. When recorded wave signals were not clear enough to detect t_{pk}, it was employed to trigger by a single sinusoidal wave at a frequency that is equal to the predominant one of the first half cycle of the input wave triggered by the single pulse.

3.2 Comparison between static and dynamic vertical Young's moduli

The three specimens were subjected to isotropic consolidation (and isotropic unloading).In order to evaluate dynamic vertical Young's moduli, at different stress levels, compression waves were generated while keeping the stress states constant, followed by application of small vertical unload / reload cycles in order to evaluate statically the quasi-elastic vertical Young's moduli (E_v). As the top cap touched the whole cross-section of the specimen that was with free side boundaries, we assumed that unconstrained compression waves were generated. Its wave velocity V_p is directly related to the small-strain vertical Young's modulus, E_d, of the material by the dynamic measurement as:

$$E_d = \rho V_p^2 \qquad (1)$$

where ρ is the mass density of the specimen, and the compression wave velocity V_p was calculated by

$$V_p = L/t \qquad (2)$$

Here t is the travel time of the wave to cover the distance L. The distance L was computed by using the following equations considering the trigger as a planar source to generate the compression waves as shown in Fig. 3.

$$L = L_2 - L_1 \qquad (3)$$

As shown in Fig.8, vertical Young's moduli obtained by static and dynamic methods were found repeatable in all three tests conducted at similar densities. The values of the dynamic vertical Young's moduli were by 15-25% larger than the static values. The latter difference is due possibly to the effects of inhomogenity of the specimen. As given in detail in Maqbool et al. (2004b), the larger grain size and larger uniformity coefficient of soil would lead to larger structural or microscopic heterogeneity inside the specimen. With such heterogeneity, the

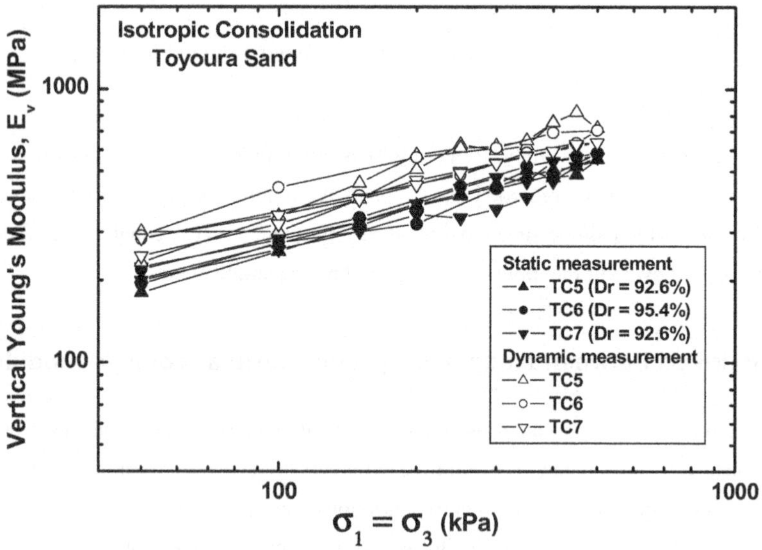

Fig.8 - Comparison of vertical Young's moduli during isotropic consolidation for Toyoura sand

Young's moduli by dynamic method would become relatively larger, since the dynamic wave tends to transmit through stiffer regions of the specimen.

It should be noted that, if the compression waves are assumed to be under constrained condition, the above difference between the dynamic and static vertical Young's moduli would decrease. It would become 10-20 % when the Poisson's ratio in the following equation is set equal to 0.15, which was measured statically by vertical and horizontal LDTs during the small unloading/reloading cycles under otherwise the isotropic stress states.

$$M = \rho Vc^2 = \frac{E(1-\upsilon)}{(1-2\upsilon)(1+\upsilon)} \qquad (4)$$

Here V_c is compression wave velocity under constrained condition, E is vertical Young's modulus and υ is Poisson's ratio. Further investigations are required to clarify this issue quantitatively.

4- CONCLUSIONS

1- Compression wave velocities obtained by single pulse wave were consistent with those by single sinusoidal waves especially at a frequency that is equal to the predominant one for the first half cycle of the measured wave triggered by the single pulse.

2- Wave velocities obtained by continuous sinusoidal wave were found largely inconsistent with those obtained by the single pulse wave except at the predominant frequency as mentioned in 1.

3- Vertical Young's moduli obtained by static and dynamic methods were repeatable in all three tests conducted at similar densities. When the assumption of unconstrained condition in computing the dynamic moduli is adopted, the values of the dynamic vertical Young's moduli were by 15-25% larger than the static values.

REFERENCES

1- AnhDan, L.Q., Koseki, J. and Sato, T. (2006), "A Large Scale True Triaxial Apparatus Developed for Gravel", submitted for possible publication in Geotechnical Testing Journal, ASTM.

2- Jiang, G.L., Tatsuoka, F., Flora, A. and Koseki, J. (1997), "Inherent and stress-state-induced anisotropy in very small strain stiffness of a sandy gravel", Geotechnique, Vol. 47, No.3, pp. 509-521.

3- Maqbool, S., Sato, T. and Koseki, J. (2004a), "Measurement of Young's moduli of Toyoura sand by static and dynamic methods using large scale prismatic specimen", Proceedings of the Sixth International Summer Symposium, JSCE, Saitama, Japan, pp. 233-236.
4- Maqbool, S., Koseki, J. and Sato, T., (2004b), "Effects of Compaction on Small Strain Young's Moduli of Gravel by Dynamic and Static Measurements", Bulletin of ERS (Earthquake Resistant Structure Research Center), IIS, The University of Tokyo, No.37, pp.41-50
5- Maqbool, S. (2005), "Effect of compaction on strength and deformation properties of gravel in triaxial and plane strain compression tests", PhD thesis, The University of Tokyo, Japan.
6- Sato, T., AnhDan, L.Q. and Koseki, J. (2001), "Development of true triaxial testing system for large scale apparatus", Proceedings of the Thirty-Sixth Japan National Conference on Geotechnical Engineering, Tokushima, Vol.1, pp.545-546. (in Japanese)

Soil Stress-Strain Behavior: Measurement, Modeling and Analysis
Geotechnical Symposium in Roma, March 16 & 17, 2006

SAMPLE DISTURBANCE IN RESONANT COLUMN TEST MEASUREMENT OF SMALL-STRAIN SHEAR-WAVE VELOCITY

Nicola Chiara
Department of Civil Engineering
Columbia University, New York, NY 10027, USA
e-mail: nc2112@columbia.edu
and
K.H. Stokoe II
Department of Civil Engineering,
The University of Texas, Austin, TX 78712, USA
e-mail: k.stokoe@mail.utexas.edu

ABSTRACT
The accurate assessment of dynamic soil properties is a crucial step in the solution process of geotechnical earthquake engineering problems. The resonant column test is one of the ordinary procedures for dynamic characterization of soil. In this paper, the impact of sample disturbance on the resonant column test measurement of small-strain S-wave velocity is examined. Sample disturbance is shown to be a function of the ratio of the laboratory to field S-wave velocities: $V_{s,lab}/V_{s,field}$. *The influence of four parameters - soil stiffness, soil plasticity index, in-situ sample depth and in-situ effective mean confining pressure - on sample disturbance is investigated both qualitatively and quantitatively. The relative importance of each parameter in predicting the small-strain field S-wave velocity from the resonant column test values is illustrated and predictive equations are presented.*
Keywords: resonant-column testing; wave velocity; sample disturbance.

1. INTRODUCTION

The measurement of dynamic soil properties is a crucial element in the solution process of geotechnical earthquake engineering problems. Evaluation of soil response to earthquake shaking requires knowledge of the shear modulus, G, and the material damping, D, of the soil and how these properties vary with shearing strain amplitude.

At small strains, i.e. shearing strains less than about 0.001%, shear modulus and shear-wave velocity are related by the following:

$$V_s = \sqrt{\frac{G_{max}}{\rho}} \qquad (1)$$

where V_s = small-strain shear-wave velocity of the soil,
ρ = unit mass density of soil, and
G_{max} = small-strain shear modulus.

Hoe I. Ling et al. (eds.), Soil Stress-Strain Behavior: Measurement, Modeling and Analysis, 605–613.

Hence, small-strain shear-wave velocity, V_s, is used as an indicator of the soil stiffness G_{max}. Small-strain-shear-wave velocities of soil can be assessed using laboratory tests and/or seismic test methods in the field. Laboratory tests have some advantages relative to field seismic tests because they permit the parameters that affect V_s and G_{max} to be studied. In particular, factors that control the soil stiffness such as the magnitude and duration of the effective confining pressure and the amplitude of the shear straining, γ, can be investigated.

The resonant column test is one of the common tests for determining laboratory small-strain S-wave velocity of a soil sample. The test is typically conducted while the specimen is confined isotropically [1]. The soil specimen is tested by exciting it in first-mode torsional resonance. Determinations of the resonant frequency, f_r, and maximum amplitude of vibration at resonance, A_r, are made from the response curve, as illustrated in Figure 1. These values are then combined with equipment characteristics and specimen size to compute the shear-wave velocity, the shear modulus and the shearing strain amplitude.

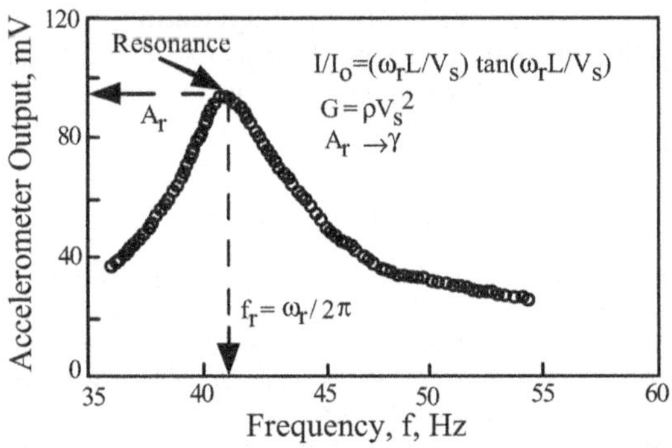

Figure 1 - Frequency response curve measured in the RC test (from Stokoe *et al.*, 1999)

The value of small-strain-shear-wave velocity determined from the resonant column test is based on the theory of elastic wave propagation in prismatic rods. The fixed-free resonant column apparatus is modeled as a column of soil with a mass attached to the free end. The frequency equation for the fundamental torsional mode of such a system has been presented by Richart, Hall, and Woods [2] as:

$$\frac{I}{I_0}=\frac{\omega_n L}{V_s}\tan\left(\frac{\omega_n L}{V_s}\right) \tag{2}$$

where V_s = shear-wave velocity of soil,

I = mass polar moment of inertia of the sample,

I_0 = mass polar moment of inertia of the attached mass,

ω_n = natural circular frequency, and

L = length of sample.

In conventional resonance tests, the damped first-mode resonant circular frequency of the sample (ω_r in Figure 1) is measured instead of ω_n. The difference between ω_n and ω_r is negligible in small-strain tests since material damping is less than 10 percent and causes ω_r to be within 1 percent of ω_n. Hence, ω_n and ω_r are assumed to be interchangeable in calculating V_s.

Field seismic tests, on the one hand, are presently used to determine only small-strain dynamic soil properties due to the limited power of the seismic source in use. On the other hand, they can be used to sample large volumes of undisturbed soil.

Field small-strain-shear-wave velocities are usually measured using the suspension logging test, a.k.a. OYO test [3], the downhole test [4], and the crosshole test [5].

2. DATABASE

Data from a total of 90 undisturbed soil specimens and from 90 field companion measurements were collected in a database. The distribution of soil samples according to soil type and soil plasticity is listed in Table 1 and 2, respectively.

Table 1 – Database: distribution of soil samples according to soil type

N. samples	Soil type
20	CH (25<PI<57)
22	CL (9<PI<30)
3	CL – ML (7<PI<22)
3	MH (17<PI<32)
4	ML (0<PI<9)
6	SC (13<PI<34)
1	SC-SM (PI = 5)
21	SM (0<PI<1)
1	SP (PI = 0)
5	SP-SM (PI = 0)
4	SW-SM (PI = 0)
Total 90	

Table 2 – Database: distribution of soil samples according to soil plasticity

N. samples	Soil Plasticity
33	Non-Plastic
3	0-5%
8	5-10%
18	10-20%
10	20-30%
12	30-40%
4	40-50%
2	>50%
Total 90	

After field S-wave velocity tests were performed, several soil specimens were recovered from different depths, and these specimens were tested at their estimated mean effective confining pressure. The field S-wave velocity data comes from two distinct sources: suspension logging tests from recent projects and downhole tests as well as crosshole tests from past projects [6]. The field S-wave velocity at a certain depth was computed by averaging the measurement recorded at that depth and the two immediately adjacent (above and below) depths.

3. SAMPLE DISTURBANCE

It is well known that the values of S-wave velocities measured in the field by seismic method differ from the values determined in the laboratory with intact soil samples recovered from the field sites. Such a difference is usually ascribed to the disturbance that a soil specimen undergoes during the sampling, handling, and testing procedures. Even high-quality "undisturbed" specimens have a certain degree of disturbance. In this study, sample disturbance is computed as the ratio between the laboratory and the field small-strain shear modulus; hence, $G_{max,lab}/G_{max,field}$. Accordingly, the sample disturbance can also be expressed as a function of the laboratory and field S-wave velocity ratio

$$\frac{V_{s,lab}}{V_{s,field}} = \sqrt{\frac{G_{max,lab}}{G_{max,field}}} \tag{3}$$

and thus, sample disturbance increases as the ratio $V_{s,lab}/V_{s,field}$ moves away from unity. The degree of sample disturbance can be represented by the coefficient of disturbance, I_d, which is function of the measured small-strain-shear moduli

$$I_d = \left| 1 - \frac{G_{max,lab}}{G_{max,field}} \right| \tag{4}$$

The degree of disturbance has been qualitatively defined as low, medium or high and then related to values of $V_{s,lab}/V_{s,field}$ ratio (Table 3).

Table 3- Ranges of Vs,lab/Vs,field and degrees of disturbance

Qualitative Degree of Disturbance	I_d	$\frac{G_{max,lab}}{G_{max,field}}$	$\frac{V_{s,lab}}{V_{s,field}}$
Low	0 to 0.2	0.8 to 1.2	0.90 to 1.10
Medium	0.2 to 0.6	(0.4 to 0.8)or (1.2 to 1.6)	(0.65 to 0.90) or (1.10 to 1.25)
High	Above 0.6	(0 to 0.4)or above 1.6	(0 to 0.65) or above 1.25

This study examines the effects of four parameters that may affect soil disturbance: in-situ soil stiffness, soil plasticity index, in-situ sample depth and in-situ effective mean confining pressure. As shown in the subsequent discussions, soil stiffness, is the most important of these parameters.

The influence of soil stiffness on sample disturbance is illustrated in Figure 2, which represents the S-wave velocity ratios from 90 undisturbed samples versus the field S-wave velocities. It is shown a clear trend that sample disturbance increases as soil stiffness increases.

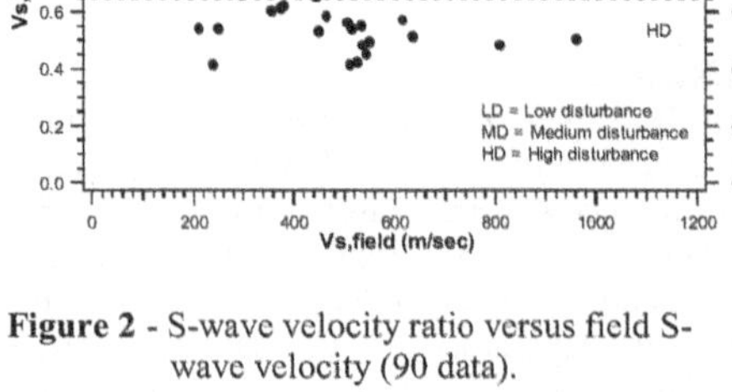

Figure 2 - S-wave velocity ratio versus field S-wave velocity (90 data).

The influence of soil plasticity index on sample disturbance has been examined by considering data with PI greater than 0 (55 data). A slight trend implying that sample disturbance is inversely proportional to the PI is shown in Figure 3. However, this trend has a few outliers.

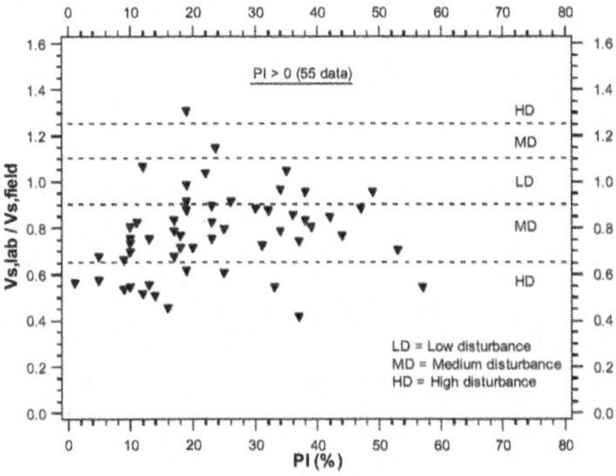

Figure 3 - S-wave velocity ratio versus soil plasticity index (55 data).

The influence of the in-situ sample depth has been investigated by considering the complete data set versus sample depth (Figure 4). As seen in the figure, there is a general trend showing that the disturbance increases as the sample depth increases. However, this trend is not as strong as the trend shown in Figure 2.

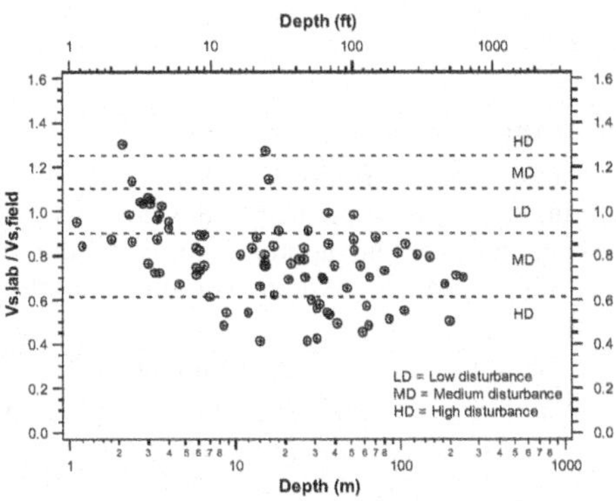

Figure 4 - S-wave velocity ratio versus in-situ sample depth (90 data).

Finally, the influence of in-situ effective mean confining pressure has been studied by taking into consideration the complete database versus effective mean confining pressure (Figure 5). Figure 5 shows a general trend of an increasing sample disturbance as the effective mean confining pressure increases. It is worth noticing that this trend is similar to the one shown in Figure 4. This relationship is expected because as depth increases so does the effective mean confining pressure.

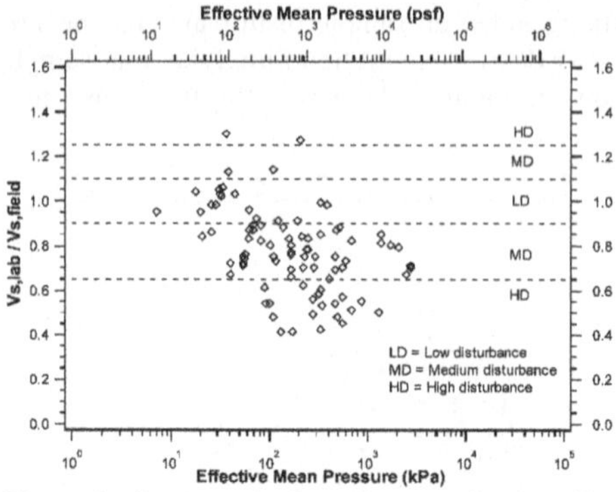

Figure 5 - S-wave velocity ratio versus in-situ effective mean confining pressure (90 data).

4. PREDICTION OF FIELD S-WAVE VELOCITY FROM LABORATORY MEASUREMENTS

The general situation in which geotechnical engineers are placed is that, once the laboratory value of S-wave velocity is measured, they are required to estimate the corresponding field value. In this section, the complete database is used to evaluate the key parameters in predicting the field S-wave velocity from the laboratory value. As before, the parameters studied are soil stiffness, PI, depth and mean confining pressure. As one would expect, soil stiffness is the most important parameter while the other three parameters play a secondary role.

4.1 Relative Importance of the Various Parameters

The weight of each variable (soil stiffness, soil plasticity index, in-situ sample depth and in-situ effective mean confining pressure) in predicting the field S-wave velocity is investigated using multiple linear regression analysis. One approach to assess the relative importance of the various predictor variables is to look at their coefficients in the multiple regression equation once the variables are standardized.

The standardized variables z_i have a mean of zero and a unitary standard deviation.

Since each standardized variable has exactly the same standard deviation and mean, the absolute value of the coefficients of the standardized multiple linear regression equation

$$z_{Vs,field} = coefficient_1 * z_1 + ... + coefficient_n * z_n \quad (5)$$

indicate the rank order of the predictor variables [7].

The regression equations and results from the analyses are presented in Table 4. From Table 4, it can be inferred that:

- the laboratory S-wave velocity is the most important parameter in predicting the field S-wave velocity because its coefficient (0.731 to 0.905) is generally one or more orders of magnitude greater than the coefficients of the other predictor variables.

- an effective prediction of $V_{s,field}$ can be carried out using $V_{s,lab}$ as the only predictor variable. In fact, after comparing the values of the squared correlation coefficient, R^2, it is apparent that adding another predictor variable to $V_{s,lab}$ can only yield a tiny increase in R^2.

Table 4 - Linear regression equations

n.	Standardized Linear Regression Equation	R^2
1	$z_{Vs,field} = 0.870 * z_{Vs,lab}$	76.4%
2	$z_{Vs,field} = 0.781 * z_{Vs,lab} + 0.116 * z_{Depth}$	76.9%
3	$z_{Vs,field} = 0.905 * z_{Vs,lab} - 0.041 * z_{Mean\ pressure}$	76.5%
4	$z_{Vs,field} = 0.860 * z_{Vs,lab} - 0.034 * z_{PI}$	76.5%
5	$z_{Vs,field} = 0.731 * z_{Vs,lab} + 0.142 * z_{Depth} - 0.060 * z_{PI}$	77.1%
6	$z_{Vs,field} = 0.881 * z_{Vs,lab} - 0.030 * z_{Mean\ pressure} - 0.028 * z_{PI}$	76.5%

4.2 Quantitative Relationship

Three different equation forms were utilized to analyze the predictive relationship between $V_{s,field}$ and $V_{s,lab}$: linear, power and exponential. The results of this analysis are summarized in Table 5. These results show that the best fit $\left(R^2 = 80.12\%\right)$ can be obtained using a power equation form

$$V_{s,field} = 0.56 * V_{s,lab}^{1.16} \tag{6}$$

The mean and $\pm 1\sigma$ and $\pm 2\sigma$ lines are shown in Figure 6.

Table 5 - Least squares regression analysis

Type	Regression Equation	R^2
Linear	$V_{s,field} = 1.44 * V_{s,lab}$	76.06%
Power	$V_{s,field} = 0.56 * V_{s,lab}^{1.16}$	80.12%
Exponential	$V_{s,field} = 92.3e^{0.0048 V_{s,lab}}$	75.62%

4.3 Quantitative Relationship: "Non-cohesive" and "Cohesive" Database

Two least-squares regression analyses were performed to find the best fit of "non-cohesive" data of the database (35 data with PI = 0 to 5%) and "cohesive" data of the database (55 data with PI > 5%). Eventually, a power regression equation was the best fit for the "non-cohesive" data

$$V_{s,field}^{non-cohesive} = 0.42 * V_{s,lab}^{1.21} \tag{7}$$

and the "cohesive" data

$$V_{s,field}^{cohesive} = 0.64 * V_{s,lab}^{1.13} \quad (8)$$

The three predictive equations (6,7 and 8) indicate that the laboratory small-strain-shear modulus, $G_{max,lab}$, may underestimate the actual field small-strain-shear modulus, $G_{max,field}$, by a "correction" factor that is proportional to the soil stiffness (Table 6). As shown in Table 6, this "correction" factor may range from 1.2 to 2.4.

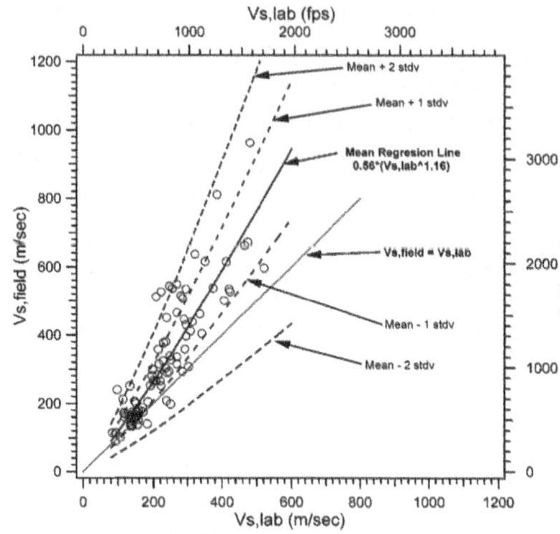

Figure 6 - Regression analysis plot: database (90 data)

Table 6 – Soil Disturbance: predicted correction factors

$V_{s,lab}$ (m/sec)	$\frac{G_{max,field}}{G_{max,lab}}$ Complete database	$\frac{G_{max,field}}{G_{max,lab}}$ "Non-cohesive"	$\frac{G_{max,field}}{G_{max,lab}}$ "Cohesive"
100	1.4	1.2	1.4
150	1.6	1.4	1.5
200	1.7	1.6	1.6
250	1.8	1.8	1.7
300	1.9	1.9	1.8
350	2.0	2.1	1.9
400	2.1	2.2	1.9
450	2.2	2.3	2.0
500	2.3	2.4	2.1

5. CONCLUSIONS

A database with small-strain-shear wave velocity data from a total of 90 undisturbed soil specimens and from the 90 field companion measurements was build. This database was then used to investigate the impact of sample disturbance on the resonant column test measurement of small-strain-S-wave velocity. The influence of four parameters (soil stiffness, soil plasticity, in-situ depth and in-situ effective mean confining pressure) on sample disturbance was analyzed. The importance of various parameters in predicting the field S-wave velocity was also illustrated. The key findings of this study are summarized as follows:

- Sample disturbance increases as soil stiffness increases.
- Sample disturbance seems to be inversely proportional to the soil plastic index; though this trend is not well defined.
- Sample disturbance increases as the sample depth or the effective mean confining pressure increases. However, this trend is not as strong as the trend with respect to the soil stiffness.
- The laboratory S-wave velocity is the most important parameter in predicting the field S-wave velocity.
- The resonant column test may underestimate the actual field small-strain shear modulus due to sample disturbance. The predicted correction factor, increasing with soil stiffness, ranges from 1.2 to 2.4.

REFERENCES

[1] Stokoe, K.H., II and Darandeli, M.B., Andrus, R. D. and Brown, L.T. (1999). Dynamic soil properties: laboratory, field and correlation studies, *Proceeding of the Second Int. Conf. on Earthquake Geotechnical Engineering*, Lisbon. 21-25 June 1999, (3): pp. 811-845, Rotterdam, Balkema.

[2] Richart, J.E., Jr., Hall, J.R., Jr., and Woods, R.O. (1970). *Vibrations of Soils and Foundations*, Prentice-Hall Inc., Englewood Cliffs, New Jersey.

[3] Nigbor, R.L. and Imai, Y. (1994). *The Suspension P-S Velocity Logging Method: Geophysical Characterization of Sites*, by TC10 for XIII ICSMFE, New Delhi, pp. 57-61.

[4] Stokoe, K.H., II and Hoar, R.J. (1978). Variables Affecting In Situ Seismic Measurements, *Proceeding of the Conference on Earthquake Engineering and Soil Dynamics*, ASCE, Vol. II, June, pp. 919-939.

[5] Stokoe, K.H., II and Woods, R.D. (1972). In Situ Shear Wave Velocity by Crosshole Method, *Journal of Soil Mechanics and Foundation Engineering Div.*, ASCE, Vol. 98 No. SM5, May, pp 443-460.

[6] Chiara, N. (2001). Investigation of Small-Strain Shear Stiffness Measured in Field and Laboratory Geotechnical Studies , *M.S. Thesis*, University of Texas at Austin.

[7] Kachigan, S.K. (1986). *Statistical Analysis*, Radius Press, New York, pp.260-263.

Soil Stress-Strain Behavior: Measurement, Modeling and Analysis
Geotechnical Symposium in Roma, March 16 & 17, 2006

CYCLIC BEHAVIOR OF NONPLASTIC SILTY SAND UNDER DIRECT SIMPLE SHEAR LOADING

El-Mamlouk, H.H[1], Hussein, A.K. [2], and Hassan, A.M.[3]

Abstract: The cyclic behavior of clean sands pertaining to earthquake loading have been studied extensively. However, natural sands may contain significant amounts of silt or clay, which have long been thought to affect the cyclic resistance of such soils. In fact, the literature contains what appears to be conflicting results. There was a need to clarify the effect of fines content on the cyclic resistance of sand which was addressed in this paper. A series of constant volume stress-controlled cyclic tests using the DSS apparatus was performed on reconstituted specimens of sand with fines content of 0%, 20%, and 40%. The samples were prepared at relative densities of 40% and 80%; consolidated at vertical stresses of 50, 100, 200, and 400kPa; and sheared at cyclic stress ratios of 0.05, 0.10, and 0.15. Test results indicate that failure criterion should be identified based on whether the sand behavior is contractive or dilative. The cyclic resistance is found to be controlled by the relative density of the soil.

Introduction

It has been understood since the 1960's that the presence of nonplastic fine particles may affect the resistance of sand to liquefaction (Seed and Lee, 1966). However, a review of studies published in the literature shows that no clear conclusions can be drawn as to in what manner altering the fines content affects the liquefaction resistance of sand under cyclic loading. Numerous laboratory studies have been performed, and have produced what appear to be conflicting results. Some studies have reported that increasing the silt content in sand increases its liquefaction resistance (e.g., Seed and Idriss, 1971), while others reported that increasing silt decreases the liquefaction resistance of the sand (e.g., Shen et al., 1977; Troncoso and Verdugo, 1985; Finn et al., 1994; Vaid, 1994; Lade and Yamamuro, 1997; Yamamuro and Lade, 1997; Zlatovic and Ishihara, 1997; Carraro et al., 2003), or decreases the liquefaction resistance until some minimum cyclic resistance is reached, and then increases its resistance (Law and Ling, 1992; Koester, 1994; Polito, 1999). Additionally, several studies have shown that the liquefaction resistance of silty sand is more closely related to its sand skeleton void ratio than to its silt content (e.g. Thevanayagam et al. 2002).

Following a brief description of the experimental program, analysis of results, conclusions and, study implications will be presented.

[1]Prof., Soil Mechanics and Foundations Div., Faculty of Engineering, Cairo University.
[2]Assoc. Prof., Soil Mechanics and Foundations Div., Faculty of Engineering, Cairo University.
[3]M.Sc. Candidate, Soil Mechanics and Foundations Div., Faculty of Engineering, Cairo University.

Hoe I. Ling et al. (eds.), Soil Stress-Strain Behavior: Measurement, Modeling and Analysis, 615–624.

Experimental Program

The main material chosen for this research is poorly-graded silica sand. The tested sand has a coefficient of uniformity, $C_u = 2.13$, coefficient of curvature, $C_c = 0.74$, and specific gravity, $G_s = 2.65$. The fines were prepared by milling a part of the tested clean sand, sieving the milled powder on sieve #200. This was done to be sure that the used fines are nonplastic. The grain size distributions for both clean sand and the fines are shown in Figure 1. Three combinations of sand and silt were created using fines contents of 0%, 20%, and 40%. Additional tests were performed for several sand-silt mixtures to determine the maximum and minimum void ratios. A plot of the maximum and minimum void ratios versus silt content is presented in Figure 2.

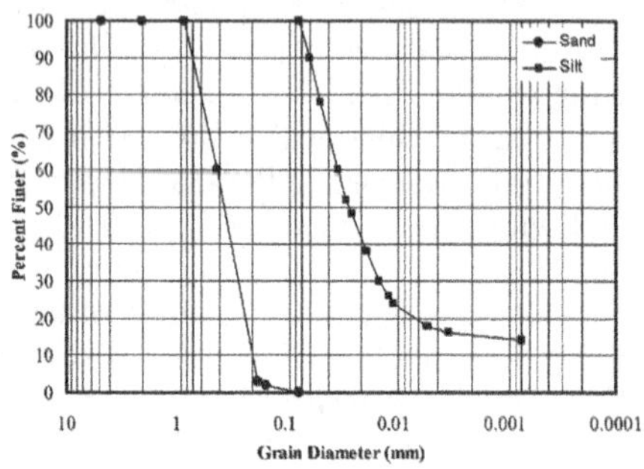

Figure 1. Grain Size Distributions for Sand and Silt

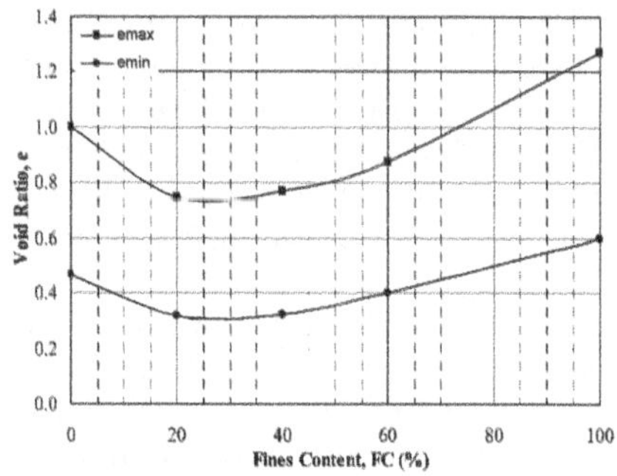

Figure 2. Variation in Maximum and Minimum Void Ratios with Fines Content

The cyclic resistance of the soils tested was determined using cyclic direct simple shear apparatus manufactured by Geonor. The reconstituted specimens tested were 66.8 mm in diameter and 16 mm in height. Each specimen was prepared at a specified initial relative density in order to obtain the target postconsolidation relative density of 40 or 80%. Although constant volume tests often are performed using dry specimens, water was circulated through the specimen to minimize concerns about potential water lubrication effects. Then, the specimens were consolidated at a vertical stress of 50, 100, 200, or 400 kPa. After consolidation, a constant volume stress-controlled cyclic test was conducted at cyclic shear stress amplitude corresponding to a cyclic stress ratio of 0.05, 0.10, or 0.15. The frequency of cyclic stress was 0.4 Hz. The shape of the loading function was sinusoidal.

Analysis of Results

The aim of this research is to study the cyclic resistance of silty sand. The main parameters investigated were cyclic shear stress amplitude, vertical consolidation stress, cyclic stress ratio, and fines content. When analyzing test results, the significance of the failure criterion is delineated based on whether contractive or dilative behavior of sand. The effect of fines content on cyclic resistance of sand is examined via various density measures such as relative density (D_r), gross void ratio (e), and sand skeleton void ratio (e_s). An explanation of what appears to be conflicting results reported in the literature as previous laboratory tests and cyclic resistance charts based on field measurements is provided in the light of the findings of this study.

Failure Criteria

Typical cyclic direct simple shear test results are presented in Figures 3. and 4. In these figures the pore pressure and shear strain amplitude are each plotted against number of loading cycles. Figure 3. displays the response of a specimen of postconsolidation relative density of 40%. The specimen behavior is characterized by the very small strains throughout the loading until just before the onset of initial liquefaction (Figure 3.a.) at which a large sudden shear strain occurs (Figure 3.b.). It should be pointed out that the magnitude of the strain is restricted by the limits of the apparatus, not by the behavior of the soil specimen. This flow liquefaction failure is typical for contractive behavior. Therefore, the failure criterion is defined as the number of cycles required to cause initial liquefaction condition (N_L). (Peacock and Seed, 1968)

On the other hand, Figure 4. displays the response of a specimen of postconsoliadtion relative density of 80%. The specimen behavior is characterized by the nearly uniform development of the shear strain amplitude throughout the cyclic loading, and the pore pressure never reaches the initial liquefaction condition, Figure 4.a. The specimen continues to deform gradually, Figure 4.b. This cyclic mobility failure is typical for dilative behavior. Therefore, the failure criterion is defined as the number of cycles (N_f) required to cause a specified shear strain amplitude (Lee and Seed, 1967; Vaid and Sivathayalan, 1996) of 3.75% (corresponding to axial strain amplitude of 2.5%).

Finally, the term *"cyclic resistance"* is defined as the cyclic stress ratio required to cause initial liquefaction in 15 cycles, for loose specimens (D_r = 40%). Whereas, for dense specimens (D_r = 80%), the cyclic resistance is defined as the cyclic stress ratio required to cause 3.75% single shear strain amplitude in 15 cycles.

Effect of Cyclic Shear Stress Amplitude

The effect of altering the shear stress amplitude (τ_{cy}) while holding fines content (FC) and vertical consolidation stress (σ'_{vc}) constant was first examined. For both loose and dense specimens, number of cycles required to cause failure (N_L or N_f) decreases as cyclic shear stress amplitude increases. It was found also that the rate of decrease in N_L or N_f becomes lower as τ_{cy} increases. Figure 5. gives an example of this behavior which is a typical trend (Seed and Lee; 1966; Lee and Seed, 1967; Peacock and Seed, 1968; Finn et al., 1971).

Effect of Vertical Consolidation Stress

Next, the effect of holding fines content (FC) and cyclic shear stress amplitude (τ_{cy}) constant while altering the applied vertical consolidation stress (σ'_{vc}) was examined. For both loose and dense specimens, number of cycles required to cause failure (N_L or N_f) increases as applied vertical consolidation stress increases. It was found also that the rate of increase in N_L or N_f becomes lower as τ_{cy} increases. Figure 6. illustrates this typical effect of vertical consolidation stress (σ'_{vc}) on N_L or N_f of sands.

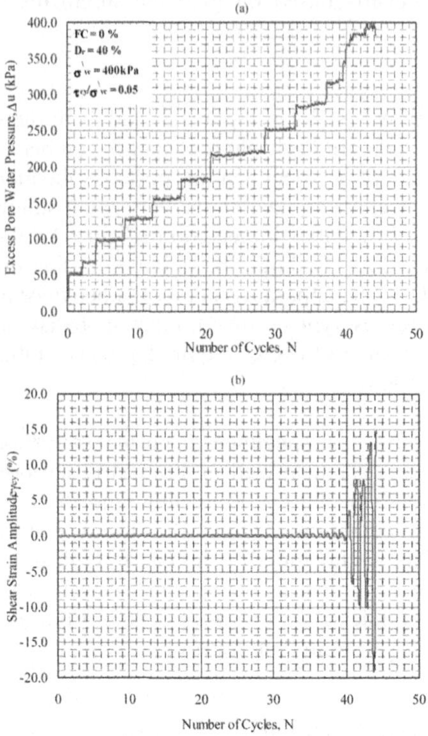

Figure 3. Typical Behavior of a Loose Sand Specimen Susceptible to Flow Liquefaction

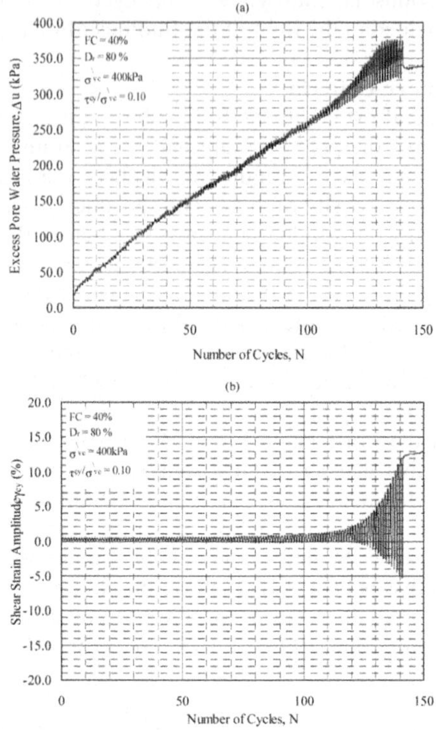

Figure 4. Typical Behavior of a Dense Sand Specimen Susceptible to Cyclic Mobility

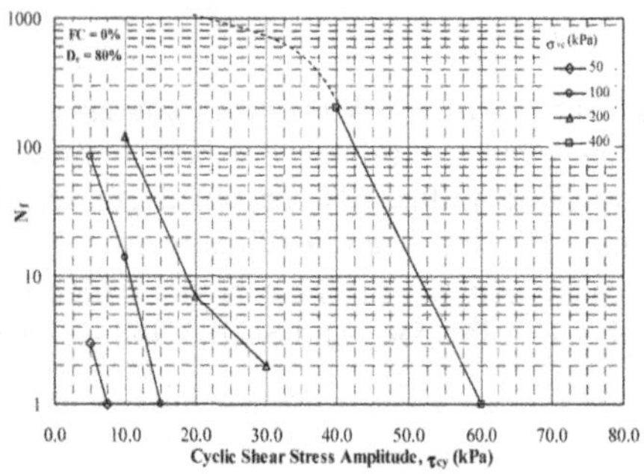

Figure 5.Cycles to Failure versus Cyclic Shear Stress Amplitude at Various Vertical Consolidation Stresses, FC = 0%, Dr = 80%

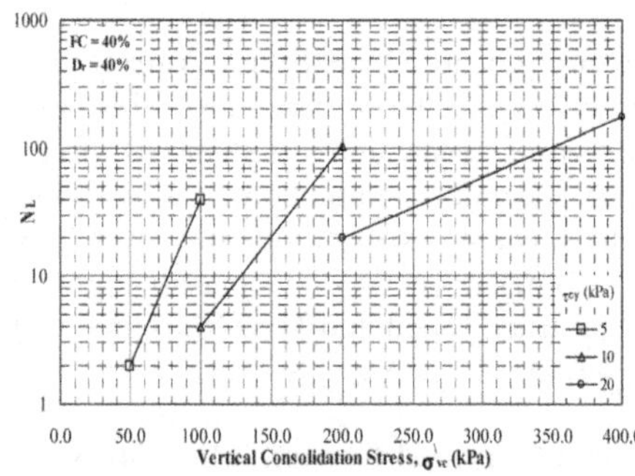

Figure 6. Cycles to Failure versus Vertical Consolidation Stress at Various Cyclic Shear Stress Amplitudes, FC = 40%, Dr = 40%

Effect of Cyclic Stress Ratio

The effect of holding fines content (FC) and cyclic shear stress amplitude (τ_{cy}) constant while altering the cyclic shear stress ratio (τ_{cy}/σ'_{vc}) was examined. For both loose and dense specimens, number of cycles required to cause failure (N_L or N_f) decreases as cyclic shear stress ratio increases, Figure 6. It should be pointed out that when the data are plotted versus cyclic stress ratio instead of cyclic shear stress amplitude, the cyclic resistance is no longer dependent on vertical consolidation stress either for dense or loose specimens. This observation has been acknowledged by some researchers (e.g., Lee and Seed, 1967; Finn et al., 1971).

Vaid and Sivathayalan (1996) based on simple shear tests reported the same observation for saturated sand but only at the loosest state, where vertical consolidation stress does not have much influence on cyclic resistance. At denser states, however, they showed that the cyclic stress ratio required to cause liquefaction decreases with increasing vertical consolidation stress. However, as the testing results of Vaid and Sivathayalan (1996) were scrutinized, it was found that for the densest tested sand (D_r = 72%), this decrease in cyclic resistance is less than 15%, when vertical consolidation stress increases from 50 kPa to 400 kPa, which practically has little significance.

Effect of Fines Content

The effect of fines content (FC) was examined through the cyclic resistance concept. For example, Figure 8. shows cyclic resistance plotted versus vertical consolidation stress (σ'_{vc}) at fines contents (FC) of 0, 20, and 40% and relative density of 40%. It is noted that all points plotted fall in a narrow band. This figure indicates that no clear trend is evident in the testing results with the increase of fines content, which implies that for a given relative density, the number of loading cycles required to cause failure of sand does not depend on the fines content.

This conclusion was approved by Polito and Martin (2001). They stated that if the silt content of the soil is below the limiting silt content, the soil could be described as having silt contained in a sand matrix. The cyclic resistance of the soil is then controlled by the relative density of the specimen and is independent of vertical consolidation stress or silt content. The limiting silt content (LSC) depends on maximum void ratio and specific gravity of host sand ($e_{max, HS}$, G_{ss}) and of pure silt ($e_{max, HF}$, G_{sf}).

$$\mathbf{LSC} = \frac{\mathbf{G_{sf} e_{max,HS}}}{\mathbf{G_{ss}(1 + e_{max,HF})}} \mathbf{(\%)} \tag{1}$$

By applying the above equation in this study, the limiting silt content is about 44%. As the maximum fines content used was 40% (less than LSC), the results conform to the behavior of silty sand below the limiting silt content ,however, the behavior of above this zone cannot be detected.

Gross Void Ratio Concept

In this study, the concept of constant gross void ratio was not applied. However, by studying the available testing results, a conclusion regarding the effect of silt content on cyclic behavior of sand while gross void ratio is constant can be deduced. Figure 9. shows that the cyclic resistance value determined at a relative density of 40% for each of the three sand-silt mixtures is constant since relative density is constant. For the same relative density, the gross void ratio decreases from 0.786 for clean sand to a value of 0.572 at 20% fines, then gross void ratio slightly increases to 0.591 again at 40% fines. For the same sand-silt mixture, one would expect that if the gross void ratio increases the cyclic resistance decreases, presumably due to the decrease in relative density. On the contrary, if gross void ratio decreases cyclic resistance increases. If this presumption is applied for each sand-silt mixture, a contour line could be drawn, as in Figure 10., to represent the cyclic resistance as a function of fines content for a constant gross void ratio. The deduced contour lines in Figure 9. indicates that at a constant gross void ratio, as fines content increases, cyclic resistance decreases then increases again which is consistent with the reported trends in the literature (Law and Ling, 1992; Koester, 1994; Polito, 1999). This decrease and then increase of cyclic resistance with increasing silt content can be explained in terms of the relative density of the sand-silt mixture. At the same gross void ratio, as the silt content of the soil increases, its relative density first decreases and then increases as the maximum and minimum void ratios vary in Figure 2.

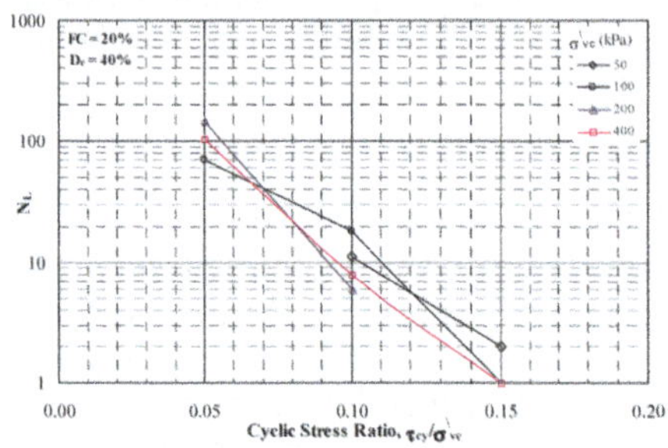

Figure 7. Cycles to Failure versus Cyclic Stress Ratio at Various Vertical Consolidation Stresses, FC = 20%, Dr = 40%

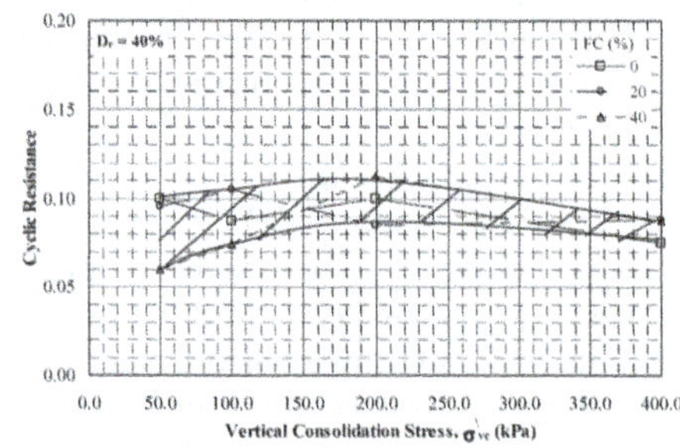

Figure 8. Cyclic Resistance versus Vertical Consolidation Stress at Various Fines Contents (N = 15), Dr = 40%

Other investigators (e.g., Carraro et al. ,2003) have reported a general decrease in cyclic resistance with increasing fines content for specimens prepared at a constant gross void ratio. The decrease of cyclic resistance with increasing silt content can be attributed to the change in the relative density. The specimens tested were all prepared at constant gross void ratios, creating soil relative densities that varied with varying silt content. As the silt content of the specimens increased, their relative density decreased. In their testing program, since the maximum silt content used was 15%, the limiting silt content was not reached and the maximum and minimum void ratios, and as a result the relative density, did not increase.

Sand Skeleton Void Ratio Concept

The concept of constant sand skeleton void ratio was not used in this study; however the reported trends may still be explained via testing results obtained herein. If specimens are prepared at a constant sand skeleton void ratio, their gross void ratio decreases as silt content increases. The maximum and minimum void ratios may decrease less rapidly than the decrease in gross void ratio. As a result, although sand skeleton void ratio is constant, soil relative density increases with increasing silt content. So the tested soil is expected to exhibit some increase in cyclic resistance as reported in several studies (e.g., Shen et al., 1977; Kubris et al., 1988; Vaid, 1994; Polito and Martin, 2001; Thevayangayam et al., 2002).

Such trend can still be deduced from the testing results of this study. Figure 10. illustrates that at the same relative density of 40%, sand skeleton void ratio increases from 0.786 for clean sand to 0.952 at 20% fines, and to 1.652 at 40% fines. For the same sand-silt mixture, one would argue that if sand skeleton void ratio increases, cyclic resistance decreases as discussed above. On the contrary, if sand skeleton void ratio decreases, cyclic resistance increases. As this presumption is applied for each sand-silt mixture, a contour line is plotted to represent the cyclic resistance as a function of fines content for a constant gross void ratio. As can be seen in Figure 10. the deduced contour lines indicate that at a constant sand skeleton void ratio, cyclic resistance increases with increasing fines content.

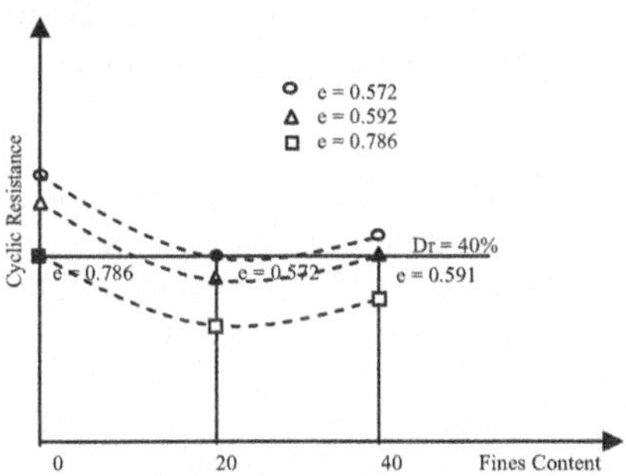

Figure 9. Cyclic Resistance Contour Lines Deduced at Various Gross Void Ratios

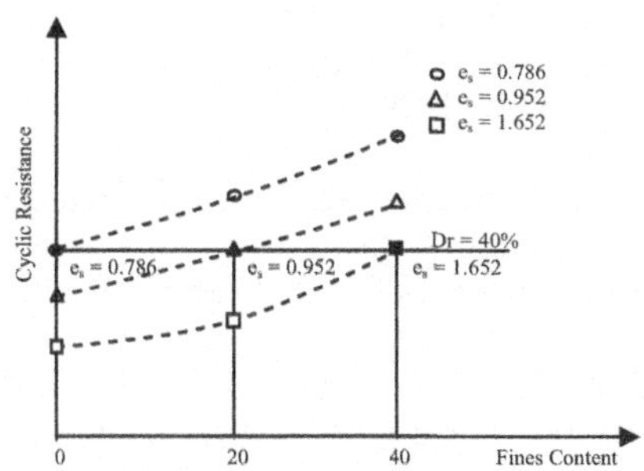

Figure 10. Cyclic Resistance Contour Lines Deduced at Various Sand Skeleton Void Ratios

Cyclic Resistance Based on SPT Results

According to Seed et al. (1983) the presence of silt lowers the penetration resistance (i.e. $(N_1)_{60}$-value) of sand as it prevents full drainage of the soil. According to this study, the cyclic resistance does not depend on the silt content but only does depend on the relative density, so the correction factor for a specific silt content should presumably cause the cyclic resistance curve of silty sand to collapse on top of the clean sand curve. This, of course, could only be presumed for the case when the soil under investigation has a limiting silt content greater than the silt content for which the cyclic resistance curve is intended. Some researchers have proposed various correction factors to take into account the effect of silt content on penetration resistance (i.e., Seed et al., 1983; NCEER, 1997; Polito, 1999). It was noted that these corrections tends towards approving the above inference by getting the cyclic resistance curves of silty sand closer to the clean sand until the curves become coincidental with the clean sand curve.

It is of great interest to compare cyclic resistance values obtained in this experimental study with those obtained based on SPT results. Cyclic resistance values were obtained in this study as cyclic stress ratios required to cause failure (either liquefaction or γ_{cy} =

3.75% for loose or dense specimens, respectively) in 15 cycles in order to be compared to data which are based on earthquakes with a magnitude of 7.5 and an equivalent number of cycles of 15. For each sand-silt mixture, the value of cyclic resistance determined only at a vertical consolidation stress of 100 kPa. Moreover, the corresponding $(N_1)_{60}$-value for each relative density was determined using the following relation (Skempton, 1986):

$$(N_1)_{60}/D_r^2 = 60 \qquad (2)$$

Values of cyclic resistance via this study are plotted in Figure 11. as a function of $(N_1)_{60}$ for clean sand. It is clear in the figure that, for D_r = 40%, and FC = 0 and 20% ($(N_1)_{60}$ = 9.6), the laboratory data falls in the vicinity of the clean sand curve. This confirms that the cyclic resistance of silty sand is independent of the fines content if the fines content is below the limiting silt content (in this study is around 44%). On the other hand, at D_r = 40% and FC = 40% the cyclic resistance obtained from the clean sand curve is overestimate. This could be explained as followed; the fines content of 40% is in the vicinity of the limiting silt content where the cyclic resistance is expected to be much lower than that below the limiting silt content. This reveals the dangerous of using the available methods of analysis if the fines content exceeds the limiting silt content. On the contrary, at D_r = 80% ($(N_1)_{60}$ = 38.4), the laboratory data falls below the recommended curve for all sand-silt mixtures.

According to Seed et al. (1985) the evidence of liquefaction takes very different forms, so, a condition of "liquefaction" for a sand with $(N_1)_{60}$-value of larger than 25 (dense soil) involves a different form of behavior than that for a sand with an $(N_1)_{60}$-value of less than 15 (loose soil). That what was differentiated in laboratory testing by referring to first type of behavior as "cyclic mobility" with limiting strain potential and the second type as "liquefaction" involving very large strains. Since the laboratory criterion used for cyclic resistance of dense sand was the development of γ_{cy} = 3.75%, the cyclic resistance recorded would clearly be underestimate if the field data represents an occurrence of much larger strain levels. These results are in agreement with the conclusions found by Vaid and Sivathayalan (1996).

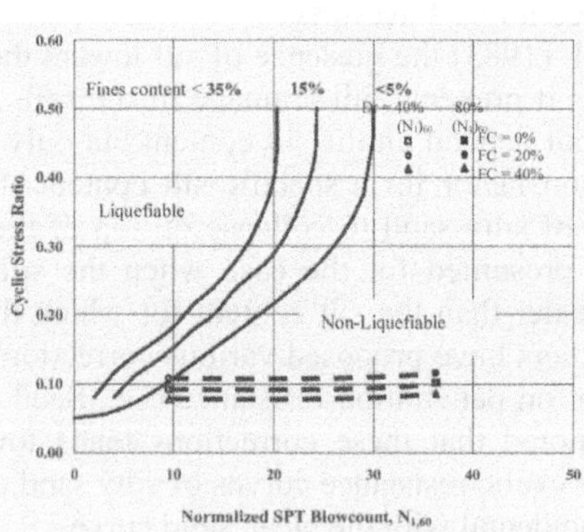

Figure 11. Liquefaction Chart based on SPT Results (after Seed and Idriss, 1971)

Conclusions and Study Implications

An experimental study has been conducted to clarify the effect of nonplastic fines on the cyclic resistance of sand. The results were evaluated based on the relative density concept. The chosen criterion for failure used in the analysis was found to be dependent on the soil behavior whether is contractive or dilative. A series of constant volume stress-controlled tests performed using the DSS apparatus. The tests were consolidated at four different vertical stresses and sheared at three various cyclic stress ratios. The effects of cyclic shear stress amplitude, vertical consolidation stress, and cyclic stress ratio on cyclic behavior of silty sand are similar to those of clean sand. The cyclic resistance is uniquely related to cyclic stress ratio regardless of vertical consolidation stress value.

Contradictory trends reported in the literature related to effect of fines content on cyclic behavior of sand are attributed to the various density measures adopted when analyzing test results. However, at a specified void ratio, as fines content increases, relative density decreases then increases as a result of analogous changes in maximum and minimum void ratios. So, cyclic behavior of silty sand is better characterized in light of relative density and limiting silt content concepts. When proper fines correction is applied to penetration resistance ($(N_1)_{60}$-value), constant cyclic resistance should be predicted for soils with a constant density regardless of fines content.

A study of the effect of silt content on steady state line via monotonic and cyclic testing should be carried out. In the same time, the cyclic behavior of entire spectrum: from clean sand to pure silt should be studied. The research may be extended to include the effect of plastic fines on cyclic behavior of sand. Finally, a better understanding of the exact relationship between penetration resistance and fines content is needed.

List of Symbols and Abbreviations

Symbol	Description	Symbol	Description
σ'_{vc}	Vertical consolidation stress	FC	Fines content (%)
γ_{cy}	Cyclic shear strain amplitude	e	Gross void ratio
τ_{cy}	Cyclic shear stress amplitude	e_s	Sand skeleton void ratio
Δu	Pore pressure	LSC	Limiting silt content
D_r	Relative density	G_{sf}	Specific gravity of pure silt
G_{ss}	Specific gravity of host sand	$e_{max,\,HS}$	Maximum void ratio of host sand
$e_{max,\,HF}$	Maximum void ratio of pure silt		
N_L	Number of cycles required to cause initial liquefaction		
N_f	Number of cycles required to cause single shear strain amplitude of 3.75%		
$e_{max,\,HF}$	Maximum void ratio of pure silt		
$(N_1)_{60}$-value	Measured number of blowcounts based on SPT test		

References

Carraro, J.A.H., Bandini, P., and Salgado, R. (2003). "Liquefaction resistance of clean sand and nonplastic silty sands based on cone penetration resistance." *J. Geotech. Engrg.*, ASCE, 129 (11), 965-976.

Finn, W.D.L., Ledbetter, R.H., and Wu, G. (1994). " Liquefaction in silty soils: Design and analysis." *Ground failures under seismic conditions, Geotech. Engrg.*, ASCE, 110 (8), 1091-1105.

Finn, W.D.L., Pickering,D.J., and Bransby, P.L. (1971). "Sand liquefaction in triaxial and simple shear tests." *J. Geotech. Engrg.*, ASCE, 97 (4), 639-659.

Hassan, A.M. (2005). "Cyclic behavior of nonplastic silty sand under direct simple shear loading." M. Sc. thesis. Public Works Department, Faculty of Engineering, Cairo University.

Lade and Yamamuro. (1997). "Effects of nonplastic fines on static liquefaction sands." *Can. Geotech. J.*, Ottawa. 34, 918-928.

Lee, K.L., and Seed, H.B. (1967). "Cyclic Stress Conditions Causing Liquefaction of Sand." *J. Geotech. Engrg.*, ASCE, Vol. 93 (1), 47-70.

Peacock, W.H. and Seed, H.B. (1968). "Sand liquefaction under cyclic loading simple shear conditions." *J. Geotech. Engrg.*, ASCE, 94 (3), 689-708.

Polito, C.P. (1999). "The effects of non-plastic and plastic fines on the liquefaction of sandy sands." Ph.D. thesis. Blacksburg, *Virginia Polytechnic Institute and State University.*

Polito, C.P., and Martin, J.R. (2001). "The effects of non-plastic and plastic fines on the liquefaction of sandy sands." *J. Geotech. Engrg.*, ASCE, 127 (5), 408-415.

Seed, H.B. (1968). "Landslides during earthquakes due to soil liquefaction." *J. Geotech. Engrg.*, ASCE, 94 (5), 1055-1122.

Seed, H.B., and Idriss, I.M. (1971). "Simplified procedure for evaluation soil liquefaction potential." *J. Geotech. Engrg.*, ASCE, 97 (9), 1249-1273.

Seed, H.B., and Lee, K.L. (1966). "Liquefaction of saturated sands during cyclic loading." *J. Geotech. Engrg.*, ASCE, 92 (6), 105-134.

Seed, H.B., Idriss, I.M., and Arango, I. (1983). "Evaluation of liquefaction potential using field performance data." *J. Geotech. Engrg.*, ASCE, 109 (3), 458-482.

Seed, H.B., Tokimatsu, K., Harder, L., and Chung, R. (1985). "Influence of SPT procedures in soil liquefaction resistance evaluations." *J. Geotech. Engrg.*, ASCE, 111 (12), 1425-1445.

Shen, C.K., Vrymoed, J.L., and Uyeno, C.K. (1977). "The effects of fines on liquefaction of sands." *Proc., 9th Int. Conf. on Soil Mech. and Found. Engrg. Div.*, ASCE, 102 (5), 511-523.

Skempton, A.W. (1986). "Standard penetration test procedures and the effects in sands of overburden pressure, relative density, particle size, ageing, and overconsolidation." *Geotechnique*, 36 (3), 425 – 447.

Thevanayagam, S., Shenthan, T., Mohan, S., and Liang, J. (2002). "Undrained fragility of clean sands, silty sands, and sandy silts." *J. Geotech. Engrg.*, ASCE, 128 (10), 849-859.

Troncoso, J.H., and Verdugo, R. (1985). "Silt content and dynamic behavior of tailing sands." *Proc., 12th Int. Conf. on Soil Mech. and Found. Engrg.*, 1311-1314.

Vaid, V.P. (1994). "Liquefaction of silty soils." Ground failures under seismic conditions, *Geotech. Spec. Publ. No. 44*, ASCE, New York, 1-16.

Vaid, Y. P., and Sivathayalan, S. (1996). "Static and cyclic liquefaction potential of Fraser Delta sand in simple shear and triaxial tests." *Can. Geotech. J.*, 33 (4), 281-289.

Yamamuro, J.A., and Lade, P.V. (1997). "Static liquefaction of very loose sands." *Can. Geotech. J.*, Ottawa, 34, 905-917.

Youd, L., and Idriss, I.M. (1997). "Summary Report" *Proceedings of the NCEER Workshop on Evaluation of Liquefaction Resistance of Soils*, Technical Report NCEER-97-0022, 1-40.

Zlatovic, S., and Ishihara, K. (1997). "Normalized behavior of very loose nonplastic soil: Effects of fabric." *Soils and Found.*, Tokyo, 37 (4), 47-56.

Soil Stress-Strain Behavior: Measurement, Modeling and Analysis
Geotechnical Symposium in Roma, March 16 & 17, 2006

MODELLING OF STRESS-STRAIN RELATIONSHIP OF TOYOURA SAND IN LARGE CYCLIC TORSIONAL LOADING

Nguyen Hong Nam
Lecturer, Division of Geotechnical Engineering, Hanoi Water Resources University
175 Tay Son street, Dong Da district, Hanoi, Vietnam. E-mail: hongnam@wru.edu.vn

Junichi Koseki
Professor, Institute of Industrial Science, The University of Tokyo
Komaba 4-6-1, Meguro-ku, Tokyo 153-8505, Japan. E-mail: koseki@iis.u-tokyo.ac.jp

ABSTRACT
The relationships between normalized shear stress and plastic shear strain of air-dried, dense Toyoura sand measured during large amplitude cyclic torsional loading with using local strain measurement could be well simulated numerically by the proportional rule combined with the drag rule. The proportional rule is an extended version of the Masing's second rule and can account for unsymmetrical stress strain behavior about neutral axis. The drag rule can account for strain hardening in cyclic loadings. Use of the newly proposed hypoelastic model for the quasi-elastic properties, the backbone curve using general hyperbolic equation or newly proposed lognormal equation for monotonic loading behavior, and the combination of the proportional rule and the drag rule for cyclic loading behavior would enable more precise simulation of deformation properties than before.

1. INTRODUCTION
In certain conditions, soils in the field are frequently subjected to cyclic loads that are caused by, for example, earthquakes, sea waves, traffic loads, and the subsequent loads during the construction of civil structures. Although present studies on the monotonic stress-strain relationship of soils have approached an advanced level, those on cyclic one are still limited. This is, possibly, because the elastic deformation properties of soils are usually assumed to be constant in several elasto-plastic soil models like Cam clay family (Schofield and Wroth, 1968). Nowadays, it has been well known that inside the conventional yield region, the behaviour of soil is not purely elastic (Jardine, 1992; Tatsuoka et al., 1997). Given any stable stress state before reaching peak, on the other hand, strain caused by performing several unload/reload cycles at very small amplitude, say, 0.001 % could be reversed, exhibiting quasi-elastic behavior. The development of hypoelastic model for sand (Tatsuoka et al., 1999) has enabled to deduce accurately such elastic strains. Therefore, if creep effect is neglected, the plastic modelling of stress-strain relationship of sand during cyclic loading with using modified Masing rule (Masing, 1926) can be a possible task. Following this approach, Masuda (1998) and Tatsuoka et al.

Hoe I. Ling et al. (eds.), Soil Stress-Strain Behavior: Measurement, Modeling and Analysis, 625–636.

(2003) have simulated stress-strain relationship of Toyoura sand in large cyclic plane strain tests; Balakrishnaiyer (2000) simulated deformation characteristics of Chiba gravel subjected to large cyclic loading in triaxial tests. However, modelling of stress-strain relationship of sand in torsional shear tests which allow the rotation of principal stress axes, following the above-mentioned approach, has never been studied. Therefore, this study also aims at plastic modelling of the stress-strain behavior of sand in torsional cyclic loading employing a newly proposed hypoelastic model (HongNam and Koseki, 2005) for evaluating elastic components.

2. TEST APPARATUS, MATERIAL AND PROCEDURES

Test apparatus

The hollow cylinder torsional shear apparatus developed at Institute of Industrial Science, The University of Tokyo, is employed. Refer to HongNam and Koseki (2005) for the description of this apparatus. The capacity of vertical loading is 8 kN. The capacity of torque is 0.15 kN.m. These loading systems can be controlled independently and automatically by a personal computer. Throughout the tests, inner and outer cell pressures are kept equal to each other, and a high capacity E/P transducer having a maximum capacity of 980 kPa is used to control them through the D-A converters. Thus, the effective radial stress σ'_r was kept equal to the effective circumferential stress σ'_θ throughout the tests. The vertical load F_z and the torque T were measured by a two-component loadcell placed at the top cap of the specimen, inside the pressure cell.

Test material and procedures

Test material was air-dried, dense Toyoura sand, a uniform sand with sub-angular particles (D_{50} = 0.18 mm, U_c = 1.6, G_s = 2.635, e_{max} = 0.966 and e_{min} = 0.600). The hollow cylindrical specimens (outer diameter D_o = 20, inner diameter D_i = 12 and height H = 30 cm) were prepared by air-pluviation method. The initial void ratio e_{ini}, at confining pressure σ'_c = 30 kPa, was in the range between 0.697 and 0.740 (D_{rini} = 73.5 to 61.7%).

Among the different tests employed, the results of tests TOYOG19 (e_{ini} = 0.708) and TOYOG20 (e_{ini} = 0.705) measured during the following stress paths will be analyzed herein.

- IC (isotropic consolidation). This path was conducted firstly, in which σ'_c were increased from 30 kPa up to 400 kPa, then down to 100 kPa.
- Cyclic TSI (torsional shear from isotropic stress state) with $-0.8 < \tau_{z\theta}/\sigma'_\theta < 0.8$ at constant and the same vertical and horizontal stresses $\sigma'_z = \sigma'_r = \sigma'_\theta$ =100 kPa.

Note that in test TOYOG19, the TSI path was conducted after IC and triaxial loading/unloading, while in test TOYOG20, it was conducted right after IC.

During the test, at several stable stress states, after a creep period of about 10 to 20 minutes, three small unload/reload cycles were applied in the vertical and torsional directions with single strain amplitude of 0.001 and 0.0015%, respectively, to evaluate the quasi-elastic properties such as vertical Young's modulus E_z, shear modulus $G_{z\theta}$, Poisson's ratios $\nu_{z\theta}$ and ν_{zr}.

Strain measurement

Figure 1 shows the layout of the local strain transducers used. Two sets of pin-typed local deformation transducers (P-LDTs) were arranged symmetrically, over the central regions of the specimen to avoid bedding error and end-restraint effect. Each set consisted of outer right-triangular P-LDTs (the free length of the horizontal, vertical, and diagonal P-LDT is 69.75, 69.50, and 98.30 mm, respectively) and an inner horizontal P-LDT (the free length is 56.80 mm).

Working principle of P-LDT (HongNam et al., 2001) is similar to that of conventional LDT (Goto et al. 1991). Strain of soil can be determined via the bending strain of the heat-treated phosphor bronze (HTPB) strip on which the electric resistant strain gages are cemented to form a Wheatstone bridge. The distinct features of P-LDT in comparison to conventional LDT are structures of the two pin-shaped ends of HTPB strip and the hinge which bears the conical hole(s), in order to maintain a free rotation of the HTPB strip at two contact points (ends of conical holes) during the test. The hinges, which were made of HTPB, were glued to the surfaces of the inner and outer rubber membranes of the specimen.

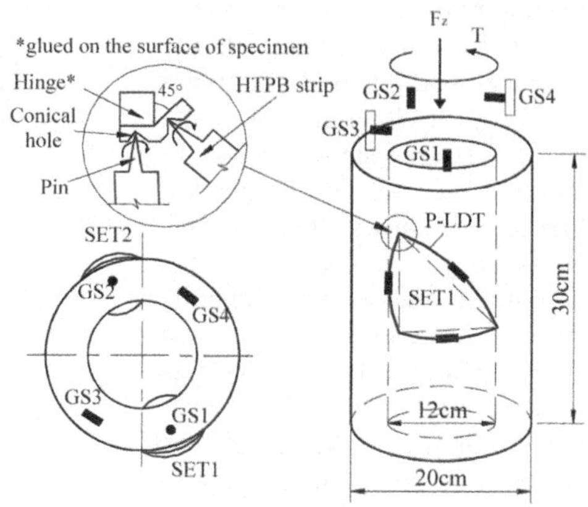

Fig. 1. Layout of transducers

Four average strain components of the specimen (ε_z, ε_r, ε_θ and $\gamma_{z\theta}$) were calculated using two assumptions; i) the central angle made of two ends of the horizontal P-LDTs and the intersection of the horizontal plane (containing them) and the symmetrical vertical axis of the specimen is constant, and ii) the specimen remains right hollow cylinder in shape. Refer to HongNam et al. (2005) for the details of strain calculation.

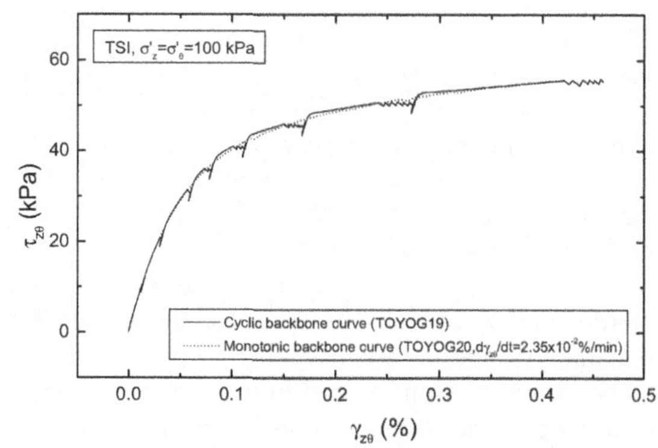

Fig. 2. Comparison between the cyclic and the monotonic backbone curves

In addition, four gap sensors (GS1 through GS4) with capacity of 4 mm were employed to measure ε_z and $\gamma_{z\theta}$ via the vertical and rotational displacements, respectively, of the top cap.

3. MODELLING OF BACKBONE CURVE

Modelling of the monotonic stress-strain relationship of geomaterials during original loading, which is known as the backbone curve, is necessary to simulate large amplitude cyclic loading. The backbone curve, considered in the present study, is composed of two kinds. The first one, as above-mentioned, is *monotonic backbone curve* of the purely monotonic loading at constant loading rate, and the other one is *cyclic backbone curve* of the primary loading with applying small cyclic loadings at some certain stress state levels to measure quasi-elastic properties. Figure 2 shows the overall consistency between the monotonic and cyclic backbone curves measured in tests TOYOG20 and TOYOG19, respectively, with the similar values of e_{ini}. Therefore, both curves can be simulated by the same equation as shown later.

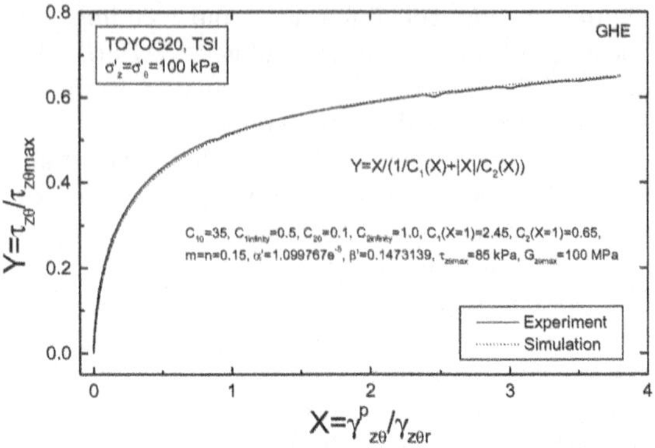

Fig. 3. Modelling of backbone curve by a general hyperbolic equation

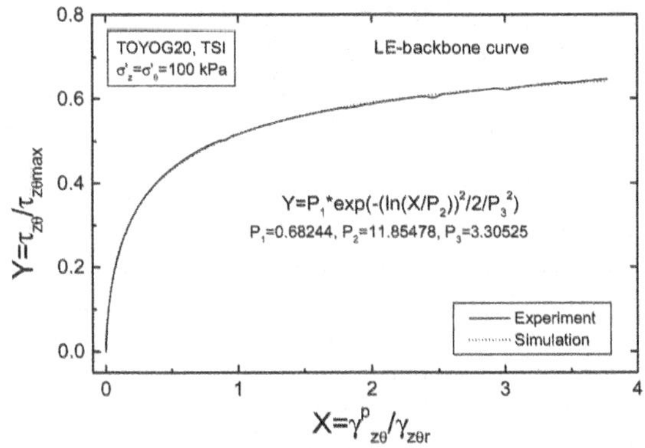

Fig. 4. Modelling of backbone curve by a lognormal equation

Modelling of backbone curve with the hyperbolic equation

For a wide range of strain, simulating the backbone curve by using the classical hyperbolic equation (Kondner, 1963) is difficult as stated by Tatsuoka and Shibuya (1991). Among many hyperbolic equations (e.g. Duncan and Chang, 1970; Hardin and Drnevich, 1972; Tatsuoka and Shibuya, 1991; Hayashi et al., 1994), the general hyperbolic equation (GHE) proposed by Tatsuoka and Shibuya (1991) for the stress-strain relation of sand in triaxial testing was employed in this study to simulate the backbone curve during torsional shear test. Note that the modifications of the equation by Tatsuoka and Shibuya (1991) are implemented in terms of the definition of parameters and power values as shown below.

$$Y = X/\left(1/C_1(X)+|X|/C_2(X)\right) \quad (1)$$

in which, $X = \gamma^p_{z\theta}/\gamma_{z\theta r}$ and $Y = \tau_{z\theta}/\tau_{z\theta\max}$ (2)

$$\gamma^p_{z\theta} = \gamma_{z\theta} - \gamma^e_{z\theta} \quad (3) \text{ (neglecting the time effects)}$$

where $\gamma^e_{z\theta}$ and $\gamma^p_{z\theta}$ denote the elastic and plastic shear strains, respectively.

The elastic shear strain $\gamma^e_{z\theta}$ can be evaluated by a newly proposed hypo-elastic model (HongNam and Koseki, 2005; as briefly described in Appendix).

$\gamma_{z\theta r}$ denotes the reference shear strain that is defined by $\gamma_{z\theta r} = \tau_{z\theta\max}/G_{z\theta\max}$ (4)

where $\tau_{z\theta\max}$ denotes the maximum shear stress (shear strength) and $G_{z\theta\max}$ denotes the initial shear modulus (at infinitesimally small strain).
Two coefficients $C_1(X)$ and $C_2(X)$ are defined as below.

$$C_1(X) = (C_1(0) + C_1(\infty))/2 + (C_1(0) - C_1(\infty))/2\cos\left(\pi/\left(|\alpha'/X|^{m_t} + 1\right)\right) \quad (5)$$

$$C_2(X) = (C_2(0) + C_2(\infty))/2 + (C_2(0) - C_2(\infty))/2\cos\left(\pi/\left(|\beta'/X|^{n_t} + 1\right)\right) \quad (6)$$

Refer to Tatsuoka and Shibuya (1991) and Tatsuoka et al. (2003) for the determination of $C_1(0)$, $C_1(\infty)$, $C_2(0)$, $C_2(\infty)$, α', β', m_t and n_t for the plane strain compression (PSC) tests.

Modelling of backbone curve with the newly proposed lognormal equation

A new lognormal equation (LE) is proposed as shown below for simulating the backbone curve of sand during torsional shear test.

$$Y = (X/|X|)P_1\, e^{-[\ln(|X|/P_2)]^2/2P_3^2} \quad (7)$$

in which $\boldsymbol{e}$ is the Euler number $\boldsymbol{e} = 2.718281828\ldots$, P_1, P_2 and P_3 are the three parameters. They can be determined based on the experimental data by using the non-linear least squares fitting technique in which the non-linear regression method based on the Levenberg-Marquardt algorithm is employed (Press et al., 1992). P_2 is the scale parameter and P_3 is the shape parameter.

Simulation results and discussion

Both GHE and LE were employed to simulate the backbone curves of Toyoura sand in torsional shear path TSI of tests TOYOG19 and 20 at $\sigma'_z = \sigma'_\theta = 100$ kPa. The values of parameters of the hypo-elastic model (HongNam and Koseki, 2005) for calculating the elastic shear strain are follows:

For test TOYOG19: $E_o = 212.5$ MPa, $\nu_o = 0.174$, $a = 1.067$, $m = n = 0.5$, $k = 0.3$, $\sigma'_o = 100$ kPa.

For test TOYOG20: $E_o = 212.5$ MPa, $\nu_o = 0.113$, $a = 1.141$, $m = n = 0.5$, $k = 0.3$, $\sigma'_o = 100$ kPa.

<u>Simulation by a general hyperbolic equation</u>

The values of parameters employed for the simulation of backbone curve using general hyperbolic equation are: $\tau_{z\theta max} = 85$ kPa, $G_{z\theta max} = 100$ MPa, $C_1(0) = 35$, $C_1(\infty) = 0.5$, $C_2(0) = 0.1$, $C_2(\infty) = 1.0$, $\alpha' = 1.099767\text{e}^{-5}$, and $\beta' = 0.1473139$ (C_1 ($X = 1$) = 2.45, C_2 ($X = 1$) = 0.65).

Figure 3 shows that the simulation result and experimental data of test TOYOG20 match very well.

Note that Tatsuoka et al. (2003) and Balakrishnaiyer (2000) used $m_t = n_t = 1$ for the backbone curves of Toyoura sand and Chiba gravel, respectively, in triaxial tests. However, in the present study modified values of $m_t = n_t = 0.15$ had to be employed for both backbone curves in tests TOYOG19 and TOYOG20.

<u>Simulation by a lognormal equation</u>

The values of parameters employed for the simulation of backbone curve using the lognormal equation are: $\tau_{z\theta max} = 85$ kPa, $G_{z\theta max} = 100$ MPa, $P_1 = 0.68244$, $P_2 = 11.85478$

and $P_3 = 3.30525$. Figure 4 shows a very good consistency between simulation result and experimental data of test TOYOG20.

Equation (7) shows that when X, P_1, P_2, $P_3 > 0$, $Y = Y_{max} = P_1$ at $X = P_2$ and $Y < P_1$ when $X > P_2$. This suggests that $(X, Y) = (P_2, P_1)$ corresponds to the peak state, and Eq. (7) should be applied in the range of $X \leq P_2$. Therefore, when modelling the (normalized) stress-strain relationship of sand, the value of P_2 should be carefully compared with the value of the (normalized) strain parameter at the peak stress state. Fortunately, the data of test TOYOG20 on Toyoura sand during TSI at $\sigma'_z = \sigma'_\theta = 100$ kPa shows that the calculated value of $P_2 = 11.85478$ was much greater than the largest measured value of $X = \gamma^{p}_{z\theta max} / \gamma_{z\theta r}$ (within the range of the stress path employed) that is approaching the peak stress state. Kawakami (1999) obtained $\tau_{z\theta max}$ (at peak stress state) = 70.5 kPa at $\gamma_{z\theta max}$ = 8.8 % during TSI at $\sigma'_z = \sigma'_\theta = 98$ kPa for dense Toyoura sand (D_{rini} = 64.7 %) using the small hollow cylindrical specimen ($D_o = 10$, $D_i = 6$, $H = 20$ cm). Using his data, with the same assumptions of $\tau_{z\theta max}$ = 85 kPa and $G_{z\theta max}$ = 100 kPa as employed with test TOYOG20, we get $X = \gamma^{p}_{z\theta max} / \gamma_{z\theta r} < \gamma_{z\theta max} / \gamma_{z\theta r} = 8.8/(85/100) = 10.4 < P_2$. Therefore, the newly proposed equation can simulate the stress-strain relationship of dense sand during torsional loading up to peak stress state.

It is reasonable to set the scale parameter $P_2 = \gamma_{z\theta max} / \gamma_{z\theta r}$ (the normalized total shear strain at peak stress state). Since the data obtained from torsional loading tests on Toyoura sand in this study is limited, more parametric and experimental studies are required on this issue.

4. MODELLING OF LARGE AMPLITUDE TORSIONAL CYCLIC LOADING

It has been well known that the stress-strain relationship of sand depends on various factors such as initial void ratio, stress history, loading rate, loading type, drainage condition etc. In order to make the problem simpler, in this study, we concentrated on the modelling of the stress-strain relationship of sand in purely large amplitude torsional cyclic loading before reaching peak stress state without small cyclic loadings applied.

In order to model cyclic loading behavior of soils, the Masing rule (Masing, 1926; Ohsaki, 1980) has been used widely. However, since the observed behaviours of soil subjected to cyclic loading do not always follow original Masing's rule (Pyke, 1979; Tatsuoka et al., 1997), several modified versions of Masing's rule have been proposed. Among them, we employed the model proposed by Tatsuoka et al. (2003) for sand in cyclic plane strain tests in which the backbone curve is simulated by a GHE and the hysteretic curve is simulated by the proportional rule with drag. Following are brief descriptions of the proportional and drag rules.

Proportional rule

The proportional rule (Tatsuoka et al., 2003), which is an extended version of the Masing's second rule (Masing, 1926; Ohsaki, 1980), can account for unsymmetrical stress strain behavior about neutral axis (Fig. 5). It consists of external and internal rules, while the same principle is adopted to evaluate the hysteretic curve in both rules.

Suppose that the backbone curve in the compression side is represented by the equation $y = f(X)$ and that in the extension side by the equation $y = g(X)$, the hysteretic curve, which consists of unloading and reloading curves, passing through the initial point (X_o, Y_o) can be simulated by the following equation.

$$(Y - Y_o)/n_p = h((X - X_o)/n_p) \quad (8)$$

where $h(X) = g(X)$ for unloading and $h(X) = f(X)$ for reloading; n_p denotes the scaling factor, which is calculated by

$n_p = -(Y_A - Y_C)/Y_C$

or $n_p = -(X_A - X_C)/X_C$ (9)

The locations of stress and strain states A and C are shown in Fig. 5. In general, $n_p > 2$ and changes during cyclic loading. Note that when $f(X) = -g(X)$ as is the case with the Masing's rule, the n_p value to be used in the external rule is always equal to 2.

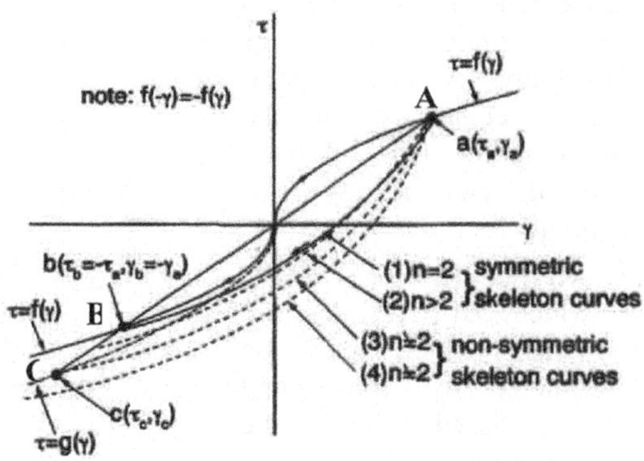

Fig. 5. Proportional rule
(After Tatsuoka et al., 2003)

Drag rule

The drag rule considers the rearrangement of sand particles; therefore, it could reflect the effect of strain hardening in cyclic loadings. The crucial points of the drag rule is that due to the rearrangement of soil particles, two branches of the original backbone curve in the compression and extension sides should be dragged to newly different positions in opposite directions of the X-axis, i.e., the two subsequently dragged backbone curves in either compression or extension sides during loading history cannot be coincided. Suppose that the stress-strain curve starts from the origin (Fig. 6), during loading the backbone curve in the opposite loading direction (i.e. unloading) is dragged (translated) along the X-axis in the positive side and vice versa. The currently dragged backbone curve can be expressed by

$$Y = h(X - \beta) \quad (10)$$

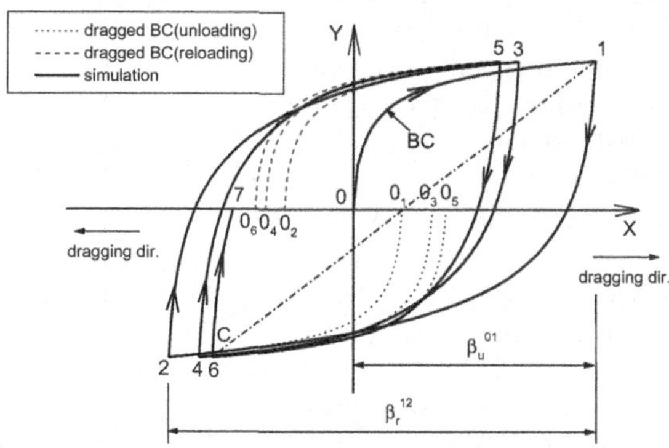

Fig. 6. Drag rule

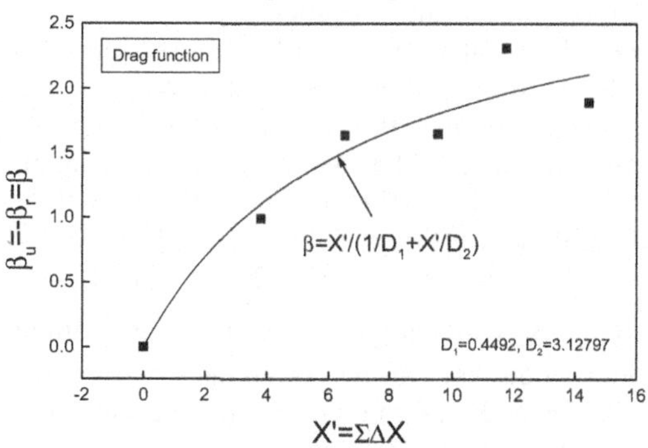

Fig. 7. A drag function

In which, β denotes the amount of drag that can be evaluated by a drag function of the accumulation of the increment of normalized plastic strain in one direction (loading or unloading). Among several possible equations for the drag function, the following hyperbola can be used for simplicity.

$$\beta = X'/(1/D_1 + X'/D_2) \quad (11)$$

in which D_1 and D_2 are constants, which can be determined by trials and errors based on the experimental data, and

$$X' = \sum \Delta X \quad (12)$$

where ΔX denotes the increment of normalized plastic strain in one direction (loading or unloading).

The dragged hysteretic curve (Fig. 6) passing the initial point (X_o, Y_o) is calculated by

$$(Y - Y_o)/n_p = h((X - X_o)/n_p) \quad (13)$$

where $n_p = -(Y_1 - Y_C)/Y_C$ or $n_p = -(X_1 - X_C)/(X_C - X_{o1})$ (14)

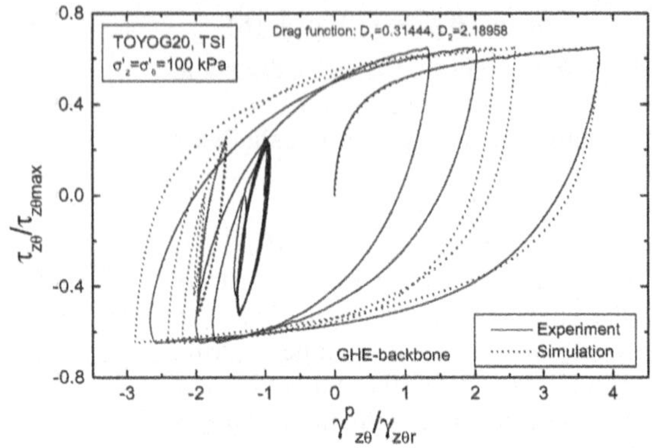

Fig. 8. Modelling of large cyclic loading with 70% of initial drag values (GHE backbone; Test TOYOG20)

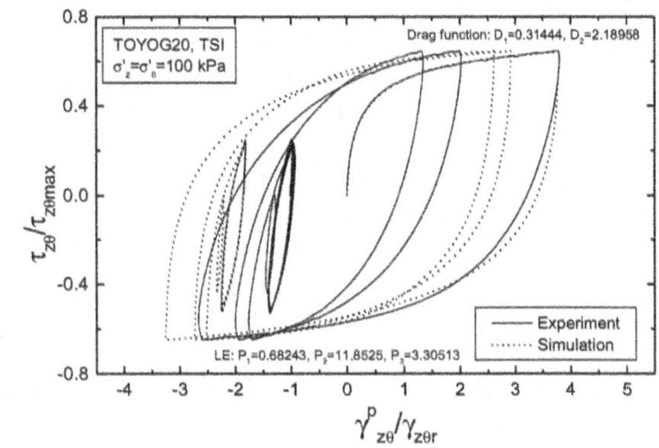

Fig. 9. Modelling of large cyclic loading with 70% of initial drag values (LE backbone; Test TOYOG20)

Newly proposed assumption relating to drag rule

With regard to the proportional rule combined with drag, Tatsuoka et al. (2003) proposed subrules with two special cases for the internal rule. However, they were not enough to treat the problem of tests TOYOG19 and TOYOG20 since the location of the current stress-strain curve was so far from the currently dragged backbone curve that the application of external rule to find the outmost curve was very difficult. Thus, one additional assumption related to these subrules was simplified as follows. Refer to HongNam (2004) for more detailed explanations.

$$(Y - Y_1)/n_p = f((X - X_1 - \Delta\beta)/n_p) \quad (15)$$

where $\Delta\beta$ denotes the increment of drag amount: $\Delta\beta = \beta - \beta_1$ in which β is the current drag value that can be determined by the drag function; β_1 is the drag value, which is calculated using the previously normalized plastic strain in unloading or reloading accumulated up to the turning point 1 (X_1, Y_1) from unloading to reloading or vice versa.

Based on the proportional rule combined with drag rule (including internal and external rules; see Tatsuoka et al., 2003), in general, the large amplitude cyclic loading could be simulated effectively.

Simulation results and discussion

In the present study, the proportional rule with drag (Tatsuoka et al., 2003) was extended to simulate the torsional cyclic loading (TSI) in which the backbone curve was simulated by a GHE and a LE. The simulation was implemented with both stress control and strain control.

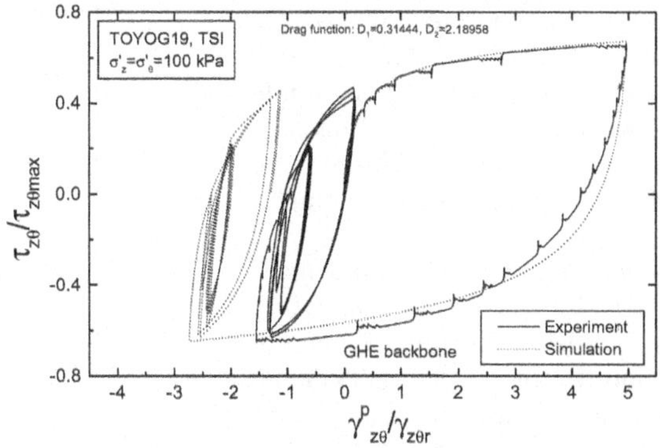

Fig. 10. Modelling of large cyclic loading with 70% of initial drag values (GHE backbone; Test TOYOG19)

Simulation by the proportional rule with drag (stress control)

Figure 7 shows the drag function $\beta = X'/(1/D_1+X'/D_2)$ ($D_1 = 0.4492$, $D_2 = 3.12797$) for unloading and reloading that was obtained from experimental data of test TOYOG20 by trials and errors. However, as described later, it was found that some reduction of the drag amount would result in better simulations. Simulations of large amplitude cyclic loadings were implemented with using two backbone curves simulated by a GHE and a LE. Note that the backbone curves in the two torsional directions were simulated by the same equation as mentioned above.

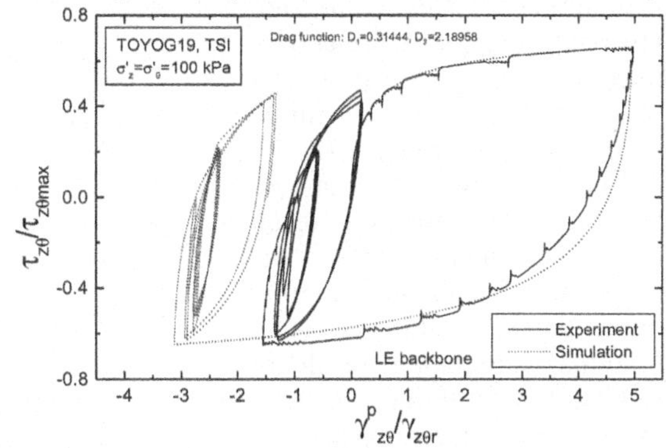

Fig. 11. Modelling of large cyclic loading with 70% of initial drag values (LE backbone; Test TOYOG19)

Comparisons between the simulation results using the amount of 70% of the initial drag ($D_1 = 0.31444$, $D_2 = 2.18958$) and experimental data of test TOYOG20, in which the backbone curves were simulated by a GHE and LE, are shown in Figs. 8 and 9, respectively. It can be seen from these figures that the simulation results and the experimental data of test TOYOG20 were consistent to each other.

Similar comparisons for test TOYOG19 are shown in Figs. 10 and 11, respectively. It can be seen from these figures that the simulation results and the experimental data of test TOYOG19 were generally consistent in shape to each other. However, the measured normalized strain values were remarkably smaller than the simulated ones at the end of the first large amplitude unloading. This could be due to effect of creep strains, which were generated by repeated processes to control stable stress states to measure quasi-elastic properties during the first large amplitude cyclic loading, resulting in changing the soil structure. Nevertheless, in test TOYOG20, this phenomenon could be negligible since the small cyclic loadings were not applied during the large amplitude cyclic loading.

Simulation by the proportional rule with drag (strain control)

Simulation result by strain control of test TOYOG20 during TSI with the same stress path as mentioned above in which the backbone curve was simulated by a GHE was plotted in Fig. 12. The simulation procedure employed the above-mentioned drag function that was deduced from the initial drag function (Fig. 7) by reducing the initial drag amount to 70%. The simulation result was consistent with the experimental data, in particular during the third unloading cycle.

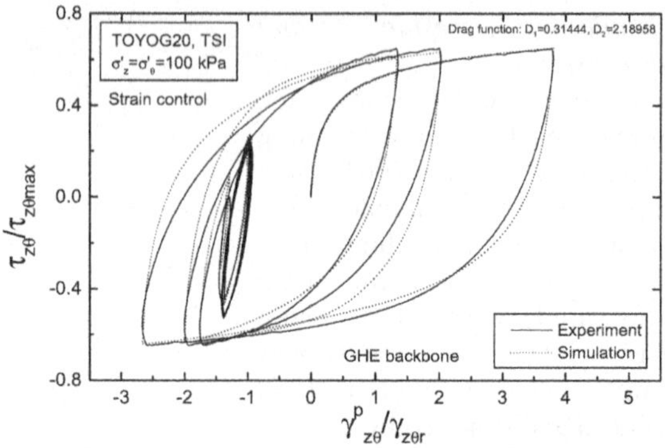

Fig. 12. Modelling of large cyclic loading with 70% of initial drag values (GHE backbone; Test TOYOG20), strain control

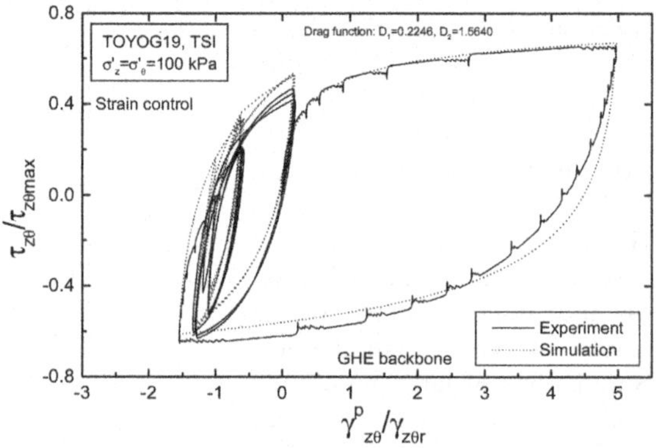

Fig. 13. Modelling of large cyclic loading with 50% of initial drag values (GHE backbone; Test TOYOG19), strain control

Similar simulation result of test TOYOG19 during TSI with the same stress path as mentioned above in which the backbone curve was simulated by a GHE was plotted in Fig. 13. Note that the employed drag values were reduced to 50% of the initial ones (Fig. 7). The simulation result was consistent with the experimental data of test TOYOG19. Note that the simulation by reducing the initial drag amount to 70% for test TOYOG19 was not consistent with the experimental data. This could be due to the effect of creep as mentioned above.

Although all tests in this study were conducted at the constant loading rate, the effect of loading rate on the stress-strain behavior of sand should be carefully considered in the future.

5. CONCLUSIONS

Both the general hyperbolic equation (GHE) and the newly proposed lognormal equation (LE) could well simulate the backbone curve of air-dried, dense Toyoura sand. Simulation by using GHE required a modification of parameters (m_t = n_t = 0.15).

Simulation by using LE requires fewer parameters in comparison to that by using GHE, while LE has a certain applicable range for the strain parameter (up to peak stress state).

Large amplitude cyclic torsional loading (without small cyclic loadings) conducted from isotropic stress state could be well simulated by the combination of the proportional rule, an extension of the well-known Masing's rule, and the drag rule by selecting a proper drag function, while the backbone curve was simulated by either GHE or LE. The combined rules could reflect the rearrangement of sand particles and densification of air-dried, dense sand. An additional assumption related to the drag rule was proposed and was proved to be effective in the simulations of tests TOYOG19 and TOYOG20.

REFERENCES

1) Balakrishnayer, K. (2000). Modelling of deformation characteristics of gravel subjected to large cyclic loading. *Ph.D. thesis*, Dept. of Civil Engineering, The University of Tokyo, Japan.

2) Duncan, J. M. and Chang, C. Y. (1970). Nonlinear analysis of stress and strain in soils. *Journal of Soil Mech. Fdns Div.*, ASCE, Vol. 96, No. SM 5, pp. 1629-1653.

3) Goto, S., Tatsuoka, F., Shibuya, S., Kim, Y. S. and Sato, T. (1991). A simple gauge for local small strain measurements in the laboratory. *Soils and Foundations*, Vol. 31, No. 1, pp. 169-180.

4) Hardin, B. O. and Drnevich, V. P. (1972). Shear modulus and damping in soils: Design equations and curves. *Journal of Soil Mechanics and Foundation Division*, ASCE, Vol. 98, No. SM7, pp. 667-692.

5) Hayashi, H., Honda, M., Yamada, T. and Tatsuoka, F. (1994). Modeling of nonlinear stress strain relations of sands for dynamic response analysis. *In Proc. of the Tenth World Conference on Earthquake Engineering*, Balkema, Rotterdam, pp. 6819-6825.

6) HongNam, N. (2004). Locally measured deformation properties of Toyoura sand in cyclic triaxial and torsional loadings and their modelling, PhD Thesis, Dept. of Civil Engineering, The Univ. of Tokyo, Japan.

7) HongNam, N., Sato, T. and Koseki, J. (2001). Development of triangular pin-typed LDTs for hollow cylindrical specimen. *Proc. of 36th annual meeting of JGS*, pp. 441-442.

8) HongNam, N. and Koseki, J. (2005). Quasi-elastic deformation properties of Toyoura sand in cyclic triaxial and torsional loadings, *Soils and Foundations*, Vol. 45, No. 5, pp. 19-38.

9) HongNam, N., Koseki, J. and Sato, T. (2005): Effect of specimen size on quasi-elastic properties of Toyoura sand in hollow cylinder triaxial and torsional shear tests (*submitted for possible publication in Geotechnical Testing Journal, ASTM*).

10) Jardine, R. J. (1992). Some observations on the kinematic nature of soil stiffness. *Soils and Foundations*, Vol. 32, No. 2, pp. 111-124.

11) Kawakami, S. (1999). Deformation characteristics of sand during liquefaction process using hollow-cylindrical torsional shear tests. *Master of Engineering Thesis*, Department of Civil Engineering, The University of Tokyo, Japan (in Japanese).

12) Kondner, R. L. (1963). Hyperbolic stress-strain response: Cohesive soils. *Journal of Soil Mechanics and Foundation Division*, ASCE, Vol. 89, No. SM1, pp. 115-143.

13) Masing, G. (1926). Eiganspannungen und verfestigung beim messing. *Proceedings of the Second International Conference of Applied Mechanics*, pp 332-335.

14) Masuda, T. (1998). Study on the effect of pre-load on the deformation of excavated

ground. *Doctor of Engineering Thesis*, Dept. of Civil Engineering, The University of Tokyo, Japan (in Japanese).
15) Ohsaki, Y. (1980). Some notes on Masing law and non-linear response of soil deposits. *Journal of the Faculty of Engineering*, The University of Tokyo, Vol. XXXXV, No. 4, pp. 513-536.
16) Press, W., Teukolsky, S., Vetterling, W., and Flannery, B. (1992). *Numerical Recipes in C* (2nd Edition). Cambridge University Press, Cambridge.
17) Pyke, R. 1979. Non-linear soil models for irregular cyclic loading. *Journal of Geotechnical Engineering Division*, ASCE, Vol. 105, No. GT6, pp. 715-726.
18) Schofield, A. N. and Wroth, C. P. (1968). *Critical state soil mechanics*. McGraw Hill, London.
19) Tatsuoka, F. and Shibuya, S. (1991). Modelling of non-linear stress-train relations of soils and rocks- Part 2: New equation. *Seisan-kenkyu*, Journal of IIS, The University of Tokyo, Vol. 43, No. 10, pp. 435-437.
20) Tatsuoka, F., Jardine, R. J., Lo Presti, D., Di Benedetto, H. and Kodaka, T. (1997). Characterising the Pre-Failure Deformation Properties of Geomaterials, *Proc. of XIV IC on SMFE,* Hamburg, Vol. 4, pp. 2129-2164.
21) Tatsuoka, F., Ishihara, M., Uchimura, T. and Gomes Correia, A. (1999). Non-linear resilient behaviour of unbound granular materials predicted by the cross-anisotropic hypo-quasi-elasticity model. *Unbound Granular Materials*, Gomes Correia (ed.), Balkema, pp. 197-204.
22) Tatsuoka, F., Masuda, T., Siddiquee, M. S. A. and Koseki, J. (2003). Modeling the stress-strain relations of sand in cyclic plane strain loading. *Journal of Geotechnical and Geoenvironmental Engineering*, ASCE, Vol. 129, No. 6, pp. 450-467.

APPENDIX. CALCULATION OF ELASTIC SHEAR STRAIN COMPONENT BY A NEWLY PROPOSED HYPO-ELASTIC MODEL (IIS MODEL)

In the material axes (*z, r, θ*), the stress and strain increments of an elastic material can be formulated by the generalized Hooke's law as

$$[d\varepsilon_z^e\ d\varepsilon_r^e\ d\varepsilon_\theta^e\ d\gamma_{\theta z}^e]^T=[M][d\sigma'_z\ d\sigma'_r\ d\sigma'_\theta\ d\tau_{\theta z}]^T \quad \text{(A-1)}$$

Similarly, in the principal stress axes (ξ,ρ,η), the stress and strain increments can be written as $[d\varepsilon_\xi^e\ d\varepsilon_\rho^e\ d\varepsilon_\eta^e\ d\gamma_{\eta\xi}^e]^T=[\overline{M}][d\sigma'_\xi\ d\sigma'_\rho\ d\sigma'_\eta\ d\tau_{\eta\xi}]^T$ (A-2)

The compliance matrix in the material axes can be calculated by the following transformation law. $[M]=[T_\sigma]^T[\overline{M}][T_\sigma]$ (A-3)

The compliance matrix in the principal stress axes (ξ,ρ,η) can be proposed as follows.

$$[\overline{M}]=\begin{bmatrix} 1/E_\xi & -\nu_{\rho\xi}/E_\rho & -\nu_{\eta\xi}/E_\eta & -\alpha_{1o}/E_{zo} \\ -\nu_{\xi\rho}/E_\xi & 1/E_\rho & -\nu_{\eta\rho}/E_\eta & -\alpha_{2o}/E_{zo} \\ -\nu_{\xi\eta}/E_\xi & -\nu_{\rho\eta}/E_\rho & 1/E_\eta & -\alpha_{3o}/E_{zo} \\ -\alpha_{1o}/E_{zo} & -\alpha_{2o}/E_{zo} & -\alpha_{3o}/E_{zo} & 1/G_{\eta\xi} \end{bmatrix} \quad \text{(A-4)}$$

Elastic shear strain can be calculated by $\gamma^e{}_{z\theta}=\int_0^{\tau_{z\theta}} M_{44}\,d\tau_{z\theta}$ (A-5)

Refer to HongNam and Koseki (2005) for more details about this model.

Soil Stress-Strain Behavior: Measurement, Modeling and Analysis
Geotechnical Symposium in Roma, March 16 & 17, 2006

EFFECT OF LOADING CONDITION ON LIQEFACTION STRENGTH OF SATURATED SAND

Shun-ichi Sawada
Group Manager, Earthquake Geotechnical Engineering Group
OYO CORPORATION, 43 Miyukigaoka, Tsukuba, Ibaraki, 305-0841, JAPAN
e-mail:sawada-shun@oyonet.oyo.co.jp

ABSTRACT

Liquefaction strength is, obtained by means of a cyclic triaxial test in the engineering practice. This loading procedure of a cyclic triaxial test is equivalent to the shear stress condition acting on the horizontal plane in the ground during an earthquake. This method has a disadvantage, however, that effective mean stress changes. When simulating the behavior of the level ground, the cyclic shear stress must be applied while inhibiting lateral deformation as a torsional shear test. These tests were performed. The following conclusions were obtained from a series of test result.

(1) The result of these laboratory tests shows that the liquefaction strength defined as 7.5% shear strain in double amplitude is generally agreement between triaxial and torsional test.

(2) The stress condition appears to affect the behavior of excess pore pressure and shear strain up to liquefaction strength. In particular, the behavior of the cyclic triaxial tests is different stress condition between in-situ and laboratory.

(3) It was noteworthy that the minimum cyclic shear strength was observed at cyclic triaxial test that is ordinary used in small shear strain up to 7.5% in double amplitude. The main reason for this behavior is that the effective confining pressure is decrease, when the cyclic axial stress direction is extension.

1. INTRODUCTION

A major destruction of soil structures during earthquake occurs due to saturated sandy soil liquefaction. To achieve an accurate assessment of the dynamic performance of soil structures due to seismic excitations, an estimation of liquefaction strength is of pivotal importance. In this study, the effect of K_0-condition on liquefaction behavior is mainly discussed.

It is generally accepted in the practice to measure liquefaction strength by triaxial cyclic loading test rather than torsional cyclic test. In triaxial cyclic loading test, an axial loading is applied with the isotropic initial stress condition while a cyclic torsion is applied with the K_0-initial stress condition in torsional cyclic test. In-situ stress condition tends to exhibit K_0-condition, which is different from the stress condition of triaxial cyclic loading tests. Based on such issues, it is worth to investigate the effect of K_0-condition on liquefaction behavior.

In numerical analysis, material parameters are usually determined based on laboratory test results. Considering the fact that an accuracy of numerical solution is considered to

Hoe I. Ling et al. (eds.), Soil Stress-Strain Behavior: Measurement, Modeling and Analysis, 637–644.

rely on test data, test condition should duplicate in-situ soil conditions. In liquefaction analysis, the parameters to dominant liquefaction behavior are basically determined from laboratory liquefaction test such as triaxial cyclic loading test or torsional cyclic loading test. It is important, therefore, to make sure the effect of initial stress condition (K_0-condition) on numerical result in determining the material parameters.
Additional major damage of earth structure occurs due to the dissipation of pore water pressure after the earthquake. In this study, effect of K_0-condition on volumetric strain due to dissipation of pore water pressure is also discussed.

2. EXPERIMENTAL STUDY ON THE EFFECT OF K_0-CONDITION ON LIQUEFACTION STRENGTH

There are several laboratory tests to estimate liquefaction strength; triaxial cyclic loading test is widely used in the practice and torsional cyclic loading test. Triaxial cyclic loading test is usually conducted with the initial isotropic stress the condition, while torsional cyclic loading test is conducted with several stress conditions such as K_0-stress condition. It is generally accepted that in-situ stress condition tends to exhibit K_0 state. It is important, therefore, to consider stress condition for the estimation of liquefaction strength in the laboratory test. In this chapter, comparisons of the behavior from the above two tests results are carried out in order to discuss the effect of K_0 condition on liquefaction strength for silty sand.

Test samples and test condition

The test samples are silty sand with approximately 10 *N*-value from SPT. This sample is obtained from Tokyo bay area. The fine content of this sample is approximately 60%. The following three series of tests are carried out with 0.1Hz sinusoidal wave.

(1) Triaxial cyclic loading test with initial isotropic stress condition is widely used in the practice
(2) Torsional cyclic loading test with initial isotropic stress condition
(3) Torsional cyclic loading test with K_0 stress condition

Discussion on liquefaction strength and liquefaction behavior

Fig.1 illustrates stress ratio and number of cycles *Nc* relationship, whose shear strain corresponds to 1.5, 3, 7.5, and 10% in double amplitude, obtained from the series of tests.

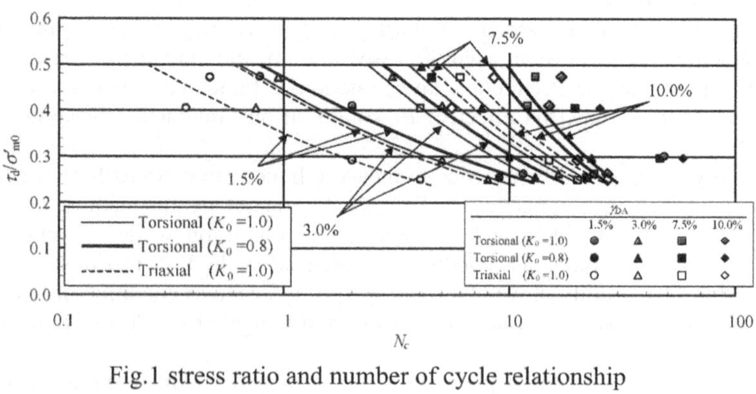

Fig.1 stress ratio and number of cycle relationship

The above relations of 7.5% shear strain which generally defined as initial liquefaction exhibit quite similar behavior regardless of initial stress condition. The behavior of other tests, which correspond to shear strain less than 7.5%, for example, exhibits different relation and seems to be dependent on initial stress condition.

Shear strain and number of cycles relationship, and pore water pressure ratio and number of cycles relationship are shown in Fig.2(a) and Fig.2(b), respectively. Comparison of three different tests yields that the amplitude of pore water pressure ratio in triaxial test exhibits larger than that of torsional test.

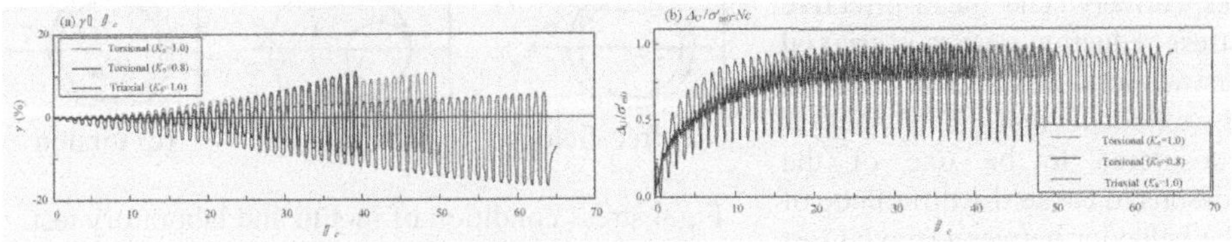

Fig.2 (a) shear strain and number of cycles relationship,
(b) pore water pressure ratio and number of cycles relationship

Fig.3(a) and Fig.3(b) illustrates shear strain and number of cycles ratio relationship, and pore water pressure ratio and number of cycles ratio relationship, respectively. Number of cycle's ratio is defined as number of cycles normalized by number of cycles at 7.5% shear strain. This figure also shows the difference of shear strain development especially at the lower number of cycles.

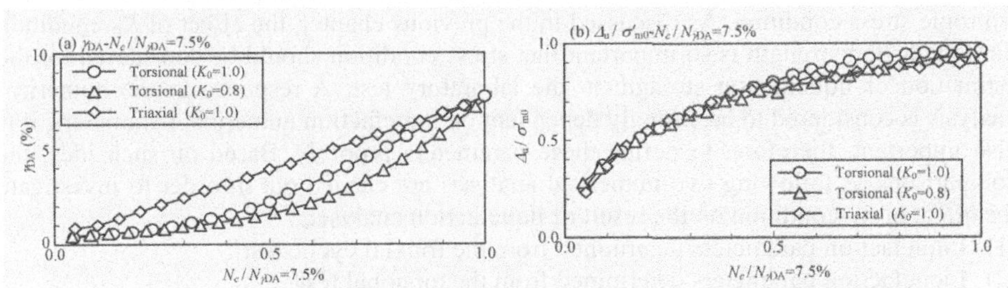

Fig.3 (a) shear strain and number of cycle ratio relationship,
(b) pore wter pressure ratio and number of cycle ratio relationship

Fig.4 (a), Fig.4 (b) and Fig.4(c) explain the difference of the behavior mentioned above. Fig.4 (a) illustrates the stress condition of free-field (in-situ condition). A cyclic shear stress is applied in the earthquake. The stress conditions of triaxial test and torsional test is illustrated in Fig.4 (b) and Fig.4(c), respectively. The mean effective stress reduction occurs in triaxial test while it seems to be constant in torsional test. This fact is considered to be one of the reasons to cause the difference of the behavior between triaxial test and torsional test, especially at the lower shear strain level for silty sand.

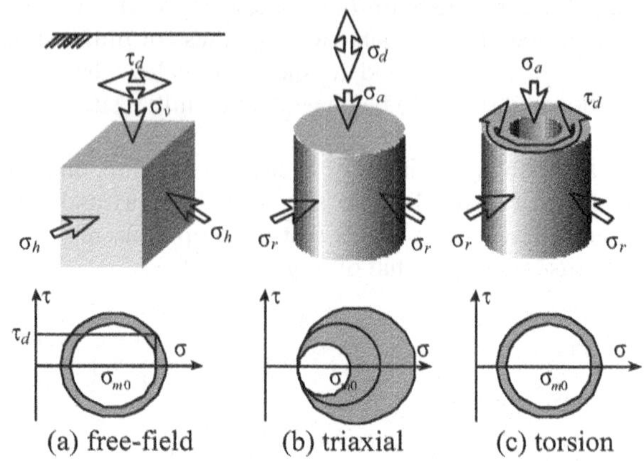

Fig.4 stress condition of in-situ and laboratory test

3. NUMERICAL STUDY ON THE EFFECT OF K_0-CONDITION ON LIQUEFACTION PROPERTIES

It is commonly accepted to use the liquefaction strength obtained from triaxial cyclic loading tests in the liquefaction analysis. In-situ stress condition, which exhibits K_0 state, is different from the stress condition of triaxial cyclic loading tests, which assumes isotropic stress condition. As discussed in the previous chapter, the effect of K_0-condition on liquefaction strength is so important that stress condition should be considered for the estimation of liquefaction strength in the laboratory test. A response in the numerical analysis is considered to be strongly dependent on liquefaction numerical parameters, it is also important, therefore, to define these parameters properly. Based on such idea, the comparisons of following two numerical analyses are carried out in order to investigate the effect of K_0-condition on the result of liquefaction analysis.

(1) Liquefaction parameters determined from the triaxial cyclic test
(2) Liquefaction parameters determined from the torsional test

Analysis model

Liquefaction analysis with 1-dimensional soil model is carried out in order to investigate the effect of numerical parameters defined from different K_0-condition on the process to generate pore water pressure. The soil model and parameters illustrated in Fig.5 was in Kawagishi-cho, Niigata which was affected by Niigata earthquake in 1964. No2, 3, and 4 layers are sandy soil which may become an object of liquefaction. The earthquake motion to use the analysis was observed in Akita prefecture office in the Niigata earthquake. The amplitude of the earthquake motion is amplified to 0.6m/s^2 in the analysis. The finite element program code "FLIP (finite element liquefaction program)" is used.

Fig.6 illustrates stress ratio and number of cycles relationship to simulate "elemental

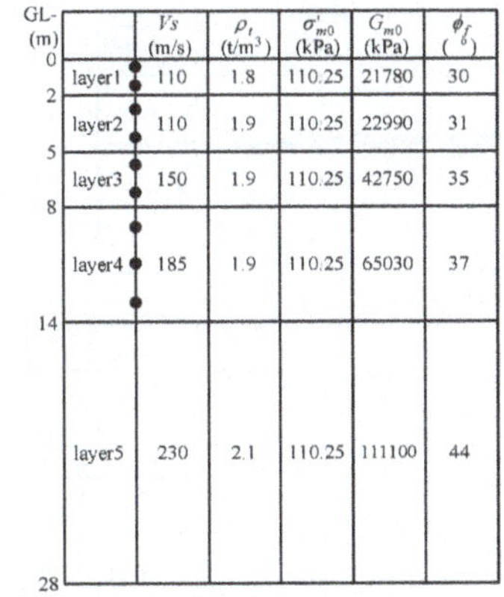

Fig.5 Soil model

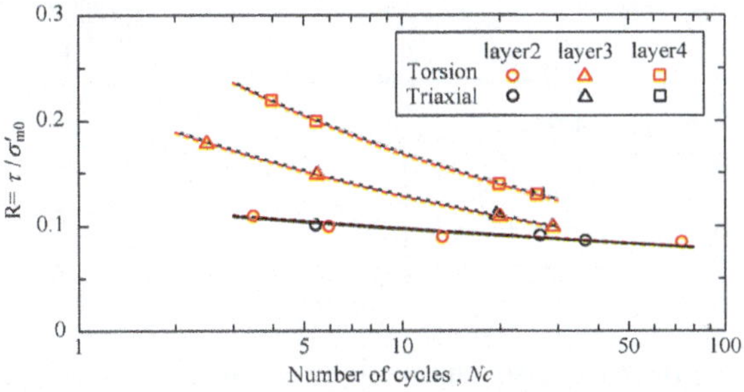

Fig.6 stress ratio and number of cycles relation

behavior" for both triaxial cyclic test result and torsional cyclic test result. Shear strain and number of cycles relationship, and pore water pressure ratio and number of cycles relationship in the same simulation are shown in Fig.7(a) and Fig.7(b), respectively. Both simulations have quite good agreement in stress ratio and number of cycle's relationship, however, the generation of pore water pressure exhibit different development especially at the lower number of cycles.

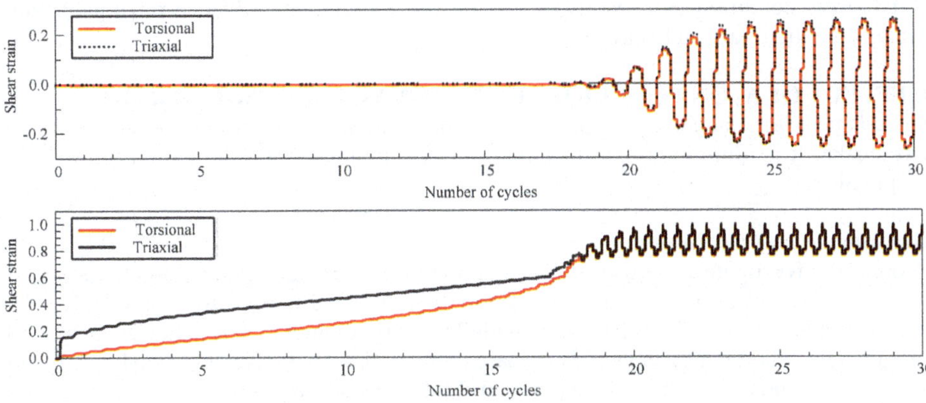

Fig.7 (a) shear strain and number of cycles relationship,
(b) pore water pressure ratio and number of cycles relationship

Discussion of the result

Fig.8 illustrates the comparisons of maximum acceleration, pore water pressure ratio and maximum strain distribution along the depth. The time history of acceleration at the crest and the time history of pore water pressure ratio at No3 layer is shown in Fig.9(a) and Fig.9(b), respectively. The maximum acceleration distribution exhibits quite similar result, on the other hand, there are some difference in pore water pressure ratio distribution. The

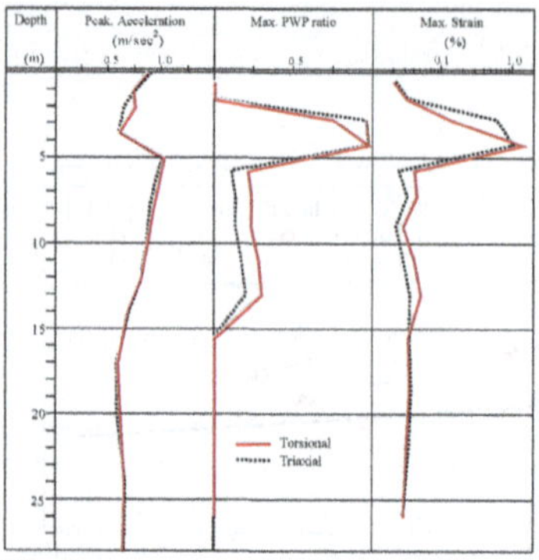

Fig.8 response distribution along the depth

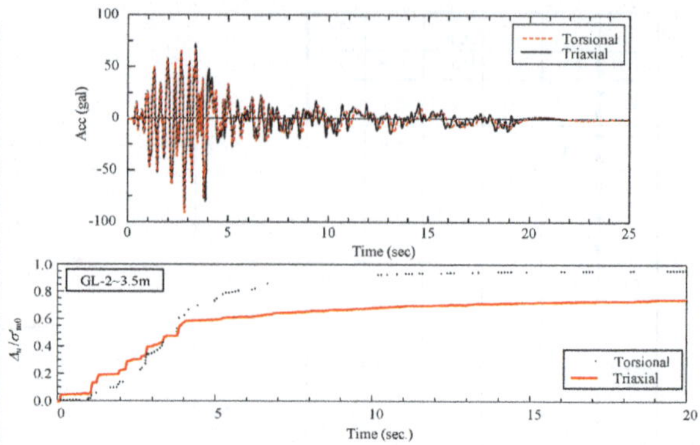

Fig.9 (a) Acceleration time history at the crest
(b) Pore water pressure ratio time historyat No3 layer

comparison of acceleration time history as shown in Fig.9(a) also illustrates good agreement. This comes from the fact that the maximum acceleration appears before the initial liquefaction which defines as pore water pressure to be 0.95. The comparison of pore water pressure ratio time history as shown in Fig.9(b) illustrates the difference of pore water pressure generation development. This is due to the difference of the liquefaction parameters explained previously. Throughout the analysis, it was found that initial stress condition and loading condition had significant influence on liquefaction strength and liquefaction behavior.

4. EFFECT OF K_0 ON POST-LIQUEFACTRION VOLUMETRIC STRAIN

There is post-liquefaction deformation due to dissipation of pore water pressure after the earthquake. It is quite important to figure out post-liquefaction deformation properly in the practice.

Test samples and test condition

In order to investigate the effect of K_0-condition on post-liquefaction volumetric strain, a series of torsional cyclic test are carried out. Toyoura sand is used for test sample. The test sample is hollow cylindrical shape with 7cm outer diameter, 3cm inner diameter and 14cm height. The relative density is prepared 50% and 80%. 0.1Hz sinusoidal wave with undrained boundary condition is applied until the shear strain double amplitude becomes 15%, then boundary condition is changed so that volume change due to pore water dissipation is allowed.

Discussion on effect of K_0-condition on post-liquefaction volumetric strain

Fig.10(a) and Fig.10(b) illustrate post-liquefaction volumetric strain and number of cycles after it reaches the initial liquefaction (7.5% double amplitude shear strain) relationship. In both cases, the more number of cycles, the more volumetric strain tends to be appeared. This tendency seems to be dependent on K_0-condition and relative density. Fig.11(a) and Fig.11(b) illustrate post-liquefaction volumetric strain and consolidation stress ratio relationship. In the relative density 50% sample, the post-liquefaction

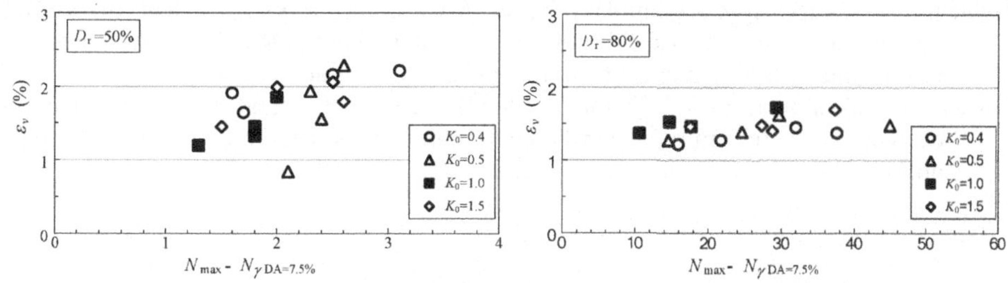

Fig.10 (a) volumetric strain and number of cycles relationship for Dr=50%
(b) volumetric strain and number of cycles relationship for Dr=80%

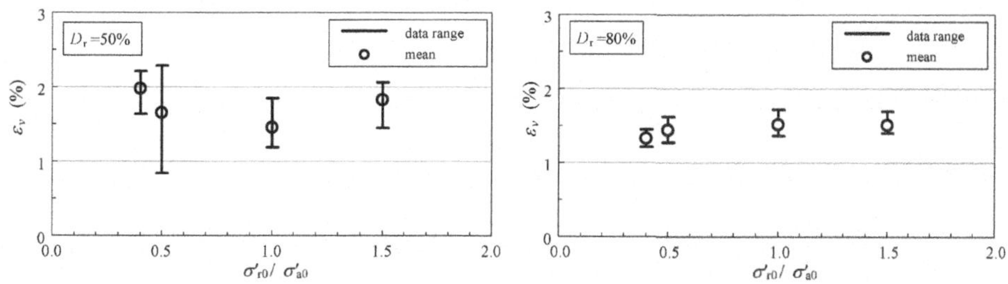

Fig.11 (a) volumetric strain and stress ratio relationship for Dr=50%
(b) volumetric strain and stress ratio relationship for Dr=80%

volumetric strain has the minimum value when the stress ratio is equal to one, while the post-liquefaction volumetric strain has the maximum value when the stress ratio is equal to one in the relative density 80% sample. The relative density 50% sample seems to have more sensitive effect of K_0-condition than the relative density 80% sample has.

5. CONCLUSIONS

Throughout this study, the following conclusions are derived.

(1) A liquefaction behavior of silty sand seems to be dependent on initial stress condition at shear strain less than 7.5%.

(2) In liquefaction analysis, a development of pore water pressure generation seems to be different in torsional test and triaxial test even if shear stress and number of cycles relationship exhibits similar behavior.

(3) A post-liquefaction volumetric strain seems to be dependent on K_0-condition and relative density.

6. REFERENCES

Ishihara, K. & Yasuda S. [1975]. "Sand liquefaction in hollow cylinder torsion under irregular excitation" Soils and Foundations, JSSMFE, Vol.15, No.1, pp.45-59.

Ishihara, K. & Takatsu H. [1979]. "Effects of overconsolidation and Ko conditions on the liquefaction characteristics of sands" Soils and Foundations, JSSMFE, Vol.19, No.4, pp.59-68.

Peiris, T. A. and Yoshida, N. [1996]: Modeling of volume change characteristics of sand under cyclic loading, Proc., 11th WCEE, Acapulco, Mexico, Paper No. 1087

Sawada, S., Sakuraba, R., Ohmukai, N. & Mikami, T. [2001]. "Effect of Ko on Liquefaction Strength of Silty Sand" 4th Inter. Conf. On Recent Advances in Geotechnical Earthquake Engineering and Soil Dynamics.

Sawada, S., Takeshima, Y. & Mikami, T. [2003]"Effect of K_0-condition on liquefaction characteristics of saturated sand" Deformation Characteristics of Geomaterials. pp.511-517.

Soil Stress-Strain Behavior: Measurement, Modeling and Analysis
Geotechnical Symposium in Roma, March 16 & 17, 2006

EXPERIMENTAL STUDY ON THE BEHAVIORS OF SAND-GRAVEL COMPOSITES LIQUEFACTION

Xianjing Kong, Bin Xu, Degao Zou
Department of Civil Engineering
Dalian University of Technology, Dalian, 116024, PRC
e-mail: kongxj@dlut.edu.cn

ABSTRACT

By use of medium scale dynamic triaxial apparatus(ϕ 200×500mm) the development of axial strain and pore water pressure of sand-gravel composites are studied in cyclic loading. Adopting same relative density, a series of substituted material specimens gained by eliminating the oversized (>5mm) gravel particles are studied. The results show that with isotropic consolidation, the development of excess pore water pressure and axial strain in sand-gravel composites differs from that in substituted material. A series of undrained cyclic triaxial tests were performed on sand-gravel composites specimens with relative density of 50%, 55%, and 60%. Test results showed that the increase of relative density may delay the development of pore pressure of sand-gravel composites.

1. INTRODUCTION

In 1964 the Niigata and Alaska earthquakes inflicted huge damage to buildings, bridges and other structures founded on saturated sand deposits by liquefying loose-sandy soils. Since then, extensive researches have been carried out to elucidate liquefaction mechanisms (Seed and Lee, 1966; Ishihara et al., 1975). After these earthquakes, for the purpose of evaluating dynamic properties and liquefaction behavior of gravelly soils, many studies on the liquefaction of sandy soils have been conducted by laboratory cyclic shear tests, shaking table tests, site investigations and analyses. Most of these researches aimed at sand or silty sand. As a natural foundation and filled material, sand-gravel composites has the characteristic of low compressibility and high shear strength. Due to its high hydraulic conductivity, gravels and gravelly soils were once thought to be unliquefiable. However, cases have occasionally been reported where liquefaction-associated damage took place in gravelly soil. For instance, at the time of the Haicheng earthquake of February 4, 1975 and Tangshan earthquake of July 28, 1976 in China, signs of disastrous liquefaction were observed in sand-gravel composites filled dam foundation. During the 1983 Borah Peak earthquake in the US, liquefaction was reported to have occurred in gravelly soil deposits at several sites, causing lateral spreading over the gently sloping hillsides. While the drainage conditions surrounding gravelly deposits may exert some influence on the dissipation of pore water pressure and hence on the liquefiability, it is of prime importance to clarify the resistance of gravelly sand itself to cyclic loading. One of the earlier endeavors in this context was made by Wong et al. (1975) who performed a series of cyclic triaxial tests on reconstituted specimens of gravelly soils with different gradation by means of a large-size triaxial test apparatus. The results of these tests indicated somewhat

Hoe I. Ling et al. (eds.), Soil Stress-Strain Behavior: Measurement, Modeling and Analysis, 645–651.

higher cyclic strength as compared to the strength of clean sands. However, whether the result of such tests reflects the cyclic strength of in situ deposits remained open to question. These liquefaction-induced failures in gravel and gravelly soils prompted a critical reevaluation of the behavior of gravelly soils subjected to dynamic loading.

Previous laboratory testing has yielded much knowledge in this research area (Ishihara 1985; Evans and Seed 1992; Evans and Harder 1993), however, most of these research focused on the effect of membrane compliance on the liquefaction of uniformly graded gravel, concerned specifically with the liquefaction behavior of sand-gravel composites are limited (Evans and Zhou 1995). Considering this problem, a series of undrained cyclic triaxial tests was performed by Evans and Zhou (1995) to quantify the effect of gravel content on the liquefaction resistance of sand-gravel composites. They prepared soil specimens by mixing poorly graded sand and poorly graded gravel with different ratios to make gap-graded specimens with different gravel contents. They found that gravely soils showed evidently larger liquefaction resistance than sand with the save relative density. In P. R. China, some recent researches were done by Chang Y. P. (1998), Liu II. S. (1998) and Wang K. Y. (2002). Most of these studies are limited since large-scale triaxial apparatus is not available in most laboratories. Even in tests performed by Evans and Zhou (1995), the maximum grain size is 10 mm.

In this study, a series of cyclic triaxial tests were conducted under undrained conditions to demonstrate the difference between sand-gravel composites and sand on liquefaction characteristics. Pore water pressure and axial strain development were examined with special attention. Several factors influence liquefaction resistance of soils, including soil density, soil composition and grain characteristics was studied. A middle-scale triaxial apparatus is the primary testing device used in this study.

2. MATERIALS AND APPARATUS

Materials Description and Maximum and Minimum Density Test. Nierji Dam (Heilongjiang Province, in China) foundation sand-gravel composites was used to investigate liquefaction characteristics in this study. Scalping the oversized particles and taking similar grade method, sand was got for parallel tests. Grain size distributions are shown in Fig. 1. For the triaxial specimen diameter of 200 mm used in this study, the ratio of the specimen diameter to the maximum particle size is about 5. A ratio of 5 to 6 is generally considered necessary for meaningful test results.

To determine the relative density, maximum and minimum dry density tests were performed for the sand-gravel composites and sand first. The maximum density was determined by vibratory compaction method. Specimens were vibrated in a cylindrical mold 100 mm inner diameter and 150 mm high. The minimum density was determined using a cylinder 50mm inner diameter and 400 mm high. The measuring cylinder was filled sand-gravel or sand no more than 1/3 to 1/2, capped with hand, then upset and uprighted carefully 3 to 4 times to achieve a very loose state, and the volume could be got. In Table 1, the characteristics for the density tests and fine content tests are presented.

Table 1. Characters of sand-gravel composites and sand tested

Material	Grain size /mm									ρ_{dmax} g/cm^3	ρ_{dmin} g/cm^3	D_{50} /mm	G_s
	>40	40~20	20~10	10~5	5~2	2~1	1~0.25	0.25~0.1	<0.1				
Sand-gravel	0	8.6	28.66	12.94	10.96	8.5	24	5.53	0.77	2.205	1.808	5.1	2.64
Sand	0	0	0	0	19.5	20.6	20.3	22.7	16.9	2.150	1.642	0.53	2.75

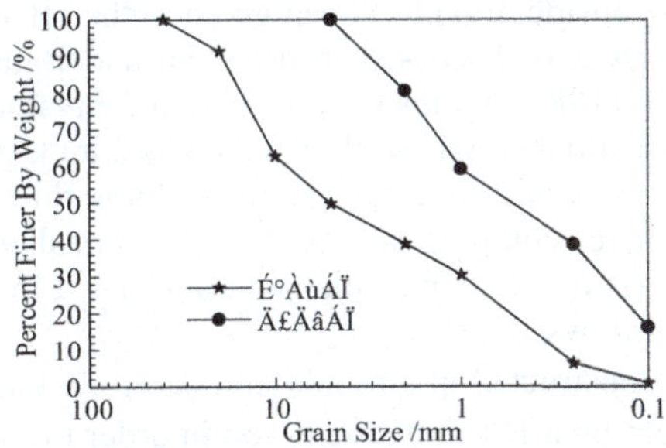

Figure 1 Grain Size Distribution Curves for Tested Materials

Relative density D_r is a pertinent parameter to evaluate undrained cyclic strength of granular soils of different particle gradations and defined by the maximum and minimum dry density ρ_{max} and ρ_{min}, respectively, as

$$D_r = \frac{1/\rho_{min} - 1/\rho}{1/\rho_{min} - 1/\rho_{max}} \times 100\% \tag{1}$$

Here, ρ_{max} and ρ_{min} were determined as discussed above, and ρ_d is actual density tested.

Experimental Setup. The axial apparatus used in the tests is illustrated in Fig.2. An inner load transducer was equipped in the chamber, and three deformation transducer, LDT, Gap sensor and external deformation transducer, were used for precise measurements of strain. Gap sensor consists of two small non-contacting discs each encasing an electromagnetic coil. The setup of this type of equipment was developed by Kokusho (1980). As an alternative technique, a device called a local deformation transducer (LDT) was developed by Goto et al. (1991) for measuring small strains of sedimentary rocks tested in the triaxial chamber. A pair of LDTs sit around the specimen shown in Fig. 2. This device is claimed to permit precise measurement of shear strain to be made even to an infinitesimally small strain on the order of 10^{-6}.

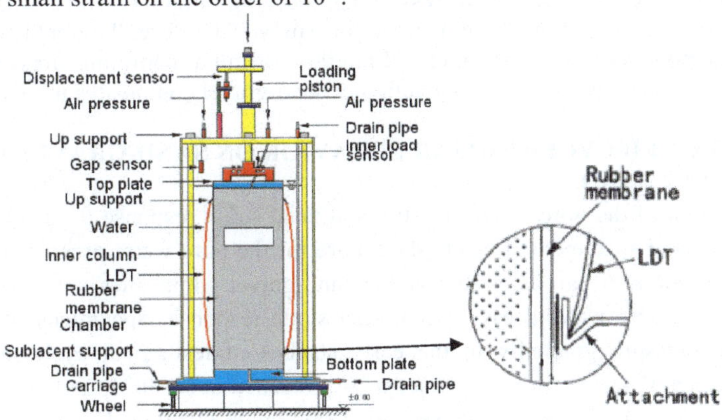

Figure 2 Triaxial test device

Test Method. No special consideration has been given to the effect of membrane penetration is this study, although it is discussed by Ramana and Raju (1981), Vaid and Negussey (1984), Seed et al. (1989), Evans et al. (1992) and Sivathayalan and Vaid (1998). A thicker membrane made of latex, whose thickness was 2 mm, was used for sand-gravel instead of the membrane, also made of latex, having 0.2 mm thickness used for sand. The specimen size of sand-gravel composites was d = 200 mm diameter and h = 500 mm height and that of sand was d = 61.8 mm diameter and h = 125 mm height. All the test procedure is outlined as follows:

1. The specimen was constructed in a mould and carefully tamped in five equal-mass layers 100 mm height by a 10 cm diameter rod in order to control specimen density.
2. The specimen was fully saturated and consolidated isotropically under a certain confining pressure, $p_c^{'}$.
3. In the undrained cyclic loading triaxial tests, the axial stress was cyclically controlled by sinusoidal waves with frequency of 0.5 Hz based on the fact that undrained strength of noncohesive soils is almost independent of the loading frequency. The cyclic loading was continued until the water pore pressure reached its maximum value (almost 90% of confining pressure for sand-gravel composites).

3. DEFINITION OF LIQUEFACTION

The basic mechanism of onset of liquefaction is elucidated from observation of the behavior of a sand sample undergoing cyclic stress application in the laboratory triaxial test apparatus. When the axial stress σ_d is applied undrained, the shear stress induced on the 45° plane is $\sigma_d/2$. The normal stress of $\sigma_d/2$ which is mostly transmitted to pore water without inducing any change in the existing effective confining stress $\sigma_0^{'}$.

The pore water pressure approaches a value equal to the initially applied confining pressure or producing an axial strain of about 5% in double amplitude. Such a stated has been referred to as initial liquefaction or simply liquefaction of sand or silty sand. Because of the stiffness of sand-gravel composites being generally high, the double amplitude of 5% and the pore water pressure approaching initial confining stress was difficult to be induced in the test specimens in this study. Therefore, the double amplitude of 2% and the pore water pressure approaching 90% of initial confining stress is taken as a criterion to identify the state of liquefaction of sand-gravel composites in this study.

4. UNDRAINED CYCLIC SHEAR BEHAVIOR ON SAND-GRAVEL COMPOSITES AND SAND

Fig. 3 is a plot of the pore water pressure and axial strain response for sand-gravel composites and sand specimens. Results shown are for the pore water pressure in specimens approaching 90%. It can be seen that the sand-gravel composites and sand specimens have different pore water pressure and axial strain response. Specimens of sand developed pore pressure gradually in the early stage s of testing, at cycle number ratio, N/N_f reached 0.4, pore pressure ratio, u/u_f, reached about 0.25, and after cycle number ratio reached 0.8, the pore pressure developed quickly and lead to initial liquefaction.

No appreciable axial strain developed in sand specimens until pore pressure ratio reached almost 1.0. Once initial liquefaction occurred, double axial strain suddenly

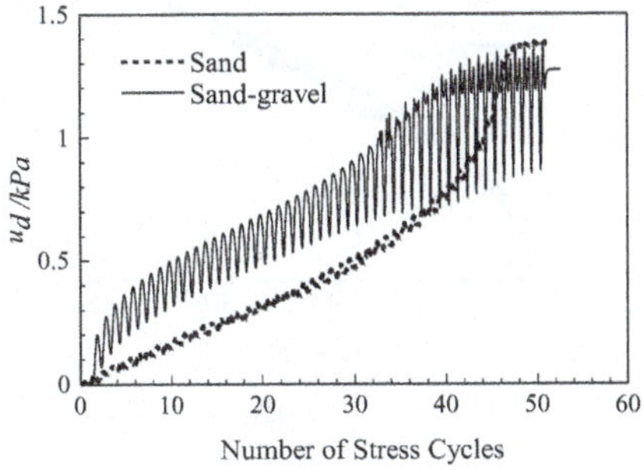

(a)

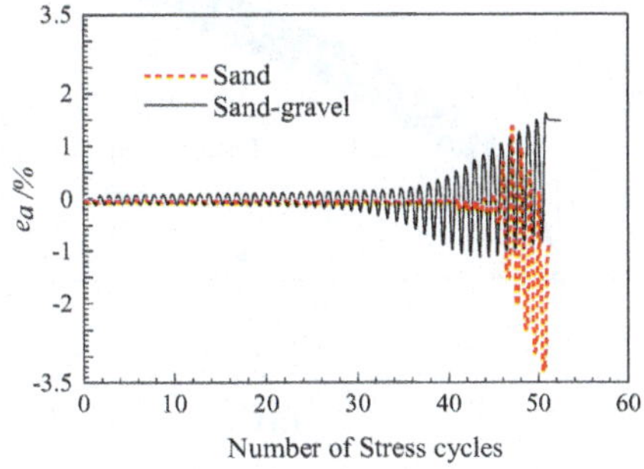

(b)

Figure 3 Pore pressure and axial strain responses of sand and sand-gravel composites

reached 4~5% in 1 to 2 additional stress cycles. The specimens of sand-gravel composites, however, show different pore pressure and axial strain responses. Axial strain of sand-gravel composites accumulated gradually from the first stress cycle, and pore pressure developed quickly in the beginning several stress cycles. Amplitude of pore pressure in a stress cycle is obviously larger than that of sand specimens.

Proposed Pore pressure Development Model. As discussed above, the difference of pore pressure development between sand and sand-gravel specimens is obvious. Seed (1976) has established an experiential model of sand, as

$$u/u_f = \frac{2}{\pi}\arcsin\left(\frac{N}{N_f}\right)^{\frac{1}{2\theta}} \tag{2}$$

The residual pore pressure responses for sand-gravel composites with various relative densities are shown in Fig.4 (a) in terms of residual pore pressure versus the number of stress cycles. With a relative density of 50%, the residual pore pressure responses of different stress ratio under $\sigma_{30}^{'} = 100kPa$ are illustrated in Fig. 4 (b). The value of N_f、u_f of all sand-gravel composites specimens tested are summarized in Table 2.

Table 2. N_f、u_f of all sand-gravel composites specimens tested

Relative density	Confine pressure /kPa	Number of initial liquefaction	Pore water pressure /kPa
0.50	100	6	0.910
		30	0.927
		50	0.893
	150	47	1.332
		16	1.364
0.55	100	26	0.913
		32	0.896
	150	16	0.913
0.60	100	17	0.878
		36	0.931

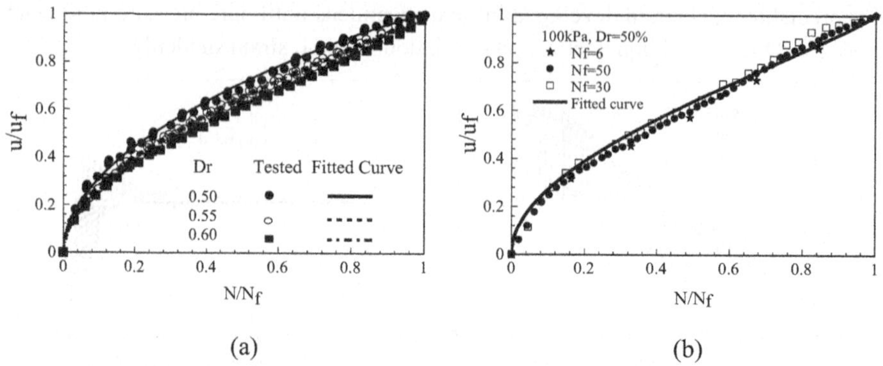

Figure 4 The residual pore pressure in sand-gravel composites

Fitted residual pore pressure curve of sand-gravel composites, it can be developed as follows:

$$u/u_f = 1 - \frac{2}{\pi}\arcsin(1 - \frac{N}{N_f})^{\frac{1}{2\theta}} \tag{3}$$

From the test results in this study, it was found that the increase of relative density may delay the development of residual pore pressure of sand-gravel composites.

5. CONCLUSIONS

Undrained cyclic triaxial tests on sand and sand-gravel composites have been carried out to investigate the difference of the development of axial strain and pore pressure. These experimental studies have yielded the following major findings:

1. The response of axial strain in sand-gravel composites differs from that in substituted material and it increases steadily during cyclic stress action. The development of axial strain in sand specimens is not appreciable before initial liquefaction.
2. The development of excess pore water pressure in sand-gravel composites differs from that in sand and its curve can be fitted by corrected arcsine. Whereas, the pore water pressure curve in substituted material can be fitted by arcsine function established by Seed.
3. The increase of relative density may delay the development of residual pore pressure of sand-gravel composites.

The corrected pore pressure development model in this study was evaluated based on limited data for sand-gravel composites specimens. The effect of gravel content and coarser particles crushed were still not investigated. These factors will be researched in future to complete the corrected pore pressure development model presented in this paper.

ACKNOWLEDGEMENTS

This study was funded through several arrangements: National Natural Science Foundation of China (50278009), National Natural Science Foundation of China (50578029), The Foundation of the State Key Laboratory of Coastal and Offshore Engineering (LP0505).

REFERENCES

Seed, H. B. and Lee, K. L. (1966). "Liquefaction of saturated sands during cyclic loading." *Journal of Soil Mechanics and Foundation Division*, ASCE, 92(SM6), 105-134.

Ishihara, K., Tatsuoka, F. and Yasuda, S. (1975). "Undrained deformation and liquefaction of sand under cyclic stresses." *Soils and Foundations*, 15(1), 29-44.

Wong, R. T., Seed, H.B. and Chan, C.K. (1975). "Cyclic loading liquefaction of gravelly soils." *Journal of Geotchnical Engineering*, ASCE, GT6, 571-583.

Ishihara, K. (1985). "Stability of natural deposits during earthquakes." Proc., *11th Int. Conf. on Soil Mechanics and Foundation Engineering*, Vol. 1, A. A. Balkema, Rotterdam, The Netherlands.

Evans, M. D., Seed, H. B., and Seed, R. B. (1992). "Membrane compliance and liquefaction of sliced gravel specimens." *Journal of Geotechnical Engineering*, ASCE, 118(6), 856-872.

Wang W. S., Chang Y. P., Zuo X. H., (1986). "Liquefaction characteristics of saturated sand-gravels under cyclic stress." *Papers Institute of Water Conservancy and Hydroelectric Power Research,* 23, 195-203 (in Chinese).

Chang Y. P.,Wang K. Y., Chen N., (1998). "Experimental Study on Liquefaction Behavior of a Saturated Gravel.", *Soil Dynamics from Theory to Practice*, Dalian, 1998, 161-166(in Chinese).

Liu H. S., (1998). "On Liquefaction of Gravel Soils.", *Soil Dynamics from Theory to Practice*, Dalian, 1998,155-160(in Chinese).

Wang K. Y., Chang Y. P., Chen N., (2002). "Experimental study on liquefaction characteristics of saturated sandy gravel", *Journal of Hydraulic Eng.*, 2, 37-41(in Chinese)

Evans, M. D., Zhou S. P., (1995). "Liquefaction behavior of sand-gravel composites." *Journal of Geotechnical Engineering.*, ASCE, 121(3), 287-298.

Kokusho, T. (1980). "Cyclic triaxial test of dynamic soil properties for wide strain range." *Soils and Foundations*,20(2), 45-60.

Goto, S., Tatsuoka, F., Shibuya, S., Kim, Y.S., and Sato, T. (1991). "A simple gauge for local small strain measurements in the laboratory." *Soils and Foundations*, 31(1), 169-180.

Ishihara (1996). "*Soil Behavior in Earthquake Geotechnic.*" Oxford University Press Inc., New York.

Soil Stress-Strain Behavior: Measurement, Modeling and Analysis
Geotechnical Symposium in Roma, March 16 & 17, 2006

ACCUMULATED DEFORMATION OF SAND IN ONE-WAY CYCLIC LOADING UNDER UNDRAINED CONDITIONS

Goran Arangelovski
[1] National Institute for Rural Engineering,
2-1-6 Kannondai, Tsukuba-shi, Ibaraki 305-8609.
e-mail: goran@nkk.affrc.go.jp

Ikuo Towhata
The University of Tokyo
7-3-1, Bunkyo-ku, Tokyo 113-8656
e-mail: towhata@geot.t.u-tokyo.ac.jp

ABSTRACT

The problem of development of the deformations in natural slopes, slopes of embankments and dams and quay walls is influenced by the presence of initial shear stress in the soil. Therefore, in this paper an experimental study focuses on accumulation of residual deformation in soil subjected to static shear stress and cyclic stress components under undrained conditions. Furthermore, in such structures during earthquake shaking forces acting in the shear plane are generally one-way cyclic loading without reversing its direction (torsional stress loading only on positive or negative side). Series of torsional shear tests on hollow cylindrical specimens of Toyoura sand with relative density of 40% and 60%, and isotropic stress condition under effective pressure of 98kPa, 196kPa and 294kPa were made in order to investigate the incremental shear strain per cycle during cyclic loading. The influence of confining pressure, isotropic conditions, cyclic stress ratio and void ratio on the accumulation of shear strain are studied.

The experimental results shows a linear relationship between accumulated shear stress and the number of cycles when they are plotted in logarithmic scale. The experimental data shows that shear strain increment during one cycle is dependent on the level of accumulated shear strain. Moreover, the shear strain increment during one cycle when unloading is constant and not depends on the level of accumulated shear strain. Based on these findings and by using the experimental data a relatively simple model for prediction of accumulated deformation of sand can be proposed.

1. INTRODUCTION

In many engineering structures the soil is subjected to initial shear stresses. Such conditions can take place in the soil element in structures like under quay walls, natural slopes, embankments and similar structures that are subjected to gravity loads under static conditions. The undrained behavior of sand with zero static shear stress in a soil element has been well investigated, which is characteristic for a soil element under the free surface. Therefore, the undrained behavior of sand with static shear stress will give a

Hoe I. Ling et al. (eds.), Soil Stress-Strain Behavior: Measurement, Modeling and Analysis, 653–662.

better understanding of the liquefaction phenomena and deformation development. The presence of initial shear stress under gravity condition combined with the seismic cyclic loading results in one-way cyclic loading without reversing its direction. As a result, during undrained cyclic loading the effective stress state is located near the failure conditions and accumulation of deformations occurs. Therefore accumulation of deformation during one-way cyclic loading was studied.

2. TESTING APPARATUS, MATERIAL AND PROCEDURE

During the experimental study hollow torsional shear apparatus was used (Fig.1). The specimens were with height of 19.5cm, outer diameter of 10.0cm and inner diameter of 6.0cm. Toyoura sand with specific gravity G_s = 2.65 was used. Mean particle size of tested material was D_{50}=0.16mm with the minimum and maximum void ratios of e_{min}=0.60 and e_{max}=0.98, respectively. The relative densities of around 40% and 60% were used during testing, isotropically consolidated under the pressures of 98kPa, 196kPa and 294kPa.

The specimens were prepared by air-pluviation of air-dried sand. The density of the specimen was controlled by changing the height of fall and the rate of pluviation of the sand. The specimen was then filled with carbon dioxide gas (CO_2) and de-aired water was percolated through the specimen in order to ensure a high level of saturation. A saturation backpressure of 100kPa was applied in order to achieve a B value exceeding 0.98. Then a specimen was isotropically consolidated to the desired level of stress.

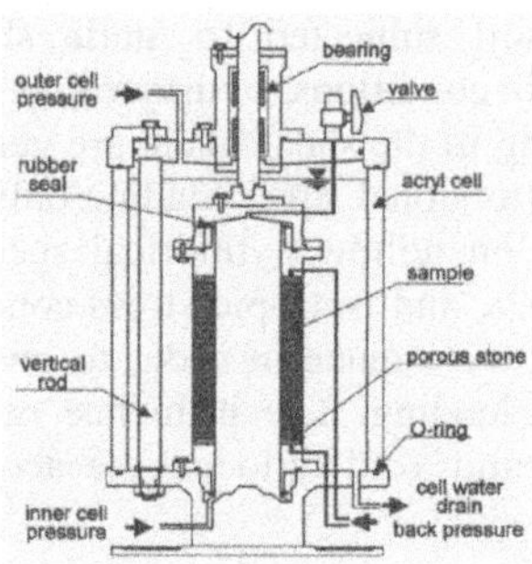

Fig. 1 Torsional shear apparatus

In this study several parameters were varied such as amplitude of cyclic torsional loading, relative density and initial isotropic stress conditions. At first, the specimens were monotonically sheared in a torsional manner under drained conditions up to 50% of the ultimate shear strength under drained conditions (factor of safety FS=2). Afterwards one-way cyclic torsional shear loading was initiated under undrained conditions where cyclic stress loading had a double amplitude of either 50% (CSR=0.5) or 100% (CSR=1.0) of the ultimate shear strength under drained conditions. Since CSR ≤1.0, the direction of shear stress was not reversed. CSR stands for Cyclic Stress Ratio and is defined as the ratio between double amplitude of cyclic stress and the drained ultimate strength. It should be noted that vertical and horizontal stresses were kept constant throughout the testing. Cyclic loading had a sinusoidal time history with period of 20 minutes. The long period of the cyclic loading was chosen in order to provide good stress control during loading.

2. EXPERIMENTAL PROGRAM AND RESULTS

The experimental program was focused on investigating the influence of initial static shear stress on undrained strength as well as on the accumulation of shear strain during one-way cyclic loading.

2.1 Influence of static shear stress

The influence of initial static shear stress was investigated on loose Toyoura sand specimens with dry density Dr=40% isotropically consolidated under 98kPa and 196kPa effective pressures. A comparison was made of Toyoura sand samples sheared under (i) drained, (ii) undrained with zero static shear stress, and (iii) undrained with initial static shear stress (initial static shear stress was achieved under drained conditions). In all cases, the total vertical, radial and tangential stresses were kept constant during the loading.

The influence of initial shear stress was observed in undrained behavior of sand with Dr=40% and isotropically consolidated at 98kPa and 196kPa with different magnitudes of initial static shear stress. The magnitude of initial static shear stress was placed at 50% of the ultimate strength under drained condition ($\tau_{initial}$=30kPa, τ_{ult}=60kPa, FS=2) in the case with samples isotropically consolidated at 98kPa, while the samples isotropically consolidated at 196kPa the initial shear stress was placed at 25% (τ_{static}=30kPa, FS=4), 50% (τ_{static}=60kPa, FS=2) and 75% (τ_{static}=90kPa, FS=1.33) of the ultimate strength under drained condition (τ_{ult}=120kPa) (Fig.2).

In general the shear stress-strain relationship showed that after the initial shear stress state was achieved under drained conditions and undrained loading was initiated, an increase in the shear stress at peak was observed and then softening started (Table 1).

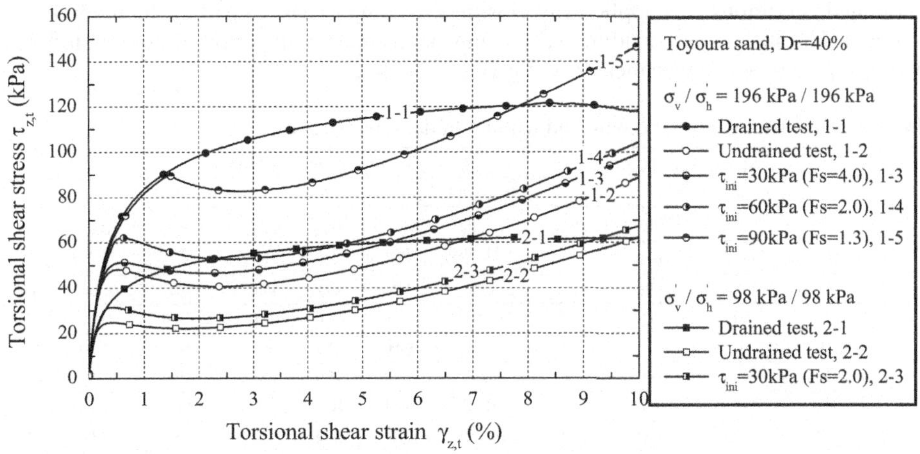

Fig. 2. Effect of static shear stress for series of tests with Dr=40% and isotropic condition σ'_m=98kPa and 196kPa. Torsional shear stress - shear strain relationship.

Exception was noticed in the case when the initial static shear stress was rather high (τ_{static}=90kPa, σ_v'/σ_h'=196kPa) where instantaneous softening occurred when undrained conditions were applied.

Table 1. Shear strength of sand with and without initial shear stress

Isotopic consolidation pressure (kPa)	Initial static shear stress (% of ultimate strength)	Void ratio e (density): After consolidation, the start of the test (drained shearing)	Void ratio e (density): At start of undrained shearing	Peak shear stress, undrained condition (kPa)	Shear stress at softening, undrained condition (kPa)
196	0		0.821 (41.3%)	48.2	40.6
	25	0.820 (41.5%)	0.819 (41.6%)	51.0	47.6
	50	0.820 (41.5%)	0.818 (42.1%)	62.1	52.5
	75	0.810 (44.2%)	0.807 (44.9%)	90.0	82.8
98	0		0.822 (41.0%)	24.8	22.2
	50	0.832 (38.5%)	0.830 (38.8%)	31.5	26.6

2.2 Accumulated shear strain during cyclic loading

The cyclic undrained behavior of sand with static shear stress was observed in terms of the shear strain increment per cycle as well as the accumulated shear strain. Also, some details will be given on the shear strain increment during loading and unloading.

The behavior of sand with initial shear stress was investigated on a soil specimens with dry densities of Dr=40% and Dr=60%. The specimens were isotropic consolidated at σ_v'/σ_h'=98kPa, 196kPa and 294kPa.

The loading part was divided in two parts. The starting part of the loading was monotonic loading under drained conditions. When desired initial shear stress was achieved, an undrained conditions was applied (by closing the drainage valve) and cyclic loading was initiated. The shear cyclic loading have a sinusoidal shape with period of 20 minutes. The details on the twelve experiments were given in Table 2.

Table 2. Shear strength of sand with and without initial shear stress

Dry density	σ_v'/σ_h'	Monotonic drained loading		Undrained cyclic loading							
				CSR=0.5				CSR=1.0			
		Initial shear stress	Ultimate shear stress	Mean value	Cyclic component	Max stress	Min Stress	Mean value	Cyclic component	Max stress	Min Stress
		kPa	kPa	kPa	kPa	kPa	kPa	kPa	kPa	kPa	kPa
≈40	98	30	60	30	±15.0	45.0	15.0	30	±30.0	60.0	0.0
	196	60	120	60	±30.0	90.0	30.0	60	±60.0	120.0	0.0
	294	90	180	90	±45.0	135.0	45.0	90	±90.0	180.0	0.0
≈60	98	33	66	33	±16.5	49.5	16.5	33	±33	66.0	0.0
	196	65	130	65	±32.5	97.5	32.5	65	±65.0	130.0	0.0
	294	94	188	94	±47.0	141	47.0	94	±94.0	188.0	0.0

Typical stress-strain relationships (Dr=40%, σ_v'/σ_h'=98kPa) are given in Fig. 3(a) with the stress path in Fig. 3(b). It was observed that during the one-way cyclic test with cyclic stress ratio CSR=0.5, the sample stayed on the failure line for a short period of time during the increase of shear stress. When higher cyclic stress ratio was applied (CSR=1.0) the loading mainly stayed on the failure line (Fig. 3(b)). At the same time the torsional shear stress ratio ($\tau_{z,t}$ / σ'_m) shows development of shear strain when shear stress ratio reaches the ultimate stress ratio (Fig. 3(a)). In all cases the development of shear strain was associated with the loading when effective stress state lying near the failure conditions.

The focus of the analysis was the shear strain increment per cycle, defined as a shear strain increment between two consecutive peak stresses ($\Delta\gamma_{z,t\ (N)}$). The shear strain increment per cycle was analyzed in terms of total shear strain increment per cycle as well as shear strain increment per cycle during loading and unloading. Moreover, the analysis on the accumulated shear strain was made. Only the cases with dry density of Dr=60% will be presented, however the case with Dr=40% showing similar behavior.

The data shows that the shear strain increment during the first cycle was considerably greater than the increment during the following cycles (Fig. 4a), mainly due to the plastic behavior of the sand during the first cycle.

During the cyclic loading the shear stress continuously increases with the number of cycles. When accumulated shear strain was plotted versus number of cycles, a nearly linear relationship in logarithmic manner was found (Figs. 4.a1 and 4.b1). It can be noticed that at lower densities and higher isotropic consolidation pressures, the accumulated shear stress was higher.

When shear strain increment per cycle ($\Delta\gamma_{z,t}$) was plotted versus the number of cycles, a logarithmical linear relationship between $\Delta\gamma_{z,t}$ and the number of cycles was observed, excluding $\Delta\gamma_{z,t}$ in the first cycle, which is considerable higher as mention previously (Figs. 4.a2 and 4.b2). The shear strain increment per cycle tends to decrease with the number of cycles.

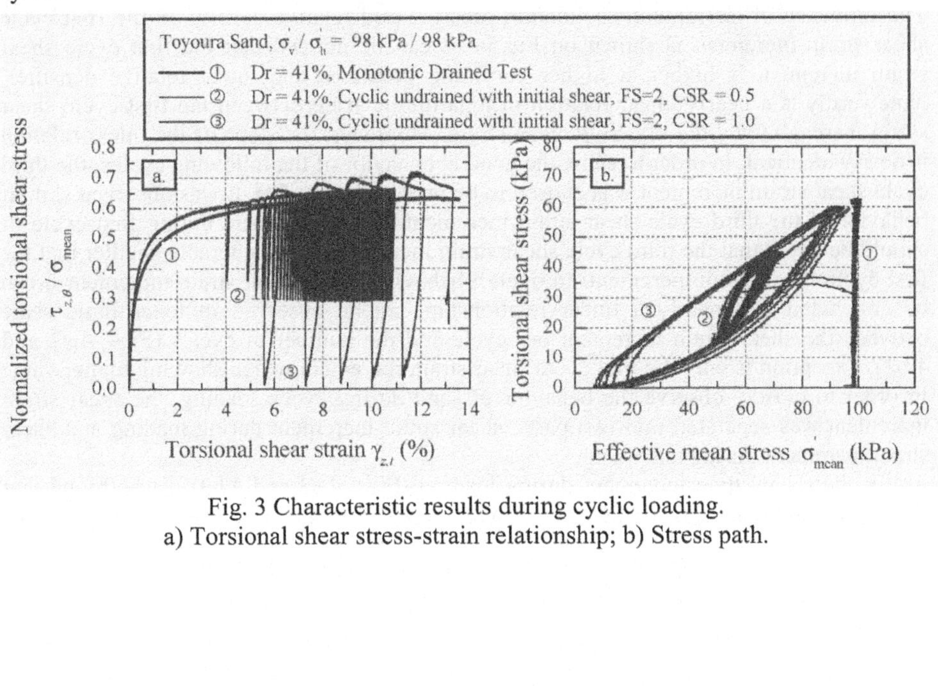

Fig. 3 Characteristic results during cyclic loading.
a) Torsional shear stress-strain relationship; b) Stress path.

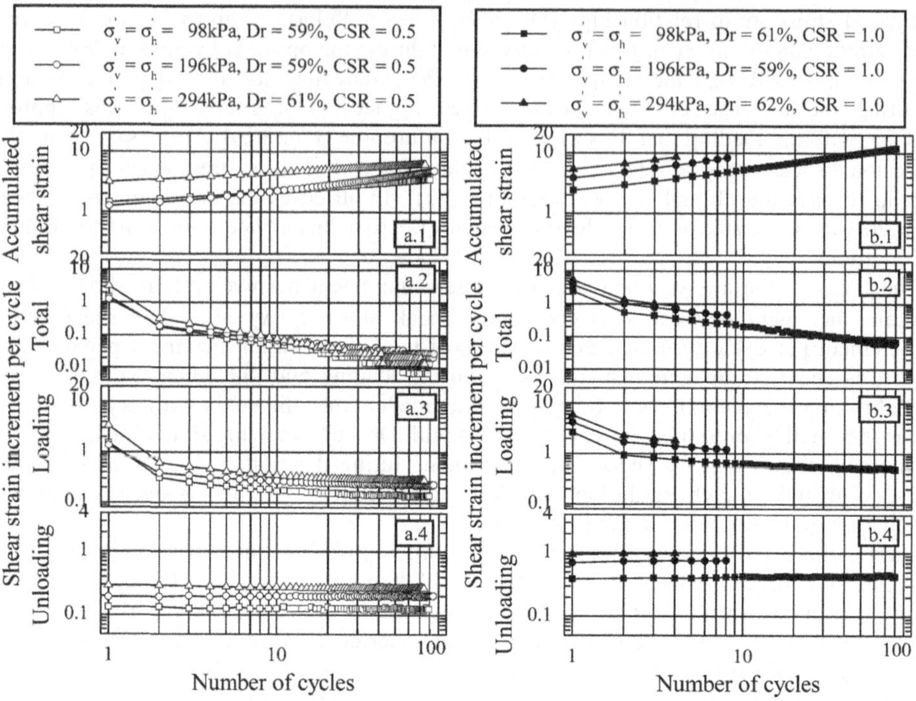

Fig. 4 Influence of different isotropic consolidation condition and relative density on torsional shear strain increment. a) Dr≈60% and CSR=0.5, b) Dr≈60% and CSR=1.0.

The influence of isotropic consolidation pressure and relative density on the first cycle shear strain increment is shown on Fig.5a. It can be noticed that the first cycle shear strain increment is higher at higher confining pressures and lower relative densities. Noteworthy is a nearly linear relation in logarithmic scale between the first cycle shear strain increment and the effective mean stress. Moreover, the slope of the linear relation is nearly identical. In order to show the general behavior of the following cycles, the third cycle shear strain increment was chosen to be presented (Fig.5b). It was observed similar behavior of the third cycle shear strain increment as the increment during first cycle. It should be noted that the third cycle shear strain increment is considerable smaller that the first cycle shear strain increment. In overall behavior of the shear strain increment it can be concluded that nearly a linear relationship can be observed in logarithmic scale between the shear strain increment per cycle and the number of cycles (Figs. 4.a2 and 4.b2). Exception is only the first cycle shear strain increment which showing higher value. In order to closely observe the behavior of sand during cyclic loading, the shear strain increment was separated into two parts, shear strain increment during loading and shear strain increment during unloading.

As for the shear strain increment during loading (Figs. 4.a3 and 4.b3) it was found that the shear strain during loading decreases with the number of cycles. It is observed that the shear strain increment is higher in looser specimens. Moreover, when cyclic amplitude and isotropic confining pressure increases, the shear strain during loading is higher.

In the case of shear strain increment during unloading it was found that the shear strain increment is independent from the number of cycles and remains nearly constant throughout the cyclic loading (Figs. 4.a4 and 4.b4). The shear strain increment during unloading was higher at higher amplitude of cyclic loading as well as at higher isotropic confining pressure. Should be noted that it remained nearly similar for different relative densities (Dr=40% and Dr=60%).

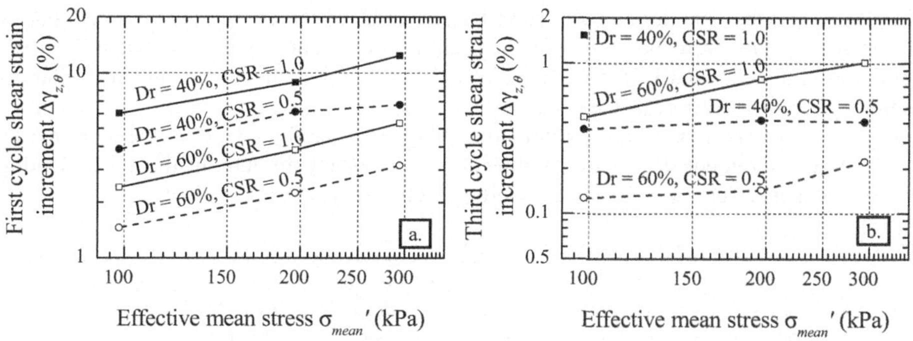

Fig. 5 Influence of different isotropic consolidation condition and relative density on torsional shear strain increment. a) First cycle, b) Third cycle.

2.3 Influence of non-uniform cyclic loading

The nature of earthquake loading is very random and constant amplitude loading can be used only for experimental purposes. In order to investigate the effect of different cyclic loading additional test with non uniform cyclic loading were performed. The different cycle loading was composed of the previously investigated cyclic stress ratios (CSR=0.5 and 1.0). Only relative density Dr=60% was used due to the shear strain capacity of the apparatus. The non uniform cyclic loading was performed with 5 cycles of CSR=0.5 followed by 5 cycles of CSR=1.0 alternatively. A case that starts with 5 cycles of CSR=1.0 and followed by 5 cycles of CSR=0.5 was also performed. The cases with isotropic confining pressure of 196kPa and 294kPa were performed with 3 cycles instead of 5 cycles.

From the experimental data it can be observed that the accumulated shear strain in the case of the non-uniform cyclic loading have values between the accumulated shear strain during uniform cyclic loading with CSR=0.5 and CSR=1.0 (Figs. 6.a1 and 6.b1). Moreover it can be noted that the accumulated shear strain during non-uniform cyclic loading is identical after completing same number of cycles with different cyclic amplitude.

The good repeatability of the experiment was observed on the graph between the shear strain increment per cycle and the number of cycles (Figs. 6.a2 and 6.b2). The first cycles which correspond to the same cycle stress ratio showing good correlation in the amount of shear strain increment per cycle. In the case when the cyclic loading started with CSR=0.5, after the initial group of cycles (five or three) were completed and cyclic stress ratio was changed to CSR=1.0, the shear strain increment was higher than in the case with constant CSR=1.0. When loading started with the higher cyclic stress ratio CSR=1.0, and cyclic stress ratio changed to CSR=0.5 the shear strain increment was smaller than the in the case with constant CSR=0.5. This behavior is due to the fact that the accumulated shear strain was different at the time when the cyclic stress ratio was changed.

The relation between the shear strain increment per cycle during loading and number of cycles showing similar behavior (Figs. 6.a3 and 6.b3). It can be noticed that the difference is rather smaller and a linear relation can be observed. Should be noted that the in-between points corresponds to the shear strain increment at the cycle when the stress ratio was changed. This is due to the reason that when a cyclic stress ratio was changed actual cyclic stress ratio was CSR=0.75.

As for the shear increment during unloading (Figs. 6.a4 and 6.b4) it can be seen that the shear strain increment during unloading remains constant throughout the cyclic loading and changes its intensity only when different cyclic stress ratio is applied.

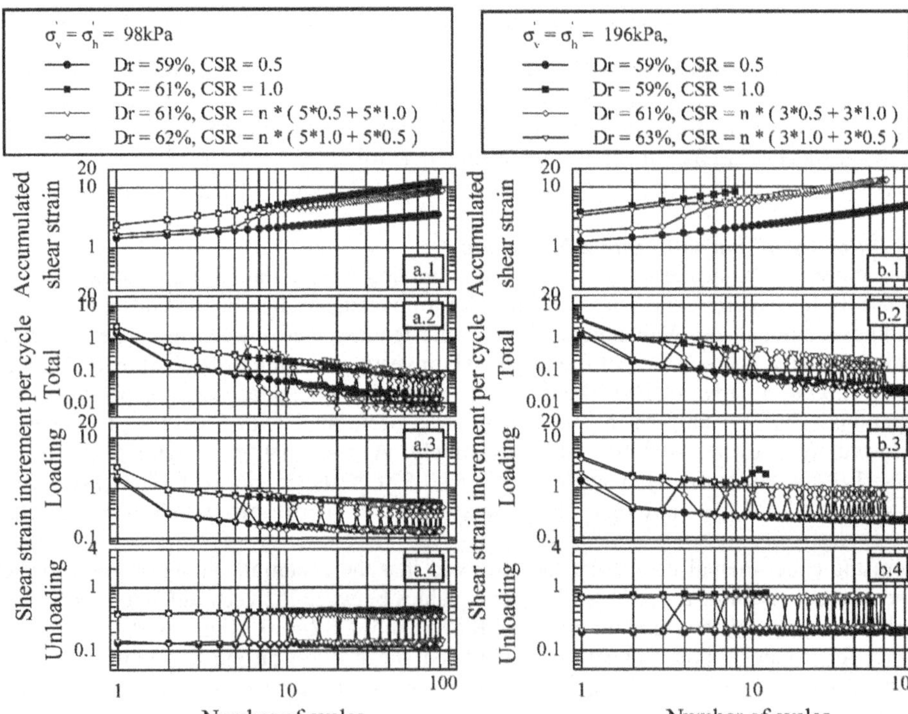

Fig. 6 Effect of non-uniform cyclic loading on torsional shear strain increment relationships, Dr≈60%. a) $\sigma_v'=\sigma_h'=98$kPa, b) $\sigma_v'=\sigma_h'=196$kPa.

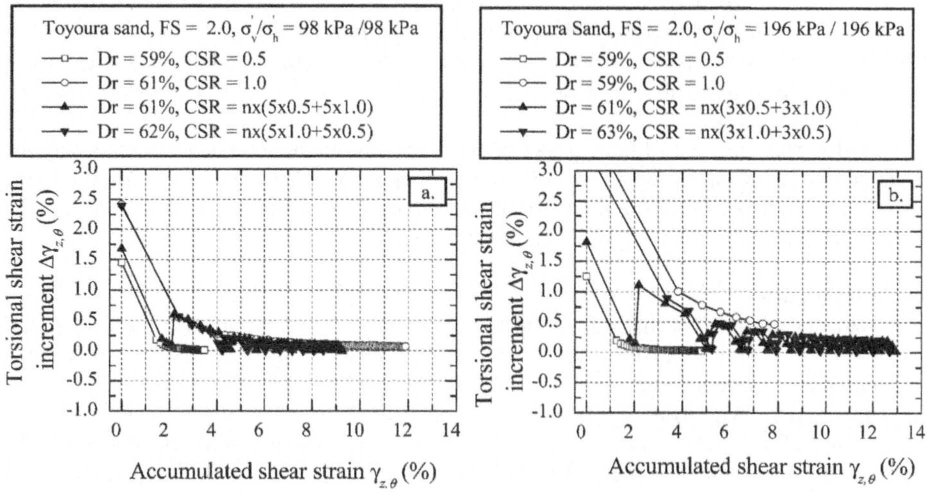

Fig. 7 Dependency of shear strain increment on accumulated shear strain, Dr=60%. a) Isotropic condition under 98kPa; b) Isotropic condition under 196kPa.

In order to investigate the finding in Figs. 6.a2 and 6.b2, that the difference in the shear strain increment is due to the accumulated shear strain, the shear strain increment per cycle was plotted against the accumulated shear strain (Fig. 7.a and 7.b). On the graphs the relationships for constant cyclic stress ratio CSR=0.5 and CSR=1.0 together with the cases with non-uniform cyclic loading were presented, for a case with isotropic condition under 98kPa and 196kPa. On the graph the relation between the shear strain increment per cycle and the accumulated shear strain for constant cycle stress ration CSR=0.5 and CSR=1.0 were considered as a reference. From the presented data in can be observed that when a cyclic loading was changed from CSR=0.5 to SCR=1.0 or vice versa, the shear strain increment is joining the reference line for a given cyclic stress ratio. Similar behavior was observed for isotropic condition under 196kPa. As mention before the intermediate points refers to a cyclic stress ratio grater than 0.5 and lower than 1.0 due to the change in cycle stress ratio.

3. CONCLUSIONS

Experimental results presented above could be summarized as follows:

1) The presence of initial shear stress in the soil element, achieved under drained conditions, increases the peak shear stress as well as the shear stress at the initiation of and after the softening under monotonic undrained loading (Fig. 2). The presence of initial shear stress in the soil can be considered as favorable by increasing the undrained shear strength. At the same time it can have a negative effect in the cases with high initial shear stress where sudden undrained conditions will lead to the softening in the soil and reducing the shear strength.

2) The relationship between the accumulated shear strain and the number of cycles showing a linear relationship when plotted in logarithmic manner.
3) The shear strain increment per cycle during unloading, have a constant value for a given cycle stress ratio, and does not depend on the accumulated shear strain or number of cycles.
4) Good correlation was found between the shear strain increment per cycle and the accumulated shear strain for a given cyclic stress ratio (Fig. 7.a and 7.b). With this relation a prediction of the shear strain increment for a given cyclic stress ratio can be predicted based on the accumulated shear strain.

REFERENCES

Arangelovski G. and Towhata, I. (2004): Accumulated deformation of sand with initial shear stress and effective stress state lying near failure conditions, *Soils and Foundations, 44* (6), 1-16.

Ghalandarzadeh, A., Orita, T., Towhata, I. and Yun, F. (1998): Shaking table tests on seismic deformation of gravity quay walls, *Soils and Foundations,* Special Issue on Geotechnical Aspects of the January 17 1995 Hyogoken-Nambu Earthquake, No.2, pp. 115-132.

Ishihara, K. and Okada, S. (1978): Effects of stress history on cyclic behavior of sand, *Soils and Foundations*, Vol. 18, No. 4, pp. 31-45.

Seed, H.B., Mori, K. and Chan, C,K. (1977): Influence of seismic history on liquefaction of sands, *Proc., ASCE*, Vol. 103, GT4, pp.257-270.

Tatsuoka, F., Sonoda, S, Hara, K., Fukushima, S. and Pradhan, T.B.S (1986): Failure and deformation of sand in torsional shear, *Soils and Foundations*, Vol.26, No.4, pp.79-97.

Towhata, I. and Ishihara, K. (1985): Undrained strength of sand undergoing cyclic rotation of principal stress axes, *Soils and Foundations*, Vol. 25, No. 2, pp. 135-147.

Towhata, I. and Ishihara, K. (1985): Shear work and pore water pressure in undrained shear, *Soils and Foundations*, Vol. 25, No. 3, pp. 73-84.

Vaid, Y.P. and Chern, J.C. (1983): Effect of static shear on resistance to liquefaction", *Soils and Foundations*, Vol. 23, No. 1, pp. 47-60.

Soil Stress-Strain Behavior: Measurement, Modeling and Analysis
Geotechnical Symposium in Roma, March 16 & 17, 2006

ANALYSIS FOR THE DEFORMATION OF THE DAMAGED EMBANKMENTS DURING THE 2004 NIIGATAKEN-CHUETSU EARTHQUAKE BY USING STRESS-STRAIN CURVES OF LIQUEFIED SANDS OR SOFTENED CLAYS

Susumu Yasuda[1),] Motohiro Inagaki[2)], Kazuyuki Nagao[2)],
Shin-ichi Yamada[3)] and Keisuke Ishikawa[3)]
[1)] Department of Civil and Environmental Engineering
Tokyo Denki University, Saitama, 3500394, JAPAN
e-mail: yasuda@g.dendai.ac.jp
[2)] Central Nippon Expressway Co., Ltd., JAPAN
[3)] Kiso-jiban Consultants Co., Ltd., JAPAN

ABSTRACT

Stress-strain curves of liquefied sands and softened clays were obtained by cyclic torsional shear tests. In the tests, 20 cycles of loadings were applied at first then a monotonic loading was applied. Shear modulus of the liquefied sands or softened clays, G_1, was evaluated based on the stress-strain curves during the monotonic loading. Then, a relationship among cyclic stress ratio to cause 7.5 % of double amplitude of shear strain, $R_L(\gamma_{DA}=7.5\ \%,\ N_L=20)$, severity of liquefaction, F_L and $G_1/\sigma_c{}'$ were derived. Based on the relationship, analysis for the deformation of damaged road embankment during the 2004 Niigataken-chuetsu earthquake, was conducted. Analyzed deformation was fairly coincided with the measured deformation.

1. INTRODUCTION

Loose sandy grounds liquefy during earthquakes and large settlement of embankments occurs due to the liquefaction. For example, the levee of Yodo River settled 2.7 m during the 1995 Kobe earthquake. On the contrary, soft clayey grounds do not liquefy during earthquakes. However, some settlement of embankments occurs due to softening of the clayey grounds. During the 2004 Niigatake-chuetsu earthquake, highway embankments on clayey grounds settled several ten cm in Ojya City.

In the seismic design of embankments, it is desired to evaluate not only stability but also settlement of the embankment. Yasuda et al. (1999) proposed a static analysis method named "ALID (Analysis for Liquefaction-Induced Deformation)" to estimate large deformation of liquefied sandy ground. In the method, stress-strain curves of liquefied soils are used. Yasuda et al. (1999) conducted cyclic torsional shear tests to obtain the stress-strain curves of sandy soils. Test results were applied to evaluate the settlement of river dikes, and showed the adaptability of the method for the estimation of the settlement of embankments (Yasuda et al., 2003, Yasuda, 2004). However, it was not clear whether this method can be applied for clayey soils or not. Then, cyclic torsional shear tests were carried out for silty, clayey and peaty soils, to derive a unified relationship (Yasuda et al., 2004).

Hoe I. Ling et al. (eds.), Soil Stress-Strain Behavior: Measurement, Modeling and Analysis, 663–672.

In this paper, test procedure to obtain stress-strain curves of liquefied sands or softened clays is introduced at first together with the derived unified relationship. Then, the unified relationship was applied to analyse the settlements of the damaged highway embankments during the Niigataken-chuetsu earthquake.

2. CYCLIC TORSIONAL TESTS TO OBTAIN STRESS-STRAIN CURVES OF LIQUEFIED SANDS OR SOFTEND CLAYS

Yasuda et al. (1999) conducted cyclic torsional tests to obtain stress-strain curves of liquefied sands. In the test, three reconstituted samples and a series of undisturbed samples were used. In addition, many undisturbed samples of sandy soils and clayey soils were taken at five sites (Yasuda et al., 2004). Tested samples are listed in Table 1.

In the tests, samples were trimmed to become hollow cylindrical specimens. Outer diameter, inner diameter and height of the specimen were 7 cm, 3 cm and 7 cm, respectively. Then the specimens were saturated and isotropically consolidated. The isotropic confining pressure, σ_c' was adjusted as in-situ effective overburden pressure. After the consolidation, 20 cycles of cyclic loading was applied to the specimens in undrained condition with a frequency of 0.1 cycles/sec., as shown in Fig.1. Then, under

Table 1 Summary of test results (Yasuda et al., 2004)

Site	Sample No.	F_C(%)	I_P	σ_C' (kPa)	R_{L20}	F_L=0.9		
						Type	G_1(kPa)	G_1/σ'_C
Tokusima	1-1	23.2	0	98	0.223	A	98.8	1.01
	1-2	77.7	3.41	128	0.238	A	180	1.41
	1-3	75.2	6.55	157	0.233	B	120	0.76
	1-4	92.4	7.75	177	0.258	A	170	0.96
	1-5	98.4	12.63	177	0.249	A	76.2	0.43
	8-1	24.1	0	69	0.279	A	55.0	0.80
	8-2	33.4	0	108	0.228	A	19.7	0.18
	8-3	84.7	4.11	137	0.233	A	150	1.09
	8-4	80.8	13.23	177	0.243	B	190	1.07
	4-1	87.1	12.6	49	0.325	A	129	2.63
	4-2	27.3	0	88	0.259	A	56.2	0.64
	5-1	97.0	11.28	49	0.285	A	76.2	1.56
	5-2	13.0	0	78	0.302	A	53.1	0.68
	5-3	63.0	3.07	128	0.308	B	120	0.82
Yasuda et al.(1999)	M-1	5.8		98.1	0.217	A	40.0	0.41
	M-2	5.8		98.1	0.228	A	70.0	0.71
	T-1	0		98.1	0.201	A	6.00	0.06
	T-2	0		98.1	0.238	A	11.0	0.11
	T-3	0		98.1	0.466	B	83.3	0.82
	AK-1	35		49.1	0.285	A	100	2.04
	AK-2	23		114.8	0.183	A	40.0	0.35
	MK-1	47		53.0	0.28	A	52.6	0.99
	MK-2	22		99.1	0.209	A	240	2.42
	MS-1	27		54.9	0.323	A	180	3.28
	MS-2	50		75.5	0.383	A	110	1.46
	MS-3	92		89.3	0.319	A	500	5.60
	AR-1	32		49.1	0.22	A	140	2.85
	AR-2	24		65.7	0.219	A	142	2.16
	AR-3	35		104.0	0.197	B	40.0	0.38
Yasuda et al.(2003)	TK-1	7		9.8	0.208	B	0.91	0.09
	TK-2	7		9.8	0.323	B	1.50	0.15
	TY-1	0		9.8	0.225	B	12.2	1.24
	TY-2	0		9.8	0.393	B	23.0	2.35
Iwamizawa	1	100.0	302.9	186	0.463	B	3160	16.9
	2	100.0	312	29.4	0.623	B	708	24.0
Kanda	1	100.0	86.6	186	0.435	B	2800	15.0
	2	100.0	94	118	0.354	B	1510	12.8
Takeo	1	100.0	50	137	0.346	B	2130	15.5
	2	98.0	42.8	29.4	0.461	B	489	1.55
Ohsawago	1	100.0	112.9	186	0.423	B	3200	17.2
	2	100.0	206.4	19.6	0.610	B	332	16.9

the same isotropic confining pressure, a monotonic loading was applied in undrained condition with a speed of 10 % of shear strain/minute.

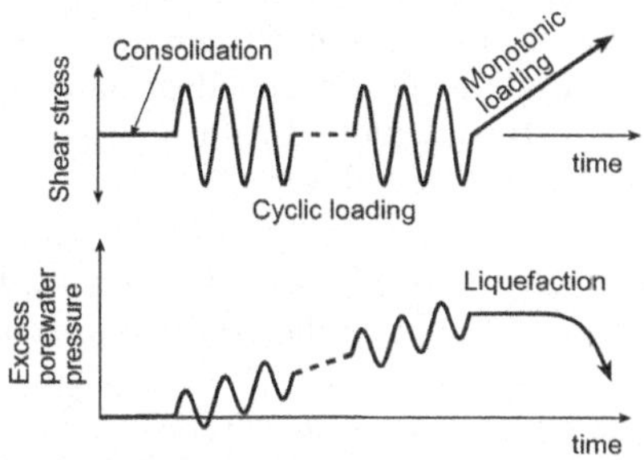

Fig.1. Procedure of cyclic and monotonic loading

Time histories of shear stress, shear strain and pore water pressure during the monotonic loading were measured. About 4 to 8 specimens were used in one sample. Different amplitude of cyclic loading was applied to each specimen to control safety factor against liquefaction, F_L, which implies severity of liquefaction or failure, mentioned later. In addition, a static test by applying a monotonic loading only was carried out. This test is called as "static" in this paper.

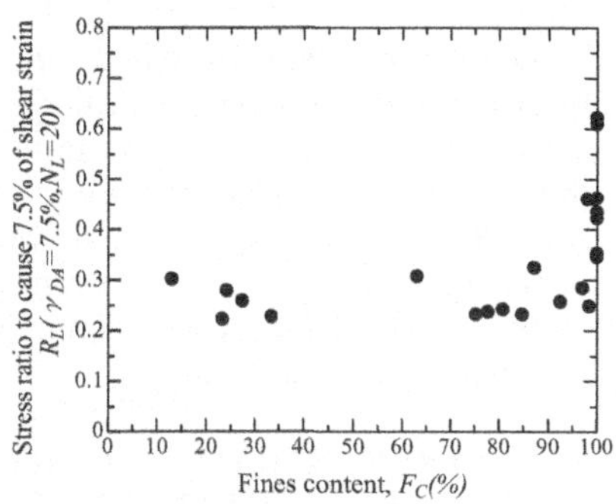

Fig.2. Relationship between R_L and F_c

As mentioned above, different amplitude of cyclic loadings were applied to each specimen. Then, relationships between cyclic stress ratio, τ_d/σ_c' and double amplitude of shear strain at 20th cycle, γ_{DA} $(N=20)$ were plotted. And, the stress ratio to cause 7.5 % of shear strain by 20 cycles, $R_L(\gamma_{DA}=7.5\ \%,\ N_L=20)$ was estimated. This stress ratio is same as the stress ratio to cause liquefaction, $R_L(\varepsilon_{DA}=5\ \%,\ N_L=20)$ in cyclic triaxial tests for sandy soils. Therefore, this stress ratio means liquefaction strength in sandy soils. In clayey soils, this stress ratio implies the stress ratio to cause a kind of failure. So, it must be said as undrained cyclic shear strength. The $R_L(\gamma_{DA}=7.5\ \%,\ N_L=20)$ for each sample are shown in Table 1.

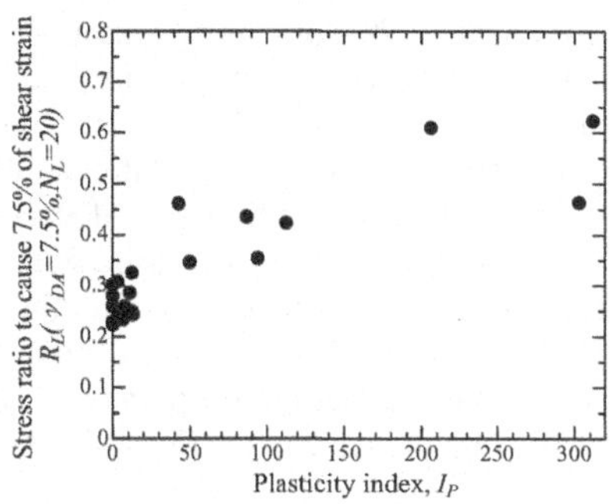

Fig.3. Relationship between R_L and I_P

Figure 2 shows relationships between $R_L(\gamma_{DA}=7.5\ \%,\ N_L=20)$ and fines content, F_c. The $R_L(\gamma_{DA}=7.5\ \%,\ N_L=20)$ increased with F_c. Figure 3 shows relationships between $R_L(\gamma_{DA}=7.5\ \%,\ N_L=20)$ and Plasticity Index, *Ip*. The $R_L(\gamma_{DA}=7.5\ \%,\ N_L=20)$ increased with I_P also. These two relations mean that the $R_L(\gamma_{DA}=7.5\ \%,\ N_L=20)$ of clayey soils is greater than the $R_L(\gamma_{DA}=7.5\ \%,\ N_L=20)$ of sandy soils. However, if a very strong earthquake hit a clayey ground, large stress ratio such as 0.5 must be induces in the ground. This means that some failure may occur even in clayey grounds, as same as liquefaction in sandy grounds.

To clarify what kind of failure occurs in clayey grounds, excess pore water pressure ratios at 20th cycle of just liquefied or failed specimens (F_L=1.0) were plotted with F_c in Fig.4. Here F_L was defined as follows:

$$F_L=R_L(\gamma_{DA}=7.5\ \%,\ N_L=20)/(\tau_d/\sigma_c') \quad (1)$$

where, τ_d/σ_c': applied shear stress ratio.

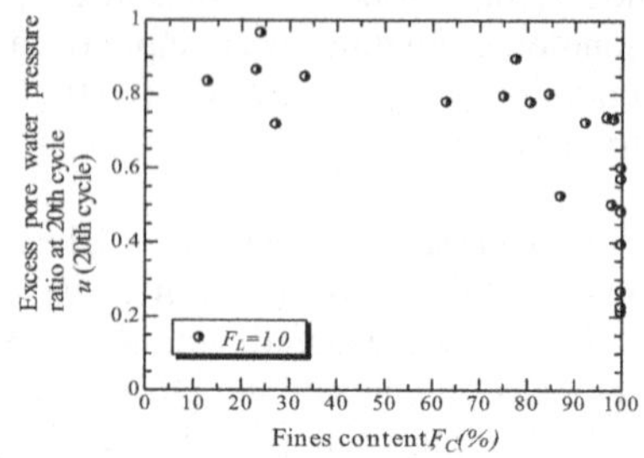

Fig.4 Relationship between F_c and excess pore water pressure ratio of liquefied or failed specimen

The excess pore water pressure ratio decreased with F_c as shown in Fig.4. In Fig.5, the excess pore water pressures ratio was plotted with Plasticity Index, I_p. It is clear that the pore water pressure ratio decreased with I_p, up to 0.2. This means that effective stress in clayey soil does not decease to zero due to cyclic loading, even though large shear strain occurs. Therefore, some resistance or shear modulus would remain in clayey ground even if very strong earthquake hit the ground. This phenomenon must be called not "failure" but "softening".

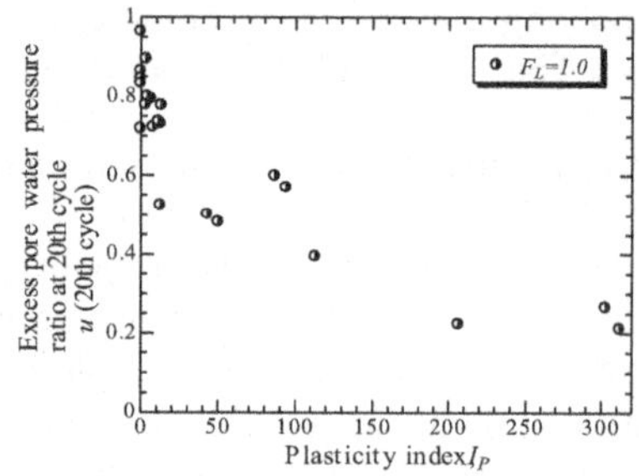

Fig.5 Relationship between I_P and excess pore water pressure ratio of liquefied or failed specimen

3. STRESS-STRAIN CURVES OF LIQUEFIED OR SOFTENED SOILS

Typical time histories of shear stress and excess pore water pressure during monotonic loading for a silty sand, a sandy silt and a peat are shown in Figs.6 to 8. Scales of axes in (c) are enlarged ones of (a). In the silty sand, stress-strain curve under static loading was normal. Namely shape of the curve was convex. However, the shapes of stress-strain curves of liquefied specimen (F_L <1.0) were different. Shear strain increased with very low shear stress up to large strain. Then, after a resistance transformation point, the shear stress increased comparatively rapidly with shear strain, following rapid decrease of pore water pressure. The shear strain up to the resistance transformation point increased with the decrease of F_L. This behavior is similar as the one in clean sands (Yasuda et al., 1998, 1999).

In the peat, shear stress at certain shear strain slightly decreased with F_L as shown in Fig.8. However, stress-strain curves of the peat were normal even F_L is less than 1.0. And, stress strain curves of softened specimens were not so different from the curve of the specimen tested under static loading only. This behavior is different from the behavior of liquefied sands. Stress-strain curves for the sandy silt were intermediate between those for the silty sand and the peat.

Then the authors classified the shape of the stress-strain curve into two types as shown in Fig.9. Type A corresponds to the stress–strain curve for liquefied sandy soils. Type B is the stress-strain curves for a softened clays or peats. In type A, stress-strain curves before

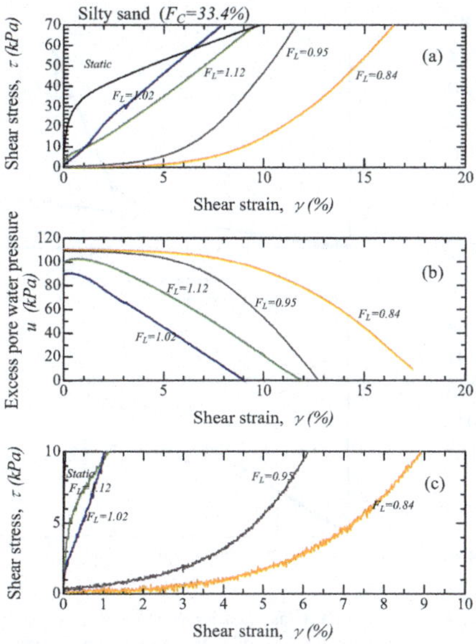

Fig.6. Time histories of shear stress and pore water pressure for a silty sand

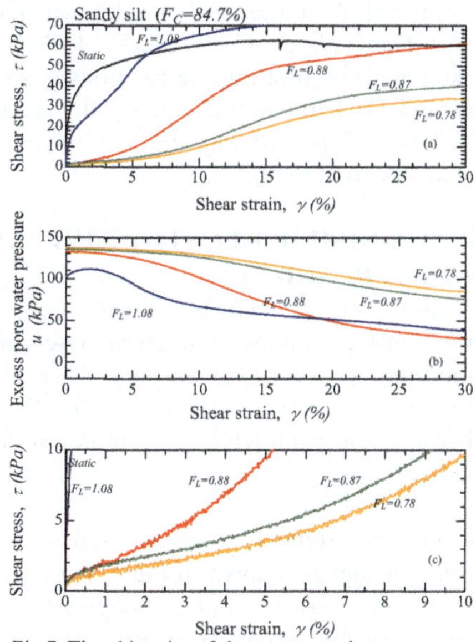

Fig.7. Time histories of shear stress and pore water pressure for a sandy silt

and after the resistance transformation point can be presented approximately by a bilinear model with G_1, G_2 and γ_L as same as for clean sands (Yasuda et al., 1999). The G_1 means shear modulus of a liquefied soil. The γ_L is influenced by grain size, density, F_L and other factors. When F_L=0.9, the γ_L was about 5 to 20 % for loose sands. In type B, it is necessary to select reference strain to define shear modulus. In clayey ground, it seems that large shear strain does not induce during earthquakes because some amount of resistance remains though cyclic loadings are applied, mentioned before. Therefore the authors defined shear modulus of softened soil, G_1 as the secant modulus at 1% of shear strain as shown in Fig.9 (b).

Then the liquefied or softened shear moduli, G_1, were estimated for all stress-strain curves. In sandy silts, two values of G_1 defined by Type A and Type B were

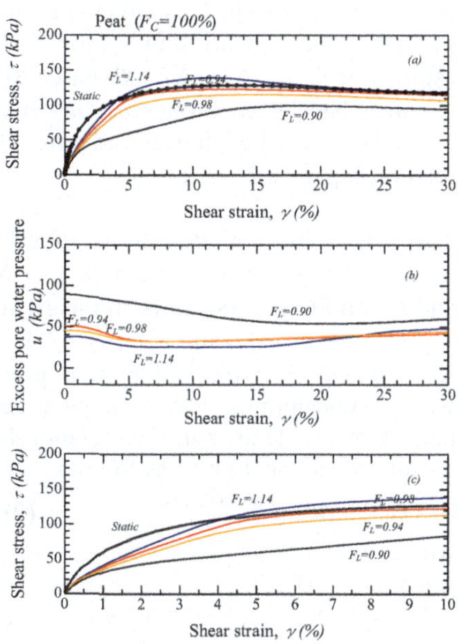

Fig.8. Time histories of shear stress and pore water pressure for a peat

estimated. And, the smaller value was judged as G_1. Then G_1 at F_L=0.8, 0.9, 1.0, 1.1 were estimated, by plotting the relationships between G_1 and F_L. Estimated G_1 for F_L=0.9 are listed in Table 1. Types of the stress-strain curves are also shown in the table.

4. A UNIFIED RELATIONSHIP AMONG G_1/σ_c', R_L AND F_L

Yasuda et al. (1998, 1999) proposed a relationship among reduction rate of shear modulus G_1/G_0, F_L, and F_c. However, this relationship cannot be extended to clayey soils because the parameter of F_c is not suitable for clayey soils.

Then the authors tried to find out a new relationship. As shown in Fig.5, $R_L(\gamma_{DA}=7.5\ \%, N_L=20)$ increased with I_P. Though a figure is not shown in this paper, G_1 increased with I_P also. Then, it was judged that the parameter $R_L(\gamma_{DA}=7.5\ \%, N_L=20)$ must be introduced in the new relationship. This value is easy to evaluate because many simple formulae to estimate $R_L(\gamma_{DA}=5\ \%, N_L=20)$ based SPT N-value, have been proposed. One more parameter to be introduced in the relationship must be σ_c'.

Then, the authors plotted the relationship between G_1/σ_c', $R_L(\gamma_{DA}=7.5\ \%, N_L=20)$ and F_L, on Fig.10. As shown in this figure, G_1/σ_c' increased with $R_L(\gamma_{DA}=7.5\ \%, N_L=20)$ and F_L. And, all plotted points were concentrated in comparatively narrow bands. Thus, a unified relationship for all soil can be derived as follows:

$$G_1/\sigma_c' = ae^{(-\exp(-b(R_L-c)))} \qquad (2)$$

where,

$a = 23.6F_L + 0.98$,

$b = 9.32F_L^3 - 10.8F_L^2 + 13.27F_L - 0.806$, $\quad c = -1.40F_L^3 + 3.87F_L^2 - 4.14F_L + 1.95$

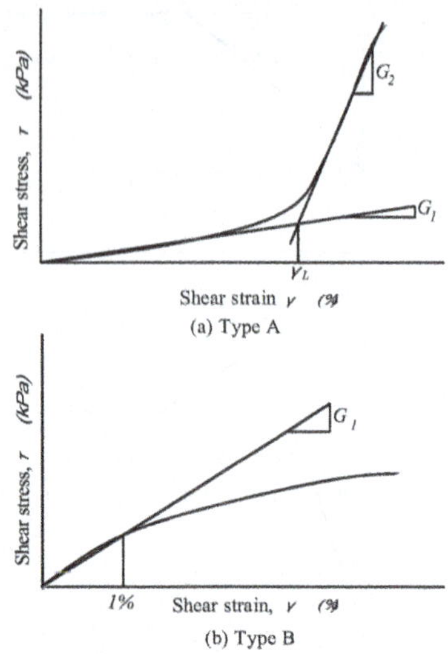

Fig.9. Classification of stress-strain curves of liquefied sandy soils and softened clayey soils

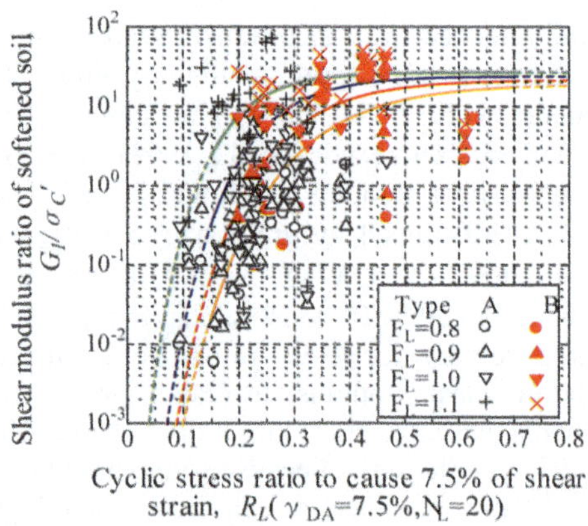

Fig.10 Relationship between G_1/σ_c' and R_L for F_L=0.8, 0.9, 1.0 and 1.1 (Yasuda et al., 2004)

5. DAMAGE OF HIGHWAY EMBANKMENTS DURING THE 2004 NIIGATAKEN-CHUETSU EARTHQUAKE

On October 23, 2004, the Niigataken-chuetsu earthquake, of Magnitude 6.8, occurred in

Japan. The maximum surface acceleration recorded was 1722 gals. Many railways including Shinkansen, roads, houses, pipelines and other structures were severely damaged. Moreover, huge number of landslides, more than 1700, occurred and hit many towns and villages. The slid soils dammed up rivers and made natural dams.

In Kan-Etsu Expressway, highway embankments between Horinouchi IC and Nagaoka IC were severely damaged. The highway between Horinouchi IC and Kagaguchi IC was constructed by cut and fill method on the slope of hill. Sliding of filled embankments occurred at several sites. On the contrary, the highway between Kawaguchi IC and Nagaoka IC was constructed by filling soils on lowland plain. Surface soil of the plane is gravel, sand or clay. Heights of the embankments are 5 to 9 m. Many underpasses cross the road embankments. The underpasses were constructed by concrete culvert boxes. Due to the Niigataken-chuetsu earthquake, the highway embankments settled several ten cm. Both side of toes of the road embankment spread in lateral direction. Culvert boxes settled and separated at joints, and embankment soils fell down through the opened joints. Due to the spread, the grounds adjacent the settled embankments moved laterally.

6. DETAILED SOIL INVESTIGATION AND ANALYSES FOR SETTLEMENT

To demonstrate the mechanism of the settlement of the embankments between Kawagchi IC and Nagaoka IC, the authors selected two sites; Ojiya No.2 and Kawaguchi No.22, for detailed soil investigation. Figure 11 shows locations of two sites, together with K-net Ojiya Site, where accelerograph is installed. The maximum surface acceleration recorded at K-net Ojiya Site was 1314 cm/s^2 in EW direction. Surface soil conditions at Ojiya No.2 and Kawaguchi No.22 sites, investigated after the earthquake, are shown in Figs.12 and 13. Soils for the embankments at two sites are clayey soils with 70 % of fines. Heights of the embankments at two sites are 5.3 to 5.6 m and 5.6 to 6.8 m, respectively. A thin soft silt layer with 2 m in thickness is deposited under the embankment at Ojiya No.2. Then, silty sand, silt, sandy silt and silt layers, with 10 to 20 of SPT *N*-values, underlayed to the depth of 24 m. At Kawaguchi No.22, thick soft silty layers, with about 5 of SPT *N*-values, are deposited to the depth of 16 m. Fig.14 shows a schematic diagram of the deformation of damaged road embankments. Surface of road settled several ten cm, both side of toes of the road embankments spread in lateral direction, and culvert boxes settled and separated at joints. The settlement of surface of road at Ojiya No.2 and Kawaguchi No.22 were 65 cm and 70 cm, respectively. Opening of the joints of culvert boxes at Ojiya No.2 and Kawaguchi No.22 were 75 cm and 30 cm, respectively.

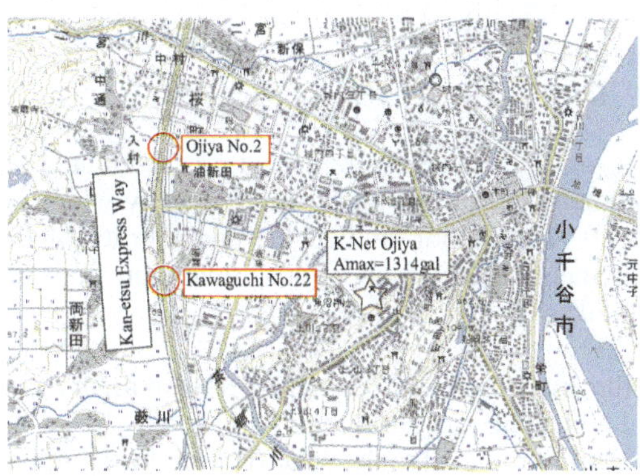

Fig.11 Location of detail investigation sites

Based on the soil conditions, analyses for deformation due to the earthquake were conducted by using analytical code "ALID (Yasuda et. al., 1999)". In this code, finite element method is applied in the following steps: i) In the first step, the deformation of the ground and the embankment before earthquake is calculated by using the stress-strain relationships of not-liquefied or softened soils. ii) The deformation of the ground and the embankment due to liquefaction or softening of soils is calculated in the second step, by using Fig.10. iii) Finally, deformation of the ground due to the dissipation of excess pore pressure is calculated based on the simple relationships among volumetric strain, F_L and relative density which was proposed by Ishihara & Yoshimine (1992). The author studied the adaptability of the ALID to various structures, such as settlement of raft foundations, floatation of buried structures and flow of the ground behind sea walls.

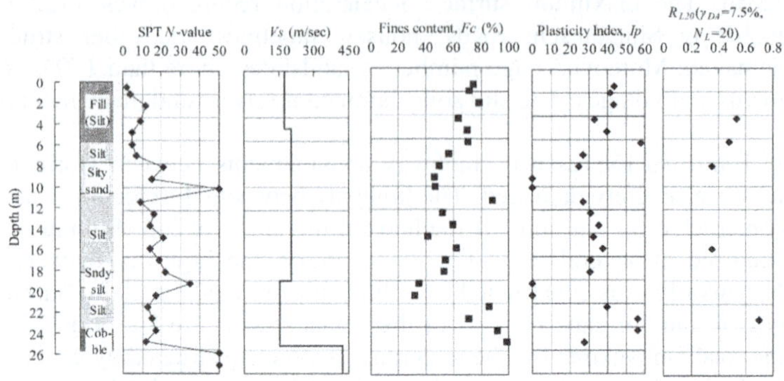

Fig.12 Soil profile and tests results at Ojiya No.2

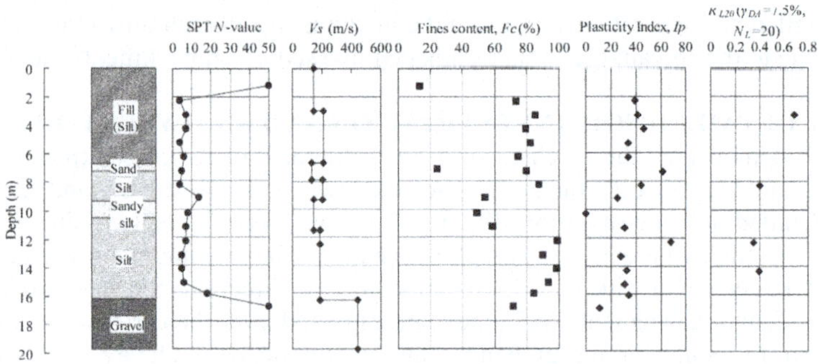

Fig.13 Soil profile and tests results at Kawaguchi No.22

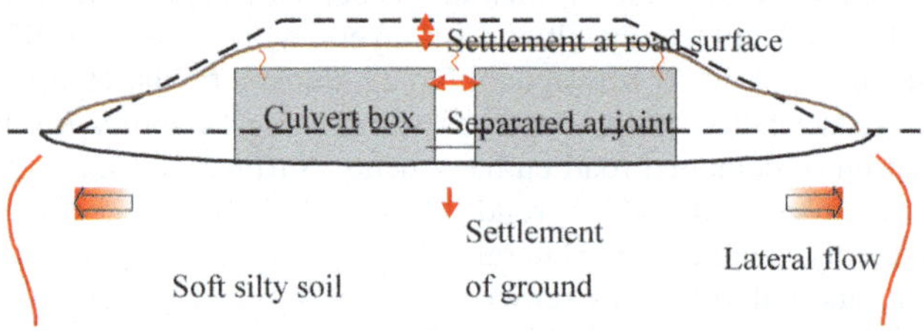

Fig.14 Schematic diagram of deformation of damaged road embankments

In the analyses for the damaged embankments, cyclic shear stress induced in the grounds and embankments during the Niigataken-chuetu earthquake were estimated by the simple method introduced in the specification for Highway Bridge (2002), based on the recorded maximum surface acceleration. As the grounds at two sites are silty soil, stress ratio to cause failure, R_L was estimated from unconfined compressive strength q_u by using the relationship between R_L and $(q_u/2)/\sigma_v'$ as show in Fig.15, which was derived by the authors. Then the distribution of F_L values in the ground was estimated for two sites.

Reduction rate of shear modulus due to shaking for the embankments and unsaturated layers, were assumed as 1/40 and 1/10, respectively base on the previous study (Yasuda et al., 2003).

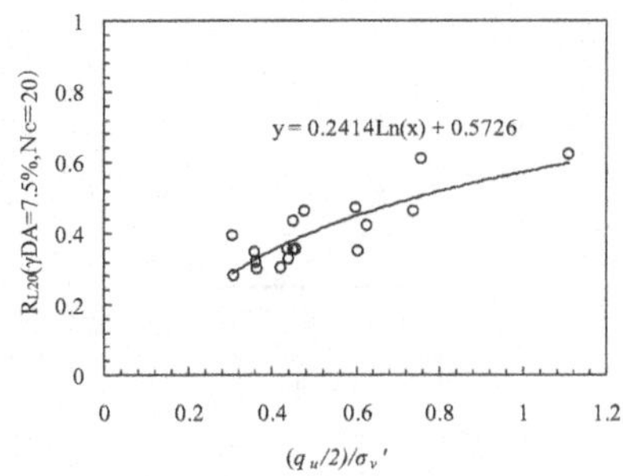

Fig.15 Relationship between R_L and $(q_u/2)/\sigma_v'$

Deformations of embankments and grounds before and after earthquake are shown in Figs 16(a) and 16(b). Grounds under the embankments were spread laterally, and embankments were stretched and settled, as same as the actual deformation shown in Fig.14. Analyzed settlement at the center of surface of roads at Ojiya No.2 and Kawaguchi No.22, were 26.6 cm and 52.1 cm, respectively. Figures 17 (a) to 17(c) compare analyzed settlements and horizontal displacements at both toes, with the measured ones, respectively. As shown in these figures, analyzed settlements and

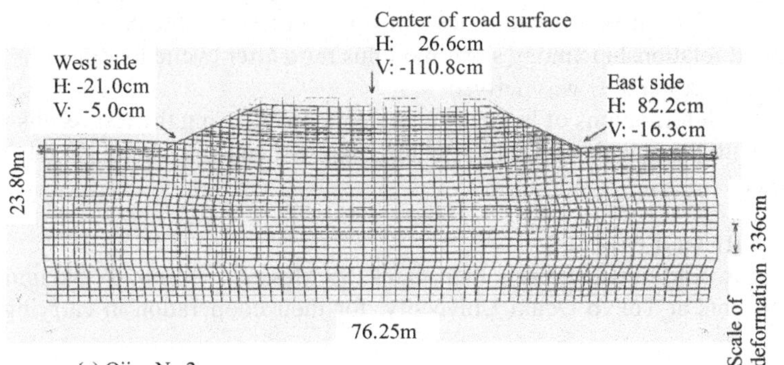

(a) Ojiya No.2

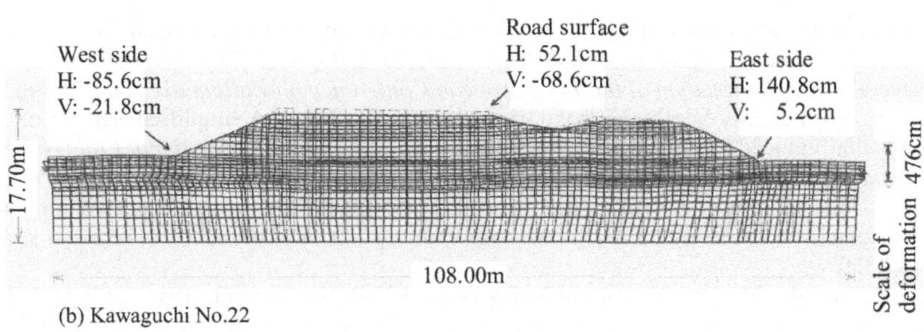

(b) Kawaguchi No.22

Fig.16 Analyzed deformation of embankments and grounds (only central zones are show in these figures)

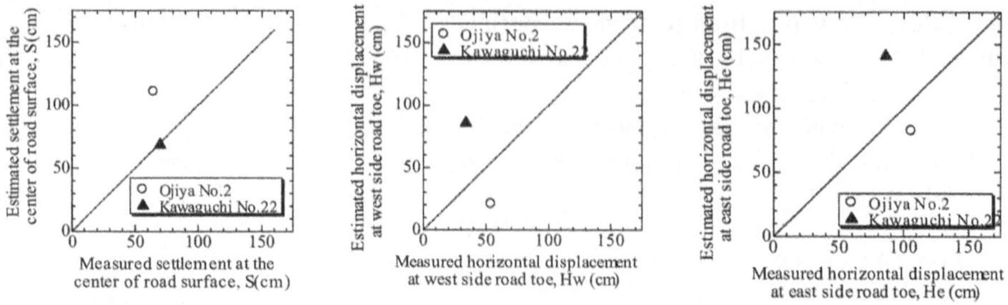

(a) Surface of road (b) West side toe (c) East side toe

Fig.17 Comparison of analyzed and measured displacements

horizontal displacements were fairly coincided with the measured values.

CONCLUSIONS

Cyclic torsional tests were carried out to demonstrate the stress-strain curves of liquefied sands or softened clays. Analyses for deformation of embankments were conducted by using the test results. The following conclusions were derived through this study:

(1) Cyclic stress ratio to cause 7. 5 % of shear strain by 20 cycles of cyclic loading, $R_L(\gamma_{DA}=7.5\ \%,\ N_L=20)$, increases with fines content and plasticity index.

(2) A unified relationship among shear modulus ratio after cyclic loading, G_1/σ_c', $R_L(\gamma_{DA}=7.5\ \%,\ N_L=20)$ and F_L was derived.

(3) Analyzed deformations of highway embankments by using the test results were fairly coincided with the actual deformation of the embankments during the Niigataken-chuetsu earthquake.

ACKNOWREDGEMENTS

The authors would like to express their thanks to Messrs T. Ideno, Y. Sakurai, S. Nakao, former students at Tokyo Denki University, for their cooperation in carrying laboratory tests.

REFERENCES

The Japan Road Association (2002): Specification for Highway Bridges. (in Japanese)

Yasuda, S., Terauchi, T., Morimoto, H., Erken, A. and Yoshida, N. (1998): Post liquefaction behavior of several sands, *Proceeding of the 11th European Conference on Earthquake Engineering.*

Yasuda, S., Yoshida, N., Adachi, K., Kiku, H. and Gose, S. (1999): A simplified analysis of liquefaction-induced residual deformation, *Proceedings of the 2nd International Conference on Earthquake Geotechnical Engineering*, pp.555-560.

Yasuda, S., Ideno, T., Sakurai, Y., Yosida, N. and Kiku, H. (2003): Analyses of liquefaction-induced settlement of river levees by ALID, *Proceedings of the 12th Asian Regional Conference on SMGE*, pp.347-350.

Yasuda, S., Inagaki, M., Yamada, S. and Ishikawa, K.(2004): Stress-strain curves of liquefied sands and softened clays, *Proceedings of the International Symposium on Engineering Practice and Performance of Soft Deposits*, pp.337-342.

Yasuda, S. (2004): Evaluation of liquefaction-induced deformation of structures, Chapter 6, *Recent Advances in Earthquake Geotechnical Engineering and Microzonation*, Kluwer Academic Publishers, pp.199-230, 2004.

Soil Stress-Strain Behavior: Measurement, Modeling and Analysis
Geotechnical Symposium in Roma, March 16 & 17, 2006

NUMERICAL SIMULATION OF SEIMIC BEHAVIOR OF PIPELINE IN LIQUEFIABLE SOIL

Degao Zou, Xianjing Kong, Bin Xu
Department of Civil Engineering
Dalian University of Technology, Dalian 116023, PRC
e-mail: zoudegao@dlut.edu.cn

ABSTRACT

This study focused on the behavior of a burial pipe with special reference to its stability against floatation subject to soil liquefaction. The excess pore water pressure response behaviors of soil foundations, and the effectiveness of different types of drainage or reinforcement measures were investigated using Finite Element Method (FEM). FEM numerical model is a coupled stress-flow finite element procedure, based on u-p formulation of dynamic Biot's equations (Zienkiewicz, 1982). The hyperbolic stress and strain relationship was used in the numerical model, which takes into account the stiffness and strength degradation. Pore pressure generation due to earthquake loading was calculated via the pore pressure model (Seed et al, 1979). Performance of the numerical models was studied by simulating a series of shake table tests. Excess pore pressures predicted by numerical models were compared with the pore pressure transducer records during experiments. Also, the effectiveness of different drainage measures against uplifting of pipelines was compared. It was demonstrated that the models were able to provide results in agreement with experiments.

1. INTRODUCTION

Pipelines are used in daily utilities, such as communications, natural gas, water supply and sewage systems. They are also used in the agricultural and industrial instruments, for irrigation, drainage and transport of petroleum. Pipelines are usually constructed by the open-cut method, in which excavation is backfilled with sandy soil. This backfill soil is potentially liquefiable if it is loose and saturated below the ground water level. Actually, pipes suffered major uplift Niigata Earthquake and Alaska Earthquake (Hall and O'Yourke, 1991). Increased attention has been paid to pipeline damages following recent major earthquakes, such as the 1993 Nansei-Oki Earthquake (Mohri et al. 1995), the 1993 Kushiro-Oki earthquake and the 1994 Hokkaido-Toho-Oki earthquake (Koseki et al. 2000).

Based on these case histories, the significance of liquefaction on the uplift stability of pipelines has been recognized. In order to evaluate the uplift mechanism and also develop appropriate mitigation technique, a series of shaking table tests (Koseki et al., 1997; Mohri et al., 1999; Kong et al., 2002) and centrifugal shaking table tests (Ling et al., 2003; Sun, 2001) have been conducted. Shake table tests contributed significantly toward understanding the behavior of pipeline buried in liquefiable soil, but they are more expensive to perform compared to numerical analysis. The development of robust Finite Element Method procedures improved the analysis of liquefaction problem (Zienkiewicz et al., 1993; Zienkiewicz et al., 1999).

Hoe I. Ling et al. (eds.), Soil Stress-Strain Behavior: Measurement, Modeling and Analysis, 673–682.

Very little works have been accomplished in liquefaction analysis of burial pipeline in the Liquefiable soil. Wang et al. (1990) adopted a nonlinear elastic approach to analyze pipeline flotation. More works should be conducted to evaluate the uplift mechanism of pipelines. Thus, in this paper, FEM were performed to investigate the stability of pipelines in a liquefiable soil deposit.

2. FEM MODEL

Governing equation. Saturated soil behavior is governed by following u-p equation (Zienkiewicz et al. 1999)

$$\mathbf{M}\ddot{\overline{\mathbf{u}}} + \int_{\Omega}\mathbf{B}^{\mathrm{T}}\sigma' \mathrm{d}\Omega - \mathbf{Q}\overline{\mathbf{p}} - \mathbf{f}^{(1)} = \mathbf{0} \tag{1}$$

$$\mathbf{Q}^{T}\dot{\overline{\mathbf{u}}} + \mathbf{S}\dot{\overline{\mathbf{p}}} + \mathbf{H}\overline{\mathbf{p}} - \mathbf{f}^{(2)} = 0 \tag{2}$$

where $\mathbf{M}$ is mass matrix; σ' is effective stress vector; $\mathbf{Q}$ is coupling matrix; $\mathbf{S}$ is compressibility matrix; $\mathbf{H}$ is penetrability matrix; $\mathbf{f}^{(1)}$ and $\mathbf{f}^{(2)}$ are the applied forces in the solid and fluid phases, respectively.

The general procedures of the Finite Element discretization of above equations were described in detail by Zienkiewicz et al. (1999).

System damping is considered by adding a matrix of the form $\mathbf{C}\dot{\overline{\mathbf{u}}}$ to the dynamic equation. That is, equation (1) is rewritten as

$$\mathbf{M}\ddot{\overline{\mathbf{u}}} + \mathbf{C}\dot{\overline{\mathbf{u}}} + \int_{\Omega}\mathbf{B}^{\mathrm{T}}\sigma' \mathrm{d}\Omega - \mathbf{Q}\overline{\mathbf{p}} - \mathbf{f}^{(1)} = \mathbf{0} \tag{3}$$

Rayleigh damping (Clough and Penzien, 1993) is usually used in dynamic analysis.

$$\mathbf{C} = \alpha\mathbf{M} + \beta\mathbf{K} \tag{4}$$

where $\mathbf{K}$ is stiffness matrix. α and β are coefficient written as (Idriss, et al.)

$$\beta = \lambda/\omega \tag{5}$$

$$\alpha = \lambda\omega \tag{6}$$

where λ is damping ratio; ω is the first frequency of foundation.

Time integration. The time integration was conducted using the Generalized Newmark GNpj scheme (Newmark, 1959; Katona and Zienkiewicz, 1985). Using GN22 for the displacement parameters $\overline{\mathbf{u}}$ and GN11 for the pore pressure parameter $\overline{\mathbf{p}}$, Equation (3) and (2) are rewritten as

$$\mathbf{M}\ddot{\overline{\mathbf{u}}}_{n+!} + \mathbf{C}\dot{\overline{\mathbf{u}}}_{n+1} + \int_{\Omega}\mathbf{B}^{\mathrm{T}}\sigma'_{n+1} \mathrm{d}\Omega - \mathbf{Q}\overline{\mathbf{p}}_{n+1} - \mathbf{f}_{n+1}{}^{(1)} = \mathbf{0} \tag{7}$$

$$\mathbf{Q}^{T}\dot{\overline{\mathbf{u}}}_{n+1} + \mathbf{S}\dot{\overline{\mathbf{p}}}_{n+1} + \mathbf{H}\overline{\mathbf{p}}_{n+1} - \mathbf{f}_{n+1}{}^{(2)} = 0 \tag{8}$$

The quantities at time t_{n+1} are expressed as

$$\ddot{\mathbf{u}}_{n+1} = \ddot{\mathbf{u}}_n + \Delta\ddot{\mathbf{u}}_n \tag{9a}$$

$$\dot{\mathbf{u}}_{n+1} = \dot{\mathbf{u}}_n + \ddot{\mathbf{u}}_n \Delta t + \beta_1 \Delta\ddot{\mathbf{u}}_n \Delta t \tag{9b}$$

$$\overline{\mathbf{u}}_{n+1} = \overline{\mathbf{u}}_n + \dot{\mathbf{u}}_n \Delta t + \ddot{\mathbf{u}}_n \Delta t^2/2 + \beta_2 \Delta\ddot{\mathbf{u}}_n \Delta t^2/2 \tag{9c}$$

$$\dot{\overline{\mathbf{p}}}_{n+1} = \dot{\overline{\mathbf{p}}}_n + \Delta\dot{\overline{\mathbf{p}}}_n \tag{10a}$$

$$\overline{\mathbf{p}}_{n+1} = \overline{\mathbf{p}}_n + \dot{\overline{\mathbf{p}}}_n \Delta t + \overline{\beta}_1 \Delta\dot{\overline{\mathbf{p}}}_n \Delta t \tag{10b}$$

The parameter β_1, β_2 and $\overline{\beta}_1$ are usually chosen in the range 0.0~1.0. For unconditionally stable of the recurrence scheme we require that $\beta_2 \geq \beta_1 \geq 1/2$ and $\overline{\beta}_1 \geq 1/2$. In this study, $\beta_1 = 0.6$, $\beta_2 = 0.605$, $\overline{\beta}_1 = 0.6$.

Soil constitutive model. Soil behavior was modeled using a hyperbolic stress-strain relationship as shown in Figure 1.

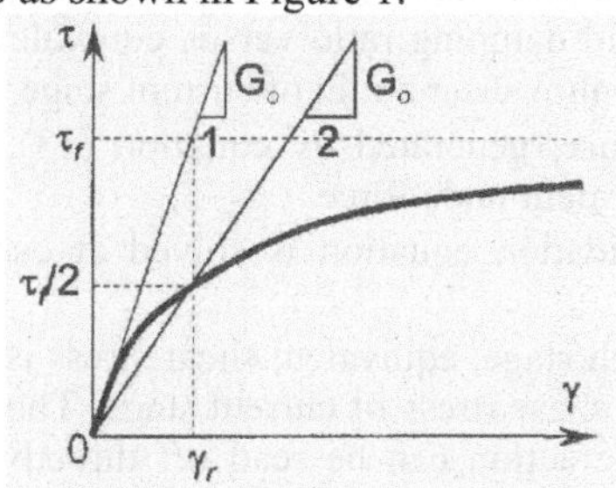

Figure 1. Hyperbolic model

$$\tau = \frac{G_0 \gamma}{1 + G_0 \gamma / \tau_f} \tag{11}$$

where τ_f is maximum shear strength. G_0 is the initial maximum tangent modulus is expressed as.

$$G_0 = K \cdot P_a \cdot (\sigma_0 / P_a)^n \tag{12}$$

where K is modulus coefficient, P_a is the engineering atmospheric pressure, σ_0 is the mean effective stress, n is modulus exponent.
G/G_0 and the damping ratio can be read off directly from the experimental curve of G/G_0 and damping ratio versus equivalent shear strain.

Pore pressure generation The rate of pore pressure generation during earthquake shaking was calculated using the following equation proposed by Seed et al. (1976).

$$p_g = \sigma_v'/2 + \sigma_v' \arcsin[2(N/N_l)^{1/\alpha} - 1]/\pi \tag{13}$$

where p_g is the excess pore pressure generated due to earthquake loading, α is a coefficient depends the test condition and soil properties, σ'_v is the initial effective vertical stress, N_l is the number of cycles required for liquefaction and N/N_l is the rate at which the cyclic shear stress is applied to the soil.

The constitutive laws can be written as (Zienkiewicz et al., 1978)

$$d\boldsymbol{\sigma}' = \mathbf{D}_T(d\boldsymbol{\varepsilon} - \mathbf{m}d\varepsilon_v^0/3) = \mathbf{D}_T d\boldsymbol{\varepsilon} + \mathbf{m}dp_g \tag{14}$$

where $d\varepsilon_v^0$ is defined as the autogenous volumetric strain increment to reflect the compaction of grain configuration due to cyclic loading, $\mathbf{D}_T$ is the drained tangential constitutive matrix, $\mathbf{m}$ is a vector written as $\{1\ 1\ 0\}^T$ for two dimension problem. Therefore, equation (7) is rewritten as.

$$\mathbf{M}\ddot{\bar{\mathbf{u}}}_{n+1} + \mathbf{C}\dot{\bar{\mathbf{u}}}_{n+1} + \int_\Omega \mathbf{B}^T\boldsymbol{\sigma}'_{n+1}d\Omega + \int_\Omega \mathbf{B}^T\mathbf{p}^g_{n+1}d\Omega - \mathbf{Q}\bar{\mathbf{p}}_{n+1} - \mathbf{f}_{n+1}^{(1)} = \mathbf{0} \tag{15}$$

Calculation steps. The calculation steps involved in the analysis can be summarized as follows:

The whole earthquake wave history is divided into a few stages of time, in each stage:

(1) G/G_0 and the damping ratio are read off directly from the experimental curve of G/G_0 and damping ratio versus equivalent shear strain, which is 65% of the maximum shear strain of current stage.

(2) The pore pressure, generated by equation (13), is added into dynamic equation as equivalent node force.

(3) Dynamic consolidation equation is solved at each time step of current stage.

(4) At the end of each stage, equivalent shear stress is obtained, which is 65% of the maximum shear stress of current stage. The number of stress cycles required for liquefaction can be read off directly from the experimental curve of cyclic shear stress ratio versus number of cycles required for liquefaction, based on the equivalent shear stress ratio of the soil at current stage. Equivalent cyclic number of current stage is estimated By Martin's method (1976). Pore pressure increment within current stage can be calculated using equation (13).

(5) Estimate if all elements are convergent. The convergence criterion is $|G^{n+1}-G^n)/G^n| \le 0.01$ and $|(P_g^{n+1} - P_g^n)/P_g^n| \le 0.01$. Return (1) for iterating until convergence. In this study, the iteration number is usually 4~6.

(6) Repeat (1)~(5) within next stage until the end of earthquake.

3. FEM ANALYSIS ON SIMULATING SHAKE TABLE TESTS

The physical model and instrumentation for the two shake table tests are shown in Figure 3. Case 1 is soil deposit. One pipe without any measure and another pipe with rectangle drainage measure were buried in its left and right, respectively. Case 2 is soil deposit. One pipe without any measure and another pipe with U shape drainage measure were buried in its left and right, respectively. In this study, the gravel is used for drainage measure. The FEM mesh is shown in Figure 4. The composite isoparametric quadrilateral elements having four solid nodes and four fluid nodes were used. The two sides and bottom were impermeable. The top of the model is traction free with zero reference fluid pressure. The solid nodes at two sides and bottom were subject to input motion of acceleration of 0.6g shown in Figure 5. The wave history is divided into 14 stages, and each stage is one second.

A static analysis is first conducted to determine the initial stress states in soil foundation. Then a dynamic analysis is performed to simulate earthquake excitation. The time step of each stage in the dynamic analysis is 0.01 second.

Model parameters. Sand deposit: Dry density $\rho = 1590 kg/m^3$; Porosity ratio $n = 0.4$; Shear modulus coefficient $k = 352$; Shear modulus exponent $n = 0.36$; Poisson ratio $\nu = 0.33$; Coefficient of hydraulic conductivity $k_x = k_y = 1.2\times10^{-5}$ m/s; $G/G_0\sim\gamma$ and $\lambda\sim\gamma$ curve are shown in Figure 6; Shear stress ratio curve is shown in Figure 7. Gravel: Dry density $\rho = 1550\text{kg/m}^3$; Porosity $n = 0.43$; Shear modulus coefficient $k = 898.5$; Shear modulus exponent $n = 0.5$; Poisson ratio $\nu = 0.33$; Coefficient of hydraulic conductivity $k_x = k_y = 5.0\times10^{-3}$ m/s; $G/G_0\sim\gamma$ and $\lambda\sim\gamma$ curve are shown in Figure 6. Pipeline: Young's modulus $E = 5\times10^9$Pa; Density $\rho = 2400\text{kg/m}^3$; Poisson ratio $\nu = 0.286$, Damping ratio $\lambda = 0.05$. Fluid (water): Density $\rho = 1000\text{kg/m}^3$; Bulk modulus $Ks = 2.24\times10^9$Pa.

Excess pore pressure responses. Figure8~Figure16 show the excess pore pressure responses of the analysis and the test.

Case 1: pore pressure transducer P1 was located at the bottom of the pipe. A higher excess pore pressure was obtained in the analysis compared to the test result. The result was due to the fact that separation between the pipe and soil was not accounted for in the analysis. For P2, p4 and p5, the excess pore pressures obtained in the analysis were qualitatively with the experimental result. For P3, which was located at the top of the pipe, the excess pore pressure obtained from the experiment showed a decrease not long after shaking. The result was due to the fact that the sand surface crack with the uplift of the pipe results in the pore pressure dissipating quickly. However, this was not observed in the analysis. In P6, which was located in the gravel at the top of pipe, the pore pressure appeared negative. The result was due to the transducer uplift with the floatation of pipe results in the change of hydrostatic pressure.

Case 2: For P4 and P5, the excess pore pressure increased after shaking, and then dissipated quickly soon. But it not appeared in analysis. This result maybe was due to the pore pressure model was not used in the gravel. For P6, a higher excess pore pressure was obtained in the analysis compared to the test result. But the general trend of response was quite similar.

Safety factor against uplift. The equilibrium of vertical forces acting on the pipe and the overburden soil block (Koseki et al., 1997) is schematically shown in Figure 2. The safety factor F_s of pipes against uplift defined with an assumption of vertical slip surfaces in the overlying soil, as indicated in the same figure.

$$F_s = \frac{W+Q}{U_s + U_d + F} \tag{16}$$

where W is the dead weight of pipe and overburden soil block; Q is frictional resistance along slip surfaces; U_s is buoyant force due to hydrostatic; U_d is uplift force due to excess pore pressure; F is uplift force due to flowing soil, which is generally zero or negligibly small.

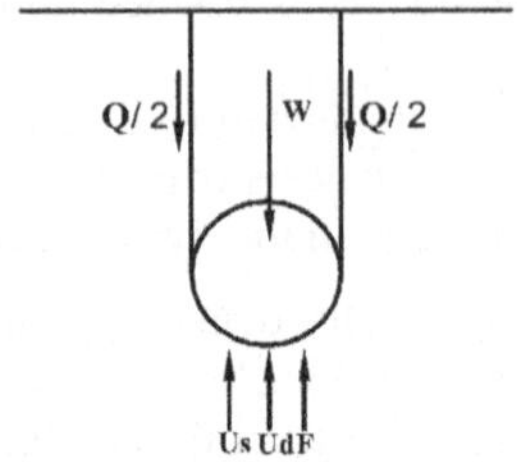

Figure 2. Assumed vertical slip surfaces in overburden soil

The minimum safety factors were obtained using equation 16 in the analysis. The uplift displacements were obtained from model tests. The safety factors and the uplift displacements are listed in Table 1. It was demonstrated that the uplift displacement decreased when the safety factor increased. The effectiveness against floatation of the "U" shape drainage measure was the best in two cases.

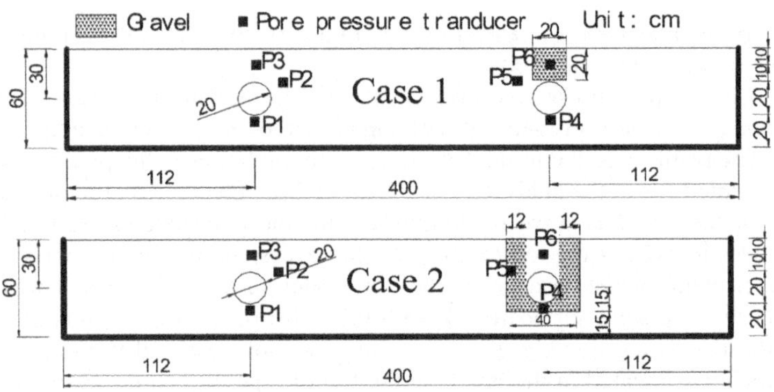

Figure 3. Shake table model

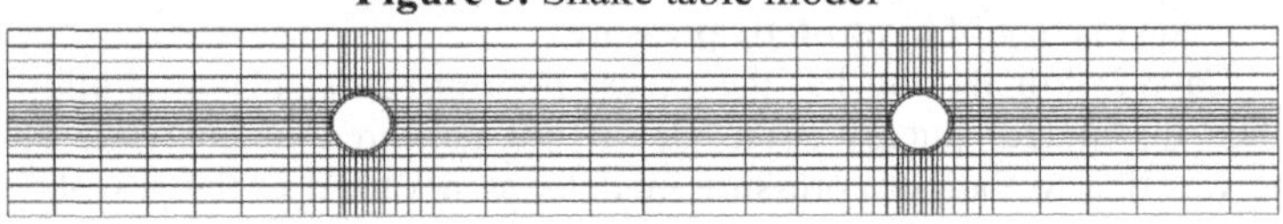

Figure 4. FEM mesh

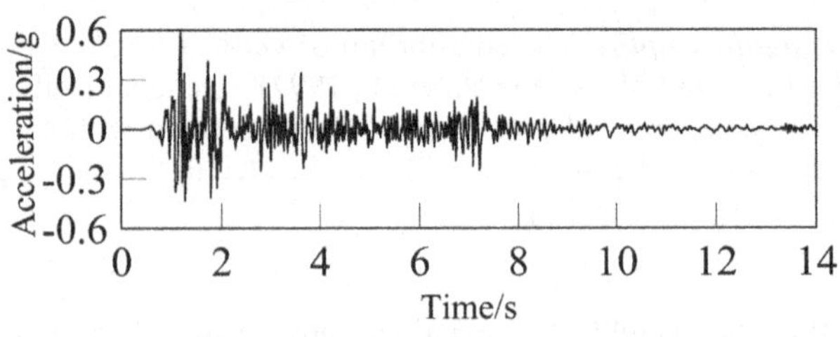

Figure 5. Earthquake input

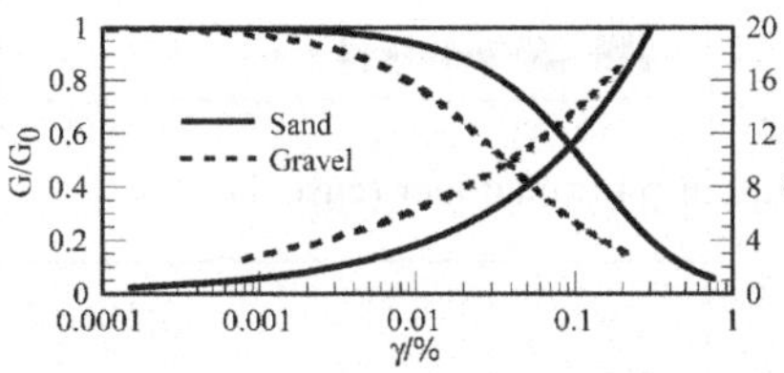

Figure 6. G/G$_0$ and λ curve of sand and gravel

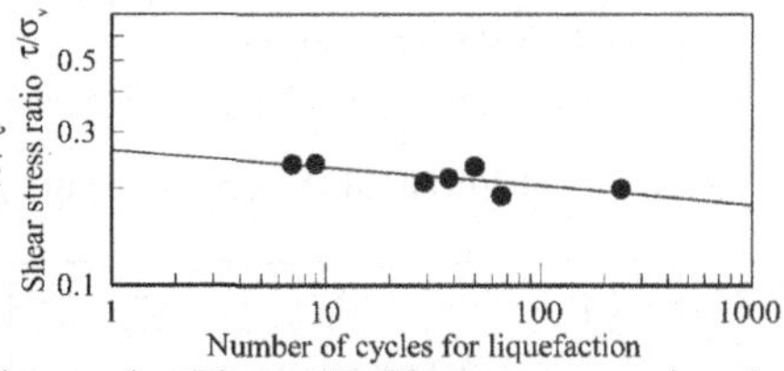

Figure 7. Shear stress ratio of sand

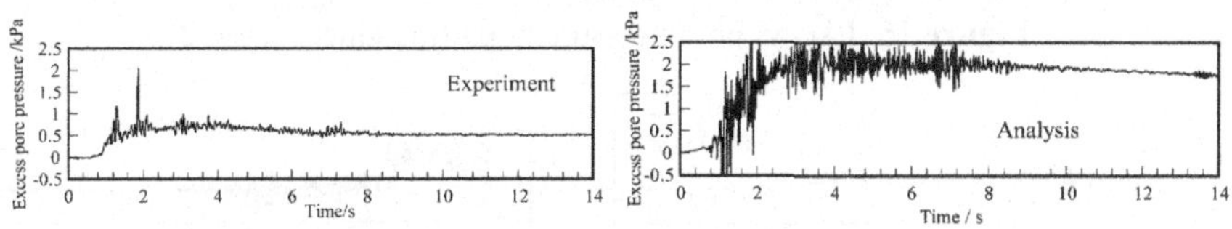

Figure 8. Excess pore pressure at p1 transducer (case 1)

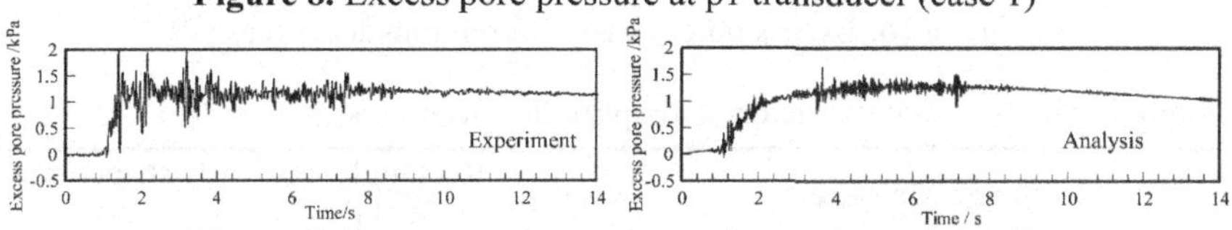

Figure 9. Excess pore pressure at p2 transducer (case 1)

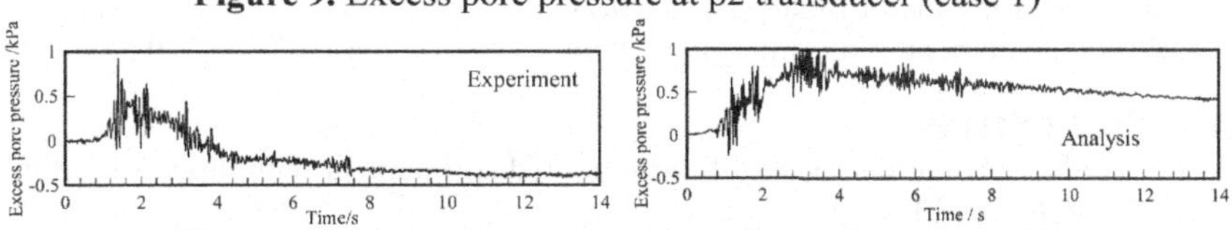

Figure 10. Excess pore pressure at p3 transducer (case 1)

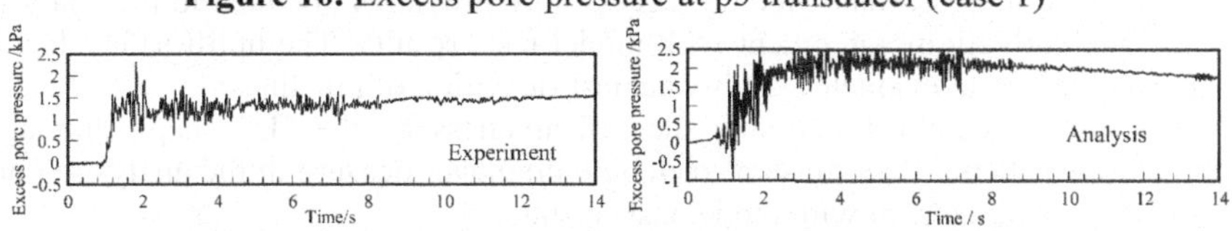

Figure 11. Excess pore pressure at p4 transducer (case 1)

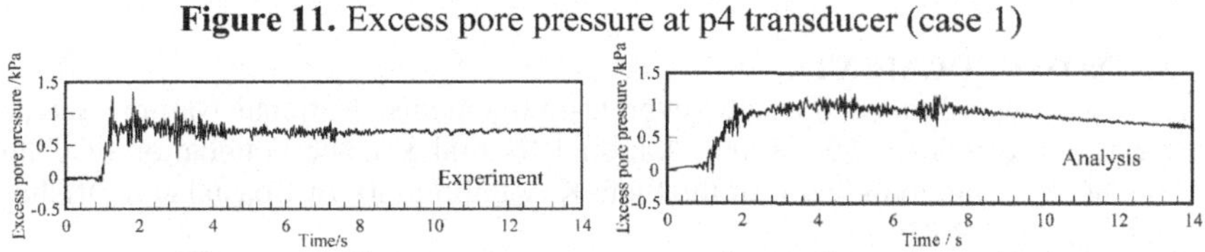

Figure 12. Excess pore pressure at p5 transducer (case 1)

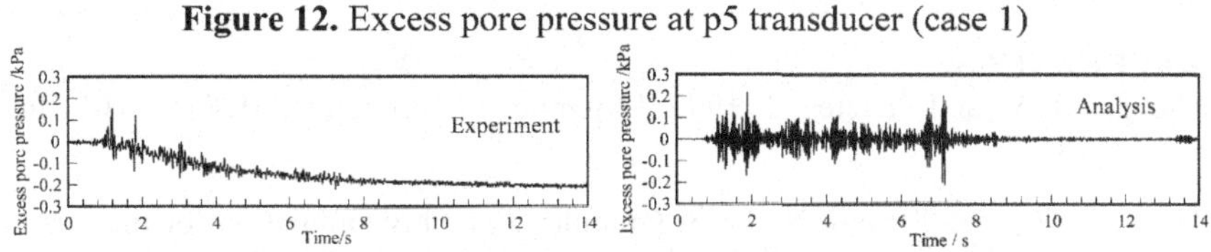

Figure 13. Excess pore pressure at p6 transducer (case 1)

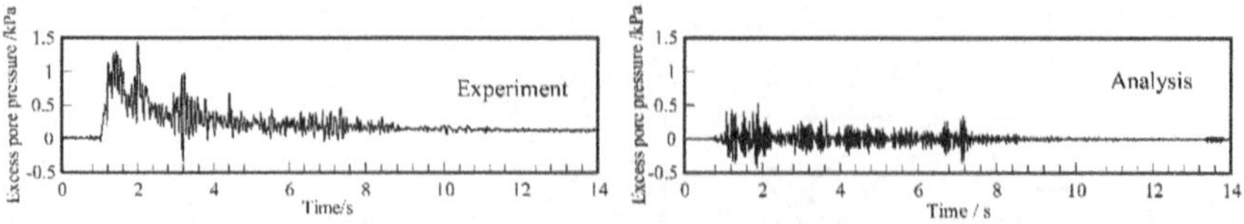

Figure 14. Excess pore pressure at p4 transducer (case 2)

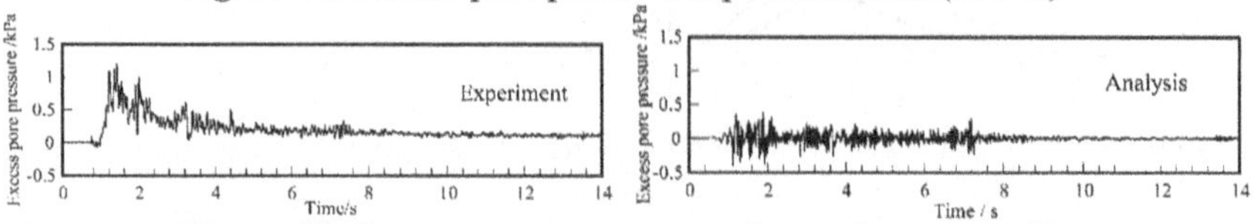

Figure 15. Excess pore pressure at p5 transducer (case 2)

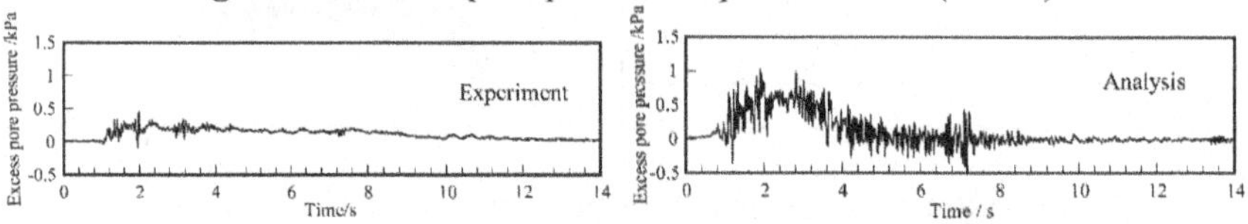

Figure 16. Excess pore pressure at p6 transducer (case 2)

Table 1. Minimum safety factors and uplift displacements

Drainage Measure	None	Rectangle(case 1)	U shape(case 2)
Minimum safety factor (analysis)	0.74	0.99	1.44
Uplift displacement (experiment)	8cm	2.6cm	0.35cm

4. CONCLUSIONS

In this paper, seismic responses of the saturated soil deposit with pipes buried in it were simulated by FEM.

By Finite Element Method, excess pore pressures obtained in the analysis were qualitatively in agreement with model tests results. The uplift safety factor can be obtained to evaluate stability against floatation of pipelines.

It was demonstrated that the effectiveness of the "U" shape drainage measure was better than those of rectangle drainage measure in the analysis. This result was in agreement with model tests results.

ACKNOWLEDGMENTS

This study was funded through several arrangements: National Natural Science Foundation of China (50278009), National Natural Science Foundation of China (50578029), The Foundation of the State Key Laboratory of Coastal and Offshore Engineering (LP0505).

REFERENCES

Clough, R.W. and Penzien, J. 1993. Dynamics of Structures, McGraw-Hill, Inc., New York.

Hakuno, M., and Tarumi, Y. 1988. Granular assembly simulation for the seismic liquefaction of sand. Proceedings of JSCE, No. 398/I-10, 129-138.

Hall, W.J. and O'Rourke 1991. Seismic behavior and vulnerability of pipeline. Lifeline earthquake Engineering, Cassaro, (eds.), ASCE, 761-773.

Idriss, I.M. QUAD-4, A Computer Program For Evaluating the Seismic Response of Soil Structure by Variable Damping Finite Element Procedures, Report No.UCB/EERC-73-16.

Katona, M.G. and Zienkiewicz, O.C. 1985. A unified set of single step algorithms part 3: the Beta-m method, a generalization of the Newmark scheme. Int. J. Num. Eng., 21, 1345~1359.

KONG, X.J., ZOU, D.G., Ling, H.I., and MA, H.C. 2002. Experiment study on the seismic behavior of pipeline buried in saturated sand foundation. Proceedings of the 4th China-Japan-USA Trilateral Symposium on Lifeline Earthquake Engineering, Hu, Y. X., Takada S., and Kiremidjian, A. S., eds., Qingdao, China, 229-234.

Koseki, J., Matsuo, O. and Hayashi, Y. 1997. Model tests on uplifts of sewer manholes and pipes accompanying deformation of liquefied backfill soil. Proceeding of International Symposium on deformation and Progressive Failure in Geomechanics, Nagoya, Japan, 593~598.

Koseki, J., Matsuo, O., Sasaki, T., Saito, K., Yamashita, M. 2000. Damage to sewer pipes during the 1993 Kushiro-Oki and the 1994 Hokkaido-Toho-Oki earthquakes. Soils and Foundations, 40(1): 99~111.

Ling, H.I., Mohri, Y, Kawabata, T., Liu, H.B., Burke, C. and Sun, L.X. 2003. Centrifugal Modeling of Seismic Behavior of Large-Diameter Pipe in Liquefiable Soil. Journal of Geotechnical and Geoenvironmental Engineering. ASCE, Vol. 129(12):1092~1101

Martin P.P., Seed H.B. 1976. Simplified procedure for effective stress analysis of ground response. Journal of the Geotechnical Engineering division, ASCE, 105(GT6): 739~758.

Mohri Y., Yasunaka M., Tani S. 1995. Damage to buried pipeline due to liquefaction induced performance at the ground by the Hokkaido-Nansei-Oki Earthquake in 1993. Proceedings of First International Conference on Earthquake Geotechnical Engineering. Rotterdam: Balkema, 31-36.

Mohri Y., Kawabata T., and Ling H.I. 1999. Experimental study on the effects of vertical shaking on the behavior of underground pipelines. Proceedings of Second International Conference on Earthquake Geotechnical Engineering, Lisbon, 489-494.

Newmark N.M. 1959 A method of computation for structural dynamics. Proc. ASCE, 8, 67~94.

Seed H.B., Martin P.P., Lysmer, J. 1976. Pore-water pressure changes during soil liquefaction. Journal of the Geotechnical Engineering Division, ASCE, 102(GT4):323–346.

Sun L.X. 2001. Centrifuge modeling and fem element analysis of pipeline buried in liquefiable soil. Ph. D. Dissertation of Columbia University, 2001.

Wang L.R.L, Shim J.S., Ishibashi L. and Wang Y. 1990. Dynamic response of buried pipelines during liquefaction process. Soil Dynamic and Earthquake Engineering, 9(1), 44~50.

Zienkiewicz, O.C., Chang, C.T. 1978. Non-Linear Seismic Response and Liquefaction. International Journal of Numerical and Analytical Methods in Geomechanics. Vol. 2:381~404.

Zienkiewicz, O.C. 1982. Basic formulation of static and dynamic behavior of soil and other porous media. In Numerical Methods in Geomechanics. J. B. Martin (ed.). D.Reidl.

Zienkiewicz O.C., Huang M., Pastor M. 1993. Numerical prediction for Model No. 1. In: Arulanandan K, Scott R F eds. Proceeding International Conference on the Verification of Numerical Procedures for the Analysis of Soil Liquefaction Problems, Vol 1, Rotterdam: Balkema.

Zienkiewicz O.C., Chan A.H.C., Pastor M., Schrefler B.A. and Shiomi T. 1999. Computational geomechanics with special reference to earthquake engineering. Wiley, London.

Soil Stress-Strain Behavior: Measurement, Modeling and Analysis
Geotechnical Symposium in Roma, March 16 & 17, 2006

DEFORMATION ANALYSIS OF LIQUEFIED GROUND BY PARTICLE METHOD

Kobayashi Yoshikazu
Department of Civil Engineering,
College of Science and Technology, Nihon University
1-8-14, Kanda-Surugadai, Chiyodaku, Tokyo, 101-83-8, Japan
e-mail: kobayasi@civil.cst.nihon-u.ac.jp

ABSTRACT

Quantitative estimation of liquefied ground deformation has been required for performance design of foundations, embedded structures, embankments and others in recent years. This paper evaluates the applicability of MPS (Moving Particle Semi-implicit) for the estimation of liquefied ground deformation by performing simple example computation. The results suggest that MPS is potentially capable to be applied for the estimation of liquefied ground deformation.

1. INTRODUCTION

Performance based design requires quantitative estimation of ground deformation after complete liquefaction. This estimation has been challenged by many research projects. Towhata et al. presented energy principle based method that is modeling liquefied sand as Newtonian fluid or Bingham Fluid for two dimensional problem[1)]. This method was extended to three dimensional problem, and applied to estimate the deformation of liquefied ground that was caused by large earthquakes in past[2)]. Finite element analysis has also been applied to the estimation of liquefied ground deformation[3)]. This conventional method is able to analyze behavior of liquefied ground from pre-liquefaction to post-liquefaction; however, there are difficulties to consider the effect of large deformation since it originally aims prediction of possibility of liquefaction at a target site.

On the other hand, the moving boundary problem has been studied in the field of computational fluid dynamics due to the importance of the interaction of fluid and solids. A series of particle method has been developed to overcome this problem. Among them, MPS(Moving Particle Semi-implicit)[4)] are recently applied to many fields of engineering. Hence, in this paper, a method of deformation analysis of liquefied ground on MPS is presented to take the advantage of MPS into account. Then, a simple example analysis is conducted to asses the feature of the presented method.

2. MPS(Moving Particle Semi-implicit) METHOD

MPS was developed by Koshizuka for the field of computational fluid dynamics[4)]. This method

Hoe I. Ling et al. (eds.), Soil Stress-Strain Behavior: Measurement, Modeling and Analysis, 683–689.

discritizes continuum by a group of particles as illustrated in Fig.1. MPS discritizes the continuum by defining relationships between each particle, and the relationships are determined by contribute of a particle to the other particles. This contribute is defined based on following weight function that is illustrated in Fig.2.

$$w(r)=\begin{cases}\frac{r_e}{r}-1 & (0\le r<r_e)\\ 0 & (r_e\le r)\end{cases} \tag{1}$$

Generally, gradient, divergence and Laplacian are necessary to numerically solve engineering problems. These are modeled by Eq.2, Eq.3 and Eq.4 based on the weight function, respectively.

$$\langle\nabla\phi\rangle_i=\frac{d}{n^0}\sum_{j\neq i}\left[\frac{\phi_j-\phi_i}{\left|r_j-r_i\right|^2}\left(r_j-r_i\right)w\left(\left|r_j-r_i\right|\right)\right] \tag{2}$$

$$\langle\nabla\bullet u\rangle_i=\frac{d}{n^0}\sum_{i\neq j}\frac{\left(u_j-u_i\right)\bullet\left(r_j-r_i\right)}{\left|r_j-r_i\right|^2}w\left(\left|r_j-r_i\right|\right) \tag{3}$$

$$\langle\nabla^2\phi\rangle_i=\frac{2d}{\lambda n^0}\sum_{j\neq i}\left[\left(\phi_j-\phi_i\right)w\left(\left|r_j-r_i\right|\right)\right] \tag{4}$$

In which, d is a dimension number, ϕ is a state value of the particle i, r_i and r_j are coordinate vectors of particle i and j, respectively. n_0 is a density of the particle i that is defined as following equation.

$$n_0=\sum_{i\neq j}w\left(\left|r_j-r_i\right|\right) \tag{5}$$

λ is a parameter that adjusts the contribution of ϕ at particle i to the other particles in the Laplacian model described in Eq.4. λ is defined as follow.

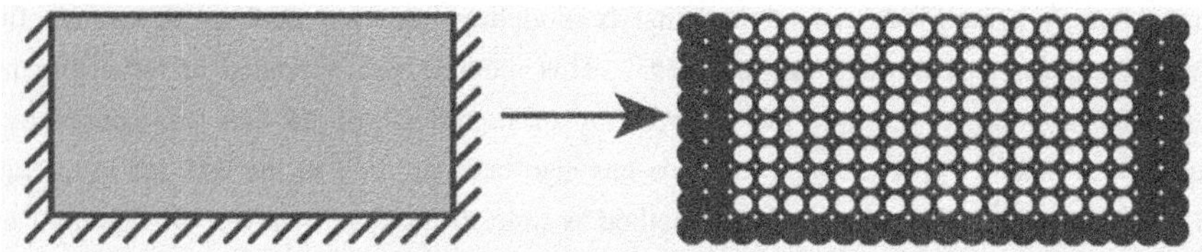

Fig.1 Discritization by MPS

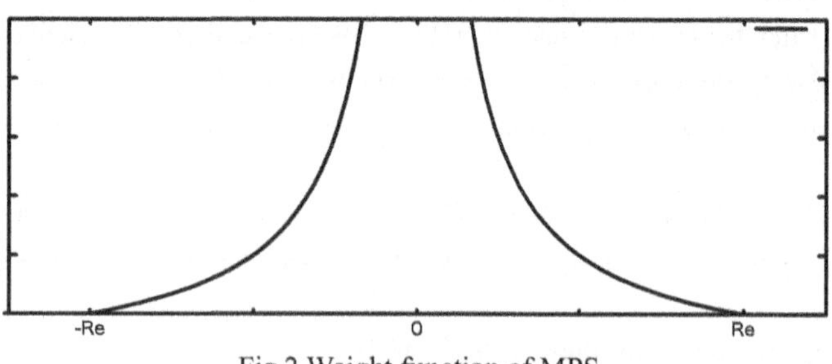

Fig.2 Weight function of MPS

$$\lambda = \frac{\sum_{j \neq i} |r_j - r_i|^2 w(|r_j - r_i|)}{\sum_{j \neq i} w(|r_j - r_i|)} \tag{6}$$

Dynamic analysis of MPS for elastic material was presented by Koshizuka[5]. Governing equation for elastic material is described by using Lame's constant.

$$\rho \frac{\partial^2 u_i}{\partial t^2} = \frac{\partial}{\partial x_j}\left(\lambda \varepsilon_v \delta_{ij} + 2\mu\varepsilon_{ij}\right) + \rho K_i \tag{7}$$

In which, u_i and K_i are displacement and body force in direction of i, respectively. ε_v is volumetric strain defined by

$$\varepsilon_v = \varepsilon_{11} + \varepsilon_{22} + \varepsilon_{33} \tag{8}$$

For solid analysis of MPS, each particle has three degree of freedom in two dimensional model as described in Eq.9.

$$u = \{u_1 \quad u_2 \quad \theta\}' \tag{9}$$

u_1 and u_2 are translations of a particle in x and y direction, and θ is a rotation of the particle. MPS defines the volumetric strain as

$$\varepsilon_v = \nabla \bullet u = \frac{d}{n_0} \sum_{i \neq j} \frac{u_{ij} \bullet r_{ij}}{|r_{ij}^0||r_{ij}|} w(|r_{ij}^0|) \tag{10}$$

r_{ij}, r_{ij}^0 and u_{ij} are relative coordinate, initial relative coordinate and relative displacement between particle i and j. These are defined as follows.

$$r_{ij} = r_j - r_i \tag{11}$$

$$r_{ij}^0 = r_j^0 - r_i^0 \tag{12}$$

$$u_{ij} = r_{ij} - R r_{ij}^0 \tag{13}$$

In which, R is a rotation matrix that is denoted as

$$R = \begin{bmatrix} \cos\frac{\theta_i + \theta_j}{2} & -\sin\frac{\theta_i + \theta_j}{2} \\ \sin\frac{\theta_i + \theta_j}{2} & \cos\frac{\theta_i + \theta_j}{2} \end{bmatrix} \tag{14}$$

First term of Eq.7 is calculated from the volumetric strain ε_v which is obtained by Eq.10.

$$p_i = -\lambda \varepsilon_v \tag{15}$$

Second and third term of Eq.10 are components on Lame's constant μ. These are obtained from stress components σ_{ij}^n which direction is the same with r_{ij} and σ_{ij}^s .which direction is orthogonal to r_{ij} between particle i and j.

$$\sigma_{ij}^{n} = 2\mu\varepsilon_{ij}^{n} = 2\mu\frac{u_{ij}^{n}}{\left|r_{ij}^{0}\right|} \tag{16}$$

$$\sigma_{ij}^{s} = 2\mu\varepsilon_{ij}^{s} = 2\mu\frac{u_{ij}^{s}}{\left|r_{ij}^{0}\right|} \tag{17}$$

By substituting Eq.15, 16, 17 to Eq.7, acceleration of particle i is computed as follow.

$$\left[\frac{\partial v_i}{\partial t}\right] = \left[\frac{\partial v_i}{\partial t}\right]_n + \left[\frac{\partial v_i}{\partial t}\right]_s + \left[\frac{\partial v_i}{\partial t}\right]_p + K_i \tag{18}$$

$$\left[\frac{\partial v_i}{\partial t}\right]_n = \frac{d}{\rho n_0}\sum_{i\neq j}\frac{2\sigma_{ij}^{n}}{\left|r_{ij}^{0}\right|}w\left(\left|r_{ij}^{0}\right|\right) \tag{19}$$

$$\left[\frac{\partial v_i}{\partial t}\right]_s = \frac{d}{\rho n_0}\sum_{i\neq j}\frac{2\sigma_{ij}^{s}}{\left|r_{ij}^{0}\right|}w\left(\left|r_{ij}^{0}\right|\right) \tag{20}$$

$$\left[\frac{\partial v_i}{\partial t}\right]_n = -\frac{d}{\rho n_0}\sum_{i\neq j}\frac{2p_{ij}r_{ij}}{\left|r_{ij}^{0}\right|\left|r_{ij}\right|}w\left(\left|r_{ij}^{0}\right|\right) \tag{21}$$

$$\left[\frac{\partial \theta_i}{\partial t}\right]_n = -\frac{m_i}{\rho I}\frac{2d}{n_0}\sum_{i\neq j}\frac{\sigma_{ij}^{s}}{\left|r_{ij}^{0}\right|}w\left(\left|r_{ij}^{0}\right|\right) \tag{22}$$

Time integration is conducted explicitly. Velocity and displacement at time step number k+1 is calculated from the acceleration that is indicated in Eq.18 at time step number k.

$$v_i^{k+1} = v_i^{k} + \Delta t\left[\frac{\partial v_i}{\partial t}\right]^{k} \tag{23}$$

$$r_i^{k+1} = r_i^{k} + \Delta t\, v_i^{k+1} \tag{24}$$

$$\theta_i^{k+1} = \theta_i^{k} + \Delta t\left[\frac{\partial \theta_i}{\partial t}\right]^{k} \tag{25}$$

MPS has been applied to multi-phase analysis. The analyses modeled the multi-phase medium as a mixture of solid particles and liquid particles in one plane. This modeling assumes that solid phase and liquid phase are not superimposed each other. This assumption is sufficient if the permeability of solid particle can be negligibly small. However, liquefiable sand deposit typically has voids in it and the void can be filled with pore water. Furthermore, the permeability can not be negligible since the lateral spread of liquefied ground mainly occurs after complete liquefaction, and the duration is enough long to break the undrain condition. This fact implies that superimposition of solid phase and liquid phase should be considered for deformation analysis of liquefied ground. Therefore, liquefied sand is modeled as a mixture of solid phase and liquid phase in a method that is proposed in this paper. The excess pore water pressure rising during strong motion is represented by decreasing the lame's constants. When the lame's constants are enough small, behavior of liquefied sand is changed from solid material to liquid material. This is achieved by changing the constitutive model of the solid material from elastic to fluid model that satisfies only mass-conservation. Mass conservation is fulfilled by applying the idea of contact analysis that was proposed by Koshizuka.

The proposed method identifies the contact from the particle density in Eq.5. Pressure resulted in contact is calculated from the variation of the particle density n_i.

$$p = \lambda \frac{n_i - n_0}{n_0} \tag{26}$$

If p in Eq.26 is larger than the pressure obtained by Eq.15, it is identified that the particle contacts with the other particles. The present method adopts this idea to fulfill the conservation of mass during analyses. When the model reaches complete liquefaction, constitutive equation of all particles switched from linear elastic model to the mass conservation model presented in Eq.26. This process represents large deformation of the liquefied ground since only particle density controls only the normal direction force between particles and shear force is ignored while after complete liquefaction.

3. COMPUTATIONAL EXAMPLE

The presented method is applied to a simple example that is illustrated in fig.3. The size of the soil container is 2.0m in width and 0.6m in height. A slope of loose sand is installed on the soil container. The material property of the slope is shown in Table.1. These parameters are not real parameters since this analysis aims to only check the feature of the presented method. The stiffness of the loose sand is linearly decreased, and 90% of the stiffness is reduced in 10000 computational time steps. After 10000 steps, constitutive model is switched from linear elastic medium to mass conservation model, and computation is continued until 150000 steps. Some particles are installed at the bottom and right side of the soil container that is reproduced by fixed particles. These particles prevent that soil particles pass through the soil container under high pressure state.

4. RESULT OF THE EXAMPLE CALCULATION

Fig.4-6 shows variation of the slope shape in time domain. These figures indicate large movement of particles is not observed during the period of pre-liquefaction even if 90% of its stiffness is reduced. However, after switching to mass conservation model, the slope started to flow from upstream side to downstream side, and ultimate equilibrium is roughly achieved after enough elapsed time as illustrated in Fig.6. Fig.7 shows time history of elevation of a particle at the top of the slope. This result also shows that the large deformation of the slope starts after cutting off the stiffness of the slope.

The surface of the slope is not completely horizontal in this computed result as shown in Fig.6.

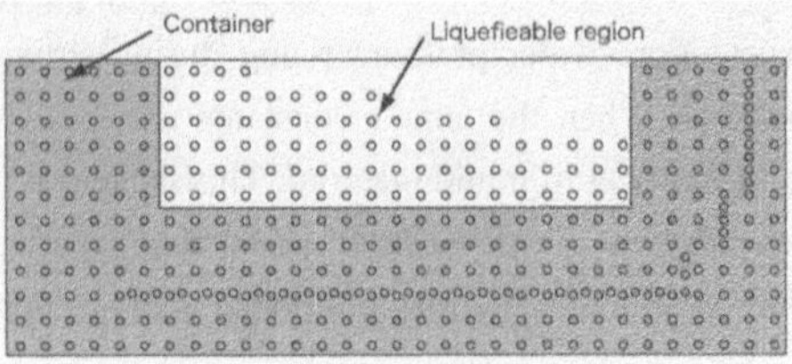

Fig.3 Example model

Table.1 Material property of the example model

Mass density	1000kg/m^3
Viscosity	5000Pas
Time interval	0.0002[s]
λ	50000000[Pa/m^2]
μ	30000000[Pa/m^2]

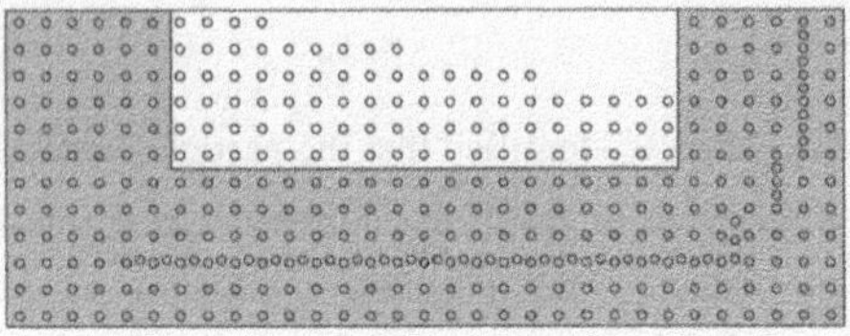

Fig.4 Deformation after 9900 steps (Approx. 2.0sec)

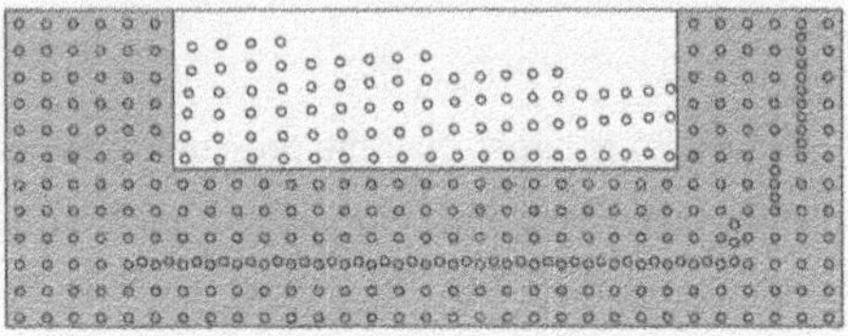

Fig.5 Deformation after 14850 steps (Approx. 3.0sec)

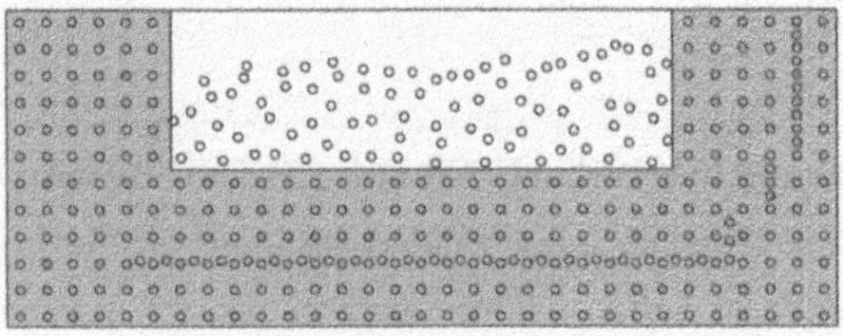

Fig.6 Deformation after 148500 steps (Approx. 30ses)

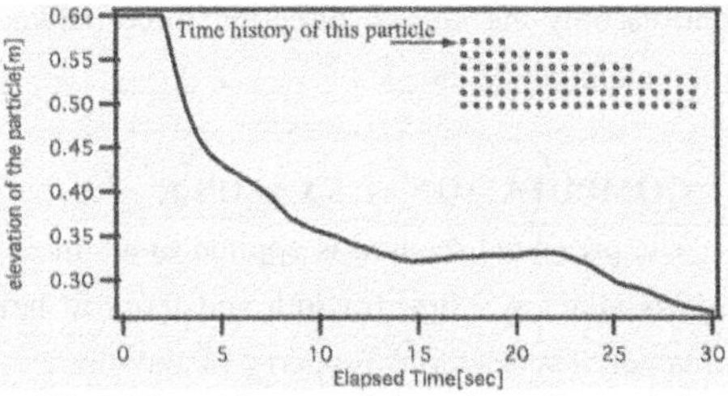

Fig.7 Time history of elevation of a particle at the top of the slope

The shape of the ground should be horizontal in the theoretical view, however, this is not due to the error of discritization caused by Re. For this example, the Re is almost 30% of the height of slope; therefore the elevation of the ground surface easily changes if one particle is stacked on a particle around the surface. This is the error of discritization on the series of particle methods and avoidable if more small particle is adopted and more large number of particles is used to build up the example model.

5. CONCLUSIONS AND REMARKS

This paper presented a method of deformation analysis of liquefied ground based on MPS. The phase change of loose sand from solid to liquid was represented by switching the constitutive equation from linear elastic model to mass conservation model. The result of example analysis shows that the phase change of the sand was simulated by the presented method. Implementation of adequate constitutive model that reproduces the generation of pore-water pressure and others is required for practical application; however, the potential of the application of MPS was suggested by the result of simple example.

As future works, constitutive equation of soil material will be implemented for solid phase due to the representation of the process of excess pore water pressure rising during strong motion. This is significant for quantitative estimation of lateral spread of liquefied ground. Furthermore, soil particle and pore water will be separated to individual plane each other to take the pore water pressure change by movement of soil particles into account.

REFERENCES

1) Ikuo Towhata, Roland P. Orense and Hirofumi Toyota, Mathematical Principles in prediction of lateral ground displacement induced by seismic liquefaction, Soils and Foundations, Vol.39, No.2, pp,1-19, 1996

2) Y. Kobayashi, S. Nishimura , 2 and 3-Dimensional Dynamic Analysis of Large Deformation of Liquefied Ground, Proceedings of the Twelfth Asian Regional Conference on Soil Mechanics and Geotechnical Engineering, pp.303-306, 2003.8

3) Masoud Mohajeri, Yoshikazu Kobayashi, Kazuhiro Kawaguchi, Masayoshi Sato, Numerical Study on Lateral Spreading of Liquefied Ground behind a Sheet pile Model in a Large Scale Shake Table Test, Thirteenth World Conference on Earthquake Engineering, CD, 2004.8

4)Koshizuka, S. and Oka, Y., Moving-Particle Semi-implicit Method for Fragmentation of Incompressible Fluid, Nucl. Sci. Eng., 123, pp.421-434, 1996

5) Koshizuka, S., Chikazawa, Y. and Oka, Y., Particle Method for Fluid and Solid Dynamics, Computational Fluid and Solid Mechanics, 2, pp.1269-1271, 2001

Soil Stress-Strain Behavior: Measurement, Modeling and Analysis
Geotechnical Symposium in Roma, March 16 & 17, 2006

EFFECTS OF CONSTITUTIVE PARAMETERS ON SHEAR BAND FORMATION IN GRANULAR SOILS

Marte S. Gutierrez
Department of Civil and Environmental Engineering
Virginia Polytechnic Institute and State University
200 Patton Hall, Blacksburg, VA 24061-0105
e-mail: magutier@vt.edu

ABSTRACT

This paper presents an analysis of the effects of constitutive response on bifurcation and shear band formation in granular soils. A simple elastoplastic constitutive model for biaxial loading condition is introduced which adequately captures the monotonic response of granular soils particularly the variation of dilatancy and mobilized friction with shear deformation. Using the constitutive model, a strain localization criterion expressed in terms of constitutive parameters is developed. Both the model and the strain localization criterion are verified against experimental data. The model and the bifurcation criterion are then applied in a parametric study to determine the sensitivity of strain localization in granular soils to constitutive parameters. Strain localization is analyzed in terms of Vermeer's compliance approach which is re-derived and generalized in this paper.

1. INTRODUCTION

Localization of deformation, a phenomenon commonly observed during loading of geomaterials, has been extensively studied theoretically and experimentally in recent years. The localization of deformation was treated as a bifurcation problem by Mandel (1966), Rudnicki and Rice (1975), and Vardoulakis (1980). According to bifurcation theory, a material which undergoes homogeneous deformation can reach a bifurcation point at which the material experiences instability and deformation becomes non-homogenous. Since the predominant yielding mechanism for geomaterials is shearing, the zone of strain localization is called a shear band.

Extensive experimental studies have been conducted on strain localization in granular materials. Numerous analytical and numerical studies have been performed to simulate various phenomena observed during strain localization. Many of the studies on strain localization in granular soils have been expressed in terms of fundamental quantities such as friction and dilation angles. However, these quantities are not constant but vary with stress and deformation level in the soil. The main objective of this paper is to analyze bifurcation and shear band formation in granular soils in terms of their full constitutive response. A simple elastoplastic constitutive model is introduced which adequately captures the monotonic response of granular soils under biaxial loading condition. Using the constitutive model, a strain localization criterion expressed in terms of constitutive parameters is developed. Another objective of the paper is to study the

Hoe I. Ling et al. (eds.), Soil Stress-Strain Behavior: Measurement, Modeling and Analysis, 691–706.

sensitivity of strain localization in granular soils to constitutive parameters. Strain localization is analyzed in terms of Vermeer's (1982) compliance approach which is re-derived and generalized in this paper. The study will be done in terms of two-dimensional plane strain and axisymmetric triaxial loading conditions.

2. LOCALIZATION CRITERION

Bifurcation theory is concerned with the prediction of how instability leads to localized deformations in elastoplastic materials. Mandel's (1966) definition of instability, which is related to discontinuities in the strain rate due to incremental non-uniqueness in material constitutive relations, is used in this paper. This definition is different from the instability criterion proposed by Drucker (1951, 1956) and Hill (1958) known as the maximum energy principle based on energy considerations. It is to be noted that unlike Drucker's and Hills criteria, Mandel's criterion is a necessary and sufficient criterion to detect the onset of instability in materials. Mandel stability criterion is based on the assumption that a stable material is able to propagate small perturbations in the forms of waves. Instability and strain localization occur when a small perturbation in the form of a wave cannot propagate across a material in a given direction n_i ($i = 1,2,3$). This condition appears when the acoustic tensor B_{ik} (defined below) has a zero or negative determinant, becomes non-positive definite, and produces negative eigenvalues. This condition may be stated for the case in which co-rotational terms are neglected as:

$$|B_{ik}| = |n_j D^{p}_{ijlk} n_l| \leq 0 \tag{1}$$

The acoustic tensor B_{ij} is a function of the elastoplastic constitutive matrix D^{p}_{ijkl} and of the direction of wave propagation n_i. Rudnicki and Rice (1975) showed that non-symmetry of D^{p}_{ijkl} causes B_{ik} to become non-positive definite under hardening conditions, in which peak strength has not yet been mobilized. Non-symmetry occurs in case of non-associated flow rule or when the plastic potential is different from the yield surface.

For an elastoplastic material, the constitutive matrix can be expressed in terms of a yield criterion $f = f(\sigma_{ij})$, plastic potential function $g = g(\sigma_{ij})$ and a hardening function :

$$D^{ep}_{ijkl} = D^{e}_{ijkl} - \frac{1}{h_p}\left(D^{e}_{ijpq} \frac{\partial g}{\partial \sigma_{pq}} \frac{\partial f}{\partial \sigma_{rs}} D^{e}_{rskl} \right) \tag{2}$$

where D^{e}_{ijkl} is the elasticity tensor, and h_p is the plastic modulus defined as:

$$h_p = h + \frac{\partial f}{\partial \sigma_{ij}} D^{e}_{ijkl} \frac{\partial g}{\partial \sigma_{kl}} \tag{3}$$

and h is the plastic hardening parameter.

An easier way to understand and formulate the static bifurcation problem is via Vermeer's compliance approach (Vermeer 1982). A re-derivation and generalization of this approach is presented below. Consider an element subjected to the principal stresses σ_1 and σ_3 and let the yet to be determined shear band direction be parallel to the x_1-axis, which is oriented at an angle θ from the σ_3-axis (Fig. 1). Shear band formation implies non-uniform stress and strain rate distributions, with stresses and strain rates different inside and outside the band. Due to equilibrium requirements, some of these differences are zero:

$$\Delta\varepsilon_{11} = \Delta\sigma_{22} = \Delta\sigma_{12} = 0 \tag{4}$$

where the symbol Δ denotes difference in quantity inside and outside the band. The only admissible discontinuity is that of the stress along the shear band direction:

$$\Delta\sigma_{11} \neq 0 \tag{5}$$

Using the stress-strain relation:

$$\Delta\varepsilon_{11} = C_{11}\Delta\sigma_{11} + C_{12}\Delta\sigma_{22} + C_{13}\Delta\sigma_{13} \tag{6}$$

where C_{ij} are the components of the compliance matrix, gives

$$\Delta\varepsilon_{11} = C_{11}\Delta\sigma_{11}, \text{ or } C_{11} = 0 \tag{7}$$

Vermeer's approach simply requires that the compliance matrix C_{ij} be formulated using the axis coinciding with the yet unknown shear band direction, and then setting the component C_{11} parallel to the shear band orientation to zero. This gives the condition for the onset of instability and shear banding.

For elastoplastic materials, the compliance C_{11} is equal to:

$$C_{11} = \frac{(1-\nu^2)}{E} + \frac{1}{h}\frac{\partial f}{\partial\sigma_{11}}\frac{\partial g}{\partial\sigma_{11}} \tag{8}$$

where h = plastic hardening modulus, f = yield function or failure criterion, g = plastic potential function, E =Young's modulus and ν = Poisson's ratio. In two dimensions, the yield and plastic potential functions can be written in terms of the two-dimensional MIT stress invariants s and t, which are equal to the coordinate of the center and the radius of the Mohr's circle for stress, respectively (Fig. 2):

$$s = \frac{1}{2}(\sigma_{11} + \sigma_{22}) \;,\; t = \sqrt{\left(\frac{\sigma_{11} - \sigma_{22}}{4}\right)^2 + \sigma_{12}^2} \tag{9}$$

Using chain-rule of differentiation:

$$\frac{\partial f}{\partial\sigma_{11}} = F_s\frac{\partial s}{\partial\sigma_{11}} + F_t\frac{\partial t}{\partial\sigma_{11}} = \frac{1}{2}F_s + \frac{1}{2}F_t\frac{(\sigma_{11} - \sigma_{22})}{2t} \tag{10}$$

where $F_s = \partial f/\partial s$ and $F_t = \partial f/\partial t$. Note that the term $(\sigma_{11} - \sigma_{22})/2t$ is equal 2θ as illustrated in Fig. 2, where θ is the shear band orientation measured from the σ_3-axis. Thus, Eq. (9) can be written as:

$$\frac{\partial f}{\partial\sigma_{11}} = \frac{1}{2}F_s + \frac{1}{2}F_t\cos 2\theta \tag{11}$$

A similar expression can be derived for $\partial g/\partial\sigma_{11}$:

$$\frac{\partial g}{\partial\sigma_{11}} = \frac{1}{2}G_s + \frac{1}{2}G_t\cos 2\theta \tag{12}$$

where $G_s = \partial g/\partial s$ and $G_t = \partial g/\partial t$. Substituting Eqs. (11) and (12) in Eq. (8), setting $C_{11} = 0$ and solving for the plastic hardening modulus h gives:

$$h = -\frac{E}{4(1-\nu^2)}\left(F_s + F_t\cos 2\theta\right)\left(G_s + G_t\cos 2\theta\right) \tag{13}$$

Since the hardening modulus decreases gradually as failure is approached during loading, detection of the onset of strain localization consists of finding the critical orientation θ

that will maximize the value of the hardening modulus h in Eq. (13). The critical orientation θ gives the shear band orientation, while the corresponding maximum value of h gives the critical hardening modulus h_B at bifurcation. Differentiating the right-hand side of Eq. (13) with respect to θ, equating the derivative to 0 and solving for θ gives the orientation of the shear band at critical hardening modulus:

$$\cos 2\theta = \frac{1}{2}\left(F_s / F_t + G_s / G_t\right) \tag{14}$$

Substituting back to Eq. (13) gives the critical hardening modulus at bifurcation h=h_B:

$$h = h_B = \frac{E(F_s / F_t - G_s / G_t)^2}{16(1-\nu^2)} \tag{15}$$

For cohesionless granular soils, the yield criterion and the plastic potential function required are usually represented by the Mohr-Coulomb model as:

$$f = t - s\sin\phi \ , \quad g = t - s\sin\psi \tag{16}$$

where ϕ = mobilized friction angle and ψ= mobilized dilation angle. Hence, the derivatives are:

$$F_s = \frac{\partial f}{\partial s} = -\sin\phi \ , \ F_t = \frac{\partial f}{\partial t} = 1 \ , \ G_s = \frac{\partial g}{\partial s} = -\sin\psi \ , \ G_t = \frac{\partial g}{\partial t} = 1 \tag{17}$$

Substituting the above derivatives in Eq. (15) gives the critical hardening modulus h_B at which bifurcation may occur:

$$h = h_B = \frac{E\left(\sin\phi - \sin\psi\right)^2}{16(1-\nu^2)} \tag{18}$$

This critical hardening modulus was derived by Vardoulakis (1980), and Vermeer (1982). This equation implies that for a strain hardening material (i.e., h>0) with non-associated flow rule (i.e., $\phi \neq \psi$), bifurcation can occur in the strain hardening regime before classical failure. For associated flow ($\phi \neq \psi$), bifurcation coincides with classical failure at h=0 and at peak friction angle.

Substituting Eqs. (17) in Eq. (18) gives the following expression for the shear band orientation:

$$\theta = \theta_{AV} = 45° + \frac{\phi}{4} + \frac{\psi}{4} \tag{19}$$

This orientation was first observed experimentally by Arthur et al. (1977) and was later derived analytically by Vardoulakis (1980). For $h = 0$, two shear band orientations are possible: The Coulomb orientation:

$$\theta = \theta_C = 45° + \frac{\phi}{2} \tag{20}$$

and the Roscoe (1970) orientation:

$$\theta = \theta_R = 45° + \frac{\psi}{2} \tag{21}$$

The Coulomb and the Roscoe solutions are, respectively, the upper and lower bound values of the shear band angle measured from the σ_3-axis. Experimentally measured shear band orientations in soils lie between these two values (Bardet 1991). Note that the Arthur-Vardoulakis orientation θ_{AV} is an average of the Coulomb and Roscoe solutions.

Experimental verification of the different shear band orientations has been exhaustively performed by Tatsuoka et al. (1990).

3. SIMPLE MODEL FOR BIAXIAL LOADING OF GRANULAR SOILS

It is noted that yield and plastic potential functions given in Eqs. (16) with constant values of ϕ and ψ are only valid at failure. For granular soils, the mobilized friction and dilation angles vary during loading and, therefore, their variations must be known in order to predict the bifurcation point. To apply Eqs. (18) and (19) to predict strain localization, a complete constitutive model is required. In this regard, a simple constitutive model for the response of the sand is first constructed. The constitutive model for biaxial loading of granular soils has the following components:

1) A strain-hardening Mohr-Coulomb yield surface with the yielding function f given in Eq. (16).

2) A strain hardening function based on a hyperbolic relation between mobilized friction angle $\sin\phi = t/s$ and plastic shear strain $\gamma^p = \int \dot{\gamma}^p$:

$$\sin\phi = \sin\phi_p \frac{\gamma^p}{A+\gamma^p} \tag{22}$$

where ϕ_p = the theoretical peak friction angle that would have been achieved if the granular soil stress-strain response did not bifurcate, and $1/A$ is equal to the initial slope of the $\sin\phi$ vs. γ^p curves. The plastic shear strain increment $\dot{\gamma}^p$ is defined as:

$$\dot{\gamma}^p = \sqrt{\left(\dot{\varepsilon}_{11}^p - \dot{\varepsilon}_{22}^p\right)^2 + 4\dot{\varepsilon}_{12}^{p\,2}} \tag{23}$$

3) As noted above, dilation in granular soils is not constant but varies with shear deformation, thus the plastic potential g given in Eq. (16) cannot be directly used. Instead, a plastic potential $g = g(s,t)$ function is derived from the stress-dilatancy relation:

$$\frac{\dot{v}^p}{\dot{\gamma}^p} = \sin\phi_c - \frac{s}{t} \tag{24}$$

where ϕ_c = the friction at phase transformation or zero dilation point where the soil changes from contractive to dilative response during shear loading (Ishihara et al. 1975). The plastic volumetric strain increment $\dot{v}^p$ is defined as:

$$\dot{v}^p = \dot{\varepsilon}_{11}^p + \dot{\varepsilon}_{22}^p \tag{25}$$

Eq. (24) can be integrated as follows to obtain the plastic potential function g:

$$\frac{\dot{v}^p}{\dot{\gamma}^p} = \frac{\lambda(\partial g/\partial s)}{\lambda(\partial g/\partial t)} = \sin\phi_c - \frac{s}{t} \rightarrow g = t - s\sin\phi_c \ln\left(\frac{s}{s_o}\right) \tag{26}$$

where s_o=the intersection of the plastic potential and s-axis.

4) Elastic shear strains are calculated using constant shear modulus G and Poisson's ratio ν. The elastic shear modulus G is estimated from Hardin and Black (1968):

$$G = 3200\frac{(2.97-e)^2}{(1+e)}\sqrt{s_o} \tag{27}$$

where s_o is the initial confining pressure and e is initial the void ratio.

Using the above equations, the hardening parameter h can be derived from the consistency condition:

$$df = \frac{\partial f}{\partial s}ds + \frac{\partial f}{\partial t}dt + \frac{\partial f}{\partial \sin\phi}d\sin\phi = 0$$

$$h = s\sin\phi_p \frac{A}{(A+\gamma^p)^2} = \frac{s\sin\phi_p}{A}\left(1-\frac{\sin\phi}{\sin\phi_p}\right)^2 \tag{28}$$

It can be shown that $A = \sin\phi_p/(G_p)$, where G_p=initial plastic shear modulus equal to the initial slope of the shear stress t vs. shear strain γ^p curve. Thus, the hardening modulus can also be written as:

$$h = G_p s\left(1-\frac{\sin\phi}{\sin\phi_p}\right)^2 \tag{29}$$

Based on the definitions of the mobilized dilatancy and friction angles (i.e., $\sin\psi = -\dot{v}^p/\dot{\gamma}^p$ and $\sin\phi = t/s$), the stress-dilatancy relation (Eq. 24) can also be written as:

$$-\sin\psi = \sin\phi_c - \sin\phi \tag{30}$$

The minus sign before $\sin\psi$ in the above equation implies that dilation is negative and volumetric contraction is positive. Substituting Eqs. (29) and (30) in the bifurcation criterion (Eq. 18), yields the following equation:

$$G_p s\left(1-\frac{\sin\phi}{\sin\phi_p}\right)^2 = \frac{E\sin\phi_c^{\,2}}{16(1-\nu^2)} \tag{31}$$

Using $E = 2G(1+\nu)$, the above equation can be re-written as:

$$\left(1-\frac{\sin\phi}{\sin\phi_p}\right)^2 = \frac{1}{(1-\nu)}\frac{G}{G_p}\frac{\sin\phi_c^{\,2}}{8s} \tag{32}$$

This equation can be solved for the friction angle at bifurcation $\phi=\phi_B$:

$$\sin\phi_B = \sin\phi_p\left(1-\sin\phi_c\sqrt{\frac{1}{8(1-\nu)s_B}\frac{G}{G_p}}\right) \tag{33}$$

where s_B=mean stress at the point of bifurcation. Substituting $\sin\phi_B$ in Eq. (22) gives the corresponding shear strain γ_B at the point of bifurcation:

$$\gamma_B = \frac{A\sin\phi_B}{\sin\phi_p - \sin\phi_B} \tag{34}$$

Also, the theoretical shear band orientation can be shown to be equal to:

$$\theta = 45^\circ + \frac{2\phi_B - \phi_c}{4} \tag{35}$$

4. EXPERIMENTAL VERIFICATION

To show the validity of the above formulation, model predictions are compared against experimental data. The validations are performed for two sands: Nevada sand and Hostun sand. Data for Nevada sand are from drained triaxial tests, while data for Hostun sand are taken from drained biaxial tests performed by Desrues and Hammad (1989). Comparisons will be made in terms of the predicted stress-strain response from the constitutive model and from the predictions from the localization criterion. The stress-strain prediction will be performed only until the bifurcation point and no attempt will be made to model the post-bifurcation response.

For Nevada sand, test results are presented in terms of the Cambridge invariants for triaxial conditions which are defined as: $p = (\sigma_1 + 2\sigma_3)/3$, $q = \sigma_1 - \sigma_3$, $\dot{\varepsilon}_v = \dot{\varepsilon}_1 + 2\dot{\varepsilon}_3$ and $\dot{\varepsilon}_s = 2(\dot{\varepsilon}_1 - \dot{\varepsilon}_3)/3$. Although the bifurcation criterion and the constitutive equations presented above are for two-dimensional conditions, they will be applied to triaxial compression condition by neglecting the effect of the intermediate principal stress (equal to the minor principal stress in triaxial compression condition). To apply the equations developed above to triaxial compression conditions, the mobilized friction angle ϕ at any level of deformation is related to the shear stress ratio $\eta = q/p$ by the relationship $\sin\phi = 3\eta/(6+\eta)$. A similar expression is used for the dilation angle. For Hostun sand, the constitutive model and localization criterion are directly applicable to the biaxial test results.

The triaxial test results for Nevada sand from three samples isotropically consolidated at initial mean stresses of p_o=40, 80 and 120 kPa are shown in Fig. 3. The tests were carried out under p-constant conditions. The results are shown in Fig. 3 in terms of shear stress q vs. shear strain ε_s, and volumetric strain ε_v vs. ε_s. The plane strain biaxial test results for Hostun sand from four samples isotropically consolidated at s_o=100, 200, 400 and 800 kPa are shown in Fig. 4. The tests were carried out under σ_3-constant conditions. The results are shown in Fig. 4 in terms of shear stress t vs. shear strain γ, and volumetric strain v vs. γ. For both sands, the test results are compared with the simulated response using the constitutive model described above. Both sands, which were sheared up to large shear strains, exhibit well-defined peak shear stress and strongly dilatant response at large strains. Nevada sand showed very small volumetric contraction at the initial part of the loading. Both sands show clear strain softening due to bifurcation and formation of well-developed shear bands.

As can be seen, the simulated and experimental results are in good agreement for the two sands up to the bifurcation points. The predicted bifurcation points in the stress-strain curves are very close to the measured peak shear stresses. For Hostun sand, the predicted bifurcation points are slightly above the experimental peak points. Because the stress-strain curves become more flat near the bifurcation point, a slight difference between the predicted and measure peak shear stress will cause a much larger difference in the predicted and experimental bifurcation shear strain. However, overall the model appears to satisfactorily predict the stress-strain curves and the locations of the bifurcation points. After the bifurcation point, the predicted stress-strain curve, which assumes homogeneous response, continue to show strain hardening and deviate from the experimental data, which show strain softening. The predicted volumetric vs. shear strain curves start to deviate from the experimental data after bifurcation. After bifurcation, the model predictions show linear relationship between volumetric strain and shear strain indicating constant dilatancy at large strains. The experimental data deviate from this linear plot and show less dilatancy after bifurcation.

Fig. 5 summarizes the predicted and measured friction angles at bifurcation ϕ_B and shear band orientations θ for the three tests on Nevada sand. Fig. 6 shows similar plots for the four tests on Hostun sand. For Nevada sand, good agreement between predicted and measured data is obtained for both friction angles at bifurcation ϕ_B and shear band orientations θ. For Hostun sand, good agreement is obtained between predicted and measured values of ϕ_B. However, the predicted values of the shear band orientation θ are significantly smaller than the measured data. The results indicate that correct prediction of the bifurcation point may not necessarily provide correct prediction of the shear band orientations. Desrues and Hammad (1989) showed that shear band orientations for Hostun sand do not follow any of the three established equations for shear band orientations discussed above, which are the Coulomb (Eq. 20), Roscoe (Eq. 21) and Arthur-Vardoulakis (Eq. 19) orientations. Instead, shear band orientations in Hostun sand were found to be between the Coulomb and Arthur-Vardoulakis orientations. Bardet (1991) has previously shown that the shear band orientations from Coulomb, Roscoe and Vardoulakis do not always provide accurate predictions of actual shear band orientations from experiments. Bardet suggested that there is a need to augment classical elastoplasticity theory and introduce vertices in the yield and plastic potential functions to accurately predict experimental shear band orientations.

An interesting result from the model prediction is the dependency of friction angle at bifurcation ϕ_B and shear band orientation θ with the level of stress confinements. Both parameters decrease with increasing confinement. These results are consistent with the experimental observations of Desrues and Hammad (1989). The dependency of the bifurcation friction angle ϕ_B and shear band orientation θ with the level of stress confinements appears to be properly accounted in the constitutive model. In particular, the hardening modulus h in the model is linearly dependent on the mean stress s (Eq. 29), and the mean stress s provides a stabilizing effect on the stress-strain response. As a result, the bifurcation friction angle ϕ_B is also dependent on the mean stress $s = s_B$ as shown in Eq. (33). Further discussions on the effects of the stress confinement on bifurcation are given below.

5. EFFECTS OF CONSTITUTIVE PARAMETERS ON STRAIN LOCALIZATION

It has been shown above that prediction of localization in granular soils can be directly related to the parameters required to model the constitutive response. This can be seen in the strain localization criterion given in Eq. (33) which is directly expressed in terms of constitutive parameters. This relationship provides a convenient means to perform parametric investigation on the sensitivity of bifurcation predictions without the need to perform multiple simulations of the stress-strain response. The validity of this criterion has been validated against experiments. Eq. (33) gives the friction angle at bifurcation ϕ_B as a function of the theoretical peak friction angle ϕ_p that would have been achieved if the soil does not localize, the friction angle at phase transformation ϕ_c, the ratio of the elastic to plastic shear modulus G/G_p, the Poisson's ratio ν and the mean stress at bifurcation s_B. These parameters can be obtained by fitting the model through pre-bifurcation experimental stress-strain curves.

Figs. 7a to 7d show plots of friction angle at bifurcation in terms of $\sin\phi_B/\sin\phi_p$ as function of the different constitutive parameters involved in the localization criterion in Eq. (33). The ratio $\sin\phi_B/\sin\phi_p$ gives a measure of the reduction in frictional strength of granular soils due to localization. A ratio equal to 1 indicates classical failure. Fig. 7a illustrates the effect of the ratio of the elastic to plastic shear modulus G/G_p, Fig. 7b the effect of Poisson's ratio ν, Fig. 7c the effect of the friction angle at phase transformation ϕ_c, and Fig. 7d the effect of mean stress at bifurcation s_B on the bifurcation point. The variation of $\sin\phi_B/\sin\phi_p$ is expressed in terms of typical range of values of the different constitutive parameters.

Figs. 7a to 7c indicate that the friction angle at bifurcation ϕ_B decreases with increasing values of G/G_p, ν and ϕ_c, and increases with increasing s_B. An increase in G/G_p means that plastic strains become more dominant than the elastic strains, and the increased plastic deformation enhances localization. On the other hand, an increase in stress confinement results in more stable response, and ϕ_B decreases with increasing s_B. From the parametric analyses, it was found that G/G_p has the most significant effect on the friction angle at bifurcation, and the ratio $\sin\phi_B/\sin\phi_p$ varies from close to 1.0 for G/G_p=1 to about 0.8 for G/G_p=100. Due to the importance of the G/G_p ratio, three values of G/G_p=1, 10 and 100 are used in the parametric analyses for ν, ϕ_c and s_B. For a given value of G/G_p, the changes in $\sin\phi_B/\sin\phi_p$ are less than about 10% points by increasing ν from 0 to 0.5, by increasing ϕ_c from 10 to 50°, and by increasing s_B from 100 to 1000 kPa. An exception is the variation in $\sin\phi_B/\sin\phi_p$ in terms of ϕ_c for G/G_p=10, which shows a reduction in $\sin\phi_B/\sin\phi_p$ from about 0.93 for ϕ_c=10° to about 0.7 for ϕ_c=50°. The effect of G/G_p is most pronounced for G/G_p>10, while the effect of s_B is most pronounced for s_B<100 kPa.

Of the different parameters required in the constitutive model, the elastic shear modulus G is possibly the most difficult to determine experimentally. Ideally, experimental determination of the elastic shear modulus would require loading and unloading in shear to establish an elastic response. At the same time, the elastic shear modulus G in relation to the plastic shear modulus G_p has the strongest effect on the predicted bifurcation point.

6. CONCLUSIONS

Bifurcation and shear band formation in granular soils was analyzed in terms of a simple constitutive model for biaxial loading. The constitutive model uses well-established empirical relationships, and is able to adequately capture important features of the monotonic response of granular soils particularly the variations of the mobilized friction and dilatancy in granular soils during shear loading. Using Vermeer's (1982) compliance approach, which was re-derived and extended, and the constitutive model, a strain localization criterion was developed in terms of the materials parameters involved in the constitutive model.

The constitutive model and the strain localization criterion were shown to satisfactorily simulate the stress-strain response and predict the bifurcation points of Nevada and Hostun sands. For Nevada sand, satisfactory prediction of the shear band orientations were also made. For Hostun sand, the predicted shear band orientations were smaller than the measured values. The discrepancy in predicted and measured shear band orientations in Hostun sand is consistent with the observations by Desrues and Hammad (1989), and Bardet (1991) that shear band orientations in sand do not always obey existing analytical expressions for shear band orientation. The strain localization criterion was shown to adequately capture the effect of confining stress on the friction angle at localization and the shear band orientations, as first observed by Desrues and Hammad (1989).

An advantage of linking the localization criterion directly to the constitutive parameters is that parametric analyses of the localization conditions can be made without full simulation of the stress-strain response. It was found that the ratio of the elastic to plastic shear modulus G/G_p, particularly for G/G_p >10, and the mean stress at bifurcation s_B, particularly for s_B>100 kPa, have the strongest effect on the friction angle at bifurcation ϕ_B. An increase in G/G_p results in more plastic deformation in relation to the elastic deformation and lowers ϕ_B. On the other hand, because hardening modulus h increases with the mean stress s, the mean stress was shown to have a stabilizing effect on the stress-strain response and an increase in mean or confining stress results in higher ϕ_B. It is noted that of the different parameters required in the model, the elastic shear modulus G is probably the most difficult to determine, yet in combination with the plastic shear modulus G_p, the elastic modulus has a strong influence on the bifurcation point.

7. REFERENCES

Arthur, J.F.R., Dunstan, T., Al-Ani, Q.A.J. and Assadi, A. (1977). "Plastic deformation and failure in granular material." *Géotechnique*, 27(1), 53-74.

Bardet, J.P. (1991). "Orientation of shear bands in frictional soils." *J. Eng. Mech., ASCE*, 117, 1466-1484.

Desrues, J. and Hammad, W. (1989). "Experimental study of the localization of deformation on sand: Influence of mean stress." *Proc. 12th Intl. Conf. Soil Mech. Fdn. Eng.*, Rio de Janeiro, 1, 31-21.

Drucker, D.C. (1951). "A more fundamental approach to stress-strain relations." *Proc. First U.S. Natl. Cong. Appl. Mech.*, 487-491.

Drucker, D.C. (1956). "On uniqueness in the theory of plasticity." *Quart. Appl. Math.*, 14, 35-42.

Hardin, B.O. and Black, W.L. (1968). Vibration modulus of normally consolidated clay. *J. Soil Mech. Fnds. Div.* ASCE, 94 (2), 353-369

Hill, R. (1958). "A general theory of uniqueness and stability in elastic-plastic solids." *J. Mech. Phys. Solids*, 6, 236-249.

Mandel, J. (1966). "Conditions de stabilité et postulat de Drucker." *Proc. IUTAM Symp. Rheology and Soil Mechanics,* Springer, Berlin, 58–68.

Roscoe, K.H. (1970). "The influence of strain in soil mechanics." *Géotechnique*, 20(2), 129-170.

Rudnicki, J.W. and Rice J.R. (1975). "Conditions for the localization of deformation in pressure-sensitive dilatant materials." *J. Mech. Phys. Solids*, 23, 371-394.

Tatsuoka, F., Nakamura, S., Huang, C.C. and Tani, K. (1990). "Strength anisotropy and shear band direction in plane strain tests of sand." *Soils and Foundations*, 30(1), 35-54.

Vardoulakis, I. (1980). "Shear band inclination and shear modulus of sand in biaxial tests." *Intl. J. Num. Analy. Meth. Geomech.*, 12, 155-168.

Vermeer, P.A. (1982). "A simple shear-band analysis using compliances." *Proc. IUTAM Conf. Deform. Failure Granular Matls.*, Delft, Netherlands, 493-499.

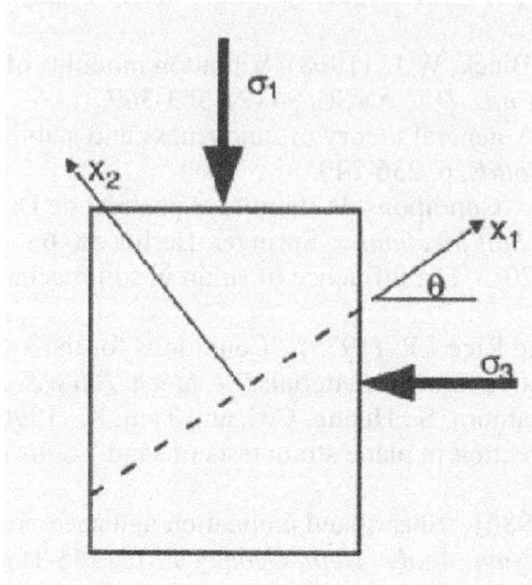

Fig. 1 – Soil sample under biaxial loading conditions with a shear band. Axis x_1 is parallel to the shear band and θ is the shear band orientation measured from the σ_3-axis.

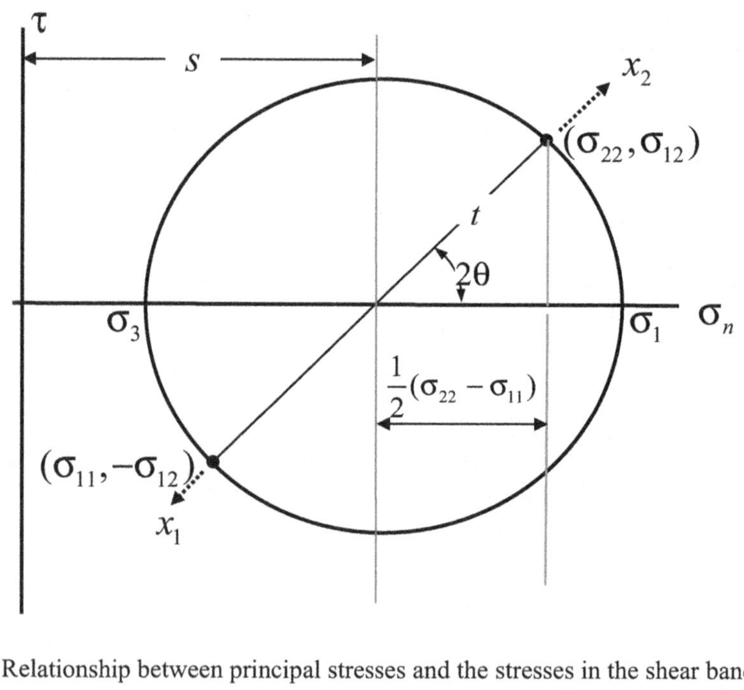

Fig. 2 – Relationship between principal stresses and the stresses in the shear band.

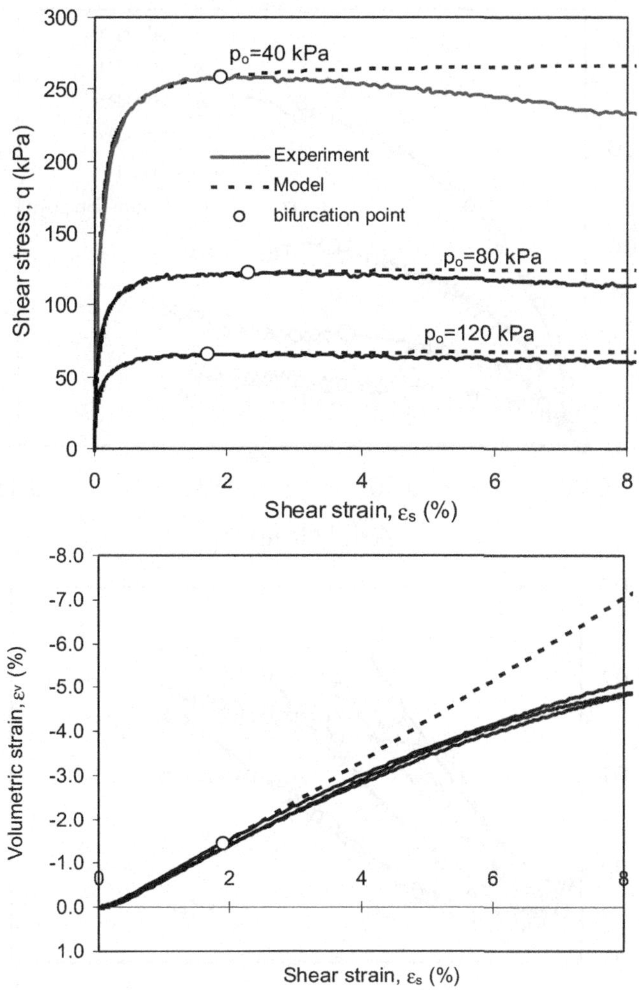

Fig. 3 – Comparison between simulated and experimental stress-strain data for Nevada sand under drained *p*-constant triaxial tests.

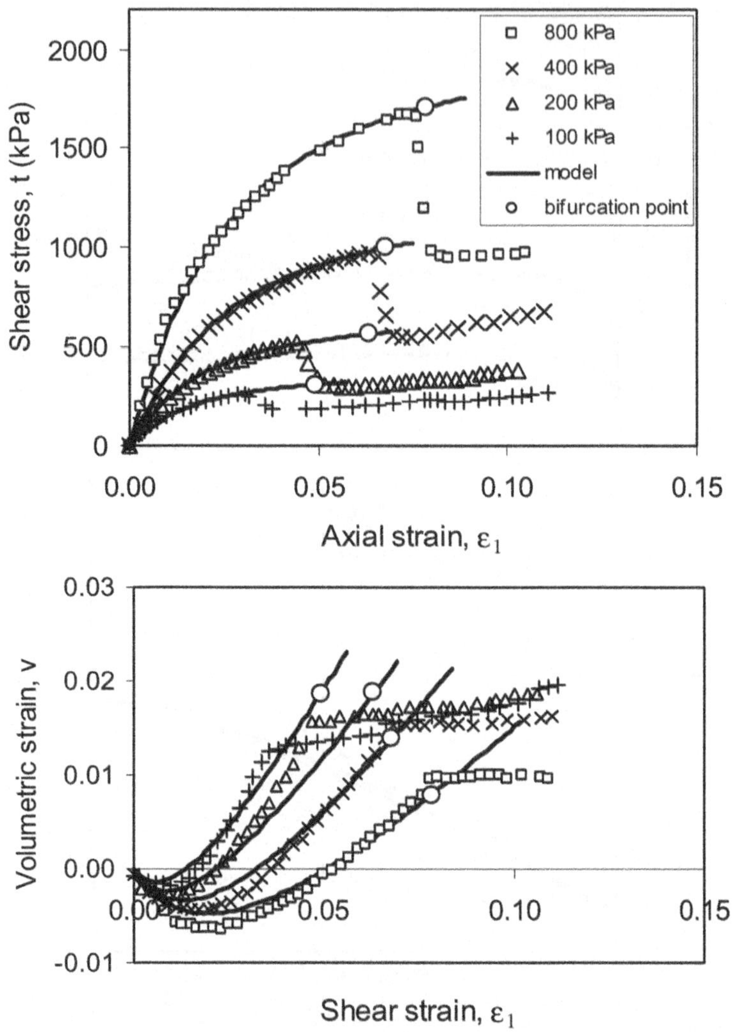

Fig. 4 – Comparison between simulated and experimental stress-strain data for Hostun sand under drained σ_3-constant and plane strain biaxial tests. Values shown in the legend are the confining stress used in each tests.

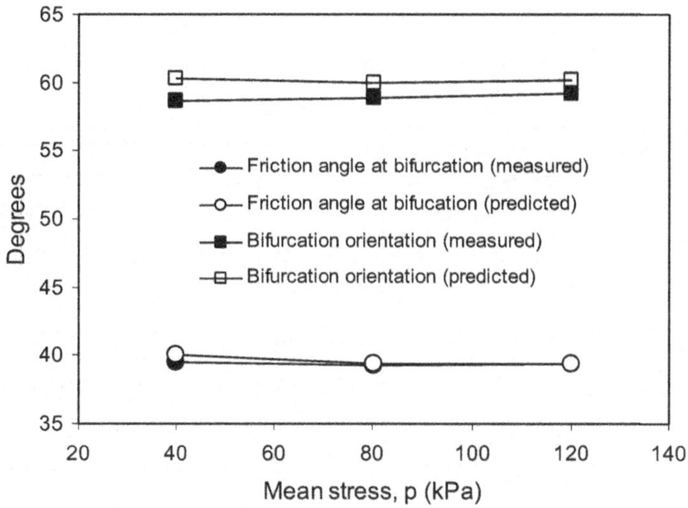

Fig. 5 – Comparison between predicted and measured friction angles at bifurcation and shear band orientations for Nevada sand.

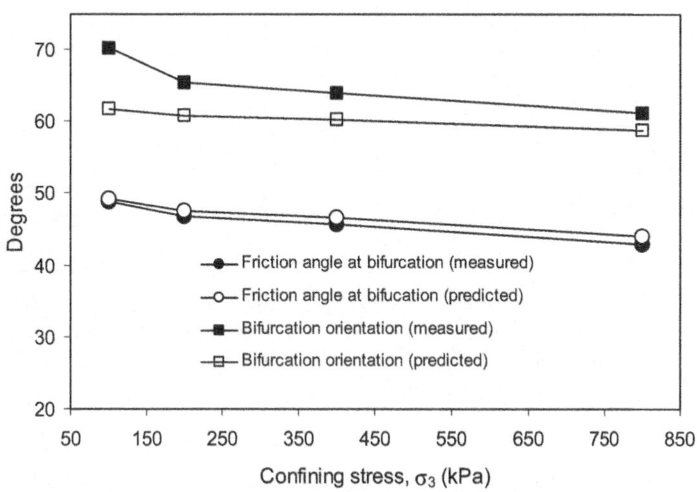

Fig. 6 – Comparison between predicted and measured friction angles at bifurcation and shear band orientations for Hostun sand.

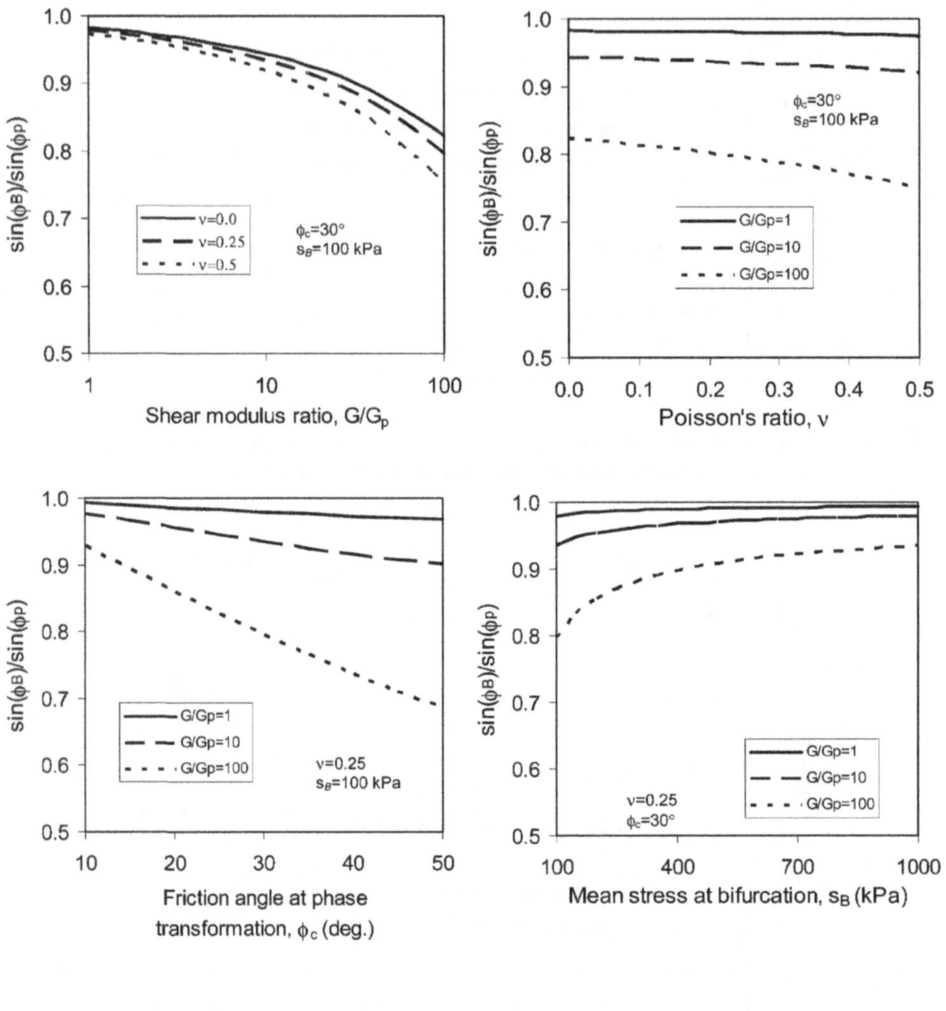

Fig. 7 – Effect of constitutive parameters on friction angle at bifurcation.

THE BEHAVIOUR OF A NORMALLY LOADED CLAYEY SOIL AND ITS SIMULATION

Georgios Belokas
Civil Engineering Faculty - Geotechnics Division
National Technical University of Athens, Iroon Polytechniou 9, Zografou GR-15780, Greece
e-mail: gbelokas@central.ntua.gr

Angelo Amorosi
Department of Civil and Environmental Engineering
Technical University of Bari, Via E. Orabona 4, 70120 Bari, Italy
e-mail: a.amorosi@poliba.it

Michael Kavvadas
Civil Engineering Faculty - Geotechnics Division
National Technical University of Athens, Iroon Polytechniou 9, Zografou GR-15780, Greece
e-mail: kavvadas@central.ntua.gr

ABSTRACT
The paper investigates the behaviour of reconstituted Vallericca clay under radial, axisymmetric, compressive stress paths and subsequent undrained triaxial shear. As reconstituted soils have no bonding, their behaviour under such stress paths is controlled by the initial stress state and specific volume. The simulation of volumetric compression is explored within the recently presented framework for radial stress paths of unbonded soils. Then the anisotropic, plasticity-based constitutive model MSS-2 is used to simulate the experimental results. The model predictions compare well with the corresponding test results.

1. INTRODUCTION
The behaviour of reconstituted soils under axisymmetric radial compression paths has been examined by Kavvadas & Belokas (2001) and Belokas *et al* (2005), which lead to their proposed behavioural framework for radial stress paths of normally loaded unbonded soils. These paths correspond to drained stress paths resulting from a proportional increase of the cell pressure and the axial stress in a standard triaxial cell. Normally loaded reconstituted soils lack any bonding and their state corresponds to their maximum precompression pressure (i.e. OCR=1). Therefore, their behaviour under radial stress paths is controlled by the initial stress state and specific volume, which means that, according to Leroueil & Vaughan (1990) definition, they are in a structureless state.

In this paper the behaviour of reconstituted Vallericca clay is being presented and simulated. The first aim is to show how the experimental results fit in the behavioural

Hoe I. Ling et al. (eds.), Soil Stress-Strain Behavior: Measurement, Modeling and Analysis, 707–718.

framework for radial stress paths. Then the variance of the constitutive model MSS-2 (Kavvadas & Belokas, 2001) for unbonded soils presented by Belokas *et al* (2005) is being used to simulate the experimental results.

2. SOIL EXAMINED – EXPERIMENTAL PROGRAM

Reconstituted specimens of Vallericca clay, a Plio-Pleistocene marine clay from a site a few kilometres north of Rome, were tested in the laboratory. This soil material is a stiff, overconsolidated, medium plasticity and activity clay (liquid limit 55.5%, plastic limit 25.3%, plasticity index 30.3%) characterised by a calcium carbonate content of about 30% (Rampello *et al*, 1993; Amorosi, 1996). Samples were reconstituted at moisture content equal to about 75% of the liquid limit and consolidated one-dimensionally in a large diameter consolidometer under a vertical stress of σ_v=100kPa. Then four cylindrical specimens having typical diameter 38.2mm and height 76.4mm where prepared from the batch. As specimen preparation involves undrained unloading (zero isotropic total stress), negative pore pressures (suction) developed.

The loading sequence in the triaxial cell involved four stages. An initial isotropic stress (σ_v=σ_h) equal to about 50 - 100kPa was applied to overcome the negative pore pressure caused by specimen preparation. The second stage involved a p=ct stress path up to the desired stress ratio η=q/p (except test 1), while the third one a radial stress path with η=ct (Table 1 and Figure 1). Two of the specimens (Tests 1, 3) were subsequently sheared under undrained conditions starting from the final stress level reached during radial consolidation (fourth stage).

Table 1. Testing program on Vallericca clay.

Test	1	2	3	4
Stress ratio of the radial stress path : $\eta = q/p$	0.0	0.3	0.5	0.7
Maximum mean effective stress : p (kPa)	2650	700	700	600
Subsequent undrained shear :	yes	no	yes	no

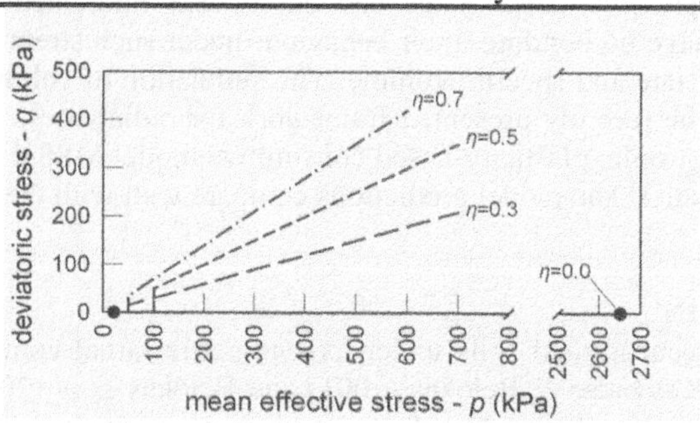

Figure 1. Radial compression paths.

Test 1 was performed in a standard triaxial apparatus equipped with a plexiglass cell. The hydrostatic cell pressure was applied in 10 increments, each maintained for 24 hours. At that time, full dissipation of excess pore water pressures (pwp) was achieved. Drainage of the specimen was permitted from top and bottom only, without lateral filter strips. Volumetric strain was measured by monitoring the volume of water squeezed out of the specimen.

Tests 2, 3 and 4 were performed in a high pressure triaxial apparatus equipped with a steel cell and a very stiff steel frame to practically eliminate compliance of the load application system. During the *p=ct* and *η=ct* stress paths, the axial force was increased at a sufficiently slow rate to prevent excess pore water pressure build-up in the specimen. Drainage of the specimens was accelerated by adding filter strips at the perimeter. The applied axial force and vertical displacements were measured both internally (using an internal load cell and *LVDT* transducers) and externally. External and internal measurements were in very good agreement, proving that the compliance of the loading system was indeed very small. Volumetric strain was measured by monitoring the volume of water squeezed out of the specimen. Figures 1 and 2 plot the four radial stress paths in the *p–q* and *v–p* spaces, respectively.

Following radial consolidation to the maximum mean effective stress (Table 1), Tests 1 and 3 were sheared in an undrained mode at a constant rate of axial displacement of 0.1 mm/min, a value sufficiently slow to ensure pore water pressure equilibration through the specimen. In Test 1, shearing started shortly after the end of radial consolidation, while in Test 3 some creep was allowed under the maximum consolidation pressure prior to undrained shearing. The resulting stress path is presented in Figure 3.

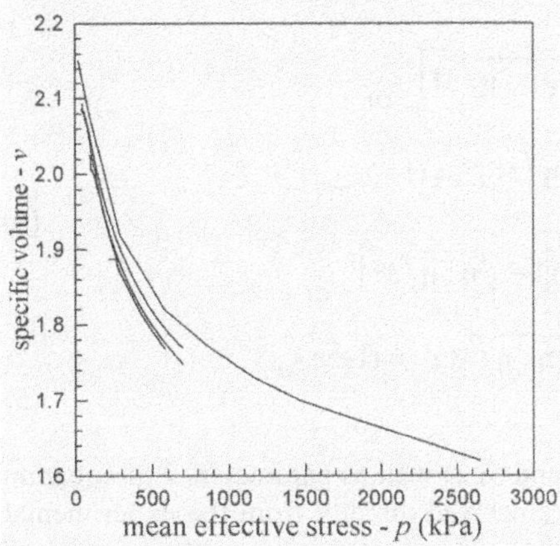

Figure 2. Experimental compression curves.

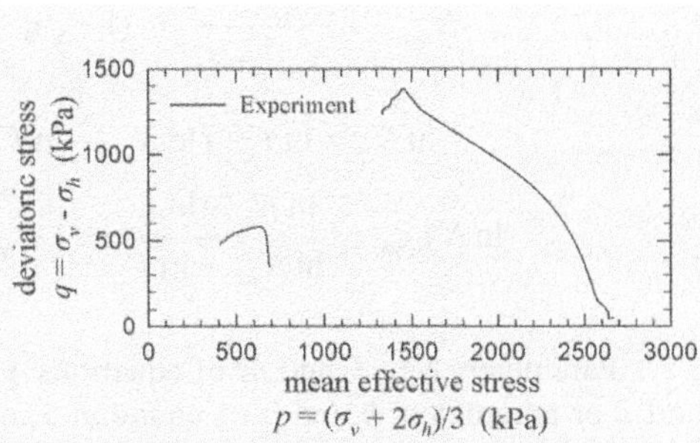

Figure 3. Undrained stress paths.

3. SIMULATION OF VOLUMETRIC BEHAVIOUR

Various approaches have been proposed and used for the simulation of the volumetric behaviour of soils, especially under radial compression. The most widespread used is the linear $v - \ln p$ (equation 1), which although it proves satisfactory for the usual range of applied stresses it has the theoretical disadvantage that in extremely high stresses it can predict negative specific volume. This problem has been overcome by the introduction of the linear $\ln v - \ln p$ (equation 2) relation (e.g. Butterfield, 1979, Hashiguchi 1995), while more recently the linear $\ln v - p^{\alpha}$ (equation 3) relation has been proposed (Wan & Guo (1999) used it to describe critical state of coarse grained soils).

$$v = N_\eta - \lambda \ln p \quad (1)$$

$$\ln v = \ln N_\eta - \lambda \ln p \quad (2)$$

$$\ln v = \ln N_\eta - (p/h)^a \quad (3)$$

where in equations 1 and 2 N_η is a parameter that depends on the stress ratio (η=q/p) of the applied radial loading and corresponds to the specific volume for p=1kPa, while is λ a material constant controlling the volumetric strain rate. For isotropic compression it is N_η=N_{iso}, while for critical state it is N_η=Γ. In equation 3 N_η it corresponds to the specific volume for p=0kPa (i.e. the loosest possible state) and if we assume a unique loosest possible state, then it is N_η=ct=N_o (i.e. material constant). In this case parameters a and h may also depend on stress ratio.

According to the framework for radial compression of Belokas *et al* (2005) and Belokas & Kavvadas (2006), when equation 1 (linear v – ln$p$) is used, equation 4 can be used for the estimation of N_η. This equation includes the assumption that λ=ct and has been based on an extensive literature review, which provided enough data. For linear ln$v$ – ln$p$ representation equation 5 is proposed in this paper (it also implies that λ of equation 2 is constant), while the case of linear ln$v$ – p^a is discussed later in this paragraph. In equations 4 and 5, M is the critical slope and $\boldsymbol{\eta}$=$\mathbf{s}/\sigma$ is the stress ratio tensor.

$$N_\eta = \Gamma + (N_{iso} - \Gamma)\left[1 - \sqrt{\boldsymbol{\eta}:\boldsymbol{\eta}}/M\right]^n \text{ or}$$
$$N_{norm} = \frac{N_\eta^{\ *} - \Gamma^*}{N_{iso}^{\ *} - \Gamma^*} = (1 - \sqrt{\boldsymbol{\eta}:\boldsymbol{\eta}}/M)^n = (1 - \eta_{norm})^n \quad (4)$$

$$\ln N_\eta = \ln \Gamma + (\ln N_{iso} - \ln \Gamma)\left[1 - \sqrt{\boldsymbol{\eta}:\boldsymbol{\eta}}/M\right]^n \text{ or}$$
$$(\ln N)_{norm} = \frac{\ln N_\eta - \ln \Gamma}{\ln N_{iso} - \ln \Gamma} = (1 - \sqrt{\boldsymbol{\eta}:\boldsymbol{\eta}}/M)^n = (1 - \eta_{norm})^n \quad (5)$$

Parameters N_{iso}, Γ and M of equations 4 and 5, as well as parameters λ of equation 1 and 2 or parameters h and a of equation 3, are derived directly from the experimental data. In order to estimate parameter n, a straightforward approach is to use the normalized plots of N_{norm} vs η_{norm} (when equation 4 is used) or lnN_{norm} vs η_{norm} (when equation 5 is used). The analysis of the data of the reconstituted Vallericca clay leads to the parameters of Tables 2a and 2b when equations 1 and 2 are used. In Figures 4a and 4b the normalized plots are presented from which parameter n was obtained.

Table 2a. Parameters of *ICC* curves of equation 1 derived directly from the experimental data.

η=	0	0.3	0.5	0.7	0.87 (=M)
N_η=	2.810	2.772	2.752	2.742	2.7100
λ=	0.153	0.153	0.153	0.153	0.153

Table 2b. Parameters of *ICC* curves of equation 2 derived directly from the experimental data.

η=	0	0.3	0.5	0.7	0.87 (=M)
$\ln N_\eta$=	1.138	1.117	1.106	1.100	1.075
λ=	0.0835	0.0835	0.0835	0.0835	0.0835

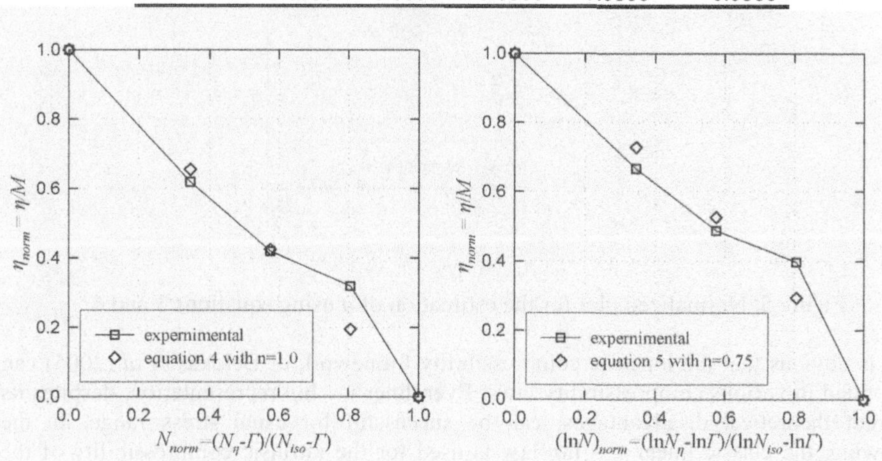

(a) (b)

Figure 4. Normalized plot for the estimation of *n* using equations: (a) 1 and 4 and (b) 2 and 5.

When attempting to simulate the experimental data of the reconstituted Vallericca clay with equation 3 (i.e. linear $\ln v - p^a$) we come up to the conclusion that, besides keeping $N_\eta=N_o=ct$, we can also keep $a=ct$ (it controls the shape of the *ICC* curves). A set of parameters that simulates well the data is given in Table 3. Parameter h can be normalized to $h_{norm}=(h_\eta-h_{cs})/(h_{iso}-h_{cs})$, where the subscript denotes the state or the stress ratio. The normalized data are presented in Figure 5, where a simulation using the newly proposed equation 6 is made.

Table 3. Parameters of *ICC* curves of equation 3 derived directly from the experimental data

η=	0	0.3	0.5	0.7	0.87 (=M)
$\ln N_\eta=\ln N_o$=	3.350	3.350	3.350	3.350	3.350
h=	25000	19500	17000	16000	12000
a=	0.135	0.135	0.135	0.135	0.135

$$h_\eta = h_{cs} + (h_{iso} - h_{cs})\left[\sqrt{\boldsymbol{\eta}:\boldsymbol{\eta}}/M\right]^n \quad \text{or}$$

$$h_{norm} = \frac{h_\eta - h_{cs}}{h_{iso} - h_{cs}} = \left(\sqrt{\boldsymbol{\eta}:\boldsymbol{\eta}}/M\right)^n = (\eta_{norm})^n \qquad (6)$$

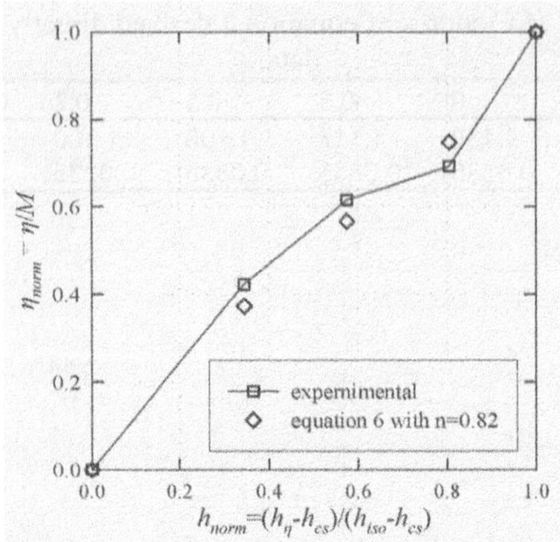

Figure 5. Normalized plot for the estimation of *n* using equations 3 and 6.

It is obvious that the intrinsic compressibility framework of Belokas *et al* (2005) can be applied in various compressibility laws. Even liner $v - \ln p$ representation, despites its inherent theoretical disadvantages, can be successful for usual stress range. In the following, the classic linear $v - \ln p$ law is used for the intrinsic compressibility of the MSS-2 model.

4. FORMULATION OF THE MSS-2 MODEL FOR UNBONDED SOILS

The present formulation is a variance for unbonded soils, of the model proposed by Kavvadas & Belokas (2001) for anisotropic and structured soils. It is a two–surface elasto–plastic model formulated in the context of bounding surface plasticity (Dafalias & Popov, 1975). The model is expressed in a general tensorial stress ($\boldsymbol{\sigma}$) and strain ($\boldsymbol{\varepsilon}$) space, decomposed in isotropic components (σ, ε) and deviatoric components (**s**, **e**) :

$$\boldsymbol{\sigma} = \mathbf{s} + \sigma \mathbf{I} \quad , \quad \boldsymbol{\varepsilon} = \mathbf{e} + (1/3)\varepsilon \mathbf{I} \qquad (7), (8)$$

In the present case, as the laboratory tests were performed in a triaxial cell (axisymmetric conditions), experimental results and model predictions are presented and evaluated in the classical (p, q) effective stress space, where $p=\sigma=(\sigma_v+2\sigma_h)/3$, and $q=\sqrt{3/2(\mathbf{s}:\mathbf{s})}=\sigma_v-\sigma_h$, and conjugate strain quantities: volumetric $\varepsilon_{vol}=\varepsilon=\varepsilon_v+2\varepsilon_h$ and deviatoric $\varepsilon_q=\sqrt{2/3(\mathbf{e}:\mathbf{e})}=\varepsilon_v-\varepsilon_h$.

The model has two characteristic surfaces in stress space, an Intrinsic Strength Envelope (*ISE*) and a Plastic Yield Envelope (*PYE*) (Figure 6). The *ISE*, described by equation 9, is a bounding envelope enclosing all possible states of the unbonded material. It is a rotated and distorted ellipsoid, characterised by its size along the isotropic axis (hardening variable α), its size in the deviatoric space (equal to $c\alpha$, where c is a material constant) and the location of its centre (K) (tensor $\boldsymbol{\sigma}_K$). The deviatoric part ($\mathbf{s}_K$) of $\boldsymbol{\sigma}_K = \mathbf{s}_K + \sigma_K \boldsymbol{I}$, gives the offset of the *ISE* from the isotropic axis and thus is a measure of material strength anisotropy.

$$F(\boldsymbol{\sigma};\boldsymbol{\sigma}_K,\alpha)\equiv\frac{1}{c^2}(\mathbf{s}-\frac{\sigma}{\sigma_K}\mathbf{s}_K):(\mathbf{s}-\frac{\sigma}{\sigma_K}\mathbf{s}_K)+(\sigma-\sigma_K)^2-\alpha^2=0 \quad (9)$$

The Plastic Yield Envelope *PYE,* described by equation 10, encloses the region where material behaviour is elastic (i.e., it is a classical yield surface). It is a scaled-down version of the ISE (by the constant $\xi<1$), fully enclosed in the *ISE*, with its centre (*L*) given by the tensor $\boldsymbol{\sigma}_L = \mathrm{s}_L + \sigma_L \boldsymbol{I}$. Stress states ($\boldsymbol{\sigma}$) can be either inside the *PYE* (elastic states) or on the *PYE* (plastic states).

$$f(\boldsymbol{\sigma};\boldsymbol{\sigma}_K,\boldsymbol{\sigma}_L,\alpha)\equiv\frac{1}{c^2}(\mathbf{s}-\frac{\sigma-\sigma_L}{\sigma_K}\mathbf{s}_K-\mathbf{s}_L):(\mathbf{s}-\frac{\sigma-\sigma_L}{\sigma_K}\mathbf{s}_K-\mathbf{s}_L)+(\sigma-\sigma_L)^2-(\xi\alpha)^2=0 \quad (10)$$

Both envelopes may harden isotropically (by varying their sizes α and $\xi\alpha$) and kinematically (by varying the location of their centres *K* and *L*). As the two envelopes are similar in shape, when the stress state reaches *ISE*, the *PYE* comes in contact with *ISE* at conjugate points (i.e., no crossing).

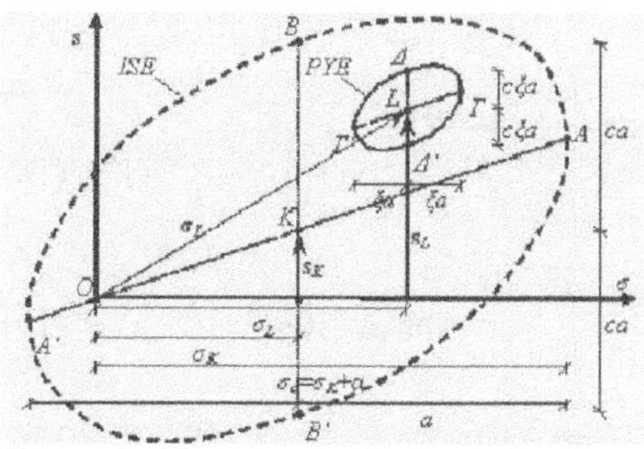

Figure 6. The two – surface envelopes

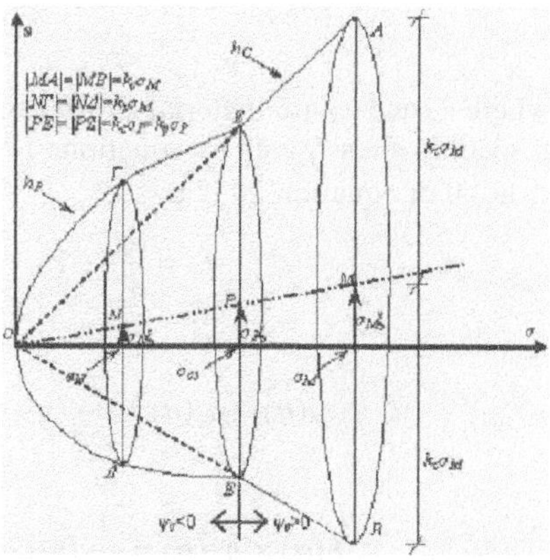

Figure 7. Representation of functions h_c and h_p

Plastic strains are given by a classical incrementally linear non–associated flow rule:

$$\dot{\boldsymbol{\varepsilon}}^p = \dot{\Lambda}_f \cdot \mathbf{P}_f \quad (11)$$

where:

$$\dot{\Lambda}_f = \frac{1}{H_f}\left(\mathbf{Q}_f : \dot{\boldsymbol{\sigma}}\right) = \frac{1}{H_f}\left(Q_f \cdot \dot{\sigma} + \mathbf{Q}'_f : \dot{\mathbf{s}}\right) \quad (12)$$

is a scalar measure of the magnitude of plastic strains, $\mathbf{P}_f$ is the plastic potential tensor representing the direction of the plastic strains, H_f is the plastic hardening modulus and

$\mathbf{Q}_f$=$\mathbf{Q}_f'$+ $Q_f\mathbf{I}$ is the gradient of the *PYE* given by equations 13 (isotropic component) and 14 (deviatoric component):

$$Q_f = \frac{\partial f}{\partial \sigma} = 2(\sigma - \sigma_L) - \frac{2}{c^2}\frac{1}{\sigma_K}(\mathbf{s}\text{-}\frac{\sigma - \sigma_L}{\sigma_K}\mathbf{s}_K - \mathbf{s}_L) : \mathbf{s}_K \tag{13}$$

$$\mathbf{Q}_f = \frac{\partial f}{\partial \boldsymbol{\sigma}} = \frac{\partial f}{\partial \boldsymbol{\sigma}} - \frac{1}{3}Q_f\mathbf{I} = \frac{\partial f}{\partial \boldsymbol{\sigma}} - \frac{1}{3}\left(\frac{\partial f}{\partial \mathbf{s}} : \mathbf{I}\right)\mathbf{I} = \frac{2}{c^2}(\mathbf{s}\text{-}\frac{\sigma - \sigma_L}{\sigma_K}\mathbf{s}_K - \mathbf{s}_L) \tag{14}$$

The deviatoric ($\mathbf{P}_f'$) and isotropic (P_f) components of the plastic potential tensor ($\mathbf{P}_f$= $\mathbf{P}_f'$ + $P_f\mathbf{I}$) are:

$$P_f \equiv -\lambda_1 \psi_\sigma \left[\frac{h(\boldsymbol{\sigma})}{\sigma_o^2}\right]\sigma \tag{15}$$

$$\mathbf{P}_f' \equiv (\mathbf{s} - \lambda_2 \frac{\sigma}{\sigma_K}\mathbf{s}_K) \tag{16}$$

where λ_1 and λ_2 are material constants, σ_o=σ_K+α (Figure 5), ψ_σ is a state parameter with respect to stress given by equations 17 and 18 and $h(\boldsymbol{\sigma})$ is a function given by equation 19 if $\psi_\sigma \geq 0$ or equation 20 if $\psi_\sigma < 0$.

$$\psi_\sigma = \frac{\sigma}{\sigma_{cs}} - 1 \quad \text{and} \quad \sigma_{cs} = \exp\left[\frac{\Gamma - v_{current}}{\lambda^*}\right] \tag{17), (18}$$

$$h(\boldsymbol{\sigma}) \equiv h_c(\boldsymbol{\sigma}) = \frac{1}{k^2}(\mathbf{s} - \sigma\boldsymbol{\xi}) : (\mathbf{s} - \sigma\boldsymbol{\xi}) - \sigma^2 = 0 \quad \text{for } \psi_\sigma \geq 0 \tag{19}$$

$$h(\boldsymbol{\sigma}) \equiv h_p(\boldsymbol{\sigma}) = \frac{1}{k^2}(\mathbf{s} - \sigma\boldsymbol{\xi}) : (\mathbf{s} - \sigma\boldsymbol{\xi}) - \sigma\sigma_{cs} = 0 \quad \text{for } \psi_\sigma < 0 \tag{20}$$

Equation 18 defines an equivalent stress on the critical state curve (*CSC*) for the current specific volume and expresses the phase transformation curve. Equation 19 represents a rotated and distorted conical surface, while equation 16 a rotated and distorted parabolic cone (Figure 7). Scalar k and deviatoric tensor $\boldsymbol{\xi}$ are material constants describing these conical surfaces (Figure 7).

The model possesses a Cam–clay type isotropic hardening rule controlling the size α of the *ISE* (equation 21). The hardening parameter α corresponds to that derived from equations 1 and 4 for σ=σ_o and $\boldsymbol{\eta}$=$\boldsymbol{\eta}_K$=$\mathbf{s}_K/\sigma_K$.

$$\dot{\alpha} = \alpha\left(\frac{v}{\lambda - \kappa}\right)\dot{\varepsilon}_v^p \tag{21}$$

The model includes kinematic hardening of both *ISE* and *PYE*. The type of kinematic hardening depends on the position of material state with respect to *ISE*. For states inside the *ISE* (but on the *PYE*), kinematic hardening ensures that *ISE* does not rotate in stress space:

$$\dot{\boldsymbol{\sigma}}_K = (\dot{\alpha}/\alpha)\boldsymbol{\sigma}_K \tag{22}$$

For states on the *ISE* (and on the *PYE*), the *ISE* may rotate by the quantity: $\psi[\mathbf{s}-\chi(\sigma/\sigma_K)\mathbf{s}_K]$, where ψ and χ are material constants :

$$\dot{\boldsymbol{\sigma}}_K = (\dot{\alpha}/\alpha)[\boldsymbol{\sigma}_K + \psi(\mathbf{s} - \chi(\sigma/\sigma_K)\mathbf{s}_K)] \tag{23}$$

Kinematic hardening of the *PYE* aims at ensuring that when *PYE* comes in contact with *ISE*, they contact at conjugate points thus avoiding intersection. This is achieved by tensor $\boldsymbol{\beta}$ which connects the current stress state (on the *PYE*) with its conjugate point on the *ISE*:

$$\boldsymbol{\beta} = (\boldsymbol{\sigma}\text{-}\boldsymbol{\sigma}_L)/\xi - (\boldsymbol{\sigma}\text{-}\boldsymbol{\sigma}_K) \tag{24}$$

$$\dot{\boldsymbol{\sigma}}_L = (\dot{\alpha}/\alpha)\boldsymbol{\sigma}_L + \dot{\mu}\cdot\boldsymbol{\beta} \tag{25}$$

The magnitude of kinematic hardening is defined by the scalar $\dot{\mu}$ which is evaluated from the consistency condition on the *PYE*:

$$\dot{\mu} = \frac{\mathbf{Q}_f : \dot{\boldsymbol{\sigma}} - \frac{\dot{\alpha}}{\alpha}\left(\mathbf{Q}_f : \boldsymbol{\sigma}_L + 2(\xi\alpha)^2\right)}{\mathbf{Q}_f : \boldsymbol{\beta}} \tag{26}$$

The model includes conventional isotropic poro-elasticity with stress-sensitive bulk modulus $K=v\sigma/\kappa$ and proportional shear modulus $G=(½)(2G/K)K$. Scalars κ and $(2G/K)$ are constants :

$$\dot{\sigma} = K\dot{\varepsilon}_v^e \qquad \dot{\mathbf{s}} = 2G\dot{\mathbf{e}}^e \tag{27), (28}$$

For material states on the *ISE*, the plastic hardening modulus is given by equation 29, where scalars R_f (equation 31) and T_f (equation 32) are determined by using the consistency condition and the plastic flow rule. For states inside the *ISE,* the interpolation mapping rule proposed by Kavvadas & Amorosi (2000) is used to achieve a smooth transition between elasticity and states on the *ISE*, which makes use of two parameters γ and μ (see Kavvadas & Amorosi, 2000).

$$H_{fo} \equiv H_f = 2\cdot\xi\cdot R_f\cdot T_f \tag{29}$$

$$R_f = \frac{v}{\lambda - \kappa} P_f \tag{30}$$

$$T_f = (\sigma - \sigma_K)\sigma + \frac{1}{c^2}\left(\mathbf{s} - \frac{\sigma}{\sigma_K}\mathbf{s}_K\right) : \left[\mathbf{s} - \frac{\sigma}{\sigma_K}\mathbf{s}_K + \frac{\sigma}{\sigma_K}\psi\left(\mathbf{s} - \chi\frac{\sigma}{\sigma_K}\mathbf{s}_K\right)\right] \tag{31}$$

5. SIMULATION WITH MSS-2 MODEL

MSS-2 plasticity based model was used for the numerical simulation of the tests performed. Concerning the volumetric simulation, it was shown that for the examined stress range all three alternatives can be used equally successful. In the current paper linear v versus ln$p$ behaviour was chosen for the volumetric law of the MASS-2 model.

The values of constants κ, λ, k, N_{iso}, Γ and n are obtained directly from the experimental compression curves. The value of ξ is arbitrarily set equal to a small value (0.05) as the elastic domain of normally consolidated unbonded clays is very small (e.g. Jardine, 1992; Smith *et al.*, 1992). The values of constants $2G/K$, ψ, χ, λ_1, λ_2, c, γ and μ are selected indirectly during the model calibration process. The value of λ_I is best determined from the results of isotropic and one-dimensional compression radial stress paths. Material constants γ and μ are best determined from unloading (rebound) tests and undrained shearing tests on overconsolidated specimens; such tests were not available in the present database.

Table 3: Values of material constants used in the simulation of Vallericca clay

κ	λ	$2G/K$	N_{iso}	Γ	n	λ_1	λ_2	ξ	k	c	ψ	χ	γ	μ
0.035	0.153	1.5	2.81	2.70	1	1.4	0.5	0.05	0.71	0.80	1.2	1	3	1

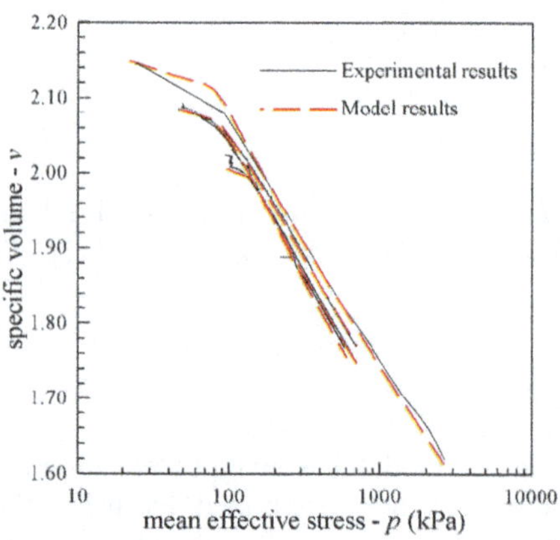

Figure 8. Compression curves simulation

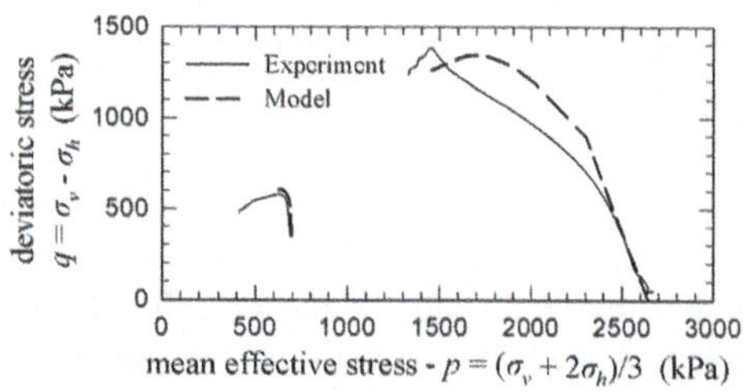

Figure 9. Undrained stress paths simulation

Figure 8 compares the radial compression curves in v–ln$p$ space with the corresponding model predictions. The laboratory curves are practically parallel for stress levels exceeding about 100 kPa verifying the basic assumption of the proposed framework. As several material constants are obtained from these curves, the excellent match with model predictions should be anticipated. Figures 9 to 11 compare the results of the two undrained shearing tests with the corresponding model predictions and show that the proposed model can simulate successfully both compression and shearing tests, despite the fact that the model calibration database was very small.

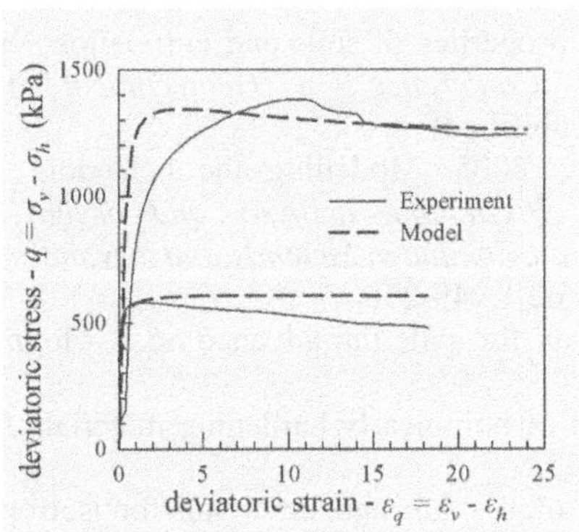

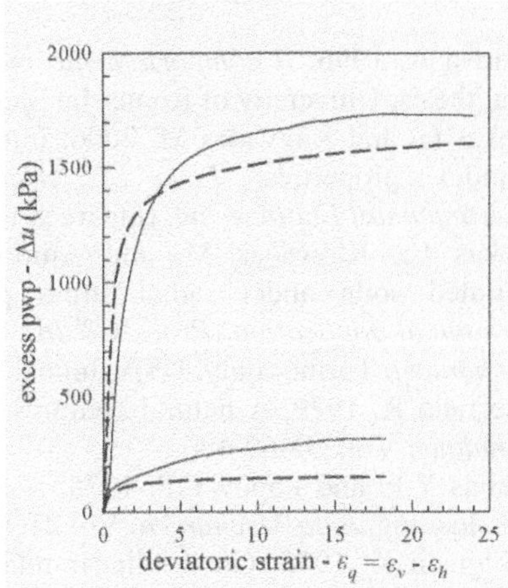

Figure 10. Deviatoric stress – strain curves Figure 11. Excess pwp vs deviatoric strain

6. CONCLUSIONS

The paper investigates the behaviour of the normally loaded reconstituted Vallericca clay and the application of the intrinsic compressibility framework (Belokas *et al*, 2005, Belokas & Kavvadas, 2006) and of the constitutive model MSS-2.

The intrinsic compressibility framework describes the behaviour of unbonded clays under radial stress paths. This framework is applied successfully to the normally loaded reconstituted Vallericca clay and it is exhibited that different intrinsic compressibility laws fit well into the framework.

A variance of the plasticity-based constitutive model MSS-2 is applied for the numerical analyses of unbonded anisotropic soils. The model is used to simulate the observed behaviour of the reconstituted normally consolidated specimens of Vallericca clay compressed under four radial stress paths and two undrained shearing paths (starting at the end of the radial compression). Several of the model parameters are obtained directly from the radial and shearing compression paths while the rest are selected indirectly during the calibration process (using both compression and shearing test results). Comparison between observed behaviour and model predictions shows that the proposed model can simulate successfully both compression and shearing tests, despite the fact that the model calibration database was very small.

7. ACKNOWLEDGEMENTS

The laboratory tests on Vallericca clay were carried out by the first author at the Geotechnical Laboratory of the University of Rome "La Sapienza" under the Socrates/Erasmus program. The authors wish to express their gratitude to Prof. G. Calabresi of the University of Rome "La Sapienza" for the use of the laboratory equipment and to Prof. S. Rampello, Dr A. Sciotti and Dr L. Callisto for their assistance with the testing procedures. Research work carried out by the doctoral student G. Belokas is funded by the National Technical University of Athens.

8. REFERENCES

Amorosi A. 1996. *Il comportamento meccanico di una argilla naturale consistente*, Doctoral thesis, University of Rome 'La Sapienza'.

Belokas G. and Kavvadas M. 2006. Intrinsic properties of soils and correlation with their index properties. *Proc. 5th Hellenic Conference on Geotechnical and Geoenvironmental Engineering*. (in Greek to be published)

Belokas G., Kavvadas M. and Amorosi A. 2005. Modelling the behaviour of reconstituted soils under radial stress paths. *Prediction, analysis and design in geomechanical application (Proc. 11th Int. Conf. on Computer Methods and Advances in Geomechanics)*. Torino, Italy, USA, June 2005. Vo2 1:249-256.

Butterfield R. 1979. A natural compression law for soils (an advance on e – logp'). *Géotechnique*. Vol.29:469-480.

Dafalias Y.F. and Popov E.P. 1975. A model of nonlinearly hardening materials for complex loading. *Acta Mechanica*. Vol.21:173-192.

Hashiguchi K. 1995. On the linear relations of v – ln$p$ and ln$v$ – ln$p$ for isotropic consolidation of soils. Short communication. *International Journal for Numerical and Analytical Methods in Geomechanics*. Vol.19:367-376.

Jardine R.J. 1992. Some observations on the kinematic nature of soil stiffness. *Soils and Foundations*. Vol.32(2):111-124.

Kavvadas M. and Belokas G. 2001. An Anisotropic Elastoplastic Constitutive Model for Natural Soils. *Proc. 10th Int. Conf. on Computer Methods and Advances in Geomechanics (IACMAG)*. Tucson, Arizona, USA, January 2001. Vol 1:335-340.

Kavvadas M. and Amorosi A. 2000. A Constitutive Model for Structured Soils. *Géotechnique*. 50(3): 263-274.

Leroueil S. and Vaughan P.R. 1990. The general and congruent effects of structure in natural soils and weak rocks. *Géotechnique*. Vol.40(3):467-488

Rampello, S., Georgiannou, V.N. and Viggiani G. 1993. Strength and dilatancy of natural and reconstituted Vallericca clay, *Proc. Int. Symp. on Hard Soils - Soft Rocks*, Athens, Vol. 2, pp 761-768.

Smith P.R., Jardine R.J. and Hight D.W. 1992. The yielding of Bothkennar clay. *Géotechnique*. Vol.42, No2:257-274.

Wan R.G. and Guo P.J. 1999. A pressure and density dependent dilatancy model for granular materials. *Soils and Foundations*. Vol.39(6):1-11.

Soil Stress-Strain Behavior: Measurement, Modeling and Analysis
Geotechnical Symposium in Roma, March 16 & 17, 2006

A FAST IMPLICIT INTEGRATION SCHEME TO SOLVE HIGHLY NONLINEAR SYSTEM

Md. Saiful Alam Siddiquee
Professor, Civil Engineering Department,
Bangladesh University of Engineering and Technology (BUET), Dhaka, Bangladesh
e-mail: sid@ce.buet.ac.bd

ABSTRACT

Now-a-days researchers are formulating new generation of soil-models based on combined theory. That means researchers are trying to put forward a unified material model, which would predict at least the behaviour of all types of soils under all types of stress and time paths. So the solution techniques so far being used by the nonlinear Finite Element packages no longer can meet the huge demand of computational speed created by those models. It was necessary to develop a new type of solution scheme for the sophisticated models. Usually material nonlinearity makes it difficult to create a robust solution technique. So it is important to develop a solution scheme which will be very robust at the same time. That means the solution scheme should not break-down even for a notoriously complicated unified model. In this paper, we have developed an implicit solution scheme, which solves the resulting nonlinear equations of motion by implicit dynamic relaxation. There are a myriad number of implicit schemes for the use. Here a relatively less used method -- called "Houbolt's integration scheme" has been used. It is very similar to the central difference scheme only difference is the use of the higher-order terms in the definition of velocity and acceleration. In order to make it faster, sparse-matrix solution scheme is used with partial pivoting and reordering of matrix elements to minimize the fill-ins. The combined effect is quite dramatic. It provides the main two traits of a good nonlinear solution technique--i.e., speed and robustness of solution. The solution scheme is applied to trace the full loading path of an elasto-visco-plastically defined material behaviour of a Plane Strain Compression (PSC) test sample. There is a huge gain in speed and robustness compared to the other techniques of solution.

Hoe I. Ling et al. (eds.), Soil Stress-Strain Behavior: Measurement, Modeling and Analysis, 719–726.

1. INTRODUCTION

Due to the advent of modern computational facilities, there is an outpour of combined soil models encompassing monotonic, cyclic, visco-elasto-plastic and muti-axial loading with inherent, induced anisotropies and aging effect. These apparently complex models pose a big challenge to the computational community to solve a boundary value problem of soil comprised of such model with in a reasonable time. The biggest problem of this kind of computation is the robustness of the solution scheme. As the resulting systems of equations are highly nonlinear both in space and time, it is easy to be in a flat plateau of solution, where the solution does not converge anymore. So far, there is a single kind of solution technique, which is very robust in its tenacity to trace the solution, that is, Dynamic Relaxation technique. This falls within the group of Conjugate Gradient type of solution, which are of explicit type of integration. Although this type of solution technique is very robust, it suffers from the limitation of time step. That means it has stability limits. So in order to achieve a highly accurate solution of a boundary value problem with very fine mesh can take unusually very high computer time, which is only comparable with Discrete Element Method (DEM).

On the other hand, there are some implicit types of integrations techniques, which are free from any stability limit. But these implicit types of stable integrators suffer from lack of robustness, especially in case of highly nonlinear material constitutive equations. Due to its nature of computational scheme, it needs to construct a global stiffness matrix and solve it every time in its force or dynamic equilibrium iteration, which is also time consuming. The objective of this paper is to describe an implicit type of integration technique, which is (1) free from stability limit (2) robustly tenacious and (3) computationally fast.

2. BASIC IMPLICIT INTEGRATION SCHEME

In this research, some of the widely used Implicit Integration Schemes are tested to find out a suitable candidate, which might fulfill the above three desired traits. It has been stated before that Dynamic Relaxation (Siddiquee *et al.*, 2001, 2006) possesses the robustness of solution, which is based on explicit central difference formula. Following the holy grail of central difference technique, Houbolt (Houbolt, 1950) integration scheme is found, which is very similar to central difference scheme according to its construction. The following finite difference expansions of velocity and accelerations are employed in the Houbolt integration method:

$$^{t+\Delta t}\ddot{U} = \frac{1}{\Delta t^2}\left(2\,^{t+\Delta t}U - 5\,^{t}U + 4\,^{t-\Delta t}U - 2\,^{t-2\Delta t}U\right) \tag{1}$$

and

$$^{t+\Delta t}\dot{U} = \frac{1}{6\Delta t}\left(11\,^{t+\Delta t}U - 18\,^{t}U + 9\,^{t-\Delta t}U - 2\,^{t-2\Delta t}U\right) \tag{2}$$

which are two backward difference formulas with errors of order $(\Delta t)^2$. In order to obtain the solution at time $t+\Delta t$, equilibrium equation can be updated as follows:

$$\left(\frac{2}{\Delta t^2}M+\frac{11}{6\Delta t}C+K\right)^{t+\Delta t}U = {}^{t+\Delta t}R+\left(\frac{5}{\Delta t^2}M+\frac{3}{\Delta t}C\right)^{t}U-\left(\frac{4}{\Delta t^2}M+\frac{3}{2\Delta t}C\right)^{t-\Delta t}U+\left(\frac{1}{\Delta t^2}M+\frac{1}{3\Delta t}C\right)^{t-2\Delta t}U \tag{3}$$

As shown in Eq. (3), the solution of $^{t+\Delta t}U$ requires the knowledge of ^{t}U , $^{t-\Delta t}U$ and $^{t-2\Delta t}U$. Although the knowledge of displacement, velocity and acceleration at the start is useful to start the Houbolt integration scheme, it may be more accurate to calculate $^{\Delta t}U$ and $^{2\Delta t}U$ by some other means.

In case of Nonlinear Analysis the Houbolt's Scheme may be re-written in the following format:

1. Formation of Global Jacobian Matrix and diagonal mass matrix.
2. Determination of critical damping parameter, by Raleigh damping as follows:

$$\alpha = \frac{\Delta u^T K \Delta u}{\Delta u^T M \Delta u} \tag{4}$$

3. Factorization or inversion of the global Jacobian (stiffness) matrix
4. Determination of the total external force as follows:

$$^{t+\Delta t}\overline{R} = \left(M\,^{t+\Delta t}\ddot{U} + \alpha M\,^{t+\Delta t}\dot{U}\right) + \left(F^{ext} - F^{\text{int}} + F^{init}\right) \tag{5}$$

5. Solution of the following incremental equation:

$$\overline{K}\,^{t+\Delta t}\Delta U = {}^{t+\Delta t}\overline{R} \tag{6}$$

6. Increasing the robustness of the algorithm by including a simple line search scheme:

$$s = \frac{^{t+\Delta t}\overline{R}\,^{t+\Delta t}\Delta U}{^{t+\Delta t}\overline{R}\,^{t+\Delta t}\Delta U + \left(F^{ext} - F^{\text{int}} + F^{init}\right)^{t+\Delta t}\Delta U} \tag{7}$$

$$^{t+\Delta t}U = {}^{t}U + s \times {}^{t+\Delta t}\Delta U$$

7. Applying any existing displacement constraint on the computed displacement field ${}^{t+\Delta t}U$
8. Computed displacement field ${}^{t+\Delta t}U$ is used to calculate the internal force vector at each node of the finite element mesh and the residual force is calculated as follows:

$${}^{t+\Delta t}R = \left(F^{ext} - F^{\text{int}} + F^{init}\right) \tag{8}$$

9. The acceleration and velocity are updated using Eq. (1) and Eq. (2).
10. Norm of displacement and out-of-balance force are calculated. Displacement norm is checked with specified tolerance for the dynamic equilibrium. If it is not satisfied, then iteration continues from step 4.

3. FINITE ELEMENT IMPLEMENTATION

The finite element implementation of the above code is carried out in an explicit Dynamic Relaxation code named "GNA", which stands for Geotechnical Nonlinear Analysis (GNA). In the FEM implementation of this code, three variety of solution technique is implemented. The solution techniques are (1) Ordinary Gaussian Elimination with full pivoting (2) Sparse Gaussian Elimination with re-ordering to reduce the fill-in and (3) Inverting-once and multiply-many technique.

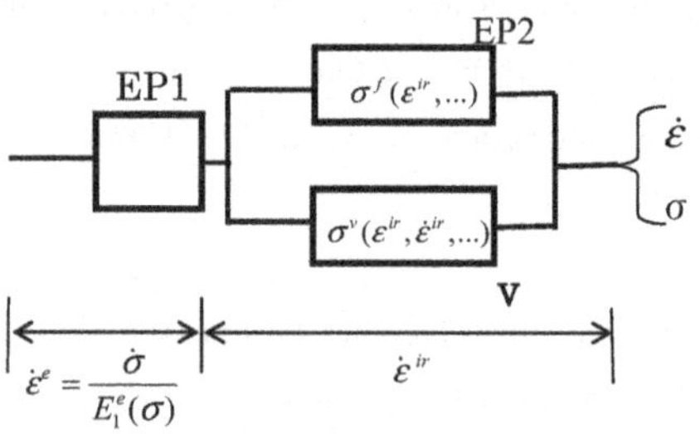

Fig. 1 General non-linear three-component model .

Within the framework of the general non-linear three-component model (Fig. 1), Tatsuoka et al. (2002) proposed a set of stress-strain models to simulate the effects of material viscosity on the stress-strain behaviour of geomaterial (i.e., clay, sand, gravel and sedimentary softrock). They showed that the viscous property of clean sand (i.e., uniform sand) is different from that of clay in that the viscous effect decays with an increase in the irreversible strain and proposed a specific model to describe the above. Here, material is elasto-visco-plastic. And all the three components are nonlinear and follow the conventional visco-plastic framework (Siddiquee *et al.*, 2006).

4. ANALYSIS AND RESULTS

In order to perform numerical simulations, the 10 cm x 10 cm one-element configuration was devised (Fig. 2A). To confirm whether the present FEM simulation can be applied to general boundary value problems, multi-element PSC test simulations were also performed using the mesh presented in Fig. 2B for a PSC test and the results were compared with those by one-element simulations. The top and bottom ends of the domain are supported with rollers to simulate a high quality of lubrication in the physical PSC tests.

In each simulation case, the reference stress-strain curve was first simulated with some trial and error by the FEM simulation without including any viscous component. The reference stress-strain relation thus obtained was used as the inviscid stress-strain relation in the FEM simulation taking into account the viscous effects. The FEM simulation was made at a very slow strain rate (10^{-6} %/second).

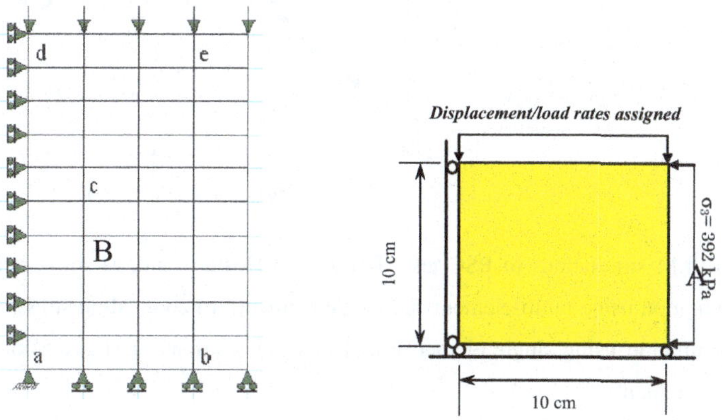

Fig. 2. FEM mesh for multi-element PSC test simulation (not to the scale, one-fourth of the whole domain, 8 cm x 20 cm, is presented here); local stress-strain relations at points *a* through *e* obtained by the analysis of PSC test "Hsd03" are presented in Fig. 3.

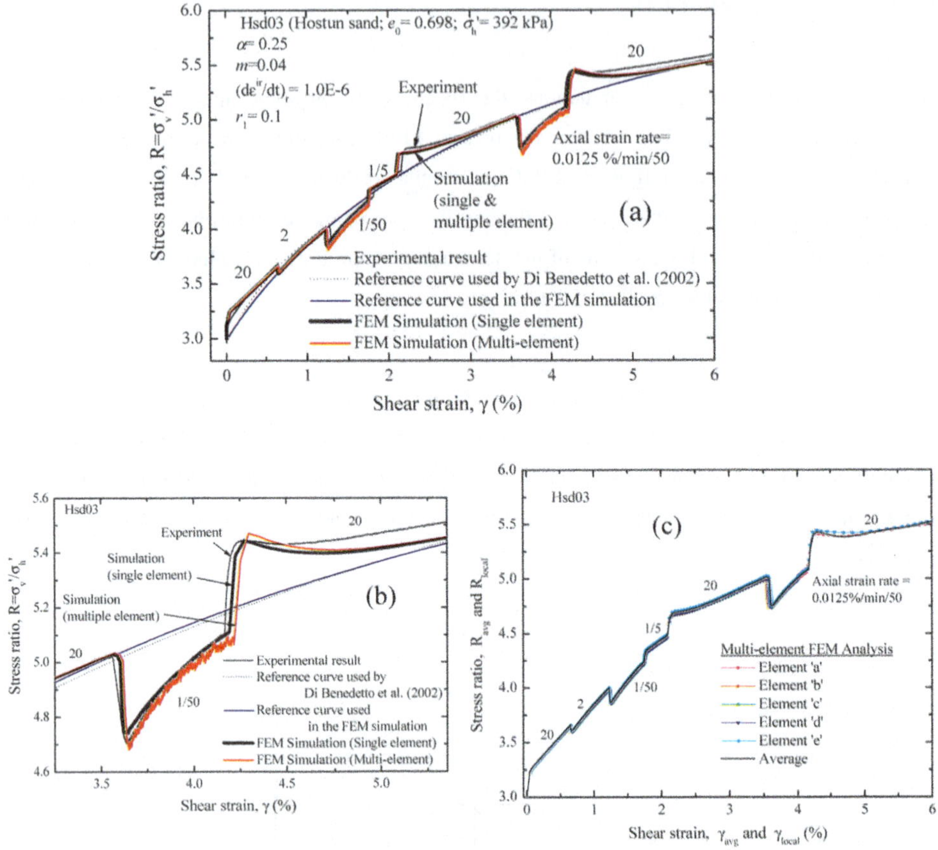

Fig. 3 Test results and FEM simulations of PSC test "Hsd03" on Hostun sand; a) stress ratio - shear strain relation (simulation using multi-element and single element); b) zoomed-up stress ratio – shear strain ration (simulation using single element); and c) local relations of stress ratio and shear strains in a multiple-element

The results from the FEM simulations carried out on a single-element and a multi-element are essentially the same (Fig. 3). Only a small difference can be seen in the average stress – average strain relations from the single-element and multi-element analyses in Fig. 3b. The difference is due to different numerical errors between the single and multiple element analysis. In Fig. 3c, local stress-strain relations in several representatives local elements (indicated in Fig. 2) in the multiple-element analysis are presented. It may be seen that the local stress-strain relations are very similar.

Table 1. Comparison of solution time required to solve the PSC simulation problem.

Type of solution techniques used	Dynamic Relaxation (minutes)	Implicit Integration by Houbolt's method (minutes)
Gaussian Elimination with full pivoting	623.21	123.3
Sparse Gaussian Elimination with re-ordering	623.21	72.34
Invert-Once and multiply-many	623.21	58.31

Table 1 shows the time of analysis of the multi-element case. It has been observed from the table that the time of solution for the case of Dynamic Relaxation is about ten times larger than the time taken by the Implicit method with stiffness matrix inverted-once technique. The sparse Gaussian Elimination with re-ordering is also competitive. But this method may be very competitive for large global stiffness matrix with higher percentage of sparsity. The results of the analysis shown here did not contain any creep. In case of analysis with long creep, Dynamic Relaxation takes very long time due to its inherent time-step limitation arising from the stability restriction. But Houbolt's implicit integration can perform creep analysis very fast by increasing the time steps in large segments after the primary part of the creep. As practical analysis requires mostly creep behaviour of any geotechnical structure, this implicit method of integration can be very useful.

5. CONCLUSIONS

The following inferences can be drawn from the above analysis:

1. Implicit Integration technique can be fast and robust at the same time.
2. Inverting the stiffness matrix once and multiply many times with residual vector to compute the incremental displacement can be very fast.
3. Houbolt's method of implicit integration can solve boundary value problem composed of very complex material model.

REFERENCE

Houbolt, J. C. (1950), "A Recurrence Matrix Solution for the Dynamic Response of Elastic Aircraft", Journal of the Aeronautical Sciences, Vol. 17, pp. 540-550.

Siddiquee, M. S. A. and Tatsuoka, F. (2001), "Modeling time-dependent stress-strain behaviour of stiff geomaterials and its applications", *Proc. 10th International Conference on Computer Methods and Advances in Geomechanics (IACMAG)*, Tucson, Arizona on January 7-12.

Siddiquee, M. S. A., Tatsuoka, F. and Tanaka, T. (2006), "FEM simulation of the viscous effects on the stress-strain behaviour of sand in plane strain compression", in press.

Tatsuoka,F., Ishihara,M., Di Benedetto,H. and Kuwano,R. (2002), "Time-dependent shear deformation characteristics of geomaterials and their simulation", *Soils and Foundations*, Vol. 42, No.2, pp.103-129.

Soil Stress-Strain Behavior: Measurement, Modeling and Analysis
Geotechnical Symposium in Roma, March 16 & 17, 2006

ANISOTROPIC BEHAVIOUR OF SAND IN THE SMALL STRAIN DOMAIN. EXPERIMENTAL MEASUREMENTS AND MODELLING.

A. Ezaoui*, H. Di Benedetto*, D. Pham Van Bang**
* Département Génie Civil et Bâtiment, CNRS URA 1652
Ecole Nationale des Travaux Publics de l'Etat (ENTPE)
rue Maurice Audin, 69518 Vaulx en Velin cedex
e-mail: ezaoui@entpe.fr, herve.dibenedetto@entpe.fr
** Laboratoire National d'Hydraulique et d'Environnement (LNHE)
6 quai Watier, 78401 Chatou Cedex

ABSTRACT

This paper deals with the initial and loading path induced anisotropy for a sub angular granular material, Hostun sand. The "quasi" elastic properties observed in the small strain domain ($<10^{-5}$ m/m) are considered. A "static and dynamic" triaxial device is used for the experimental campaign. First, the five parameters of the transverse isotropic elastic compliance tensor are experimentally obtained. The experimental investigations consist in applying small axial cyclic loadings (strain amplitude cycle $\varepsilon_{sa} \cong 10^{-5}$ m/m) and four types of dynamic wave propagations, generated by piezoelectric sensors (compressive and shear waves in axial and radial directions). The followed isotropic and deviatoric stress path underlines the effects of respectively inherent and induced anisotropy. A rheological hypoelastic model, called DBGS model, which takes into account the stress induced anisotropy, is firstly described. This model is not sufficient to properly describe experimental results at isotropic stress state as well as thus obtained during deviatoric stress path for medium and large strain. Then, an extension of the model is proposed, called DBGSP model, where strain induced anisotropy is taken into account. The concept of virtual strain induced anisotropy is introduced in this rheological hypoelastic model developed at ENTPE, and the ability of the model to foresee experimental behaviour is checked.

Keywords: granular media, sand, small strain, hypoelasticity, anisotropy, wave propagations, local measurements, triaxial device.

1. INTRODUCTION

The effects of anisotropy have been illustrated on a wide kind of granular unbound materials (sand, gravel, glass bead, two dimensions material, etc.) and for a large range of loadings (Oda 1972; Hardin 1989; Ibrahim 1991; Tatsuoka 1997; Hoque 1998; Kuwano 1999; Anhdan 2005). These studies showed the influence of the shape of particles, of history parameters such as stress and strain states, and of the considered fabric methods.

Hoe I. Ling et al. (eds.), Soil Stress-Strain Behavior: Measurement, Modeling and Analysis, 727–742.

In the present study, some investigations have been performed in the small strain elastic (or "quasi" elastic) domain ($\varepsilon_{sa}<10^{-5}$ m/m) such as dynamic wave propagations or small quasi static cyclic loadings. Indeed, with the advances of laboratory testing systems, as local stress and strain measurements, high precision sensors, as LDT or non contact sensors, and more recently with the wide spreading of non-destructive wave propagation measurements (Viggiani 1995; Belloti 1996; Brignoli 1996; Fioravante 2001), the existence of a "quasi elastic" domain for geomaterials and more particularly for sand and granular materials is now accepted for very small loadings ($\varepsilon_{sa}<10^{-5}$ m/m) (Tatsuoka 1991; Cazacliu 1996; Di Benedetto 1997; Tatsuoka 1997; Duttine 2005).

In this context, ENTPE has developed, since 1991, accurate laboratoring testing systems to reach this small strain level ($\varepsilon_{sa}<10^{-5}$ m/m), by avoiding bedding and compliance system errors (Cazacliu 1996; Sauzéat 2003; Pham Van Bang 2004; Duttine 2005). A new triaxial device, presented in the next paragraph, which takes into account the requirements in terms of local strain measurements and dynamic loadings (Pham Van Bang 2004), has been recently improved to allow wave propagations in vertical and horizontal directions.

Triaxial compression and extension tests have been carried out on air dried Hostun sand, a sub angular granular material presented in paragraph 3. Procedures and experimental campaign are also clearly defined in this latter to characterize the influence of strain and stress state as well as the fabric method on material anisotropy. Experimental results provided by cyclic static loadings at very small strain amplitudes and dynamic loadings (wave propagations) are presented in paragraph 4. The interpretations of dynamic loadings are conducted in paragraph 5, using inverse analysis with the assumption of a transverse isotropic elastic behaviour (Di Benedetto 2005). Finally, a hypoelastic model called "*DBGSP*" and which is based on a hypoelastic approach developed at ENTPE ("*DBGS*" model (Di Benedetto 2001)), is proposed in paragraph 6. This new formulation takes into account the inherent anisotropy of the specimen as well as the one induced by strain evolution during loading. The new anisotropic history parameter is the deviatoric strain.

2. TESTING APPARATUS: TRIAXIAL "STADY"

2.1 Static measuring device

The apparatus used in the present study, called Triaxial "StaDy", is a prototype developed recently at ENTPE and able to apply Static and Dynamic loadings on a specimen (Pham Van Bang 2004).

This triaxial device, presented figure 1, has internal tie bars. A 10 kN load cell is placed inside the cell. Two displacement measuring systems were designed in order to obtain locally axial and radial displacements in the central part of the sample. The shape of the sample is cylindrical with the following dimensions: diameter = 7 cm and height = 14 cm. Four axial displacement sensors (non contact type, 1 mm range) are fixed on mobile support and aim at four aluminium targets. These targets are fixed on hung rings that are glued at 3 points at 120° on the specimen (cf. figure1). The displacement sensor's supports are moved from outside the cell by micrometric screws crossing the top plate. This allows all the axial sensors to always remain inside their measuring range (1mm). Two radial displacement sensors (also, non contact type, 1 mm range) are fixed on movable supports and aim at sheets of aluminium paper placed on the inner side of the (0.4mm thick) neoprene membrane. Mobility of radial sensors is assured by micro-motor piloted from outside the cell. In total, seven sensors, six identical non-contact displacement transducers and one load cell, are used for local and static accurate measurements. The signals provided from these sensors are filtered at low frequencies by a four order low pass analogical filter with a cut-off frequency fixed at 7 Hz. The accuracy for axial and radial strain measurements is estimated to some 10^{-6} m/m for axial strain and to 0.3 kPa for the axial stress deviator.

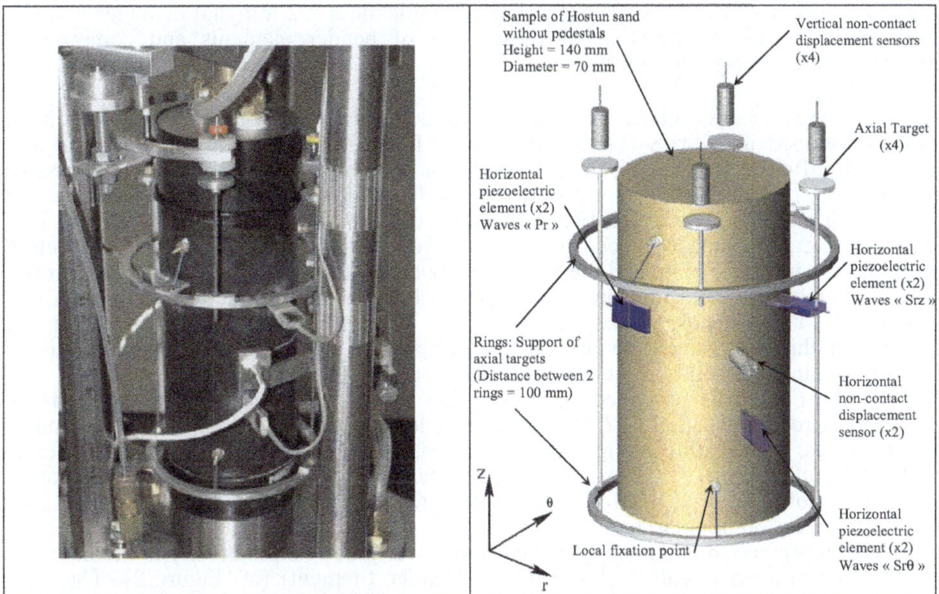

Figure 1. Picture of the instrumented sample (left) and schematic view of static and dynamic measurement system (right).

2.2 Static Loading system

The axial loading system consists in an electromechanical test machine (ref. MTS DY36) which is controlled by closed loop feedback schemes. The system imposes rate of loading from either stress or strain controlled modes.

Air pressure applied in the cell and/or vacuum inside the sample is used for lateral stress. Recording of pressure data is assured by a pressure transducer. The accuracy of the pressure transducer is estimated to 0.3kPa.

2.3 Dynamic testing system

2.3.1 Description of Transducers

The dynamic testing system is composed of five couples of piezoelectric transducers developed at ISMES (Bergamo, Italy): 3 pairs for horizontal propagation and 2 couples for vertical propagation.

The two last couples of transducers, already diffused in some geotechnical laboratories, are embedded into the pedestals (1 bender element and 1 compressive transducer on each pedestal) (cf. figure 5). The bender elements are used to generate a wave, horizontally polarized, propagating along vertical direction. This dynamic shear mode gives the wave propagation velocity, noted V_{zr}^S (*S* for shear mode, *z* for the direction of propagation and *r* for the direction of polarization). These elements have the following dimensions: height = 20 mm, width = 10 mm, and thickness = 1.3 mm. The compressive transducers generate compression wave, vertically polarized, propagating along the vertical direction. This compressive mode gives the wave propagation velocity, noted V_z^P (*P* for compressive mode, *z* for the direction of propagation). These transducers have a cylindrical shape whose dimensions are: thickness = 2 mm and diameter = 8 mm (For more technical details about these kind of bender elements and compressive transducers (cf. (Brignoli 1996)).

The 3 pairs of piezoelectric transducers used for horizontal propagation waves have been developed at ISMES (Fioravante 2001). These transducers are all identical and placed in different ways along the membrane in order to generate shear or compressive radial waves. These arrangements are described in figure 2 and 5. Each transducer is glued on a metallic plate which fits and touches the lateral surface of the specimen. This interface metallic plate is placed between the soil and the membrane. The connection between the interface plate and the transducer is ensured by buttonhole (2 × 5mm) made on the membrane. The transducer is directly glued to the interface plate through this buttonhole. These plates, whose dimensions are 10 mm in width, 17 mm in height and 0.1mm in thickness, avoid to disturb the sample and transmit the impulse from the transducers into the soil by friction or compression.

The shear mode is produced by transducers glued perpendicularly to the interface plate: this process is called *FBE* (Frictional Bender Element) (cf. Figure 2). Two kind of shear waves, propagating along radial direction, are generated: radially and vertically polarized (cf. figure 5). These dynamic shear modes give the wave propagation velocities, noted $V_{r\theta}^S$ or V_{rz}^S (*S* for shear mode, *r* for the direction of propagation and *z* or θ for the direction of polarization).

The compressive mode is produced by transducer glued longitudinally to the interface plate. This process is called *PBE* (Pulsate Bender Element) (cf. Figure 2). The *PBE* generates compressive waves, horizontally polarized, propagating along horizontal direction (cf. figure 5). This dynamic compressive mode gives the wave propagation velocity, noted V_r^P (*P* for shear mode, *r* for the direction of propagation).

These horizontal dynamic systems (*FBE* and *PBE*), placed on the specimen, are also presented in figure 1.

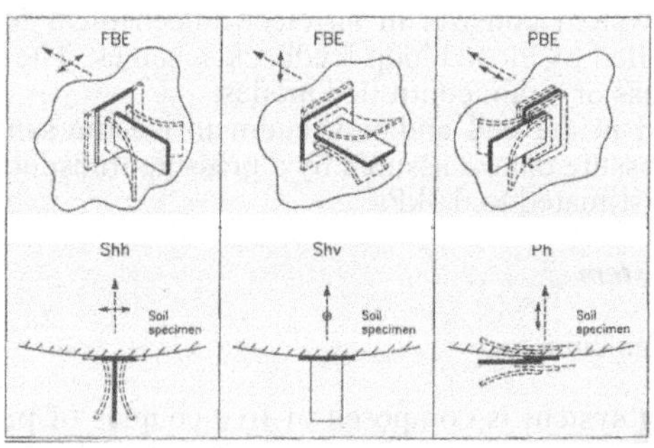

Figure 2. Scheme of the horizontal transducer arrangements used to measured $V_{r\theta}^S, V_{rz}^S, V_r^P$ respectively from left to right (Fioravante 2001).

2.3.2 Procedure for dynamic tests

The two types of transducers presented before (vertically and horizontally propagating waves), are used within a dynamic testing chain. The first element of this chain is a function generator (ref. HP 33120A) which produced the excitation signal. The generated signal is a single sinus, with amplitude of 20 Volts peak to peak, and frequency ranging from 5 to 25 kHz (depending of the sensors). The single sinus is repeated at a frequency of 10Hz. After crossing the sample, the signal which activates the corresponding receiver is amplified (ref. power amplifier Bruel & kjaer 2713) (for the compressive transducers embedded into the pedestals, the amplification is made just after the transmitter sensor).

The signal is visualised and recorded by mean of an oscilloscope whose sampling rate is of 10^6 samples per second. An average is made on 256 impulsions. Finally the acquisition of the data is ensured by a computer

During this experimental campaign, 2 or 3 frequencies are used for each kind of wave, in order to improve the determination of the corresponding wave travel time. This method allow to avoids some errors due to near field effects, compressive wave perturbations, or principal mode of vibration influence.

3. TESTED MATERIALS AND TESTING PROGRAM

3.1. Tested material and sample preparation

The tested material is an air dried poor graded sand, called Hostun sand from its original location. This sand is a quartz dominated sub angular shape whose grading curve and view of particle shape are reported on figure 3. The analyses of this grading curve are summarized in table 1.

One type of granular packing is tested in the present study. This granular arrangement corresponds to a rather dense sample whose the initial void ratio *e* (after fabrication at an initial depression of 25kPa) is close to 0.81 (Relative Density Dr ≈ 55%). In this case, the preparation method consists in deposit sand by pluviation in air. The drop height is kept constant (close to zero) to ensure the homogeneity and reproducibility of the specimen.

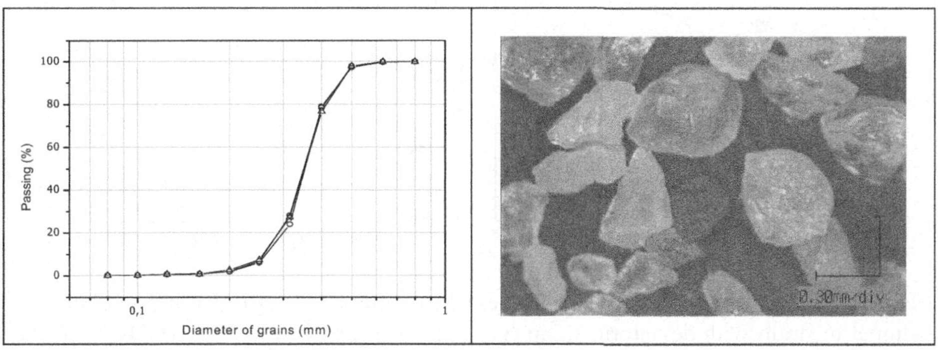

Figure 3. Grading of Hostun sand batches used in the present study and view of particle shapes.

	Passing Diameter (mm)			Coefficients		Void Ratios	
	D_{10}*	D_{30}*	D_{60}*	Cu**	CC**	e_{min}***	e_{max}***
Hostun S28	0.26	0.32	0.37	1.42	1.06	0.648	1.041

* D_x defined by x% passing particle size

** Coefficient of uniformity: $Cu=D_{60}/D_{10}$ and coefficient of curvature: $Cc=(D_{30})^2/(D_{10}D_{60})$

*** Hostun: after Flavigny (1990)

Table 1. Grading and void ratio characteristics of Hostun sand.

After the fabrication of the specimen (at 25kPa), the mould is removed and the sample is fully instrumented, then isotropically consolidated from 25kPa to 400kPa. During this step of consolidation and during the deviatoric loading steps, different loading investigations, resumed in the next paragraph, are performed.

3.2 Experimental campaign

One triaxial extension and one triaxial compression tests with a large loading and unloading stress paths are considered in this paper (cf. figure 4). These tests are part of a more general experimental campaign performed on the device and whose nomenclature obeys to the following convention: "TE/TC" stands for Triaxial Compression or Triaxial Extension test, "H" for Hostun sand, "400" for the confining pressure in kPa, ".795" for void ratio $e_0=0,795$ at initial state (25kPa) and "p" for the fabric method used: pluviation. The figure 4 presents the $(\varepsilon_z-\varepsilon_r,q)$ (deviatoric strain versus deviatoric stress) and the $(\varepsilon_z-\varepsilon_r, \varepsilon_{Vol})$ (deviatoric strain versus volumetric strain) curves for the two tests.

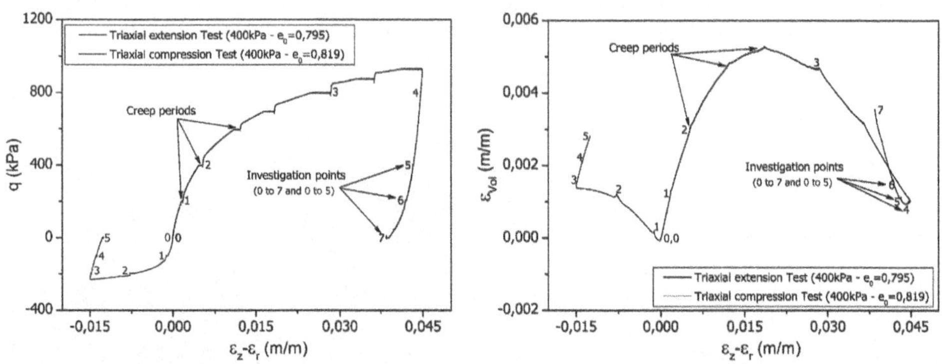

Figure 4. Deviatoric stress (q) versus deviatoric strain $(\varepsilon_z-\varepsilon_r)$ (left) and evolution of volumetric strain with deviatoric strain $(\varepsilon_z-\varepsilon_r)$ (right) for the tests "TE_H400.795p" and "TC_H400.819p".

Investigation points consist in applying loadings on the sample, in the small strain domain ($\varepsilon_{sa}<10^{-5}$ m/m). Two types of loading (dynamic and static) are performed successively:

- Vertical and horizontal wave propagations, illustrated in figure 5.
- Small vertical quasi static cyclic loadings.

The 2 types of loading are performed after large loading or unloading stress path until the whished stress state is reached, and after a creep period (2 hours) is imposed in order to eliminate the viscous behaviour effects of the material (Pham Van Bang 2006). These steps are successively repeated for each investigation point.

NB: The viscous properties of sand characterized by creep effect (cf. figure 4), are out of the scope of this paper. Creep is imposed to reach a stable point called "investigation point", where viscous effects are negligible and dimension of "quasi" elastic domain is the largest.

In order to study the influence of strain value on the "elastic properties" of Hostun sand, different investigation points having different strain histories but identical stress levels have been considered. It is illustrated in figure 4 (for example the couples of point (2-5) for "TC_H400.819p" or (0-5) for "TE_H400.795p").

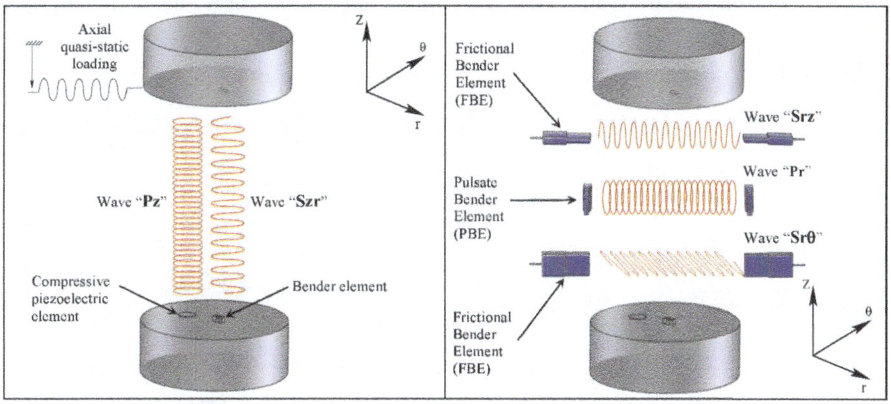

Figure 5. Schematic views of vertical (left) and horizontal (right) dynamic loadings applied in the small strain domain for each investigation point.

4. EXPERIMENTAL RESULTS IN THE SMALL STRAIN DOMAIN

Figure 6 shows a small vertical static loading in the axes (ε_z,q) (axial strain versus deviatoric stress) and (ε_z, ε_r) (axial strain versus radial strain). The linearity of the curves and the good superposition between loading and unloading points confirms the "quasi" elastic behaviour for that range of loading. The slopes of the curves gives directly access to the axial Young modulus (E_z) and to the Poisson ratio (ν_{rz}) in direction "r" and "z" (cf. paragraph 5.1) whose values are given in the figure 6.

Dynamic loadings provide velocity for each kind of waves, deduced from corresponding wave travel time as indicated in figure 7. These velocities are then linked to mechanical behaviour of the soil, using transverse isotropic assumption described in the next paragraph.

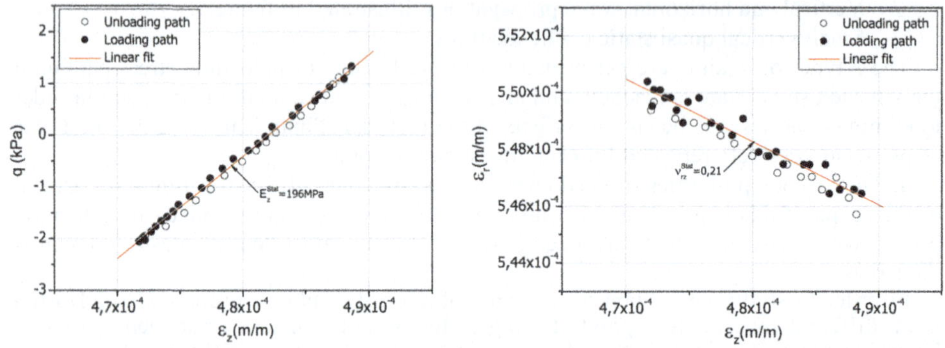

Figure 6. Example of very small unload /reload quasi static cyclic loading (1 cycle) in the axes (ε_z,q) (left) and (ε_z,ε_r) (right) to evaluate respectively vertical Young modulus "E_z" and Poisson ratio "ν_{rz}". (Test TE_H400.795p – isotropic investigation point σ_z=σ_r=100kPa).

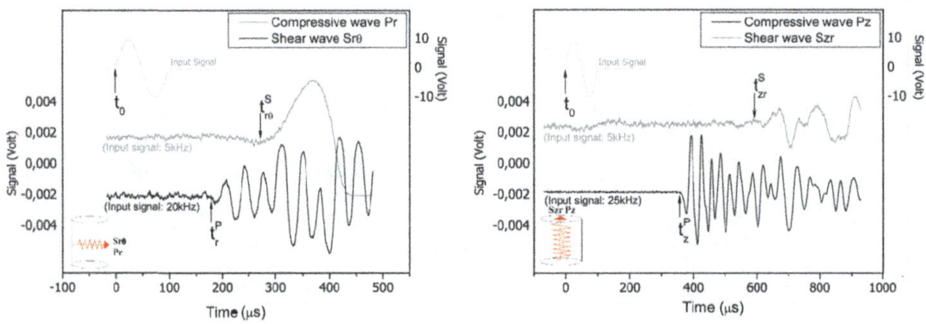

Figure 7. Example of horizontal (left) and vertical (right) wave signals recorded with the oscilloscope. The arrows indicate the start and the arrival time of the corresponding waves. (Test TE_H400.795p – isotropic investigation point σ_z=σ_r=100kPa).

5. TRANSVERSE ISOTROPIC ANALYSIS

The assumption of a transverse isotropic elastic behaviour is considered here. In the small strain domain (ε_{sa}<10^{-5} m/m), this assumption appears as a relevant approximation for triaxial test specimens. Then, the tensor $\underline{\underline{M}}$, considered as symmetrical, linking the strain increment $\underline{\Delta\varepsilon}$ to the stress increment $\underline{\Delta\sigma}$, is written in the axes of the sample (r,θ,z) (cf. figure 1) (symmetry around the vertical axis "z") as mentioned in equation 1, and the two last relations $\Delta\varepsilon_{r\theta} = \frac{1+\nu_{rr}}{E_r}\Delta\sigma_{r\theta}$ and $\Delta\varepsilon_{z\theta} = \frac{1}{2G}\Delta\sigma_{z\theta}$ are defined with the same set of parameters used in relation 1.

This tensor $\underline{\underline{M}}$ is then completly characterized by five independent elastic parameters: $\{ E_r, E_z, \nu_{rz}, \nu_{rr}, \overline{G} \}$. The purpose of this paragraph is to explain the adopted procedure to determine all of these elastic parameters, using both static and dynamic experimental results (cf. §4).

$$\underline{\Delta\varepsilon} = \underline{\underline{M}}.\underline{\Delta\sigma} \qquad \text{with} \qquad \begin{pmatrix} \Delta\varepsilon_r \\ \Delta\varepsilon_r \\ \Delta\varepsilon_z \\ \sqrt{2}.\Delta\varepsilon_{rz} \end{pmatrix} = \begin{pmatrix} \frac{1}{E_r} & \frac{-\nu_{rr}}{E_r} & \frac{-\nu_{rz}}{E_z} & 0 \\ \frac{-\nu_{rr}}{E_r} & \frac{1}{E_r} & \frac{-\nu_{rz}}{E_z} & 0 \\ \frac{-\nu_{rz}}{E_z} & \frac{-\nu_{rz}}{E_z} & \frac{1}{E_z} & 0 \\ 0 & 0 & 0 & \frac{1}{2.G} \end{pmatrix} . \begin{pmatrix} \Delta\sigma_r \\ \Delta\sigma_r \\ \Delta\sigma_z \\ \sqrt{2}.\Delta\sigma_{rz} \end{pmatrix} \tag{1}$$

5.1 Interpretation of static results

The static cyclic loadings, which consist in applying vertical stress increment $\Delta\sigma_z$ ($\Delta\sigma_r = 0$), gives directly the 2 following elastic parameters (cf. figure 6):

$$E_z = \frac{\Delta\sigma_z}{\Delta\varepsilon_z} \qquad \text{and} \qquad \nu_{rz} = -\frac{\Delta\varepsilon_r}{\Delta\varepsilon_z} \tag{2\&3}$$

These 2 elastic parameters, noted E_z^{stat} and ν_{rz}^{stat} ("stat" stand for the elastic moduli infered from static loadings) which do not require any assumptions, are used hereafter as reference data in order to improve the determination of all the elastic parameters using dynamic results.

5.2 Interpretation of dynamic results

The interpretation of dynamic results is based on wave propagations theory in an elastic continuum media. Within this framework, the following equations can be written for vertical (eq. 4 and 5) and horizontal (eq. 6 and 7) propagations of plane waves in a semi-infinite transverse isotropic elastic field:

$$\rho.(V_z^P)^2 = \frac{E_z^2(\nu_{rr}-1)}{(\nu_{rr}-1)E_z + 2\nu_{rz}^2 E_r} \qquad \rho.(V_{zr}^S)^2 = G_{rz} \tag{4\&5}$$

$$\rho.(V_r^P)^2 = E_r \frac{\nu_{rz}^2 E_r - E_z}{(\nu_{rr}^2 - 1)E_z + 2\nu_{rz}^2(1+\nu_{rr})E_r} \qquad \rho.(V_{r\theta}^S)^2 = \frac{E_r}{2(1+\nu_{rr})} \tag{6\&7}$$

The experimental set of dynamic wave measurements $\{ V_z^P, V_r^P, V_{zr}^S, V_{r\theta}^S \}$ is not sufficient to determine the

5 unknown parameters noted hereafter $E_r^{dyn}, E_z^{dyn}, \nu_{rz}^{dyn}, \nu_{rr}^{dyn}, G_{rz}^{dyn}$ ("dyn" stand for the elastic moduli infered from dynamic loadings). The results from quasi static measurements are additionally used to solve the global problem. The system is solved by modifying the value of the parameter ν_{rr} (from 0.1 to 0.3) by an optimization process. The chosen ν_{rr} value (ν_{rr}^{dyn}) minimizes the difference between static and dynamic parameters E_z and ν_{rz} ($\left|\frac{E_z^{dyn} - E_z^{stat}}{E_z^{stat}}\right|$ and $\left|\frac{\nu_{rz}^{dyn} - \nu_{rz}^{stat}}{\nu_{rz}^{stat}}\right|$ are minimized).

Figure 8 presents the comparison between static and dynamic moduli obtained from the optimization method. The chosen parameter ν_{rr} for the best optimization calculus is plotted for each investigation point in figure 8. A well accordance can be noted on figures 8 as well as a few dispersion and a constant value close to 0.2 for the Poisson ratio ν_{rr}.

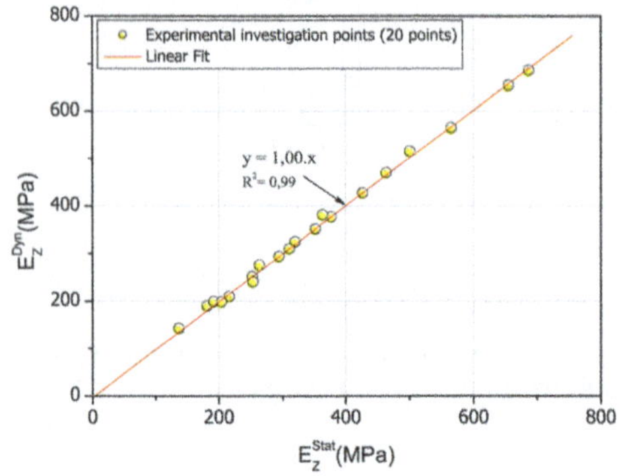

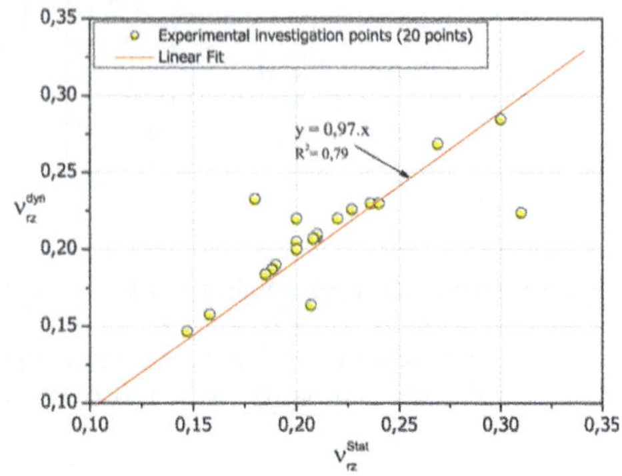

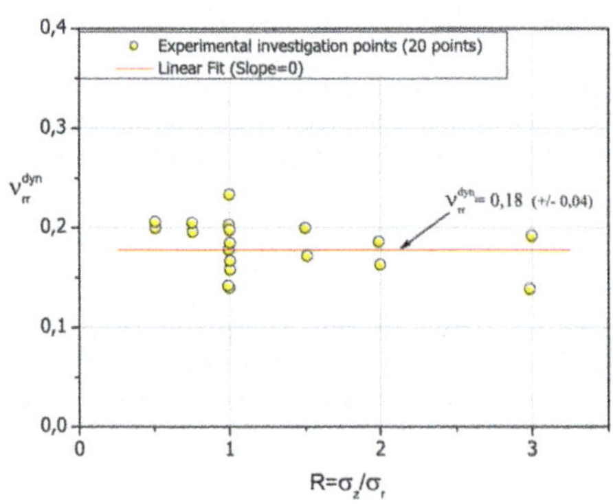

Resolution of equations 4 to 7:

- ν_{rr} is fixed (between 0.1 and 0.3)
- Resolution of the system (eq. 4 to 7)
- Calculus of the difference between static and dynamic moduli (Ez and νrz)
- Choice of the ν_{rr} for the best optimization (the 5 corresponding "dynamic" elastic parameters are kept).

Figure 8. Dynamic and static elastic parameters obtained after the resolution of the system (eq. 4 to 7) and optimized values of ν_{rr} plotted versus stress ratio R (tests "TE_H400.795p" and "TC_H400.819p").

In the following, the presented experimental dynamic elastic parameters are obtained from the procedure presented in this section 5.

6. HYPOELASTIC MODELLING

6.1 Hypoelastic formulation: *DBGS* Model

DBGS model is based on a hypoelastic formulation linking stress and strain objective increment (respectively $\underline{\delta\varepsilon}$ and $\underline{\delta\sigma}$):

$$\underline{\delta\varepsilon} = \underline{\underline{M}}(h).\underline{\delta\sigma} \tag{9}$$

where "h" is a set of history parameters and $\underline{\underline{M}}$ the rheological compliance tensor.

From relation (9), Di Benedetto et al. (2001) suggested the following anisotropic expression of the rheological tensor $\underline{\underline{M}}$:

$$\underline{\underline{M}}^{DBGS} = \frac{1}{f(e)} \frac{\underline{\underline{S}}_v . \underline{\underline{\Sigma}}_p + {}^t(\underline{\underline{S}}_v . \underline{\underline{\Sigma}}_p)}{2} \quad (10)$$

with "f(e)" a function of void ratio and $\{ \underline{\underline{S}}_v, \underline{\underline{\Sigma}}_p \}$ the tensors originally defined by (Hardin 1989):

$$S_v = \begin{bmatrix} 1 & -\nu_0 & -\nu_0 & 0 & 0 & 0 \\ -\nu_0 & 1 & -\nu_0 & 0 & 0 & 0 \\ -\nu_0 & -\nu_0 & 1 & 0 & 0 & 0 \\ 0 & 0 & 0 & 1+\nu_0 & 0 & 0 \\ 0 & 0 & 0 & 0 & 1+\nu_0 & 0 \\ 0 & 0 & 0 & 0 & 0 & 1+\nu_0 \end{bmatrix} \quad \Sigma_p = P^n_{ref} \cdot \begin{bmatrix} \frac{1}{\sigma_1^n} & 0 & 0 & 0 & 0 & 0 \\ 0 & \frac{1}{\sigma_2^n} & 0 & 0 & 0 & 0 \\ 0 & 0 & \frac{1}{\sigma_3^n} & 0 & 0 & 0 \\ 0 & 0 & 0 & \frac{1}{(\sigma_2 . \sigma_3)^{n/2}} & 0 & 0 \\ 0 & 0 & 0 & 0 & \frac{1}{(\sigma_1 . \sigma_3)^{n/2}} & 0 \\ 0 & 0 & 0 & 0 & 0 & \frac{1}{(\sigma_1 . \sigma_2)^{n/2}} \end{bmatrix} \quad (11\&12)$$

where $\{n, \nu_0\}$ are two constants (ν_0 stands for the Poisson ratio at isotropic stress state), P_{ref} is a reference pressure and $\{\sigma_1, \sigma_2, \sigma_3\}$ are the principal stress values.

The expression proposed for the tensor $\underline{\underline{M}}^{DBGS}$ is valid only in the principal axes of stress. The assumption of the symmetrical expression of the tensor $\underline{\underline{M}}$ (based on the recent experimental results obtained successively by (Cazacliu 1996; Sauzéat 2003; Duttine 2005) is imposed by relation (10). Finally, the set of history parameters "h" gathers the void ratio "e" (an isotropic parameter) and the stress state (anisotropic parameter). Expressions of the 3 elastic moduli can be deduced from relation (10), (11) and (12) and written for our triaxial condition test as follow:

$$E_r = \frac{f(e)}{P_{ref}^{n-1}} . \sigma_r^n \qquad E_z = \frac{f(e)}{P_{ref}^{n-1}} . \sigma_z^n \qquad G = \frac{f(e)}{2(1+\nu_0) P_{ref}^{n-1}} . (\sigma_r \sigma_z)^{n/2} \quad (13\&14\&15)$$

The experimental results presented in figure 9 show for isotropic consolidation stress state a good accordance between dynamic moduli $\{ E_r^{dyn}, E_z^{dyn}, G^{dyn} \}$ and the corresponding evolution given by equation 13, 14 and 15. The determination of the rheological parameter "n" is clearly defined figure 9 for Young moduli E_z and E_r (n=0.44 and 0.45) and for shear modulus G (n/2=0.22). The chosen value for simulation is n=0.445.

Nevertheless the model does not take into account the difference between the moduli E_z and E_r at isotropic stress state involved by the fabric method. This anisotropic behaviour is visible in figure 9 where "E_r" is different of "E_z" ($E_r/E_z \approx 1.1$). This initial anisotropy has been observed in other studies by measuring directly the static moduli

(Anhdan 2005), by the use of dynamic loadings (wave propagation) (Belloti 1996), or by measuring directly the repartition of intergranular contacts (Ibrahim 1991).

Strain history induced anisotropy which corresponds to some modifications into the granular structure appears from medium and large strains. This phenomenon involves consequent modifications in terms of Young Moduli (E_z and E_r) compared with those predicted by a stress induced anisotropy (*DBGS* model) as shown in figure 10. The *DBGSP* model represented in figure 10 and which takes into account the both fabric and strain induced anisotropy, is detailed in the next paragraph 6.2.

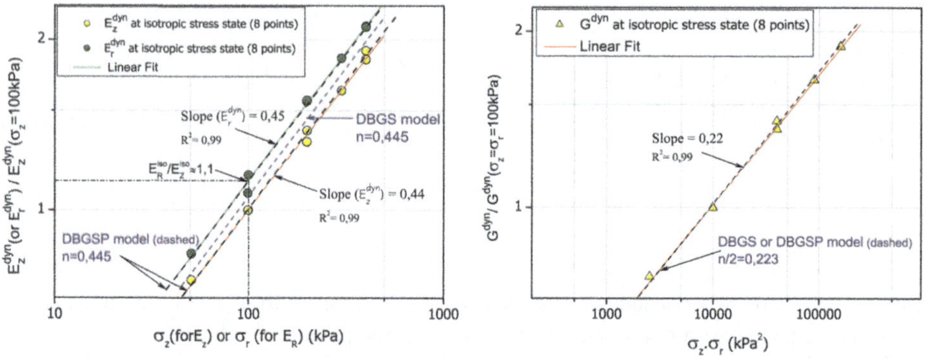

Figure 9. Stress state dependency of dynamic Young Moduli (E_Z and E_r) (left) normalized by vertical Young modulus at 100kPa as well as of shear modulus (G) (right) normalized by the shear modulus at 100kPa, experimental results obtained from investigation points during isotropic consolidation loading, tests "TE_H400.795p" and "TC_H400.819p".

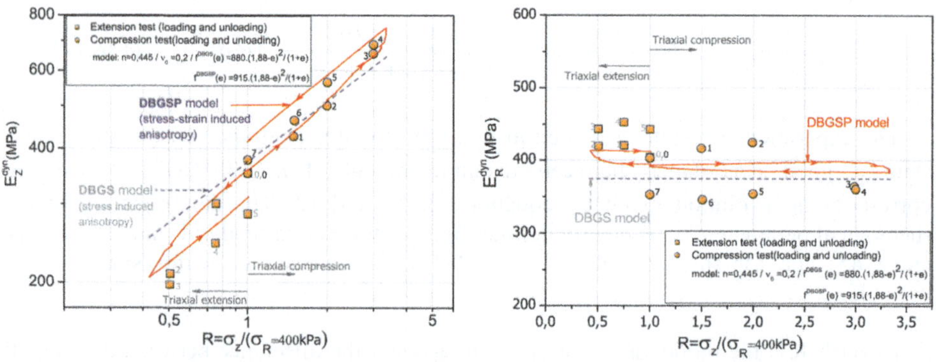

Figure 10. Stress-strain state dependency of Young Moduli "E_z" (left) and "E_r" (right) obtained from investigation points, (n°0 to 5) and (n°0 to 7) (cf. figure 4) for respectively tests "TE_H400.795p" and "TC_H400.819p".

6.2 Modification of DBGS law: introduction of fabric and strain induced anisotropy (*DBGSP* model)

The modification of the *DBGS* model into *DBGSP* model (Pham Van Bang 2004) consists from relation (9) and (10) in adding a new history parameter, the total irreversible deviatoric strain $\underline{\underline{\bar{\varepsilon}}}_{ir}^{Tot}$. The introduction of such a parameter is realised by the mean of the new tensor $\underline{\underline{\Gamma}}$:

$$\underline{\underline{\Gamma}} = \begin{pmatrix} \frac{1}{\chi(\bar{\varepsilon}_{ir,1}^{Tot})} & 0 & 0 & 0 & 0 & 0 \\ 0 & \frac{1}{\chi(\bar{\varepsilon}_{ir,2}^{Tot})} & 0 & 0 & 0 & 0 \\ 0 & 0 & \frac{1}{\chi(\bar{\varepsilon}_{ir,3}^{Tot})} & 0 & 0 & 0 \\ 0 & 0 & 0 & \frac{1}{\tilde{\chi}(\bar{\varepsilon}_{ir,2}^{Tot},\bar{\varepsilon}_{ir,3}^{Tot})} & 0 & 0 \\ 0 & 0 & 0 & 0 & \frac{1}{\tilde{\chi}(\bar{\varepsilon}_{ir,1}^{Tot},\bar{\varepsilon}_{ir,3}^{Tot})} & 0 \\ 0 & 0 & 0 & 0 & 0 & \frac{1}{\tilde{\chi}(\bar{\varepsilon}_{ir,1}^{Tot},\bar{\varepsilon}_{ir,2}^{Tot})} \end{pmatrix} \qquad \underline{\underline{\bar{\varepsilon}}}_{ir}^{Tot} = \underline{\underline{\bar{\varepsilon}}}_{ir}^{Mes} + \underline{\underline{\bar{\varepsilon}}}_{ir}^{Vir} \tag{16\&17}$$

where $\underline{\underline{\Gamma}}$ is written in the principal axes of $\underline{\underline{\bar{\varepsilon}}}_{ir}^{Tot}$ ($\underline{\underline{\bar{\varepsilon}}}_{ir} = \underline{\underline{\varepsilon}}_{ir} - \frac{tr(\underline{\underline{\varepsilon}}_{ir})}{3}.\underline{\underline{\delta}}$) ("ir" stands for the irreversible part the deviatoric strain tensor) with principal strain values noted $\{\bar{\varepsilon}_{ir,1}^{Tot}, \bar{\varepsilon}_{ir,2}^{Tot}, \bar{\varepsilon}_{ir,3}^{Tot}\}$. The term $\underline{\underline{\bar{\varepsilon}}}_{ir}^{Tot}$ introduces the concept of virtual initial state. χ and $\tilde{\chi}$ are 2 functions, only χ is presented in this paper. $\underline{\underline{\bar{\varepsilon}}}_{ir}^{Tot}$ corresponds to the deviatoric total strain that should be applied to the sample from a virtual isotropic intial state under isotropic stress, to reach the considered stress and material state. This tensor is the sum of the virtual deviatoric strain tensor $\underline{\underline{\bar{\varepsilon}}}_{ir}^{Vir}$ which stands for initial (fabric) anisotropy, and the deviatoric strain tensor applied to the sample during the test $\underline{\underline{\bar{\varepsilon}}}_{ir}^{Mes}$ (from the isotropic stress state of 25kPa). The strain $\underline{\underline{\bar{\varepsilon}}}_{ir}^{Mes}$ is measured during the test.

The concept of virtual initial strain is able to take into account the initial anisotropy as a virtual strain path induced anisotropy. This virtual strain $\underline{\underline{\bar{\varepsilon}}}_{ir}^{Vir}$ which is unknown is evaluated from the data of E_z and E_r at the beginning of the test, when the stress state is isotropic. For example, if "E_z" is equal to "E_r", then $\underline{\underline{\bar{\varepsilon}}}_{ir}^{Vir} = 0$ and $\underline{\underline{\bar{\varepsilon}}}_{ir}^{Tot} = \underline{\underline{\bar{\varepsilon}}}_{ir}^{Mes}$ (the strain applied to the sample during the test). In the *DBGSP* model, equation 10 of *DBGS* model is extended as follow:

$$\underline{\underline{M}}^{DBGSP} = \frac{1}{f(e)} \frac{\underline{\underline{S}}_v.\underline{\underline{\Gamma}}.\underline{\underline{\Sigma}}_p + {}^t(\underline{\underline{S}}_v.\underline{\underline{\Gamma}}.\underline{\underline{\Sigma}}_p)}{2} \tag{18}$$

Therefore, considering our triaxial test conditions, the formulation of $\underline{\underline{M}}^{DBGSP}$ can be written by relation (19):

$$\underline{\underline{M}}^{DBGSP}=\frac{P^n_{ref}}{f(e)}\begin{pmatrix} \frac{1}{\chi(\bar{\varepsilon}^{Tot}_{ir,r}).\sigma_r^n} & \frac{-\nu_0}{\chi(\bar{\varepsilon}^{Tot}_{ir,r}).\sigma_r^n} & \frac{-\nu_0}{2}\left(\frac{1}{\chi(\bar{\varepsilon}^{Tot}_{ir,r}).\sigma_r^n}+\frac{1}{\chi(\bar{\varepsilon}^{Tot}_{ir,z}).\sigma_z^n}\right) & 0 \\ \frac{-\nu_0}{\chi(\bar{\varepsilon}^{Tot}_{ir,r}).\sigma_r^n} & \frac{1}{\chi(\bar{\varepsilon}^{Tot}_{ir,r}).\sigma_r^n} & \frac{-\nu_0}{2}\left(\frac{1}{\chi(\bar{\varepsilon}^{Tot}_{ir,r}).\sigma_r^n}+\frac{1}{\chi(\bar{\varepsilon}^{Tot}_{ir,z}).\sigma_z^n}\right) & 0 \\ \frac{-\nu_0}{2}\left(\frac{1}{\chi(\bar{\varepsilon}^{Tot}_{ir,r}).\sigma_r^n}+\frac{1}{\chi(\bar{\varepsilon}^{Tot}_{ir,z}).\sigma_z^n}\right) & \frac{-\nu_0}{2}\left(\frac{1}{\chi(\bar{\varepsilon}^{Tot}_{ir,r}).\sigma_r^n}+\frac{1}{\chi(\bar{\varepsilon}^{Tot}_{ir,z}).\sigma_z^n}\right) & \frac{1}{\chi(\bar{\varepsilon}^{Tot}_{ir,z}).\sigma_z^n} & 0 \\ 0 & 0 & 0 & \frac{2(1+\nu_0)}{\chi(\bar{\varepsilon}^{Tot}_{ir,r},\bar{\varepsilon}^{Tot}_{ir,z}).(\sigma_r\sigma_z)^{n/2}} \end{pmatrix} \quad (19)$$

The determination and the expression of the function χ is presented in figure 11, from the comparisons of Young moduli obtained from different investigation points having different strain histories but identical stress levels. Each experimental point in figure 11 is the result of the ratio of a couple of Young moduli, corrected by void ratio "e" (for E_z cf. couples figure 4 at the same stress level and for E_r, each investigation point is compared with isotropic stress state at 400kPa because radial stress σ_r is constant during deviatoric stress loading), plotted versus deviatoric strain measured from the beginning of the test. Function χ, calibrated in figure 11, is then used into "*DBGSP* model" and presented in figure 10.

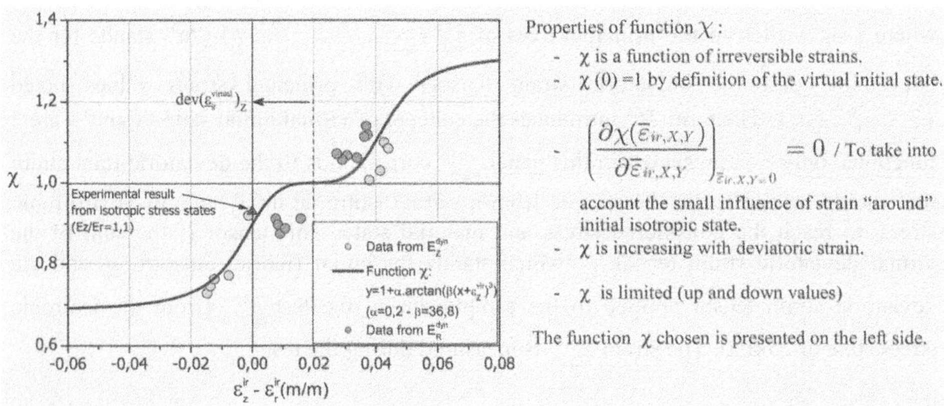

Figure 11. Calibration (from investigation points) and properties of χ function, tests "TE_H400.795p" and "TC_H400.819p".

7. CONCLUSION

Vertical and horizontal investigations using static and dynamic loadings have been performed in the small strain domain with a triaxial apparatus on a dry poor graded and sub angular Hostun sand. Vertical and horizontal wave propagation results as well as quasi static loadings and the assumptions of a transverse isotropic elastic behaviour have led to the determination of the 5 terms of the compliance elastic tensor for the triaxial conditions.

This campaign is in agreement with various findings concerning the inherent anisotropy induced by the fabric method. Moreover, load path induced anisotropy, observed experimentally on the elastic tensor for medium and large strains (from 1% of axial strain), cannot be simulated only with void ratio and stress state. A model introducing stress and strain induced anisotropy in a hypoelastic formulation, called *DBGSP* is presented. Simulations with the *DBGSP* model are in a good agreement with experimental observations.

8. BIBLIOGRAPHY

Anhdan, L., Koseki, J. (2005). "Small strain behaviour of dense granular soils by true triaxial tests." Soils and Foundations **45**(3): 21-38.

Belloti, R. J., M., Lo Presti, D.C.F, O'neill, D.A. (1996). "Anisotropy of small strain stiffness in Ticino sand." Geotechnique **46**(1): 115-131.

Brignoli, E., Gotti, M., Stokoe, K.H. (1996). "Measurements of shear waves in laboratory specimen by means of piezoelectric transducers." Geotechnical Testing Journal **19**(4): 384-397.

Cazacliu, B. (1996). Comportement des sables en petites et moyennes déformations prototype d'essai de torsion compression sur cylindre creux. Lyon, Ecole doctorale MEGA, INSA: 241.

Di Benedetto, H. (1997). Viscous effect and anisotropy for sand. (Panel discussion). Proc. of the 15th Int. Conf. of Soils Mechanics and Foundation Engineering, Hamburg.

Di Benedetto, H., Geoffroy H., Duttine, A., Sauzéat, C. (2005). Anisotropic behaviour of soils and site investigation based on wave propagation tests. (In french). 16th Int. Conf. on Soils Mechanics and Geotechnical Engineering., Osaka.

Di Benedetto, H., Sauzéat C., Geoffroy H. (2001). Hollow cylinder test and modelling of prefailure behaviour of sand. Proc. of the 2nd Int. Conference Albert Caquot, Paris.

Duttine, A. (2005). Comportement des sables et des mélanges sable/argile sous sollicitations statiques et dynamiques avec et sans "rotations d'axes". Lyon, Ecole doctorale MEGA, INSA: 317.

Fioravante, V., Capoferri, R. (2001). "On the use of multi-directional piezoelectric transducers in triaxial testing." Geotechnical Testing Journal **24**(3): 243-255.

Hardin, B. O., Blandford, G.E. (1989). "Elasticity of particulate materials." Journal of Geotechnical Engineering **115**(6): 788-805.

Hoque, E., Tatsuoka, F. (1998). "Anisotropy in the elastic deformation of granular materials." Soils and Foundations **38**(1): 163-179.

Ibrahim, A. A., Kagawa, T. (1991). "Microscopic measurement of sand fabric from cyclic tests causing liquefaction." Geotechnical Testing Journal **14**(4): 371-382.

Kuwano, R., Connoly, T.M., Jardine, R.J. (1999). "Anisotropic stiffness measurements in a stress-path triaxial cell." Geotechnical Testing Journal **23**(2): 141-157.

Oda, M. (1972). "Initial fabrics and their relations to mechanical properties of granular materials." Soils and Foundations **12**(1): 17-36.

Pham Van Bang, D. (2004). Comportement instantané et différé des sables des petites aux moyennes déformations: expérimentation et modélisation. Lyon, Ecole doctorale MEGA, INSA: 238.

Pham Van Bang, D., Di Benedetto H., Duttine, A., Ezaoui, A. (2006). "Viscous behaviours of dry sand." IJANMG / to be published.

Sauzéat, C. (2003). Comportement des sables dans le domaine des petites et moyennes déformations. Lyon, Ecole doctorale MEGA, INSA: 331.

Tatsuoka, F., Jardine, R.J., Lo Presti, D., Di Benedetto, H., Kodata, T. (1997). Characteristing the Pre-Failure Deformation Properties of Geomaterials. XIV ICSMFE, Hamburg.

Tatsuoka, F., Shibuya, S. (1991). Deformation characteristics of soils and rocks from field and laboratory test. Proc. of the 9th Asian Regional Conf. on SMFE, Bangkok.

Viggiani, G., Atkinson, J.H. (1995). "Stiffness of fine grained at very small strains." Geotechnique **45**(2): 249-265.

Soil Stress-Strain Behavior: Measurement, Modeling and Analysis
Geotechnical Symposium in Roma, March 16 & 17, 2006

ADAPTING A GENERALIZED PLASTICITY MODEL TO REPRODUCE THE STRESS-STRAIN RESPONSE OF SILTY SOILS FORMING THE VENICE LAGOON BASIN

Simonetta Cola
Dpt. IMAGE
University of Padova, ITALY
e-mail: simonetta.cola@unipd.it

Laura Tonni
Dpt. DISTART
University of Bologna, ITALY
email: laura.tonni@unibo.it

ABSTRACT

A Generalized Plasticity model, originally developed for the analysis of sandy soil behaviour, is modified in order to properly simulate the stress-strain response of a wide class of non-active natural soils, forming the upper profile of the Venice Lagoon basin. The main modification consists in introducing a state-dependent dilatancy, which allows proper modelling of granular soils over a wide range of pressures and densities, fulfilling at the same time basic premises of critical state soil mechanics. Moreover, according to recent developments on the isotropic compression of granular soils, few adjustments are introduced in the plastic modulus expression. The approach is validated by comparing the model predictions with experimental data obtained from drained triaxial compression tests on natural and reconstituted samples of soils having different fine contents.

1. INTRODUCTION

Over the last twenty years the subsoil of the historical city of Venice and the surrounding lagoon has been extensively studied in order to formulate a reliable geotechnical model of the ground, useful for the design of the submersible barriers intended to protect the area against flooding.

Due to a very complex depositional history, the upper 100m of the Venetian basin profile are mainly composed of silts, always combined with sand or clay or both together in an endless, chaotic alternation of stratified sediments.

A comprehensive study carried out by Cola and Simonini (2000, 2002) showed that the mechanical behaviour of such soils is strongly dependent on the stress level and seems to be mostly controlled by inter-particle friction rather than electrochemical action.

Recently, Tonni *et al.* (2003) presented an attempt to model the experimental stress-strain response of Venetian soils through a Generalized Plasticity approach, using a constitutive model for sands developed by Pastor *et al.* (1990) within such versatile theoretical framework.

The preliminary calibration work, based on few drained triaxial tests, showed a rather

Hoe I. Ling et al. (eds.), Soil Stress-Strain Behavior: Measurement, Modeling and Analysis, 743–758.

satisfactory applicability of the model for predicting the behaviour of these soils, but suggested at the same time a more sensitive calibration study and few minor corrections to be introduced into the constitutive equations in order to get unified modelling over a wide range of densities and stress levels.

Therefore, following recent studies on the internal-state dependence of granular soil behaviour (Been & Jefferies, 1985; Muir Wood *et al.*, 1994; Manzari & Dafalias, 1997; Li & Dafalias, 2000; Wang *et al.*, 2002) and according to the available experimental data, modifications in the plastic flow rule and in the plastic modulus expressions have been introduced.

In this paper, after a brief introduction to the original formulation of the constitutive model and a review of the basic properties of Venetian soils, results of the calibration studies are discussed, with particular reference to those material parameters appearing in the modified relationships of dilatancy and plastic modulus.

Finally, the approach is validated by comparing experimental data of drained triaxial compression tests with the model predictions.

2. GENERALIZED PLASTICITY

2.1. Basic theory

The basic idea of Generalized Plasticity (*GP*), introduced by Zienkiewicz & Mroz (1984) and later extended by Pastor & Zienkiewicz (1986), is to allow for plastic deformations irrespective of the direction of the stress increment $d\boldsymbol{\sigma}$, i.e. in both loading and unloading conditions. Moreover, plastic deformations are introduced without specifying any yield or plastic potential surfaces: the gradients to these surfaces are explicitly defined, instead of the functions themeselves.

The elastoplastic behaviour of the material is described by the general incremental relationship:

$$d\sigma = \mathbf{D}_t : d\varepsilon \tag{1}$$

in which the tangent elastoplastic stiffness tensor $\mathbf{D}_t$ depends on the current stress state, on the direction of the stress increment $d\boldsymbol{\sigma}$ and on a set of internal variables.

The dependence of $\mathbf{D}_t$ on the direction of $d\boldsymbol{\sigma}$ is expressed by simply distinguishing between two different loading classes, namely *loading* (*L*) and *unloading* (*U*). As shown in Pastor & Zienkiewicz (1986), loading/unloading conditions are determined by considering the sign of the dot product between the stress increment $d\boldsymbol{\sigma}$ and a normalized direction $\mathbf{n}$ defined in the stress space.

Correspondingly the tangent elastoplastic stiffness tensor $\mathbf{D}_t$ (represented with subscript *L* or *U* whether *loading* or *unloading* is occurring respectively), is expressed as follows:

$$\left(\mathbf{D}_{t,L/U}\right)^{-1} = \left(\mathbf{D}_t^e\right)^{-1} + \frac{1}{H_{L/U}}\left[\mathbf{m}_{L/U} \otimes \mathbf{n}\right] \tag{2}$$

where $\mathbf{m}_{L/U}$ is a direction of unit norm, $H_{L/U}$ stands for the plastic modulus and $\mathbf{D}_t^e$ is the tangent elastic stiffness tensor.

The limit case, called *neutral loading*, corresponds to stress increments for which the material behaviour is elastic and no plastic deformations occur.

Since the plastic modulus $H_{L/U}$ and the directions $\mathbf{n}$ and $\mathbf{m}_{L/U}$ are fully determined without reference to any yield surface or plastic potential function, different expressions can be selected for them depending on whether *loading* or *unloading* is occurring.

As usual, the strain rate $d\boldsymbol{\varepsilon}$ in eq. (1) can be decomposed into an elastic component $d\boldsymbol{\varepsilon}^e$ and a plastic component $d\boldsymbol{\varepsilon}^p$:

$$d\boldsymbol{\varepsilon} = d\boldsymbol{\varepsilon}^e + d\boldsymbol{\varepsilon}^p \tag{3}$$

with:

$$d\boldsymbol{\varepsilon}^e = (\mathbf{D}_t^e)^{-1} : d\boldsymbol{\sigma} \tag{4}$$

$$d\boldsymbol{\varepsilon}^p = d\boldsymbol{\varepsilon}_{L/U}^p = \frac{1}{H_{L/U}}[\mathbf{m}_{L/U} \otimes \mathbf{n}] : d\boldsymbol{\sigma} \tag{5}$$

Hence it follows that a constitutive model developed in the framework of *GP* is fully determined by specifying three directions (the loading direction $\mathbf{n}$ and the plastic flow directions $\mathbf{m}_L$ and $\mathbf{m}_U$), two scalars (the plastic moduli H_L and H_U) and the elastic stiffness tensor $\mathbf{D}_t^e$. The above "main ingredients" will be next expressed with reference to a specific constitutive model for granular soils - the *PZ model* (Pastor *et al.*, 1990) - which is able to simulate various features of loose and dense sand behaviour under monotonic and cyclic loading, in drained as well as in undrained conditions.

2.2. A Generalized Plasticity model for sands

The *PZ* model assumes an isotropic material response in both elastic and plastic ranges, hence the constitutive equations can be written in terms of the three stress invariants p', q, ϑ and the work-conjugate strain invariants $\boldsymbol{\varepsilon}_v$ and $\boldsymbol{\varepsilon}_s$.

In what follows the model is presented only in the $q - p'$ formulation, since in this work the validation of the constitutive equations is restricted to triaxial tests.

For *loading* stress increments, the plastic flow direction $\mathbf{m}_L^T = (m_{Lv}, m_{Ls})$ is given by:

$$m_{L,v} = \frac{d_g}{\sqrt{1+d_g^2}}; \qquad m_{L,s} = \frac{1}{\sqrt{1+d_g^2}} \tag{6a, b}$$

where the soil dilatancy d_g is a linear function of the stress ratio $\eta = q/p'$:

$$d_g = (1+\alpha_g)\cdot(M_g - \eta) \tag{7}$$

being M_g the slope of the critical state line in the q-p' plane and α_g a material parameter.

In unloading conditions irreversible strains are contractive and the m_v-component of the plastic flow direction $\mathbf{m}_U$ changes as follows:

$$m_{U,v} = -|m_{L,v}| \tag{8}$$

The model assumes a non-associated flow rule, thus the loading direction $\mathbf{n}^T = (n_v, n_s)$ is different from $\mathbf{m}$, but with similar expressions for its components which are given by:

$$n_v = \frac{d_f}{\sqrt{1+d_f^2}}; \qquad n_s = \frac{1}{\sqrt{1+d_f^2}} \tag{9a, b}$$

where:

$$d_f = (1+\alpha_f)(M_f - \eta) \tag{10}$$

with M_f and α_f material parameters.

In order to model the main features of sand behaviour (i.e. softening of dense sands, failure at the critical stress ratio M_g for all densities, liquefaction and cyclic mobility), Pastor & Zienkiewicz proposed for the plastic moduli H_L and H_U the following relationships:

$$H_L = H_0 \cdot p' \cdot H_f \cdot (H_v + H_s) \cdot H_{Dm} \tag{11}$$

with:

$$H_f = \left(1 - \frac{\eta}{M_f} \frac{\alpha_f}{1+\alpha_f}\right)^4; \qquad H_v = \left(1 - \frac{\eta}{M_g}\right) \tag{12a, b}$$

$$H_s = \beta_0 \beta_1 \exp(-\beta_0 \xi); \qquad H_{Dm} = \left(\frac{\zeta_{MAX}}{\zeta}\right)^\gamma \tag{12c, d}$$

and

$$H_U = H_{u0} \quad \text{for} \quad \left|\frac{M_g}{\eta_u}\right| \le 1; \qquad H_U = H_{u0}\left(\frac{M_g}{\eta_u}\right)^{\gamma_u} \quad \text{for} \quad \left|\frac{M_g}{\eta_u}\right| > 1 \tag{13a, b}$$

Constants H_0, β_0, β_1, γ, H_{u0} and γ_u appearing in eqs. (11)-(13) are constitutive parameters, ξ is the accumulated deviatoric plastic strain and ζ_{MAX} is the maximum value of the *mobilized stress function* ζ, accounting for the soil stress history. Finally, η_u stands for the stress ratio from which unloading takes place.
The material has a non-linear elastic response, according to the following relationships:

$$dp' = K_t \cdot d\varepsilon_v^e \qquad dq = 3G_t \cdot d\varepsilon_s^e \tag{14a, b}$$

The tangent bulk and shear moduli K_t and G_t are assumed to be dependent only on the hydrostatic part of the stress tensor:

$$K_t = K_o \cdot \frac{p'}{p'_0}; \qquad G_t = G_o \cdot \frac{p'}{p'_0} \tag{15a, b}$$

being K_o and G_o the bulk and shear moduli at the reference mean effective stress p'_0, respectively.
As shown, the *PZ* model requires the definition of 12 material parameters, which can be calibrated from tests routinely performed in geotechnical laboratories, such as drained and undrained monotonic triaxial tests or undrained cyclic triaxial tests (Zienkiewicz *et al.*, 1999). Moreover, the number of material parameters that need calibration is dependent on the stress path under consideration: parameters such as γ, γ_u and H_{u0} were introduced in eqs. (12d)-(13a,b) for describing soil behaviour in unloading/reloading processes, hence they don't require any definition in case of monotonic loading.

3. THE VENICE LAGOON SOILS

3.1. Basic properties

The Venice lagoon lies on 800m of Quaternary deposits, alternatively originated from continental and marine sedimentation, as a consequence of marine regressions and transgressions over the last 2 million years. The shallowest deposits, from ground level to 5-10 m below m.s.l., are due to the present lagunar cycle (Holocene).
Below Holocene sediments, down to a depth of 100 m, there is a chaotic interbedding of

continental deposits, occurred in the last glacial Pleistocenic period (Würm).
The 95% of sediments can be grouped into 3 classes: medium-fine sands (SP-SM) with sub-angular grains, silts (ML) and very silty clays (CL). The remaining 5% may be classified as organic clay and peat.
Coarser sediments are mainly composed of silicate and carbonate minerals (dolomite, calcite, quartz and feldspars) while silts and silty clays, which originated from mechanical degradation of sands, have a higher content of clay minerals (illite with minor quantities of chlorite, kaolinite and smectite), though never exceeding 20% in weight. Generally, cohesive soils are slightly overconsolidated due to aging or exsiccation.

3.2. Main features of soil behaviour

The mechanical behaviour of Venetian soils was recently analysed by Cola & Simonini (2002), who examined laboratory test results of a rather detailed investigation carried out at the Malamocco Test Site (MTS), located in one of the Lagoon inlets.
Such a study showed that, due to the high silty content and to the low activity of clay minerals, the mechanical behaviour of these soils is mainly controlled by inter-particle friction. Moreover, geotechnical parameters gradually vary as the grain size distribution changes from medium-fine sands (SP-SM) to silty clays (CL), with very few exceptions concerning organic samples.
According to these remarks, Cola and Simonini expressed a number of intrinsic parameters, such as the critical state parameters λ, e_{ref}, M or the maximum shear stiffness G_{max}, as functions of the grain size composition through the grain size index I_{GS}. Such index, defined as the ratio between the non-uniformity coefficient U and the mean particle diameter D_{50}, seemed to be particularly suitable to express the dependence of intrinsic parameters on the grain-size characteristics.
A number of relationships proposed by Cola and Simonini were successfully used in a preliminary attempt to model the Venetian soils behaviour (Tonni *et al.*, 2003) using a Generalized Plasticity approach.
More recently, Biscontin *et al.* (2006) examined the non-linear compression behaviour of Venetian soils using the unified approach proposed by Pestana & Whittle (1995) for sandy and silty soils, which assumes that at very high pressures soils reach a unique Limiting Compression Curve (LCC), independently of the initial density.
In the log$e$-logσ'_v plane the LCC is well fitted by a line whose slope is intrinsically related to the soil mineralogy, being dependent only on the grain resistance to crushing.
Since in the LCC regime the coefficient of earth pressure K_0 can be considered constant, the isotropic and one-dimensional compression curves are parallel: hence, Biscontin *et al.* estimated the Pestana – Whittle model parameters using experimental results of a number of 1D compression tests

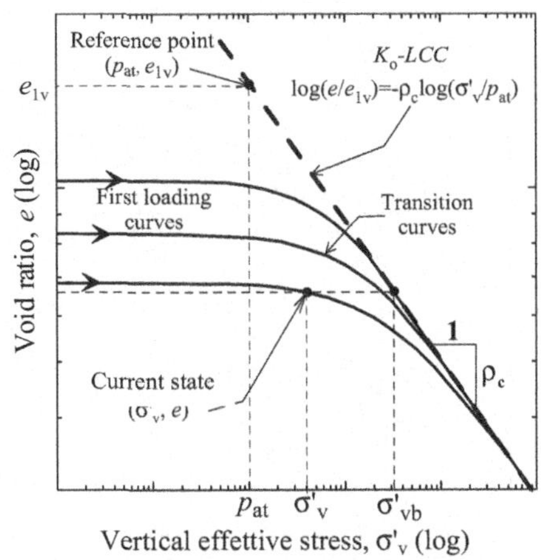

Figure 1. Compression curves for granular soils (adapted from Pestana & Whittle, 1995).

on Venetian soils, pushed up to a maximum pressure of about 30 MPa.
Moreover, according to Mitchell theory on mixtures (1976), Biscontin *et al.* demonstrated that Venetian soils can be considered as mixtures of two fractions, namely a coarse-grained fraction composed of medium-fine sands and a fine-grained fraction composed of particles smaller than 5 μm.
The position of the LCC in the loge-logσ'$_v$ plane – i.e. the void ratio e_l at the reference pressure σ'$_v$= 100 kPa – can be expressed as a function of the fine content FF, while the LCC slope is an intrinsic parameter which can be assumed as constant for all Venetian soils, except for few organic clays.

4. CALIBRATION AND MODIFICATION OF THE ORIGINAL MODEL

4.1. Experimental database

In Tonni *et al.* (2003), a preliminary attempt to apply the *PZ* model for analysing the mechanical behaviour of Venice soils was proposed, even though using a few tests on samples coming from the Malamocco Test Site (Cola & Simonini, 2002).
The triaxial testing programme on MTS soils, performed with a standard equipment, provided experimental evidence that sandy and silty samples tended to form shear bands as soon as the peak in the deviatoric stress was overcome, thus making it difficult to detect the critical state condition.
In order to dispose of a more reliable set of experimental data for the current calibration, four samples from MTS, 140 mm high and 70 mm diameter, were tested using an advanced triaxial cell available at the University of Padova, fully controlled through a personal computer and provided with local displacement transducers. According to suggestions of Rowe & Barden (1964) and Head (1982), the cell was outfitted with lubricated heads (San Vitale, 2004) in order to delay the development of shear bands. The main characteristics of the samples are listed in Table 1.
Table 1 also contains data of two more tests, SP-LD300 and SP-HD300, performed on 100 mm-diameter reconstituted samples of a medium-fine, highly uniform sand coming from MTS and prepared at different relative densities (Dalla Vecchia, 2002). Unlike the

Table 1. List of tests and basic properties of the samples.

Test	SP-200	SM-200	SM-480	ML-200	SP-HD300	SP-LD300
Soil type	Fine sand	Sandy silt	Sandy silt	Silt	Uniform fine sand	Uniform fine sand
Type of sample	Natural	Natural	Natural	Natural	Reconst.	Reconst.
Depth[(1)] (m below MSL)	40,5	33,1	33,1	18,5	40,8	40,8
D_{50} (mm)	0,170	0,045	0,045	0,034	0,170	0,170
$U=D_{60}/D_{10}$	1,98	2,68	2,68	7,08	1,64	1,64
I_{GS}	0,086	0,017	0,017	0,0043	0,104	0,104
LL, IP	-	-	-	29, 9	-	-
Fine Fraction, FF (d<5μm) (%)	0,0	6,0	6,0	10,5	0,0	0,0
Cell pressure, σ_c (kPa)	200	200	480	200	300	300
Void ratio at consolidation, e_c	0,676	0,787	0,748	0,726	0,712	0,820

[(1)] All the samples were collected from MSgM2 bore, located in MTS at a point with bottom sea at 10,2 m from MSL.

first group of tests, the latter was performed using a standard triaxial cell. The samples were prepared by dry pluviation in a mould placed on the triaxial base.

In Figure 2 deviatoric stress and volumetric strain of both sets of tests are plotted against the axial strain. Note that although lubricated ends were used, localization (characterized by a sudden drop in the deviatoric stress) was only delayed but not completely avoided. Except for the sample tested in ML-200, containing few quantities of fines (FF = 10,5%), all the other samples show a dilatant behaviour that becomes less evident as the cell pressure increases (compare the volumetric response of SM-200 and SM-480 tests) or the density decreases (compare SP-HD300 and SP-LD300 tests). Such result confirms what observed by Cola and Simonini, i.e. that sandy and silty soils dilate at medium pressures and that dilatancy disappears when the mean stress goes over 1000 kPa. Contractive behaviour can be observed only when soils have a significant clay fraction.

In what follows, a calibration study of *PZ* model parameters, based on such new experimental data, will be presented. Since only drained triaxial tests in monotonic compression were considered, attention was merely focused on those features of the model accounting for monotonic loading conditions. As a result, the validation work was restricted to the dilatancy relationship and to those components of the plastic modulus such as H_0, H_f, H_v *and* H_s. Components H_U and H_{Dm}, describing the unloading-reloading behaviour, were not examined in this work.

4.2. *Critical state parameters*

As mentioned before, Cola and Simonini found that for Venice soils the material constants characterizing the CSL in the *e*-ln*p*' and *q*-*p*' plane can be related to the grain size composition through the index I_{GS}. Such relationships are:

$$\phi'_c = 38.0 + 1.55 \cdot \log I_{GS} \quad (°) \tag{16}$$

$$\lambda_c = 0.066 - 0.016 \cdot \log I_{GS} \tag{17}$$

$$e_{ref} = 1.13 + 0.10 \cdot \log I_{GS} \tag{18}$$

which apply in the range $8 \cdot 10^{-5} \leq I_{GS} \leq 0.12$.

In Tonni *et al.* (2003), only eq.(16) was used for calibrating the *PZ* model parameters. In this context, according to modifications introduced in the constitutive equations, the whole set of equations (16)-(17)-(18) had to be used in order to evaluate the state parameter ψ. Values of ϕ'_c, λ_c, e_{ref}, referring to the six

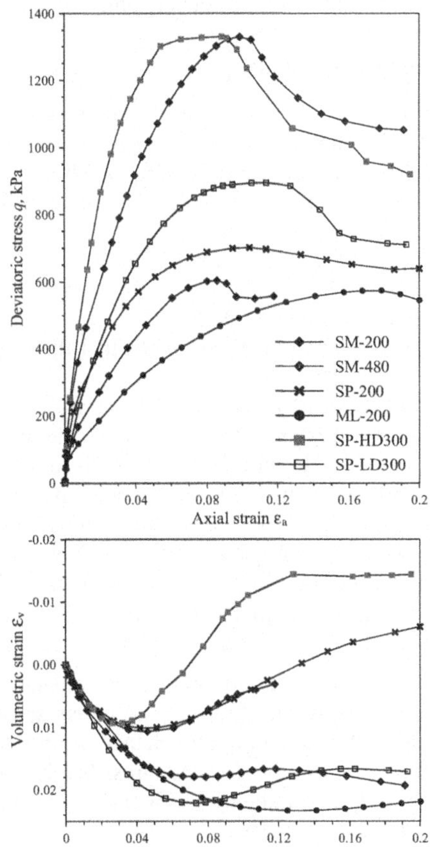

Figure 2. Deviatoric stress and volumetric strain vs. axial strain in tests used for calibration.

available samples, are summarized in Table 2.

4.3. Elastic moduli

As shown in section 3.2, in the *PZ* model the elastic behaviour of the material is assumed non-linear and is described by eqs. (14)-(15).

In this work, K_o and G_o in eqs. (14)-(15) were estimated using a method successfully adopted in Tonni *et al.* (2003). According to such procedure, the initial shear modulus G_o can be estimated from G_{max}, which can be in turn determined by means of the Hardin & Drenevich (1972) relationship:

$$\frac{G_{\max}}{p'_{\text{ref}}} = D\frac{(2.97-e)^2}{(1+e)}\left(\frac{p'}{p'_{\text{ref}}}\right)^n \quad (19)$$

being D and n material constants and p'_{ref} a reference mean effective pressure assumed equal to 100 kPa. Parameters D and n were determined using the experimental fitting procedure recently proposed by Cola and Simonini for MTS soils. As a result, n can be assumed equal to 0.6 independently of the material, while D can be related to the grain size index as follows:

$$D = 470 + 60{,}4 \cdot \log I_{GS} \quad (20)$$

The modulus G_{max} in eq. (19) describes the response at very small strains (ε_s <0.001%) while G_0 refers to larger strain levels; hence G_0 was calculated reducing G_{max} by a factor of 2.5, as also suggested by Gajo & Muir Wood (1999).

The bulk modulus K_0 was then calculated using the well-known relationship:

$$K_0 = \frac{2(1+\nu)}{3(1-2\nu)} \cdot G_0. \quad (21)$$

in which ν is equal to 0.15.

Values of D, G_0 and K_0 are listed in Table 2.

4.4. Dilatancy of Venetian soils: a state dependent relationship

As known, dilatancy plays a crucial role in the modelling of the mechanical behaviour of granular soils. In what follows we discuss some issues related to the dilatancy expression originally adopted in the *PZ* model. Then we will examine the effect on the stress-strain response induced by using two different flow rules.

Following the pioneristic work of Rowe and according to many later contributions on sand modelling (e.g. Nova & Wood, 1979), the *PZ* model assumes that dilatancy $d_g = d\varepsilon_v^p / d\varepsilon_s^p$ is a unique function of the stress ratio $\eta = q/p'$, irrespective of the material internal state.

One of the major shortcomings of considering dilatancy as uniquely related to η is that different sets of constitutive parameters are needed for a single sand with different initial densities, thus avoiding any chance of unified modelling of the mechanical response over a wide range of densities and stress levels. As pointed out by Li & Dafalias (2000), a sand model with its dilatancy following equation (7) works well only when the change in the internal state is minor.

In recent years attempts (Wan & Guo, 1998; Li *et al.*, 1999; Gajo & Wood, 1999; Li & Dafalias, 2000) have been made to treat dilatancy as a state-dependent quantity, with the

Table 2. Parameters obtained from calibration.

Test	SP-200	SM-200	SM-480	ML-200	SP-HD300	SP-LD300
Soil type	Fine sand	Sandy silt	Sandy silt	Silt	Uniform fine sand	Uniform fine sand
ϕ'_c	36.3	35.3	35.3	34.3	36.5	36.5
λ	0.083	0.095	0.095	0.104	0.082	0.082
e_{ref}	1.023	0.953	0.953	0.893	1.032	1.032
M_g	1.48	1.43	1.43	1.39	1.48	1.48
M_f	0.70	0.81	0.81	0.79	1.10	0.88
K_0 (kPa)	84580	64300	115160	63360	103550	88310
G_0 (kPa)	77220	58710	105150	57850	94540	80630
α_f	0.45	0.45	0.45	0.45	0.45	0.45
m_d	0.05	1	1	0.2	0.05	0.05
D_0	1	1	1	1	1	1
$k_d^{(*)}$	-	1.1	1.1	-	-	-
$A_d^{(*)}$	-	1	1	-	-	-
β'_0	0.33	0.80	0.80	0.33	0.35	0.35
β'_1	10	10	10	10	10	10

(*) Parameters k_d and A_d refer to the dilatancy expression of eq. (23), which was only used for comparison purposes with eq. (25) in tests SM-200 and SM-400.

concept of critical state as basis: in these contributions indeed dilatancy is expressed in terms of void ratio-dependent parameters which measure the deviation of the current state from the critical one.

According to such developments, in this work a modified expression of the plastic flow rule was introduced in the original constitutive equations, in order to address the deficiencies of the model in capturing the evolution of sand behaviour due to pressure and density changes.

In particular we focused our attention on two existing relationships, recently proposed by Li & Dafalias (2000) and Gajo & Muir Wood (1999) respectively:

$$d_g = \frac{d_o}{M_g}\left[M_g \exp(m_d \psi) - \eta\right] \tag{22}$$

$$d_g = A_d[M_g(1 + k_d \psi) - \eta] \tag{23}$$

in which d_0, m_d, A_d and k_d are material parameters while ψ is the well-known state parameter (Been & Jefferies, 1985) measuring the difference between the current and the critical void ratios at the same mean effective pressure p':

$$\psi = e - e_{crit} \tag{24}$$

In this work eq. (22) was expressed in an equivalent form, in which the ratio d_0/M_g was replaced by the parameter D_0:

$$d_g = D_0\left[M_g \exp(m_d \psi) - \eta\right] \tag{25}$$

Eqs. (23) and (25), obtained from eq. (7) by introducing in it an exponential and a linear

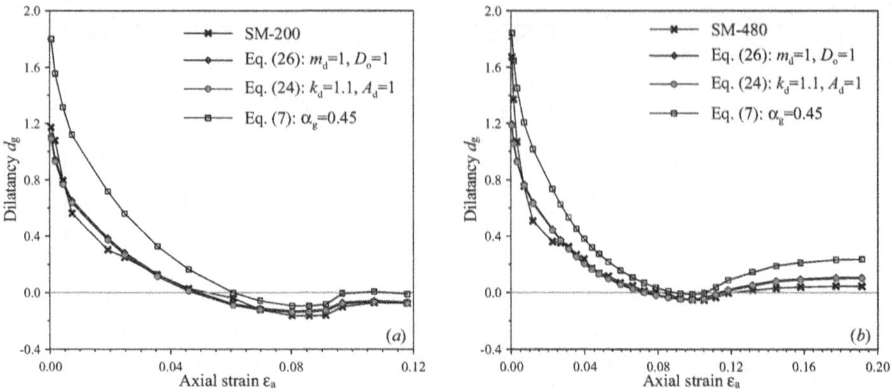

Figure 3. Comparison among experimental and predicted dilatancy in tests SM-200 (a) and SM-480 (b), using different flow rules.

dependence on ψ respectively, still guarantee basic premises of critical state soil mechanics and embed at the same time both pycnotropy and barotropy through the state parameter ψ that changes during the deformation process.

The calibration of parameters appearing in the above equations can be performed fitting the experimental ε_v-ε_a curve, by a trial and error procedure, taking into account that constants m_d and k_d can be determined at the phase transformation state, where the soil behaviour changes from contractive to dilatative. In that state, being dilatancy equal to zero, parameter k_d of eq. (23) is given by:

$$k_d = \frac{1}{\psi_{PTP}} \cdot \left(\frac{\eta_{PTP}}{M_g} - 1 \right) \tag{26}$$

while m_d of eq. (25) is given by:

$$m_d = \frac{1}{\psi_{PTP}} \cdot \ln\left(\frac{\eta_{PTP}}{M_g} \right) \tag{27}$$

being η_{PTP} and ψ_{PTP} the stress ratio and the state parameter at the phase transformation point, respectively.

According to these remarks, a preliminary estimate of m_d and k_d was determined for every single test, based on the phase transformation point; then, in order to avoid as much as possible multiple sets of parameters, mean values of m_d and k_d were considered. Particularly, a unique value was adopted for samples having the same I_{GS}. Final values of such parameters are listed in Table 2, together with the other material constants.

As an example, in Figures (3a) and (3b) experimental data of $\mathrm{d}\varepsilon_v/\mathrm{d}\varepsilon_d \approx \mathrm{d}\varepsilon_v^p/\mathrm{d}\varepsilon_d^p = d_g$ are compared with dilatancy predictions obtained from eqs. (23), (25) and (7), the latter being the expression originally proposed in the model. Such curves refer to SM-200 and SM-480 tests performed on a sandy silt consolidated at two different stress levels.

It can be noted that both eqs. (23) and (25) can successfully simulate the experimental data; moreover, the best fit of experimental curves gives similar responses for both formulations.

Hence, the two expressions seem to be both suitable to describe dilatancy of Venetian soils and appear almost equivalent in their predictions. Moreover SM-200 and SM480 were reproduced without changing the material parameters.

4.5. A convenient plastic modulus expression for venetian soils

4.5.1. Plastic modulus at constant stress ratio compression

There are a number of constitutive models assuming that sands, compressed at constant stress ratios, do not experience any plastic deformation before reaching the crushing regime. Such an assumption implies that in isotropic compression (which is a special case of constant stress ratio compression) the plastic modulus is infinite as long as the material remains within a "cap" limiting the elastic behaviour.

In the original *PZ* model, there is no need to introduce any cap in order to allow for plastic deformations in isotropic compression, since the volumetric plastic strain is inversely proportional to the material parameter H_0, as evident from eqs. (11)-(13) when η is assumed equal to zero.

According to the Cam-Clay model, plastic deformations in clay occur when the stress path moves along the normal compression line (NCL); hence, H_0 can be determined as:

$$H_0 = \frac{1+e}{\lambda - \kappa} \tag{28}$$

in which e is the initial void ratio while λ and κ are the slopes of the isotropic normal compression and swelling lines respectively. Using eq. (28), H_0 may vary in clay within 5-200, with the lowest values only for very soft clays: as an example, in Bangkok clay H_0 is equal to 6.6 (Pastor *et al.*, 1990).

As regards granular soils, Zienkiewicz *et al.* (1999) observed that such parameter can be determined by fitting the experimental curves p-ε_a or q-ε_a: estimations of H_0 reported in Pastor *et al.* (1990) vary between 350 and 16000, with higher values generally for dense sands.

In Tonni *et al.* (2003), the calibration of H_0, performed by fitting the q- ε_a curves of three drained triaxial tests, gave values equal to 800, 1000 and 2800 for CL, ML and SP-SM samples respectively. Although the estimated values seemed to be in good agreement with those reported in Pastor *et al.*, the procedure did not appear sufficiently reliable. In order to address this issue, alternative formulations for the isotropic compression of sands were examined, with particular reference to studies of Pestana & Whittle (1995) and Jefferies & Been (2000). The unified approach proposed by Pestana & Whittle (1995) overcomes the limitations of some constitutive models assuming that soils exhibit irreversible volumetric strains for η-constant paths, also at low pressures: plastic strain gradually increases as long as the distance from the LCC decreases.

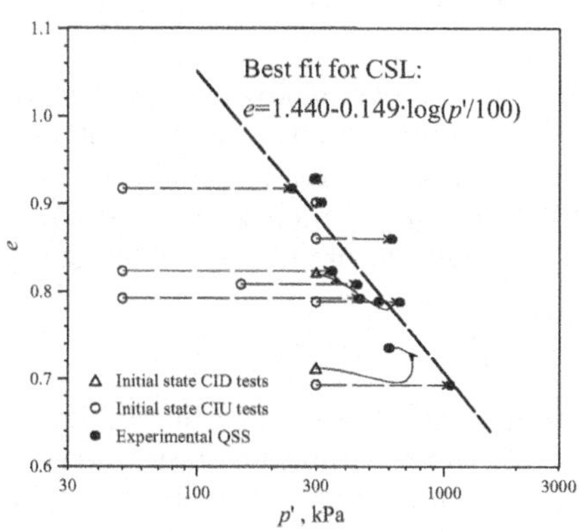

Figure 4. CSL for sand of tests SP-HD300 and SP-LD300 (from Dalla Vecchia, 2002).

As observed by Biscontin *et al.*, such a response characterizes also the 1D

Table 3. Values of H_0 from Pestana-Whittle and Jefferies-Been formulations.

Test	SP-200	SM-200	SM-480	ML-200	SP-HD300	SP-LD300
Pestana-Whittle	640	207	109	220	332	169
Jefferies-Been	1017	226	159	214	611	258

compression behaviour of Venetian soils, so that the Pestana-Whittle formulation can be successfully used to predict experimental data from normal compression tests on these soils. In such model, the volumetric plastic strain in isotropic compression is given by:

$$d\varepsilon_v^p = \frac{e}{1+e}\rho_c\left(1-\delta_b^{\vartheta}\right)\frac{dp'}{p'} \tag{29}$$

being:

$$\delta_b = 1 - p'/p'_b \tag{30}$$

a normalized parameter varying in the range of [0-1], accounting for the distance from the limiting compression curve (LCC), and p'_b the octahedral stress on the LCC at the same void ratio. The exponent θ in eq. (29) governs the compression curve curvature when it comes close to the LCC: the higher θ, the higher the curvature.
Comparing eqs. (29)-(30) with eq. (11), it follows that H_0 can be expressed as:

$$H_0 = \frac{1+e}{e}\frac{1}{\rho_c(1-\delta_b^{\vartheta})} \tag{31}$$

Values of H_0, obtained from eq. (31) using data from Biscontin *et al.*, range within 109-640 (see Table 3).
On the other hand, the alternative approach on the isotropic compression of sands, recently proposed by Jefferies & Been (2000), combines premises of CSSM (as the uniqueness of critical state line) with the idea that there is a single, unique LCC only when grain crushing becomes prevalent.
Based on several isotropic compression tests on Erksak sand, the authors found that the plastic bulk modulus K^p could be expressed as follows:

$$K^p = 0{,}3\cdot K^e\cdot\exp(-6{,}5\psi)\left(1+1{,}3\frac{p'}{\sigma_\chi}\right) \tag{32}$$

where ψ is the well-known state parameter and σ_χ is the apparent grain crushing pressure in shear, corresponding to a discontinuity of the CSL slope in the *e*-ln*p*' plane (Verdugo, 1992). Indeed, the critical void ratio is proportional to the logarithm of p' when $p'<\sigma_\chi$ and becomes proportional to the mean pressure for $p'>\sigma_\chi$.
As K^p coincides with H_L, H_0 can be estimated as:

$$H_0 = \frac{K^p}{p'} \tag{33}$$

To estimate the apparent grain crushing pressure in shear σ_χ of Venetian soils, required in eq. (32), two experimental observations were taken into account.
The first one is that sandy and silty samples did not exhibit any dilatancy only when triaxial tests were performed at mean confining pressures higher than 1-1.2 MPa. Moreover, results (Dalla Vecchia, 2002) from drained and undrained triaxial tests

performed on the same soil tested in SP- DH300 and SP-LD300, showed that in a semilogarithmic plot the critical state line is linear up to 1 MPa (Figure 4).

Secondly, sandy and silty fractions in Venetian soils are a mixture of carbonates and silicates with very fine grains, the grain size distribution varying from uniform to well graded.

Since McDowell & Bolton (1998) remark that fine well-graded soils crush at higher pressures than coarse-grained soils, it follows that the apparent grain crushing pressure in shear for Venetian soils should be higher than the value (700 kPa) adopted for Erksak sand, being the latter a coarse, uniform soil. All considered, a final value of 1,2 MPa was adopted for σ_χ.

For comparison purposes, value of H_0 obtained through both Pestana-Whittle and Jefferies-Been methods, at the same initial confining pressures, are listed in Table 3: note that the two procedures seem to be alternative as they give values that are similar or, in some cases, within the same order of magnitude. Such a result confirms the reliability of the Been & Jefferies approach.

Eq. (32) was obtained by fitting experimental tests on Erksak sand samples, thus a more sensitive calibration work would be necessary in order to properly apply the relationship to other granular soils. However, the formulation proposed by Jefferies & Been has the advantage of describing the isotropic compression behaviour of sands in terms of the same parameter ψ which was introduced in the modified dilatancy equation, thus allowing a unified modelling within the state parameter framework. As a result, in this work H_0 was determined using eq. (32), according to Been & Jefferies formulation.

4.5.2. Other components of plastic modulus

The components of the plastic modulus defined in eqs. (12a,c) account for the decrease of H_L as the shear plastic deformation increases and the critical state is reached. In what follows we briefly discuss some issues related to the calibration of parameters appearing in the expression of H_L.

Since no particular tests had been performed for determining α_f and M_f, reliable estimations of such parameters were obtained according to suggestions of Zienkiewicz *et al.* (1999). Hence a constant value equal to 0.45 was adopted for α_f while a preliminary estimation of M_f was obtained by considering the following relationship:

$$M_f = M_g \cdot D_R \tag{34}$$

where the relative density D_r was assumed equal to 50%, being this a reliable value for Venetian soils. M_f was then adjusted in

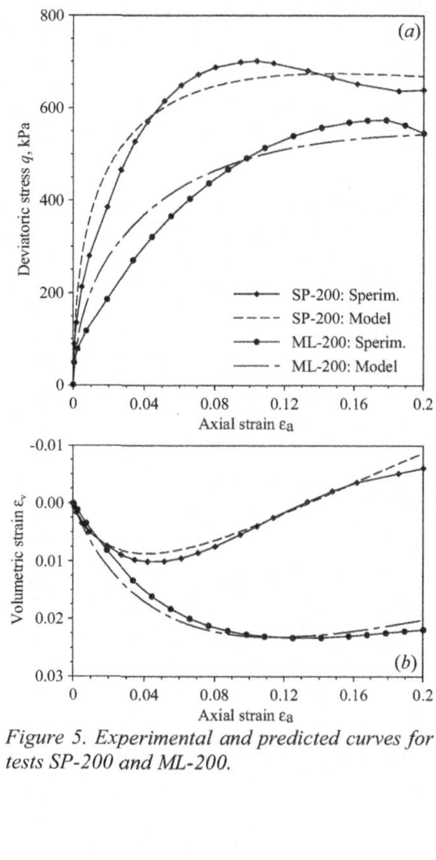

Figure 5. Experimental and predicted curves for tests SP-200 and ML-200.

order to get a better fit of the stress ratio at which the soil behaviour changes from contractive to dilatative.

The expression of H_s was slightly modified as follows:

$$H_s = \beta'_0 \exp(-\beta'_1 \xi) \qquad (35)$$

being β_0' e β_1' two material parameters corresponding to $\beta_0 \cdot \beta_1$ and β_0 of eq. (12c) respectively. The modification was introduced in order to make the calibration of such parameters easier: indeed, keeping fixed all the other material parameters, experimental values of H_s can be determined from test results and fitted in the $H_s - \log\xi$ plot.

From calibration, parameter β_1' turned out equal to 10 for all the tests, while β_1' varied from 0.3 to 0.8, as summarized in Table 2.

5. VALIDATION OF THE MODEL AND FINAL REMARKS

Table 2 summarizes the parameter values obtained through the calibration work, which was performed according to the procedure described in previous sections.

Correspondingly, in Figures 5, 6 and 7 model predictions of soil response, plotted in terms of deviatoric stress and volumetric strain versus axial strain, are compared with experimental data.

Such predictions, obtained using the Li-Dafalias flow rule, show that as long as the strains do not localize, the volumetric behaviour is successfully modelled in all the tests.

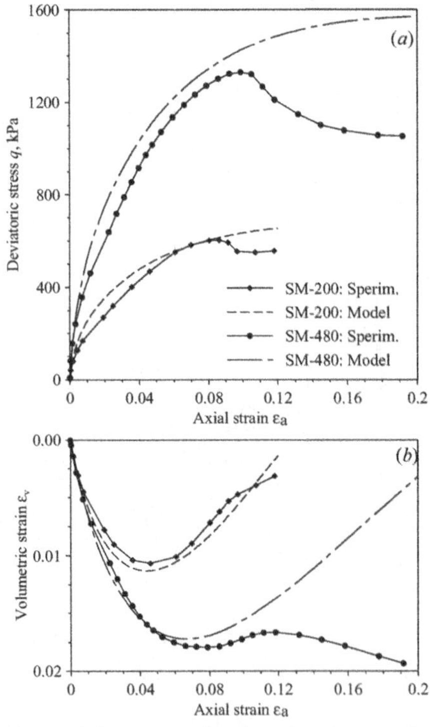

Figure 6. Experimental and predicted curves for tests SM-200 and SM-480 on sandy silt.

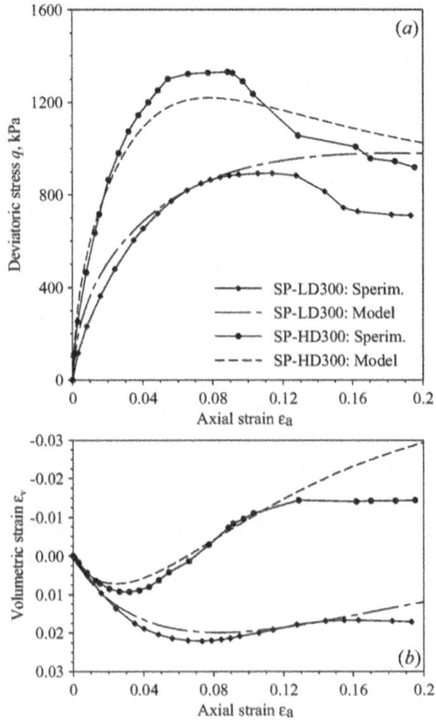

Figure 7. Experimental and predicted curves for tests SP-HD300 and SP-LD300.

As regards parameter m_d appearing in dilatancy equation (25), it must be noted that different values had to be considered with varying the tests. Nevertheless, a unique value could be adopted when considering two soils having the same fine content. Moreover, the small number of available tests did not allow to explain m_d oscillations: difficulties in determining a reliable value of such parameter could be maybe related to its dependence on the state parameter ψ, which was in turn calculated from empirical relationships having some degree of uncertainty.

Observing curves of deviatoric stress vs. axial strain, it can be noted that there is a good agreement between model predictions and experimental data, particularly at small and medium strain levels.

It's worth remarking that the model predictions match fairly well the experimental stress-strain curves for pre-peak deformations, while in post-peak regime they show a lower rate of softening. Nevertheless it is well known that, due to localization phenomena, an inhomogeneous distribution of stresses and strains occurs in the specimens, leading to potentially unreliable data. As a result, the steepness of the post-peak load-displacement curve can be significantly overestimated in laboratory tests and the softening, as experimentally observed, cannot be regarded as a material property. According to this remarks and considering that the model in use cannot describe localization of shear bands, the calibration of the parameters was performed so as to achieve the best fit of the experimental curves before reaching the post-peak regime, although numerical analyses were in general pushed up to an axial strain of about 20%. It's obvious that the comparison is no longer meaningful.

Finally, although high heterogeneity makes difficult proper modelling of Venetian soils, it must be noted that the proposed calibration procedure provides a rather simple tool for evaluating with a reasonable accuracy the model parameters, for all the Venetian soil classes.

Moreover, the introduction of a state dependent dilatancy into the original *PZ* model resulted in reliable predictions of the volumetric response using a unique set of constitutive parameters over a wide range of pressures.

On the other hand, the new expression of the plastic modulus component H_0 allowed embedding recent developments on the isotropic compression behaviour of sands within a Generalized Plasticity approach.

Further improvements should be introduced in the constitutive equations, in order to get a completely unified modelling of such natural soils over a full range of densities and stress levels. Modifications should concern those components of the plastic modulus different from H_0, by introducing in them a further dependence on the state parameter ψ.

REFERENCES

Been, K. & Jefferies, M.G., 1985. A state parameter for sands. *Géotechnique*, 35(2): 99-112.

Biscontin, G., Cola, S., Pestana, J.M. & Simonini, P. 2006. A unified compression model for the Venice lagoon natural silts. In printing.

Cola, S. & Simonini, P. 2000. Geotechnical characterization of Venetian soils: basic properties and stress history. *Memorie e Studi dell'Istituto di Costruzioni Marittime e di Geotecnica*, Università di Padova (in Italian).

Cola, S. & Simonini, P. 2002. Mechanical behaviour of silty soils of the Venice lagoon as a function of their grading characteristics. *Canadian Geotechnical Journal* 39: 879-893.

Dalla Vecchia, P. 2002. Taratura di un modello costitutivo per le sabbie di Venezia. Degree thesis at University of Padova. (In Italian)

Gajo, A. & Muir Wood, D. 1999. Severn-Trent sand: a kinematic-hardening constitutive model: the *q-p* formulation. *Géotechnique* 49(**5**), 595-614.

Hardin, B.O. & Drnevich, V.P. 1972. Shear modulus and damping in soils: design equations and curves. *J. of SMFE Div., Proc. ASCE* 98, SM7, 667-692.

Head, K.H. 1982. Manual of soil laboratory testing, vol.2, Pentech Press, London.

Ishihara, K., Tatsuoka, F., & Yashuda, S. 1975. Undrained strength and deformation of sand under cyclic stresses. *Soils and Foundations*, 15(1), 29-44.

Jefferies, M. & Been, K. 2000. Implications for critical state theory from isotropic compression of sand. *Géotechnique* 50(**4**), 419-429.

Li, X.S. & Dafalias, Y.F, 2000. Dilatancy for cohesionless soils. *Geotechnique*, 50(**4**), 449-460.

Manzari, M.T. & Dafalias, Y.F., 1997. A two-surface critical plasticity model for sand. *Géotechnique* 47(**2**), 255-272.

Mc Dowell, G.R., & Bolton, M.D. 1998. On the micromechanics of crushable aggregates. *Géotechnique*, Vol. 8(**5**), 667-679.

Mitchell, J.K., 1976. Fundamental of soil behaviour. J. Wiley, New York.

Muir Wood, D., Belkheir, K. & Liu, D.F. 1994. Strain softening ans state parameter for sand modeling. *Géotechnique* 44(*2*), 335-339.

Nova, R. & Muir Wood, D., 1979. Constitutive model for sand in triaxial compression. *International Journal for Numerical and Analytical Methods in Geomechanics*, 3(3), 255-278.

Rowe, P.W. & Barden, L. 1964. Importance of free ends in triaxial testing. *J. Soil Mech. and Found. Div. ASCE, 90*, SMI, 1-27.

Pastor, M. & Zienkiewicz, O.C., 1986. A generalized plasticity, hierarchical model for sand under monotonic and cyclic loading. In G.N. Pande & W.F. Van Impe (eds), *Proc. 2nd Int. Symp. on Numerical Models in Geomechanics, Ghent*, Belgium: 131-150. M. Jackson and Son Pub.

Pastor, M., Zienkiewicz, O.C. & Chan, A.H.C., 1990. Generalized plasticity and the modelling of soil behaviour. *Int. J. Numer. and Anal. Methods in Geomechanics* 14, 151-190.

Pestana, J.M. & Whittle, A.J., 1995. Compression model for cohesionless soils. *Geotechnique*, Vol. 45, No. 4, pp. 611-632.

San Vitale, N., 2004. Taratura di un modello costitutivo per i terreni di Venezia. *Degree thesis at University of Padova.* (In Italian)

Tonni, L., Gottardi, G., Cola, S., Simonini, P., Pastor, M. & Mira, P., 2003. Use of Generalized Plasticity to describe the behaviour of a wide class of non-active natural soils. *ISLyon 2003 Deformation Characteristics of Geomaterials*, 1145-1153, Di Benedetto et al. (eds), Swets & Zeitlinger, Lisse.

Verdugo, R., 1992. The critical state af sands (discussion). *Géotechnique* 42(*4*), 655-663.

Wan, R.G. & Guo, P.J., 1998. A Simple constitutive model for granular soils: modified stress-dilatancy approach. Computers and Geotechnics, 22(*2*), 109-133.

Wang, Z.L., Dafalias, Y.F., Li, X.S. & Makdisi, F.I., 2002. State pressure index for modelling sand behaviour. *J. Geotech. Geoenviron. Eng.*, 128(6), 511-519.

Zienkiewicz, O.C. & Mroz, Z. 1984. Generalized plasticity formulation and applications to geomechanics. In C.S. Desai & R.H. Gallagher (eds), *Mechanics of Engineering Materials*, 655- 679. Wiley.

Soil Stress-Strain Behaviour: Measurement, Modeling and Analysis
Geotechnical Symposium in Roma, March 16 & 17, 2006

VALIDATION OF A NEW SOIL CONSTITUTIVE MODEL FOR CYCLIC LOADING BY FEM ANALYSIS

G. Abate, C. Caruso, M. R. Massimino, M. Maugeri
Department of Geotechnical Engineering
University of Catania, 95125 Ct, ITALY
e-mail: mmassimi@dica.unict.it

ABSTRACT

Nowadays, there is an increasing need to understand the behaviour of geotechnical structures during earthquakes. The damage caused by the recent earthquakes has shown that the local geology and the geotechnical characteristics of the foundation soil can influence significantly the seismic response of structures. So, in order to correctly predict the behaviour of a structure subjected to an earthquake, it is necessary to focus attention on the dynamic soil behaviour. In general, very simple soil constitutive models are implemented in commercial codes. Several studies have shown that when shear strains in the soil are small, it is possible to use the elastic-linear model; for medium strains it is convenient to use equivalent linear or nonlinear models. Elastic-plastic models, incrementally nonlinear models or hypoplastic models can more accurately capture response for sites that experience higher strains. A recent elasto-plastic constitutive model including both isotropic and kinematic hardening has been implemented in a FEM code. The numerical results achieved by the new version of the FEM code, are discussed and validated by means of the comparison with laboratory experimental results involving sands of different relative densities. An interesting parametric analysis is also presented, in order to investigate the effects of the implemented constitutive model parameter variation in the soil cyclic behaviour.

1. INTRODUCTION

The damage caused by recent earthquakes (Kobe, 1995; Umbria-Marche, 1997; Kocaeli, 1999; Athens, 1999) has shown that local geology and geotechnical characteristics of foundation soil can influence significantly the seismic response of structures. So, in order to correctly predict the behaviour of structures subjected to earthquakes, it is necessary to focus attention on the dynamic soil behaviour.

Numerical codes for geotechnical site response analyses have developed significantly over the past 30 years, starting from linear elastic analyses codes, SHAKE (Schnabel et al., 1972) and QUAD-4 (Idriss et al., 1973), developed at the beginning of the 1970s. Nonlinear site response analyses have been available since the second part of the 1970s, by means of codes such as DESRA (Lee and Finn., 1978). A significant number of numerical codes, which can be used for one-, two-, and three-dimensional problems, are now available for site response analyses, as well as for the study of any geotechnical structure. General analysis codes, such as FLAC (1996), PLAXIS (1998), STRAUS

Hoe I. Ling et al. (eds.), Soil Stress-Strain Behavior: Measurement, Modeling and Analysis, 759–768.

(1999) and ADINA (Bathe, 1996), now have dynamic analysis capabilities and are becoming increasingly popular for such applications.

Nevertheless, very simple soil constitutive models (elastic-linear, elastic-perfectly plastic Mohr-Coulomb or Drucker-Prager, Cam-Clay, etc.) are implemented in these commercial codes. Actually, geotechnical materials show a great variety of behaviour when subjected to different conditions. In particular, sands respond differently due to density and confining pressure.

Experimental investigation has shown that when a sand is sheared under high confining pressure, a large reduction in the volume and angle of internal friction may result (Vesic and Clough 1968; Banks and MacIver, 1969; Miura et al. 1984; Bolton 1986; Yamamuro and Lade 1996), because of the crushing of soil particles, which is however very difficult to predict theoretically and/or numerically. The behaviour of sand under very low confining pressure has also been studied by several researchers (Ponce and Bell 1971; Fukushima and Tatsuoka 1984; Tatsuoka et al. 1986; Maeda and Miura 1999). In this last case, the test results show a dilatant deformation behaviour and the angle of internal friction increases markedly as the confining pressure is reduced. Other test results, covering a wide range of pressures, lead to similar conclusions (Lee and Seed 1967; Verdugo and Ishihara 1996).

Nevertheless, no mathematical model can completely describe the complex behaviour of real soils under all conditions. Thus, drastic idealizations are essential in order to develop simple mathematical models for practical applications. Each soil model is aimed at a certain class of phenomena, captures their essential features, and disregards what is considered to be of minor importance in that class of applications.

Studies have shown that when shear strains in the soil are small (which typically occurs when the ground motions are weak or the site consists of stiff soils), it is possible to use the elastic-linear model; for small to medium strains it is convenient to use equivalent linear or nonlinear models (Kodner and Zelasko, 1963; Desai, 1971; Breth et al., 1973; Daniel et al., 1975). Elastic-plastic models (Drucker et al., 1957; Roscoe et al., 1958; Schofield and Wroth, 1968; Lade and Dancan, 1975; Nova and Wood, 1979) can more accurately capture response for sites that experience medium to high strains. For high to very high strains (strong motions affecting soft soil sites) it is necessary to use isotropic-kinematic hardening constitutive models (Gajo and Muir Wood, 1999a and 1999b), incrementally nonlinear models (Darve, 1978, 1990) or hypoplastic models (Chambon et al., 1994; di Prisco et al., 2003).

According to what has been said up to this point, the implementation of a greater number of more appropriate and realistic soil constitutive models in numerical codes should be encouraged.

The present paper deals with the analysis of the behaviour of sand samples subjected to cyclic load conditions by means of a quite recent elasto-plastic constitutive model, including isotropic-kinematic hardening, implemented in a finite element code. The analysis is performed by comparing laboratory experimental results with the numerical ones, involving loose and dense sand samples subjected to cyclic triaxial tests under drained conditions. Finally, a parametric study on the implemented constitutive model parameters in the field of cyclic loads is discussed.

2. AN ISOTROPIC-KINEMATIC HARDENING CONSTITUTIVE MODEL FOR SANDS IMPLEMENTED IN A FEM CODE

In the field of elasto-plasticity the Authors focus their attention particularly on the Gajo and Muir Wood model (1999a and 1999b). It is formulated within the framework of kinematic hardening and bounding surface plasticity and it is capable of describing the general multiaxial stress-strain behaviour of granular material over a wide range of densities and mean pressures.

The implemented model combines the Drucker-Prager yielding, the critical state, the dependence of strength and stiffness on the state parameter ψ, introduced by Been and Jefferies (1985), a hyperbolic law for plastic stiffness degradation and a non associated flow role derived from that of the Cam-Clay model. The model is characterized by three surfaces: the yield surface, which can both change in size and rotate, while remaining inside the strength surface; the strength surface, which can change in size and degenerate into the critical state surface at infinite shear strains ($\psi = 0$); the critical state surface (CLS), which is a reference surface. The ratio between the yield surface size and the strength surface size is a constitutive parameter. The problem of softening are avoided by formulating the elasto-plastic relations in a "normalized" stress space, in which both the yield and strength surfaces have constant size.

The model has the great advantage of being based on only ten parameters: two elastic (E and ν) and eight plastic (v_λ, λ, φ_{cv}, A, k_d, B, k, R) determinable by simple triaxial tests. Particularly, v_λ and λ define the local position of the critical-state line, φ_{cv} defines the critical-state strength, A and k_d define the flow rule, B defines the hyperbolic plastic stiffness relationship and, finally, k and R define the yield and strength surfaces.

The model has been recently implemented by the Authors in the ADINA commercial code; the new version of this code has already been validated successfully in the field of monotonic loads (Abate et al., 2005).

The code can perform an enormous quantity of static and dynamic numerical FEM analyses. It, like many other commercial codes, is a very powerful code, with a very easy-to-use and complete interface and with a large number of well-supplied libraries, which regard element types, fixity and loading conditions. As geotechnical constitutive models the user can choose from the following materials: Cam Clay material, elastic-perfectly plastic Drucker-Prager material modified to include an elliptical cap hardening, elastic-perfectly plastic Mohr-Coulomb material and elastic linear material. The above constitutive models are those generally present in a lot of commercial codes. But, unfortunately, none of these constitutive models include, for example, kinematic hardening and simultaneously the effects of density and mean pressure, which are generally of crucial importance for the study of granular materials, in particular under dynamic loading. Nevertheless, the code allows the user to introduce other material constitutive models, using the User Supplied Material Model command, and so that has been done. For more details on the implementation of the model in the ADINA code see Caruso, 2006.

3. NUMERICAL MODELLING OF CYCLIC TRIAXIAL TESTS

3.1. NUMERICAL MODEL

To investigate the predictive capability of the implemented model in simulating behaviour of sands under cyclic loading, experimental results of cyclic triaxial tests, reported in the geotechnical literature, are compared with the numerical ones.
The comparison between numerical and experimental results is presented in terms of stress ratio *q/p'* versus shear strain γ, and in terms of deviatoric stress *q* versus axial strain ε_a. The comparison regards the cyclic behaviour of: i) loose Fuji River sand (Tatsuoka and Ishihara, 1974); and ii) dense Fountainebleau sand (Ling and Liu; 2003). Moreover, a parametric analysis is shown on Fuji River sand, in order to investigate the effects of soil parameters requested by the implemented constitutive model.
The numerical model (fig. 1) consists of half a cylinder subdivided in 80 3-D solid isoparametric elements, due to the symmetry of triaxial apparatus samples. Each element is characterized by means of 20 nodes and 2x2x2 Gauss integration points. In order to reproduce the boundary conditions of the triaxial apparatus using half a cylinder, the boundary fixities shown in Fig. 1 are applied to the mesh. Load conditions are varied in relation to the different cyclic triaxial tests to be simulated.

3.2.COMPARISON BETWEEN NUMERICAL AND EXPERIMENTAL RESULTS

All of the simulated triaxial tests are stress-controlled drained tests on loose and dense sands under cyclic load conditions.
The first and second simulations deal with conventional triaxial tests regarding two loose samples of Fuji River sand, performed by changing the axial load, while keeping the cell pressure constant throughout the tests (σ_r = 200 kPa). In particular, in the first test, only one cycle of triaxial compression and extension is performed, reaching the maximum value of *q/p'* equal to 0.9; in the second test, different cycles of triaxial compression and extension with increasing in *q/p'* amplitude are applied. Particularly, the following maximum absolute values of *q/p'* are applied: 0.5, 0.75, 1.0, 1.25 and 1.4.

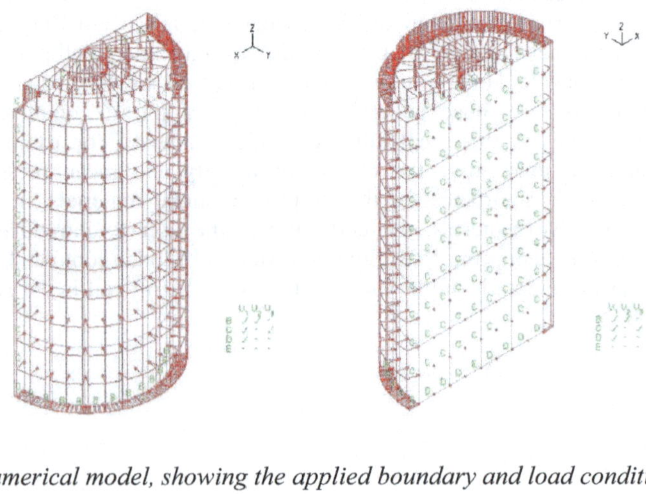

fig. 1- Numerical model, showing the applied boundary and load conditions.

The tested Fuji River sand is sub-angular in grain shape and has the effective particle size D_{10} of 0.22 mm, the uniformity coefficient is 2.21, the specific gravity of 2.68, the maximum and minimum void ratios of 1.08 and 0.53, respectively (Tatsuoka and Ishihara, 1974). The parameters of the implemented soil constitutive model for this sand are reported in Tables 1 and 2, respectively for the first test and for the second one. They are established considering the data reported in Tatsuoka and Ishihara (1974) and applying a trial and error procedure, to fit as well as possible the experimental curves. Figures 2 and 3 report the comparison between the experimental and numerical results in terms of γ vs q/p and q/p vs γ, respectively for the first test and the second one. A good agreement between experimental and numerical curves can be observed.

Table 1. Implemented soil constitutive model parameters for Fuji River sand under a triaxial compression and extension cycle.

parameter	E	ν	v_λ	λ	ϕ_{cv}	A	k_d	B	k	R
value	30000	0.35	1.939	0.04	30°	0.75	1.0	0.0018	2.0	0.10

Table 2. Implemented soil constitutive model parameters for Fuji River sand under different cycles of triaxial compression and extension with increasing in stress amplitude.

parameter	E	ν	v_λ	λ	ϕ_{cv}	A	k_d	B	k	R
value	40000	0.35	1.939	0.04	30°	0.90	1.0	0.0010	2.0	0.01

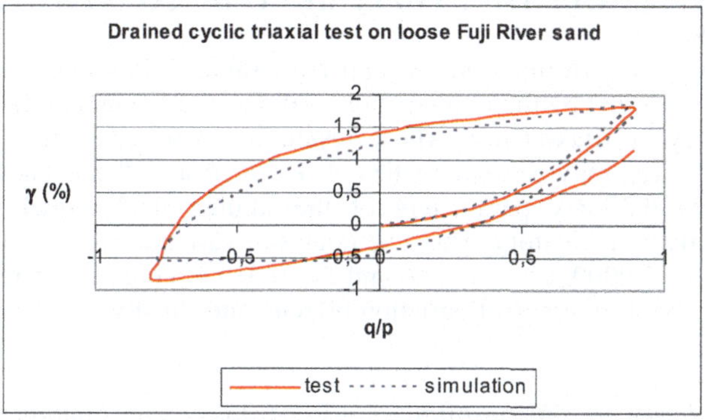

fig.2- Comparison between experimental and numerical results for Fuji River sand under a triaxial compression and extension cycle.

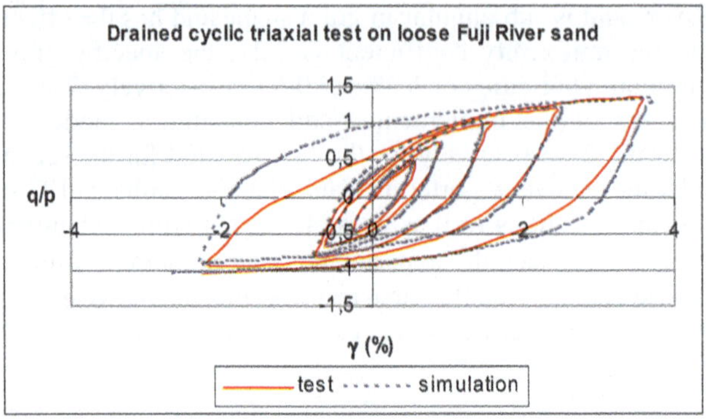

fig.3- Comparison between experimental and numerical results for Fuji River san under different cycles of triaxial compression and extension with increasing in stress amplitude.

As far as the third simulation is concerned, the behaviour of a constant-p' ($p' = 200$ kPa) triaxial test of dense Fountainebleau sand is reproduced, changing axial and radial stresses simultaneously, in order to keep the effective mean stress p' constant. The Fontainebleau sand is a standard fine siliceous uniform sand, which is characterized by a relative density D_r of 65%, and by a void ratio e_0 equal to 0.633 (Ling and Liu, 2003). The parameters required by the implemented soil constitutive model are reported in Table 3; they are established considering the data reported in Ling and Liu (2003) and applying a trial and error procedure. Figure 4 reports the comparison between the experimental and numerical results in terms of q vs ε_a; a good capability of the implemented model in simulating experimental sand behaviour can be, once more, observed.

3.3. SENSITIVITY OF THE IMPLEMENTED CONSTITUTIVE MODEL PARAMETERS

A parametric study is performed on the Fuji river sand, to investigate the influence of some parameters required by the implemented soil constitutive model. Due to the lack of space, only the cyclic triaxial test, whose results are reported in fig. 3, is taken into account. In particular, only the parameters A, k_d, B, k and R are taken into account, because they are the only parameters of the implemented model that cannot be determined directly by laboratory tests. The ranges of variation are: $A = 0.9 \pm 0.2$, $k_d = 1 \pm 0.5$, $B = 0.0010 \pm 0.0009$, $k = 2.0 \pm 1.0$ and $R = 0.010 \pm 0.005$. Fig. 5 refers to A and k_d variation effects; fig. 6 refers to B variation effects; and, finally, fig. 7 refers to k and R variation effects.

Table 3. Implemented soil constitutive model parameters for Fontainebleau sand.

parameter	E	ν	ν_λ	λ	ϕ_{cv}	A	k_d	B	k	R
value	80000	0.35	1.939	0.04	31°	0.75	1.0	0.0004	2.0	0.50

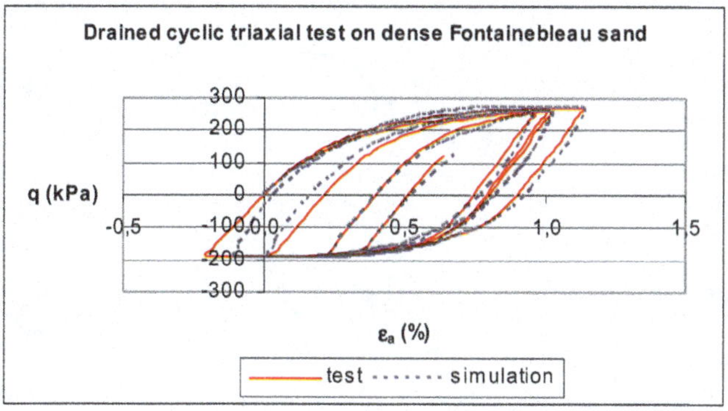

fig.4 - Comparison between experimental and numerical results for Fontainebleau sand.

In general, for loose sands subjected to monotonic loads under drained conditions the variation of k_d, B and k has negligible effects, while the increase of A can significantly increase the volumetric compressibility. For dense sands subjected to monotonic loads under drained conditions an increase of A leads to a more rapid softening and to greater volumetric strains; the increase of k_d can cause lower volumetric strains; while the increase of B leads to a significant reduction of the stiffness and of the peak strength; finally the increase of k increases the peak strength (Gajo and Muir Wood, 1999a).

The present analysis, which is performed per each parameter considering the value of Table 2 and to other values, one higher and the other lower than the value of Table 2, leads for the examined sand to the following conclusions. The increase of A causes an increase of the shear strain, which is evident with the increasing of the number of cycles (fig. 5). The variation of k_d does not show any significant effects (fig. 5). The greatest sample sensitivity is related to parameter B variation (fig. 6); in this case low values of B lead to a quite coincidence of all the cycles and to very low shear strains. Great values of B lead to the expansion of each cycle area and so to greater shear strains. The increase of k causes some reduction of the shear strains (fig. 7). Finally, very low values of R cause the contraction of all the cycle areas (fig. 7).

4. CONCLUSIONS

Numerical codes to study geotechnical problems have developed greatly over the past 30 years; nevertheless, very simple soil constitutive models are implemented in these commercial codes. The implementation of a greater number of more appropriate and realistic soil constitutive models in commercial numerical codes should be encouraged. The behaviour of sand samples subjected to cyclic load conditions is investigated by means of a recent elasto-plastic constitutive model, including isotropic-kinematic hardening, implemented by the Authors in a FEM code. Drained cyclic triaxial tests on sands with different relative densities are taken into account. The comparison between experimental and numerical curves shows generally a good predictive capability of the implemented constitutive model in reproducing cyclic sand behaviour.

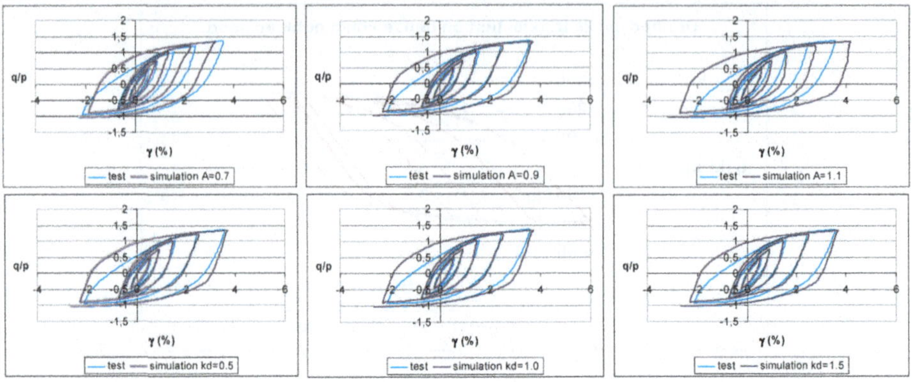

fig. 5 - Parametric analysis of the effects induced by A and k_d parameters (Fuji River sand).

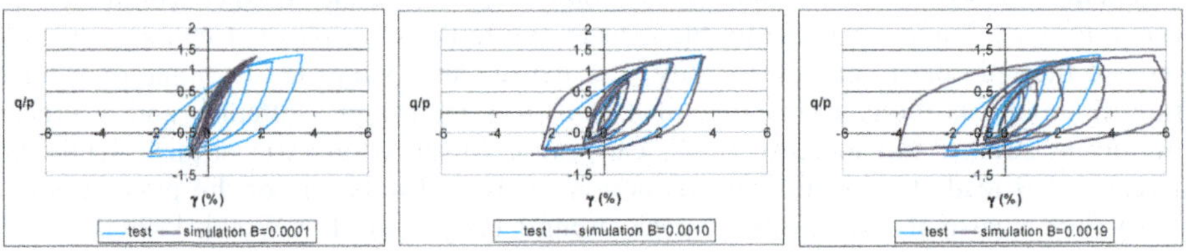

fig. 6 - Parametric analysis of the effects induced by B parameter (Fuji River sand).

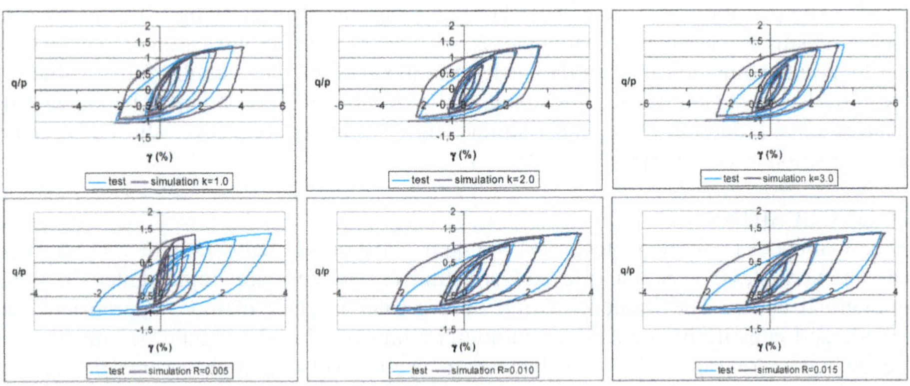

fig. 7 - Parametric analysis of the effects induced by k and R parameters (Fuji River sand).

A parametric analysis is also performed for the implemented constitutive model parameters, that have been determined by means of a trial and error procedure on the basis of triaxial tests. Generally these parameters show significant effects on the cyclic sand behaviour, so they need a very accurate trial and error procedure to be deduced before using the constitutive model to predict full-scale geotechnical structure behaviour.

REFERENCES

Abate G., Caruso C., Massimino M.R., Maugeri M. (2005). "A kinematic-isotropic hardening constitutive model for granular soils: implementation in a FEM code and application to finite problems". Symposium on "Granular Matter: Mathematical Modelling and Physical Instance", Reggio Calabria, June 26-29, 2005 (oral presentation).

Banks D. C., Maclver B. N. (1969). "Variation in angle of internal friction with confining pressure". *Rep. to U.S. Army Engineer Nuclear Cratering Group.*

Bathe K.J. (1996). "Finite Element Procedures". Prentice Hall, Englewood Cliffs, NJ.

Been K. and Jefferies M.G. (1985). "A state parameter for sands". *Géotecnique*, **35**(1), 99-112.

Bolton M. D. (1986). "The strength and dilatancy of sands". Geotechnique, **36** (1), 65–78.

Breth H, Schuster E., and Pise P. (1973). "Axial stress-strain characteristics of sand". JGED, ASCE, 99: 617-632.

Caruso C. (2006) "Implementazione e validazione di un modello costitutivo per terreni a grana grossa in un codice di calcolo agli elementi finiti". Ph. D. Thesis, Catania University.

Chambon R., Desrues J., Hammad W., Charlier R. CloE. (1994). "A new rate type constitutive model for geomaterials. Theiretical basis and implementation". *Int. J. Num. Anal. Meth. Geomech., Vol. 18, 53-78.*

Daniel A.W.T., Harvey R.C., and Burley E. (1975). "Stress-strain characteristics of sand". JGED, ASCE, 101: 508-512.

Darve F. (1978). "Une formulation incrementale des lois rhéologiques. Application aux sols". Thèse docteur es – Sciences physiques, INPG Grenoble.

Darve F. (1990). "Incrementally non-linear constitutive relationships. Geomaterials Constitutive Equations and Modelling". F. Darve ed., Elsevier Applied Science, 213-238.

Desai C.S. (1971). "Non linear analysis using spline functions". JGED, ASCE, 97: 1461-1480.

di Prisco C., Nova R., Sibilia A. (2003). "Shallow footing under cyclic loading: experimental behaviour and constitutive modeling". Geotechnical analysis of the seismic vulnerability of historical monuments. Maugeri M. & Nova R. Editors, Pàtron editore, Bologna.

Drucker D. C., Gibson R. E., Henkel D. J. (1957) "Soil Mechanics and Work-Hardening Theories of Plasticity". Trans., ASCE, **122**, 338-346.

FLAC-3D (1996). "FLAC. Fast Lagrangian Analysis of Continua". Itasca Consulting Group, Inc.(1996). Minneapolis, Minnesota, USA.

Fukushima S., Tatsuoka F. (1984). "Strength and deformation characteristics of saturated sand at extremely low pressures". *Soils Found., 24 (4), 30–48.*

Gajo A., Muir Wood D. (1999.a). "Severn-Trent sand: a kinematic-hardening constitutive model: the q-p formulation". *Géotechnique*, **49**(5), 595-614.

Gajo A., Muir Wood D. (1999.b). "A kinematic hardening consitutive model for sands: the multiaxial formulation". *Int. Journal Numer. Anal. Meth. Geomech.*, **23**, 925-965.

Idriss I.M., Lysmer J., Hwang R., Seed H.B. (1973) QUAD-4: a computer program for evaluating the seismic response of soil structures by variable damping finite element procedures. Department of Civil Engineering, University of California, Berkeley.

Kondner R.L., Zelasko J.S. (1963). "A hyperbolic stress-strain formulation for sands". Proc. 2nd Pan-American Conference on Soil Mechanics and Foundations Engineering, Brazil, I: 289-324.

Lade P.V. and Dancan J.M. (1975). "Elasto-plastic stress-strain theory for cohesionless soil". *J. Soil Mech. Found. Div.*, ASCE, **92**(SM2), 353-369.

Lee M.K.W., Finn W.D.L. (1978). "DESRA-2, dynamic effective stress response analysis of soil deposits with energy transmitting boundary including assessment of liquefaction potential', Soil Mech. Series No. 38, Dept. of Civil Eng., Univ. of British Columbia, Vancouver B.C.

Lee K.L., Seed H.B. (1967). "Drained characteristics of sands". *J. Soil Mech. and Found. Div.*, ASCE, **93**(6), 117-141.

Ling Hoe I. and Huabei Liu. (2003). "Pressure level dependency and densification behaviour of sand through generalized plasticity model". *J. Eng. Mech., ASCE, **129**, no 8, 851-860.*

Maeda K., Miura, K. *(1999)*. "Confining stress dependency of mechanical properties of sands". *Soils Found., 39 (1), 53–67.*

Miura N., Murata H., Yasufuku N. (1984). "Stress-strain characteristics of sand in a particle-crushing region". *Soils Found., **24** (1), 77–89.*

Nova R., Wood D. M. (1979). "A constitutive model for sand in triaxial compression". *Int. J. Numer. Analyt. Meth. Geomech., 3, 255–278.*

PLAXIS (1998). "Plaxis Manual". Version 8, Balkema, Rotterdam.

Ponce V. M., Bell J. M. (1971). "Shear strength of sand at extremely low pressures". *J. Soil Mech. Found. Div., Am. Soc. Civ. Eng., 97 (4), 625–638.*

Roscoe P.W., Schofield A.N. and Wroth C.P. (1958). "On the yielding of soils". *Géotechnique*, 1, 22-52.

Schanabel P. B., Lysmer J., Seed H. B. (1972). „A computer program for earthquake response analysis of horizontally layered sites". Earthquake Engineering Research Center, Report N. EERC 72-12, University of California, Berkeley.

Schofield A.N. and Wroth C.P. (1968). "Critical state soil mechanics". *McGray-Hill*, London.

STRAUS-7 (1999). "Guida all'uso di Straus-7. Concezione e sviluppo: G+D Computing", Sydney NSW 2000 Australia. Distribuzione e consulenza: HSH s.r.l., Padova .

Tatsuoka F. and Ishihara K. (1974). "Drained deformation of sand under cyclic stresses reversing direction". *Soils and Foundations, **14**, no. 3, 51-65.*

Tatsuoka F., Sakamoto M., Kawamura T., Fukushima S. (1986). "Strength and deformation characteristic of sand in plane strain compression at extremely low pressure". *Soils and Foundations, **26**, no. 1, 65-84.*

Tatsuoka F. et al. (2003). "Modeling the stress-strain relations of sand in cyclic plane strain loading". *J. of Geot. and Geoenv. Eng. , ASCE, **129**, no. 6, 450-467.*

Verdugo R., Ishihara K. (1996). "The steady state of sandy soils". *Soils Found., 36 (2), 81–91.*

Vesic A. S., Clough G. B. (1968). "Behaviour of granular materials under high stresses". *J. Soil Mech. Found. Div., Am. Soc. Civ. Eng., **94**(3), 661-668.*

Yamamuro J. A., Lade P. V. (1996). "Drained sand behaviour in axisymmetric tests at high pressures". *J. Geotech. Eng., **122** (2), 109–119.*

Soil Stress-Strain Behavior: Measurement, Modeling and Analysis
Geotechnical Symposium in Roma, March 16 & 17, 2006

VISCOPLASTICITY OF GEOMATERIALS AND FINITE ELEMENT ANALYSIS

T. Tanaka
Department of Biological & Environmental Engineering, University of Tokyo, Bunkyo-Ku, Tokyo, JAPAN, atanak@mail.ecc.u-tokyo.ac.jp

Abstract

The elasto-plastic and voscoplastic constitutive relations with kinematic strain hardening-softening model for geomaterials are applied to boundary value problems. The constitutive models are based on experimental findings about inherent anisotropies and strain rate dependent properties involved in soils. A generalized return-mapping algorithm is applied to solution methods of the problems. The dynamic relaxation method for static problems and the dynamic analysis for earthquake responses are solved based on finite element methods.

1. Introduction

The dynamic relaxation method combined with the generalized return-mapping algorithm is applied to the integration algorithms of viscoplstic constitutive relations including the effect of the shear band. The explicit type dynamic relaxation method (Tanaka and Kawamoto, 1988) is applied to the element tests of soils.

The viscoplastic kinematic hardening model (modified and extended soil model of isotropic strain-hardening-softening) is then developed. The TESRA model developed for soils is applied to the boundary value problems such as plane strain compression test and dynamic response analysis of embankment dam.

2. Dynamic Relaxation Method

Solution to systems of nonlinear equations involving the governing non-linear equation is obtained as

$$P - P^{init} = F \quad \text{and} \quad P = \sum_N \int_{vol} B^T \sigma dv \tag{1}$$

where P is the internal force vector, P^{init} is the nodal forces due to initial stresses, F is the external force vector, B^T is the strain-displacement transformation matrix, N is the number of elements in finite element discretization, σ is the stresses at Gauss points in each element, and *vol* is the volume of each element. The solution to the above governing equation can be obtained by achieving the steady state response of the following dynamic equation of motion.

$$M_D a + Cv + P - P^{init} = F \tag{2}$$

where M_D is the diagonalized mass matrix, C is the damping matrix, which is a vector for critically damped dynamic relaxation, v is the velocity vector, and a is the acceleration vector.

Then, applying the central difference method to Eq. (2) and replacing the damping by the following relation;

$$C = \alpha M_D \tag{3}$$

the following relaxation equation can be derived.

Hoe I. Ling et al. (eds.), Soil Stress-Strain Behavior: Measurement, Modeling and Analysis, 769–778.

$$q_{n+1} = \frac{1}{1+0.5\alpha\Delta t}\left[\frac{\Delta t^2}{M_D}(F - P + P^{init})' + 2q_n - (1-0.5\Delta t)q_{n-1}\right] \quad (4)$$

Here, q_n is the displacement vector at time n, Δt is the time increment and α is the damping ratio which is the most critical value to be determined.

A number of methods can estimate a reasonable value of the critical damping parameter. We employ the Rayleigh's quotient to determine the approximate damping in an adaptive way using the current solution parameters.

3. Constitutive Model for Plasticity

A simplified and generalized version of mesh size-dependent softening modulus method is used in this study. A material model for a real granular material (i.e., Toyoura sand) used with the features of nonlinear pre-peak, pressure-sensitivity of the deformation and strength characteristics of sand, non-associated flow characteristics, post-peak strain softening, and strain-localization into a shear band with a specific width.

The yield function (f) and the plastic potential function (Φ) are given by:

$$f = \alpha I_1 + \frac{\overline{\sigma}}{g(\theta_L)} = 0 \quad (5)$$

$$\Phi = \alpha' I_1 + \overline{\sigma} = 0 \quad (6)$$

where

$$\alpha = \frac{2\sin\phi}{\sqrt{3}(3-\sin\phi)} \qquad \alpha' = \frac{2\sin\psi}{\sqrt{3}(3-\sin\psi)} \quad (7)$$

where I_1 is the first invariant (positive in tension) of deviatoric stresses and $\overline{\sigma}$ is the second invariant of deviatoric stress. With the Mohr-Coulomb model, $g(\theta_L)$ takes the following form:

$$g(\theta_L) = \frac{3-\sin\phi}{2\sqrt{3}\cos\theta_L - 2\sin\theta_L \sin\phi} \quad (8)$$

ϕ is the mobilized friction angle and θ_L is the Lode angle. The frictional hardening-softening functions expressed as follows were used:

$$\alpha(\kappa) = \left(\frac{2\sqrt{\kappa\varepsilon_f}}{\kappa+\varepsilon_f}\right)^m \alpha_p \quad (\kappa \le \varepsilon_f) \; : \; \text{hardening-regime} \quad (9)$$

$$\alpha(\kappa) = \alpha_r + (\alpha_p - \alpha_r)\exp\left\{-\left(\frac{\kappa-\varepsilon_f}{\varepsilon_r}\right)^2\right\} (\kappa \ge \varepsilon_f)$$

$$: \; \text{softening-regime} \quad (10)$$

where m, ε_f and ε_r are the material constants (Yoshida et al. 1995) and α_p and α_r are the values of α at the peak and residual states. The residual friction angle (ϕ_r) and Poisson's ratio (ν) were chosen based on the data from the test of air-dried dense Toyoura sand. The peak friction angle (ϕ_P) was estimated from the empirical relations based on the plane strain compression test on dense Toyoura sand. ψ is dilatancy angle. The introduction of shear banding in the numerical analysis was achieved by introducing a strain localization parameter s in the following additive decomposition of total strain increment as follows:

$$d\varepsilon_{ij} = d\varepsilon_{ij}^{e} + sd\varepsilon_{ij}^{p} \quad , \quad s = F_b / F_e \tag{11}$$

where F_b is the area of a single shear band in each element; and F_e is the area of the element

The jump kinematic hardening model considering the cumulative deformation from cyclic loading is developed. This is a modified and extended soil model of strain-hardening-softening property in order to take into account the cyclic behavior.

Within bounding surface (or normal surface), plastic behavior is assumed and hardening modulus is much greater comparing the plastic behavior outside the bounding surface.

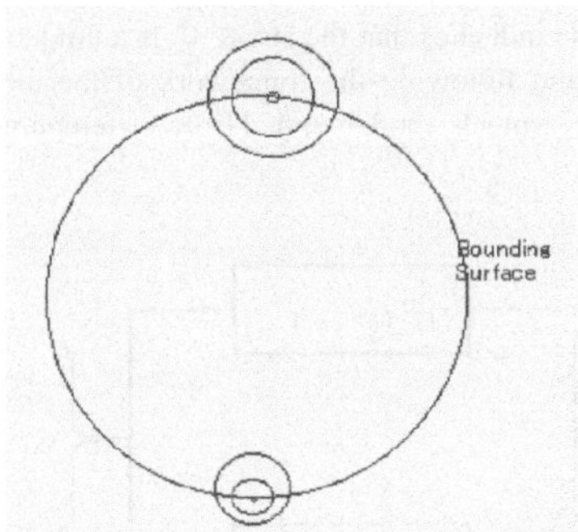

Fig.1 Jump kinematic model on π plane (Mohr-Coulomb model takes pyramid shape)

$$\alpha\ (\kappa) = (\alpha_p - \alpha_0)\left(\frac{2\sqrt{\kappa\varepsilon_f}}{\kappa + \varepsilon_f}\right)^m \quad \text{or} \quad \alpha'\ (\kappa) = (\alpha_p + \alpha_0)\left(\frac{2\sqrt{\kappa\varepsilon_f}}{\kappa + \varepsilon_f}\right)^m \tag{16}$$

where α_0 indicates the reversing point.

4. Constitutive Model for Viscoplasticity

The constitutive equations of rate-dependent plasticity originally proposed by Duvaut-Lions are as follows.

$$\sigma^{v} = \eta C\dot{\varepsilon}^{vp} = \sigma - \overline{\sigma} \tag{17}$$

$$\dot{q} = \frac{-1}{\eta}[q - \overline{q}] \tag{18}$$

where η fluidity parameter, q is internal variables, $\overline{\sigma}$ and $\overline{q}$ are rate independent solution, C is elastic modulus.

Eq. (17) can be rewritten in incremental form as follows.

$$\Delta\varepsilon^{ir} = \frac{\Delta t_{n+1}}{\eta}(\sigma_{n+1} - \overline{\sigma}_{n+1})C^{-1} \tag{19}$$

We can obtain the following equation.

$$\sigma_{n+1}=\frac{\eta C\Delta\varepsilon_{n+1}+\eta\sigma_n+\Delta t_{n+1}\overline{\sigma}_{n+1}}{\Delta t_{n+1}+\eta}$$
$$=\frac{\eta\sigma_{n+1}^{trial}+\Delta t_{n+1}\overline{\sigma}_{n+1}}{\Delta t_{n+1}+\eta} \tag{20}$$

Similarly, eq. (18) is transformed to the following equation.

$$q_{n+1}=\frac{q_n+\Delta t/\eta\overline{q}_{n+1}}{1+\Delta t/\eta} \tag{21}$$

A great deal of experimental results indicates that the stress σ is a unique function of irreversible strain and its rate (Tatsuoka et al. 2002) and following the framework of the three component model (Fig.2), Tatsuoka et al. proposed the New isotach model and TESRA (temporary effect of strain rate and acceleration) model.

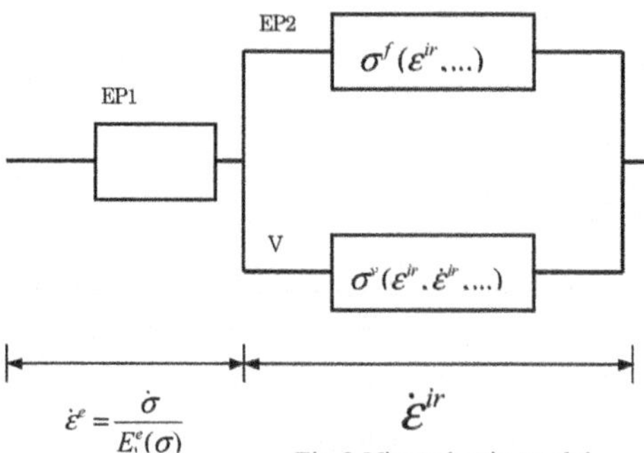

Fig.2 Viscoplastic model

$$\sigma=\sigma^f(\varepsilon^{ir})+\sigma^v(\varepsilon^{ir},\dot{\varepsilon}^{ir}) \tag{22}$$

where ε^{ir} is irreversible strain, $\dot{\varepsilon}^{ir}$ is irreversible strain rate, σ^f is time independent stress, σ^v is time dependent stress.

The New Isotach model takes the following form.

$$\sigma^v=\sigma^f\cdot g(\dot{\varepsilon}^{ir}) \tag{23}$$

$$g(\dot{\varepsilon}^{ir})=\alpha[1-\exp\{1-(\frac{\dot{\varepsilon}^{ir}}{\dot{\varepsilon}_r^{ir}})^m\}] \qquad \text{(viscosity function)}$$

From this equation, we can obtain the equivalent fluidity parameter as follow.

$$\eta=\frac{\sigma^f\alpha[1-\exp\{1-(\frac{\dot{\varepsilon}^{ir}}{\dot{\varepsilon}_r^{ir}}+1)^m\}]}{C\dot{\varepsilon}^{ir}} \tag{24}$$

Introducing the following decay function, Tatsuoka et al. obtained TESRA model

$$g_{decay}(\varepsilon^{ir}-\tau)=r_1^{(\varepsilon^{ir}-\tau)} \tag{25}$$

These New Isotach and TESRA model for one dimensional viscous stress of soils are extended to three dimensional analyses by applying following relation.

$$\sigma = \sigma_{\alpha\alpha} n_\alpha \otimes n_\alpha \tag{26}$$

where n_α principal directions. In this computation, the dynamic relaxation method combined with the generalized return mapping algorithm is applied to obtain the three dimensional stresses by the elasto-plastic constitutive relations. The obtained stresses are transformed to principal stresses. Then the nonlinear viscous stresses in each direction are computed using equivalent fluidity parameter and thus $\sigma_{\alpha\alpha}$ are computed. Applying eq.(26) we can obtain the normal three dimensional stresses in the case of isotropic materials.

5. Cyclic Analyses of Plane Strain Tests

The simulation of plane strain tests by the finite element method using one element was carried out with cyclic elasto-plastic constitutive model. The dynamic relaxation method was applied to this problem. The material constants of Toyoura sand used for calculation are as follow: D_r= 60%, ν = 0.3, φ_r= 34 (deg), ε_r= 0.6, ε_f = 0.1, m = 0.3, shear band thickness = 0.3cm. Fig.3 shows the relationship between undrained mean stress and stress difference and Fig.4 shows the relationship undrained shear strain and stress difference.

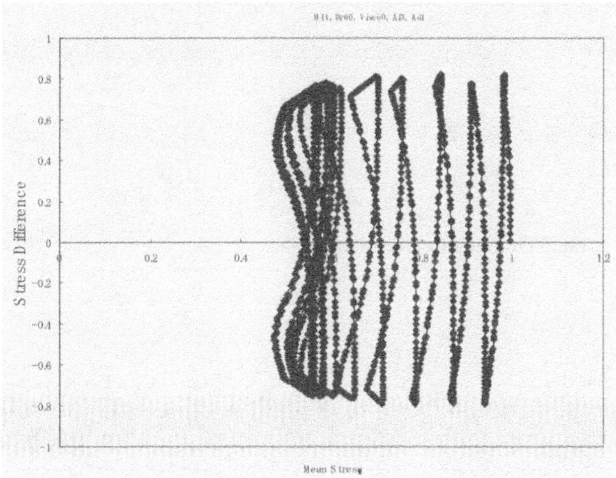

Fig.3 Relationship between undrained mean stress and stress difference

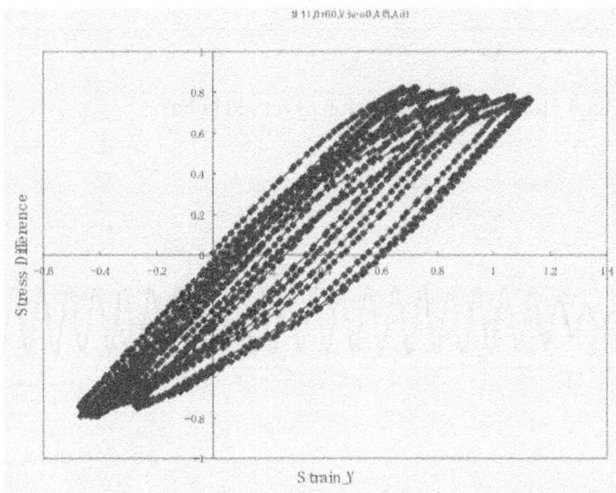

Fig.4 Relationship between undrained shear strain and stress difference

6. Dynamic Progressive Failure of Embankment Dams

The two-dimensional model dams were prepared with a plane strain condition. Models were constructed on the shaking table using the Toyoura sand with water content 5 %. The sand with predetermined weight was thrown on the shaking table every 5 cm, and compacted so that the relative density might become 50 % using the vibrator. In order to measure the response acceleration by simulated earthquake, the accelerometer was embedded in the predetermined layer on the central section. The size of a model dam was 255 cm wide, 80 cm high, 277 cm long, and the slope was 1:1.5 (Fig.5). The experiments were carried out at National Research Institute of Agricultural Engineering. The simulated earthquake was horizontal acceleration (Fig. 6, 7). Fig.8 shows finite element mesh of embankment dam used for the analysis using kinematic hardening-softening constitutive model.

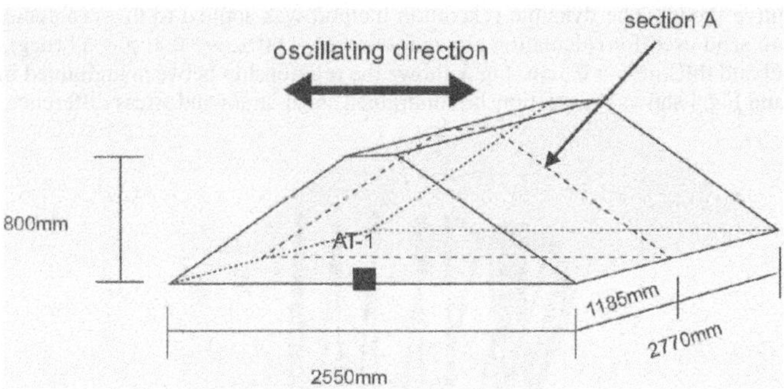

Fig.5 Outline of model dam

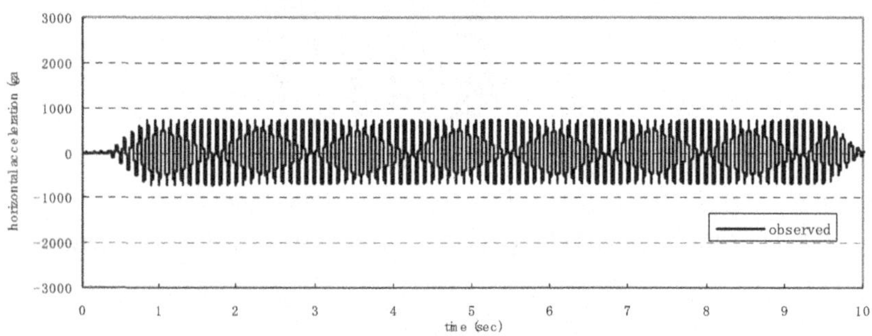

Fig.6 Horizontal acceleration observed at base

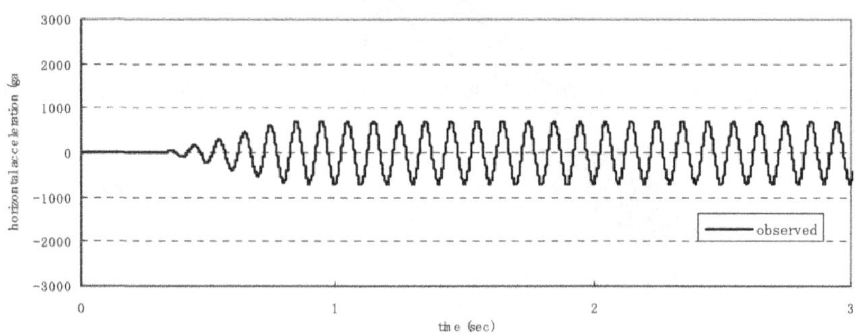

Fig.7 Horizontal acceleration observed at base (first 3 seconds)

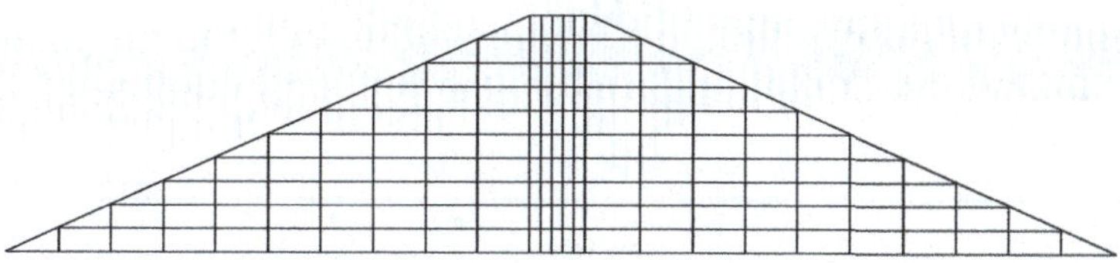

Fig. 8 Finite element mesh of model dam

Fig.9 shows sectional view of model dams after dynamic tests and also displacements and accelerations of dam crest.

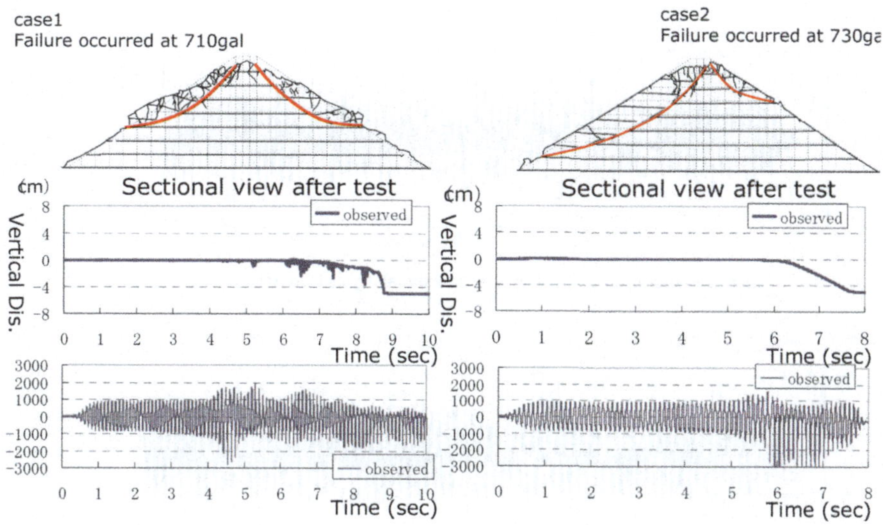

Fig.9 Sectional view after model experiment and vertical displacements and accelerations of dam crest

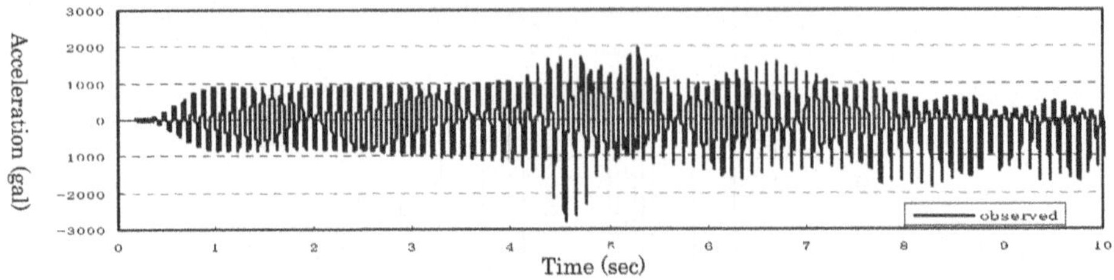

Fig.10 Acceleration at crest by experiment

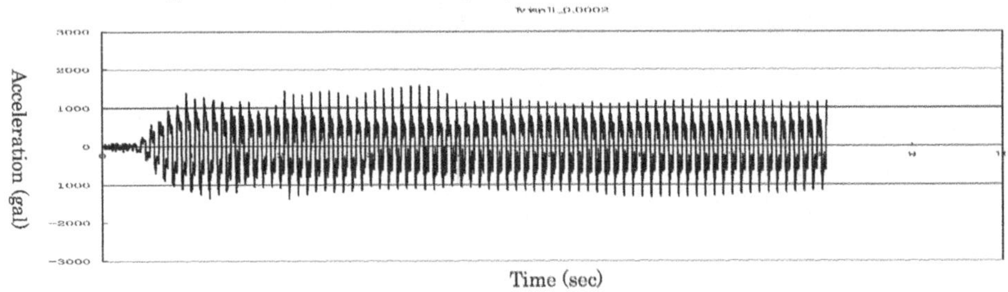

Fig.11 Acceleration at crest by viscoplastic analysis (constant viscosity: η=0.0002)

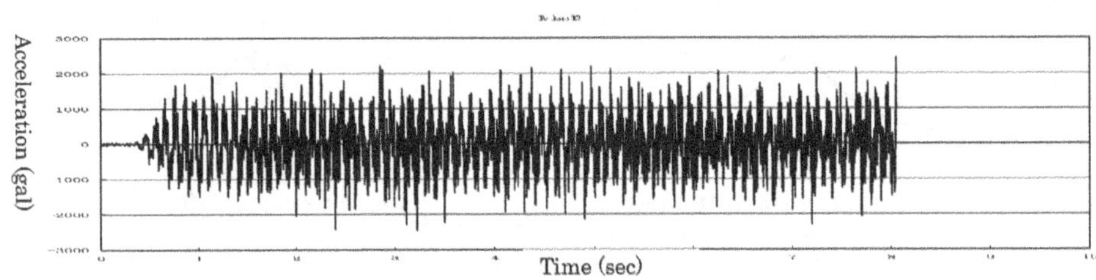

Fig.12 Acceleration at crest by viscoplastic analysis (Isotach)

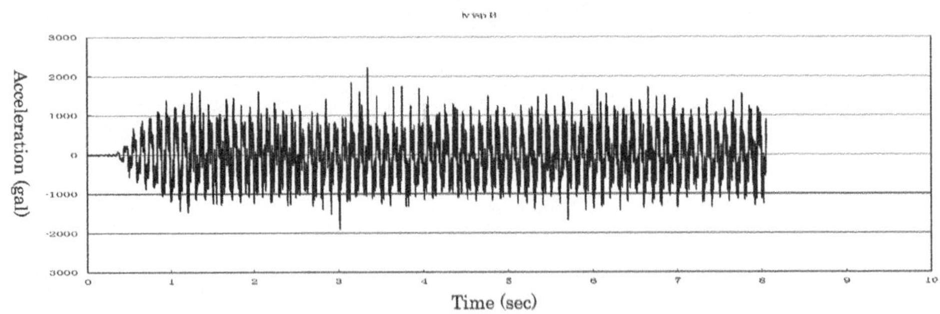

Fig.13 Acceleration at crest by viscoplastic analysis (TESRA)

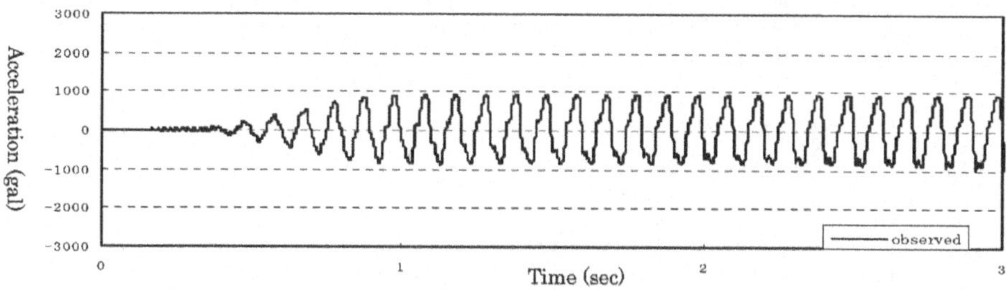

Fig.14 Acceleration at crest by experiment (first 3 sec)

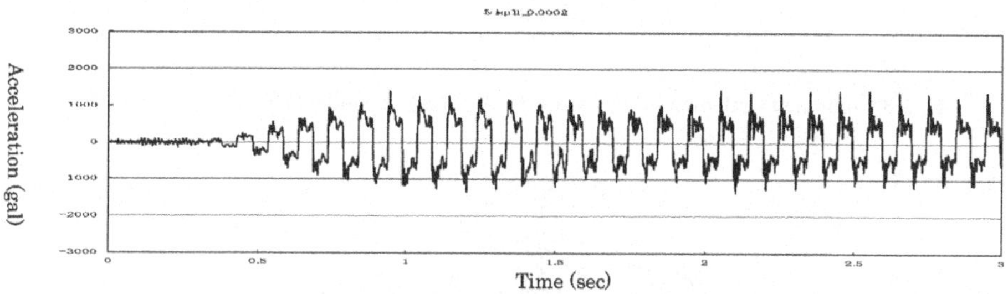

Fig. 15 Acceleration at crest by viscoplastic analysis (constant viscosity: η=0.0002: first 3 sec)

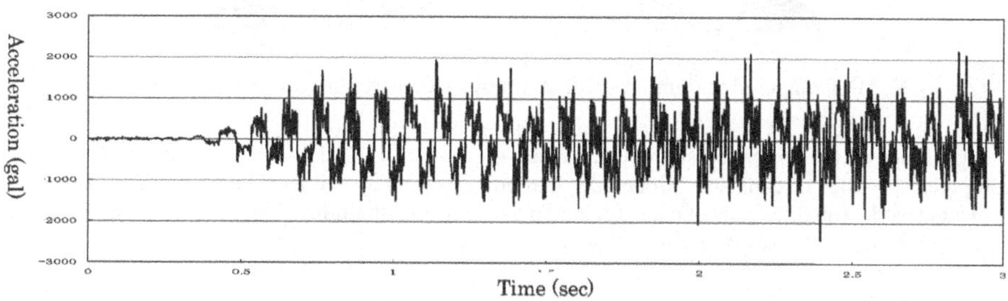

Fig.16 Acceleration at crest by viscoplastic analysis (Isotach: first 3 sec)

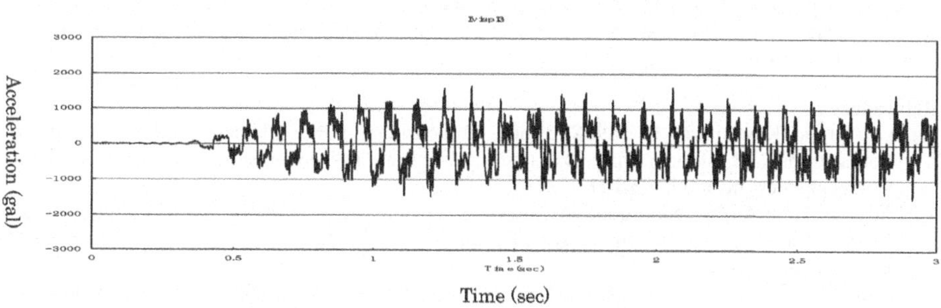

Fig.17 Acceleration at crest by viscoplastic analysis (TESRA: first 3 sec)

Fig.10-17 show accelerations at crest by experiments and analyses with by viscoplastic kinematic hardening model (constant viscosity, Isotach and TESRA). The computed accelerations by the viscoplastic model are approximately identical to the experimental results. Especially TESRA which can realistically represent the behavior of element test of sand is promising also for dynamic response analysis of embankment dam model.

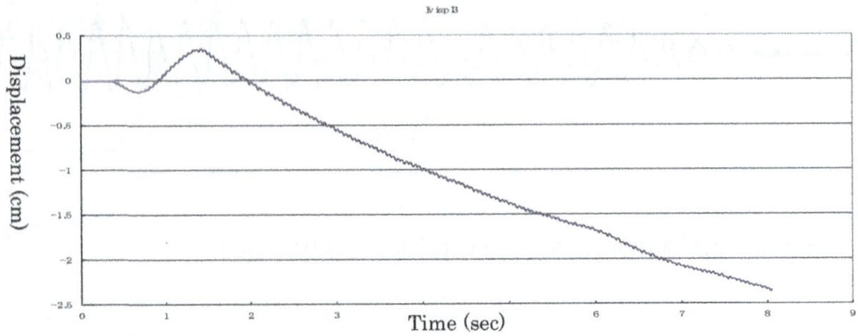

Fig.18 Computed settlement of dam crest by viscoplastic model (TESRA)

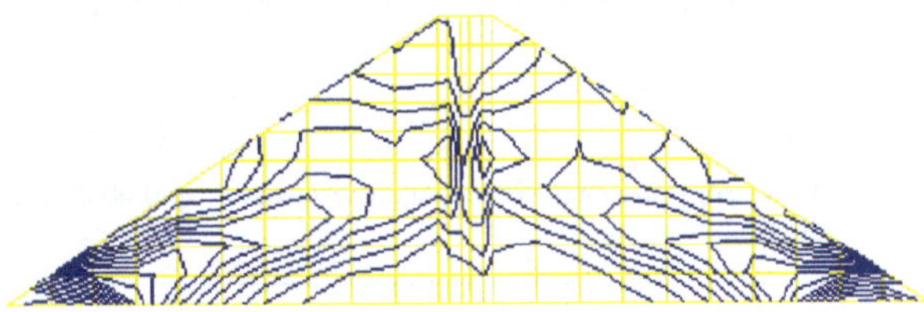

Fig.19 Computed maximum shear strain by viscoplastic model (TESRA: at 9 sec)

Fig. 18 shows the computed settlement of dam crest and Fig.19 shows the maximum shear strain computed by the viscoplastic kinematic hardening-softening model (TESRA). The kinematic hardening model (modified and extended soil model of isotropic strain-hardening-softening) is promising for the prediction of cumulative deformation of soil structures.

6. Conclusions

The viscoplastic kinematic hardening model is developed. The Duvaut-Lions viscoplastic model is used to solve the problem and TESRA model of constitutive relation for soils is applied to the boundary value problems such as plane strain compression tests and dynamic response analysis of embankment dam models successfully.

References

Tanaka, T. and Kawamoto, O. (1988): Three Dimensional Finite Element Collapse Analysis for Foundations and Slopes using Dynamic Relaxation, Proceedings of Numerical Methods in Geomechanics, Insbruch, pp.1213-1218.

Yoshida, T., Tatsuoka, F., Siddiquee, M. S. A. and Kamegai, Y. (1995): Shear Banding in Sands Observed in Plane Strain Compression, Localization and Bifurcation Theory for Soils and Rocks, Balkema, pp.165-179.

Tatsuoka, F., Ishihara, M., Di Benedetto, H. and Kuwano, R. (2002): Time-dependent Shear Deformation Characteristics of Geomaterials and Simulation, Soils and Foundations, Vol.42, No.2, pp.103-129.

Soil Stress-Strain Behavior: Measurement, Modeling and Analysis
Geotechnical Symposium in Roma, March 16 & 17, 2006

NUMERICAL MODELING OF WAVE PROPAGATION IN BOGOTÁ SOFT SOILS

D.K. Reyes
Universidad de los Andes, Bogotá, Colombia
e-mail: dian-rey@uniandes.edu.co
C. Grandas
Universidad de los Andes, Bogotá, Colombia
e-mail: cgrandas@uniandes.edu.co
A. Lizcano
Universidad de los Andes, Bogotá, Colombia
e-mail: alizcanos@uniandes.edu.co

ABSTRACT
Visco-hypoplastic constitutive models reproduce properly the non-linear soil dynamic behavior, carrying effective stress analysis out. That is, viscohypoplastic models can simulate pore pressure build up, stiffness degradation, permanent deformation, non-linear viscous effects (time and rate dependence) and some non-linear effects observed in the wave propagation in undrained soft soils. In this work dynamic equations are derived and analyzed using this model. The results of numerical simulations of shear wave propagation in Bogotà soft soils as well as the influence of initial and boundary conditions are presented.

1. INTRODUCTION
Seismological measurements (Beresnev et al., 1994; Beresnev et al., 1995, Beresnev et al., 1998; Cultrera et al., 1999; and Yang et al., 2000) and lab measurements (Tatsuoka et al, 1999) show a non-linear soil behavior under dynamic loads. Additionally, the strain-stress relationship of soils is influenced by the current stress state, the stress conditions and the load process. During seismic wave propagation the shear stiffness changes sudden and continuously because effective stress state, strain direction and strain rate change fast too.

Ossinov and Gudehus (1996), Ossinov (1997) and Ossinov (1998) has been studied the plane shear wave propagation through saturated granular media using a hypoplastic constitutive theory to model the soil response.

The visco-hypoplastic constitutive model proposed by Niemunis (2003) is able to describe non-linear time-rate effects of soft soils under monotonic and cyclic loads for drained and undrained conditions. The model simulates properly stiffness dependence on initial effective stress, initial void ratio, strain level, strain rate, number of cycles and effective stress history. Pore pressure build up and permanent deformations can be obtained from the visco-hypoplastic constitutive model.

The equations for the wave propagation in soft soils using the visco-hypoplastic law after Niemunis is derived and analyzed in this work. Simulations of the influence of the initial stress and the boundary conditions on the wave propagation velocity are showed.

Hoe I. Ling et al. (eds.), Soil Stress-Strain Behavior: Measurement, Modeling and Analysis, 779–789.

Nonlinear changes in the length wave and analyses of the amplification level are illustrated. For the numerical simulations were used the visco-hypoplastic parameter of Bogotá soft soil obtained from oedometric and monotonic and cyclic triaxial tests..

2. MOVEMENT EQUATION

The soft soil is supposed as a simple skeleton without cementation, macro-pores, structure, nor cohesion. The soil is composed by two incompressible fractions: a solid skeleton and a liquid with the volumes V_s and V_l respectively. Each pore is fully occupied by the liquid.

The void ratio e establishes the portion of the two fractions $e = V_l/V_s$. To describe the state of de body, the stresses in both fractions, liquid pressure, p_l, and stress in skeleton, $\mathbf{T}$, have to be known. The total stress $\mathbf{T}_{\text{total}}$ is the sum of the stresses in the two fractions

$$\mathbf{T}_{\text{total}} = \mathbf{T} - p_l \mathbf{1}, \tag{0.1}$$

where $\mathbf{1}$ is the second order unit tensor. The permeability of the soft soil and the time interval are assumed small enough to avoid seepage. Thus, no volume change of the body can occur during wave propagation, $e = const.$, or the velocity of the fluid $\mathbf{v}_l$ and the velocity of the solid fraction $\mathbf{v}_s$ coincide in all points, $\mathbf{v} = \mathbf{v}_l = \mathbf{v}_s$. From the mass balance

$$\text{Div}\,\mathbf{v} = 0 \tag{0.2}$$

the movement can be described by the velocity field vector $\mathbf{v}(\mathbf{x},t)$. This velocity is a function of the current position $\mathbf{x}$ (spatial description) and time t. When the soil is at rest, (initial state), the liquid fraction stress $p_l^0(\mathbf{x})$ and the solid fraction $\mathbf{T}^0(\mathbf{x})$ must satisfy the equilibrium condition:

$$\text{Div}\,\mathbf{T}^0 - \text{grad}\,p_l^0 + \rho\mathbf{f} = 0, \tag{0.3}$$

where $\rho = (\rho_l e + \rho_s)/(1+e)$ is the density of the body, $\mathbf{f}$ is the body force vector (time independent), and ρ_l and ρ_s are the liquid and solid density respectively. Note that the stresses in both fractions depend on the position $\mathbf{x}$ only.

During the wave propagation, the impulse balance equation is

$$\text{Div}\,\mathbf{T}_{\text{total}} + \rho\mathbf{f} = \rho\frac{d\mathbf{v}}{dt} \tag{0.4}$$

where $d\mathbf{v}/dt$ is the material time derivate with respect to time. The equation (0.4) can be rewritten in terms of solid and liquid stresses:

$$\text{Div}\,\mathbf{T} - \text{grad}\,p_l = \rho\frac{d\mathbf{v}}{dt} \tag{0.5}$$

The equation (0.5) relates the stress of solid skeleton $\mathbf{T}$, the liquid pressure ρ_l with the velocity field $\mathbf{v}$. To solve the equation of motion (0.5) it is necessary a stress-strain relationship to describe de soil behavior and to connect the stress $\mathbf{T}$ with $\mathbf{v}$. In this work, the visco-hypoplastic equation is used.

3. VISCO-HIPOPLASTIC MODEL

The effective stresses-rate $\overset{\circ}{\mathbf{T}}$ is determined depending on the current effective stress $\mathbf{T}$, the void ratio e, the history of deformations $\mathbf{D}^{vis}$ (related with *OCR*), the strain-rate $\mathbf{D}$, and the recent deformation history via intergranular strain tensor $\mathbf{h}$. Niemunis (2003) introduces the intergranular strain concept to improve the behavior at small cycles of load. This modification considers the immediate stiffness caused by a change in the strain direction (Atkinson et al., 1990). The effective stress-rate depends on

$$\overset{\circ}{\mathbf{T}} = \mathbf{H}(\mathbf{T}, \mathbf{D}, e, \mathbf{h}). \tag{0.1}$$

The constitutive model can be described as the tensorial expression

$$\overset{\circ}{\mathbf{T}} = \mathcal{M} : \mathbf{D} - \mathsf{L} : \mathbf{D}^{vis}, \tag{0.2}$$

with the fourth order tensor $\mathcal{M}(\mathbf{T}, \mathbf{h})$ (that depends on stress and strain direction)

$$\mathcal{M} = m f_b \hat{\mathsf{L}} = \left[\rho^{\chi} m_T + (1-\rho^{\chi}) m_R\right] \mathsf{L} + \begin{cases} \rho^{\chi}(1-m_T)\mathsf{L} : \vec{\mathbf{h}}\vec{\mathbf{h}} & \text{for } \vec{\mathbf{h}} : \mathbf{D} > 0 \\ \rho^{\chi}(m_R - m_T)\mathsf{L} : \vec{\mathbf{h}}\vec{\mathbf{h}} & \text{for } \vec{\mathbf{h}} : \mathbf{D} \leq 0 \end{cases}, \tag{0.3}$$

where $\rho = \|\mathbf{h}\|/R$, R is the elastic range, m_T and m_R are the increments of stiffness during a direction change of 90° and 180° respectively, and χ is a interpolation constant. The barotropy function f_b depends on stress via:

$$f_b(\mathbf{T}) = -\frac{\operatorname{tr}\mathbf{T}}{(1 + a^2/3)\kappa}, \tag{0.4}$$

where κ is a material parameter (the unloading-reloading slope). The intergranular strain evolution depends on the stretching tensor $\overset{0}{\mathbf{h}}(\mathbf{D})$:

$$\overset{0}{\mathbf{h}} = \begin{cases} \left(\mathsf{I} - \vec{\mathbf{h}}\vec{\mathbf{h}}\rho^{\beta_r}\right) : \mathbf{D} & \text{for } \vec{\mathbf{h}} : \mathbf{D} > 0 \\ \mathbf{D} & \text{for } \vec{\mathbf{h}} : \mathbf{D} \leq 0 \end{cases}. \tag{0.5}$$

In (3.3) the four order tensor (that depends on stress) $\mathsf{L}(\mathbf{T})$ is defined as

$$\mathsf{L} = f_b \hat{\mathsf{L}} = f_b a^2 \left[\left(\frac{F}{a}\right)^2 \mathsf{I} + \hat{\mathbf{T}}\hat{\mathbf{T}}\right], \tag{0.6}$$

with the four order unit tensor I and the dimensionless stress $\hat{\mathbf{T}} = \mathbf{T}/\operatorname{tr}\mathbf{T}$. $a = \sqrt{3}(3 - \sin\varphi_c)/(2\sqrt{2}\sin\varphi_c)$ and $F(\mathbf{T})$ are related with Matsuoka-Nakai yield surface and depends on stress. The objective stress tensor $\overset{\circ}{\mathbf{T}}$ is defined by:

$$\overset{\circ}{\mathbf{T}} = \dot{\mathbf{T}} + \mathbf{T}\cdot\mathbf{W} - \mathbf{T}\cdot\mathbf{W}, \tag{0.7}$$

where the dot over the tensor $\mathbf{T}$ indicates the material time derivate of stress. The spin tensor $\mathbf{W}$ is defined by:

$$W_{ij} = \frac{1}{2}\left(\frac{\partial v_i}{\partial x_j} - \frac{\partial v_j}{\partial x_i}\right). \tag{0.8}$$

The stretching tensor $\mathbf{D}$ is composed by an elastic and a viscous part:

$$\mathbf{D} = \mathbf{D}^{elas} + \mathbf{D}^{vis} \tag{0.9}$$

and is defined by:

$$D_{ij} = \frac{1}{2}\left(\frac{\partial v_i}{\partial x_j} + \frac{\partial v_j}{\partial x_i}\right). \tag{0.10}$$

The viscous strain rate $\mathbf{D}^{vis}(\mathbf{T}, e)$ depends on the current stress $\mathbf{T}$ via *OCR* and the void ratio, e. $\mathbf{D}^{vis}$ can be discomposed in a direction of flow $\vec{\mathbf{B}}(\mathbf{T})$ and a intensity of creep:

$$\mathbf{D}^{vis} = -D_r\vec{\mathbf{B}}\left(\frac{1}{OCR}\right)^{\frac{1}{I_v}}. \tag{0.11}$$

In (3.11) D_r is a reference velocity of creep and I_v is the viscosity index. The *OCR* is now defined as the "distance" between the current mean stress and the equivalent stress that corresponds to the current void ratio in a reference isotach. The direction of the viscous flow is given by:

$$\vec{\mathbf{B}} = \frac{\left(F/a\right)^2\left(\hat{\mathbf{T}}+\hat{\mathbf{T}}^*\right)+\hat{\mathbf{T}}:\hat{\mathbf{T}}\hat{\mathbf{T}}^*-\hat{\mathbf{T}}\hat{\mathbf{T}}:\hat{\mathbf{T}}^*}{\left\|\left(F/a\right)^2\left(\hat{\mathbf{T}}+\hat{\mathbf{T}}^*\right)+\hat{\mathbf{T}}:\hat{\mathbf{T}}\hat{\mathbf{T}}^*-\hat{\mathbf{T}}\hat{\mathbf{T}}:\hat{\mathbf{T}}^*\right\|}, \tag{0.12}$$

where $\hat{\mathbf{T}}^* = \overset{\circ}{\mathbf{T}} - \frac{1}{3}\mathbf{1}$ is the dimensionless stress deviator.

4. THE INITIAL-BOUNDARY VALUE PROBLEM

In this section the motion equation (0.5) is combined with the stress strain relationship (0.2) to describe the propagation of plane shear wave through a fully saturated soft soil. A soil deposit is idealized as a semi-infinite half-space. The dimension of the soil extends to infinity in x_2 and x_3 horizontal directions. In the positive x_1 vertical direction, the soil has a finite extension and bounds with the air (**Figure 1**). The particles move only in the x_2 direction and the wave travels in the x_1 direction, so the velocity field has the components

$$v_1 = 0, v_3 = 0 \text{ and } v_2 = v_2(x_1, t) \quad (0.1)$$

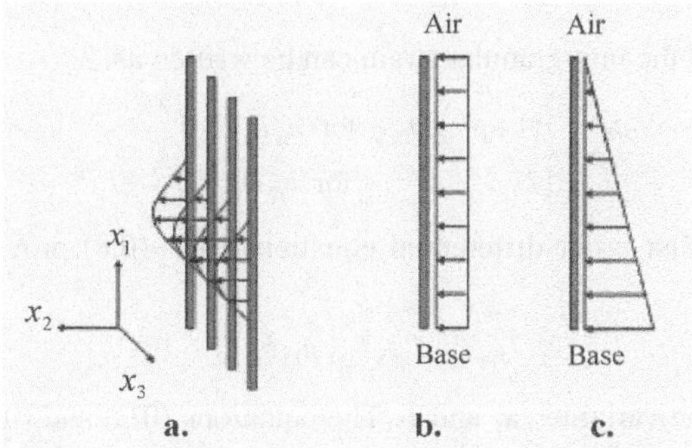

Figure 1: **a.** Plane waves traveling in the x_1 direction. **b.** A constant stress distribution with respect to x_1. **c.** A geostatic distribution of stress with respect to x_1.

With those restrictions the stretching tensor $\mathbf{D}$, equation (0.10), has only two non cero components, $D_{12} = D_{21} = 1/2(\partial v_2/\partial x_1)$. In consequence, during propagation tr **D**=0 and volume changes do not occur. The intergranular strain tensor $\mathbf{h}$, that depends on $\mathbf{D}$, has only two non cero components, $h_{12} = h_{21}$ too.

The stress tensor components $T^0_{13}, T^0_{23}, T_{13}$ and T_{23} are equal to zero. All others variables are function of x_1 and t only. With this restriction on the velocity and stress field the equation of motion (0.5) takes the form:

$$\frac{\partial T_{11}}{\partial x_1} - \frac{\partial p_l}{\partial x_1} = 0 \quad (0.2)$$

$$\frac{\partial T_{12}}{\partial x_1} - \rho \frac{\partial v_2}{\partial t} = 0 \quad (0.3)$$

Using the equation (0.7) the constitutive equation (0.1) can be written as:

$$\frac{\partial T_{11}}{\partial t} - H_{11}\left(\mathbf{T}, \frac{\partial v_2}{\partial x_1}, e, h_{12}\right) + (T_{12})\frac{\partial v_2}{\partial x_1} = 0, \quad (0.4)$$

$$\frac{\partial T_{12}}{\partial t} - H_{12}\left(\mathbf{T}, \frac{\partial v_2}{\partial x_1}, e, h_{12}\right) + \frac{1}{2}(T_{22} - T_{11})\frac{\partial v_2}{\partial x_1} = 0, \quad (0.5)$$

$$\frac{\partial T_{22}}{\partial t} - H_{22}\left(\mathbf{T}, \frac{\partial v_2}{\partial x_1}, e, h_{12}\right) - (T_{12})\frac{\partial v_2}{\partial x_1} = 0 \text{ and} \quad (0.6)$$

$$\frac{\partial T_{33}}{\partial t} - H_{33}\left(\mathbf{T}, \frac{\partial v_2}{\partial x_1}, e, h_{12}\right) = 0. \tag{0.7}$$

The evolution of the intergranular strain can be written as:

$$\frac{\partial h_{12}}{\partial t} = \begin{cases} \left(1-\rho^{\beta_r}\right) D_{12} & \text{for } h_{12}D_{12} > 0 \\ D_{12} & \text{for } h_{12}D_{12} \leq 0 \end{cases}. \tag{0.8}$$

The system of first order differential equations (0.2)-(0.8) provides the solution for the variables

$$v_2, T_{11}, T_{12}, T_{22}, T_{33}, p_l, h_{12} \tag{0.9}$$

as function of the variables x_1 and t. The equations (0.3) and (0.5) can be combined to obtain a "visco-hypoplastic wave equation", as function of the unknown variable v_2 only. The resulting differential equation can be solved using a finite difference approach. In order to minimize the approximation error is convenient to transform the first order differential equation into a second order one. This can be performed using the displacement in the direction x_2, u_2, instead v_2 with the expression:

$$\frac{\partial v_2}{\partial x_1} = \frac{\partial^2 u_2}{\partial x_1 \partial t}. \tag{0.10}$$

The combination of equations (0.3) and (0.5) conduces to

$$\left(m \cdot f_b \cdot \hat{L}_{1212} - \frac{1}{2}\left(T_{22} - T_{11}\right)\right)\frac{\partial^2 u_2}{\partial x_1^2} - f_b \cdot \hat{L}_{12kl} \cdot D_{kl}^{vis}\frac{\partial t}{\partial x_1} - \rho\frac{\partial^2 u_2}{\partial x_1} = 0 \quad \text{with } k,l = 1,2,3. \tag{0.11}$$

This equation can be solved for u_2, from which $\partial v_2/\partial x_1$ can be obtained and replaced in the other equations to obtain all the unknown functions.

The initial boundary value problem consist in to solve the system of equations (0.2), (0.4)-(0.8) and (0.11) in the domain $0 \leq x_1 \leq l$, $t \geq 0$ subject to the initial conditions

$$u_2(x_1, 0) = 0, \qquad \mathbf{T}(x_1, 0) = \mathbf{T}^0 \tag{0.12}$$

and to the boundary conditions:

$$u_2(0, t) = b(t), \qquad \mathbf{T}(l, t) = 0. \tag{0.13}$$

The function $b(t)$ is an harmonic signal introduced at the base. The perturbation propagates in the positive x_1 direction, goes to the soil-air boundary and then reflects in the negative x_1 direction.

The problem has been solved using an explicit difference finite scheme programmed in a MATLAB script. To maintain the algorithm stability, the velocity of propagation should be a little lower than the mesh ratio dx/dt (spatial step / time step).

5. THE WAVE VELOCITY

Supposing an initial stress state where $\mathbf{T}^0$ is a linear function of x_1, so a geostatic stress distribution is obtained (**Figure 1**c), and the stress $T_{22}^0 = T_{33}^0 = K_0 T_{11}^0$, where $K_0 < 1$. The stiffness of the visco-hypoplastic media depends on the direction of strain respect to the recent strain history (the sing of the product $h_{12}D_{12}$). From the equation (0.3) the scalar m can be rewritten as:

$$m = k_1 + \begin{cases} k_2 & \text{for } h_{12}D_{12} > 0 \\ k_3 & \text{for } h_{12}D_{12} \leq 0 \end{cases}, \tag{0.1}$$

with:

$$k_1 = \rho^{\chi} m_T + \left(1-\rho^{\chi}\right) m_R, \quad k_2 = \rho^{\chi}\left(1-m_T\right), \quad k_3 = \rho^{\chi}\left(m_R - m_T\right) \ . \tag{0.2}$$

The equation (0.11) can be rewritten as:

$$\left[\left(k_1+k_2\right) f_b \hat{L}_{1212} - \frac{1}{2}\left(T_{22}-T_{11}\right)\right]\frac{\partial^2 u_2}{\partial x_1^2} - f_b \hat{L}_{12kl} D_{kl}^{vis} \frac{\partial t}{\partial x_1} - \rho \frac{\partial^2 u_2}{\partial x_1} = 0 \quad \text{for } h_{12}D_{12} > 0 \tag{0.3}$$

$$\left[\left(k_1+k_3\right) f_b \hat{L}_{1212} - \frac{1}{2}\left(T_{22}-T_{11}\right)\right]\frac{\partial^2 u_2}{\partial x_1^2} - f_b \hat{L}_{12kl} D_{kl}^{vis} \frac{\partial t}{\partial x_1} - \rho \frac{\partial^2 u_2}{\partial x_1} = 0 \quad \text{for } h_{12}D_{12} \leq 0 \tag{0.4}$$

Consequently, there are two wave velocity c_1 and c_2 instead a single velocity c. The two cases are:

$$c_1 = \sqrt{\frac{\left(k_1+k_2\right) f_b \hat{L}_{1212} - \left(T_{22}-T_{11}\right) - f_b \hat{L}_{12kl} D_{kl}^{vis}}{\rho}} \quad \text{for } h_{12}D_{12} > 0, \tag{0.5}$$

$$c_2 = \sqrt{\frac{\left(k_1+k_3\right) f_b \hat{L}_{1212} - \left(T_{22}-T_{11}\right) - f_b \hat{L}_{12kl} D_{kl}^{vis}}{\rho}} \quad \text{for } h_{12}D_{12} \leq 0. \tag{0.6}$$

The velocity c_1 corresponds to the situation where the point in the body moves in the same direction in which it has been recently deformed. The velocity c_2 is obtained when the motion induces a sudden change in the direction of the deformation. When the second case occurs, the stiffness of the body increases immediately. If the deformation is big enough, the stiffness evolves towards the first case. In both cases, if K_0 decreases, the velocity decreases too. The maximum wave velocity is reached for the isotropic case, K_0=1.

The equation (0.5) and (0.6) shows that velocity depends on stress level. When the disturbance travels from the base (when the stress is bigger) to the surface (when stress tends to zero), the wave velocity decrease in the positive direction of x_1. This fact generates a change in the wave length λ: it decreases when approximates to the free surface.

6. RESULTS OF WAVE PROPAGATION

This section shows the results of the numerical simulations of plane shear wave propagations in Bogotá soft soil using the above described viscohypoplastic model (Niemunis, 2003). The Table 1 presents the parameters of Bogotá soft soils obtained from oedometer and monotonic and cyclic triaxial tests (Grandas, 2006).

Table 1 Material parameters used in the numerical simulations

φ_c	λ	κ	I_v	β	OCR	e_{100}	m_R	m_T	R_{max}	β_x	χ	D_r	$\delta\varepsilon/\delta t$
[°]	[-]	[-]	[-]	[-]	[-]	[-]	[-]	[-]	[-]	[-]	[-]	[-/s]	[-/s]
20	0.2	0.02	0.06	0.95	1.0	2.5	5.0	2.0	1.0E-4	0.12	6.0	1.15E-5	1.1E-5

As perturbation for the wave propagation analysis was used a sine signal restricted to x_1 direction. **Figure 2** shows the wave propagation velocity for two different stress conditions. The velocity increases with the confining pressure and is major for isotropic than for oedometric condition.

Figure 3 presents the results of the numerical simulation of the propagation in the positive x_1 direction of a sine wave with a frequency of 10 *Hz* through a 100 m one layer soil profile. The signal is captured in some time instants during the propagation. The length wave, and consequently the velocity, diminishes in the considered wave propagation direction. This nonlinear effect is a consequence of the confining pressure.

To fulfill the boundary condition, the wave shows an amplification of the input signal at the soil surface (see Figure 3). The amplification level depends on the stiffness of the soil layers. In order to analyze this effect, numerical simulations of sine wave propagations (frequency = 10Hz) through a soil profile with two 50 m thick layers (**Figure 4**) were performed. **Figure 4**a shows the results of the simulations when both layers have the same stiffness. In **Figure 4**b the stiffness of the upper layer is major than the lower layer. In **Figure 4**c the stiffness of the upper layer is minor than the lower layer.

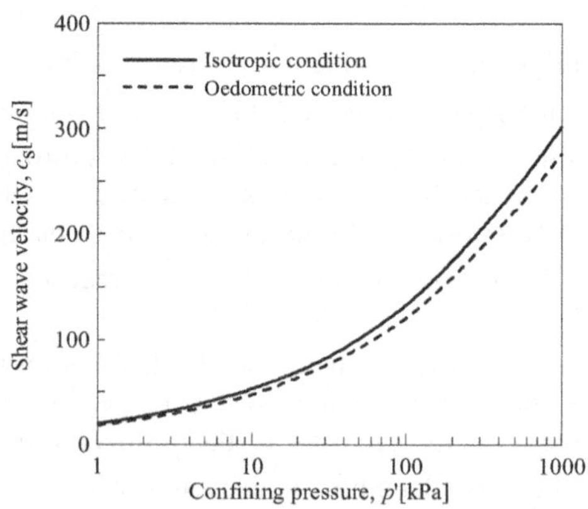

Figure 2: Dependence of the wave velocity from confining pressures for isotropic and oedometric conditions

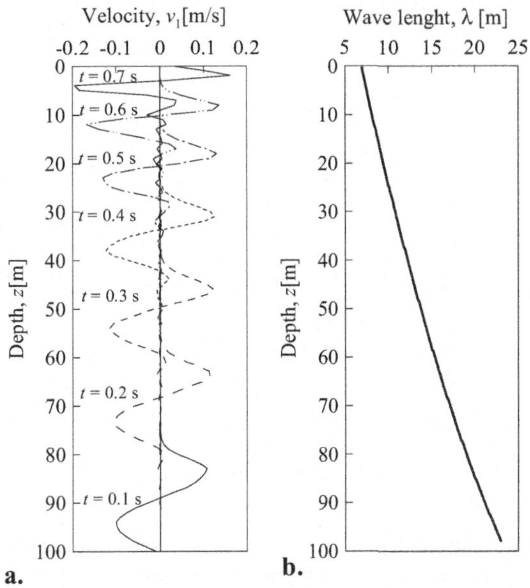

Figure 3: Variation of the length wave with the depth

Additionally, the **Figure 4**d shows the amplification of the maximum velocity for the previous cases. When a soft soil layer of bigger stiffness is over the soft soil layer of smaller stiffness, the amplification level diminishes. In this case, the waves arrive in smaller time than in the opposite case.

7. CONCLUSIONS

1) The equations that govern the wave propagation in a visco-hypoplastic medium can be described by the expressions (0.2)-(0.8).
2) The wave propagation velocity depends on the initial stress state and the initial void ratio. Further, the wave propagation depends on the stress condition: if K_0 approaches to 1 (isotropic state) the velocity increases; if K_0 is move away of 1 (anisotropic state) the velocity diminishes.
3) The propagation wave velocity depends on the strain direction and the sudden change in a time instant.
4) In order to satisfy the boundary conditions, amplification must be appear when the wave approaches the surface
5) The wave length and the wave velocity must vary in the depth as consequence of the geostatic stress state.

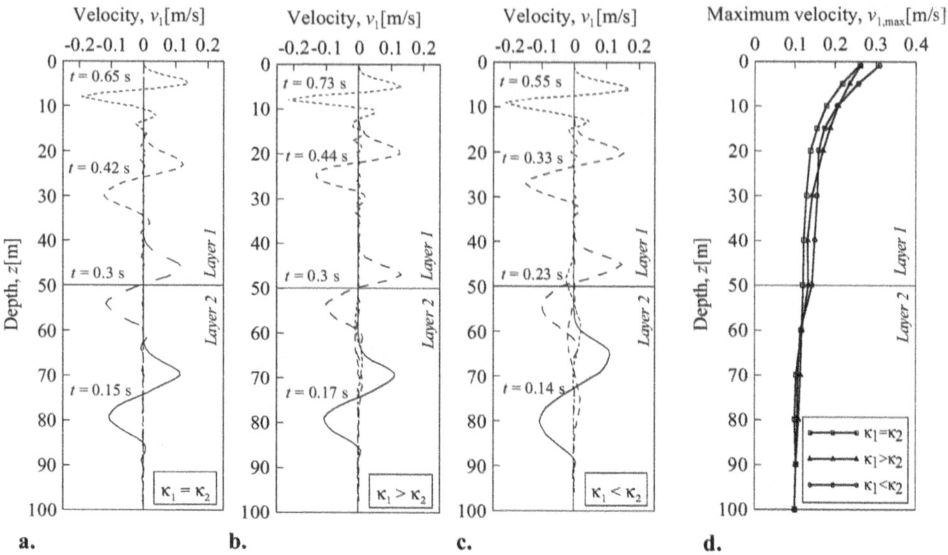

Figure 4: Wave propagation results for two soil layers with different stiffness. **a.** Two layer with equal stiffness. **b.** Stiffer layer over softer layer. **c.** Softer layer over stiffer layer. **d.** Profile of maximum velocity values for previous cases.

6) When a layer of great stiffness lies over a layer of smaller stiffness, the amplification level diminishes. In this case, the waves arrive in smaller time than in the opposite case.

8. REFERENCES

Atkinson J.H., Richardson D. and Stallebrass S.E. (1990). *Effect of recent stress history on the stiffness of overconsolidated soil.* Geothecnique, 40, (4) 531-540.

Beresnev, I.A., Wen, K.-L. and Yeh, Y.T. (1994). *Seismological evidence for nonlinear elastic ground behavior during large earthquakes.* Soil Dyn Earthquake Eng. 14, 103-114.

Beresnev, I. A., K.-L. Wen, and Y. T. Yeh (1995). *Nonlinear soil amplification: its corroboration in Taiwan.* Bull. Seism. Soc. Am. 85, No. 2,496 -515

Beresnev, I.A. and Atkinson G.M. (1998). *Stochastic finite-fault modeling of ground motions from the 1994 Northridge, California, earthquake.* Bull. Seism. Soc. Am. 88, 1392-1401

Cultrera, G., Boore, D.M., Joyner, W.B. and Dietel, C.M. (1999). *Nonlinear soil response in the vicinity of the Van Norman Complex following the 1994 Northridge, California, earthquake.* Bull. Seism. Soc. Am. 89, 1214–1231

Grandas, C. (2006). *Wave propagations in soft soils using a viscohipoplastic constitutive model (in spanish).* Master of Sciene Thesis, *to be publisher in march of 2006*

Niemunis, A. (2003). *Extended hypoplastic models for soils.* Heft 34 des Institutes für Grundbau und Bodenmechanik der Ruhr-Universität Bochum

Osinov, V. A. & Gudehus, G, (1996). *Plane shear waves and loss of stability in a saturated granular body.* Mechanics of Cohesive-frictional Materials, 1,25-44

Osinov V. A. (1997). *Plane waves and dynamic ill-possedness in granular media.* Powders and Grains. Behringer and Jenkins. 363-366, Rotterdam, Balkema

Osinov V. A. (1998). *On the formation of discontinuities of wave fronts in a saturated granular body*. Continium Mechanics and Thermodynamic. 10, 253-268.

Tatsuoka F. Modoni, G. Anh Dan, L.Q. Flora A. Matsuhita M and Koseki J *(1999). Stress-Strain behavior at small strains of unbound granular materials and its laboratory test, Keynote lecture,* Procc. of Workshop on modelling and advanced testing for unbound granular materials. 17-6*1*

Yang, J., Sato, T., and Xiang-Song, L. (2000). *Nonlinear site effects on strong ground motion at a reclaimed island.* Can. Geotech. J. 37, 26 -39.

Soil Stress-Strain Behavior: Measurement, Modeling and Analysis
Geotechnical Symposium in Roma, March 16 & 17, 2006

A CONSTITUTIVE MODEL FOR SOFT ROCKS

Dr. Mohammed Kabirul Islam, Associate Professor
Mr. Mohammed Ibrahim, Post Graduate Student
Department of Civil Engineering
Bangladesh University of Engineering & Technology, Dhaka-1000, Bangladesh
e-mail: raja4bd@yahoo.com

ABSTRACT
An isotropic, volumetric strain hardening, non-associated plasticity model, based on the critical state framework and similar to the Modified Cam Clay model, is proposed for simulating the stress-strain and volume change response of soft rocks. The model is relatively simple and has four parameters in addition to those of the Modified Cam Clay. A numerical study was conducted to investigate the effects of three of the four additional parameters of the proposed model on the predicted stress-strain and volume change response. The numerical predictions were also compared with the experimental data available for natural Corinth marl, which is a soft rock. It was concluded that the values of the parameters of the proposed model could be adjusted to give a reasonable approximation of the stress-strain and volume change behaviour of soft rocks.

1. MODEL DESCRIPTION
The proposed model for soft rocks is an isotropic volumetric strain-hardening model within the critical state framework. The yield locus of the model is elliptical and similar to the yield locus of the Modified Cam Clay (MCC) model. However it is scaled down in $p'-q$ space by a scaling factor termed as the spacing ratio parameter r. The mean effective pressure p' the deviator stress q is defined as done in the standard literature and is given as below:

$$p' = \frac{(\sigma_1' + \sigma_2' + \sigma_3')}{3}$$

$$q = \frac{1}{\sqrt{2}}\sqrt{(\sigma_1' - \sigma_2')^2 + (\sigma_2' - \sigma_3')^2 + (\sigma_3' - \sigma_1')^2}$$

In the above equations, $\sigma_1', \sigma_2', \sigma_3'$ are the mean effective pressure in the three principal directions. These directions are the axial and radial directions in a conventional triaxial test. The spacing ratio r is the ratio of the mean effective pressure at the Normal Consolidation Line (NCL) to the mean effective pressure at the Critical State Line (CSL). These mean effective pressures are obtained from the intersection of an elastic rebound line with the NCL and CSL in the $e - \ln p'$ space, where e is the void ratio of the soil.

Hoe I. Ling et al. (eds.), Soil Stress-Strain Behavior: Measurement, Modeling and Analysis, 791–800.

The equation for the yield locus for the proposed soft rock model is given as follows:

$$\left[\frac{q}{M(p'+p_t')}\right]^n = \frac{\dfrac{p_o'+p_t'}{p'+p_t'}-1}{r-1} \tag{1}$$

In equation (1), $r = \dfrac{p_o'+p_t'}{p_c'+p_t'}$, where p_c' may be defined as the mean effective pressure at the intersection of the CSL with elastic rebound line in $e-\ln p'$ space. Alternatively p_c' may also be defined as the mean effective pressure at the intersection of the CSL with the yield locus in $p'-q$ space. Figure 1 shows the shape and size of yield locus of the proposed model in $p'-q$ space for various values of the exponent parameter *n*. Figure 2 shows the shape and size of yield locus of the proposed model in $p'-q$ space for various values of the spacing parameter *r*.

A spacing ratio value greater than 2 is observed for many soft rocks. In equation (1), *M* is the slope of the critical state line, p_o' is the preconsolidation pressure and p_t' is the effective tensile strength. The parameter p_t' shifts the axis of origin of the yield locus to the left on the mean effective pressure axis by an amount equal to the value of tensile strength of the soil. A value of the parameter *n* greater than 2 expands the size of yield locus in the deviator direction. This may predict higher initial yield stresses observed during the shearing of soft rocks. Constant elasticity may be assumed within the yield locus. This is observed in many soft rocks prior to initial yield. However, in this paper pressure dependent elasticity has been assumed.

Elastic and plastic consolidation of the soil is defined respectively by equation (2) and (3) as given below:

$$e = \Gamma - \lambda \ln p' \tag{2}$$

$$e = e_k - \kappa \ln p' \tag{3}$$

In equation (2) and (3), λ and κ are respectively the slopes of the plastic consolidation line and elastic rebound line in $e-\ln p'$ space, *e* is the void ratio of the soil, p' is the mean effective pressure and Γ and e_κ are constants. Γ and e_κ are respectively the void ratio at the plastic consolidation line and the elastic rebound line, at unit pressure. Equations (2) and (3), along with the Poisson's ratio μ, may be used to define the elastic bulk and shear modulus, in case of pressure dependent elasticity. In case the soft rock is assumed to have a constant elastic bulk modulus, the Poisson's ratio μ is used to compute the corresponding constant elastic shear modulus. Equations (2) and (3) are also used to define the strain hardening associated with plastic volumetric strains.

The plastic flow rule for the model is in general non-associated as observed in many soft rocks. It is defined by a stress-dilatancy equation identical to that of the Modified Cam Clay model and is given as below:

$$\frac{d\varepsilon_q^p}{d\varepsilon_v^p} = \frac{2\eta}{M^2 - \eta^2} \tag{4}$$

In equation (4), $d\varepsilon_v^p$ and $d\varepsilon_q^p$ are the incremental plastic volumetric and deviator strains respectively and η is the stress ratio given by $\eta = \frac{q}{p'}$. Equation (4) appears to be a good approximation of the stress-dilatancy behaviour observed in many soft rocks.

When the critical state is reached, the soft rock will undergo continuous plastic shearing strains at constant volume and at constant stress. This is the fundamental assumption of any soil model within the critical state framework.

2. DETERMINATION OF MODEL PARAMETERS

The proposed model for soft rocks has 10 material parameters. 6 of the 10 model parameters are identical to those of the Modified Cam Clay. These 6 parameters are respectively the slope of the normal consolidation line λ, the slope of elastic rebound line κ, the preconsolidation pressure p_o', the slope of the critical state line M, the elastic Poisson's ratio μ and the initial void ratio e of the soft rock. The four additional parameters of the model are the Young's modulus of elasticity E (in case constant elasticity assumption is used), exponent of the yield locus equation n, the spacing ratio parameter r and the tensile strength p_t'.

The parameters λ, κ, p_o' and e of the proposed model may be determined from high pressure isotropic consolidation tests. The parameter M may be determined from drained triaxial shear of soft rock. The elastic Young's modulus E may be computed as the ratio of the initial or elastic part of the deviator stress to the initial or elastic part of the axial strain. The elastic Poisson's ratio μ may be indirectly computed from the initial axial and volumetric strains experienced by the rock sample during triaxial shear. The tensile strength p_t' may be determined indirectly from unconfined compression test or directly from tensile strength test of the rock. The exponent n has to be determined numerically by curve fitting triaxial test data of the soft rock with model predictions.

3. SIGNIFICANCE OF ADDITIONAL MODEL PARAMETERS

The significance of the four additional parameters of the proposed model are now discussed. The exponent n determines the size of the yield locus in the deviator direction. Figure 1. shows that as the value of the exponent is gradually increased from 2, the yield locus expands in the deviator direction. At the same time, the yield locus loses it's symmetrical shape. An increased value of n will generally predict increased initial yield stresses. Figure 2. shows the effect of the value of the spacing ratio parameter r on the shape and size of the yield locus. As the value of r is increased, a flatter yield locus, which is smaller in size in the deviator direction, is obtained. This is similar to the effect of the parameter β proposed by Lagioa and Nova (1995). A higher value of r shifts the point of intersection of the CSL with the yield locus to the left, when measured along the mean pressure axis. This implies that for higher values of spacing ratio r, points on the yield locus with relatively low mean pressure values, will be to the right or wet side of the critical state line. This may result in compressive volume changes for soft rocks with high spacing ratios, even when sheared at low cell pressures. The tensile strength parameter p_t' in equation (1), shifts the origin of the yield locus to the left of the zero point, along the mean effective pressure axis. Thus the model can predict shear strength at zero mean effective pressure. When a constant value for the Young's modulus of elasticity E and elastic Poisson's μ ratio is assumed, a linear elastic stress-strain response prior to initial yield will be predicted. This is observed in many soft rocks.

4. DRAINED PREDICTIONS FOR NATURAL CORINTH MARL

This paper shows drained predictions of the soft rock model for natural Corinth Marl. The predictions of the model were compared with the experimental data of Anagnostopoulos et al, 1991. Values of critical state parameters for natural Corinth marl were obtained from the experimental data presented by Anagnostopoulos et al, 1991. A combination of three among the four additional parameters were selected by the authors for generating numerical predictions.

Table 1.

Parameter	λ	κ	p_o' **(kPa)**	Γ	**M**	μ	r	p_t' **(kPa)**	n	E
Case										
Ref	**0.04**	**0.008**	**3800**	**0.775**	**1.38**	**0.25**	**2**	**0.0**	**2**	**NA**
1.1									3	
1.2									4	
2.1							3			
2.2							4			
3.1								190		
3.2								380		

The first six of the reference set of parameters were obtained from the experimental data presented by Anagnostopoulos et al, 1991 and Liu and Carter (2002). In case of the parameter set termed as "Ref" in Table 1, standard values were chosen for the additional four parameters. For comparison purposes, predictions were generated each time using the reference set of parameters and in addition, two sets of parameters for each case (Case 1, 2 and 3). In each case, only one of the 3 additional parameters was varied relative to the reference set.. Predictions were obtained separately for two different cell pressures (1500kPa and 294 kPa), one being relatively high and the other being relatively low, compared to the preconsolidation pressure of the soil. Thus for each cell pressure, and for each case, three set of predictions were generated, which were compared with each other as well as with the experimental data presented by Anagnostopoulos at al, 1991. The results of this numerical exercise are shown in Figures 3. to Figures 14.. The figures presented show the effect of each of the three of four additional parameters of the proposed soft rock model on the stress-strain and volume change response of natural Corinth marl.

5. CONCLUSIONS

The results of the numerical study clearly show that the proposed model is quite capable of predicting the stress-strain response of soft rocks. In case of volume change response, the model can approximate reasonably well the volume change behaviour at cell pressures which are higher relative to the preconsolidation pressure of the soil. However, at cell pressures which are lower relative to the preconsolidation pressure of the soil, significantly larger expansive volume change behaviour is predicted. The model may be used in finite element programs and boundary value problems may be solved. It can then be investigated how the limitations of the proposed model in volume change predictions at relatively low pressures affect the prediction of the overall response of structures resting on soft rocks.

6. REFERENCES

Anagnostopoulos, A. G., Kalteziotis, N. and Tsiambaos, G. K. and Kavvadas, M. (1991). "Geotechnical properties of the Corinth Canal marls", *Geotechnical and Geological Engineering*, Vol. 9(1), pp. 1-26

Lagioia, R. and Nova, R. (1995). " An experimental and theoretical study of the behaviour of a calcerinite in triaxial compression", *Geotechnique*, Vol 45(4), pp. 633-648.

Liu, M. D. and Carter, J. P. (2002). "A structured Cam Clay model", *Research Report No R814*, Centre for Geotechnical Research, University of Sydney.

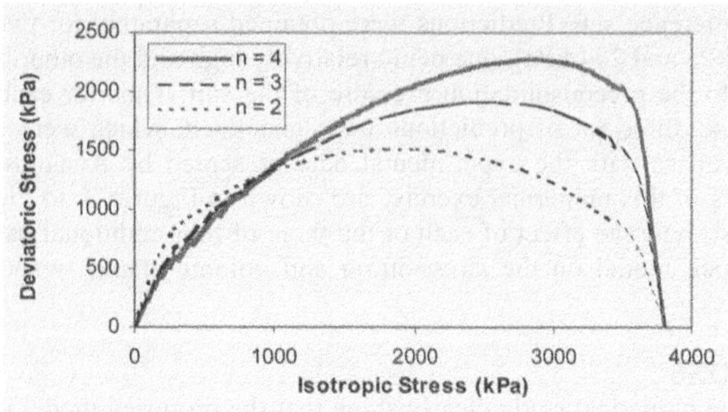

Figure 1. Effect of the exponent parameter on the shape of the yield locus of the soft rock model

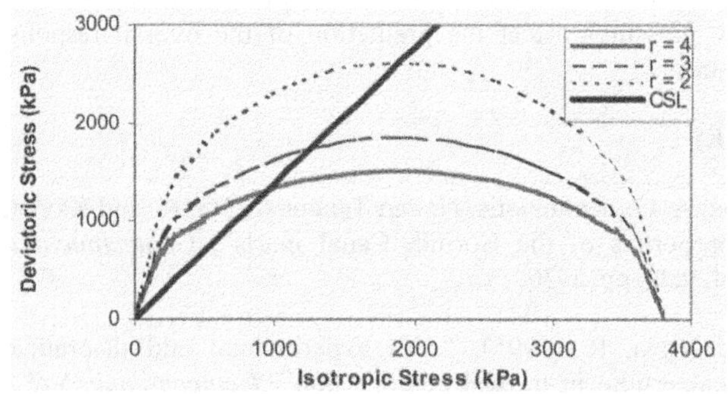

Figure 2. Effect of the spacing ratio parameter on the shape of the yield locus of the soft rock model

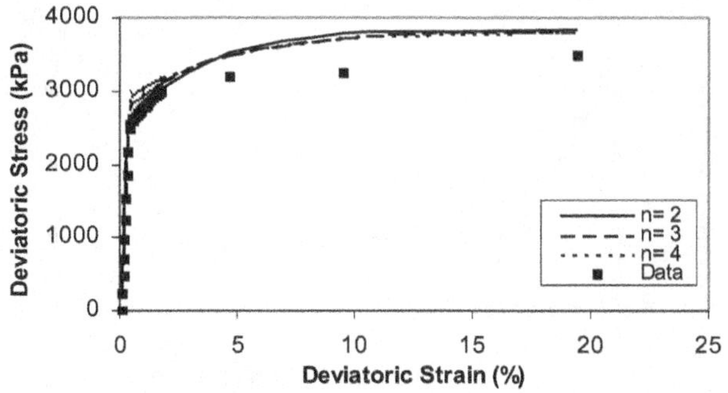

Figure 3. Effect of exponent parameter on stress-strain response of natural Corinth marl, cell pressure 1500 kPa

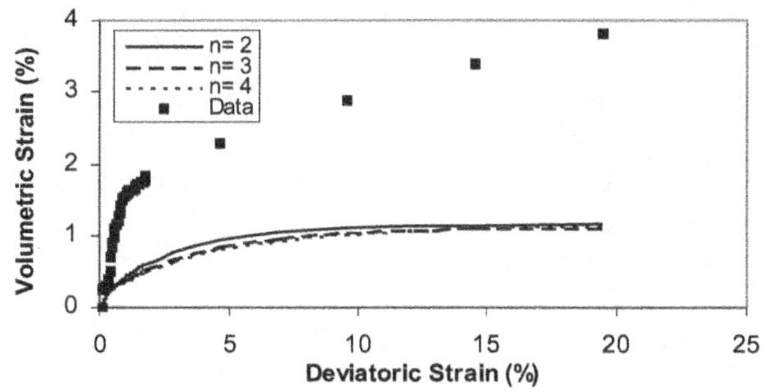

Figure 4. Effect of exponent parameter on volume change response of natural Corinth marl, cell pressure 1500 kPa

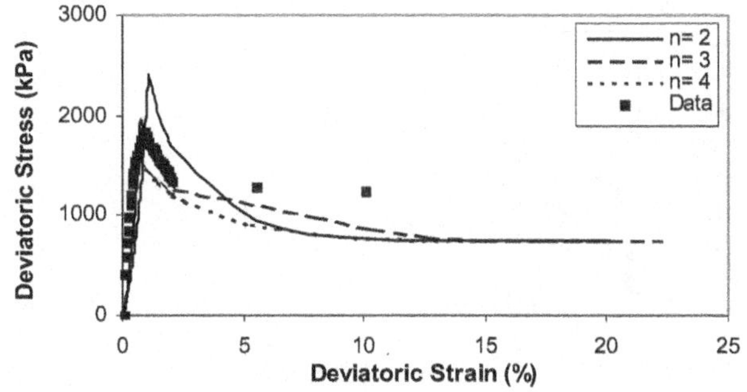

Figure 5. Effect of exponent parameter on stress-strain response of natural Corinth marl, cell pressure 294 kPa

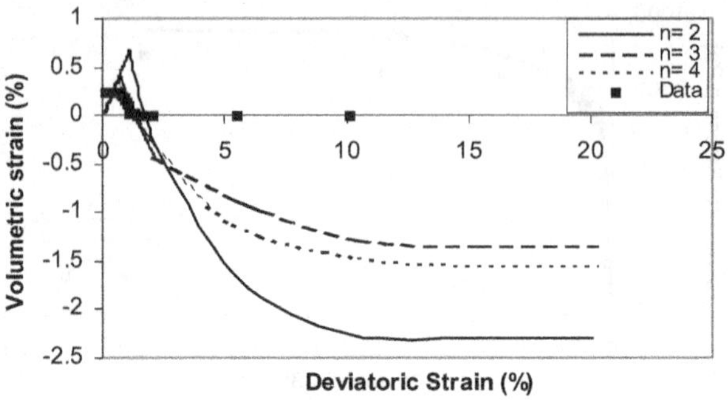

Figure 6. Effect of exponent parameter on volume change response of natural Corinth marl, cell pressure 294 kPa

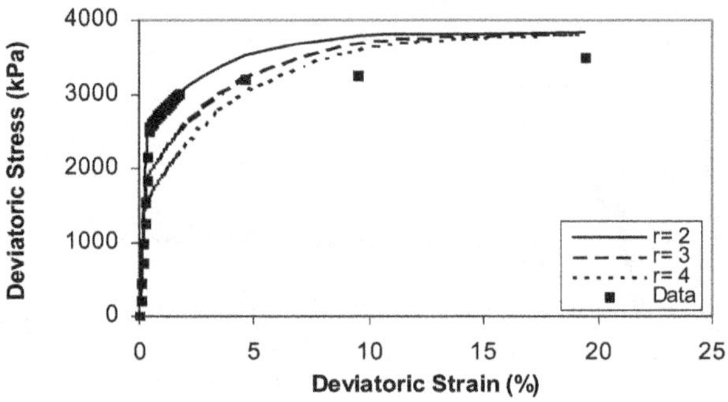

Figure 7. Effect of spacing ratio on stress-strain response of natural Corinth marl, cell pressure 1500 kPa

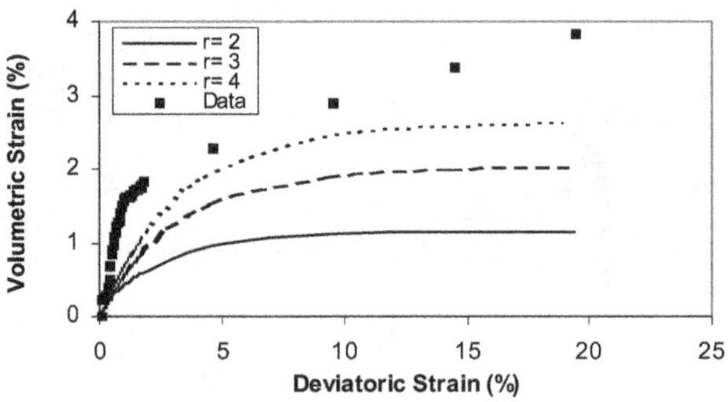

Figure 8. Effect of spacing ratio on volume change response of natural Corinth marl, cell pressure 1500 kPa

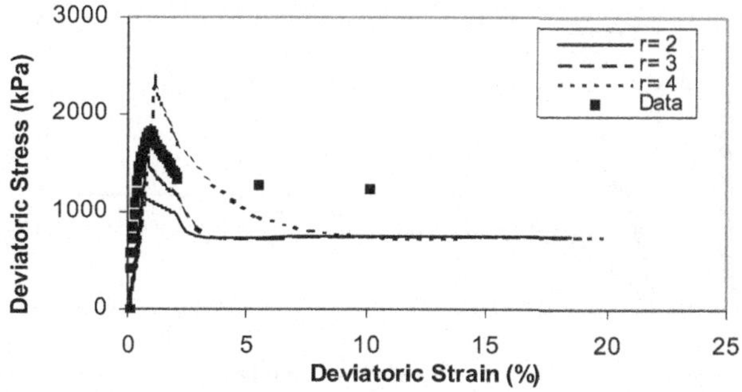

Figure 9. Effect of spacing ratio on stress-strain response of natural Corinth marl, cell pressure 294 kPa

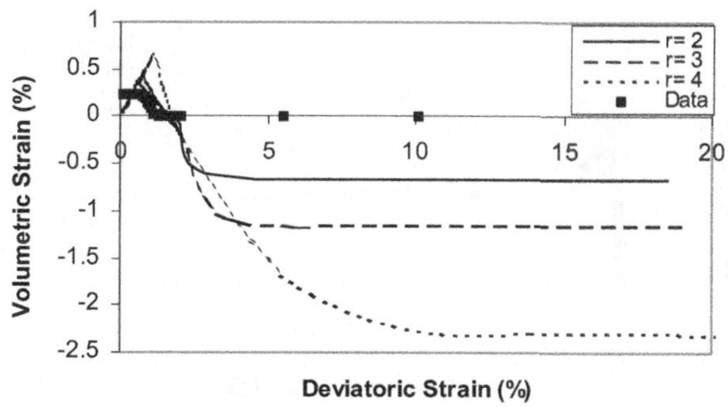

Figure 10. Effect of spacing ratio on volume change response of natural Corinth marl, cell pressure 294 kPa

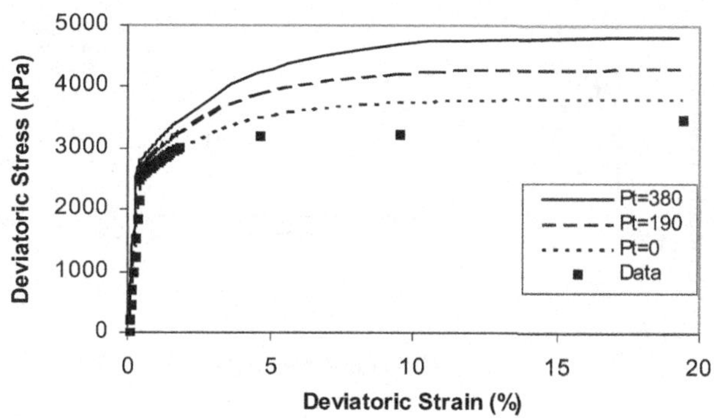

Figure 11. Effect of tensile strength on stress-strain response of natural Corinth marl, cell pressure 1500 kPa

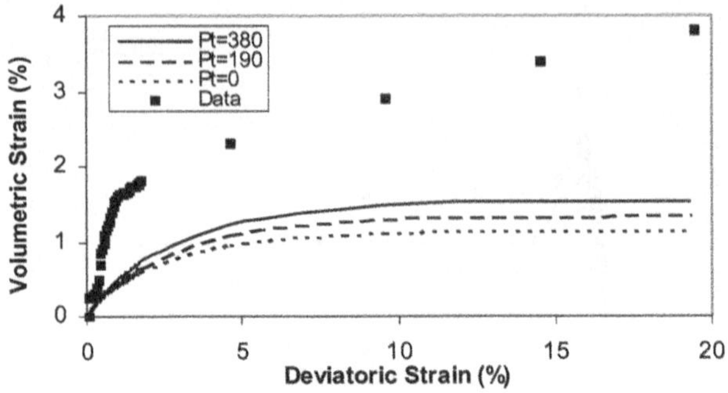

Figure 12. Effect of tensile strength on volume change response of natural Corinth marl, cell pressure 1500 kPa

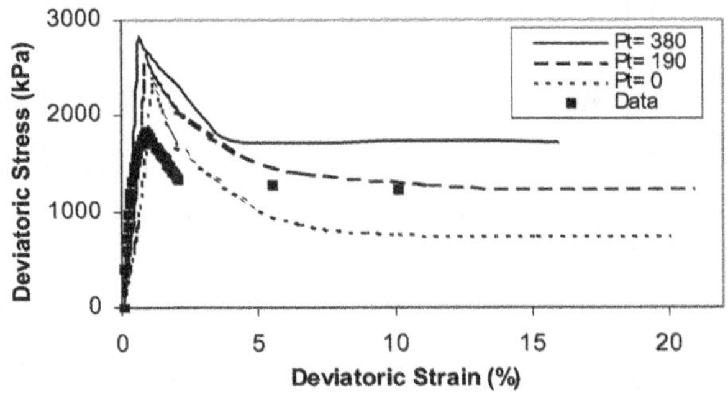

Figure 13. Effect of tensile strength on stress-strain change response of natural Corinth marl, cell pressure 294 kPa

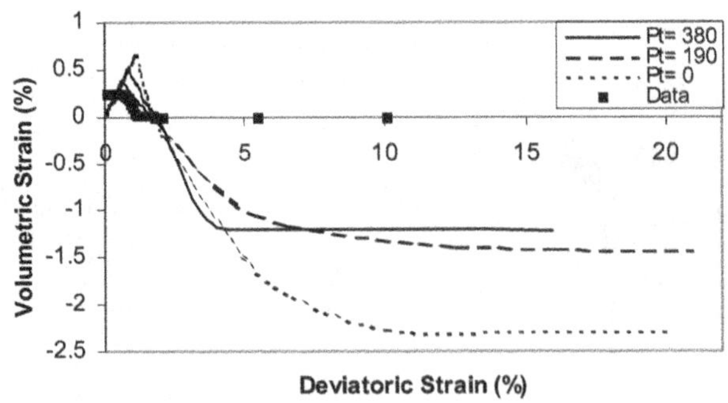

Figure 14. Effect of tensile strength on volume change response of natural Corinth marl, cell pressure 294 kPa

Soil Stress-Strain Behavior: Measurement, Modeling and Analysis
Geotechnical Symposium in Roma, March 16 & 17, 2006

EMBEDDED TEMPORARY PROP FOR BALLAST BED RENEWAL IN RAILWAYS

Tomokazu ISE
Fiber and Industrial Materials Company, Kuraray Co., Ltd. JAPAN
1-1-3 Otemachi, Chiyoda-ku, Tokyo 100-8115 JAPAN
e-mail: tomokazu_ise@kuraray.co.jp

ABSTRACT: In the railroad maintenances, the ballast bed renewal with temporary wooden props is tough and troublesome work. Then, we have developed a brand-new method in which no removal of the temporary props is needed by using the water-soluble synthetic fibers.

1. Introduction

In Japan, present temporary props are made of some wooden pieces. They are put under the rails or sleepers to support the rail tracks at the ballast bed renewal. Though they must be removed from the new ballast after finishing the renewal, it's a very tough and troublesome work. Therefore, the removal of the temporary props is the bottleneck of the ballast bed renewal process. We have been trying to extend the ballast bed renewal length per one night by omitting the removal of the temporary props.

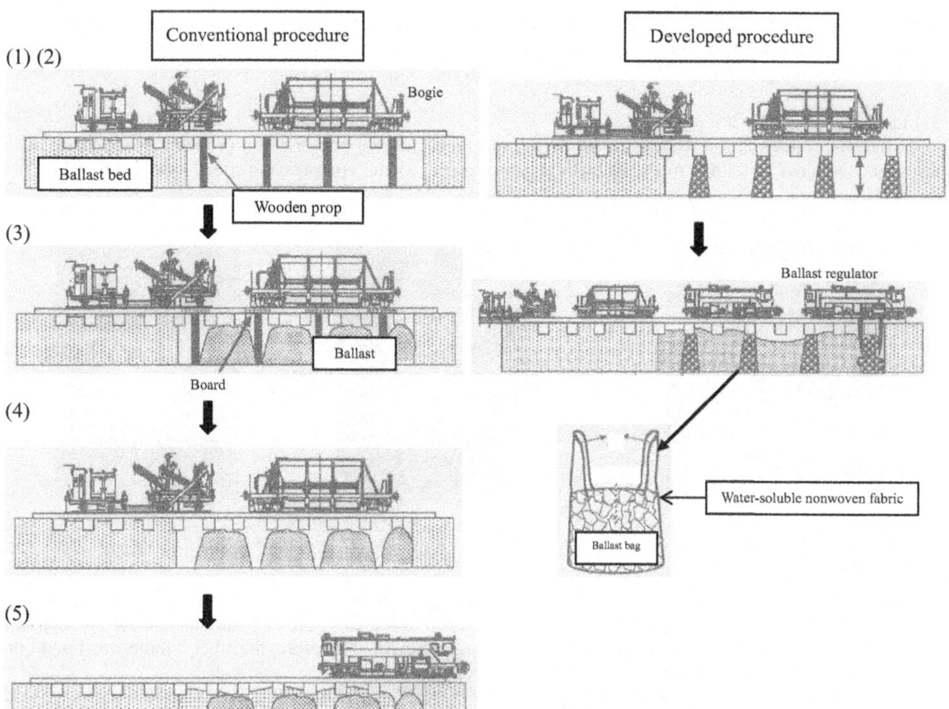

Fig.1 Ballast bed renewal procedures

Hoe I. Ling et al. (eds.), Soil Stress-Strain Behavior: Measurement, Modeling and Analysis, 801–805.

2. Design and preparation of embedded temporary prop

In order to make the temporary props embeddable, the props should be mainly made of ballasts. Since we have water-soluble synthetic fibers, we planed to prepare water-soluble bags from the fiber and filled ballasts into the bags to build the temporary props. The chemical component of the water-soluble fiber is polyvinyl alcohol (Fig.2), which contains no harmful elements and shows biodegradability after dissolving into water. The physical properties of the water-soluble bag are shown in Table 1. This fiber has enough tensile strength in dry condition and is easy to handle to convert them to woven fabrics or nonwoven fabrics.

Table 1 Physical properties of the water-soluble bag

Chemical component	PVA	
Weight	350<	g/m^2
Thickness	4.0<	mm
Tensile strength (Warp)	100<	N/cm
Tensile strength (Weft)	160<	N/cm
Tear strength (Warp)	350<	N
Tear strength (Weft)	400<	N
Breaking strength	2.5<	MPa
Seam strength	100<	N/cm

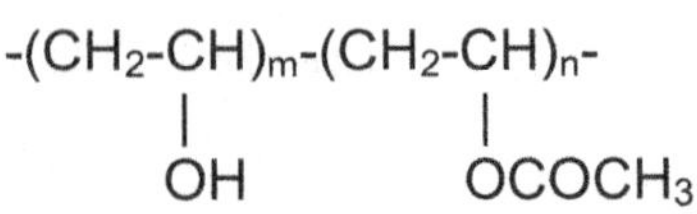

Fig.2 Chemical formula of PVA (Polyvinyl Alcohol)

From various constructions of fabrics, we chose the needle-punched nonwoven fabric as the material of the bags for the embedded temporary props because the bag needs certain elongation and high resistance against edges of the ballast stones.

The shape of the bag is column and continuous fiber bundle is bound after filling ballast in order to restrain the increasing of the column diameter when it receives the load of the bogie wagon. The fiber bundle is made of the same water-soluble PVA. The diameter of the bag is less than 30cm to fit the width of the sleepers and the height of the bag is controllable by choosing the amount of the filling ballast and normally settled to 20 ~ 30 cm.

Bogie wagon (type 931) weighs 48.7 ton when filled with ballast. Since the bogie has four axes, the load for each wheel is estimated to 6.1 ton. When the safety factor is set to 0.4, the withstand load of the prop is about 9.0 ton force (88kN).

3. Indoor test

At the first, compression tests were conducted to investigate the compression property of the props. The test was conducted by 980kN full scale testing apparatus(Fig.3) and compression speed was 1cm/min(Fig.4).

Fig.3 Test apparatus

Fig.4 Compression test

Even though the fiber bundle was not bound on the bag, both the nonwoven fabric and seam area didn't break when 88kN of load was added to the temporary prop at the first test. But the diameter of the bag was extended and the height went down to the half of its initial height. Therefore, the fiber bundle was bound on the side surface of the bags afterwards.

For the 25cm diameter bag, the axial displacements were 50mm at the 88kN load and 40mm at the 59kN load (the initial height of the bag was 250mm). These test results, listed in Table 2, are estimated to be enough performance for the yard test.

Table 2 Compression test result

Bag Diameter	Height	Weight	59kN(6tf) load		88kN(9tf) load	
			Displacement	Strain	Displacement	Strain
20cm	25cm	12kg	4.8cm	21%	6.0cm	26%
25cm	25cm	20kg	3.7cm	16%	4.7cm	20%

Secondly, water-solubility test was conducted. As shown in Fig.5, the bags dissolved completely by simply sprinkling ordinary temperature water on the bags.

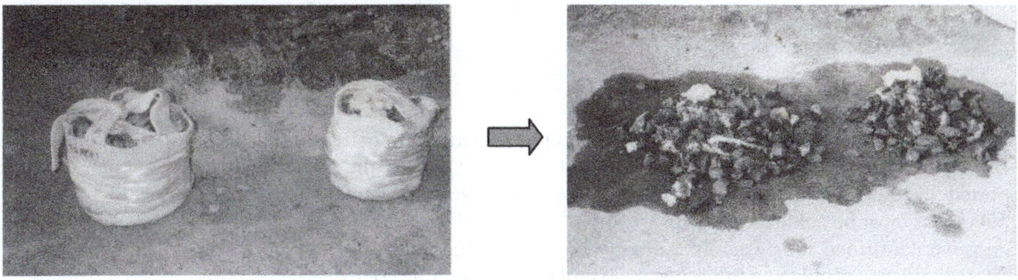

Fig.5 Water-solubility test

4. Yard test

Since we got the good indoor test result, yard test was conducted at the marshalling yard. The ballast bed renewal was conducted by using the embedded temporary props made of water-soluble bag and ballast. Diameter of the props was 20cm and the props were set at every two sleepers (Fig. 6~9).

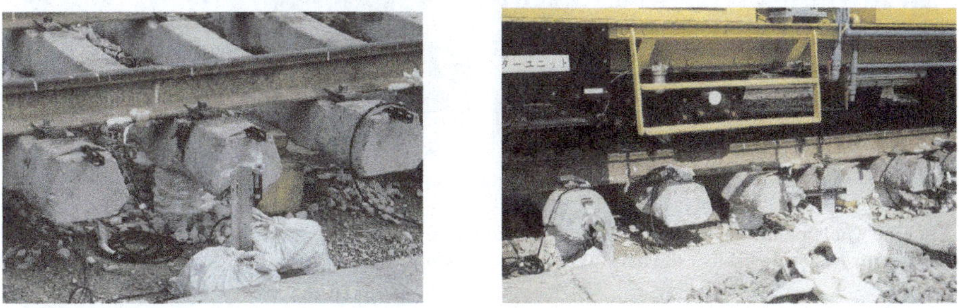

Fig.6 Temporary props put under the rails

The load and displacement of the props were measured when the bogie goes on the rails. As shown in Figure 7 and 8, the bigger load on the rails, the bigger displacements.

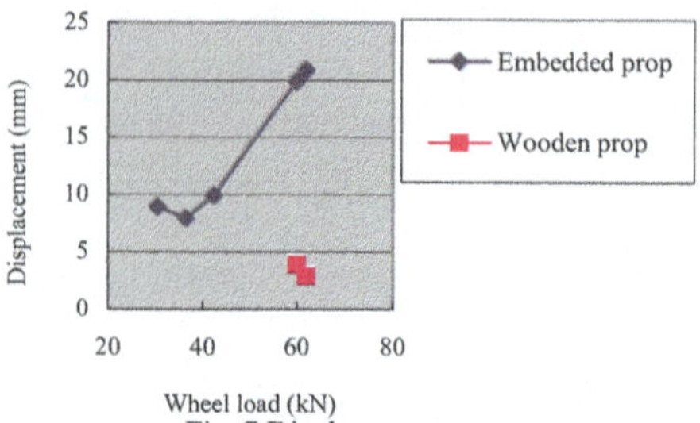

Fig. 7 Displacement

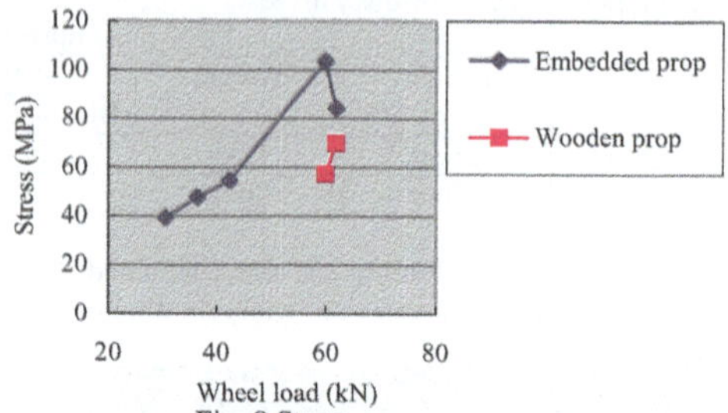

Fig. 8 Stress

The most important parameter is the rail stress. The maximum rail stress in this test was about one third of the rail resistance (400MPa). This indicates that it's possible to pass the bogie wagons on the rails, which are supported by the temporary props. But to keep the same displacement as present wooden props, some improvement must be achieved.

Since we confirmed the possibility, the wagon filled with ballast was led on the rails. The ballasting was conducted with no trouble and the props were embedded. (see Fig. 9)

Fig. 9 Ballasting from the wagon

After the whole ballasting procedures, enough quantity of water was sprinkled over the rails to dissolve the bag. (Fig. 10) To check whether the bags were dissolved or not, the new ballast were removed again. (Fig.11)

Fig. 10 Ballast leveling and sprinkling

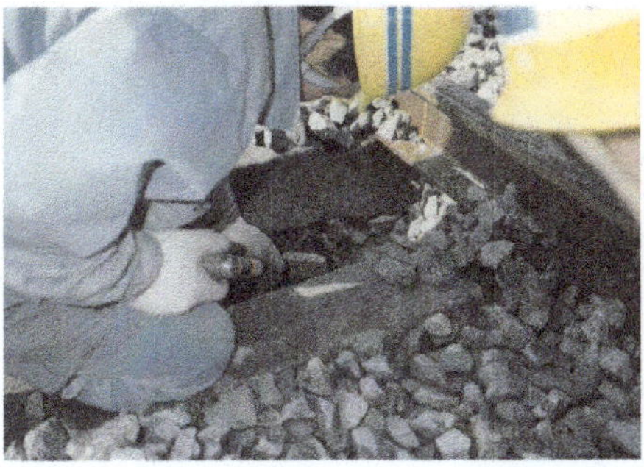

Fig. 11 Digging out the ballast

As a result, the bags completely dissolved and the inside ballast have become united with the outside ballast. This indicates that the props are embeddable and no need to dig out the props from the new ballast bed after the ballast bed renewal.

5. Conclusion

The embedded temporary props for the ballast bed renewal can be produced from the water-soluble bag and ballast. The performance of the props is enough to practical use. But further improvement is required to apply this technology to the railways of the high-speed railways like Shinkansen (Japanese bullet train).

REFERENCE

Hiroki Shiozaki, Tamaki Shiomi, Tomokazu Ise, 1997. Proceedings of 52th JSCE Annual Conference.

Soil Stress-Strain Behavior: Measurement, Modeling and Analysis
Geotechnical Symposium in Roma, March 16 & 17, 2006

BEHAVIOUR OF SAND REINFORCED WITH FIBRES

Erdin Ibraim & Stephane Fourmont
Department of Civil Engineering
University of Bristol, Bristol, BS8 1TR, UK
e-mail: erdin.ibraim@bristol.ac.uk

ABSTRACT

This study focuses on the effects of randomly oriented discrete crimped polypropylene fibres on the mechanical response of very fine sand. Compaction and direct shear tests were performed on sand specimens of different densities unreinforced and reinforced with fibres in different proportions. The presence of reinforcement provides an extra resistance to the compaction, causing a less dense packing as the quantity of fibres is increased. The results of the direct shear tests indicate that inclusion of fibres increases the peak shear strength and the strain required to reach the peak. The post-peak strength at large strains was also higher when fibres were included. The presence of fibres leads to more dilative behaviour. In this study, for the range of the effective normal stresses employed, a linear failure envelope has been recorded for all densities and fibre concentrations. The increase of the peak shear strength was almost linear for all densities at low effective normal stress and approached a limiting value for higher normal stresses. For the loosest specimens reinforced with the highest percentage of fibres that could be employed in the laboratory using a moist tamping fabrication method, the relative increase of the peak shear strength was more than 50%.

1. INTRODUCTION

The influence and the contribution of fibre reinforcement to the shear strength of sand have been examined by various investigators. Several parameters such as confining stress, fibre type (natural and synthetic), volume fraction, density, length, aspect ratio, modulus of elasticity, orientation, and soil characteristics including particle size, shape, gradation have been studied using monotonic loading in direct shear tests, consolidated drained triaxial tests or unconfined compression tests.

Experimental results have shown that fibre reinforcement causes significant improvement in strength. Strength increase of the reinforced sand seems to be linear with the amount of fibres at high confining stress. Nevertheless, above some limiting content and for lower values of the confining stress, this increase seems to approach an asymptotic upper limit (Gray and Al-Refeai 1986, Ranjan et al. 1996, Murray et al. 2000). For a given weight fraction, strength, as expressed by the major principal stress at failure, varies linearly with aspect ratio. The slope of this curve increases with larger fibre concentrations. It has been noted that for the same confining stress, the strength of the reinforced sand increases with reducing average grain size D_{50} (Maher and Gray 1990, Gray and Al-Refeai 1986). Also, a better gradation - increase in coefficient of uniformity, C_u - and a smaller D_{50} result in higher contribution to strength. Increasing the length of

Hoe I. Ling et al. (eds.), Soil Stress-Strain Behavior: Measurement, Modeling and Analysis, 807–818.

fibre reinforcements increases the shear strength of the fibre-sand composite but only up to a point beyond which any further increase in fibre length has no effect on shear strength (Al-Refeai 1991). For some types of fibres, like nylon, Kumar and Tabor (2003) showed that the relative increase of the residual strength was much higher than the relative increase of the peak shear strength.

Sand reinforced with randomly distributed discrete fibres exhibits either curved-linear (for uniform, rounded sand) or bilinear (for well-graded or angular sands) failure envelopes (Gray and Ohashi 1983, Maher and Gray 1990, Bailey and Knox 1997). Above a threshold confining stress, referred to as a critical confining stress σ_{crit}, failure envelopes for the reinforced sand are parallel to the unreinforced sand envelope. Below the critical confining stress, it is considered that the reinforcing mechanism is not fully mobilised and with the shearing process the fibres tend to slip or pull out. It was also observed experimentally that σ_{crit} increases with sphericity index of sand and with reduction of the fibre aspect ratio and decreases with C_u, whereas it could be relatively unaffected by the amount of fibre content and D_{50}.

Presented in this paper are direct shear test results for a very fine sand, Hostun RF sand, reinforced with randomly distributed discrete crimped polypropylene fibres. These fibres are currently used in 'marginal' applications such as foundations for sport pitches, horse racing tracks, access for secondary roadways. The objective of this research is to examine the mechanical of the mixed material and provide insight into the suitability of this particular fibre reinforcement for use in other geotechnical applications including earth retaining structures, embankments and tailing dams, subgrade stabilisation beneath shallow footings, pavements.

2. EXPERIMENTAL TESTS

The effect of the polypropylene fibres on the mechanical behaviour of the reinforced sand was investigated mainly by conducting direct shear tests using a recently improved apparatus (Dietz, 2001, Lings and Dietz, 2004). Different combinations of sand densities, effective normal stresses and fibre contents have been used. Preliminary compaction tests to examine the influence of moisture content and fibre concentration have also been performed.

2.1 MATERIALS

The sand used in this study was Hostun RF (S28) sand. The main characteristics of the Hostun RF sand are presented in the Table 1. Crimped polypropylene fibres were used as reinforced material. Fibre lengths (ℓ) range from 30 to 35 mm and diameter is approximately 0.1mm (Figure 1). The physical properties of the polypropylene fibres are given in Table 2.

2.2 COMPACTION TESTS

Compaction tests using a Proctor compaction apparatus were performed to determine the moisture density relationships of the Hostun RF sand without fibres first and then reinforced at fibre contents of 0.1, 0.3 and 0.5 percent by dry weight of soil (Vijayasingam and Heng 2003). The compaction curves obtained are shown in Figure 2. For all compaction tests, the specimens had the same amount of dry sand. The pre-mixed sand with water and fibres placed into the compaction mould as a single layer experienced 20 blows of a 4.5 kg hammer. At the end of the test, as sand grains tended to stick to the sides of the mould, a light manual compaction using a tamper was also necessary in order to level off the top layer, and 15 light blows were employed for this. As the quantity of dry sand and compaction energy was identical, the change of the dry density with the water and fibre contents was the result of a different final volume occupied by the soil. As also observed by different authors, for a given compactive energy the presence of the reinforcement provides higher resistance to the compaction. A less dense packing is obtained once the quantity of fibres is increased. The variation of the maximum dry density with the fibre content is presented in Figure 2. Fibres up to 0.5% of dry weight of sand seemed to have no significant influence on the optimum moisture content (OMC) which appears to be around 10%. The extremely low moisture absorption characteristic of polypropylene fibres could explain this result.

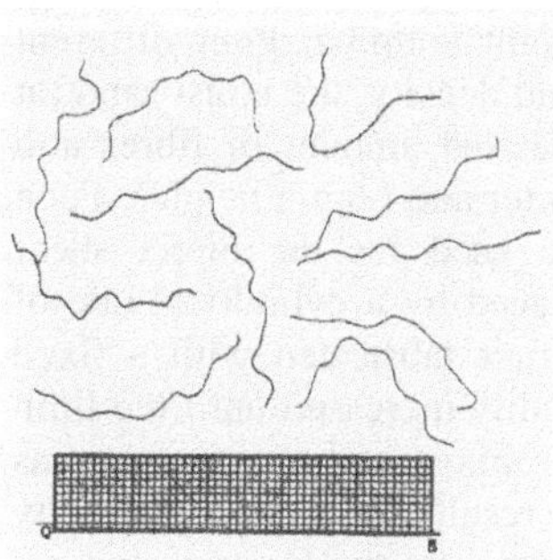

Figure 1. Individual polypropylene fibres.

Table 1. Main characteristics of Hostun RF (S28) sand

D_{50}	C_u	C_g	e_{min}	e_{max}	G_s
0.32	1.7	1.1	0.62	1.0	2.65

D_{50} = mean grain size, C_u = coefficient of uniformity, C_g = coefficient of gradation, G_s = specific gravity, e_{max}, e_{min} = maximum and minimum void ratio

Table2. Main characteristics of the reinforcement

Tensile strength (MPa)	Young's Modulus (MPa)	Specific gravity	Diameter, *d* (mm)	Aspect ratio, λ, $\lambda = l/d$
300	4,000	0.91	0.1	350

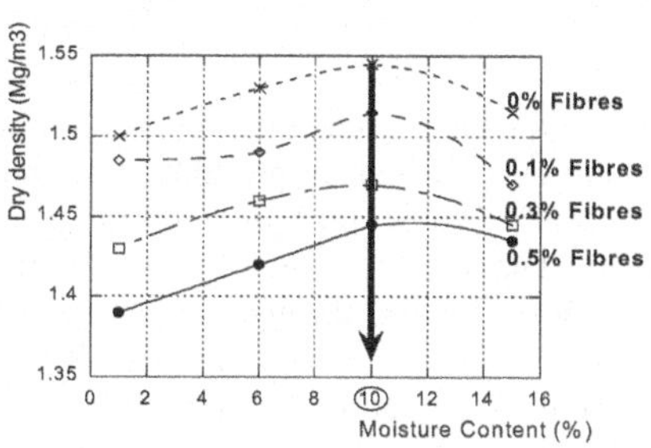

Figure 2. Compaction test results.

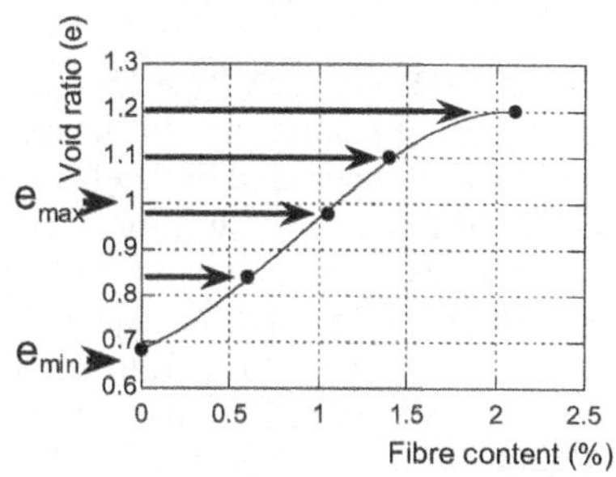

Figure 3. Maximum amount of fibres that could be mixed with a fixed amount of sand without leading to a change in specimen volume using the moist light tamping.

2.3 SPECIMEN FABRICATION METHOD

There are two critical stages when fibre-reinforced specimens are prepared in the laboratory: mixing and formation. Initial mixing tests showed that some amount of water is required to mix more efficiently the sand with the fibres and also to prevent the fibre/sand segregation. In this study, the amount of water used for the mixing stage was chosen to be equal to the (OMC). The mixing of the soil and water was done manually using a small spoon until the grains became wet and it was continued by adding progressively small amounts of fibres. The mixing process between the soil, water and fibres was stopped when, by visual examination, it was considered that the fibres are relatively well distributed throughout the soil mass. Three layers of mixed soil of different heights were used for the shear box specimens. Each layer was delicately deposited into the box to ensure a zero drop height and minimal disturbances and each layer was compacted to the predetermined height using a small drop light rectangular hammer. The presumably shear plane locates in the central part of the middle layer. The specimen dimensions were carefully measured and recorded in order to estimate the initial void ratio, e_o. In this study, as a convention, the void ratio represents the intergranular void ratio; the voids are occupied by air, water and fibres.

Before starting the experimental programme, it was considered necessary to estimate the maximum dosage of fibres that can be mixed with a given amount of sand, placed into a given volume and compacted using moist light tamping. Four different densities of sand have been chosen for this test. For each sand density, the moist sand (at OMC) divided in three equal parts was mixed with the desired amount of fibres and placed in three layers into a cylindrical mould (72mm diameter and 66mm height). Each layer was compacted using the same compaction device used for the direct shear specimen fabrication but with the rectangular hammer replaced by a cylindrical one of 50mm diameter. For a given density, the specimen was first fabricated with a fixed amount of fibres. Then, the fibre dosage was each time steadily increased until the limit beyond which it was practically impossible to compact the mixed soil any further was reached. This limit of the fibre content was recorded and the results for each sand density are shown in Figure 3. The minimum and maximum void ratios of the Hostun sand are also plotted in the graph. A value of 2% of fibre content seems to be a limit beyond which sand reinforced specimens cannot be obtained.

2.4 DIRECT SHEAR BOX TESTS

A Direct Shear Apparatus has been used for the purposes of this research. Important features previously upgraded by Dietz (2001) and have recently been presented by Lings and Dietz (2004). A pair of swan-neck type yokes attached to the mid points of the upper frame's side-walls brings the point of the shear load transmission to the specimen centre. The shear load is applied in equal measure to each wing through an arm. The transmission of the shear load from each arm to each wing is done through a ball race. The load pad is clamped to the upper frame prior to the application of any shear displacement.

The apparatus tests 100x100x45mm specimens. Sheets of flexible neoprene membranes (1mm thickness) are attached to the internal walls of the shear box (two split sheets spanning the divide) to prevent excessive specimen extrusion via this opening. Silicone grease was used to attach the shielding to the internal wall of the shear box.

The initial opening between the upper and lower frames of the shear box was installed prior to specimen deposition. This gap was fixed according to Dietz (2001) and in this study, irrespective of the presence of fibres, it was taken equal to five times the mean grain size of the Hostun RF sand (D_{50}), which represents around 1.5mm. It should be noted that no research was carried out to confirm that this value actually represents the optimum value for this particular type of sand and fibre mixture.

The measurement of the test parameters was done by a load cell for the shear force and by six LVDTs, five for monitoring the vertical displacement of the load pad (one in each corner of the upper frame and one on the midpoint of the apparatus' normal load hanger) and one monitoring the carriage's horizontal displacement. The evolution of the vertical displacement shown in this paper is an average of all four corner placed transducers values. These four transducers also enabled the rotation of the upper frame to be recorded.

The rate of the horizontal displacement employed was around 1.2mm/min.

2.4.1 PRELIMINARY TESTS

It was found useful to know whether the direct shear tests can be conducted under saturated specimen conditions or simply unsaturated at the OMC (10%). Under shear, the moisture content of an unsaturated specimen could provide an extra interaction between grains by introducing suction forces. Santoni et al. (2001) reported that the unconfined compression strength of polypropylene reinforced sand increased with the moisture content up to a limit of approximately 9.0%. Beyond this moisture content, the specimen's unconfined compressive strength was progressively less beneficial.

Five series of direct shear tests have been conducted: one on pluviated and saturated sand specimens, two on moist tamped (one fully saturated, one unsaturated) and two on fibre reinforced specimens (one fully saturated, one unsaturated). Only the failure lines resulted from direct shear tests are presented (Figure 4). The normal stress ranges between 50kPa and 350kPa and all the specimens reinforced or unreinforced had an initial void ratio of 0.9. The failure envelopes have been obtained by linear regression; excellent values of the regression coefficients (R^2) have been found for all series of tests. As can be observed, 10% of moisture content and the moist tamping fabrication technique give a failure line that lies significantly above that obtained for pluviated but fully saturated specimens. However, the slopes of the failure lines seem to remain identical. An apparent cohesion of around 3kPa probably due to the suction but also to the moist tamping compaction seems to be present.

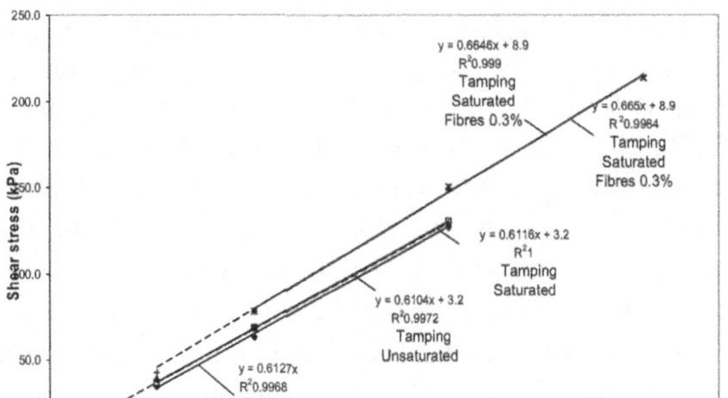

Figure 4. Influence of the moisture content.

Another series of tests performed on moist tamped specimens but this time fully saturated before the shearing test revealed an identical failure line with those specimens fabricated also by tamping but sheared under unsaturated conditions (fixed 10% moisture content). These results show that, apparently, 10% of moisture content has the same effect on the shear stress as that recorded for fully saturated specimens (or, at least, 10% of water content has no measurable effect on the stress deformation properties of the sand). In view of these observations and due to the lack of experimental evidence to the contrary, it is considered that the differences between the pluviated/saturated and tamped/unsaturated failure lines as well as the existence of the apparent cohesion are due to the tamping compaction. The same conclusions can be drawn when fibres are added to the sand and the specimens are moist tamped. Identical failure line envelopes have been obtained for 0.3% of fibre inclusions into unsaturated and fully saturated specimens. A higher cohesion intercept is recorded in this case, which can be the effect of the fibres but also the tamping compaction.

Taking into account these results, it was decided to use specimens under unsaturated conditions at 10% moisture content.

Prior to the experimental programme, a few series of direct shear tests were performed in order to assess the ability of the apparatus to reproduce test results for specimens with similar compositions. Scatter in the test results can also be attributed to inconsistencies of the specimen fabrication method. Together with the moist tamping fabrication technique, a dry pluviation of sand using a funnel under 'zero' drop height has also been used. The experimental results are not presented here; however, an excellent repeatability of the shear stress and volumetric responses has been obtained for either dry pluviated or moist tamped specimens. A good repeatability of the results was also obtained for specimens reinforced with fibres, especially the pre-peak and peak stages. The initial structure of the soil given by the fabrication method and the distribution of the fibres throughout the soil seem to be repeatable; however, no indication on the quality of the fibre distribution throughout the specimen can be made.

2.4.2 Experimental test results

Three different specimen void ratio values of 0.8, 0.9 and 1.0 have been chosen for the experimental direct shear box testing programme, which in terms of the relative density represent approximately 60%, 30% and 0%, respectively. The applied normal stresses ranged from 55.3 to 310.6kPa. Typical direct shear test results under a normal stress of 55.3kPa are presented in the Figure 5; the percentages of fibres used for each density are also specified in the figure. The figure presents the variation of the shear stress, vertical displacement (v_y) and dilation angle (ψ) with the horizontal displacement (v_x). The dilation angle (ψ) represents the ratio of incremental volume change and incremental shear strain, which for the direct shear test is given by $\tan\psi = \delta v_y/\delta v_x$, where δv_x is the horizontal incremental displacement and δv_y the vertical incremental displacement.

The shear stress-strain response of a medium/dense unreinforced sand (e_o=0.8) shows a pronounced peak, followed by softening and stabilisation of the shear stress with large shear strains (bold line, Figure 5). Under the same effective normal stress, the peak shear stress decreases and a reduced softening post-peak behaviour is observed for lower densities. The shear strain needed to reach the peak shear stress is smaller for a denser specimen. The vertical displacement shows a reduction in specimen's height at the earliest stages of the test which is followed by steady increase at different rates, depending on the specimen density and effective normal stress. The dilation angle, ψ, shows an increase up to a maximum value, ψ_{max}, and then a decrease and stabilization around zero degrees. Table 3 shows the values of ψ_{max} for all the unreinforced specimens. For a constant effective normal stress, the dilation increases with the density, whereas for a constant void ratio, the tendency for dilation is inhibited by a higher effective normal stress.

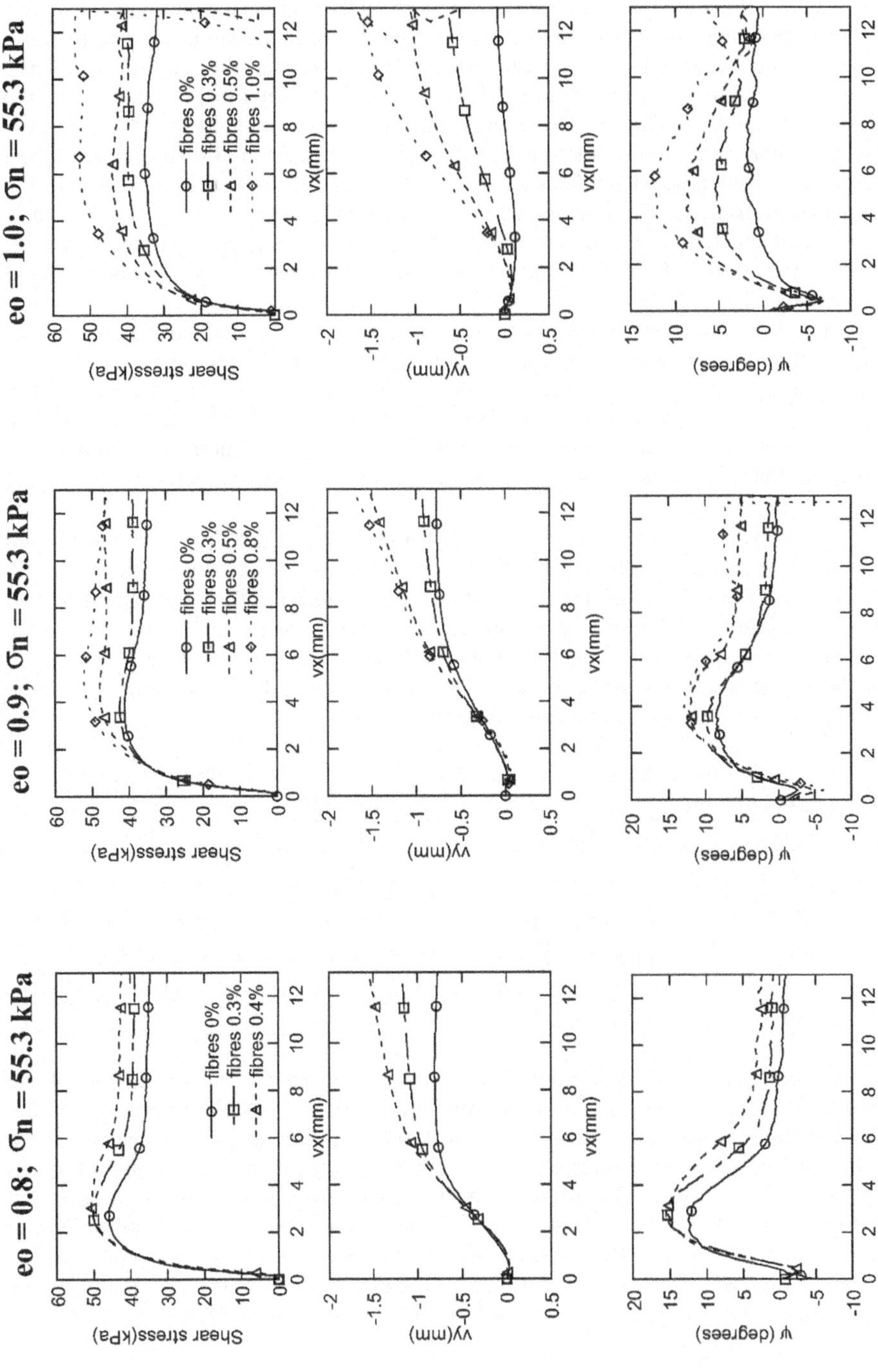

Figure 5. Typical direct shear test results: influence of density and fibre content.

As shown by Figure 5, the experimental shear stress - displacement and the volume change behaviours are clearly affected by the polypropylene fibre inclusions. The shear stress-horizontal displacement curves for specimens at the same void ratio show that the reinforced sand exhibits a greater peak shear stress than the unreinforced sand. The stress-strain relationships seem to be similar at low horizontal displacements, but the reinforced soil is able to absorb more energy and therefore producing higher peak shear stresses. Increasing the amount of fibres leads to a larger horizontal displacement being required to reach the peak shear failure. The residual strength is also increased with the fibre concentrations and reductions in the post peak shearing resistance are systematically observed. As expected, a higher confining pressure gives higher shear strength and deformation values at failure; the bond which develops between the fibres and the sand grains is enhanced.

For all the specimen densities considered in this study, the shape of the volumetric curve of reinforced sand showed, as for the plain sand, an initial contraction followed by dilation. However, the amount of vertical deformation was higher and the volumetric response became more dilative with the presence of fibres and the fibre content. The addition of fibres to the sand does not inhibit the dilatancy. For liquefaction studies, this dilative behaviour can result in an increase in the stress increment required to initiate pore-pressure built up.

For identical densities and fibre concentrations, the higher the effective normal stresses, the smaller the dilatancy. The maximum values of the dilation angle, ψ_{max}, are presented in the Table 3. A non linear variation of ψ_{max} with the fibre concentration can be observed in Figure 6 where, for example, results for specimens with a void ratio of 1.0 tested under three effective normal stresses 55.3, 106.4 and 208.5kPa are shown. In general for all densities and fibre contents, the dilation angle does not stabilise at large displacements and does not reach a zero value (Figure 5).

Table 3 Maximum dilation angle for all densities, stress and fibre conditions.

Maximum Dilation angle, ψ_{max} (°)	Void ratio										
	0.8			0.9				1.0			
	Fibre content (%)										
Normal stress (kPa)	0	0.3	0.4	0	0.3	0.5	0.8	0	0.3	0.5	1.0
55.3	12.5	15.6	15.6	8.6	9.9	12.3	13.0	1.9	5.5	8.7	12.7
106.4	11.3	13.5	14.4	4.7	7.5	9.0	11.0	1.7	4.2	5.8	10.7
208.5	9.2	11.7	13.4	3.2	6.5	7.5	8.5	2.1	4.0	5.0	7.4
310.6	-	10.2	11.0	-	4.7	6.6	-	-	-	-	-

Although the use of a scanning electron microscope for the inspection of the polypropylene fibres would be more recommended for a possible indication of the deformation process (Michalowski and Cermak 2003), at this stage of the research, only a post-test visual inspection was carried out. No fibre damages by the deformation process were observed in any test. The limit of the resistance afforded by the reinforcement seems, therefore, to be due mainly to a bond failure rather than a tension failure within the reinforcement. In these conditions, and for the range of the effective normal stresses used in this study, the strength of the mixed soil appears to be controlled mainly by the friction between the soil and reinforcement and it seems not dependent on the ultimate strength characteristics of the polypropylene fibres. These observations could explain the absence of bilinear or curved-linear peak strength envelopes for all specimen densities and fibre contents. As an example, the peak shear strength envelopes for reinforced sand specimens at 1.0 void ratio are shown in Figures 7. The strength envelope of unreinforced sand is also been plotted for comparison. The friction angle value is also depending on the confining stresses under which the specimens are tested and, therefore, the whole envelope is expected to be nonlinear. However, for the limited range of the effective normal stresses used in this study and for an easier quantification of the effects of fibre reinforcement, a linear envelope found by linear regression would be a reasonable approximation of the state of stress at failure. The extrapolation of this line towards the zero effective normal stress gives a cohesion intercept. As a system of uncemented particles, the sand should not have any true cohesion in terms of effective stresses. However, the inter-particle interactions (interlocking) induced by the tamping fabrication technique seem to be important and this can explain the existence of an apparent cohesion that increases slightly with the addition of fibres. Although further research should be done to explain the origin of this apparent cohesion, in this study the failure envelopes are defined in terms of the angle of friction, ϕ, and the cohesion intercept, c. The shear strength parameters obtained from direct shear tests on unreinforced and reinforced Hostun RF sand are also summarised in the Table 4.

Table 4 Angle of friction and cohesion intercept for all series of tests.

Void ratio	Fibre content (%)											
	0		0.3		0.4		0.5		0.8		1.0	
	c (kPa	ϕ°	c (kPa	ϕ°	c (kPa	ϕ°	c (kPa	ϕ°	c (kPa)	ϕ°	c (kPa	ϕ°
0.8	8.5	34.4	8.6	37.4	11.5	37.6	-	-	-	-	-	-
0.9	4.5	31.4	8.9	33.6	-	-	9.1	35.7	10.18	37.1	-	-
1.0	0.8	30.9	5.1	32.8	-	-	3.9	35.3	-	-	7.0	39.4

The peak shear strength variation with the amounts of fibres is presented in Figure 8. The values of the peak shear stress are given by the equations of the failure lines obtained by linear regression. The evolution of the peak shear strength is almost linear for all specimen densities at low effective normal stresses and, particularly, for specimens with 1.0 void ratio at higher values of the normal stress. The slope of these lines increases with the effective normal stress. For all the normal stresses, much closer trends could be noted for the specimens with 0.9 and 1.0 void ratio. For higher densities and normal stresses greater than 200kPa, the evolution of the peak shear stress approaches a limiting level. As a consequence, it can be very clearly observed that a fibre concentration of 0.8% gives practically the same peak shear stress for specimens with 0.9 and 1.0 void ratios. A similar trend was observed by Gray and Al-Refeai (1986), Rajan et al. (1996) and Murray et al. (2000) but for lower confining stress levels.

It can also be noted that for all densities, the loosest specimens (e_o=1.0) reinforced with 0.5% of fibres give a similar peak shear strength as the densest but unreinforced specimens (e_o=0.8); the loosest fibre reinforced specimens further offer the advantage of a ductile behaviour (Figure 5). This suggests that fibres may be used in shallow foundations, sand fills, and other earthworks that may suffer excessive deformations.

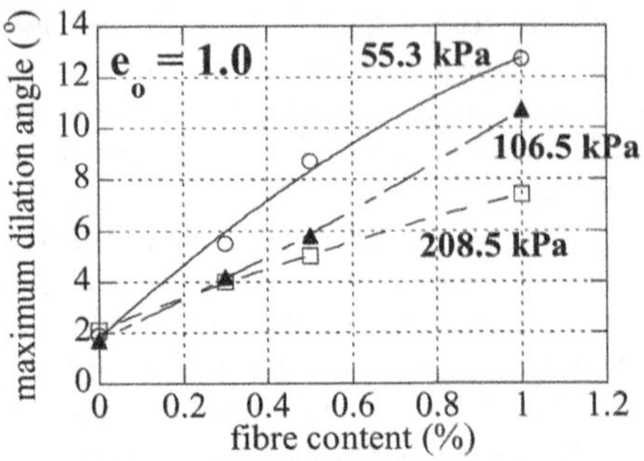

Figure 6. Variation of maximum dilation angle with fibre content at e_o=1.0.

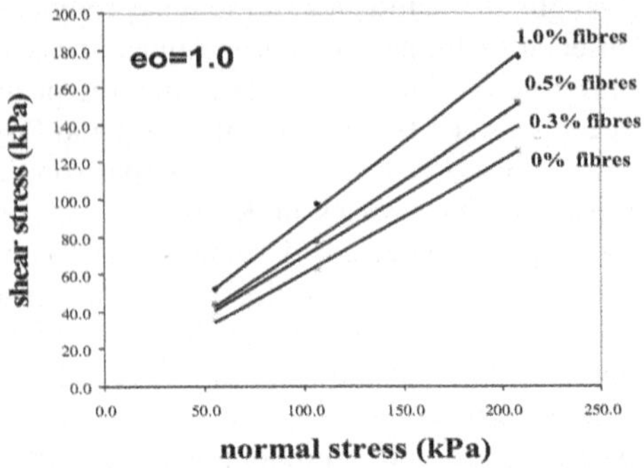

Figure7. Failure envelopes of reinforced and unreinforced sand.

If the relative increase in peak shear strength due to fibre content is represented by the parameter F_{peak}, the ratio of the peak shear strength of the fibre-reinforced specimen to the peak strength of the unreinforced specimen, for all densities and normal stresses used in this study, values of F_{peak} between 15 and 20% can be recorded for a limit of 0.4% of fibres. For 0.8% of fibre content, the increase is higher and ranges between 30 and 40%, while at the highest fibre content of 1.0% and specimens with 1.0 void ratio, the relative gain in strength increases significantly up to 60%.

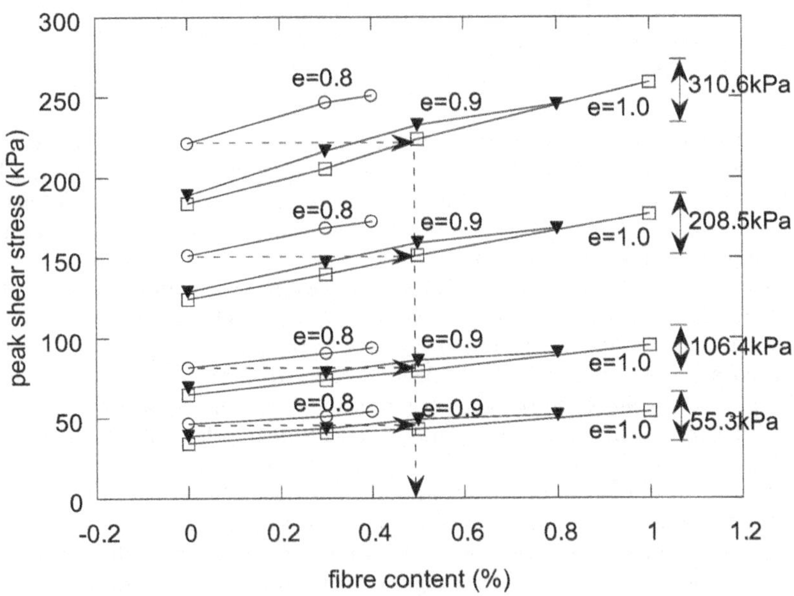

Figure 8. Evolution of the peak shear stress with fibre content, density and effective normal stress

3. CONCLUSIONS

An experimental programme test was undertaken to investigate the individual effect of randomly oriented crimped polypropylene fibres on the mechanical behaviour of the Hostun RF sand. Compaction and direct shear tests were conducted on unreinforced and reinforced sand.

The results of the compaction tests indicated that the maximum dry density of reinforced sand decreases with increasing fibre content. An optimum moisture content of 10%, independent of the amount of fibres, was also recorded.

In direct shear test, the randomly oriented polypropylene fibre inclusions increase the failure peak shear strength as well as the corresponding horizontal displacement. The specimens become more dilative with the increasing amount of fibres, the dilation is not inhibited by the presence of fibres. This result suggests that the use of mixed materials can improve the static liquefaction behaviour of loose sands. The residual strength is also affected by the fibre content. The trend in change of the apparent cohesion and angle of shearing resistance seems to be consistent with the fibre inclusions. A linear failure line for reinforced specimens has been obtained for all densities and fibre concentrations and this seems to be in agreement with the post-test observations of the state of fibres. The failure mechanism seems to be due to the slippage or pull out rather than stretching or breaking of fibres.

The increase of the peak shear strength was almost linear for all specimen densities at low effective normal stresses. For higher normal stresses, the evolution of the peak shear stress with the fibre content approaches a limiting level, while for specimens with 1.0 void ratio the trend still remains linear. The loosest specimens reinforced with 0.5% of fibres showed the same peak shear strength as densest plain specimens.

When the relative increase of the peak shear stress was considered, for all densities and normal stresses used in this study, qualitatively, a limit of 0.4% of fibres gives between 15 and 20% increase in shear strength. A higher increase, between 30 and 40%, was recorded for 0.8% of fibre content, while at the highest fibre content (1.0%) and specimens with 1.0 void ratio, the gain in strength was, around 60%.

These first experimental results have led to encouraging conclusions concerning the potential use of flexible random fibres for the reinforcement of fine granular materials in applications including stronger and liquefaction resistant earthfills, foundations for buildings, heavy trafficked pavements, and slope stabilisation. The experimental programme using a triaxial and a hollow cylindrical apparatus is in progress and the laboratory study will be also accompanied by the development of analytical model developments.

REFERENCES

Al-Refeai, T.O. – Behaviour of granular soils reinforced with discrete randomly oriented inclusions, Geotextiles and Geomembranes, 10, pp. 319-333, 1991

Bailey, R. and Knox W.R.A. – The strength properties of fibre-reinforced sand, Proc. of 3rd Int. Conf. On ground improvement geosystems, London, pp. 349-357, 1997

Dietz, M.S. – Developing an holistic understanding of interface friction using sand within the direct shear appartus, PhD thesis, University of Bristol, 2001

Gray, D.H. and Ohashi, H. – Mechanics of fibre reinforcement in sand, Journal of Geotechnical Engineering, vol 109, no. 3, 1983

Gray, D.H. and Al-Refeai, T. O. - Behaviour of fabric - versus fiber-reinforced sand, Journal of Geotechnical Engineering, vol 112, no. 8, 1986

Kumar, S. and Tabor, E. – Strength characteristics of silty clay reinforced with randomly oriented nylon fibres, EJGE, paper 10, 2003

Lings, M.L., Dietz, M.S. – an improved direct shear apparatus for sand. Géotechnique, 54, no. 4, p. 245-256, 2004

Maher, M.H.and Woods, R.D. – Dynamic response of sand reinforced with randomly distributed fibres, Journal of Geotechnical Engineering, vol 116, no. 7, 1990

Maher, M.H and Gray, D.H. – Static response of sands reinforced with randomly distributed fibres, Journal of Geotechnical Engineering, vol 116, no. 11, 1990

Michalowschi, RL and Cermak, J. – Triaxial compression of sand reinforced with fibres. Journal of Geotechnical and Geoenvironmental Engineering, Vol. 129, No. 2, February 2003

McGown, A., Andrawes, K.Z., Al-Hasani – Effect of inclusion properties on the behavior of sand, Géotechnique, 28, no. 3, p. 327-346, 1978

Murray, J.J, Frost, J.D., Wang, Y. – Behaviour of s sandy silt reinforced with discontinuous recycled fibre inclusions, Transportation Research record, 1714, 9-17, 2000

Nataatmadja, A. and Parkin, A.K. - Axial Deformation Measurement in Repeated Load Triaxial Testing, *Geotechnical Testing Journal*, Vol. 13, no. 1, p. 45-48, 1990

Ranjan, G., Vasan, R.M., Charan, H.D. – Probabilistic analysis of randomly distributed fiber-reinforced soil, Journal of Geotechnical Engineering, vol 122, no. 6, 1996

Santoni, R.L., Tingle, J.S., Webster, S.L. – Engineering properties of sand-fiber mistures for road construction, Journal of Geotechnical and Geoenvironmental Engineering, Vol. 127, No. 3, March 2001

Vijayasingam, B., Heng, G.Y. – The laboratory study of granular soils reinforced with randomly oriented distributed flexible fibres. Major Research Project, University of Bristol, 45p, 2003

DEFORMATION BEHAVIORS OF GEOSYNTHETIC REINFORCED SOIL WALLS ON SHALLOW WEAK GROUND

You-Seong Kim
Professor, Dept. of Civil Engineering, Chonbuk National University, Korea
e-mail: yusung@moak.chonbuk.ac.kr

Myoung-Soo Won
Researcher, Research Center of Industrial Technology, Chonbuk National University, Korea
e-mail: wondain@kunsan.ac.kr

ABSTRACT
In this study, the fifteen-month behavior of two geosynthetic reinforced soil walls, which was constructed on the shallow weak ground, was measured and analyzed. The walls were backfilled with clayey soil obtained from the construction site nearby, and the safety factors obtained from general limit equilibrium analysis were less than 1.3 in both wall. To compare with the measured data from the real GRS walls and unreinforced soil mass, a series of finite element method (FEM) analyses on two field GRS walls and unreinforced soil mass were conducted. The FEM analysis results showed that failure plane of unreinforced soil mass was consistent with the Rankine active state, but failure plane did not occur in GRS walls. In addition, maximum horizontal displacements and shear strains in GRS walls were 50% smaller than those found in unreinforced soil mass. Modeling results such as the maximum horizontal displacements, horizontal pressure, and geosynthetic tensile strengths in GRS wall have a god agreement with the measured data. Based on this study, it could be concluded that geosynthetic reinforcement are effective to reduce the displacement of the wall face and/or the deformation of the backfill soil even if the mobilized tensile stress after construction is very small.
Keywords: geosynthetic, geosynthetic reinforced soil wall, reinforced soil, finite element method analysis, shear strain.

1. Introduction
The limited equilibrium technique has been used for the design and the analysis of reinforced soil wall since the reinforced earth was commercially used at the first time (Schlosser and Vidal, 1969; Vidal, 1969). In the limited equilibrium design, the force applied to the top of the wall is used to calculate the horizontal pressure, which is resisted by the reinforcement. Although these forces are easily applied to the limited equilibrium design, they cannot be simply incorporated to the prediction of deformation. Finite element technique was applied to analyze the behavior of the reinforced earth in the middle of 70's (Romstad, et al., 1976; Shen, et al., 1976; Hermann and Al Yassin, 1978; Al-Hussaini and Johnson, 1978; Schlosser and Elias, 1978). FEM has been used for the study of numerous parameters and for the analysis of the GRS wall. However, it is not generally used to design the GRS wall. In the research for the GRS wall, FEM has been mainly applied to predict the reinforcing strains and the deformation of the wall face (Bathurst

Hoe I. Ling et al. (eds.), Soil Stress-Strain Behavior: Measurement, Modeling and Analysis, 819–830.

and Koerner, 1988; Ling and Tatsuoka, 1992; Christopher, B.R., 1993; Chew and Mitchell, 1994; Boyle, 1995; Kim and Won, 1999). FEM has been also applied to analyze the parameters such as the length (Christopher, et al., 1990; Xi, 1992; Ho and Rowe, 1993; Chew and Mitchell, 1994), the strength, the spacing, the stiffness, and the arrangement of the reinforcement (Schmetmann, et al., 1989; Chew, et al., 1990; Yoo, 2001), facing material and facing construction (Allen, et al., 1992; Tatsuoka, 1993), compaction stress (Mitchell, 1987; Schmertmann, et al., 1989), and friction at the interface between the soil and the reinforcement and the relative motion (Hermann and Al-Yassin, 1978; Xi, 1992; Boyle, 1995).
The conclusion obtained from these studies is presented as below:

(1) The FEM is an appropriate tool for the investigations of the deformation and the behavior of the GRS wall.
(2) The deformation of the GRS wall is very sensitive to the ratio of L/H when the height (H) of wall/ the length (L) of wall is less than 0.7. Lateral deformation is increased when the ratio of L/H is decreased. This ratio has been identified as an important parameter for the total deformation.
(3) Increasing the vertical space of layers or decreasing the strength of the reinforcement for a given space increases the deformation.
(4) The greater rigidity of the face element, the smaller deformation of the wall.
(5) Compaction energy and compaction induce the deformation of the soil stress during the construction and influence the stress of the reinforcement after the construction.
(6) The deformation of the wall face cannot be exactly estimated by integrating the strains of the reinforcement since these strains do not include the external factors (e.g., foundation settlement or global wall rotation).

In this study, horizontal and vertical stresses and horizontal and shear displacements working on the wall face with or without geosynthetic reinforcement at the backfill will be compared and analyzed by FEM based on measured data from the real GRS wall. In addition, effect of reinforcement installation will be shown. Measured and computed tensile strength data will be evaluated in order to analyze the internal/external behavior and the safety.

2. GRS wall construction and FEM modeling

2.1 The method of the GRS wall construction

According to the results of boring log, profile of soil layer at the site of GRS wall is the indicated fill, the gravel, and the soft rock layers in order. The thickness of each layer and SPT blow counts are shown in Table 1. Two GRS walls with the dimension of 8.0 × 3.9 × 5.0 (length × width × height) (m) as shown in Table 2 were constructed on the shallow weak ground. Average N value is between 3 and 4. Clays with low or medium plasticity from the construction site were used as the backfill of the GRS wall. Non-woven geotextile, woven geotextile, and geogrid were used as reinforcement. Table 3 shows the properties of reinforcement.

Table 1. Results of the field boring log.

	Fill	Silty Clay	Gravel	Soft Rock
Thickness (m)	1.4	3.4	0.7	2.5
Depth (m)	0~1.4	1.4~4.8	4.8~5.5	5.5~8.0 (End boring)
SPT N Value	-	3~4	50	-

Table 2. Geosynthetic lengths placed in GRS walls (SECTION I·II).

Layer No.	Reinforcement (Geosynthetics)	Overlap length (m)	Vertical spacing (m)	Inside length (m)
17	Nonwoven	0.5	0.2	8.0
16	Woven (or Geogrid)	0.5	0.3	1.5
15	Nonwoven	0.5	0.3	1.5
14	Woven (or Geogrid)	0.5	0.3	1.5
13	Nonwoven	0.5	0.3	1.5
12	Nonwoven	0.5	0.3	8.0
11	Nonwoven	0.5	0.3	1.5
10	Woven (or Geogrid)	0.5	0.3	1.5
9	Nonwoven	0.5	0.3	1.5
8	Woven (or Geogrid)	0.5	0.3	1.5
7	Nonwoven	0.5	0.3	8.0
6	Woven or Geogrid	0.5	0.3	1.5
5	Nonwoven	0.5	0.3	1.5
4	Woven or Geogrid	0.5	0.3	1.5
3	Nonwoven	0.5	0.3	1.5
2	Woven (or Geogrid)	0.5	0.3	1.5
1	Nonwoven	0.5	0.3	2.7
Note 1) Geothynthetics which are shown in ( ) in the column of reinforcement classification indicate the reinforcement in SECTION II. Note 2) The length of the geosynthetic length column is included as the gabion part.				

Table 3. Geosynthetic reinforcement properties.

Products	Materials	Description	Thickness (mm)	Tensile strength (tf/m)	
				Manufacturer (KS K 0520)	Researchers (ASTM D 4595)
KOLON P5100	Polyester	Nonwoven needle-punched	5	10	8.97
KOLON KM5001	Polyester	Woven multi-filament	0.25	5	5.11
AKILEN GRID5/3	Polyester	Geogrid coated with PVC resin	0.5	5	4.43

The GRS wall is divided into two sections (Section I; Non-woven and woven & Section II; Non-woven and geogrid) according to the combination of the reinforcement as shown in Fig. 1 and Table 2. Combining non-woven and woven in Section I and non-woven and geogrid in Section II constructed reinforcement installed in the GRS wall. These constructions optimize the reinforcement effects by combining stiff woven and geogrid geosynthetics with non-woven with drainage capacity. The length of geosynthetic reinforcements (except for layers 1, 7, 12, and 17 from the bottom of the wall) is 1.5m, which is 30% of the wall height (5m) and is the minimum reinforcement length as shown in Figs. 1 and 2. This is done in order to verify if the GRS wall (geosynthetic reinforcement) can be safely constructed on shallow weak ground and to analyze the GRS wall deformation caused by non-uniform settlement.

Measuring instruments fixed at the GRS wall have been installed at Section I and II as shown in Fig. 2 and automatically monitored by a data logger. Fig.2 shows the measuring instruments used in this study.

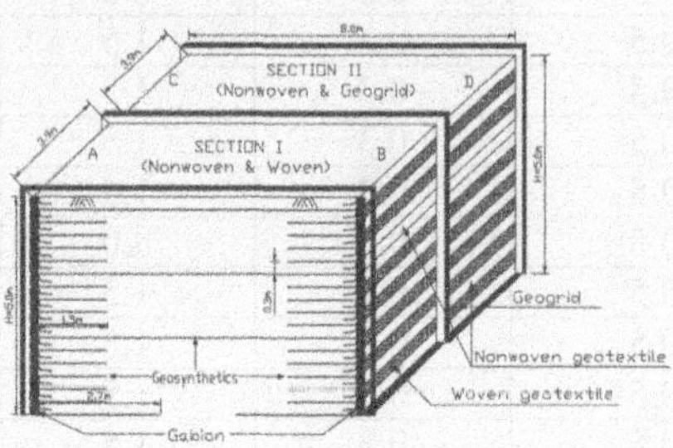

Fig. 1. GRS wall design

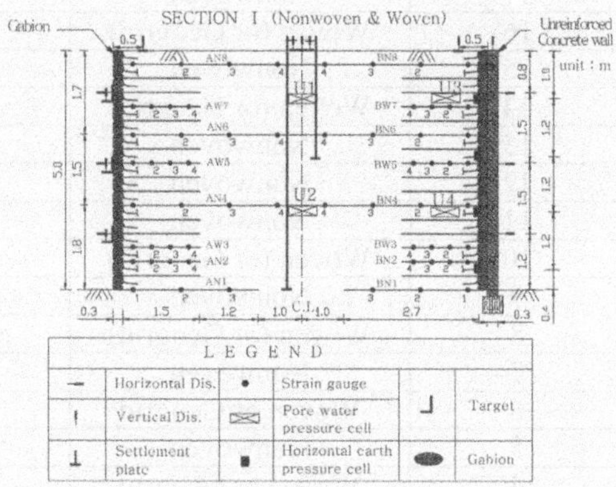

Fig.2. Installed Instruments at SECTION I

Table 4. Instruments used to measure the deformation behavior of GRS walls

Instruments	Quantities	SECTION			
		SECTION I		SECTION II	
		A	B	C	D
Strain gages	124	31	31	31	31
Horizontal earth pressure cells	4	-	4	-	-
Pore water pressure cells	4	-	4	-	-

2.2 FEM Mesh

REA that is two-dimensional FEM program capable of mobilizing construction process was used to simulate the deformation of the GRS wall in this study. Finite element meshes are horizontally laid at 2H (H=wall height) and vertically laid at 1H for the foundation of the GRS wall to investigate the foundation's stress-strain induced by the construction process. Fig. 3 shows the assumed boundary conditions and distinguished layers according to the representative materials. The base of the foundation has been fixed at the boundary condition. Boundary conditions for right and left sides of the foundation are considered to be rollers allowing only vertical deformation. The geosynthetics in the GRS wall are considered as a bar element to allow the both vertical and horizontal deformation and not to allow the rotation.

Table 5. Analysis parameters of the foundation and the backfill.

Parameter	Foundation		Backfill	Concrete	Facing (Gabion)
	Upper	Lower			
Unit weight, γ(tf/m³)	1.7	1.7	1.85	2.4	-
Young's modulus number, K	50	30	60	-	-
Unload-reload modulus number, K_{ur}	75	45	90	-	-
Failure ratio, R_f	0.7	0.7	0.7	-	-
Young's modulus exponent, n	0.6	0.45	0.45	-	-
Poisson's ratio, ν	0.33	0.33	0.33	0.167	0.33
Cohesion, c (tf/m²)	2.0	2.0	3.8	-	-
Friction angle, ø(Deg)	25	20	24.5	-	-
Modulus of elasticity, E (tf/m²)	-	-	-	2,100,000	15,000
Yield stress, σ_a (tf/m²)	-	-	-	-	1,500
Atmospheric pressure, P_a (tf/m²)	10.33	10.33	10.33	-	
Unified classification	SM~SC	CL	CL	-	

Soil parameters of the backfill are determined by lab test and by the report of Duncan et al. (1980). The mechanical properties of the geosynthetics shown in Table 3 are determined by wide-width tensile tests. Parameters of the foundation are determined by feedback analysis based on the measured data from the GRS wall. Table 5 shows the analyzed parameters of the backfill and the foundation. The total number of nodal points is 692. The backfill and the foundation has 629 elements of 4-node rectangular, geosynthetic reinforcement has 132 elements of 2-node bar, and the GRS wall face has 17 elements of beam.

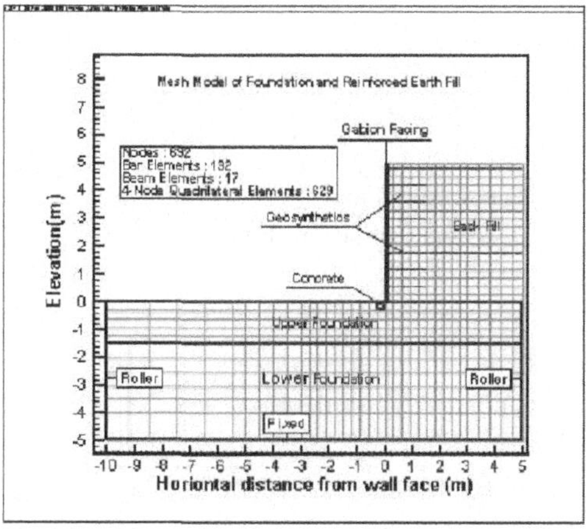

Fig. 3. FE mesh for the GRS wall.

3. Comparison and analysis of measured data and the results of FEM analysis

FEM analysis for the GRS wall and the unreinforced soil mass has been conducted to investigate the effect of reinforcement. The initial data collected within 10 days after the construction of the reinforced soil wall have been used to minimize any external influences such as rainstorm or typhoons.

3.1 Horizontal and vertical stress

Figs. 4 and 5 show the horizontal and vertical stress distributions behind the GRS wall faces. The measured distributions of horizontal stress is less than Rankine's active earth pressure and greater than the analytic value and is similar to the results of FEM analysis. The measured data are larger than the results of FEM analysis because the backfill used in the GRS wall is highly compacted Clay. In addition, the GRS wall face is constructed by wrapping reinforcement around gabion at each layer. This construction led to increase in horizontal stress brought on by reinforcement's increased tensile stress as the backfill material settle unevenly from the compaction. In the GRS wall analysis, vertical stress nearby the wall face is much less than the theoretical results computed by FEM analysis. In this case, vertical stress is transformed to horizontal stress due to interaction between soil and reinforcement due to friction between the soil and the wall face as noted in Pal (1997)' s report. This phenomenon is precisely illustrated in Fig. 6, which shows the vertical stress distribution at the bottom of the wall.

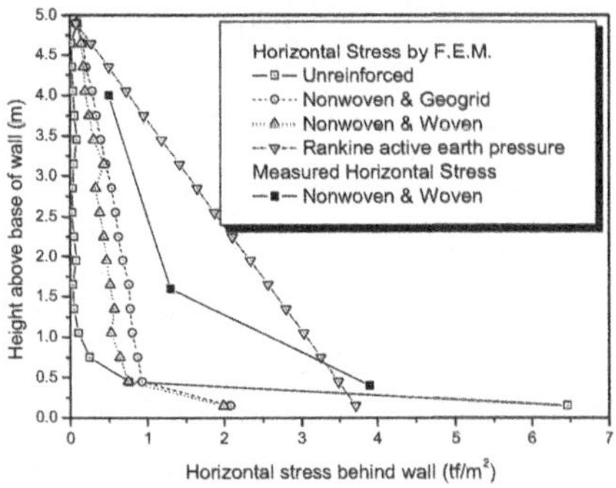

Fig. 4. Horizontal stress distribution at the GRS wall face

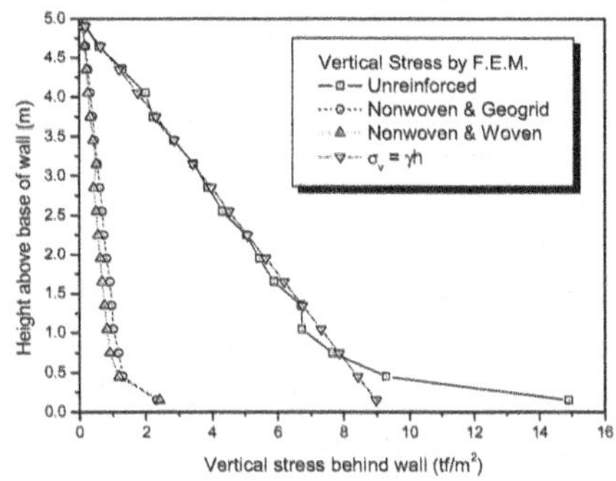

Fig. 5. Vertical stress distribution at the GRS wall face.

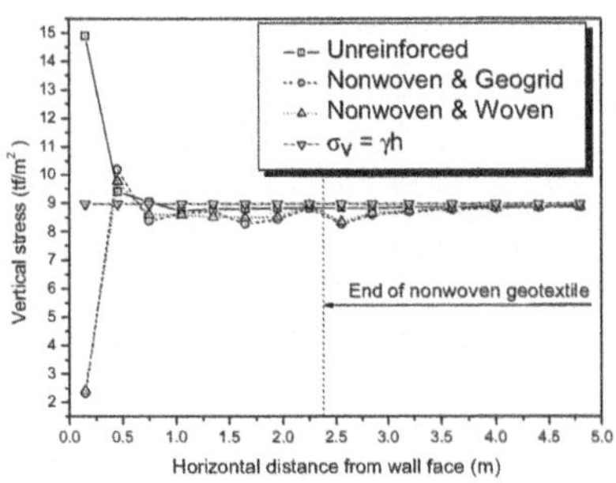

Fig. 6. Vertical stress distribution at the bottom of the wall

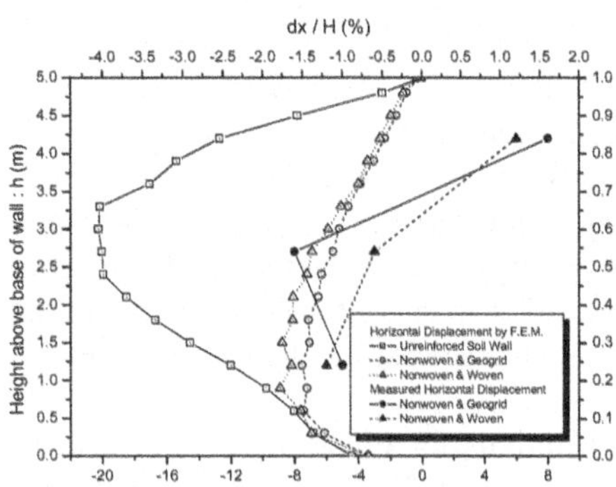

Fig. 7. Horizontal displacement at the GRS wall face.

3.2 Horizontal deformation

Fig. 7 shows a comparison of measured horizontal displacement at the GRS wall face and the computation results. The measured maximum horizontal displacements at Section I and II of the wall faces are 1.25 % and 1.6 % of the wall height, respectively. In addition, the FEM results of the maximum horizontal displacements at Section I (non-woven and woven) and II (non-woven and geogrid) of the wall faces are 1.75 % and 1.5 % of the wall height. The results of FEM analysis show a good agreement with the measured data. For the typical GRS wall after the construction, the maximum displacement at the wall face is approximately between 0.9 % and 1.5 % of the wall height (Christopher et al., 1990). The maximum horizontal displacement by FEM analysis for the unreinforced soil mass is about 4 % of the wall height. This means that horizontal displacement of the backfill is considerably restrained by the tensile strength of the reinforcement material and by the interaction between the soil and the reinforcement. Unlike the Section I, the measured data of Section II exhibit maximum horizontal displacement, which is even, less than the measured data of Section I in the middle section of the wall. This result may be due to the constructional error since the GRS wall is built by wrapping geosynthetic on gabion.

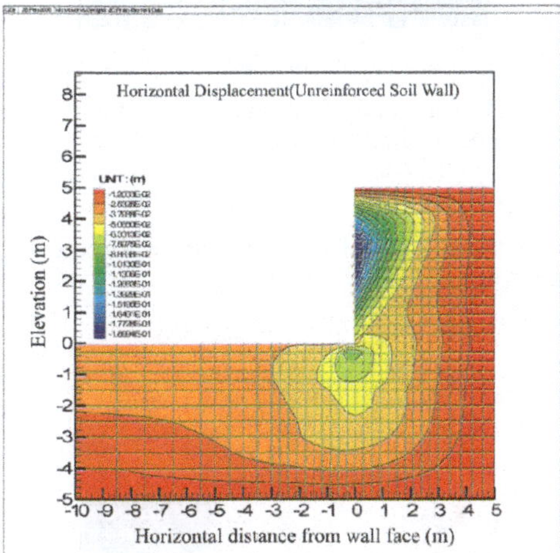

Fig. 8. Horizontal displacement contour and displacement vector in unreinforced soil mass.

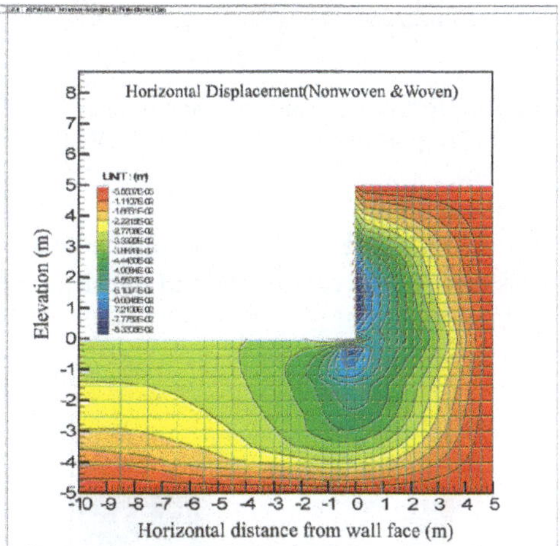

Fig. 9. Horizontal displacement contour and displacement vector at SECTION I

Figs. 8, 9, and 10 show the contour illustrating the horizontal deformation behavior of the unreinforced soil mass and the GRS wall according to the FEM analysis. In these figures, small arrows show the horizontal and vertical collective vectors representing the deformation behavior of a structure. The maximum horizontal deformation in the GRS wall (Figs. 9 and 10) is mainly concentrated at H/3 from the bottom of the wall. On the other hand, the maximum displacement in the unreinforced soil mass is occurred at H/2 (Fig. 8). It is seen that the horizontal displacement in measured data is smaller than that in results of analysis because the top of the GRS wall moved slightly toward the backfill as it is settled due to heavy rain. Based on the results of analysis, it could be inferred that the maximum deformation of the GRS wall occurred in the lower area than the unreinforced soil mass because of reinforcing effects derived from friction between the soil and the reinforcement. In addition, the horizontal displacement of the toe area in the GRS wall is quite similar to those of the unreinforced soil mass.

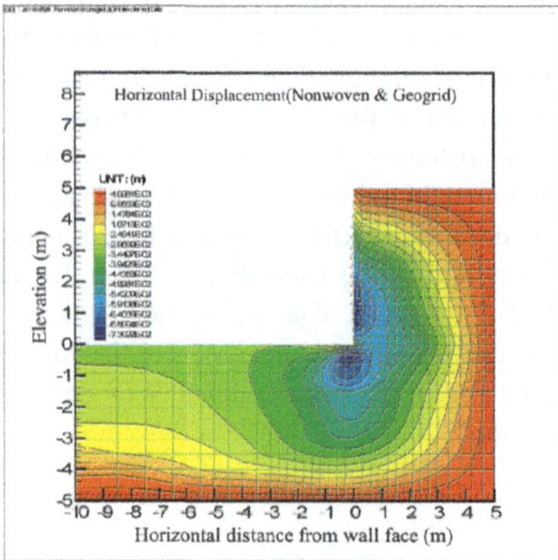

Fig. 10. Horizontal displacement contour and displacement vector at SECTION II.

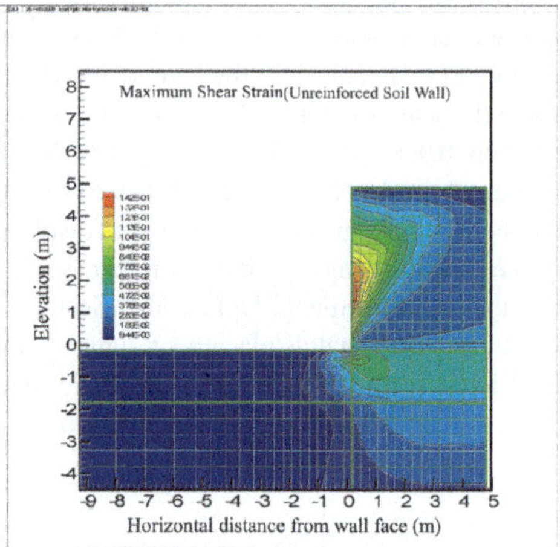

Fig. 11. Maximum shear strain contour at unreinforced soil mass.

3.3 Shear strain

Figs. 11, 12, and 13 show the failure planes of the GRS wall and the unreinforced soil mass through the examination of the shear strain.

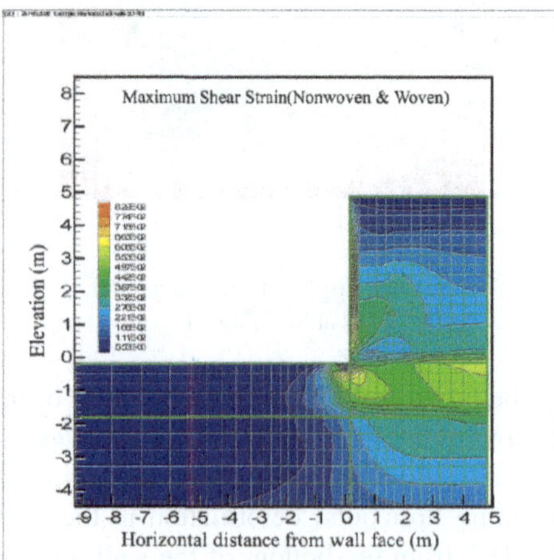

Fig. 12. Maximum shear strain contour at SECTION I.

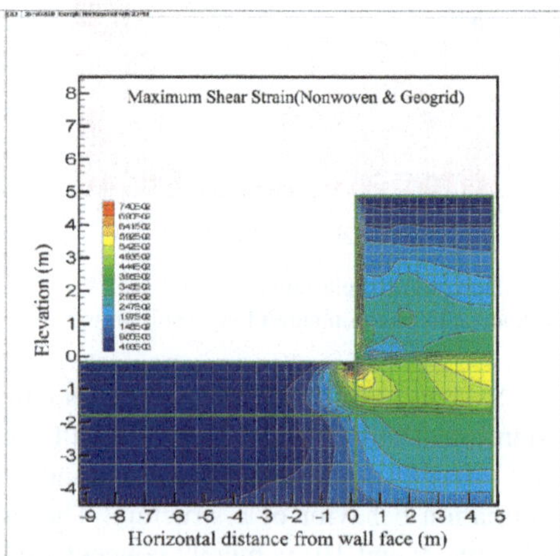

Fig. 13. Maximum shear strain contour at SECTION II.

Figs. 12 and 13 indicate that the maximum shear strain is concentrated at the toe of the GRS wall. This means that the shear stress is concentrated at the toe of the GRS wall since the reinforced soil mass is likely to be overturned. The unreinforced soil mass reaches its maximum shear strain at H/3 caused by Rankine active pressure exerted at H/3 as shown in Fig. 11. In the unreinforced soil mass, the failure plane shows Rankine's type of active failure while the reinforced soil shows no clear type of failure plane. In the GRS wall (non-woven and geogrid) with slightly greater geosynthetic stiffness, the deformation appears to be slightly smaller when compared to less stiff materials (non-woven and woven). The shear strain at the GRS wall is also about 1.5 times less than that at the unreinforced soil mass. It demonstrates the effectiveness of geosynthetic reinforcement.

3.4 Tensile stress of geosythetic

Figs. 14 and 15 show the tensile stress of geosynthetic reinforcements induced by friction between the geosynthetic and the backfill within the GRS wall.

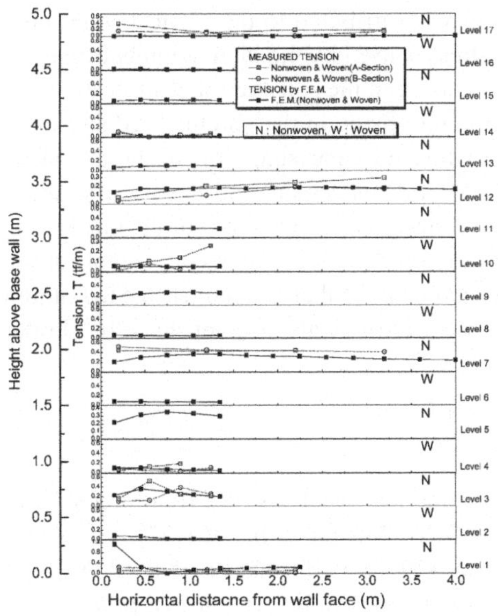

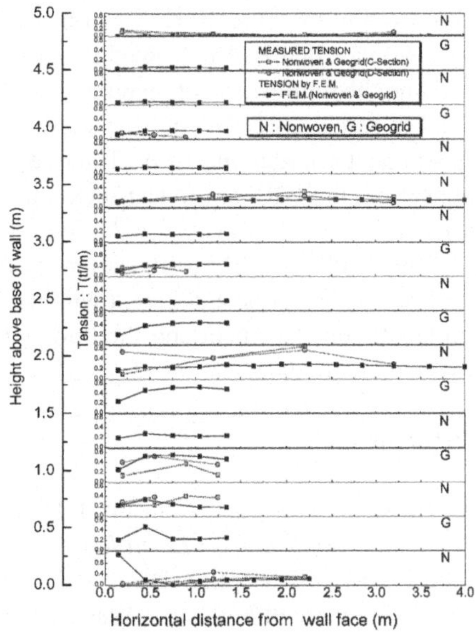

Fig. 14. Tensile strength of geosynthetics at SECTION I. Fig. 15. Tensile strength of geosynthetics SECTION II.

In these figures, the maximum values collected from the data measured within 10-day period are 0.665t f/m in non-woven, 0.260t f/m in woven, and 0.524t f/m in geogrid. These values are 7.4 %, 5.1 %, and 12.1 % of the maximum tensile strength in wide-width tensile test conducted for this study. In the design of the reinforcement soil structure, the allowable tensile strength of the reinforcement is generally 30 % ~ 50 % of the maximum tensile strength that indicates that the tensile stresses obtained from this study are safe values. The results of FEM analysis have a good agreement with the measured data. However, the results of the toe analysis are considerably greater than the measured data. In addition, the measured data at the top 17 layers of the reinforcement material are greater than the results of FEM analysis. In the FEM analysis, a force is trying to overturn the wall from the toe as shown in shear strain while the measured data doesn't show such tendency. The data measured in the top reinforcement layer are greater than the results of analysis because it reflects external influences such as compaction and non-uniform settlement. Despite many external influences (underground water level, foundation materials, settlement, construction method, etc.), the measured data are generally in good agreement with the results of FEM analysis for the GRS wall. This suggests that FEM could be a useful tool to understand and predict the deformation behavior ofthe GRS wall.

4. Conclusion

(1) Results of FEM analysis show the greater horizontal stress at the GRS wall face compared to the unreinforced soil mass. This can be attributed to the vertical stress near the wall face that is transformed to horizontal stress caused by the interaction between the soil and the reinforcement materials and friction between the soil and the wall face.
(2) The results of FEM analysis show that the failure plane in the unreinforced soil mass is similar to a Rankine type of active failure while no clear shape of the failure plane appears in the GRS wall. The maximum horizontal displacement and the shear strain at the unreinforced soil mass are 2.5 times and 1.5 times greater than those of the GRS wall.
(3) The results of FEM analysis for the tensile stress of the reinforcement, except for the toe area, have a good agreement with the measured data for that. Moreover, the maximum horizontal displacement and the horizontal pressure at the wall face appear to be similar to the measure data.

References

Al-Hussaini, M.M. and Johnson, L.D. (1978) Numerical Analysis of a Reinforced Earth Wall. *Symposium on Earth Reinforcement*, ASCE, Pittsburgh, pp. 351-379.

Allen, T.M., Chrisopher, B.R., and Holtz, R.D. (1992) Performance of a 12.6m high geotextile wall in Seattle, Washington. *Geosynthetic-Reinforced Soil Retaining Walls, Proceedings of the International Symposium on Geosynthetic-Reinforced Soil Retaining Walls*, Denver, J.T.H. Wu edl, A.A. Balkema Publ., Rotterdam, pp. 81-100.

Bathurst, R.J., and Koerner, R.M. (1988) *Results of Class A Predictions for the RMC Reinforced Soil Wall Trials.* The Application of Polymeric Reinforcement in Soil Retaining Structures, P.M. Jarrett and A. McGown eds., NATO ASI Series, Series E: Applied Sciences – Vol. 147, Kluwer Academic Publishers, Boston, pp. 127-171.

Boyle, S.R. (1995) *Deformation Prediction of Geosynthetic Reinforced Soil Retaining Walls*. PhD. dissertation, University of Washington, U.S.A.

Chew, S.H. and Mitchell, J.K. (1994) Deformation Evaluation Procedure for Reinforced Soil Walls. *Fifth International Conference on Geotextiles, Geomembranes and Related Products*, Singapore, Vol. 1, pp. 171-176.

Chew, S.H., Schmertmann, G.R. and Mitchell, J.K. (1990) Reinforced soil wall deformations by finite element method. *Performance of Reinforced Soil Structures, Proceedings of International Reinforced Soil Conference*, A. McGown, K.C. Yeo, and K.Z. Andrawes, eds., British Geotechnical Society, Glasgow, pp. 35-40.

Christopher, B.R. (1993) *Deformation Response and Wall Stiffness in Relation to Reinforced Soil Wall Design.*. PhD. Dissertation, Purdue University, U.S.A.

Christopher, B.R., Gill, S.A., Giroud, J.P., Juran, I., Mitchell, J.K., Schlosser, F. and Dunnicliff, J. (1990) *Reinforced Soil Structures Volume I. Design and Construction Guidelines and Reinforced Soil Structures Volume II. Summary of Research and Systems Information.* Federal Highway Adminstration, FHWARD-89-043, Washington, D.C., Vol. 1, 283 p., Vol. 2, 158 p.

Duncan, J.M., Byrne, P., Wong, K.S., and Mabry, P. (1980) *Strength, Stress-Strain and Bulk Modulus Parameters for Finite Element Analysis of Stresses and Movements in Soil Masses.* University of California, Berkeley, Geotechnical Engineering Report No. UCB/GT/80-01.

Kim, Y.S. (in Korean) and Won, M.S. (in Korean) (1999) Behavior Analysis of Geosynthetics Reinforced Earth Walls using Analytical Method. *Chonbuk National University Engineering Research Institute, Engineering Research*, Vol. 30, pp. 103-110.

Herrmann, L.R., and Al-Yassin, Z. (1978) Numerical Analysis of Reinforced Soil Systems. *Symposium on Earth Reinforcement*, ASCE, Pittsburgh, pp. 428-457.

Ho, S.K. and Rowe, R.K. (1993) Finite Element Analysis of Geosynthetics-Reinforced Soil Walls. *Geosynthetics '93*, Vancouver, 1993, Vol. 1, pp. 189-201.

Ling, H.I., and Tatsuoka, F. (1992) Nonlinear analysis of reinforced soil structures by Modified CANDE (M-CANDE). *Geosynthetic-Reinforced Soil Retaining Walls, Proceedings of the International Symposium on Geosynthetic-Reinforced Soil Retaining Walls*, Denver, J.T.H. Wu ed., A.A. Balkema Publishers, Rotterdam, pp. 279-296.

Mitchell, J.K. (1987) Reinforcement for Earthwork Construction and Ground Stabilization. *Proceedings of the Eigth Pan American Conference on Soil Mechanics and Foundations*, Vol. 1, pp. 147-153.

Pal, S. (1997) *Numerical simulation of geosynthetic reinforced earth structures using finite element method.* PhD. dissertation, Louisiana State University, U.S.A.

Romstad, K.M., Herrmann, L.R., and Shen, C.K. (1976) Integrated Study od Reinforced Eearth - I: Theoretical Formulation. **Journal of Geotechnical Engineering**, ASCE, Vol. 102, No. GT5, pp. 457-472.

Schlosser, F. and Elias, V. (1978) Friction in Reiforced Earth. *Symposium on Earth Reinforcement*, ASCE, Pittsburgh, pp. 735-763.

Schlosser, F. and Vidal H. (1969) *Reinforced Earth.* Bulletin de Liasison Des Laboratories Routiers-Ponts et Chaussees, No. 41 (English Translation).

Schmertmann, G.R., Chew, S.H. and Mitchell, J.K. (1989) *Finite Element Modeling of Reinforced Soil Wall Behaviour.* University of California, Berkeley, Geotechnical Engineering Report No. UCB/GT/89-01, 220 p.

Tatsuoka, F. (1993) Keynote Lecture: Roles of Facing Rigidity in Soil Reinforcing. *Earth Reinforcement Practice, Proceedings of the International Symposium on Earth Reinforcement Ractice*, Fukuoka, H. Ochiai, S. Hayashi and J. Otani, eds., Balkema, Rotterdam, Vol. 2, pp. 831-870.

Vidal, H. (1969) *The Principle of Reinforced Earth.* Highway Research Record 282: Soil Theries: Reinforced Earth, Displacements, Bearing and Seepage, Highway Research Board, Washington, D.C., pp. 1-16.

Yoo, C.S. (in Korean) (2001) Seismic Response of Soil-Reinforced Segmental Retaining Walls by Finite Element Analysis, **Journal of the Korean Geotechnical Society**, Vol. 17, No. 4, pp. 15-25.

Xi, F. (1992) *Finite Element Analysis of Geostynthetically Reinforced Walls: A Parametric Study.* Masters thesis, University of Delaware, U.S.A.

Soil Stress-Strain Behavior: Measurement, Modeling and Analysis
Geotechnical Symposium in Roma, March 16 & 17, 2006

EFFECTS OF CUSHION ON THE INDUCED EARTH PRESSURE BY ROLLER COMPACTION

H.S. Roh
Korea Highway Corporation, SeongNam, Korea
e-mail: hsroh@freeway.co.kr
H.J. Lee
Department of Civil and Environmental Engineering,
Sejong University, Seoul, Korea

ABSTRACT

The high earth pressure induced by dynamic loading frequently results in cracks in the culvert walls. In this study, expanded polystyrene (EPS) panel as a cushion material is applied on the culvert walls in backfill areas to reduce the compaction-induced dynamic earth pressure as well as to improve the characteristics of compacted soils. The result of numerical analysis shows that a cushion material with low elastic modulus and high damping ratio can effectively reduce the dynamic earth pressure. Field tests are also performed to validate the performance of the EPS panel. It is found that the EPS panel successfully reduces the dynamic earth pressure acting on the culvert wall.

1. INTRODUCTION

High-capacity vibratory rollers have been widely used for the construction of road to maximize the compaction effects. However, the use of vibratory roller has been limited in near the culvert walls. Because the high-frequency dynamic loading induced by the roller is highly possible to create cracks in the culvert walls. This structural vulnerability of backfill causes the differential settlement between backfill and culvert.

Applying a cushion material to the wall surface of concrete culvert was effective to reduce the compaction-induced dynamic earth pressure as well as to improve the characteristics of compacted soils (Roh et al., 2000). The effect of retaining wall movement on the earth pressure was studied (Duncan and Seed, 1986; Duncan et al., 1992). If the wall moves out from the backfill, the earth pressure reduces. On the other hand, the wall moves into the backfill, the earth pressure increases. When a cushion material, which has low stiffness and high damping ratio compared to soils, is installed between the culvert wall and backfill, the dynamic earth pressure due to the vibratory roller could be reduced similarly.

One of the major objectives of this study is to evaluate the effects of the cushion material on the compaction-induced dynamic earth pressure. A series of numerical analyses was conducted for a typical concrete culvert with and without the cushion material. A sensitivity analysis was also conducted to investigate how the changes in mechanical properties of the cushion material affect the dynamic earth pressure. In addition to the numerical analysis, two field experiments were performed to verify the results of the numerical analysis.

Hoe I. Ling et al. (eds.), Soil Stress-Strain Behavior: Measurement, Modeling and Analysis, 831–836.

NUMERICAL ANALYSIS

Analyses Conditions

Both static and dynamic analyses in time domain are performed using FLAC2D program. For the numerical analyses, a typical concrete culvert is selected from the standard types of concrete culverts for highways (Korea Highway Corporation, 1994) and modeled as shown in Fig. 1. The two dimensional numerical analysis with plane strain condition is performed to model the stage construction of backfill soils. It is assumed that the lift thickness of each backfill layer is 20 cm and the soils are homogeneous, isotropic, and elastic. The analysis using direct time integration method is performed in time domain to examine the earth pressure after maximum stress is produced. It is assumed that vertical stress and strain do not occur in the undisturbed soil and there is no horizontal displacement at the boundary in lateral direction. The vertical and horizontal dimensions of the culvert model are 10 m in both directions.

The compaction loading generated by a high-capacity vibratory roller is simply modeled as dynamic loading with amplitude of 700kN/m^2 and frequency of 40Hz. The dynamic loading is applied at a distance of 1 m from the culvert wall as shown in Fig. 1.

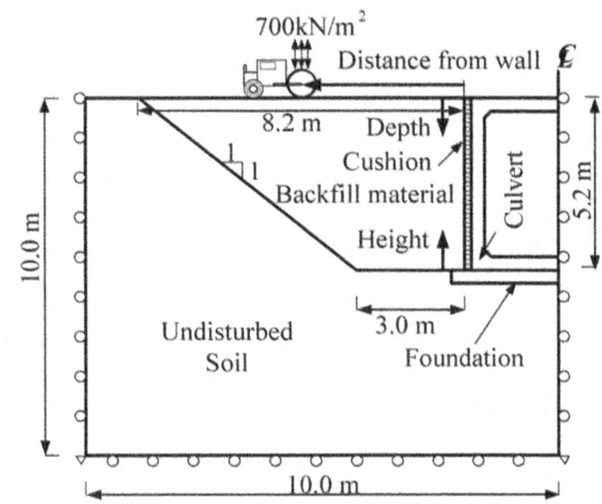

Figure 1. Schematic diagram of model

Material properties used in the analysis are described in Table 1. As can be seen in the table, typical values for elastic modulus, Poisson's ratio, and unit weight of the concrete culvert, foundation, undisturbed and backfill soils are selected. The elastic modulus and damping ratio of the cushion material are dynamic characteristics of the material depending on the rate of applied loading. Therefore, the value of elastic modulus of the cushion material is assumed 10 % of the value of backfill soil. The damping ratio of the cushion is assumed 20. In the sensitivity analysis, however, varying values of elastic modulus and damping ratio for the cushion material are used.

Table 1. Material Properties Used in Numerical Analysis

Materials	Elastic Modulus (kN/m^2)	Poisson's Ratio	Unit Weight (kN/m^3)	Remark
Concrete Culvert	201.1 x 10^6	0.2	24.5	
Culvert Foundation	2.94 x 10^6	0.2	22.6	
Undisturbed Soil	290.0 x 10^3	0.3	19.6	
Backfill Soil	98.1 x 10^3	0.3	19.6	Damping Ratio 5%
Cushion Material	10.0 x 10^3	0.2	0.2	Damping Ratio 20%

Results of the Sensitivity Analysis of Elastic Modulus and Damping Ratio

The distribution of horizontal earth pressure induced by the dynamic loading is examined. To investigate the effects of the stiffness (i.e., elastic modulus) of the cushion material on the dynamic earth pressures, a sensitivity analysis is performed using varying stiffness values of the cushion. For the backfill soil, a constant stiffness value is used in the analysis. The various stiffness values of the cushion is determined based on the stiffness ratio defined as follows:

$$RE = E_{CUSHION}/E_{BACKFILL} \quad (1)$$

where RE is the stiffness ratio, $E_{CUSHION}$ and $E_{BACKFILL}$ are the stiffness of the cushion and backfill soil, respectively. RE values ranging from 1 % to 100 % are used in the analysis. Fig. 2 shows the horizontal earth pressures acting on the culvert wall that are estimated at the location of 0.05 m apart from the culvert wall. The static and the total earth pressures (i.e., sum of static and dynamic earth pressures) for the different stiffness ratios are presented in this figure. Earth pressures at rest (Ko-line) are also presented in this figure. The static earth pressures, which are left side of Ko-line in Fig. 2, continuously increase up to a certain depth and then decreases as the stiffness ratios decrease (i.e., as stiffness values of the cushion decrease). On the other hand, the dynamic earth pressures and the total earth pressures decrease when the stiffness values of the cushion decrease. For example, the peak value of horizontal earth pressures for 100% of SR (i.e., without cushion) are significantly reduced to two third and one third of the values for 10 % and 1 % of SR, respectively. It can be concluded from this observation that the cushion materials having lower stiffness values compared to backfill soils are more effective to reduce the compaction-induced horizontal earth pressure.

Another sensitivity analysis is conducted to evaluate the effects of damping ratio on the dynamic earth pressure. Damping ratio used in the analysis ranges from 5 % to 80 %. Fig. 3 shows the distribution of total horizontal earth pressures estimated at the distance of 0.05m from the culvert wall. As shown in this figure, the peak value of the horizontal earth pressure is reduced up to 30 % as the damping ratio of the cushion increases from 5 % to 80 %. It can be said from this result that higher damping ratio of the cushion is desirable for reducing the compaction-induced horizontal earth pressure.

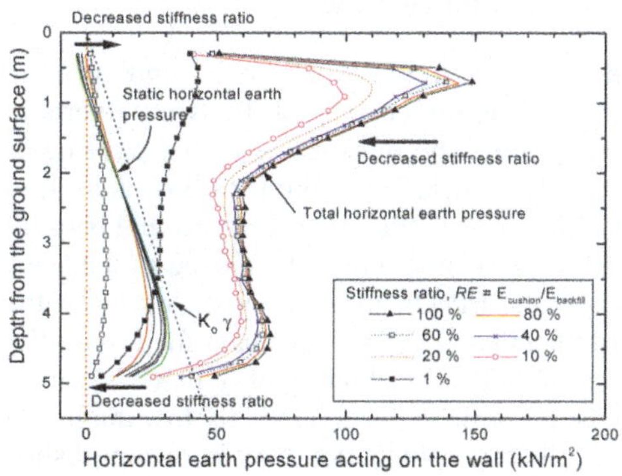

Figure 2. Effect of elastic modulus of cushion on the horizontal earth pressure

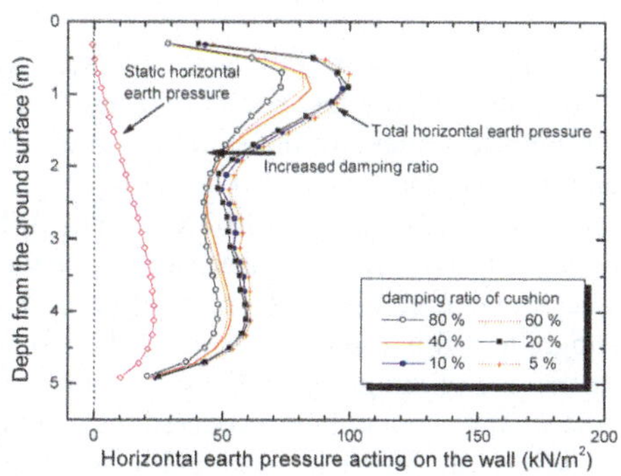

Figure 3. Effect of damping ratio of cushion on the horizontal earth pressure

FIELD EXPERIMENTS

Materials and Constructions

Two field experiments are conducted to validate the results of the numerical analysis as well as to evaluate the performance of a cushion material. An Expanded PolyStyrene (EPS) board is selected as the cushion material in this study. Before the field experiments, a uniaxial compressive repeated loading tests are carried out for the EPS specimen to measure the stiffness and damping ratio of the material. Test specimens of EPS are prepared by cutting of EPS block. The area and length of EPS specimen for the tests are 8836 mm^2 and 199 mm, respectively. A servo hydraulic closed-loop testing system is used for the repeated loading tests. A haversine loading with a amplitude of 10 kg and a frequency of 5 Hz is continuously applied to the specimen for 20 seconds under the stress-controlled mode of loading. The results obtained from the last four cycles (97th ~100th cycles) are used in the estimation of the stiffness and damping ratio. It is observed from the tests results that the EPS specimen shows a linear elastic behavior under the applied loading. The damping ratio of the EPS is negligible. The stiffness of EPS specimens measured from the tests is 9,387 kN/m^2 that is approximately 10 % of backfill soil used in the numerical analysis as shown in Table 1.

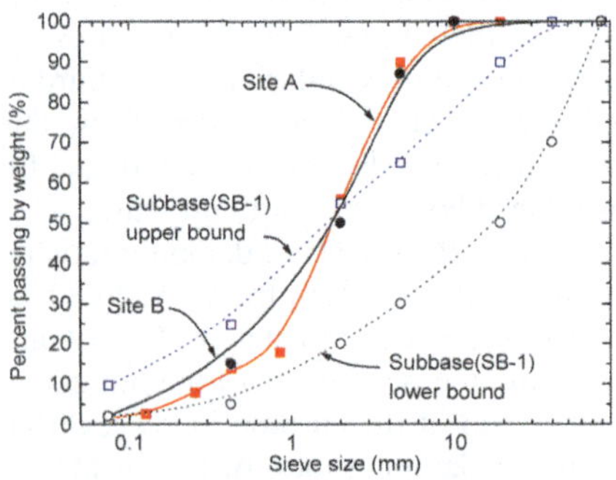

Figure 4. Grain size distribution of backfill material

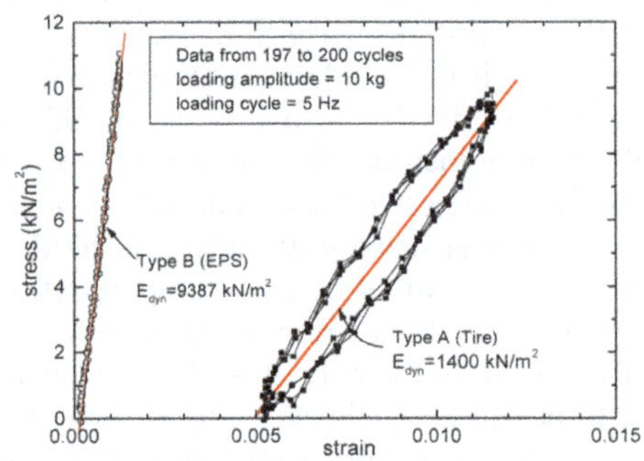

Figure 5. Results of cyclic loading test for cushions

Two concrete culverts are selected for the field experiments and general information of the test site is presented in Fig. 6. The dimensions of the two culverts in the both test sites are the same but the differences are the thicknesses of the pavements and the road materials used in the two test sites. The thickness of the EPS boards used in this study is 10 cm. Two different types of backfill soils including a conventional backfill material (KHC typically uses subbase materials) and a subgrade soil are used in the tests. The conventional backfill material is used where the EPS board is not installed while the subgrade soil is used in case of the EPS board installed. This is to evaluate the possibility of the use of subgrade soils as a backfill material instead of high quality materials for a cost saving. The authors expected that it may be possible to use a relatively low quality of backfill material such as subgrade soils since a better compaction can be accomplished due to the application of the cushion material.

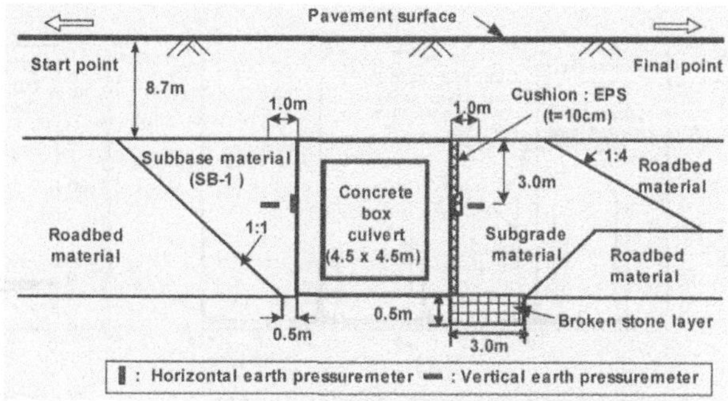

Figure 6. Typical Profile of Test Site

The field density requirement of the backfill soils is that in-place density should be greater than 95 % of maximum dry density of the material. In addition, the coefficient of subgrade reaction ($k_{0.3}$) should be greater than <u>15kg/cm</u>. Physical properties of the subgrade soils used in the backfill areas are presented in Table 2. The subgrade soil used in the backfill of the site A is classified as SM based on the unified soil classification system (USCS). In-place water content is slightly higher than the optimum water content and percent passing of No. 200 sieve is in the range of 1.4 to 6.0 %. The backfill soil of the site B is classified as SC based on USCS. The maximum dry density is 2.0g/cm3, and optimum water content is 9.0 % that is higher than the in-place water content of 7.8 %.

Table 2. Soil properties in sites

Soil classification	Water content, ω_n (%)	Max. dry density, $\gamma_{d,max}$ (t/m^3)	OMC (%)	Plastic index	Specific gravity	CBR (%)
SM	8.3	1.951	7.0	N.P	2.647	6.41
SC	7.8	2.01	9.0	N.P	2.64	-

Field Measurements

Electrical dynamic pressure cells are embedded near the culvert walls to measure the compaction-induced dynamic earth pressures in real-time shown in Fig. 8. The soil pressures are measured by moving a vibratory compaction roller back and forth. The vibration speed of the roller is around 2,400 rpm (40 Hz).

Results of Field Experiments

Typical dynamic earth pressures measured during the compaction of the conventional backfill material (i.e., subbase material) are presented in Fig. 7. The earth pressures presented in Fig. 7(a) and 7(b) are obtained from the same pressure cell under the similar loading conditions but the difference is the fill heights like 3.0 m of fill height for Fig. 7(a), 3.6 m of fill height for Fig. 7(b). As shown in Fig. 7(a), the peak value of vertical earth pressure is approximately 230kN/m^2. When the fill height increased, the peak vertical pressure decreased to 150kN/m^2 because of the decreased transfer of the compaction-induced stresses.

The effects of EPS board on the dynamic earth pressure is analyzed and presented in Fig. 7. The data presented in this figure is measured at the depth of 0.55 m from the wall distance of 0.4 ~ 0. 5 m. It can be seen from Fig. 7(a) that the peak horizontal earth pressure is decreased from 13 to 10 kN/m^2 using the EPS board. The peak earth pressures with and without the EPS board develops at different times because the two pressure cells are differently located in horizontal direction.

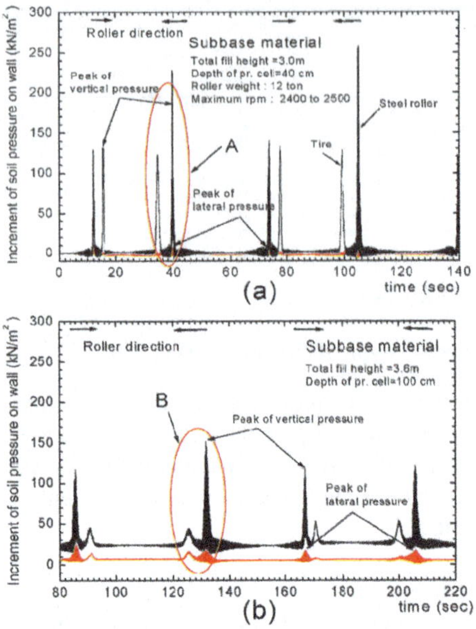

Figure 7. Compaction-induced soil pr.

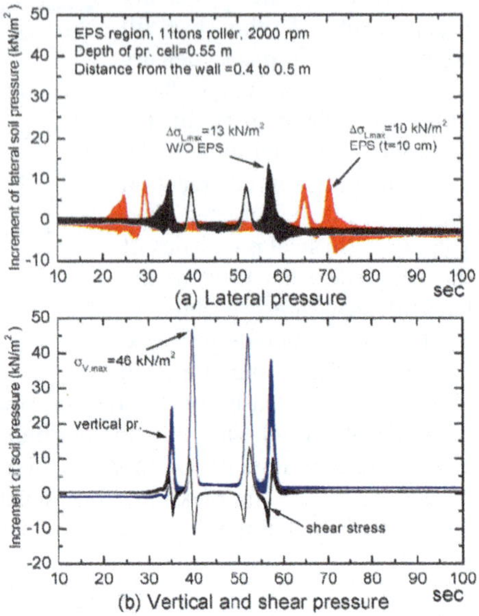

Figure 8. Compaction-induced soil pr.

CONCLUSIONS

Using FEA and field measurements, the reduction of earth pressure associated with the cushion on the concrete culvert induced by dynamic compaction was investigated. According to the numerical analysis results, the peak horizontal earth pressure near to the culvert wall induced by stepwise backfill of 20cm substantially decreases with the cushion, which has one tenth of the soil stiffness and damping ratio of 20 %. As the elastic modulus of the cushion relative to the backfill material decreases, the horizontal earth pressure acting on the culvert wall induced by compaction loading decreases. It could be said that the cushion having the low values of the elastic modulus ratio is recommendable to be used. And also, the horizontal earth pressure decreases to about one third when damping ratio of the cushion is changed 5 % to 80 %. The effect of damping ratio of the cushion is also substantially important during the compaction. It is also observed from the field experiments that the magnitudes of the dynamic vertical and horizontal earth pressures are depending on the backfill materials, fill height, and the distance of the compacting roller.

REFERENCES

Roh, H.S. and Choi, Y.C. Standardized Selection of Backfill Materials for Embedded Structures, Korea Highway Corporation (KHC), 2000, pp. 119-139

Duncan, J.M. and Seed, R.B. Compaction-Induced Earth Pressure under Ko-Conditions. Journal of Geotechnical Engineering, ASCE, 1986, 112, No. 1, pp. 1-22

Duncan, J.M., Williams G..W., Sehn A.L., and Seed, R.B. Estimation Earth Pressure Due to Compaction. Journal of Geotechnical Engineering, ASCE, 1992, 117, No. 12, pp. 1833-1847

Roh, H.S., Choi Y.C., and Kim S.H., Earth Pressure on Culvert during Compaction of Backfill. GeoEng 2000, International Society for Rock Mechanics, Melbourne, Australia, No. UW0775, 2000, Vol. 2

Seed, R.B. and Duncan J.M., FE Analyses: Compaction-Induced Stresses and Deformations, Journal of Geotechnical Eng.. ASCE, 1986, 112, No. 1, 1986, pp. 23-43

Soil Stress-Strain Behavior: Measurement, Modeling and Analysis
Geotechnical Symposium in Roma, March 16 & 17, 2006

MECHANICAL BEHAVIOR OF REINFORCED SPECIMEN USING CONSTANT PRESSURE LARGE DIRECT SHEAR TEST

Matsushima, K., Mohri, Y. and Aqil, U.
National Institute for Rural Engineering, Ibaraki Pref. Japan
e-mail: kenichim@affrc.go.jp
Yamazaki, S.
Mitsui Chemicals Industrial Products, Ltd, Tokyo, Japan
Tatsuoka, F.
Department of Civil Engineering
Tokyo University of Science, Noda City, Chiba, Japan

ABSTRACT

To investigate the deformation and strength characteristics of reinforced soil, a series of large scale direct shear tests (LDST) were performed on reinforced and unreinforced specimens of Toyoura sand using eight different types of reinforcements and rubber sheet. Two layers of reinforcement, 600mm × 500mm in cross section were installed perpendicular to shear direction in a shear box of size L×W×H 800mm×500mm× 600mm. The tests results confirmed the accuracy of data obtained from LDST. Also, it was found that the tensile strain developed in the reinforcement was only about 2 ~3% at peak state and much less than the rupture strain. The reinforced specimens exhibited high peak stress ratio and more dilative behavior provided mobilization of tensile stress was large enough to compensate the reduction of strength of sand, resulting in the spreading of the shear strain development area along with reinforcement. The pressure dependency of reinforced specimen was found to be higher than that of unreinforced specimen. Also the role of the mobilized tensile stress to cater the reduction of soil strength caused by strain softening became higher with lower constant pressures. I hope to provide a simple guideline about the paper.

1. INTRODUCTION

The use of geosynthetics reinforced soil (GRS) structure has gained popularity in recent years because they are proved to be more feasible than conventional earth retaining structures both in terms of cost effectiveness and performance. The increased use of GRS has provided the basis to have a deep insight into the shear characteristics and failure mechanism of reinforced soil. In reinforced soil, the tensile force of reinforcement is mobilized along with the shear deformation of the adjacent soil. As the shear strength of soil varies significantly with strain levels because of their nonlinear stress strain behavior, it is important to know the strain compatibility between tensile strain of reinforcement and strain of infill material in reinforced soils. For this reason, many researchers have used direct shear test apparatus to study the interaction between the backfill soil and the reinforcement. Direct shear test has advantage over the other testing

Hoe I. Ling et al. (eds.), Soil Stress-Strain Behavior: Measurement, Modeling and Analysis, 837–847.

method in term of shear plane crossing the reinforcement. However, most of the previous studies were limited to small and medium size direct shear tests (Qiu. et al 2000 and Wu 2003). Moreover, in small and medium size direct shear test apparatuses, reinforcements are subjected to pullout during shearing and may not truly simulate the field conditions, and somehow mask the true behavior of reinforced soil. Therefore, in the present study, at first a large scale direct shear test apparatus (LDST) was developed based on original framework and feedback control system by Qiu. et al. 2000, in National Institute for Rural Engineering (NIRE). Later, a series of tests were conducted with unreinforced and reinforced specimens with eight different types of reinforcements and rubber sheet in commercial-use under different vertical pressures, to investigate their shear characteristics and pressure dependency.

2. DESCRIPTION OF LARGE DIRECT SHEAR TEST

Large direct shear test device and its procedure were followed in Qiu. et al. They mentioned some techniques about middle direct shear test and reported the precision of this device. Large direct shear test procedure was followed up their procedure. The precision of large direct shear test also was confirmed.

2.1 *Large direct shear test (LDST)*

Figure 1 (a) and (b) show features of LDST apparatus (a), and description of measurement system (b). Analogous to the medium size direct shear test devise, the test results also confirmed the accuracy and precision of the LDST not only at low confining pressure but also at high confining pressure, and will be discussed in details later in the paper. Some of the features of the LDST are as follows:

(i) The shear box was 800mm in length, 500 in width and 600 mm in height. The loading platen and side-wall of the upper shear box were fixed to apply constant vertical pressure on potential shear plane.

(ii) Sponge tape was glued to the periphery of the upper and lower shear boxes to prevent sand from the opening during shearing. A thin but stiff phosphor-bronze channel member was placed over the passive side sponge tape to sustain some shear load on the passive vertical plane for the thickness of the opening, while preventing damage to the sponge.

(iii) The horizontal and vertical friction forces acting on linear guides, as indicated by ③ and ⑪ were measured by two directional load cells ⑥, ⑩, and designated respectively as Load10, Load11, Load12 and Load13 for horizontal friction force and Load14 and Load 15 for vertical friction force in Figure 1 (a), (b).

(vi) To investigate the shear characteristics of reinforced soil at low and high vertical stresses, dual vertical loading system consisting of air and oil cylinders were used. It is because air cylinders limit an applied vertical pressure range of around 70kPa. On the contrary, high vertical loading level such as over 120kPa can be applied by oil cylinder using the pneumatic booster, which can amplify a pressure. However, oil cylinders are not good at controlling low loading level due to a friction between cylinder piston in the pneumatic booster. Therefore, both air and oils were installed to facilitate the application of wide range of vertical pressure, i.e., σ_v=30 to 500kPa.

2.2 *Procedure*

First of all, the upper and lower shear boxes were set together and spacers were inserted in the gap between them to keep the constant opening of 10mm. They were then fixed by bolts to avoid any disturbance until the start of consolidation. Specimens were prepared by pluviating air-dried Toyoura sand (ρ_S=2.64g/cm^3, D_{50}=0.21mm, Uc=1.2) from the top through multiple sieves hopper. The sand in the top sieve was adjusted so as to make nearly flat surface of specimen during air pluviation. This method achieved an initial void ratio, e_i from 0.6226 to 0.6559 in all the specimens.

The shear box was placed on the base of the apparatus ⑲ and fixed by bolts. The loading platen ⑦ was then lowered until it touched the top of upper shear box, and both were fixed together using bolts. The inserted spacers were then removed and prescribed vertical pressure for consolidation was applied so that each axial rod (Axial rod 0,1,2 and 3 in Figure 1 (b)) achieved equal target load.

Four vertical displacement transducers (LVDTs) were set to measure the vertical displacements at the four corners of the top loading platen and were designated as Disp0, Disp1, Disp2, and Disp3 respectively. To maintain the upper shear box parallel with the lower shear box under the constant pressure, a compensating balance moment was applied by two sets of air/oil cylinders through the feedback system. For this purpose, the vertical displacements of four LVDTs were reset to be zero after the completion of consolidation. The degree of the tilt of the top loading platen in shear and its orthogonal direction were within approximately 0.2mm respectively and will be discussed in the next section. Two transducers for measuring the shear displacements were attached to the side-wall of the upper shear box and respectively named as Disp4 and Disp5. Shear load was applied by means of two screw jacks activated by a pair of servomotors. The two screw jacks were operated to move at the same speed by the feed back control system. Shearing was performed at constant rate of about 0.23mm/min in all the tests.

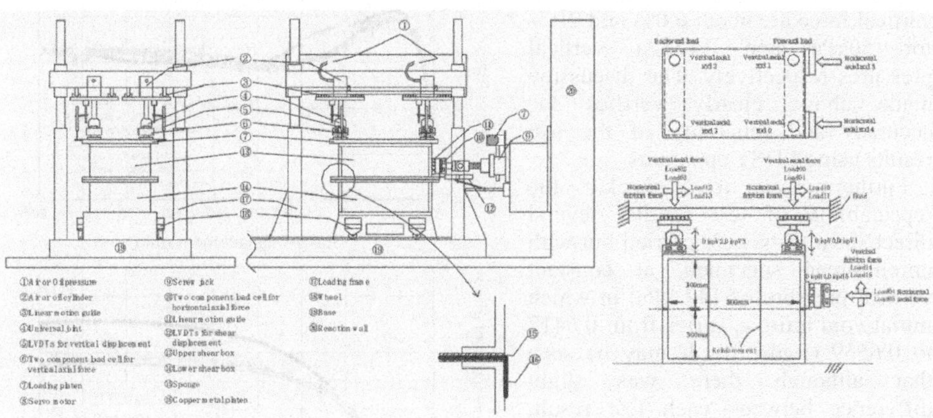

(a) Features of LDST apparatus (b) Description of measurement system

Fig.1 Schematic diagram of LDST apparatus

2.3 Computations and reliability of LDST

Vertical and shear stresses, taking into account the friction forces, were calculated as follows (refer also Figure 1 (b)):

$$\sigma_{vave} = (Weight_{uppershearbox} + (Load00 + Load01 + Load02 + Load03) - (Load14 + Load15)) / Area$$

$$\tau_{have} = ((Load04 + Load05) - (Load10 + Load11 + Load12 + Load13)) / Area$$

The vertical displacement of the specimen was obtained by averaging the readings of four displacement transducers Disp0, Disp1, Disp2, and Disp3, while the shear displacement of the specimen was obtained by averaging the readings of two displacement transducers Disp4 and Disp5.

Figure 2 shows results from two typical constant pressure tests conducted at an average vertical stresses of $\sigma_{v\,ave}$=50kPa and 147kPa using air and oil cylinders respectively on un-reinforced specimens of Toyoura sand. It may be seen from figure 2 (c) and (d), which show the difference of vertical displacements in the shear and its orthogonal direction, that the inclinations of the top platen were approximately within 0.05mm and 0.2mm with air and oil cylinders respectively. These values reflect that the feedback system could maintain the upper shear box parallel with the lower shear box. Consequently due to parallel control, forward load was increased and backward load was decreased with shear displacement to compensate the moment caused by applied shear force as seen from figure 2 (e). Simultaneously, the prescribed vertical pressures were almost kept constant with both air and oil cylinders, however, variation of vertical pressure using oil cylinder was slightly larger than with air cylinder due to pneumatic booster friction as shown in Figure 2 (f). Figure 2 (g), and (h) show total frictional forces in horizontal and vertical directions respectively. The ratio of total horizontal frictional force to shear force are about 6.3% and 3.0%, and the ratio of total vertical friction force to vertical force are about 6.0% and 2.1% for 50kPa and 147kPa vertical pressures respectively. The discussion made above clearly verifies the accuracy and reliability of the test results using LDST apparatus.

Furthermore, to check the repeatability of test results, several direct shear tests were carried out with unreinforced specimen, at constant vertical pressure of 147 kPa, in which initial void ratio, e_i varies from 0.6412 to 0.6559 (figure 3). It may be seen that although there was slight difference between each test result, however the degree of variation in peak stress ratio and in vertical displacement at residual state were 0.80 to 0.90 and –6.1mm to –4.6mm respectively. These variation values appear to be reasonable with large direct shear apparatus.

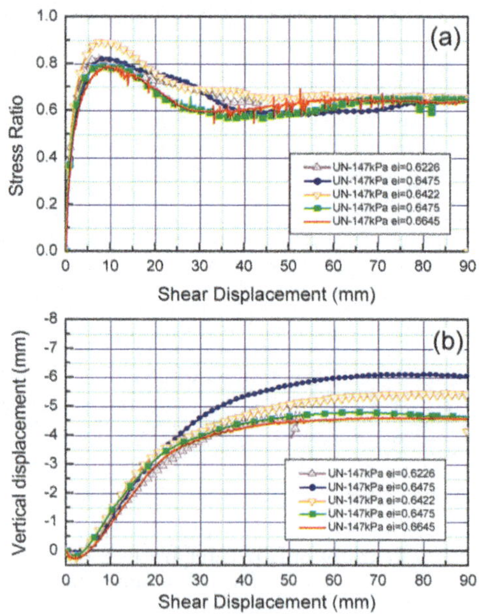

Fig. 3 Results from unreinforced sand tests with a constant pressure of 147kPa

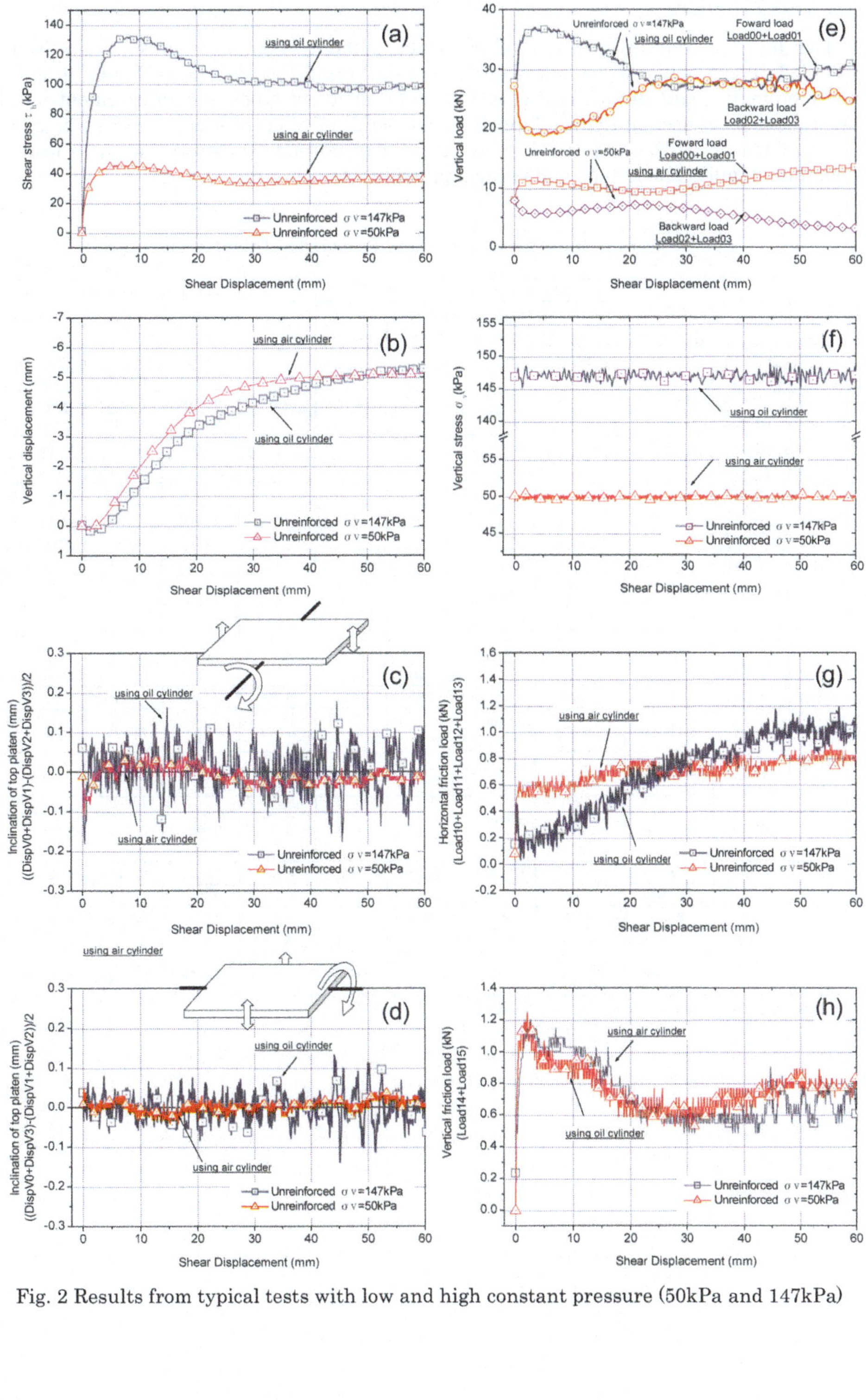

Fig. 2 Results from typical tests with low and high constant pressure (50kPa and 147kPa)

2.4 *Test Program*

Two layers of reinforcement or rubber sheet, 600mm×500mm in cross section were installed perpendicular to the shear direction in all reinforced specimen. Figure 4 and 5 show properties and stress-strain curves of eight different types of reinforcements A-H and rubber sheet used in LDST. Alphabetical legends of reinforcements were arranged in order of tensile stiffness. Table 1 shows the list of constant pressure tests with testing condition, reinforcement type and vertical pressure, etc.

Table 1 Testing condition of LDST and Properties of reinforcements

Test type	Test name	Reinforcements				Cover ratio	Initial void ratio	Constan t pressure	Remarks
		Tensile Stiffness*	Roughness, Thickness, Aperture; Transverse × Longitudinal size		Texture shape				
Unrein-forced series	UN-30	-				-	0.6559	30 kPa	Monotonic
	UN-50	-				-	0.6475	50 kPa	Cyclic
	UN-50	-				-	0.6649	50 kPa	Monotonic
	UN-50	-				-	0.6412	50 kPa	Monotonic
	UN-80	-				-	0.6538	80 kPa	Monotonic
	UN-147	-				-	0.6226	147 kPa	Monotonic
	UN-147	-				-	0.6475	147 kPa	Monotonic
	UN-147	-				-	0.6422	147 kPa	Monotonic
	UN-147	-				-	0.6645	147 kPa	Monotonic
	UN-147	-				-	0.6475	147 kPa	Monotonic
High tensile stiffnes s series	A-147	152.7 kN/m	Very smooth	0.5mm, 22mm	Reed-shaped	0.272	0.6433	147 kPa	Monotonic
	B-147	21.7 kN/m	Rough	1.1mm, 20mm×20mm	Grid	0.252	0.6464	147 kPa	Monotonic
	B-147	21.7 kN/m	Rough	1.1mm, 20mm×20mm	Grid	0.252	0.8712	147 kPa	Monotonic
	C-147	17.7 kN/m	Rough	1.1mm, 20mm×20mm	Grid	0.23	0.6401	147 kPa	Monotonic
	D-30	8.4 kN/m	Smooth	3.8mm, 20mm×20mm	Grid	0.384	0.6538	30 kPa	Cyclic
	D-50	8.4 kN/m	Smooth	3.8mm, 20mm×20mm	Grid	0.384	0.6517	50 kPa	Monotonic
	D-50	8.4 kN/m	Smooth	3.8mm, 20mm×20mm	Grid	0.384	0.6475	50 kPa	Cyclic
	D-147	8.4 kN/m	Smooth	3.8mm, 20mm×20mm	Grid	0.384	0.6443	147 kPa	Monotonic Strain gauges
	E-147	5.7 kN/m	Rough	1.3mm, 20mm×20mm	Grid	0.225	0.6443	147 kPa	Monotonic
Low tensile stiffnes s series	F-147	5.0 kN/m	Smooth	2.0mm, 28mm×33mm	Grid	0.222	0.6515	147 kPa	Monotonic
	G-147	3.8 kN/m	Rough	1.5mm, 28mm×33mm	Grid	0.232	0.6412	147 kPa	Monotonic
	H-147	2.2 kN/m	Rough	0.9mm, 9mm×9mm	Grid	0.289	0.6412	147 kPa	Monotonic
	Rubber-147	1.0 kN/m	Relatively smooth	1.5mm	Sheet	1.00	0.6518	147 kPa	Monotonic

*Secant modulus of tensile stiffness at tensile strain 2.5%

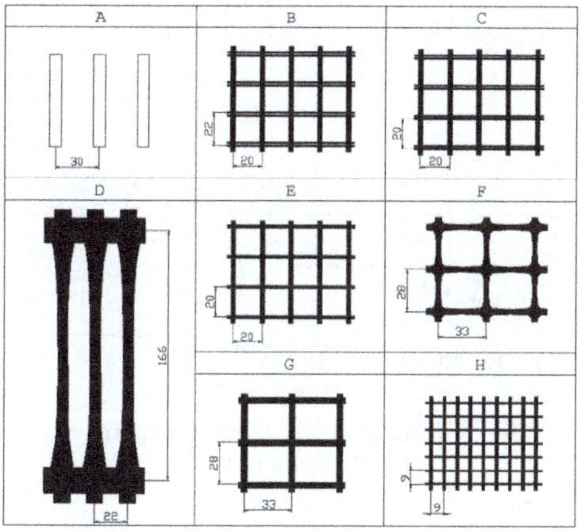

Fig. 4 Geometrical shapes of reinforcements

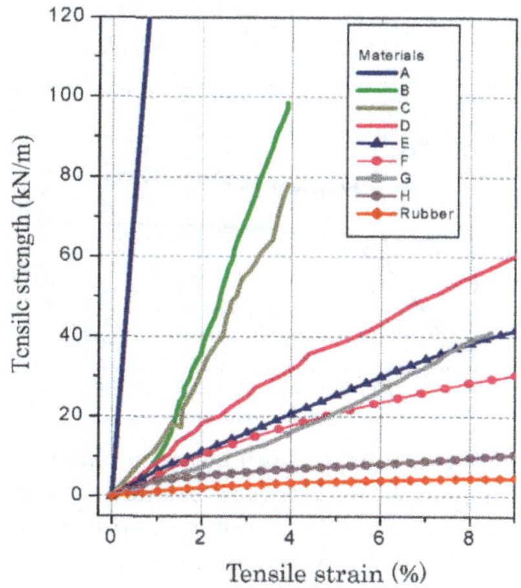

Fig. 5 Relationship between tensile stress and strain

3. RESULT

3.1 *Tensile strain of reinforcement*

In test D-147, the reinforcement D was equipped with strain gauges as shown in Figure 6. Figure 7 shows the relationships between shear displacement, stress ratio, and shear displacement-vertical displacement. The softening behavior of unreinforced specimen starts at shear displacement of 9mm. Similarly the shear stiffness of reinforced specimen decreases after shear displacement of 10mm. It is said that peak shear displacement of reinforced specimen may be larger than that of the unreinforced specimen due to development of wider shear zone. However in case using Toyoura sand, peak shear displacement of reinforced specimen may be 30mm by conservative estimate, because the gradient of vertical displacement gradually became smaller after about shear displacement 26mm in Figure 7. Therefore, it seems that this reduction of shear stiffness is caused by softening behavior of Toyoura sand.

Figure 8 shows the relationship between shear displacement and incremental tensile strain. It was found that at shear displacement of 30mm the tensile strain developed in the reinforcement near the shear plane was only 2～3% and much less than the rupture strain. This reveals that in addition to tensile rupture strength, the tensile stiffness should also needs to be considered while

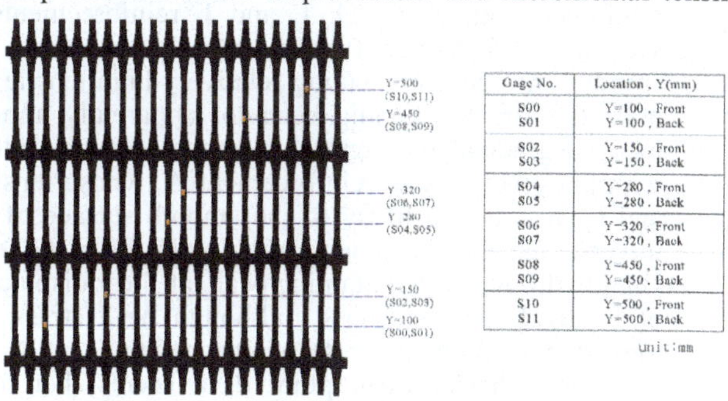

Fig. 6 Position of strain gauges in reinforcement D as used in test D-147

working with reinforced soil design.

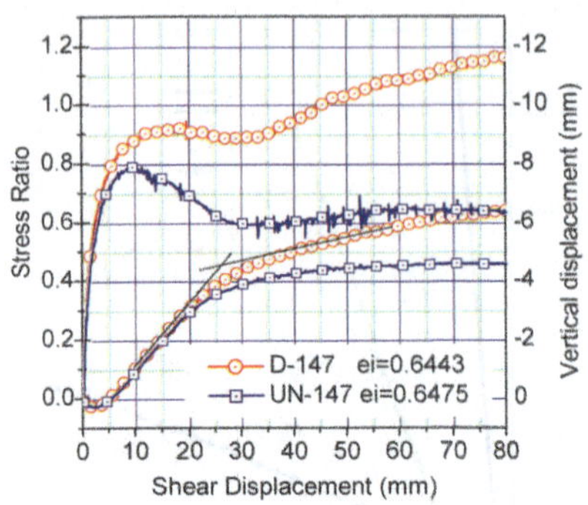

Fig. 7 Relationship between shear displacement, stress ratio and vertical displacement for unreinforced and reinforced specimens

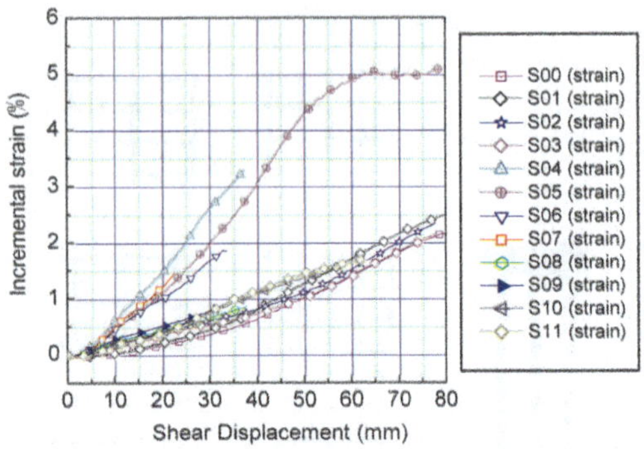

Fig. 8 Tensile strain behavior using reinforcement D

3.2 *Effect of reinforcement*

Figure 9 and 10 show the relationships between overall shear displacement and stress ratio (a), the zoom up shear displacement and stress ratio at small shear displacement (b) and shear displacement and vertical displacement (c) for sand specimen reinforced with high and low tensile stiffness reinforcements respectively. It is evident from these figures that the peak strength of all the reinforced specimens are higher than the unreinforced specimen. Furthermore, to specifically evaluate tensile stiffness effect of reinforcement, on peak strength, volume change and post-peak stress-strain behavior, various shear displacement points a, b, c and d were considered in these figures. The following trend of behavior may be seen:

a. For low tensile stiffness reinforcements, G and H, although the peak stress ratio was higher than unreinforced specimen, the peak shear displacement was almost similar to the unreinforced specimen, i.e., at point a (fig. 10(b)). On the other hand, for the other reinforced specimens (having relatively high tensile stiffness), the peak shear displacements were higher than that of the unreinforced specimen, i.e., at point b. (refer to fig. 9(b) and 10 (b)).

b. The volume change behavior was almost similar among reinforced specimens until point c, after which the vertical displacement becomes larger with specimens reinforced with B, C, D, E and F reinforcements in order of tensile stiffness specimen (refer to fig. 9(c) and 10 (c)).

c. With reinforcements having relatively high tensile stiffness, considering B and C only, the strain softening behavior apparently almost disappeared and started to increase gradually strength after point c. However, this behavior was not consistent with phosphor-bronze reinforcement A, whose tensile stiffness was higher than B and C. It was due likely to the smooth surface of reinforcement and absence of traverse rib. On the other hand, for D, E, F, G, and H reinforcements, strength slightly dropped after attaining peak, at shear displacement point c. For rubber sheet of relatively low stiffness , although peak stress was not improved, trend was toward maintaining peak stress after point a.

d. At shear displacement point d, it seems that the pull out of reinforcements occurred even using the large shear box, therefore tensile stiffness effect were not evaluated precisely. However, it was found that strength continues to increases from point c to point d.

From the discussion made above, it appears that the maintained high stress ratio and more dilative behavior of specimens reinforced with high tensile stiff reinforcement were due to mobilization of large tensile stress enough to compensate the reduction of strength of sand, resulting in the spreading of the shear strain development area along with reinforcement. Compared with unreinforced specimen, toughness of reinforced specimens were also improved considerably with all reinforced specimens, including rubber sheet of quite low stiffness rubber sheet.

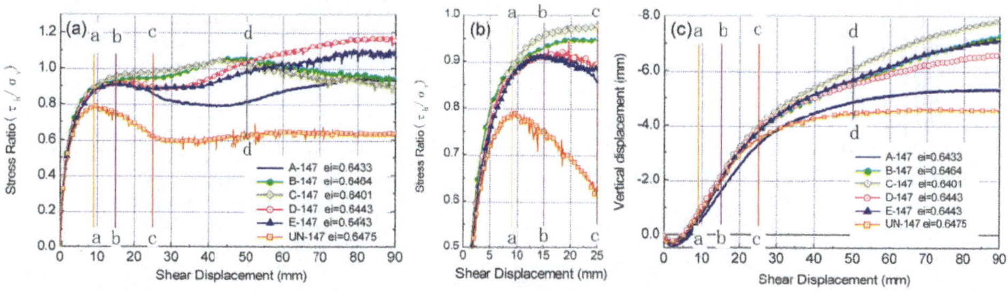

Fig. 9 (a), (b) and(c) Reinforcement A-E series:(a) shear displacement and stress ratio relationship (b) Zoom up shear displacement and stress ratio relationship at small shear displacement (c) shear displacement and vertical displacement relationship

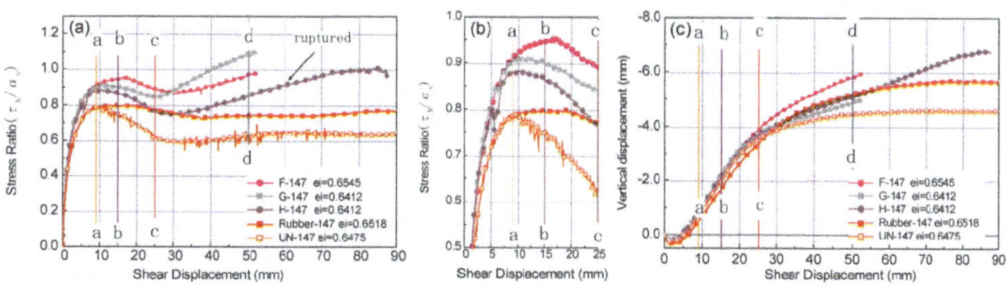

Fig. 10 (a), (b) and(c) Reinforcement F-H and Rubber sheet series:(a) shear displacement and stress ratio relationship (b) Zoom up shear displacement and stress ratio relationship at small shear displacement (c) shear displacement and vertical displacement relationship

3.3 *Pressure dependency*

Figure 11 shows relationship between shear displacement, stress ratio and vertical displacement for different constant vertical pressures tests, for unreinforced specimens and specimens reinforced with reinforcement D, performed to evaluate their pressure dependency. It was found that with the increase of constant vertical pressure, the peak stress ratio and the pre-peak stiffness of the stress ratio and shear displacement relation of both unreinforced and reinforced specimens decrease. Furthermore, the vertical pressure dependency was more with reinforced specimens. This may be attributed to the presence of tensile reinforcements in reinforced specimen.

On other hand, the effect of vertical pressure on vertical displacement was much less with reinforced specimens suggesting that there was not much difference between the shear strain developed areas.

As the tensile force of reinforcement may not increase with the change in confining pressure (in present case vertical pressure), it is likely that the tensile reinforced effect relatively increases with the smaller confining pressure because the reduction of soil strength also became smaller.

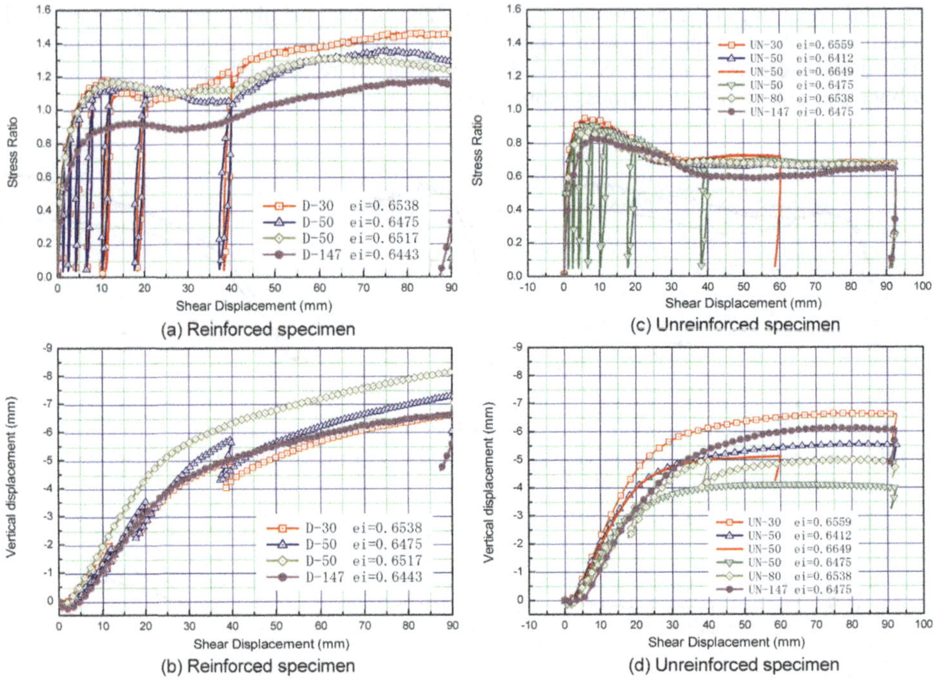

Fig. 11 Pressure dependency of unreinforced and reinforced specimen

4. CONCLUSIONS

The following conclusions can be derived from the test results presented in this paper.

1. The mobilized peak stress ratio (peak shear stress) of reinforced specimen was strongly affected by tensile stiffness of reinforcement. With higher stiffness reinforcement, after peak state, shear strength could maintain well due to the mobilization of high tensile force to cater the reduction of infill soil strength. While, with low stiffness reinforcement or reinforcement having smooth surface and no transverse rib, shear strength temporary dropped down.
2. The reinforced specimens were found to be more pressure dependent than the unreinforced specimen. Also the role of the mobilized tensile stress to cater the reduction of soil strength caused by strain softening became higher with lower vertical pressures.

In the current design codes, performance of reinforcement is only determined by their tensile strength without taking into account the tensile stiffness. However, from the present study it is apparent that peak shear stress is strongly affected by tensile stiffness. Therefore, it is recommended that strain compatibility between reinforcement and soil should also be considered while designing the reinforced soil.

REFERENCES

1) Jewell, R. A. and Wroth, C. P. (1987): "Direct shear tests on reinforced sand", Geotechnique, Vol. 37, No.1, pp53-69.
2) Matsushima, K., Wu, P., Uchimura, T., Tatsuoka, F., and Mohri, Y., (2002): "The Shear Zone Pattern of Reinforced Soil in Direct Shear Test", 37th annual conference of Japanese Geotechnical Society, Niigata, pp765-766. (in Japanese)
3) Public Work Research Centre (2000): "Design and Construction Manual for Reinforced soil structure using Geotextile (revised edition)", pp76-78.
4) Qiu, J., Tatsuoka, F., and Uchimura, T. (2000): "Constant Pressure and Constant Volume Direct Shear Tests on Reinforced sand", Soils and Foundations, Vol. 40, No. 4, pp.1-17.
5) Wu, P. (2003): "Particle Size Effects on Soil-Reinforcement Interaction in Direct Shear Tests", PhD. Thesis, University of Tokyo.
6) Wu, P., Matsushima, K., Tatsuoka, F., Uchimura, T., (2002): "Shear Zone Formation in Reinforced Soil Subjected to Direct Shear", 37th annual conference of Japanese Geotechnical Society, Niigata, pp767-768.

REFERENCES

[illegible]

Soil Stress-Strain Behavior: Measurement, Modeling and Analysis
Geotechnical Symposium in Roma, March 16 & 17, 2006

INELASTIC DEFORMATION OF SAND REINFORCED WITH DIFFERENT REINFORCING MATERIALS

Warat Kongkitkul & Fumio Tatsuoka
Department of Civil Engineering
Tokyo University of Science, 2641 Yamazaki, Noda, Chiba, Japan 278-8510
e-mail: warat@rs.noda.tus.ac.jp & tatsuoka@rs.noda.tus.ac.jp

ABSTRACT

Geomaterial as well as polymer geosynthetic reinforcement are known to have significant visco-plastic property. A series of plane strain compression tests were performed on air-dried Toyoura sand reinforced with four grid types of reinforcement to evaluate the effects of the strength and deformation characteristics (inextensible or extensible; and nearly elastic or highly non-linear elasto-viscoplastic), surface conditions (smooth or rough) and the degree of in-plane dispersion of reinforcement. The effect of reinforcement stiffness on the pre-peak stiffness and peak strength of reinforced sand was not significant, while the effects of the surfaced roughness and the degree of in-plane dispersion were more significant. Significant viscous effects on the vertical stress - vertical strain behaviour of reinforced sand were observed, which were due mainly to the viscous property of sand. The tensile load in the geogrid reinforcement arranged in the sand subjected to sustained constant vertical load decreased with time, indicating that the current design method to evaluate the long-term tensile strength of geosynthetic assuming that the constant tensile load is maintained for life time is on the safe side, perhaps overly. The residual strain of reinforced sand can be made very small by preloading and further sustained loading at the preloaded state, making the effects of reinforcement stiffness negligible.

1. INTRODUCTION

The strength and stiffness of reinforced soil increases with an increase in the confining pressure that develops by the restraining effects of reinforcement, affected by the stress-strain property (e.g., stiffness, elasto-viscoplastic property and rupture strength), the surface condition and the structure of reinforcement (e.g., Tatsuoka et al., 2004; Kongkitkul & Tatsuoka, 2004, Kongkitkul et al., 2005, 2006; Kongkitkul & Tatsuoka, 2005). It is necessary for the performance-based design of reinforced soil structure to evaluate the deformation rather than the rupture load. In this respect, it is usually more important to predict the residual deformation by sustained loading as well as cyclic loading than the deformation by primary loading during construction. However, the evaluation of the deformation (in particular residual one) of a given reinforced soil structure is usually very complicated and therefore very difficult, as it is controlled by not

Hoe I. Ling et al. (eds.), Soil Stress-Strain Behavior: Measurement, Modeling and Analysis, 849–864.

only the elastic properties but also the inelastic or visco-plastic (i.e., irreversible and rate-dependent) properties of both sand and reinforcement (in particular in the case of polymer geosynthetic reinforcement), while these properties as well as their interactions are not well-understood. For the same reason, it is not known how the tensile load in the reinforcement arranged in the backfill that is subjected to fixed sustained load under typical working load conditions changes with time. In the conventional design, the design long-term tensile strength of geosynthetic reinforcement is evaluated by assuming that the tensile load in the reinforcement is kept constant for life time despite that this assumption has not been warranted.

It is known that the plane strain compression (PSC) test is one of the relevant performance test methods of reinforced soil simulating typical field plane strain conditions of reinforced soil structure with tensile reinforcement layers arranged in only one direction. Moreover, in the PSC tests, it is possible to observe the deformation on the intermediate principal stress plane, from which the local strains in not only the backfill but also the reinforcement can be evaluated.

In the present study, a series of PSC tests were performed on air-dried Toyoura sand either unreinforced or reinforced with four different grid type reinforcements. Effects of the strength and deformation characteristics, surface roughness and structure of reinforcement on the strength and deformation of reinforced sand as well as the reinforcement force, in particular their inelastic behaviour, were evaluated.

2. TEST MATERIALS AND TEST PROCEDURES

PSC specimens (96 mm-wide x 62 mm-deep x 120 mm-high; Fig. 1) were prepared by pluviating Toyoura sand (D_{50}= 0.2 mm) through air to obtain a relative density about 84 - 88 %. The specimens were either unreinforced or reinforced with two layers of reinforcement grid located at ¼ and ¾ of the specimen height. The following four grid type reinforcements were used (Fig. 2): 1) a Polyester (PET) geogrid; 2) a PVA

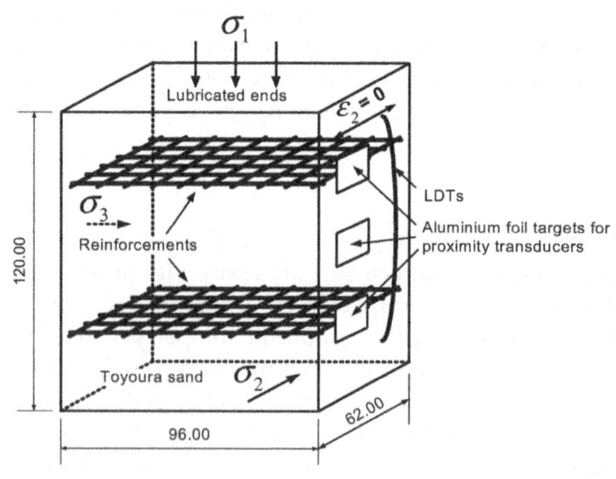

Figure 1. PSC specimen of Toyoura sand reinforced with two reinforcement layers at ¼ and ¾ of the height

(Polyvinyl alcohol) geogrid; 3) a smooth phosphor bronze (PB) grid; and 4) a rough PB grid. The PB grids were prepared to have the same geometry as the PET geogrid, which had a covering ratio (CR)= 22.2 %. On the other hand, the PVA geogrid had a CR equal to 25 % with a smaller degree of in-plane dispersion (i.e., the reinforcement is more concentrated). Constant confining pressure of 30 kPa was applied by means of partial vacuum. The average axial and horizontal strains were obtained by measuring, respectively, the vertical displacement of the loading piston with a LVDT and the lateral expansion of specimen with three pairs of gap sensor (see Fig. 1). The local axial strains were measured also with a pair of LDTs, which were used to measure the vertical strains at small strain range. The average horizontal strains shown in this paper are equal to [(two values at the two levels of reinforcement layers) + (2 times the value at the centre between the two reinforcement layers)]/4.

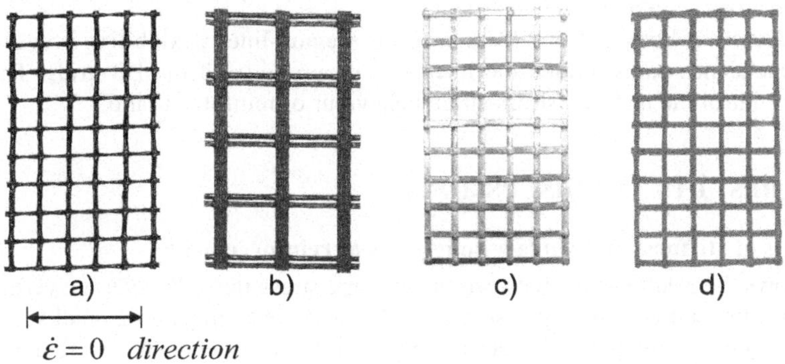

Figure 2. a) PET geogrid; b) PVA geogrid; c) smooth PB grid; and d) rough PB grid used to reinforce Toyoura sand

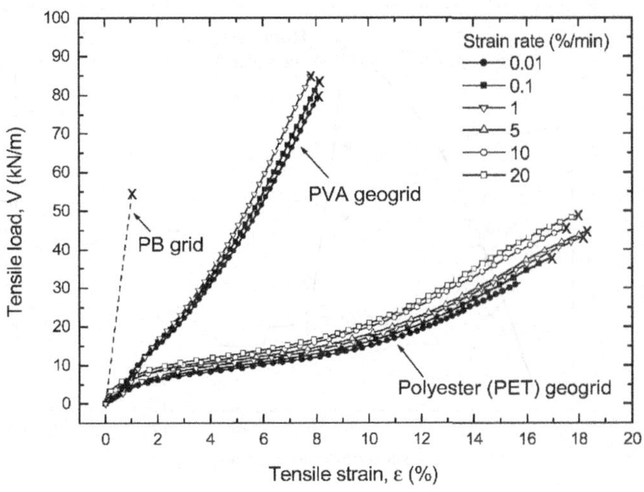

Figure 3. Tensile load - tensile strain relations of PET geogrid, PVA geogrid and PB grid

The tensile load - tensile strain relations of above-mentioned four types of reinforcement are presented in Fig. 3. The relations of the PET and PVA geogrids were obtained by performing tensile loading tests at different strain rates (Hirakawa et al., 2003). On the other hand, the relation together with the rupture strength of the PB grids was obtained from the known Young's modulus and yield stress of phosphor bronze. It is considered that the sand paper glued on the surface of the PB grid to make rough has no effect on its strength and stiffness. The followings can be seen from Fig. 3:

1. The rupture strength of the PB grids and the PET geogrid are similar, while the PVA geogrid is much stronger.
2. The pre-peak stiffness and rupture strain are substantially different among the three materials. Despite that this factor may largely affect the strength and stiffness of reinforced soil, the effects of other factors (e.g., surface roughness and structure of reinforcement) could also be important.
3. The relations of both PET and PVA geogrids are non-linear, exhibiting irreversible and rate-dependent strains, unlike the linear-elastic property of the PB grids. This factor should control the inelastic stress-strain behaviour of reinforced sand.

3. TEST RESULTS AND DISCUSSIONS

3.1 Effects of stiffness and surface roughness of reinforcement

Fig. 4 shows the relationships between the average stress ratio, R= average vertical stress divided by constant confining pressure (30 kPa), and the average horizontal strain, $\varepsilon_{h,avg}$, from monotonic loading (ML) tests on an unreinforced specimen and four reinforced ones. Significant effects of the stiffness, partly rupture strength, surface condition and structure of reinforcement, as described below, may be seen:

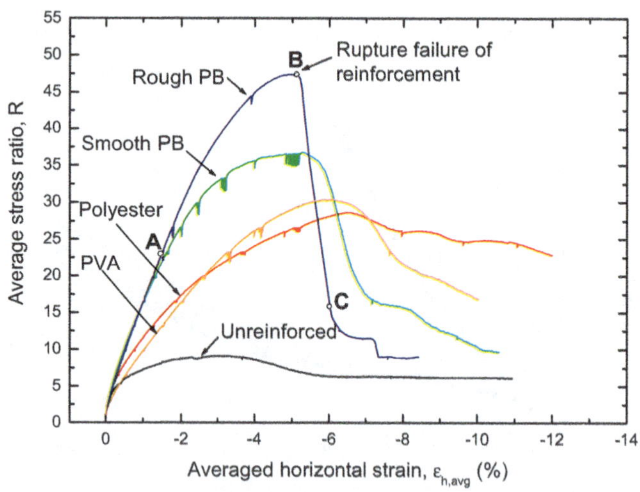

Figure 4. Averaged stress ratio - averaged horizontal strain relations from continuous ML tests on unreinforced and reinforced Toyoura sand

1. The specimen reinforced with a stiffer reinforcement generally exhibits a stiffer response. However, the difference in the pre-peak stiffness between the specimens reinforced with a rough PB grid and a PET geogrid, which have nearly the same structure, is much less significant than the difference in the stiffness between these two reinforcements. Moreover, the pre-peak stress-strain relations and peak strengths of the specimens reinforced with the PVA and PET geogrids are nearly the same despite large differences in the pre-peak stiffness and peak strength of the two geogrids (Fig. 3). This surprising trend of behaviour is likely a result of balancing between the effects of pre-peak stiffness and structure of reinforcement. That is, when compared with the PET geogrid, the reinforcing effects of the PVA geogrid are larger due to a larger pre-peak stiffness while smaller due to a smaller degree of in-plane dispersion. The effects of the latter factor has been observed in the direct shear tests on Toyoura sand reinforced with PB strips by Qiu et al. (2000). In their tests, for the same total amount and covering ratio of reinforcement, the shear strength of reinforced sand decreased with a decrease in the number of PB strip (i.e., with a decrease in the degree of the in-plane dispersion of reinforcement).
2. The specimens reinforced with the rough and smooth PB grids, having the same stiffness, exhibit similar stress-strain relations until a certain stress ratio (Fig. 4). Subsequently, the difference becomes larger with an increase in the stress revel, resulting into a large difference in the peak strength. This result indicates that the effects of surface conditions of reinforcement become more significant after slipping at the surface of reinforcement has started.

3.2 Inelastic response of reinforced sand

Figs. 5a through 5d show the R-$\varepsilon_{h,avg}$ relations from four special PSC tests applying stepwise changes in the strain rate, one or two sustained loading(s), a stress relaxation and a set of cyclic loading performed on Toyoura sand reinforced with the four types of reinforcement shown in Fig. 2. The average vertical stress exhibits a jump, $\Delta\sigma_{v,avg}$, upon a stepwise change in the average horizontal strain rate, $\dot{\varepsilon}_{h,avg}$, which represents the viscous property of the reinforced sand (Fig. 6). Treating a reinforced PSC specimen as a single homogenous element, the respective $\Delta\sigma_{v,avg}$ value observed for a given ratio of the $\dot{\varepsilon}_{h,avg}$ values after and before a step change, was divided by the instantaneous average vertical stress, $\sigma_{v,avg}$, and plotted against the logarithm of $\left(\dot{\varepsilon}^{ir}_{h,avg}\right)_{after}/\left(\dot{\varepsilon}^{ir}_{h,avg}\right)_{before}$ (Fig. 7). As $\varepsilon_{h,avg}$ changes stepwise according to a step change in the average vertical strain, $\varepsilon_{v,avg}$, the value of $\left(\dot{\varepsilon}^{ir}_{h,avg}\right)_{after}/\left(\dot{\varepsilon}^{ir}_{h,avg}\right)_{before}$ is always essentially the same as the corresponding ratio, $\left(\dot{\varepsilon}^{ir}_{v,avg}\right)_{after}/\left(\dot{\varepsilon}^{ir}_{v,avg}\right)_{before}$. Note also that the ratio, $\Delta\sigma_{v,avg}/\sigma_{v,avg}$ is equal to $\Delta R/R$. The slope of the respective linear relation is defined as the rate-sensitivity coefficient, β (Tatsuoka, 2004; Di Benedetto et al., 2005).

Despite largely different stress-strain relations (e.g., Figs. 4 & 5) due to significantly different material properties, surface conditions and structures of these different reinforcements, it may be seen from Fig. 7 that the β values of the reinforced sand specimens are nearly the same while these β values are very similar to the one of the unreinforced sand specimen. It appears, therefore, that the viscous response seen in the vertical stress of reinforced sand is controlled by the viscous property of sand. That is, the viscous property of reinforcement arranged in the lateral direction, at a right angle relative to the axial direction (i.e., the direction of loading), has indirect effects on the rate-dependent behaviour in the vertical stress. On the other hand, the viscous properties of both reinforcement and sand have significant effects on the time-dependent behaviour of tensile load in the reinforcement, as shown below.

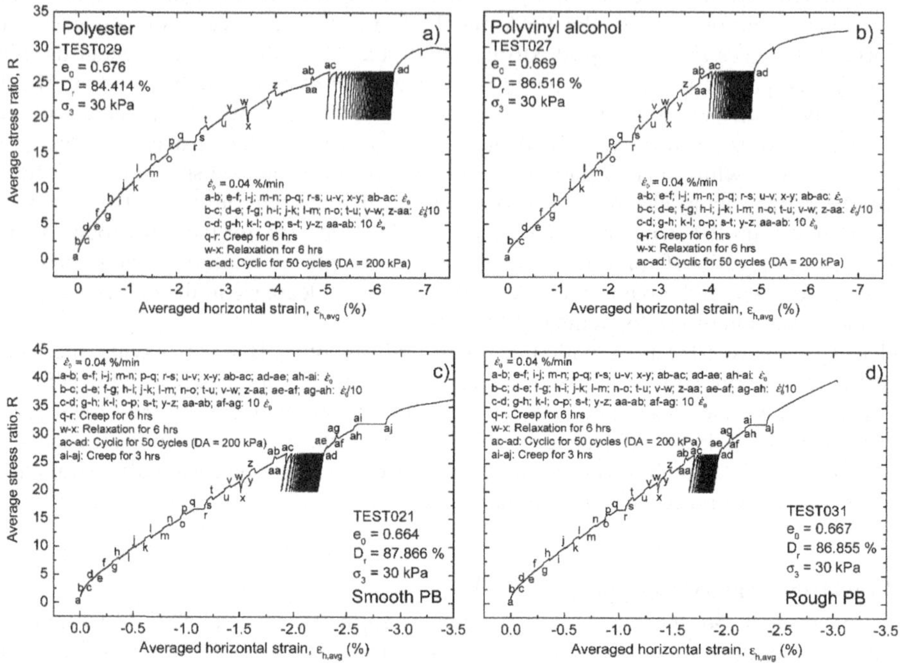

Figure 5. Averaged stress ratio - averaged horizontal strain relation from ML with stepwise changes in the strain rate, sustained loading and cyclic loading on Toyoura sand reinforced with: a) PET geogrid; b) PVA geogrid; c) smooth PB grid; and d) rough PB grid

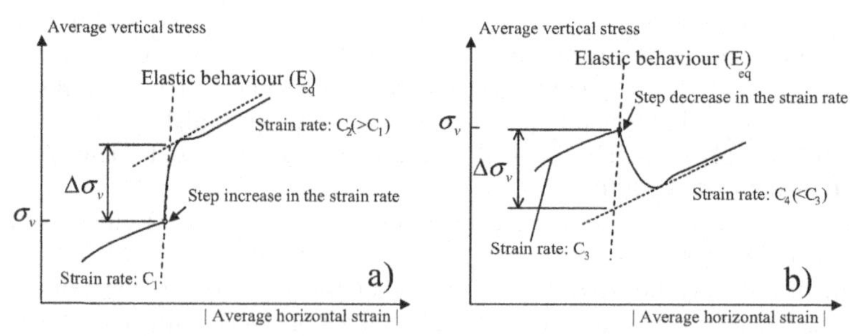

Figure 6. Definition of stress jump upon a step increase/decrease in the strain rate

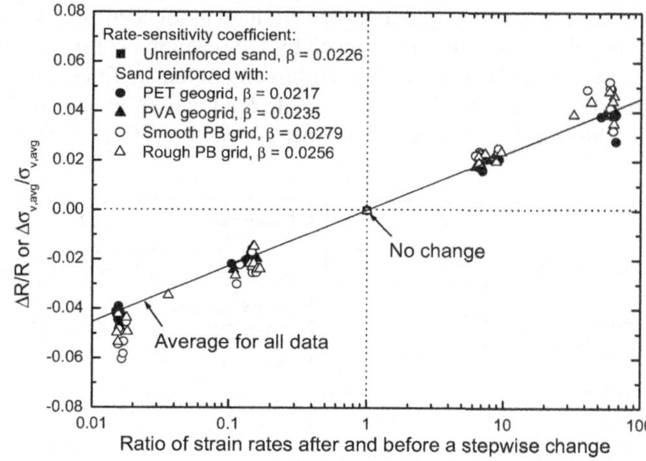

Figure 7. Rate-sensitivity coefficients for unreinforced and reinforced Toyoura sand

3.3 Relaxation of tensile load during sustained loading of reinforced sand

Figs. 8a, 8b and 8c show the R-$\varepsilon_{h,avg}$ and R-$\varepsilon_{v,avg}$ relations of sand without and with reinforcement from ML tests with sustained loading and stress relaxation stages. Note that the R-$\varepsilon_{h,avg}$ relation is relevant when analysing the tensile reinforcing mechanism while the R-$\varepsilon_{v,avg}$ relation is relevant when analysing the engineering performance of reinforced soil. The trend of stress-strain behaviour upon the restart of ML at the original strain rate after a sustained loading stage is markedly different among the three specimens. That is, with the unreinforced sand (Fig. 8a), an overshooting in the stress following yielding is noticeable along the stress-strain relation after the restart of ML at a constant strain rate from the ends of sustained loading and stress relaxation stages. On the other hand, the stress-strain relation of the PET geogrid-reinforced sand specimen exhibits a yield stress that is noticeably below the primary curve by continuous ML at a constant strain rate, and the subsequent stress-strain relation only gradually rejoins the primary curve (Fig. 8b). This large drop in the yield stress indicates that the confining pressure in the sand, therefore the reinforcement tensile load, decreased during the sustained loading and stress relaxation stages. Also with the sand specimen reinforced with the smooth PB grid (Fig. 8c), the trends of behaviour observed with the PET-reinforced sand (Fig. 8b) are noticeable, but they are much weaker. This difference should be attributed to the fact that the PB grid has essentially elastic property while the PET geogrid has markedly viscous property.

These results indicate that the time-dependency of the tensile load of reinforcement arranged in sand and the corresponding local confining pressure of sand are controlled by the interactions of the following three viscous mechanisms (Fig. 9):

a) an increase with time in the reinforcement tensile load associated with an increase in the reinforcement tensile strain imposed by an increase in the viscous horizontal tensile strain of sand caused by creep deformation of sand subjected to constant vertical load;

b) a decrease with time in the reinforcement tensile load associated with a decrease of the reinforcement tensile strain with time caused by the development of viscous horizontal compressive strain of sand caused by the reinforcement tensile force; and
c) a decrease with time in the reinforcement tensile load caused by the load relaxation of reinforcement that would take place even under constant reinforcement tensile strain.

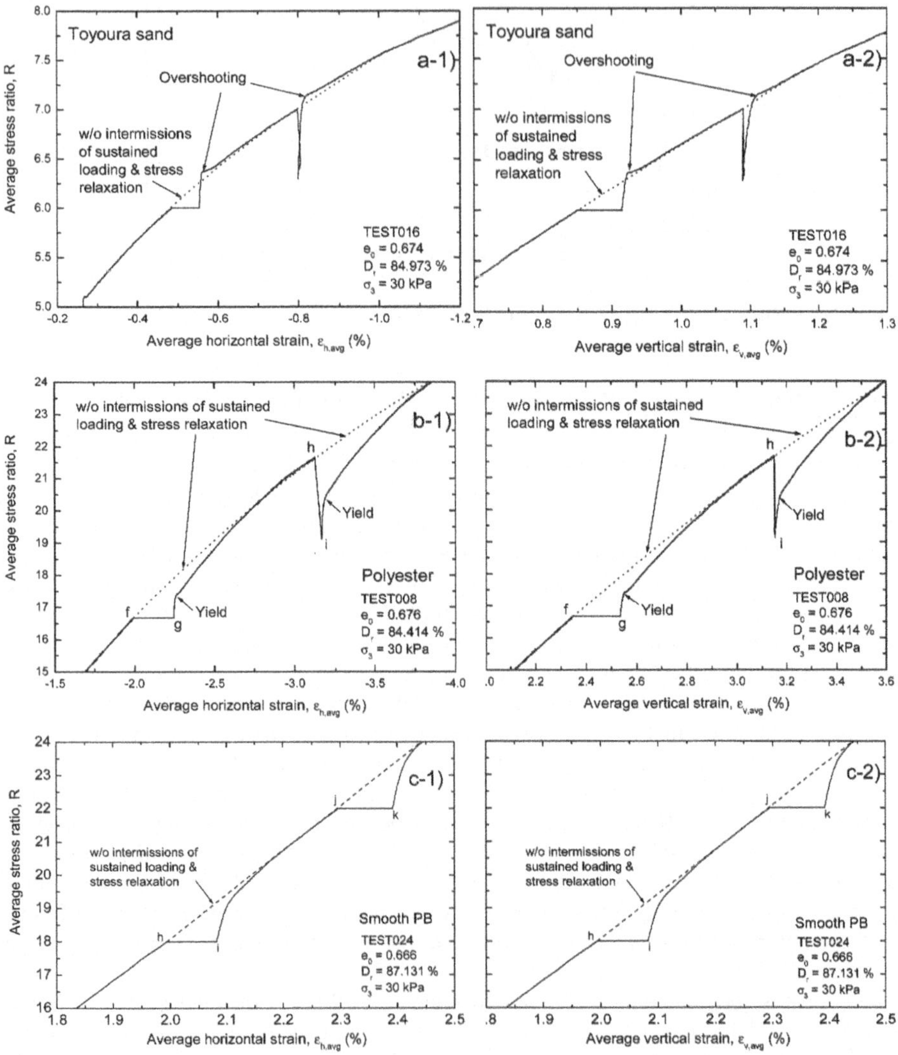

Figure 8. Close-up of R-$\varepsilon_{h,avg}$ and R-$\varepsilon_{v,avg}$ relations including sustained loading and stress relaxation stages for: a) unreinforced Toyoura sand; b) Toyoura sand reinforced with PET geogrid; and c) Toyoura sand reinforced with smooth PB grid

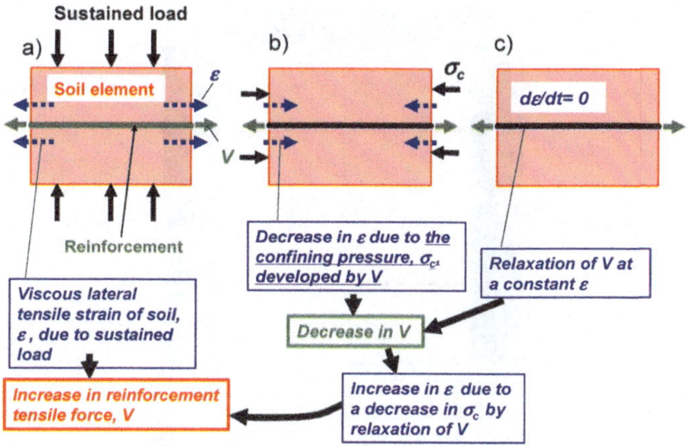

Figure 9. Illustration of interactions between deformations of reinforcement and sand during sustained loading of reinforced sand

All these mechanisms are relevant to the PET geogrid-reinforced sand, while only mechanisms ***a*** & ***b*** are relevant to the PB grid-reinforced sand. When sand is reinforced with a polymer geogrid having noticeably viscous property, the geogrid tensile load can decrease with time during the sustained loading of reinforced sand when the effects of mechanisms ***b*** & ***c*** are larger than those of mechanism ***a*** (i.e., the case in Fig. 8b). On the other hand, when reinforced with a reinforcement having negligible viscous property, the effect of mechanism ***c*** becomes negligible while the effect of mechanism ***a*** could be somehow balanced with that of mechanism ***b***, resulting into a small interaction between the viscous properties of sand and reinforcement (i.e., the case in Fig. 8c).

3.4 Local strain distribution in reinforced sand

A photogrametric analysis of a number of photographs of the σ_2-surface of the PSC specimen, as typically shown in Fig. 10, was performed. Fig. 11 shows the contours of local horizontal strain in the rough PB grid-reinforced specimen at stages **A**, **B** and **C** denoted along the stress-strain relation presented in Fig. 4. That is, point **A** represents the stress state at R= 23 during primary loading, far before the peak state, point **B** represents the peak state, and point **C** represents the state immediately after the stress had started dropping due to the rupture failure of the rough PB grid. Restraining effects of reinforcement on the sand strain are noticeable at stage **A** while they become much more obvious at state **B** (the peak state). Namely, the local horizontal strains around the PB grid layers are noticeably smaller than those in the other zones of the specimen. It may be seen that the horizontal strains started localising from both ends of both lower and upper PB grids. At stage **C**, upon a sudden rupture at the middle of the lower PB grid, the horizontal strain has been localised into two shear bands (Fig. 11c). This rupture mode of reinforcement indicates that the tensile force was largest at the central part of the reinforcement.

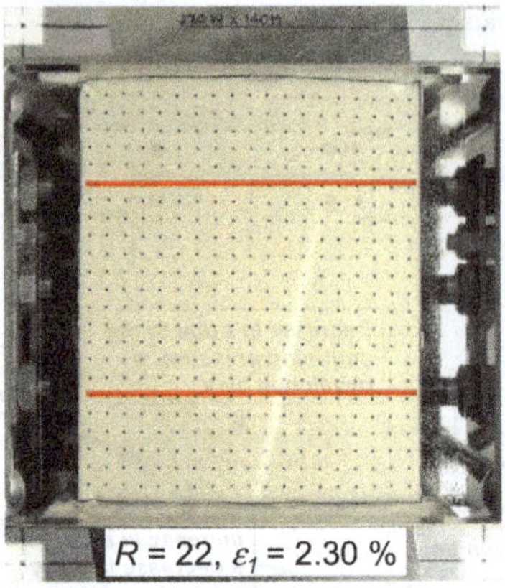

Figure 10. σ_2-view of PSC specimen

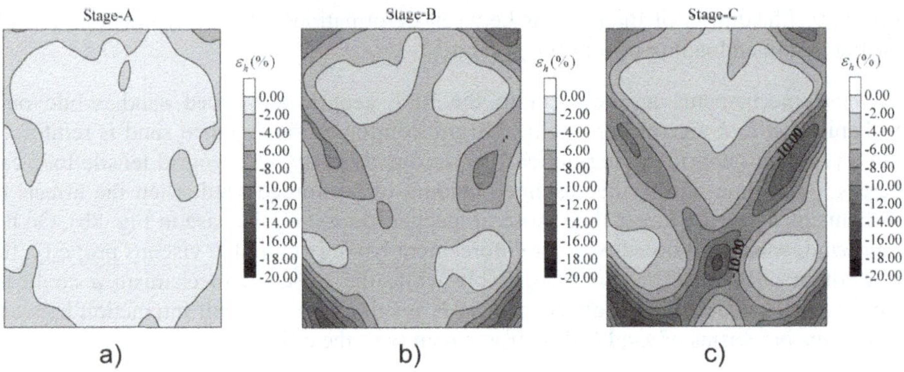

Figure 11. Horizontal strain contours on rough PB grid-reinforced Toyoura sand (n.b., stages A, B and C are denoted in Fig. 4)

3.5 Tensile load-strain relation of geogrid arranged in sand

Figs. 12a and 12b compares with the R-$\varepsilon_{h,avg}$ and R-$\varepsilon_{v,avg}$ relations from two ML tests at a constant strain rate (i.e., 0.04 %/min) on PET geogrid-reinforced PSC specimens with and without a long-term sustained loading test (i.e., for 30 days) at a stress level of about a half of the peak strength, R = 16.67. The peak strength of the reinforced sand increased noticeably by the 30 day-long sustained loading. This result indicates that the creep deformation of the reinforced Toyoura sand, which is due to creep deformation of PET geogrid and sand, had no deteriorating effect on the subsequent stress-strain behaviour of the reinforced sand. The ultimate rupture strength of the PET geogrid does not increase by a sustained loading for 30 days performed during otherwise ML at a constant strain rate (Fig. 13). Therefore, the observed increase in the peak strength of the reinforced sand by the long-term sustained loading was not due to an increase in the ultimate strength of the reinforcement during the sustained loading, but it might be due to the development of a more stable interlocking between the reinforcement and the sand during the sustained loading.

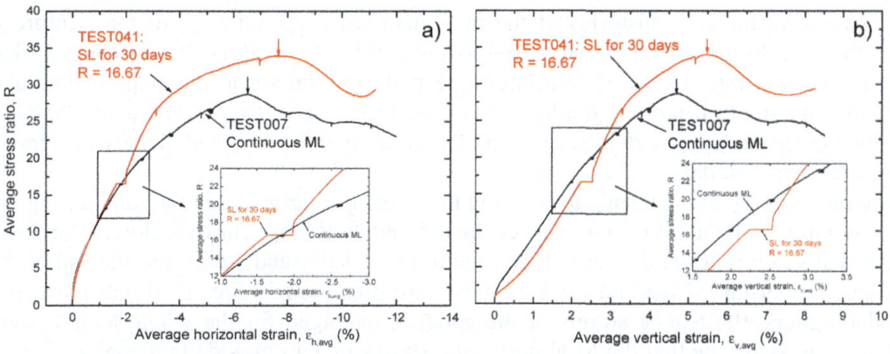

Figure 12. a) R-$\varepsilon_{h,avg}$ and b) R-$\varepsilon_{v,avg}$ relations from PSC tests with and without a 30 day-long sustained loading stage on Toyoura sand reinforced with a PET geogrid

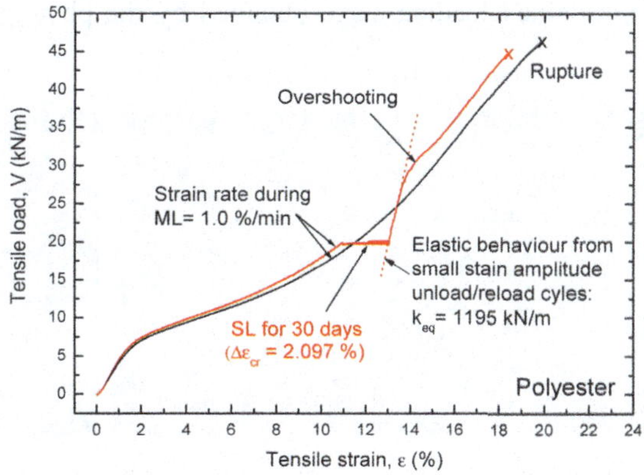

Figure 13. Tensile load - tensile strain relations from tensile tests of PET geogrid with and without a 30 day-long sustained loading stage

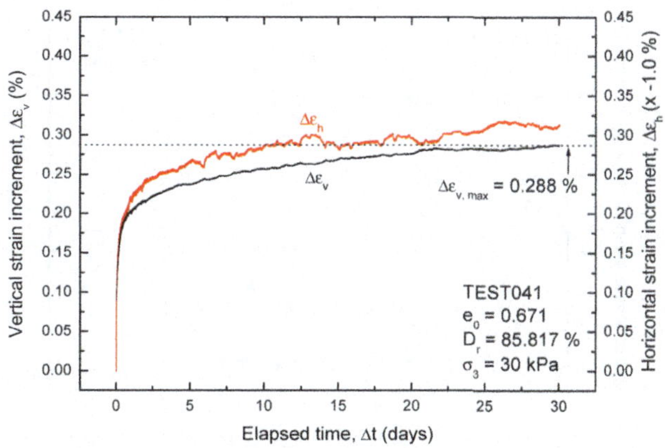

Figure 14. Time histories of averaged vertical and horizontal strain increments of PET geogrid-reinforced Toyoura sand during 30 day-long sustained loading

Fig. 14 shows the time histories of the increments of $\varepsilon_{h,avg}$ and $\varepsilon_{v,avg}$ of the reinforced sand that developed during the sustained loading stage for 30 days. It may be seen that the creep strain rate decreased at a large rate with an increase in the elapsed time and became very small, if not negligible, at an elapsed time of 30 days, while about 70 % of the total creep strain increment seen at an elapsed time of 30 days had already developed during the first one day.

The time history of the tensile load acting in the geogrid arranged in the sand during the 30 day-long sustained loading was evaluated indirectly as shown below. The local horizontal strain fields at the start of, elapsed times of 1 day and 3 days and the end of the 30 day-long sustained loading (at R = 16.67) are presented in Fig. 15. From these and similar figures, the tensile strains of the geogrid, averaged for the whole length, were obtained by assuming that the local horizontal strains in a 1 cm-wide thin zone including the respective geogrid layer are the same as those of the geogrid layers. The data points shown in Fig. 16 are the averaged horizontal strain increments of the geogrid reinforcement at the sustained loading stage obtained by the photogrametric technique

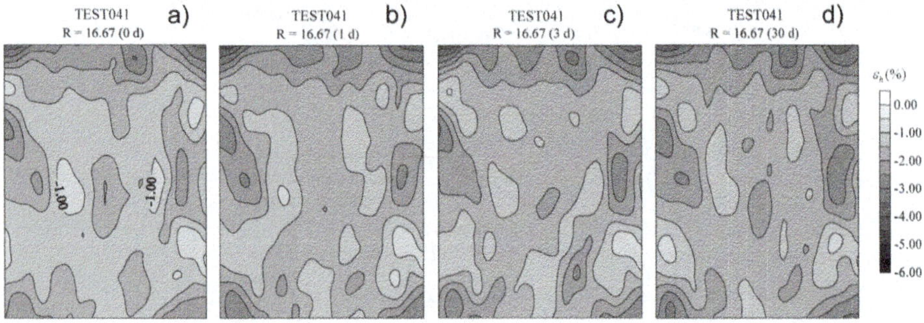

Figure 15. Local horizontal strain contours from 30 day-long sustained loading at R = 16.67: a) at the start of sustained loading; and after b) 1 day; c) 3 days; and d) 30 days

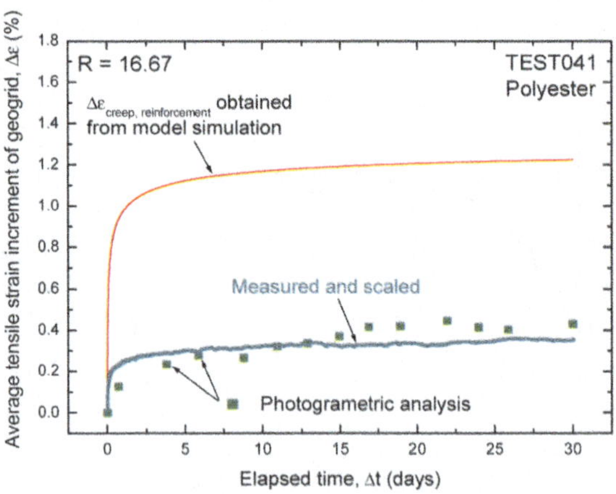

Figure 16. Comparison of average tensile strain in the geogrid between during sustained loading of reinforced sand and during sustained loading of a geogrid alone

described above. These strain values are not very different from the average horizontal strain increments of specimen measured with external sensors (i.e., gap sensors) located at the levels of two geogrid layers, shown in Fig. 14. Then, the time history of average geogrid tensile strain was obtained by scaling the time history of the average horizontal strain increment of specimen presented in Fig. 14 to be fitted to the data points obtained by the photogrametric analysis (Fig. 16).

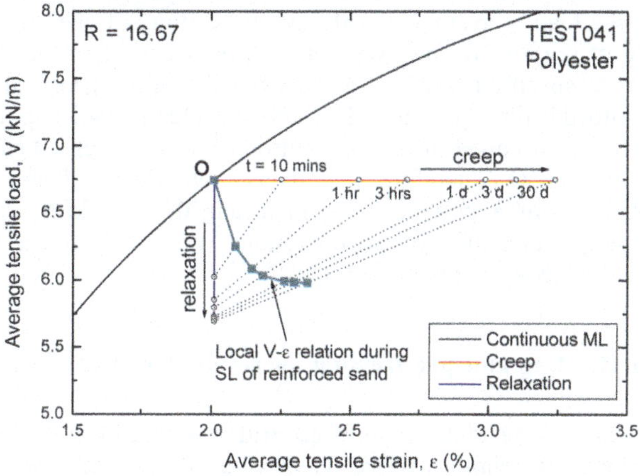

Figure 17. Estimated average tensile load - average tensile strain relation of PET geogrid during the 30 day-long sustained loading of reinforced sand at $R = 16.67$

The relation presented in Fig. 17 is the average tensile load - average tensile strain relation of the PET geogrid during continuous ML. The relation was obtained by a non-linear three-component model (Fig. 18) for a constant strain rate equal to the average value during the continuous ML that was obtained by the average horizontal strain increment based on the photogrametric method described above. The model is described in detail in Hirakawa et al. (2003) and Kongkitkul et al. (2004). The relation during the 30 day-long sustained loading of the reinforced sand that is also presented with data points in this figure was obtained as follows. First, the average load - average strain - time relations of the PET geogrid during sustained loading at a constant load level and load relaxation at a constant strain, both fixed to the values (i.e., at point **O**), were obtained by a numerical simulation based on the model. The time history of average tensile strain of the geogrid during the sustained loading at a fixed tensile load obtained

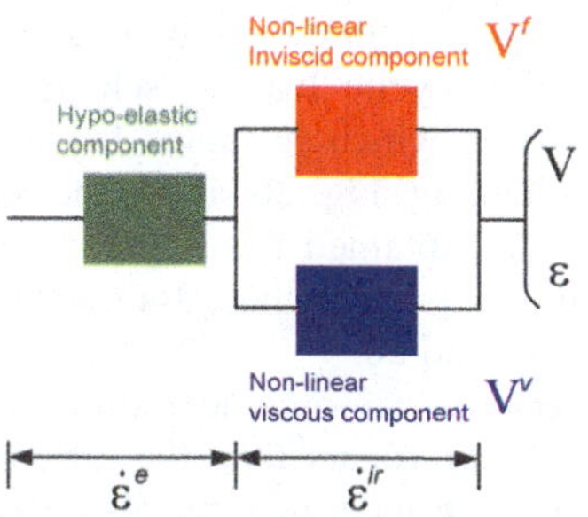

Figure 18. Non-linear three-component model modified for geosynthetic reinforcement (Hirakawa et al., 2003)

from the numerical simulation is presented in Fig. 16. It may be seen that, if it is assumed that the tensile load in the geogrid is kept constant, the creep tensile strain in the geogrid placed in the sand subjected to sustained loading at a constant load is over-estimated, likely largely with reinforced soil structures under usual working loads. Then, in Fig. 17, time-contours are also depicted from these two relations for sustained loading and load relaxation assuming that the contours are linear. This assumption has been found relevant according to numerical simulations based on the model (Kongkitkul et al., 2006). Finally, the average tensile load - average tensile strain relation of the geogrid during the sustained loading of reinforced sand was obtained by substituting the strain increments (starting from **O**) at specified times obtained from the time history of geogrid tensile strain (Fig. 16) into the time-contours. This relation clearly indicates that the geogrid tensile load during the sustained loading of reinforced sand decreased substantially from the initial value at point **O**. Therefore, it is likely that, when evaluating the long-term design tensile strength of a polymer geogrid arranged in backfill of reinforced soil structure under the typical static working conditions, it is overly conservative to assume that the tensile load in geogrid remains constant.

3.6 Effects of preloading and prestressing on residual vertical strain of reinforced sand

A method to decrease the residual strain of geogrid-reinforced soil structure is discussed below. Fig. 19a shows the relations between the averaged stress ratio, R, and the averaged vertical strain obtained from two PSC tests on Toyoura sand specimens reinforced with either a PET geogrid or a rough PB grid. The specimens were first monotonically loaded at a constant axial strain rate equal to 0.04 %/min up to $R = 26$ (herein denoted as preload, PL), followed by an immediate start of unloading toward $R = 10$. A cyclic loading scheme (described in the figure) was then applied, followed by further unloading toward $R = 1$, where another cyclic loading scheme was applied. Fig. 19b shows results from two similar tests, in which sustained loading was applied for 1,440 minutes at $R = 26$ (PL) prior to unloading. Fig. 20 shows the residual vertical strains plotted against the number of load cycles obtained from cyclic loading tests with a stress amplitude $\Delta q = 150$ kPa performed at $R = 10$ in these four tests. The following trends of behaviour may be seen from Figs. 19 and 20:

1. The effects of reinforcement stiffness on the R - averaged vertical strain relations are much less significant than the difference in the reinforcement stiffness. Furthermore, this trend is stronger than the R - averaged lateral strain relations presented in Fig. 4.
2. In both cases with and without pre-sustained loading at $R = 26$, the residual vertical strains developed by cyclic loading at $R = 10$ are generally very small, while the effects of the stiffness of reinforcement are negligible despite a large difference. This result indicates that, when designed based on performance during service, the use of polymer reinforcement, which has relatively much lower stiffness when compared with the metal reinforcement, becomes warranted by applying such preloading as this. Moreover, by sustained loading at $R = 26$, the residual vertical strains that developed by cyclic loading at unloaded condition became nearly zero, resulting in nearly elastic behaviour of reinforced specimen, equally with both types of reinforcement.

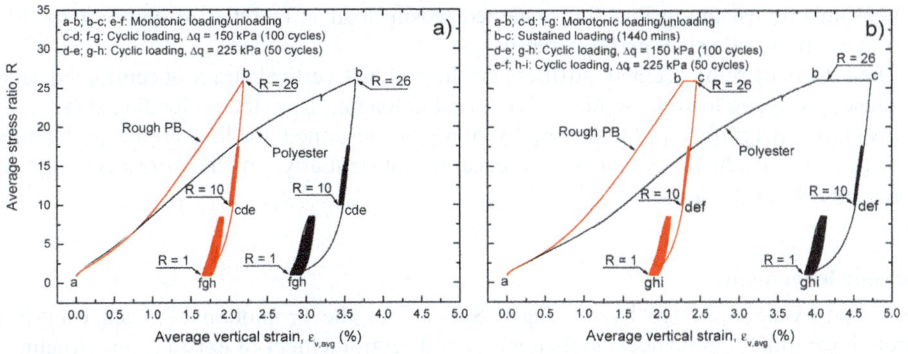

Figure 19. Average stress ratio - average vertical strain relations during loading and unloading: a) without sustained load at the maximum load; and b) with sustained load at the maximum load

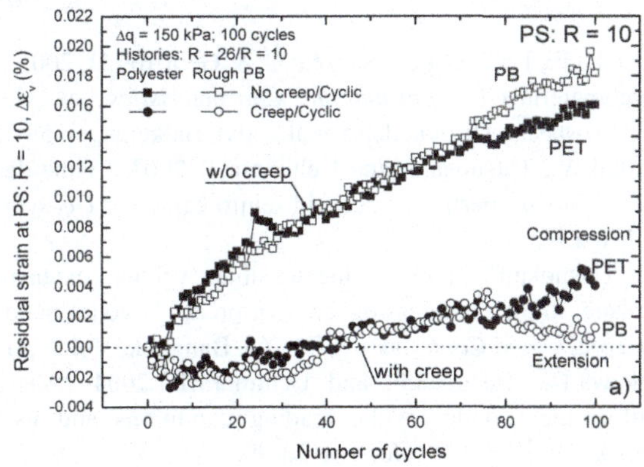

Figure 20. Residual averaged vertical strain increments developed by cyclic loading at R = 10 (unloaded stage)

4. CONCLUSIONS

The following conclusions can be derived:

1. PSC sand specimens reinforced with a stiffer reinforcement exhibited a stiffer stress-strain relation during primary loading in PSC tests. However, the difference in the stiffness was substantially smaller than the difference in the reinforcement stiffness.
2. Significant viscous effects on the stress-strain behaviour were also observed in reinforced PSC tests: a) when the strain rate was changed stepwise; b) during sustained loading or stress relaxation stages; and c) immediately after loading was restarted following a sustained loading stage. There is an apparent similarity in the viscous response in the vertical stress - vertical strain relation between unreinforced sand and sand reinforced with different types of reinforcement.
3. Tensile load activated in the geogrid reinforcement arranged in the sand specimen subjected to sustained loading at a fixed vertical load decreased with time, becoming substantially lower than the initial value at the start of sustained loading. This result indicates that the current method to evaluate the long-term design tensile strength of

geosynthetic assuming that the sustained tensile load is maintained constant for life time is on the safe side, perhaps overly.

4. The effects of reinforcement stiffness on the residual vertical strain of reinforced sand caused by cyclic loading applied after partial unloading from the preloading stage were generally insignificant. Moreover, by applying sustained loading at the preloading stage, the residual vertical strain became substantially smaller irrespectively of reinforcement stiffness.

Acknowledgement

This study was supported by the Japan Society for the Promotion of Science (JSPS) through the grant: "Advanced application of soil reinforcement of existing soil structures and construction of highly-earthquake resistant and environment-friendly soil structures".

Reference:

Di Benedetto,H., Tatsuoka,F., Lo Presti,D., Sauzéat,C. & Geoffroy,H. 2005. "Time effects on the behaviour of geomaterials", *Deformation Characteristics of Geomaterials: Recent Investigations and Prospects*, Di Benedetto et al. (eds), Balkema, pp.59-123.

Hirakawa,D., Kongkitkul,W., Tatsuoka,F. and Uchimura,T. 2003. "Time-dependent stress-strain behaviour due to viscous properties of geogrid reinforcement", *Geosynthetics International*, Vol.10, No.6, pp.176-199.

Kongkitkul,W. and Tatsuoka,F. 2004. "Inextensible versus extensible reinforcements; performance in plane strain compression on reinforced Toyoura sand", *Proc. of the 15th Southeast Asian Geotechnical Conference (SEAGC)*, Bangkok, Vol.1, pp.503-508.

Kongkitkul,W., Hirakawa,D., Tatsuoka,F. and Uchimura,T. 2004. "Viscous deformation of geosynthetic reinforcement under cyclic loading conditions and its model simulation", *Geosynthetics International*, Vol.11, No.2, pp.73-99.

Kongkitkul,W. and Tatsuoka,F. 2005. "Viscous deformation of geogrid-reinforced sand in plane strain compression", *Proc. of the 16th International Conference on Soil Mechanics and Geotechnical Engineering*, Osaka, Vol.2, pp.1071-1074.

Kongkitkul,W., Hirakawa,D. and Tatsuoka,F. 2005. "Behaviour of geogrid-reinforced sand subject to sustained loading in PSC", *Geosynthetics and Geosynthetic-Engineered Soil Structures; Contributions from the Symposium Honoring Prof. Robert M. Koerner*, Ling et al. (eds), McMat2005, Baton Rouge, Louisiana, USA, pp.251-280.

Kongkitkul,W., Tatsuoka,F. and Hirakawa,D. 2006. "Rate-dependent load-strain behaviour of geogrid arranged in sand under plane strain compression", *Soils and Foundations* (submitted).

Qiu,J.-Y., Tatsuoka,F. and Uchimura,T. 2000. "Constant pressure and constant volume direct shear tests on reinforced sand", *Soils and Foundations*, Vol.40, No.4, pp.1-17.

Tatsuoka,F. 2004. "Effects of viscous properties and ageing on the stress-strain behaviour of geomaterials." *Geomechanics Testing, Modeling and Simulation, Proc. of the 1st GI-JGS workshop*, Boston, ASCE SPT No. 143, Yamamuro & Koseki (eds), pp.1-60.

Tatsuoka,F., Hirakawa,D., Shinoda,M., Kongkitkul,W. and Uchimura,T. 2004. "An old but new issue; viscous properties of polymer geosynthetic reinforcement and geosynthetic-reinforced soil structures", Keynote Lecture, *Proc. of the 3rd Asian Regional Conference on Geosynthetics (GeoAsia 2004)*, Seoul, pp.29-77.

Soil Stress-Strain Behavior: Measurement, Modeling and Analysis
Geotechnical Symposium in Roma, March 16 & 17, 2006

RESIDUAL EARTH PRESSURE ON A RETAINING WALL WITH SAND BACKFILL SUBJECTED TO FORCED CYCLIC LATERAL DISPLACEMENTS

Daiki HIRAKAWA[1)], Minehiro NOJIRI, Hiroyuki AIZAWA and Fumio TATSUOKA
Department of Civil Engineering, Tokyo University of Science, Japan
e-mail[1)]: dhirakaw@rs.noda.tus.ac.jp

Takashi SUMIYOSHI
Institute of Civil Engineering, Tokyo Metropolitan Government, Japan

Taro UCHIMURA
Department of Civil Engineering, the University of Tokyo, Japan

ABSTRACT

A pair of about 11 m-high soil retaining walls of an U-shaped underground reinforced concrete (RC) structure in Tokyo exhibited a large residual inward (i.e., toward the active side) displacement with potential structural damage, which became 18 cm between the tops of the two walls about three years after its completion. Noticeable settlements of the backfill were observed behind the walls. A series of small-scale model tests was performed in the laboratory to understand this field behaviour. The results from in-situ investigation and model tests showed that this wall behaviour can be attributed to a gradual increase in the residual lateral earth pressure, resulting from cyclic lateral displacements of the walls caused by a small number of relatively large seasonal thermal cyclic displacement of the RC wall facing and bottom slab of the structure, not by a great number of relatively small daily displacement. Three factors for the mechanism of this wall behaviour (i.e., ratcheting, cyclic hardening and cyclic loading-induced residual deformation of the backfill) were identified and analyzed based the model test results. The settlement in the backfill observed in the model tests is consistent with the field behaviour.

1. INTRODUCTION

An U-shaped underground reinforced concrete (RC) structure that was constructed as an open-cut to accommodate over-passing roads has two about 11 m–high soil retaining walls on the opposite sides with the backfill of sandy soil (**Figure 1**; Sugimoto et al., 2003; Sumiyoshi et al., 2005). Two side roads were constructed on the backfill immediately behind the RC walls. During the construction of the structure, the two walls were supported with horizontal steel struts. After the backfill was filled and then the steel struts were removed, the 5th section of the walls started exhibiting overturning displacements toward the active side, which gradually increased with time. The wall displacement was actually rotation about the bottom of the facing. **Figure 2a** shows the time-histories of the lateral displacement measured between the tops of the two walls (at

Hoe I. Ling et al. (eds.), Soil Stress-Strain Behavior: Measurement, Modeling and Analysis, 865–874.

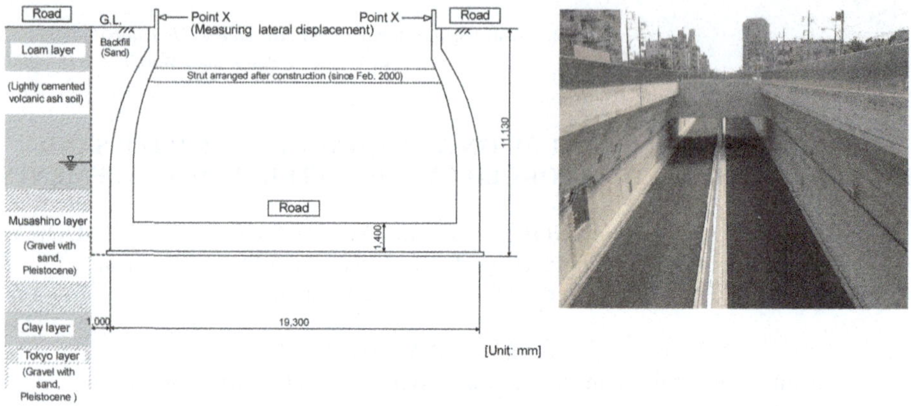

Figure 1. Cross-section of the RC U-shaped soil retaining wall at the 5th section (Sugiyama et al., 2003; Sumiyoshi et al., 2005)

point X in Fig.1) and the surrounding temperature. The displacements toward the active side between the tops of the two walls became 18 cm about three years after the completion of the structure with a possibility of structural damage if the residual displacement would continue increasing. Moreover, the backfill behind the walls exhibited noticeable settlements that increased as approaching the back of the walls. The in-situ investigation and numerical back analysis of the deformation of the walls indicated that the residual lateral earth pressure continued increasing at a decreasing rate with time after the removal of the struts and the earth pressure coefficient, K, acting on the facing reached 0.72 (Sugimoto et al., 2003). To prevent the structural damage, a permanent strut was installed February 2000, as shown in Fig. 1.

Figures 2b and **2c** show the relationships between the lateral displacement between the tops of the two walls and the surrounding temperature (presented in Fig. 2a) during a single day and about two years (Sumiyoshi et al., 2005). It may be seen that the wall top cyclically displaced in the lateral direction corresponding to daily and seasonally temperature changes. The average double amplitude (*DA*) of seasonal cyclic displacement of a single wall, δ, was about 20 mm × 0.5= 10 mm compared to the wall height, H, equal to 11 m; i.e., $\delta(DA)/H$ = about 0.09 %, which was much larger than that of daily cyclic displacement for a single wall equal to about 1.6 mm × 0.5= 0.8 mm; i.e., $\delta(DA)/H$ = about 0.07%. Based on these facts, it was considered that this wall behaviour was caused by cyclic thermal deformation (i.e., contraction and expansion) of the RC wall facing and bottom slab of the structure due to daily or seasonal changes in the temperature. It was not known however whether a great number of relatively small daily cyclic wall displacement or a small number of relatively large seasonal cyclic wall displacement or both is (are) the cause for the gradual increase in the residual earth pressure. Furthermore, the mechanism of the increase in the residual earth pressure by forced cyclic lateral wall displacement with relatively small amplitude was not known.

In view of the above, a series of model loading tests on a small-scale retaining wall was performed in the laboratory to investigate the following issues:

1) The effects of the amplitude of cyclic wall displacement and the number of cyclic loading on the development of residual earth pressure and associated residual active wall displacement.
2) The mechanism of the development of residual earth pressure by cyclic lateral wall displacements; and
3) The comparison of the wall behaviour between when subjected to forced cyclic lateral displacements and when subjected to monotonic active and passive displacements.

2. MODEL TEST PROCEDURES

A 505 mm-high model wall made of a full-height rigid facing was set up in a plane strain sandbox (1,800 mm-long × 400 mm-wide × 800 mm-high: **Figure 3**). The bottom of the facing was placed on a pair of hinge structures, which were the center of wall rotation. The back face of the model facing was made rough by gluing sandpaper #150. The model wall was equipped with nine two-component (shear and normal) load cells to measure the distribution of the earth pressure, which will be reported in the near future. The facing was cyclically displaced laterally at a constant rate of 0.4 mm/min at a hinge located 115 mm below the top of the facing. The model backfill (1,295 mm-long × 595 mm-high × 400 mm-wide) was prepared by pluviating air-dried particles of Toyoura sand throughout air using multiple sieves. The target initial relative density, D_r, was 90 % (γ_d = about 1.60 g/cm^3). Horizontal thin layers of colored Toyoura sand were arranged in the model backfill to observe the deformation of the backfill (including shear bands) through the transparent Acrylic sidewall of the sand box.

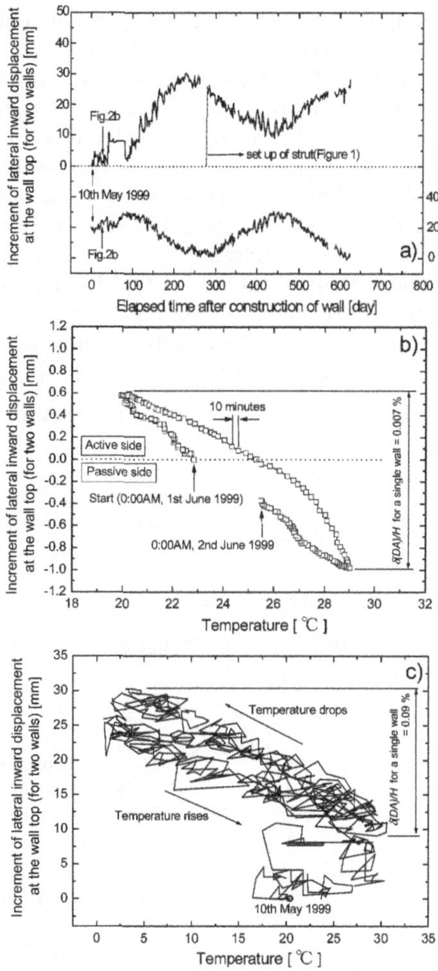

Figure 2. a) Time-histories of lateral displacement between the wall tops at point X in Fig.1 (for two sides of wall) and surrounding temperature; and their relations during: b) a single day & c) for about two years.

Several amplitudes of cyclic lateral wall displacement for a single wall in a range of $\delta(DA)/H$ from 0.02 % to 0.5 %, which nearly covers the daily and seasonal values of the prototype wall, were applied to the model wall from the K_0-state. The following two types of wall rigidity were assumed:

Rigid wall: The model facing was cyclically loaded with zero residual displacement, which means that the model wall is rigid against changes in the earth pressure. The results in this case are reported in this paper.

Elasto-plastic wall: Actual prototype walls exhibit residual active displacements when subjected to a residual increase in the lateral earth pressure (as shown in Fig. 2a). The results from the model tests assuming a non-rigid wall having elasto-plastic displacement characteristics will be reported in the near future.

The settlements of the crest of the backfill were measured at in total five locations with laser displacement transducers (Fig. 3). The lateral load acting to the facing was measured with two load cells arranged at the top and bottom hinges (Fig. 3). The earth pressure coefficient, $K= 2\cdot Q/(\gamma_d \cdot H^2)$, where Q is the total earth pressure per wall width; γ_d is the unit weight of the backfill (= about 1.60 g/cm^3); and the wall height (= 505 mm).

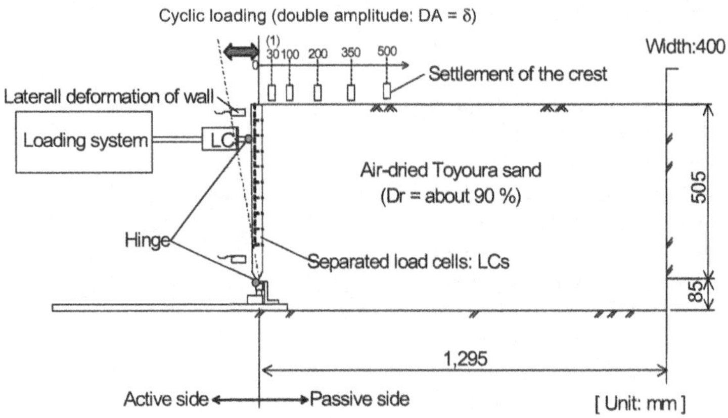

Figure 3. Model retaining wall.

3. TEST RESULTS AND DISCUSSIONS

Overall behaviour of the model wall when subjected to forced lateral cyclic displacements

Figures 4a and **4b** show two typical time-histories of total earth pressure coefficient, K, when $\delta(DA)/H$ = 0.02 % and 0.08 %, more-or-less simulating the average daily and seasonal cyclic displacements of the prototype wall. The values of K were obtained from the measurements of the load cells at the top hinge where cyclic lateral displacements were applied and the bottom one on which the facing was placed. **Figures 5a** and **5b** show the relationships between the K value and δ/H, corresponding to Figs. 4a and 4b. **Fig. 5c** is a close-up of the major part of Fig. 5b. In these figures, the results from two

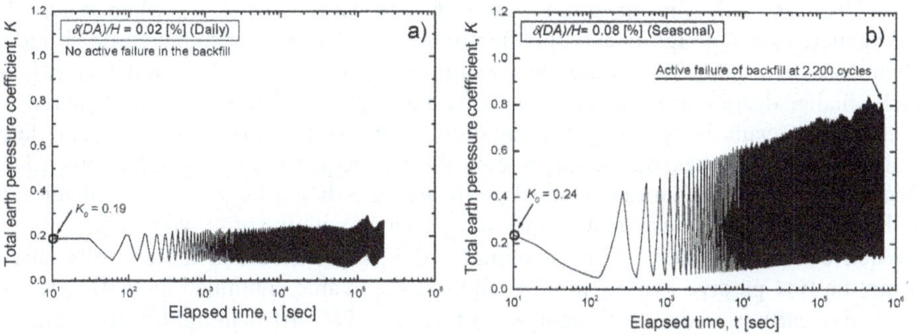

Figure 4. Time-histories of total earth pressure coefficient K: a) $\delta(DA)/H = 0.02$ % (simulating daily cyclic loading) as double of daily cyclic loading, and b) $\delta(DA)/H = 0.08$ % as seasonally cyclic loading (simulating seasonal cyclic loading).

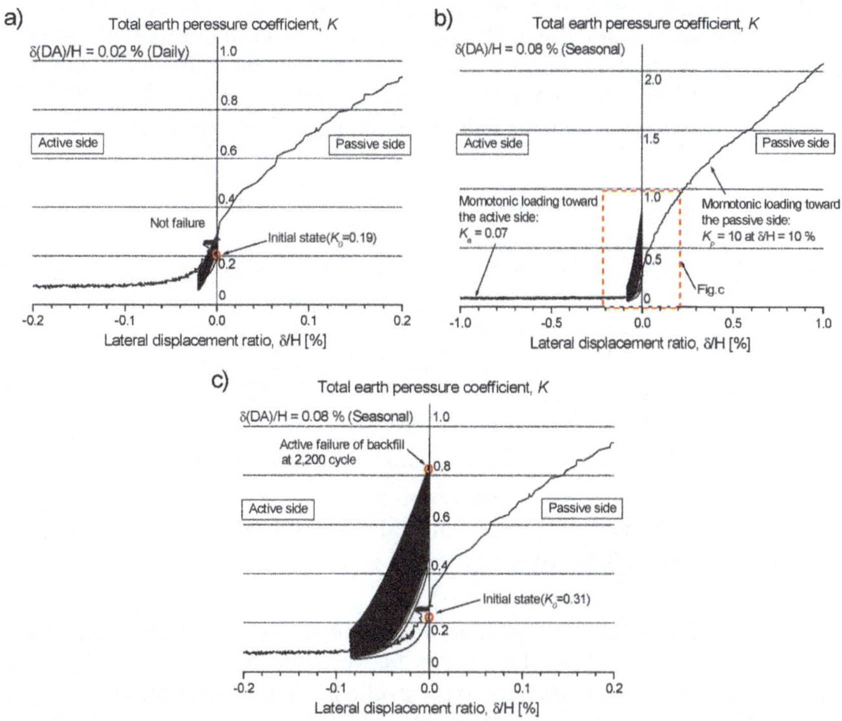

Figure 5. Relationships between K and δ/H (positive at the passive side): a) $\delta(DA)/H = 0.02\%$ (simulating daily cyclic loading), and b) & c) $\delta(DA)/H = 0.08\%$ (simulating seasonal cyclic loading).

monotonic loading (ML) tests that were continued towards the active and passive failure states are also presented. It may be seen from Figs. 4 and 5 that the development of residual earth pressure by cyclic wall displacements depends on the amplitude of cyclic lateral displacement. The maximum value of K in each cycle, K_{max}, when $\delta(DA)/H$ = 0.08 % increased at a high rate, while the minimum value, K_{min}, also increased noticeably. When $\delta(DA)/H$ = 0.02 %, on the other hand, the increase in the K_{max} and K_{min} values is much smaller despite a larger number of cyclic loading for a given period (about four times). This result indicates that a smaller number of relatively large cyclic lateral displacement, as the seasonal changes with the prototype wall (Fig. 2), has much larger effects on the development of residual earth pressure than a larger number of relatively small cyclic lateral displacements, as the daily change with the prototype wall.

When $\delta(DA)/H$ = 0.08 %, the maximum of K, K_{max}, exceeded the K_0 value already during the first passive loading process and the K_{max} value continued increasing towards the passive earth pressure coefficient, K_p, attained at δ/H= 10 % in the ML test. Although it is far below K_p, the K_{max} reached even 1.0 after many cycles. This result suggests that the earth pressure coefficient, K, of the prototype wall (Fig. 1) may reach 1.0 within its lifetime. When subjected to such large earth pressure, walls designed based on the active earth pressure either exhibits large, perhaps intolerably large, active displacements when the wall is non-rigid, or is structurally damaged when the wall is rigid. Indeed, the prototype wall (Fig. 2) was designed based on K= 0.3 – 0.4 (Sugiyama et al., 2003).

Figure 6 shows the time-histories of the settlement of the backfill crest 30 mm from the back of the model facing (point 1 in Fig. 3). The settlement when $\delta(DA)/H$ = 0.08 % was very large and is consistent with the field observation (i.e., Sumiyoshi et al., 2005).

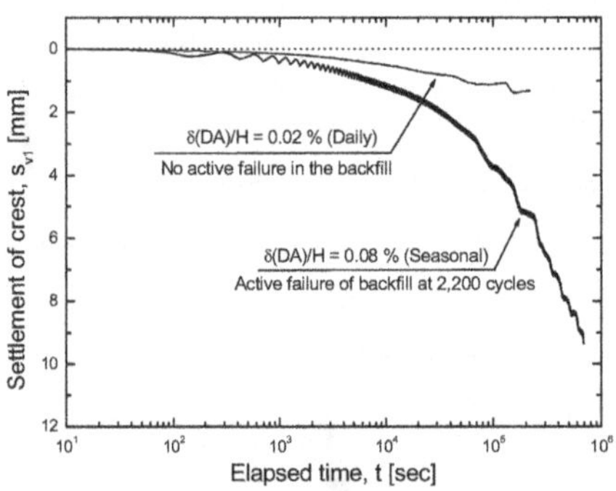

Figure 6. Time-histories of settlement at the backfill crest (30 mm from the back of model facing), the rigid facing.

The effect of cyclic amplitude of wall deformation on the increase of earth pressure

Figure 7 summarises of the relationships between the K_{max} value at the respective number of cycle (N) and $\delta(DA)/H$ from the cyclic loading model tests on the rigid wall. The value of K_0 is also indicated in this figure. The solid data points indicate the moment when the active failure plane developed in the backfill was noted. It may be seen that the values of K_{max} increases by cyclic loading at a rate that increases with an increase in $\delta(DA)/H$. In Fig. 7, the results from the 1g model tests by England et al. (2000), similar to those performed in the present study, and those from centrifuge tests by Ng et al. (1998), both using Leighton Buzzard silica sand with particle sizes between 90 – 150 μm,

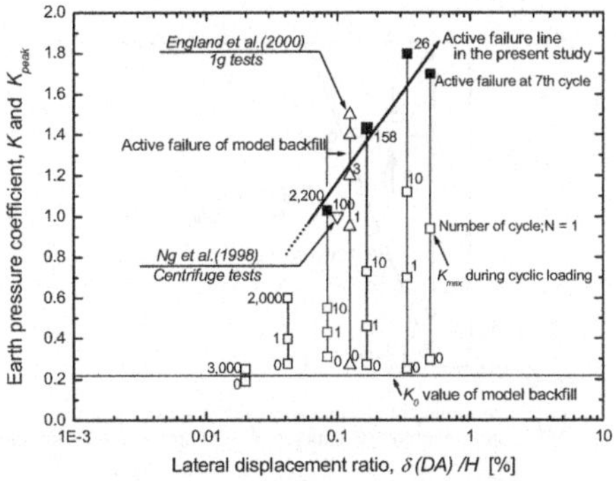

Figure 7. Relationships between K and $\delta(DA)/H$: comparison between the present study model loading tests and the previous studies for rigid wall.

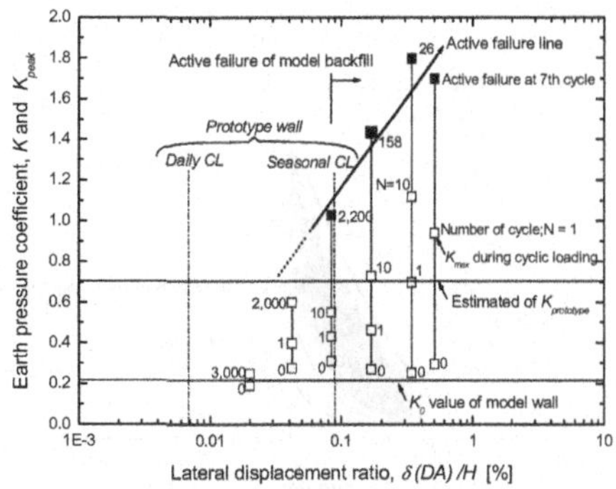

Figure 8. Relationships between K and $\delta(DA)/H$: comparison between the present study model loading tests for rigid wall and the prototype wall.

are also plotted. These previous studies were performed linked to the thermal loading problem of integrated bridge. The results from the present study (1g on Toyoura sand) and the previous studies are consistent with each other.

Figure 8 compares the results from the present study with the behaviour of the prototype wall, for which the estimated maximum residual value, $K_{prototype}$, is equal to 0.72, and the average daily and seasonal values of $\delta(DA)/H$ for single wall are equal 0.007 % and 0.09 %, are indicated. In the model tests on the rigid wall, the development of residual earth pressure was very small even after many cycles when $\delta(DA)/H$ was less than about 0.02, while the development when $\delta(DA)/H =$ about 0.1 % was significant, similar to the prototype wall. Therefore, considering also the trends of the settlement of the backfill (Fig. 6), it can be concluded that the development of relatively large residual active displacement and associated settlement with the prototype wall can be attributed to seasonal thermal displacements of the wall.

Mechanism of the development of residual earth pressure by cyclic wall displacement

The increase in the residual earth pressure by cyclic lateral wall displacements, which results into the development of residual active wall displacements when the wall is not rigid, are due to a mechanism consisting of the following three factors (**Figure 9**):

1) ratcheting in the backfill deformation;
2) cyclic hardening of backfill; and
3) cyclic loading induced residual deformation of backfill.

The first factor, the ratcheting in the backfill deformation, is illustrated in **Figure 10**, namely:

1) The facing displacement toward the active side results into a settlement of the active zone in the backfill (Fig. 10b). The first event of this process is denoted by relation **a** → **b** in Fig. 9.
2) When the wall is subsequently forced to displace toward the passive side, the active zone in the backfill cannot return to the original location due to different mechanisms

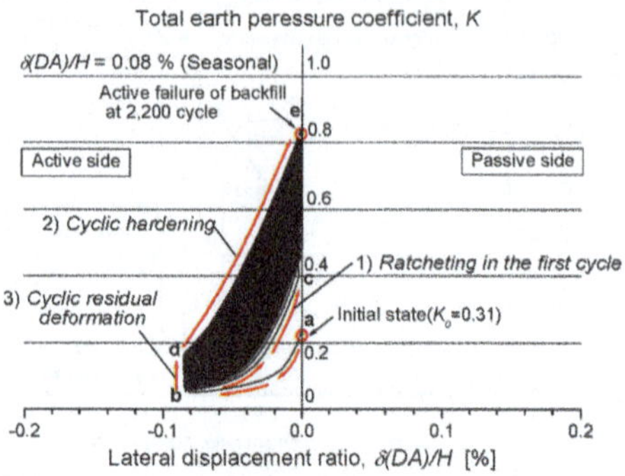

Figure 9. Three factors of the mechanism for an increase in the residual earth pressure when subjected to cyclic lateral displacements, illustrated using the test results presented in Fig. 5c.

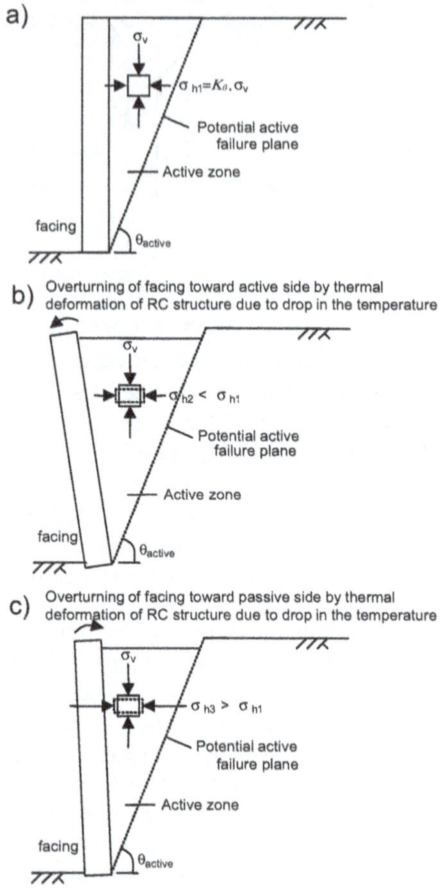

Figure 10. Ratcheting mechanism in the wall subjected to cyclic lateral displacements.

between the active and passive earth pressure developments, which results in an increase in the lateral earth pressure (Fig. 10b). The first event of this process is relation **b** → **c** in Fig. 9.

3) The ratcheting process described above is repeated during subsequent cycle loading at a rate that decreases with an increase in the number of loading cycles.

It seems that most of the passive wall displacement, which takes place subsequently to the preceding active wall displacement is absorbed by the deformation of the backfill outside the active shear band (having a thickness of order of ten times the mean diameter), while large part of the active wall displacement is absorbed by the deformation of the active shear band. Then, cyclic displacements of wall either in the fixed range of displacement when the wall is rigid, or in a range shifting toward the active side when the wall is non-rigid result into a gradual increase in the active displacement of the active zone. This means that, even if the maximum displacement during the cyclic loading is smaller than the displacement when active failure takes place in a continuous active ML test, the maximum earth pressure can exceed the K_0 value increasing towards the passive value while the active failure takes place in the backfill. In fact, in the cyclic loading tests with a displacement $\delta(DA)/H$ ranging between 0.08 and 0.5 %, an active failure shear band, as observed in the continuous active ML test, developed despite that the maximum earth pressure observed in the cyclic loading tests was far larger than the active earth pressure observed in the continuous active ML test. The angle of active shear band that developed in these cyclic loading tests was independent of $\delta(DA)/H$, and close to the one observed in the continuous active ML test.

The second factor, cyclic hardening of backfill, can be noted by a significant increase in the stiffness of the backfill by cyclic loading (i.e., a change from relation **b** → **c** to relation **d** → **e** in Fig. 9). By this factor, the maximum earth pressure coefficient, K_{max}, increases by cyclic loading even when the minimum value, K_{min}, remains constant. This factor is due to such material property that the stress-strain behaviour of unbound granular material becomes more elastic, thus the stiffness increases, when subjected to continuous cyclic straining for a fixed range of strain (Tatsuoka et al., 2003).

The last factor, cyclic loading-induced residual deformation of backfill, can be noted from an increase in the minimum value of K, K_{min}, by cyclic loading (i.e., from point **b** to pint **d** in Fig. 9). It is considered that, if the wall is subjected to cyclic earth pressure for a fixed range of K below 1.0, the active residual displacement increases during cyclic loading. It seems that this trend of behaviour results from such property of soil that the residual shear strain increases when subjected to cyclic shear stresses in the direction of currently acting neutral shear stress.

4. CONCLUSIONS

From the full-scale behaviour of a prototype structure and the results of model loading tests, the following conclusions can be derived:

1) The earth pressure can increase gradually when a RC soil retaining wall is subjected to cyclic lateral displacement caused by thermal deformation of the wall structure due to cyclic changes in the temperature even if the cyclic wall displacement is relatively small and remains on the active side. In the case of the prototype structure reported in this paper, a small number of relatively large seasonal temperature change was responsible for the development of relatively large residual active displacement of the wall while the effects of a great number of daily relatively small temperature changes can be considered negligible.
2) Even when the wall is subjected to cyclic lateral displacements that remain on the active side, the earth pressure can exceed the K_0 value while increasing towards the passive value while the backfill can exhibit active failure.
3) The mechanism for the development of residual earth pressure, and also residual active wall displacements when the wall is not rigid, by cyclic lateral wall displacements consists of three factors; namely, ratcheting in the backfill deformation, cyclic hardening of the backfill and cyclic loading-induced residual deformation of the backfill.

REFERENCES

1) England, G, L., Neil, C, M. and Bush, D, I. Integral Bridges, (2000), A fundamental approach to the time-temperature loading problem, *Thomas Telford.*
2) Ng, C., Springman, S. and Norrish,A. (1998), "Soil-structure interaction of spread-base integral bridge abutments", *Soils and Foundations*, Vol.38, No.1, pp.145-162.

3) Nojiri,M., Kasahara,T., Hirakawa,D., Tatsuoka,F., Sumiyoshi,T., Uchimura,T. and Arai,T. (2005), "Residual deformation and earth pressure of reraining wall subjected to multiple horizontal cyclic loading", *Proc. 40th Japan National Conf. on Geotechnical Engineering, Hakodate*, pp.1753-1754 (in Japanese).
4) Sugimoto, T., Sumiyoshi, T., Sasaki, S., Hiroshima, M. and Yamamura, H. (2003), "Increase in the earth pressure by cyclic displacement of wall", *Proc. of Geotechnical Engineering in Urban Construction*, The Sino-Japanese Symposium on Geotechnical Engineering, pp.189-196.
5) Sumiyoshi, T. (2005). Behaviour of U-type retaining wall subjected to cyclic displacement", *Annual report, Institute of Civil Engineering, Tokyo Metropolitan Government* 2005 (in Japanese), pp.69-78.
6) Tatsuoka,F., Masuda,T. and Siddiquee,M.S.A. (2003): Modelling the stress-strain behaviour of sand in cyclic plane strain loading, Geotechnical and Environmental Engineering, *Journal of Geotechnical and Environmental Engineering, ASCE,* Vol.129, No.6, June 1, pp.450-467.

Soil Stress-Strain Behavior: Measurement, Modeling and Analysis
Geotechnical Symposium in Roma, March 16 & 17, 2006

EXPERIMENTAL ESTIMATION OF ADFREEZE SHEAR REINFORCEMENT AT JOINT BETWEEN FROZEN SOIL AND UNDERGROUND STRUCTURES

Katsuhiro Uemoto, Teru Yoshida & Jeawoo Lee
Kajima Technical Research Institute
Kajima Corporation, Tokyo, Japan
e-mail: uemoto-k@kajima.com

ABSTRACT

Recently artificial ground freezing has become more frequently used for construction of underground facilities in urban areas because it makes underground excavation possible without lowering the groundwater level. To use artificial ground freezing in a underground excavation project, the adfreeze shear strength at the joint between an underground structure and frozen soil should be large enough to ensure structural stability. For that reason, the authors have previously proposed a method for enhancement of adfreeze shear strength by the use of stud connectors. This paper addresses a series of small-scale laboratory tests and large-scale verifying experiments that help formulate reinforcing effects of the proposed stud connectors on the adfreeze shear strength.

1. INTRODUCTION

As constructing underground facilities in urban areas, the application of dewatering or open-cut method is rarely beneficial due to their negative influence on the ground (e.g., traffic restraint and ground settlement). Moreover, the deeper the construction depth, the less economical these methods become. Artificial ground freezing has been recognized as an effective substitute to provide temporary support and water cutoff so that deep excavation is possible without concern on treatment of water inflow. Additionally it has no long-term effect on the subsurface environment because once the refrigeration unit is turned off, the frozen area gradually melts restoring the groundwater and the soil to their original states.

Fig. 1 shows an illustrative application of artificial ground freezing to underground excavation in which arch-shaped water tight frozen earth barriers are formed to protect the shield tunnels for the construction of diverging or junction lines between parallel road tunnels. In this case, particular care should be paid to adfreeze shear strength between the frozen soil and tunnel segments rather than the strength of frozen soil itself which

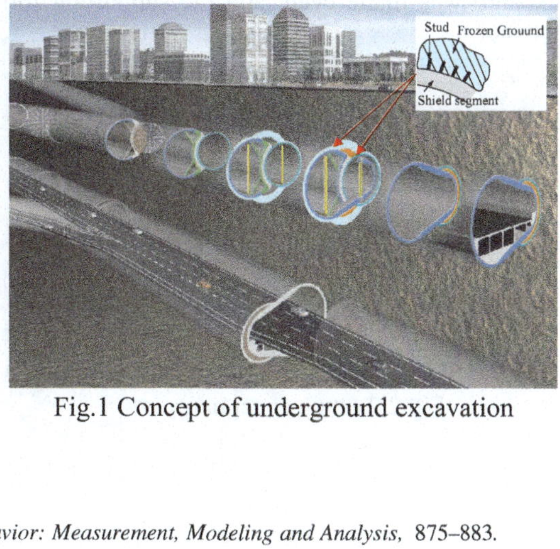

Fig.1 Concept of underground excavation

Hoe I. Ling et al. (eds.), Soil Stress-Strain Behavior: Measurement, Modeling and Analysis, 875–883.

usually exhibits high strength and water tightness almost as good as those of weak concrete. The adfreeze shear strength can be sufficiently achieved provided that the frozen soil is adherently joined to the underground structure [1]. However, the existent of greasy material or separation of rustproof coating on the outer surface of the underground structure may cause significant reduction of the adfreeze shear strength. In addition, the bending tensile failure along the joint between the frozen soil and the surface of structure may occur due to either of the stress redistribution in the frozen soil or the relative movement of underground structure during excavation. These lead the authors to propose a reinforcing method using stud connectors [2)-4)], hereinafter "studs", that have been used for composite structures of steel and concrete.

This paper presents the results of experimental studies to evaluate the reinforcing effect of studs. A simple equation for assessment of the resistance of studs is suggested based on a series of small-scale laboratory tests[5)]. And large-scale load tests exhibit validation of the suggested equation in practice.

2. Quantitative estimation of reinforcing effect of studs through small-scale laboratory tests

To examine the reinforcing effect of the proposed adfreeze shear reinforcement method and formulate the stud arrangement (penetration depth and installation spacing) as well as its reinforcement effect, a series of laboratory load tests have been carried out using the small-scale specimens made of steel plates with studs combined with the frozen soil.

2.1 Brief description of testing procedure

As shown in Fig. 2, the cubic specimens with a length of 100 mm for each side were formed between two restraint steel plates 19 mm thick with steel studs attached on the inner side of them. After assembling the restraint plates and installing a form between them that supported the cooling surface and its neighboring two surfaces (See Photo 1), the wet soil with its water content controlled to a target value was put into the matrix as the cooling surface faced downward on the cooling bed in a refrigerator, and then every surface except the cooling surface were insulated so that the freezing proceeded in the direction from the cooling surface toward the opposite surface. The specimens were kept in the refrigerator at –25 ℃ for 24 hours until they were completely frozen. After that, the surface facing opposite to the cooling surface (the one which allowed the frozen soil to expand) was made even, and the supporting form between the restraint plates was removed.

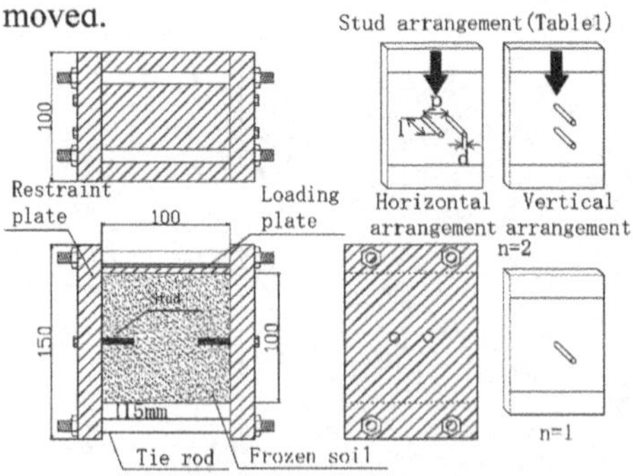

Fig.2 Testing specimen

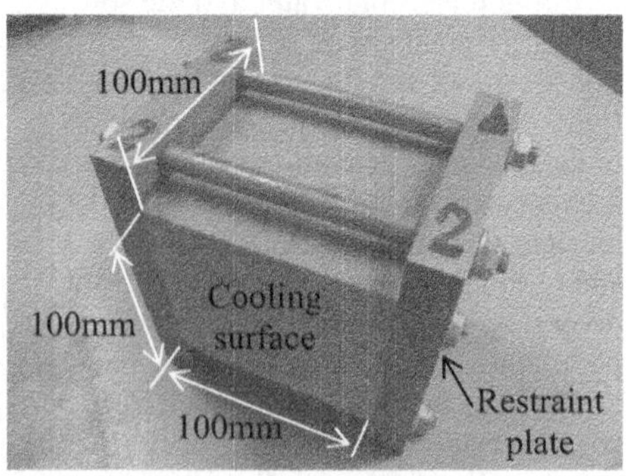

Photo. 1 Testing specimen

With an additional 24-hour curing at the experiment temperature, a loading plate (5-mm-thick steel plate) was placed on the top surface of the specimen and the downward loading was activated at a constant speed of 1 mm/min. At this time, the adfreeze resistance between the restraint plate and frozen soil was readily removed using a lubricant to simulate the actual behavior of frozen soil on greasy materials or separable rustproof coatings attached to the outer surface of underground structure. The studs were physically modeled by fixing ready-made metal hexangular bolts (full-threaded type) into the female threads readily created on the inner side of each restraint plate.
As illustrated in Fig. 2, one or two bolts were equally placed for both the inner sides of restraint plates so as to raise the reliability of testing results by averaging the measured values. The arrangement parameters and mechanical properties of studs are given in Table 1 through 2. Regarding the frozen soil, two testing temperatures were adapted for two different types of soil as shown in Table 3, which indicated that four types of soil properties exhibiting different levels of unconfined compressive strength had been tested (See Fig. 3).

Table 1. Testing conditions for stud model

	Setting condition
Number per side, n	1, 2
Diameter, d (mm) (external diameter of screw)	5, 8, 16
Arranging direction (when n=2)	Horizontal, vertical (refer to Fig. 2)
Center to center spacing, p (mm) (when n=2)	10, 25, 45
Length, l (mm)	2.5, 5, 10, 15, 20, 25, 40

Table 2. Mechanical properties of stud model

Standard	Tensile strength (N/ mm^2)	breakdown point (N/ mm^2)
JIS B 1180	400	240(d=5) to 320(d=8)

Table 3. Physical properties of frozen soil

Soil material	*Toyoura* sand		*Fujinomori* clay	
Testing temperature, T (℃)	-10	-20	-10	-20
Unconfined compressive strength, q_u (N/mm^2)	12.72	16.93	4.25	7.17
Water content, w (%)	20.4		44.9	
Moisture density, γ_t (g/cm^2)	1.80		1.67	
Specific gravity, Gs (g/cm^3)	2.64		2.69	
Degree of saturation, S_r (%)	67.0		90.9	

2. 2 Result from the small-scale laboratory tests

2.2.1 Mechanism of load transfer between the studs and frozen soil

During the series of small-scale laboratory tests, the deformation of studs appeared as three patterns as shown in Fig. 4 (Photo. 2 and 3, Fig. 7). All patterns (I -III) were observed for the frozen *Toyoura* sand and pattern I & II for the frozen *Fujinomori* clay (Fig. 7). Considering that these three deformation patterns looked similar to those of a horizontally-loaded pile with its head fixed[6], the resistance of stud can be recognized as being affected by its bending stiffness and strength, diameter, length as well as the deformation and strength characteristics of the frozen soil.

As an example for the relation between the relative displacement of frozen soil and load acting on stud, Fig. 5 shows the results obtained from the tests on a pair of studs whose diameter was 5mm with changing the stud length l under the fixed testing temperature T of -10 ℃ and the horizontal spacing, p, of 25 mm. (Although the displacement of the frozen soil herein indicated the displacement of the piston of the loading apparatus, it could be regarded as the relative displacement between the studs and frozen soil because the size of openings taking place in the back side of the studs nearly equaled the piston displacement). The frozen *Fujinomori* clay showed no definite peak load against the relative displacement that continually increased even until the body of frozen soil partially reached the lower tie rod. Regarding the frozen *Toyoura* sand, a distinct peak load was observed in the cases with Patterns II or III, and the load appeared to gradually grow as the displacement had increased in the case of Pattern I. By the way, the test result performed without stud exhibited extremely small increase of the load, which infers that the frictional resistance between the frozen soil and plates had rarely developed owing to the lubricant.

2.2.2 Effect of spacing of studs on resistance

Similar to the group pile effect in which a pile group exhibits smaller resistance than that of the sum of each individual pile in it, the shearing resistance for each stud in a pair of them is supposed to decrease as the spacing between them becomes narrower. From the point of view, a series of load tests were carried out to investigate the threshold spacing of studs that could cause the interference of multiple studs in a way that the maximum

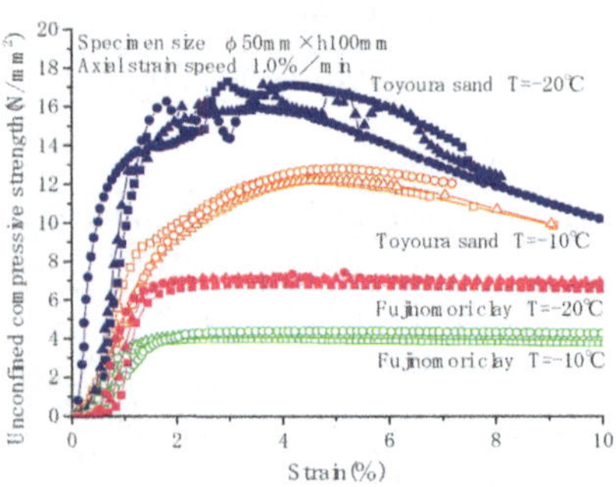

Fig. 3 Results of unconfined compressive strength test of the frozen soil

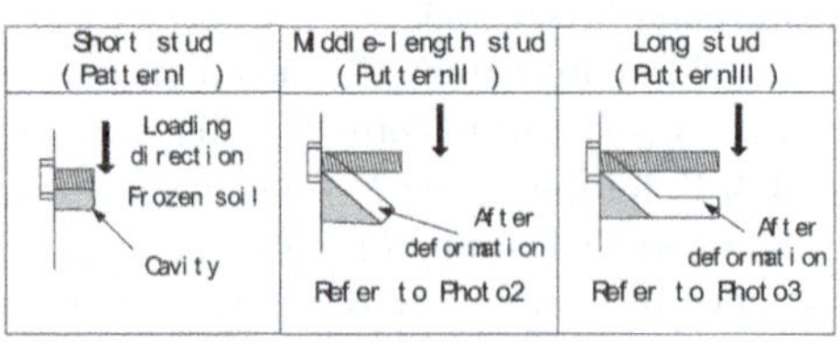

Fig. 4 Deformation types of studs

Photo. 2 Studs after experiment (*Fujinomori* clay)

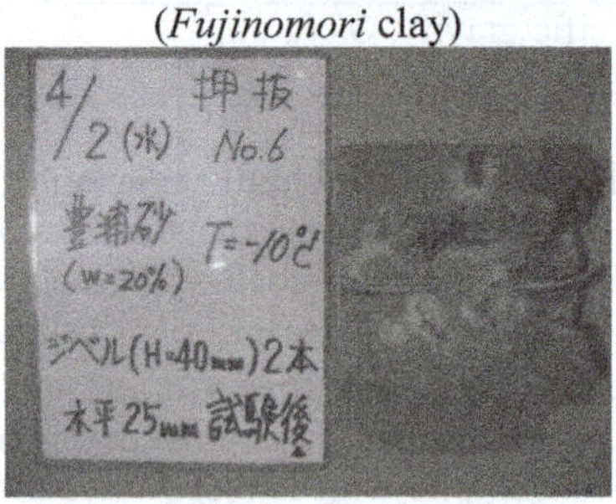

Photo .3 Studs after experiment (*Toyoura sand*)

resistances of a pair of studs, Q_u, in the frozen *Toyoura* sand formed at -10 ℃ had been compared in accordance with change of the horizontal or vertical spacing, p(herein the diameter of each stud, d, is 5 mm) (Fig. 6).

With the horizontal arrangement, at right angle with respect to the loading direction, the resistance of studs hardly appeared to decrease with decreasing p provided that the spacing, p, was no less than 10 mm, which was equivalent to two times the diameter (net distance between the studs was 1d when p = 2d). It can be then said that this arrangement can produce the resistance same as the simple sum of that of each stud without any influence of the interference. Under the vertical arrangement, along the loading direction, the resistance was affected by the stud length l, which is similar to the case of a composite structure of steel and concrete [2) through 4)]. Namely, in the range that the spacing, p, was no less than 25 mm (equivalent to 1*l*) for the stud with a length of 25 mm, and the range p ≧ 10 mm (equivalent to 1*l*) with *l* = 10 mm, a reduction of resistance was hardly observed. However, the resistance appeared to become lessened for p = 10 mm (0.4*l*) with *l* = 25 mm. It is thus necessary to leave a spacing of at least 1.0*l* for the two studs arranged along the loading direction to avoid the reduction of resistance due to their interference.

2.2.3 Effect of stud diameter d, length *l*, and shear strength of frozen soil (unconfined compressive strength, q_u)

In order to establish the equation for determining the design resistance of studs, a series of load tests had been carried out according to change of the stud length under the condi-

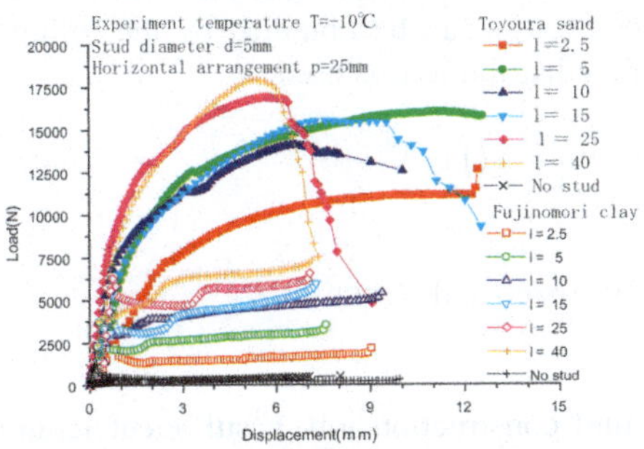

Fig. 5 Relation between load and displacement

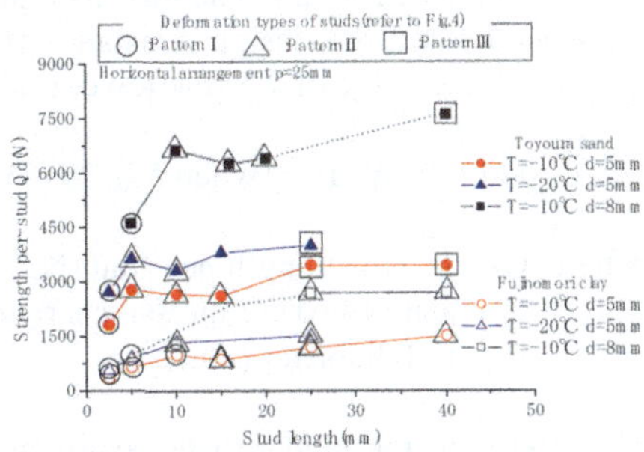

Fig. 7 Relation between strength per-stud and stud length

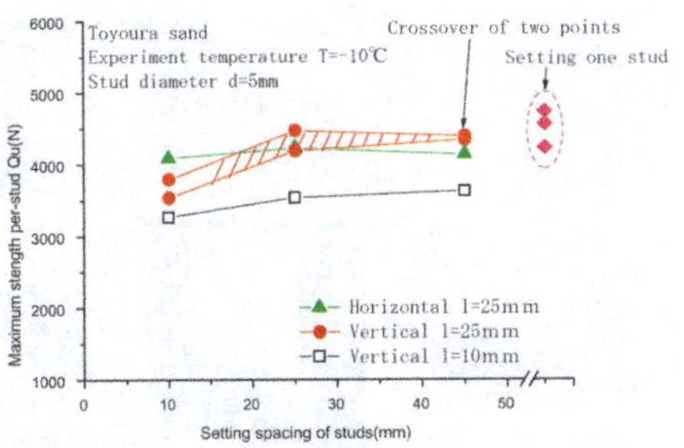

Fig. 6 Relation between maximum strength and setting spacing of studs

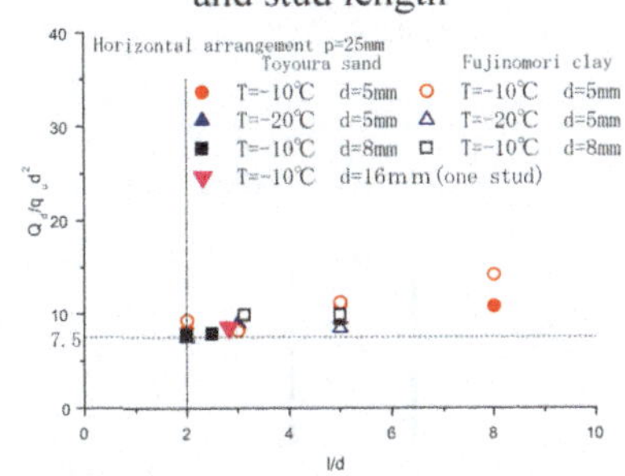

Fig.8 Relation between dimensionless strength and normalized stud length

tion that a pair of horizontally arranged studs whose diameter, d varied 5 to 8 mm were respectively placed in the four types of frozen soil, summarized in Table 3, with its fixed horizontal spacing, p of 25 mm. At this time, the allowable resistance Q_d was selected corresponding to the relative displacement, δ , of 0.5d on the load-displacement curve (Fig. 5) which was decided on the assumption of the safety factor of 2.0 for the ultimate resistances that had been observed at δ = 1.0d from the majority of previous load tests with the frozen *Toyoura* sand. Besides, even the stud with a diameter, d, differing from that in this study is expected to produce the strain distribution in the frozen soil near the connection between the stud and restraint plate similar to the one from this study within δ = 0.5d(supposing that the relative displacement δ is measured as a function of the diameter, d). In other words, the deformation of both frozen soil and stud seems to occur in a very similar fashion in spite of the varying diameter. This leads to the evaluation of design resistance at this point.

As shown in Fig. 7, the resistance per stud, Q_d, generally increased with the length *l* under the condition where the unconfined compressive strength of the frozen soil, q_u, and diameter, d, were fixed. (It appeared that Q_d did not increase monotonically with the length *l* in the frozen *Toyoura* sand, which might be owing to the alteration of deformation patterns as shown in section 2.2.1). Meanwhile, the increase associated with the length *l* was gentle for $l \geqq 2d$ and the values were almost constant. Also, for the same stud length *l*, the larger the value of q_u or d, the greater the resistance Q_d. The dimensionless values divided by $q_u d^2$ which has the same dimension as Q_d, are shown in Fig. 8 for the range of $1 \geqq 2d$. The dimensionless resistance $Q_d / q_u d^2$ generally shows a constant value, and the design resistance resistance, Q_d, can be obtained by the following formula if the value of 7.5 (the lower limit of its distribution) is used:

$$Qd = 7.5\, q_u d^2 \quad (\text{when } l \geqq 2d) \qquad (1)$$

where, Q_d : design strength per-stud (N)
q_u : unconfined compressive strength of frozen soil (N/mm2)
d : stud diameter (mm).

To determine the stud resistance for an actual construction job, a sufficient length is needed (depth of at least 2d). In addition, one should consider the unevenness of shield tunnel segments and any oil and fat on the surface, and add a safety factor according to

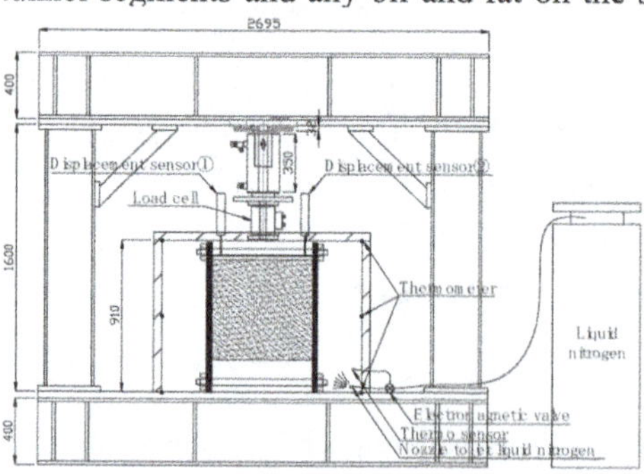

Fig. 9 Outline of experiment

Photo.4 Situation of experiment

any uncertain factor. Moreover, the applicability of formula (1) is limited to steel studs with at least the strength of those in Table 2.

3.Tests of the large-scale shear reinforcement effect for actual construction applications

The above tests have verified the reinforcing effect of the proposed adfreeze shear reinforcement method, which is followed by the investigation into its application to an underground excavation as shown in Fig. 1. The shearing stress acting on the adfreeze surface during the underground excavation is so large, therefore large-scale studs are needed according to the equation (1). The proposed equation (1) for design resistance was, however, achieved from the studs with diameters d of 5, 8, and 16 mm, and the verification of the equation for a stud diameter d =48 mm is thus required for its application to an actual construction site. For that purpose, a verification experiment of the larger-scale shear reinforcement effect has been performed.

3.1 Test method

The specimen used for the experiment consisted of two restraint plates with studs mounted on inner one side of them and the same frozen soil samples as used for the previous small-scale verification tests. A cubic frozen soil with a length of 600 mm for each side was prepared in a way similar to the small-scale tests. The specimen was frozen in a

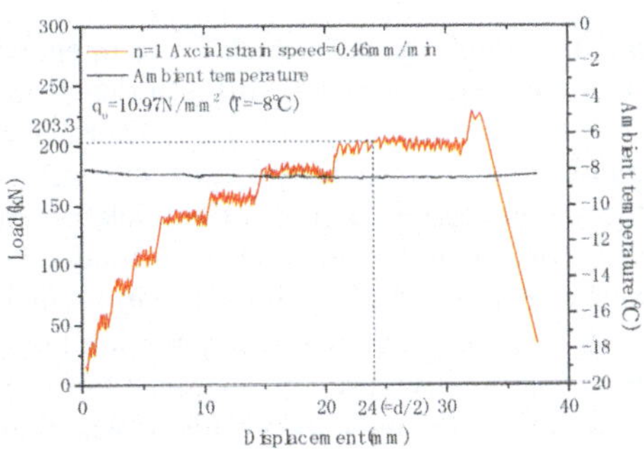

Fig.10 Relation between load and displacement

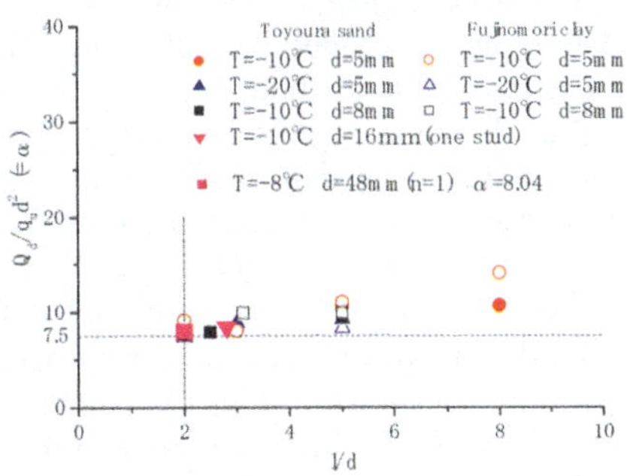

Fig.11 Relation between dimensionless strength and normalized stud length

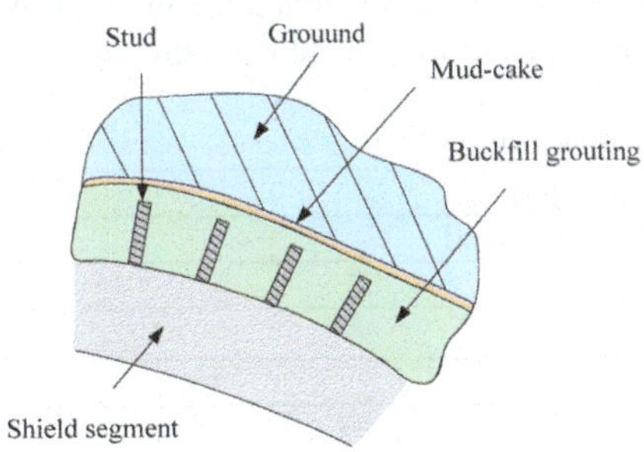

Fig.12 Ground back of shield tunnel segments

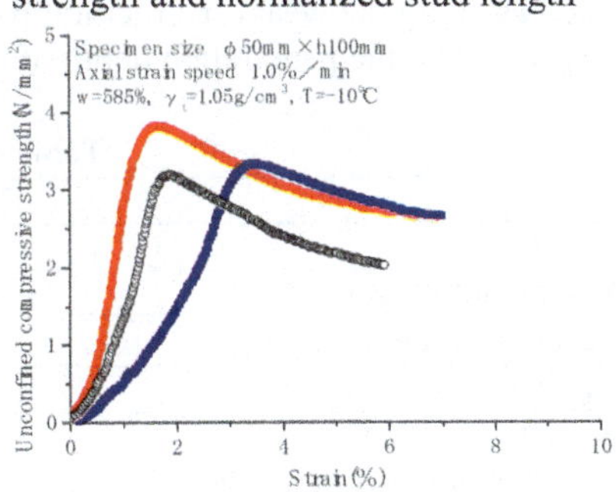

Fig.13 Result of uniaxial compression tests of the frozen slurry

large refrigerator at T=−8.0 ℃, and after forming the surface of the frozen soil, static loading tests were implemented as shown in Fig. 9 and Photo. 4. During the experiment, the specimen was covered with a heat-insulating material. To cool the specimen to the desired temperature, liquid nitrogen was sprayed with the control of a thermal sensor. The deformation of the specimen was measured at two points on the loading board. There was one test case. To evaluate the resistance per stud, a test with merely one stud was performed as described in Table 4.

3.2 Results from the experiment

Fig. 10 shows the relations between the displacement and load as well as the ambient temperature during the experiment. It presents that the ambient temperature was maintained as assumed. Also, it was observed that the peak resistance developed when the displacement was about 0.7d (=32.1mm), as was smaller than the case where the stud diameter was smaller (1.0d). Fig. 11 shows the dimensionless resistance ($Q_d/q_u\ d^2$) of shear resistance Q_u corresponding to the displacement of d /2, together with previous test results. The results obtained certainly appeared to exceed the lower limit of 7.5 that is adopted in the proposed equation for practical use. Accordingly it is recognized that the equation for the design resistance was applicable to the large stud diameters that are used in actual construction sites.

4. Penetration depth, *l*, of studs

The surrounding ground behind lining segments in a shield tunnel, which is the intended application of this method, is apt to involve over-cut areas where backfill grouting is injected as shown in Fig. 12. Those are likely to include a mud-cake area which is prone to become excessively deformable. When the mud-cake area is subject to excessive deformation, the penetration depth of studs from the shield segment requires the thickness of over-cut area plus 2d, which is very deep. Therefore, as a examination for a mud-cake area, bentonite was adjusted with 585% of water content (1.05g/cm^3 in its density) which is quite thin against a mud-cake and frozen at -10℃ to run unconfined compression tests. The results are shown in Fig. 13. The unconfined compressive strength appeared to reach 3.0 to 4.0 N/mm^2, which is similar to that for common silts and clayey soils. It was thus found that the structure of the mud-cake area was no weaker than the surrounding soil. Accordingly, for the stable structure, the application is limited to a joint, which is regarded as being the weakest structural part, between the frozen soil and the underground-structure. Also, the penetration depth of the studs is determined as 2d from the outer sur-

Table 4. Test conditions

	Conditions
Sample	*Toyoura* sand
Water content (%)	20
Freezing temperature (℃)	−8.0
Number per side (n)	1
Diameter d (mm) (external diameter of screw)	48
Length *l* (mm)	96 (=2d)

face of the underground structure. Regarding the mud-cake area, the stability will be studied as a part of the frozen soil body when studs are not penetrated, and it will be evaluated within q_u in the proposed formula (1) when studs are penetrated. Further studies on these topics are planned.

Conclusion

The following findings were obtained through the experiments of the reinforcement effect from the method of adfreeze shear reinforcement using studs.

1) The proposed reinforcement method improves the adfreeze shear reinforcement between frozen soil and an underground structure.
2) When the spacing of two adjacent studs is given as 1*l* or more (same as the stud length) along the direction of shear loading and 2d or more (twice the stud diameter) at right angle to the direction, the each stud in the pair is capable of carrying essentially the same resistance as that carried by single stud.
3) When the stud diameter d is under 48 mm, the design resistance per stud Q_d, can be expressed as $Q_d = 7.5\, q_u\, d^2$ for stud diameter, d and unconfined compressive strength of the frozen soil, q_u, provided that no interference between adjacent two studs is assured.

For the applications to practical use, further evaluation will be implemented regarding frozen cohesive soil, spacing of studs for practical scale, influence of strain speed, and different strength levels of frozen soils (experiment temperature at T=−20 ℃).

References

1) Teru Yoshida, Katushiro Uemoto and Tadashi Yoshikawa (2003), Test Method of Adfreeze Sheer Strength, Proceedings of the 38th Annual Conference of JGS, pp.339 to 340.
2) Japan Road Association, Specifications for Highway Bridges and Description, II Steel Bridge, pp.331 to 340, 2002.
3) Japan Society of Civil Engineers, Guideline for the Steel-Concrete Sandwich Structure Design (draft), pp.17 to 20, 1992.
4) Railway Technical Research Institute, Standards for Railway Structures Design and Description, Steel Composite Structures, pp.326 to 328, 1992.
5)Teru Yoshida, Katsuhiro Uemoto, Tadashi Yoshikawa and Yasuyuki Hayakawa(2004), Reinforcing Effects by Stud Connectors between Frozen Soils and Underground Structures Tsuchi-to-Kiso,Vol.52, No.7, pp.17 to 19.
6) Architectural Institute of Japan, Standards for Structure Design of Foundations, pp.262 to 296, 2001.

STRESSED AND DEFORMED CONDITION OF THE GROUNDS AROUND DRIVEN PILES

A. Zh. Zhusupbekov & A. A. Zhusupbekov
Geotechnical Institute
L. N. Gumilev Eurasian National University
Astana, Kazakhstan
e-mail: askarz@nets.kz
A. S. Zhakulin
Karaganda State Technical University
Karaganda, Kazakhstan
e-mail: zhakulina_aida@mail.ru
T. Tanaka, K. Okajima
Tokyo University,
Tokyo, Japan
e-mail: atanaka@mail.ecc.u-tokyo.as.jp
A.J. Belovitch, G.A. Sultanov
"Corporation Bazis-A, ltd,"
Almaty ,Kazakhstan
e-mail: basis_astana@mail.ru

ABSTRACT

The paper highlights the soil characteristics on the territory of the Republic of Kazakhstan along with the basic theses of the elastic and plastic soil models elaborated by the authors through solving the physical non-linear mixed problem of the elasticity and plasticity theory to describe the initial stressed and deformed condition of water saturated grounds with consideration of the pore pressure.

The elastic and plastic model was taken as a base for the axle-symmetrical problem solution to estimate the stressed and deformed condition in the areas around driven piles sectioned in 30×30 cm and 40×40 cm for the Atyrau soils condition.

1. INTRODUCTION

Taking into consideration of the emission and stratification properties of the grounds which will consequently serve as the base for buildings and constructions is obligatory both in industrious and public construction.

According to the principles of zone segregation concerning the engineering and hydogeological conditions the territory of Kazakhstan is divided into the following zones:

- Central Kazakhstan (Central Kazakhstan Low Hills, Betpak-Dala Plateau, intermountain, foothill lowlands and plains of theTengiz, Zhezkazgan, Balkhash as well as estuary valleys of the Ishim, Nura and Irtysh);
- The Corrugational formations of South Kazakhstan (the Thyan-Shyan, the Zaisan, intermountain, foothill lowlands and valleys of the Alykol and Ili);
- The Russian Platform and Skiff Plate (Pre-Caspian lowlands and the region of West Kazakhstan);
- The Turan Plate (eolithic sandy deserts of the Muyunkum, Kyzylkum, Pre-Aral; eluvial and eluvial-deluvial plains of the Torgay and Kostanay Plateau, foothill plains of the Chu Cavity and valleys of Shimkent and Caratay; valleys and deltas of rivers, sea and lake-alluvial lowlands of the Chu and Syrdarya).

Hoe I. Ling et al. (eds.), Soil Stress-Strain Behavior: Measurement, Modeling and Analysis, 885–894.

The complex zonal geoengineering surveys of the territories and eventual estimation have revealed that the bases of buildings are originally continental sedimentary non-cemented rocks of the Quatenary System.

To evaluate the properties of the Quatenary system sedimentary rocks as those of the buildings base, the conditions of their genesis have been defined. Genetically the grounds of the Quatenary system are divided into marine, eolithic, eluvial, deluvial, lake, alluvial and etc.

Eluvial grounds of the Quatenary system are more frequently used as the base for buildings on the territory of Kazakhstan, SNiP2.02.01-83 (CIS Republics Standard)stipulates the specifics of the building base designing in the eluvial grounds. Eluvial argillaceous soil is composed of weak loam and rarely of clay, both characterized by comparatively low resistance. The base structured in eluvial grounds should be designed with consideration of their particularities:

- compositional non-homogeneity and wide-ranged resistance and deformation terms;
- sandy loam and sand may get quick while water-saturated;
- sandy loam and fine sand may easily settle down while soaked;
- high mutability of physical and mechanical properties of clay and loamy grounds while water-saturated (Ter-Martirosyan 1990, Zaretskyi, 1988).

The analysis reveals that most of buildings and constructions in the Kazakhstan industrious regions are erected mainly on the grounds of the Quaternary period. The soil of the Quaternary period is distinguished by its physical mutability, complexity and variability of mechanical properties. Therefore, within the scientific object (geotechnical system) the Quaternary water-saturated loam and sandy loam of West and Central Kazakhstan were taken as a research subject.

The major factors leading to water saturation of the base are considered to be technogenic (technical and technological) and climatic (atmosphere precipitation infiltration).

Water-saturated grounds could be met in all towns of Kazakhstan without exceptions, as well as in the oil fields region along the Caspian coastline. Water-saturated grounds of the bases possessing low deforming and resistance features require further investigation to prevent accidents and structural defects of buildings and constructions.

The pore pressure is an excessive pressure of the pore water in the water-saturated ground, exceeding the hydrostatic one and found as a component of the base stressed and deformed condition. The pore pressure as a component of the stressed condition determines the character of the filtrational consolidation process and influences the stability and resistance of the water-saturated bases as buildings and constructions are being erected and exploited.

The pore pressure being one of the components of the stressed condition in the water-saturated ground principally determines the process of consolidation and influences the stability and resistance of the building and construction base. Therefore, spreading and change of the pore pressure in the ground is one of the significant geomechanical problems as to reliably design the foundations on the water-saturated base of the Quartery system.

The prognosis for the ground deformation process in the water-saturated base depends primarily on the formation of the stressed condition subsequent to the interaction with the building. At that, the sought for stressed condition is induced in the water-saturated base which considerably differs from the initial one.

We suggest exploring the pre-marginal stressed condition of the water-saturated base at its four stages: initial, primary, intermediate, final.

The primary stressed (momentary) condition in the water-saturated base forms quickly enough and is, as a rule, followed by the intermediate stage. At the given stage, the interaction between the pore water and the ground carcass does not lead to significant change in their qualitative correlation. It is not necessary to apply the equation of the consolidation theory to describe the primary stressed condition.

2. THE BASIS OF ELASTO-PLASTICAL MODEL OF SATURATED SOIL GROUND

As to research for the primary stressed and deformed condition in the water-saturated base with consideration of the pore pressure impact, the axle-symmetrical problem has been solved within the elastic and plastic set in the deformational ground model (Tatsuoka, Shibuya & Kumano 2001).

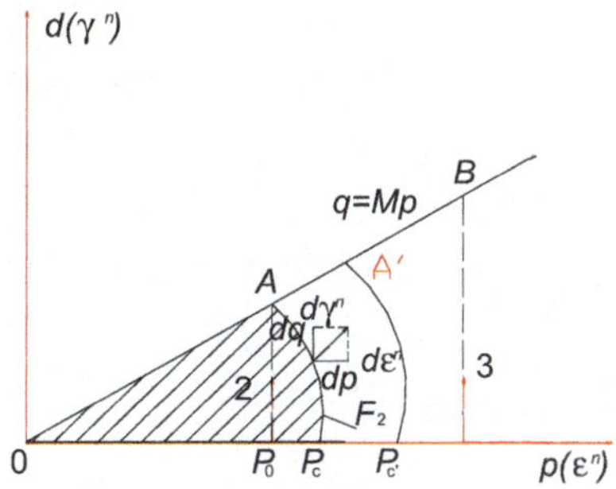

Figure 1. The diagram of elasto-plastic model of soil ground

Marquee (deformational) models are the most applicable and well-elaborated to describe the properties of the normally compressed loam and clay (Fadeev 1987). The incremental links between the stress and deformation have been considered within this model during the ground tests by the triaxle compression device. The analysis of the results is conveyed through the effective stress:

$$\eth = \frac{1}{3}\left(\sigma_1^d + 2\sigma_2^d\right) \text{ and } q = \sigma_1^d - \sigma_2^d \tag{1}$$

Under the isotropic consolidation the dependence of the pore coefficient on the pore pressure logarithm is linear. Therefore the dimensional plastic deformation is determined by the formula:

$$\begin{aligned} &\varepsilon_v = \varepsilon_{11} + \varepsilon_{22} + \varepsilon_{33} \\ &-\varepsilon_{ii}^n = -\varepsilon_v^n = \frac{\lambda - k}{1 + e_0}\log\left(\frac{-p}{-p_0}\right) \end{aligned} \tag{2}$$

in which e_0 is a basic pore coefficient; p_0 is the pore pressure.

Surmising that the total increase in the dimensional deformation constitutes of the elastic and plastic parts, the total deformation is determined as follows:

$$d\varepsilon_{ii} = d\varepsilon_v = d\varepsilon_{ii}^y + d\varepsilon_{ii}^n = d\varepsilon_v^y + d\varepsilon_v^n\ , \tag{3}$$

The growth of the elastic deformation component is connected by the linear correlation of the generalized Huke Law with the stress growth:

$$d\varepsilon_{ij}^{y} = \left[(1+v)d\sigma_{ij} - 3v\delta_{ij}d\sigma_{cp}\right]/E, \tag{4}$$

in which δ_{ij} is the Cronenera symbol, $\delta_{ij} = 1$, at $i = j$ and $\delta_{ij} = 0$ at $i \neq j$.
The plastic deformations in the experimental graphics are determined as:

$$\varepsilon_{ii}^{n} = \frac{\lambda - k}{1+e_0}\frac{1}{2.3}\ln\left(\frac{-3p}{-3p_0}\right), \tag{5}$$

in which: $-3p = P_c$
And consequently:

$$\varepsilon_{ii}^{n} = -A\ln(-P_c) + B,$$

In which

$$A = \frac{1}{2.3}\frac{\lambda - k}{1+e_0} \text{ and } B = -A\ln(-3p_0)$$

The coefficients A and B are constant.
Consequently:

$$P_c = -\exp\left(\frac{B - \varepsilon_{ii}^{n}}{A}\right). \tag{6}$$

In case of the elastic and plastic environment, the function of the surface loading is equal to the function of the surface fluidity $\Phi = f$, consequently:

$$d\varepsilon_{ii}^{p} = \lambda\frac{\partial f}{\partial \sigma_{ii}} \tag{7}$$

The associated surface of the fluidity and plasticity potential for normally compressed ground ($F_2 = f$ or AP_c) given in the model is a section of the ellipses produced through the equation:

$$\frac{(I_1 - p_0)^2}{a^2} + \frac{J_2}{b^2 g(\theta)^2} = 1, \tag{8}$$

in which a and b are geometrical parameters of the ellipses, $a/b = R = const$; p_0 is the ellipses center.
Proceeding from it the fluidity surface is calculated through the equation:

$$f = F_2 = b^2(P_c - p_0)^2 + \frac{a^2 q}{(g(\theta))^2} - a^2 b^2 = 0. \tag{9}$$

Basing on the Moor-Coulon Law, we may determine as follows:

$$g(\theta)=\frac{\cos\frac{\pi}{6}-\frac{1}{\sqrt{3}}\sin\frac{\pi}{6}\sin\varphi}{\cos\theta-\frac{1}{\sqrt{3}}\sin\theta\sin\varphi}=\frac{3-\sin\varphi}{2\sqrt{3}\left(\cos\theta-\frac{1}{\sqrt{3}}\sin\theta\sin\varphi\right)}, \quad (10)$$

in which θ is the Lode parameter;

The ellipses position is defined through two constants a and b and the function:

$$P_c=F_2\left(\varepsilon_v^n\right), \quad (11)$$

linking the marquee position with the dimensional plastic deformation - $\varepsilon_v{}^n$.

We should note that the theoretical increase in stress is defined as:

$$dp=dp^y-Kd\varepsilon_v^n \text{ and } dq=dq^v-2Gdq^n; \quad (12)$$

in which dq^y and dp^y are the stress elastic growth;
K, G are secant modules of the dimensional deformation and shearing.
The advantage of the model for undercompressed grounds is that the loading adapted for the pore pressure follows the trajectory approximate to the normal one towards the spreading marquee of the fluidity surface, the model in this case reflects properly the link between the stress and deformation. Its basic positive feature is a small number of the experimental parameters: λ, K and M, three magnitudes determined during the ground test through the compressing devices and triaxle compression device. The additional pressure will constitute of the additional effective stress in the carcass $\Delta\sigma^d$ and pore pressure growth - ΔP_w:

$$\Delta P_n=\Delta\sigma^d+m\Delta P_w, \quad (13)$$

in which $m=(1,\ 1,\ 1,\ 0)^T$

The stress mutation in the carcass results from its deformation mutation through the tangent module of the dimensional compression:

$$\Delta\sigma^d=K_k\varepsilon_v=\frac{p(1+e)}{\lambda}\varepsilon_v, \quad (14)$$

in which

$$K_k=\frac{p(1+e)}{\lambda},$$

The pore pressure is determined as follows:

$$\Delta P_w=K_w\Delta\varepsilon_{v(w)}=K_w m^T\Delta\varepsilon, \quad (15)$$

in which $\Delta\varepsilon_{v(w)} = \Delta\varepsilon_v = m^T \Delta\varepsilon$; $K_w \approx 2 \cdot 10^2$ МПа is the module of the water dimensional shrinking.
If the ground was in the normally compressed condition by the moment of the quick loading the tangent module of the dimensional compression will be equal to the tangent module of the initial consolidation. Practically all pressures in the normally compressed ground being under quick compression are imposed upon the pore water.
In accordance with the Terzagi effective stress principle, general stress s σ_{ij} in the water saturated base is equal to the sum of stress in the ground carcass $\sigma_{ij}{}^d$ and in the pore water p_w :

$$\sigma_{ij} = \sigma_{ij}^d + \delta_{ij} p_w, \tag{16}$$

Consequently the balance equation is as follows:

$$\sigma_{ij}^d + \delta_{ij} p_{w,j} = F_{\gamma i}, \tag{17}$$

in which $F_{\gamma i}$ is the dimensional power determined by the ground's own weight.
Calculating for stress p and q in the equation (9):

The expression may be produced in the matrix form as:

$$\{\sigma\} = \{\sigma^d\} + \{g\} p_w, \tag{18}$$

in which

$$\{\sigma^d\} = \{\sigma_r^d \quad \sigma_z^d \quad \sigma_\theta^d \quad \tau_{rz}^d\}^T, \ \{g\} = \{1, \ 1, \ 1, \ 0\}^T.$$

3. THE ANALYSIS of RESULTS OF NUMERICAL CALCULATION BY FEM

To make a complete description of the problem it is necessary to define border and primary conditions. Apparently, the equation system is self-starting, as by the time moment $t = 0$ of the loading application, the primary conditions are automatically realized on the account of the fact that the joint displacements and pore pressure components for the previous moment are equal to zero.
To estimate the primary stress condition with consideration of the pore pressure in the water-saturated base around a single pile and to determine its bearing capacity the axle-symmetrical problem within the elastic-plastic set was solved on the basis of the non-linear deformational model applying the principle of the effective stress of the "water-saturated base - pile" system by the final elements method according to the program "NONSOLAN" (Japan). The theoretical solutions were compared with the data of the static pile test of the vertically pressing loading carried out on the test field in Atyrau.
The ground base properties determined through the test results on the triaxle compression device (the construction by Tokyo University, Japan) are represented in the table 1 below.
Certainly, the element scheme for the single pile calculation takes only a half of the calculated zone due to the pile section symmetry. The scheme of the pile cast in the water-saturated base clarifies that relatively to axle, Z is an absolute symmetry.
The calculated area is conditionally divided in 13 zones, the joint points are marked and placed in the proper locations. To provide the complete fading of the local

disturbance the boundaries of the area in question are expanded for the distance not shorter than the piles length from each side.
In the calculation scheme during the pile cast in the water-saturated base the modeling was carried out in the form of the non-linear problem within the axle-symmetrical set generated as a result of a pace-out shearing of the field 3-4 between the joint points 4, 5 and 6 corresponding to the contact surface "water-saturated base – pile" up to the position 4', 5' and 6' as a rigid integer. The fluidity condition of the water-saturated ground is reflected through the deformation model equation. The calculation of the proper weight at the different depth levels was made in the form of the basic stress s.
The mash zone in the horizontal direction has been divided in 4 zones. zones 5 and 13 have been divided in six equal parts of 15cm, zones 6 and 12 in four equal parts of 40cm, zones 7 and 11 in three equal parts of 75cm, zones 8 and 10 in three equal parts of 150cm.
In the vertical direction the division was made in 5 character zone, zone 1 in 8 equal parts of 31.8cm, zone 2 in 36 equal parts of 16cm, zone 3 in 16 equal parts of 16cm, zone 4 in 14 equal parts of 31.8cm, and zone 9 in 24 equal parts of 63.6cm.

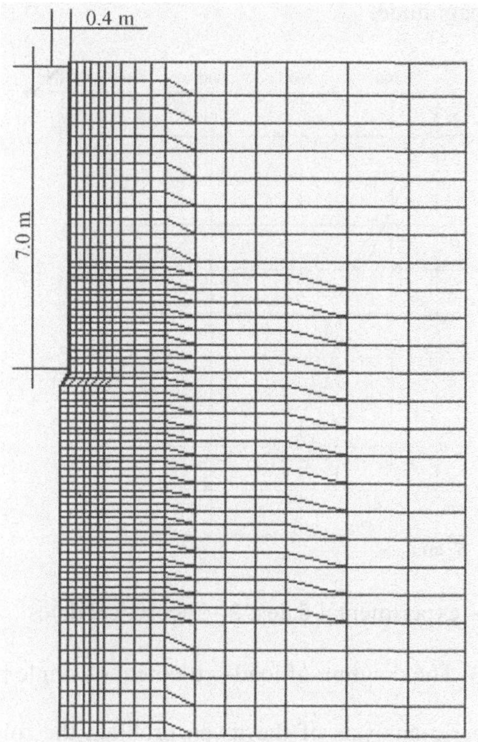

Figure 2. The calculation mesh by FEM

Table 1. The ground properties accepted while calculating.

Name of strata	*E, MPa*	ν	γ, kN/m^3	α_{cs},	e_0	λ	k	γ_{wH}, *MPa*
Loam	5.0	0.30	17.5	20	0.70	0.312	0.10	0.49

The problem was solved through the numeric method of the final elements under the following primary boundary conditions:

$1 \div 3$ joints – $\varepsilon_r = \varepsilon_2 = 0$; $4 \div 6$ joints – $\varepsilon_r = \varepsilon_2 =$ 15cm or 20cm;
$7 \div 10$ joints – $\varepsilon_r = \varepsilon_2 = 0$; $10 \div 11$ joints – $\varepsilon_2 = 0$; $11 \div 14$ joints – $\varepsilon_r = \varepsilon_2 = 0$.

Thus, the upper area is free. The right side along the axle *r* and *z* is fixed. The left side along the axle *z* is fixed, and along the axle *r* it is fixed for the pile. The boundary conditions of the pile surface on the left side are denoted by the displacement $\varepsilon_r=\varepsilon_2=0{,}15$cm or $\varepsilon_r=\varepsilon_2=20$cm. The capacity of the studied water-saturated ground block is as follows: the depth H=15m, the breadth L=20m. The piles cross section is 30×30 cm and 40×40 cm, their length is 7.0m. It is supposed that there is complete adhesion along the contact line of the lateral surface "pile - ground". The physical nature of FEM permits to consider the system "Water-saturated base - pile" together. The rigidity is defined through its geometrical characteristics specification, the ground specific gravity is replaced by the all-sided hydrostatic stress tensor which is summed with the factual stress magnitude.

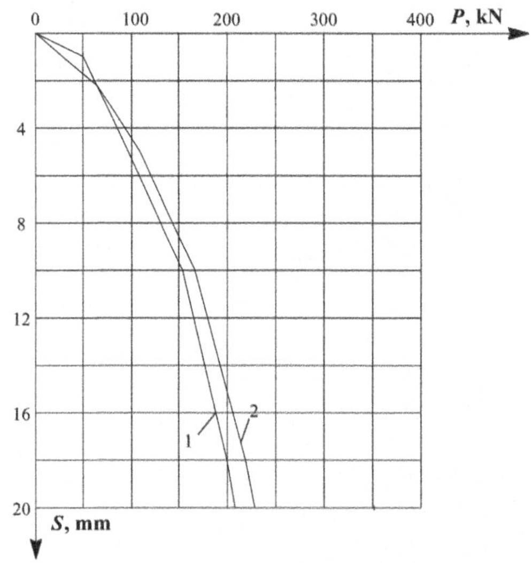

1 – experimental dates; 2 – theoretical dates

Figure 3. The diagram of load-settlement of single pile

As a result of the numeric analysis of the given problem the following epures and contour lines of the stress , deformation and pore pressure around a single pile in the water-saturated base for the standard and propagated cross sizes of 0.3×0.3m and 0.4×0.4m:

contour lines of the zero deformations radial – ε_r,, the axle– ε_z,, tangential - ε_θ;
contour lines of the maximal deformations tangent (shearing) -γ_{rz};

spreading of the stresses radial - σ_r, axle - σ_z, tangential - σ_θ;
contour lines of the stress tangent (shearing) - τ_{rz} (Figure 4);
zero contour lines of the pore pressure – P_w (Figure 5).

The contour lines of the zero radial – ε_r, axle - ε_z, and tangential - ε_θ for the piles of 30×30cm and 40×40cm are similar both in form and character.

The largest interest is attracted by the contour lines of the maximal tangent deformations around the piles, conditioned by the shearing deformations. The spreading zone of the shearing deformations for the piles of 40×40cm encompasses more base ground than the pile of 30×30cm.

Spreading of the radial - σ_r, axle - σ_z, and tangential - σ_θ stress s around the pile discovers that the stress zone generates under the tip and their spreading is similar. No the ground surface rise is observed from outside (from the upper ground surface) which is proper to the water-saturated base. The contour lines of the tangent (shearing) stress (Figure 4) indicate distinctly that two compressed zones are forming around and under the pile tip with the nucleus with the stress value of 0.1MPa and 0.05MPa which in their turn spread around up to 2.0m from the pile edges. Altogether with this, the stress spreading zone for the piles of 40×40cm is wider than that for the piles of 30×30cm.

The contour lines of the pore water pressure – Pw (Figure 5) indicate at the pore pressure spreading around the piles. Three zones of the pore pressure spreading are forming around the piles: by the tip, along the lateral surface, under the tip. The zone by the tip is up to 1.0m and takes a circle form. The second zone along the lateral surface is 2.2m in breadth and 4.3m in height. The third zone under the tip reaches 4.0m in depth for the piles of 30×30cm and 7-8m for the piles of 40×40cm.

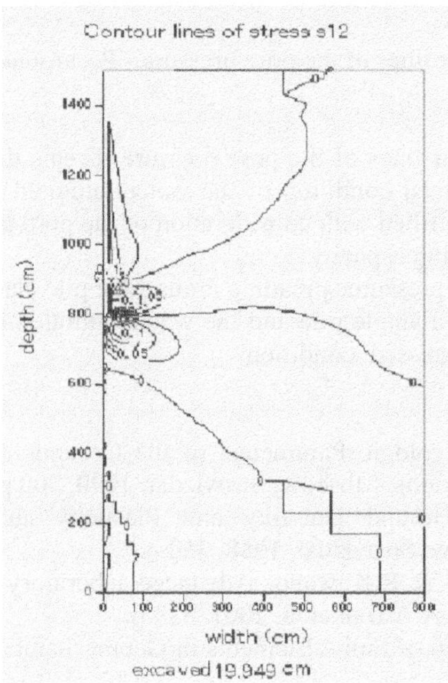

Figure 4. Contour lines of shearing stresses – τ_{rz} around the pile of 40×40 cm

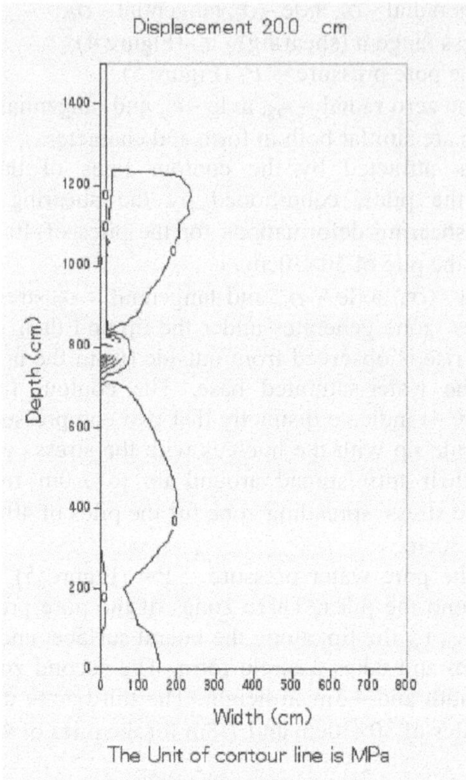

Figure 5. Zero contour lines of the pore pressure - P_w around the pile of 40×40cm

4 CONCLUSIONS

Spreading of the contour lines of the pore pressure reveals that the determination of the stressed and deformed condition of the water-saturated ground around the pile should necessarily be fulfilled with consideration of the pore pressure as it, in its turn, determines the pile bearing capacity.

The process of the pore pressure spreading around the pile clarifies the mechanism of the interaction between a single pile and the water-saturated ground base and allows to estimate its primary stressed condition.

REFERENCE

Z.G.Ter-Martirosyan. Reologic Parameters of the Grounds and Calculations of the Construction Foundations, Moscow, Stroyizdat, 1990, 200 p.

Yu.K.Zaretskyi. The Ground Ductility and Plasticity and Calculations of the Construction, Moscow, Stroyizdat, 1988, 352 p.

F. Tatsuoka, S.Shibuya & R.Kuwano. Advanced laboratory stress-strain testing of goematerials, Tokyo, A.A.Balkema, 2001, 329 p.

A.B.Fadeev. The Method of Finite Elements in Geomechanics, Moscow, Nedra, 1987, 221 p.

Soil Stress-Strain Behavior: Measurement, Modeling and Analysis
Geotechnical Symposium in Roma, March 16 & 17, 2006

CENTRIFUGE MODELING OF PILES SUBJECTED TO LATERAL LOADS

Brant, Logan and Ling, Hoe I.
Department of Civil Engineering
Columbia University, New York, NY 10027, USA
e-mail: lcb2002@columbia.edu

ABSTRACT

There are many applications where piles are employed to absorb and deflect lateral impact loads. Structural elements of this type are used to protect infrastructure and are commonly found at marine sites. A series of model tests have been conducted using Columbia University's centrifuge facility to better understand the performance of piles subjected to these loading conditions. A device was designed to install and laterally load single model piles during centrifuge flight. This device uniquely contains two lateral loading systems, one which allows static testing and another appropriate for dynamic tests. This research examines the behavior of tubular steel piles embedded within dry or saturated soil and subjected to varied rates of lateral loading. This investigation provides insight on the contribution of lateral loading rates to the behavior of piles.

1. INTRODUCTION

This paper summarizes the experimental observations obtained while studying the behavior of laterally loaded piles using Columbia University's centrifuge facility. An unusual component of this work involved the dynamic manner in which lateral loading was applied during many of these tests. This feature allowed models to simulate conditions occurring when a single free-head pile is subjected to a large horizontal impact force. In addition, many of the models were constructed using fully saturated soil. Throughout these tests design parameters were varied allowing opportunities for comparisons.

The primary motivation for conducting this work was to investigate the response of piles subjected to lateral impact loading, an area of research which has not been extensively explored despite a critical need. Current design practices such as those recommended by the American Petroleum Institute (API 1993) focus on the design of deep foundations subjected to lateral loads applied in a static manner and to a lesser extent wave induced cyclic loading.

Applications of lateral impact loaded piles commonly occur in a marine environment making it important that this research also investigate the contribution from saturated soils in this soil-structure interaction. The analysis of saturated soils introduces complications especially when involving dynamic deformations. Rapid changes to saturated soil may cause pressure within the pore fluid to increase. Soils with low hydraulic conductivity obstruct the flow of pore fluids causing un-drained conditions.

Hoe I. Ling et al. (eds.), Soil Stress-Strain Behavior: Measurement, Modeling and Analysis, 895–907.

2. SCALING LAWS & PROPERTIES

Centrifuge modeling in conjunction with limited full-scale field validation can provide a cost effective alternative to testing only full scale structures in situations where reliable soil-structure interactions are uncertain. This tool allows the capability to tailor experiments to specific design criteria. In geotechnical engineering, the body forces within the soil are important when defining how an underground structure will perform. In order to create representative soil models it is critical that body forces within the soil be scaled appropriately. The use of a geotechnical centrifuge allows scaling of stresses imposed by the soils own weight by varying the acceleration field in which the model is located. When properly constructed, reduced scale centrifuge models represent conditions existing in full scale prototype structures.

During centrifuge modeling conflicting scaling relationships exists when water is used as a pore fluid. To correct for this discrepancy centrifuge modeling of water saturated prototypes requires the use of a substitute pore fluid with density similar to water, but with a viscosity increased proportionally with the scaled centrifugal acceleration. The use of a substitute pore fluid compensates for the difference in scaling relations which then allows dynamic time and diffusion time events to occur at a similar rate, with model speeds occurring N times faster then those found in the prototype.

Table 1. Relevant Centrifuge Scaling

Quantity	Prototype	Model
Length	1	1/N
Area	1	$1/N^2$
Volume	1	$1/N^3$
Acceleration	1	N
Force	1	$1/N^2$
Stress	1	1
Strain	1	1
Dynamic Time	1	1/N
Diffusion Time	1	$1/N^2$

Replacing prototype soils with scaled model soils would require the reduction of soil grain diameters, however that would likely result in soils with very different physical properties. The same soil types found in the prototype are typically used when constructing centrifuge models. This presents challenges when shear banding or soil dilation cause changes which are not proportional to the scaled dimension of the model. Generally if the ratio of the pile diameter divided by the mean soil diameter is kept large particle size effects are minimized during this type of soil-structure interaction. In these experiments the ratio of D_{PILE} to D_{50} was equal to 85.

3. SOIL, PILE & FLUID PROPERTIES

In the field piles are constructed from a range of materials including, structural steel, reinforced concrete, timber and plastics. These materials each have their own distinct characteristics. A model pile should match closely the properties of the prototype pile material being studied, which in this case was structural steel. T316L stainless steel was used as the material for the model piles because of its similarities to the properties of A36 structural steel. In addition, stainless steel is non-corrosive and is available in a wide range of sizes.

Nevada Sand was selected as the soil used in these models because of its well researched material properties. No. 120 Nevada Sand is relatively fine poorly grated sand. Published results report the hydraulic permeability of water through Nevada Sand as 2.3×10^{-5}, 5.6×10^{-5} and 6.6×10^{-5} m/sec for soils with relative densities of 91.0, 60.1 and 40.2 percent respectively. (Arulmoli et al., 1992)

Table 2. Nevada Sand Soil Properties (Arulmoli et al., 1992)

Specific Gravity	2.67
Max. Dry Unit Weight	17.33 kN/m^3
Min. Dry Unit Weight	13.87 kN/m^3
Max. Void Ratio	0.887
Min. Void Ratio	0.511
D_{50}	0.00015 m

Table 3. Typical Scaled Pile Properties

Model (40g)	Prototype (1g)
T316 Stainless Steel	A36 Structural Steel
E = 193 GPa	E = 200 GPa
σ_{YIELD} = 290 MPa	σ_{YIELD} = 250 MPa
σ_{ULT} = 580 MPa	σ_{ULT} = 400 MPa
D = 0.0127 m	D = 0.508 m
L_{EMB} = 0.2032 m	L_{EMB} = 8.13 m
EI = 93.2 Nm^2	EI = 247.2 MNm^2

A fluid mixture containing Metolose and water was used as a substitute pore fluid because it has properties when scaled to prototype conditions that simulate water saturated soil within the model. This fluid solution has density similar to water (1000 kg/m^3) but with viscosity of 40 cps, a value which is 40 times greater than that of water.

4. EXPERIMENTAL PROCEDURES

To facilitate this work an elaborate testing device was created allowing experiments to be conducted on the arm of the centrifuge. The purpose of this equipment was to install and laterally load single model piles during centrifuge flight. This device uniquely contains two lateral loading systems, one which allowed static testing and another which created dynamic impact loads. Numerous challenges were encountered during the design of this system, including the requirement that all components be capable of operating while subjected to large accelerations. The system was controlled remotely with the operator safely removed from the centrifuge chamber. Video signals, AC power and electrical controls were transmitted to and from the centrifuge using electrical slip rings. Fluid joints allowed pressurized hydraulic fluid to enter the rotating centrifuge. An onboard data logger was used to collect signals from numerous sensors while a wireless router mounted on the centrifuge transmitted this information to a computer located outside of the centrifuge chamber. The wireless system ensured experimental data of the highest quality was obtained in real time.

The piles were typically assembled from stainless steel tubes (D = 1.27 cm and T = 0.71 mm) cut to a length of 26.0 cm. Along one side of the model pile 8 quarter bridge strain gages were attached at equally spaced intervals. The lowest was placed 2.54 cm above the pile tip and the highest at a distance of 20.32 cm, which corresponded to the fully embedded piles depth. A coating of epoxy 1 mm in thickness was applied to the exterior surface of the pile to provide the strain gages with protection from water and mechanical damage. Each experiment required the construction of a new model pile.

Homogenous soil specimens were created using an automatic sand hopper. This machine moved back and forth at a constant rate raining sand into the soil box while an operator raised the hopper to maintain a constant drop height. This device has been used to prepare uniform horizontal soil deposits in previous centrifuge studies (Ling et al., 2003). To achieve fully saturated soil specimens pore fluid was added to the soil while a vacuum pressure of 70 kPa was maintained within the covered soil box. This step was necessary for the removal of trapped gases within the soil, an important part of the process required to obtain fully saturated soil specimens.

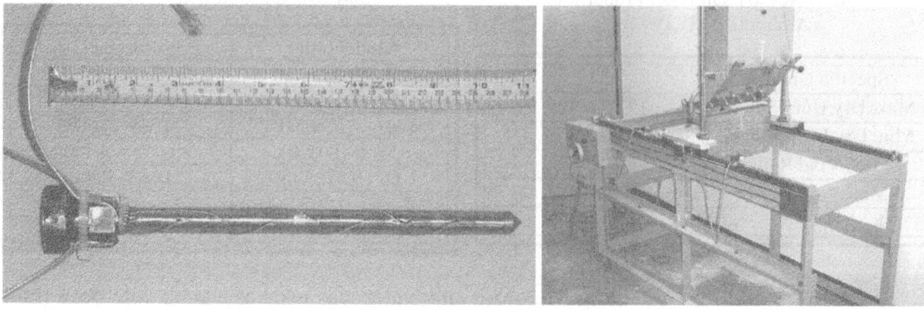

Figure 1. Example of instrumented model pile with closed tip (left) and automatic sand hopper (right)

The testing procedure itself consisted of two phases. The first utilized a motor driven displacement controlled system to uniformly push the model pile vertically into the soil at a constant rate of 0.3 mm/s. During installation sensors including a load cell, a displacement transducer and strain gages recorded the response of the pile. When the pile was fully embedded to a depth of 20.32 cm the mechanism driving the pile was stopped. The pile was released by disconnecting the electromagnet which had previously held the head of the pile during installation. The driving mechanism was then raised separating it from the pile and allowing the next phase of the experiment to begin.

This second phase was the most critical portion of this research. Vertically oriented model piles were subjected to horizontal loads applied at a height of 5.1 cm above the soil. The applied force acting on the pile or similarly the equal but opposite resistance supplied by the pile was measured by a force transducer mounted on the lateral loading mechanism. Two displacement transducers were employed to measure the rotation and displacement of the free head pile during loading. Several experiments were equipped with pore pressure transducers placed within the soil to record pore fluid pressure changes that occurred at specific locations as a result of the dynamic deformations.

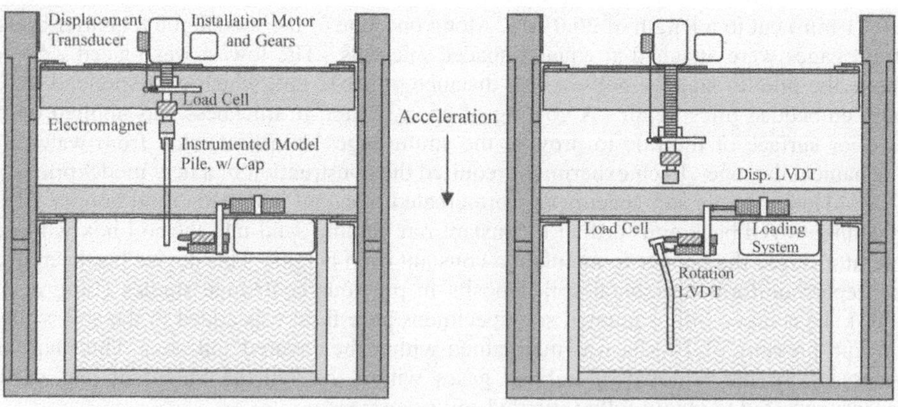

Figure 2. Testing Device

A unique feature of this device was that the applied horizontal force could be supplied from two different loading systems. One system involved a motor driven displacement controlled mechanism which applied loads extremely slowly at $1.7x10^{-5}$ m/s, simulating static loading conditions. The other was driven by a hydraulic piston capable of completing the loading within tens of milliseconds, creating a rapid impact condition with a displacement rate of 0.8 m/s. Using this dynamic system the maximum displacement of the pile head was controlled by the placement of the pile relative to the hydraulic piston with a maximum stroke limited to 8 cm. Both of these mechanisms were displacement controlled systems despite differences in their mechanical design and rate of displacement.

5. RESULTS & OBSERVATIONS

Extensive instrumentation was used to measure behaviors occurring during these laterally loaded model pile experiments. This paper presents a number of observations presented using graphs describing lateral pile resistance as a function of the pile head deflection. Other examples highlight interesting behaviors and demonstrate the variety of measurements obtained during these centrifuge tests. The scope of this paper restricts the quantity of material that may be presented. Values described in this paper reflect conditions measured during the model tests and have not been scaled to represent prototype values.

Table 4. Testing Summary

Test No.	Dr %	Pore Fluid	Duration sec	Disp m	Rate m/s	Lateral Max, N	Lateral Res, N	% Decrease	Install Type	Install Max, N	Tip Cond.	Pile T mm
4	80	Dry	0.062	0.0495	0.80	680	480	29.4	40g	N/A	O	0.71
5	80	Dry	0.064	0.0515	0.80	700	500	28.6	***1g***	N/A	O	0.71
6	80	Dry	1420	0.0245	1.7E-5	540	-	-	40g	N/A	O	0.71
7	80	Dry	1750	0.0300	1.7E-5	550	-	-	40g	N/A	O	0.71
8	80	Dry	2100	0.0360	1.7E-5	550	-	-	40g	N/A	C	0.71
9	80	Dry	2100	0.0360	1.7E-5	530	-	-	40g	N/A	O	0.71
20	80	Met	1450	0.0235	1.6E-5	560	-	-	40g	3600	O	1.24
21	80	Met	1400	0.0225	1.6E-5	510	-	-	40g	3400	O	0.71
23	***50***	Met	1550	0.0265	1.7E-5	440	-	-	40g	1120	C	0.71
24	***50***	Met	1500	0.0260	1.7E-5	410	-	-	40g	1110	O	0.71
26	80	Met	0.048	0.0350	0.73	N/A	N/A	N/A	40g	2650	O	0.71
27	80	Met	0.050	0.0375	0.75	940	750	20.2	40g	N/A	O	1.24
28	80	Met	0.052	0.0380	0.73	870	720	17.2	40g	2650	C	0.71
29	80	Met	1230	0.0225	1.8E-5	N/A	-	-	40g	3200	O	0.71

Rate of Loading - These experiments demonstrated that increases in the rate of applied loading caused the model piles to provide greater lateral resistance. Tests conducted using Nevada Sand with relative density, D_r of 80 percent showed increased resistance of 10 percent for dry sands and an increase of 35 percent when using fully saturated soil. Models constructed using saturated soil, but with piles 50 percent stiffer ($EI = 143.1\ Nm^2$) than the previous piles showed an increased resistance of 60 percent when comparing impact loading to static loading. Differences in soil response which may have contributed to these rate dependent behaviors are discussed later.

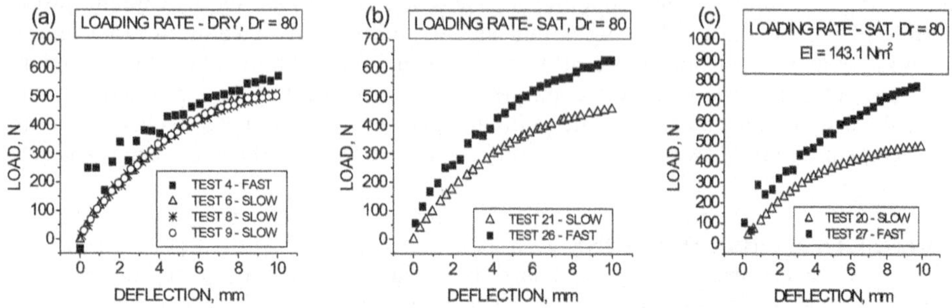

Figure 3. Pile response subjected to varied loading rates

When measuring conditions occurring during dynamic loading the presence of inertial forces can create a challenge. Inertial forces occurred during impact loading when the pile was accelerated from rest. These forces are different from those caused by the resistance from the soil-structure interaction. Distortions caused by the presence of inertial forces were intentionally minimized by applying the dynamic loading at a nearly constant rate. Horizontal accelerations remained near zero during most of the loading period. Large accelerations occurred for relatively short lengths of time at the beginning and at the end of the impact loading while producing negligible interference during the remainder of the loading. Unavoidable inertial forces were responsible for inconsistencies immediately following the initial impact. This is shown in Figure 3. Inertial resistance created a large spike at the time of impact followed by several decaying oscillations. Increased damping of these vibrations were observed in the tests which involved fully saturated soil (Figures 3b and 3c) as opposed to dry soil (Figure 3a).

Pore Fluid - Tests were conducted in Nevada sand with D_r of 80 percent in soil that was either dry or fully saturated using a substitute pore fluid. When subjected to static loading conditions the dry soil provided lateral resistance 15 percent greater than that provided by the fully saturated soil. This could be explained by the decreased effective unit weight of the saturated sand which subsequently caused decreased passive earth pressure to act horizontally against the pile. With γ_{dry} = 16.5 kN/m^3 and γ' = 10.3 kN/m^3 it might be expected that this variation in capacity would be even greater, however other factors also contribute to the passive earth pressure within the soil. Model piles subjected to dynamic impact loading provided a lateral resistance 10 percent larger when embedded in saturated soil compared to models constructed using dry soil.

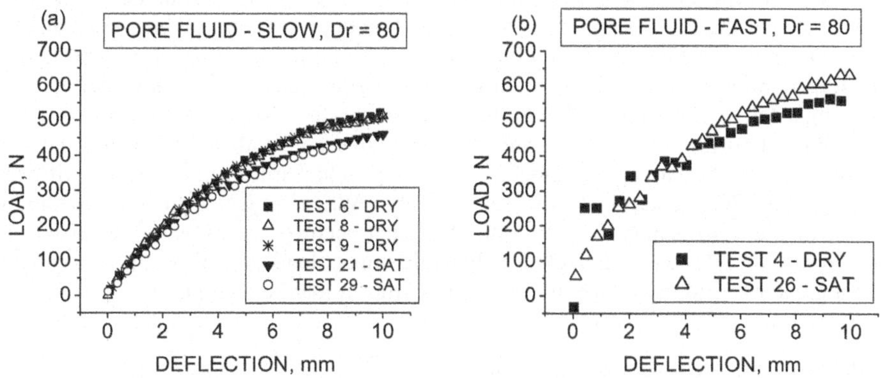

Figure 4. Pile response subjected to varied loading rates in dry and saturated soil deposits

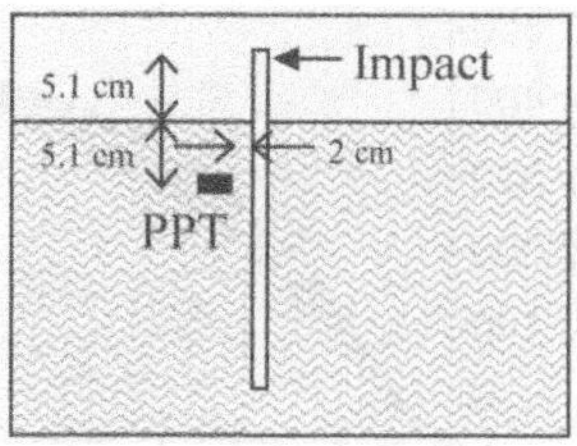

Figure 5. PPT Location

Pore pressure transducers were incorporated into several of the models to directly measure changes occurring in the pore fluid pressure during impact loading. Figure 6 shows measurements obtained during test number 15 which demonstrate changes in the pore pressure occurring during lateral impact loading. This transducer was located within the fully saturated soil with D_r of 80 percent at a depth of 5.1 cm and 2 cm in front of the instrumented pile (shown in Figure 5). A temporary pore fluid pressure decrease of 15 kPa occurred at this location. The dense sand likely underwent some degree of dilation when subjected to these loading conditions. The at-rest hydrostatic pressure in this location was 20 kPa and the total vertical stress was 40.5 kPa.

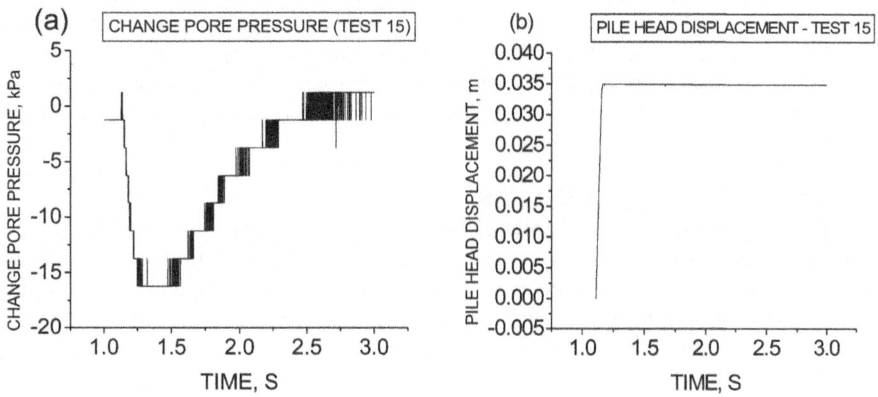

Figure 6. Relationships between changes in pore pressure & pile cap displacement vs. time.

Tip Condition - Most model piles were constructed from tubular steel pipe which was left open at the tip. As an alternative several model piles were given a conical insert that was tapered at 45 degrees providing a solid closed pile tip. It was observed that the open tipped piles became plugged during installation and at a shallow depth began producing a behavior similar to that of the solid cone shaped tip. (Figure not shown)

Pile Stiffness - Tests involving fully saturated soil with Dr of 80 percent were used to compare the lateral resistance provided by model piles with bending stiffness's (E_PI_P) of 93.2 Nm^2 and 143.1 Nm^2 respectively. When subjected to static loading an increase in lateral resistance of only 5 to 10 percent occurred when using the stiffer pile. When the model piles were subjected to dynamic impact loads the stiffer pile provided 20 percent greater lateral resistance under dynamic impact loading conditions.

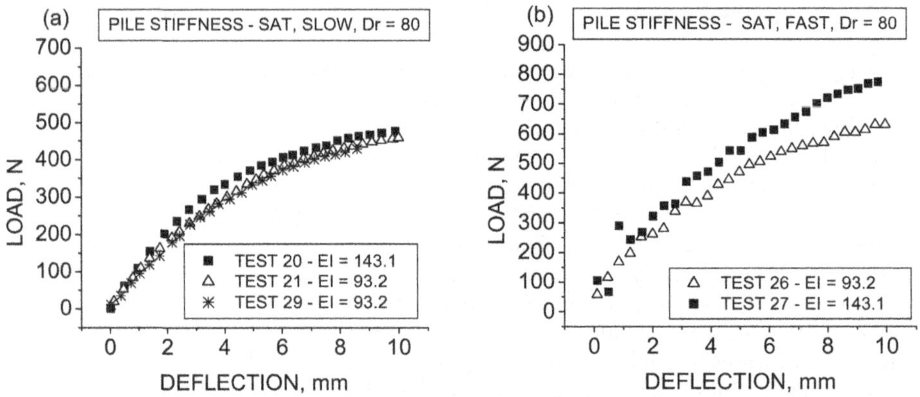

Figure 7. Graphs showing effects of varied pile stiffness (EI)

Installation Acceleration - Similar responses were observed comparing two tests with dry soil having D_r of 80 percent that were subjected to lateral impact loading when one pile was installed under an acceleration of 40g during centrifuge flight and the other was installed at 1g prior to centrifuge spinning. This comparison showed that in-flight pile installation may not be necessary when studying lateral impact loading. (Figure not shown)

6. SAMPLING OF OTHER RESULTS

Bending Moment vs. Depth (Elastic) - These graphs provide snapshots of bending moment distributions at successive increments of time while the internal forces within the pile remained within the elastic stress range. Bending moment measurements provide valuable descriptions of a pile's response and may be used to create load-transfer functions.

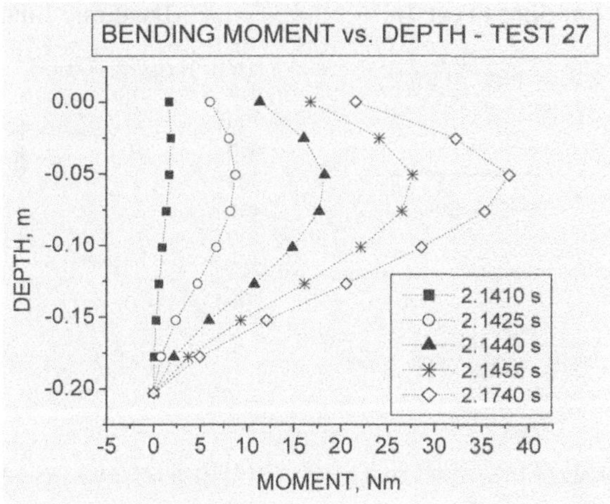

Figure 8. Bending Moment vs. Depth

Deformed Pile Shape - The deformed shape of a pile may help to determine the location within the pile where the plastic hinge developed. This understanding is relevant when investigating the behavior of this type of soil-structure interaction. Table 5 provides a summary of the locations where plastic hinges developed on the pile due to extreme loading. In general model piles subjected to impact loading and embedded in saturated soil develop a plastic hinge located closer to the surface of the soil. In dry soils this characteristic did not significantly vary regardless of the loading type.

Table 5. Deformed Pile Shape Summary

Test No.	4	7	26	29	27	20	24	28
Parameter Type	Dry Fast	Dry Slow	Sat Fast	Sat Slow	Sat Fast EI = 143	Sat Slow EI = 143	Sat Slow Dr = 50	Sat Fast Closed
Depth to Hinge, m	0.066	0.057	0.051	0.072	0.051	0.074	0.076	0.050
Depth to Hinge, %	32.5	28.1	25.0	35.6	25.0	36.3	37.5	24.4

Force vs. Time - There are interesting behaviors not clearly portrayed by graphs showing the force vs. displacement measurements recorded during these experiments. Piles subjected to lateral impact forces showed an increased horizontal resistance relative to the response when subjected to static loads. During the dynamic portion of the impact loading increased resistance was observed. When the displacement causing the dynamic loading was stopped the resistance provided by the pile fell by 20 to 30 percent. In the period that followed while the lateral force was maintained but the displacement rate was equal to zero the loading type became static. At this stage the lateral resistance provided by the pile became approximately consistent with the ultimate lateral capacity measured during the static load tests. This behavior was observed in both dry and fully saturated soils.

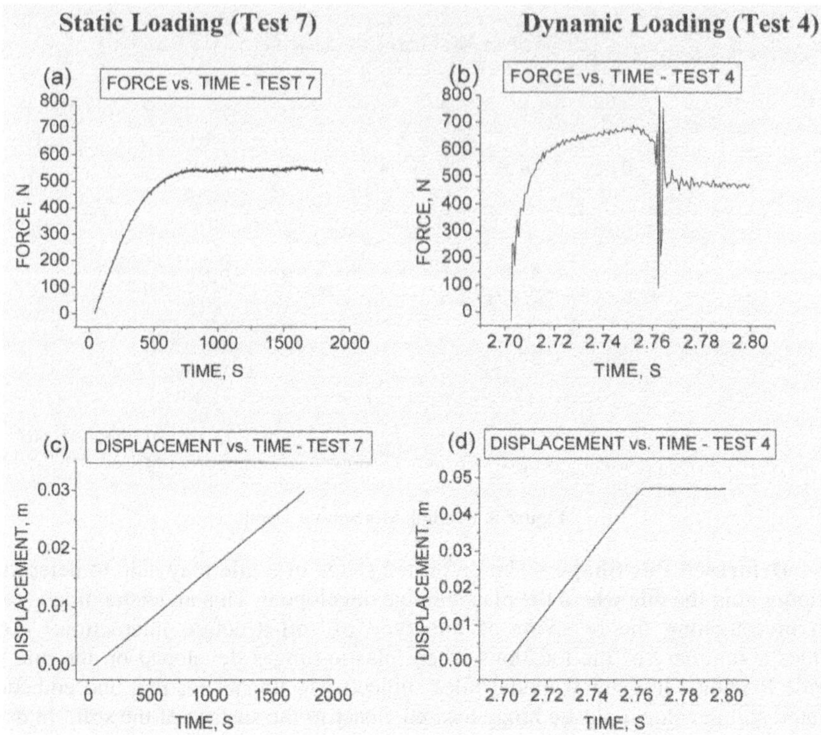

Figure 9. Examples of Force vs. Time & Disp. vs. Time Graphs

Figure 9 presents force vs. time and displacement vs. time graphs showing results from experiments numbers 4 and 7. These tests were conducted using dry Nevada sand with D_r of 80 percent. The model properties used during these tests only varied in the manner with which the lateral load was applied to the head of the pile.

There were contrasts observed during these experiments resulting from differing responses due to changes in the rate of applied load. Densely packed sand such as that used in these models generally underwent dilation. Large vertical and horizontal pressures constraining this soil made it difficult for the soil volume to increase. The soils attempt to increase volume was resisted causing increased pressures within the soil. This pressure over time would dissipate and redistribute within the surrounding soil. Soil deformed at a slow rate had sufficient time for pressures within the soil to equilibrate by deforming the surrounding material. When piles were rapidly displaced localized pressures within the soil were not able to redistribute. These pressures subsided only after the dynamic loading was stopped. It has been observed that piles subjected to impact loads provide greater lateral resistance while dynamic loading occurred, however when the displacement stopped this measured resistance significantly dropped. This behavior may explain the different responses which occurred when comparing the static and dynamic loaded piles embedded in both dry and fully saturated soils.

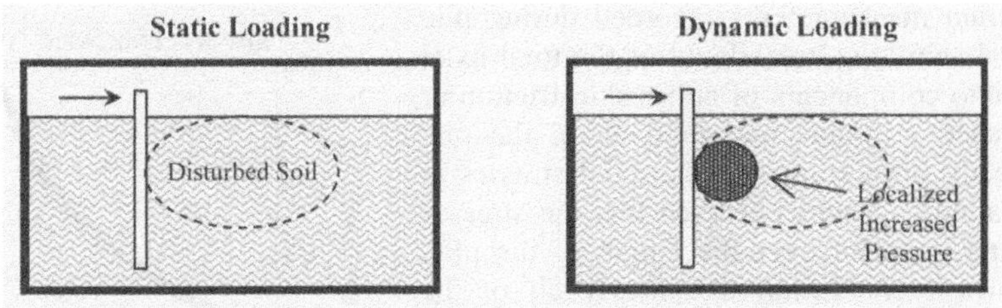

Figure 10. Differences in effected soil region caused by loading conditions

Axial Resistance - A displacement controlled mechanism which provided a constant rate of embedment of 0.3 mm/s was used to drive the model piles. Throughout these series of tests parameters were varied including; soil density, pore fluid and pile tip condition. A comparison was made relating the axial resistance provided when the model was under Earth's gravity (1g) and when it was exposed to large centrifuge imposed accelerations (40g). Figures 11a-d compare measurements obtained during installation and provide insight into the contribution from each studied parameter on the effect it had on the axial capacities provided by the pile. This information may be important because the tip resistance measured during installation is likely related to other properties contributing to the behavior of laterally loaded piles.

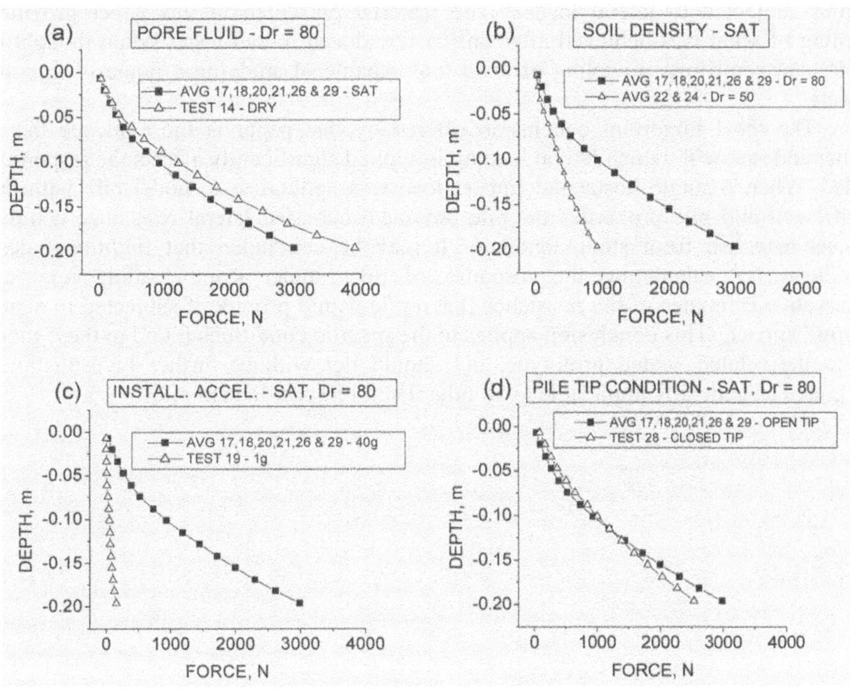

Figure 11. Axial Pile Resistance vs. Embedment Relationships

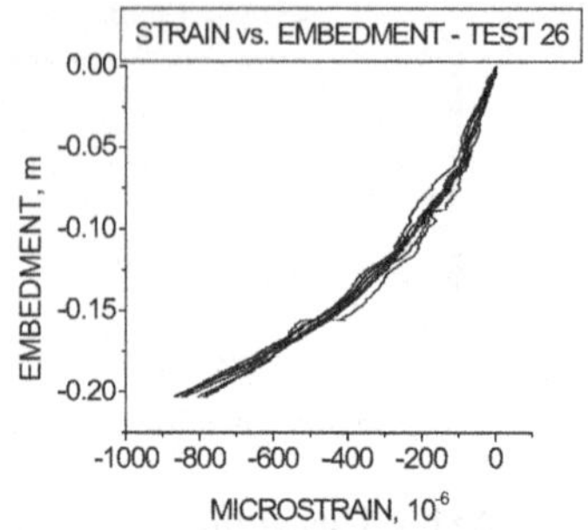

Figure 12. Strain vs. Embedment

Strain measurements recorded during pile installation show the breakdown of the total axial capacity into components of either skin friction or tip resistance. Strains measured along the pile were directly proportional to the axial stresses at those locations. It was observed that the stresses were nearly constant over the length of the pile. This allows a conclusion that nearly all of the capacity provided by the pile resulted from tip resistance. Figure 12 shows strain vs. embedment measurements recorded at 8 locations spaced over the length of a pile. The purpose of this figure is to show that these 8 measurements were similar. A greater contribution from skin friction would have resulted in stresses along the pile decreasing with depth. Skin friction did not contribute to the capacity of these piles because the sand used within these models offered little cohesive strength and the smooth epoxy coating on the pile provided a low angle of friction between the epoxy and the sand.

7. CONCLUSIONS

There is a wealth of information available through the use of centrifuge modeling of piles subjected to lateral loads. The material presented in this paper provides a sampling of what was achieved after only a few dozen model test. When thoughtfully constructed centrifuge modeling offers a tool capable of studying a range of important subjects.

The most important conclusion offered by this paper is the evidence that the manner and rate with which lateral loading is applied significantly affects the response of a pile. When a single horizontal impact load was applied to a model pile with these specific soil and pile properties the pile provided increased lateral resistance compared with its response from static loading. It may be concluded that traditional design procedures for calculating the response of piles under static loading may offer conservative estimates of the resistance that a pile would provide if subjected to a single dynamic impact. This conclusion applies to the specific conditions found in these models and to the related scaled prototype, but should not without further investigation be interpreted as generally applicable to all other lateral impact loaded piles.

8. REFERENCES

API RP 2A-WSD (1993), *Recommended Practice for Planning, Designing and Constructing Fixed Offshore Platforms - Working Stress Design,* 20th Ed., Washington, American Petroleum Institute, 65-67.

Arulmoli, K., Muraleetharan, K. K., Hossain, M. M., and Fruth, L. S. (1992). "VELACS: Verification of liquefaction analyses by centrifuge studies, laboratory testing program, soil data report." *Project No. 90-0562*, The Earth Technology Corporation, Irvine, Calif.

Chang, Y.L. (1937). "Chang on Lateral Pile-Loading Tests." Transaction of the American Society of Civil Engineers, v.102, 272-278.

Grundhoff, T., Latotzke, J., & Laue, J. (1998) "Investigation of Vertical Piles Under Horizontal Impact." *Centrifuge 98*, Rotterdam: Balkema, 569-574.

Hetyenyi, M.I. (1946). *Beams on Elastic Foundation; Theory with Applications in the Fields of Civil and Mechanical Engineering.* Ann Arbor: The University of Michigan press.

King, G.J.W. (1994). "The Interpretation of Data from Tests on Laterally Loaded Piles." *Centifuge 94*, Rotterdam: Balkema, 515-520.

Kusakabe, O. (1995). "Foundations.", *Geotechnical Centrifuge Technology*, Ed. R.N. Taylor, Blackie Academic & Professional, 118-167.

Ling, H.I., Mohri, Y., Kawabata, T., Liu, H., Burke, C., and Sun, L. (2003). "Centrifugal Modeling of Seismic Behavior of large-Diameter Pipeline in Liquefiable Soil." *Journal of Geotechnical and Geoenvironmental Engineering* , ASCE, 129(12), 1092-1101.

Matlock, H. & Reese, L.C. (1960) "Generalized Solutions for Laterally Loaded Piles." *Journal of the Soil Mechanics and Foundations Division*, ASCE, 86(5), 63-91.

Phillips, R. (1995). "Centrifuge modelling: practical considerations." *Geotechnical Centrifuge Technology*, Ed. R.N. Taylor, Blackie Academic & Professional, 34-60.

Poulos, H. G. & Davis, E. H. (1980). *Pile foundation analysis and design.* New York: Wiley.

Reese, L.C. (1977). "Laterally loaded piles: Program documentation." Journal Geotechnical Engineering Division, ASCE, 103(4), 287-305.

Reese, L.C. (1983). "Behavior of piles and pile groups under lateral load." *Rep. to the U.S. Dept. of Transportation*, Federal Highway Administration, Office of Research, Development, and Technology, Washington, D.C.

Taylor, R.N. (1995). "Centrifuge modelling: principles and scale effects." *Geotechnical Centrifuge Technology*, Ed. R.N. Taylor, Blackie Academic & Professional, 19-33

Tsinker, G.P. (1997). *Handbook of Port and Harbor Engineering: Geotechnical and Structural Aspects.* New York: Chapman & Hall

Soil Stress-Strain Behavior: Measurement, Modeling and Analysis
Geotechnical Symposium in Roma, March 16 & 17, 2006

A SIMPLIFIED SOIL-STRUCTURE INTERACTION BASED METHOD FOR CALCULATING DEFLECTION OF BURIED PIPE

Ashutosh Sutra Dhar
Assistant Professor
Department of Civil Engineering
Bangladesh University of Engineering and Technology, Dhaka-1000, Bangladesh
E-mail:ashutoshdhar@ce.buet.ac.bd

And

Md. Aynul Kabir
Graduate Student
Department of Civil Engineering
Bangladesh University of Engineering and Technology, Dhaka-1000, Bangladesh
E-mail: aynulce@yahoo.com

ABSTRACT

Soil-pipe interaction analysis was performed using the continuum theory solution and the finite element method to develop simplified equations for deflection of buried flexible pipes. The hoop and bending components of pipe deflections were studied extensively to determine the influence of different soil and pipe parameters on deflection calculations. Then, two separate simplified equations were developed for the hoop and bending components of the pipe deflection. Two factors were incorporated in the equation for bending deflection to capture the effects of different parameters. Values of those factors were determined for steel and thermoplastic pipes. The proposed simplified equations logically incorporate the hoop and bending stiffness of the soil-pipe interaction.

1. INRODUCTION

Underground pipe has been used to transport potable water to city dwellers and to remove wastewater from cities perhaps since the beginnings of modern civilization. Flexible pipes were introduced for these applications at the turn of the twentieth century, and their use increased steadily. Performance limits used in the design of buried flexible pipes include excessive deflection, wall crushing, and global buckling. The limits were derived empirically from pipe tests and failure data of pipes in service.

The first aspect of the design of flexible pipe is to limit the deflection of the pipe under overburden and live loads. Excessive deflections of the pipe may affect integrity of the joints and/or cause excessive ground settlements. A semi-empirical deflection equation developed at the Iowa State University has generally been used to calculate pipe deflections. Spangler [1] developed the equation, known as the "Iowa Formula", using assumptions based on his observations during field-loading experiments on corrugated metal pipe culverts. Spangler [1] expressed horizontal deflection as a function of the

Hoe I. Ling et al. (eds.), Soil Stress-Strain Behavior: Measurement, Modeling and Analysis, 909–919.

vertical load and the bending resistance provided by the pipe and the surrounding soil. The vertical deflection was generally assumed as equal and opposite to the horizontal deflection. Watkins later performed model studies and examined the Iowa formula, from which some modifications were made to incorporate a soil parameter with dimensions equivalent to modulus in the equation [2]. The Modified Iowa Formula has been the principal tool for estimating deflection for the past 50 years.

While the Modified Iowa Formula may work reasonably well under some circumstances, a method based on soil-structure interaction analysis would provide more reliable results. McGrath [3] proposed a simplified equation to calculate the vertical deflection of flexible thermoplastic pipes. McGrath's equation was based on the continuum theory (Burn and Richard, 1964), finite element analysis, and field observations. McGrath's equation is expressed in similar form as the Modified Iowa Formula [2], but incorporates the deflection contributions from both the hoop stiffness and the bending stiffness of the soil-pipe interaction distinctively. The Spangler equation does not include the hoop stiffness in the deflection calculation. However, Dhar [4] demonstrated that the simplified equation of McGrath [3] over-estimate the hoop component of deflection. Bending deflections also appeared to depend on number of parameters, which are not included in the proposed equation of McGrath, [4]. Therefore, there is a need to study extensively the basic principles that model the soil-structure interaction appropriately in order to develop an understanding of the behavior of buried pipe and to develop simplified design equations. Moore [5] adapted the two-dimensional elastic continuum theory of Hoeg [6] to calculate pipe deflections and strains. The continuum theory is rigorous and permits the development of unified design methods that cover metal, concrete, and polymer pipes. However, the equations derived from the theory are too rigorous for use in codified design procedures. A design procedure based on simplified equations would be preferred by the Engineers. The objective of this paper is to develop simplified design equations for deflection of buried pipe based on a rigorous study of the solution from the continuum theory and the finite element method.

2. CONTINUUM THEORY

Burns and Richard [7] analyzed the idealized post-stressed soil-pipe system subjected to vertical and horizontal pressures using the principle of continuum mechanics and developed a number of closed form solutions. The analysis presents the interaction of an elastic, circular cylinder embedded in a linearly elastic, isotropic, and homogeneous continuum under plain strain condition. Hoeg [6] derived the formulation for the more generalized case of $\sigma_h = K\sigma_v$, where σ_v, σ_h are vertical and lateral earth pressures, respectively and K is the coefficient of earth pressure at rest.

Moore [5] adopted Hoeg's solution for the analysis of buried circular pipes and culverts. However, to simplify the interpretation, the two-dimensional load system was divided into uniform (σ_m) and non-uniform components (σ_d) of pressures as shown in Fig. 1, [5]. The uniform (σ_m) and non-uniform components (σ_d) of pressures were expressed by Moore [5] in the pattern of Eq. (1).

$$\sigma_m = \frac{\sigma_v + \sigma_h}{2}$$
$$\sigma_d = \frac{\sigma_v - \sigma_h}{2} \quad (1)$$

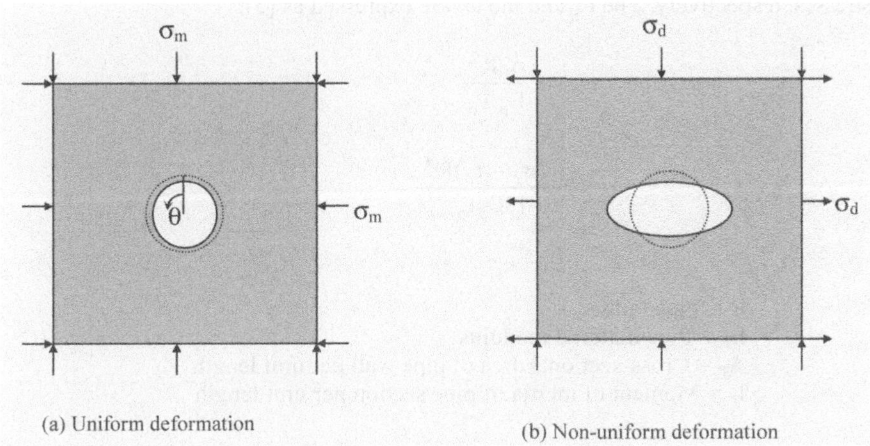

Figure 1: Uniform and non-uniform deformation of buried pipe
(After Moore, 2000)

Distributions of interface radial and shear stress on the external boundary of the pipe were then defined as:

$$\sigma = \sigma_o + \sigma_2 \cos 2\theta$$
$$\tau = \tau_2 \sin 2\theta \quad (2a)(2b)$$

Where θ is measured from the vertical axis, and

$$\left.\begin{aligned} \sigma_o &= A_m \sigma_m \\ \sigma_2 &= A_{d\sigma} \sigma_d \\ \tau_2 &= A_{d\tau} \sigma_d \end{aligned}\right\} \quad (3a)\ (3b)\ (3c)$$

Factors A_m, $A_{d\sigma}$ and $A_{d\tau}$ are called arching coefficients. These provide stresses on the pipe in terms of the mean and deviatoric components of field stresses. The coefficients are defined further in a subsequent section.

The uniform component of pressure (σ_m) produces hoop compression (circumferential shortening) and the non-uniform component of pressure (σ_d) generates out-of-round (ovaling) deflections to the pipe. These two components of stress were calculated separately and the principle of superposition was used to obtain the combined effect. Thus, the radial deflection of the pipe (ω) was expressed as [5]:

$$\omega = \omega_o + \omega_2 \cos 2\theta \tag{4}$$

Here, ω_o and ω_2 are the radial deflections due to the uniform and non-uniform component of stresses, respectively. The ω_o and the ω_2 are expressed as [5]:

$$\omega_o = -\frac{\sigma_o R^2}{E_p A_p} \tag{5}$$

$$\omega_2 = -\frac{(2\sigma_2 - \tau_2) R^4}{18 E_p I_p} \tag{6}$$

Here,

R = Pipe radius
E_P = Pipe material modulus
A_P = Cross-sectional area of pipe wall per unit length
I_P = Moment of inertia of pipe section per unit length

The hoop and bending components of pipe deflections (diametral) were studied individually, and are discussed in the following sections. The two deflections are defined as the changes in pipe diameter due to the uniform and the non-uniform components of earth pressure, respectively, and can be obtained from the radial deflections ω_o and ω_2 (Eqs. 5 and 6) as $W_o = 2\omega_o$ and $W_2 = 2\omega_2$, respectively.

3. HOOP COMPONENT OF DEFLECTION

In the continuum theory solution, the hoop component of deflection (ω_o) is as shown in Eq. (5). The term σ_o of Eq. (5) is expressed in terms of A_m, Eq. (3a), where A_m is defined as [5]:

$$A_m = \frac{2(1-\nu_s)}{1 + C(1-2\nu_s)} \tag{7}$$

Here,

$$C = \frac{E_s D}{2(1+\nu_s)(1-2\nu_s) E_p A_p} \tag{8}$$

Substituting C of Eq. (8) into Eq. (7),

$$A_m = \frac{2(1-\nu_s) E_p A_p}{E_p A_p + R \dfrac{E_s}{(1+\nu_s)}} \tag{9}$$

Now, substitution of σ_o (and A_m) in Eq. (5) yields,

$$\omega_o = \frac{2R(1-\nu_s)\sigma_m}{E_s\left(\frac{E_pA_p}{E_sR} + \frac{1}{(1+\nu_s)}\right)} \tag{10}$$

Thus, the hoop component of pipe deflections ($W_o = 2\omega_o$) is obtained as (replacing σ_m using Eq. 1):

$$W_o = \frac{2R(1-\nu_s)(1+K)\sigma_v}{E_s\left(\frac{E_pA_p}{E_sR} + \frac{1}{(1+\nu_s)}\right)} \tag{11}$$

Eq. (11) can be expressed in a non-dimensional form:

$$\frac{W_oE_s}{D\sigma_v} = \frac{(1-\nu_s)(1+K)}{\left(\frac{E_pA_p}{E_sR} + \frac{1}{(1+\nu_s)}\right)} \tag{12}$$

In Eq. (12), $D\sigma_v/(W_0E_s)$ represents a non-dimensional hoop deflection. The term $E_sR/(E_pA_p)$ is generally defined as hoop stiffness (S_h) of soil-pipe interaction [5]. Bending stiffness of the interaction is also defined as: $S_b=E_sR^3/(E_pI_p)$, [5].

Thus, hoop deflection is expressed in terms of the soil modulus (E_s), Pipe modulus (E_p), Poisson's ratio of the soil (ν_s) and the coefficient of lateral earth pressure (K). However, the parameters ν_s and K are not included in the simplified equation of McGrath [3] for hoop deflection. The influences of these parameters on hoop deflection have been examined here.

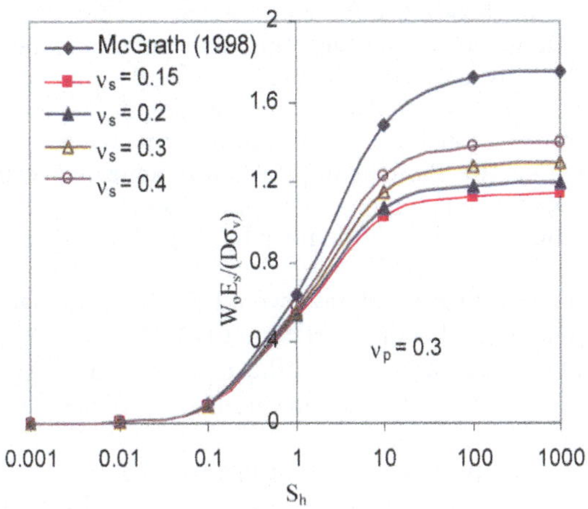

Figure 2: Hoop Component of Deflections

Figure 2 shows hoop deflections plotted as a function of hoop stiffness, [$S_h = E_sR/(E_pA_p)$]. Deflections in Fig. 2 are expressed in a non-dimensional form defined as $W_0E_s/(D\sigma_v)$. Equation (12) does not include any term for the bending stiffness (S_b) and therefore not considered. Hoop deflection calculated using the McGrath [3] equation is also plotted in the figure. Figure 2 shows that the simplified equation proposed by McGrath always overestimates the deflection. The solution from the continuum theory solution for different values of ν_s is included in the figure, which reveals that Poisson's ratio of soil significantly influences the hoop deflection. The greater the Poisson's ratio, the greater the pipe deflection.

The discussion presented above demonstrates that a simplified equation, expressed in terms of hoop stiffness only, may not estimate the hoop deflection appropriately. The contribution of the lateral earth pressure and Poisson's ratio should be included in the equation. Rearranging the terms in Eq. (12), the equation for estimating hoop deflection can be expressed as:

$$\frac{W_o}{D} = \frac{(1+K)(1-\nu_s)\sigma_v}{\left(\frac{E_pA_p}{R} + \frac{E_s}{(1+\nu_s)}\right)} \tag{13}$$

This equation provides a simplified form of the continuum theory solution for hoop deflection.

4. BENDING COMPONENT OF DEFLECTION

Equation (6) shows the bending component of deflection according to the continuum theory solution. The parameters in Eq. (6) are expressed as functions of σ_s, S_h and S_b [7]. Figure 3 compares the bending deflections based on the Iowa Formula and on the continuum theory. Deflections from the continuum theory solution with different values of hoop stiffness (S_h) are compared in Fig. 3. It shows that for low bending stiffness (S_b<10), each of the methods calculate similar bending deflections. The bending stiffness is less for ductile iron and long-span reinforced concrete pipes. However, for flexible and compressible pipes (S_b>10), the bending deflection appeared to be influenced by other factors of the pipe-soil interaction. Continuum solution provides higher or lower deflections for flexible pipes than those given by Spangler's equation.

While the continuum theory solution would provide rational modeling of soil-pipe interaction, the equation for bending deflection is rather complicated (Eq. 6). The non-uniform component of radial stress σ_2 and shear stress τ_2 of Eq. (6) are expressed in term of deviatoric stress components, through the arching coefficients $A_{d\sigma}$ and $A_{d\tau}$, where

$$A_{d\sigma} = 4(1-\nu_S)(4+3C(1-2\nu_S)-2F)/\Delta \tag{14}$$

For a bonded interface

$$A_{d\sigma} = 12(1-\nu_S)/(2F+5-6\nu_S) \tag{15}$$

For a smooth surface

$$A_{d\tau} = 16(1-\nu_S)(F+1)/\Delta \qquad (16)$$

For a bonded surface

$$A_{d\tau} = 0 \qquad (17)$$

For a smooth surface

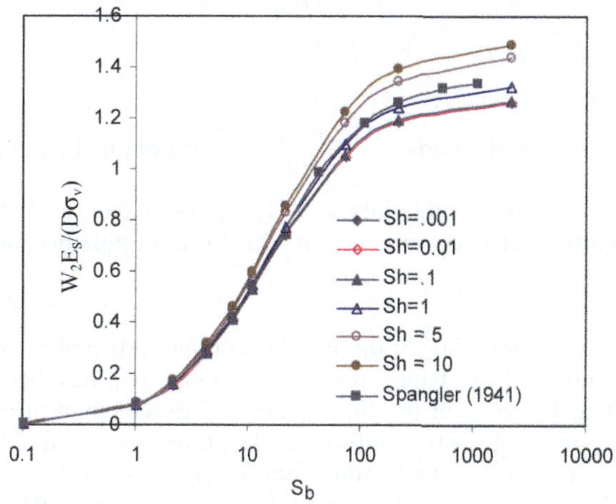

Figure 3: Bending Deflections

The denominator in Eqs. (14) and (16) is given by

$$\Delta = C(1-2\nu_S)(5-6\nu_S+2F)+2F(3-2\nu_S)+4(3-4\nu_S) \qquad (18)$$

The two stiffness parameters C and F are defined as [6]:

$$C = E_S D/[2(1+\nu_S)(1-2\nu_S)E_P A_P \qquad (19)$$

$$F = E_S D^3/[48(1-\nu_S)E_P I_P] \qquad (20)$$

Dhar [4] expressed the complicated bending component of deflection in a simplified form as in Spangler [1], but incorporating rationally the effects of parameters involved in the continuum theory solution. The equation of bending deflection was expressed as [4]:

$$\frac{W_2}{D} = \left(\frac{(1-K)A.\sigma_v}{\frac{EI}{R^3} + B.E_s} \right) \qquad (21)$$

Where 'A' and 'B' are arbitrary constants, accounting for the effects of the parameters of the continuum solution, and W_2 is the bending deflection. In the Iowa equation a factor of 0.061 was used instead of 'B' [1].

To determine the value of the respective constants of Eq. (21), the equation is expressed a linear from:

$$\frac{E_P I_P}{E_S R^3} = A\left(\frac{(1-K)D.\sigma_v}{E_S.W_2}\right) - B \tag{22}$$

Then, $\frac{E_P I_P}{E_S R^3}$ is plotted against $\left(\frac{(1-K)D.\sigma_v}{E_S.W_2}\right)$ as obtained from the continuum theory solution. Pipes of different materials namely thermoplastic and steel were considered separately to determine the constant 'A' and 'B' for each pipe product.

Figure 4 shows the relation for thermoplastic pipes. All typical range of thermoplastic pipe materials and profile shapes reported in Dhar [8] has been considered in this investigation. Dhar [8] studied the limit state of different pipe products and determined the sectional properties of profile wall pipes. The cross-sectional area per unit length was reported as from 10 mm^2/mm to 16 mm^2/mm for common profile wall pipes. Moment of inertia per unit length was ranged between 420 mm^4/mm and 10500 mm^4/mm. Modulus of elasticity for the pipe material varied from 150 MPa (long-term modulus of HDPE) to 3000 MPa (PVC pipe).

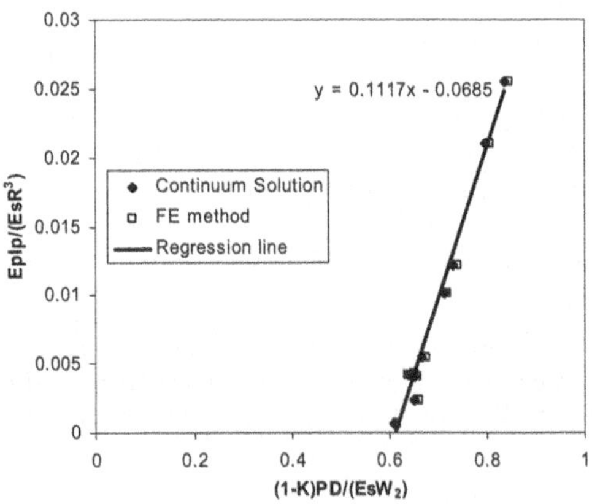

Figure 4: Study of bending deflection for thermoplastic pipe

Finite element Analysis was also used to calculate bending deflection in order to evaluate those obtained from the continuum theory. Figure 4 includes the results of the finite element analysis. Data points with open symbol correspond to finite element results, while those with closed symbol correspond to continuum solution. It reveals from the figure that the non-dimensional deflections obtained from the two methods are almost the same. Figure 5 shows the finite element mesh used, along with the idealization of the non-uniform component of stress (Fig. 1). Smooth rigid boundary was used along both horizontal and vertical lines of symmetry. Material and pipe's sectional parameters were the same as those used in the continuum theory solution. A general purpose finite element software ANSYS was used in the finite element analysis.

Figure 4 reveals that for the typical ranges of sectional parameters with the typical range of the modulus of elasticity of pipe materials, a linear relationship can be obtained between $E_PI_P/(E_SR^3)$ and $(1\text{-}K)D\sigma_v/(E_SW_2)$. Equation of the regression line is also shown in the figure. The regression line shows a slope of 0.112 and an intercept of -0.069, indicating the values of the factors 'A' and 'B' of eq. (21) as A=0.112 and B=0.069. It appears that the factor 'B' is similar to the factor of Iowa equation (i.e. 0.061).

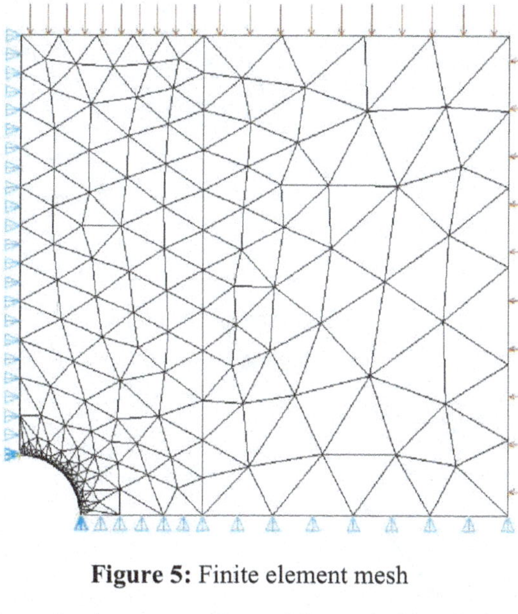

Figure 5: Finite element mesh

A similar investigation was performed for typical steel pipes to determine the factors 'A' and 'B'. Cross-sectional area and moment of inertia per unit length considered were 1.06 mm^2/mm to 5.59 mm^2/mm and 5.46 mm^4/mm to 546 mm^4/mm, respectively.

Figure 6 shows $\frac{E_PI_P}{E_SR^3}$ versus $\left(\frac{(1-K)D\sigma_v}{E_SW_2}\right)$ relation for the steel pipe. Solution from Finite Element Analysis and Continuum theory is included in Fig. 6. Both of the methods appeared to provide similar results as in the case of thermoplastic pipe. Equation of regression line shown in Fig. 6 reveals A=0.125 and B=0.08 for steel pipe. This is interesting to note that both of the factors are higher for the steel pipes (pipes with higher modulus of elasticity) compared to those for the thermoplastic pipes. However, the Iowa formula does not consider the effect.

5. PROPOSED SIMPLIFIED EQUATION

Study presented in this paper reveals that simplified deflection equation can be developed from the continuum theory solution if different pipe products are considered separately. Hoop and bending component of deflection were studied here extensively for thermoplastic and steel pipe. It was revealed that Eq. (13) and Eq. (20) can be used to calculate the hoop and bending components of pipe deflections, respectively. Factors 'A' and 'B' of Eq. (20) has been obtained for thermoplastic and steel pipes. Substitution of the values of the factors in Eq. (20) lead to the simplified equations for bending deflection as follow:

For steel pipe, $$\frac{W_2}{D} = 0.125\left(\frac{(1-K)\sigma_v}{\frac{EI}{R^3} + 0.08E_s}\right) \tag{23}$$

For thermoplastic pipe, $$\frac{W_2}{D} = 0.112\left(\frac{(1-K)\sigma_v}{\frac{EI}{R^3} + 0.069E_s}\right) \tag{24}$$

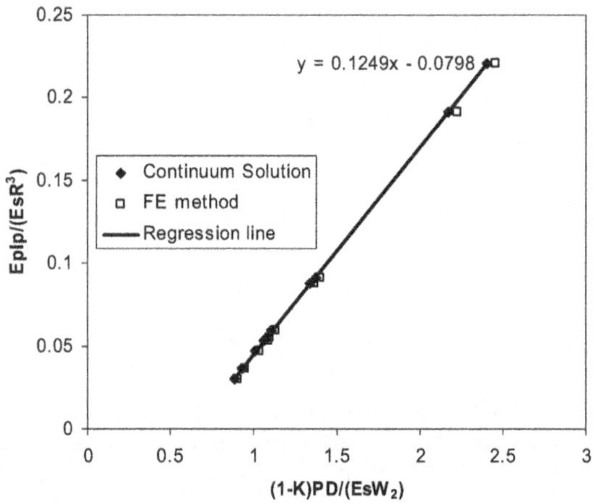

Figure 6: Study of bending deflection for steel pipe

6. CONCLUSION

Deflections of buried flexible pipes were investigated in this research using the continuum theory solution and the finite element analysis. The study revealed that simplified forms of equation can be developed to calculate pipe deflections when pipes of different material are considered separately. Two different pipe materials namely thermoplastic and steel were considered in this research to develop the simplified deflection equations.

Deflection of pipes was expressed in its two components: the hoop and the bending components, to develop the simplified equations. The hoop component of deflection could be expressed in a simplified form from the continuum theory solution. However, the bending component in the continuum theory solution is complicated. The bending component was found to depend on number of parameters including the hoop stiffness and the bending stiffness of the soil-pipe interaction, and the coefficient of lateral earth pressure, particularly for pipes with $S_b>10$.

The bending component was also expressed here in a simplified form incorporating two factors 'A' and 'B' to account for the effects of the parameters involved in the continuum theory solution. The factors were determined for typical thermoplastic and steel pipes. The study showed A=0.112 and B=0.069 for thermoplastic pipes, and A=0.125 and B=0.08 for steel pipes.

REFERENCES

[1] Spangler, M.G. (1941) "The Structural Design of Flexible pipe Culverts." Bulletin 153, Iowa Engineering Experiment Station, Ames, Iowa.

[2] Watkins, R.K. and Spangler, M.G. (1958) "Some Characteristics of Modulus of Passive Resistance of Soil: A Study in Similitude", Proc HHRB, Vol. 37, pp: 576-583..

[3] McGrath, T.J. (1998) "Design method for flexible pipe", A report to the AASHTO Flexible Culvert Liaison Committee, Simpson Gumpertz & Heger Inc., Arlington, MA.

[4] Dhar, A.S. (2003) "The Development of a Simplified Equation for Deflection of Buried Pipe", ASCE International Pipeline 2003 conference, Baltimore, Maryland, July 13-16. pp. 1096-1105.

[5] Moore, I.D. (2000) "Culverts and Buried Pipelines" Chapter 18, Geotechnical and Geoenvironmental Handbook, Edited by R.K. Rowe, Kluwer publisher, pp. 541-568.

[6] Hoeg, K. (1968) "Stress Against Underground Cylinder", Journal of Soil Mechanics and Foundation Engineering, ASCE, Vol. 94, SM4, 833-858.

[7] Burns, J.Q. and Richard, R.M. (1964) "Attenuation of stresses for Buried Cylinders", Proceedings of Symposium on soil-structure interaction, University of Arizona, pp: 379-392.

[8] Dhar, A.S. (2002). "Limit States of Profiled Thermoplastic Pipes under Deep Burial" Ph.D. Thesis, Department of Civil and Environmental Engineering. University of Western Ontario. London. Canada. 285 p.

Soil Stress-Strain Behavior: Measurement, Modeling and Analysis
Geotechnical Symposium in Roma, March 16 & 17, 2006

ALTERNATIVE REMEDIAL TECHNIQUES FOR SHEET-PILED EARTH EMBANKMENTS

Ahmet Pamuk
Department of Civil Engineering & Engineering Mechanics, Columbia University
New York, NY 10027, USA
pamuk@civil.columbia.edu

Korhan Adalier
Department of Civil & Environmental Engineering, Florida State University
Panama City, FL 32405, USA
kadalier@pc.fsu.edu

ABSTRACT

This paper proposes two different remediation techniques, slightly different from conventional-type sheet-piled earth embankments, to reduce adverse effects induced by foundation liquefaction. A total of four centrifuge tests were conducted without and with countermeasure techniques, all involving model sheet-piles. The effectiveness of each countermeasure were compared and discussed based on the recorded displacements, accelerations, pore water pressure measurements and post-earthquake deformations. The tests showed that conventional sheet pile retrofitting method may not be fully adequate to reduce the distress in the embankment cased by liquefaction. However, the utilization of proposed countermeasures was found to be more significant in reducing the embankment settlement, deformation and cracking. Besides, they are practical and can be easily applied with less expense to existing structures.

1. INTRODUCTION

Liquefaction-induced ground displacements resulting from earthquake shaking are a major cause of damage to earth structures underlain by loose saturated granular soils. Many liquefaction induced failures of structures such as highway/railway embankments, river dikes, and earth dams have been reported during recent earthquakes. Damage to embankments have been particularly destructive when the foundation soils liquefied, resulting in cracking, settlement, lateral spreading, and slumping of the embankments as shown in Figure 1. Such earthquake liquefaction hazard increased the demand to develop appropriate remediation countermeasures as well as to predict the embankment damage against earthquakes (Marcuson et al., 1996; JGS, 1996 and 1998; Adalier et al., 1998; Sasaki et al., 2004).

At present, various soil treatment methods (e.g., densification, solidification and dissipation of pore water pressure) and structural remediation measures (e.g., counterweight berms and sheet-piles) are used to reduce the effects of foundation liquefaction induced damage for earth embankments. Most of these soil treatment

Hoe I. Ling et al. (eds.), Soil Stress-Strain Behavior: Measurement, Modeling and Analysis, 921–930.

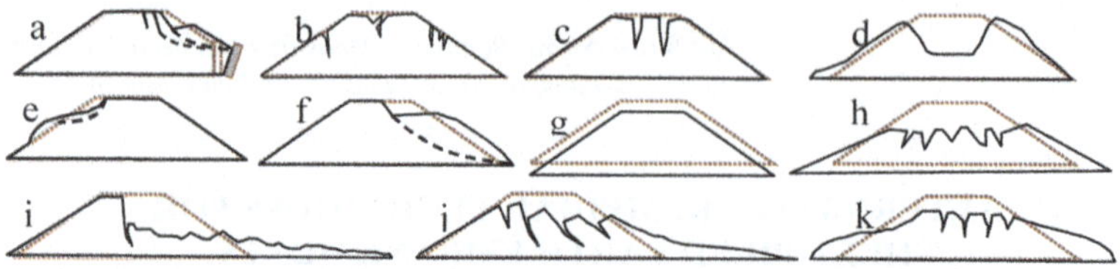

a	Failures induced by retaining walls	e	Shoulder failure	i	Flow-type slide
b	Cracks on the top and slope	f	Slope failure	j	Multiple sliding surfaces
c	Parallel cracks on the top	g	Bearing capacity failure	k	Cracks and settlement due to bulging
d	Slumping of the bank	h	Failure with foundation ground		

Figure 1. Schematic of various embankment failures based on case histories (after Yamada, 1966; Mishima and Kimura, 1970; Sasaki, 1980; Tani, 1993).

techniques have limited application to existing embankment/dikes, while sheet-pile walls can be applied to the foundations of existing structures.

Construction of sheet-pile walls are simple, and can be easily driven into soft saturated ground at both sides of the embankments. However, the rigidity of the walls is very low so as to sustain large lateral deformations in weak soils without any lateral structural support. Thus, a row of tie rods (Figure 2) is installed on the top of both walls to reduce large lateral movements. For this reason, a small excavation/trench is made along the walls to install tie rods. In practice, the tie-rods are usually placed above ground level to ease their installation.

The use of sheet-piles to enclose liquefiable foundations has been increasingly considered as a remediation technique, especially in transportation facilities (highway, railway embankments) resting on soft ground in areas prone to earthquakes in Japan (JGS, 1998). However, there is no documented case of a rehabilitated earth embankment subjected to strong earthquake shakings, together with a very limited amount of centrifuge model testing the effectiveness of sheet-pile enclosures used as a liquefaction countermeasure method (Zheng et al., 1995; Adalier et al., 1998). Hence, there is a great need for further understanding regarding the potential for reducing ground and embankment deformations to acceptable levels, as well as further understanding of the factors affecting new remedial measure designs and performance.

For this purpose, the effectiveness of three different sheet-pile enclosure methods to protect highway embankments against foundation liquefaction induced deformation and failure is evaluated through a series of instrumented dynamic centrifuge model tests. The characteristics of the dynamic response of both embankment and saturated loose foundation soil under a base motion are presented. The tests (Figure 3) included an

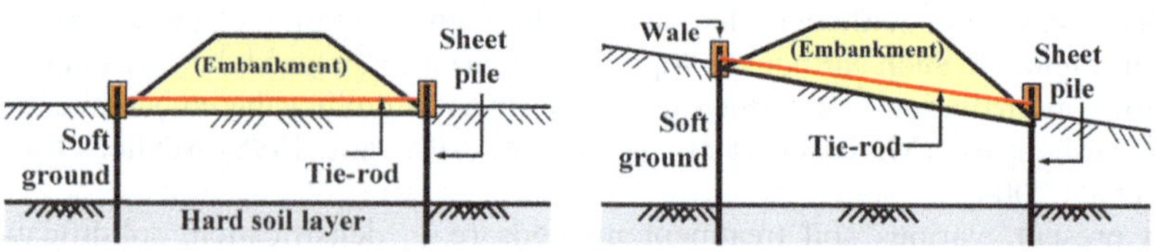

Figure 2. Typical applications of sheet pile-tie rod retrofitted earth embankments constructed on soft ground.

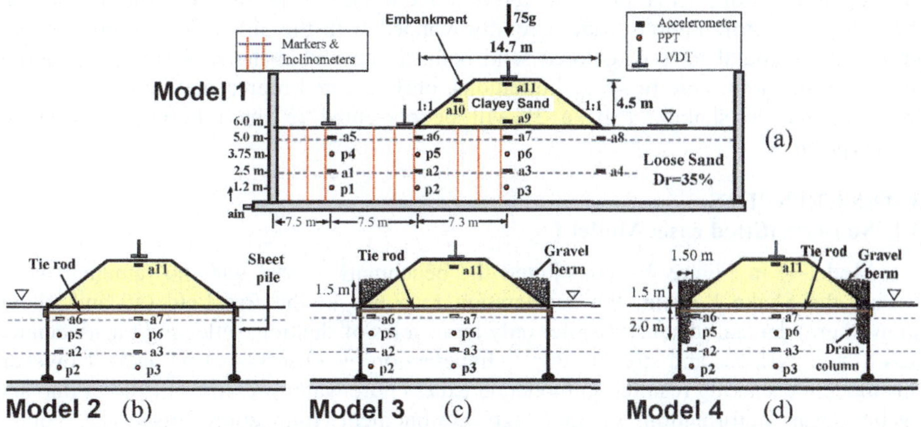

Figure 3. Setups used in centrifuge testing (dimensions in prototype units).

embankment with (a) no countermeasure, (b) sheet pile (conventional method), (c) sheet pile with toe area gravel berm, and (d) sheet pile with toe area gravel berm extending into foundation. The remedial effects as well as deformation characteristics of the models are compared in details.

2. MODEL PREPARATION AND TESTING PROGRAM

The seismic performance of a 4.5 m high (prototype) clayey sand earth embankment sitting on a 6.0 m thick saturated loose foundation soil was studied with and without liquefaction hazard mitigation measures (Figure 3). Nevada No. 120 sand was used as the foundation soil in all tests. Extensive data on the engineering properties of this sand has been documented by Arulmoli et al. (1992). Nevada sand foundation layer was poured into the testing box by dry pluviation with D_r = 35%. The model embankment was prepared using Nevada 120 sand mixed with kaolin clay, utilizing a mixing ratio of 4:1. The berm material in Models 3 and 4 was gravel with a grain size of 2.8 to 4.75 mm. In Models 2-4 (Figure 3), two steel sheet piles (1.6 mm thick in the model; 0.12 m thick prototype) were tied to each other with two steel tie-rods and located in the foundation layer. The bending stiffness of the model pile was selected to be quite high; therefore, lateral deformations due to deflection of the pile were negligible. The tests were carried out using a 3.0-meter radius, 100-g ton geotechnical centrifuge. In all tests, an in-flight shaker was employed to impart the model base shaking.

The foundation of the models were built at 1g and then saturated with de-ionized/de-aired water/glycerin solution (10 times the viscosity of water) under vacuum. Thereafter, the model containers were placed on the centrifuge, and spun to a 75g gravitational field. Each model was subjected to two successive shake-earthquake events of a uniform harmonic base input motion with a prototype frequency at 1.6-hz. The first shaking event (herein referred to as Shake-1) of 10-cycles at 0.12g amplitude simulated a moderate earthquake event. The second shaking event (referred to as Shake-2) of 20-cycles at 0.3g amplitude simulated a moderate to strong earthquake event.

As seen in Figure 3, model response was measured by miniature transducers, including horizontal accelerometers, pore pressure transducers, and linear variable differential transformers (LVDT). In addition, the models were dissected after testing and the deformed configurations were carefully mapped with the aid of soft vertical marker lines and horizontal lines of colored sand both placed within the foundation soil during model construction. Due to space limitations, only selected-representative response data from the second shaking (Shake 2) will be presented in the following sections in prototype units.

3. TEST RESULTS

3.1. Non-retrofitted case: Model 1

Model 1, in Figure 3a, constituted the benchmark model with no remedial work. During the Shake-1 event, the acceleration response in the free-field (a5 and a1-not shown here) decreased severely after only two cycles of shaking, reflecting the associated loss of soil stiffness and strength due to liquefaction (as observed at p4 data). However, this moderate shaking resulted in moderate excess pore water pressure (EPP) buildup and cyclic shear deformation in the high confinement foundation areas (i.e., under embankment). Consequently, a large portion of the overall soil strength and stiffness was preserved, reflected in strong foundation soil and embankment accelerations (i.e., essentially motion similar to input prevailed). Relatively moderate embankment crest settlement (0.38 m) was measured. Small tension cracks were observed at the surface of the embankment; however, the overall stability remained intact (i.e., lateral extension was small).

Figure 4a shows model response at selected transducer locations during the Shake-2 event. The free-field reached a fully liquefied state in a cycle or two of shaking (see a5). Actually during all tests (Models 1-4), the free-field reached full liquefaction (r_u=1.0) within a few cycles of the base input motion. Significant attenuation of input motion was observed starting from the free-field and propagating to the foundation due to high EPP induced softening and strength loss. Likewise, the embankment accelerations (a11) were gradually attenuated to about 1/3 of the input motion after 2 cycles of shaking, as the foundation soils became softer due to EPP development. However, a noticeably stiffer acceleration response prevailed within the foundation soil compared to the free-field as the foundation EPP fell short of initial liquefaction. Due to large shear and normal strains (as indicated by the post-test displacement vectors, Figure 6) induced in the foundation and the associated dilation effect the foundation EPP (p6) did not reach initial liquefaction values. The estimated r_u in the foundation was in the range of 0.6~0.7.

A large embankment crest settlement of 1.17 m was measured (Figure 4a) during Shake-2. The embankment crest settled 1.55 m during the two shaking events, and some heaving occurred in the free-field and toe areas due to the outflow of foundation soils in Model 1 (Figure 5).

Figure 6 shows the post-test displacement vectors of Models 1-4. Post-test investigation revealed that most of the embankment settlement in Model 1 was due to lateral spreading of EPP softened (severe loss of shear strength) foundation soil towards the free-field. The asymmetric behavior of the measured accelerating (a6) under the toe was indicative of significant lateral deformation (Elgamal et al., 1996; Adalier and Aydingun, 2003) towards the free-field at this location. In contrast, the lack of appreciable acceleration asymmetry below the embankment centerline demonstrates the absence of lateral strains at this locality. Lateral deformation in the foundation soil below the embankment attained its maximum value near the ground surface, decreased with depth, and become negligible at the base. The deformation was continuous, though not linear, with depth, and maximum lateral strains occurred at shallow foundation zones. This deformation may be associated with an average accumulated normal lateral tensile strain of about 20% along the embankment base. Indeed, this tensile strain was clearly manifested in the form of major cracks (as much as 1.3 m wide and 2.5 m deep) that propagated longitudinally throughout the embankment body (Figure 7). Some minor tensile cracks were also occurred at the slopes. The embankment strained excessively both in vertical and horizontal directions, and its integrity and functionality were practically lost, indicating a catastrophic failure to the highway embankment.

The test results suggested that the main mechanism responsible for the settlement of the embankment was a form of repeated bearing-capacity failure in the foundation due to EPP-induced soil softening. Essentially, some vertical compression accompanied a major lateral expansion that was superimposed on a simple shear type straining. The lateral outflow of the foundation soil was the main contributor to the observed embankment lateral and vertical deformations. Therefore, utilizing a pair of underground sheet-pile walls can be an effective method in reducing embankment deformations by restricting the lateral outflow of the foundation soil, although this obviously will not eliminate seismic-induced foundation liquefaction. Models 2, 3 and 4 (Figure 3) attempt to investigate the efficiency of the sheet-pile retrofitted embankments in various forms.

3.2. Conventional retrofitting with sheet piles: Model 2

Sheet piles were installed at both sides of the embankment (Figure 1). As shown in Figure 3b, Model 2 represents a typical case of a sheet pile enclosure application, a conventional method that was frequently used in engineering practice, as depicted in Figure 2. The model evaluates the effectiveness of using sheet piles to contain foundation soil and mitigate the adverse effects of foundation liquefaction on a cohesive highway embankment. In this model, the top of the pile was located slightly (0.15 m) above the original ground surface. This model represents a typical case of a sheet-pile enclosure application, a conventional method that was used in several projects in Japan.

Figure 4b shows the Model 2 response at selected transducer locations during the Shake-2 event. Accelerations, near the sheet pile (a6), are characterized by pronounced spikes of large amplitudes and strong overall response. This is mainly due to the rigid boundary effect (Adalier et al., 1998) caused by the nearby sheet pile, despite the large EPP buildup. Overall, the recorded accelerations displayed two important differences compared to those of the non-retrofitted model (Model 1, Figure 3a): i) lack of asymmetric acceleration response associated with lateral deformations, and ii) significant decay in acceleration amplitudes in the foundation and the embankment. This decay was associated with high values of EPP (r_u of more than 0.8) within the entire contained foundation domain (significantly larger than those of the benchmark case). The absence of significant lateral deformations and driving shear stresses below the embankment had contributed to these large EPP values. In view of the resulting significant reduction in foundation stiffness, the embankment was more efficiently isolated from the induced base excitation, as documented by the decay of the accelerometer-a11 response. This isolation is believed to have contributed significantly to the observed high degree of embankment integrity.

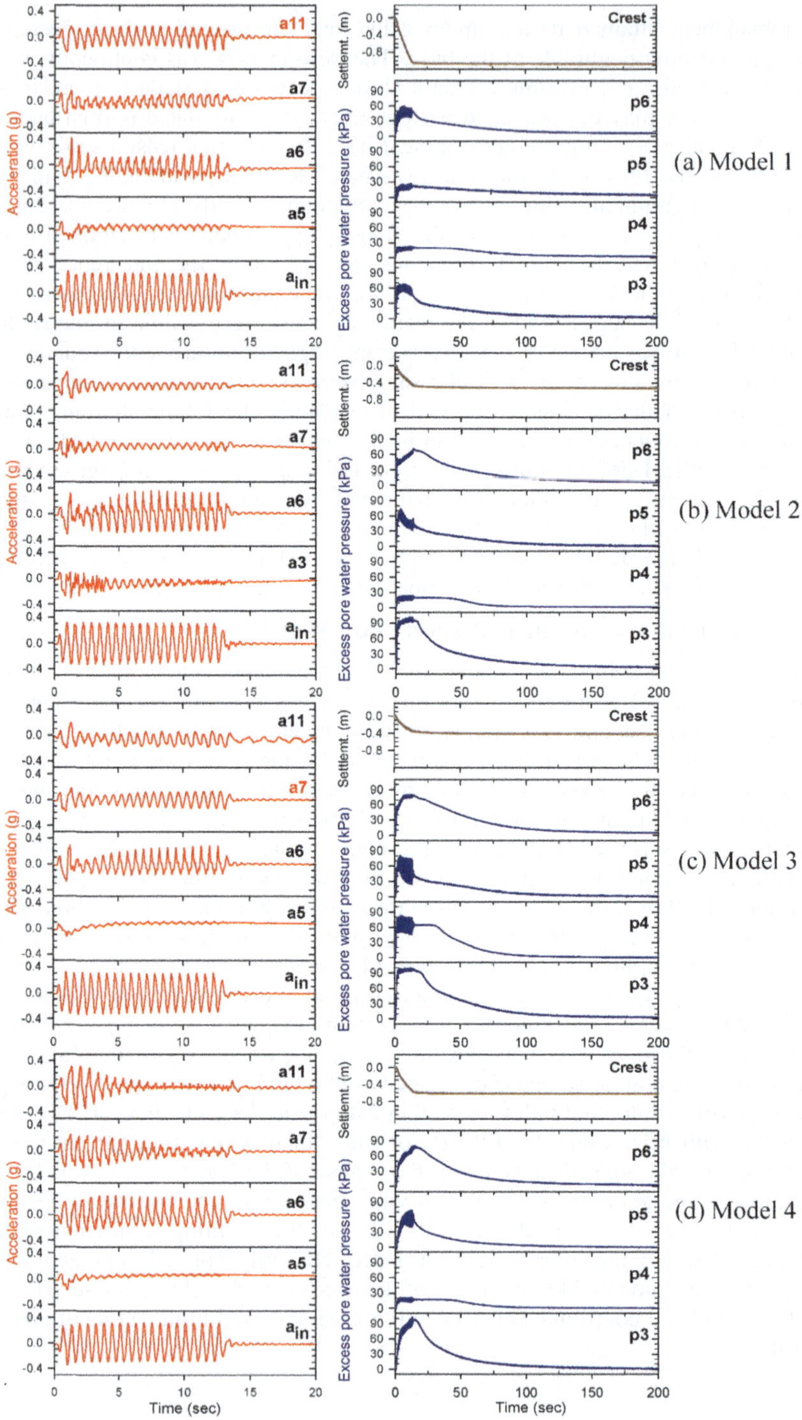

Figure 4. Selected accelerations and excess pore pressures in the free-field and under the foundation, and crest settlements during shake-2 events for Models 1~4.

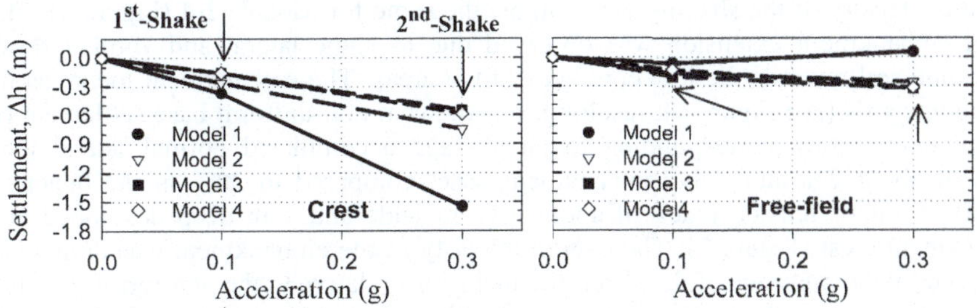

Figure 5. Accumulated settlements (after shake 1 and 2) measured at crest and free-field.

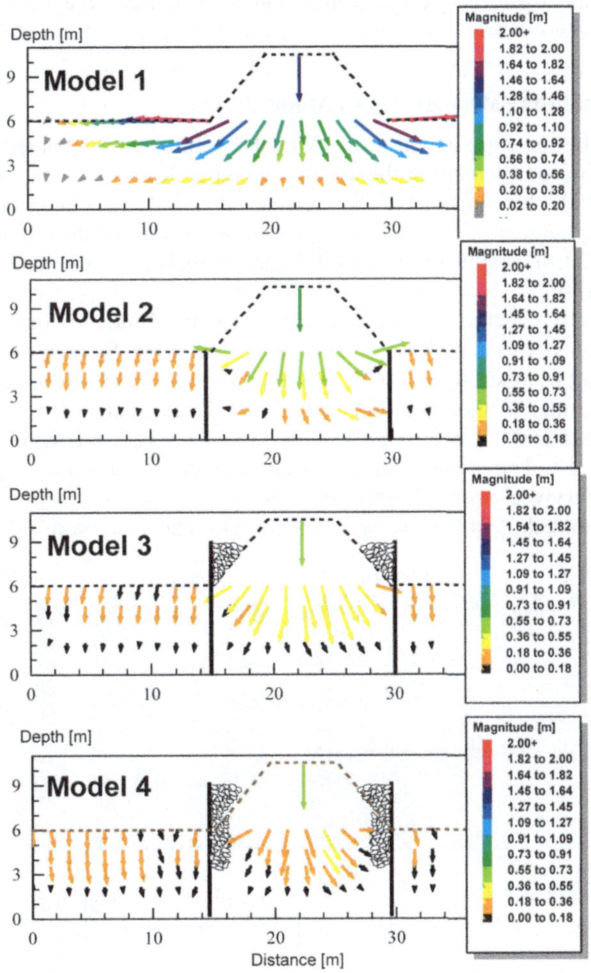

Figure 6. Displacement vectors obtained from dissected models at the end of testing.

After two simulated earthquake events, the embankment crest settled about 0.75 m, about a 50% reduction relative to Model 1 (Figure 5). The free-field settlement increased with the intensity of the shaking were almost the same for Models 2-4 (Figure 5). Minor lateral embankment extension was observed due to some lateral and vertical flow of foundation soil towards the low confinement toe areas. The embankment toe areas were pushed upwards (see Figure 6), resulting in somewhat non-uniform base settlement (i.e., convex deformation corresponding to an average accumulated normal lateral tensile strain of about 7% along the embankment base; compared to 20% in the benchmark model). A minor tension crack of about 0.15 m wide and 1 m deep developed at the embankment crest (Figure 7). The overall integrity of the embankment was dramatically improved by the presence of the sheet-pile inclusions relative to the non-retrofitted model (Model 1), however, the vertical settlement and lateral extension was still high enough to cause considerable damage to superstructures. It was not investigated herein, but a slight extension of the sheet pile above the foundation level could have further reduced the lateral straining by providing additional lateral confinement, with a minimal additional project cost.

3.3. Proposed retrofitting strategy No. I : Model 3

Small gravel berms were added to the Model 2 setup (with extended sheet piles) in order to increase the confinement in the toe areas (Figure 3c). This extra confinement was intended to reduce the heaving of the toe areas and thus the associated lateral extension (although relatively minor) of the embankment that was observed during the Model 2 test. In addition, the berm might offer added lateral support to the embankment, thus increasing its stability.

Figure 4c shows selected recorded responses for the Model 3-Shake 2 event. The addition of gravel berms resulted in somewhat stronger foundation soil behavior (manifested as somewhat stronger model acceleration records, a7 and a11). EPP records were essentially similar to those recorded in Model 2. A total of 0.55 m embankment crest settlement (0.15 and 0.4 m during the Shake 1 and Shake 2 events, respectively) was measured (Figure 5). The added berms further helped to preserve the embankment integrity, which survived with no apparent cracks (Figure 7). Normal lateral tensile strains along the embankment base were less than 1%. The vast majority of the observed

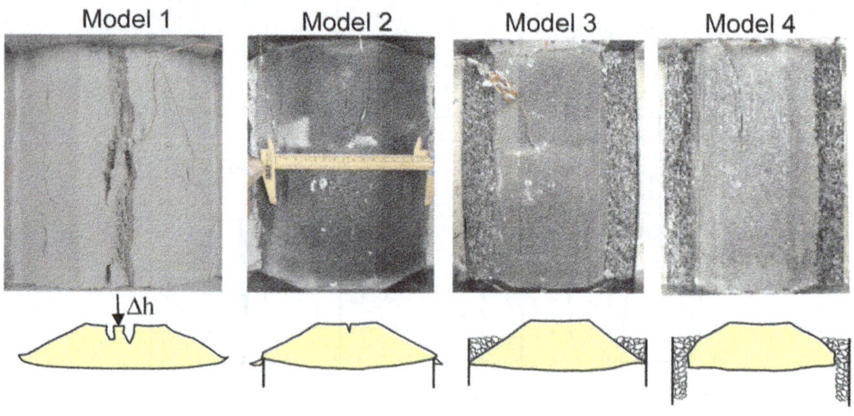

Figure 7. Top view of embankments after testing digitized side views of shapes of embankments.

embankment and foundation settlements can be attributed to post liquefaction densification and consolidation settlements. The only practical ways to reduce these kinds of settlements are either to: i) reduce volume change potential (e.g., densify the soil), and/or ii) reduce EPP buildup below the embankment model.

In an attempt to reduce EPP buildup, the gravel berms used in Model 3 were extended under the toe area about 2 m down into the foundation soil in the next model test (Model 4, Figure 3d). This additional step was chosen since it is easily applicable in the field, with minimal additional cost.

3.4. Proposed Retrofitting Strategy No. II: Model 4

Figure 4d shows selected recorded responses during Shake-2 event for Model 4. The addition of limited gravel drains under the toe initially caused a noticeable delay in the EPP buildup and the soil strength degradation (i.e., decay of accelerations) in the foundation. However, excess pore pressures (EPP) eventually reached the level that was observed in Models 2 and 3. After two shaking events the embankment crest settled 0.59 m (Figure 5). Normal lateral tensile strains along the embankment base were less than 1%. No tensile cracks were observed and the embankment integrity was good (Figure 7). The addition of such limited size gravel drains under the toes did not significantly improve the overall remediation effectiveness, compared to Model 3. This was attributed to the size of drains that were not adequate enough compared to the volume of soil to be drained during shaking.

4. CONCLUSIONS

In this paper, three simple derivations of the sheet-pile enclosure method applied to typical earth embankment configurations are evaluated through a series of dynamic centrifuge model testing simulations. Liquefaction countermeasures with sheet piles involved (a) sheet-pile extending to the foundation surface (Model 2, conventional method of retrofitting with sheet piles), (b) sheet-pile with toe area gravel surcharge berm (Model 3, proposed retrofitting strategy No. I), and (c) sheet-pile with toe area gravel surcharge berm extending into foundation (Model 4, proposed retrofitting strategy No. II).

In the plain embankment model case (Model 1), base shaking induced high excess pore pressures (or liquefaction) resulted in significant reduction in stiffness and shear strength in the loose sandy embankment foundations. This resulted in major lateral spreading or outward migration of the foundation soil leading to large lateral strains in the embankment (tensile cracking caused by embankment base stretching) and excessive settlement of the embankment base. Implemented countermeasures (Models 2, 3 and 4) were found to largely eliminate the lateral spreading of the foundation soils and reduce the embankment vertical deformations. Moreover, liquefaction within the contained foundation (i.e., between sheet piles) was observed to create a base isolation mechanism reducing dynamic shear excitation within the embankment body. However, conventional-type sheet-pile retrofitting (Model 2) may not be fully adequate in reducing the distress (i.e., in terms of settlement, deformation and cracking) caused by shaking and foundation liquefaction. Benefits of the proposed retrofitting strategies No. I and II were more or less the same in reducing the measured embankment deformations. The presence of limited size gravel drains at the toe areas in strategy No. II seemed to have insignificant remedial effects in terms of increasing the overall drainage capacity below the embankment. Thus, the proposed retrofitting method No. I (Model 3), suggesting the sheet-piles extending out of the foundation with small gravel berms added at the embankment toe areas, will be highly effective in preserving the overall integrity of the embankment against soil liquefaction. This method is cost-effective, and can be easily applied for the remediation of the existing conventional-type sheet-piled embankments, as well as for the new embankments to be retrofitted by sheet piles enclosures.

REFERENCES

Adalier, K., Elgamal, A.W., and Martin, G.R. (1998). "Foundation Liquefaction Countermeasures for Earth Embankments," *Journal of Geotechnical and Geoenvironmental Engineering, ASCE*, 123(6), 500-517.

Adalier, K., and Aydingun, O. (2003). "Numerical Analysis of Seismically Induced Liquefaction in Earth Embankment Foundations. Part II, Application of Remedial Measures," *Canadian Geotechnical Journal*, 40(4): 766-779.

Arulmoli, K, Muraleetharan, K.K., Hossain, M.M., and Fruth, L.S. (1992). "Verification of Liquefaction Analysis by Centrifuge Studies Laboratory Testing Program Soil Data," *Technical Report*, EarthTech Corp., Irvine, CA.

Elgamal, A.W., Zeghal, M., Taboada, V., and Dobry, R. (1996). "Analysis of Site Liquefaction and Lateral Spreading using Centrifuge Testing Records," *Soils and Foundations*, 36(2), 111-121.

JGS (1996). "Geotechnical Aspects of the January 17 1995 Hyogoken-Nambu Earthquake," Special Issue of *Soils and Foundations*, 359 p.

JGS (1998). Remedial Measures against Soil Liquefaction: from Investigation and Design to Implementation. Japanese Geotechnical Society (ed.), Balkema, Rotterdam, 443p.

Marcuson, W.F., Hadala, P.F., and Ledbetter, R.H. (1996). "Seismic Rehabilitation of Earth Dams," *Journal of Geotechnical Engineering*, ASCE, 122(1), 7-20.

Mishima, S., and Kimura, H. (1970). "Characteristics of Landslide and Embankment Failures during the Tokachioki Earthquake," *Soils and Foundations*, 10(2), 39-51.

Sasaki, Y (1980). "Earthquake Damage of River Dikes," *Tsuchi-to-kiso*, 28(8), 25-30.

Sasaki, Y., Kano, S., and Matsuo, O. (2004). "Research and Practices on Remedial Measures for River Dikes against Soil Liquefaction," *Journal of Japan Assoc. for Earthquake Engineering*, 4(3), 312-335.

Tani, S. (1993). "Earth Damage to Fill Dams," *Proceedings of the 3rd Int. Conference on Case Histories in Geotechnical Engineering*, No. 3.13, 595-598.

Yamada, G. (1966). "Damage to Earth Structures and Foundations by the Niigata Earthquake." *Soils and Foundations*, 6(2), 1-13.

Zheng, J., Ohbo, N., Suzuki, K., Mishima, N., and Nagao, K. (1995). "Analysis of Results of Centrifuge Tests on Seismic Behavior of Embankment," *Proc. of the 1st Int. Conference on Earthquake and Geotechnical Eng*ineering, 105-110.

Soil Stress-Strain Behavior: Measurement, Modeling and Analysis
Geotechnical Symposium in Roma, March 16 & 17, 2006

RESEARCH ACTIVITIES OF GEOTECHNICAL RESEARCH GROUP OF NIIS FROM THE PAST TO PRESENT

N. Horii, Y. Toyosawa, S. Tamate & K. Itoh
Construction Safety Research Group
National Institute of Industrial Safety (NIIS)
Incorporated Administrative Agency
e-mail: horii@s.jniosh.go.jp

ABSTRACT
In this paper, firstly the memories of Prof. Tatsuoka's laboratory and research works carried out when the first author visited Prof. Tatsuoka's laboratory as a visiting researcher from May 1986 for about 1 year are described. Secondly, the research activities of Geotechnical Research Group of NIIS are introduced. Main emphasis is given on the research activities conducted using old geotechnical centrifuge (NIIS Mark-I centrifuge) and newly developed geotechnical centrifuge (NIIS Mark-II centrifuge).

1. INTRODUCTION
In May 1986, the first author had carried out some research works in Prof. Tatsuoka's laboratory (Tatsuoka Laboratory) when he was there as a visiting researcher for about one year. Geotechnical Symposium in Roma will be held for the celebration of Prof. Tatsuoka's 60th birthday. At first, the memory of Tatsuoka laboratory is introduced briefly along with some research works carried out during that time. Secondly, the research activities of Geotechnical Research Group of NIIS are introduced.

2. MEMORIRES AND RESEARCH WORKS OF TATSUOKA LABORATORY

2.1 Staffs of Tatsuoka laboratory in those day
When I joined Tatsuoka Laboratory's research project team, many promising staffs were there who were conducting and supporting the research works of that laboratory. Late Dr. Pradhan, T.B.S(Associate Prof. of Yokohama National University), Mr. Sato, T. and Miss Tarimitu, M were there as a research assistant, talented technical staff and a smart secretary. Among the students, Mr. Goto, S. (Associate Prof. of Yamanashi University), Mr. Yamanouchi, H. (Domi Environmental Ltd., Japan), Mr. Valerio, Guttierrez, Mr. Im, Jong Chull (Prof. of Pusan National University, Korea) and Mr. Mucabi, F. Codjoe, Mr. Tani. K. (Prof. of Yokohama National University, Japan) were doctoral students whereas Mr. Nakamura, S. (Japan Construction Mechanization Association, Japan), Mr. Ching-Chuan Haung (Prof. of National Cheng Kung University, Taiwan), Mr. Samuel, I. Kofi Ampadu (Kwame Nkrumah University of Science and Technology, Ghana) and Mr. Yoshida, K. (Advanced Construction Technology Center, Japan), etc. were Master course students. While Dr. Lo Presti, D. (University of Naples Federico II, Italy) was there as a foreign visiting researcher and Mr. Tateyama, M(RTRI, Japan) and MR. Sato, Y(Nishimatsu Const. LTD) were visiting research engineers. I would like to apologize for not mentioning the names of other colleagues who were there during that time as my memory did not allow me to recall everyone of them.

2.2 Research works in Tatsuoka laboratory
In Tatsuoka laboratory, I had conduced element tests using torsional shear apparatus. In sliding phenomena of slope or bearing capacity problem, deformation mode is very important to estimate resisting force. Especially in slope failure, major part of the sliding surface shows simple shear deformation mode as shown in Fig. 1 and Fig. 2 (Pradhan et. al, 1988a). So understanding of stress-strain characteristics under simple shear

Hoe I. Ling et al. (eds.), Soil Stress-Strain Behavior: Measurement, Modeling and Analysis, 931–938.

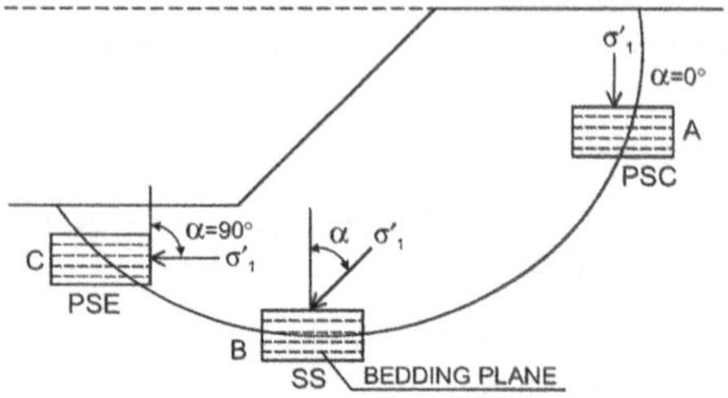

Fig. 1 Three representative soil elements deforming in plane strain.

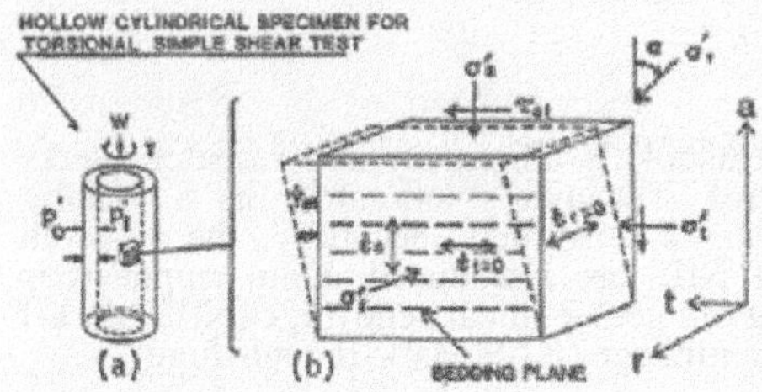

Fig. 2 Torsional simple shear(TSS) deformation.

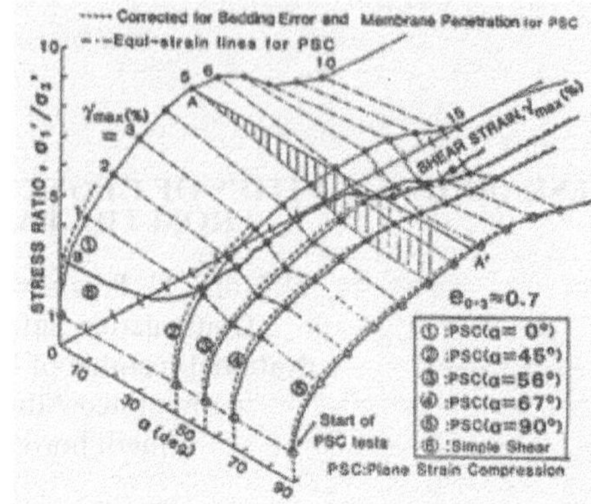

Fig. 3 Comparison of σ_1'/σ_3', $\gamma_{max}=\varepsilon_1-\varepsilon_3$ and α relation between TSS and PSC for dense specimens.

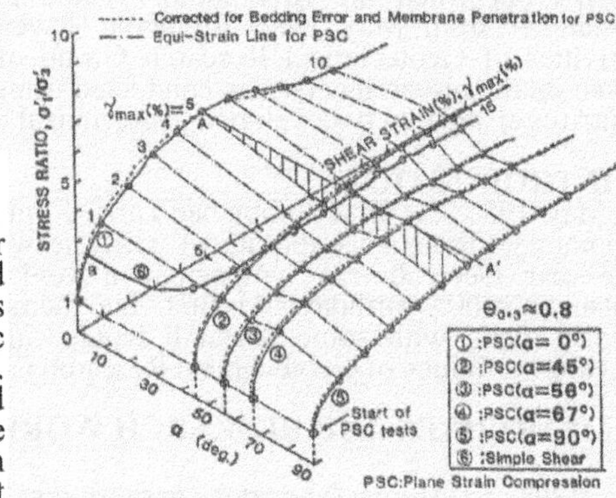

Fig. 4 Comparison of σ_1'/σ_3', $\gamma_{max}=\varepsilon_1-\varepsilon_3$ and α relation between TSS and PSC for loose specimens.

deformation is very important. Together with late Dr. Pradhan, I developed torsional simple shear testing apparatus where personal computer and pneumatic air pressure controlling devices were used. In this apparatus, the testing method used was fundamentally modified from the automated controlling of stress and strain paths in laboratory shear tests developed at Tatsuoka laboratory (Tastsuoka, 1988).

Simple shear test is one of the plane strain shear tests in which all parallel planes in the direction of shear move in parallel without changing their original shapes i.e. keeping horizontal shape constant during the shear (Fig. 2(b)). For this element, all horizontal normal strains are kept zero, without the rotation of all the horizontal planes. In Torsional simple shear test, drained simple shear condition is achieved by changing different outer and inner pressures (i.e., $p_o' \neq p_i'$) and keeping the effective axial stress constant. Typical results are shown in Fig. 3 and Fig. 4 (Pradhan et. al, 1988b) for a normally consolidated air-pluviated specimen of Toyoura sand (G_s=2.64, D_{50}=0.16mm, U_c=1.46). The relationships between σ_1'/σ_3', $\gamma_{max}=\varepsilon_1-\varepsilon_3$ in TSS and α for dense ($e \fallingdotseq 0.7$) and loose ($e \fallingdotseq 0.8$) specimens are compared with those in PSC by expressing them in three-dimensional space shown in Fig. 3 and Fig. 4, respectively. The void ratios of specimens are varied in the range between 0.702 and 0.716 for dense and between 0.781 and 0.810 for loose specimens, respectively. Although no correction for these variations has been made, the effect in relationship mentioned earlier due to change in the void ratios is very small. Each PSC tests were performed at a constant σ_1'/σ_3'=1.0 at σ_3'=49kPa (Tatsuoka et. al, 1986). A vertical plane shaded with thin solid lines in each figure at γ_{max}=5% illustrates the anisotropy observed by PSC tests.

It may be seen from Fig. 3 and Fig. 4, the curve of TSS constructed by σ_1'/σ_3', $\gamma_{max}=\varepsilon_1-\varepsilon_3$ and α is located very close to the surface of PSC tests. So, the relationship between σ_1'/σ_3', $\gamma_{max}=\varepsilon_1-\varepsilon_3$ and α in TSS are rather independent of continuous

change of α during loading. One of the possible reasons might be that the major part of rotation of α was occurred in early stage of shearing where $\gamma_{max} < 1\%$.
It can be concluded that the effects of principal stress rotation on the strength and deformation characteristics of normally consolidated sand in simple shear under monotonic loading condition in torsional simple shear test is very small.

3 INTORODUCTION OF RESEARCH ACTIVITIES OF GEOTECHNICAL RESEARCH GROUP OF NIIS

3.1 Research works using Mark-I geotechnical centrifuge

In 1988, Centrifuge called Mark-I was introduced into the National Institute of Industrial Safety (NIIS) with an effective radius of 2.3 m (Toyosawa et al., 1994). Mark-I centrifuge is shown in Photo. 1.

Photo. 1 NIIS Mark I Centrifuge (since 1988).

Since then, NIIS has conducted many centrifuge model tests to investigate the failure mechanism of soil structures which are mainly related to the labor accidents. The results of fundamental and practical research projects have been applied and reflected in some safety guidelines given out by several construction safety groups. Typical research works are demonstrated below.

3.1.1 Deformation and Failure characteristics of trench excavation in centrifuge model tests

Trench failure frequently occurred as a result of collapse of un-supported or insufficiently supported walls of trench. To study deformation characteristics of un-supported trench and failure mechanisms due to excavation, centrifuge model tests on

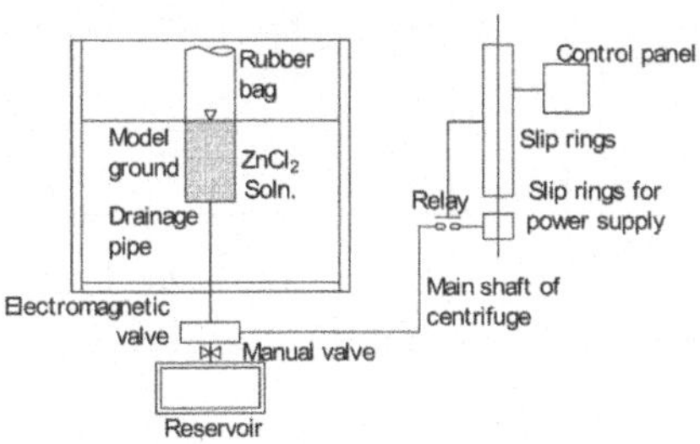

Fig. 5 Simulation system of trench excavation process.

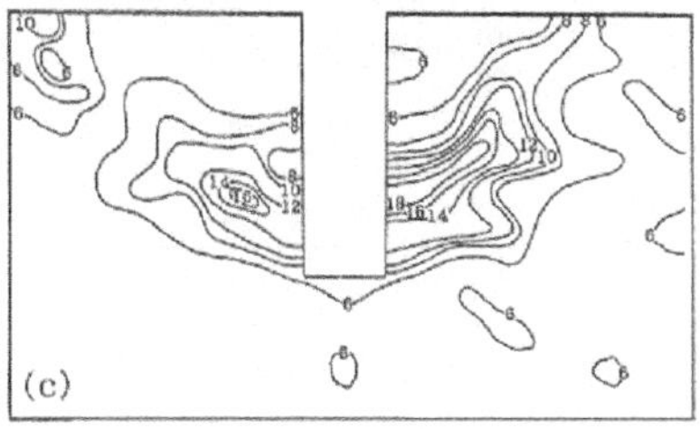

Fig. 6 Cumulative maximum shear strain γ_{max} just before failure.

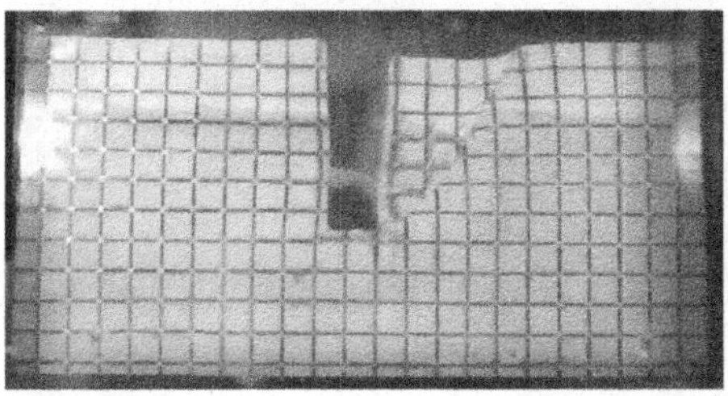

Photo. 2 Photograph of trench just after the collapse.

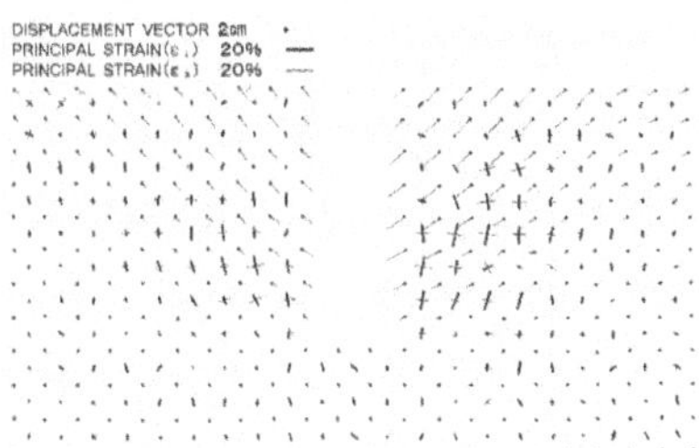

Fig.7 Principal strain ε_1, ε_3 and displacement vector.

preconsolidated kaolin were carried out (Toyosawa et al., 1994). In the centrifuge model tests, trench excavation was simulated by draining away a large amount of liquid from the bottom of the trench at the predetermined acceleration. Simulation system is shown in Fig. 5. The progress of deformation during excavation was analyzed from the deformation of targets on the model measured from as a sequence of photographs taken at intervals of about 0.3 seconds until the trench were collapsed. From the calculated strains, the development of γ_{max} up to collapse was evaluated. Based on the results of centrifuge tests simulating trench excavation, both wedge type failure and circular arc failure were observed. Photographic measurements showed that prior to failure, the strains were observed over a large area behind the trench wall(Fig. 6). Principal strain ε_1, ε_3 and Displacement vector are shown in Fig. 7. At the on set of failure, the strains were concentrated around the toe of the trench, then the shear band appeared in this area first shown in Photo. 2.

3.1.2 Stability of composite ground improved by Deep Mixing Method

Failures or unacceptable large deformations sometimes occur during constructing embankments or other soil structures on soft ground. Recently, Deep soil Mixing Method (DMM) has been widely used as a ground improving method to prevent these problems. However, there are some problems in design of DMM, which includes: the nonuniformity of the strength of the improved soil column, how to evaluate the strength of the composite ground and so on. Especially, in the case of adopting a low improvement ratio, it is considered that the strength of the composite ground would not be much higher than that of original ground because of the low improvement ratio. In this study (Horii et al., 1998), centrifuge tests and numerical analysis were carried out to investigate the effect of the DMM at the low improvement ratio. Five kinds of tests were carried out and are listed in Table.1. Improvement ratio is defined as the ratio of the total cross sectional area of the columns to the improved area. The improvement ratio was about 10% for all test cases

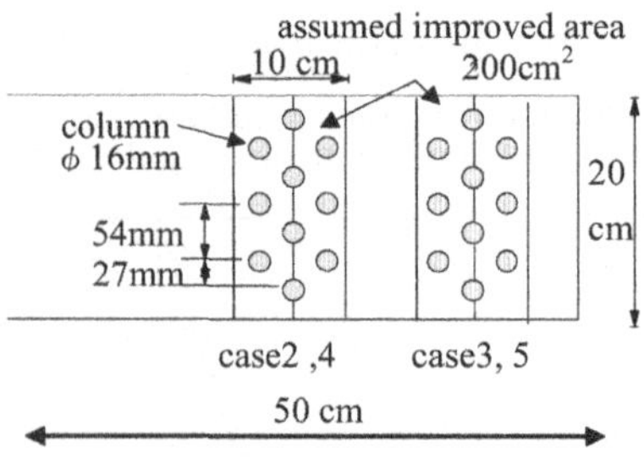

Fig.8 The Location of the Column.

Table 1 Test conditions.

	case 1	case 2	case 3	case 4	case 5
Improved area(below of a embank ment)	-----	toe	shou lder	toe	shou lder
Improveme nt ratio(%)	0	10	10	10	10
Column's angle (degrees)	-----	0	0	20	20

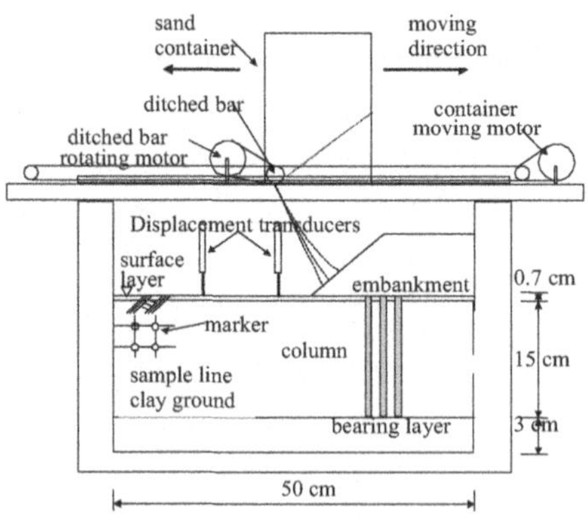

Fig. 9 Model ground and in-flight sand hopper.

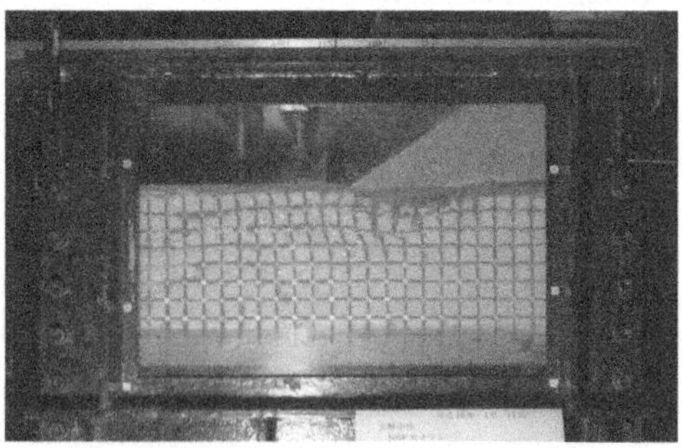

Photo.3 Failure of model ground improved by DMM.

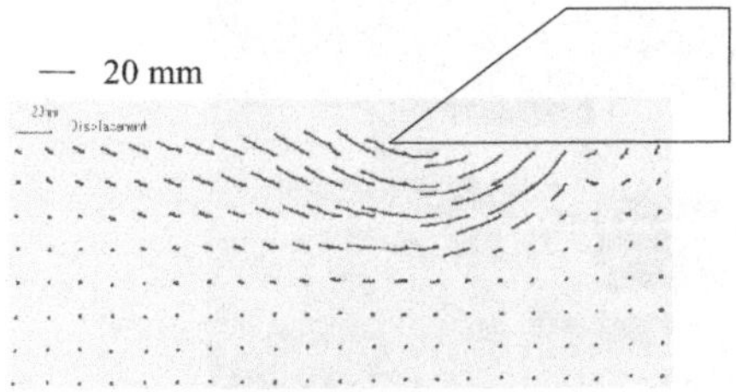

Fig.10 Case 1 (non-improvement).

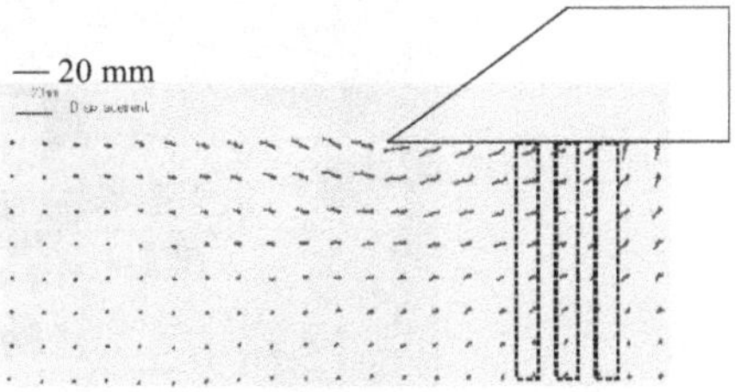

Fig.12 Case 3 (shoulder improvement).

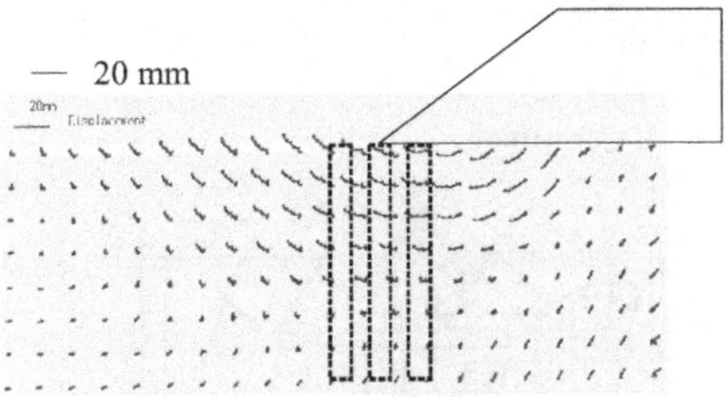

Fig.11 Case 2 (toe improvement).

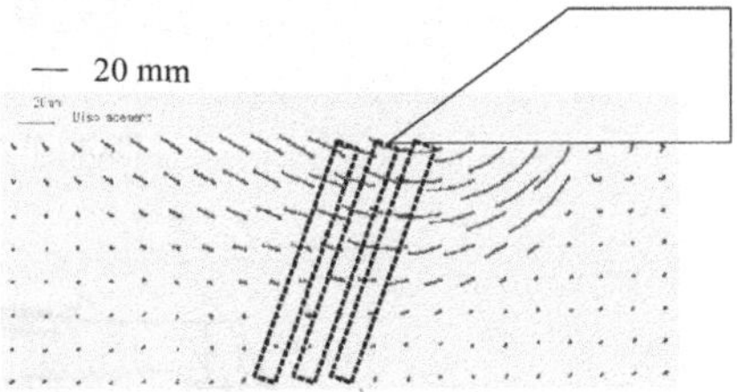

Fig.13 Case 4 (toe improvement inclined).

except case1. The column of case4 and case5 is inclined at an angle of 20 degrees clockwise. The location of the column is shown in Fig.8. However, the level of column's top of these cases is same as case2, 3. Toyoura sand was used for the bearing layer, surface layer and embankment. NSF clay (w_l=66.8%, w_p=26.4%, I_p=40.3) was used for the soft ground. Model ground and In-flight sand hopper are shown in Fig. 9 and Photo. 3 is typical result of in-flight construction of embankment. The behavior of the clay ground deformation is shown in Fig10-14. The deformations like a circular arc were observed in clay ground except case 3. In case3, the horizontal displacement surpassed vertical one. The clear circular sliding surfaces were observed in clay ground except case3. The clear circular sliding surface located from below of the shoulder of slope to below of the toe of slope in case1, 2 and 4, however, in case5 it located from the front of improved area to below of toe of slope. The depth of the deformed area in case 2 and 4 was deeper than case1.

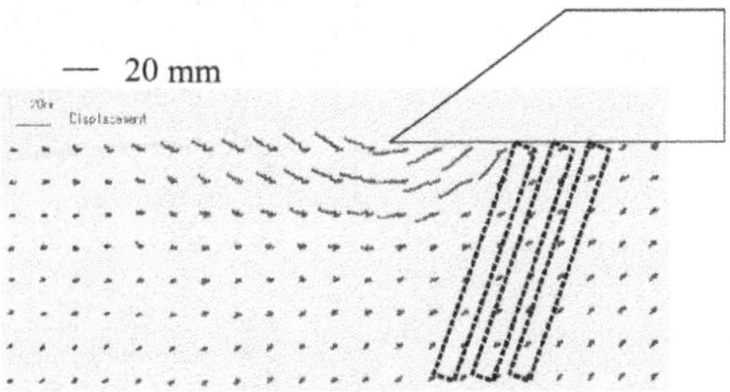

Fig.14 Case 5(shoulder improvement inclined).

4 NEW CENTRIFUGE FACILITES

4.1 AGEMENTS OF NIIS MARK II CENTRIFUGE

In 2003, NIIS started to reconstruct its Mark-I centrifuge and it was completed in 2004. A new centrifuge "NIIS Mark-II" was named for this reconstructed centrifuge. This is of medium size and this centrifuge is believed to be one of the most effective and multi-functional type of centrifuge in the world. This chapter describes the specifications, structural and mechanical design, facilities, data acquisition system, and safety components of this centrifuge. As to reconstruct new centrifuge, machine and control system have improved to accommodate the 10 years operations and experiments into the new facilities. Especially, authors mainly put emphasis on the following problems.

1. Improvement of working environment
2. Expansion of control device spaces
3. Expansion of platforms

NIIS Mark-II Centrifuge are shown in Photo. 4 and Fig. 15. As in other centrifuge, it also

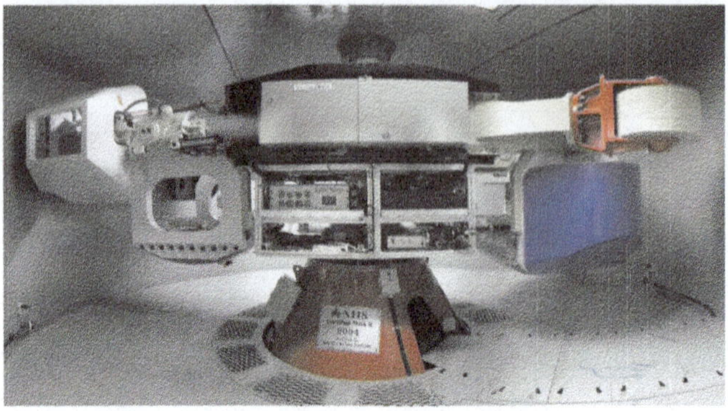

Photo 4. NIIS Mark- II Centrifuge.

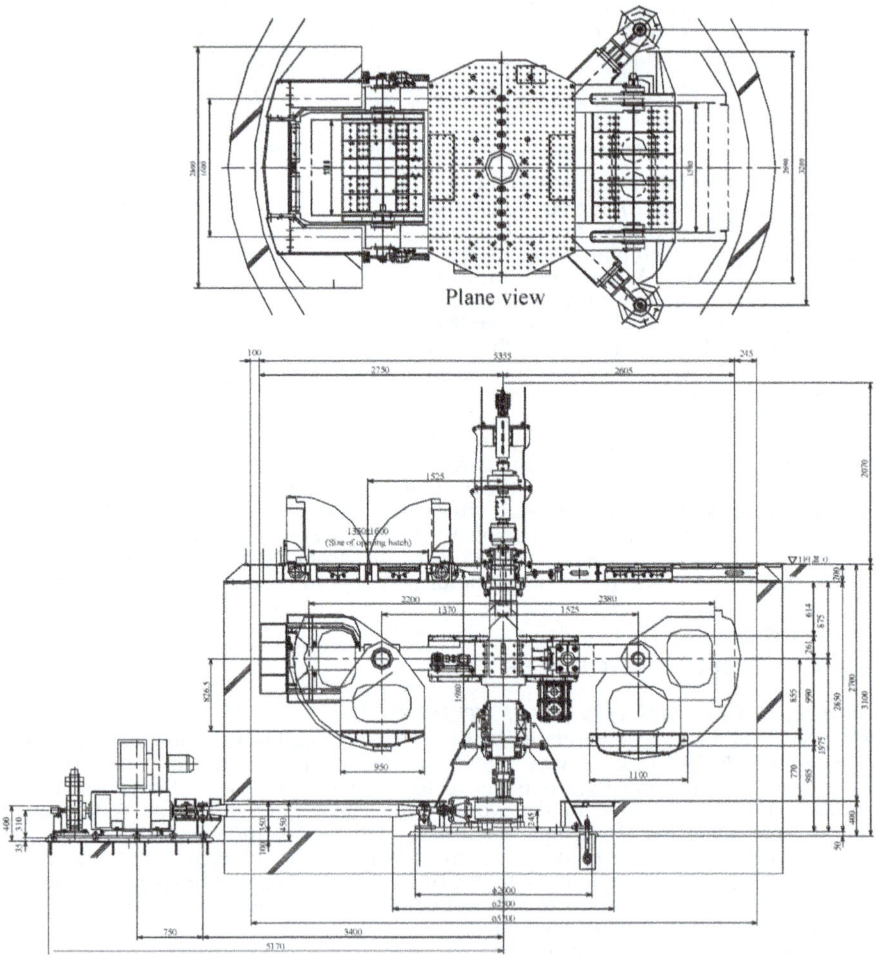

Fig.15 NIIS Mark- II Centrifuge.

has a main shaft, a drive unit, two arms, two swinging platforms. But its arms (forms) are asymmetric. One side of the arm is provided with a bridge plate where the swinging platform which could be fixed to its bridge plate using a pair of hydraulic suspension jacks. This type of system is called "Touch-down system" (Nagura et. al, 1994). This facilitates the simulation of strong earthquake motions and mounting of shaking table on to the platform (Dynamic platform). Other side of the arm is not provided with the bridge plate and hence no end plate. In order to balance this portion (weight of the end plate)) in this side of the arm while the swinging platform is moved up, two counter-weights are overhung on the two sides of this arm. This platform is used for non-shaking or static tests (Static platform). Major specification of NIIS Mark-II is listed in Table 2.

4.2 Ancillary equipment

4.2.1 Digital data acquisition system

Digital data acquisition system, which is installed in the arm, consists of amplifiers, and A/D converters (MCA, Kyowa Co Ltd.), wireless LAN, and control PC which are diagrammatically shown in Fig. 16. The A/D converters and amplifiers are simultaneously controlled through the GPIB system which are further controlled by PC of operation room adjacent to the centrifuge through an Ethernet (10/100 Base-T) network. Conditions of data recording such as gains, scanning time and duration time can be set up during the rotation. The wireless LAN which works between the main shaft and the operation room is of direct sequence Spread Spectrum (DSSS) type. And it has the transmission rate of 11 MB/sec with transmitting frequency of 2.4 GHz.

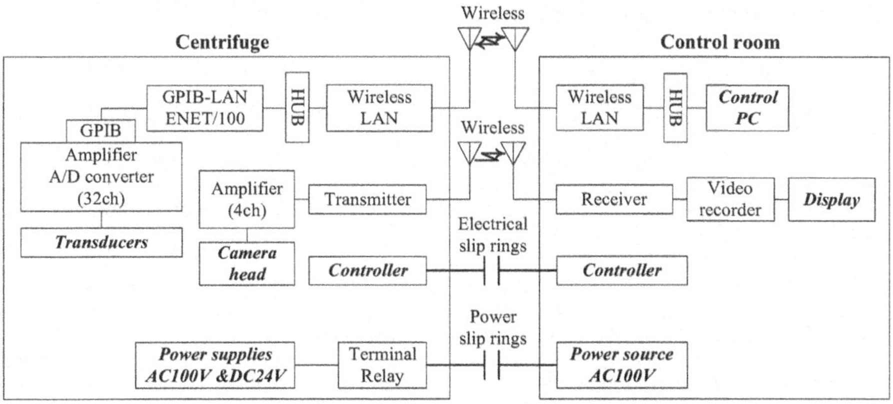

Fig. 16 Control and Acquisition flow.

4.2.2 Rotary joints and electric slip rings

The NIIS Mark-II centrifuge is equipped with two rotary joints which permit fluids to flow through the central axis of the machine to the platform. These rotary joints contain a total of five passages; two are designed to accept high pressure hydraulic fluid; two are designed for the water unit; and the remaining one is designed for carrying air. The rated capacity of the fluid passages is as shown in Table 3.The centrifuge is equipped with 46 electrical slip rings which are reused from the old centrifuge. Forty four of these are signal lines, and remaining 2 are power lines. The power electricity is transformed to 100V and DC 24V, and each electric phase can be controlled individually by means of the terminal relay. Some specifications for the assemblies are shown in Table 3.

5 SUMMARY

The memories and research works of Tatsuoka laboratory and the introduction of research works conducted by geotechnical research group of the NIIS were described in this paper. New geotechnical centrifuge facility established at the NIIS was introduced. Research works with this new centrifuge have been already started in the area of slope

Table 2 Major specification.

Maximum acceleration	Dynamic	50 g
	Static	100 g
Maximum number of rotation	Dynamic	143 rpm
	Static	194 rpm
Maximum payload	Dynamic	1000 kg
	Static	500 kg
Payload capacity	Dynamic	50 G – tons
	Static	
Effective radius	Dynamic	2.20 m
	Static	2.38 m
Payload size	Dynamic	1.1 m x 0.95 m
	Static	1.1 m x 1.5 m
Main motor	440V	
	90 kW (DC)	

Table 3 Ancillary equipment.

Rotary joints		
Oil	Maximum pressure	21.0 MPa
	Number of ports	2
Air	Maximum pressure	0.7 MPa
	Number of ports	1
Water	Maximum pressure	0.5MPa
	Number of ports	2
Electric	Capacity	AC100V,30A
	Number of channel	2 channels
Slip ring	Capacity	AC100V, 1A
	Number of channel	44 channels
Rotary joints		

failure (Tamrakar et al., 2006 and Itoh et al, 2006) and soil/structure interaction (Ichikawa et al, 2006 and Arai et al, 2006). In addition, future research works will be commenced in the area of wave propagation due to ground vibration.

ACKNOWLEDGMENTS

The authors are grateful to the support of the administrator of the National Institute of Industrial Safety, for decision making and funding to construct the centrifuge. They are also grateful to the staff of NIIS geotechnical research group for their support. Construction of the machine by Sumikin Kansai Industries, Ltd. is also acknowledged with thanks. The guidance and advice from other public and private centrifuge centers are gratefully acknowledged.

REFERENCES

Arai, F., Suemasa, N., Katada, T., Itoh, K., and Tamate, S. 2006. Dynamic stability of tower crane with pile foundation. The 6th Int. Conf. Physical Modelling in Geotechnics.

Ichikawa, S., Suemasa, N., Katada, T., and Toyosawa, Y. 2005. Centrifuge model tests on seismic stability of reinforced retaining wall. The 5th Workshop on Safety and Stability of Infrastructure against Environmental Impacts, CD-ROM.

Itoh, K., Toyosawa, Y., and Kusakabe, O. 2006. Centrifugal modelling of rockfall simulation. The 6th Int. Conf. Physical Modelling in Geotechnics.

Horii, N., Toyosawa, Y., Tamate, S., Hashizume, H. and Okochi, Y. 1998, Stability of composite ground improved by Deep Mixing Method. Proc. of 2nd Int. Conf. on Ground Improvement Techniques.

Nagura, K., Tanaka, M., Kawasaki, K., and Higuchi, Y. 1994. Development of an earthquake simulation for the TAISEI centrifuge. Proc. Int. Conf. on Centrifuge Modelling-Centrifuge 94, Balkema, Singapore, 151-156.

Pradhan, T.B.S.,Tatsuoka, F. and Horii, N. 1988a: Simple shear testing on sand in a torsional shear apparatus, Soils and Foundations, Vol.28, No.2, 95-112.

Pradhan, T.B.S.,Tatsuoka, F. and Horii, N. 1988b: Strength and deformation characteristics of sand and in a torsional simple shears, Soils and Foundations, Vol.28, No.3, 131-148.

Tamrakar, S. B., Toyosawa, Y., Itoh, K. and Timpong, S. 2006. Comparison of failure heights during excavation of slope using In-flight Excavator. The 6th Int. Conf. Physical Modelling in Geotechnics.

Tatsuoka, F., Sakamoto, M., Kawamura, T., and Fukushima, S. 1986, Failure and deformation of sand in a torsional shears, Soils and Foundations, Vol.26, No.4, 79-97.

Tatsuoka, F. 1988, Some recent developments in triaxial testing systems for cohesionless soils, ASTM, STP, No.977, Advanced Triaxial Testing of Soil and Rock, 7-67.

Toyosawa, Y., Horii, N., Tamate, S., Hanayasu, S. and Ampadu, S. K. 1994. Deformation and failure characteristics of vertical cuts and excavations in clay. Proc. Int. Conf. on Centrifuge Modelling-Centrifuge 94, Balkema, Singapore, 663-668.

Soil Stress-Strain Behavior: Measurement, Modeling and Analysis
Geotechnical Symposium in Roma, March 16 & 17, 2006

ENGINEERING IMPLICATIONS OF GROUND MOTIONS ON WELDED STEEL MOMENT RESISTING FRAME BUILDINGS

Daniel Pradel
Department of Civil and Environmental Engineering
UCLA, Los Angeles, CA 90095-1593, USA
e-mail: dpradel@ucla.edu

ABSTRACT
Although tall buildings are relatively few in the Los Angeles area, more than 150 buildings with welded steel moment resisting frames were significantly damaged during the 1994 Northridge Earthquake. The number of damaged buildings and the level of damage surprised many engineers and building officials. This study summarizes an analysis of free-field ground motions at the location of 228 buildings for which post-earthquake engineers' reports, as well as weld inspections, were available. The study shows how vulnerable the pre-Northridge type welds are during seismic events, and how typical geotechnical ground motion parameters such as PGA and PGV are useful in assessing damage levels. The study may be used to estimate average damage levels and/or probabilities of building damage, for the Northridge Earthquake in adjoining localities and may also be used to estimate damage thresholds for future earthquakes, or in zones affected by previous earthquakes, including the 1987 Whittier, and 1989 Loma Prieta earthquakes.

1. INTRODUCTION
The Northridge earthquake of January 17, 1994 had a moment magnitude $M_w = 6.7$ and is generally considered a moderate earthquake. Although tall buildings are relatively few in the Los Angeles area, more than 150 buildings with welded steel moment resisting frames were significantly damaged during the earthquake. The number of damaged buildings and the level of damage surprised many engineers and building officials and was highly publicized (EERI, 1996). Recognizing the severity and potential hazard of the situation the City of Los Angeles issued a special ordinance requiring the inspection and testing of most steel moment resisting frame buildings within its jurisdiction. By August of 1998, engineering reports were prepared and submitted for most of the 247 buildings affected by the ordinance. These reports represent an unprecedented database of information regarding earthquake damage to steel moment resisting frame buildings. The purpose of this paper is to summarize our assessment and interpretation of the relationship between the damage and geotechnical earthquake ground motion parameters.

Hoe I. Ling et al. (eds.), Soil Stress-Strain Behavior: Measurement, Modeling and Analysis, 939–947.

1.1 NORTHRIDGE EARTHQUAKE

The Northridge Earthquake of January 17, 1994, occurred near the central portion of the northern San Fernando Valley. The rupture began about 18 km below the surface, and propagated up to a depth of about 5 km. A moment magnitude of $M_w = 6.7$ was reported by Caltech for the Earthquake (Stewart et al., 1994). Many organizations produced papers and reports after the Northridge Earthquake discussing the spacial variability of the ground motions and the effects the Earthquake had throughout the greater Los Angeles area (e.g., Stewart et al. 1994). The nature and overall distribution of the damage in the Los Angeles area has been reported in various reports, including Dewey et al. (1995).

1.2 PRE-NORTHRIDGE EARTHQUAKE PRACTICE

Prior to the Northridge Earthquake, steel moment frame connections were often welded with E70T4 type electrodes in most of Southern California and many other areas of North America. This type of weld behaved in a particularly brittle manner during the Northridge Earthquake, with numerous buildings throughout the Los Angeles area experiencing considerable damage to the welded moment frame connections (EERI, 1996).

1.3 CITY OF LOS ANGELES RESPONSE AFTER THE EARTHQUAKE

Immediately following the Northridge Earthquake, damage was observed in welded steel moment resisting frame buildings that were under construction in the communities of Universal City and Brentwood (about 18 km and 20 km from the epicenter, respectively). Within the first two weeks, inspections of existing buildings revealed damage as far north as the City of Santa Clarita (about 25 km from the epicenter). In accordance with local practices, the damage to these buildings was generally reported by structural engineers to building officials. By late March, 1994, 17 to 20 damaged buildings were identified, and by early April, 1994, there were 35 to 39 buildings (Gates and Morden, 1995).

The City of Los Angeles, Department of Building and Safety responded quickly to these reports, and on April 11, and May 11, 1994, two directives were issued regarding the repair of cracked connections, and for the construction of new buildings. By early June, 1994, there were 77 damaged steel moment resisting frame buildings identified within the City of Los Angeles. Building officials decided that it was necessary to inspect all buildings with welded steel moment frames within the most severely damaged zones of the City, and prepared the first draft of an ordinance mandating weld inspections in June, 1994. This draft was never adopted; according to Gates and Morden (1995), two factors were of particular significance:

- Full scale tests performed in 1990 showed that welded connections were not reliable (Engelhart, 1993). Additional tests by Professor Engelhart in March, 1994, confirmed his earlier findings, and showed that the repair procedures proposed by the City of Los Angeles for damaged connections were not reliable.

- It was estimated that the total cost for inspection alone was as high as $10,000 per connection, and therefore building owner groups strongly opposed the ordinance.

On July 21, 1994, the City Council passed an emergency ordinance giving the Department of Building and Safety authority for emergency actions, and on August 19, 1994, an ordinance was passed prohibiting the use of the typical pre-Northridge Earthquake standard welded connection. By October 1994, 100 damaged buildings were identified.

1.4 ORDINANCE FOR INSPECTION AND REPAIRS

The draft ordinance prepared in June, 1994: (1) made no distinction between residential and commercial buildings; (2) covered the entire City of Los Angeles; and (3) required the inspection of an estimated 1000 buildings. Hence, it was considered politically impossible to support. In an effort to salvage the ordinance:

- Building officials analyzed the available data, and mapping showed that nearly all the damaged buildings were in West Los Angeles and in the San Fernando Valley. The area of required inspections was thus reduced to these damaged areas.
- Condominiums are especially costly to inspect, and the Department of Building and Safety dropped the requirement for inspection of residential buildings.

For several months, building officials met with Council members and emphasized that in the event of a major earthquake, there would be the potential for collapse and of more severe damage than from the Northridge Earthquake. Shortly after the Kobe Earthquake the ordinance was schedule for vote. The ordinance passed 12 to 0 with no discussion on February 22, 1995.

The ordinance requires building owners to hire an engineer to conduct an investigation. It also requires that reports and plans be submitted to the Department of Building and Safety before a repair permit is issued, and that all damaged connections be repaired. The hundreds of engineering reports and thousands of inspection reports of welded connections submitted to the City constitute an unprecedented database of information regarding cracked welds in steel frame buildings.

2. BACKGROUND INFORMATION

2.1 INTRODUCTION

The aim of our investigation was to establish the approximate relationship between realistic earthquake ground motion parameters and weld damage in steel moment frame buildings. In other words, this research describes the correlation between regional damage surveys with site-specific free-field ground motions at the location of the individual buildings.

2.2 PREVIOUS STUDIES

To determine the range of ground motions capable of causing damage on pre-Northridge Earthquake welds previous studies generally used early damage surveys, and simplified synthetic ground motion maps. These studies generally concentrated on regional damage patterns and often excluded buildings in certain areas, such as where the Peak Ground Accelerations (PGA) were less than 0.2g. Using the 89-building survey prepared by Bonowitz and Youssef (1995), and the synthetic ground motion maps with smoothed contours by Somerville et al. (1995), Gates (1998) found that no connection damage occurred in areas were the PGA was less than 0.24g, and the Peak Ground Velocity (PGV) less than 15 cm/s. A similar study by Durkin (1995) concluded that a relatively high proportion of the damaged buildings were exposed to a PGA between 0.25 and 0.35g, and a PGV of 20 to 35 cm/s.

2.3 EARTHQUAKE GROUND MOTIONS

In order to obtain realistic free-field ground motion parameters at the location of each steel moment frame building for this study, we needed a source that would provide accurate and consistent estimates of ground motion parameters. Ideally this source should be based on actual recordings, and not synthetic ground motion studies. Chang et al. (1996) published a comprehensive set of maps that show the distribution of ground motion parameters in the greater Los Angeles area, including, PGA, PGV, and Spectral Accelerations (Sa) at periods of 0.3 and 1.0 seconds. Since the maps by Chang et al. (1996) used actual free-field recordings they were ideally suited to estimate site-specific ground motion parameters at each of the locations of investigated buildings.

2.4 DAMAGE SURVEYS

As part of this research we received much cooperation from the City of Los Angeles Department of Building and Safety and were able to review numerous reports submitted to the City between 1994 and 1998. Cooperation from neighboring cities was not as forthcoming and we were only able to obtain few documents and data. As a result, we decided to only use the comprehensive data obtained from the City of Los Angeles. It should be noted that Northridge is located within the City of Los Angeles.

The result of our investigation culminated in the creation of a database which contained our findings for each building, including PGA, PGV, Spectral Accelerations, number of welds, weld inspection data, weld damage data, number of joints repaired, etc. The database includes weld performance data for all but 19 of the 247 buildings affected by the 1995 City of Los Angeles ordinance and is more comprehensive than the early surveys used in previous studies. For our analyses we calculated the Damage Ratio (DR), for each surveyed building where:

$$\text{Damage Ratio:}\quad DR = \frac{\text{Number of damaged welds}}{\text{Number of inspected welds}}$$

2.5 LEVELS OF DAMAGE

During the repairs different engineers evaluated the nature of the damage differently. For some minor root cracks, were acceptable. According to Gates et al. (1995), certain engineers considered some cracking tolerable and used a rule of thumb where minor root cracks were acceptable if they were present in less than 10% of the moment resisting connections. Since even minor root cracks can result in a fractured joints during an earthquake, we considered them as damage in this study. We also decided that it was important to correlate ground motions with levels of damage, i.e., with the Damage Ratio defined above.

3. RELATION BETWEEN GROUND MOTIONS AND WELD DAMAGE

3.1 PEAK GROUND ACCELERATION (PGA)

Historically, Modified Mercalli Intensities and geotechnical earthquake damage have often been correlated with PGA. Thus we compared the site-specific free-field PGA for each of the buildings in our database with the level of damage to the building. Figure 1, shows the result of such a correlation. Please note that the "Average Damage Ratio" in Figure 1 was established by first obtaining the individual Damage Ratio (DR) for each building, and then averaging them within each category (e.g., for the 0.2-0.3g category). This averaging process provides equal weight for all the buildings and reduces the influence that large buildings with numerous joints could have had on the database. As expected, Figure 1 shows that as the PGA increases the percentage of damaged joints, i.e., the damage level, increases. More importantly it shows that even for low PGA values, (in the range of 0.1 to 0.2g) significant weld damage was found. It is interesting to note that this damage is below the threshold of 0.2g used in some previous studies and suggests that it may be prudent to inspect certain buildings outside the limits of the City of Los Angeles ordinance. Figure 1 also shows that for a PGA > 0.3g the average Damage Ratio remained near constant at about 15%. Please note, that the relative drop for PGA > 0.5g is most certainly related to the limited number of buildings which experienced such high accelerations (only 15 out of 228 buildings) and is not considered statistically significant.

The relation between the percentage of damaged buildings within a PGA group and the PGA is shown on Figure 2, for DR values between 0 and 30%. A ratio DR > 0 means that some damage was found during inspection. Figure 2 is interesting as it exemplifies the probability of building damage as a function of the PGA. Figure 2 shows that in areas with PGA > 0.3g more than two thirds of the buildings surveyed suffered some weld damage and that about 17% had damage to more than 30% of the inspected welds. Thus damage was prevalent for this type of building during the Northridge Earthquake. It also shows that depending on the value of DR, the corresponding damage starts at PGAs, typically in the range of 0.1 to 0.3g.

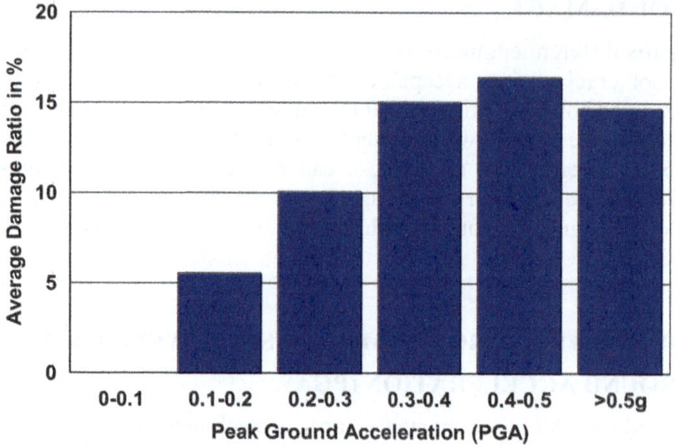

Fig. 1: Relation between damaged joints and PGA

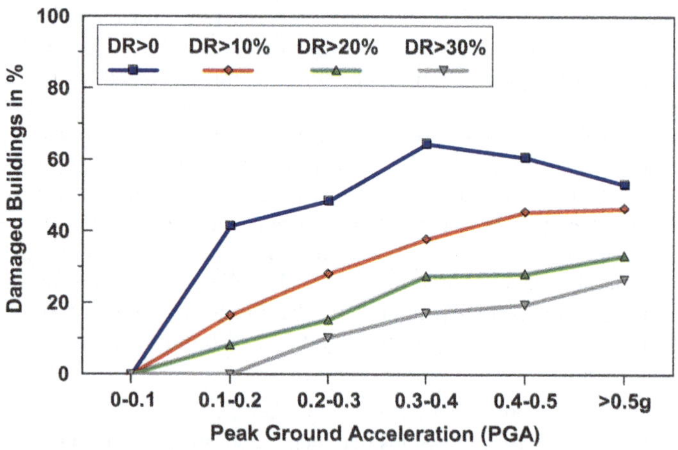

Fig. 2: Relation between damage level and PGA

3.2 PEAK GROUND VELOCITY (PGV)

When analyzing brittle type failures, the PGV is often preferred to the PGA, as it is considered a better quantifier of the large pulse that earthquakes with strong rupture directivity effects can provide. Thus we compared the relation between the PGV at each of the buildings in our database with the Damage Ratio. The results are included in Figures 3 and 4. Generally, Figures 3 and 4 show that as the PGV increases the damage level, increases. Although Figure 4 shows that damage starts at a PGV in the range of 10 to 20 cm/s, considerable damage becomes apparent only when the PGV exceeds 30 cm/s.

The data in Figures 3 and 4 is more scattered than the data in Figures 1 and 2, and thus we consider the PGV a less reliable predictor of damage than the PGA. Nevertheless, as shown in Figure 5 when used in combination with PGA, the PGV becomes a useful quantifier of damage.

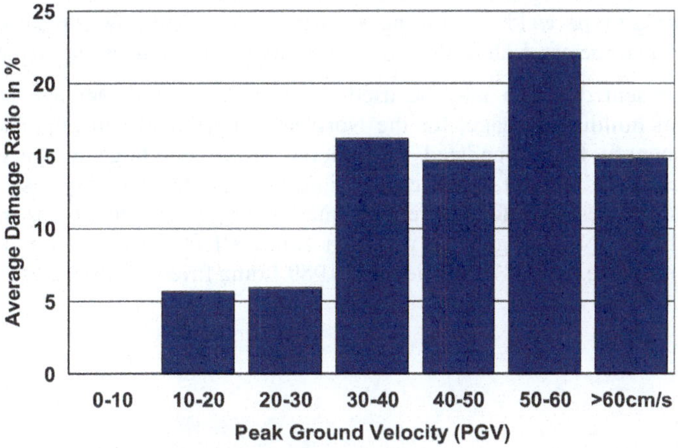

Fig.3: Relation between damaged joints and PGV

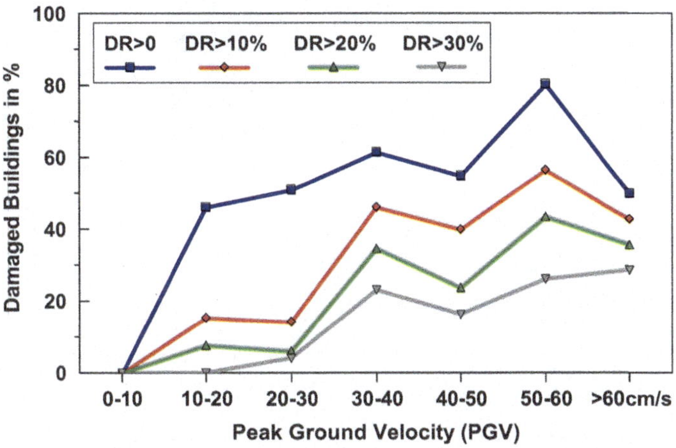

Fig.4: Relation between damage level and PGV

4. CONCLUSIONS

The ordinance passed by the City of Los Angeles after the Northridge Earthquake revealed widespread damage in welded steel moment frames. The study herein summarizes an analysis of free-field ground motions at the location of 228 buildings for which post-earthquake engineers' reports, as well as weld inspections, were submitted to the City of Los Angeles regarding cracked welds. The study herein shows how vulnerable the pre-Northridge type welds are during a seismic event, and how typical geotechnical ground motion parameters such as PGA and PGV are useful in assessing damage levels.

The figures presented herein may be used to estimate average damage levels and/or probabilities of building damage, for the Northridge Earthquake in adjoining localities that did not benefit from mandated inspections, or for buildings in the City of Los Angeles which were exempt, e.g. condominiums. The figures may also be used to estimate damage thresholds for future earthquakes, or as a guideline for inspection of buildings with pre-Northridge type welds in zones affected by previous earthquakes, including for example the 1987 Whittier, and 1989 Loma Prieta earthquakes

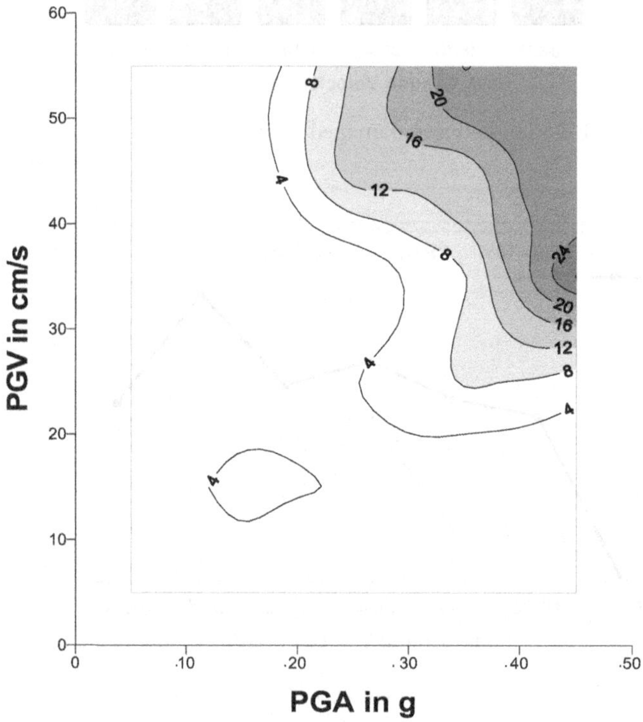

Fig.5: Contours of average Damage Ratio in percent versus PGA and PGV

5. REFERENCES

Dewey, J.W., et al. (1995): "Intensity Distribution and Isoseismal Maps for the Northridge, CA, Earthquake of Jan. 17, 1994", USGS Open-File Report No. 95-92, Denver, CO, USA

Durkin, M.E. (1995): "Inspection, Damage and Repair of Steel Frame Buildings following the Northridge Earthquake", SAC Report No. 95-06, Sacramento, CA, USA.

Chang, S. W., Bray, J.D., and Seed, R.B. (1996), "Engineering Implications of Ground Motions from the Northridge Earthquake", Bulletin of the Seismological Society of America, Vol. 86, No. 1B, pp. S270-S288.

EERI (1996): "Northridge Earthquake Reconnaissance Report, Vol.2", Earthquake Spectra, Supplement C to Vol. 11, pp. 33-47.

Engelhardt, M.D., and Husain, A.S. (1993): "Cyclic-Loading Performance of Welded Flange - Bolted Web Connections" ASCE Journal of Structural Engineering, Vol. 119, No. 12, pp. 3537–3550.

Gates, W. E., and Morden, M. (1995): "Lessons from inspection, evaluation, repair and construction of welded steel moment frames following the Northridge Earthquake", SAC Report No. 95-06, Sacramento, CA, USA.

Gates, W. E. (1998): "Summary Interpretation of SAC Survey Data on Damaged Welded Steel Moment Frames", ASCE Journal of Performance of Constructed Facilities, Vol. 12, No. 4, November 1998, pp. 180-185.

Somerville et al. (1995): "SAC Task 4: Characterize Ground Motion at sites of Subject Buildings", SAC Report No. 95-03, Sacramento, CA, USA.

Stewart, J.P., et al. (1994). "Preliminary Report on the Principal Geotechnical Aspects of the January 17, 1994 Northridge earthquake," Earthquake Engineering Research Center Report. No. UCB/EERC-94/08, Berkeley, CA, USA.

Soil Stress-Strain Behavior: Measurement, Modeling and Analysis
Geotechnical Symposium in Roma, March 16 & 17, 2006

INVERSE STABILITY ANALYSIS OF THE ST. MORITZ LANDSLIDE

Alexander M. Puzrin
Institute for Geotechnical Engineering
ETH Zurich, Zurich CH-8093, Switzerland
e-mail: alexander.puzrin@igt.baug.ethz.ch

Ivo Sterba
Institute for Geotechnical Engineering
ETH Zurich, Zurich CH-8093, Switzerland
e-mail: ivo.sterba@igt.baug.ethz.ch

ABSTRACT
This paper deals with long-term stability analysis of the Brattas-St. Moritz landslide in Switzerland. The landslide has "nowhere" to go and its downhill movement is slowing in time, which intuitively implies its long-term stability. However, exactly because the landslide is slowing down, the shear strength on the sliding surface may decrease, leading to increase in compressive stresses at the landslide foot and, ultimately, to a failure. The inverse long-term stability analysis procedure proposed by Puzrin and Sterba (2006) allows for the safety factor to be determined solely by curve fitting the observed displacement data. The time of failure can be also predicted using additional earth pressure measurements in the sliding layer. The proposed procedure helps to identify, which additional observation data is required to determine the long-term stability of the Brattas-St. Moritz landslide.

1. INTRODUCTION
Long-term stability of slowly moving landslides has been a subject of early interest in Soil Mechanics (e.g., Terzaghi, 1936; Skempton, 1964; Bjerrum,1967). In these and later studies, the analysis mainly focused on progressive failure in over-consolidated clays, where the failure is delayed in time by the development of the negative excess pore water pressure caused by shearing. As soon as this excess pore water pressure dissipated, the landslides, which did not have any kinematic constraints, accelerated and failed. But what would happen if these landslides did have a constraint, e.g. a natural (rock outcrop) or manmade (landslide protection wall) obstacle at its bottom end?

This question is not a purely academic one: one of such landslides is, in fact, rather famous – the Brattas Landslide in St. Moritz, Switzerland – and its stability is of the great concern for the community. The Leaning Tower of St. Moritz (Figure 1a) is probably the most striking evidence of this landslide. It used to be a part of St. Mauritius Church, which was built in 12th century and had to be demolished in 1893 in a view of inevitable collapse due to enormous differential settlements. The continuing displacements of the Tower has been monitored over almost a century (in 1984 the tower has been partially propped up). The detailed geological and geotechnical description of the landslide and the history of the Tower are presented elsewhere (Schluechter, 1988;

Hoe I. Ling et al. (eds.), Soil Stress-Strain Behavior: Measurement, Modeling and Analysis, 949–955.

Sterba et al., 2000). What is of particular interest here, is that the landslide is constrained by the rock outcrop along the Via Maistra (Figure 1b), where zero displacements are observed. Also the monitoring of the Tower displacements before 1984 (Sterba et al., 2000) as well as more recent measurements of the slide displacements (Lang and Sterba, 2002) indicate that the slide is gradually slowing down. Intuitively, it is tempting to conclude that the landslide will eventually stop. But is this really the case?

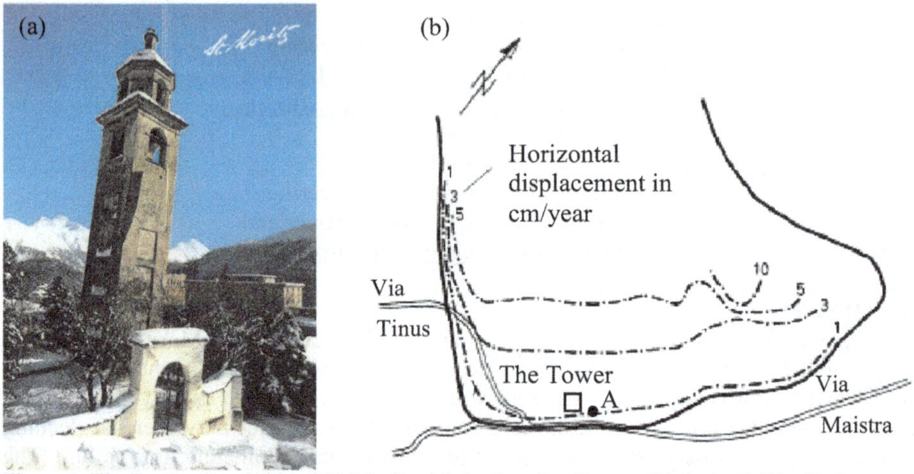

Figure 1. The Brattas Landslide of St. Moritz: (a) the Leaning Tower; (b) the landslide displacements in the lower 200m – long-term survey until 1982.

Puzrin and Sterba (2006) suggested the mechanism of the *failure scenario* of a constrained landslide evolution. In this scenario, the landslide will keep slowing down till the earth pressure at its bottom exceeds the soil resistance and the slope fails catastrophically. In this case, it is necessary to predict the time of the future failure. These objectives are achieved with the help of the inverse analysis procedure developed by Puzrin and Sterba (2006). It allows for the safety factor for long-term slope stability and final displacements to be determined solely from the observed slope displacements. This helps to reduce uncertainties caused by spatial variability of soil properties. The time of the failure can then be calculated using some additional field measurements. In this paper the proposed procedure is illustrated using selected observation data from the St. Moritz landslide. The proper long-term stability analysis of this landslide, however, will be performed after additional monitoring and field tests data is acquired.

2. THE CONSTRAINED LANDSLIDE MODEL

The schematic layout of the boundary value problem of a slowing constrained landslide is given in Figure 2. Equilibrium of the sliding layer relates the shear stress τ on the sliding surface to the average effective normal stress in the layer p' and the effective active earth pressure p'_a acting at the top of the layer:

$$p'(x,t)H + \int_x^L \tau(x,t)dx = \gamma_t H(L-x)\sin\alpha + p'_a H \tag{1}$$

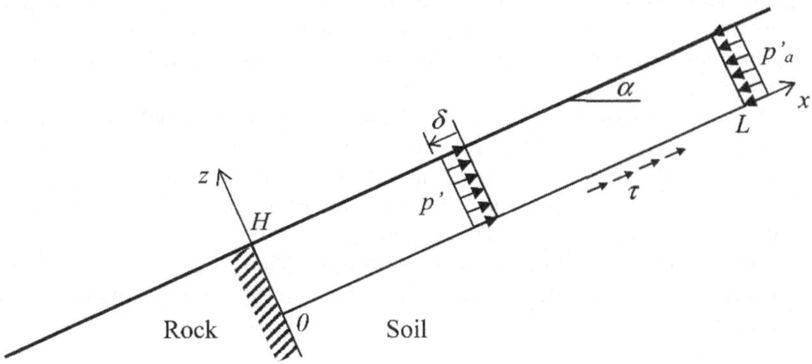

Figure 2. Schematic layout of the constrained landslide model.

Here γ_t is the total unit weight of soil; α is the slope inclination; L and H are the landslide length and thickness, respectively. We use in (1) the effective earth pressures assuming that the average pore water pressure is constant along the slope: $u(x,t) = u(t)$, i.e. either that there is a flow parallel to the slope surface, or that there is no connected water.

In a forward boundary value problem, we would supplement equation (1) with constitutive equations and solve them together in order to obtain displacements $\delta(x,t)$ and earth pressures $p'(x,t)$ and predict the landslide behavior. Because the processes in a constrained landslide are slower than in the one which is free to slide, we assume that the excess pore water pressure has enough time to dissipate. Therefore, the time dependency of displacements is solely due to viscous properties of soil. A number of possible constitutive models can then be suggested (Figure 3). As is seen from equation (1) and Figure 2, the weight and the active force in the layer are resisted by the earth pressure in the layer and the shear stress on the sliding surface. These are schematically represented in Figure 3 by the elastic spring (with elastic modulus *E*) and the slip element (with the slip stress τ_r), respectively. In order to introduce time dependency we include two dashpot elements: one to describe viscous processes within the soil layer, another - on the sliding surface (Skempton, 1985, showed the viscous character of the residual strength).

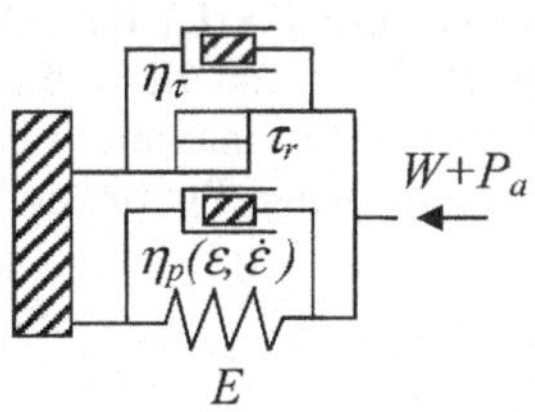

Figure 3. Schematic layout of the visco-elasto-visco-plastic (VEVP) constitutive behavior.

Disadvantage of the forward approach is that it does not take into account the observed slope displacements. Spatial variability of the soil properties results in high levels of indeterminacy in constitutive models and their parameters obtained in laboratory tests. This often causes large discrepancies between the calculated and observed behavior. In contrast to the forward approach, the inverse analysis of the problem would allow for the material properties to be back calculated directly from the observed displacements. This would account for the global slope behavior, as opposed to the behavior of the locally extracted soil samples, and would provide a more reliable basis for the future predictions. The purpose of this paper is to develop such an approach.

In order to simplify the analysis, Puzrin and Sterba (2006) proposed the following analytical function to fit the observed normalized displacements data $\bar{\delta}(\bar{x},t)=\delta(x/L,t)/L$:

$$\bar{\delta}(\bar{x},t)=\bar{\delta}_x(\bar{x})\bar{\delta}_t(t)=\bar{x}(a-b\bar{x})\left(1-e^{-ct+d}\right), \text{ where } \quad 0\leq b/a<0.5;\ c>0 \tag{2}$$

where $\bar{x}=x/L$. This function describes displacements that are zero at the landslide foot and increase monotonically (when $0\leq b/a<0.5$) along the slope towards its crest (Figure 4a), while slowing with time and approaching an asymptotic value (Figure 4b).

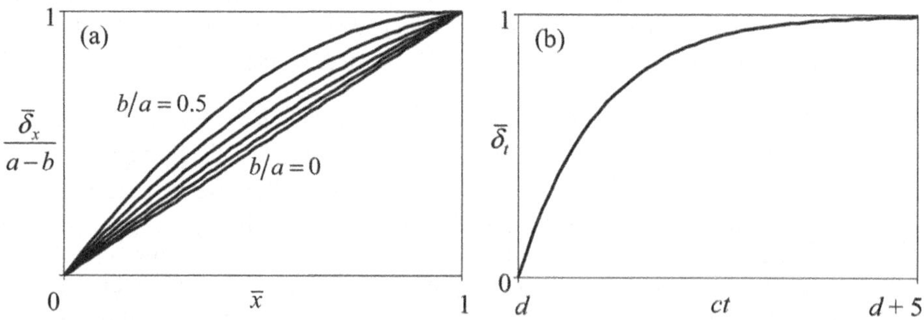

Figure 4. Normalized functions for curve fitting of slope displacements: (a) in space; (b) in time.

The function (2) is simple and yet provides sufficient flexibility to fit the observation data both in the space (parameters a and b) and time domains (parameters c and d). In fact, as shown by Puzrin and Sterba (2006), this function also has a certain theoretical background.

3. THE SAFETY FACTOR, TIME OF FAILURE AND FINAL DISPLACEMENTS

The safety factor for the slope stability can be defined as the ratio between the soil resistance (passive earth pressure) and the maximum earth pressure that can develop at the landslide foot in time (defined, using Puzrin and Sterba, 2006, inverse analysis procedure):

$$F_s=\frac{p_p'}{p'(0,\infty)}=\frac{1-2b/a}{p_a'/p_p'} \tag{3}$$

This definition identifies the *failure scenario* with $F_s < 1$. In this case the time before the slope failure is given as (Puzrin and Sterba, 2006):

$$t_f = t_0 + \frac{1}{c}\ln\frac{\Delta p'/p'_p}{\left(1-e^{-c\Delta t}\right)\left(1/F_s - 1\right)} \tag{4}$$

where Δp is the increase in the earth pressure at the slope bottom, measured over the period of time Δt, t_0 is the time of the pressure transducer installation. Note, that the entire stability analysis can be performed using only equations (3) and (4), which utilize solely the observed data and the values of the effective active and passive earth pressures in the slope found from:

$$\begin{Bmatrix} p'_a \\ p'_p \end{Bmatrix} = \frac{1}{2}\gamma' H \cos\alpha\left[1 + 2\tan^2\varphi'_p \mp 2\sqrt{\left(1+\tan^2\varphi'_p\right)\left(\tan^2\varphi'_p - \tan^2\alpha\right)}\right] \tag{5}$$

where φ'_p and γ' are the effective peak angle of internal friction and effective unit weight of the soil in the sliding layer, respectively.

If $F_s > 1$, the slope will eventually stop sliding, and the final displacement increment for the point x on the slope are defined by:

$$\delta_\infty(x) = \frac{\delta_M(x)}{1-\exp(-c(t_M - t_1))} \tag{6}$$

where $\delta_M(x) = \delta(x, t_M)$; $\delta(x, t_1) = 0$.

4. BRATTAS-ST. MORITZ LANDSLIDE

In the following landslide stability analysis we utilize selected observation data from the St. Moritz landslide. It is important to emphasize that these data were not found sufficient to definitely conclude about the landslide stability, but will allow for the missing data to be identified.

The following landslide parameters are adopted here (Sterba et al, 2000): $L = 1500$ m and $\alpha = 20^\circ$. The peak effective angle of internal friction is assumed (after Vermeer, 1997) to be within the range of $\varphi'_p = 28^\circ - 35^\circ$, so that from equation (5): $p'_a/p'_p = 0.28 - 0.15$. The best fit of the analytical curve (2) to the normalized displacement data monitored at different points in the lower 200 m of the landslide (Figure 1b) is achieved at $b/a = 0.39$ (Figure 5a). Substituting these parameters into equation (3), we obtain the range for the safety factor: $F_s = 0.78 - 1.46$. As is seen, this analysis does not exclude a possibility of the future failure. But let us first focus on the *safe scenario*. The best fit to the normalized displacement data (Lang and Sterba, 2002) monitored between 1979 and 1999 at the point A on the slope located 15 m east from the Tower (Figure 1b) is achieved using the analytical curve (2) with $c = 0.045$ (Figure 5b). The total downhill displacement of the point A between 1979 and 1999 was 177 mm, therefore in the *safe scenario* case, according to formula (6), the final displacement of this point will be 298 mm.

Of a larger concern here is that in this preliminary stability analysis of the Brattas Landslide the possibility of the failure scenario is not excluded. However, it is still too early to panic: as mentioned above, the data used in this example is far from being reliable. First of all, the data in Figure 5a is obtained by monitoring only a few points along the slide, all of them within the lower 200 m (out of total of 1500 m) of the landslide. When the curve (2) is fitted in such a small portion of the range, its ability to represent the entire range is very limited, and the parameter b/a becomes very sensitive to the measurement accuracy. The proper stability analysis of the St. Moritz landslide requires additional displacement measurements along the entire landslide length. Also, the data in Figure 5b is obtained by monitoring the displacements of one point on the slope over a relatively short period of time. The proper time related predictions can be only achieved by monitoring the long term displacements of soil along the entire landslide and measuring the earth pressure changes in time in the area of high compression at the landslide bottom. Finally, effects of the climatic changes and ground water conditions should be also studied and incorporated in the analysis. These and other measurements will become a part of the extensive additional field observation and testing program, which will be carried out by the Institute of Geotechnical Engineering, ETH Zurich commencing in the summer 2006.

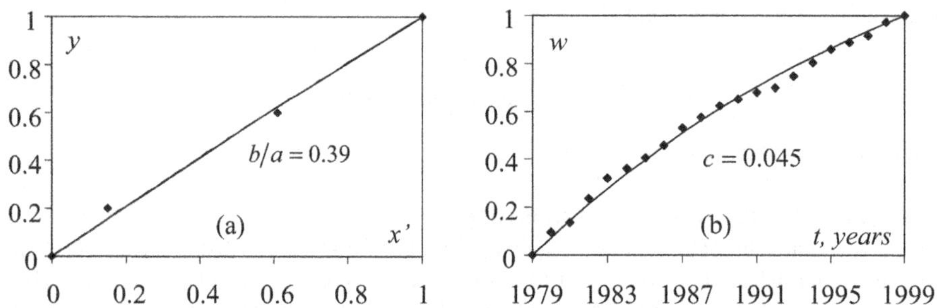

Figure 6. Curve fitting using equation (2) of the normalized displacement data monitored for the St. Moritz landslide: (a) in space (before 1983); (b) in time (1979-1999).

5. ACKNOWLEDGEMENT

The work has been partially supported by the ASTRA/VSS grant VSS 2005/502 "Landslide-Road-Interaction".

6. REFERENCES

Bjerrum, L. 1967. Progressive failure in slopes of overconsolidated plastic clay and clay shales. *ASCE Journal of the Soil Mechanics and Foundation Engineering Division,* 93(5), 1-49.

Lang, H.-J. and Sterba, I. 2002. Schiefer Turm, St. Moritz – Stellungnahme zu den Messungen. *Bericht 3676/10, Institut für Geotechnik, ETH Zürich*, November 1999, 10 p.

Puzrin, A. and Sterba, I. (2006). Inverse long-term stability analysis of a constrained landslide. *Geotechnique,* in press.

Schluechter, Ch. 1988. Instabilities in the area of St. Moritz, Switzerland – Geology, chronology, geotechnology. In: *Proceedings of the Fifth International Symposium on Landslides*, Lausanne, 10-15 July, 1988, Ed. C. Bonnard, Vol 1.

Skempton, A.W. 1964. Long-term stability of clay slopes. *Geotechnique*, 14, 77-102.

Skempton, A.W. 1985. Residual strength of clays in landslides, folded strata and the laboratory. *Geotechnique*, 35(1), 3-18.

Sterba, I., Lang, H.-J., and Amann, P. 2000. The Brattas Landslide in St. Moritz. In: *Proceedings of International Conference on Geotechnical & Geological Engineering: GeoEng 2000*, Melbourne, 12-24 November 2000.

Terzaghi, K. 1936. Stability of slopes of natural clay. In: *Proceedings of First International Conference on Soil Mechanics and Foundation Engineering*, 1, 161-165.

Vermeer, P.A. 1997. PLAXIS Practice I: The leaning Tower of St. Moritz. *PLAXIS Bulletin*, No. 4, pp. 4-7.

Soil Stress-Strain Behavior: Measurement, Modeling and Analysis
Geotechnical Symposium in Roma, March 16 & 17, 2006

TWO-DIMENSIONAL SLOPE FAILURE IN THE CENTRIFUGAL FIELD

Wu, Min-Hao*, Ling, Hoe. I.*, Pamuk, Ahmet.*, and Dov Leshchinsky**
* Department of Civil Engineering and Engineering Mechanics, Columbia University, New York, NY 10027, USA
email: mw2010@columbia.edu
** Department of Civil Engineering, University of Delaware, Newark, Delaware

ABSTRACT
Slope failures occurred frequently that resulted in the loss of lives and properties. Slope performed differently with respect to different slope angles, heights and soil properties. In this study, the centrifuge facility was used to simulate slope failure under two dimensional (2-D) conditions. Clean Nevada sand and its mixtures with different percentages of fines (up to 30% or so) were used. The slope angles were 60, 75 and 90 degrees. Slope failure was generated by increasing the gravity. A laser displacement transducer was used to measure the settlement at the top of the slopes that indicated initiation of failure. A video camera was used in front of the slope to trace failure and movement of failure soil mass. At the end of testing, the slope was cut to obtain the configurations of failure surface. The results showed a normalized behavior of slope failure surface. The normalized behavior tended to drift for less steep slope. Vertical slopes also showed shallower failure surface compared to 75- and 60-degree slopes.

1. INTRODUCTION

Slopes fail frequently that resulted in the loss of many lives and properties. The techniques, such as conventional retaining structures and slope reinforcement methods, are developed to prevent the slope failure by increasing its stability. Slopes perform differently with respect to different slope angles, heights and soil properties. The understanding of slope failures involving clay and shale slopes and seams has increased largely due to the original work by Bishop (1966), Bjerrum (1966), and Skempton (1964). The engineering behaviors of compacted soils and man-made slopes have not been studied as extensively as that of clays and sands, but they have received more attention in recent years.

The limit equilibrium methods used for evaluating the stability of slopes require an accurate and reliable estimate of the in situ shear strength of the slope materials. However, the shear strength parameters, c and ϕ are strongly influenced by many conditions, including the in situ stress, drainage, overconsolidation ratio, and loading rate, etc. Different laboratory methods are available to measure the strength parameters. Common laboratory tests for determining the shear strength of soils include the triaxial test, direct shear test and unconfined compression test. Of these, the triaxial test is one of the most commonly used methods for determining the shear strength of soil (Saada and Townsend, 1981; Tatsuoka, 1988). However, the major problem inherent in this testing technique is its lack of capability in simulating plain strain conditions and limited to certain stress

Hoe I. Ling et al. (eds.), Soil Stress-Strain Behavior: Measurement, Modeling and Analysis, 957–968.

paths. Small scale model tests on the other hand do not simulate the stress level realistically (Taylor 1995). To minimize the scale effects, centrifuge modeling technique has been used.

The scale effect in simulating the behavior of full-scale problems has been recognized by some researchers in the early century. Bucky (1931) developed the world first centrifuge at Columbia University, and since then, the testing method has gained wide acceptance in simulating the behavior of full-scale structures. In the centrifuge modeling technique, a reduced-scale model *n* times smaller than the full-scale is used, whereas the acceleration is increased by *n* times to preserve the stress and strain behavior (Taylor 1995). Scaling laws can be derived by making use of dimensional analysis (for example, Langhaar, 1951) or from a consideration of the governing differential equations. Relevant scaling rules between the centrifuge model and full-scale are given in Table 1.

In this study, slopes of three different angles and heights were tested in the centrifuge. Soils of three different properties were used. The slopes were brought to failure and the results are discussed.

Table 1.Scaling Rules in Centrifuge Testing

Quantity	**Centrifuge model/prototype**
Length	1/n
Area	$1/n^2$
Volume	$1/n^3$
Time (dynamics)	1/n
Time (diffusion)	$1/n^2$
Velocity	1/n
Acceleration	n
Mass	$1/n^3$
Force	$1/n^2$
Energy	$1/n^3$
Stress, Strain, Pressure	1
Frequency (dynamics)	n
Unit weight	n
Permeability	1/n

2. TESTING PROGRAM AND PROCEDURES

Three types of soil were prepared for the slope failure tests: pure sand, sand with 15% fines, and sand with 30% fines. Pure sand was made up of clean Nevada sand; sand with 15% fines sample was made up of 87 percent by weight of Nevada sand and 13 percent by weight of Kaolinite clay; sand with 30% fines sample was made up of 77 percent by weight of Nevada sand and 23% by weight of Kaolinite clay; Each sample was mixed with water to reach its optimum water content.

The slopes were made up of soil compacted at the maximum dry unit weight and optimum water content. As a result, standard compaction tests were first conducted. Figure 1 shows the compaction curves for the three types of soil. The maximum dry unit weights and optimum water contents for different samples of interest were obtained from the compaction curves as summarized in Table 2.

The triaxial compression tests were used to investigate the stress-strain behavior of the three types of soil (Figures 2a and b). From the stress-strain curves obtained under 16.93 kPa confining pressure (Figure 2a), the Young's Modulus of pure sand, sand with 15% fines and sand with 30% fines were 3400, 5670 and 8500 kPa, respectively. The strength parameters obtained from the triaxial compression tests are summarized in Table 3.

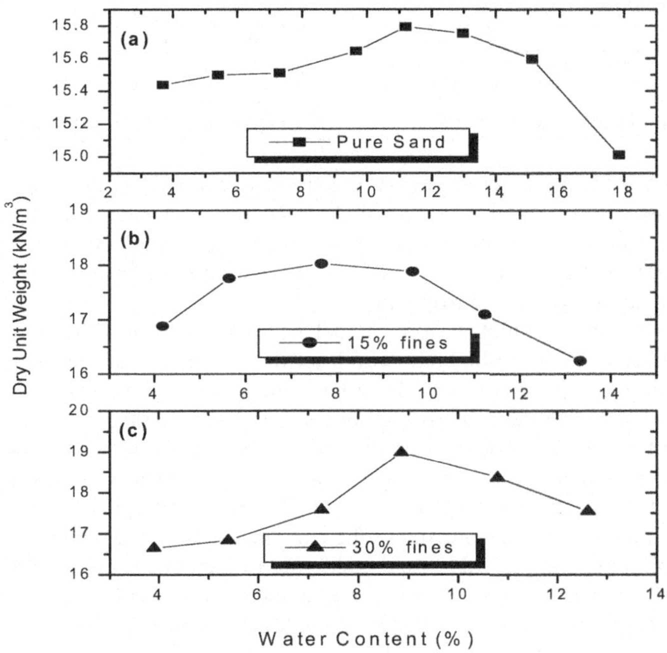

Figure 1. Compaction Curves:
(a) Pure sand, (b) Sand with 15% fines, (c) Sand with 30% fines

The soil sample was compacted to form slopes of different heights and angles. Three different heights: 10, 15 and 20 cm and three different angles: 60°, 75° and 90° were selected for the slopes.

The setup for the 2D slope test is illustrated in Figure 3. The slope model was created in an aluminum box and compacted using a rectangular steel rammer into layers of 2.5 cm thick. This compaction was continued until the thickness of bottom layer reached 5 cm. After the bottom layer was compacted, a frame was secured to the box. The frame (Figure 4) was made up of two parts—a bottom part and a slope surface.

These two parts were connected by a hinge as long as the slope width. It could be adjusted to form different angles for the slope. After fixing the angle of the frame, a thin layer (less than 1 mm) of dyed Nevada sand was spread above the bottom layer. The soil was placed behind the frame and compacted to 2.5 cm in thickness on top of the color layer. This process was repeated until the slope height reached the targeted values (10, 15, or 20 cm).

Table 2. Properties of Soil Samples

Samples	Pure sand	15% fines	30% fines
Sand fraction by weight (%)	100	87	77
Clay fraction by weight (%)	0	13	23
Maximum dry unit weight, γ_d (kN/m^3)	15.79	18.02	18.98
Optimum water content, w_{opt} (%)	11.19	7.66	8.87
Total unit weight, γ_t (kN/m^3)	17.56	19.40	20.67

The purpose of using colored sand in the slope layers was to allow the slip surface to be traced more easily. An attempt was made to decrease the side friction between the aluminum wall and slope by placing a rubber membrane smeared with grease. The membrane was pre-cut into a number of thin rubber strips in order to reduce the tensile force that may restrain the slope movement.

The box with the completed slope model was mounted onto the centrifuge. The frame was then removed from the box. The laser displacement transducer was attached to the box above the slope top surface. In order to measure the largest displacement, the sensor's position was adjusted within the sensitive measuring range. Then, the camera was installed in front of the slope. During the test, all the data from the laser displacement transducer was recorded by the data acquisition system located at the axis of the centrifuge and transferred to the computer outside the centrifuge chamber through a wireless router fixed above the centrifuge. The camera recorded the whole test procedure starting from spinning, and the video was transferred to the outside computer through the slip ring. From the video, the tests were monitored and the instant of failure was captured.

The tests were conducted under increasing acceleration (g). During the tests, the rotation speed of centrifuge was increased at the rate of 1 rpm every 10 seconds, until the slope failed. When the slope failed, the motor of the centrifuge was turned off. The centrifuge then slowed down to reach a complete stop.

Table 3. Strength Parameters - Triaxial Compression Tests

	Angle of Internal Friction φ (degrees)	Cohesion c (kPa)
Pure Sand	40.5	2
Sand with 15 % fines	41	5
Sand with 30 % fines	39	20

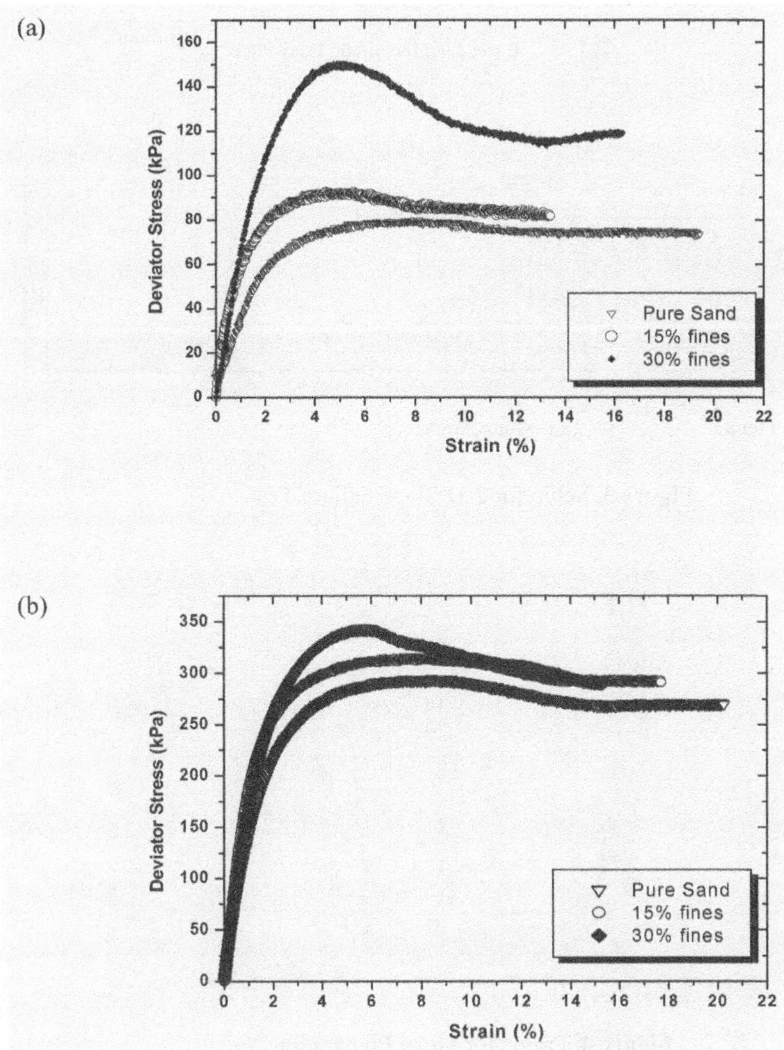

Figure 2. Stress-Strain Behavior of Soil Samples Obtained from Triaxial Compression Test: (a) 16.93 kPa, (b) 84.65 kPa Confining Pressure

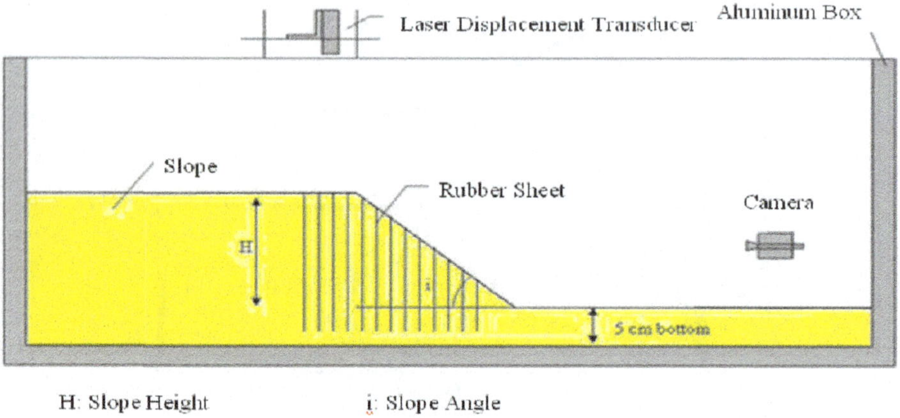

Figure 3. Setup for 2-D Slope Failure Test

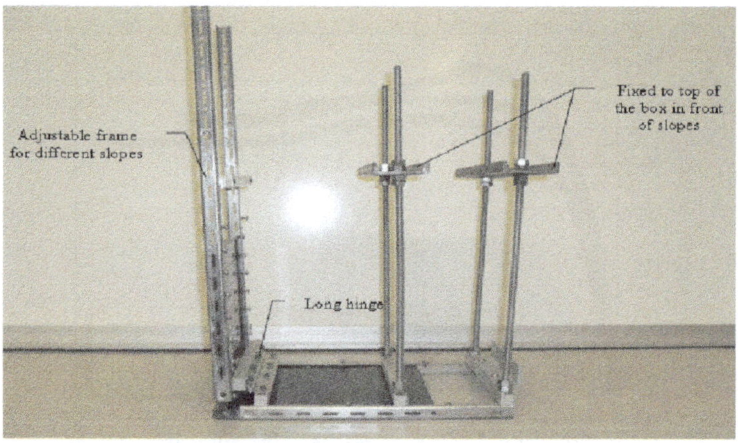

Figure 4. Frame for Slope Preparation

At the end of testing, one of the side walls of the test box was removed. The slope was cut longitudinally to obtain the configuration of its two-dimensional failure surface. The slip surface was carefully traced onto a transparency sheet and then digitized. The data measured from the laser displacement transducer was analyzed to obtain the slope settlement. The settlement was divided by the slope height to obtain the normalized settlement. The normalized settlement under increasing gravity was also used to judge the instant of failure. Note however that in some of the tests, the failure surface lied outside the target of the laser transducer and thus failure was not captured.

3. RESULTS AND DISCUSSION

Figure 5 showed an example of a series of failure pictures captured by the camera at a speed of 30 frames per second. Failure occurred in a very short period of time.

Figures 6 to 8 show the normalized slip surfaces obtained from the tests of pure sand, sand with 15% fines and 30% fines, respectively. The results show that the normalized slip surfaces could be obtained for different slope angles. For vertical slopes of the same type of soil, the failure surfaces superimposed. The normalized slip surfaces for other slope angles were close when superimposed. The results show that shallow failure surface occurred in soil of the lower fraction of fines compared to the more cohesive slopes. As the fraction of fines increased, the failure surface became deeper. This was consistent with the log-spiral analysis (e.g., Leshchinsky and Boedeker, 1989).

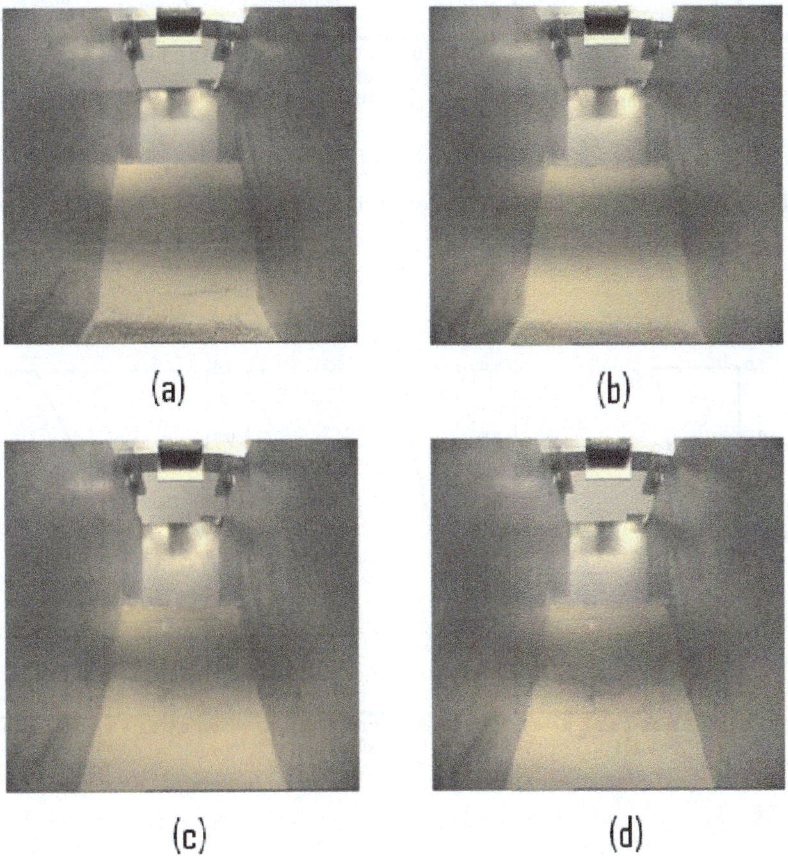

Figure 5. Pictures of Slope Failure viewed from the front
(a) Crack formed, (b) Initiation of failure, (c) Slope collapsed, (d) Failure

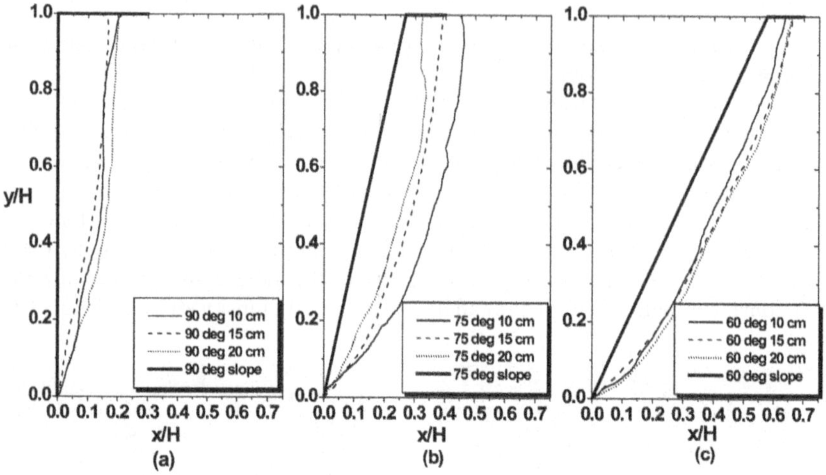

Figure 6. Normalized Slip Surfaces of Pure Sand Slopes: (a) 90°, (b) 75°, (c) 60°

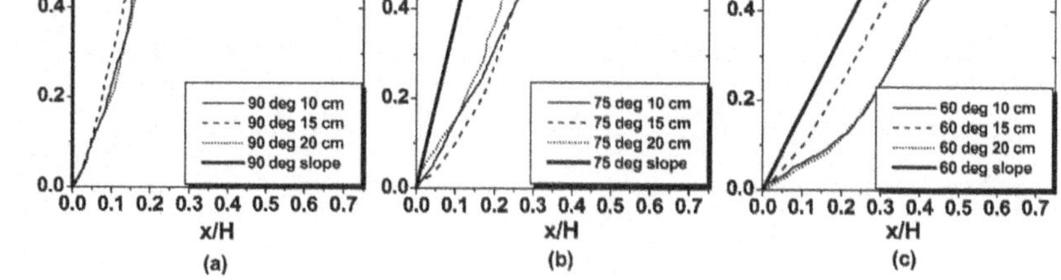

Figure 7. Normalized Slip Surfaces of Sand Slopes with 15% Fines:
(a) 90°, (b) 75°, (c) 60°

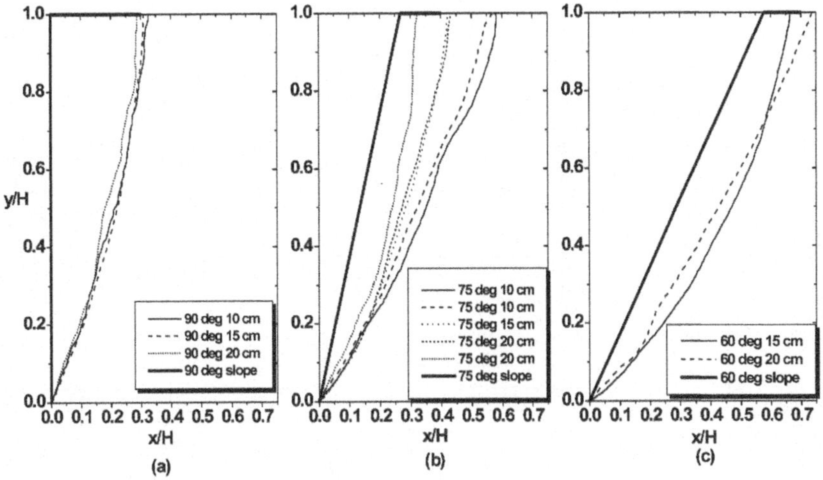

Figure 8. Normalized Slip Surfaces of Sand Slopes with 30% Fines: (a) 90°, (b) 75°, (c) 60°

Table 4 shows the experimental results of the 2-D slope failure tests. For the slopes made with the same type of soil and of the same slope angle, the higher slopes failed at lower acceleration field (gravity) than the lower slopes. It proves that lower slopes were more stable than higher ones. For the slopes with the same type of soil and of the same slope height, the steeper slopes failed at lower gravity than the less steep slopes. For the slopes having the same slope height and angle, higher fines content contributed to failure at higher gravity than those of lower fines contents. It shows that the cohesive slopes were more stable than the sandy or less cohesive slopes provided that excess pore pressure will not be an issue.

The normalized settlement of the slopes due to increasing gravity was small for all the tests. The range of normalized displacement varied from 0.002 to 0.012. From the analysis, the 90° slopes were more sensitive to the settlement than 75° and 60° slopes under increasing gravity. For example, of pure sand slopes (Figure 9), when the gravity increased up to 19g, the normalized settlements of 90°, 75° and 60° slopes were 0.0027, 0.0015 and 0.0008, respectively. As is well known, steeper slopes are less stable than less steep ones. The reasoning is that the vertical displacement and lateral displacement concurred in the tests of this study. When the vertical displacement occurred, it created lateral force to the slope front. If the deformation became large, it may induce failure. As a result, for slopes of similar mixtures, steeper slopes failed at lower gravity compared to less steep ones. The deformation was related to the stiffness of soils as presented in Figures 2a and b.

Table 4. Experimental Results of 2-D Slope Failure Tests

Sample	Total Unit Weight (kN/m^3)	Slope Angle, i (degrees)	Height at 1g (cm)	Prototype Slope Height, H (m)	Acceleration at Failure, n (g)
Pure Sand	17.56	90	20	2.226	10.96
		90	15	2.966	19.07
		75	20	2.757	14.28
		75	15	3.172	21.01
		75	10	1.965	19.94
		60	20	6.850	33.72
		60	15	6.758	45.37
		60	10	4.823	47.15
Sand with 15% Fines	19.40	90	20	3.861	19.29
		90	15	3.656	24.15
		90	10	3.249	31.62
		75	20	5.240	26.06
		75	15	5.740	37.11
		75	10	3.486	35.97
		60	20	11.538	56.32
		60	15	14.576	92.73
		60	10	8.826	87.63
Sand with 30% Fines	20.67	90	20	5.874	29.34
		90	15	8.518	56.78
		90	10	6.594	66.04
		75	20	16.524	81.78
		75	15	13.868	89.84
		75	10	10.856	105.25
		75	10	11.480	113.51
		60	20	30.876	149.93
		60	15	27.840	181.83

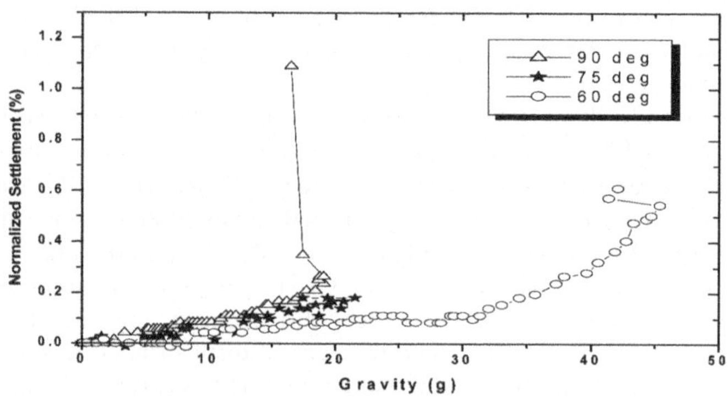

Figure 9. Pure Sand: Normalized Settlement under Increasing Gravity

Comparing the acceleration field between the slopes with the same slope angle and under the same normalized settlement, pure sand slopes required smaller acceleration than 15%, 30% fines and so on to reach the same settlement. For example of 60° slopes (Figure 10), to achieve the normalized settlement of 0.005 , the gravity was increased up to 45, 70 and 82 g for slope of pure sand, sand with 15% fines, and sand with 30% fines, respectively, which means slope of sand with 30% fines were stiffer than pure sand and sand with 15% fines. This result conformed to the stress-strain behavior of the three testing samples.

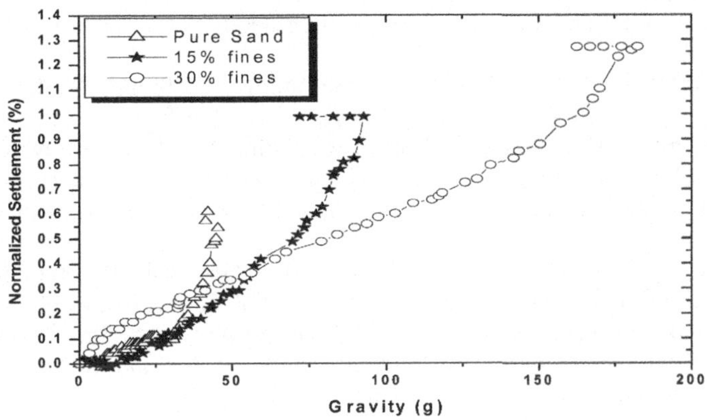

Figure 10. 60° -Slopes: Normalized Settlement under Increasing Gravity

4. CONCLUSIONS

A series of 2-D slope tests were conducted in the centrifugal field for three types of soil - pure sand, sand with 15% fines and 30% fines. The slope models were prepared by compaction to optimum water content and maximum dry unit weight. Based on the test results, the following conclusions were drawn:

The higher and steeper slopes failed at lower gravity compared to the lower and less steep slopes. The slopes with more fines had deeper failure surface compared to less cohesive ones. In addition, the more cohesive slopes failed at higher gravity. The same conclusions applied to the settlement. Pure sand slopes settled more than slopes of sand-clay mixture.

The normalized slip surfaces for each soil type could be obtained for each slope angle. That means the normalized slip surface is unique for the slopes with the same slope angle but different slope heights.

Note that the study as presented did not include the effects of pore water pressure.

5. ACKNOWLEDGMENTS

The authors wish to thank the efforts of many individuals in the Carleton Laboratory who have contributed to this study. The first author was supported by a scholarship from the Ministry of Education, Taiwan, which is supplemented by the teaching assistantship from the Department of Civil Engineering and Engineering Mechanics, Columbia University.

6. REFERENCES

Bishop, A.W. (1966), "The Strength of Soils as Engineering Materials," *Geotechnique*, 16(2), 91-128.

Bucky, P.B. (1931), "Use of Models for the Study of Mining Problems," *AIMME Technical Publications. Am. Inst. Min. Met. and Petroleum Eng.* No. 425.

Bjerrum, L. (1966), "Mechanism of Progressive Failure in Slopes of Overconsolidated Plastic Clay Shales" Preprint, *ASCE Structural Engineering Conference,* Miami.

Langhaar, H.L. (1951), *Dimensional Analysis and Theory of Models*. John Wiley, New York.

Leshchinsky, D., and Boedeker, R.H. (1989), "Geosynthetic Reinforced Earth Structures." *Journal of Geotechnical Engineering*, ASCE, 115(10), 1459-1478.

Saada, A.S., and Townsend, F.C. (1981), "State of the Art: Laboratory Strength Testing of Soils," *ASTM Special Technical Publication 740, R.N. Young and F.C. Townsend, Eds. Philadelphia, Pennsylvania: ASTM*, pp. 7-77.

Skempton, A.W. (1964), "Long-Term Stability of Clay Slopes," *Geotechnique*, Vol. 14, No. 2, pp. 77-101.

Tatsuoka, F. (1988). "Some recent developments in triaxial testing systems for cohesionless soils." *Advanced Triaxial Testing of Soil and Rock*, ASTM STP 977, Donaghe, R.T., Chaney, R.C., and Silver, M.L., Editors, American Society for Testing and Materials, 7-67.

Taylor, R.N. (1995), "Centrifuges in Modeling: Principles and Scale Effects," *Geotechnical Centrifuge Technology* (ed Taylor, R.N.), Blackie Academic and Professional, Glasgow, pp. 19-33.

Soil Stress-Strain Behavior: Measurement, Modeling and Analysis
Geotechnical Symposium in Roma, March 16 & 17, 2006

GEOTECHNICAL AND STRUCTURAL FAILURES DUE TO HEAVY RAINFALL OF MINDULLE TYPHOON IN TAIWAN

Jieh-Jiuh Wang* Hoe I. Ling**

* Department of Civil Engineering and Engineering Mechanics, Columbia University, New York, NY 10027, USA
e-mail: jw2162@columbia.edu
** Department of Civil Engineering and Engineering Mechanics, Columbia University, New York, NY 10027, USA
e-mail: hil9@columbia.edu

ABSTRACT
On July 2, 2004, a 2000-kilometer southwest air current, following the Mindulle Typhoon, caused serious harms to Taiwan. Though this event did not trigger off the most serious damages, the disaster resulted in extensive geological and structural failures. Some of the observations and causes of failures were identified. This case history showed significant implications to future disaster prevention and management works with the hope that it will attract new challenges for geotechnical engineering in solving rainfall-induced failures.

MINDULE TYPHOON (July 2, 2004)

The Mindulle Typhoon passed through Taiwan in the period of June 29 to July 2, 2004. It left a 2000-kilometer tail following the storm that continued to attack Taiwan and wreak havoc for 2 more days. The heavy rainfall induced flooding, debris flows, landslides, slope failures, among various geological failures. For 7 days under the influences of the typhoon and the southwest current, 2 rain gauge stations measured over 2000 mm of rainfall, 10 stations measured over 1600 mm and 45 stations measured over 1000 mm. The heavy rainfall caused 134 landslides and 14 main bridges were destroyed, whereas 15 additional main bridges were damaged.

Losses due to the disaster include 26 deaths and 11 missing, estimated losses of US$ 123 million in agriculture and US$ 1 billion loss in tourism. Up to 210,000 households were without electricity and 440,000 households were without water for couple of weeks after the typhoon.

The recovery cost for such rainfall-induced failures has been extremely high. The failures have been repeating themselves at many of the locations during the events of Typhoon.

Characteristics of Mindulle Typhoon

There were certain special features of Mindulle Typhoon that led to the catastrophic events:

Hoe I. Ling et al. (eds.), Soil Stress-Strain Behavior: Measurement, Modeling and Analysis, 969–978.

1. Unusually heavy rainfall: the average annual rainfall in Taiwan is about 2500mm; however, for Mindulle Typhoon, 2 rain gauge stations measured over 2000 mm of rainfall, and 10 measured over 1600 mm.
2. Soils, rocks and debris from upstream that accumulated in the river channels resulted in the rise of riverbeds of 10 to 20 meters above adjacent roadways at some locations.
3. Stones and debris buried villages. Communities located on the alluvial fans were most seriously damaged.
4. Riverbanks were destroyed by flooding and erosions. Communities located along riversides were also heavily damaged.
5. Erosion was particularly damaging in the areas which were converted into agricultural lands.

TYPES OF FAILURE

This section focuses on the issues of geological/ geotechnical and structural failures.

Geological/ Geotechnical Failures

Mindulle typhoon-triggered geological/geotechnical failures can be categorized as follows:

- The failure of foundations (Figure 1) : There were many factors that led to the failure of foundations, and accordingly resulted in the damages of the superstructures. For instance, the foundations were vulnerable to landslides, debris flows, etc. Meanwhile, because of the increase in excess pore pressure (due to rainfall infiltration), the strength of the soil at the foundation decreased. The large settlement and uneven settlement were seen resulting from the lost of bearing capacity. The extent of damage depended on the types of soil and locations.

Figure 1. The failure of foundation

Figure 2. Translational landslide

- Landslide : The rainfall-induced landslides in this case included that of translational and rotational landslides. Some of the failures were surficial and shallow.
  Translational landslide (Figure 2): The slides of the dip slope, often occurring

Figure 3. Rotational landslide

Figure 4. Surficial failure

Figure 5. Debris flow

Figure 6. Rock fall

among the layers of rocks, has caused many tragedies in Taiwan recently, especially after heavy downpours. In this case, the translational landslides were very common since the rainfall was not only torrential in intensity but also lasted for extended period of time.

Rotational landslide (Figure 3): This failure involved large scale of soil movement.

- Surficial Failures (Figure 4): Numerous small scale slope surface failures were seen. These failures were shallow.
- Debris Flow (Figure 5) : The massive movement of mixed solids and water moved downward as a viscous fluid. Debris flow may be very dense, so flows occurred slower than water. It is perceived that debris flow is the destructive disaster in Taiwan since there have been up to 1420 documented potential debris flow areas.
- Rock Fall (Figure 6) : The downward movement of rocks that fell off after separating from the bedrock. Geological structures in Taiwan are considering young and thus have a high risk of rock falls and possibly providing sufficient sources to debris flow.

Major Factors to Geological/Geotechnical Failures

The main factors of this disaster are discussed as follows:

1. Erosion

The main factors of flow that affected erosion included the discharge, flow rate and sections of the channel. After the Mindulle typhoon, heavy rainfall intensity raised the water level rapidly, and then the erosion accompanied inasmuch as the soaring flow rate. Erosion increased the angle of the slope while the shear resistance of soil was reduced in the presence of flow, thus creating unstable slope (Figure 7). Besides, erosion washed out the materials of slope roots. The soil strength was reduced and thus the factor of safety of the slope decreased drastically.

Figure 7. The erosion increased the angle of slope.

2. Sediment Deposition

The variation of flow rate dominates its shifting capability. The distribution of the flow rates varies with depths. Even though the higher flow rate is at the surface, the river bottom possesses slower flow rate due to frictional shear. That is why the seen sediments were deposited at the river bottom.

3. Excess Pore Water Pressure

The low soil permeability and high infiltration rate of rain resulted in a temporary rise of water table. The cohesive soil beneath the water table became saturated; hence the pore water pressure increased. Besides, the increasing of unit weight of soil, due to the

Figure 8. The scouring and failure of foundations

Figure 9. Roadway has failed due to scouring of the road base.

Figure 10. Bridge and road collapsed.

Figure 11. School buildings damaged by impacts from tree debris carried by the flooding river.

permeation of rainfall, promoted the sliding force and reduced the factor of safety.

4. Chemical Reactions (Leaching)

When large amount of rain permeated into the joints of bedrocks, the increasing sliding moment, owing to lateral and bottom up water pressure, raised the risk of rock falls. Furthermore, clay minerals existed in the joints of rocks. The chemical reactions between water and minerals may lead to the separation of stone from the bedrock and thus the risk of rock fall increased.

Many geological materials disintegrated after reacting with water. In Taiwan, there are many tiny particle materials, such as shale, schist and clay, which internal strength are reduced quickly after reacting with water. Moreover, the volume of clay changed (swelled) after mixing with water. The pressure resulted from swelling led to the geological failures as well.

Structural Failures

Figure 12. Building damage - Debris flow has buried buildings and transported large objects.

Figure 13. Bridge damage - Road surface on the bridge has been destroyed, although much of the structure remained intact.

The main natural factors and mechanism that caused damages to physical structures are similar to that mentioned in the geological section. Generally speaking, the types of failure and possible factors contributing to the failures are classified as follows:

- Types of building damage: Impact and/or burial by debris, lack of bearing capacity.
- Types of roadway damage: Erosion and lack of bearing capacity, rock impact, burial by debris and landslides, failure of retaining structure.
- Types of bridge damage: Lateral impact, lack of bearing capacity.
- Types of natural effect: Debris flow, surficial and deep-seated failures, sedimentations.

Major Factors Contributing to Structural Failures

1. Riverside Roads and Erosion

Many expressways and villages are located at riverine areas, some of which are vulnerable to the attacks. After downpour, water level rose quickly and flow rate increased dramatically. Gravels and boulders carried by floods led to erosion, thus the soil under the roads, bridges and building foundations lost the bearing capacity, and, furthermore, collapsed (Figures 11, 12, 13).

Figure 14. Debris-flow deposits up to the bottom of the new arch-bridge roadway.

Figure 15. Slope failure damaged the road.

2. Slope Failure

The failure of slope would harm the physical structure sitting upon it, especially when the structure was constructed on the failure surface (Figure 15). Similarly, the slope became less stable if the earth, which supported the physical structure, was damaged by the debris of structure.

3. Debris Deposits, Lateral impact and Clearance

The river deposited the debris at certain areas. Sediments of gravels and boulders piled up and blocked the waterways and increased water level (Figures 14, 16, 17).

In this case, several constructions were not able to withstand the lateral impact from boulders or giant woods. Many riverside buildings were destroyed by the lateral forces of floods and theirs hauling solids.

The sediment of solids from debris flow raised the riverbed go up. Insufficient buffer room or clearance beneath the bridge is harmful to the structural safety of bridges. Otherwise, when the rivers overflowed, the bridges

Figure 16. Debris piled up and blocked the waterway.

Figure 17. River channel filled with debris overflowed into adjacent roadway.

Figure 18. The bridge could not withstand the lateral force. Only three piers remained.

not only lost the function but also threatened the communities nearby, stepping up the regional vulnerability (Figures 14, 18).

4. The Scouring of Bridge Foundation

The scouring of bridge is a common problem in Taiwan (Figure 19). The effects of washout and erosion might be underestimated. Except for improving design and monitoring, it revealed the needs of deepening the foundations if the situations keep occurring.

Figure19. Approximately 1 meter of bedrock (shale) removed beneath the foundation

5. Lacks of Lateral Resisting Force of Retaining Walls

In many failure cases, RC retaining walls either lacked for lateral resisting force or reinforcements (Figure 20), so that the retaining walls were unable to resist against strong lateral earth pressure.

Several reinforced soil retaining walls with geosynthetic reinforcements were found to have performed well.

Figure 20. Detail of failed horizontally unreinforced retaining wall

6. Engineering Design Problems and/or Alternative Solutions

Open tunnels, used for protecting the slope and prevention from the falling rocks, are often located at the potential areas of falling rocks or sliding zones (Figure 21, 22, 23). Even though the constructions were often destroyed by debris flows or landslides, it may not be the problem of construction quality but the design and site selection. Alternative construction methods need to be sought in avoiding repeated failures.

Figure 21. The open tunnel often destroyed by landslides is the problem of site selection or construction methods.

CONCLUSIONS

This paper summarizes the types and possible causes of geological/ geotechnical and structural failures in the events of heavy rainfall.

For geological/geotechnical failures, the followings were observed:

- Increasing discharge and flow rate resulted from typhoon-triggered heavy rainfall accelerated the erosion.
- Both the angle and washout of slope resulted from erosion led to slope instability.
- The chemical reaction between water, soil and rocks (especially shale, schist and clay), not only reduced the strength of materials but also induced excess pore pressure that rendered slope unstable.
- The pressure of ground water due to rainfall permeation multiplied the risk of geological and geotechnical failures.

Figure 22. An example of RC withstanding debris flow. The approach has been washed away.

Figure 23. Abandoned pier focused debris on unreinforced retaining wall

For structural failures, the followings were observed:

- Reinforced concrete structures performed exceptionally well, though unreinforced concrete structures performed poorly.
- Thoughtful placement of structures is important in protecting bridge foundations or reducing the clearance with riverbed.
- Inexpensive / Temporary structures may be good choices for high risk locations with less critical applications.

Eventually, the aspects of geological/ geotechnical and structural failures may lead to some suggestions for better future structural and geotechnical designs and/or construction methodologies.

Different from traditional theories, high disaster frequency, unimaginable disaster seriousness and diverse disaster characteristics posed a new research field for Taiwan. Meanwhile, the experiences from Taiwan also provide the chance for better future development of both disaster prevention researches and practical works.

ACKNOWLEDGMENTS

This paper is based on the disaster investigation works, sponsored by the Public Construction Commission, Executive Yuan, Taiwan, following the Mindulle Typhoon of 2004. This support and arrangement for the site visit are gratefully acknowledged.
Many individuals participated in this works. We would like to express our appreciation to the following individuals from Columbia University who served as members of the Investigation Team: Prof. Andrew W. Smyth who served as co-host of the team, Logan Brant, Min-Hao Wu and Jui-Pin Wang. Without the dedicated assistance of Investigation Team members this study would not have been completed.

REFERENCES

Department of Civil Engineering and Engineering Mechanics, Columbia University (2004) *Reconnaissance Report— Disaster Investigation following the Mindulle (July 2) Typhoon.* Taipei: Public Construction Commission, Taiwan.

Central Weather Bureau, Taiwan: www.cwb.gov.tw

National Fire Administration, MOI, Taiwan: www.nfa.gov.tw

AUTHOR INDEX

Mechanics

SOLID MECHANICS AND ITS APPLICATIONS

Series Editor: G.M.L. Gladwell

Aims and Scope of the Series

The fundamental questions arising in mechanics are: *Why?*, *How?*, and *How much?* The aim of this series is to provide lucid accounts written by authoritative researchers giving vision and insight in answering these questions on the subject of mechanics as it relates to solids. The scope of the series covers the entire spectrum of solid mechanics. Thus it includes the foundation of mechanics; variational formulations; computational mechanics; statics, kinematics and dynamics of rigid and elastic bodies; vibrations of solids and structures; dynamical systems and chaos; the theories of elasticity, plasticity and viscoelasticity; composite materials; rods, beams, shells and membranes; structural control and stability; soils, rocks and geomechanics; fracture; tribology; experimental mechanics; biomechanics and machine design.

1. R.T. Haftka, Z. Gürdal and M.P. Kamat: *Elements of Structural Optimization*. 2nd rev.ed., 1990 ISBN 0-7923-0608-2
2. J.J. Kalker: *Three-Dimensional Elastic Bodies in Rolling Contact*. 1990 ISBN 0-7923-0712-7
3. P. Karasudhi: *Foundations of Solid Mechanics*. 1991 ISBN 0-7923-0772-0
4. *Not published*
5. *Not published.*
6. J.F. Doyle: *Static and Dynamic Analysis of Structures*. With an Emphasis on Mechanics and Computer Matrix Methods. 1991 ISBN 0-7923-1124-8; Pb 0-7923-1208-2
7. O.O. Ochoa and J.N. Reddy: *Finite Element Analysis of Composite Laminates*. ISBN 0-7923-1125-6
8. M.H. Aliabadi and D.P. Rooke: *Numerical Fracture Mechanics*. ISBN 0-7923-1175-2
9. J. Angeles and C.S. López-Cajún: *Optimization of Cam Mechanisms*. 1991 ISBN 0-7923-1355-0
10. D.E. Grierson, A. Franchi and P. Riva (eds.): *Progress in Structural Engineering*. 1991 ISBN 0-7923-1396-8
11. R.T. Haftka and Z. Gürdal: *Elements of Structural Optimization*. 3rd rev. and exp. ed. 1992 ISBN 0-7923-1504-9; Pb 0-7923-1505-7
12. J.R. Barber: *Elasticity*. 1992 ISBN 0-7923-1609-6; Pb 0-7923-1610-X
13. H.S. Tzou and G.L. Anderson (eds.): *Intelligent Structural Systems*. 1992 ISBN 0-7923-1920-6
14. E.E. Gdoutos: *Fracture Mechanics*. An Introduction. 1993 ISBN 0-7923-1932-X
15. J.P. Ward: *Solid Mechanics*. An Introduction. 1992 ISBN 0-7923-1949-4
16. M. Farshad: *Design and Analysis of Shell Structures*. 1992 ISBN 0-7923-1950-8
17. H.S. Tzou and T. Fukuda (eds.): *Precision Sensors, Actuators and Systems*. 1992 ISBN 0-7923-2015-8
18. J.R. Vinson: *The Behavior of Shells Composed of Isotropic and Composite Materials*. 1993 ISBN 0-7923-2113-8
19. H.S. Tzou: *Piezoelectric Shells*. Distributed Sensing and Control of Continua. 1993 ISBN 0-7923-2186-3
20. W. Schiehlen (ed.): *Advanced Multibody System Dynamics*. Simulation and Software Tools. 1993 ISBN 0-7923-2192-8
21. C.-W. Lee: *Vibration Analysis of Rotors*. 1993 ISBN 0-7923-2300-9
22. D.R. Smith: *An Introduction to Continuum Mechanics*. 1993 ISBN 0-7923-2454-4
23. G.M.L. Gladwell: *Inverse Problems in Scattering*. An Introduction. 1993 ISBN 0-7923-2478-1

Mechanics

SOLID MECHANICS AND ITS APPLICATIONS

Series Editor: G.M.L. Gladwell

24. G. Prathap: *The Finite Element Method in Structural Mechanics*. 1993 ISBN 0-7923-2492-7
25. J. Herskovits (ed.): *Advances in Structural Optimization*. 1995 ISBN 0-7923-2510-9
26. M.A. González-Palacios and J. Angeles: *Cam Synthesis*. 1993 ISBN 0-7923-2536-2
27. W.S. Hall: *The Boundary Element Method*. 1993 ISBN 0-7923-2580-X
28. J. Angeles, G. Hommel and P. Kovács (eds.): *Computational Kinematics*. 1993 ISBN 0-7923-2585-0
29. A. Curnier: *Computational Methods in Solid Mechanics*. 1994 ISBN 0-7923-2761-6
30. D.A. Hills and D. Nowell: *Mechanics of Fretting Fatigue*. 1994 ISBN 0-7923-2866-3
31. B. Tabarrok and F.P.J. Rimrott: *Variational Methods and Complementary Formulations in Dynamics*. 1994 ISBN 0-7923-2923-6
32. E.H. Dowell (ed.), E.F. Crawley, H.C. Curtiss Jr., D.A. Peters, R. H. Scanlan and F. Sisto: *A Modern Course in Aeroelasticity*. Third Revised and Enlarged Edition. 1995 ISBN 0-7923-2788-8; Pb: 0-7923-2789-6
33. A. Preumont: *Random Vibration and Spectral Analysis*. 1994 ISBN 0-7923-3036-6
34. J.N. Reddy (ed.): *Mechanics of Composite Materials*. Selected works of Nicholas J. Pagano. 1994 ISBN 0-7923-3041-2
35. A.P.S. Selvadurai (ed.): *Mechanics of Poroelastic Media*. 1996 ISBN 0-7923-3329-2
36. Z. Mróz, D. Weichert, S. Dorosz (eds.): *Inelastic Behaviour of Structures under Variable Loads*. 1995 ISBN 0-7923-3397-7
37. R. Pyrz (ed.): *IUTAM Symposium on Microstructure-Property Interactions in Composite Materials*. Proceedings of the IUTAM Symposium held in Aalborg, Denmark. 1995 ISBN 0-7923-3427-2
38. M.I. Friswell and J.E. Mottershead: *Finite Element Model Updating in Structural Dynamics*. 1995 ISBN 0-7923-3431-0
39. D.F. Parker and A.H. England (eds.): *IUTAM Symposium on Anisotropy, Inhomogeneity and Nonlinearity in Solid Mechanics*. Proceedings of the IUTAM Symposium held in Nottingham, U.K. 1995 ISBN 0-7923-3594-5
40. J.-P. Merlet and B. Ravani (eds.): *Computational Kinematics '95*. 1995 ISBN 0-7923-3673-9
41. L.P. Lebedev, I.I. Vorovich and G.M.L. Gladwell: *Functional Analysis*. Applications in Mechanics and Inverse Problems. 1996 ISBN 0-7923-3849-9
42. J. Menčik: *Mechanics of Components with Treated or Coated Surfaces*. 1996 ISBN 0-7923-3700-X
43. D. Bestle and W. Schiehlen (eds.): *IUTAM Symposium on Optimization of Mechanical Systems*. Proceedings of the IUTAM Symposium held in Stuttgart, Germany. 1996 ISBN 0-7923-3830-8
44. D.A. Hills, P.A. Kelly, D.N. Dai and A.M. Korsunsky: *Solution of Crack Problems*. The Distributed Dislocation Technique. 1996 ISBN 0-7923-3848-0
45. V.A. Squire, R.J. Hosking, A.D. Kerr and P.J. Langhorne: *Moving Loads on Ice Plates*. 1996 ISBN 0-7923-3953-3
46. A. Pineau and A. Zaoui (eds.): *IUTAM Symposium on Micromechanics of Plasticity and Damage of Multiphase Materials*. Proceedings of the IUTAM Symposium held in Sèvres, Paris, France. 1996 ISBN 0-7923-4188-0
47. A. Naess and S. Krenk (eds.): *IUTAM Symposium on Advances in Nonlinear Stochastic Mechanics*. Proceedings of the IUTAM Symposium held in Trondheim, Norway. 1996 ISBN 0-7923-4193-7
48. D. Ieşan and A. Scalia: *Thermoelastic Deformations*. 1996 ISBN 0-7923-4230-5

Mechanics

SOLID MECHANICS AND ITS APPLICATIONS

Series Editor: G.M.L. Gladwell

49. J.R. Willis (ed.): *IUTAM Symposium on Nonlinear Analysis of Fracture*. Proceedings of the IUTAM Symposium held in Cambridge, U.K. 1997 ISBN 0-7923-4378-6
50. A. Preumont: *Vibration Control of Active Structures*. An Introduction. 1997 ISBN 0-7923-4392-1
51. G.P. Cherepanov: *Methods of Fracture Mechanics: Solid Matter Physics*. 1997 ISBN 0-7923-4408-1
52. D.H. van Campen (ed.): *IUTAM Symposium on Interaction between Dynamics and Control in Advanced Mechanical Systems*. Proceedings of the IUTAM Symposium held in Eindhoven, The Netherlands. 1997 ISBN 0-7923-4429-4
53. N.A. Fleck and A.C.F. Cocks (eds.): *IUTAM Symposium on Mechanics of Granular and Porous Materials*. Proceedings of the IUTAM Symposium held in Cambridge, U.K. 1997 ISBN 0-7923-4553-3
54. J. Roorda and N.K. Srivastava (eds.): *Trends in Structural Mechanics*. Theory, Practice, Education. 1997 ISBN 0-7923-4603-3
55. Yu.A. Mitropolskii and N. Van Dao: *Applied Asymptotic Methods in Nonlinear Oscillations*. 1997 ISBN 0-7923-4605-X
56. C. Guedes Soares (ed.): *Probabilistic Methods for Structural Design*. 1997 ISBN 0-7923-4670-X
57. D. François, A. Pineau and A. Zaoui: *Mechanical Behaviour of Materials*. Volume I: Elasticity and Plasticity. 1998 ISBN 0-7923-4894-X
58. D. François, A. Pineau and A. Zaoui: *Mechanical Behaviour of Materials*. Volume II: Viscoplasticity, Damage, Fracture and Contact Mechanics. 1998 ISBN 0-7923-4895-8
59. L.T. Tenek and J. Argyris: *Finite Element Analysis for Composite Structures*. 1998 ISBN 0-7923-4899-0
60. Y.A. Bahei-El-Din and G.J. Dvorak (eds.): *IUTAM Symposium on Transformation Problems in Composite and Active Materials*. Proceedings of the IUTAM Symposium held in Cairo, Egypt. 1998 ISBN 0-7923-5122-3
61. I.G. Goryacheva: *Contact Mechanics in Tribology*. 1998 ISBN 0-7923-5257-2
62. O.T. Bruhns and E. Stein (eds.): *IUTAM Symposium on Micro- and Macrostructural Aspects of Thermoplasticity*. Proceedings of the IUTAM Symposium held in Bochum, Germany. 1999 ISBN 0-7923-5265-3
63. F.C. Moon: *IUTAM Symposium on New Applications of Nonlinear and Chaotic Dynamics in Mechanics*. Proceedings of the IUTAM Symposium held in Ithaca, NY, USA. 1998 ISBN 0-7923-5276-9
64. R. Wang: *IUTAM Symposium on Rheology of Bodies with Defects*. Proceedings of the IUTAM Symposium held in Beijing, China. 1999 ISBN 0-7923-5297-1
65. Yu.I. Dimitrienko: *Thermomechanics of Composites under High Temperatures*. 1999 ISBN 0-7923-4899-0
66. P. Argoul, M. Frémond and Q.S. Nguyen (eds.): *IUTAM Symposium on Variations of Domains and Free-Boundary Problems in Solid Mechanics*. Proceedings of the IUTAM Symposium held in Paris, France. 1999 ISBN 0-7923-5450-8
67. F.J. Fahy and W.G. Price (eds.): *IUTAM Symposium on Statistical Energy Analysis*. Proceedings of the IUTAM Symposium held in Southampton, U.K. 1999 ISBN 0-7923-5457-5
68. H.A. Mang and F.G. Rammerstorfer (eds.): *IUTAM Symposium on Discretization Methods in Structural Mechanics*. Proceedings of the IUTAM Symposium held in Vienna, Austria. 1999 ISBN 0-7923-5591-1

Mechanics

SOLID MECHANICS AND ITS APPLICATIONS

Series Editor: G.M.L. Gladwell

69. P. Pedersen and M.P. Bendsøe (eds.): *IUTAM Symposium on Synthesis in Bio Solid Mechanics.* Proceedings of the IUTAM Symposium held in Copenhagen, Denmark. 1999 ISBN 0-7923-5615-2
70. S.K. Agrawal and B.C. Fabien: *Optimization of Dynamic Systems.* 1999 ISBN 0-7923-5681-0
71. A. Carpinteri: *Nonlinear Crack Models for Nonmetallic Materials.* 1999 ISBN 0-7923-5750-7
72. F. Pfeifer (ed.): *IUTAM Symposium on Unilateral Multibody Contacts.* Proceedings of the IUTAM Symposium held in Munich, Germany. 1999 ISBN 0-7923-6030-3
73. E. Lavendelis and M. Zakrzhevsky (eds.): *IUTAM/IFToMM Symposium on Synthesis of Nonlinear Dynamical Systems.* Proceedings of the IUTAM/IFToMM Symposium held in Riga, Latvia. 2000 ISBN 0-7923-6106-7
74. J.-P. Merlet: *Parallel Robots.* 2000 ISBN 0-7923-6308-6
75. J.T. Pindera: *Techniques of Tomographic Isodyne Stress Analysis.* 2000 ISBN 0-7923-6388-4
76. G.A. Maugin, R. Drouot and F. Sidoroff (eds.): *Continuum Thermomechanics.* The Art and Science of Modelling Material Behaviour. 2000 ISBN 0-7923-6407-4
77. N. Van Dao and E.J. Kreuzer (eds.): *IUTAM Symposium on Recent Developments in Non-linear Oscillations of Mechanical Systems.* 2000 ISBN 0-7923-6470-8
78. S.D. Akbarov and A.N. Guz: *Mechanics of Curved Composites.* 2000 ISBN 0-7923-6477-5
79. M.B. Rubin: *Cosserat Theories: Shells, Rods and Points.* 2000 ISBN 0-7923-6489-9
80. S. Pellegrino and S.D. Guest (eds.): *IUTAM-IASS Symposium on Deployable Structures: Theory and Applications.* Proceedings of the IUTAM-IASS Symposium held in Cambridge, U.K., 6–9 September 1998. 2000 ISBN 0-7923-6516-X
81. A.D. Rosato and D.L. Blackmore (eds.): *IUTAM Symposium on Segregation in Granular Flows.* Proceedings of the IUTAM Symposium held in Cape May, NJ, U.S.A., June 5–10, 1999. 2000 ISBN 0-7923-6547-X
82. A. Lagarde (ed.): *IUTAM Symposium on Advanced Optical Methods and Applications in Solid Mechanics.* Proceedings of the IUTAM Symposium held in Futuroscope, Poitiers, France, August 31–September 4, 1998. 2000 ISBN 0-7923-6604-2
83. D. Weichert and G. Maier (eds.): *Inelastic Analysis of Structures under Variable Loads.* Theory and Engineering Applications. 2000 ISBN 0-7923-6645-X
84. T.-J. Chuang and J.W. Rudnicki (eds.): *Multiscale Deformation and Fracture in Materials and Structures.* The James R. Rice 60th Anniversary Volume. 2001 ISBN 0-7923-6718-9
85. S. Narayanan and R.N. Iyengar (eds.): *IUTAM Symposium on Nonlinearity and Stochastic Structural Dynamics.* Proceedings of the IUTAM Symposium held in Madras, Chennai, India, 4–8 January 1999 ISBN 0-7923-6733-2
86. S. Murakami and N. Ohno (eds.): *IUTAM Symposium on Creep in Structures.* Proceedings of the IUTAM Symposium held in Nagoya, Japan, 3-7 April 2000. 2001 ISBN 0-7923-6737-5
87. W. Ehlers (ed.): *IUTAM Symposium on Theoretical and Numerical Methods in Continuum Mechanics of Porous Materials.* Proceedings of the IUTAM Symposium held at the University of Stuttgart, Germany, September 5-10, 1999. 2001 ISBN 0-7923-6766-9
88. D. Durban, D. Givoli and J.G. Simmonds (eds.): *Advances in the Mechanis of Plates and Shells The Avinoam Libai Anniversary Volume.* 2001 ISBN 0-7923-6785-5
89. U. Gabbert and H.-S. Tzou (eds.): *IUTAM Symposium on Smart Structures and Structonic Systems.* Proceedings of the IUTAM Symposium held in Magdeburg, Germany, 26–29 September 2000. 2001 ISBN 0-7923-6968-8

Mechanics

SOLID MECHANICS AND ITS APPLICATIONS

Series Editor: G.M.L. Gladwell

90. Y. Ivanov, V. Cheshkov and M. Natova: *Polymer Composite Materials – Interface Phenomena & Processes*. 2001 ISBN 0-7923-7008-2
91. R.C. McPhedran, L.C. Botten and N.A. Nicorovici (eds.): *IUTAM Symposium on Mechanical and Electromagnetic Waves in Structured Media*. Proceedings of the IUTAM Symposium held in Sydney, NSW, Australia, 18-22 Januari 1999. 2001 ISBN 0-7923-7038-4
92. D.A. Sotiropoulos (ed.): *IUTAM Symposium on Mechanical Waves for Composite Structures Characterization*. Proceedings of the IUTAM Symposium held in Chania, Crete, Greece, June 14-17, 2000. 2001 ISBN 0-7923-7164-X
93. V.M. Alexandrov and D.A. Pozharskii: *Three-Dimensional Contact Problems*. 2001 ISBN 0-7923-7165-8
94. J.P. Dempsey and H.H. Shen (eds.): *IUTAM Symposium on Scaling Laws in Ice Mechanics and Ice Dynamics*. Proceedings of the IUTAM Symposium held in Fairbanks, Alaska, U.S.A., 13-16 June 2000. 2001 ISBN 1-4020-0171-1
95. U. Kirsch: *Design-Oriented Analysis of Structures*. A Unified Approach. 2002 ISBN 1-4020-0443-5
96. A. Preumont: *Vibration Control of Active Structures*. An Introduction (2nd Edition). 2002 ISBN 1-4020-0496-6
97. B.L. Karihaloo (ed.): *IUTAM Symposium on Analytical and Computational Fracture Mechanics of Non-Homogeneous Materials*. Proceedings of the IUTAM Symposium held in Cardiff, U.K., 18-22 June 2001. 2002 ISBN 1-4020-0510-5
98. S.M. Han and H. Benaroya: *Nonlinear and Stochastic Dynamics of Compliant Offshore Structures*. 2002 ISBN 1-4020-0573-3
99. A.M. Linkov: *Boundary Integral Equations in Elasticity Theory*. 2002 ISBN 1-4020-0574-1
100. L.P. Lebedev, I.I. Vorovich and G.M.L. Gladwell: *Functional Analysis*. Applications in Mechanics and Inverse Problems (2nd Edition). 2002 ISBN 1-4020-0667-5; Pb: 1-4020-0756-6
101. Q.P. Sun (ed.): *IUTAM Symposium on Mechanics of Martensitic Phase Transformation in Solids*. Proceedings of the IUTAM Symposium held in Hong Kong, China, 11-15 June 2001. 2002 ISBN 1-4020-0741-8
102. M.L. Munjal (ed.): *IUTAM Symposium on Designing for Quietness*. Proceedings of the IUTAM Symposium held in Bangkok, India, 12-14 December 2000. 2002 ISBN 1-4020-0765-5
103. J.A.C. Martins and M.D.P. Monteiro Marques (eds.): *Contact Mechanics*. Proceedings of the 3rd Contact Mechanics International Symposium, Praia da Consolação, Peniche, Portugal, 17-21 June 2001. 2002 ISBN 1-4020-0811-2
104. H.R. Drew and S. Pellegrino (eds.): *New Approaches to Structural Mechanics, Shells and Biological Structures*. 2002 ISBN 1-4020-0862-7
105. J.R. Vinson and R.L. Sierakowski: *The Behavior of Structures Composed of Composite Materials*. Second Edition. 2002 ISBN 1-4020-0904-6
106. Not yet published.
107. J.R. Barber: *Elasticity*. Second Edition. 2002 ISBN Hb 1-4020-0964-X; Pb 1-4020-0966-6
108. C. Miehe (ed.): *IUTAM Symposium on Computational Mechanics of Solid Materials at Large Strains*. Proceedings of the IUTAM Symposium held in Stuttgart, Germany, 20-24 August 2001. 2003 ISBN 1-4020-1170-9

Mechanics

SOLID MECHANICS AND ITS APPLICATIONS

Series Editor: G.M.L. Gladwell

109. P. Ståhle and K.G. Sundin (eds.): *IUTAM Symposium on Field Analyses for Determination of Material Parameters – Experimental and Numerical Aspects*. Proceedings of the IUTAM Symposium held in Abisko National Park, Kiruna, Sweden, July 31 – August 4, 2000. 2003 ISBN 1-4020-1283-7
110. N. Sri Namachchivaya and Y.K. Lin (eds.): *IUTAM Symposium on Nonlnear Stochastic Dynamics*. Proceedings of the IUTAM Symposium held in Monticello, IL, USA, 26 – 30 August, 2000. 2003 ISBN 1-4020-1471-6
111. H. Sobieckzky (ed.): *IUTAM Symposium Transsonicum IV*. Proceedings of the IUTAM Symposium held in Göttingen, Germany, 2–6 September 2002, 2003 ISBN 1-4020-1608-5
112. J.-C. Samin and P. Fisette: *Symbolic Modeling of Multibody Systems*. 2003 ISBN 1-4020-1629-8
113. A.B. Movchan (ed.): *IUTAM Symposium on Asymptotics, Singularities and Homogenisation in Problems of Mechanics*. Proceedings of the IUTAM Symposium held in Liverpool, United Kingdom, 8-11 July 2002. 2003 ISBN 1-4020-1780-4
114. S. Ahzi, M. Cherkaoui, M.A. Khaleel, H.M. Zbib, M.A. Zikry and B. LaMatina (eds.): *IUTAM Symposium on Multiscale Modeling and Characterization of Elastic-Inelastic Behavior of Engineering Materials*. Proceedings of the IUTAM Symposium held in Marrakech, Morocco, 20-25 October 2002. 2004 ISBN 1-4020-1861-4
115. H. Kitagawa and Y. Shibutani (eds.): *IUTAM Symposium on Mesoscopic Dynamics of Fracture Process and Materials Strength*. Proceedings of the IUTAM Symposium held in Osaka, Japan, 6-11 July 2003. Volume in celebration of Professor Kitagawa's retirement. 2004 ISBN 1-4020-2037-6
116. E.H. Dowell, R.L. Clark, D. Cox, H.C. Curtiss, Jr., K.C. Hall, D.A. Peters, R.H. Scanlan, E. Simiu, F. Sisto and D. Tang: *A Modern Course in Aeroelasticity*. 4th Edition, 2004 ISBN 1-4020-2039-2
117. T. Burczyński and A. Osyczka (eds.): *IUTAM Symposium on Evolutionary Methods in Mechanics*. Proceedings of the IUTAM Symposium held in Cracow, Poland, 24-27 September 2002. 2004 ISBN 1-4020-2266-2
118. D. Ieşan: *Thermoelastic Models of Continua*. 2004 ISBN 1-4020-2309-X
119. G.M.L. Gladwell: *Inverse Problems in Vibration*. Second Edition 2004 ISBN 1-4020-2670-6
120. J.R. Vinson: *Plate and Panel Structures of Isotropic, Composite and Piezoelectric Materials, Including Sandwich Construction*. 2005 ISBN 1-4020-3110-6
121. *Forthcoming*.
122. G. Rega and F. Vestroni (eds.): *IUTAM Symposium on Chaotic Dynamics and Control of Systems and Processes in Mechanics*. Proceedings of the IUTAM Symposium held in Rome, Italy, 8–13 June 2003. 2005 ISBN 1-4020-3267-6
123. E.E. Gdoutos: *Fracture Mechanics. An Introduction*. 2nd edition. 2005 ISBN 1-4020-2863-2
124. M.D. Gilchrist (ed.): *IUTAM Symposium on Impact Biomechanics from Fundamental Insights to Applications*. 2005 ISBN 1-4020-3795-3
125. J.M. Huyghe, P.A.C. Raats and S.C. Cowin (eds.): *IUTAM Symposium on Physicochemical and Electromechanical Interactions in Porous Media*. 2005 ISBN 1-4020-3864-X
126. H. Ding, W. Chen and L. Zhang: *Elasticity of Transversely Isotropic Materials*. 2005 ISBN 1-4020-4033-4
127. W. Yang (ed.): *IUTAM Symposium on Mechanics and Reliability of Actuating Materials*. Proceedings of the IUTAM Symposium held in Beijing, China, 1–3 September 2004. 2005 ISBN 1-4020-4131-6

Mechanics

SOLID MECHANICS AND ITS APPLICATIONS

Series Editor: G.M.L. Gladwell

128. J.-P. Merlet: *Parallel Robots*. 2006 ISBN 1-4020-4132-2
129. G.E.A. Meier and K.R. Sreenivasan (eds.): *IUTAM Symposium on One Hundred Years of Boundary Layer Research*. Proceedings of the IUTAM Symposium held at DLR-Göttingen, Germany, August 12–14, 2004. 2006 ISBN 1-4020-4149-7
130. H. Ulbrich and W. Günthner (eds.): *IUTAM Symposium on Vibration Control of Nonlinear Mechanisms and Structures*. 2006 ISBN 1-4020-4160-8
131. L. Librescu and O. Song: *Thin-Walled Composite Beams. Theory and Application*. 2006 ISBN 1-4020-3457-1
132. G. Ben-Dor, A. Dubinsky and T. Elperin: *Applied High-Speed Plate Penetration Dynamics*. 2006 ISBN 1-4020-3452-0
133. X. Markenscoff and A. Gupta (eds.): *Collected Works of J. D. Eshelby*. Mechanics of Defects and Inhomogeneities. 2006 ISBN 1-4020-4416-X
134. R.W. Snidle and H.P. Evans (eds.): *IUTAM Symposium on Elastohydrodynamics and Microelastohydrodynamics*. Proceedings of the IUTAM Symposium held in Cardiff, UK, 1–3 September, 2004. 2006 ISBN 1-4020-4532-8
135. T. Sadowski (ed.): *IUTAM Symposium on Multiscale Modelling of Damage and Fracture Processes in Composite Materials*. Proceedings of the IUTAM Symposium held in Kazimierz Dolny, Poland, 23–27 May 2005. 2006 ISBN 1-4020-4565-4
136. A. Preumont: *Mechatronics. Dynamics of Electromechanical and Piezoelectric Systems*. 2006 ISBN 1-4020-4695-2
137. M.P. Bendsøe, N. Olhoff and O. Sigmund (eds.): *IUTAM Symposium on Topological Design Optimization of Structures, Machines and Materials*. Status and Perspectives. 2006 ISBN 1-4020-4729-0
138. A. Klarbring: *Models of Mechanics*. 2006 ISBN 1-4020-4834-3
139. H.D. Bui: *Fracture Mechanics*. Inverse Problems and Solutions. 2006 ISBN 1-4020-4836-X
140. M. Pandey, W.-C. Xie and L. Xu (eds.): *Advances in Engineering Structures, Mechanics and Construction*. Proceedings of an International Conference on Advances in Engineering Structures, Mechanics & Construction, held in Waterloo, Ontario, Canada, May 14–17, 2006. 2006 ISBN 1-4020-4890-4
141. G.Q. Zhang, W.D. van Driel and X. J. Fan: *Mechanics of Microelectronics*. 2006 ISBN 1-4020-4934-X
142. Q.P. Sun and P. Tong (eds.): *IUTAM Symposium on Size Effects on Material and Structural Behavior at Micron- and Nano-Scales*. Proceedings of the IUTAM Symposium held in Hong Kong, China, 31 May–4 June, 2004. 2006 ISBN 1-4020-4945-5
143. A.P. Mouritz and A.G. Gibson: *Fire Properties of Polymer Composite Materials*. 2006 ISBN 1-4020-5355-X
144. Y.L. Bai, Q.S. Zheng and Y.G. Wei (eds.): *IUTAM Symposium on Mechanical Behavior and Micro-Mechanics of Nanostructured Materials*. Proceedings of the IUTAM Symposium held in Beijing, China, 27–30 June 2005. 2007 ISBN 1-4020-5623-0
145. L.P. Pook: *Metal Fatigue*. What It Is, Why It Matters. 2007 ISBN 1-4020-5596-6
146. H.I. Ling, L. Callisto, D. Leshchinsky and J. Koseki (eds.): *Soil Stress-Strain Behavior: Measurement, Modeling and Analysis*. A Collection of Papers of the Geotechnical Symposium in Rome, March 16–17, 2006. 2007 ISBN 978-1-4020-6145-5

The manufacturer's authorised representative in the EU is Springer Nature Customer Service Centre GmbH, Europaplatz 3, 69115 Heidelberg, Germany. If you have any concerns regarding our products, please contact ProductSafety@springernature.com

Printed and bound by CPI Group (UK) Ltd, Croydon, CR0 4YY

30/09/2026

02225441-0003